Ultimate Objective Neet Physics

For Class 11

Vol. 1

by

Susamoy Chakraborty

Double9 BOOKS

Ultimate Objective Neet Physics
For Class 11
Vol. 1

by Susamoy Chakraborty

ISBN: 978-93-62205-45-2

Published by

DOUBLE 9 BOOKS

2/13-B, Ansari Road
Daryaganj, New Delhi – 110002
info@double9books.com
www.double9books.com
Tel. 011-40042856

ABOUT THE AUTHOR

My Self susamoy Chakraborty. I have done my graduation from Tamralipta Mahavidyalaya and Masters from Medinipur Day College. And completed B.ed degree from Tamralipta Mahavidyalaya. I have nearly 6 years' experience in this teaching field. I observed that Actual NEET physics book is not in our books so perpouse to write the book is guide the students properly. I teach about 500 students and many of them are doctors. So in from my view point this book helps you properly. I teach also free of cost students and they are also established and achieve their goals. I am a school teacher under WBCHSE board. If you practice this book you will able to solve all kinds of problems, all kinds of numericals and thought base questions.

CONTENTS

2
UNITS AND MEASUREMENTS
SINGLE CORRECT ANSWER TYPE

1. The velocity of water waves v may depend upon their wavelength λ, the density of water ρ and the acceleration due to gravity g.The method of dimensions gives the relation between these quantities as
a) $v^2 \propto \lambda g^{-1} \rho^{-1}$ b) $v^2 \propto g\lambda\rho$ c) $v^2 \propto g\lambda$ d) $v^2 \propto g^{-1}\lambda^{-3}$
2. In a vernier callipers, one main scale division is x cm and n division of the vernier scale coincide with $(n-1)$ divisions of the main scale. The least count (in cm) of the callipers is
a) $\left(\frac{n-1}{n}\right)x$ b) $\frac{nx}{(n-1)}$ c) $\frac{x}{n}$ d) $\frac{x}{(n-1)}$
3. The work done by a battery is $W = \varepsilon\Delta q$, where Δq change transferred by battery, ε =emf of the battery. What are dimensions of emf of battery?
a) $[M^0L^0T^{-2}A^{-2}]$ b) $[ML^2T^{-3}A^{-2}]$ c) $[M^2L^0T^{-3}A^0]$ d) $[ML^2T^{-3}A^{-1}]$
4. The dimensions of universal gravitational constant are
a) $M^{-2}L^2T^{-2}$ b) $M^{-1}L^3T^{-2}$ c) $ML^{-1}T^{-2}$ d) ML^2T^{-2}
5. The dimension of $\frac{1}{\sqrt{\varepsilon_0\mu_0}}$ is that of
a) Velocity b) Time c) Capacitance d) Distance
6. $[ML^2T^{-3}]$ is the dimension of
a) Work b) Power c) Force d) Momentum
7. Crane is British unit of volume (one crane = 170.4742). convert crane into SI units.
a) 0.170474 m^3 b) 17.0474m^3 c) 0.00170474m^3 d) 1704.74m^3
8. What are the units of $K = 1/4\pi\varepsilon_0$
a) $C^2N^{-1}m^{-2}$ b) Nm^2C^{-2} c) Nm^2C^2 d) Unitless
9. If the speed of light (c), acceleration due to gravity (g) and pressure (p) are taken as the fundamental quantities, then the dimension of gravitational constant is
a) $c^2g^0p^{-2}$ b) $c^0g^2p^{-1}$ c) cg^3p^{-2} d) $c^{-1}g^0p^{-1}$
10. The unit of reduction factor of tangent galvanometer is
a) *Ampere* b) *Gauss* c) *Radian* d) None of these
11. The constant of proportionality $\frac{1}{4\pi\varepsilon_0}$ in Coulomb's law has the following dimensions
a) $C^{-2}Nm^2$ b) $C^2N^{-1}m^{-2}$ c) C^2Nm^2 d) $C^{-2}N^{-1}m^{-2}$
12. A resistor of 4 kΩ with tolerance 10% is connected in parallel with a resistor of 6 kW with tolerance 100%. The tolerance of the parallel combination is nearly
a) 10% b) 20% c) 30% d) 40%
13. An object is moving through the liquid. The viscous damping force acting on it is proportional to the velocity. Then dimension of constant of proportionality is
a) $ML^{-1}T^{-1}$ b) MLT^{-1} c) M^0LT^{-1} d) ML^0T^{-1}
14. If $v = \frac{A}{t} + Bt^2 + Ct^3$ where v is velocity, t is time and A, B and C are constants, then the dimensional formula of B is
a) $[M^0LT^0]$ b) $[ML^0T^0]$ c) $[M^0L^0T\ ]$ d) $[M^0LT^{-3}]$
15. The specific resistance ρ of a circular wire of radius r. Resistance R and and length l is given by $\rho = \frac{\pi r^2 R}{l}$.
Given, $r = 0: (24 \pm 0.02)$cm, $R = (30 \pm 1)\Omega$ and $l = (4.80 \pm 0.01)$cm. The percentage error in ρ is nearly
a) 7% b) 9% c) 13% d) 20%

16. The fundamental physical quantities that have same dimensions in the dimensional formulae of torque and angular momentum are

a) Mass, time b) Time, length c) Mass, length d) Time, mole

17. In an experiment, to measure the height of a bridge by dropping stone into water underneath, if the error in measurement of time is $0.1s$ at the end of $2s$, then the error in estimation of height of bridge will be

a) $0.49\ m$ b) $0.98\ m$ c) $1.96\ m$ d) $2.12\ m$

18. The radius of the proton is about 10^{-15}m. The radius of the observable universe is 10^{26}m. identify the distance which is half-way between these two extremes on a logarithmic scale.

a) 10^{21}m b) 10^{6}m c) 10^{-6}m d) 10^{0}m

19. Which of the following pairs is wrong

a) Pressure-Barometer b) Relative density-Pyrometer

c) Temperature-Thermometer d) Earthquake-Seismograph

20. The temperature of a body on Kelvin scale is found to be X K. When it is measured by a Fahrenheit thermometer, it is found to be X^0F. Then X is

a) 301.25 b) 574.25 c) 313 d) 40

21. A plate has a length (5 ± 0.1)cm and breadth (2 ± 0.01) cm. Then the area of the plate is

a) (10 ± 0.2)cm^2 b) (10 ± 0.01)cm^2 c) (10 ± 0.001)cm^2 d) (10 ± 1)cm^2

22. One yard in SI units is equal

a) 1.9144 *metre* b) 0.9144 *metre* c) 0.09144 *kilometre* d) 1.0936 *kilometre*

23. The dimensional formula for Boltzmann's constant is

a) $[ML^2T^{-2}\theta^{-1}]$ b) $[ML^2T^{-2}]$ c) $[ML^0T^{-2}\theta^{-1}]$ d) $[ML^{-2}T^{-1}\theta^{-1}]$

24. $Erg - m^{-1}$ can be the unit of measure for

a) Force b) Momentum c) Power d) Acceleration

25. Which physical quantities have the same dimension

a) Couple of force and work b) Force and power

c) Latent heat and specific heat d) Work and power

26. The frequency of vibration f of a mass m suspended from a spring of spring constant k is given by relation of the type $f = cm^xk^y$, where c is a dimensionless constant. The values of x and y are

a) $1/2, 1/2$ b) $-1/2, -1/2$ c) $1/2, -1/2$ d) $-1/2, 1/2$

27. What is the unit of k in the relation $U = \frac{ky}{y^2+a^2}$ where U represents the potential energy, y represents the displacement and a represents the maximum displacement *ie*, amplitude?

a) m s^{-1} b) m s c) J m d) J s^{-1}

28. The pair having the same dimensions is

a) Angular momentum, work b) Work, torque

c) Potential energy, linear momentum d) Kinetic energy, velocity

29. Which one has the dimensions different from the remaining three

a) Power b) Work c) Torque d) Energy

30. Planck's constant has the dimensions (unit) of

a) Energy b) Linear momentum c) Work d) Angular momentum

31. Dimension of R is

a) ML^2T^{-1} b) $ML^2T^{-3}A^{-2}$ c) $ML^{-1}T^{-2}$ d) None of these

32. Which of the following sets have different dimensions?

a) Pressure, Young's modulus, Stress b) Emf, Potential difference, Electric potential

c) Heat, Work done, Energy

d) Dipole moment, Electric flux, Electric field

33. Choose the incorrect statement out of the following

a) Every measurement by any measuring instrument has some error

b) Every calculated physical quantity that is based on measured values has some error

c) A measurement can have more accuracy but less precision and vice versa

d) The percentage error is different from relative error

34. A public park, in the form of a square, has an area of $(100 \pm 0.2)\mathrm{m}^2$. The side of park is

a) (10 ± 0.01)m b) (10 ± 0.1)m c) (10.0 ± 0.1)m d) (10.0 ± 0.2)m

35. Which of the following is dimensionless?

a) $\frac{v^2}{rg}$ b) $\frac{v^2 g}{r}$ c) $\frac{vg}{r}$ d) $v^2 rg$

36. If E =energy, G = gravitational constant, I = impulse and M = mass, the dimensions of $\frac{GIM^2}{E^2}$ are same as that of

a) Time b) Mass c) Length d) Force

37. The unit of percentage error is

a) Same as that of physical quantity

b) Different from that of physical quantity

c) Percentage error is unit less

d) Errors have got their own units which are different from that of physical quantity measured

38. The S.I. unit of gravitational potential is

a) J b) $J - kg^{-1}$ c) $J - kg$ d) $J - kg^{-2}$

39. Dimensions of charge are

a) $M^0L^0T^{-1}A^{-1}$ b) $MLTA^{-1}$ c) $T^{-1}A$ d) TA

40. $S = A(1 - e^{-Bxt})$, where S is speed and x is displacement. The unit of B is

a) $m^{-1}s^{-1}$ b) $m^{-2}s$ c) s^{-2} d) s^{-1}

41. A cube has numerically equal volume and surface area. The volume of such a cube is

a) 216 *units* b) 1000 *units* c) 2000 *units* d) 3000 *units*

42. The physical quantity having the dimensions $[M^{-1}L^{-3}T^3A^2]$ is

a) Resistance

b) Resistivity

c) Electrical conductivity

d) Electromotive force

43. If $x = at + bt^2$, where x is the distance travelled by the body in kilometre while t is the time in second, then the units of b are

a) km/s b) $km - s$ c) km/s^2 d) $km - s^2$

44. ‘Torr’ is the unit of

a) Pressure b) Volume c) Density d) Flux

45. A suitable unit for gravitational constant is

a) kg-m sec^{-1} b) $N\,m^{-1}\mathrm{sec}$ c) $N\,m^2\,kg^{-2}$ d) $kg\,m\mathrm{sec}^{-1}$

46. Dimensional formula for force is

a) $[\mathrm{ML^2T^{-2}}]$ b) $[\mathrm{MLT^{-2}}]$ c) $[\mathrm{ML^{-1}T^{-2}}]$ d) $[\mathrm{ML^2T^{-2}}]$

47. If L denotes the inductance of an inductor through which a current I is flowing, then the dimensional formula of LI^2 is

a) $[\mathrm{MLT^{-2}}]$

b) $[\mathrm{ML^2T^{-2}}]$

c) $[\mathrm{M^2L^2T^{-2}}]$

d) Not expressible in terms of M, L, T

48. The dimensions of electric dipole moment are

a) $[\mathrm{L^2I}]$ b) $[\mathrm{LI}]$ c) $[\mathrm{LTI}]$ d) $[\mathrm{T^{-2}}]$

49. The velocity of a freely falling body changes as $g^p h^q$ where g is acceleration due to gravity and h is the height. The values of p and q are

a) $1, \frac{1}{2}$ b) $\frac{1}{2}, \frac{1}{2}$ c) $\frac{1}{2}, 1$ d) 1,1

50. $ML^{-1}T^{-2}$represents

a) Stress

b) Young's Modulus

c) Pressure

d) All of the above three quantities

51. The unit of Wien's constant b is

a) $Wm^{-2}K^{-4}$ b) $m^{-1}K^{-1}$ c) Wm^2 d) MK

52. The SI unit of gravitational potential is

a) J b) Jkg^{-1} c) Jkg d) Jkg^2

53. A physical quantity is measured and its value is found to be nu where n = numerical value and u = unit. Then which of the following relations is true

a) $n \propto u^2$ b) $n \propto u$ c) $n \propto \sqrt{u}$ d) $n \propto \frac{1}{u}$

54. Frequency is the function of density (ρ), length (a) and surface tension (T). Then its value is

a) $k\rho^{1/2}a^{3/2}/\sqrt{T}$ b) $k\rho^{3/2}a^{3/2}/\sqrt{T}$ c) $k\rho^{1/2}a^{3/2}/T^{3/4}$ d) None of these

55. The units of modulus rigidity are

a) $N-m$ b) N/m c) $N-m^2$ d) N/m^2

56. The circular divisions of shown screw gauge are 50. It moves 0.5 mm on main scale in one rotation. The diameter of the ball is

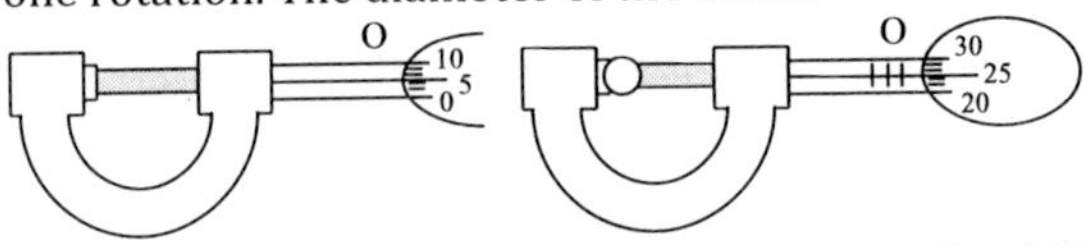

a) 2.25 mm b) 2.20 mm c) 1.20 mm d) 1.25 mm

57. Which of the following pairs has same dimensions?

a) Current density and charge density

b) Angular momentum and momentum

c) Spring constant and surface energy

d) Force and torque

58. The Vander Waal's equation of state for real gases is given as $\left(P+\frac{a}{V^2}\right)(V-b)=nRT$ which of the following terms has dimensions different from that of energy

a) PV b) $\frac{a}{V^2}$ c) $\frac{ab}{V^2}$ d) bP

59. In the relation $p=\frac{\alpha}{\beta}e^{-\frac{\alpha z}{k\theta}}$, p is the pressure, z the distance, k is Boltzmann constant and θ is the temperature, the dimensional formula of β will be

a) $[M^0L^2T^0]$ b) $[ML^2T]$ c) $[ML^0T^{-1}]$ d) $[ML^2T^{-1}]$

60. A dimensionally consistent relation for the volume V of a liquid of coefficient of viscosity η flowing per second through a tube of radius r and length l and having pressure p across its end, is

a) $V=\frac{\pi pr^4}{8\eta l}$ b) $V=\frac{\pi\eta l}{8pr^4}$ c) $V=\frac{8p\eta l}{\pi r^4}$ d) $V=\frac{\pi p\eta}{8lr^4}$

61. "Pascal-Second" has dimension of

a) Force

b) Energy

c) Pressure

d) Coefficient of viscosity

62. In the equation $X=3YZ^2$, X and Z have dimensions of capacitance and magnetic induction respectively. In MKSQ system, the dimensional formula of Y is

a) $[M^{-3}L^{-2}T^{-2}Q^{-4}]$ b) $[ML^{-2}]$ c) $[M^{-3}L^{-2}Q^4T^8]$ d) $[M^{-3}L^{-2}Q^4T^4]$

63. The dimensional formula for impulse is same as the dimensional formula for

a) Momentum

b) Force

c) Rate of change of momentum

d) Torque

64. If E, M, L and G denote energy, mass, angular momentum and gravitational constant respectively, then the quantity (EL^2/M^5G^2) has the dimensions of

a) Angle b) Length c) Mass d) Time

65. Unit of electric flux is

a) Vm b) Nm/C^{-1} c) Vm^{-1} d) CNm^{-1}

66. Dimensions of frequency are

a) $M^0L^{-1}T^0$ b) $M^0L^0T^{-1}$ c) M^0L^0T d) MT^{-2}

67. Students I, II and III perform an experiment for measuring the acceleration due to gravity (g) using a simple pendulum. They use different lengths of the pendulum and/or record time for different number of oscillations. The observations are shown in the table

Least count for length = 0.1 cm

Least count for time = 0.1 s

Student	Length of the pendulum (cm)	Number of oscillation (n)	Total time for (n) oscillations (s)	Time period (s)
I	64.0	8	128.0	16.0
I	64.0	4	64.0	16.0
III	20.0	4	36.0	9.0

If $E_{\text{I}}, E_{\text{II}}$ and E_{III} are the percentage errors in g, $i.e.$, $\left(\frac{\Delta g}{g} \times 100\right)$ for students I, II and III, respectively

a) $E_{\text{I}} = 0$ b) E_{I} is minimum c) $E_{\text{I}} = E_{\text{II}}$ d) E_{II} is maximum

68. The unit of the coefficient of viscosity in S.I. system is

a) $m/kg - s$ b) $m - s/kg^2$ c) $kg/m - s^2$ d) $kg/m - s$

69. The physical quantity angular momentum has the same dimensions as that of

a) Work b) Force c) Momentum d) Planck's constant

70. The physical quantity having the dimensions $[M^{-1}L^{-3}A^2]$ is

a) Resistance b) Resistivity

c) Electrical conductivity d) Electromotive force

71. The dimensional formula of $\frac{1}{\varepsilon_0}\frac{e^2}{hc}$ is

a) $[M^0L^0T^0A^0]$ b) $[M^{-1}L^3T^2A]$ c) $[ML^3T^{-4}A^{-2}]$ d) $[M^{-1}L^{-3}T^4]$

72. In the formula, $a = 3bc^2$, a and c have dimensions of electric capacitance and magnetic induction respectively. What are dimensions of b in MKS system?

a) $[M^{-3}L^{-2}T^4Q^4]$ b) $[M^{-3}T^4Q^4]$ c) $[M^{-3}T^3Q]$ d) $[M^{-3}L^2T^4Q^{-4}]$

73. A unit of area, often used in measuring land areas, is the hectare defined as $10^4 m^2$. An open-pit coal mine consumes 75 hectares of land, down to a depth of 26m, each year. What volume of earth, in cubic kilometre, is removed in this time?

a) 0.01 b) 0.02 c) 0.03 d) 0.04

74. If 3.8×10^{-6} is added to 4.2×10^{-5} giving due regard to significant figures, then the result will be

a) 458×10^{-5} b) 4.6×10^{-5} c) 4.5×10^{-5} d) None of the above

75. Let us choose a new unit of length such that the velocity of light in vacuum is unity. If light takes 8 min and 20 sec to cover the distance between sun and earth, this distance in terms of the new unit is

a) 5 b) 50 c) 500 d) 3×10^8

76. If L, C and R denote inductance, capacitance and resistance respectively, then which of the following combination has the dimension of time?

a) $\frac{C}{L}$ b) $\frac{1}{RC}$ c) $\frac{L}{R}$ d) $\frac{RL}{C}$

77. In an experiment the angles are required to be measured using an instrument. 29 divisions of the main scale exactly coincide with the 30 divisions of the vernier scale. If the smallest division of the main scale is half-a-degree ($= 0.5°$) then the least count of the instrument is
a) One minute b) Half minute c) One degree d) Half-degree

78. The position of a particle at time t is given by the relation $x(t) = \left(\frac{v_0}{\alpha}\right)(1 - e^{\alpha t})$, where v_0 is constant and $\alpha > 0$. The dimensions of v_0 and α are respectively
a) $M^0L^1T^{-1}$ and T^{-1} b) $M^0L^1T^0$ and T^{-1} c) $M^0L^1T^{-1}$ and LT^{-2} d) $M^0L^1T^{-1}$ and T

79. Out of the following which pair of quantities do not have same dimensions
a) Planck's constant and angular momentum
b) Work and energy
c) Pressure and Young's modulus
d) Torque and moment of inertia

80. The unit of physical quantity obtained by the line intergral of electric field is
a) NC^{-1} b) Vm^{-1} c) JC^{-1} d) $C^2N^{-1}m^{-2}$

81. Write dimensional formula for the intensity of radiation
a) $M^1L^0T^3$ b) $M^1L^0T^{-3}$ c) $M^1L^2T^{-2}$ d) $M^1L^2T^{-3}$

82. Two quantities A and B have different dimensions. Which mathematical operation given below is physically meaningful
a) A/B b) $A + B$ c) $A - B$ d) None

83. The percentage errors in the measurement of a mass and speed are 2% and 3% respectively. How much will be the maximum error in the estimate of kinetic energy obtained by measuring mass and speed?
a) 11% b) 8% c) 5% d) 1%

84. The concordeis the fastest airlines used for commercial service. It can cruise at 1450 mile per hour (about two times the speed of sound or in other words mach 2). What is it in m/s?
a) 644.4m/s b) 80 m/s c) 40 m/s d) None of these

85. The dimensions of electric potential are
a) $[ML^2T^{-2}Q^{-1}]$ b) $[MLT^{-2}Q^{-1}]$ c) $[ML^2T^{-1}Q]$ d) $[ML^2T^{-2}Q]$

86. The internal and external diameters of a hollow cylinder are measured with the help of avernier calipers. Their values are 4.23 ± 0.01 cm and 3.87 ± 0.01 cm respectively. The thickness of the wall of the cylinder is
a) 0.36 ± 0.02 cm b) 0.18 ± 0.02 cm c) 0.36 ± 0.01 cm d) 0.18 ± 0.01 cm

87. The velocity v (in cm/sec) of a particle is given in terms of time t(in sec) by the relation $v = at + \frac{b}{t+c}$; the dimensions of a, b and c are
a) $a = L^2, b = T, c = LT^2$
b) $a = LT^2, b = LT, c = L$
c) $a = LT^2, b = L, c = T$
d) $a = L, b = LT, c = T^2$

88. The dimensions of kinetic energy are
a) $[M^2L^2T]$ b) $[ML^2T]$ c) $[ML^2T^{-2}]$ d) $[ML^2T^{-1}]$

89. What is the SI unit of permeability
a) *Henryper metre*
b) Tesla *metre* per *ampere*
c) *Weber* per *ampere metre*
d) All the above units are correct

90. A student measures the distance traversed in free fall of a body, initially at rest in a given time. He uses this data to estimate g, the acceleration due to gravity. If the maximum percentage errors in measurement of the distance and the time are e_1 and e_2 respectively, the percentage error in the estimation of g is
a) $e_2 - e_1$ b) $e_1 + 2e_2$ c) $e_1 + e_2$ d) $e_1 - 2e_2$

91. Which of the following is the smallest unit
a) *Millimetre* b) *Angstrom* c) *Fermi* d) *Metre*

92. The dimensions of a rectangular block measured with calipers having least count of 0.01 cm are 5 mm × 10 mm × 5 mm. The maximum percentage error in the measurement of the volume of the block is

a) 5% b) 10% c) 15% d) 20%

93. If E, m, J and G represent energy, mass, angular momentum and gravitational constant respectively, then the dimensional formula of EJ^2/m^5G^2 is

a) $[MLT^{-2}]$ b) $[M^0L^0T]$ c) $[M^0L^2T^0]$ d) Dimensionless

94. The dimensions of couple are

a) ML^2T^{-2} b) MLT^{-2} c) $ML^{-1}T^{-3}$ d) $ML^{-2}T^{-2}$

95. The dimension of quantity $(L/RCV)$is

a) $[A]$ b) $[A^2]$ c) $[A^{-1}]$ d) None of these

96. The dimensional formula of relative density is

a) ML^{-3} b) LT^{-1} c) MLT^{-2} d) Dimensionless

97. The dimensions of emf in MKS is

a) $ML^{-1}T^{-2}Q^{-2}$ b) $ML^2T^{-2}Q^{-2}$ c) $MLT^{-2}Q^{-1}$ d) $ML^2T^{-2}Q^{-1}$

98. If the length of rod A is (3.25 ± 0.01)cm and that of B is (4.19 ± 0.01) cm, then the rod B is longer than rod A by

a) (0.94 ± 0.00) cm b) (0.94 ± 0.01) cm c) (0.94 ± 0.02) cm d) (0.94 ± 0.005) cm

99. The length l, breadth b and thickness t of a block are measured with the help of a metre scale. Given $l = 15.12 \pm 0.01$cm, $b = 10.15 \pm 0.01$cm, $t = 5.28 \pm 0.01$cm. The percentage error in volume is

a) 0.64% b) 0.28% c) 0.37% d) 0.48%

100. The mass and volume of a body are found to be $500 \pm 0.05\ kg$ and $1.00\ \pm 0.05 m^3$ respectively. Then the maximum possible percentage error in its density is

a) 6% b) 3% c) 10% d) 5%

101. What is the dimensional formula of $\frac{\text{planck's constant}}{\text{linear momentum}}$?

a) $[M^0L^0T^0]$ b) $[M^0L^0T]$ c) $[M^0LT^0]$ d) $[MLT^{-1}]$

102. The dimensional formula of coefficient of permittivity for free space (ε_0) in the equation $F = \frac{1}{4\pi\varepsilon_0}\frac{q_1q_2}{r^2}$, where symbols have their usual meanings, is

a) $[ML^3A^{-2}T^{-4}]$ b) $[M^{-1}L^{-3}T^4A^2]$ c) $[M^{-1}L^{-3}A^{-2}T^{-4}]$ d) $[ML^3A^2T^{-4}]$

103. The surface tension of a liquid is 70 $dyne/cm$, in MKS system value is

a) 70 N/m b) $7 \times 10^{-2}\ N/m$ c) $7 \times 10^3\ N/m$ d) $7 \times 10^2\ N/m$

104. If unit of length, mass and time each be doubled, the unit of work done is increased by

a) 4 times b) 6 times c) 8 times d) 2 times

105. Unit of impulse is

a) *Newton* b) $kg - m$ c) $kg - m/s$ d) *Joule*

106. The constant of proportionality $\frac{1}{4\pi\varepsilon_0}$ in Coulomb's law has the following units

a) $C^{-2}Nm^2$ b) $C^2N^{-1}m^{-2}$ c) C^2Nm^2 d) $C^{-2}N^{-1}m^{-2}$

107. From the dimensional consideration, which of the following equation is correct

a) $T = 2\pi r\sqrt{\frac{R^3}{GM}}$ b) $T = 2\pi\sqrt{\frac{GM}{R^3}}$ c) $T = 2\pi\sqrt{\frac{GM}{GR^2}}$ d) $T = 2\pi\sqrt{\frac{R^2}{GM}}$

108. The length, breadth and thickness of a block are given by $l = 12cm, b = 6\ cm$ and $t = 2.45cm$

The volume of block according to the idea of significant figures should be

a) $1 \times 10^2\ cm^3$ b) $2 \times 10^2\ cm^3$ c) $1.763 \times 10^2\ cm^3$ d) None of tehse

109. A force F is given by $F = at + bt^2$, where t is time. What are the dimensions of a and b

a) MLT^{-3} and ML^2T^{-4} b) MLT^{-3} and MLT^{-4} c) MLT^{-1} and MLT^0 d) MLT^{-4} and MLT^1

110. Which of the following is not a unit of energy
a) W–s b) $kg-m/sec$ c) N-m d) $Joule$

111. In a new system of units, unit f mass is 10 kg, unit of length is 1 km and unit of time is 1 min. The value of 1 joule in this new hypothertical system is
a) 3.6×10^{-4} new units b) 6×10^{7}new units c) 10^{11} new units d) 1.67×10^{4} new units

112. In an experiment, the following observation's were recorded: $L = 2.820\ m$, $M = 3.00kg$, $l = 0.087\ cm$, diameter $D = 0.041\ cm$. Taking $g = 9.81 m/s^2$ using the formula,
$Y = \frac{4MgL}{\pi D^2 l}$, the maximum permissible error in Y is
a) 7.96% b) 4.56% c) 6.50% d) 8.42%

113. Which relation is wrong
a) $1\ calorie = 4.18\ joule$ b) $1\ \text{Å} = 10^{-10} m$
c) $1\ MeV = 1.6 \times 10^{-13} joule$ d) $1\ newton = 10^{-5} dyne$

114. The resistance $R = \frac{V}{i}$ where $V = 100 \pm 5\ volts$ and $i = 10 \pm 0.2\ amperes$. What is the total error in R
a) 5% b) 7% c) 5.2% d) $\frac{5}{2}\%$

115. The quantity$X = \frac{\varepsilon_0 LV}{t}$: ε_0 is the permittivity of free space,$L$is length, V is potential difference and t is time. The dimensions of X are same as that of
a) Resistance b) Charge c) Voltage d) Current

116. $Ampere - hour$ is a unit of
a) Quantity of electricity b) Strength of electric current
c) Power d) Energy

117. 1 Wb/m^2 is equal to
a) 10^4 gauss b) $4\pi \times 10^{-3}$ gauss c) 10^2 gauss d) 10^{-4} gauss

118. Unit of magnetic moment is
a) $Ampere - metre^2$ b) $Ampere - metre$ c) $Weber - metre^2$ d) $Weber/metre$

119. Surface tension has the same dimensions as that of
a) Coefficient of viscosity b) Impulse
c) Momentum d) Spring constant

120. The dimensions of physical quantity X in the equation
Force $= \frac{X}{\text{Density}}$ is given by
a) $M^1L^4T^{-2}$ b) $M^2L^{-2}T^{-1}$ c) $M^2L^{-2}T^{-2}$ d) $M^1L^{-2}T^{-1}$

121. Given that $r = m^2 \sin pt$, where t represents time. If the unit of m is N, then the unit of r is
a) N b) N^2 c) N s d) N^2s

122. The dimensions of faradare
a) $M^{-1}L^{-2}T^2Q^2$ b) $M^{-1}L^{-2}TQ$ c) $M^{-1}L^{-2}T^{-2}Q$ d) $M^{-1}L^{-2}TQ^2$

123. One light year is defined as the distance travelled by light in one year. The speed of light is $3 \times 10^8 ms^{-1}$. The same in metre is
a) 3×10^{12}m b) 9.461×10^{15}m c) 3×10^{15}m d) None of these

124. The mean time period of second's pendulum is $2.00s$ and mean absolute error in the time period is $0.05s$. To express maximum estimate of error, the time period should be written as
a) $(2.00 \pm 0.01)s$ b) $(2.00 + 0.025)s$ c) $(2.00 \pm 0.05)\ s$ d) $(2.00 \pm 0.10)\ s$

125. A vernier callipers has 1 mm marks on the main scale. It has 20 equal divisions on the Vernier scale which match with 16 main scale divisions. For this Vernier callipers , the least count is
a) 0.02 mm b) 0.05 mm c) 0.1 mm d) 0.2 mm

126. The relative density of material of a body is found by weighing it first in air and then in water. If the weight in air is (5.00 ± 0.05) *newton* and weight in water is (4.00 ± 0.05) *newton* .Then the relative density along with the maximum permissible percentage error is

a) $5.0 \pm 11\%$ b) $5.0 \pm 1\%$ c) $5.0 \pm 6\%$ d) $1.25 \pm 5\%$

127. The dimensions of coefficient of self inductance are

a) $[ML^2T^{-2}A^{-2}]$ b) $[ML^2T^{-2}A^{-1}]$ c) $[MLT^{-2}A^{-2}]$ d) $[MLT^{-2}A^{-1}]$

128. If force (F), length (L) and time (T) are assumed to be the fundamental units, then the dimensional formula of the mass will be

a) $[FL^{-1}T^2]$ b) $[FL^{-1}T^{-2}]$ c) $[FL^{-1}T^{-1}]$ d) $[FL^2T^{-2}]$

129. Inductance Lcan be dimensionally represented as

a) $ML^2T^{-2}A^{-2}$ b) $ML^2T^{-4}A^{-3}$ c) $ML^{-2}T^{-2}A^{-2}$ d) $ML^2T^4A^3$

130. Linear momentum and angular momentum have the same dimensions in

a) Mass and length b) Length and time
c) Mass and time d) Mass, length and time

131. Which unit is not for length

a) Parsec b) Light year c) Angstrom d) Nano

132. SI unit of electric intensity is

a) Coulomb b) Coulomb/m^2 c) Newton d) Newton/ coulomb

133. The thrust developed by a rocket-motor is given by $F = mv + A(p_1 - p_2)$, where m is the mass of the gas ejected per unit time, v is velocity of the gas, A is area of cross-section of the noszzle, $p_1 \cdot p_2$are the pressures of the exhaust gas and surrounding atmosphere. The formula is dimensionally

a) Correct b) Wrong
c) Sometimes wrong, sometimes correct d) Data is not adequate

134. The dimensional formula for impulse is

a) MLT^{-2} b) MLT^{-1} c) ML^2T^{-1} d) M^2LT^{-1}

135. If Prepresents radiations pressure, c represents speed of light and Q represents radiation energy striking a unit area per *second*, the non-zero integers x, y and z such that $P^xQ^yc^z$ is dimensionless, are

a) $x = 1, y = 1, z = -1$ b) $x = 1, y = -1, z = 1$ c) $x = -1, y = 1, z = 1$ d) $x = 1, y = 1, z = 1$

136. Which is not a unit of electric field

a) NC^{-1} b) Vm^{-1} c) JC^{-1} d) $JC^{-1}m^{-1}$

137. The dimensional formula of self-inductance is

a) $[MLT^{-2}]$ b) $[ML^2T^{-1}A^{-2}]$ c) $[ML^2T^{-2}A^{-2}]$ d) $[ML^2T^{-2}A^{-1}]$

138. The unit of self-inductance is

a) Weber ampere b) Weber^{-1} ampere c) Ohm second d) Farad

139. Consider a new system of units in which c(speed of light in vacuum), h(Planck's constant) and G(gravitational constant) are taken as fundamental units. Which of the following would correctly represent mass in this new system?

a) $\sqrt{\frac{hc}{G}}$ b) $\sqrt{\frac{Gc}{h}}$ c) $\sqrt{\frac{hG}{c}}$ d) $\sqrt{hGc}$

140. The dimensional formula for the modulus of rigidity is

a) ML^2T^{-2} b) $ML^{-1}T^{-3}$ c) $ML^{-2}T^{-2}$ d) $ML^{-1}T^{-2}$

141. Which of the following is not represented in correct unit

a) $\frac{\text{Stress}}{\text{Strain}} = N/m^2$ b) Surface tension $= N/m$
c) Energy $= kg - m/sec$ d) Pressure $= N/m^2$

142. If 1 g cm s^{-1}= x newton-second, then the number x is equal to

a) 1×10^{-3} b) 3.6×10^{-3} c) 1×10^{-5} d) 6×10^{-4}

143. The time taken by an electron to go from ground state to excited state is one shake (one shake = 10^{-8}s). this time in nanosecond will be
a) 10 ns b) 4 ns c) 2 ns d) 25 ns

144. The dimensional formula of angular velocity is
a) $M^0L^0T^{-1}$ b) MLT^{-1} c) $M^0L^0T^1$ d) ML^0T^{-2}

145. The physical quantity which has the dimensional formula M^1T^{-3} is
a) Surface tension b) Solar constant c) Density d) Compressibility

146. The dimensions of $e^2/4\pi\varepsilon_0 hc$, where e, ε_0, h and c are electronic charge, electric permittivity, Planck's constant and velocity of light in vacuum respectively
a) $[M^0L^0T^0]$ b) $[M^1L^0T^0]$ c) $[M^0L^1T^0]$ d) $[M^0L^0T^1]$

147. Farad is not equivalent to
a) $\frac{q}{V}$ b) qv^2 c) $\frac{q^2}{J}$ d) $\frac{J}{V^2}$

148. Given $X = (Gh/c^3)^{1/2}$, where G, h and c are gravitational constant, Planck's constant and the velocity of light respectively. Dimensions of X are the same as those of
a) Mass b) Time c) Length d) Acceleration

149. The velocity of a particle v at an instant t is given by $v = at + bt^2$ the dimension of b is
a) [L] b) $[LT^{-1}]$ c) $[LT^{-2}]$ d) $[LT^{-3}]$

150. Dimensions of luminous flux are
a) ML^2T^{-2} b) ML^2T^{-3} c) ML^2T^{-1} d) MLT^{-2}

151. The ratio of the dimension of Planck's constant and that of moment of inertia is the dimension of
a) Frequency b) Velocity c) Angular momentum d) Time

152. The equation of a wave is given by

$$Y = A\sin\omega\left(\frac{x}{v} - k\right)$$

where $\omega$is the angular velocity and v is the linear velocity.
The dimension of $k$is
a) LT b) T c) T^{-1} d) T^2

153. The dimensional formula for impulse is
a) $[MLT^{-1}]$ b) $[ML^{-1}T]$ c) $[M^{-1}LT^{-1}]$ d) $[ML^{-1}T^{-1}]$

154. The radius of the sphere I (4.3 ± 0.1)cm. The percentage error in its volume is
a) $\frac{0.1}{4.3} \times 100$ b) $3 \times \frac{0.1 \times 100}{4.3}$ c) $\frac{1}{3} \times \frac{0.1 \times 100}{4.3}$ d) $3 + \frac{0.1 \times 100}{4.3}$

155. Size of universe is about
a) Ten million light years b) Million light years
c) Hundred million light years d) 10 million light years

156. A cube has a side of length 1.2×10^{-2}m. Calculate its volume.
a) $1.7 \times 10^{-6}m^3$ b) $1.73 \times 10^{-6}m^3$ c) $1.70 \times 10^{-6}m^3$ d) $1.732 \times 10^{-6}m^3$

157. The SI unit of momentum is
a) $\frac{kg}{m}$ b) $\frac{kg.m}{sec}$ c) $\frac{kg.m^2}{sec}$ d) $kg \times newton$

158. A calorie is a unit of heat and equal 4.2 J. Suppose we employ a system of units in which the unit of mass is αkg, the unit of length is β metre and the unit of time is γ sec. In this new system. 1 calorie =
a) $\alpha^{-1}\beta^{-2}\gamma^2$ b) $4.2\alpha\beta^2\gamma^2$ c) $\alpha\beta^2\gamma^2$ d) $4.2\alpha^{-1}\beta^{-2}\gamma^2$

159. Number of base SI unit is
a) 4 b) 7 c) 3 d) 5

160. Dimensions of bulk modulus are
a) $[M^{-1}LT^{-2}]$ b) $[ML^{-1}T^{-2}]$ c) $[ML^{-2}T^{-2}]$ d) $[M^2L^2T^{-1}]$

161. Position of body with acceleration 'a' is given by $x = Ka^m t^n$, here t is time. Find dimensions of m and n

a) $m = 1, n = 1$ b) $m = 1, n = 2$ c) $m = 2, n = 1$ d) $m = 2, n = 2$

162. A pressure of 10^6 dyne cm^{-2} is equivalent to

a) $10^5 N\,m^{-2}$ b) 10^4 $N\,m^{-2}$ c) $10^6 N\,m^{-2}$ d) $10^7 N\,m^{-2}$

163. The unit of Planck's constant is

a) *Joule* b) *Joule/s* c) *Joule/m* d) *Joule-s*

164. *Candela* is the unit of

a) Electric intensity b) Luminous intensity c) Sound intensity d) None of these

165. The dimensions of gravitational constant G and the moment of inertia are respectively

a) ML^3T^{-2}; ML^2T^0 b) $M^{-1}L^3T^{-2}$; ML^2T^0 c) $M^{-1}L^3T^{-2}$; $M^{-1}L^2T$ d) ML^3T^{-2}; $M^{-1}L^2T$

166. What will be the unit of time in that system in which the unit of length is metre, unit of mass is kg and unit of force is kg wt?

a) $(9.8)^2$sec b) 9.8 sec c) $\sqrt{9.8}$ sec d) $\frac{1}{\sqrt{9.8}}$ sec

167. If pressure P, velocity V and time T are taken as fundamental physical quantities, the dimensional formula of force is

a) PV^2T^2 b) $P^{-1}V^2T^{-2}$ c) PVT^2 d) $P^{-1}VT^2$

168. If I is the moment of inertia and ω the angular velocity, what is the dimensional formula of rotational kinetic energy $\frac{1}{2}I\omega^2$?

a) $[ML^2T^{-1}]$ b) $[M^2L^{-1}T^{-2}]$ c) $[ML^2T^{-2}]$ d) $[M^2L^{-1}T^{-2}]$

169. The dimensional formula for Planck's constant (h) is

a) $ML^{-2}T^{-3}$ b) ML^2T^{-2} c) ML^2T^{-1} d) $ML^{-2}T^{-2}$

170. The physical quantities not having same dimensions are

a) Speed and $(\mu_0\varepsilon_0)^{-1/2}$ b) Torque and work

c) Momentum and Planck's constant d) Stress and Young's modulus

171. Pressure gradient has the same dimension as that of

a) Velocity gradient b) Potential gradient c) Energy gradient d) None of these

172. The damping force on an oscillator is directly proportional to the velocity. The units of the constant of proportionality are

a) $kgms^{-1}$ b) $kgms^{-2}$ c) kgs^{-1} d) kgs

173. Dimensional formula of capacitance (or farad) is

a) $M^{-1}L^{-2}T^4A^2$ b) $ML^2T^4A^{-2}$ c) $MLT^{-4}A^2$ d) $M^{-1}L^{-2}T^{-4}A^{-2}$

174. *Joule-second* is the unit of

a) Work b) Momentum c) Pressure d) Angular momentum

175. If $L = 2.331$ cm, $B = 2.1$ cm, then $L + B$ is equal to

a) 4.431 cm b) 4.43 cm c) 4.4 cm d) 4 cm

176. If the radius of the sphere is $(5.3 \pm 0.1)cm$. Then percentage error in its volume will be

a) $3 + 6.01 \times \frac{100}{5.3}$ b) $\frac{1}{3} \times 0.01 \times \frac{100}{5.3}$ c) $\left(\frac{3 \times 0.01}{5.3}\right) \times 100$ d) $\frac{0.1}{5.3} \times 100$

177. The dimension of magnetic field in M, L, T and C (coulomb) is given as

a) MT^2C^{-2} b) $MT^{-1}C^{-1}$ c) $MT^{-2}C^{-1}$ d) $MLT^{-1}C^{-1}$

178. In a system of units if force (F), acceleration (A), and time (T) are taken as fundamental units then the dimensional formula of energy is

a) FA^2T b) FAT^2 c) F^2AT d) FAT

179. In the equation S_{nth} the $= u + \frac{a}{2}(2n - 1)$, the letters have their usual meanings. The dimensional formula of S_{nth} is

a) $[ML^0T]$ b) $[ML^{-1}T^{-1}]$ c) $[M^0LT^{-1}]$ d) $[M^0LT^0]$

180. Which of the following system of units is not based on units of mass, length and time alone

a) SI b) MKS c) FPS d) CGS

181. The dimensions of permittivity ε_0 are

a) $A^2T^2M^{-1}L^{-3}$ b) $A^2T^4M^{-1}L^{-3}$ c) $A^{-2}T^{-4}ML^3$ d) $A^2T^{-4}M^{-1}L^{-3}$

182. Which one of the following is not a unit of Young's modulus?

a) Nm^{-1} b) Nm^{-2} c) Dyne cm^{-2} d) Mega pascal

183. The unit of surface tension in SI system is

a) $Dyne/cm^2$ b) $Newton\ /m$ c) $Dyne/cm$ d) $Newton/m^2$

184. The speed (v) of ripples on the surface of water depends on surface tension (σ), density (ρ) and wavelength (λ). The square of speed (v) is proportional to

a) $\frac{\sigma}{\rho\lambda}$ b) $\frac{\rho}{\sigma\lambda}$ c) $\frac{\lambda}{\sigma\rho}$ d) $P\lambda\sigma$

185. The dimension of $\frac{1}{2}\epsilon_0E^2$, where ϵ_0 is permittivity of free space and E is electric field, is

a) MLT^1 b) ML^2T^{-2} c) $ML^{-1}T^{-2}$ d) ML^2T^{-1}

186. If the dimensions of a physical quantity are given by $M^aL^bT^c$, then the physical quantity will be

a) Pressure if $a = 1, b = -1, c = -2$ b) Velocity if$a = 1, b = 0, c = -1$

c) Acceleration if$a = 1, b = 1, c = -2$ d) Force if $a = 0, b = -1, c = -2$

187. The dimensional formula of modulus of rigidity is

a) $[ML^{-2}T^{-2}]$ b) $[ML^{-3}T_2]$ c) $[ML^2T^{-2}]$ d) $[ML^{-1}T^{-2}]$

188. SI unit of pressure is

a) *Pascal* b) $dynes/cm^2$ c) *cmofHg* d) *Atmosphere*

189. The dimensional formula of capacitance in terms of M, L, T and I is

a) $[ML^2T^2I^2]$ b) $[ML^{-2}T^4I^2]$ c) $[M^{-1}L^3T^3I]$ d) $[M^{-1}L^{-2}T^4I^2]$

190. Which of the following groups have different dimensions

a) Potential difference, EMF, voltage b) Pressure, stress, young's modulus

c) Heat, energy, work-done d) Dipole moment, electric flux, electric field

191. The fundamental unit, which has the same power in the dimensional formulae of surface tension and viscosity is

a) Mass b) Length c) Time d) None of these

192. Two full turns of the circular scale of a screw gauge cover a distance of 1 mm on its main scale. The total number of divisions on the circular scale is 50. Further, it is found that the screw gauge has a zero error of -0.03 mm. While measuring the diameter of a thin wire, a student notes the main scale reading of 3 mm and the number of circular scale divisions in line with the main scale as 35. The diameter of the wire is

a) 3.32 mm b) 3.73 mm c) 3.67 mm d) 3.38 mm

193. One million electron *volt* (1 *MeV*) is equal to

a) $10^5\ eV$ b) $10^6\ eV$ c) $10^4\ eV$ d) $10^7\ eV$

194. Given, Force $= \frac{\alpha}{\text{density} + \beta^3}$

What are the dimensions of α, β?

a) $[ML^2T^{-2}], [ML^{-1/3}]$ b) $[M^2L^4T^{-2}], [M^{1/3}L^{-1}]$ c) $[M^2L^{-2}T^{-2}], [M^{1/3}L^{-1}]$ d) $[M^2L^{-2}T_2], [ML^{-3}]$

195. Hertz is the unit for

a) Frequency b) Force c) Electric charge d) Magnetic flux

196. If $S = \frac{1}{3}ft^3$, f has the dimensions of

a) $[M^0L^{-1}T^3]$ b) $[MLT^{-3}]$ c) $[M^0L^1T^{-3}]$ d) $[M^0L^{-1}T^{-3}]$

197. If $x = a - b$, then the maximum percentage error in the measurement of x will be

a) $\left(\frac{\Delta a + \Delta b}{a - b}\right) \times 100\%$ b) $\left(\frac{\Delta a}{a} - \frac{\Delta b}{b}\right) \times 100\%$

c) $\left(\frac{\Delta a}{a-a}+\frac{\Delta b}{a-b}\right)\times 100\%$ d) $\left(\frac{\Delta a}{a-a}-\frac{\Delta b}{a-b}\right)\times 100\%$

198. If the error in the measurement of radius of a sphere is 2%, then the error in the determination of volume of the sphere will be

a) 8% b) 2% c) 4% d) 6%

199. Which pair has the same dimensions

a) Work and power b) Density and relative density
c) Momentum and impulse d) Stress and strain

200. The value of Planck's constant is

a) 6.63×10^{-34} *J-sec* b) 6.63×10^{34} *J-sec* c) $6.63\times 10^{-34}\ kg\text{-}m^2$ d) 6.63×10^{34} *kg-sec*

201. Which of the following is the unit of specific heat?

a) $\mathrm{Jkg\,°C^{-1}}$ b) $\mathrm{Jkg^{-1}\,°C^{-1}}$ c) $\mathrm{kg\,°CJ^{-1}}$ d) $\mathrm{J/kg^{-1}\,°C^{-2}}$

202. A body of mass $m = 3.513$ kg is moving along the x −axis with a speed of $5.00\ \mathrm{ms^{-1}}$. The magnitude of its momentum is recorded as

a) $17.6\ \mathrm{kg\,ms^{-1}}$ b) $17.565\ \mathrm{kg\,ms^{-1}}$ c) $17.56\ \mathrm{kg\,ms^{-1}}$ d) $17.57\ \mathrm{kg\,ms^{-1}}$

203. A sextant is used to measure

a) Area of hill b) Height of an object
c) Breadth of a tower d) Volume of the building

204. In the context of accuracy of measurement and significant figures in expressing results of experiment, which of the following is/are correct

(1) Out of the two measurements 50.14 *cm*and 0.00025 *ampere*, the first one has greater accuracy

(2) If one travels 478 *km* by rail and 397 *m* by road, the total distance travelled is 478 *km*

a) Only (1) is correct b) Only (2) is correct
c) Both are correct d) None of them is correct

205. Which one of the following pair of quantities has same dimension?

a) Force and work done b) Momentum and impulse
c) Pressure and force d) Surface tension and stress

206. If error in radius is 3%, what is error in volume of sphere?

a) 3% b) 27% c) 9% d) 6%

207. Which of the following relation is wrong

a) 1 ampere × 1 ohm = 1 volt b) 1 watt × 1 sec = 1 joule
c) 1 × newton per coulo = 1 volt per meter d) 1 coulomb × 1 volt = 1 watt

208. Let L denote the self-inductance of coil which is in series with a capacitor of capacitance C. Which of the following has the unit second?

a) $\sqrt{LC}$ b) C/L c) CL d) L^2/C^2

209. An important milestone in the evolution of the universe just after the Big Bang is the Planck time t_P, the value of which depends on three fundamental constants-speed c of light in vacuum, gravitational constant G and Planck's constant h. Then, $t_P \propto$

a) Ghc^5 b) $\frac{c^5}{Gh}$ c) $\frac{Gh}{c^5}$ d) $\left(\frac{Gh}{c^5}\right)^{1/2}$

210. One is equivalent to 931 MeV energy. The rest mass of electron is 9.1×10^{-31}kg. the Mass equivalent energy is

(1 amu= 1.67×10^{-17}kg)

a) 0.5073 MeV b) 0.693 MeV c) 4.0093 MeV d) None of these

211. Out of the following four dimensional quantities, which one qualifies to be called a dimensional constant?

a) Acceleration due to gravity b) Surface tension of water
c) Weight of a standard kilogram mass d) The velocity of light in vacuum

212. The unit of Stefan's constant is

a) $Wm^{-2}K^{-1}$ b) $Wm\,K^{-4}$ c) $Wm^{-2}K^{-4}$ d) $Nm^{-2}K^{-4}$

213. The correct value of $0°C$ on the Kelvin scale is

a) $273.15\,K$ b) $272.85\,K$ c) $273\,K$ d) $273.2\,K$

214. The velocity v of water waves may depend on their wavelength (λ), the density of water (ρ) and the acceleration due to gravity (g). The method of dimensions gives the relation between these quantities as

a) $v^2 \propto \lambda^{-1}\rho^{-1}$ b) $v^2 \propto g\lambda$ c) $v^2 \propto g\lambda\rho$ d) $g^{-1} \propto \lambda^3$

215. Which of the following is not the unit of time

a) Micro *second* b) Leap year c) Lunar month d) Parallactic *second*

216. The values of two resistors are $R_1 = (6 \pm 0.3)k\Omega$ and $R_2 = (10 \pm 0.2)k\Omega$. The percentage error in the equivalent resistance when they are connected in parallel is

a) 5.125% b) 2% c) 10.125% d) 7%

217. SI unit of permittivity is

a) $C^2m^2N^2$ b) $C^2m^{-2}N^{-1}$ c) $C^2m^2N^{-1}$ d) $C^{-1}m^2N^{-2}$

218. The speed (v) of ripples on the surface of water depends on surface tension (σ), density (ρ) and wavelength (λ). The square of speed (v) is proportional to

a) $\frac{\sigma}{\rho\lambda}$ b) $\frac{\rho}{\sigma\lambda}$ c) $\frac{\lambda}{\sigma\rho}$ d) $\rho\lambda\sigma$

219. The dimensions of $1/2\varepsilon E^2$ are same as

a) Energy density (energy per unit volume) b) Energy
c) Power d) None of the above

220. Which one of the following pairs of quantities and their unit is proper match?

a) Electric field-coulomb/m b) Magnetic flux-weber
c) Power-farad d) Capacitance-henry

221. The physical quantity which is not a unit of energy is

a) Volt-coulomb b) MeV-sec c) Henry $(\text{ampere})^2$ d) Farad-$(\text{volt})^2$

222. If the constant of gravitation (G), Plank's constant (h) and the velocity of light (c) be chosen as fundamental units. The dimension of the radius of gyration is

a) $h^{1/2}c^{-3/2}G^{1/2}$ b) $h^{1/2}c^{3/2}G^{1/2}$ c) $h^{1/2}c^{-3/2}G^{-1/2}$ d) $h^{-1/2}c^{-3/2}G^{1/2}$

223. Which of the following quantities is dimensionless

a) Gravitational constant b) Planck's constant
c) Power of a convex lens d) None

224. If the units of mass, length and time are doubled unit of angular momentum will be

a) Doubled b) Tripled
c) Quadrupled d) Eight times the original value

225. $\frac{h}{2\pi}$ is the dimension of

a) Velocity b) Momentum c) Energy d) Angular momentum

226. What is the dimensional formula of mc^2, where the letters have their usual meanings?

a) $[MLT^{-1}]$ b) $[ML^0T^0]$ c) $[ML^2T^{-2}]$ d) $[M^{-1}L^3T^6]$

227. Which is different from others by units

a) Phase difference b) Mechanical equivalent
c) Loudness of sound d) Poisson's ratio

228. Which of the following five physical parameters have the same dimensions
(A) Energy density (B) Refractive index (C) Dielectric constant (D) Young's modulus (E) Magnetic field

a) (A) and (D) b) (A) and (E) c) (B) and (D) d) (C) and (E)

229. Which of the following is not equal to *watt*

a) *joule/second* b) *ampere × volt* c) *(ampere)² × ohm* d) *ampere/volt*

230. The dimensions of $\frac{a}{b}$ in the equation $P = \frac{a-t^2}{bx}$, where P is pressure, x is distance and t is time, are

a) MT^{-2} b) M^2LT^{-3} c) ML^3T^{-1} d) LT^{-3}

231. Dimensions of impulse are same as that of

a) Force b) Momentum c) Energy d) Acceleration

232. $Dyne/cm^2$ is not a unit of

a) Pressure b) Stress c) Strain d) Young's modulus

233. Given $\pi = 3.14$. the value of π^2 with due regard for significant figures is

a) 9.86 b) 9.859 c) 9.8596 d) 9.85960

234. Which is the correct unit for measuring nuclear radii

a) *Micron* b) *Millimetre* c) *Angstrom* d) *Fermi*

235. The dimensional formula of electrical conductivity is

a) $[M^{-1}L^{-3}T^3A^2]$ b) $[ML^3T^3A^2]$ c) $[M^2L^3T^{-3}A^2]$ d) $[ML^3T^3A^{-2}]$

236. If the time period (T) of vibration of a liquid drop depends on surface tension (S), radius (r) of the drop and density (ρ) of the liquid, then the expression of T is

a) $T = k\sqrt{\rho r^3/S}$ b) $T = k\sqrt{\rho^{1/2}r^3/S}$ c) $T = k\sqrt{\rho r^3/S^{1/2}}$ d) None of these

237. The period of oscillation of a simple pendulum in the experiment is recorded as 2.63 s, 2.56 s, 2.42 s, 2.71 s and 2.80 s respectively. The average absolute error is

a) 0.1 s b) 0.11 s c) 0.01 s d) 1.0 s

238. Dimensional formula of Stefan's constant is

a) $MT^{-3}K^{-4}$ b) $ML^2T^{-2}K^{-4}$ c) ML^2T^{-2} d) $MT^{-2}L^0$

239. The dimensions of universal gas constant is

a) $[ML^2T^{-2}\theta^{-1}]$ b) $[M^2LT^{-2}\theta]$ c) $[ML^3T^{-1}\theta^{-1}]$ d) None of these

240. In the determination of Young's modulus $\left(Y = \frac{4\,MLg}{\pi\,\ell d^2}\right)$ by using Searle's method, a wire of length $L = 2m$ and diameter $d = 0.5mm$ is used. For a load $M = 2.5\ kg$, an extension $\ell = 0.25mm$ in the length of the wire is observed. Quantities d and ℓ are measured using a screw gauge and a micrometer, respectively. They have the same pitch of $0.5\ mm$. The number of divisions on their circular scale is 100. The contributions to the maximum probable scale is 100. The contributions to the maximum probable error of the Y measurement

a) Due to the errors in the measurements of d and ℓ are the same

b) Due to the error in the measurement of d is twice that due to the error in the measurement of ℓ

c) Due to the error in the measurement of ℓ is twice that due to the error in the measurement of d

d) Due to the error in the measurement of d is four time that due to the error in the measurement of ℓ

241. A screw gauge gives the following reading when used to measure the diameter of a wire.

Main scale reading : 0 mm

Circular scale reading : 52 divisions

Given that 1 mm on main scale corresponds to 100 divisions of the circular scale.

The diameter of wire from the above data is

a) 0.052 cm b) 0.026 cm c) 0.005 cm d) 0.52 cm

242. Out of the following pairs, which one does not have identical dimensions

a) Moment of inertia and moment of force b) Work and torque

c) Angular momentum and Planck's constant
d) Impulse and momentum

243. Out of following four dimensional quantities , which one quantity is to be called a dimensional constant
a) Acceleration due to gravity
b) Surface tension of water
c) Weight of a standard kilogram mass
d) The velocity of light in vacuum

244. Given, potential difference $V=(8 \pm 0.5)$ volt and current $I = (2 \pm 0.2)$A. The value of resistance R is
a) $4 \pm 16.25\%$
b) $4 \pm 6.25\%$
c) $4 \pm 10\%$
d) $4 \pm 8\%$

245. The density of a material in CGS system of units is $4g/cm^3$, In a system of units which unit of length is 10 cm and unit of mass is 100 g , the value of density of material will be
a) 400
b) 0.04
c) 0.4
d) 40

246. The density of the material of a cube is measured by measuring its mass and length of its side. If the maximum errors in the measurement of mass and the length are 3% and 2% respectively. The maximum error in the measurement of density is
a) 1%
b) 5%
c) 7%
d) 9%

247. The equation of state of some gases can be expressed as $\left(P + \frac{a}{V^2}\right) = \frac{R\theta}{V}$. Where P is the pressure, V the volume, θ the absolute temperature and a and b are constants. The dimensional formula of a is
a) $[ML^5T^{-2}]$
b) $[M^{-1}L^5T^{-2}]$
c) $[ML^{-1}T^{-2}]$
d) $[ML^{-5}T^{-2}]$

248. The only mechanical quantity which has negative dimension of mass is
a) Angular momentum
b) Torque
c) Coefficient of thermal conductivity
d) Gravitational constant

249. The dimensions of shear modulus are
a) MLT^{-1}
b) ML^2T^{-2}
c) $ML^{-1}T^{-2}$
d) MLT^{-2}

250. The values of two resistors are $R_1 = (6 \pm 0.3)$kΩ and $R_2 = (10 \pm 0.2)$kΩ. The percentage error in the equivalent resistance when they are connected in parallel is
a) 5.125%
b) 2%
c) 3.125%
d) 10.125%

251. In which of the following system of units, *weber* is the unit of magnetic flux
a) CGS
b) MKS
c) SI
d) None of these

252. A physical quantity is given by $X = [\mathrm{M}^a\mathrm{L}^b\mathrm{T}^c]$. The percentage error in measurement of M, Land T are $\alpha, \beta$and γ respectively. Then, the maximum % error in the quantity X is
a) $a\alpha + b\beta + c\gamma$
b) $a\alpha + b\beta - c\gamma$
c) $\frac{a}{\alpha} + \frac{b}{\beta} + \frac{c}{\gamma}$
d) None of these

253. If the velocity of light c, gravitational constant G and Planck's constant h are chosen as fundamental units, the dimensions of length L in the new system is
a) hcG^{-1}
b) $[h^{1/2}c^{1/2}G^{-1/2}]$
c) $[hc^{-3}G^1]$
d) $[h^{1/2}c^{-3/2}G^{1/2}]$

254. Which of the following physical quantities has neither dimensions nor unit?
a) Angle
b) Luminous intensity
c) Coefficient of friction
d) Current

255. The modulus of elasticity is dimensionally equivalent to
a) Strain
b) Force
c) Stress
d) Coefficient of viscosity

256. What is the area of a disc of radius 1.1 cm?
a) 3.8028571 cm^2
b) 3.8029 cm^2
c) 3.803 cm^2
d) 3.8 cm^2

257. The length of a cylinder is measured with a meter rod having least count 0.1 cm. Its diameter is measured with vernier callipers having least count 0.01cm. Given that length is 5.0 cm. and radius is 2.0cm. The percentage error in the calculated value of the volume will be
a) 1%
b) 2%
c) 3%
d) 4%

258. If C and R represent capacitance and resistance respectively, then the dimensions of RC are

a) $M^0L^0T^2$ b) $M^0L^0\,T$ c) ML^{-1} d) None of these above

259. In the following list, the only pair which have different dimensions, is

a) Linear momentum and moment of a force
b) Planck's constant and angular momentum
c) Pressure and modulus of electricity
d) Torque and potential energy

260. Dimensions of $\frac{1}{\mu_0\varepsilon_0}$, where symbols have their usual meaning, are

a) $[L^{-1}T]$ b) $[L^2T^2]$ c) $[L^2T^{-2}]$ d) $[LT^{-1}]$

261. In the relation $y = r\sin(\omega t - kx)$, the dimensions of ω/k are

a) $[M^0L^0T^0]$ b) $[M^0L^1T^{-1}]$ c) $[M^0L^0T^1]$ d) $[M^0L^1T^0]$

262. Electron – volt is the unit of energy ($1\text{ eV} = 1.6 \times 10^{-19}$J). in H-atom, the binding energy of electron in first orbit is 13.6 eV. The same in joule (J) is

a) 10×10^{-19}J b) 21.76×10^{-19}J c) 13.6×10^{-19}J d) None of these

263. The dimensions of pressure is equal to

a) Force per unit volume b) Energy per unit volume
c) Force d) energy

264. For the equation $F \propto A^a v^b d^c$, where F is the force, A is the area v is the velocity and d is the density, the value of a, b and c are respectively

a) 1,2,1 b) 2,1,1 c) 1,1,2 d) 0,1,1

265. The SI unit of universal gas constant (R) is

a) $Watt\ K^{-1}mol^{-1}$ b) $Newton\ K^{-1}mol^{-1}$ c) $Joule\ K^{-1}mol^{-1}$ d) $Erg\ K^{-1}mol^{-1}$

266. If E = energy, G = gravitational constant, I = impulse and M = mass, then dimensions of $\frac{GIM^2}{E^2}$ are same as that of

a) Time b) Mass c) Length d) Force

267. Dimensional formula for force is

a) $[M^1L^2T^{-2}]$ b) $[M^1L^1T^{-2}]$ c) $[M^1L^{-1}T^{-2}]$ d) $[M^1L^{-2}T^{-2}]$

268. Dimensions of the following three quantities are the same

a) Work, energy, force b) Velocity, momentum, impulse
c) Potential energy, kinetic energy, momentum d) Pressure, stress, coefficient of elasticity

269. A gas bubble from an explosion under water oscillates with a time period T, depends upon static pressure p, density of water ρ and the total energy of explosion E. Find the expression for the time period T.(where, k is a dimensionless constant)

a) $T = kp^{-5/6}\rho^{1/2}E^{1/3}$ a b) $T = kp^{-4/7}\rho^{1/2}E^{1/3}$ c) $T = kp^{-5/6}\rho^{1/2}E^{1/2}$ d) $T = kp^{-4/7}\rho^{1/3}E^{1/2}$

270. The dimensional formula of magnetic flux is

a) $[MLT^{-2}A^{-1}]$ b) $[ML^2T^{-1}A^{-1}]$ c) $[ML^2T^{-1}A^{-2}]$ d) $[ML^2T^{-2}A^{-1}]$

271. The dimensions of K in the equation $W = \frac{1}{2}Kx^2$ is

a) $M^1L^0T^{-2}$ b) $M^0L^1T^{-1}$ c) $M^1L^1T^{-2}$ d) $M^1L^0T^{-1}$

272. *Kilowatt – hour*is a unit of

a) Electrical charge b) Energy c) Power d) Force

273. If $L = 2.331\ cm, B = 2.1\ cm$, then $L + B =$

a) 4.431 cm b) 4.43 cm c) 4.4 cm d) 4 cm

274. The dimensions of coefficient of thermal conductivity is

a) $ML^2T^{-2}K^{-1}$ b) $MLT^{-3}K^{-1}$ c) $MLT^{-2}K^{-1}$ d) $MLT^{-3}K$

275. Which of the following is a derived unit

a) Unit of mass b) Unit of length c) Unit of time d) Unit of volume

276. The density of a solid ball is to be determined in an experiment. The diameter of the ball is measured with a screw gauge, whose pitch is 0.5 mm and there are 50 divisions on the circular scale. The reading on the main scale is 2.5 mm and that on the circular scale is 20 divisions. If the measured mass of the ball has a relative error of 2%, the relative percentage error in the density is

a) 0.9% b) 2.4% c) 3.1% d) 4.2%

277. Density of a liquid in CGS system is $0.625\ g/cm^3$. What is its magnitude in SI system

a) 0.625 b) 0.0625 c) 0.00625 d) 625

278. The dimensions of magnetic field in M, L, T and C (coulomb) is given as

a) $[MLT^{-1}C^{-1}]$ b) $[MT^2C^{-2}]$ c) $[MT^{-1}C^{-1}]$ d) $[MT^{-2}C^{-1}]$

279. Which one of the following represents the correct dimensions of the coefficient of viscosity?

a) $[ML^{-1}T^{-2}]$ b) $[MLT^{-1}]$ c) $[ML^{-1}T^{-1}]$ d) $[ML^{-2}T^{-2}]$

280. When a wave traverses a medium, the displacement of a particle located at x at a time t is given by $y = a\sin(bt - cx)$, where a, b and c are constants of the wave. Which of the following is a quantity with dimensions

a) $\frac{y}{a}$ b) bt c) cx d) $\frac{b}{c}$

281. Energy per unit volume represents

a) Pressure b) Force c) Thrust d) Work

282. The unit of self inductance of a coil is

a) *Farad* b) *Henry* c) *Weber* d) *Tesla*

283. The length, breadth and thickness of a metal block is given by $l = 90$ cm, $b = 8$ cm, $t = 2.45$ cm. The volume of the block is

a) $2 \times 10^2 cm^3$ b) $1.8 \times 10^2 cm^3$ c) $1.77 \times 10^2 cm^3$ d) $1.764 \times 10^2 cm^3$

284. One slug is equivalent to 14.6 kg. A force of 10 pound is applied on a body of 1 kg. The acceleration of the body is

a) $44.5\ ms^{-2}$ b) $4.448\ ms^{-2}$ c) $44.4\ ms^{-2}$ d) None of these

285. A physical quantity A is related to four observations a, b, c and d as follows, $A = \frac{a^2b^3}{c\sqrt{d}}$. The percentage error of measurement in a, b, c and d are 1%, 3%, 2% and 2% respectively. What is the percentage error in the quantity A

a) 12% b) 7% c) 5% d) 14%

286. The SI unit of electrochemical equivalent is

a) Kg C b) $C\ kg^{-1}$ c) $Kg\ C^{-1}$ d) kg^2C^{-1}

287. The position of a particle at time t is given by the equation $x(t) = \frac{v_0}{A}(1 - e^{At})$, $v_0 =$ constant and $A > 0$. Dimensions of v_0 and A respectively are

a) $[M^0LT^0]$ and $[M^0L^0T^{-1}]$ b) $[M^0LT^{-1}]$ and $[M^0LT^{-2}]$

c) $[M^0LT^{-1}]$ and $[M^0L^0T]$ d) $[M^0LT^{-1}]$ and $[M^0L^0T^{-1}]$

288. Dimensions of magnetic field intensity is

a) $[M^0L^{-1}T^0A^1]$ b) $[MLT^{-1}A^{-1}]$ c) $[ML^0T^{-2}A^{-1}]$ d) $[MLT^{-2}A]$

289. If K denotes coefficient of thermal conductivity, d the density and c the specific heat, the unit of X, where $X = K/dc$ will be

a) $cm\ sec^{-1}$ b) $cm^2\ sec^{-2}$ c) cm sec d) $cm^2\ sec^{-1}$

290. The period of oscillation of a simple pendulum is given by $T = 2\pi\sqrt{\frac{l}{g}}$ where l is about 100 cm and is known to have 1mm accuracy. The period is about 2s. The time of 100 oscillations is measured by a stop watch of least count 0.1 s. The percentage error in g is

a) 0.1% b) 1% c) 0.2% d) 0.8%

291. A weber is equivalent to

a) $A\ m^{-2}$ b) $A\ m^{-1}$ c) $A\ m^2$ d) $T\ m^2$

292. How many wavelengths of the Kr^{89} are there in one metre?
a) 658189.63 b) 2348123.73 c) 1650763.73 d) 1553164.12

293. If L and R are respectively the inductance and resistance, then the dimensions of $\frac{L}{R}$ will be
a) $M^0L^0T^{-1}$
b) M^0LT°
c) $M\ {}^0L^0T$
d) Cannot be represented in terms of M, L and T

294. If energy (E), velocity (v) and force (F) be taken as fundamental quantity, then what are the dimensions of mass
a) Ev^2 b) Ev^{-2} c) Fv^{-1} d) Fv^{-2}

295. The physical quantities not having same dimensions are
a) Torque and work
b) Momentum and Planck's constant
c) Stress and Young's modules
d) Speed and $(\mu_0\varepsilon_0)^{-1/2}$

296. The Martians use force (F), acceleration (A) and time (T) as their fundamental physical quantities. The dimensions of length on Martians system are
a) FT^2 b) $F^{-1}T^2$ c) $F^{-1}A^2T^{-1}$ d) AT^2

297. In the gas equation $\left(p+\frac{a}{V^2}\right)(V-b)=RT$, the dimensions of a are
a) $[ML^3T^{-2}]$ b) $[M^{-1}L^3T^{-1}]$ c) $[ML^5T^{-2}]$ d) $[M^{-1}L^{-5}T^2]$

298. The dimension of the ratio of angular to linear momentum is
a) $M^0L^1T^0$ b) $M^1L^1T^{-1}$ c) $M^1L^2T^{-1}$ d) $M^{-1}L^{-1}T^{-1}$

299. If the acceleration due to gravity is 10 ms^{-2} and the units of length and time are changed in kilometer and hour respectively, the numerical value of the acceleration is
a) 360000 b) 72,000 c) 36,000 d) 129600

300. The dimensions of $\frac{R}{L}$ are
[here, R =electric resistance, L =self inductance]
a) $[T^{-2}]$ b) $[T^{-1}]$ c) $[ML^{-1}]$ d) [T]

301. The number of significant figures in all the given numbers 25.12, 2009, 4.156 and 1.217×10^{-4} is
a) 1 b) 2 c) 3 d) 4

302. The unit of momentum is
a) N s b) Ns^{-1} c) N m d) N m^{-1}

303. The maximum static friction on a body is $F=\mu N$. Here, N = normal reaction force on the body μ = coefficient of static friction. The dimensions of μ are
a) $[MLT^{-2}]$ b) $[M^0L^0T^0\theta^{-1}]$ c) Dimensionless d) None of these

304. The speed of light (c), gravitational constant (G) and Planck's constant (h) are taken as the fundamental units in a system. The dimension of time in this new system should be
a) $G^{1/2}h^{1/2}c^{-5/2}$ b) $G^{-1/2}h^{1/2}c^{1/2}$ c) $G^{1/2}h^{1/2}c^{-3/2}$ d) $G^{1/2}h^{1/2}c^{1/2}$

305. Dimensions of kinetic energy are
a) ML^2T^{-2} b) M^2LT^{-1} c) ML^2T^{-1} d) ML^3T^{-1}

306. The air bubble formed by explosion inside water performed oscillation with time period T that is directly proportional to $p^ad^bE^c$, where p is the pressure, d is the density and E is the energy due to explosion. The values of a, b and c will be
a) -5/6, 1/2, 1/3 b) 5/6, 1/3, 1/2 c) 5/6, 1/2, 1/3 d) None of these

307. Identify the pair whose dimensions are equal
a) Torque and work b) Stress and energy c) Force and stress d) Force and work

308. Which of the following cannot be regarded as an essential characteristic of a unit of measurement?
a) Inaccessibility
b) Indenstructibility
c) Invariability
d) Reproductibility

309. What is the power of a 100 W bulb in CGS units?
a) $10^6 ergs^{-1}$ b) $10^7 ergs^{-1}$ c) $10^9 ergs^{-1}$ d) $10^{11} ergs^{-1}$

310. The pressure on a square plate is measured by measuring the force on the plate and the length of the sides of the plate by using the formula $p = \frac{F}{l^2}$. If the maximum errors in the measurement of force and length are 4% and 2% respectively, then the maximum error in the measurement of pressure is
a) 1% b) 2% c) 8% d) 10%

311. How many wavelengths of Kr^{86} are there in one *metre*
a) 1553164.13 b) 1650763.73 c) 652189.63 d) 2348123.73

312. One femtometre is equivalent to
a) 10^{15} m b) 10^{-15} m c) 10^{-12} m d) 10^{12} m

313. A rectangular beam which is supported at its two ends and leaded in the middle with weight w sags by an amount δ such that $\delta = \frac{wl^3}{4Yd^3}$, where l, d and Y represent length, depth and elasticity respectively.
Guess the unknown factor using dimensional considerations
a) Breadth b) $(\text{breadth})^2$ c) $(\text{breadth})^3$ d) Mass

314. The unit of L/R is (where L =inductance and R = Resistance)
a) *Sec* b) Sec^{-1} c) *Volt* d) *Ampere*

315. The dimensions of *calorie* are
a) ML^2T^{-2} b) MLT^{-2} c) ML^2T^{-1} d) ML^2T^{-3}

316. The dimensions of potential are the same as that of
a) Work b) Electric field per unit charge
c) Work per unit charge d) Force per unit charge

317. If the units of M and L are increased three times, then the unit of energy will be increased by
a) 3 times b) 6 times c) 27 times d) 81 times

318. Dimensions of resistance in an electrical circuit, in terms of dimension of mass M, of length L, of time T and current I, would be
a) $[ML^2T^{-3}I^{-1}]$ b) $[ML^2T^{-2}]$ c) $[ML^2T^{-1}I^{-1}]$ d) $[ML^2T^{-3}I^{-2}]$

319. An athletic coach told his team that muscle times speed equals power. What dimensions does he view for muscle
a) MLT^{-2} b) ML^2T^{-2} c) MLT^2 d) L

320. The following observations were take for determining surface tension of water by capillary tube method. Diameter of capillary, $D = 1.25 \times 10^{-2}$ m and rise of water in capillary. $h = 1.46 \times 10^{-2}$m
Taking $g = 9.80 ms^{-2}$ and using the relation $T = (rgh/2) \times 103 Nm^{-1}$, what is the possible error in surface tension$T$?
a) 2.4% b) 15% c) 1.6% d) 0.15%

321. $X = 3YZ^2$ find dimension of Y in (MKSA) system, If X and Z are the dimensions of capacity and magnetic field respectively
a) $M^{-3}L^{-2}T^{-4}A^{-1}$ b) ML^{-2} c) $M^{-3}L^{-2}T^4A^4$ d) $M^{-3}L^{-2}T^8A^4$

322. Student I, II and III perform an experiment for measuring the acceleration due to gravity (g) using a simple pendulum. They use different lengths of the pendulum and/or record time for different number of oscillations. The observations are shown in the table..
Least count for length = 0.1 cm.
Least count for time = 0.1 s.

Student	Length of the pendulum (cm)	Number of oscillations(n)	Total time for (n) oscillations (s)	Time period (s)
I	64.0	8	128.0	16.0
II	64.0	4	64.0	16.0
III	20.0	4	36.0	9.0

If E_1, E_{II} and E_{III} are the percentage errors in g, *ie*, $\left(\frac{\Delta g}{g} \times 100\right)$, for students I, II and III respectively.

a) $E_I = 0$
b) E_I is minimum
c) $E_I = E_{II}$
d) E_{II} is maximum

323. The physical quantity which has dimensional formula as that of $\frac{\text{Energy}}{\text{Mass} \times \text{Length}}$ is

a) Force b) Power c) Pressure d) Acceleration

324. Young's modulus of a material has the same units as

a) Pressure b) Strain c) Compressibility d) Force

325. The dimensional formula for $r.m.s.$ (root mean square) velocity is

a) M^0LT^{-1} b) $M^0L^0T^{-2}$ c) $M^0L^0T^{-1}$ d) MLT^{-3}

326. Dimensions of potential energy are

a) MLT^{-1} b) ML^2T^{-2} c) $ML^{-1}T^{-2}$ d) $ML^{-1}T^{-2}$

327. The unit of $e.m.f.$ is

a) $Joule$ b) $Joule - coulomb$ c) $Volt - coulomb$ d) $Joule/coulomb$

328. A student performs an experiment for determination of $g = \frac{4\pi^2 l}{T^2}$ and he commits an error of Δl. For that he takes the time of n oscillations with the stop watch of least count ΔT and he commits a human error of $0.1\ sec$. For which of the following data, the measurement of g will be most accurate

$\Delta l \Delta T n$ Ampli. of oscill.

a) $5\ mm 0.2 sec 10 5 mm$
b) $5\ mm 0.2 sec 20 5 mm$
c) $5\ mm 0.1 sec 20 1 mm$
d) $1\ mm 0.1 sec 50 1 mm$

329. If C is capacitance and q is charge, then the dimension of q^2/C is same as that of

a) Work b) Angular momentum c) Force d) Torque

330. Error in the measurement of radius of sphere is 2%. The error in the measurement of volume is

a) 1% b) 5% c) 3% d) 6%

331. Curie is a unit of

a) Energy of γ-rays b) Half life c) Radioactivity d) Intensity of γ-rays

332. Which has not the same unit as other?

a) Watt-sec b) Kilowatt-hour c) eV d) Js

333. The length of a cube is 2.1×10^{-2}m. the volume in significant figures will be

a) $9.2 \times 10^{-6}\text{m}^3$ b) $9.3 \times 10^{-6}\text{m}^3$ c) $9.26 \times 10^{-6}\text{m}^3$ d) $9.261 \times 10^{-6}\text{m}^3$

334. Position of a body with acceleration a is given by $x = ka^m t^n$.Here t is time. Find the dimensions of m and n.

a) $m = 1, n = 1$ b) $m = 1, n = 2$ c) $m = 2, n = 1$ d) $m = 2, n = 2$

335. Which one of the following units is not that of mutual inductance?

a) Henry
b) $(\text{Weber})^{-1}$
c) Ohm second
d) Volt second $(\text{ampere})^{-1}$

336. Two quantities A and B are related by the relation $\frac{A}{B}$=m, where m is linear mass den sity and A is force. The dimensions of B will be
a) Mass as that of latent heat
b) Same a that of pressure
c) Same as that of work
d) Same as that of momentum

337. The dimensions of electric potential are
a) $[ML^2T^{-2}Q^{-1}]$
b) $[MLT^{-2}Q^{-1}]$
c) $[ML^2T^{-1}Q]$
d) $[ML^2T^{-2}Q]$

338. The dimensions of gravitational constant G and the moment of inertia are respectively
a) $[ML^3T^{-2}]$; $[ML^2T^0]$
b) $[M^{-1}L^3T^{-2}]$; $[ML^2T^0]$
c) $[M^{-1}L^3T^{-2}]$; $[M^{-1}L^2T]$
d) $[ML^3T^{-2}]$; $[M^{-1}L^2T]$

339. The unit of absolute permittivity is
a) Fm $(farad - metre)$
b) Fm^{-1} $(farad/metre)$
c) Fm^{-2} $(farad/metre^2)$
d) F (farad)

340. $[ML^2L^{-2}]$ are dimensions of
a) Force
b) Moment of force
c) Momentum
d) Power

341. the dimensional formula of latent heat is
a) $[M^0L^2T^{-2}]$
b) $[MLT^2]$
c) $[ML^2T^{-2}]$
d) $[MLT^{-1}]$

342. Which of the following is not the unit of energy
a) *Calorie*
b) *Joule*
c) *Electron volt*
d) *Watt*

343. The unit of nuclear dose given to a patient is
a) *Fermi*
b) Rutherford
c) Curie
d) Roentgen

344. The least count of a stop watch is 0.2 s. The time of 20 oscillations of a pendulum is measured to be 25 s. The percentage error in the measurement of time will be
a) 8%
b) 1.8%
c) 0.8%
d) 0.1%

345. The dimensions of resistance are same as those of .... Where h is the Planck's constant and e is the charge.
a) $\frac{h^2}{e^2}$
b) $\frac{h^2}{e}$
c) $\frac{h}{e^2}$
d) $\frac{h}{e}$

346. The dimensional formula of magnetic permeability is
a) $[M^0L^{-1}T]$
b) $[M^0L^2T^{-1}]$
c) $[M^0L^2T^{-1}A^2]$
d) $[MLT^{-2}A^{-2}]$

347. The dimensional formula for young's modulus is
a) $ML^{-1}T^{-2}$
b) M^0LT^{-2}
c) MLT^{-2}
d) ML^2T^{-2}

348. $ML^3T^{-1}Q^{-2}$ is dimensions of
a) Resistivity
b) Conductivity
c) Resistance
d) None of these

349. The relative density of the material of a body I the ratio of its weight in air and the loss of its weight in water. By using a spring balance, the weight of the body in air in measured to be 5.00 $\pm$0.05N. The weight of the body in water is measured to be 4.00 $\pm$ 0.05N. Then the maximum possible percentage error in relative density is
a) 11%
b) 10%
c) 9%
d) 7%

350. If the unit of length and force be increased four times, then the unit of energy is
a) Increased 4 times
b) Increased 8 times
c) Increased 16 times
d) Decreased 16 times

351. The physical quantity which has the dimensional formula $[M^1T^{-3}]$ is
a) Surface tension
b) Density
c) Solar constant
d) Compressibility

352. Unit of surface tension is
a) Nm^{-1}
b) Nm^{-2}
c) N^2m^{-1}
d) Nm^{-3}

353. The magnetic force on a point moving charge is $\vec{F} = q(\vec{V} \times \vec{B})$.
Here, q =electric charge
$\vec{V}$ = velocity of the point charge
$\vec{B}$ =magnetic field
The dimensions of $\vec{B}$
a) $[MLT^{-1}A]$
b) $[MLT^{-2}A^{-1}]$
c) $[MT^{-1}A^{-1}]$
d) None of these

354. IF L, C and R denote the inductance, capacitance and resistance respectively, the dimensional formula for C^2LR is

a) $[ML^{-2}T^{-1}I^0]$ b) $[M^0L^0T^3I^0]$ c) $[M^{-1}L^{-2}T^6I^2]$ d) $[M^0L^0T^2I^0]$

355. The dimensions of $e^2/4\pi\varepsilon_0 hc$, where e, ε_0, h and c are electronic charge, electric permittivity, Planck's constant and velocity of light in vacuum respectively, are

a) $[M^0L^0T^0]$ b) $[ML^0T^0]$ c) $[M^0LT^0]$ d) $[M^0L^0T^1]$

356. The period of a body under SHM is represented by $T = P^aD^bS^c$; where P is pressure, D is density and S is surface tension. The value of a, b and c are

a) $-\frac{3}{2}, \frac{1}{2}, 1$ b) $-1, -2, 3$ c) $\frac{1}{2}, -\frac{3}{2}, -\frac{1}{2}$ d) $1, 2, \frac{1}{3}$

357. *Faraday* is the unit of

a) Charge b) Emf c) Mass d) Energy

358. If $X = A \times B$ and ΔX and ΔA and ΔB are maximum absolute errors in X, A and B respectively, then the maximum relative error in X is given by

a) $\Delta X = \Delta A + \Delta B$ b) $\Delta X = \Delta A - \Delta B$ c) $\frac{\Delta X}{X} = \frac{\Delta A}{A} - \frac{\Delta B}{B}$ d) $\frac{\Delta X}{X} = \frac{\Delta A}{A} + \frac{\Delta B}{B}$

359. The dimensions of Planck's constant are

a) $[M^2L^2T^{-2}]$ b) $[MLT^{-2}]$ c) $[ML^2T^{-2}]$ d) $[ML^2T^{-1}]$

360. A student performs an experiment to determine the Young's modulus of a wire, exactly 2 m long, by Searle's method. In a particular reading, the student measures the extension in the length of the wire to be 0.8 mm with an uncertainty of $\pm$ 0.05 mm at a load of exactly 1.0 kg. The student also measures the diameter of the wire to be 0.4mm with an uncertainty of $\pm$ 0.01 mm. Take $g = 9.8$ ms^{-2} (exact). The Young's modulus obtained from the reading is

a) $(2.0 \pm 0.3) \times 10^{11}$Nm^{-2} b) $(2.0 \pm 0.2) \times 10^{11}$Nm^{-2}
c) $(2.0 \pm 0.1) \times 10^{11}$Nm^{-2} d) $(2.0 \pm 0.05) \times 10^{11}$Nm^{-2}

361. Tesla is a unit for measuring

a) Magnetic moment b) Magnetic induction
c) Magnetic intensity d) Magnetic pole strength

362. The mass of a box is 2.3 g. Two gold pieces, each of mass 0.035 g, are placed in it. The total mass of the box and gold pieces is

a) 2.3 g b) 2.4 g c) 2.37 g d) 2.370 g

363. The dimensions of $\frac{a}{b}$ in the equation $p = \frac{a - t^2}{bx}$ where p is pressure, x is distance and t is time, are

a) $[M^2LT^{-3}]$ b) $[MT^{-2}]$ c) $[LT^{-3}]$ d) $[ML^3T^{-1}]$

364. If velocity v, acceleration A and force F are chosen a fundamental quantities, then the dimensional formula of angular momentum in terms of v, A and F would be

a) $FA^{-1}v$ b) Fv^3A^{-2} c) Fv^2A^{-1} d) $F^2v^2A^{-1}$

365. $[ML^2T^{-3}A^{-2}]$ is the dimensional formula of

a) Electric resistance b) Capacity c) Electric potential d) Specific resistance

366. The dimensions of CV^2 matches with the dimensions of

a) L^2I b) L^2I^2 c) LI^2 d) $\frac{1}{LI}$

367. To determine the Young's modulus of a wire, the formula is $Y = \frac{F}{A} \times \frac{L}{\Delta L}$; where L = length, A = area of cross-section of the wire, ΔL = change in lengthof the wire when stretched with a force F. The conversion factor to change it from CGS to MKS system is

a) 1 b) 10 c) 0.1 d) 0.01

368. If C and L denote capacitance and inductance respectively, then the dimensions of LC are

a) $M^0L^0T^0$ b) $M^0L^0T^2$ c) $M^2L^0T^2$ d) MLT^2

369. The potential energy of a particle varies with distance x from a fixed origin as $U = \left(\frac{A\sqrt{X}}{x+B}\right)$; where A and B are constants. The dimensions of AB are

a) $[ML^{5/2}T^{-2}]$ b) $[ML^2T^{-2}]$ c) $[M^{3/2}L^{3/2}T^{-2}]$ d) $[ML^{7/2}T^{-2}]$

370. Dimensional formula for torque is

a) L^2MT^{-2} b) $L^{-1}MT^{-2}$ c) L^2MT^{-3} d) LMT^{-2}

371. Dimensions of CR are those of

a) Frequency b) Energy c) Time period d) Current

372. The dimensions of inter atomic force constant are

a) MT^{-2} b) MLT^{-1} c) MLT^{-2} d) $ML^{-1}T^{-1}$

373. The ratio of 1 kWh to 1 MeV is

a) 2.25×10^{17} b) 2.25×10^{19} c) 2.25×10^{23} d) $2.25 \times 4.4 \times 10^9$

374. The physical quantity that has no dimensions is

a) Angular Velocity b) Linear momentum c) Angular momentum d) Strain

375. Out of the following pairs, which one does not have identical dimensions?

a) Angular momentum and Planck's constant b) Impulse and momentum

c) Moment of inertia and moment of a force d) Work and torque

376. A physical quantity is given by $X = M^aL^bT^c$. The percentage error in measurement of M, L and T are α, β and γ respectively. The maximum percentage error in the quantity X is

a) $a\alpha + b\beta + c\gamma$ b) $a\alpha + b\beta - c\gamma$ c) $\frac{a}{\alpha}+\frac{b}{\beta}+\frac{c}{\gamma}$ d) None of these

377. $[ML^{-2}T^{-2}]$represents dimensional formula of which of the following physical quantities?

a) Energy b) pressure c) Torque d) Pressure gradient

378. Which of the following pairs does not have similar dimensions

a) Stress and pressure b) Angle and strain

c) Tension and surface tension d) Planck's constant and angular momentum

379. Density of liquid in CGS system is $0.625\ \text{g cm}^{-3}$. What is its magnitude in SI system?

a) 0.625 b) 0.0625 c) 0.00625 d) 625

380. Taking frequency f, velocity v and density ρ to be the fundamental quantities, then the dimensional formula for momentum will be

a) $[\rho v^4 f^{-3}]$ b) $[\rho v^3 f^{-1}]$ c) $[\rho v f^2]$ d) $[\rho^2 v^2 f^2]$

381. Oersted is a unit of

a) Dip b) Magnetic intensity c) Magnetic moment d) Pole strength

382. One nanometre is equal to

a) $10^9 mm$ b) $10^{-6} cm$ c) $10^{-7} cm$ d) $10^{-9} cm$

383. If the length of rod A is $3.25 \pm 0.01\ cm$ and that of B is $4.19 \pm 0.01\ cm$ then the rod B is longer than rod A by

a) $0.94 \pm 0.00\ cm$ b) $0.94 \pm 0.01\ cm$ c) $0.94 \pm 0.02\ cm$ d) $0.94 \pm 0.005\ cm$

384. If force (F), length (L) and time (T) are assumed to be fundamental units, then the dimensional formula of the mass will be

a) $FL^{-1}T^2$ b) $FL^{-1}T^{-2}$ c) $FL^{-1}T^{-1}$ d) FL^2T^2

385. The dimensions of surface tension are

a) $ML^{-1}T^{-2}$ b) MLT^{-2} c) $ML^{-1}T^{-1}$ d) MT^{-2}

386. Unit of power is

a) *Kilowatt* b) *Kilowatt-hour* c) *Dyne* d) *Joule*

387. Which of the following is dimensionally correct

a) Pressure = Energy per unit area
b) Pressure = Energy per unit volume
c) Pressure = Force per unit volume
d) Pressure = Momentum per unit volume per unit time

388. The sides of a rectangle are 6.01 m and 12m. taking the significant figures into account, the area of the rectangle is

a) 7.2 m^2 b) 72.1 m^2 c) 72.00 m^2 d) 72.12 m^2

389. The unit of angular acceleration in the SI system is

a) $N\,kg^{-1}$ b) ms^{-2} c) $rad\ s^{-2}$ d) $m\,kg^{-1}K$

390. If the unit of force is 1 kN, the length is 1 km and time is 100 s, what will be the unit of mass?

a) 1 kg b) 100 kg c) 1000 kg d) 10000 kg

391. Two full turns of the circular scale of a screw gauge cover a distance of 1 mm on its main scale. The total number of divisions on the circular scale is 50. Further, it is found that the screw gauge has a zero error of −0.03 mm. While measuring the diameter of a thin wire, a student notes the main scale reading of 3 mm and the number of circular scale divisions in line with the main scale as 35. The diameter of the wire is

a) 3.73 mm b) 3.67 mm c) 3.38 mm d) 3.32 mm

392. $[ML^3T^{-1}Q^{-2}]$ is the dimensional formula of

a) Resistance b) Resistivity c) Conductance d) Conductivity

393. The damping force of an oscillating particle is observed to be proportional to velocity. The constant of proportionality can be measured in

a) Kg s^{-1} b) Kg s c) Kg ms^{-1} d) Kg $m^{-1}s^{-1}$

394. A physical quantity is represented by $X = M^a L^b T^{-c}$. If percentage errors in the measurements of M, L and T are $\alpha\%, \beta\%$and $\gamma\%$ respectively, then total, percentage error is

a) $(\alpha a + \beta b - \gamma c)\%$ b) $(\alpha a + \beta b + \gamma c)\%$ c) $(\alpha a - \beta b - \gamma c)\%$ d) 0%

395. The value of 0.99−0.989 is

a) 0.001 b) 0.010×10^{-1} c) 0.01×10^{-1} d) 0.1×10^{-3}

396. The dimensional formula for areal velocity is

a) $[M^0L^{-2}T]$ b) $[M^0L^{-2}T^{-1}]$ c) $[M^0L^2T^{-1}]$ d) $[M^0L^2T]$

397. Temperature can be expressed as a derived quantity in terms of any of the following

a) Length and mass b) Mass and time
c) Length, mass and time d) None of these

398. Unit of stress is

a) N/m b) $N-m$ c) N/m^2 d) $N-m^2$

399. If $f = x^2$, then the relative error in f is

a) $\frac{2\Delta x}{x}$ b) $\frac{(\Delta x)^2}{x}$ c) $\frac{\Delta x}{x}$ d) $(\Delta x)^2$

400. If the velocity of light (c), gravitational constant (G) and Planck's constant (h) are chosen as fundamental units, then the dimensions of mass in new system is

a) $c^{1/2}G^{1/2}h^{1/2}$ b) $c^{1/2}G^{1/2}h^{-1/2}$ c) $c^{1/2}G^{-1/2}h^{1/2}$ d) $c^{-1/2}G^{1/2}h^{1/2}$

401. The dimensions of resistivity in terms of $M, L, T$and$Q$where Qstands for the dimensions of charge, is

a) $ML^3T^{-1}Q^{-2}$ b) $ML^3T^{-2}Q^{-1}$ c) $ML^2T^{-1}Q^{-1}$ d) $MLT^{-1}Q^{-1}$

402. A spectrometer gives the following reading when used to measure the angle of a prism
Main scale reading : 58.5 degree
Vernier scale reading : 09 divisions
Given that 1 division on main scale corresponds to 0.5 degree. Total divisions on the vernier scale is 30 and match with 29 divisions of the main scale. The angle of the prism from the above data

a) 58.59 *Degree* b) 58.77 *Degree* c) 58.65 *Degree* d) 59 *Degree*

403. Ampere-hour is the unit of

a) Quantity of charge b) Potential c) Energy d) Current

404. A physical quantity u is given by the relation $u = \frac{B^2}{2\mu_0}$

here, B =magnetic field strength

μ_0 =magnetic permeability of vacuum.

a) Energy b) Energy density c) Pressure d) None of these

405. Which does not have the same unit as others

a) *Watt-sec* b) *Kilowatt-hour* c) *eV* d) *J-sec*

406. The radius of a wire is 0.24 mm. Then its area of cross section by taking significant figures into consideration is

a) 0.1 mm^2 b) 0.2 mm^2 c) 0.18 mm^2 d) 0.180 mm^2

407. The expression $[ML^{-1}T^{-1}]$ represents

a) Momentum b) Force

c) Pressure d) Coefficient of viscosity

408. In an experiment the angles are required to be measured using an instrument. 29 divisions of the main scale exactly coincide with the 30 divisions of the vernier scale. If the smallest division of the main scale is half-a-degree (= 0.5°), then the least count of the instrument is

a) One minute b) Half minute c) One degree d) Half degree

409. The length, breadth and thickness of a block is measured to be 50 cm, 2.0 cm and 1.00 cm. The percentage error in the measurement of volume is

a) 0.8 % b) 8% c) 10% d) 12.5%

410. Given that v is speed, r is the radius and g is the acceleration due to gravity. Which of the following is dimensionless

a) v^2/rg b) v^2r/g c) v^2g/r d) v^2rg

411. The focal length of a mirror is given by $\frac{1}{f} = \frac{1}{u} + \frac{1}{v}$ where u and v represent object and image distances respectively. The maximum relative error in f is

a) $\frac{\Delta f}{f} = \frac{\Delta u}{u} + \frac{\Delta v}{v}$ b) $\frac{\Delta f}{f} = \frac{1}{\Delta u/u} + \frac{1}{\Delta v/v}$

c) $\frac{\Delta f}{f} = \frac{\Delta u}{u} + \frac{\Delta v}{v} - \frac{\Delta(u+v)}{u+v}$ d) $\frac{\Delta f}{f} = \frac{\Delta u}{u} + \frac{\Delta v}{v} + \frac{\Delta u}{u+v} + \frac{\Delta v}{u+v}$

412. The equation of alternating current is $I = I_0 e^{-t/CR}$, where t is time, C is capacitance and R is resistance of coil, then the dimensions of C R is

a) $[MLT^{-1}]$ b) $[M^0LT]$ c) $[M^0L^0T]$ d) None of these

413. The refractive index of a material is given by the equation $n = \frac{A+B}{\lambda^2}$, where A and B are constant. The dimensional formula for B is

a) $[M^0L^2T^{-1}]$ b) $[M^0L^{-2}T^0]$ c) $[M^0L^2T^{-2}]$ d) $[M^0L^2T^0]$

414. There are atomic clocks capable of measuring time with an accuracy of 1 part in 10^{11}. If two such clocks are operated with precision, then after running for 5000 years, these will record

a) A difference of nearly 1 s b) A difference of 1 day

c) A difference of 10^{11} s d) A difference of 1 year

415. R, L and Crepresent the physical quantities resistance, inductance and capacitance respectively. Which one of the following combination has dimension of frequency?

a) $\frac{1}{\sqrt{RC}}$ b) $\frac{R}{L}$ c) $\frac{1}{LC}$ d) $\frac{C}{L}$

416. Wavelength of ray of light is 0.00006 m. It is equal to

a) 6 *micron* b) 60 *micron* c) 600 *micron* d) 0.6 *micron*

417. Dimensions of permeability are

a) $A^{-2}M^1L^1T^{-2}$ b) MLT^{-2} c) ML^0T^{-1} d) $A^{-1}MLT^2$

418. If Ldenotes the inductance of an inductor through which a current i is flowing, the dimensions of Li^2 are

a) ML^2T^{-2}
b) Not expressible in MLT
c) MLT^{-2}
d) $M^2L^2T^{-2}$

419. Ins is defined as

a) 10^{-9} s of Kr – clock of 1650763.73 oscillations
b) 10^{-9} s of Kr – clock of 6521389.63 oscillations
c) 10^{-9} s of Cs – clock of 1650763.73 oscillations
d) 10^{-9} s of Cs – clock of 9192631770 oscillations

420. Density of wood is 0.5 gm/cc in the CGS system of units. The corresponding value in MKS units is

a) 500
b) 5
c) 0.5
d) 5000

421. Unit of moment of inertia in MKS system

a) $kg \times cm^2$
b) kg/cm^2
c) $kg \times m^2$
d) $Joule \times m$

422. The percentage error in the above problem is

a) 7%
b) 5.95%
c) 8.95%
d) 9.85%

423. Force constant has the same dimensions as

a) Coefficient of viscosity
b) Surface tension
c) Frequency
d) Impulse

424. A physical quantity P is given by $P = \frac{A^3 B^{\frac{1}{2}}}{C^{-4} D^{\frac{3}{2}}}$. The quantity which brings in the maximum percentage error in P is

a) A
b) B
c) C
d) D

425. The surface tension is $T = \frac{F}{l}$, then the dimensions of surface tension are

a) $[MLT^{-2}]$
b) $[MT^{-2}]$
c) $[M^0L^0T^0]$
d) None of these

426. What is dimensional formula of thermal conductivity?

a) $[MLT^{-1}\theta^{-1}]$
b) $[MLT^{-3}\theta^{-1}]$
c) $[M^2LT^{-3}\theta^{-2}]$
d) $[ML^2T^{-2}\theta]$

427. Universal time is based on

a) Rotation of the earth on its axis
b) Earth's orbital motion around the Sun
c) Vibrations of cesium atom
d) Oscillations of quartz crystal

428. If F denotes force and t time, then in equation $F = at^{-1} + bt^2$, the dimensions of a and b respectively are

a) $[LT^{-4}]$ and $[LT^{-1}]$
b) $[LT^{-1}]$ and $[LT^{-4}]$
c) $[MLT^{-4}]$ and $[MLT^{-1}]$
d) $[MLT^{-1}]$ and $[MLT^{-4}]$

429. Identify the pair which has different dimensions

a) Planck's constant and angular momentum
b) Impulse and linear momentum
c) Angular momentum and frequency
d) Pressure and Young's modulus

430. If voltage $V = (100 \pm 5)$ volt and current $I = (10 \pm 0.2)$ A, the percentage error in resistance R is

a) 5.2%
b) 25%
c) 7%
d) 10%

431. Dimensions of ohm are same as (h –Planck's constant, e –charge)

a) h/e
b) h^2/e
c) h/e^2
d) h^2e^2

432. The respective number of significant figures for the numbers 23.023, 0.0003 and 2.1×10^{-3} are

a) 4, 4, 2
b) 5, 1, 2
c) 5, 1, 5
d) 5, 5, 2

433. The difference in the lengths of a mean solar day and a sidereal day is about

a) 1 min
b) 4 min
c) 15 min
d) 56 min

434. Electron *volt* is a unit of

a) Charge
b) Potential difference
c) Momentum
d) Energy

435. The measured mass and volume of a body are 23.42 g and 4.9 cm^3 respectively with possible error 0.01 g and 0.1 cm^3. The maximum error in density is nearly

a) 0.2% b) 2% c) 5% d) 10%

436. In the relation $y = a\cos(\omega t - kx)$, the dimensional formula for k is

a) $[M^0L^{-1}T^{-1}]$ b) $[M^0LT^{-1}]$ c) $[M^0L^{-1}T^0]$ d) $[M^0LT]$

437. A quantity X is given by $\varepsilon_0 L\frac{\Delta V}{\Delta t}$, where ε_0 is the permittivity of free space, L is a length, ΔV is a potential difference and Δt is a time interval. The dimensional formula for X is the same as that of

a) Electrical resistance b) Electric charge c) Electric voltage d) Electric current

438. Dimension of electric current is

a) $[M^0L^0T^{-1}Q]$ b) $[ML^2T^{-1}Q]$ c) $[M^2LT^{-1}Q]$ d) $[M^2L^2T^{-1}Q]$

439. The velocity of a particle (v) at an instant t is given by $v = at + bt^2$ the dimension of b is

a) L b) LT^{-1} c) LT^-2 d) LT^{-3}

440. The dimensions of emf in MKS is

a) $[ML^{-1}T^{-2}Q^{-2}]$ b) $[ML^{-2}T^{-2}Q^{-2}]$ c) $[MLT^{-2}Q^{-1}]$ d) $[ML^2T^{-2}Q^{-1}]$

441. Number of particles is given by $n = -D\frac{n_2-n_1}{x_2-x_1}$crossing a unit area perpendicular to X-axis in unit time, where $n_1$and n_2 are number of particles per unit volume for the value of x meant to x_2 and x_1.Find dimensions of D called as diffusion constant

a) M^0LT^2 b) $M^0L^2T^{-4}$ c) M^0LT^{-3} d) $M^0L^2T^{-1}$

442. If the velocity v(is cms^{-1}) of a particle is given in terms of t (in second) by the relation

$v = at + \frac{b}{t+c}$

then, the dimensions of a, $b$and c are

abc

a) $[L][LT][T^2]$ b) $[L^2][T][LT^{-2}]$ c) $[LT^2][LT][L]$ d) $[LT^{-2}][L][T]$

443. A current of 2..34 A flows in a resistance of 11.11111Ω. The potential difference across the given resistance with due regard for significant figure is

a) 26.000 V b) 26.00 V c) 26.0 V d) 26 V

444. The dimensions of Planck's constant is same as that of

a) Angular momentum b) Linear momentum
c) Work d) Coefficient of viscosity

445. Solar constant is defined as energy received by earth per cm^2 per minute. The dimensions of solar constant are

a) $[ML^2T^{-3}]$ b) $[M^2L^0T^{-1}]$ c) $[ML^0T^{-3}]$ d) $[MLT^{-2}]$

446. Dimensional formula for volume elasticity is

a) $M^1L^{-2}T^{-2}$ b) $M^1L^{-3}T^{-2}$ c) $M^1L^2T^{-2}$ d) $M^1L^{-1}T^{-2}$

447. If p represents radiation pressure, C represents speed of light and q represents radiation energy striking a unit area per second, then non-zero integers a, b and c are such that $p^aq^bC^c$ is dimensionless, then

a) $a = 1, b = 1, c = -1$ b) $a = 1, b = -1, c = 1$
c) $a = -1, b = 1, c = 1$ d) $a = 1, b = 1, c = 1$

448. The unit of reactance is

a) *Ohm* b) *Volt* c) *Mho* d) *Newton*

449. A wire has a mass 0.3 $\pm$ 0.003 g, radius 0.5 $\pm$ 0.005 mm and length 6 $\pm$ 0.06 cm. The maximum percentage error in the measurement of its density is

a) 1 b) 2 c) 3 d) 4

450. A highly rigid cubical block A of small mass M and side L is fixed rigidly on to another cubical block of same dimensions and of low modulus of rigidity η such that the lower face of A completely covers the upper face of B. The lower face of B is rigidly held on a horizontal surface. A small force F is applied perpendicular to one of the side faces of

A. After the force is withdrawn, block *A* executes small oscillations, the time period o which is given by

a) $2\pi\sqrt{M\eta L}$ b) $2\pi\sqrt{\frac{M\eta}{L}}$ c) $2\pi\sqrt{\frac{ML}{\eta}}$ d) $2\pi\sqrt{\frac{M}{\eta L}}$

451. Dimensions of coefficient of viscosity are
a) ML^2T^{-2} b) ML^2T^{-1} c) $ML^{-1}T^{-1}$ d) MLT

452. Which physical quantities have same dimensions
a) Force and power b) Torque and energy c) Torque and power d) Force and torque

453. If L, C and R represent inductance, capacitance and resistance respectively, then which of the following does not represent dimensions of frequency
a) $\frac{1}{RC}$ b) $\frac{R}{L}$ c) $\frac{1}{\sqrt{LC}}$ d) $\frac{C}{L}$

454. Which of the following is/are the units of strength of magnetic field at a point?
a) $\mathrm{NA\ m^{-1}}$ b) NA m c) $\mathrm{NA^{-1}m^{-1}}$ d) $\mathrm{NA^{-2}m^{-2}}$

455. A screw gauge gives the following reading when used to measure the diameter of a wire
Main scale reading : 0 *mm*
Circular scale reading : 52 divisions
Given that 1 *mm* on main scale corresponds to 100 divisions on the circular scale.
The diameter of wire from the above data is
a) 0.52 *cm* b) 0.052 *cm* c) 0.026 *cm* d) 0.005 *cm*

456. The energy (E), andgular momentum (L) and universal gravitational constant (G) are chosen as fundamental quantities. The dimensions of universal gravitational constant in the dimensional formula of Planck's constant (h) is
a) Zero b) -1 c) $\frac{5}{3}$ d) 1

457. A thin copper wire of length $lmetre$ increases in length by 2% when heated through 10^oC. What is the percentage increase in area when a square copper sheet of length $lmetre$ is heated through 10^oC
a) 4% b) 8% c) 16% d) None of the above

458. Which one of the following is not a fundamental SI unit?
a) Ampere b) Candela c) Newton d) Kelvin

459. Find the dimensions of electric permittivity
a) $[A^2M^{-1}L^{-3}T^4]$ b) $[A^2M^{-1}L^{-3}T^0]$ c) $[AM^{-1}L^{-3}T^4]$ d) $[A^2M^0L^{-3}T^4]$

460. The respective number of significant figures for the numbers 23.02310.0003 and 2.1×10^{-3} are
a) 5, 1, 2 b) 5, 1, 5 c) 5, 5, 2 d) 4, 4, 2

461. A body travels uniformly a distance of $(13.8 \pm 0.2)\ m$ in a time (4.0 ± 0.3) s. The velocity of the body within error limits is
a) $(3.45 \pm 0.2)\ ms^{-1}$ b) $(3.45 \pm 0.3)\ ms^{-1}$ c) $(3.45 \pm 0.4)\ ms^{-1}$ d) $(3.45 \pm 0.5)\ ms^{-1}$

462. The force Fon the sphere of radius 'a' moving in a medium with velocity 'v' is given by $F = 6\pi\eta av$. The dimensions of $\eta$are
a) $ML^{-1}T^{-1}$ b) MT^{-1} c) MLT^{-2} d) ML^{-3}

463. Given that $2l\sqrt{\frac{m}{T}}$, where l is the length of a string of linear density m, under tension T ha the same dimensional formula as that of
a) Mass b) Time c) Length d) Mole

464. In C.G.S. system the magnitude of the force is 100 *dyne*. In another system where fundamental physical quantities are kilogram, *metre* and minute, the magnitude of the force is

a) 0.036 b) 0.36 c) 3.6 d) 36

465. The length of a simple pendulum is about 100 cm known to an accuracy of 1 mm. Its period of oscillation is 2s determined by measuring the time for 100 oscillations using a clock of 0.1 s resolution. What is the accuracy in the determined value of g?

a) 0.2% b) 0.5% c) 0.1% d) 2%

466. Which one of the following pairs of quantities and their units is a proper match

a) Electric field-*coulomb*/*m* b) Magnetic flux-*weber*
c) Power-*farad* d) Capacitance-*henry*

467. Which of the following units denotes the dimensions ML^2/Q^2, where Q denotes the electric charge

a) Henry (H) b) H/m^2 c) Weber (Wb) d) Wb/m^2

468. The dimensional formula for the magnetic field is

a) $[MT^{-2}A^{-1}]$ b) $[ML^2T^{-1}A^{-2}]$ c) $[MT^{-2}A^{-2}]$ d) $[MT^{-1}A^{-2}]$

469. The dimensions of stress are equal to

a) Force b) Pressure c) Work d) $\frac{1}{\text{Pressure}}$

470. If C the restoring couple per unit radian twist and I is the moment of inertia, then the dimensional representation of $2\pi\sqrt{\frac{I}{C}}$ will be

a) $[M^0L^0T^{-1}]$ b) $[M^0L^0T\ ]$ c) $[M^0L\ T^{-1}]$ d) $[M\ L^2T^{-2}]$

471. The density of a solid ball is to be determined in an experiment. The diameter of the ball is measured with a screw gauge, whose pitch is 0.5 *mm* and there are 50 divisions on the circular scale. The reading on the main scale is 2.5 *mm* and that on the circular scale is 20 divisions. If the measured mass of the ball has a relative error of 2%, the relative percentage error in the density is

a) 0.9% b) 2.4% c) 3.1% d) 4.2%

472. Which one of the following does not have the same dimensions

a) Work and energy b) Angle and strain
c) Relative density and refractive index d) Planck constant and energy

473. Dimensional formula of heat energy is

a) ML^2T^{-2} b) MLT^{-1} c) $M^0L^0T^{-2}$ d) None of these

474. Given that : $y = A\sin\left[\left(\frac{2\pi}{\lambda}\right)(ct - x)\right]$ where, y and x are measured in metre. Which of the following statements is true?

a) The unit of λ is same as that of x and A b) The unit of λ is same as that of x but not of A
c) The unit of c is same as that of $\frac{2\pi}{\lambda}$ d) The unit of $(ct - x)$ is same as that of $\frac{2\pi}{\lambda}$

475. Length is measured in metre and time in second as usual. But a new unit of mass is so chosen that $G = 1$. This new unit of mass is equal to

a) 1.5×10^7kg b) 1.5×10^{10}kg c) 6.67×10^{-11}kg d) 6.67×10^{-8}kg

476. The dimensional formula for entropy is

a) $[MLT^{-2}K^{-1}]$ b) $[ML^2T^{-2}]$ c) $[ML^2T^{-2}K^{-1}]$ d) $[ML^{-2}T^{-2}K^{-1}]$

477. Which of the following represents a *volt*

a) *Joule*/*second* b) *Watt*/*ampere* c) *Watt*/*coulomb* d) *Coulomb*/*joule*

478. The equation $\left(P + \frac{a}{V^2}\right).(V - b) =$ constant. The unit of a is

a) $Dyne \times cm^5$ b) $Dyne \times cm^4$ c) $Dyne \times cm^3$ d) $Dyne \times cm^2$

479. The unit of potential energy is

a) $g(cm/\sec^2)$ b) $g(cm/\sec)^2$ c) $g(cm^2/\sec)$ d) $g(cm/\sec)$

480. If C be the capacitance and V be the electric potential, then the dimensional formula of CV^2 is

a) $[ML^{-3}TA]$ b) $[K^0LT^{-2}A^0]$ c) $[ML^1T^{-2}A^{-1}]$ d) $[ML^2T^{-2}A^0]$

481. The mass and volume of a body are found to be 5.00 ± 0.05 kg and 1.00 ± 0.05 m^3 respectively. Then the maximum possible percentage error in its density is

a) 6% b) 3% c) 10% d) 5%

482. The number of significant figures in all the given numbers 25.12, 2009, 4.156 and 1.217×10^{-4} is

a) 1 b) 2 c) 3 d) 4

483. The unit of permittivity of free space ε_0 is

a) $Coulomb/newton - metre$
b) $Newton - metre^2/coulomb^2$
c) $Coulomb^2/(newton - metre)^2$
d) $Coulomb^2/newton - metre^2$

484. If the units of mass, length and time are doubled, unit of angular momentum will be

a) Doubled
b) Tripled
c) Quadrupled
d) 8 times the original value

485. $1kWh =$

a) $1000\ W$ b) $36 \times 10^5\ J$ c) $1000\ J$ d) $3600\ J$

486. A wire has a mass $0.3 \pm 0.003g$, radius $0.5 \pm 0.005\ mm$ and length $6 \pm 0.06\ cm$. The maximum percentage error in the measurement of its density is

a) 1 b) 2 c) 3 d) 4

487. Coefficient of thermal conductivity has the dimensions

a) $[MLT^{-3}K^{-1}]$ b) $[ML^3T^3K^2]$ c) $[ML^3T^{-3}K^{-2}]$ d) $[M^2L^3T^{-3}K^2]$

488. If C, R, L and I denote capacity, resistance, inductance and electric current respectively, the quantities having the same dimensions of time are

(1) CR
(2) $\frac{L}{R}$
(3) $\sqrt{LC}$
(4) LI^2

a) (1) and (2) only
b) (1) and (3) only
c) (1) and (4) only
d) (1), (2) and (3) only

489. In the relation $P = \frac{\alpha}{\beta} e^{\frac{\alpha Z}{k\theta}}$ P is pressure, Z is the distance, k is Boltzmann's constant and θ is the temperature. The dimensional formula of β will be

a) $[M^0L^2T^0]$ b) $[M^1L^2T^1]$ c) $[M^1L^0T^{-1}]$ d) $[M^0L^2T^{-1}]$

490. Which physical quantities have same dimensions?

a) Force and power b) Torque and energy c) Torque and power d) Force and torque

491. In the equation $y = a \sin(\omega t + kx,)$ the dimensional formula of ω is

a) $[M^0L^0T^{-1}]$ b) $[M^0LT^{-1}]$ c) $[ML^0T^0]$ d) $[M^0L^{-1}T^0]$

492. Resistance of a given wire is obtained by measuring the current flowing in it and the voltage difference applied across it. If the percentage errors in the measurement of the current and the voltage difference are 3% each, then error in the value of resistance of the wire is

a) 6% b) Zero c) 1% d) 3%

493. Electric displacement is given by $D = \varepsilon E$,
Here, ε =electric permittivity
E =electric field strength
The dimensions of electric displacement are

a) $[ML^{-2}TA]$ b) $[L^{-2}T^{-1}A]$ c) $[L^{-2}TA]$ d) None of these

494. The dimension of k in the equation $W = \frac{1}{2}kx^2$ is

a) $[ML^0T^{-2}]$ b) $[M^0LT^{-1}]$ c) $[MLT^{-2}]$ d) $[ML^0T^{-1}]$

495. In$S = a + bt + ct^2$.S is measured in metre and t in *second*. The unit of c is

a) None b) m c) ms^{-1} d) ms^{-2}

496. Dimensional formula for the universal gravitational constant G is

a) $[M^{-1}L^2T^{-2}]$ b) $[M^0L^0T^0]$ c) $[M^{-1}L^3T^{-2}]$ d) $[M^{-1}L^3T^{-1}]$

497. Select the pair whose dimensions are same

a) Pressure and stress b) Stress and strain c) Pressure and force d) Power and force

498. A resistor of 10 kΩ having tolerance 10% is connected in series with another resistor of 20kΩ having tolerance 20%. The tolerance of the combination will be approximately

a) 10% b) 13% c) 17% d) 20%

499. R and L represent respectively resistance and self inductance, which of the following combinations has the dimensions of frequency

a) $\frac{R}{L}$ b) $\frac{L}{R}$ c) $\sqrt{\frac{R}{L}}$ d) $\sqrt{\frac{L}{R}}$

500. The unit of specific resistance is

a) Ohm/cm^2 b) Ohm/cm c) $Ohm - cm$ d) $(Ohm - cm)^{-1}$

501. If C is the capacitance and V is the potential, the dimensional formula for CV^2 is

a) $[ML^2T^{-1}]$ b) $[ML^{-2}T^{-3}]$ c) $[ML^2T^{-2}]$ d) $[ML^{-2}T^{-2}]$

502. The equation of state of some gases can be expressed as $\left(P + \frac{a}{V^2}\right)(V - b) = RT$. Here P is the pressure, $V$is the volume, T is the absolute temperature and a, b, R are constants. The dimensions of 'a' are

a) ML^5T^{-2} b) $ML^{-1}T^{-2}$ c) $M^0L^3T^0$ d) $M^0L^6T^0$

503. With the usual notations, the following equation $S_t = u + \frac{1}{2}a(2t - 1)$is

a) Only numerically correct b) Only dimensionally correct c) Both numerically and dimensionally correct d) Neither numerically nor dimensionally correct

504. *Volt/metre* is the unit of

a) Potential b) Work c) Force d) Electric intensity

505. Length cannot be measured by

a) *Fermi* b) *Debye* c) *Micron* d) *Light year*

506. *Henry/ohm* can be expressed in

a) *Second* b) *Coulomb* c) *Mho* d) *Metre*

507. 1 $a.m.u.$ is equivalent to

a) $1.6 \times 10^{-27} kg$ b) 934 MeV c) 1.6×10^{-24} gm d) All above

508. The unit of Stefan's constant $\sigma$is

a) $W\,m^{-2}K^{-1}$ b) $W\ m^2\ K^{-4}$ c) $W\,m^{-2}K^{-4}$ d) $W\,m^{-2}K^4$

509. The dimensions of time constant are

a) $[M^0L^0T^0]$ b) $[M^0L^0T]$ c) $[MLT]$ d) None of these

510. The wavelength associated with a moving particle depends upon power p of its mass m, qth power of its velocity v and rth power of planck's constant h. Then the correct set of values of p, q and r is

a) $p = 1, q = -1, r = 1$ b) $p = 1, q = 1, r = 1$ c) $p = -1, q = -1, r = -1$ d) $p = -1, q = -1, r = 1$

511. Parsec is a unit of

a) Distance b) Velocity c) Time d) Angle

512. A student performs an experiment for determination of g$\left(= \frac{4\pi^2 l}{T^2}\right)$, $l \approx$ 1m, and he commits an error of Δl. For T he takes the time of n oscillations with the stop watch of

least countΔT and he commits a human error of 0.1 s. For which of the following data, the measurement of g will be most accurate?

a) $\Delta L = 0.5, \Delta T = 0.1, n = 20$
b) $\Delta L = 0.5, \Delta T = 0.1, n = 50$
c) $\Delta L = 0.5, \Delta T = 0.01, n = 20$
d) $\Delta L = 0.5, \Delta T = 0.05, n = 50$

513. Dimensional formula of magnetic flux is

a) $ML^2T^{-2}A^{-1}$
b) $ML^0T^{-2}A^{-2}$
c) $M^0L^{-2}T^{-2}A^{-3}$
d) $ML^2T^{-2}A^3$

514. The SI unit of length is the metre. Suppose we adopt a new unit of length which equal x metre. The area of 1 m^2 expressed in terms of the new unit has a magnitude

a) x
b) x^2
c) x^{-1}
d) x^{-2}

515. A physical quantity A is related to four observables a, b, c and d as follows

$$A = \frac{a^2b^3}{c\sqrt{d}}$$

The percentage errors of measurement in a, b, c and d are 1%, 3%, 2% and 2% respectively. What is the percentage error in the quantity A?

a) 12%
b) 7%
c) 5%
d) 14%

516. The percentage errors in the measurement of length and time period of a simple pendulum are 1% and 2% respectively. Then the maximum error in the measurement of acceleration due to gravity is

a) 8%
b) 3%
c) 4%
d) 5%

517. If the acceleration due to gravity is $10ms^{-2}$ and the units of length and time are changed in kilometer and hour respectively, the numerical value ofacceleration is

a) 360000
b) 72000
c) 36000
d) 129600

518. The square root of the product of inductance and capacitance has the dimension of

a) Length
b) Mass
c) Time
d) No dimension

519. The number of particles given by $n = -D\frac{n_2-n_1}{x_2-x_1}$ are crossing a unit area perpendicular to x-axis in unit time, where n_1 and n_2 are the number of particles per unit volume for the values x_1 and x_2 of x respectively. Then the dimensional formula of diffusion constant D is

a) $[M^0LT^0]$
b) $[M^0L^2T^{-4}]$
c) $[M^0LT^{-3}]$
d) $[M^0L^2T^{-1}]$

520. The expression for centripetal force depends upon mass of body, speed of the body and the radius of circular path. Find the expression for centripetal force

a) $F = \frac{mv^2}{2r^3}$
b) $F = \frac{mv^2}{r}$
c) $F = \frac{mv^2}{r^2}$
d) $F = \frac{m^2v^2}{2r}$

521. A small steel ball of radius $r$is allowed to fall under gravity through a column of a viscous liquid of coefficient of viscosity η.After sometime the velocity of the ball attains a constant value known as terminal velocity v_T. The terminal velocity depends on (i) the mass of the ball m,(ii)η , (iii)r and (iv) acceleration due to gravity g. Which of the following relations is dimensionally correct

a) $v_T \propto \frac{mg}{\eta r}$
b) $v_T \propto \frac{\eta r}{mg}$
c) $v_T \propto \eta rmg$
d) $v_T \propto \frac{mgr}{\eta}$

522. If $F = 6\pi\eta^a r^b v^c$,
Where F =viscous force
η = coefficient of viscosity
r = radius of spherical body
v = terminal velocity of the body.
Find the values of $a, b$and c.

a) $a = 1, b = 2, c = 1$
b) $a = 1, b = 1, c = 1$
c) $a = 2, b = 1, c = 1$
d) $a = 2, b = 1, c = 2$

523. A physical parameter a can be determined by measuring the parameters b, c, d and e using the relation $a = b^\alpha c^\beta / d^\gamma e^\delta$. If the maximum errors in the measurement of b, c, d and e are b_1%, c_1%, d_1% and e_1%, then the maximum error in the value of a determined by the experiment is

a) $(b_1 + c_1 + d_1 + e_1)\%$
b) $(b_1 + c_1 - d_1 - e_1)\%$
c) $(\alpha b_1 + \beta c_1 - \gamma d_1 - \delta e_1)\%$
d) $(\alpha b_1 + \beta c_1 + \gamma d_1 + \delta e_1)\%$

524. The frequency f of vibration of mass m suspended from a spring of spring constant k is given by

$$f = cm^x k^y$$

Where c is dimensionless constant.The values of x and y are respectively

a) 1/2, 1/2
b) -1/2, 1/2
c) 1/2, -1/2
d) -1/2, -1/2

525. The dimensional formula of universal gas constant is

a) $[ML^2T^{-2}\theta^{-1}]$
b) $[M^2LT^{-2}\theta]$
c) $[ML^3T^{-1}\theta^{-1}]$
d) None of these

526. Dimensions of $\frac{1}{\mu_0 \epsilon_0}$, where symbols have their usual meaning, are

a) $[lT^{-1}]$
b) $[L^{-1}T]$
c) $[L^{-2}T^2]$
d) $[L^2T^{-2}]$

527. In an experiment, we measure quantities a, $b$and c. Then x is calculated from the formula $x = \frac{ab^2}{c^3}$. The percentage errors in a, b, c are $\pm 1\%, \pm 3\%$, and $\pm 2\%$ respectively. The percentage error in x can be

a) $\pm 1\%$
b) $\pm 4\%$
c) 7%
d) $\pm 13\%$

528. The dimensions of potential are the same as that of

a) Work
b) Electric field per unit charge
c) Work per unit charge
d) Force per unit charge

529. Which of the following units denotes the dimensions $[ML^2/Q^2]$, where Q denotes the electric charge?

a) Wbm^{-2}
b) Henry (H)
c) Hm^{-2}
d) Weber (Wb)

530. The dimensional formula of the ratio of angular to linear momentum is

a) $[M^0LT^0]$
b) $[MLT]$
c) $[ML^2T^{-1}]$
d) $[M^{-1}L^{-1}T^{-1}]$

531. If force is proportional to square of velocity, then the dimensions of proportionality constant are

a) $[ML^{-1}T]$
b) $[ML^{-1}T^0]$
c) $[MLT^0]$
d) $[M^0LT^{-1}]$

532. The number of significant figures in the numbers 4.8000×10^4 and 48000.50 are respectively

a) 5 and 6
b) 5 and 7
c) 2 and 7
d) 2 and 6

533. The pressure on a square plate is measured by measuring the force on the plate and the length of the sides of the plate. If the maximum error in the measurement of force and length are respectively 4% and 2%, the maximum error in the measurement of pressure is

a) 1%
b) 2%
c) 6%
d) 8%

534. The frequency of vibration of string is given by $v = \frac{P}{2l}\left[\frac{F}{m}\right]^{1/2}$. Here p is number of segments in the string and l is the length. The dimensional formula for m will be

a) $[M^0LT^{-1}]$
b) $[ML^0T^{-1}]$
c) $[ML^{-1}T^0]$
d) $[M^0L^0T^0]$

535. The time dependence of a physical quantity P is given by $P = P_0 e^{\alpha t^2} - \alpha t^2$ where α is a constant and t is time. Then constant α is

a) Dimensionless
b) Dimensionless of T^{-2}
c) Dimensionless of P
d) Dimensionless of T^2

536. The percentage errors in the measurement of length and time period of a simple pendulum are 1% and 2% respectively. Then the maximum error in the measurement of acceleration due to gravity is

a) 8%
b) 3%
c) 4%
d) 5%

537. According to *Joule's* law of heating, heat produced $H = I^2Rt$, where I is current, R is resistance and t is time. If the errors in the measurement of I, R and t are 3%, 4% and 6% respectively then error in the measurement of H is

a) $\pm 17\%$
b) $\pm 16\%$
c) $\pm 19\%$
d) $\pm 25\%$

538. From the equation $\tan\theta = \frac{rg}{v^2}$, one can obtain the angle of banking θ for a cyclist taking a curve (the symbols have their usual meanings). Then say it is,

a) Both dimensionally and numerically correct
b) Neither numerically nor dimensionally correct
c) Dimensionally correct only
d) Numerically correct only

539. If the length of a rectangle $l = 10.5$ cm, breadth $b = 2.1$ cm and minimum possible measurement by scale = 0.1 cm, then the area is

a) 22.0 cm^2 b) 22.1 cm^2 c) 22.05 cm^2 d) 22 cm^2

540. According to Newton, the viscous force acting between liquid layers of area A and velocity gradient $\Delta v/\Delta z$ is given by $F = -\eta A \frac{\Delta v}{\Delta x}$ where η is constant called coefficient of viscosity. The dimensions of η are

a) $[ML^2T^{-2}]$ b) $[ML^{-1}T^{-1}]$ c) $[ML^{-2}T^{-2}]$ d) $[M^0L^0T^0]$

541. SI unit of intensity of wave is

a) $Jm^{-2}s^{-1}$ b) $Jm^{-1}s^{-2}$ c) $W\,m^{-2}$ d) $J\,m^{-2}$

542. The value of universal gas constant is $R = 8.3$ J/K-mol. The value of R in atmosphere litre per Kelvin mol

a) 8.12 b) 0.00812 c) 81.2 d) 0.0812

543. E, m, I and G denote energy, mass, angular momentum and gravitational constant respectively, then the dimensions of $\frac{EI^2}{m^5G^2}$ are

a) Angle b) Length c) Mass d) Time

544. The velocity of a body is given by the equation $v = \frac{b}{t} + ct^2 + dt^2$

The dimensional formula of b is

a) $[M^0LT^0]$ b) $[ML^0T^0]$ c) $[M^0L^0T]$ d) $[MLT^{-1}]$

545. If V denotes the potential difference across the plates of a capacitor of capacitance C, the dimensions of CV^2 are

a) Not expressible in MLT
b) MLT^{-2}
c) M^2LT^{-1}
d) ML^2T^{-2}

546. The dimensions of power are

a) $M^1L^2T^{-3}$ b) $M^2L^1T^{-2}$ c) $M^1L^2\,T^{-1}$ d) $M^1L^1T^{-2}$

547. Light year is a unit of

a) Time b) Mass c) Distance d) Energy

548. The surface tension of mercury is 32 dyne cm^{-1}. Its value in SI units is

a) 0.032 b) 0.32 c) 3200 d) 32000

549. Which of the following quantities has not been expressed in proper unit

a) Torque : Newton metre
b) Stress : Newton $metre^{-2}$
c) Modulus of elasticity : Newton $metre^{-2}$
d) Surface tension : Newton $metre^{-2}$

550. One femtometer is equivalent to

a) $10^{15}\,m$ b) $10^{-15}\,m$ c) $10^{-12}\,m$ d) $10^{12}\,m$

551. One side of a cubical block is measured with the help of a verniercallipers of vernier constant 0.01 cm. This side comes out to be 1.23 cm. What is the percentage error in the measurement of area?

a) $\frac{1.23}{0.01} \times 100$ b) $\frac{0.01}{1.23} \times 100$ c) $2 \times \frac{0.01}{1.23} \times 100$ d) $3 \times \frac{0.01}{1.23} \times 100$

552. Assuming the mass of Earth as $6.64 \times 10^{24} kg$ and the average mass of the atoms that make up earth as $40u$ (atomic mass unit), the number of atoms in the Earth are approximately

a) 10^{30} b) 10^{40} c) 10^{50} d) 10^{60}

553. The velocity of transverse wave in a string is $v = \sqrt{\frac{T}{m}}$,where T is the tension in the string and m is mass per unit length. If $T = 3.0$ kgf, mass of string is 2.5 g and length of string is 1.00m, then the percentage error in the measurement of velocity is

a) 0.5 b) 0.7 c) 2.3 d) 3.6

554. $Newton/metre^2$ is the unit of

a) Energy b) Momentum c) Force d) Pressure

555. The unit of magnetic moment is

a) TJ^{-1} b) JT^{-1} c) Am^{-2} d) Am^{-1}

556. Dimensional formula of Stefan's constant is

a) $[MT^{-3}K^{-4}]$ b) $[ML^2T^{-2}K^{-4}]$ c) $[ML^2T^{-2}]$ d) $[MT^{-2}L^0]$

557. In the relation $x = \cos(\omega t + kx)$, the dimensions of ω are

a) $[M^0LT]$ b) $[M^0L^{-1}T^0]$ c) $[M^0L^0T^{-1}]$ d) $[M^0LT^{-1}]$

558. $newton - second$ is the unit of

a) Velocity b) Angular momentum c) Momentum d) Energy

559. Which of the two have same dimensions

a) Force and strain b) Force and stress
c) Angular velocity and frequency d) Energy and strain

560. The initial temperature of a liquid is $(80.0 \pm 0.1)^0$C. After it has been cooled, its temperature is $(10.0 \pm 0.1)^0$C. The fall in temperature in degree centigrade is

a) 70.0 b) 70.0 ± 0.3 c) 70.0 ± 0.2 d) 70.0 ± 0.1

561. The dimensional formula of magnetic induction B is

a) $[M^0ALT^0]$ b) $[M^0AL^{-1}T^0]$ c) $[M^0AL^2T^0]$ d) $[ML^2T^{-2}A^{-1}]$

562. Which one of the following is not a unit of young's modulus

a) Nm^{-1} b) Nm^{-2} c) $Dyne\ cm^{-2}$ d) Mega Pascal

563. A student has measured the length of a wire equal to 0.04580 m. This value of length has the number of significant figures equal to

a) Five b) Four c) Six d) None of these

564. If u_1 and u_2 are the units selected in two systems of measurement and n_1 andn_2 their numerical values, then

a) $n_1u_1 = n_2u_2$ b) $n_1u_1 + n_2u_2 = 0$
c) $n_1n_2 = u_1u_2$ d) $(n_1 + u_1) = (n_2 + u_2)$

565. Universal time is based on

a) Rotation of earth on its axis
b) Oscillations of quartz crystal
c) Vibrations of cesium atom
d) Earth's orbital motion around the sun

566. The power of lens is $P = \frac{1}{f}$, where f is focal length of the lens . The dimensions of power of lens are

a) $[LT^{-2}]$ b) $[M^0L^{-1}T^0]$ c) $[M^0L^0T^0]$ d) None of these

567. If there is a positive error of 50% in the measurement of speed of a body, then the error in the measurement of kinetic energy is

a) 25% b) 50% c) 100% d) 125%

568. Dimensional formula for angular momentum is

a) ML^2T^{-2} b) $ML^2\ T^{-1}$ c) MLT^{-1} d) $M^0L^2T^{-2}$

569. If the value of the resistance is 10.845Ω and the value of the current is 3.23 A, then the potential difference is 35.02935 V. its value in correct significant figures would be

a) 35 V b) 35.0 V c) 35.03 V d) 35.029 V

570. Which of the following quantities has the same dimensions as that of energy

a) Power b) Force c) Momentum d) Work

571. Which of the following sets of quantities have same dimensional formula?
a) Frequency, angular frequency and angular momentum
b) Surface tension, stress and spring constant
c) Acceleration, momentum and retardation
d) Work, energy and torque

572. Dimensions of strain are
a) MLT^{-1} b) ML^2T^{-1} c) MLT^{-2} d) $M^0L^0T^0$

573. The circular scale of a screw gauge has 50 divisions and pitch of 0.5 mm. Find the diameter of sphere. Main scale reading is 2.

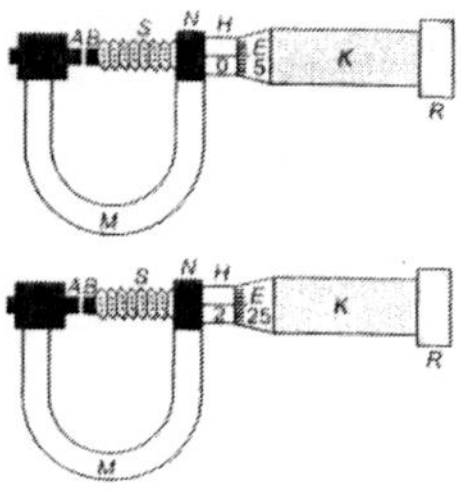

a) 1.2 b) 1.25 c) 2.20 d) 2.25

574. The random error in the arithmetic mean of 100 observations is x;then random error in the arithmetic mean of 4000 observations would be
a) $4x$ b) $\frac{1}{4}x$ c) $2x$ d) $\frac{1}{2}x$

575. Which of the following quantity is expressed as force per unit area
a) Work b) Pressure c) Volume d) Area

2
UNITS AND MEASUREMENTS

ANSWER KEY

1)	c	2)	c	3)	d	4)	b
5)	a	6)	b	7)	a	8)	b
9)	b	10)	a	11)	a	12)	c
13)	d	14)	d	15)	d	16)	c
17)	c	18)	b	19)	b	20)	b
21)	a	22)	b	23)	a	24)	a
25)	a	26)	d	27)	c	28)	b
29)	a	30)	d	31)	b	32)	d
33)	d	34)	a	35)	a	36)	a
37)	c	38)	b	39)	d	40)	a
41)	a	42)	c	43)	c	44)	a
45)	c	46)	b	47)	b	48)	c
49)	b	50)	d	51)	d	52)	b
53)	d	54)	a	55)	d	56)	c
57)	c	58)	b	59)	a	60)	a
61)	d	62)	d	63)	a	64)	a
65)	a	66)	b	67)	b	68)	d
69)	d	70)	c	71)	a	72)	a
73)	b	74)	b	75)	c	76)	c
77)	a	78)	a	79)	d	80)	c
81)	b	82)	a	83)	b	84)	a
85)	a	86)	b	87)	c	88)	c
89)	d	90)	b	91)	c	92)	a
93)	d	94)	a	95)	c	96)	d
97)	d	98)	c	99)	c	100)	a
101)	c	102)	b	103)	b	104)	d
105)	c	106)	a	107)	a	108)	b
109)	b	110)	b	111)	a	112)	c
113)	d	114)	b	115)	d	116)	a
117)	a	118)	a	119)	d	120)	c
121)	b	122)	a	123)	b	124)	c

125)	d	126)	a	127)	a	128)	a
129)	a	130)	c	131)	d	132)	d
133)	a	134)	b	135)	b	136)	c
137)	c	138)	c	139)	a	140)	d
141)	c	142)	c	143)	a	144)	a
145)	b	146)	a	147)	b	148)	c
149)	d	150)	b	151)	a	152)	b
153)	a	154)	a	155)	d	156)	a
157)	b	158)	d	159)	b	160)	b
161)	b	162)	a	163)	d	164)	b
165)	b	166)	d	167)	a	168)	c
169)	c	170)	c	171)	d	172)	c
173)	a	174)	d	175)	b	176)	c
177)	b	178)	b	179)	c	180)	a
181)	b	182)	a	183)	b	184)	a
185)	c	186)	a	187)	d	188)	a
189)	d	190)	d	191)	a	192)	d
193)	b	194)	d	195)	a	196)	a
197)	a	198)	d	199)	c	200)	a
201)	b	202)	a	203)	b	204)	c
205)	b	206)	c	207)	d	208)	a
209)	d	210)	a	211)	d	212)	c
213)	a	214)	b	215)	d	216)	c
217)	b	218)	a	219)	a	220)	b
221)	b	222)	a	223)	d	224)	c
225)	d	226)	c	227)	d	228)	a
229)	d	230)	a	231)	b	232)	c
233)	a	234)	d	235)	a	236)	a
237)	b	238)	a	239)	a	240)	a
241)	a	242)	a	243)	d	244)	a
245)	d	246)	d	247)	a	248)	d
249)	c	250)	d	251)	c	252)	a
253)	d	254)	c	255)	c	256)	d
257)	c	258)	b	259)	a	260)	c
261)	b	262)	b	263)	b	264)	a
265)	c	266)	a	267)	b	268)	d

269)	a	270)	d	271)	a	272)	b
273)	c	274)	b	275)	d	276)	c
277)	d	278)	c	279)	c	280)	d
281)	a	282)	b	283)	a	284)	a
285)	d	286)	c	287)	d	288)	c
289)	d	290)	c	291)	d	292)	c
293)	c	294)	b	295)	b	296)	d
297)	c	298)	a	299)	d	300)	b
301)	d	302)	a	303)	c	304)	a
305)	a	306)	a	307)	a	308)	a
309)	c	310)	c	311)	b	312)	b
313)	a	314)	a	315)	a	316)	c
317)	c	318)	d	319)	a	320)	c
321)	d	322)	b	323)	d	324)	a
325)	a	326)	b	327)	d	328)	d
329)	a	330)	d	331)	c	332)	d
333)	b	334)	b	335)	b	336)	a
337)	a	338)	b	339)	b	340)	b
341)	a	342)	d	343)	d	344)	c
345)	c	346)	d	347)	a	348)	a
349)	a	350)	c	351)	c	352)	a
353)	c	354)	b	355)	a	356)	a
357)	a	358)	d	359)	d	360)	b
361)	b	362)	b	363)	b	364)	b
365)	c	366)	c	367)	c	368)	b
369)	d	370)	a	371)	c	372)	a
373)	b	374)	d	375)	c	376)	a
377)	d	378)	c	379)	d	380)	a
381)	b	382)	c	383)	c	384)	a
385)	d	386)	a	387)	b	388)	a
389)	c	390)	b	391)	c	392)	b
393)	a	394)	b	395)	c	396)	c
397)	d	398)	c	399)	a	400)	c
401)	a	402)	c	403)	a	404)	b
405)	d	406)	c	407)	d	408)	a
409)	b	410)	a	411)	d	412)	c

413)	d	414)	a	415)	b	416)	b
417)	a	418)	a	419)	d	420)	a
421)	c	422)	c	423)	b	424)	c
425)	b	426)	b	427)	c	428)	d
429)	c	430)	c	431)	c	432)	b
433)	b	434)	d	435)	b	436)	c
437)	d	438)	a	439)	d	440)	d
441)	d	442)	d	443)	c	444)	a
445)	c	446)	d	447)	b	448)	a
449)	d	450)	d	451)	c	452)	b
453)	d	454)	c	455)	b	456)	a
457)	a	458)	c	459)	a	460)	a
461)	b	462)	a	463)	b	464)	c
465)	a	466)	b	467)	a	468)	a
469)	b	470)	b	471)	c	472)	d
473)	a	474)	a	475)	b	476)	c
477)	b	478)	b	479)	b	480)	d
481)	a	482)	d	483)	d	484)	c
485)	b	486)	d	487)	a	488)	d
489)	a	490)	b	491)	a	492)	a
493)	c	494)	a	495)	d	496)	c
497)	a	498)	c	499)	a	500)	c
501)	c	502)	a	503)	c	504)	d
505)	b	506)	a	507)	d	508)	c
509)	b	510)	d	511)	a	512)	d
513)	a	514)	d	515)	d	516)	d
517)	d	518)	c	519)	d	520)	b
521)	a	522)	b	523)	d	524)	b
525)	a	526)	d	527)	d	528)	c
529)	b	530)	a	531)	b	532)	b
533)	d	534)	c	535)	b	536)	d
537)	b	538)	c	539)	a	540)	b
541)	c	542)	d	543)	a	544)	a
545)	d	546)	a	547)	c	548)	a
549)	d	550)	b	551)	c	552)	c
553)	d	554)	d	555)	b	556)	b

557)	c	558)	c	559)	c	560)	c
561)	d	562)	a	563)	b	564)	a
565)	a	566)	b	567)	d	568)	b
569)	b	570)	d	571)	d	572)	d
573)	a	574)	b	575)	b		

2
UNITS AND MEASUREMENTS
HINTS AND SOLUTIONS

1 **(c)**
Let $v^x = kg^y\lambda^z\rho^\delta$. Now by submitting the dimensions of each quantities and equating the powers of M, L and T we get $\delta = 0$ and $x = 2, y = 1, z = 1$

2 **(c)**
One main scale division, 1 M. S. D. $= x\ cm$
One vernier scale division , 1 V. S. D. $= \frac{(n-1)x}{n}$
Least count = 1 M. S. D. −1 V. S. D.
$$= \frac{nx - nx + x}{n} = \frac{x}{n}\ cm$$

3 **(d)**
Work done $W = \varepsilon\Delta q$
$$\therefore \quad \varepsilon = \frac{W}{\Delta q} = \frac{[\mathrm{ML^2T^{-2}}]}{[\mathrm{AT}]}$$
$$\therefore \quad [\varepsilon] = [\mathrm{ML^2T^{-3}A^{-1}}]$$

4 **(b)**
$$F = \frac{Gm_1m_2}{d^2} \Rightarrow G = \frac{Fd^2}{m_1m_2}$$
$$\therefore [G] = \frac{[MLT^{-2}][L^2]}{[M^2]} = [M^{-1}L^3T^{-2}]$$

5 **(a)**
$\frac{1}{\sqrt{\varepsilon_0\mu_0}} = C$ =velocity of light

6 **(b)**
Power$= \frac{\text{Work}}{\text{Time}}$
$\therefore [\text{Power}] = \frac{[\text{Work}]}{[\text{Time}]} = \frac{[ML^2T^{-2}]}{[T]}$
$= [ML^2T^{-3}]$

7 **(a)**
$$n_1u_1 = n_2u_2$$
$$n_2 = \frac{n_1u_1}{u_2}$$
$$= \frac{170.474L}{M^3}$$
$$= \frac{170.474 \times 10^{-3}M^3}{M^3}$$
$= 0.170474$

8 **(b)**
Unit of $\varepsilon_0 = C^2/N - m^2$ $\therefore$ Unit of $K = Nm^2C^{-2}$

9 **(b)**
Let $[G] \propto c^xg^yp^z$
By substituting the following dimensions:
$[G] = [M^{-1}L^3T^{-2}], [c] = [LT^{-1}], [g] = [LT^{-2}]$
$[p] = [ML^{-1}T^{-2}]$
and by comparing the powers of both sides
we can get $x = 0, y = 2, z = -1$
$\therefore [G] \propto c^0g^2p^{-1}$

11 **(a)**
According to Coulombs law $F = \frac{1}{4\pi\varepsilon_0}\frac{q_1q_2}{r^2}$
$$\therefore \frac{1}{4\pi\varepsilon_0} = \frac{Fr^2}{q_1q_2} = \frac{(\text{newton})(\text{meter})^2}{(\text{coulomb})(\text{coulomb})}$$
$$= \frac{Nm^2}{C^2} = C^{-2}Nm^2$$

12 **(c)**
$$R_s = \frac{R_1R_2}{R_1 + R_2},$$
$$\frac{\Delta R_s}{R_s} \times 100$$
$$= \frac{\Delta R_1}{R_1} \times 100 + \frac{\Delta R_2}{R_2} \times 100 + \frac{\Delta(R_1 + R_2)}{R_1 + R_2}$$
$$= 100$$
Now, $\Delta R_1 = \frac{10}{100} \times 4\mathrm{k\Omega}=0.4\mathrm{k\Omega}$,
$\Delta R_2 = \frac{10}{100} \times 6\mathrm{k\Omega} = 0.6\mathrm{k\Omega}$
Again, $\frac{\Delta R_s}{R_s} \times 100 = \frac{0.4}{4} \times 100 + \frac{0.6}{6} \times 100$
$$+\frac{0.4 + 0.6}{10} \times 100$$
$= 10 + 10 + 10 = 30$

13 **(d)**
$$F \propto v \Rightarrow F = kv \Rightarrow [k] = \left[\frac{F}{v}\right] = \left[\frac{MLT^2}{LT^{-1}}\right]$$
$$= [MT^{-1}]$$

14 **(d)**
Velocity $v = Bt^2$
$$\therefore B = \frac{v}{t^2} = \frac{[\mathrm{LT^{-1}}]}{[\mathrm{T^2}]} = [\mathrm{LT^{-3}}]$$

15 **(d)**
Required percentage $= \frac{2\times0.02}{0.24} \times 100 + \frac{1}{30} \times 100 + \frac{0.01}{4.80} \times 100$
$= 16.7 + 3.3 + 0.2$
$= 20\%$

16 **(c)**
Torque = $[ML^2T^{-2}]$, Angular momentum = $[ML^2T^{-1}]$
So mass and length have the same dimensions

17 **(c)**
From $h = ut + \frac{1}{2}gt^2$
$$h = 0 + \frac{1}{2} \times 9.8 \times (2)^2 = 19.6\, m$$
$$\frac{\Delta h}{h} = \pm 2\frac{\Delta t}{t} \qquad [\because a = g = \text{constant}]$$
$$= \pm 2\left(\frac{0.1}{2}\right) = \pm\frac{1}{10}$$
$$\therefore \Delta h = \pm\frac{h}{10} = \pm\frac{19.6}{10} = \pm 1.96\, m$$

18 **(b)**
$r_1 = 10^{-15}\text{m}, r_2 = 10^{26}\text{m}$
Log $r = \frac{1}{2}[\log 10^{-15} + \log 10^{26}]$
$=\frac{1}{2}[-15 + 26] = 5.5 \approx 6 \Rightarrow r = 10^6\text{m}$

19 **(b)**
Pyrometer is used the for measurement for temperature

20 **(b)**
$$\frac{F-32}{9} = \frac{K-273}{5} \Rightarrow \frac{x-32}{9} = \frac{x-273}{5}$$
$$\Rightarrow x = 574.25$$

22 **(b)**
$1\ yard = 36\ inch = 36 \times 2.54\ cm$
$= 0.9144m$

23 **(a)**
$$k = \left[\frac{R}{N}\right] = [ML^2T^{-2}\theta^{-1}]$$

24 **(a)**
$$\text{Energy } (E) = F \times d \Rightarrow F = \frac{E}{d}$$
So $Erg/metre$ can be the unit of force

25 **(a)**
Couple of force = $|\vec{r} \times \vec{F}| = [ML^2T^{-2}]$
Work = $[\vec{F}.\vec{d}] = [ML^2T^{-2}]$

26 **(d)**
$[\text{M}^0\text{L}^0\text{T}^{-1}] = [\text{M}^x][\text{M}^y\text{T}^{2y}]$
$= [\text{M}^{x+y}\text{T}^{-2y}]$
Equating powers of M and T.
$x + y = 0, -2y = -1$
Or $y = \frac{1}{2}, x + \frac{1}{2} = 0$ or $x = -\frac{1}{2}$

27 **(c)**
The right hand side of the given relation is basically $\frac{k}{\text{metre}}$. But, since the left hand side is joule, therefore k should be J m.

28 **(b)**
Dimension of work and torque = $[ML^2T^{-2}]$

29 **(a)**
Power = $\frac{\text{Energy}}{\text{Time}}$

30 **(d)**
$[h]$ = [Angular momentum] =$[ML^2T^{-1}]$

31 **(b)**
$$R = \frac{V}{I} = \left[\frac{ML^2T^{-3}A^{-1}}{A}\right] = [ML^2T^{-3}A^{-2}]$$

32 **(d)**
Dipole momen = (charge) × (distance)

Electric flux = (electric field) × (area)

34 **(a)**
Percentage error inside = $\frac{1}{2}\left[\frac{0.2}{100} \times 100\right] =$ 0.1
Absolute error inside = $\frac{0.1}{100} \times 10 = 0.01$

35 **(a)**
The velocity of a body at highest point of vertical circle is,
$$v = \sqrt{rg}$$
Or $v^2 = rg$

Or $\frac{v^2}{rg}$ = constant

Hence, $\frac{v^2}{rg}$ is dimensionless.

36 **(a)**
$E = [ML^2T^{-2}], G = [M^{-1}L^3T^{-2}], I$
$= [MLT^{-1}]$
$$\therefore \frac{GIM^2}{E^2} = \frac{[M^{-1}L^3T^{-2}][MLT^{-1}][M^2]}{[ML^2T^{-2}]^2} = [T]$$

38 **(b)**
$V = \frac{W}{m}$ so, SI unit = $\frac{joule}{kg}$

39 **(d)**
Charge = Current × Time = $[AT]$

40 **(a)**
Bxt is unitless. ∴ Unit of B is$m^{-1}s^{-1}$

41 **(a)**
Volume of cube =a^3
Surface area of cube = $6a^2$

according to problem $a^3 = 6a^2 \Rightarrow a = 6$
$\therefore V = a^3 = 216\, units$

43 **(c)**
$[x] = [bt^2] \Rightarrow [b] = [x/t^2] = km/s^2$

45 **(c)**
$F = \frac{Gm_1m_2}{d^2};\ \therefore G = \frac{Fd^2}{m_1m_2} = Nm^2/kg^2$

46 **(b)**
Force = mass × acceleration

Or $F = ma$

$\therefore [F] = [m][a]$

$= [\text{M}][\text{LT}^2]$

$= [\text{MLT}^{-2}]$

47 **(b)**
Both $\frac{1}{2}LI^2$ and $\frac{1}{2}CV^2$ represent energy.

48 **(c)**
Electric dipole moment

$p = q(2l) = it(2l)$

The dimensions of electric dipole moment is

$[p] = [i][t][l]$

$= [\text{I}][\text{T}][\text{L}]$

$= [\text{LTI}]$

49 **(b)**
$v \propto g^p h^q$ (given)
By submitting the dimension of each quantity and comparing the powers on both sides we get $[LT^{-1}] = [LT^{-2}]^p[L]^q$
$\Rightarrow p + q = 1, -2p = -1, \therefore p = \frac{1}{2}, q = \frac{1}{2}$

51 **(d)**
According to Wien's law the product of wavelength corresponding to maximum intensity of radiation and temperature of body (in Kelvin) is constant *ie*, $\lambda_m T = b$ = constant, where b is Wien's constant and has value 2.89×10^{-3}m − K.

52 **(b)**
Gravitational potential $= \frac{\text{work}}{\text{mass}}$

Hence, SI unit gravitational potential

$= \frac{\text{unit of work}}{\text{unit of mass}}$

$= \frac{\text{J}}{\text{kg}}$
$= \text{Jkg}^{-1}$ or ms^{-2}

53 **(d)**
$P = nu \therefore n \propto \frac{1}{u}$

54 **(a)**
Let $n = k\rho^a a^b T^c$ where $[\rho] = [ML^{-3}], [a] = [L]$ and $[T] = [MT^{-2}]$
Comparing dimensions both sides we get $a = \frac{-1}{2}, b = \frac{-3}{2}$ and $c = \frac{1}{2}\ \therefore \eta = k\rho^{-1/2}a^{-3/2}T^{-1/2}$
$= \frac{K\sqrt{T}}{\rho^{1/2}a^{3/2}}$

56 **(c)**
Zero error $= 5 \times \frac{0.5}{50} = 0.05\, mm$
Actual measurement $= 2 \times 0.5\, mm + 25 \times \frac{0.5}{50} - 0.05\, mm$
$= 1\, mm + 0.25\, mm - 0.05\, mm$
$= 1.20\, mm$

57 **(c)**
Spring constant $= \frac{F}{l} = [\text{ML}^0\text{T}^{-2}]$.
Surface energy $= \frac{\text{Energy}}{\text{Area}} = [\text{ML}^0\text{T}^{-2}]$

58 **(b)**
PV =[energy]
Vander Waal's equation is $\left(P + \frac{a}{V^2}\right)(V - b) = nRT$
The dimensions of $\frac{a}{V^2}$ should be that of P and b is that of volume
Work done (or energy) should have the dimensions of PV
$\therefore \left[\frac{a}{V^2} \times b\right] = [\text{Energy}]$
$[bP] = [\text{Energy}]$
$\left[\frac{a}{V^2}\right] = [P]$ is having dimensions different from energy

59 **(a)**

In given equation, $\frac{az}{k\theta}$ should be dimensionless

$$\alpha = \frac{k\theta}{z}$$

$$\Rightarrow \quad [\alpha] = \frac{[ML^2T^{-2}K^{-1} \times K]}{[L]} = [MLT^{-2}]$$

And $\quad p = \frac{\alpha}{\beta}$

$$\Rightarrow \quad [\beta] = \left[\frac{\alpha}{p}\right] = \frac{[MLT^{-2}]}{[ML^{-1}T^{-2}]} = [M^0L^2T^0]$$

60 **(a)**

Formula for viscosity $\eta = \frac{\pi pr^4}{8Vl} \Rightarrow V = \frac{\pi pr^4}{8\eta l}$

61 **(d)**

$NSm^{-2} = Nm^{-2} \times S =$ Pascal-second

62 **(d)**

[capacitance X]=$[M^{-1}L^{-2}T^2Q^2]$

[Magnetic induction Z] = $[MT^{-1}Q^{-1}]$

$[Z^2]$=$[M^2T^{-2}Q^{-2}]$

Given, $X = 3YZ^2$ or $Y = \frac{X}{3Z^2}$ or $[Y] = \frac{[X]}{[Z]^2}$

$\therefore \quad [Y] = \frac{[M^{-1}L^{-2}T^2Q^2]}{[M^2T^{-2}Q^{-2}]} = [M^{-3}L^{-2}T^4Q^4]$

63 **(a)**

Momentum $= mv = [ML \;\; T^{-1}]$

Impulse = Force × Time $= [MLT^{-2}] \times [T] = [MLT^{-1}]$

64 **(a)**

$[E] = [ML^2T^{-2}]$

$[M] = [M]$

$[L] = [ML^2T^{-1}]$

$[G] = [M^{-1}L^3T^{-2}]$

$$\left[\frac{EL^2}{M^5G^2}\right] = \frac{[ML^2T^{-2}][ML^2T^{-1}]^2}{[M]^5[M^{-1}L^3T^{-2}]^2}$$

$$= \frac{[ML^2T^{-2}][M^2L^4T^{-2}]}{[M^5][M^{-2}L^6T^{-4}]} = \frac{[M^3L^6T^{-4}]}{[M^3L^6T^{-4}]}$$

$= [m^0L^0T^0] =$ Angle

65 **(a)**

If **E** is the intensity of electric field over a small area element **dS** and θ is angle between **E** and outdrawn normal to area element. Therefore, electric flux through this element is

$$d\phi_E = (dS)(E\cos\theta)$$

$$= E\,dS\cos\theta = \mathbf{E}.\mathbf{dS}$$

Hence, $\phi_E = E\,.S$

$$= \frac{V}{d}\,.S$$

$$\therefore \text{Unit of } \phi_E = \frac{\text{volt} \times \text{metre}^2}{\text{metre}}$$

$$= \text{volt} - \text{metre}$$

66 **(b)**

Frequency $= \frac{1}{T} = [M^0L^0\,T^{-1}]$

67 **(b)**

% error in $g = \frac{\Delta g}{g} \times 100 = \left(\frac{\Delta l}{l}\right) \times 100 + 2\left(\frac{\Delta T}{T}\right) \times 100$

$$E_I = \frac{0.1}{64} \times 100 + 2\left(\frac{0.1}{128}\right) \times 100 = 0.3125\%$$

$$E_{II} = \frac{0.1}{64} \times 100 + 2\left(\frac{0.1}{64}\right) \times 100 = 0.4687\%$$

$$E_{III} = \frac{0.1}{20} \times 100 + 2\left(\frac{0.1}{36}\right) \times 100 = 1.055\%$$

68 **(d)**

$[\eta] = ML^{-1}T^{-1}$ so its unit will be kg/m-sec

69 **(d)**

Angular momentum,

$$[J] = [I\omega] = [ML^2T^{-1}]$$

Planck's constant, $[h] = \frac{[E]}{[v]} = [ML^2T^{-1}]$

70 **(c)**

Resistivity, $\rho = \frac{m}{ne^2\tau}$

$$\therefore [\rho] = \frac{[M]}{[L^{-3}][AT][T^2]}$$

$$= [ML^3A^{-2}T^{-3}]$$

So, electrical conductivity

$$\sigma = \frac{1}{\rho}$$

$$\Rightarrow \quad [\sigma] = \frac{1}{[\rho]} = [M^{-1}L^{-3}A^2T^3]$$

71 **(a)**
The formula for fine structure constant is
$= \dfrac{e^2}{4\pi\varepsilon_0\left(\frac{h}{2\pi}\right)c}$

72 **(a)**
Given, $a = 3bc^2$
$\Rightarrow b = \dfrac{a}{3c^2}$
Writing dimensions for a and c, we have
$[b] = \dfrac{[Q/V]}{[B]^2} = \dfrac{[Q][ML^2T^{-2}Q^{-1}]}{[MT^{-1}Q^{-1}]^2}$
$= [M^{-3}L^{-2}T^4Q^4]$

73 **(b)**
Required volume $= \frac{75\times10^4\times26}{10^3\times10^3\times10^3}(\text{km})^3$

75 **(c)**
New unit of distance = speed in new units × 500 s

76 **(c)**
Dimensions of L and R

$[R] = [ML^2T^{-3}A^{-2}]$

$[L] = [ML^2T^{-2}A^{-2}]$

$$\left[\frac{L}{R}\right] = \frac{[ML^2T^{-2}A^{-2}]}{[ML^2T^{-3}A^{-2}]}$$

$$= [T]$$

77 **(a)**
$30\,VSD = 29\,MSD$
$1\,VSD = \dfrac{29}{30}\,MSD$
$L.C. = 1\,MSD - 1\,VSD$
$= \left(1 - \frac{29}{30}\right) MSD = \frac{1}{30} \times 0.5° = 1$ minute

78 **(a)**
Dimension of $\alpha t = [M^0L^0T^0] \therefore [\alpha] = [T^{-1}]$
Again $\left[\frac{v_0}{\alpha}\right] = [L]$ so $[v_0] = [LT^{-1}]$

79 **(d)**
Torque $= [ML^2T^{-2}]$, Moment of inertia $[ML^2]$

80 **(c)**
The unit of physical quantity obtained by the line intergral of electric field is JC^{-1}.

81 **(b)**
Intensity of radiation $= \frac{\text{Radiation Energy}}{\text{Area} \times \text{time}}$
$\Rightarrow I = \dfrac{[ML^2T^{-2}]}{[L^2 \times T]} = [ML^0T^{-3}]$

82 **(a)**
Quantities having different dimensions can only be divided or multiplied but they cannot be added or subtracted

83 **(b)**
We know that kinetic energy $= \frac{1}{2}mv^2$
Required percentage error is 2%+2× 3% *ie*, 8%

84 **(a)**
$n_1u_1 = n_2u_2$
$n_2 = \dfrac{n_1u_1}{u_2}$
$= \dfrac{1450\text{mile/h}}{\text{m/s}} = \dfrac{1450 \text{ s/mile}}{\text{mh}}$
$= \dfrac{1450 \text{ s} \times 1.6\text{km}}{10^{-3}\text{km } 60 \times 60\text{s}} = 644.4$
1450 mile/h=644.4m/s

85 **(a)**
$V = \dfrac{W}{Q} = [ML^2T^{-2}Q^{-1}]$

86 **(b)**
Subtract 3.87 from 4.23 and then divide by 2.

87 **(c)**
From the principle of dimensional homogenity $[v] = [at] \Rightarrow [a] = [LT^{-2}]$.
Similarly $[b] = [L]$ and $[c] = [T]$

88 **(c)**
$\text{KE} = \dfrac{1}{2}mv^2$

$\therefore [KE] = [M][LT^{-1}]^2 = [ML^2T^{-2}]$

90 **(b)**
$\ln g = \ln h - 2\ln t$
$\left(\frac{\Delta g}{g} \times 100\right)_{max} = \frac{\Delta h}{h} \times 100 + 2\frac{\Delta t}{t} \times 100$
$= e_1 + 2e_2$

91 **(c)**
1 fermi $= 10^{-15} metre$

92 **(a)**
Required percentage error
$= 2 \times \frac{0.01}{15.12} \times 0 + \frac{0.001}{10.15} \times 10 = 4 + 1$
$= 5$

93 **(d)**
$\frac{[E][J]^2}{[M]^5[G]^2} \frac{[ML^2T^{-2}][ML^2T^{-1}]^2}{[M^5][M^{-1}L^3T^{-2}]^2} = [M^0L^0T^0]$

94 **(a)**
Couple = Force × Arm length =
$[MLT^{-2}][L] = [ML^2T^{-2}]$

95 **(c)**
$\frac{L}{RCV} = \left[\frac{L}{R}\right]\frac{1}{CV} = \frac{T}{Q} = [A^{-1}]$

96 **(d)**
Relative density $= \frac{\text{Density of substance}}{\text{Density of water}} =$
$[M^0L^0T^0]$

97 **(d)**
$e = L\frac{di}{dt} \Rightarrow [e] = [ML^2T^{-2}A^{-2}]\left[\frac{A}{T}\right]$
$[e] = \left[\frac{ML^2T^{-2}}{AT}\right] = [ML^2T^{-2}Q^{-1}]$

98 **(c)**
Given, length of rod A is

$L_A = 3.25 \pm 0.01$

Of B is $L_B = 4.19 \pm 0.01$

Then, the rod B is longer than rod A by a length

$\Delta l = L_B - L_A$

$\Delta l = (4.19 \pm 0.01) - (3.25 \pm 0.01)$

$\Delta l = (0.94 \pm 0.02)cm$

99 **(c)**
$\frac{\Delta l}{l} \times 100 = \frac{0.01}{15.12} \times 100 = 0.07,$
$\frac{\Delta b}{b} \times 100 = \frac{0.01}{10.15} \times 100 = 0.1,$
$\frac{\Delta t}{t} \times 100 = \frac{0.01}{5.28} \times 100 = 0.2$
Required percentage = 0.07 + 0.1 + 0.2 = 0.37%

100 **(a)**
Here,
Mass of a body, $M = 5.00 \pm 0.05\ kg$
Volume of a body, $V = 1.00 \pm 0.05 m^3$
Density, $\rho = \frac{M}{V}$
Relative error in density is
$\frac{\Delta\rho}{\rho} = \frac{\Delta M}{M} + \frac{\Delta V}{V}$
Percentage error in density is
$\frac{\Delta\rho}{\rho} \times 100 = \frac{\Delta M}{M} \times 100 + \frac{\Delta V}{V} \times 100$
$= \left(\frac{0.05}{5} \times 100\right) + \left(\frac{0.05}{1} \times 100\right)$
$= 1\% + 5\% = 6\%$

101 **(c)**
$\frac{[\text{planck's constant}]}{[\text{linear momentum}]} = \frac{[ML^2T^{-1}]}{[MLT^{-1}]}$
$= [M^0LT^0]$

102 **(b)**
$[\varepsilon_0] = \frac{[A^2T^2]}{[ML^3T^{-2}]} = [M^{-1}L^{-3}T^4A^2]$

103 **(b)**
$1\ dyne = 10^{-5}\ newton, 1\ cm = 10^{-2}\ m$
$70\frac{dyne}{cm} = \frac{70 \times 10^{-5}}{10^{-2}}\frac{N}{m}$
$= 7 \times 10^{-2} N/m$

104 **(d)**
$n_2 = n_1\left[\frac{M_2}{M_1}\right]^1\left[\frac{L_2}{L_1}\right]^2\left[\frac{T_2}{T_1}\right]^{-2}$

Given, $M_2 = 2M_1, L_2 = 2L_1, T_2 = 2T_1$

$\therefore n_2 = n_1[2]^1[2]^2[2]^{-2} = 2n_1$

105 **(c)**
Impulse = Force × time $= (kg - m/s^2) \times s = kg\text{-}m/s$

106 **(a)**
From Coulomb's law

$$F = \frac{1}{4\pi\varepsilon_0}\frac{q_1 q_2}{r^2}$$

$$\Rightarrow \left[\frac{1}{4\pi\varepsilon_0}\right] = \frac{[F \times r^2]}{[q]^2}$$

$$= \frac{[\text{newton}][\text{metre}]^2}{[\text{coulomb}]^2}$$

$$= \text{Nm}^2\text{C}^{-2}$$

107 **(a)**

By substituting the dimensions in $T = 2\pi\sqrt{\frac{R^3}{GM}}$ we get $\sqrt{\frac{L^3}{M^{-1}L^3T^{-2}\times M}} = T$

108 **(b)**

Volume $V = l \times b \times t$
$= 12 \times 6 \times 2.45 = 176.4\ cm^3$
$V = 1.764 \times 10^2\ cm^3$
Since, the minimum number of significant figure is one in breadth, hence volume will also contain only one significant figure. Hence, $V = 2 \times 10^2\ cm^3$

109 **(b)**

From the principle of dimensional homogeneity
$[a] = \left[\frac{F}{t}\right] = [MLT^{-3}]$ and $[b] = \left[\frac{F}{t^2}\right] = [MLT^{-4}]$

110 **(b)**

Kg-m/sec is the unit of linear momentum

111 **(a)**

We know that the dimensional formula of energy is $[ML^2T^{-2}]$

$$n_2 = 1\left[\left[\frac{1\text{kg}}{10\text{kg}}\right]^1 \left[\frac{1\text{m}}{1\text{km}}\right]\left[\frac{1\text{s}}{1\ \text{min}}\right]^2\right.$$

$$= \frac{1}{10} \times \frac{1}{10^6} \times \frac{1}{(60)^{-2}} = \frac{3600}{10^7} = 3.6 \times 10^{-4}$$

112 **(c)**

$Y = \frac{4MgL}{\pi D^2 l}$ so maximum permissible error in Y

$$= \frac{\Delta Y}{Y} \times 100 = \left(\frac{\Delta M}{M} + \frac{\Delta g}{g} + \frac{\Delta L}{L} + \frac{2\Delta D}{D} + \frac{\Delta l}{l}\right) \times 100$$

$$= \left(\frac{1}{300} + \frac{1}{981} + \frac{1}{2820} + 2 \times \frac{1}{41} + \frac{1}{87}\right) \times 100$$

$$= 0.065 \times 100 = 6.5\%$$

113 **(d)**

$1\ newton\ =\ 10^{-5}\ dyne$

114 **(b)**

$$\therefore \left(\frac{\Delta R}{R} \times 100\right)_{\max} = \frac{\Delta V}{V} \times 100 + \frac{\Delta I}{I} \times 100$$

$$= \frac{5}{100} \times 100 + \frac{0.2}{10} \times 100 = (5 + 2)\% = 7\%$$

115 **(d)**

$$[\varepsilon_0 L] = [C] \therefore X = \frac{\varepsilon_0 LV}{t} = \frac{C \times V}{t} = \frac{Q}{t} = \text{current}$$

116 **(a)**

Charge = current × time

117 **(a)**

If a charge of 1 C moving with a velocity of 1 ms^{-1} perpendicular to a uniform magnetic field experiences a force of 1 N, then the magnitude of the field is 1 T. The SI unit of magnetic field is Wb m^{-2}. Thus,

$$1\ \text{T} = 1\ \text{NA}^{-1}\text{m}^{-1} = 1\ \text{Wbm}^{-2}$$

In CGS systems

$$1\ \text{tesla} = 10^4\ \text{gauss} = 1\ \text{Wbm}^{-2}$$

118 **(a)**

M =Pole strength × length
$= amp - metre\ \times metre$
$= amp - metre^2$

119 **(d)**

Surface Tension $= \frac{\text{Force}}{\text{Length}}$

$$= \frac{[MLT^{-2}]}{[L]} = [ML^0T^{-2}]$$

Spring constant $= \frac{\text{Force}}{\text{Length}}$

$$= \frac{[MLT^{-2}]}{[L]} = [ML^0T^{-2}]$$

120 **(c)**

$$[X] = [F] \times [\rho] = [MLT^{-2}] \times \left[\frac{M}{L^3}\right] = [M^2L^{-2}T^{-2}]$$

121 **(b)**

T-ratios are dimensionless. So the unit of $r$is N^2.

122 **(a)**

Farad is the unit of capacitance and

$C = \frac{Q}{V} = \left[\frac{Q}{[ML^2T^{-2}Q^{-1}]}\right] = M^{-1}L^{-2}T^2Q^2$

123 **(b)**

One light year

$= 3 \times 10^8$ m/s year

$= \frac{3\times10^8}{s} \times 365 \times 24 \times 60 \times 60\text{s}$

$= 3 \times 10^8 \times 365 \times 24 \times 60 \times 60\text{m}$

$= 9.461 \times 10^{15}\text{m}$

124 **(c)**

Mean time period $T = 2.00$ sec

& Mean absolute error $= \Delta T = 0.05$ sec

To express maximum estimate of error, the time period

should be written as $(2.00\ \pm 0.05)$sec

125 **(d)**

$20\ VSD = 16\ MSD$

$1\ VSD = 0.8\ MSD$

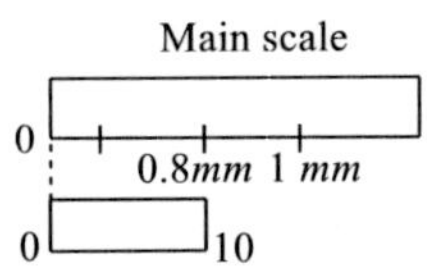

Least count $= MSD - VSD$

$= 1\ mm - 0.8\ mm = 0.2\ mm$

126 **(a)**

Weight in air $= (5.00 \pm 0.05)N$

Weight in water $=(4.00\ \pm 0.05)N$

Loss of weight in water $=(1.00 \pm 0.1)N$

Now relative density$=\frac{\text{weight in air}}{\text{weight loss in water}}$

$i.e. R.D = \frac{5.00 \pm 0.05}{1.00 \pm 0.1}$

Now relative density with max permissible error

$= \frac{5.00}{1.00} \pm \left(\frac{0.05}{5.00} + \frac{0.1}{1.00}\right) \times 100$

$= 5.0 \pm (1 + 10)\%$

$= 5.0 \pm 11\%$

127 **(a)**

Energy $U = \frac{1}{2}LI^2$

$\Rightarrow \qquad L = \frac{2U}{I^2}$

$\therefore [L] = \frac{[U]}{[I]^2} = \frac{[\text{ML}^2\text{T}^{-2}]}{[\text{A}]^2} = [\text{ML}^2\text{T}^{-2}\text{A}^{-2}]$

128 **(a)**

From Newton's second law

Force(F) = Mass$(M) \times$ acceleration

Dimensions of $[F] = [\text{MLT}^{-2}]$

$\therefore \qquad [M] = [\text{FL}^{-1}\text{T}^2]$

129 **(a)**

$E = \frac{1}{2}Li^2$ hence $L = [ML^2T^{-2}A^{-2}]$

130 **(c)**

Linear momentum $= [\text{MLT}^{-1}]$

Angular momentum $= [\text{ML}^2\text{T}^{-1}]$

132 **(d)**

$E = F/q =$ Newton/coulomb

133 **(a)**

Each of three terms in the given equation has the dimensional formula of force.

134 **(b)**

Impulse =Force × Time $= [MLT^{-2}][T] = [MLT^{-1}]$

135 **(b)**

By substituting the dimension of given quantities

$[ML^{-1}T^{-2}]^x[MT^{-3}]^y[LT^{-1}]^z = [MLT]^0$

By comparing the power of M, L, T in both sides

$x + y = 0$...(i)

$-x + z = 0$ ...(ii)

$-2x - 3y - z = 0$...(iii)

The only values of x, y, z satisfying (i), (ii) and (iii) corresponds to (b)

137 **(c)**

Energy $E = \frac{1}{2}LI^2 \Rightarrow [\text{L}] = \frac{[E]}{[I]^2} = \frac{[\text{ML}^2\text{T}^{-2}]}{[\text{A}^2]} = [\text{ML}^2\text{T}^{-2}\text{A}^{-2}]$

138 **(c)**

The magnitude of induced *emf* is

$|\varepsilon| = L\frac{dI}{dt}$ or $L = \frac{|\varepsilon|dt}{dI}$

$L = \frac{\text{volt} \times \text{second}}{\text{ampere}} = ohm$ second

139 **(a)**

$[h] = [\text{ML}^2\text{T}^{-1}]$

$[c] = [\text{LT}^{-1}]$

$[\text{G}] = [\text{M}^{-1}\text{L}^3\text{T}^{-2}]$

140 **(d)**

$$\text{Modulas of rigidity} = \frac{\text{Shear stress}}{\text{Shear strain}} = [ML^{-1}T^{-2}]$$

141 **(c)**

Unit of energy will be $kg - m^2/sec^2$

142 **(c)**

$$x = \frac{1\text{gcms}^{-1}}{\text{T}^2} = \frac{1\text{gcms}^{-1}}{1\text{kg} \times 1\text{ms}^{-1} \times 1\text{s}}$$

$$= \frac{1\text{gcms}^{-1}}{10^3\text{g} \times 10^2\text{cms}^2 \times 1\text{s}} = 10^{-5}$$

143 **(a)**

$n_1 u_1 = n_2 u_2$

$$n_2 = \frac{1 \text{ shake}}{1 \text{ ns}}$$

$$= \frac{10^{-8}\text{s}}{10^{-9}\text{s}}$$

$\therefore \; n_2 = 10$

144 **(a)**

Angular velocity $= \frac{\theta}{t}, [\omega] = \frac{[M^0L^0T^0]}{[T]} = [T^{-1}]$

145 **(b)**

Solar constant is energy received per unit area per unit time i.e. $\frac{[ML^2T^{-2}]}{[L^2][T]} = [M^1T^{-3}]$

146 **(a)**

$[e] = [AT], \epsilon_0 = [M^{-1}L^{-3}T^4A^2], [h] = [ML^2T^{-1}]$

And $[c] = [LT^{-1}]$

$$\therefore \left[\frac{e^2}{4\pi\epsilon_0 hc}\right]$$

$$= \left[\frac{A^2T^2}{M^{-1}L^{-3}T^4A^2 \times ML^2T^{-1} \times LT^{-1}}\right]$$

$= [M^0L^0T^0]$

147 **(b)**

Capacitance $C = \frac{\text{Charge}}{\text{potential}} = \frac{q}{V}$

Also potential $= \frac{\text{work}}{\text{charge}} \left(\because V = \frac{W}{q}\right)$

$\therefore \; C = \frac{q^2}{J}$ as well as $C = \frac{J}{V^2}$.

Thus, (a), (c), (d) are equivalent to farad but (b) is not equivalent to farad.

148 **(c)**

$$[X] = \left[\frac{\text{M}^{-1}\text{L}^3\text{T}^{-2} \times \text{ML}^2\text{T}^{-1}}{\text{L}^3\text{T}^{-3}}\right]^{-1/2} = [\text{L}]$$

149 **(d)**

Given, $v = at + bt^2$

Applying the law of homogeneity $[v] = [bt^2]$

Or $[\text{LT}^{-1}] = [b\text{T}^2]$

Or $[b] = [\text{LT}^{-3}]$

151 **(a)**

$$\frac{h}{I} = \left[\frac{ML^2T^{-1}}{ML^2}\right] = [T^{-1}]$$

152 **(b)**

From the principle of homogenity $\left(\frac{x}{v}\right)$ has dimensions of T

153 **(a)**

Impulse = force × time

$= [\text{MLT}^{-2}][\text{T}]$

$= [\text{MLT}^{-1}]$

154 **(a)**

Percentage error in radius is $\frac{0.1}{4.3} \times 100$.

again, $V \propto R^3$

155 **(d)**

Size of universe is about $10^{26}\text{m} = 10^6 \times (9.46 \times 10^{15})\text{m}$
$= 10^{10}\text{ly}$

156 **(a)**

Volume $V = l^3 = (1.2 \times 10^{-2}m)^3 = 1.728 \times 10^{-6}m^3$

$\because$ length l has two significance figures. Therefore, the correct answer is

$V = 1.7 \times 10^{-6}m^3$

157 **(b)**

$$mv = kg\left(\frac{m}{\text{sec}}\right)$$

158 **(d)**

$[\text{Calorie}] = [\text{ML}^2\text{T}^{-2}]$

Comparing with general dimensional formula $[M^aL^bT^c]$, we get

$a = 1, b = 2, c = -2.$

$$n_2 = 4.2\left[\frac{1\text{kg}}{\alpha\text{kg}}\right]^1\left[\frac{1\text{m}}{\beta\text{m}}\right]^2\left[\frac{1\text{s}}{\gamma\text{s}}\right]^{-2} = 4.2\alpha^{-1}\beta^{-2}\gamma^2$$

160 **(b)**

$$\text{Bulk modulus K} = \frac{\text{normal stress}}{\text{volumetric strain}}$$

$$= \frac{F/A}{-\Delta V/V}$$

$$= -\frac{FV}{A\,\Delta V}$$

Now, $\frac{F}{A} = p$

$$\therefore \quad K = \frac{pV}{\Delta V}$$

As volumetric strain is dimensionless.

$\therefore$ Dimensions of K
= dimensions of normal stress

$$\Rightarrow \quad [K] = [\mathrm{ML^{-1}T^{-2}}]$$

161 **(b)**
As $x = ka^m \times t^n$
$[M^0LT^0] = [LT^{-2}]^m[T]^n = [L^mT^{-2m+n}]$
$\therefore m = 1$ and $-2m + n = 0 \Rightarrow n = 2$

162 **(a)**
1 Newton = 10^5 dyne and 1m=100 cm

166 **(d)**
We know $[F] = [\mathrm{MLT^{-2}}]$
$T^2 = \frac{ML}{F} = \frac{1\mathrm{kg} \times 1\mathrm{m}}{1\mathrm{kg} - \mathrm{wt}} = \frac{1\mathrm{kg} \times 1\mathrm{m}}{9.8\mathrm{N}}$
$T = \frac{1}{\sqrt{9.8}}$ sec

167 **(a)**
Let $F \propto P^xV^yT^z$
By substituting the following dimensions:
$[P] = [ML^{-1}T^{-2}][V] = [LT^{-1}], [T] = [T]$
and comparing the dimension of both sides
$x = 1, y = 2, z = 2$, so $F = PV^2T^2$

168 **(c)**
Do not think in terms of I and ω.
Remember; kinetic energy is fundamentally 'work'
W =Force × distance
$= [\mathrm{MLT^{-2}}] \times [\mathrm{L}]$
$= [\mathrm{ML^2T^{-2}}]$

169 **(c)**
$E = hv \Rightarrow [ML^2T^{-2}] = [h][T^{-1}] \Rightarrow [h] = [ML^2T^{-1}]$

170 **(c)**
Momentum $[MLT^{-1}]$, Plank's constant $[ML^2T^{-1}]$

171 **(d)**
Velocity gradient $= \frac{v}{x} = \frac{[LT^{-1}]}{[L]} = [T^{-1}]$
Potential gradient $= \frac{V}{x} = \frac{[ML^2T^{-3}A^{-1}]}{[L]} = [MLT^{-3}A^{-1}]$
Energy gradient $= \frac{E}{x} = \frac{[ML^2T^2]}{[L]} = [MLT^{-2}]$
And pressure gradient $= \frac{P}{x} = \frac{[ML^{-1}T^{-2}]}{[L]} = [ML^{-2}T^{-2}]$

172 **(c)**
$F \propto v \Rightarrow F = kv$
$k = \frac{F}{v} \Rightarrow [k] = \frac{[kgms^{-2}]}{[ms^{-1}]} = kg\ s^{-1}$

173 **(a)**
$$[C] = \left(\frac{Q}{V}\right) = \left(\frac{Q^2}{W}\right) = \left[\frac{A^2T^2}{ML^2T^{-2}}\right] = [M^{-1}L^{-2}T^4A^2]$$

174 **(d)**
$\tau = \frac{dL}{dt} \Rightarrow dL = \tau \times dt = r \times F \times dt$
i.e., the unit of angular momentum is *joule-second*

175 **(b)**
In mathematical operation, involving addition the result would be correct up to minimum number of decimal places in any of the quantity involve.

Given, L=2.3331 cm and B=2.10 cm.

Taking correct up to decimal places and since, can be rounded off less than 5, the preceding digit is unaffiliated.

$\therefore L = 2.33\ cm, B = 2.10\ cm$

$L + B = 2.33 + 2.10 = 4.43\ cm$

176 **(c)**
Volume of sphere $(V) = \frac{4}{3}\pi r^3$
% error in volume $= 3 \times \frac{\Delta r}{r} \times 100 = \left(3 \times \frac{0.1}{5.3}\right) \times 100$

177 **(b)**

$$[B] = \frac{[F]}{[I][L]} = \frac{[MLT^{-2}]}{[CT^{-1}][L]} = [MT^{-1}C^{-1}]$$

178 **(b)**

$E = KF^a A^b T^c$

$[ML^2T^{-2}] = [MLT^{-2}]^a[LT^{-2}]^b[T]^c$

$[ML^2T^{-2}] = [M^a L^{a+b} T^{-2a-2b+c}]$

$\therefore a = 1, a + b = 2 \Rightarrow b = 1$

And $-2a - 2b + c = -2 \Rightarrow c = 2$

$\therefore E = KFAT^2$

179 **(c)**

S_{nth} represents the distance covered in nth sec.

180 **(a)**

Because in S.I. system there are seven fundamental quantities

181 **(b)**

$$F = \frac{1}{4\pi\varepsilon_0}\frac{q_1q_2}{r^2}$$

$$\Rightarrow \varepsilon_0 = \frac{|q_1||q_2|}{[F][r^2]} = \frac{[A^2T^2]}{[MLT^{-2}][L^2]} = [A^2T^4M^{-1}L^{-3}]$$

182 **(a)**

Young's modulus $Y = \frac{\text{stress}}{\text{strain}} = \text{N/m}^2$ or pascal (in SI system)

And $Y = \frac{\text{dyne}}{\text{cm}^2}$ (in CGS system)

Then, Nm^{-1} is not the unit of Young's modulus.

183 **(b)**

$$\text{Surface tension} = \frac{\text{Force}}{\text{Length}} = newton/metre$$

184 **(a)**

Let $\quad v \propto \sigma^a \rho^b \lambda^c$

Equating dimensions on both sides.

$$[\text{M}^0\text{LT}^{-1}] \propto [\text{MT}^{-2}]^a[\text{ML}^{-3}]^b[\text{L}]^c$$

$$\propto [\text{M}]^{a+b}[\text{L}]^{-3b+c}[\text{T}]^{-2a}$$

Equating the powers of M, L, T on both sides, we get

$$a + b = 0$$

$$-3b + c = 1$$

$$-2a = -1$$

Solving, we get

$$a = \frac{1}{2}, b = -\frac{1}{2}, c = -\frac{1}{2}$$

$$\therefore \quad v \propto \sigma^{1/2}\rho^{-1/2}\lambda^{-1/2}$$

$$\therefore \quad v^2 \propto \frac{\sigma}{\rho\lambda}$$

185 **(c)**

$$\left[\frac{1}{2}\epsilon_0 E^2\right] = [\text{Energy density}]$$

$$= \frac{ML^2T^{-2}}{L^3} = ML^{-1}T^{-2}$$

187 **(d)**

Modulus of rigidity $\eta = \frac{F/A}{\phi}$

Dimensions of $\eta = \frac{[\text{MLT}^{-2}]}{[\text{L}^2]}$

$$= [\text{ML}^{-1}\text{T}^{-2}]$$

189 **(d)**

Capacitances $C = \frac{Q}{V} = \frac{QQ}{W} = \frac{Q^2}{W} = \frac{I^2t^2}{W}$,

$$[C] = \frac{[\text{I}^2\text{T}^2]}{[\text{ML}^2\text{T}^{-2}]} = [\text{M}^{-1}\text{L}^{-2}\text{T}^4\text{I}^2]$$

191 **(a)**

[surface tension]=$[\text{ML}^0\text{T}^{-2}]$,
[viscosity]=$[\text{ML}^{-1}\text{T}^{-1}]$.
Clearly, mass has the same power in the two physical quantities.

192 **(d)**

Diameter = Main scale reading

+ Circular scale reading × LC + Zero error

$$= 3 + 35 \times \frac{1}{2 \times 50} + 0.03 = 3.38 \text{ mm}$$

193 **(b)**

$1\ MeV = 10^6 eV$

194 **(d)**

Dimensions of $\beta^3 =$ dimensions of density $= [ML^{-3}]$

$$\beta = [M^{1/3}L^{-1}]$$

Also $\alpha =$ force $\times$ density

$$= [MLT^{-2}][ML^{-3}]$$

$$= [M^2L^{-2}T^{-2}]$$

196 **(a)**

Here, $[f] = \frac{[S]}{[t^3]} = [M^0LT^{-3}]$.

197 **(a)**

Maximum absolute error is $\Delta a + \Delta b$. Now work out the relative error ad finally the percentage error.

198 **(d)**

As $v = \frac{4}{3}\pi r^3$

$\frac{dv}{v} = 3\left(\frac{dr}{r}\right)$

∴ Percentage error in determination of volume = 3

(Percentage error in measurement of radius) $= 3(2\%) = 6\%$

199 **(c)**

Impulse = change in momentum so dimensions of both quantities will be same and equal to MLT^{-1}

201 **(b)**

We know that

Specific heat $= \frac{Q}{m\,\Delta t}$

Unit of specific heat

$$= \frac{\text{unit of heat}}{\text{unit of mass} \times \text{unit of temperature}}$$

∴ Unit of specific heat $= \frac{J}{kg°C}$

$= Jkg^{-1}°C^{-1}$

202 **(a)**

Given, $m = 3.513$ kg and $v = 5.00\ ms^{-1}$

So, momentum, $p = mv = 17.565$

As the number of significant digits in m is 4 and v is 3, so, p must have 3 significant digits

$$p = 17.6\ kgms^{-1}$$

203 **(b)**

The height of a tree, building tower, hill etc, can be determined with the help of a sextant.

204 **(c)**

Since for 50.14 cm,significant number = 4 and for 0.00025, significant number=2

206 **(c)**

Volume of sphere is given by

$$v = \frac{4}{3}\pi R^3$$

Where R is radius of sphere

$$\therefore \frac{\Delta V}{V} = 3\frac{\Delta R}{R}$$

Hence, percentage error in volume

$$\frac{\Delta V}{V} \times 100 = 3\left(\frac{\Delta R}{R} \times 100\right)\%$$

$$= 3 \times 3\% = 9\%$$

208 **(a)**

Time $T = 2\pi\sqrt{LC}$.

209 **(d)**

Note carefully that every alterative has Gh and c^5.

$[Gh] = [M^{-1}L^3T^{-2}][ML^2T^{-1}] = [M^0L^5T^3]$

$[c] = [LT^{-1}]$

$\therefore \left(\frac{Gh}{c^5}\right)^{1/2} = [T]$

210 **(a)**

The mass of electron $= \frac{9.1\times10^{-31}}{1.67\times10^{-27}}$

$\therefore E = \frac{9.1\times10^{-31}}{1.67\times10^{-27}} \times 931 \text{ MeV}$
$= 0.5073 \text{ MeV}$

212 **(c)**

According to Stefan's law, the energy radiated per second or power radiated is given by

$$P = \sigma A T^4$$

$$\therefore \quad \sigma = \frac{P}{AT^4}$$

Therefore, unit of $\sigma = \frac{\text{W}}{\text{m}^2\text{K}^4} = \text{Wm}^{-2}\text{K}^{-4}$

213 **(a)**

$K = C + 273.15$

214 **(b)**

Velocity $v = k\lambda^a\rho^b g^c \Rightarrow [\text{M}^0\text{LT}^{-1}] = [\text{L}^a][\text{M}^b\text{L}^{-3b}][\text{L}^c\text{T}^{-2c}]$

Or $\quad [\text{M}^0\text{LT}^{-1}] = [\text{M}^b\text{L}^{a-3b+c}\text{T}^{-2c}]$

Equating powers of M, L and T, we get

$-2c = -1$

Again, $a - 3b + c = 1, b = 0, c = \frac{1}{2}$

$\therefore \ v = k\lambda^{1/2}\rho^0 g^{1/2}$ or $v^2 \propto g\lambda$

216 **(c)**

$$R_{\text{parallel}} = \frac{R_1R_2}{(R_1+R_2)}$$

$$\Rightarrow \frac{\Delta R_p}{R_p} = \frac{\Delta R_1}{R_1} + \frac{\Delta R_2}{R_2} + \frac{\Delta(R_1+R_2)}{R_1+R_2}$$

$$\Rightarrow \frac{\Delta R_p}{R_p} = \frac{0.3}{6} + \frac{0.2}{10} + \frac{(0.3+0.2)}{10+6}$$

$= 0.05 + 0.02 + 0.03125 = 0.10125$

$\therefore \frac{\Delta R_p}{R_p} \times 100 = 10.125$ or 10.125%

217 **(b)**

From Coulomb's law

$$F = \frac{1}{4\pi\varepsilon_0}\frac{q_1q_2}{r^2}$$

Or $\varepsilon_0 = \frac{q_1q_2}{4\pi F r^2}$

$\therefore$ Units of ε_0(permittivity)

$$= \frac{\text{C}^2}{\text{N}-\text{m}^2} = \text{C}^2\text{N}^{-1}\text{m}^{-2}$$

218 **(a)**

Let $v \propto \sigma^a\rho^b\lambda^c$

Equating dimensions on both sides,

$[M^0L^1T^{-1}] \propto [MT^{-2}]^a[ML^{-3}]^b[L]^c$

$\propto [M]^{a+b}[L]^{-3b+c}[T]^{-2a}$

Equating the powers of M, L, T on both sides, we get

$a + b = 0$

$-3b + c = 1$

$-2a = -1$

Solving, we get

$a = \frac{1}{2}, b = -\frac{1}{2}, c = -\frac{1}{2}$

$\therefore \ v \propto \sigma^{1/2}\rho^{-1/2}\lambda^{-1/2}$

$\therefore \ v^2 \propto \frac{\sigma}{\rho\lambda}$

219 **(a)**

The unit of $\frac{1}{2}\varepsilon E^2 = \frac{\text{C}^2}{\text{Nm}^2}\left(\frac{\text{N}}{\text{C}}\right)^2$

$= \frac{\text{C}^2}{\text{Nm}^2}\frac{\text{N}^2}{\text{C}^2} = \frac{\text{N}}{\text{m}^2} = \frac{\text{Nm}}{\text{m}^3}$

$= \frac{\text{J}}{\text{m}^3}$ =energy density

221 **(b)**

MeV-sec is not a unit of energy. While others are units of energy.

222 **(a)**

Let radius of gyration $[k] \propto [h]^x[c]^y[G]^z$

By substituting the dimension of $[k] = [L]$

$[h] = [ML^2T^{-1}]$

$[c] = [LT^{-1}]$

$[G] = [M^{-1}L^3T^{-2}]$

And by comparing the power of both sides

We can get $x = 1/2, y = -3/2, z = 1/2$

Therefore dimension of radius of gyration is

$[h]^{1/2}[c]^{-3/2}[G]^{1/2}$

223 **(d)**

$[G] = [M^{-1}L^3T^{-2}]; [h] = [ML^2T^{-1}]$

Power $= \frac{1}{\text{focal length}} = [L^{-1}]$

All quantities have dimensions

224 **(c)**

The dimensions of angular momentum are

$$J = [\text{ML}^2T^{-1}]$$

When units of mass, length and time are doubled, than

$J' = [2M(2L)^2(2T)^{-1}]$

$\Rightarrow \quad J' = 4[ML^2T^{-1}] = 4\,J$

Unit of angular momentum is quadrupled.

226 **(c)**

Einstein's mass-energy equivalence is $E = mc^2$.

227 **(d)**

Poisson's ratio is a unitless quantity

228 **(a)**

Energy density $= \frac{Energy}{Volume} = \frac{ML^2T^{-2}}{L^3} = [ML^{-1}T^{-2}]$

Young' s modulus $= \frac{Stress}{Strain} = \frac{ML^{-1}T^{-2}}{M^0L^0T^0} = [ML^{-1}T^{-2}]$

Both have the same dimensions

229 **(d)**

$Watt = Joule/second = Ampere \times volt$
$= Ampere^2 \times Ohm$

230 **(a)**

$[a] = [T^2]$ and $[b] = \frac{[a-t^2]}{[P][X]} = \frac{T^2}{[ML^{-1}T^{-2}][L]}$

$\Rightarrow [b] = [M^{-1}T^4]$

So $\left[\frac{a}{b}\right] = \frac{[T^2]}{[M^{-1}T^4]} = [MT^{-2}]$

231 **(b)**

The action of impulse is to change the momentum of a body or particle and the impulse of force is equal to the change in momentum.

Thus, the dimensions of impulse are same as that of momentum.

235 **(a)**

Conductivity $\left(\frac{1}{\rho}\right) = \frac{1}{aR} = \frac{[L]}{[L^2][ML^2T^{-3}A^{-2}]}$

$= [M^{-1}L^{-3}T^3A^2]$

236 **(a)**

Let $T \propto S^x r^y p^z$

By substituting the dimensions of $[T] = [T]$

$[S] = [MT^{-2}], [r] = [L], [\rho] = [ML^{-3}]$

and by comparing the power of both the sides

$x = -1/2, y = 3/2, z = 1/2$

so $T \propto \sqrt{\rho r^3/S} \Rightarrow T = k\sqrt{\frac{\rho r^3}{S}}$

237 **(b)**

Average value

$= \frac{2.63 + 2.56 + 2.42 + 2.71 + 2.80}{5}$

$= 2.62$ sec

Now $|\Delta T_1| = 2.63 - 2.62 = 0.01$

$|\Delta T_2| = 2.62 - 2.56 = 0.06$

$|\Delta T_3| = 2.62 - 2.42 = 0.20$

$|\Delta T_4| = 2.71 - 2.62 = 0.09$

$|\Delta T_5| = 2.80 - 2.62 = 0.18$

Mean absolute error

$$\Delta T = \frac{|\Delta T_1| + |\Delta T_2| + |\Delta T_3| + |\Delta T_4| + |\Delta T_5|}{5}$$

$$= \frac{0.54}{5} = 0.108 = 0.11\ sec$$

238 **(a)**

Stefan's constant

$$\sigma = \frac{\text{Energy}}{\text{Area} \times \text{Time} \times (\text{Temperature})^4}$$

$$\therefore \sigma = \frac{[ML^2T^{-2}]}{[L^2][T][K]^4} = [ML^0T^{-3}K^{-4}]$$
$$= [MT^{-3}K^{-4}]$$

239 **(a)**

$$\because R = \frac{PV}{T} = \left[\frac{ML^{-1}T^{-2} \times L^3}{\theta}\right]$$
$$= [ML^2T^{-2}\theta^{-1}]$$

240 **(a)**

Least count of both instrument

$\Delta d = \Delta \ell = \frac{0.5}{100}\ mm = 5 \times 10^{-3}\ mm$

$Y = \frac{4MLg}{\pi \ell d^2}$

$\left(\frac{\Delta Y}{Y}\right)_{\max} = \frac{\Delta \ell}{\ell} + 2\frac{\Delta d}{d}$

Error due to ℓ measurement $\frac{\Delta \ell}{\ell} = \frac{0.5/100\ mm}{0.25\ mm} = 2\%$

Error due to d measurement $2\frac{\Delta d}{d} = \frac{2\times\frac{0.5}{100}}{0.5\ mm}$

$= \frac{0.5/100}{0.25} = 2\%$

Hence due to the errors in the measurements of d and ℓ are the same

241 **(a)**

Diameter of wire,

$$d = \text{MSR} + \text{CSR} \times \text{LC}$$

$$= 0 + 52 \times \frac{1}{100}$$

$$= 0.52 \text{ mm} = 0.052 \text{ cm.}$$

244 **(a)**

$V = (8 + 0.5)$

$I = (2 + 0.2)$

$R = \frac{8}{2} = 4$

$\frac{\Delta R}{R}\% = \left(\frac{\Delta V}{V} + \frac{\Delta I}{I}\right)$

$= \left(\frac{0.5}{8} + \frac{0.2}{2}\right) \times 100 = 16.25\%$

$\therefore\ R = (4 \pm 16.25\%)$

245 **(d)**

$n_2 = n_1 \left[\frac{m_1}{m_2}\right]\left[\frac{L_1}{L_2}\right]^{-3}$

$= 4\left[\frac{1\ gm}{100\ gm}\right]\left[\frac{cm}{10\ cm}\right]^{-3} = 4 \times \frac{1}{100} \times 10^3$

$= 40$ units

246 **(d)**

Required error in density $= 3\% + 3 \times 2\% = 9\%$.

247 **(a)**

By the principle of dimensional homogenity

$[P] = \left[\frac{a}{V^2}\right] \Rightarrow [a] = [P] \times [V^2]$

$= [ML^{-1}T^{-2}][L^6]$

$= [ML^5T^{-2}]$

248 **(d)**

Dimensional formula of angular momentum $(L) = [ML^2T^{-1}]$

Torque $(\tau) = [ML^2T^{-2}]$

Coefficient of thermal conductivity

$(K) = [MLT^{-3}K^{-1}]$

Gravitational constant $(G) = [M^{-1}L^3T^{-2}]$

Thus, gravitational constant has negative dimension of mass.

249 **(c)**

Shear modulus $= \frac{\text{Shearing stress}}{\text{Shearing strain}} = \frac{F}{A\theta} = [ML^{-1}T^{-2}]$

250 **(d)**

$R_1 = (6 \pm 0.3)\text{k}\Omega, R_2 = (10 \pm 0.2)\text{k}\Omega$

$$R_{parallel} = \frac{R_1R_2}{(R_1 + R_2)}$$

Let $(R_1 + R_2) = x$

$$\Rightarrow \quad R_P = \frac{R_1R_2}{x}$$

Taking log of both sides

$$\ln R_P = \ln R_1 + \ln R_2 - \ln x$$

Differentiating,

$$\frac{\Delta R_P}{R_P} = \frac{\Delta R_1}{R_1} + \frac{\Delta R_2}{R_2} + \left(-\frac{\Delta x}{x}\right)$$

$$\Delta x_{\text{mean}} = \frac{0.3 + 0.2}{2} = 0.25\ \Omega$$

$$R_{\text{mean}} = \frac{6 + 10}{2} = 8\ \Omega$$

$$\therefore \quad x = \frac{6 + 10}{2} = 8\ \Omega$$

$$\Rightarrow \quad \frac{\Delta x}{x} = \frac{0.25}{8}$$

$$\therefore \text{Total error} = \frac{0.3}{6} + \frac{0.2}{10} + \frac{0.25}{8}$$

$$= 0.05 + 0.02 + 0.03125$$
$$= 0.10125$$

$$\therefore \quad \frac{\Delta R_P}{R_P} = 10.125\%$$

252 **(a)**

$X = [M^aL^bT^c]$

Maximum % error in $X = a\alpha + b\beta + c\gamma$

253 **(d)**

Let $L = [h^a c^b G^c]$

$\therefore$ $[L']$
$= [ML^2T^{-1}]^a [LT^{-1}]^b [M^{-1}L^3T^{-2}]^c$

$\Rightarrow \quad a = \frac{1}{2}, b = -\frac{3}{2}, c = \frac{1}{2}$

Hence, $L = \left[h^{1/2} c^{-3/2} G^{1/2}\right]$

254 **(c)**

Coefficient o friction= $\frac{\text{Applied force}}{\text{Normal reactiom}}$

$= \frac{[MLT^{-2}]}{[MLT^{-2}]}$ =no dimensions

Unit= $\frac{N}{N}$ =no unit

256 **(d)**

Express the result in two significant figures.

257 **(c)**

Volume of cylinder $V = \pi r^2 l$

Percentage error in volume

$$\frac{\Delta V}{V} \times 100 = \frac{2\Delta r}{r} \times 100 + \frac{\Delta l}{l} \times 100$$

$$= \left(2 \times \frac{0.01}{2.0} \times 100 + \frac{0.1}{5.0} \times 100 \right)$$

$$= (1+2)\% = 3\%$$

258 **(b)**

$RC = T$

$\because [R] = ML^2T^{-3}A^{-2}]$ and $[C] = [M^{-1}L^{-2}T^4A^2]$

259 **(a)**

Linear momentum = Mass $\times$ Velocity = $[MLT^{-1}]$

Moment of a force = Force $\times$ Distance = $[ML^2T^{-2}]$

260 **(c)**

Velocity is given by

$$v = \frac{1}{\sqrt{\mu_0 \varepsilon_0}}$$

$$\therefore \quad v^2 = \frac{1}{\mu_0 \varepsilon_0} = [LT^{-1}]^2$$

$$\therefore \quad \frac{1}{\mu_0 \varepsilon_0} = [L^2T^{-2}]$$

261 **(b)**

$y = r \sin(\omega t - kx)$

There ωt =angle $\therefore \quad \omega = \frac{1}{T} = T^{-1}$

Similarly kx =angle $\therefore \quad k = \frac{1}{x} = L^{-1}$

$\therefore \quad \frac{\omega}{k} = \frac{[T^{-1}]}{[L^{-1}]} = [LT^{-1}]$

262 **(b)**

Given, 1 eV=1.6$\times 10^{-19}$ J

13.6 eV=13.6$\times$ 1.6 $\times 110^{-19}$ J

$= 21.76 \times 10^{-19}$J

263 **(b)**

$\frac{\text{Energy}}{\text{Volume}} = \frac{ML^2T^{-2}}{L^3} = [ML^{-1}T^{-2}]$ = Pressure

264 **(a)**

$[MLT^{-2}] = [L^{2a}][L^bT^b][M^cL^{-3c}]$
$= [M^cL^{2a+c-3c}T^{-b}]$

Comparing powers of M, L and T, we get

$c = 1, 2a + b - 3c = 1, -b = -2$ or $b = 2$

$2a + 2 - 3(1) = 1 \quad \Rightarrow \quad 2a = 2$or $a = 1$.

265 **(c)**

$PV = nRT \Rightarrow R = \frac{PV}{nT} = \frac{joule}{mole \times kelvin}$
$= JK^{-1}mol^{-1}$

266 **(a)**

Dimensions of $E = [ML^2T^{-2}]$

Dimensions of $G = [M^{-1}L^3T^{-2}]$

Dimensions of $I = [MLT^{-1}]$

And dimension of $M = [M]$

$\therefore$ Dimensions of $\frac{GIM^2}{E^2}$

$$= \frac{[M^{-1}L^3T^{-2}][MLT^{-1}][M^2]}{[ML^2T^{-2}]^2}$$

$= [T]$

= Dimensions of time

267 **(b)**

[Force] = $[M]$[Acceleration] $\Rightarrow [F] = [M^1L^1T^{-2}]$

268 **(d)**

[Pressure]=[Stress]=[coefficient of elasticity] =$[ML^{-1}T^{-2}]$

269 **(a)**

Time period

$T \propto p^a \rho^b E^c$

Or, $T = kp^a \rho^b E^c$

k,is a dimensionless constant.

According to homogeneity of dimensions, LHS=RHS

$\therefore$ $[\text{T}] = [\text{ML}^{-1}\text{T}^{-2}]^a[\text{ML}^{-3}]^b[\text{ML}^2\text{T}^{-2}]^c$

$[\text{T}] = [\text{M}^{a+b+c}][\text{L}^{-a-3b+2c}][\text{T}^{-2a-2c}]$

Comparing the powers, we obtain

$a + b + c = 0$

$-a - 3b + 2c = 0$

$-2a - 2c = 1$

On solving, we get

$a = -\frac{5}{6}, b = \frac{1}{2}, c = \frac{1}{3}$

270 **(d)**

Dimensional formula of magnetic flux

$= [\text{ML}^2\text{T}^{-2}\text{A}^{-1}]$

271 **(a)**

$W = \frac{1}{2} kx^2 \Rightarrow [k] = \frac{[W]}{[x^2]} = \left[\frac{ML^2T^{-2}}{L^2}\right]$

$= [MT^{-2}]$

273 **(c)**

Given, $L = 2.331\ cm$

$= 2.33$ (correct upto two decimal places)

And $B = 2.1\ cm = 2.10\ cm$

$\therefore L + B = 2.33 + 2.10 = 4.43\ cm$

$= 4.4\ cm$

Since minimum significant figure is 2

274 **(b)**

$\frac{dQ}{dt} = -KA\left(\frac{d\theta}{dx}\right)$

$\Rightarrow [K] = \frac{[ML^2T^{-2}]}{[T]} \times \frac{[L]}{[L^2][K]}$

$= MLT^{-3}K^{-1}$

276 **(c)**

Least count of screw gauge $= \frac{0.5}{50}$

$= 0.01$ mm

$= \Delta r$

Diameter $r = 2.5$ mm $+ 20 \times \frac{0.5}{50} =$ 2.70 mm

$\frac{\Delta r}{r} = \frac{0.01}{2.70}$

Or$\frac{\Delta r}{r} \times 100 = \frac{1}{2.7}$

Now density $d = \frac{m}{V} = \frac{m}{\frac{4}{3}\pi\left(\frac{r}{2}\right)^3}$

Here, r is the diameter.

$\therefore \quad \frac{\Delta d}{d} \times 100 = \left\{\frac{\Delta m}{m} + 3\left(\frac{\Delta r}{r}\right)\right\} \times 100$

$= \frac{\Delta m}{m} \times 100 + 3 \times \left(\frac{\Delta r}{r}\right) \times 100$

$= 2\% + 3 \times \frac{1}{2.7}$

$= 3.11\%$

277 **(d)**

1 CGS SI

$N_1U_1 = N_2U_2$

$N_1[M_1L_1^{-3}] = N_2[M_2L_2^{-3}]$

$\therefore N_2 = N_1\left[\frac{M_1}{M_2}\right] \times \left[\frac{L_1}{L_2}\right]^{-3}$

$= 0.625\left[\frac{1g}{1kg}\right] \times \left[\frac{1cm}{1m}\right]^{-3}$

$= 0.625 \times 10^{-3} \times 10^6 = 625$

278 **(c)**

Force $\quad F = qvB$

$[\text{MLT}^{-2}] = [\text{C}][\text{LT}^{-1}][\text{B}]$

$\Rightarrow \quad [\text{B}] = [\text{MC}^{-1}\text{T}^{-1}]$

279 **c)**

By Newton's formula

$\eta = \frac{F}{A(\Delta v_x/\Delta z)}$

∴ Dimensions of η

$$= \frac{\text{dimensions of force}}{\text{dimensions of area} \times \text{dimensions of velocity gradient}}$$

$$= \frac{[MLT^{-2}]}{[L^2][T^{-1}]} = [ML^{-1}T^{-1}]$$

280 **(d)**

Given equation , $y = a\sin(bt - cx)$
Comparing the given equation with general wave equation

$y = a\sin\left(\frac{2\pi t}{T} - \frac{2\pi x}{\lambda}\right)$,

We get $b = \frac{2\pi}{T}, c = \frac{2\pi}{\lambda}$

Dimension of $\frac{b}{c}$

$= \frac{2\pi/T}{2\pi/\lambda} = [LT^{-1}]$, and other three quantity is dimensionless

281 **(a)**

$$\frac{[\text{Energy}]}{[\text{Volume}]} = \frac{[ML^2T^{-2}]}{[L^3]} = [ML^{-1}T^{-2}]$$

$[\text{pressure}] = \frac{[MLT^{-2}]}{[L^2]} = [ML^{-1}T^{-2}]$

282 **(b)**

$$L = \frac{\emptyset}{I} = \frac{Wb}{A} = Henry$$

283 **(a)**

The result has to be in one significant umber only.

284 **(a)**

Force, $F = ma$

$$\therefore a = \frac{F}{m} = \frac{10 \text{ pound}}{\text{kg}}$$

$$= 10\frac{\text{pound}}{\text{kg}} = \frac{10\text{slug ft}}{\text{kg s}^2} = 146\frac{\text{ft}}{\text{s}^2}$$

$= 146 \times 0.30 \text{ ms}^{-2}$

$= 44.5 \text{ ms}^{-2}$

285 **(d)**

Percentage error in A

$$= \left(2 \times 1 + 3 \times 3 + 1 \times 2 + \frac{1}{2} \times 2\right)\%$$

$$= 14\%$$

286 **(c)**

According to Faraday's first law of electrolysis, $m = Zq$ or $Z = \frac{m}{q}$, so, SI unit of Z is kg C^{-1}

287 **(d)**

The dimensions of x = dimensions of $\frac{v_0}{A}$

Therefore, out of the given options v_0 has dimensions equal to $[M^0LT^{-1}]$ and A has dimensions equal to $[M^0L^0T^{-1}]$

So, that $\frac{[v_0]}{[A]} = \frac{[M^0LT^{-1}]}{[M^0L^0T^{-1}]} = [L]$

= dimension of x

288 **(c)**

$$B = \frac{F}{IL} = \frac{[MLT^{-2}]}{[A][L]} = [MT^{-2}A^{-2}]$$

289 **(d)**

We know that

$$Q = \frac{KA(\theta_1 - \theta_2)t}{d}$$

$$\Rightarrow \quad K = \frac{Qd}{A(\theta_1 - \theta_2)t}$$

So, the unit of $K = \frac{\text{cal} \times \text{metre}}{\text{metre}^2 \times \text{K} \times \text{sec}}$

$$= \frac{\text{cal}}{\text{m}}.\text{K}.\text{sec} \quad \text{... (i)}$$

And density $d = \frac{m}{V}$, where m = mass, V = volume.

So, the unit of $d = \frac{\text{kg}}{\text{m}^3}$ ... (ii)

Also, we know that $Q = mc\Delta\theta$

$$\Rightarrow \quad c = \frac{Q}{m.\Delta\theta}$$

So, the unit of $c = \frac{\text{cal}}{\text{kg.K}}$ ... (iii)

Hence, the unit of $X = \frac{K}{dc}$

On putting the values of unit from Eqs. (i), (ii) and (iii), the unit of

$$X = \frac{\text{cal/mK} - \text{sec}}{\text{kg/m}^3 \times \text{cal/kg . K}}$$

$\Rightarrow = \text{m}^2/\text{s}$

So, the unit of X in CGS system is cm^2s^{-1}.

290 **(c)**

$T = 2\pi\sqrt{l/g} \Rightarrow T^2 = 4\pi^2 l/g \Rightarrow g = \frac{4\pi^2 l}{T^2}$

Here, % error in $l = \frac{1mm}{100cm} \times 100 = \frac{0.1}{100} \times 100 = 0.1\%$

And % in error in $T = \frac{0.1}{2\times100} \times 100 = 0.05\%$

$\therefore$ % error in g = % error in l + 2(% error in T)

$= 0.1 + 2 \times 0.05 = 0.2$ %

291 **(d)**

1T=1 Wbm^{-2}

293 **(c)**

$\frac{L}{R}$ = Time constant

294 **(b)**

Let $m \propto E^x v^y F^z$

By substituting the following dimensions:

$E = [ML^2T^{-2}], [v] = [LT^{-1}], [F] = [MLT^{-2}]$

and by equating the both sides

$x = 1, y = -2, z = 0$. So $[m] = [Ev^{-2}]$

295 **(b)**

Planck's constant $(h) = \text{J} - \text{s} = [\text{ML}^2\text{T}^{-2}][\text{T}] = [\text{ML}^2\text{T}^{-1}]$

Linear momentum $(p) = \text{kg} - \text{ms}^{-1}$

$= [\text{M}][\text{L}][\text{T}]^{-1}$
$= [\text{MLT}^{-1}]$

296 **(d)**

Acceleration $= \frac{\text{Distance}}{\text{time}^2} \Rightarrow A = LT^{-2} \Rightarrow L = AT^2$

297 **(c)**

Given, $\left(p + \frac{a}{V^2}\right)(V - b) = RT$

According to principle of homogeneity

Dimension of $\frac{a}{V^2}$ = dimension of p

Dimension of a = dimension of $p \times$ dimension of V^2

$= [\text{ML}^{-1}\text{T}^{-2}][\text{L}^3]^2$
$= [\text{ML}^5\text{T}^{-2}]$

298 **(a)**

$\frac{\text{Angular momentum}}{\text{Linear momentum}} = \frac{mvr}{mv} = r = [M^0L^1T^0]$

299 **(d)**

$n_2 = n_1 \left[\frac{L_1}{L_2}\right]^1 \left[\frac{T_1}{T_2}\right]^{-2} = 10 \left[\frac{m}{km}\right]^1 \left[\frac{sec}{hr}\right]^{-2}$

$n_2 = 10 \left[\frac{m}{10^3 m}\right]^1 \left[\frac{\text{sec}}{3600 \text{ sec}}\right]^{-2} = 129600$

300 **(b)**

Time constant $= \frac{L}{R}$

$\therefore \left[\frac{L}{R}\right] = [\text{T}]$

$\therefore \left[\frac{R}{L}\right] = [\text{T}^{-1}]$

301 **(d)**

The number of significant figures in all of the given numbers is 4.

302 **(a)**

$F = \frac{dp}{dt} \Rightarrow dp = Fdt$

303 **(c)**

Friction $F = \mu N$

$\mu = \frac{F}{N}$

$\therefore [\mu] = \left[\frac{F}{N}\right] = \frac{[\text{MLT}^{-2}]}{[\text{MLT}^{-2}]}$ = dimensionless

304 **(a)**

Time $\propto c^x G^y h^z \Rightarrow T = kc^x G^y h^z$

Putting the dimensions in the above relation

$\Rightarrow [M^0L^0T^1]$
$= [LT^{-1}]^x[M^{-1}L^3T^{-2}]^y[ML^2T^{-1}]^z$
$\Rightarrow [M^0L^0T^{-1}]$
$= [M^{-y+z}L^{x+3y+2z}T^{-x-2y-z}]$

Comparing the powers of M, L and T

$-y + z = 0$...(i)
$x + 3y + 2z = 0$...(ii)
$-x - 2y - z = 1$ ...(iii)
On solving equations (i) and (ii) and (iii)
$x = \frac{-5}{2}, y = z = \frac{1}{2}$
Hence dimension of time are
$[G^{1/2}h^{1/2}c^{-5/2}]$

305 **(a)**
Kinetic energy $= \frac{1}{2}mv^2 = M[LT^{-1}]^2 = [ML^2T^{-2}]$

306 **(a)**
Given that,

Time period, $T \propto p^a d^b E^c$ ... (i)

The dimensions of these quantities are given as

$$p = [\text{ML}^{-1}\text{T}^{-2}]$$

$$d = [\text{ML}^{-3}]$$

$$E = [\text{ML}^2\text{T}^{-2}]$$

In Eq. (i), on writing the dimensions on both sides.

$[\text{M}^0\text{L}^0\text{T}] \propto [\text{ML}^{-1}\text{T}^{-2}]^a[\text{ML}^{-3}]^b[\text{ML}^2\text{T}^{-2}]^c$

$\Rightarrow [\text{M}^0\text{L}^0\text{T}] \propto [\text{M}^{a+b+c}\text{L}^{-a-3b+2c}\text{T}^{-2a-2c}]$

On comparing the powers of M, L, T on both sides.

$\Rightarrow \quad a + b + c = 0$ ... (ii)

$-a - 3b + 2c = 0$ ... (iii)

$-2a - 2c = 1$ ... (iv)

Solving, we get value of a, b and c, $-\frac{5}{6}, \frac{1}{2}$ and $\frac{1}{3}$ repectively.

308 **(a)**
Indestructibility, invariability and reproductibility are essential characteristics of a unit of measurement.

309 **(c)**
100 W= 100 J $\text{s}^{-1} = 10^9$ erg s^{-1}

310 **(c)**
Maximum percentage error in $P = 4\% + 2 \times 2\%$
$= 8\%$

311 **(b)**
According to the definition

312 **(b)**
One femtometre is equivalent to 10^{-15}m

ie, $1\text{fm} = 10^{-15}$ m

313 **(a)**
$[?] = \frac{[Sl^3]}{[4yd^3\delta]} = \frac{[\text{MLT}^{-2}][\text{L}^3][\text{L}^2]}{[\text{MLT}^{-2}][\text{L}^3][\text{L}]} = [\text{L}]$

314 **(a)**
$[L/R]$ is a time constant so its unit is *second*

315 **(a)**
Calorie is the unit of heat *i.e.*, energy
So dimensions of energy $= ML^2T^{-2}$

316 **(c)**
According to definition of potential

317 **(c)**
[Energy] $= [\text{ML}^2\text{T}^{-2}]$. Increasing M and L by a factor of 3 energy is increased 27 times.

318 **(d)**
Resistance, $R = \frac{\text{Potential difference}}{\text{Current}} = \frac{V}{i} = \frac{W}{qi}$

($\because$ Potential difference is equal to work do

So, dimensions of R

$$= \frac{[\text{Dimensions of work}]}{[\text{Dimensions of charge}][\text{Dimensions of c}}$$

$$= \frac{[\text{ML}^2\text{T}^{-2}]}{[\text{IT}][\text{I}]} = [\text{ML}^2\text{T}^{-3}\text{I}^{-2}]$$

319 **(a)**
According to problem muscle × speed = power
$\therefore$ muscle $= \frac{\text{power}}{\text{speed}} = \frac{ML^2T^{-3}}{LT^{-1}} = MLT^{-2}$

320 **(c)**

Percentage error in $T = \frac{0.01}{1.26} \times 100 + \frac{0.01}{9.80} \times 100 + \frac{0.01}{1.45} \times 100$
$= 0.8 + 0.1 + 0.7 = 1.6$

321 **(d)**

$$Y = \frac{X}{3Z^2} = \frac{M^{-1}L^{-2}T^4A^2}{[MT^{-2}A^{-1}]^2} = [M^{-3}L^{-2}T^8A^4]$$

322 **(b)**

Time period $T = 2\pi\sqrt{\frac{l}{g}}$

Or $\frac{t}{n} = 2\pi\sqrt{\frac{l}{g}}$

$\therefore\ g = \frac{(4\pi^2)(n^2)l}{t^2}$

$\%\text{error in } g = \frac{\Delta g}{g} \times 100$
$= \left(\frac{\Delta l}{l} + \frac{2\Delta t}{t}\right) \times 100$

$$E_I = \left(\frac{0.1}{64} + \frac{2 \times 0.1}{128}\right) \times 100 = 0.3125\%$$

$$E_{II} = \left(\frac{0.1}{64} + \frac{2 \times 0.1}{64}\right) \times 100 = 0.46875\%$$

$$E_{III} = \left(\frac{0.1}{20} + \frac{2 \times 0.1}{36}\right) \times 100 = 1.055\%$$

Hence, E_I is minimum.

323 **(d)**

$$\frac{\text{Energy}}{\text{mass} \times \text{length}} = \frac{[ML^2T^{-2}]}{[M][L]} = [LT^{-2}]$$

324 **(a)**

$$Y = \frac{\text{Stress}}{\text{Strain}} = \frac{\text{Force/Area}}{\text{Dimensionless}} \Rightarrow Y \equiv \text{Pressure}$$

327 **(d)**

Unit of $e.m.f. = volt = joule/coulomb$

328 **(d)**

For best results amplitude of oscillation should be as small as possible and more oscillations should be taken

330 **(d)**

Volume $\times\, r^3$
So, error is $3 \times 2\% = 6\%$

331 **(c)**

$Curie = disintegration/second$

333 **(b)**

Volume $= (2.1 \times 10^{-2})^3 \text{m}^3 = 9.261 \times 10^{-6} \text{m}^3$. Rounding off two significant figures, we get $9.3 \times 10^{-6} \text{m}^3$.

334 **(b)**

Positions $x = ka^m t^n$

$$[M^0LT^0] = [LT^{-2}]^m[T]^n$$
$$= [M^0L^mT^{-2m+n}]$$

On comparing both sides

$$m = 1$$
$$-2m + n = 0$$
$$n = 2m$$
$$n = 2 \times 1 = 2$$

336 **(a)**

$[B] = \left[\frac{\text{force} \times \text{length}}{\text{mass}}\right] = \left[\frac{\text{energy}}{\text{mass}}\right] = [\text{latent heat}]$

337 **(a)**

Electric potential $V = \frac{W}{q} = \frac{\text{joule}}{\text{coulomb}} = \frac{\text{newton} \times \text{metre}}{\text{coulomb}}$

$$= \frac{(\text{kg} - \text{ms}^{-2}) \times \text{m}}{\text{coulomb}}$$

$$= \text{kg} - \text{ms}^{-2} \times \text{m} \times \text{coulomb}^{-1}$$

$$\therefore\ = [ML^2T^{-2}Q^{-1}]$$

338 **(b)**

$$F = \frac{Gm_1m_2}{d^2}$$

$$\Rightarrow\ G=\frac{Fd^2}{m_1m_2}$$

$$[G]=\frac{[MLT^{-2}][L^2]}{[M^2]}=[M^{-1}L^3T^2]$$

Moment of inertia $I=mK^2=[ML^2]$

340 **(b)**

In general, moment (M) of force (F) is

$$M=r\times F$$

$\therefore$ Dimensions of $M=[L][MLT^{-2}]$

$$=[ML^2T^{-2}]$$

341 **(a)**

$$\frac{\text{Heat}}{\text{Mass}}=\frac{[ML^2T^{-2}]}{[M]}=[M^0L^2T^{-2}]$$

342 **(d)**

Watt is a unit of power

344 **(c)**

$$\frac{0.2}{25}\times 100=0.8$$

345 **(c)**

Resistance, $R=\frac{V}{i}=\frac{W}{qi}$

$$=\frac{[ML^2T^{-2}]}{[A^2T]}$$

$$R=[ML^2T^{-3}A^{-2}]$$

$$\left[\frac{h}{e^2}\right]=\frac{[ML^2T^{-1}]}{[AT]^2}$$

$$=[ML^2T^{-3}A^{-2}]$$

346 **(d)**

From the expression $=\frac{\text{Power}}{\text{Area}}\left(\because\ \frac{\text{Energy}}{\text{time}}=\text{power}\right)$

$$\frac{W}{m^2}=Wm^{-2}$$

$$\frac{F}{l}=\frac{\mu_0I_1I_2}{2\pi r}$$

Or $[\mu_0]=\frac{[F]}{[I_1I_2]}=\frac{[MLT^{-2}]}{[A^2]}=[MLT^{-2}A^{-2}]$.

348 **(a)**

Resistivity $[\rho]=\frac{[R].[A]}{[l]}$ where $[R]=[ML^2T^{-1}Q^{-2}]$

$\therefore [\rho]=[ML^3T^{-1}Q^{-2}]$

349 **(a)**

$$RD=\frac{w_1}{w_1-w_2},\frac{\Delta(RD)}{RD}\times 100$$

$$=\frac{\Delta w_1}{w_1}\times 100+\frac{D(w_1-w_2)}{w_1-w_2}\times 100$$

$$=\frac{0.05}{5.00}\times 100+\frac{0.05+0.05}{1.00}\times 100$$

$$=1+10=11\%$$

350 **(c)**

Energy = force × distance, so if both are increased by 4 times then energy will increase by 16 times

351 **(c)**

$[MT^{-3}]=\frac{[ML^2T^{-2}]}{[L^2][T]}$ =energy /area ×time=dimensions of solar constant.

353 **(c)**

Force $\vec{F}=q\vec{v}\times\vec{B}$

Or $F=qvB\sin\theta$

$$\therefore\quad [B]=\left[\frac{F}{qv}\right]=\frac{[MLT^{-2}]}{[ATLT^{-1}]}=[MT^{-2}A^{-1}]$$

354 **(b)**

$$C^2LR=[C^2L^2]\times\left[\frac{R}{L}\right]=[T^4]\times\left[\frac{1}{T}\right]$$

$$=[T^3]$$

As $\left[\frac{L}{R}\right]=T$ and $\sqrt{LC}=T$

355 **(a)**

The dimension of $y=\frac{e^2}{4\pi\varepsilon_0hc}$

Putting the dimensions of

$$[e]=[Q]=[AT]$$

$$[\varepsilon_0]=[M^{-1}L^{-3}T^4A^2], h=[ML^2T^{-1}], c=[LT^{-1}]$$

$$y=\frac{[A^2T^2]}{[M^{-1}L^{-3}T^4A^2][ML^2T^{-1}][LT^{-1}]}$$

$$y=[M^0L^0T^0]$$

356 **(a)**

By submitting the dimensions of each quantity we get

$T = [ML^{-1}T^{-2}]^a[L^{-3}M]^b[MT^{-2}]^c$

357 **(a)**

$1\ Faraday = 96500\ coulomb$

358 **(d)**

When two quantities are multiplied, their maximum relative errors are added up.

359 **(d)**

According to Planck's hypothesis

$$E = hv$$

Or $$h = \frac{E}{v}$$

Substituting the dimensions of energy E and frequency v, we get

$$[h] = \frac{[\mathrm{ML^2T^{-2}}]}{[\mathrm{T^{-1}}]}$$

$\therefore$ $$[h] = [\mathrm{ML^2T^{-1}}]$$

360 **(b)**

Young's modulus $Y = \frac{FL}{Al} = \frac{4FL}{\pi d^2 l}$

$$= \frac{(4)(1.0 \times 9.8)(2)}{\pi(0.4 \times 10^{-3})^2(0.8 \times 10^{-3})}$$

$$= 2.0 \times 10^{11}\mathrm{Nm^{-2}}$$

Further, $$\frac{\Delta Y}{Y} = 2\left(\frac{\Delta d}{d}\right) + \left(\frac{\Delta l}{l}\right)$$

$\therefore$ $$\Delta Y = \left\{2\left(\frac{\Delta d}{d}\right) + \left(\frac{\Delta l}{l}\right)\right\} Y$$

$$= \left\{2 \times \frac{0.01}{0.4} + \frac{0.05}{0.8}\right\} \times 2.0 \times 10^{11}$$

$$= 0.2 \times 10^{11}\mathrm{Nm^{-2}}$$

Or $(Y + \Delta Y) = (2 + 0.2) \times 10^{11}\mathrm{Nm^{-2}}$

362 **(b)**

$(2.3 + 0.035 + 0.035)$g$=2.37$ g

But we have to retain only one decimal place.

So, the total mass is 2.4 g.

363 **(b)**

Given, $p = \frac{a-t^2}{bx}$ or $pbx = a - t^2$

By the law of homogeneity of dimensional equation.

Dimensions of a = dimensions of t^2 = $[\mathrm{T^2}]$

Dimensions of b =dimensions of $\frac{t^2}{px}$ = $[\mathrm{M^{-1}T^4}]$

So, dimensions of $\frac{a}{b}$ is $[\mathrm{MT^{-2}}]$.

364 **(b)**

$L \propto v^x A^y F^z \Rightarrow L = kv^x A^y F^z$

Putting the dimensions in the above relation

$[ML^2T^{-1}] = k[LT^{-1}]^x[LT^{-2}]^y[MLT^{-2}]^z$

$\Rightarrow [ML^2T^{-2}] = k[M^z L^{x+y+z} T^{-x-2y-2z}]$

Comparing the powers of M, L and T

$z = 1$ ...(i)

$x + y + z = 2$ ...(ii)

$-x - 2y - 2z = -1$...(iii)

On solving (i), (ii) and (iii) $x = 3, y = -2, z = 1$

So dimension of L in terms of v, A and f

$[L] = [Fv^3A^{-2}]$

365 **(c)**

Electric potential $V = IR, [R] = \left[\frac{V}{I}\right] = \left[\frac{\text{Work done}}{\text{Charge} \times \text{I}}\right]$

$= \frac{[\mathrm{ML^2T^{-2}}]}{[\mathrm{A^2T}]} = [\mathrm{ML^2T^{-3}A^{-2}}]$

366 **(c)**

Both are the formula of energy.$\left(E = \frac{1}{2}CV^2 = \frac{1}{2}LI^2\right)$

367 **(c)**

$$Y = \frac{F}{A} \cdot \frac{L}{\Delta L} = \frac{dyne}{cm^2} = \frac{10^{-5}N}{10^{-4}m^2} = 0.1\ N/m^2$$

368 **(b)**

$$f = \frac{1}{2\pi\sqrt{LC}} \Rightarrow LC = \frac{1}{f^2} = [M^0L^0T^2]$$

369 **(d)**

Given, $U = \frac{A\sqrt{x}}{x+B}$ ... (i)

Dimensions of U = dimensions of potential energy

$= [ML^2T^{-2}]$

From Eq. (i),

Dimensions of B = dimensions of x = $[M^0LT^0]$

∴ Dimensions of A

$= \frac{\text{dimensions of } U \times \text{dimensions of } (x+B)}{\text{dimension of} \sqrt{x}}$

$= \frac{[ML^2T^{-2}][M^0LT^0]}{[M^0L^{1/2}T^0]}$

$= [ML^{5/2}T^{-2}]$

Hence, dimensions of AB

$= [ML^{5/2}T^{-2}][M^0LT^0]$

$= [ML^{7/2}T^{-2}]$

370 **(a)**

Torque = force × distance = $[ML^2T^{-2}]$

371 **(c)**

Capacity × Resistance

$= \frac{\text{Charge}}{\text{Potential}} \times \frac{\text{Volt}}{\text{amp}}$

$= \frac{\text{amp} \times \text{second} \times \text{Volt}}{\text{Volt} \times \text{amp}} = \text{Second}$

372 **(a)**

$K = Y \times r_0 = [ML^{-1}T^{-2}] \times [L] = [MT^{-2}]$

Y=Young's modulus and r_0 = Interatomic distance

374 **(d)**

Strain has no dimensions

375 **(c)**

Moment of inertia $I = mr^2$

∴ $[I] = [ML^2]$

And τ = moment of force = $\mathbf{r} \times \mathbf{F}$

∴ $[\tau] = [L][MLT^{-2}]$
$= [ML^2T^{-2}]$

376 **(a)**

Percentage error in $X = a\alpha + b\beta + c\gamma$

377 **(d)**

$[ML^{-2}T^{-2}] = \frac{[MLT^{-2}]}{[L][L^2]}$

$= \frac{\text{Force}}{\text{distance} \times \text{area}} = \frac{\text{pressure}}{\text{distance}}$

= pressure gradient.

378 **(c)**

Tension = $[MLT^{-2}]$, Surface Tension =$[MT^{-2}]$

379 **(d)**

We know that

$$\text{density} = \frac{\text{mass}}{\text{volume}}$$

In CGS units

$d = 0.625 \text{ gcm}^{-3}$

In SI units

$$d = \frac{0.625 \times 10^{-3}\text{kg}}{10^{-6}\text{m}^3}$$
$= 625 \text{ kgm}^{-3}$

380 **(a)**

Momentum $p \propto f^a v^b \rho^c$

$[MLT^{-1}]=[T^{-1}]^a[LT^{-1}]^b[ML^{-3}]^c$

$[MLT^{-1}] = [M^cL^{b-3c}T^{-a-b}]$

$\Rightarrow c = 1.$

$b - 3c = 1 \Rightarrow b = 4$

$-a - b = -1$

$a + b = 1, a = -3$

∴ $[p] = [f^{-3}v^4\rho]$

381 **(b)**

1 Oersted = 1 Gauss = 10^{-4} *Tesla*

382 **(c)**

$1\ nm = 10^{-9}\ m = 10^{-7} cm$

384 **(a)**
Let $m = KF^aL^bT^c$
Substituting the dimension of
$[F] = [MLT^{-2}], [C] = [L]$ and $[T] = [T]$
And comparing both sides, we get $m = FL^{-1}T^2$

385 **(d)**
Surface tension $= \frac{\text{Force}}{\text{Length}} = \frac{[MLT^{-2}]}{L} = [MT^{-2}]$

387 **(b)**
$$\text{Pressure} = \frac{\text{Force}}{\text{Area}} = \frac{\text{Energy}}{\text{Volume}} = ML^{-1}T^{-2}$$

388 **(a)**
Result should have only two significant numbers (same as in 12 m).

389 **(c)**
$$\text{Angular acceleration} = \frac{\text{Angular velocity}}{\text{Time}}$$
$$= \frac{rad}{sec^2}$$

391 **(c)**
Least count $= \frac{\text{Pitch}}{\text{No.of div.in circular scale}}$
$= \frac{0.5}{50} = 0.01\ mm$
Actual reading $= 0.01 \times 35 + 3 = 3.35\ mm$
Taking error into consideration
$= 3.35 + 0.03 = 3.38\ mm$

393 **(a)**
Force, F $= kv, [k] = \frac{[F]}{[v]} = \frac{[\text{MlT}^{-2}]}{[\text{LT}^{-1}]} = [\text{MT}^{-1}]$.
So, unit is kgs^{-1}

394 **(b)**
$X = \text{M}^a\text{L}^b\text{T}^{-c}$
$$\therefore \quad \frac{\Delta X}{X} = \pm\left[\alpha\frac{\Delta M}{M} + b\frac{\Delta L}{L} + c\frac{\Delta T}{T}\right]$$
$= \pm[a\alpha + \beta b + \gamma c]\%$

395 **(c)**
The result should have two decimal places (same as 0.99)*ie*, $0.\ 010\times 10^{-1}$

396 **(c)**
Area velocity is area covered per unit time.

397 **(d)**
Because temperature is a fundamental quantity

398 **(c)**
$$\text{Stress} = \frac{\text{Force}}{\text{Area}} = \frac{N}{m^2}$$

399 **(a)**
Required relative error=power × relative error in x.

400 **(c)**
Let $m \propto C^xG^yh^z$
By substituting the following dimensions:
$[C] = LT^{-1}; [G] = [M^{-1}L^3T^{-2}]$ and $[h] = [ML^2T^{-1}]$
Now comparing both sides we will get
$x = 1/2; y = -1/2, z = +1/2$
So $m \propto c^{1/2}G^{-1/2}h^{1/2}$

401 **(a)**
$\rho = \frac{RA}{l}$ $i.e.$ dimension of resistivity is $[ML^3T^{-1}Q^{-2}]$

402 **(c)**
$30\ VSD = 29\ MSD$
$$1\ VSD = \frac{29}{30}\ MSD$$
Least count of vernier =
$1\ M.S.D. - 1\ V.S.D.$
$$= 0.5° - \frac{29}{30} \times 0.5° = \frac{0.5°}{30}$$
Reading of vernier = M.S. reading + V.S. reading $\times\ L.C.$
$$= 58.5° + 9 \times \frac{0.5°}{30} = 58.65$$

404 **(b)**
Physical quantity $u = \frac{B^2}{2\mu_0}$
Unit of $u = \frac{(\text{N/Am})^2}{\text{N/A}^2} = \frac{\text{N}^2\text{A}^2}{\text{NA}^2\text{m}^2}$
$$= \frac{\text{N}}{\text{m}^2} = \frac{\text{Nm}}{\text{m}^3} = \frac{\text{J}}{\text{m}^3}$$
= energy per unit volume = energy density

405 **(d)**
Joule-sec is the unit of angular momentum where as other units are of energy

406 **(c)**
Area of cross section $= \frac{22}{7} \times 0.24 \times 0.24\text{mm}^2 = 018\text{mm}^2$

407 **(d)**
Coefficient of viscosity $= \frac{F\times r}{A\times v} = \frac{[\text{MLT}^{-2}]\times[\text{L}]}{[\text{L}^2]\times[\text{LT}^{-1}]}$
$= [\text{ML}^{-1}\text{T}^{-1}]$

408 **(a)**
Least count
$$= \frac{\text{Value of main scale division}}{\text{No. of divisions on vernier scale}}$$

$$= \frac{1}{30} \text{MSD} = \frac{1}{30} \times \frac{1^\circ}{2} = \frac{1^\circ}{60}$$
$$= 1 \text{ min}$$

409 **(b)**

Percentage error in length $= \frac{1}{50} \times 100 = 2$

Percentage error in breadth $= \frac{0.1}{2.0} \times 100 = 5$

Percentage error in thickness$= \frac{0.1}{1.00} \times 100 = 1$

Percentage error in volume$= 2 + 5 + 1 = 8$

410 **(a)**

Angle of banking: $\tan\theta = \frac{v^2}{rg}$. i.e. $\frac{v^2}{rg}$ is dimensionless

411 **(d)**

$$f = \frac{uv}{u+v}, \frac{\Delta f}{f} = \frac{\Delta u}{u} + \frac{\Delta v}{v} + \frac{(u+v)}{u+v}$$

412 **(c)**

CR is known as time constant $CR = [T]$

413 **(d)**

The given equation is

$$n = \frac{A+B}{\lambda^2}$$

Where A and B are constants.

By homogeneity principle the dimensions of all the terms on both sides should be same

ie, $[B] = [A] = [n\lambda^2]$

$\therefore$ $[B] = [\mathrm{M^0L^0T^0}][\mathrm{L^2}]$

$= [\mathrm{M^0L^2T^0}]$

414 **(a)**

Required time $= \frac{5000 \times 86400 \times 365.25}{10^{11}}\text{s} = 1.6\text{s}$

415 **(b)**

The dimension of frequency $(f) = [\mathrm{T^{-1}}]$

The dimension of $\left(\frac{R}{L}\right) = \frac{[\mathrm{ML^2T^{-3}A^{-2}}]}{[\mathrm{ML^2T^2A^{-2}}]}$

$$= \left[\frac{1}{T}\right]$$

$$= [\mathrm{T^{-1}}]$$

416 **(b)**

$6 \times 10^{-5} = 60 \times 10^{-6} = 60\, microns$

417 **(a)**

$$F = \frac{\mu_0}{4\pi}\frac{2I_1I_2I}{r} \Rightarrow \mu_0 = [F][A]^{-2}$$
$$= [MLT^{-2}A^2]$$

418 **(a)**

$\frac{1}{2}Li^2 =$ Stored energy in an inductor$= [ML^2T^{-2}]$

419 **(d)**

The second is the duration of 9192631770 period of the radiation corresponding to the transition between the two hyperfine levels of the ground state of cesium-133 atom. Therefore, 1 ns is 10^{-9}s of Cs-clock of 9192631770 oscillations.

420 **(a)**

1 C.G.S. unit of density = 1000 M.K.S. unit of density

$\Rightarrow 0.5\, gm/cc = 500\, kg/m^3$

421 **(c)**

As $I = MR^2 = kg - m^2$

422 **(c)**

% error in velocity = % error in L + % error in t

$$= \frac{0.2}{13.8} \times 100 + \frac{0.3}{4} \times 100$$
$$= 1.44 + 7.5 = 8.94\,\%$$

423 **(b)**

Both force constant and surface tension represent force per unit length.

424 **(c)**

Quantity C has maximum power. So it brings maximum error in P

425 **(b)**

Surface tension, $T = \frac{F}{l}$

$\therefore$ $[T] = \frac{[F]}{[l]}$

$$= \frac{[\mathrm{MLT^{-2}}]}{[\mathrm{L}]} = [\mathrm{ML^0T^{-2}}] = [\mathrm{MT^{-2}}]$$

426 **(b)**

Heat ΔQ transferred through a rod of length L and area A in time Δt is

$$\Delta Q = KA\left(\frac{T_1 - T_2}{L}\right)\Delta t$$

$$\therefore \quad K = \frac{\Delta Q \times L}{A(T_1 - T_2)\Delta t} \quad \text{... (i)}$$

Substituting dimensions for corresponding quantities in Eq. (i), we have

$$[K] = \frac{[\mathrm{ML^2T^{-2}}][\mathrm{L}]}{[\mathrm{L^2}][\theta][\mathrm{T}]}$$

$$= [\mathrm{MLT^{-3}\theta^{-1}}]$$

427 **(c)**

According to the definition

428 **(d)**

By the principle of dimensions homogeneity

$$F = at^{-1}$$

$$[\mathrm{MLT^{-2}}] = a[\mathrm{T^{-1}}]$$

$$a = [\mathrm{MLT^{-1}}]$$

Similarly for $b = [\mathrm{MLT^{-4}}]$

429 **(c)**

Angular momentum $= [ML^2T^{-1}]$, Frequency $= [T^{-1}]$

430 **(c)**

Given, voltage $V = (100 \pm 5)$volt,

Current $I = (10 \pm 0.2)$A

From Ohm's law $V = IR$

$\therefore$ Resistance $R = \frac{V}{I}$

Maximum percentage error in resistance

$$\left(\frac{\Delta R}{R} \times 100\right) = \left(\frac{\Delta V}{V} \times 100\right) + \left(\frac{\Delta I}{I} \times 100\right)$$

$$= \left(\frac{5}{100} \times 100\right) + \left(\frac{0.2}{10} \times 100\right)$$

$$= 5 + 2 = 7\%$$

431 **(c)**

Dimensional formula of $ohm = [ML^2T^{-3}A^{-2}]$

and that of $\frac{h}{e^2} = \frac{[ML^2T^{-1}]}{A^2T^2} = [ML^2T^{-3}A^{-2}]$

432 **(b)**

In 23.023 number of significant figures will be 5 because all the zero's between non zero digits are significant. In 0.0003, number of significant figures will be one because all the zero's before and after decimal point are insignificant if the number is less then one. In 2.1×10^{-3} number of significant figure are to because power of 10 is not considered as significant figure

433 **(b)**

The difference in the sidereal year and solar year is about 1 day (or $24 \times 60 =$ 1440 min)

$\therefore$ Difference in sidereal day and solar day is about $\frac{1440}{365} \cong 4$ min

i.e., sidereal day is 4 min smaller than the solar day

435 **(b)**

Percentage error in mass $= \frac{0.01}{23.42} \times 100 =$ 0.04

Percentage error in volume $= \frac{0.1}{4.9} \times 100 =$ 2.04

Adding up the percentage errors, we get nearly 2%.

436 **(c)**

$[kx] =$ Dimension of $\omega t =$ (dimensionless)

Hence $K = \frac{1}{X} = \frac{1}{L} = [L^{-1}]$ $\therefore [K] = [L^{-1}]$

437 **(d)**

Dimensions of $\varepsilon_0 = [\mathrm{M^{-1}L^{-3}T^4A^2}]$

Dimensions of $L = [\text{L}]$

Dimensions of $\Delta V = [\text{ML}^2\text{T}^{-3}\text{A}^{-1}]$

Dimensions of $\Delta t = [\text{T}]$

As $X = \varepsilon_0 L \frac{\Delta V}{\Delta t}$

Dimensions of $X = \frac{[\text{M}^{-1}\text{L}^{-3}\text{T}^4\text{A}^2][\text{L}][\text{ML}^2\text{T}^{-3}\text{A}^{-1}]}{[\text{T}]}$

$= [\text{A}]$

438 **(a)**
$I = \frac{Q}{t} = \frac{[Q]}{[T]} = [M^0L^0T^{-1}Q]$

439 **(d)**
$v = at + bt^2$
$[v] = [bt^2]$ or $LT^{-1} = bT^2 \Rightarrow [b] = [LT^{-3}]$

440 **(d)**
$e = \frac{Ldi}{dt} \Rightarrow [e] = [\text{ML}^2\text{T}^{-2}\text{A}^{-2}]\left[\frac{\text{A}}{\text{T}}\right]$
$[e] = [\text{ML}^2\text{T}^{-2}\text{Q}^{-1}]$

441 **(d)**
$[n]$ = Number of particles crossing a unit area in unit time$=[L^{-2}T^{-1}]$
$[n_2] = [n_1]$ =number of particles per unit volume $= [L^{-3}]$
$[x_2] = [x_1]$= positions
$\therefore D = \frac{[n][x_2 - x_1]}{[n_2 - n_1]} = \frac{[L^{-2}T^{-1}] \times [L]}{[L^{-3}]}$
$= [L^2T^{-1}]$

442 **(d)**
Given, $v = at + \frac{b}{t+c}$

Since, LHS is equal to velocity, so at and $\frac{b}{t+c}$ must have the dimensions of velocity.

$\therefore \quad at = v$

Or $a = \frac{v}{t} = \frac{[\text{LT}^{-1}]}{[\text{T}]} = [\text{LT}^{-2}]$

Now, c = time $\quad (\because$ like quantities are added)

$\therefore \quad c = t = [\text{T}]$

Now,

$\frac{b}{t+c} = v$

$\therefore \quad b = v \times \text{time} = [\text{LT}^{-1}][\text{T}] = [\text{L}]$

443 **(c)**
$V = 2.34 \times 11.11111$volt= 26.0 volt.
[Because the final result should contain three significant figure.]

444 **(a)**
Dimensions of Planck's constant

$h = \text{Dimensions of} \frac{E}{v}$

$= \frac{[\text{ML}^2\text{T}^{-2}]}{[\text{T}^{-1}]} = [\text{ML}^2\text{T}^{-1}]$

Dimensions of angular momentum

= Dimension of mvr

$= [\text{MLT}^{-1}] . [\text{L}]$

$= [\text{ML}^2\text{T}^{-1}]$

445 **(c)**
Solar constant $= \frac{\text{energy}}{\text{cm}^2\text{min}}$
$\therefore$ The dimensions of solar constant
$= \frac{[\text{ML}^2\text{T}^{-2}]}{[\text{L}^2\text{T}]} = [\text{ML}^0\text{T}^{-3}]$

446 **(d)**
Volume elasticity $= \frac{\text{Force/Area}}{\text{Volume strain}}$
Strain is dimensionless, so
$= \frac{\text{Force}}{\text{Area}} = \frac{MLT^{-2}}{L^2} = [ML^{-1}T^{-2}]$

447 **(b)**
Here, $[\text{M}^0\text{L}^0\text{T}^0] = [\text{M} \ \text{L}^{-1}\text{T}^{-2}]^a[\text{MT}^{-3}]^b[\text{LT}^{-1}]^c$
Or $\quad [\text{M}^0\text{L}^0\text{T}^0] = [\text{M}^{a+b}\text{L}^{-a+c}\text{T}^{-2a-3b-c}]$
Comparing powers of M, L and T, we get
$a + b = 0, -a + c = 0, -2a - 3b = 0$
Solving $a = 1, b = -1, c = 1$

449 **(d)**

Density

$$\rho = \frac{m}{\pi r^2 L}$$

$$\therefore \frac{\Delta\rho}{\rho} \times 100 = \left(\frac{\Delta m}{m} + 2\frac{\Delta r}{r} + \frac{\Delta L}{L}\right) \times 100$$

After substituting the values we get the maximum percentage error in density =4%

450 **(d)**

$[\eta] = [\mathrm{ML^{-1}T^{-2}}]$or $[T]=\left[\frac{\mathrm{M}}{\mathrm{L}\eta}\right]^{1/2}$

Time period $= 2\pi\sqrt{\frac{M}{L\eta}}$

451 **(c)**

$$F = -\eta . A\frac{dv}{dx} \Rightarrow [\eta] = [ML^{-1}T^{-1}]$$

452 **(b)**

Force = Mass × acceleration

$= [M][LT^{-2}] = [MLT^{-2}]$

Torque = Force × distance

$= [MLT^{-2}][L]$

$= [ML^2T^{-2}]$

Work = Force × distance $= [MLT^{-2}][L]$

$= [ML^2T^{-2}]$

Energy $= [ML^2T^{-2}]$

$$Power = \frac{\text{Work}}{\text{Time}} = \frac{[ML^2T^{-2}]}{[T]} = [ML^2T^{-3}]$$

453 **(d)**

$$f = \frac{1}{2\pi\sqrt{LC}}$$

$\therefore \left(\frac{C}{L}\right)$ does not represent the dimensions of frequency

454 **(c)**

Magnetic field $B = \frac{F}{qv\sin\theta}$

Hence, $1\ \mathrm{T} = \frac{1\mathrm{N}}{1\mathrm{C} \times 1\ \mathrm{ms^{-1}}}$

$$= \frac{1\ \mathrm{N}}{1\ \mathrm{As} \times 1\ \mathrm{ms^{-1}}} (\because 1\mathrm{C} = 1\ \text{ampere} \times 1\ \text{second})$$

$= 1\ \mathrm{NA^{-1}m^{-1}}$

455 **(b)**

Least count of screw gauge $= \frac{1}{100} mm = 0.01\ mm$

Diameter = Divisions on circular scale × least count +

$= 52 \times \frac{1}{100} + 0 = 0.52\ mm$

Diameter $= 0.052\ cm$

456 **(a)**

Let $h \propto G^x L^y E^z$

$[\mathrm{ML^2T^{-1}}] \propto [\mathrm{M^{-1}L^3T^{-2}}]^x[\mathrm{ML^2T^{-1}}]^y[\mathrm{ML^2T^{-2}}]^z$

$[\mathrm{ML^2T^{-1}}] = k[\mathrm{M^{-1}L^3T^{-2}}]^x[\mathrm{ML^2T^{-1}}]^y[\mathrm{ML^2T^{-2}}]^z$

Comparing the powers, we get

$$1 = -x + y + z \quad \text{...(i)}$$

$$2 = 3x + 2y + 2z \quad \text{...(ii)}$$

$$-1 = -2x - y - 2z \quad \text{...(iii)}$$

On solving Eqs. (i), (ii) and (iii), we get

$$x = 0$$

∴ Gravitational constant has no dimensions

457 **(a)**

Since percentage increase in length =2%

Hence, percentage increase in area if square sheet

$=2 \times 2\% = 4\%$

459 **(a)**

From Coulomb's law, the force of attraction/repulsion between two point charges q and q separated by distance r is

$$F = \frac{1}{4\pi\varepsilon_0}\frac{q^2}{r^2}$$

$$\Rightarrow \qquad \varepsilon_0 = \frac{1}{4\pi} \cdot \frac{q^2}{Fr^2}$$

Where ε_0 is electric permittivity.

Dimensions of $\varepsilon_0 = \frac{[AT]^2}{[MLT^{-2}][L^2]}$

$[\varepsilon_0] = [A^2M^{-1}L^{-3}T^{-4}]$

460 **(a)**

The number of significant figures in 23.023 = 5.

The number of significant figures in 0.0003 = 1

(Zeros after decimal and before a non-zero number are not significant.)

The number of significant figures in $2.1 \times 10^{-3} = 2$ (zero in powers of are not counted)

(All the zeros between two non-zero numbers are significant).

461 **(b)**

Here, $S = (13.8 \pm 0.2)m$
and $t = (4.0 \pm 0.3)sec$
Expressing it in percentage error, we have,

$S = 13.8 \pm \frac{0.2}{13.8} \times 100\% = 13.8 \pm 1.4\%$

and $t = 4.0 \pm \frac{0.3}{4} \times 100\% = 4 \pm 7.5\%$

$\because V = \frac{s}{t} = \frac{13.8 \pm 1.4}{4 \pm 7.5} = (3.45 \pm 0.3)\ m/s$

462 **(a)**

$\eta = \frac{F}{av} = \frac{[MLT^{-2}]}{[L][LT^{-1}]} = [ML^{-1}T^{-1}]$

463 **(b)**

$v = \frac{1}{2l}\sqrt{\frac{T}{m}} \Rightarrow \frac{1}{v} = T = 2l\sqrt{\frac{m}{T}}$ has the dimensions of time.

464 **(c)**

$$n_2 = n_1 \left(\frac{M_1}{M_2}\right)^1 \left(\frac{L_1}{L_2}\right)^1 \left(\frac{T}{T_2}\right)^{-2}$$

$$= 100\left(\frac{gm}{kg}\right)^1 \left(\frac{cm}{m}\right)^1 \left(\frac{\sec}{\min}\right)^{-2}$$

$$= 100\left(\frac{gm}{10^3 gm}\right)^1 \left(\frac{cm}{10^2 cm}\right)^1 \left(\frac{\sec}{60 \sec}\right)^{-2}$$

$$n_2 = \frac{3600}{10^3} = 3.6$$

465 **(a)**

Time period of a simple pendulum

$$T = 2\pi\sqrt{\frac{L}{8}}$$

$$Or\ \ g = \frac{4\pi^2 L}{T^2} \qquad \ldots.(i)$$

Differentiating Eq. (i), we have

$$\frac{\Delta g}{g} = \frac{\Delta L}{L} + \frac{2\Delta T}{T} \qquad \ldots\ldots\ldots(ii)$$

Given, L=100 cm, T=2s,

$$\Delta T = \frac{0.1}{100} = 0.001s,$$

$$\Delta L = 1mm = 0.1\ cm$$

Substituting the in Eq. (ii), we have

$$\therefore \left|\frac{\Delta g}{g}\right|_{\max} = \frac{\Delta L}{L} + \frac{2\Delta T}{T}$$

$$= \frac{0.1}{100} + 2 \times \frac{0.001}{2}$$

Thus, maximum percentage eror

$$\left|\frac{\Delta g}{g}\right|_{\max} \times 100 = \left(\frac{0.1}{100} \times 100\right) + \left(\frac{2 \times 0.001}{2} \times 100\right)$$

$$= 0.1\% + 0.1\% = 0.2\%$$

467 **(a)**

Henry is a unit of coefficient of self induction (L)

Dimension of coefficient of self induction $= [ML^2T^{-2}A^{-2}] = [ML^2Q^{-2}]$

468 **(a)**

Magnetic field $= \frac{\text{Force}}{\text{Charge} \times \text{velocity}}$

$$= \frac{[\text{MLT}^{-2}]}{[\text{AT}][\text{LT}^{-1}]} = [\text{MA}^{-1}\text{T}^{-2}]$$

469 **(b)**

[Pressure]=[stress]= $[ML^{-1}T^{-2}]$

470 **(b)**

Use formula for time period in angular SHM.

471 **(c)**

Least count $= \frac{0.5}{50} = 0.01\ mm$

Diameter of ball $D = 2.5\ mm + (20)(0.01)$

$D = 2.7\ mm$

$$\rho = \frac{M}{vol} = \frac{M}{\frac{4}{3}\pi\left(\frac{D}{2}\right)^3} \Rightarrow \left(\frac{\Delta\rho}{\rho}\right)_{max} = \frac{\Delta M}{M} + 3\frac{\Delta D}{D}$$

$$\left(\frac{\Delta\rho}{\rho}\right)_{max} = 2\% + 3\left(\frac{0.01}{2.7}\right) \times 100\% \Rightarrow \frac{\Delta\rho}{\rho} = 3.1\%$$

472 **(d)**

[Planck constant] $= [ML^2T^{-1}]$ and [Energy] $= [ML^2T^{-2}]$

473 **(a)**

$Q = [ML^2T^{-2}]$ (All energies have same dimensions)

474 **(a)**

Here, $\frac{2\pi}{\lambda}(ct - x)$ is dimensionless. Hence, $\frac{ct}{\lambda}$ is also dimensionless and unit of ct is same as that of x. Therefore, unit of λ is same as that of x. Also unit of y is same as that of A, which is also the unit of x.

475 **(b)**

New unit of mass is $\frac{1}{6.67\times10^{-11}}$kg ie,1.5× 10^{10} kg.

476 **(c)**

[Entropy]=$\frac{Q}{T} = \frac{[\text{ML}^2\text{T}^{-2}]}{[\text{K}]} = [\text{ML}^2\text{T}^{-2}\text{K}^{-1}]$

477 **(b)**

$$\frac{watt}{ampere} = volt$$

478 **(b)**

Units of a and PV^2 are same and equal to $dyne \times cm^4$

479 **(b)**

Potential energy $= mgh = g\left(\frac{\text{cm}}{\text{sec}^2}\right)\text{cm} = g\left(\frac{\text{cm}}{\text{sec}}\right)^2$

480 **(d)**

$CV^2 =$ Energy

The dimensional formula is $[\text{ML}^2\text{T}^{-2}]$.

481 **(a)**

$$\frac{\Delta p}{p} = \frac{\Delta m}{m} + \frac{\Delta V}{V}$$

$$= \frac{0.05}{5} \times 100 + \frac{0.05}{1} \times 100$$

$$= 6$$

$$\frac{\Delta p}{p} = 6\%$$

482 **(d)**

The number of significant figures in all of the given number is 4

483 **(d)**

$$F = \frac{1}{4\pi\varepsilon_0}\cdot\frac{Q_1Q_2}{r^2}$$

$$\Rightarrow \varepsilon_0 \propto \frac{Q^2}{F \times r^2}$$

So ε_0 has units of $coulomb^2/newton - m^2$

484 **(c)**

$P_1 = [ML^2T^{-1}]$

$D_2 = [(2M)(2L)^2(2T)^{-1}]$

$P_2 = 4[ML^2T^{-1}] = 4P_1$

485 **(b)**

$1\ kWh = 1 \times 10^3 \times 3600\ W \times \text{sec} = 36 \times 10^5\ J$

486 **(d)**

$\because$ Density, $\rho = \frac{M}{V} = \frac{M}{\pi\gamma^2 L}$

$$\Rightarrow \frac{\Delta\rho}{\rho} = \frac{\Delta M}{M} + 2\frac{\Delta r}{r} + \frac{\Delta L}{L}$$

$= \frac{0.003}{0.3} + 2 \times \frac{0.005}{0.5} + \frac{0.06}{6}$
$= 0.01 + 0.02 + 0.01 = 0.04$
$\therefore$ Percentage error $= \frac{\Delta\rho}{\rho} \times 100$
$= 0.04 \times 100 = 4\%$

487 **(a)**

Heat $Q = \frac{KA(\theta_1 - \theta_2)t}{d}$ or $K = \frac{Qd}{A(\theta_1 - \theta_2)t}$

$[k] = \frac{[ML^2T^{-2}][L]}{[L]^2[K][T]} = [MLK^{-1}T^{-3}]$

488 **(d)**

$[C] = [M^{-1}L^{-2}T^4A^2], [R] = [ML^2T^{-3}A^{-2}]$

$[L] = [ML^2T^{-2}A^{-2}]$ and $[I] = [M^0L^0T^0A]$

1. $[CR] = [M^{-1}L^{-2}T^4A^2][ML^2T^{-3}A^{-2}]$

   $= [M^0L^0TA^0]$

2. $\frac{[L]}{[R]} = \frac{[ML^2T^{-2}A^{-2}]}{[ML^2T^{-3}A^{-2}]} = [M^0L^0TA^0]$

3. $(\sqrt{LC}) = ([ML^2T^{-2}A^{-2}] \times [M^{-1}L^{-2}A^2])^{1/2}$

   $= [M^0L^0TA^0]$

4. $[LI^2] = [ML^2T^{-2}A^{-2}][M^0L^0T^0A^1]^2$

   $= [ML^2T^{-2}A^0]$

489 **(a)**

In given equation, $\frac{\alpha z}{k\theta}$ should be dimensionless

$\therefore \alpha = \frac{k\theta}{z}$

$\Rightarrow [\alpha] = \frac{[ML^2T^{-2}K^{-1} \times K]}{[L]} = [MLT^{-2}]$

And $P = \frac{\alpha}{\beta}$

$\Rightarrow [\beta] = \left[\frac{\alpha}{P}\right] = \frac{[MLT^{-2}]}{[ML^{-1}T^{-2}]}$

$\Rightarrow [\beta] = [M^0L^2T^0]$

490 **(b)**

Force = Mass $\times$ acceleration

$\therefore$ Dimensions of force $= [M][LT^{-2}] = [MLT^{-2}]$

$$\text{Power} = \frac{\text{Work}}{\text{Time}}$$

$\therefore$ Dimensions of power $= \frac{[ML^2T^{-2}]}{[T]} = [ML^2T^{-3}]$

Torque = Force $\times$ displacement

$\therefore$ Dimensions of torque

$= [MLT^{-2}][L] = [ML^2T^{-2}]$

And dimensions of energy $= [ML^2T^{-2}]$

Hence, torque and energy have same dimensions.

491 **(a)**

$y = a \sin(\omega t + kx)$.
Here, ωt should be dimensionless

$\therefore [\omega] = \left[\frac{1}{t}\right]$

$[\omega] = [M^0L^0T^{-1}]$

492 **(a)**

$R = \frac{V}{I} \Rightarrow \pm\frac{\Delta R}{R} = \pm\frac{\Delta V}{V} \pm \frac{\Delta I}{I}$
$= 3 + 3 = 6\%$

493 **(c)**

Electric displacement, $D = \varepsilon E$

Unit of $D = \frac{C^2}{Nm^2}\frac{N}{C}$

$\therefore \quad [D] = \left(\frac{C}{m^2}\right) = \frac{[AT]}{[L^2]} = [L^{-2}TA]$

494 **(a)**

Given, $W = \frac{1}{2}kx^2$

Writing the dimensions on both sides

$[ML^2T^{-2}] = k[M^0L^2T^0]$

$\therefore$ Dimensions of $k = [MT^{-2}] = [ML^0T^{-2}]$

495 **(d)**

ct^2 must have dimensions of L
$\Rightarrow c$ must have dimensions of L/T^2 i.e. LT^{-2}

496 **(c)**

Gravitational force, $F = \frac{GM_1M_2}{R^2}$

$$\Rightarrow \quad G = \frac{FR^2}{M_1M_2}$$

$$[G] = \frac{[\mathrm{MLT^{-2}}][\mathrm{L^2}]}{[\mathrm{M^2}]}$$

$$= [\mathrm{M^{-1}L^3T^{-2}}]$$

497 **(a)**

$$\text{Pressure} = \frac{\text{Force}}{\text{Area}} = ML^{-1}T^{-2}$$

$$\text{Stress} = \frac{\text{Restoring Force}}{\text{Area}} = ML^{-1}T^{-2}$$

498 **(c)**

$$\Delta R_s = \Delta R_1 + \Delta R_2 = \left[\frac{10}{100} \times 10 + \frac{20}{100} \times 20\right] \mathrm{k\Omega} = 5\mathrm{k\Omega}$$

$$\frac{\Delta R_s}{R_s} \times 100 = \frac{5}{30} \times 100 = \frac{50}{3} = 17$$

499 **(a)**

$$\frac{R}{L} = \frac{V/I}{V \times T/I} = \frac{1}{T} = \text{Frequency}$$

500 **(c)**

$$R = \rho\frac{L}{A} \Rightarrow \rho = \frac{RA}{L} = ohm \times cm$$

501 **(c)**

$$[C] = [\mathrm{M^{-1}L^{-2}T^4A^2}]$$

$$[V] = [\mathrm{ML^2T^{-3}A^{-1}}]$$

$$\therefore [CV^2] = [\mathrm{M^{-1}L^{-2}T^4A^2}][\mathrm{ML^2T^{-3}A^{-1}}]^2$$

$$= [\mathrm{ML^2T^{-2}}]$$

502 **(a)**

By principle of dimensional homogenity

$$\left[\frac{a}{V^2}\right] = [P]$$

$$\therefore [a] = [P][V^2] = [ML^{-1}T^{-2}] \times [L^6] = [ML^5T^{-2}]$$

503 **(c)**

We can derive this equation from equations of motion so it is numerically correct

S_1 =distance trvelled in t^{th} second = $\frac{\text{Distance}}{\text{time}} = [LT^{-1}]$

u = velocity =$[LT^{-1}]$ and $\frac{1}{2}a(2t-1) = [LT^{-1}]$

As dimensions of each term in the given equation are same, hence equation is dimensionally correct also

504 **(d)**

$$E = -\frac{dV}{dx}$$

506 **(a)**

$\frac{L}{R}$ is a time constant of L-R circuit so *Henry/ohm* can be expressed as *second*

508 **(c)**

Stefan's law is $E = \sigma(T^4) \Rightarrow \sigma - \frac{E}{T^4}$

$$\text{where, } E = \frac{\text{Energy}}{\text{Area} \times \text{Time}} = \frac{\text{Watt}}{m^2}$$

$$\sigma = \frac{\text{Watt} - m^{-2}}{K^4} = Watt - m^{-2}K^{-4}$$

509 **(b)**

Time constant in an $R - C$ circuit

$\tau = R - C$

$[\tau] = [R][C]$

$= [\mathrm{ML^2T^{-3}A^{-2}}][\mathrm{M^{-1}L^{-2}T^4A^2}]$

$= [\mathrm{M^0L^0T}]$

510 **(d)**

$\lambda = m^p v^q h^r$

$[\mathrm{M^0LT^0}] = [\mathrm{M}^p][\mathrm{LT^{-1}}]^q[\mathrm{ML^2T^{-2}}]^r$

$[\mathrm{M^0LT^0}] = [\mathrm{M}^{p+r}\mathrm{L}^{q+2r}\mathrm{T}^{-q-r}]$

$\therefore \ p + r = 0, q + 2r = 1, -q - r = 0$

After solving we get

$p = -1, q = -1, r = 1$

511 **(a)**

Astronomical unit of distance

512 **(d)**

$$\frac{\Delta \mathrm{g}}{\mathrm{g}} = \frac{\Delta l}{l} + 2\frac{\Delta T}{T}$$

In option (d) error in Δg is minimum and number of observations made are maximum. Hence, in this case error in g will be minimum.

513 **(a)**

$$\phi = BA = \frac{F}{I \times L} A = \frac{[MLT^{-2}][L^2]}{[A][L]} = [ML^2T^{-2}A^{-1}]$$

514 **(d)**

$n(xm)^2 = 1\text{m}^2$ or $n = \frac{1}{x^2}$

515 **(d)**

Percentage error in

$$A = \left(2\frac{\Delta a}{a} + 3\frac{\Delta b}{b} + \frac{\Delta c}{c} + \frac{1}{2}\frac{\Delta d}{d}\right) \times 100\%$$

$$= 2 \times 1 + 3 \times 3 + 2 + \frac{1}{2} \times 2$$

$$= 2 + 9 + 2 + 1 = 14\%$$

516 **(d)**

The number of significant figures in 4.8000×10^4 is 5 (zeros on right after decimal are counted while zeros in powers of 10 are not counted).

The number of significant figures in 48000.50 is 7 (all the zeros between two non-zero digits are significant).

517 **(d)**

$$n_2 = n_1 \left[\frac{L_1}{L_2}\right]^1 \left[\frac{T_1}{T_2}\right]^{-2} = 10 \left[\frac{\text{m}}{\text{km}}\right]^2 \left[\frac{\text{sec}}{\text{hr}}\right]^{-2}$$

$$n_2 = 10 \left[\frac{\text{m}}{10^3\text{m}}\right]^1 \left[\frac{\text{sec}}{3600\text{ sec}}\right]^{-2} = 129600$$

518 **(c)**

We know, $f = \frac{1}{2\pi\sqrt{LC}}$

Or $\sqrt{LC} = \frac{1}{2\pi f} = \text{time}$

Thus, $\sqrt{LC}$ has the dimension of time.

519 **(d)**

From the given relation, $D = -\frac{n(x_2 - x_1)}{n_2 - n_1}$

Here $[n] = \left[\frac{1}{\text{area} \times \text{time}}\right] = \frac{1}{[\text{L}^2\text{T}]} = [\text{L}^{-2}\text{T}^{-1}]$

$x_2 - x_1 = [\text{L}]$ and $n_2 - n_1 = \left[\frac{1}{\text{volume}}\right] = \left[\frac{1}{\text{L}^3}\right] = [\text{L}^{-3}]$

So, $[D] = \frac{[\text{L}^{-2}\text{T}^{-1}\text{L}]}{[\text{L}^{-3}]} = [\text{L}^2\text{T}^{-1}]$.

520 **(b)**

According to question

$F \propto m^a v^b r^c$

$F = km^a v^b r^c$

k, being a dimensionless constant.

From homogeneity of dimensions,

LHS=RHS

$[\text{MLT}^{-2}] = [\text{M}]^a[\text{LT}^{-1}]^b[\text{L}]^c$

Or $[\text{MLT}^{-2}] = [\text{M}^a\text{L}^{b+c}\text{T}^{-b}]$

Comparing the powers, we obtain

$a = 1$

$b + c = 1$

$-b = -2 \Rightarrow b = 2$

$\therefore\ 2 + c = 1$

$\Rightarrow c = -1$

Therefore, $F = kmv^2r^{-1} = \frac{kmv^2}{r}$

The experimental value of k is found to be 1 here

$$\therefore\ F = \frac{mv^2}{r}$$

521 **(a)**

By submitting dimension of each quantity in R.H.S. of option (a) we get

$$\left[\frac{mg}{\eta r}\right] = \left[\frac{M \times LT^{-2}}{ML^{-1}T^{-1} \times L}\right] = [LT^{-1}]$$

This option gives the dimension of velocity

522 **(b)**

According to homogeneity principle

LHS=RHS

Or $F = 6\pi\eta^a r^b v^c$

Or $[\text{MLT}^{-2}] = [\text{ML}^{-1}\text{T}^{-1}]^a[\text{L}]^b[\text{LT}^{-1}]^c$

Or $[\text{M}^1\text{L}^1\text{T}^{-2}] = [\text{M}^0\text{L}^{-a+b+c}\text{T}^{-a-c}]$

$\therefore\ a = 1, -a + b + c = 1$ and $-a - c = -2$

After solving, we get

$a = 1, b = 1, c = 1$

523 **(d)**

$a = b^\alpha c^\beta / d^\gamma e^\delta$

So maximum error in a is given by

$$\left(\frac{\Delta a}{a} \times 100\right)_{\max} = \alpha.\frac{\Delta b}{b} \times 100 + \beta.\frac{\Delta c}{c} \times 100 + \gamma.\frac{\Delta d}{d} \times 100 + \delta.\frac{\Delta e}{e} \times 100$$

$$= (\alpha b_1 + \beta c_1 + \gamma d_1 + \delta e_1)\%$$

524 **(b)**

Frequency $f = cm^x k^y, k = \text{Force/Length}$

$$[M^0L^0T^{-1}] = [M]^x[ML^0T^{-2}]^y$$

$$= [M]^{x+y}[L]^0[T]^{-2y}$$

Comparing the powers on M, L and T

$$-2y = -1$$

$$\Rightarrow \quad y = \frac{1}{2}$$

And $x + y = 0$

$$\therefore \ x = -y = -\frac{1}{2}$$

525 **(a)**

Using the relation $R = \frac{pV}{\theta}$

Dimensions of gas constant R

$$= \frac{\text{dimensions of pressure } \times \text{ dimensions of volume}}{\text{dimension of temperature}}$$

$$= \frac{[ML^{-1}T^{-2}][L^3]}{\theta}$$

$$= [ML^2T^{-2}\theta^{-1}]$$

526 **(d)**

$$C = \frac{1}{\sqrt{\mu_0\varepsilon_0}} \Rightarrow \frac{1}{\mu_0\varepsilon_0} = c^2 = [L^2T^{-2}]$$

527 **(d)**

Percentage error in $x = 1\% + 2 \times 3\% + 3 \times 2\% = 13\%$.

The sign $\pm$ has been used because the words `maximum percentage error' have not been used.

Note percentage error is $\pm\frac{\Delta A}{A} \times 100$

Maximum percentage error is $\frac{\Delta A}{A} \times 100$

528 **(c)**

Potential can be written a potential energy per unit charge,

$$V = \frac{W}{q} = \frac{U}{q}$$

Hence, dimensions of potential are the same as that of work per unit charge.

529 **(b)**

Magnetic energy $= \frac{1}{2}Li^2 = \frac{LQ^2}{2t^2}$

[L = inductance, i = current.]

Energy has the dimensions = $[ML^2T^{-2}]$.

Equate the dimensions, we have

$$[ML^2T^{-2}] = (\text{henry}) \times \frac{[Q^2]}{[T^2]}$$

$$\Rightarrow \quad [\text{henry}] = \frac{[ML^2]}{[Q^2]}$$

530 **(a)**

$$\frac{\text{angular momentum}}{\text{linear momentum}} = \frac{[ML^2T^{-1}]}{[MLT^{-1}]} = [M^0LT^0]$$

531 **(b)**

Given, $\quad F \propto v^2$

Or $\quad F = kv^2$

$$\therefore \quad k = \frac{F}{v^2}$$

$$[k] = \frac{[F]}{[v^2]} = \frac{[MLT^{-2}]}{[L^2T^{-2}]}$$

$$= [ML^{-1}T^0]$$

532 **(b)**

Time period of simple pendulum is

$$T = 2\pi\sqrt{\frac{l}{g}}$$

Or $\frac{\Delta T}{T} = \frac{1}{2}\left(\frac{\Delta l}{l} - \frac{\Delta g}{g}\right)$

Or $\frac{\Delta g}{g} = \frac{\Delta l}{l} - \frac{2\Delta T}{T}$

∴ Maximum percentage error in equation

$$\frac{\Delta g}{g} \times 100 = \frac{\Delta l}{l} \times 100 + \frac{2\Delta T}{T} \times 100$$

$$= 1 \times 100 + 2 \times 2 \times 100$$

$$= 5 \times 100 = 5\%$$

533 **(d)**

$P = \frac{F}{A} = \frac{F}{l^2}$, so maximum error in pressure (P)

$\left(\frac{\Delta P}{P} \times 100\right)_{max} = \frac{\Delta F}{F} \times 100 + 2\frac{\Delta l}{l} \times 100$

$= 4\% + 2 \times 2\% = 8\%$

534 **(c)**

$v = \frac{P}{2l}\left[\frac{F}{m}\right]^{1/2}$

$\Rightarrow v^2 = \frac{P^2}{4l^2}\left[\frac{F}{m}\right]$

$\therefore m \propto \frac{F}{l^2v^2}$

$\Rightarrow [m] = \left[\frac{MLT^{-2}}{L^2T^{-2}}\right] = [ML^{-1}T^0]$

535 **(b)**

Here αt^2 is a dimensionless. Therefore, $\alpha = \frac{1}{t^2}$ and has the dimension of $[T^{-2}]$.

536 **(d)**

Time period of a simple pendulum is

$$T = 2\pi\sqrt{\frac{L}{g}} \Rightarrow g = \frac{4\pi^2 L}{T^2}$$

$$\therefore \frac{\Delta g}{g} \times 100 = \left(\frac{\Delta L}{L} + 2\frac{\Delta T}{T}\right) \times 100$$

$$= 1\% + 2 \times 2\% = 5\%$$

537 **(b)**

$H = I^2Rt$

$\therefore \frac{\Delta H}{H} \times 100 = \left(\frac{2\Delta I}{I} + \frac{\Delta R}{R} + \frac{\Delta t}{t}\right) \times 100$

$= (2 \times 3 + 4 + 6)\% = 16\%$

538 **(c)**

Given equation is dimensionally correct because both sides are dimensionless but numerically wrong because the correct equation is $\tan\theta = \frac{v^2}{rg}$

539 **(a)**

Area of rectangle

$A = lb$

$= 10.5 \times 2.1$

$= 22.05\ cm^2$

Minimum possible measurement of scale =0.1 cm

So, area measured by scale = $22.0 cm^2$

540 **(b)**

$F = -\eta A\frac{\Delta v}{\Delta z} \Rightarrow [\eta] = [ML^{-1}T^{-1}]$

As $F = [MLT^{-2}], A = [L^2], \frac{\Delta v}{\Delta z} = [T^{-1}]$

∴ Dimensions of $\eta = \frac{F}{A}\frac{\Delta z}{\Delta v} = \frac{MLT^{-2}}{L^2}.\frac{1}{T^{-1}} = [ML^{-1}T^{-1}]$

541 **(c)**

Intensity $(I) = \frac{\text{Energy}}{\text{Area} \times \text{time}}$

542 **(d)**

R=8.3 J/K-mol

$n_1u_1 = n_2u_2$

$\therefore n_2 = \frac{n_1u_1}{u_2}$

$= \frac{8.3 J/K - mol}{atm\ L/K - mol}$

$= \frac{8.3J/K - mol}{(1.013 \times 10^5 N/m^2)(10^{-3}m^3)/K - mol}$

$= \frac{8.12}{10^2} = 0.0812$

∴ 8.3 J/K-mol=0.0812 atm L/K-mol

543 **(a)**

$[E] = [ML^2T^{-2}], [m] = [M], [l] = [ML^2T^{-1}]$ and $[G] = [M^{-1}L^3T^{-2}]$

Substituting the dimensions of above quantities in the given formula:

$$\frac{El^2}{m^5G^2} = \frac{[ML^2T^{-2}][ML^2T^{-1}]^2}{[M^5][M^{-1}L^3T^{-2}]^2} = \frac{M^3L^6T^{-4}}{M^3L^6T^{-4}}$$
$$= [M^0L^0T^0]$$

544 **(a)**

Dimensionally. $\left[\frac{b}{t}\right] = [v]$ or $[b] = [vt] =$ [L].

545 **(d)**

$\frac{1}{2}CV^2 =$ Stored energy in a capacitor $= [ML^2T^{-2}]$

546 **(a)**

$$\text{Power} = \frac{\text{Work done}}{\text{Time}} = \left[\frac{ML^2T^{-2}}{T}\right] = [ML^2T^{-3}]$$

547 **(c)**

Light year is a distance which light travels in one year

548 **(a)**

$$T = \frac{32 \times 10^{-5}}{(10)^{-2}}$$
$$= 32 \times 10^{-3}\text{Nm}^{-1} = 0.032\text{Nm}^{-1}.$$

549 **(d)**

Surface tension is defined as the force per unit length and hence its unit is newton metre^{-1} or Nm^{-1}

551 **(c)**

Percentage error in measurement of a side

$$= \frac{0.01}{1.23} \times 100$$

Percentage error in measurement of area

$$= 2 \times \frac{0.01}{1.23} \times 100$$

552 **(c)**

$n \times 40\, amu = 6.64 \times 10^{24}$

$n \times 40 \times 1.6 \times 10^{-27} = 6.64 \times 10^{24}$

$n = 10^{50}$

553 **(d)**

$$v = \sqrt{\frac{T}{m}} = \left[\frac{m'g}{\frac{M}{t}}\right]^{1/2} = \left[\frac{m'lg}{M}\right]^{1/2}$$

It follows from here, $\frac{\Delta v}{v} = \frac{1}{2}\left[\frac{\Delta m'}{m} + \frac{\Delta l}{l} + \frac{\Delta M}{M}\right]$

$$= \frac{1}{2}\left[\frac{0.1}{3.0} + \frac{0.001}{1.000} + \frac{0.1}{2.5}\right]$$
$$= \frac{1}{2}[0.03 + 0.001 + 0.04]$$
$$= 0.036$$

Percentage error in the measurement=3.6

555 **(b)**

Magnetic moment is the strength of magnet. Its SI unit is amp $\times$ m^2 or N $-$ m/telsa or JT^{-1}.

556 **(b)**

By Stefan's law,

$$E = \sigma T^4$$

Where σ is the Stefan's constant

$$\sigma = \frac{E}{T^4}$$

$$[\sigma] = \frac{[E]}{T^4} = \frac{[\text{ML}^2\text{T}^{-2}]}{[\text{K}^4]}$$

$$= [\text{ML}^2\text{T}^{-2}\text{K}^{-4}]$$

557 **(c)**

Given, $x = \cos(\omega t + kx)$

Here, $(\omega t + kx)$ is an angle so the dimension of

$(\omega t + kx) = [\text{M}^0\text{L}^0\text{T}^0]$

Or dimensions of $\omega t = [\text{M}^0\text{L}^0\text{T}^0]$

Or dimensions of $\omega = \frac{[\text{M}^0\text{L}^0\text{T}^0]}{[\text{T}]}$

Or $= [\text{M}^0\text{L}^0\text{T}^{-1}]$

558 **(c)**

Impulse = change in momentum = $F \times t$

So the unit of momentum will be equal to *Newton-sec*

559 **(c)**

$\omega = \frac{d\theta}{dt} = [T^{-1}]$ and frequency $[n] = [T^{-1}]$

560 **(c)**

When quantities are subtracted, their maximum absolute errors are added up.

561 **(d)**

$F_m = Bqv$ or $[B] = \left[\frac{F_m}{qv}\right] = \left[\frac{F_m}{Itv}\right]$

$= \frac{[\text{MLT}^{-2}]}{[\text{A}][\text{T}][\text{LT}^{-1}]} = [\text{ML}^0\text{T}^{-2}\text{A}^{-1}]$

Alternate

$F = BIl \Rightarrow [B] = \left[\frac{F}{Il}\right] = [\text{ML}^0\text{T}^{-2}\text{A}^{-1}]$

563 **(b)**

We will use the general rule of addition by making the powers same.

ie, we will add 3.8×10^{-6} and 42×10^{-6} we get s

$$= 45.8 \times 10^{-6} = 4.58 \times 10^{-5}$$

As least number of significant figures in given values are 2, so

We round off the result to 4.6×10^{-5}.

564 **(a)**

Physical quantity(p) = Numerical value $(n) \times$ Unit (u)

If physical quantity remains constant then $n \propto 1/u$

$\therefore n_1u_1 = n_2u_2$

565 **(a)**

Time defined in terms of the rotation of the earth is called universal time (UT).

566 **(b)**

Power of lens $P = \frac{1}{f}$

$\therefore \ [P] = \frac{1}{[f]} = \frac{1}{[\text{L}]} = [\text{M}^0\text{L}^{-1}\text{T}^0]$

567 **(d)**

Kinetic energy, $E = \frac{1}{2}mv^2$

$\therefore \ \frac{\Delta E}{E} \times 100 = \frac{v'^2 - v^2}{v^2} \times 100$

$= [(1.5)^2 - 1] \times 100$

$= 125\%$

568 **(b)**

Angular momentum $= mvr$

$= [MLT^{-1}][L] = [ML^2T^{-1}]$

569 **(b)**

We have to retain three significant figures in the result.

570 **(d)**

Energy = Work done [Dimensionally]

571 **(d)**

The dimensional formula of

Work = Energy = Torque = $[\text{ML}^2\text{T}^{-2}]$

572 **(d)**

Strain is dimensionless

573 **(a)**

Least count LC

$= \frac{Pitch}{Number\ of\ divisions\ on\ circular\ scale}$

$= \frac{0.5}{50} = 0.01\ mm$

Now, diameter of ball

$= (2 \times 0.5\ mm) + (25 - 5)(0.01)$

$= 1.2mm$

3

MOTION IN A STRAIGHT LINE

SINGLE CORRECT ANSWER TYPE

1. A body, thrown upwards with some velocity reaches the maximum height of 50 m. Another body with the double the mass thrown up with double the initial velocity will reach a maximum height of
   a) 100 m b) 200 m c) 300 m d) 400 m
2. A car accelerates from rest at a constant rate α for some time, after which it decelerates at a constant rate β and comes to rest. If the total time elapsed is t, then the maximum velocity acquired by the car is
   a) $\left(\frac{\alpha t+\beta^2}{\alpha\beta}\right)t$ b) $\left(\frac{\alpha^2-\beta^2}{\alpha\beta}\right)t$ c) $\frac{(\alpha+\beta)t}{\alpha\beta}$ d) $\frac{\alpha\beta t}{\alpha+\beta}$
3. Two cars A and B are travelling in the same direction with velocities v_1 and $v_2(v_1 > v_2)$. When the car A is at a distance d behind the car B, the driver of the car A applies the brake producing uniform retardation, a. There will be no collision when
   a) $d<\left(\frac{v_1-v_2}{2a}\right)$ b) $d>\frac{v_1^2-v_2^2}{2a}$ c) $d>\frac{(v_1-v_2)^2}{2a}$ d) $d<\frac{v_1^2-v_2^2}{2a}$
4. A body of *mass* 10 kg is moving with a constant velocity of 10 m/s. When a constant force acts for 4 *seconds* on it, it moves with a velocity 2 m/sec in the opposite direction. The acceleration produced in it is
   a) $3m/\sec^2$ b) $-3m/\sec^2$ c) $0.3m/\sec^2$ d) $-0.3m/\sec^2$
5. A bullet is fired with a speed of 1000 m/sec in order to hit a target 100 m away. If $g = 10\ m/s^2$, the gun should be aimed
   a) Directly towards the target b) 5 cm above the target
   c) 10 cm above the target d) 15 cm above the target
6. The $v-t$ graph for a particle is as shown. The distance travelled in the first four second is

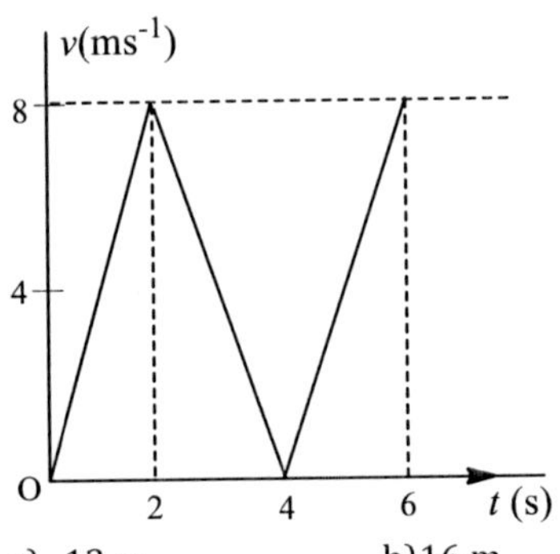

   a) 12 m b) 16 m c) 20 m d) 24 m
7. A body starts from rest, with uniform acceleration. If its velocity after n seconds is v, then its displacement in the last two seconds is
   a) $\frac{2v(n+1)}{n}$ b) $\frac{v(n+1)}{n}$ c) $\frac{v(n-1)}{n}$ d) $\frac{2v(n-1)}{n}$
8. A body, thrown upwards with some velocity reaches the maximum height of 50 m. Another body with the double the mass thrown up with double the initial velocity will reach a maximum height of
   a) 100 m b) 200 m c) 300 m d) 400 m

9. Figure given shows the distance –time graph of the motion of a car. It follows from the graph that the car is

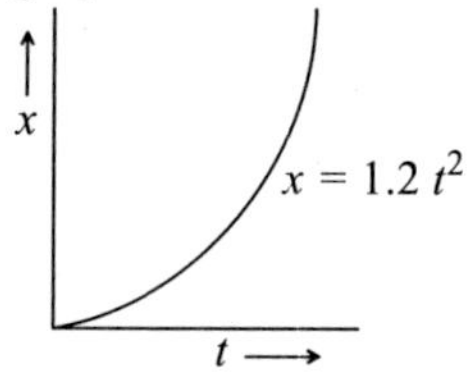

a) At rest
b) In uniform motion
c) In non-uniform acceleration
d) Uniformly accelerated

10. An object is moving with a uniform acceleration which is parallel to its instantaneous direction of motion. The displacement (s) –velocity (v) graph of this object is

a)
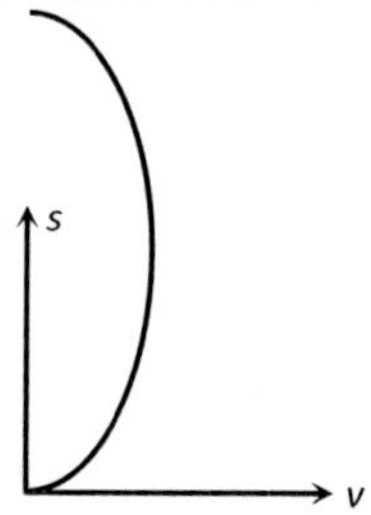

b)
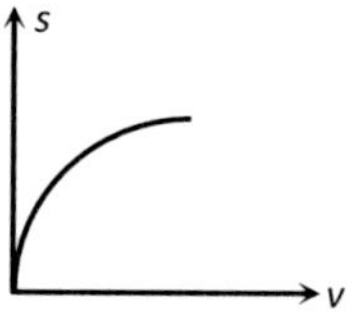

c)
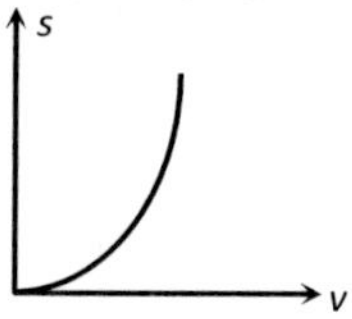

d)
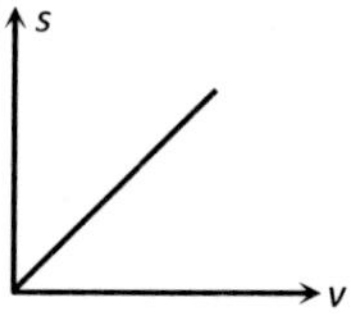

11. Water drops fall at regular intervals from a tap which is 5 m above the ground. The third drop is leaving the tap at the instant the first drop touches the ground. How far above the ground is the second drop at that instant

a) 2.50 m
b) 3.75 m
c) 4.00 m
d) 1.25 m

12. A police jeep is chasing with velocity of 45 km/h a thief in another jeep moving with velocity 153 km/h. Police fires a bullet with muzzle velocity of 180 m/s. The velocity with which it will strike the car of the thief is

a) 150 m/s
b) 27 m/s
c) 450 m/s
d) 250 m/s

13. A particle moves along a semicircle of radius 10m in 5 seconds. The average velocity of the particle is

a) $2\pi\ ms^{-1}$
b) $4\pi\ ms^{-1}$
c) $2\ ms^{-1}$
d) $4\ ms^{-1}$

14. Two balls are dropped from height h and $2h$ respectively from the earth surface. The ratio of time of these balls to reach the earth is

a) $1 : \sqrt{2}$
b) $\sqrt{2} : 1$
c) $2 : 1$
d) $1 : 4$

15. A body is projected with a velocity v and after some time it returns to the point from which it was projected. The average velocity and average speed of the body for the total time of flight are

a) $\vec{v}/2$ and $v/2$
b) 0 and $v/2$
c) 0 and 0
d) $\vec{v}/2$ and 0

16. A particle starts from rest at $t = 0$ and undergoes an acceleration a in ms^{-2} with time t in seconds which is as shown

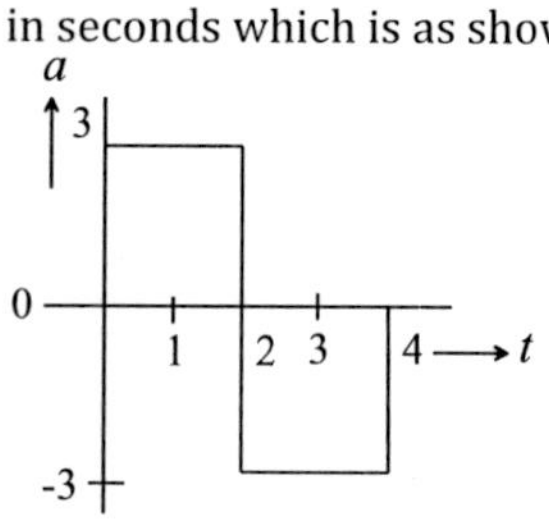

Which one of the following plot represents velocity V in ms^{-1} versus time t in seconds

a) 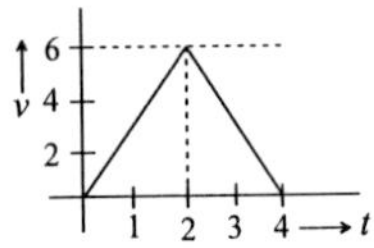

b) 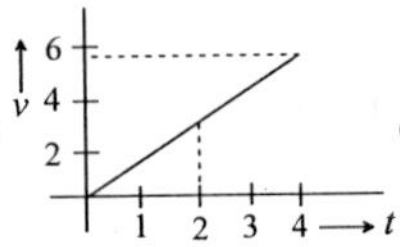

c) 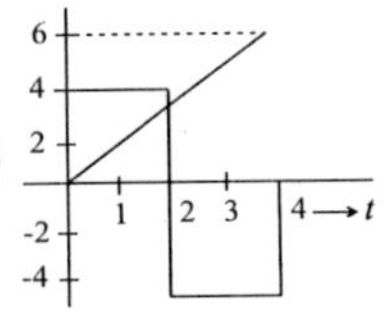

d) 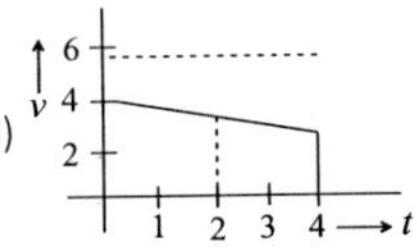

17. Which of the following speed-time graphs exist in the nature?

a) 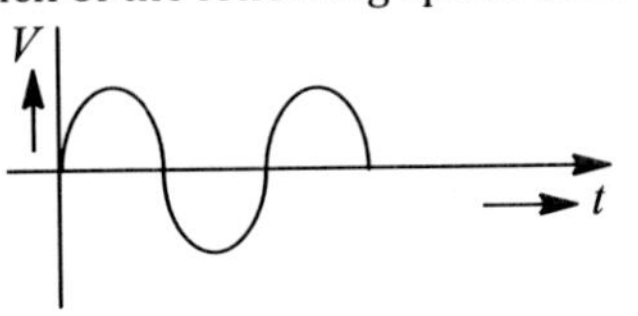

b) 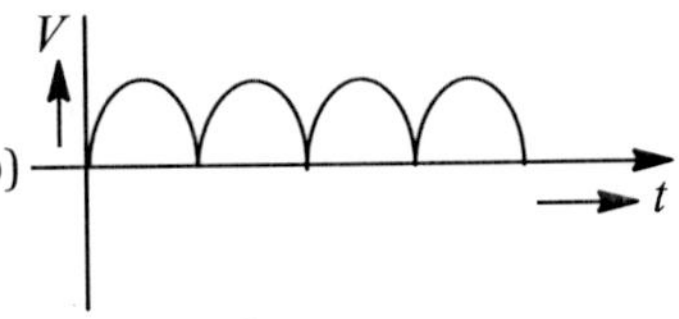

c) 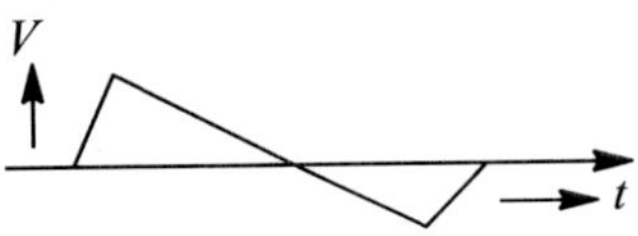

d) All of the above

18. If a body loses half of its velocity on penetrating 3 cm in a wooden block, then how much will it penetrate more before coming to rest?
a) 1 cm b) 2 cm c) 3 cm d) 4 cm

19. A body dropped from a height h with an initial speed zero, strikes the ground with a velocity 3 km/h . Another body of same mass is dropped from the same height h with an initial speed $-u' = 4km/h$. Find the final velocity of second body with which it strikes the ground
a) 3 km/h b) 4 km/h c) 5 km/h d) 12 km/h

20. A car starting from rest, accelerates at the rate f through a distance S, then continues at constant speed for time t and then decelerates at the rate $f/2$ to come to rest. If the total distance travelled is 15 S, then
a) $S = ft$ b) $S = \frac{1}{6}ft^2$ c) $S = \frac{1}{2}ft^2$ d) None of these

21. What determines the nature of the path followed by the particle
a) Speed b) Velocity c) Acceleration d) Both (b) and (c)

22. A body is thrown vertically up with a velocity u. It passes three points A, B and C in its upward journey with velocities $\frac{u}{2}, \frac{u}{3}$ and $\frac{u}{4}$ respectively. The ratio of the separations between points A and B and between B and C $i.e., \frac{AB}{BC}$ is
a) 1 b) 2 c) $\frac{10}{7}$ d) $\frac{20}{7}$

23. A particle moves along a straight line such that its position x at any time t is $x = 6t^2 - t^3$. Where x in metre ant t is in second, then
a) At $t = 0$ acceleration is 12 ms^{-2} b) $x - t$ curve has maximum at 4 s
c) Both (a) and (b) are wrong d) Both (a) and (b) are correct

24. A particle starts from rest, acceleration at 2 m/s^2 for 10 s and then goes with constant speed for 30 s and then decelerates at 4 m/s^2 till it stops. What is the distance travelled by it
a) 750 m b) 800 m c) 700 m d) 850 m

25. Select the incorrect statements from the following
S1 : Average velocity is path length divided by time interval
S2 : In general, speed is greater than the magnitude of the velocity

S3 : A particle moving in a given direction with a non-zero velocity can have zero speed
S4 : The magnitude of average velocity is the average speed

a) S2 and S3 b) S1 and S4 c) S1, S3 and S4 d) All four statements

26. Acceleration-time graph of a body is shown. The corresponding velocity-time graph of the same body is

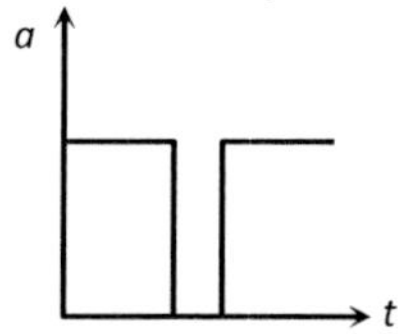

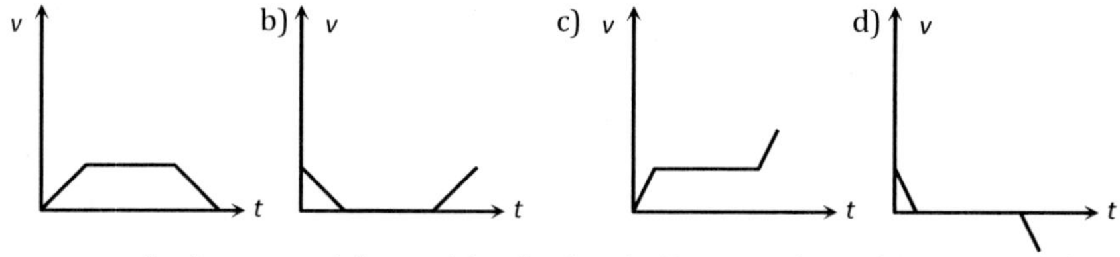

27. A person travels along a straight road for the first half time with a velocity v_1 and the next half time with a velocity v_2
The mean velocity V of the man is

a) $\frac{2}{V} = \frac{1}{v_1} + \frac{1}{v_2}$ b) $V = \frac{v_1 + v_2}{2}$ c) $V = \sqrt{v_1 v_2}$ d) $V = \sqrt{\frac{v_1}{v_2}}$

28. A body is moving along a straight line path with constant velocity. At an instant of time the distance travelled by it is s and its displacement is D, then

a) $D < s$ b) $D > s$ c) $D = s$ d) $D \leq s$

29. A packet is dropped from a balloon which is going upwards with the velocity 12 m/s, the velocity of the packet after 2 seconds will be

a) $-12\ m/s$ b) $12\ m/s$ c) $-7.6\ m/s$ d) $7.6\ m/s$

30. A body is moving in a straight line a shown in velocity-time graph. The displacement and distance travelled by in 8s are respectively

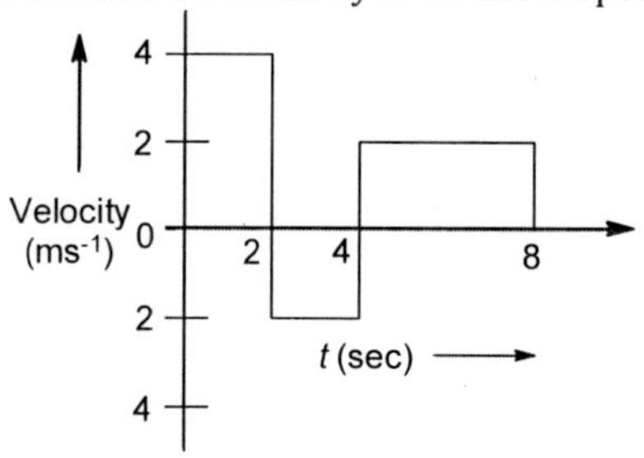

a) 12 m, 20 m b) 20m, 12 m c) 12 m, 12 m d) 20 m, 20 m

31. The displacement-time graph of a moving object is shown in figure. Which of the velocity-time graphs shown in figure could represent the motion of the same body?

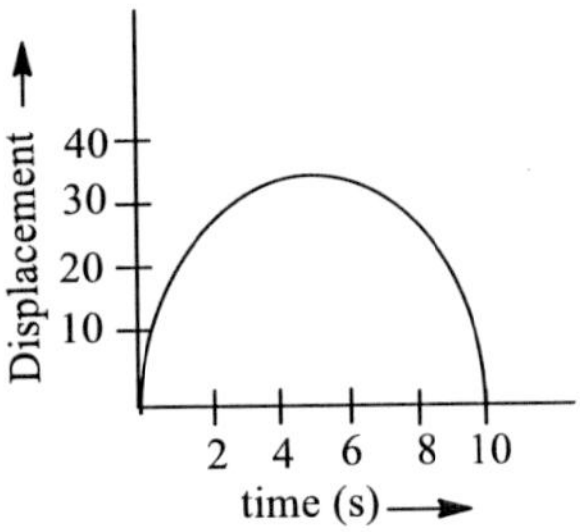

a) 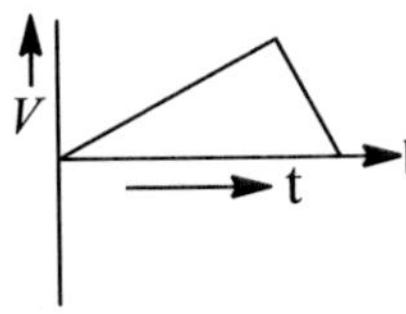

b) 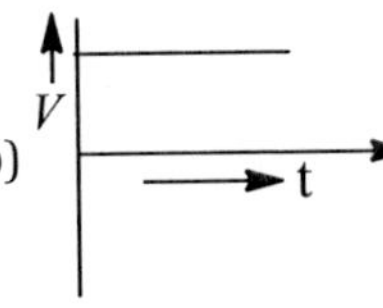

c) 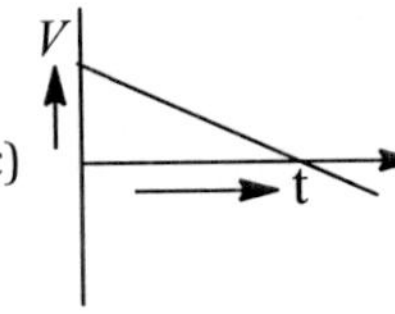

d) 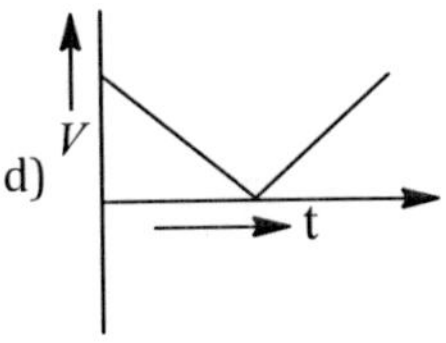

32. The displacement of the particle varies with time according to the relation $x = \frac{k}{b}[1 - e^{-bt}]$.Then the velocity of the particle is
a) $k(e^{-bt})$ b) $\frac{k}{b^2 e^{-bt}}$ c) $k\ b\ e^{-bt}$ d) None of these

33. A student is standing at a distance of 50 metres from the bus. As soon as the bus begins its motion with an acceleration of 1 ms^{-2}, the student starts running towards the bus with a uniform velocity u. Assuming the motion to be along a straight road, the minimum value of u, so that the student is able to catch the bus is
a) 52 ms^{-1} b) 8 ms^{-1} c) 10 ms^{-1} d) 12 ms^{-1}

34. A stone falls from a balloon that is descending at a uniform rate of 12ms^{-1}. The displacement of the stone from the point of release after 10s is
a) 725 m b) 610 m c) 510 m d) 490 m

35. An elevator car, whose floor to ceiling distance is equal to 2.7m, starts ascending with constant acceleration of 1.2ms^{-2}. 2 sec after the start, a bolt begins falling from the ceiling of the car. The free fall time of the bolt is
a) $\sqrt{0.54}s$ b) $\sqrt{6}\ s$ c) 0.7 s d) 1 s

36. A body thrown vertically upwards with an initial velocity u reaches maximum height in 6 seconds. The ratio of the distances travelled by the body in the first second and the seventh second is
a) 1 : 1 b) 11 : 1 c) 1 : 2 d) 1 : 11

37. A body falls from rest, its velocity at the end of first second is ($g = 32ft/sec$)
a) 16 ft/sec b) 32 ft/sec c) 64 ft/sec d) 24 ft/sec

38. A police car is travelling in a straight line with a constant speed v. A truck travelling in the same direction with constant velocity 3v/2 passes, the police car at $t = 0$. The police car starts acceleration 10 s after passing the truck, at a constant rate of 3 ms^{-2}, while truck continues to move at constant speed. If he police car takes 10 s further to catch the truck, find the value of v
a) 10 ms^{-1} b) 15 ms^{-1} c) 20 ms^{-1} d) 30 ms^{-1}

39. The displacement of a particle is given by $y = a + bt + ct^2 - dt^4$. The initial velocity and acceleration are respectively
a) $b, -4d$ b) $-b, 2c$ c) $b, 2c$ d) $2c, -4d$

40. The position of a particle x (in metres) at a time t seconds is given by the relation $\vec{r} = (3t\hat{\imath} - t^2\hat{\jmath} + 4\hat{k})$. Calculate the magnitude of velocity of the particle after 5 seconds
a) 3.55 b) 5.03 c) 8.75 d) 10.44

41. A steamer taken 12 days to reach from part x to part y. Every day only one steamer sets out from both the ports. How many steamers does each steamer meet in the open sea?
a) 23 b) 25 c) 27 d) 21.

42. A ball of mass m_1 and another ball of mass m_2 are dropped from equal height. If time taken by the balls are t_1 and t_2 respectively, then

a) $t_1 = \frac{t_2}{2}$ b) $t_1 = t_2$ c) $t_1 = 4t_2$ d) $t_1 = \frac{t_2}{4}$

43. If a car covers $2/5^{th}$ of the total distance with v_1 speed and $3/5^{th}$ distance with v_2 then average speed is

a) $\frac{1}{2}\sqrt{v_1 v_2}$ b) $\frac{v_1 + v_2}{2}$ c) $\frac{2v_1 v_2}{v_1 + v_2}$ d) $\frac{5v_1 v_2}{3v_1 + 2v_2}$

44. Which of the following 4 statements is false

a) A body can have zero velocity and still be accelerated
b) A body can have a constant velocity and still have a varying speed
c) A body can have a constant speed and still have a varying velocity
d) The direction of the velocity of a body can change when its acceleration is constant

45. Figures (i) and (ii) below show the displacement-time graphs of two particles moving along the x-axis. We can say that

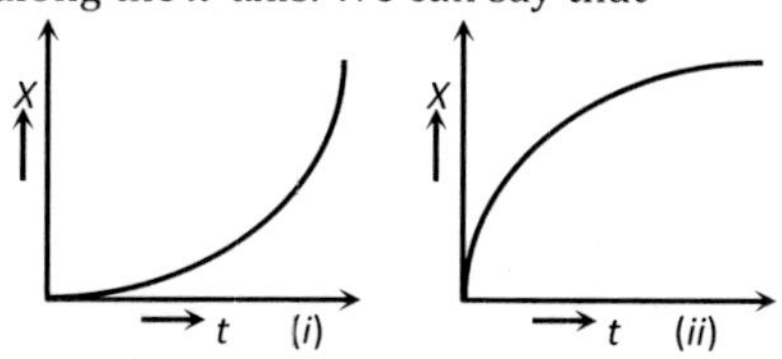

a) Both the particles are having a uniformly accelerated motion
b) Both the particles are having a uniformly retarded motion
c) Particle (i) is having a uniformly accelerated motion while particle (ii) is having a uniformly retarded motion
d) Particle (i) is having a uniformly retarded motion while particle (ii) is having a uniformly accelerated motion

46. A particle moves a distance x in time t according to equation $x = (t + 5)^{-1}$. The acceleration of particle is proportional to

a) $(\text{Velocity})^{2/3}$ b) $(\text{Velocity})^{3/2}$ c) $(\text{distance})^2$ d) $(\text{distance})^{-2}$

47. A body moves with initial velocity 10 ms^{-1}. If it covers a distance of $20m$ in $2s$, then acceleration of the body is

a) Zero b) $10ms^{-2}$ c) $5ms^{-2}$ d) $2ms^{-2}$

48. A ball is thrown vertically upwards. Which of the following plots represents the speed-time graph of the ball during its flight if the air resistance is not ignored

a)
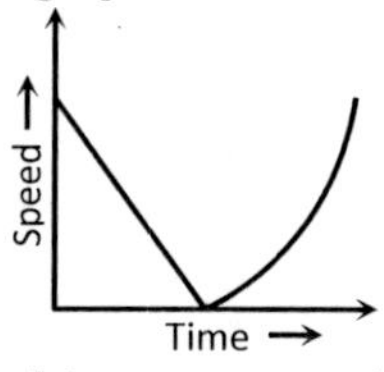

b)
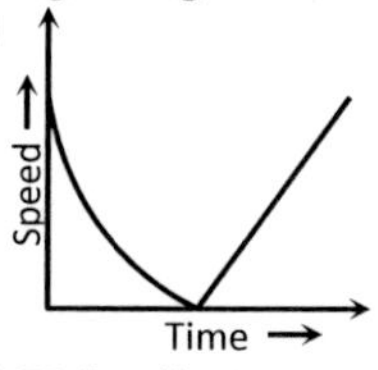

c)
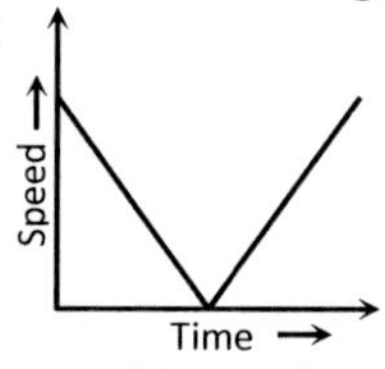

d)
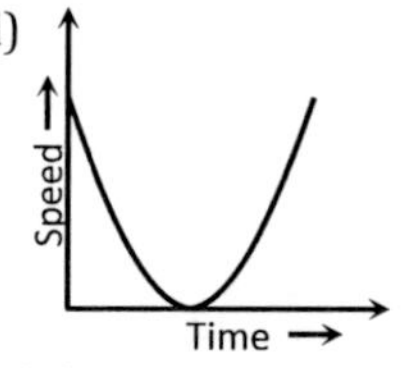

49. You drive a car at seed of 70 km/hr in a straight road for 8.4 km, and then the car runs out of petrol. You walk for 30 min to reach a petrol pump at a distance of 2 km. The average velocity from the beginning of your drive till you reach the petrol pump is

a) 16.8 km/hr b) 35 km/hr c) 64 km/hr d) 18.6 km/hr

50. Which of the following velocity-time graphs represent uniform motion

a)
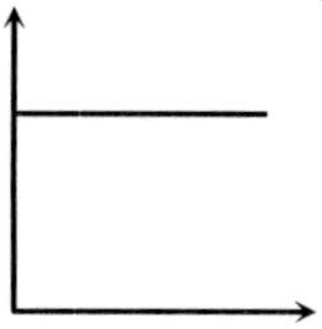

b)
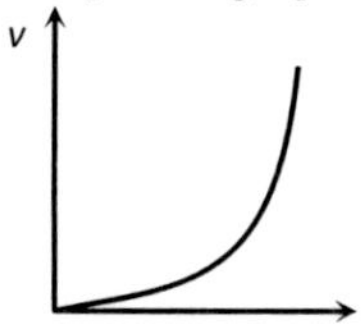

c)
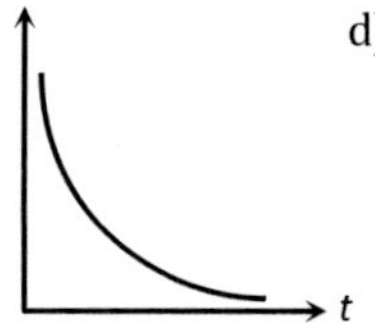

d)
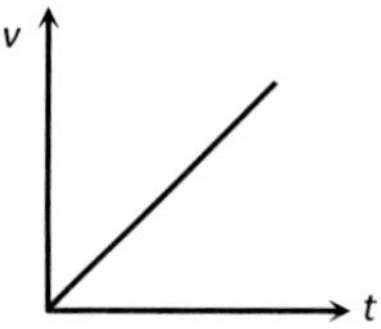

51. A boggy of uniformly moving train is suddenly detached from train and stops after covering some distance. The distance covered by the boggy and distance covered by the train in the same time has relation

a) Both will be equal b)First will be half of second
c) First will be 1/4 of second d)No definite ratio

52. A balloon starts rising from the ground with an acceleration of 1.25 m/s^2 after 8s, a stone is released from the balloon. The stone will $(g - 10\ m/s^2)$

a) Reach the ground in 4 second
b) Begin to move down after being released
c) Have a displacement of 50 m
d) Cover a distance of 40 m in reaching the ground

53. A particle starts from rest and travels a distance s with uniform acceleration, then it travels a distance 2s with uniform speed, finally it travels a distance 3s with uniform retardation and comes to rest. If the complete motion of the particle in a straight line then the ratio of its average velocity to maximum velocity in

a) 6/7 b)4/5 c)3/5 d)2/5

54. The displacement x of a particle varies with time $t, x = ae^{-\alpha t} + be^{\beta t}$, where a, b, α and β are positive constants. The velocity of the particle will

a) Go on decreasing with time b)Be independent of α and β
c) Drop to zero when $\alpha = \beta$ d)Go on increasing with time

55. A body is thrown vertically upwards. If air resistance is to be taken into account, then the time during which the body rises is

a) Equal to the time of fall b)Less than the time of fall
c) Greater than the time of fall d)Twice the time of fall

56. The motion of a particle is described by the equation $u = at$. The distance travelled by the particle in the first 4 seconds

a) $4a$ b)$12a$ c)$6a$ d)$8a$

57. Which graph represents a state of rest for an object

a) 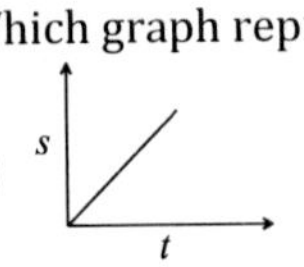

b) 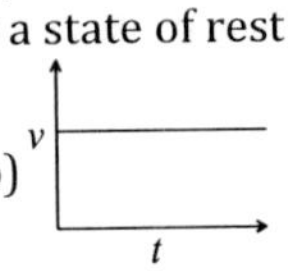

c) 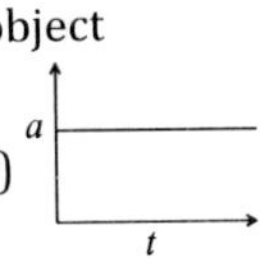

d) 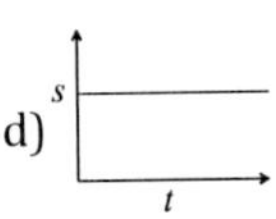

58. A stone is thrown with an initial speed of 4.9 m/s from a bridge in vertically upward direction. It falls down in water after 2 sec. The height of the bridge is

a) 4.9 m b)9.8 m c) 19.8 m d)24.7 m

59. A parachutist after bailing out falls 50 m without friction. When parachute opens, it decelerates at 2 ms^{-2}. He reaches the ground with a speed of 3 ms^{-1}. At what height, did he bail out?

a) 91 m b)182 m c) 293 m d)111 m

60. If a train travelling at 72 $kmph$ is to be brought to rest in a distance of 200 metres, then its retardation should be

a) 20 ms^{-2} b)10 ms^{-2} c) 2 ms^{-2} d)1ms^{-2}

61. A body freely falling from rest has a velocity v after it falls through distance h. The distance it has to fall down further for its velocity to become double is

a) h b)$2h$ c)$3h$ d)$4h$

62. Four marbles are dropped from the top of a tower one after the other with an interval of one second. The first one reaches the ground 4 seconds. When the first one reaches the ground the distances between the first and second, the second and third and the third and forth will be respectively

a) 35,25 and 15 m
b) 30,20 and 10 m
c) 20,10 and 5 m
d) 40,30 and 20 m

63. A particle starts from the origin and moves along the X-axis such that the velocity at any instant is given by $4t^3 - 2t$, where t is in second and velocity is in ms^{-1}. What is the acceleration of the particle when it is 2 m from the origin?

a) $10ms^{-2}$ b) $12ms^{-2}$ c) $22ms^{-2}$ d) $28ms^{-2}$

64. The ratios of the distances traversed, in successive intervals of time by a body, falling from rest are

a) $1:3:5:7:9:...$ b) $2:4:6:8:10:...$ c) $1:4:7:10:13:...$ d) None of these

65. Which of the following graphs can not possibly represent one dimensional motion of a particle

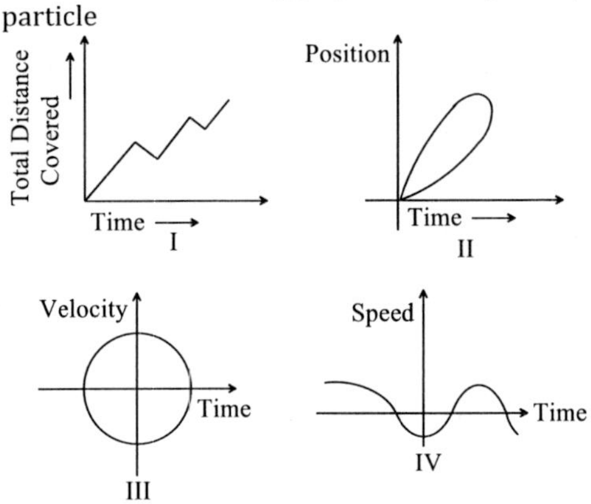

a) I and II b) II and III c) II and IV d) All four

66. A particle moves along x –axis as
$x = 4(t-2) + a(t-2)^2$ Which of the following is true?

a) The initial velocity of particle is 4 b) The acceleration of particle is $2a$
c) The particle is at origin at $t = 0$ d) None of the above

67. An object is dropped from rest. Its v-t graph is

a) b) c) d)

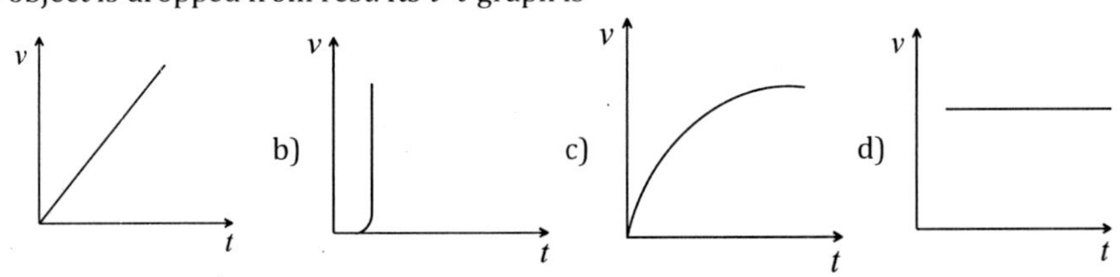

68. A parachutist after bailing out falls 50 m without friction. When parachute opens, it decelerates at 2 m/s^2. He reaches the ground with a speed of 3 m/s. At what height, did he bail out

a) 293 m b) 111 m c) 91 m d) 182 m

69. Figures (i) and (ii) below show the displacement-time graphs of two particles moving along the x-axis. We can say that

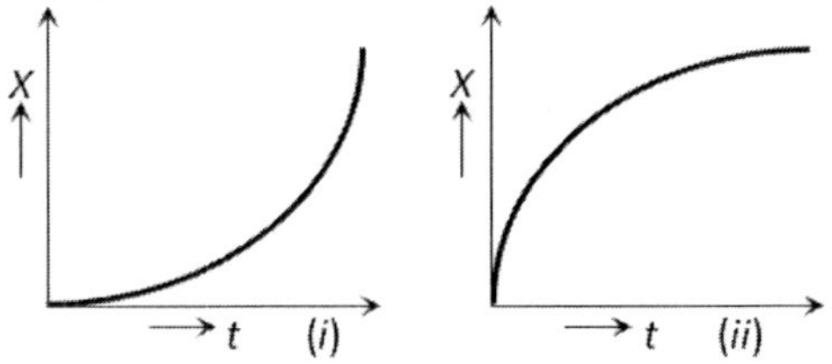

a) Both the particles are having a uniformly accelerated motion
b) Both the particles are having a uniformly retarded motion

c) Particle (i) is having a uniformly accelerated motion while particle (ii) is having a uniformly retarded motion

d) Particle (i) is having a uniformly retarded motion while particle (ii) is having a uniformly accelerated motion

70. A body is released from a great height falls freely towards the earth. Another body is released from the same height exactly a second later. Then the separation between two bodies, 2 s after the release of the second body is, nearly

a) 15 m b) 20 m c) 25 m d) 30 m

71. Which of the following curves represent the $v-t$ graph of an object falling on a metallic surface and bouncing back?

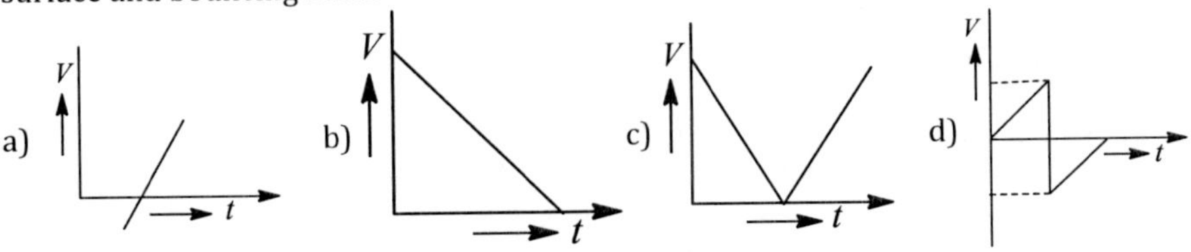

72. Velocity of a body on reaching the point from which it was projected upwards, is

a) $v=0$ b) $v=2u$ c) $v=0.5u$ d) $v=u$

73. An object accelerates from rest to a velocity 27.5 ms^{-1} in 10 s, then the distance covered after next 10 s is

a) 550 m b) 137.5 m c) 412.5 m d) 175 m

74. A body falls from a height $h=200m$ (at New Delhi). The ratio of distance travelled in each 2 sec during $t=0$ to $t=6$ second of the journey is

a) 1 : 4 : 9 b) 1 : 2 : 4 c) 1 : 3 : 5 d) 1 : 2 : 3

75. A ball is dropped from a high rise platform at $t=0$ starting from rest. After 6 seconds another ball is thrown downwards from the same platform with a speed v. The two balls meet at $t=18s$. What is the value of v? (taking $g=10\ m/s^2$)

a) 60 m/s b) 75 m/s c) 55 m/s d) 40 m/s

76. The three initial and final position of a man on the x –axis are given as

(i) $(-8m, 7m)$ (ii) $(7m, -3m)$ and (iii) $(-7m, 3m)$

Which pair gives the negative displacement

a) (i) b) (ii) c) (iii) d) (i) and (iii)

77. An object moving with a speed of 6.25 m/s, is decelerated at a rate given by $\frac{dv}{dt}=2.5\sqrt{v}$ where v is the instantaneous speed. The time taken by the object, to come to rest would be

a) $1s$ b) $2s$ c) $4s$ d) $8s$

78. A particle covers half of its total distance with speed v_1 and the rest half distance with speed v_2. Its average speed during the complete journey is

a) $\frac{v_1^2 v_2^2}{v_1^2+v_2^2}$ b) $\frac{v_1+v_2}{2}$ c) $\frac{v_1 v_2}{v_1+v_2}$ d) $\frac{2v_1 v_2}{v_1+v_2}$

79. A body starts from rest and falls vertically from a height of 19.6m. If $g=9.8ms^{-2}$, then the distance travelled by the body in the last 0.1s of its motion is

a) 1.9 m b) 0.049 m c) 17.7 m d) 19.6 m

80. For the velocity-time graph shown in figure below the distance covered by the body in last two seconds of its motion is what fraction of the total distance covered by it in all the seven seconds

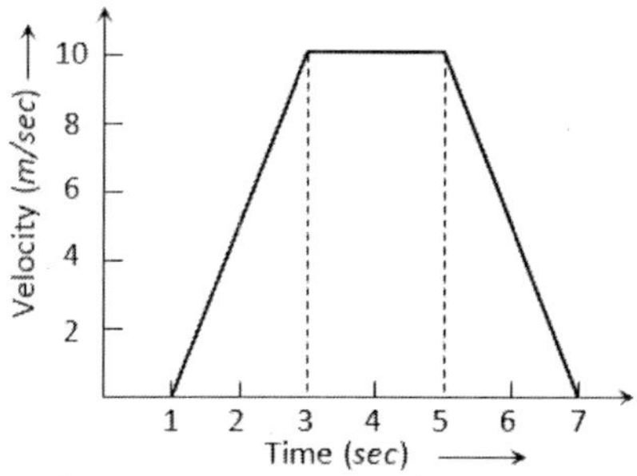

a) $\frac{1}{2}$ b) $\frac{1}{4}$ c) $\frac{1}{3}$ d) $\frac{2}{3}$

81. A truck and a car are moving with equal velocity. On applying the brakes both will stop after certain distance, then
a) Truck will cover less distance before rest b) Car will cover less distance before rest
c) Both will cover equal distance d) None

82. A stone falls freely from rest from a height h and it travels a distance $\frac{9h}{25}$ in the last second. The value of h is
a) 145 m b) 100 m c) 122.5 m d) 200 m

83. Figure given shows the distance –time graph of the motion of a car. It follows from the graph that the car is

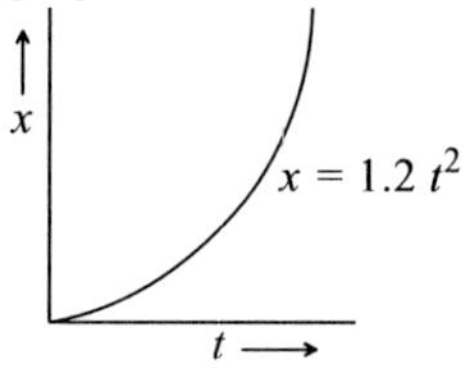

a) At rest b) In uniform motion
c) In non-uniform acceleration d) Uniformly accelerated

84. A train is moving slowly on a straight track with a constant speed of 2 ms^{-1}. A passenger in that train starts walking at a steady aped of a 2 ms^{-1} to the back of the train in the opposite direction of the motion of the train. So to an observer standing on the platform directly in front of that passenger. The velocity of the passenger appears to be
a) 4 ms^{-1} b) 2 ms^{-1}
c) 2 ms^{-1} in the opposite direction of the train d) Zero

85. A rocket is fired upward from the earth's surface such that it creates an acceleration of 19.6 $m/\sec^2$. If after 5 sec its engine is switched off, the maximum height of the rocket from earth's surface would be
a) 245 m b) 490 m c) 980 m d) 735 m

86. A ball is dropped on the floor from a height of 10 m. It rebounds to a height of 2.5m. If the ball is in contact with the floor for 0.01 sec, the average acceleration during contact is
a) 2100 $m/\sec^2$ downwards b) 2100 $m/\sec^2$ upwards
c) 1400 $m/\sec^2$ d) 700$m/\sec^2$

87. A bird files for 4s with a velocity of $|t-2|\text{ms}^{-1}$ in a straight line, where t =time in second. It covers a distance of

a) 8 m b)6 m c) 4 m d)2 m

88. A body is released from a great height and falls freely towards the earth. Another body is released from the same height exactly one second later. The separation between the two bodies , two seconds after the release of the second body is

a) 4.9 m b)9.8 m c) 19.6 m d)24.5 m

89. A 150 m long train is moving with a uniform velocity of 45 km/h. The time taken by the train to cross a bridge of length 850 m is

a) 56 sec b)68 sec c) 80 sec d)92 sec

90. If a body is thrown up with the velocity of 15 m/s then maximum height attained by the body is $(g = 10\ m/s^2)$

a) 11.25 m b)16.2 m c) 24.5 m d)7.62 m

91. A body of mass m moving along a straight line covers half the distance with a speed of 2ms^{-1}. The remaining half of distance is covered in two equal time intervals with a speed of 3 $\text{ms}^{-1}$and 5 ms^{-1} respectively. The average speed of the particle for the entire journey is

a) $\frac{3}{8}\text{ms}^{-1}$ b)$\frac{8}{3}\text{ms}^{-1}$ c)$\frac{4}{3}\text{ms}^{-1}$ d)$\frac{16}{3}\text{ms}^{-1}$

92. Equation of displacement for any particle is $s = 3t^3 + 7t^2 + 14t + 8m$. Its acceleration at time $t = 1$ sec is

a) 10 m/s^2 b)16 m/s^2 c) 25 m/s^2 d)32 m/s^2

93. A person travels along a straight road for the first half time with a velocity v_1 and the next half time with a velocity v_2
The mean velocity V of the man is

a) $\frac{2}{V} = \frac{1}{v_1} + \frac{1}{v_2}$ b)$V = \frac{v_1 + v_2}{2}$ c) $V = \sqrt{v_1 v_2}$ d)$V = \sqrt{\frac{v_1}{v_2}}$

94. The acceleration a of a particle starting from rest varies with time according to relation $a = \alpha t + \beta$. The velocity of the particle after a time t will be

a) $\frac{\alpha t^2}{2} + \beta$ b)$\frac{\alpha t^2}{2} + \beta t$ c) $\alpha t^2 + \frac{1}{2}\beta t$ d)$\frac{(\alpha t^2 + \beta)}{2}$

95. A man runs at a speed of $4\text{ms}^{-1}$to overtake a standing bus. When he is 6 m behind the door at $t = 0$, the bus moves forward and continues with a constant acceleration of 1.2ms^{-2}. The man reaches the door in time t. Then,

a) $4t = 6 + 0.6t^2$ b)$1.2t^2 = 4t$ c) $4t^2 = 1.2t$ d)$6 + 4t = 0.2t^2$

96. The graph of displacement v/s time is

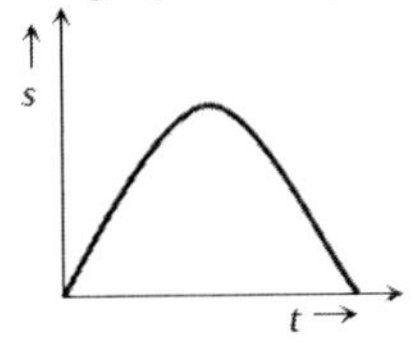

Its corresponding velocity-time graph will be

a)

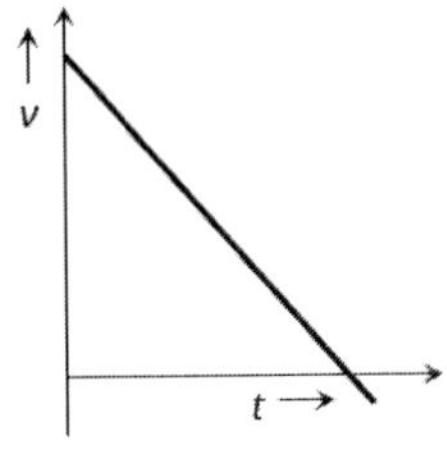

b) 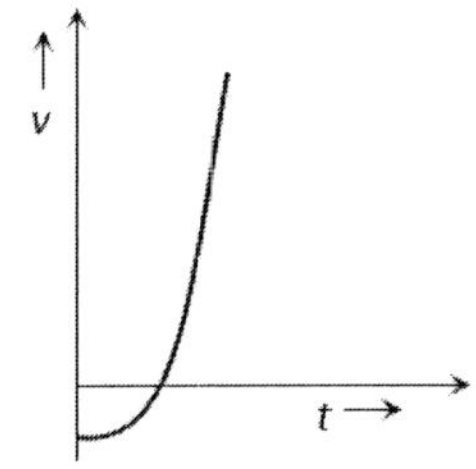

c) 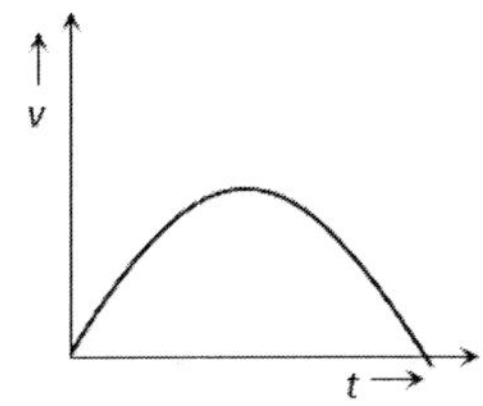

d) 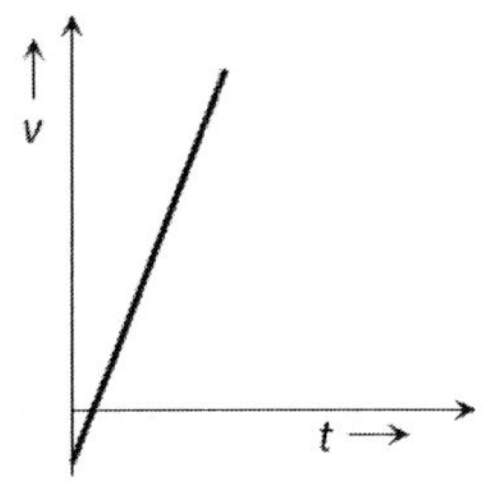

97. The time taken by a block of wood (initially at rest)to slide down a smooth inclined plane 9.8 $m$long (angle of inclination is 30°) is

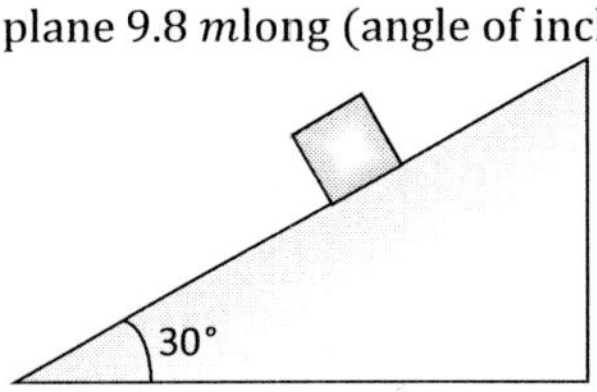

a) $\frac{1}{2} sec$ b)$2 sec$ c) $4 sec$ d)$1 sec$

98. A car travels half the distance with constant velocity of 40 $kmph$ and the remaining half with a constant velocity of 60 $kmph$. The average velocity of the car in $kmph$ is
a) 40 b)45 c) 48 d)50

99. From a balloon rising vertically upwards at $5 m/s$ a stone is thrown up at 10 m/s relative to the balloon. Its velocity with respect to ground after 2 s is (assume $g = 10 m/s^2$)
a) 0 b)20 m/s c) 10 m/s d)5 m/s

100. If a body starts from rest and travels 120 cm in the 6th second, then what is the acceleration
a) 0.20 m/s^2 b)0.027 m/s^2 c) $0.218 m/s^2$ d)$0.03 m/s^2$

101. A ball is thrown vertically upwards. It was observed at a height h twice with a time interval Δt. The initial velocity of the ball is
a) $\sqrt{8gh + g^2(\Delta t)^2}$ b) $\sqrt{8gh + \left(\frac{g\Delta t}{2}\right)^2}$ c)$\frac{1}{2}\sqrt{8gh + g^2(\Delta t)^2}$ d)$\sqrt{8gh + 4g^2(\Delta t)^2}$

102. A motion boat covers a given distance in 6 h moving down stream of a river. It covers the same distance in 10 h moving upstream. The time (in hour) it takes to cover the same distance in still water is
a) 6 b)7.5 c) 10 d)15

103. Acceleration-time graph of a body is shown. The corresponding velocity-time graph of the same body is

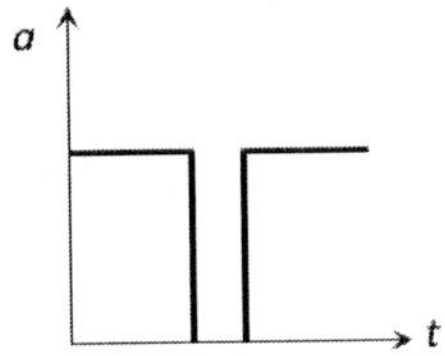

a) 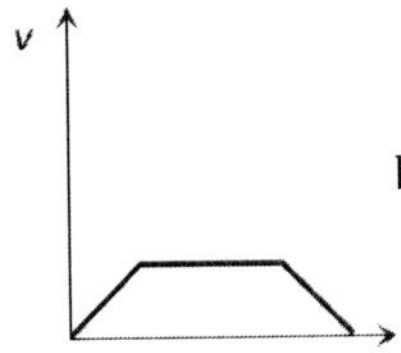 b) 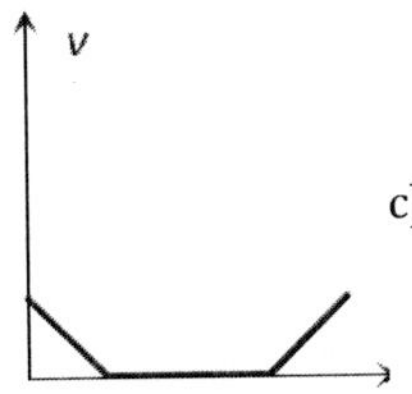 c) 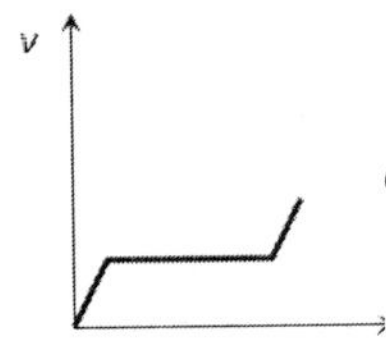 d) 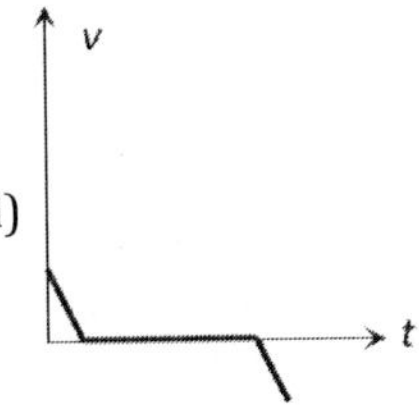

104. A body starting from rest moves with uniform acceleration. The distance covered by the body in time t is proportional to

a) $\sqrt{t}$ b) $t^{3/2}$ c) t^2 d) t^3

105. A boy released a ball from the top of a building. It will clear a window 2m high at a distance 10m below the top in nearly

a) 1 s b) 1.3 s c) 0.6 s d) 0.13 s

106. A man wants to reach point B on the opposite bank of a river flowing at a speed as shown in figure. What minimum speed relative to water should the man have so that he can reach point B? In which direction should be swim?

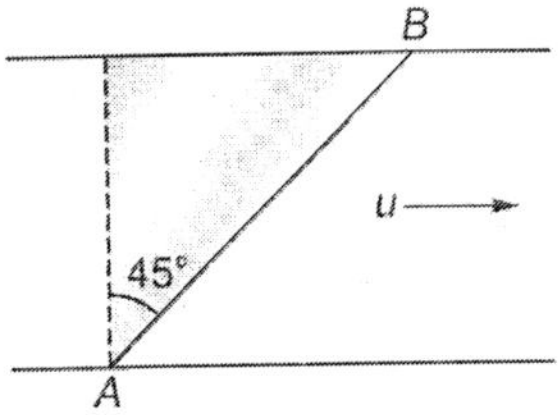

a) $u\sqrt{2}$ b) $u/\sqrt{2}$ c) 2u d) $u/2$

107. A balloon is rising vertically up with a velocity of 29 ms^{-1}. A stone is dropped from it and it reaches the ground in 10 seconds. The height of the balloon when the stone was dropped from it is ($g = 9.8\ ms^{-2}$)

a) 100 m b) 200 m c) 400 m d) 150 m

108. Acceleration velocity graph of a particle moving in a straight line is as shown in figure. The slope of velocity-displacement graph

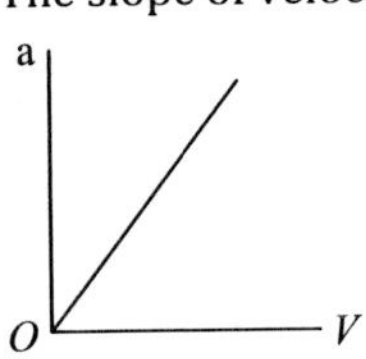

a) Increases linearly b) Decreases linearly
c) Is constant d) Increases parabolically

109. A car starts from rest and moves with uniform acceleration a on a straight road from time $t = 0$ to $t = T$. After that, a constant deceleration brings it to rest. In this process the average speed of the car is

a) $\frac{aT}{4}$ b) $\frac{3aT}{2}$ c) $\frac{aT}{2}$ d) aT

110. A stone dropped from a building of height h and it reaches after t seconds on earth. From the same building if two stones are thrown (one upwards and other downwards) with the same velocity u and they reach the earth surface after t_1 and t_2 seconds respectively, then

a) $t = t_1 - t_2$ b) $t = \frac{t_1 + t_2}{2}$ c) $t = \sqrt{t_1 t_2}$ d) $t = t_1^2 t_2^2$

111. A train is moving towards east and a car is along north, both with same speed. The observed direction of car to the passenger in the train is

a) East- north direction
b) West-north direction
c) South-east direction
d) None of these

112. The engine of a train can impart a maximum acceleration of $1 ms^{-2}$ and the brakes can give a maximum retardation of $3 ms^{-2}$. The least time during which a train can go from one place to the other place at a distance of 1.2 km is nearly

a) 108 s b) 191 s c) 56.6 s d) Time is fixed

113. A balloon is at a height of 81m and is ascending upwards with a velocity of $12 ms^{-1}$. A body of 2 kg weight is dropped from it. If $g = 10 ms^{-2}$, the body will reach the surface of the earth in

a) 1.5 s b) 4.025 s c) 5.4 s d) 6.75 s

114. A boat is sent across a river with a velocity of boat is 10 km/hr. If the resultant velocity of boat is 10 km/hr, then velocity of the river is :

a) 10 km/hr b) 8 km/hr c) 6 km/hr d) 4 km/hr

115. A 150 m long train is moving with a uniform velocity of 45 km/h. The time taken by the train to cross a bridge of length 850 m is

a) 56 sec b) 68 sec c) 80 sec d) 92 sec

116. At he instant a motor bike starts from rest in a given direction, a car overtakes the motor bike, both moving in the same direction. The speed-time graphs for motor bike and car are represented by OAB and CD respectively.
Then

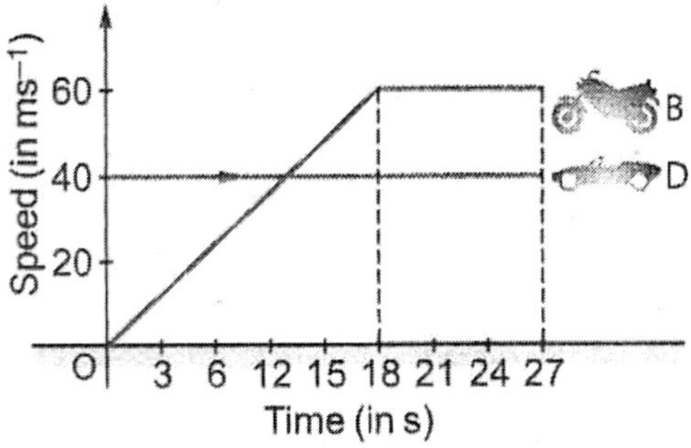

a) At $t = 18$ s the motor bike and car are 180m apart
b) At $t = 18$ s the motor bike and car are 720m apart
c) The relative distance between motor bike and car reduces to zero at $t = 27$ s and both are 1080m far from origin
d) The relative distance between motor bike and car always remains same

117. The variation of velocity of a particle with time moving along a straight line is illustrated in the following figure. The distance travelled by the particle in four seconds is

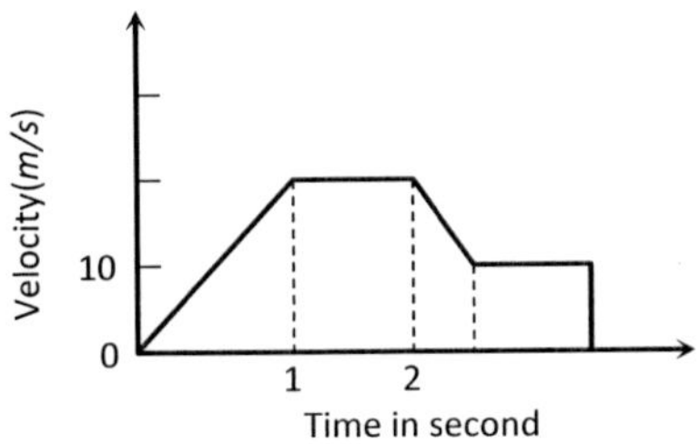

a) 60 m b) 55 m c) 25 m d) 30 m

118. Two balls of same size but the density of one is greater than that of the other are dropped from the same height, then which ball will reach the earth first (air resistance is negligible)?

a) Heavy ball b) Light ball
c) Both simultaneously d) Will depend upon the density of the balls

119. A bird flies for 4 s with a velocity of $|t-2| m/s$ in a straight line, where t is time in seconds. It covers a distance of

a) 2 m b) 4 m c) 6 m d) 8 m

120. The correct statement from the following is

a) A body having zero velocity will not necessarily have zero acceleration
b) A body having zero velocity will necessarily have zero acceleration
c) A body having uniform speed can have only uniform acceleration
d) A body having non-uniform velocity will have zero acceleration

121. A particle when thrown, moves such that it passes from same height at 2 and 10s, the height is

a) g b) $2g$ c) $5g$ d) $10g$

122. A body falling for 2 seconds covers a distance S is equal to that covered in next second. Taking $g = 10 m/s^2, S =$

a) 30 m b) 10 m c) 60 m d) 20m

123. Two boys are standing at the ends A and B of a ground where $AB = a$. The boy at B starts running in a direction perpendicular to AB with velocity v_1. The boy at A starts running simultaneously with velocity v and catches the other boy in a time t, when t is

a) $\frac{a}{\sqrt{(v^2+v_1^2)}}$ b) $\sqrt{a^2/(v^2-v_1^2)}$ c) $a/(v-v_1)$ d) $a/(v+v_1)$

124. With what velocity a ball be projected vertically so that the distance covered by it in 5th second is twice the distance it covers in its 6th second ($g = 10 m/s^2$)

a) 58.8 m/s b) 49 m/s c) 65 m/s d) 19.6 m/s

125. A juggler keeps on moving four balls in the air throws the balls in regular interval of time. When one ball leaves his hand (speed=$20 ms^{-1}$), the position of other ball will be (Take $g = 10 ms^{-2}$)

a) 10m, 20m, 10m b) 15m, 20m, 15m c) 5m, 15m, 20m d) 5m, 10m, 20m

126. An object start sliding on a frictionless inclined plane and from same height another object start falling freely

a) Both will reach with same speed b) Both will reach with the same acceleration
c) Both will reach in same time d) None of above

127. A ball is thrown vertically upwards. Which of the following plots represents the speed-time graph of the ball during its flight if the air resistance is not ignored

a) 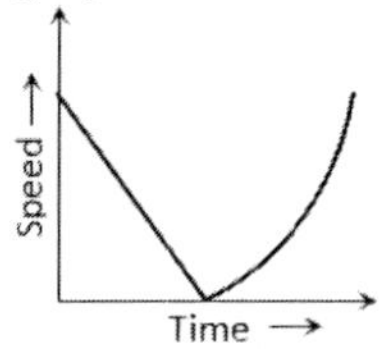

b) 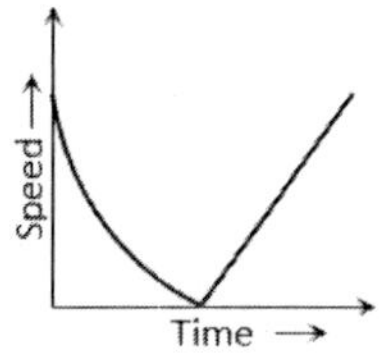

c) 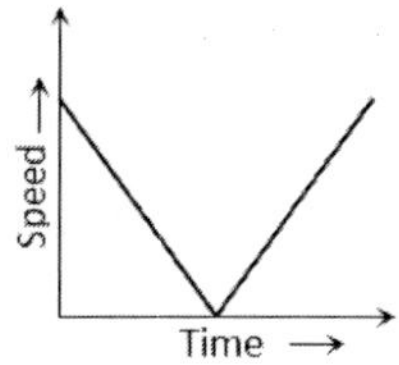

d) 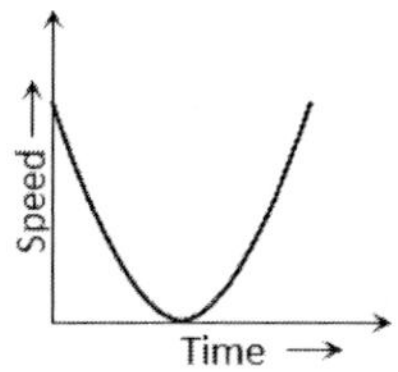

128. The variation of velocity of a particle with time moving along a straight line is illustrated in the following figure. The distance travelled by the particle in four seconds is

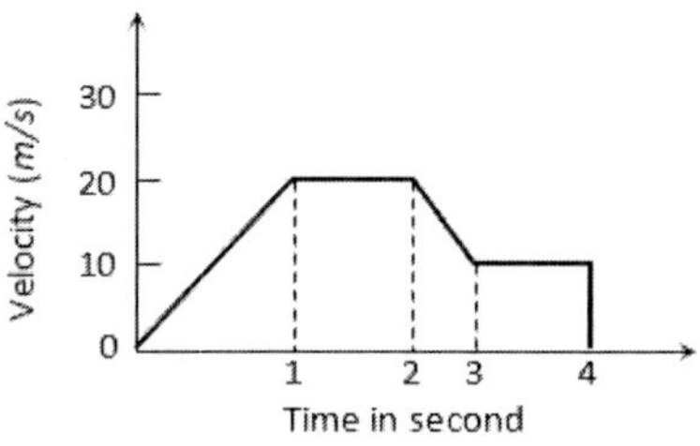

a) 60 m b) 55 m c) 25 m d) 30 m

129. 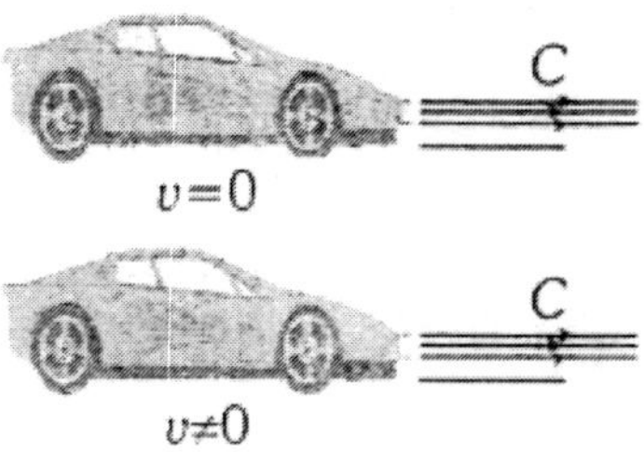

In figure, one car at rest and velocity of the light from head light is c, tehn velocity of light from head light for the moving car at velocity v, would be

a) $c + v$ b) $c - v$ c) $c \times v$ d) c

130. A ball is dropped on the floor from a height of 10m. It rebounds to a height of 2.5m. If the ball is in contact with the floor for 0.01 s, the average acceleration during contact is nearly [Take g = $10 ms^{-2}$]

a) $500\sqrt{2}$ ms^{-2}upwards b) $1800 ms^{-2}$downwards
c) $1500\sqrt{5}$ ms^{-2}upwards d) $1500\sqrt{2}$ ms^{-2}downwards

131. A ball is thrown vertically upwards from the top of a tower at $4.9 ms^{-1}$. It strikes the pond near the base of the tower after 3s. The height of the tower is

a) 29.4m b) 44.1m c) 73.5m d) 490m

132. A particle moves along with x −axis. The position x of particle with respect to time t from origin given by $x = b_0 + b_1 t + b_2 t^2$. The acceleration of particle is

a) b_0 b) b_1 c) b_2 d) $2b_2$

133. The effective acceleration of a body, when thrown upwards with acceleration a will be:

a) $\sqrt{a - g^2}$ b) $\sqrt{a^2 + g^2}$ c) $(a - g)$ d) $(a + g)$

134. The velocity-time graph of a body moving in a straight line is shown in the figure. The displacement and distance travelled by the body in 6 sec are respectively

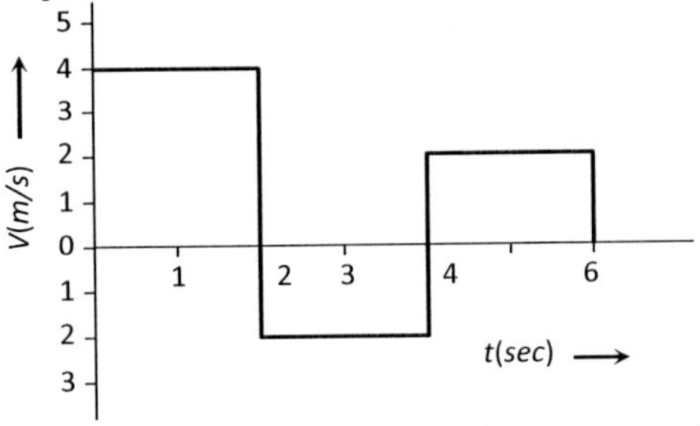

a) $8\ m, 16\ m$ b) $16\ m, 8\ m$ c) $16\ m, 16\ m$ d) $8\ m, 8\ m$

135. A ball falls from height h. After 1 second, another ball falls freely from a point $20\ m$ below the point from where the first ball falls. Both of them reach the ground at the same time. What is the value of h

a) $11.2\ m$ b) $21.2\ m$ c) $31.2\ m$ d) $41.2\ m$

136. A particle moves 200 cm in the first 2s and 220 cm in the next 4s with uniform deceleration. The velocity of the particle at the end of 7s is

a) $12\ cms^{-1}$ b) $11\ cms^{-1}$ c) $10 cms^{-1}$ d) $5\ cms^{-1}$

137. From a high tower at time $t = 0$, one stone is dropped from rest and simultaneously another stone is projected vertically up with an initial velocity. The graph of the distance S between the two stones, before either his the ground, plotted against time t will be as

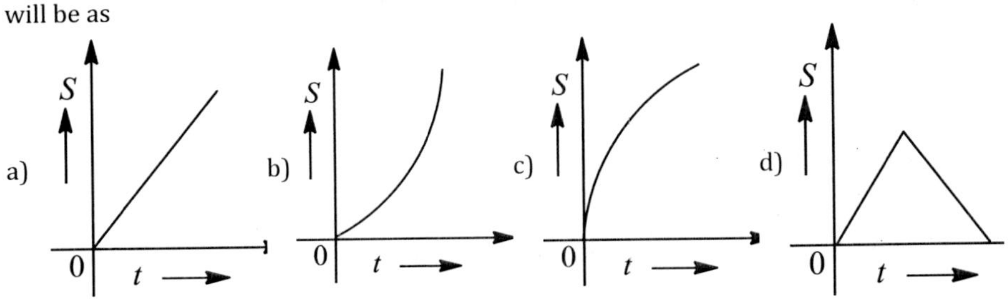

138. A ball is dropped vertically from a height d above the ground. It hits the ground and bounces up vertically to a height $d/2$. Neglecting subsequent motion and air resistance, its velocity v varies with the height h above the ground as

a) 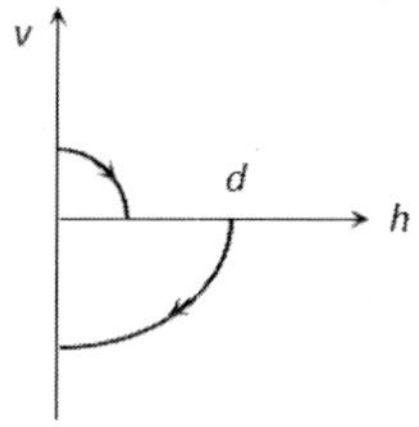

b) 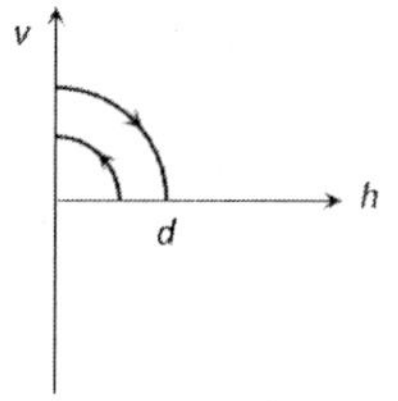

c) 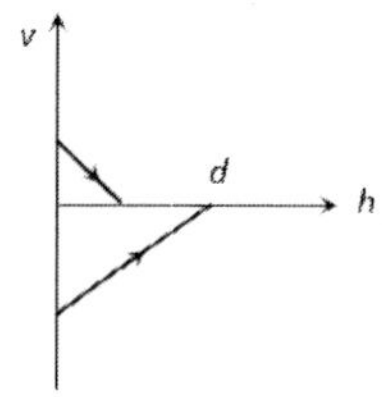

d) 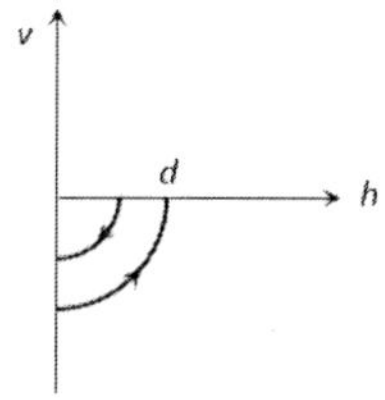

139. A ball is dropped from a bridge at a height of 176.4 m over a river. After 2 s, a second ball is thrown straight downwards. What should be the initial velocity of the second ball so that both hit the water simultaneously?
a) 2.45 ms^{-1} b) 49 ms^{-1} c) 14.5 ms^{-1} d) 24.5 ms^{-1}

A ball is thrown vertically upwards from the surface of earth with a speed of 18kmh^{-1}.
140. If $g = 10ms^{-2}$, then the maximum height attained by the ball is
a) 1.25 m b) 3 m c) 10 m d) 180 m

141. An elevator car, whose floor to ceiling distance is equal to $2.7m$, starts ascending with constant acceleration of $1.2ms^{-2}$. 2 sec after the start, a bolt begins falling from the ceiling of the car. The free fall time of the bolt is
a) $\sqrt{0.54}s$ b) $\sqrt{6}\,s$ c) 0.7 s d) 1 s

142. A boy walks to his school at a distance of $6km$ with constant speed of 2.5 $km/hour$ and walks back with a constant speed of 4 km/hr. His average speed for round trip expressed in $km/hour$, is
a) 24/13 b) 40/13 c) 3 d) 1/2

143. A car is travelling at 72 kmh^{-1} and is 20 m from a barrier when the driver puts on the brakes. The car hits the barrier 2s later. What is the magnitude of the constant deceleration?
a) 7.2 ms^{-2} b) 10 ms^{-2} c) 36 ms^{-2} d) 15 ms^{-2}

144. The acceleration of a particle is increasing linearly with time t as bt.The particle starts from the origin with an initial velocity v_0. The distance travelled by the particle in time t will be

a) $v_0 t+\frac{1}{3}bt^2$ b) $v_0 t+\frac{1}{3}bt^3$ c) $v_0 t+\frac{1}{6}bt^3$ d) $v_0 t+\frac{1}{2}bt^2$

145. The displacement-time graphs of two moving particles make angles of 30° and 45° with the x-axis. The ratio of the two velocities is

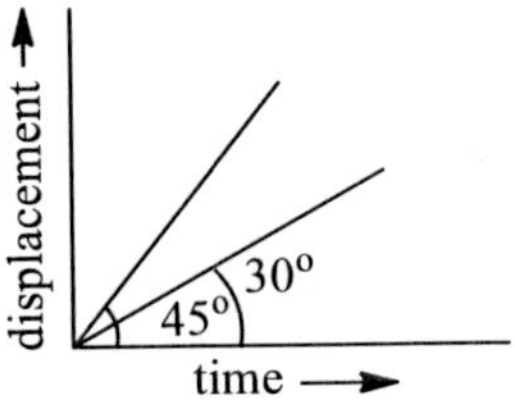

a) $\sqrt{3}:1$ b) $1:1$ c) $1:2$ d) $1:\sqrt{3}$

146. A body is released from a great height and falls freely towards the earth. Another body is released from the same height exactly one second later. The separation between the two bodies , two seconds after the release of the second body is

a) 4.9 m b) 9.8 m c) 19.6 m d) 24.5 m

147. A particle has initial velocity $(2\hat{\imath}+3\hat{\jmath})$ and acceleration $(0.3\hat{\imath}+0.2\hat{\jmath})$. The magnitude of velocity after 10 seconds will be

a) $9\sqrt{2}$ units b) $5\sqrt{2}$ units c) 5 units d) 9 units

148. Two bodies of different masses m_a and m_b are dropped from two different heights a and b. The ratio of the time taken by the two to cover these distances are

a) $a:b$ b) $b:a$ c) $\sqrt{a}:\sqrt{b}$ d) $a^2:b^2$

149. Speed of two identical cars u and $4u$ at a specific instant. The ratio of the respective distances in which the two cars are stopped from that instant is

a) 1:1 b) 1:4 c) 1:8 d) 1:16

150. A body is moving along a straight line path with constant velocity.At an instant of time the distance travelled by it is S and its displacement is D, then

a) D<S b) $D>S$ c) $D=S$ d) $D\leq S$

151. Two bodies are thrown simultaneously from a tower with same initial velocity v_0: one vertically upwards, the other vertically downwards. The distance between the two bodies after time t is

a) $2v_0 t+\frac{1}{2}gt^2$ b) $2v_0 t$ c) $v_0 t+\frac{1}{2}gt^2$ d) $v_0 t$

152. A car, starting from rest, accelerates at the rate f through a distance S, then continues at constant speed for time t and then decelerates at the rate $\frac{f}{2}$ to come to rest. If the total distance traversed is 15 S, then

a) $S=\frac{1}{2}ft^2$

b) $S=\frac{1}{4}ft^2$

c) $S=\frac{1}{72}ft^2$

d) $S=\frac{1}{6}ft^2$

153. A point initially at rest moves along x –axis. Its acceleration varies with time as $a=(6t+5)\text{ms}^{-2}$. If it starts from origin, the distance covered in 2 s is

a) 20 m b) 18 m c) 16 m d) 25 m

154. A body released from a great height falls freely towards the earth. Another body is released from the same height exactly one second later. The separation between the two bodies two second after the release of the second body is

a) 9.8 m b) 4.9 m c) 24.5 m d) 19.6 m

155. A body of mass 3 kg falls from the multi-storeyed building 100m high and buries itself 2m deep in the sand. The time of penetration will be

a) 0.09 s b) 0.9 s c) 9 s d) 10 s

156. At $t = 0$, a stone of mass 10 gm is thrown straight up from the ground level with a speed 10 m/s. After 1 s, a second stone of the same mass is thrown from the same position with a speed 20 m/s. What is the position of the first stone from the ground level at that moment? (Take g $= 10\ m/s^2$)

a) 10 m b) 1 m c) 2 m d) 5 m

157. A particle moves along x-axis as

$x = 4(t-2) + a(t-2)^2$

Which of the following is true

a) The initial velocity of the particle is 4
b) The acceleration of particle is $2a$
c) The particle is at origin at $t = 0$
d) None of these

158. A particle is constrained to move on a straight line path. It returns to the starting point after 10 sec. The total distance covered by the particle during this time is 30 m. Which of the following statement about the motion of the particle is false

a) Displacement of the particle is $zero$
b) Average speed of the particle is 3 m/s
c) Displacement of the particle is 30 m
d) Both (a) and (b)

159. A boy standing at the top of a tower of 20 m height drops a stone. Assuming $g = 10\ ms^{-2}$, the velocity with which it hits the ground is

a) 5.0 m/s b) 10.0 m/s c) 20.0 m/s d) 40.0 m/s

160. The position of a particle moving along x-axis at certain times is given below:

$t(s)$	0	1	2	3
$x(m)$	-2	0	6	16

Which of the following describes the motion correctly

a) Uniform accelerated
b) Uniform decelerated
c) Non-uniform accelerated
d) There is not enough data for generalization

161. An object dropped from rest. Its $v - t$ graph is

a) 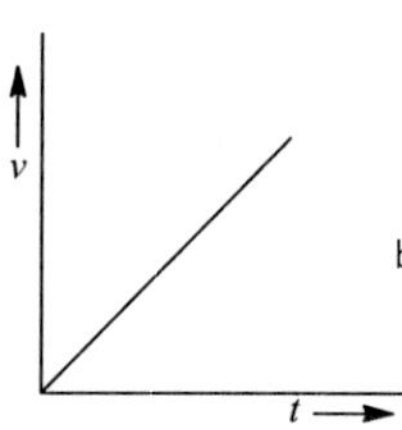

b) 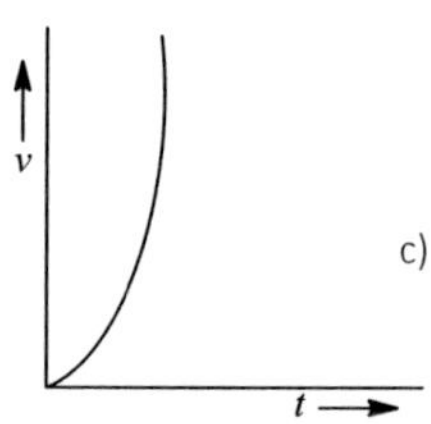

c) 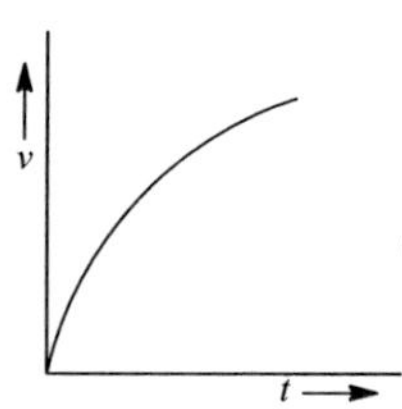

d) 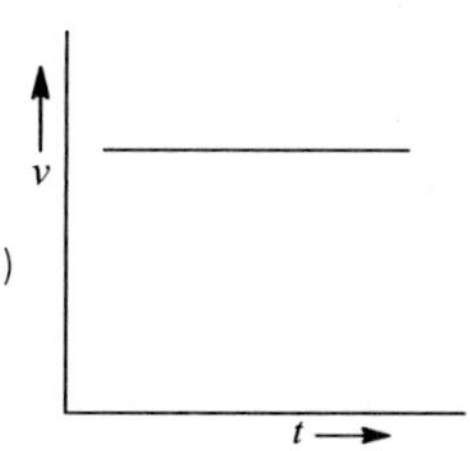

162. Which of the following velocity-time graphs represent uniform motion

a) 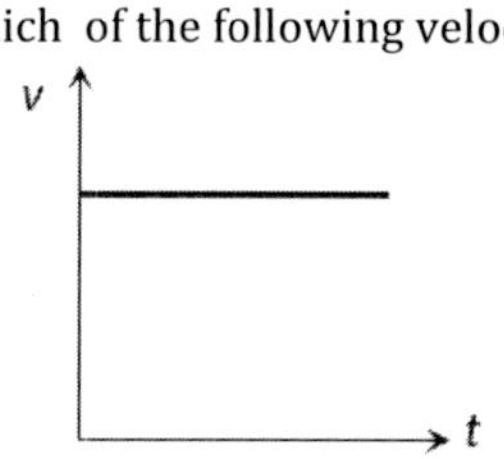

b) 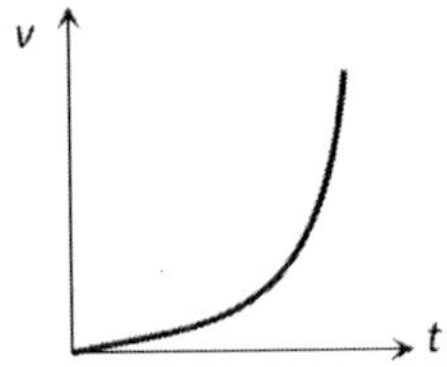

c) 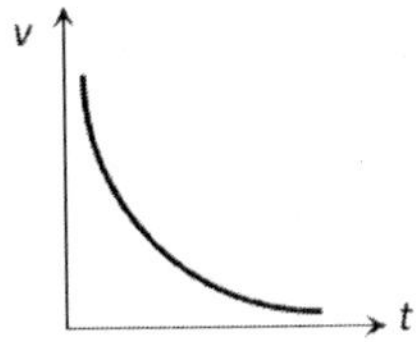

d) 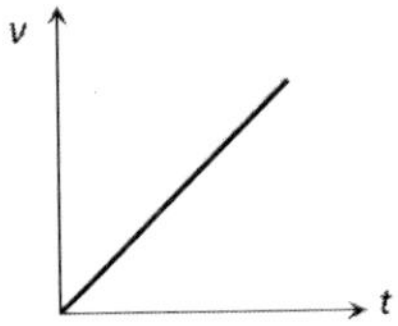

163. A balloon rises from rest with a constant acceleration $g/8$. A stone is released from it when it has risen to height h. The time taken by the stone to reach the ground is

a) $4\sqrt{\frac{h}{g}}$ b) $2\sqrt{\frac{h}{g}}$ c) $\sqrt{\frac{2h}{g}}$ d) $\sqrt{\frac{g}{h}}$

164. A Body moves 6 m north.8 m east and 10 m vertically upwards, what is its resultant displacement from initial position

a) $10\sqrt{2m}$ b) $10\ m$ c) $\frac{10}{\sqrt{2}}m$ d) $10 \times 2m$

165. A body begins to walk eastward along a street in front of his house and the graph of his position from home is shown in the following figure. His average speed for the whole time interval is equal to

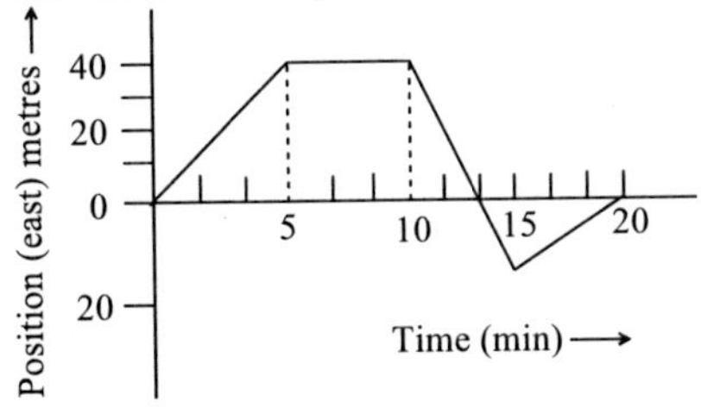

a) $8\ m/min$ b) $6\ m/min$ c) $\frac{8}{3}\ m/min$ d) $2\ m/min$

166. Speed of two identical cars u and $4u$ at a specific instant. The ratio of the respective distances in which the two cars are stopped from that instant is

a) 1: 1 b) 1: 4 c) 1: 8 d) 1: 16

167. The acceleration a of a particle starting from rest varies with time according to relation $a = \alpha t + \beta$. The velocity of the particle after a time t will be

a) $\frac{\alpha t^2}{2} + \beta$ b) $\frac{\alpha t^2}{2} + \beta t$ c) $\alpha t^2 + \frac{1}{2}\beta t$ d) $\frac{(\alpha t^2 + \beta t)}{2}$

168. A bullet is fired with a speed of 1000 m/sec in order to hit a target 100 m away. If $g = 10\ m/s^2$, the gun should be aimed

a) Directly towards the target b) 5 cm above the target
c) 10 cm above the target d) 15 cm above the target

169. A boy walks to his school at a distance of $6km$ with constant speed of 2.5 $km/hour$ and walks back with a constant speed of 4 km/hr. His average speed for round trip expressed in $km/hour$, is

a) 24/13 b) 40/13 c) 3 d) 1/2

170. A car moves from X to Y with a uniform speed v_u and returns to Y with a uniform speed v_d. The average speed for this round trip is

a) $\frac{2v_d v_u}{v_d + v_u}$ b) $\sqrt{v_u v_d}$ c) $\frac{v_d v_u}{v_d + v_u}$ d) $\frac{v_u + v_d}{2}$

171. A car moves a distance of 200 m. it covers first half of the distance at speed 60 kmh^{-1} and the second half at speed v. If the average speed is 40 kmh^{-1}, the value of v is

a) 30 kmh^{-1} b) 13 kmh^{-1} c) 60 kmh^{-1} d) 40 kmh^{-1}

172. The displacement x of a particle at the instant when its velocity v is given by $v = \sqrt{3x + 16}$. Its acceleration and initial velocity are

a) 1.5 units, 4 units b) 3 units, 4 units c) 16 units, 1.6 units d) 16 units, 3 units

173. A car moving with speed of 40 km/h can be stopped by applying brakes after atleast 2 m. If the same car is moving with a speed of 80 km/h, what is the minimum stopping distance

a) 8 m b) 2 m c) 4 m d) 6 m

174. A body is projected vertically upwards. The times corresponding to height h while ascending and while descending are t_1 and t_2 respectively. Then the velocity of projection is (g is acceleration due to gravity)

a) $g\sqrt{t_1 t_2}$ b) $\frac{gt_1 t_2}{t_1 + t_2}$ c) $\frac{g\sqrt{t_1 t_2}}{2}$ d) $\frac{g(t_1 + t_2)}{2}$

175. Two trains each 50 m long are travelling in opposite direction with velocity 10 m/s and 15 m/s. The time of crossing is

a) $2s$ b) $4s$ c) $2\sqrt{3}s$ d) $4\sqrt{3}\,s$

176. A ball is thrown vertically upwards from the top of a tower at 4.9 ms^{-1}.It strikes the pond near the base of the tower after 3 $seconds$. The height of the tower is

a) 73.5 m b) 44.1 m c) 29.4 m d) None of these

177. A train of 150 m length is going towards north direction at a speed of 10m/sec. A parrot flies at the speed of 5 m/sec towards south direction parallel to the railway track. The time taken by the parrot to cross the train is

a) 12 sec b) 8 sec c) 15 sec d) 10 sec

178. A cricket ball is thrown up with a speed of 19.6 ms^{-1}. The maximum height it can reach is

a) 9.8 m b) 19.6 m c) 29.4 m d) 39.2 m

179. A man drops a ball downside from the roof of a tower of height 400 m. At the same time another ball is thrown upside with a velocity 50 m/s. From the surface of the tower, then they will meet at which height from the surface of the tower

a) 100 m b) 320 m c) 80 m d) 240 m

180. A train is moving towards east and a car is along north, both with same speed. The observed direction of car to the passenger in the train is

a) East- north direction b) West-north direction c) South-east direction d) None of these

181. A body is moving with uniform acceleration covers 200 m in the first 2 s and 220 m in the next 4 s. find the velocity in ms^{-1} after 7 s.

a) 10 b) 15 c) 20 d) 30

182. A stone thrown vertically upward files past a window one second after it was thrown upward and after three second on its way downward. The height of the window above the ground is (Take $g = 10ms^{-2}$)

a) 20 m b) 15 m c) 10 m d) 5 m

183. A particle is projected with velocity v_0 along $x - axis$.The deceleration on the particle is proportional to the square of the distance from the origin i.e., $a = -\alpha x^2$.The distance at which the particle stops is

a) $\sqrt{\frac{3v_0}{2\alpha}}$ b) $\left(\frac{3v_0}{2\alpha}\right)^{\frac{1}{3}}$ c) $\sqrt{\frac{3v_0^2}{2\alpha}}$ d) $\left(\frac{3v_0^2}{2\alpha}\right)^{\frac{1}{3}}$

184. A boat crosses a river from port A to port B, which are just on the opposite side. The speed of the water is V_W and that of boat is V_B relative to still water. Assume $V_B = 2V_W$. What is the time taken by the boat, if It has to cross the river directly on the AB line

a) $\frac{2D}{V_B\sqrt{3}}$ b) $\frac{\sqrt{3}D}{2V_B}$ c) $\frac{D}{V_B\sqrt{2}}$ d) $\frac{D\sqrt{2}}{V_B}$

185. A particle moves along a straight line OX. At a time t (in seconds) the distance x (in metres) of the particle from O is given by $x = 40 + 12t - t^3$
How long would the particle travel before coming to rest

a) $24\ m$ b) $40\ m$ c) $56\ m$ d) $16\ m$

186. If the velocity v of a particle moving along a straight line decreases linearly with its displacement s from 20ms^{-1} to a value approaching zero $s = 30$ m, then acceleration of the particle at s = 15m is

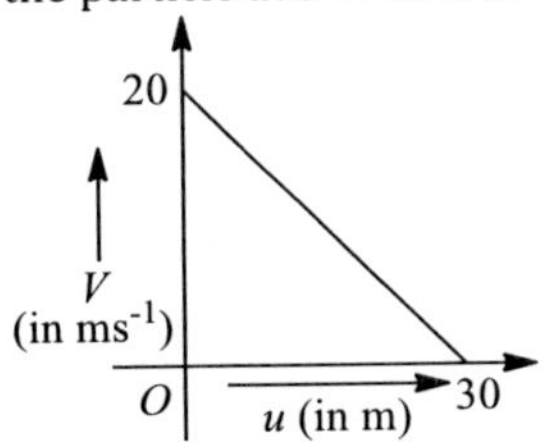

a) $\frac{2}{3}\text{ms}^{-2}$ b) $-\frac{2}{3}\text{ms}^{-2}$ c) $\frac{20}{3}\text{ms}^{-2}$ d) $-\frac{20}{3}\text{ms}^{-2}$

187. Which of the following are true?
(i) A body having constant speed can have verying velocity.
(ii) Position-time graphs for two objects with zero relative velocity are parallel.
(iii) The numerical ration of velocity to speed of an object can never be more than one.

a) (i) b) (ii) and (iii) c) All d) None of these

188. The position x of a particle varies with time t as $x = at^2 - bt^3$. The acceleration of the particle will be zero at time t equal to

a) $\frac{a}{b}$ b) $\frac{2a}{3b}$ c) $\frac{a}{3b}$ d) Zero

189. A train accelerates from rest at a constant rate α for distance x_1 and time t_1. After that it retards to rest at constant rate β for distance x_2 and time t_2. Then it is found that

a) $\frac{x_1}{x_2} = \frac{\alpha}{\beta} = \frac{t_1}{t_2}$ b) $\frac{x_1}{x_2} = \frac{\beta}{\alpha} = \frac{t_1}{t_2}$ c) $\frac{x_1}{x_2} = \frac{\beta}{\alpha} = \frac{t_2}{t_1}$ d) $\frac{x_1}{x_2} = \frac{\beta}{\alpha} = \frac{t_2}{t_1}$

190. A particle is moving with constant initial velocity $4\ \text{ms}^{-1}$ till $t = 1.5$ s. Then it accelerates at $10\ \text{ms}^{-2}$ till $t = 3$. The distance covered is (Take g = 10ms^{-2})

a) 17.25 m b) 36.25 m c) 40 m d) 23.25 m

191. Rain drops fall vertically at a speed of 20ms^{-1}. At what angle do they fall on the wind screen of a car moving with a velocity of 15ms^{-1}, if the wind screen velocity inclined at an angle of 23° to the vertical?
$\left(\cot^{-1}\left[\frac{4}{3}\right] \approx 36°\right)$

a) 60° b) 30° c) 45° d) 90°

192. Two cars leave one after the other and travel with and acceleration of $0.4 ms^{-2}$. Two minutes after the departure of the first car, the distance between them becomes 1.90 km. The time interval between their departures is

a) 50 s b) 60 s c) 70 s d) 80 s

193. A body falling from a high Minaret travels 40 m in the last 2 seconds of its fall to ground. Height of Minaret in meters is (take $g = 10\ m/s^{-2}$)

a) 60
b) 45
c) 80
d) 50

194. The graph of displacement-time for a body travelling in a straight line is given. We can conclude that

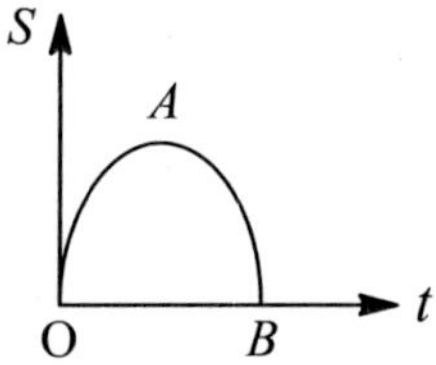

a) The velocity is constant
b) The velocity increases uniformly.
c) The body is subjected to acceleration from O to A.
d) The velocity of the body at A is zero.

195. The distance travelled by a particle starting from rest and moving with an acceleration $\frac{4}{3} ms^{-2}$, in the third second is

a) $\frac{10}{3} m$ b) $\frac{19}{3} m$ c) 6 m d) 4 m

196. Which graph represents the uniform acceleration

a) 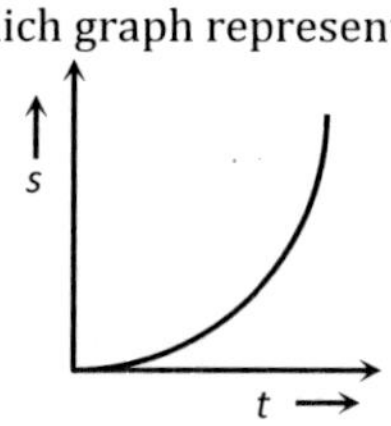

b) 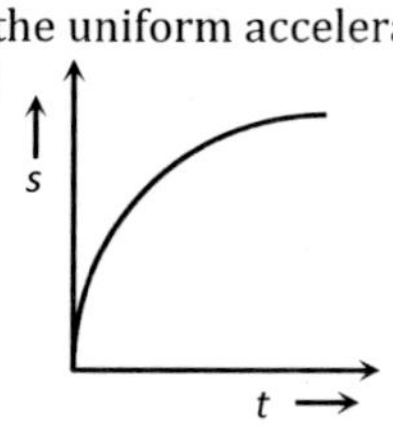

c) 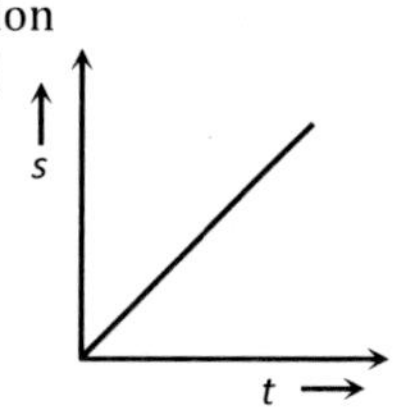

d) 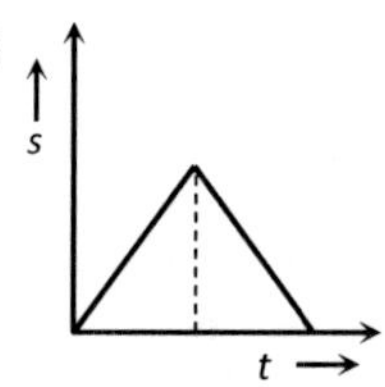

197. For a body moving with relativistic speed, if the velocity is doubled, then

a) Its linear momentum is doubled
b) Its linear momentum will be less than double
c) Its linear momentum will be more than double
d) Its linear momentum remains unchanged

198. A river is flowing from W to E with a speed of 5 m/min. A man can swim in still water with a velocity 10 m/min. In which direction should the man swim so as to take the shortest possible path to go to the south

a) 30° with downstream b) 60° with downstream
c) 120° with downstream d) South

199. A bullet comes out of the barrel of gun of length 2m with a speed 80 ms^{-1}. The average acceleration of the bullet is

a) 1.6 ms^{-2} b) 160 ms^{-2} c) 1600 ms^{-2} d) 16 ms^{-2}

200. A ball is released from the top of a tower of height h metre. It takes T sec to reach the ground. What is the position of the ball in $\frac{T}{3}$ s?

a) $h/9$ m from the ground b) $7h/9$ m from the ground
c) $8h/9$ m from the ground d) $17h/18$ m from the ground

201. A particle moves along X- axis in such a way that its coordinate X varies with time t according to the equation $x = (2 - 5t + 6t^2)m$. The initial velocity of the particle is

a) $-5\ m/s$ b) $6\ m/s$ c) $-3\ m/s$ d) $3\ m/s$

202. A body starts to fall freely under gravity. The distance covered by it in first, second and third *second* are in ratio

a) 1: 3: 5 b) 1: 2: 3 c) 1: 4: 9 d) 1: 5: 6

203. A frictionless wire AB is fixed on a sphere of radius R. A very small spherical ball slips on this wire. The time taken by this ball to slip from A to B is

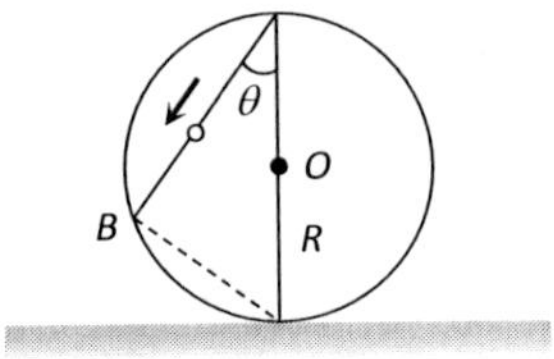

a) $\frac{2\sqrt{gR}}{g\cos\theta}$ b) $2\sqrt{gR}.\frac{\cos\theta}{g}$ c) $2\sqrt{\frac{R}{g}}$ d) $\frac{gR}{\sqrt{g\cos\theta}}$

204. If a ball is thrown vertically upwards with speed u, the distance covered during the last t seconds of its ascent is

a) $\frac{1}{2}gt^2$ b) $ut - \frac{1}{2}gt^2$ c) $(u - gt)t$ d) ut

205. The path of a particle moving under the influence of a force fixed in magnitude and direction is

a) Straight line b) Circle c) Parabola d) Ellipse

206. A body starts from rest and moves with a constant acceleration. The ratio of distance covered in the nth second to the distance covered in n second is

a) $\frac{2}{n} - \frac{1}{n^2}$ b) $\frac{1}{n^2} - \frac{1}{n}$ c) $\frac{2}{n^2} - \frac{1}{n}$ d) $\frac{2}{n} + \frac{1}{n^2}$

207. A bullet emerge from a barrel of length 1.2 m with a speed of 640 ms^{-1}. Assuming constant acceleration, the approximate time that is spends in the barrel after the gun is fired is

a) 4 ms b) 40 ms c) 400 μs d) 1 s

208. Which of the following is a one dimensional motion

a) Landing of an aircraft b) Earth revolving around the sun
c) Motion of wheels of moving train d) Train running on a straight track

209. Free fall of an object (in vacuum) is a case of motion with

a) Uniform velocity b) Uniform acceleration c) Variable acceleration d) Constant momentum

210. Three persons P, Q and R of same mass travel with same speed u such that each one faces the other always. After how much time will they meet each other?

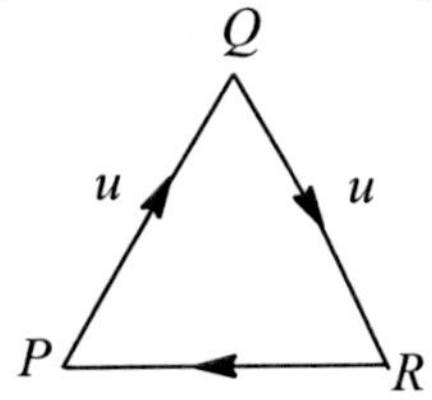

a) d/u sec b) $2d/3u$ sec c) $2d/\sqrt{(3)}u$ sec d) $d/\sqrt{3}u$ sec

211. Consider a rubber ball freely falling from a height $h = 4.9$ m onto a horizontal elastic plate. Assume that the duration of collision is negligible and the collision with the plate is totally elastic. Then the velocity as a function of time will be

a)
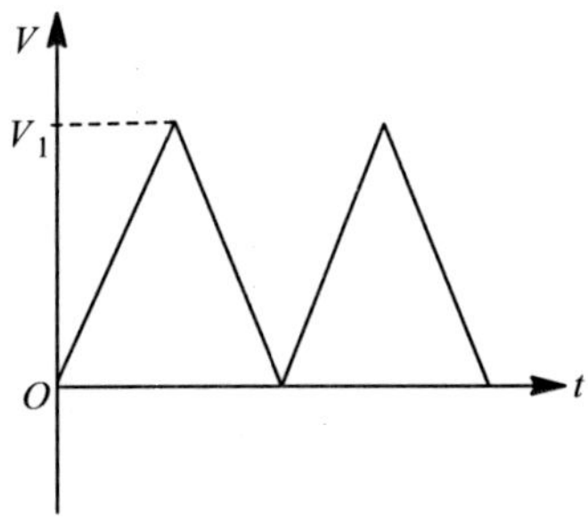

b)
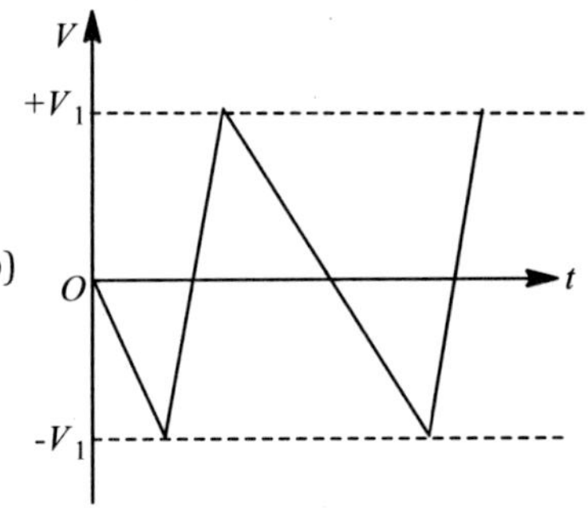

c)
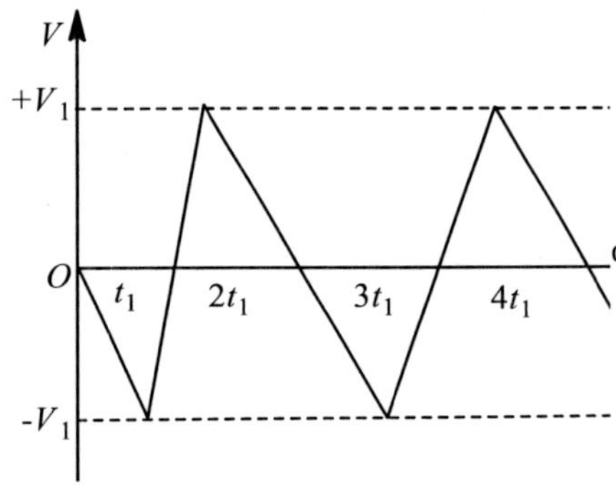

d)
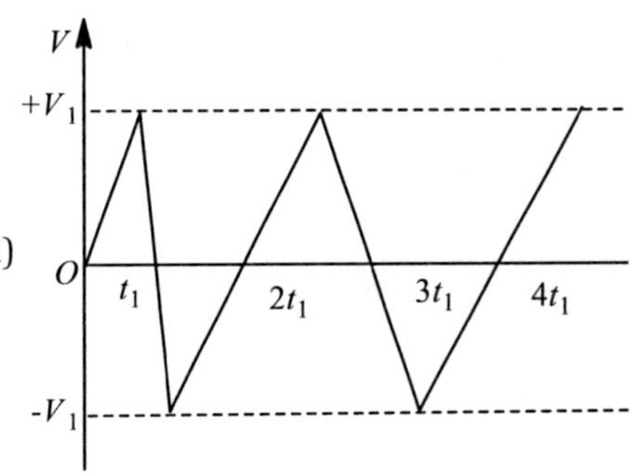

212.
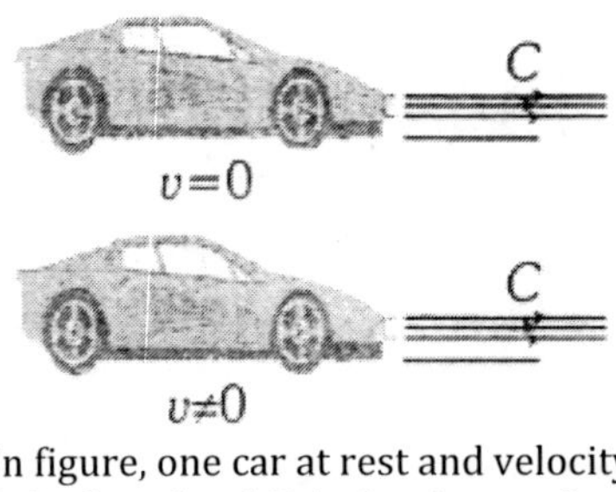

In figure, one car at rest and velocity of the light from head light is c, tehn velocity of light from head light for the moving car at velocity v, would be

a) $c + v$ b) $c - v$ c) $c \times v$ d) c

213. A packet is dropped from a balloon which is going upwards with the velocity 12 m/s, the velocity of the packet after 2 seconds will be

a) $-12\ m/s$ b) $12\ m/s$ c) $-7.6\ m/s$ d) $7.6\ m/s$

214. A man throws balls with the same speed vertically upwards one after the other at an interval of two seconds. What should be the speed of the throw so that more than two balls are in the sky at any time (Given $g = 9.8\ m/s^2$)

a) At least 0.8 m/s

b) Any speed less than 19.6 m/s

c) Only with speed 19.6 m/s

d) More than 19.6

215. Velocity-time (v-t) graph for a moving object is shown in the figure. Total displacement of the object during the same interval when there is non-zero acceleration and retardation is

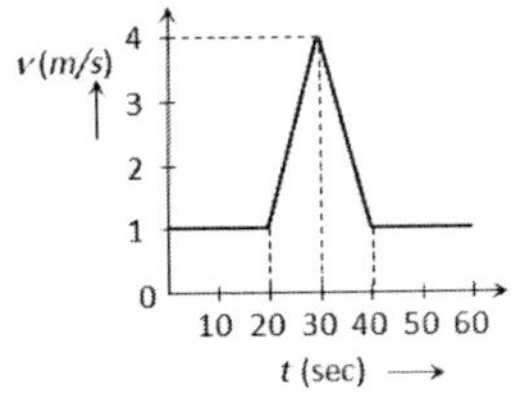

a) $60\ m$ b) $50\ m$ c) $30\ m$ d) $40\ m$

216. The displacement- time graph for two particles A and B are straight lines inclined at angles of 30° and 60° with the time axis. The ratio of velocities of $V_A : V_B$ is

a) $1:2$ b) $1:\sqrt{3}$ c) $\sqrt{3}:1$ d) $1:3$

217. A particle moves along a straight line such that its displacement at any time t is given by $S = t^3 - 6t^2 + 3t + 4\ metres$

The velocity when the acceleration is zero is

a) $3ms^{-1}$ b) $-12ms^{-1}$ c) $42ms^{-1}$ d) $-9ms^{-1}$

218. The velocity of a body of mass 20 kg decreases from 20 ms^{-1} to 5 ms^{-1} in a distance of 100 m. Force on the body is

a) -27.5 N b) -47.5 N c) -37.5 N d) -67.5 N

219. From the top of a tower two stones, whose masses are in the ratio 1: 2 are thrown one straight up with an initial speed u and the second straight down with the same speed u. Then, neglecting air resistance

a) The heavier stone hits the ground with a higher speed
b) The lighter stone hits the ground with a higher speed
c) Both the stones will have the same speed when they hit the ground
d) The speed can't be determined with the give data

220. A particle has initial velocity $(2\hat{i} + 3\hat{j})$ and acceleration $(0.3\hat{i} + 0.2\hat{j})$. The magnitude of velocity after 10 seconds will be

a) $9\sqrt{2}$ units b) $5\sqrt{2}$ units c) 5 units d) 9 units

221. A car A is travelling on a straight level road with a uniform speed of 60 km/h. It is followed by another car B which is moving with a speed of $70 km/h$. When the distance between them is 2.5 km, the car B is given a deceleration of 20 km/h^2. After how much time will B catch up with A.

a) $1\ hr$ b) $1/2hr$ c) $1/4hr$ d) $1/8hr$

222. The distance travelled by a particle starting from rest and moving with an acceleration $\frac{4}{3}ms^{-2}$, in the third second is

a) $\frac{10}{3}m$ b) $\frac{19}{3}m$ c) $6\ m$ d) $4\ m$

223. The ratio of the numerical values of the average velocity and average speed of a body is always

a) Unity b) Unity or less c) Unity or more d) Less than unity

224. A ball thrown upward from the top of a tower with speed v reaches the ground in t_1 second. If this ball is thrown downward from the top of he same tower with speed v it reaches the ground in t_2 second. In what time the ball shall reach the ground if it is allowed to falls freely under gravity from the top of the tower?

a) $\frac{t_1 + t_2}{2}$ b) $\frac{t_1 - t_2}{2}$ c) $\sqrt{t_1 t_2}$ d) $t_1 + t_2$

225. A body freely falling from the rest has a velocity 'v' after it falls through a height 'h'. The distance it has to fall down for its velocity to become double, is

a) $2h$ b) $4h$ c) $6h$ d) $8h$

226. From the top of tower, a stone is thrown up. It reaches the ground in t_1 second. A second stone thrown down with the same speed reaches the ground in t_2 second. A third stone released from rest reaches the ground in t_3 second. Then

a) $t_3 = \frac{(t_1 + t_2)}{2}$ b) $t_3 = \sqrt{t_1 t_2}$ c) $\frac{1}{t_3} = \frac{1}{t_1} - \frac{1}{t_2}$ d) $t_3^2 = t_2^2 - t_1^2$

227. The velocity-time graph of a particle in linear motion is shown. Both v and t are in SI units. What is the displacement of the particle from the origin after 8 s?

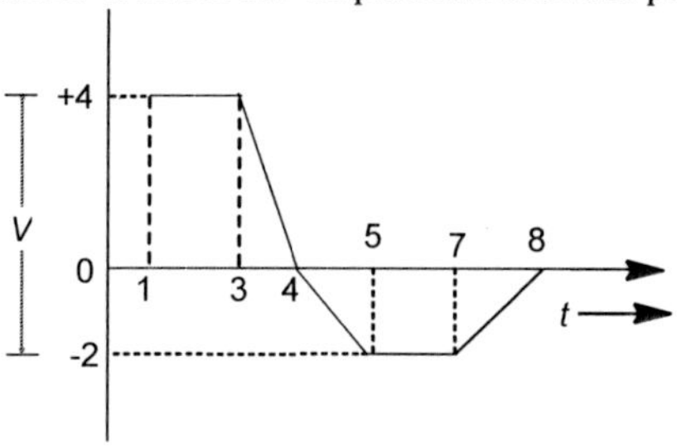

a) 6m b) 8m c) 16m d) 18m

228. The acceleration of a particle is increasing linearly with time t as bt.The particle starts from the origin with an initial velocity v_0. The distance travelled by the particle in time t will be

a) $v_0 t + \frac{1}{3}bt^2$ b) $v_0 t + \frac{1}{3}bt^3$ c) $v_0 t + \frac{1}{6}bt^3$ d) $v_0 t + \frac{1}{2}bt^2$

229. Two cars A and B are moving with same speed of 45 kmh^{-1} along same direction. If a third car C coming from the opposite direction with a speed of 36 kmh^{-1} meets two cars in an interval of 5 min, the distance of separation of two cars A and B should be (in km)

a) 6.75 b) 7.25 c) 5.55 d) 8.35

230. A boy begins to walk eastward along a street infront of his house and the graph of his displacement from home is shown in the following figure. His average speed for in the whole time interval is equal to

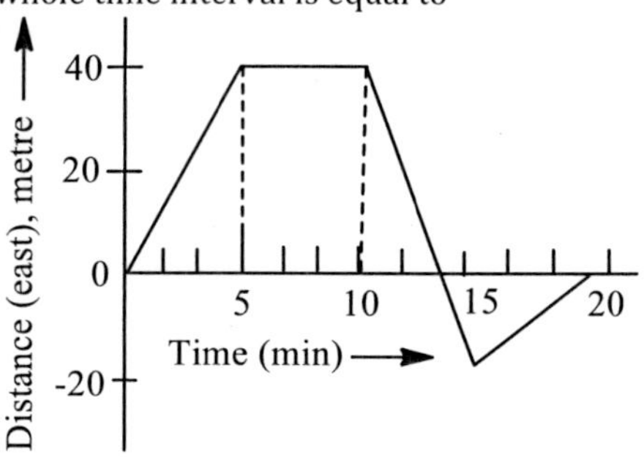

a) 8 $mmin^{-1}$ b) 6 $mmin^{-1}$ c) $\frac{8}{3}$ $mmin^{-1}$ d) 2 $mmin^{-1}$

231. A body moves for a total of nine second starting from rest with uniform acceleration and then with uniform retardation, which is twice the value of acceleration and then stops. The duration of uniform acceleration

a) 3 s b) 4.5 s c) 5 s d) 6 s

232. Two spheres of same size, one of mas 2 kg and another of mass 4 kg, are dropped simultaneously from the top of Qutub Minar (height = $72m$). When they are 1 m above the ground, the two spheres have the same

a) Momentum b) Kinetic energy c) Potential energy d) Acceleration

randomly to the left or to right by 90∘. After walking a total of 40 steps, the maximum displacement of the person from its starting point can be

a) Zero b) $8\sqrt{2}$m c) $16\sqrt{2}$m d) 32 m

234. A ball P is dropped vertically and another ball Q is thrown horizontally with the same velocities from the same height and at the same time. If air resistance is neglected, then

a) Ball P reaches the ground first
b) Ball Q reaches the ground first
c) Both reach the ground at the same time
d) The respective masses of the two balls will decide the time

235. A body has speed of V,2V and 3V in first 1/3 of distance S, seconds 1/3 of S and third 1/3 of S respectively. Its average speed will be

a) V b) $2V$ c) $\frac{18}{11}V$ d) $\frac{11}{18}V$

236. A particle moving in a straight line covers half the distance with speed of 3m/s. The other half of the distance is covered in two equal time intervals with speed of 4.5 m/s and 7.5 m/s respectively. The average speed of particle during this motion is

a) 4m/s b) 5m/s c) 5.5m/s d) 4.8m/s

237. A boat is sent across a river with a velocity of boat is 10 km/hr. If the resultant velocity of boat is 10 km/hr, then velocity of the river is :

a) 10 km/hr
b) 8 km/hr
c) 6 km/hr
d) 4 km/hr

238. Two spheres of same size, one of mass 2 kg and another of mass 4 kg, are dropped simultaneously from the top of QutabMinar (height = 72 m). When they are 1 m above the ground, the two spheres have the same

a) Momentum b) Kinetic energy c) Potential energy d) Acceleration

239. If a train travelling at 72 $kmph$ is to be brought to rest in a distance of 200 metres, then its retardation should be

a) 20 ms^{-2} b) 10 ms^{-2} c) 2 ms^{-2} d) 1ms^{-2}

240. A particle moving with a uniform acceleration travels 24 m and 64 m in the first two consecutive intervals of 4 sec each.
Its initial velocity is

a) 1 m/sec b) 10 m/sec c) 5 m/sec d) 2 m/sec

241. A body moves for a total of nine second started from rest with uniform acceleration and then with uniform retardation, which is twice the value of acceleration and then stops. The duration of uniform acceleration

a) 3 s b) 4.5 s c) 5 s d) 6 s

242. A man goes 10 m towards North, then 20 m towards east then displacement is

a) 22.5m b) 25m c) 25.5m d) 30m

243. An object moving with a speed of 6.25 m/s, is decelerated at a rate given by $\frac{dv}{dt} = 2.5\sqrt{v}$ where v is the instantaneous speed. The time taken by the object, to come to rest would be

a) 1s b) 2s c) 4s d) 8s

244. A river is flowing from W to E with a speed of 5 m/min. A man can swim in still water with a velocity 10 m/min. In which direction should the man swim so as to take the shortest possible path to go to the south

a) 30° with downstream
b) 60° with downstream
c) 120° with downstream
d) South

245. Look at the graphs (a) to (d) carefully and indicate which of these possibly represents one dimensional motion of a particle

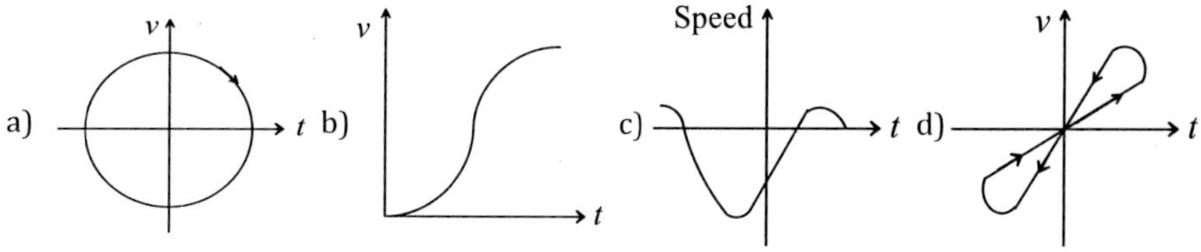

246. A particle travels $10m$ in first 5 sec and $10m$ in next 3 sec. Assuming constant acceleration what is the distance travelled in next 2 sec

a) 8.3 m b) 9.3 m c) 10.3 m d) None of above

247. A particle is thrown vertically upwards. If it velocity at half of the maximum height is 10 m/sec, then maximum height attained by it is (Take $g = 10\ m/\sec^2$)

a) 8 m b) 10 m c) 12 m d) 16 m

248. The displacement of the body along $x -$ axis depends on time as $\sqrt{x} = t + 1$. Then the velocity of body

a) Increases with time b) Decreases with time c) Independent of time d) None of these

249. A particle starts from rest and experiences constant acceleration for 6 s. if it travels a distance d_1 in the first two second, a distance d_2 in the next two seconds and a distance d_3 in the last two second, then

a) $d_1 : d_2 : d_3 = 1 : 1 : 1$ b) $d_1 : d_2 : d_3 = 1 : 2 : 3$

c) $d_1 : d_2 : d_3 = 1 : 3 : 5$ d) $d_1 : d_2 : d_3 = 1 : 5 : 9$

250. Two bodies are thrown vertically upwards with their initial speed in the ratio 2 : 3. The ratio of the maximum heights reached by then and the ratio of their time taken by them to return back to the ground respectively are

a) 4 : 9 and 2 : 3 b) 2 : 3 and $\sqrt{2} : \sqrt{3}$ c) $\sqrt{2} : \sqrt{3}$ and 4 : 9 d) $\sqrt{2} : \sqrt{3}$ and 2 : 3

251. A student is standing at a distance of 50 metres from the bus. As soon as the bus begins its motion with an acceleration of 1 ms^{-2}, the student starts running towards the bus with a uniform velocity u. Assuming the motion to be along a straight road, the minimum value of u, so that the student is able to catch the bus is

a) 52 ms^{-1} b) 8 ms^{-1} c) 10 ms^{-1} d) 12 ms^{-1}

252. Three different objects of masses m_1, m_2 and m_3 are allowed to fall from rest and from the same point 'O' along three different frictionless paths. The speeds of the three objects, on reaching the ground, will be in the ratio of

a) $m_1 : m_2 : m_3$ b) $m_1 : 2m_2 : 3m_3$ c) 1: 1: 1 d) $\frac{1}{m_1} : \frac{1}{m_2} : \frac{1}{m_3}$

253. A ball is dropped from a high rise platform at $t = 0$ starting from rest. After 6 seconds another ball is thrown downwards from the same platform with a speed v. The two balls meet at $t = 18s$. What is the value of v? (taking $g = 10\ m/s^2$)

a) 60 m/s b) 75 m/s c) 55 m/s d) 40 m/s

254. If a body is thrown up with the velocity of 15 m/s then maximum height attained by the body is ($g = 10\ m/s^2$)

a) 11.25 m b) 16.2 m c) 24.5 m d) 7.62 m

255. The position coordinates of a particle moving in $X - Y$ as a function of time t are

$x = 2t^2 + 6t + 25$

$y = t^2 + 2t + 1$

The speed of the object at $t = 10$ s is approximately

a) 31 b) 51 c) 71 d) 81

256. A body projected vertically upwards with velocity u returns to the starting point in 4 seconds. If $g = 10\ m/\sec^2$, the value of u is

a) 5 m/sec b) 10 m/sec c) 15 m/sec d) 20 m/sec

257. A body is at rest at $x = 0$. At $t = 0$, it starts moving in the positive x –direction with a constant acceleration. At the same instant another body passes through $x = 0$ moving in the positive x –direction with a constant speed. The position of the first body is given by $x_1(t)$ after time t and that of the second body by $x_2(t)$ after the same time interval. Which of the following graphs correctly describes $(x_1 - x_2)$ as a function of time t?

a) 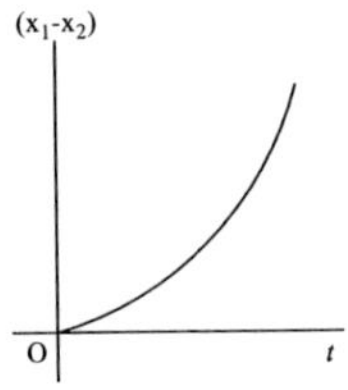

b) 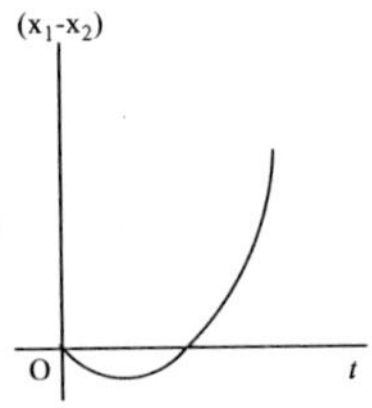

c) 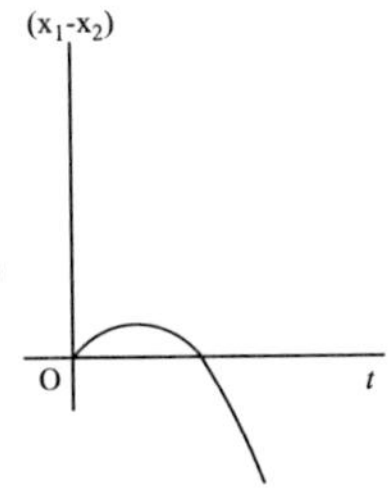

d) 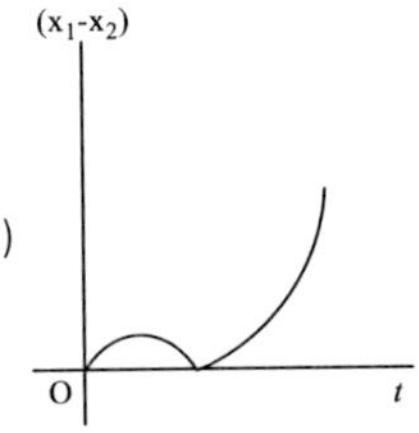

258. A particle of mass m is initially situated at the point P inside a hemispherical surface of radius r as shown in figure. A horizontal acceleration of magnitude$a_0$is suddenly produced on the particle in the horizontal direction. If gravitational acceleration is neglected, the time taken by particle to touch the sphere again is

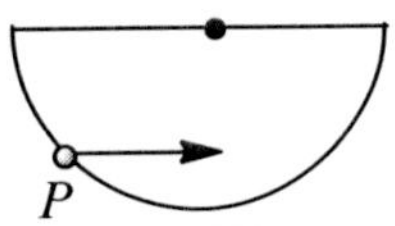

a) $\sqrt{\frac{4r\sin\alpha}{a_0}}$ b) $\sqrt{\frac{4r\tan\alpha}{a_0}}$ c) $\sqrt{\frac{4r\cos\alpha}{a_0}}$ d)None of these

259. At $t = 0$, a stone of mass 10 gm is thrown straight up from the ground level with a speed 10 m/s. After 1 s, a second stone of the same mass is thrown from the same position with a speed 20 m/s. What is the position of the first stone from the ground level at that moment? (Take g = 10 m/s^2)

a) 10 m b)1 m c) 2 m d)5 m

260. The time taken by a block of wood (initially at rest)to slide down a smooth inclined plane 9.8 m long (angle of inclination is 30°) is

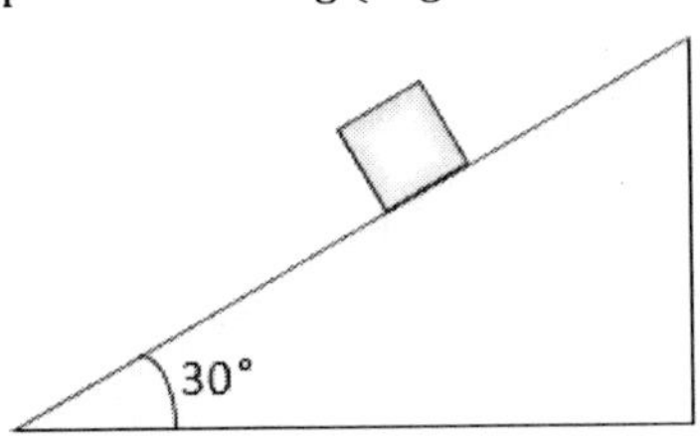

a) $\frac{1}{2}sec$ b)2sec c) 4sec d)1sec

261. A particle moves along a straight line OX. At a time t (in second) the distance x (in metre) of the particle from O is given by $x = 40 + 12t - t^3$. How long would the particle travel before coming to rest?

a) 24 m b)40 m c) 56 m d)16 m

262. The graph between the displacement x and t for a particle moving in a straight line is shown in figure. During the interval OA, AB, BC and CD, the acceleration of the particle is

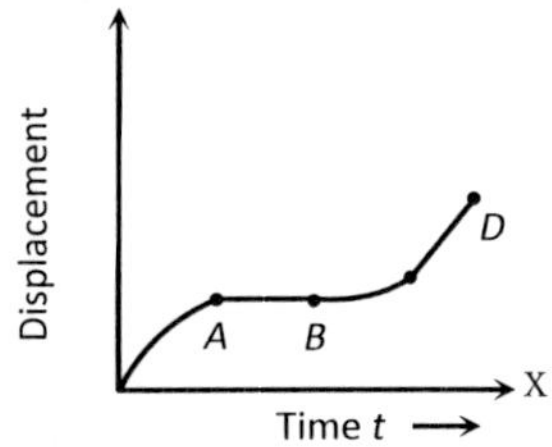

OA, AB, BC, CD

a) + 0 + + b)- 0 + 0 c) + 0 - + d)- 0 - 0

263. The particles A, B and C are thrown from the top of a tower with the same speed. A is thrown up, B is thrown down and C is horizontally. They hit the ground with speeds V_A, V_B and V_C respectively

a) $V_A = V_B = V_C$ b) $V_A = V_B > V_C$ c) $V_B > V_C > V_A$ d) $V_A > V_B = V_C$

264. A ball is thrown vertically upwards from the top of a tower of height h with velocity v. The ball strikes the ground after

a) $\frac{v}{g}\left[1+\sqrt{1+\frac{2gh}{v^2}}\right]$ b) $\frac{v}{g}\left[1-\sqrt{1+\frac{2gh}{v^2}}\right]$ c) $\frac{v}{g}\left(1+\frac{2gh}{v^2}\right)^{1/2}$ d) $\frac{v}{g}\left(1-\frac{2gh}{v^2}\right)^{1/2}$

265. A target is made of two plates, one of wood and the other of iron. The thickness of the wooden plate is 4 cm and that of iron plate is 2 cm. A bullet fired goes through the wood first and then penetrates 1 cm into iron. A similar bullet fired with the same velocity from opposite direction goes through iron first and then penetrates 2cm into wood. If a_1 and $a_2$be the retardation offered to the bullet by wood and iron plates respectively, then

a) $t_1 + t_2$ b) $a_2 = 2a_1$ c) $a_1 = a_2$ d) Data insufficient

266. A body, thrown upwards with some velocity, reaches the maximum height of $20m$. Another body with double the mass thrown up, with double initial velocity will reach a maximum height of

a) $200\ m$ b) $16\ m$ c) $80\ m$ d) $40\ m$

267. Two bodies being a free fall from rest, from the same height 2s apart. How long after the first body begins to fall the two bodies will be 40m apart? (Take $g = 10\text{ms}^{-2}$)

a) 1 s b) 2 s c) 3 s d) 4 s

268. A stone falls freely from rest from a height h and it travels a distance $\frac{9h}{25}$ in the last second. The value of h is

a) $145\ m$ b) $100\ m$ c) $122.5\ m$ d) $200\ m$

269. An aeroplane is moving with horizontal velocity u at height h.The velocity of a packet dropped from it on the earth's surface will be (g is acceleration due to gravity)

a) $\sqrt{u^2 + 2gh}$ b) $\sqrt{2gh}$ c) $2gh$ d) $\sqrt{u^2 - 2gh}$

270. A small block sides without friction down an inclined plane starting from rest. Let S_n be the distance travelled from time $t = n - 1$ to $t = n$. Then $\frac{S_n}{S_{n+1}}$ is

a) $\frac{2n-1}{2n}$ b) $\frac{2n+1}{2n-1}$ c) $\frac{2n-1}{2n+1}$ d) $\frac{2n}{2n+1}$

271. A point starts moving in a straight line with a certain acceleration. At a time t after beginning of motion the acceleration suddenly becomes retardation of the same value. The time in which the point returns to the initial point is

a) $\sqrt{2t}$

b) $(2 + \sqrt{2})t$

c) $\frac{t}{\sqrt{2}}$

d) Cannot be predicted unless acceleration is given

272. The initial velocity of a particle is u (at $t = 0$) and the acceleration f is given by at. Which of the following relation is valid

a) $v = u + at^2$ b) $v = u + a\frac{t^2}{2}$ c) $v = u + at$ d) $v = u$

273. From a balloon rising vertically upwards at $5m/s$ a stone is thrown up at $10\ m/s$ relative to the balloon. Its velocity with respect to ground after 2 s is (assume $g = 10m/s^2$)

a) 0 b) $20\ m/s$ c) $10\ m/s$ d) $5\ m/s$

274. Two bodies of different masses m_a and m_b are dropped from two different heights a and b. The ratio of the time taken by the two to cover these distances are

a) $a : b$ b) $b : a$ c) $\sqrt{a} : \sqrt{b}$ d) $a^2 : b^2$

275. A particle is constrained to move on a straight line path. It returns to the starting point after 10 sec. The total distance covered by the particle during this time is 30 m. Which of the following statement about the motion of the particle is false

a) Displacement of the particle is *zero*

b) Average speed of the particle is $3\ m/s$

c) Displacement of the particle is 30 m

d) Both (a) and (b)

276. The position x of a particle with respect to time t along x-axis is given by $x = 9t^2 - t^3$ where x is in metres and t in second. What will be the position of this particle when it achieves maximum speed along the $+x$ direction

a) 32 m b) 54 m c) 81 m d) 24 m

277. The driver of an express train moving with a velocity v_1 finds that a goods train is moving with a velocity v_2 in the same direction on the same track. He applies the brakes and produces a retardation a. The minimum time required to avoid collision is

a) $\frac{v_1 - v_2}{a}$ b) $\frac{v_1}{a}$ c) $\frac{v_2}{a}$ d) $\frac{v_1 + v_2}{a}$

278. A body is thrown vertically upwards. If air resistance is to be taken into account, then the time during which the body rises is

a) Equal to the time of fall

b) Less than the time of fall

c) Greater than the time of fall

d) Twice the time of fall

279. A particle moves along a straight line OX. At a time t (in seconds) the distance x (in metres) of the particle from O is given by $x = 40 + 12t - t^3$

How long would the particle travel before coming to rest

a) 24 m b) 40 m c) 56 m d) 16 m

280. A particle starts from rest. Its acceleration (a) *versus* time (t) is as shown in the figure. The maximum speed of the particle will be

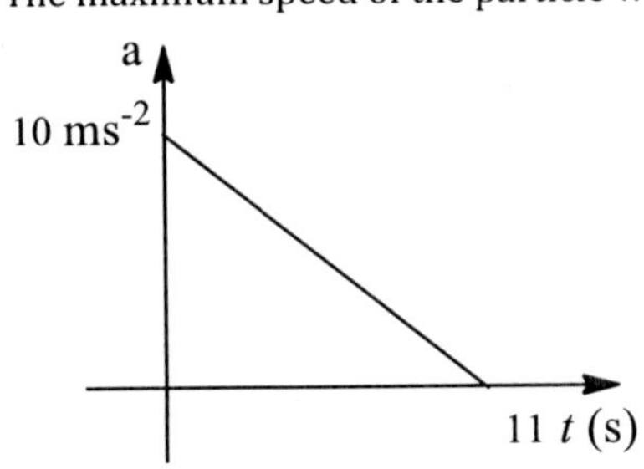

a) 110 ms^{-1} b) 55 ms^{-1} c) 550 ms^{-1} d) 660 ms^{-1}

281. A body is travelling in a straight line with a uniformly increasing speed. Which one of the plot represents the changes in distance (s) travelled with time (t)

a) 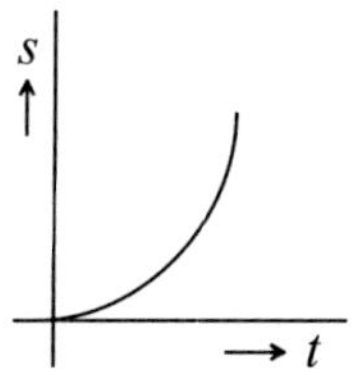

b) 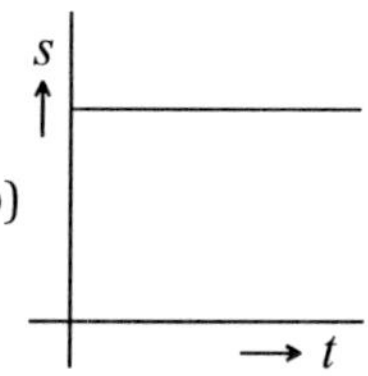

c) 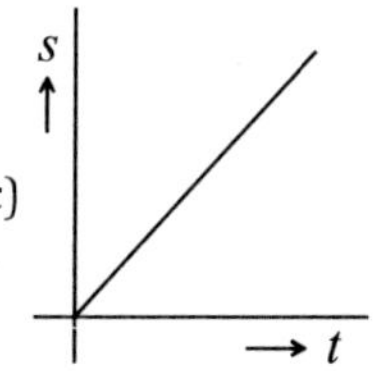

d) 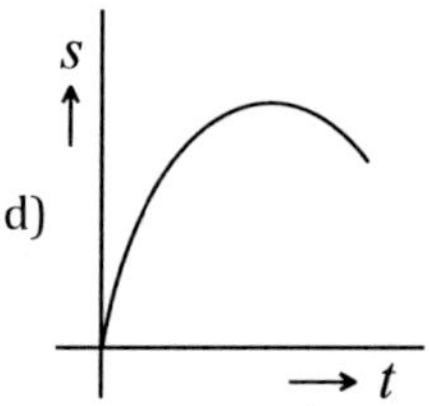

282. A body A moves with a uniform acceleration a and zero initial velocity. Another body B, starts from the same point moves in the same direction with a constant velocity v. The two bodies meet after a time$t$. The value of t is

a) $\frac{2v}{a}$ b) $\frac{v}{a}$ c) $\frac{v}{2a}$ d) $\sqrt{\frac{v}{2a}}$

283. The velocity of a body depends on time according to the equation $v = 20 + 0.1t^2$. The body is undergoing

a) Uniform acceleration b) Uniform retardation
c) Non-uniform acceleration d) Zero acceleration

284. A cat moves from X to Y with a uniform speed $v_u$and returns toX with a uniform speed v_d. The average speed for this ground trip is

a) $-\frac{2v_d v_u}{v_d + v_u}$ b) $\sqrt{v_u v_d}$ c) $\frac{v_d v_u}{v_d + v_u}$ d) $\frac{v_u+v_d}{2}$

285. Which of the following options is correct for the object having a straight line motion represented by the following graph

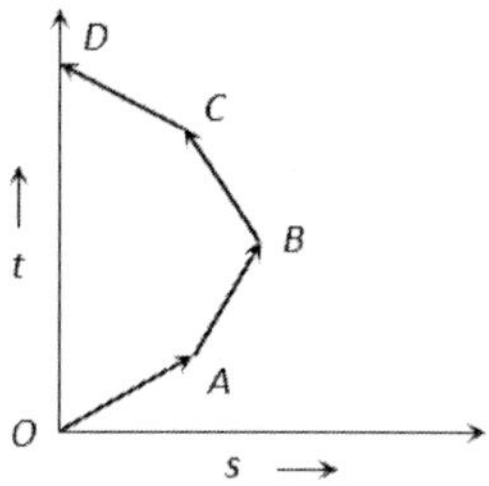

a) The object moves with constantly increasing velocity from O to A and then it moves with constant velocity
b) Velocity of the object increases uniformly
c) Average velocity is zero
d) The graph shown is impossible

286. A balloon gong upward with a velocity of 12 ms^{-1} is at a height of 65 m from the earth's surface at any instant. Exactly at this instant a ball drops from it. How much time will the ball take in reaching the surface of earth?
($g = 10\ ms^{-2}$)

a) 5 s b) 6 s c) 10 s d) None of these

287. The velocity-time graph of a body moving in a straight line is shown in the figure. The displacement and distance travelled by the body in 6 sec are respectively

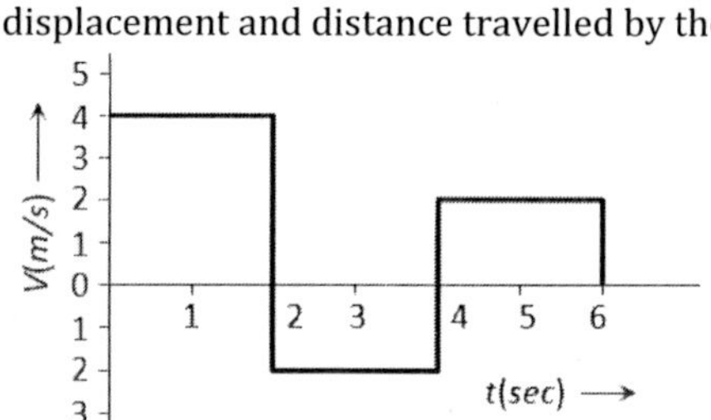

a) $8\ m, 16\ m$ b) $16\ m, 8\ m$ c) $16\ m, 16\ m$ d) $8\ m, 8\ m$

288. A train has a speed of 60 km/h.for the first one hour and 40 km/h for the next half hour. Its average speed in km/h is

a) 50 b) 53.33 c) 48 d) 70

289. Two cars A and B are moving with same speed of 45 km/hr along same direction. If a third car C coming from the opposite direction with a speed of 36 km/hr meets two cars in an interval of 5 minutes, the distance of separation of two cars A and B should be (in km)

a) 6.75 b) 7.25 c) 5.55 d) 8.35

290. A body is thrown vertically upwards with velocity u. The distance travelled by it in the fifth and the sixth seconds are equal. The velocity u is given by ($g = 9.8 \text{ ms}^{-2}$)

a) 24.5 ms^{-1} b) 49.0 ms^{-1} c) 73.5 ms^{-1} d) 98.0 ms^{-1}

291. A body is released from the top of a tower of height h. It takes $t sec$ to reach the ground. Where will be the ball after time $t/2$ sec

a) At $h/2$ from the ground b) At $h/4$ from the ground
c) Depends upon mass and volume of the body d) At $3h/4$ from the ground

292. A juggler throws balls into air. He throws one whenever the previous one is at its highest point. If the throws n balls each second, the height to which each ball will rise is

a) $\frac{g}{2n^2}$ b) $\frac{2g}{n^2}$ c) $\frac{2g}{n}$ d) $\frac{g}{4n^2}$

293. The position of a particle x (in metres) at a time t seconds is given by the relation $\vec{r} = (3t\hat{\imath} - t^2\hat{\jmath} + 4\hat{k})$. Calculate the magnitude of velocity of the particle after 5 seconds

a) 3.55 b) 5.03 c) 8.75 d) 10.44

294. A ball is dropped on the floor from a height of 10 m. It rebounds to a height of 2.5m. If the ball is in contact with the floor for 0.01 sec, the average acceleration during contact is

a) 2100 $m/\sec^2$ downwards b) 2100 $m/\sec^2$ upwards
c) 1400 $m/\sec^2$ d) 700m/sec^2

295. The motion of a particle is described by the equation $u = at$. The distance travelled by the particle in the first 4 seconds

a) $4a$ b) $12a$ c) $6a$ d) $8a$

296. A man is 45 m behind the bus, when the bus starts accelerating from rest with acceleration 2.5 ms^{-2}. With what minimum velocity should the man start running to catch the bus?

a) 12 ms^{-1} b) 14 ms^{-1} c) 15 ms^{-1} d) 16 ms^{-1}

297. A body is projected up with a speed 'u' and the time taken by it is T to reach the maximum height H.Pick out the correct statement

a) It reaches $H/2$ in $T/2$ sec b) It acquires velocity $u/2$ in $T/2sec$
c) Its velocity is $u/2$ at $H/2$ d) Same velocity at $2T$

298. A man drops a ball downside from the roof of a tower of height 400 m. At the same time another ball is thrown upside with a velocity 50 m/s. From the surface of the tower, then they will meet at which height from the surface of the tower

a) 100 m b) 320 m c) 80 m d) 240 m

299. If a freely falling body travels in the last second a distance equal to the distance travelled by it in the first three second, the time of the travel is

a) 6 sec b) 5 sec c) 4 sec d) 3 sec

300. A balloon is rising vertically up with a velocity of 29 ms^{-1}. A stone is dropped from it and it reaches the ground in 10 s. The height of the balloon when the stone was dropped from it is ($g = 9.8 \text{ ms}^{-2}$)

a) 400 m b) 150 m c) 100 m d) 200 m

301. A train starts from station with an acceleration 1ms^{-2}. A boy is 48 m behind the train with a constant velocity 10ms^{-1}, the minimum time after which the boy will catch the train is

a) 4.8 s b) 8 s c) 10 s d) 12 s

302. An aeroplane is moving with horizontal velocity u at height h. The velocity of a packet dropped from it on the earth's surface will be (g is acceleration due to gravity)

a) $\sqrt{u^2+2gh}$ b) $\sqrt{2gh}$ c) $2gh$ d) $\sqrt{u^2-2gh}$

303. What is the relation between displacement, time and acceleration in case of a body having uniform acceleration

a) $S=ut+\frac{1}{2}ft^2$ b) $S=(u+f)\,t$ c) $S=v^2-2fs$ d) None of these

304. The displacement of particle is given by $x=a_0+\frac{a_1t}{2}-\frac{a_2t^2}{3}$

What is its acceleration?

a) $\frac{2a_2}{3}$ b) $-\frac{2a_2}{3}$ c) a_2 d) Zero

305. Two trains one of 100 m and another of length 125m, are moving in mutually opposite directions along parallel lines, meet each other, each with speed 10m/s.

If their acceleration are $0.3m/s^2$ and $0.2m/s^2$ respectively, then the time taken to pass each other will be

a) 5 s b) 10 s c) 15 s d) 20 s

306. A body is moving along a straight line path with constant velocity. At an instant of time the distance travelled by it is S and its displacement is D, then

a) D<S b) $D>S$ c) $D=S$ d) $D\leq S$

307. An automobile in travelling at 50 kmh^{-1}, can be stopped at a distance of 40 m by applying brakes. If the same automobile is travelling at 90 kmh^{-1}, all other conditions remaining same and assuming no skidding, the minimum stopping distance in metre is

a) 72 b) 92.5 c) 102.6 d) 129.6

308. A ball P is dropped vertically and another ball Q is thrown horizontally with the same velocities from the same height and at the same time. If air resistance is neglected, then

a) Ball P reaches the ground first
b) Ball Q reaches the ground first
c) Both reach the ground at the same time
d) The respective masses of the two balls will decide the time

309. In the following graph, distance travelled by the body in meters is

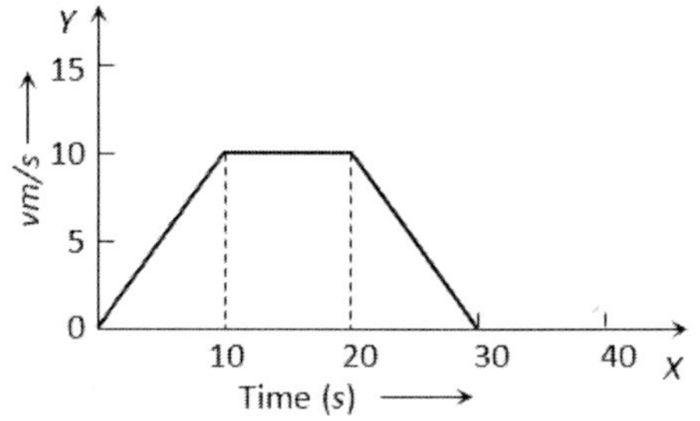

a) 200
b) 250
c) 300
d) 400

310. A car starts from rest and accelerates uniformly to a speed of 180 kmh^{-1} in 10 *seconds*. The distance covered by the car in this time interval is

a) 500 m b) 250 m c) 100 m d) 200 m

311. The velocity-time graph of a body is shown in figure. The ratio of the ... during the intervals OA and AB is....

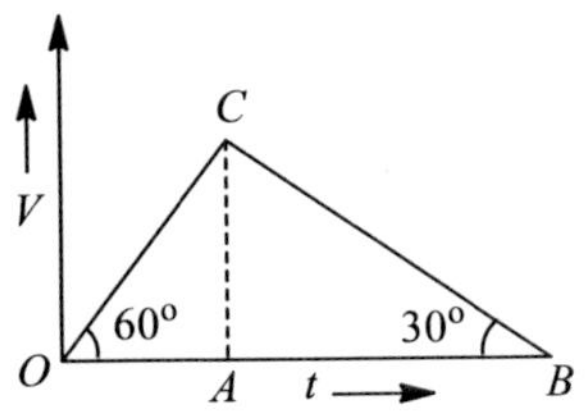

a) Average velocities:2

b) $\frac{OA}{AB}:\frac{1}{3}$

c) Average accelerations, same as distance covered

d) Distance covered : $\frac{1}{2}$

312. A bee files a line from a point A to another point B in 4 s with a velocity of $|t-2|\text{ms}^{-1}$. The distance between A and B in metre is

a) 2 b)4 c) 6 d)8

313. If a car at rest accelerates uniformly to a speed of 144 km/h in 20 s. Then it covers a distance of

a) 20 m b)400 m c) 1440m d)2880 m

314. A ball falls from height h. After 1 second, another ball falls freely from a point 20 m below the point from where the first ball falls. Both of them reach the ground at the same time. What is the value of h

a) 11.2 m b)21.2 m c) 31.2 m d)41.2 m

315. A particle moves in a straight line with a constant acceleration. It changes its velocity from 10 $ms^{-1}$to 20 ms^{-1} while passing through a distance 135 m in t second. The value of t is

a) 12 b)9 c) 10 d)1.8

316. Look at the graphs (a) to (d) carefully and indicate which of these possibly represents one dimensional motion of a particle

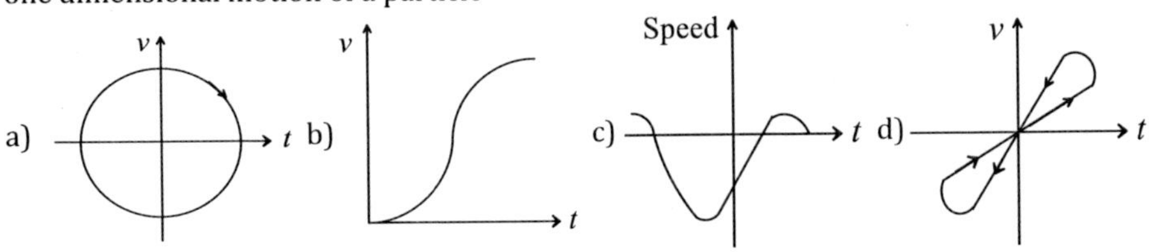

317. A bus begins to move with an acceleration of 1ms^{-2}. A man who is 48m behind the bus starts running at 10 ms^{-1} to catch the bus. The man will be able to catch the bus after

a) 6s b)5s c) 3s d)8s

318. A particle moving with a uniform acceleration travels 24 m and 64 m in the first two consecutive interval of 4 s each. Its initial velocity will be

a) 5 ms^{-1} b)3 ms^{-1} c) 1 ms^{-1} d)4 ms^{-1}

319. The driver of a car moving with a speed of 10ms^{-1} sees a red light ahead, applies breaks and stops after covering 10m distance. If the same car were moving with a speed of 20 ms^{-2}, the same driver would have stopped the car after covering 30 m distance. Within what distance the car can be stopped if travelling with a velocity of 15ms^{-1}? Assume the same reaction time and the same deceleration in each case.

a) 18.75 m b)20.75 m c) 22.75 m d)25 m

320. A particle located at $x=0$ at time $t=0$, starts moving along the positive x-direction with a velocity 'v' that varies as $v=a\sqrt{x}$. The displacement of the particle varies with time as

a) t b)$t^{1/2}$ c) t^3 d)t^2

321. A point initially at rest moves along x-axis. Its acceleration varies with time as $a = (6t + 5)m/s^2$. If it starts from origin, the distance covered in 2 s is

a) 20 m b) 18 m c) 16 m d) 25 m

322. A rocket is fired upward from the earth's surface such that it creates an acceleration of 19.6 $m/\sec^2$. If after 5 sec its engine is switched off, the maximum height of the rocket from earth's surface would be

a) 245 m b) 490 m c) 980 m d) 735 m

323. Velocity-time $(v - t)$ graph for a moving object is shown in the figure. Total displacement of the object during the time interval when there is non-zero acceleration and retardation is

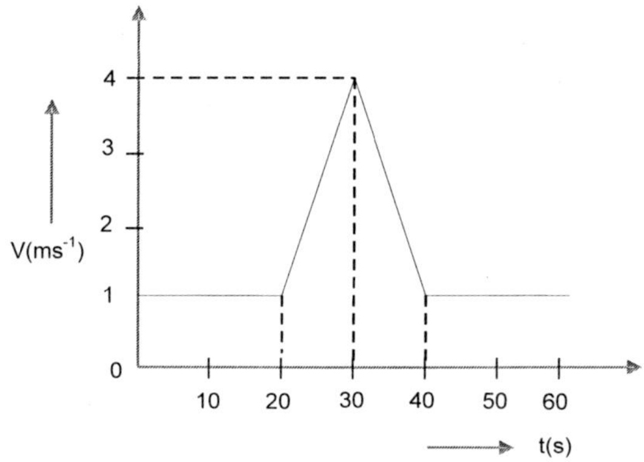

a) 60 m b) 50 m c) 30 m d) 40 m

324. A balloon is at a height of81 m and is ascending upwards with a velocity of 12 m/s. A body of $2kg$ weight is dropped from it. If $g = 10\ m/s^2$, the body will reach the surface of earth in

a) 1.5 s b) 4.025 s c) 5.4 s d) 6.75 s

325. Two cars A and B are travelling in the same direction with velocities v_A and $v_B (v_A > v_B)$. When the car A is at a distance s behind car B, the driver of the car A applies the brakes producing a uniform retardation a, there will be no collision when

a) $s < \frac{(v_A - v_B)^2}{2a}$ b) $s = \frac{(v_A - v_B)^2}{2a}$ c) $s \geq \frac{(v_A - v_B)^2}{2a}$ d) $s \leq \frac{(v_A - v_B)^2}{2a}$

326. A ball is dropped from top of a building. The ball take 0.5 s to fall past the 3 m length of window some distance from top of building with what speed does the ball pass the top of window?

a) 6 ms^{-1} b) 12 ms^{-1} c) 7 ms^{-1} d) 3.5 ms^{-1}

327. If the velocity of particle is given by $v = (180 - 16x)^{1/2}\ m/s$, then its acceleration will be

a) Zero b) 8 m/s^2 c) $-8\ m/s^2$ d) 4 m/s^2

328. A particle shows distance-time curve as given in this figure. The maximum instantaneous velocity of the particle is around the point

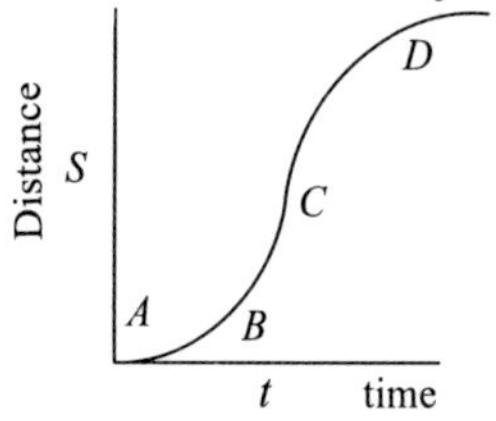

a) D b) A c) B d) C

329. Two cars A and Bat rest at same point initially. If A starts with uniform velocity of $40\ m/sec$ and B starts in the same direction with constant acceleration of $4m/s^2$, then B will catch A after how much time

a) $10\ sec$ b)$20\ sec$ c) $30\ sec$ d)$35\ sec$

330. A particle starts from rest, acceleration at $2\ m/s^2$ for $10\ s$ and then goes with constant speed for $30\ s$ and then decelerates at $4\ m/s^2$ till it stops. What is the distance travelled by it

a) $750\ m$ b)$800\ m$ c) $700\ m$ d)$850\ m$

331. The motion of a particle along a straight line is described by equation :

$x = 8 + 12t - t^3$

Where x is in metre and t in second. The retardation of the particle when its velocity becomes zero, is

a) $24ms^{-2}$ b)Zero c) $6ms^{-2}$ d)$12ms^{-2}$

332. A particle moves along the sides AB, BC, CD of a square of side $25\ m$ wuth a velocity of $15\ ms^{-1}$. Its average velocity is

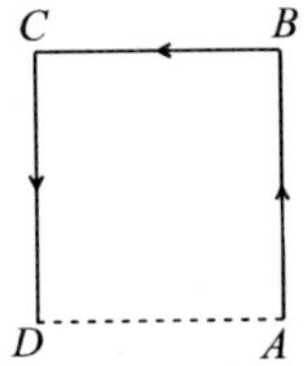

a) $15\ ms^{-1}$ b)$10ms^{-1}$ c) $7.5ms^{-1}$ d)$5ms^{-1}$

333. A stone falls freely from rest and the total distance covered by it in the last second of its motion equals the distance covered by it in the first three seconds of its motion. The stone remains in the air for

a) $6\ s$ b)$5\ s$ c) $7\ s$ d)$4\ s$

334. A constant force acts on a body of mass $0.9\ kg$ at rest for $10s$.If the body moves a distance of $250\ m$, the magnitude of the force is

a) $3N$ b)$3.5N$ c) $4.0N$ d)$4.5N$

335. A car moves a distance of $200\ m$. It covers first half of the distance at speed $60\ kmh^{-1}$ and the second half at speed v. If the average speed is $40\ kmh^{-1}$, the value of v is

a) $30\ kmh^{-1}$ b)$13\ kmh^{-1}$ c) $60\ kmh^{-1}$ d)$40\ kmh^{-1}$

336. Starting from rest, acceleration of a particle is $a = 2(t-1)$. The velocity of the particle at $t = 5s$ is

a) $15\ m/sec$ b)$25\ m/sec$ c) $5\ m/sec$ d)None of these

337. One car moving on a straight road covers one third of the distance with $20\ km/hr$ and the rest with $60\ km/hr$. The average speed is

a) $40\ km/hr$ b)$80\ km/hr$ c) $46\frac{2}{3}\ km/hr$ d)$36\ km/hr$

338. A scooterist sees a bus 1 km ahead of him moving with a velocity of 10 ms^{-1}. With what speed the scooterist should move so as to overtake the bus in 100 s?

a) 10 ms^{-1} b)15 ms^{-1} c) 20 ms^{-1} d)17 ms^{-1}

339. A stone is released from a balloon moving upwards with velocity v_0 a height $h$at $t = 0$. Which of the following graphs is best representation of velocity-time graph for the motion of stone?

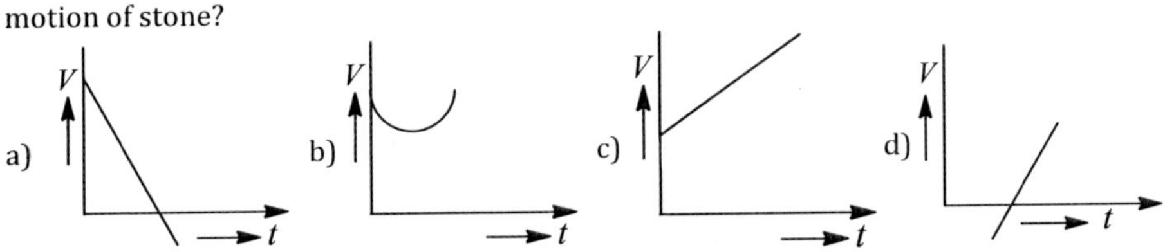

340. The displacement of a particle is proportional to the cube of time elapsed. How does the acceleration of the particle depends on time obtained

a) $a \propto t^2$ b)$a \propto 2t$ c)$a \propto t^3$ d)$a \propto t$

341. Which of the following graphs can not possibly represent one dimensional motion of a particle

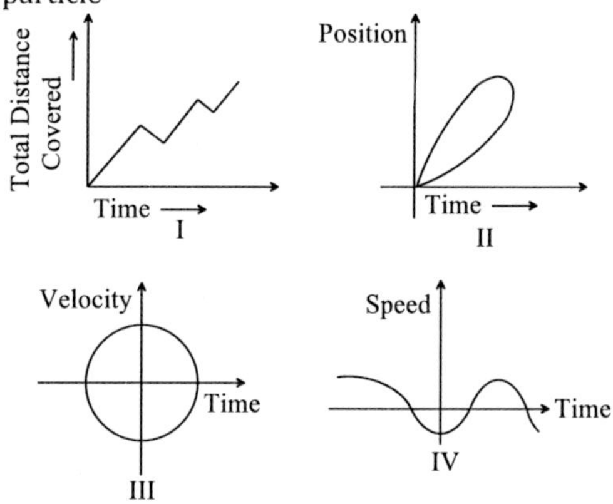

a) I and II b)II and III c) II and IV d)All four

342. One car moving on a straight road covers one third of the distance with 20 km/hr and the rest with 60 km/hr. The average speed is

a) 40 km/hr b)80 km/hr c) $46\frac{2}{3}$ km/hr d)36 km/hr

343. A body falls from a height $h = 200m$ (at New Delhi). The ratio of distance travelled in each 2 sec during $t = 0$ to $t = 6$ second of the journey is

a) 1 : 4 : 9 b)1 : 2 : 4 c) 1 : 3 : 5 d)1 : 2 : 3

344. The speed of body moving with uniform acceleration is u.This speed is doubled while covering distance S, its speed would be become

a) $\sqrt{3}u$ b)$\sqrt{5}u$ c) $\sqrt{11}u$ d)$\sqrt{7}u$

345. A ball of mass $m_1$and another ball of mass m_2 are dropped from equal height. If time taken by the balls are t_1 and t_2 respectively, then

a) $t_1 = \frac{t_2}{2}$ b)$t_1 = t_2$ c) $t_1 = 4t_2$ d)$t_1 = \frac{t_2}{4}$

346. The velocity of a bullet is reduced from 200 m/s to 100 m/s while travelling through a wooden block of thickness 10 cm. The retardation, assuming it to be uniform, will be

a) 10×10^4 m/s^2 b)12×10^4 m/s^2 c) 13.5×10^4 m/s^2 d)15×10^4 m/s^2

347. A ball is thrown vertically upwards. Which of the following graph/graphs represent velocity-time graph of the ball during its flight (air resistance is neglected)

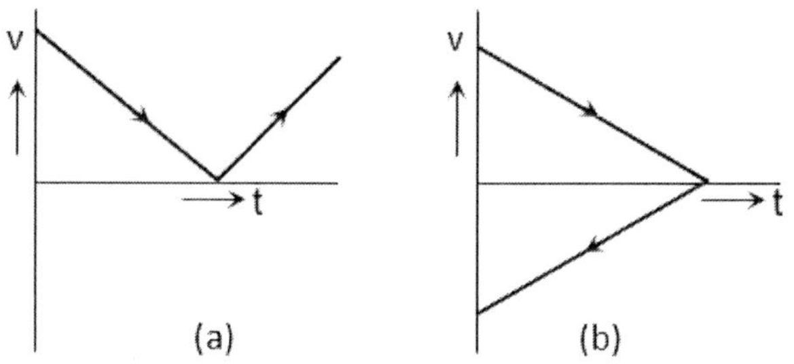

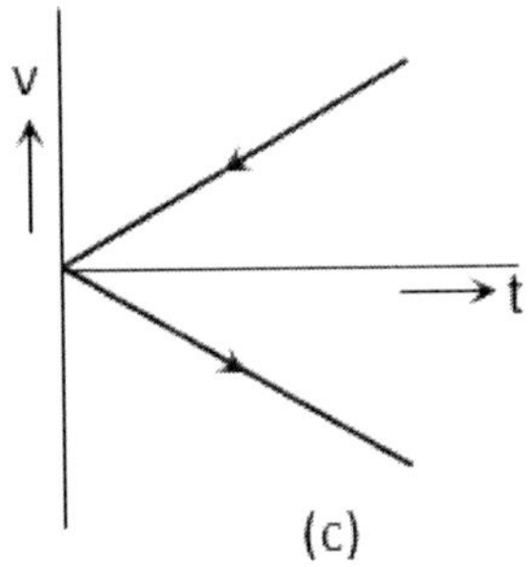

(c)

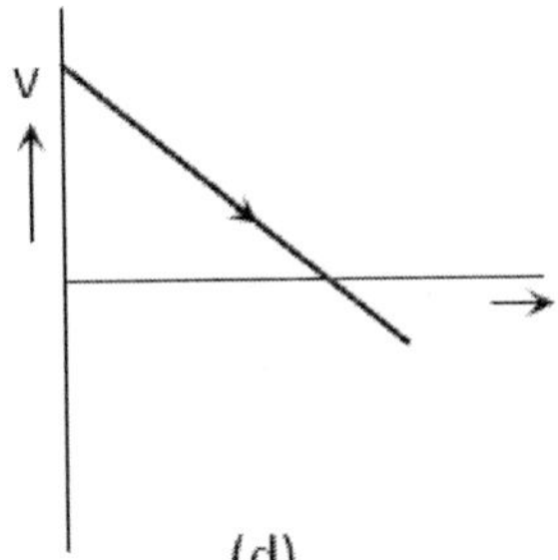

(d)

a) A
b) B
c) C
d) D

348. A car accelerates from rest at a constant rate α for some time, after which it decelerates at a constant rate β and comes to rest. If the total time elapsed is t, then the maximum velocity acquired by the car is
a) $\left(\frac{\alpha t+\beta^2}{\alpha\beta}\right)t$ b) $\left(\frac{\alpha^2-\beta^2}{\alpha\beta}\right)t$ c) $\frac{(\alpha+\beta)t}{\alpha\beta}$ d) $\frac{\alpha\beta t}{\alpha+\beta}$

349. Starting from rest, acceleration of a particle is $a = 2(t-1)$. The velocity of the particle at $t = 5s$ is
a) $15\ m/sec$ b) $25\ m/sec$ c) $5\ m/sec$ d) None of these

350. A particle moves along X- axis in such a way that its coordinate X varies with time t according to the equation $x = (2-5t+6t^2)m$. The initial velocity of the particle is
a) $-5\ m/s$ b) $6\ m/s$ c) $-3\ m/s$ d) $3\ m/s$

351. A body falling for 2 seconds covers a distance S is equal to that covered in next second. Taking $g = 10m/s^2, S =$
a) $30\ m$ b) $10\ m$ c) $60\ m$ d) $20m$

352. Two trains are moving with equal speed in opposite directions along two parallel railway tracks. If the wind is blowing with speed u along the track so that the relative velocities of the trains with respect to the wind are in the ratio 1: 2, then the speed of each train must be
a) $3u$ b) $2u$ c) $5u$ d) $4u$

353. You drive a car at seed of $70\ km/hr$ in a straight road for $8.4\ km$, and then the car runs out of petrol. You walk for $30\ min$ to reach a petrol pump at a distance of $2\ km$. The average velocity from the beginning of your drive till you reach the petrol pump is
a) $16.8\ km/hr$ b) $35\ km/hr$ c) $64\ km/hr$ d) $18.6\ km/hr$

354. A body is thrown vertically up from the ground. It reaches a maximum height of $100m$ in $5sec$. After what time it will reach the ground from the maximum height position
a) $1.2\ sec$ b) $5\ sec$ c) $10\ sec$ d) $25\ sec$

355. A body starts from rest. What is the ratio of the distance travelled by the body during the 4th and 3rd second
a) $\frac{7}{5}$ b) $\frac{5}{7}$ c) $\frac{7}{3}$ d) $\frac{3}{7}$

356. Which of the following options is correct for the object having a straight line motion represented by the following graph

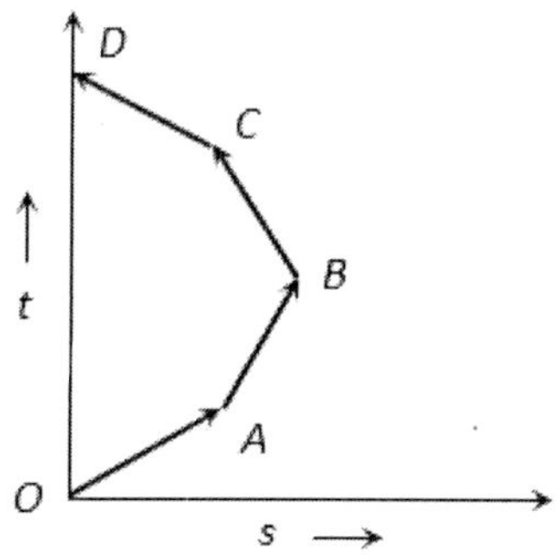

a) The object moves with constantly increasing velocity from O to A and then it moves with constant velocity
b) Velocity of the object increases uniformly
c) Average velocity is zero
d) The graph shown is impossible

357. A body is projected with a velocity v and after some time it returns to the point from which it was projected. The average velocity and average speed of the body for the total time of flight are
a) $\vec{v}/2$ and $v/2$ b) 0 and $v/2$ c) 0 and 0 d) $\vec{v}/2$ and 0

358. When a ball is thrown up vertically with velocity V_0, it reaches a maximum height of $'h'$. If one wishes to triple the maximum height then the ball should be thrown with velocity
a) $\sqrt{3}V_0$ b) $3V_0$ c) $9V_0$ d) $3/2V_0$

359. A ball is thrown vertically upwards with a velocity of $25\ ms^{-1}$ from the top of a tower of height 30 m. How long will it travel before it hits ground
a) $6s$ b) $5s$ c) $4s$ d) $12s$

360. A car moving with a velocity of 10 m/s can be stopped by the application of a constant force F in a distance of 20 m. If the velocity of the car is 30 m/s. It can be stopped by this force in
a) $\frac{20}{3}m$ b) 20 m c) 60 m d) 180 m

361. A particle moving in a straight line covers half the distance with speed of 3 m/s. The other half of the distance is covered in two equal time intervals with speed of4.5 m/s and 7.5 m/s respectively. The average speed of the particle during the motion is
a) 4.0 m/s b) 5.0 m/s c) 5.5 m/s d) 4.8 m/s

362. An aeroplane is flying horizontally with a velocity of 600 kmh^{-1} and at a height of 1960 m. When it is vertically above a point A on the ground a bomb is released from it. The bomb strikes the ground at point B. The distance AB is
a) 1200 m b) 0.33 km c) 333.3 km d) 3.33 km

363. An automobile travelling with a speed of 60 kmh^{-1}, can brake to stop within a distance of 20 m. if the car is going twice as fast, *ie.*, 120 kmh^{-1}, the stopping distance will be
a) 20 m b) 40 m c) 60 m d) 80 m

364. The particles A, B and C are thrown from the top of a tower with the same speed. A is thrown up, B is thrown down and C is horizontally. They hit the ground with speeds V_A, V_B and V_C respectively
a) $V_A = V_B = V_C$ b) $V_A = V_B > V_C$ c) $V_B > V_C > V_A$ d) $V_A > V_B = V_C$

365. A cyclist starts from the centre O of a circular park of radius one kilometre, reaches the edge P of the park, then cycles along the circumference and returns to the centre along QO as shown in figure. If the round trip takes ten minutes, the net displacement and average speed of the cyclist (in metre and kilometre per hour) is

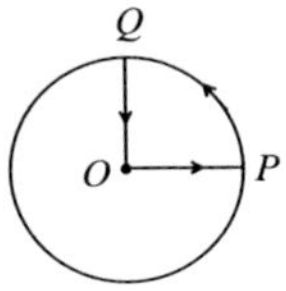

a) 0, 1 b) $\frac{\pi+4}{2}, 0$ c) $21.4, \frac{\pi+4}{2}$ d) $0, 21.4$

366. A projectile is fired vertically upwards with an initial velocity u. After an interval of T seconds a second projectile is fired vertically upwards, also with initial velocity is

a) They meet at time $t = \frac{u}{g}$ and at a height $\frac{u^2}{2G} + \frac{gT^2}{8}$

b) They meet at time $t = \frac{u}{g} + \frac{T}{2}$ and at a height $\frac{u^2}{2g} + \frac{gT^2}{8}$

c) They meet at time $t = \frac{u}{g} + \frac{T}{2}$ and at a height $\frac{u^2}{2g} - \frac{gT^2}{8}$

d) They never meet

367. The initial velocity of a body moving along a straight line is 7 m/s. It has a uniform acceleration of 4 m/s^2. The distance covered by the body in the 5^{th}second of its motion is

a) 25 m b) 35 m c) 50 m d) 85 m

368. A ball is thrown straight upward with a speed v from a height h above the ground. The time taken for the ball to strike the ground is given by

a) $-h = vt - \frac{1}{2}gt^2$ b) $h = vt - \frac{1}{2}gt^2$ c) $\frac{1}{2}gt^2$ d) $\sqrt{\frac{2g}{h}}$

369. An express train is moving with a velocity v_1. Its driver finds another train is moving on the same track in the same direction with velocity v_2. To escape collision, driver applies a retardation a on the train. The minimum time of escaping collision will be

a) $t = \frac{v_1 - v_2}{a}$ b) $t = \frac{v_1^2 - v_2^2}{2}$ c) None d) Both

370. A tiger chases a deer 30m ahead of it and gains 3m in 5s after the chase began. The distance gained by the tiger in 10 s is

a) 6m b) 12 m c) 18 m d) 20 m

371. A particle moves along x –axis in such a way that its coordinate (x) varies with time t according to the expression $x = 2 - 5t + 6t^2$m, the initial velocity of the particle is

a) 3 ms^{-1} b) 6 ms^{-1} c) -3 ms^{-1} d) -5 ms^{-1}

372. A man throws a ball vertically upward and it rises through 20 m and returns to his hands. What was the initial velocity (u) of the ball and for how much time (T) it remained in the air $[g = 10m/s^2]$

a) $u = 10m/s, T = 2s$ b) $u = 10m/s, T = 4s$ c) $u = 20m/s, T = 2s$ d) $u = 20m/s, T = 4s$

373. A particle moves with constant acceleration and v_1, v_2 and v_3 denote the average velocities in the three successive interval t_1, t_2 and t_3 of time. Which of the following relations is correct?

a) $\frac{v_1 - v_2}{v_2 - v_3} = \frac{t_1 - t_2}{t_2 + t_3}$ b) $\frac{v_1 - v_2}{v_2 - v_3} = \frac{t_1 - t_2}{t_1 - t_3}$ c) $\frac{v_1 - v_2}{v_2 - v_3} = \frac{t_1 - t_2}{t_2 - t_3}$ d) $\frac{v_1 - v_2}{v_2 - v_3} = \frac{t_1 + t_2}{t_2 + t_3}$

374. A bullet fired into a fixed wooden block loses half of its velocity after penetration 40 cm. it comes to rest after penetrating a further distance of

a) $\frac{22}{3}$ cm b) $\frac{40}{3}$ cm c) $\frac{20}{3}$ cm d) $\frac{22}{5}$ cm

375. A 210 m long train is moving due North at a speed of 25 m/s. A small bird is flying due South a little above the train with speed 5 m/s. The time taken by the bird to cross the train is

a) 6s b) 7s c) 9s d) 10s

376. A body starting from rest moves with constant acceleration. The ratio of distance covered by the body during the $5^{th} sec$ to that covered in 5 sec is

a) 9/15 b)3/5 c) 25/9 d)1/25

377. Two bodies are thrown vertically upwards with their initial speed in the ratio 2 : 3. The ratio of the maximum heights reached by then and the ratio of their time taken by them to return back to the ground respectively are

a) 4 : 9 and 2 : 3 b)2 : 3 and $\sqrt{2}$: $\sqrt{3}$ c) $\sqrt{2}$: $\sqrt{3}$ and 4 : 9 d)$\sqrt{2}$: $\sqrt{3}$ and 2 : 3

378. A body is released from the top of a tower of height h. It takes $t sec$ to reach the ground. Where will be the ball after time $t/2$ sec

a) At $h/2$ from the ground

b) At $h/4$ from the ground

c) Depends upon mass and volume of the body

d) At $3h/4$ from the ground

379. Two balls of same size but the density of one is greater than that of the other are dropped from the same height, then which ball will reach the earth first (air resistance is negligible)

a) Heavy ball b)Light ball c) Both simultaneously d)Will depend upon the density of the balls

380. A point initially at rest moves along x-axis. Its acceleration varies with time as $a = (6t + 5)$in ms^{-2}. It starts from origin, the distance covered in 1 s is

a) 3 m b)3.5 m c) 4 m d)4.5 m

381. Two particles A and B start moving simultaneously from point P with velocity 15 $ms^{-1}$and $20ms^{-1}$ respectively. They move with equal accelerations but opposite in direction. When A overtakes B at C then its velocity is $30ms^{-1}$. The velocity of B at C will lbe

a) $5ms^{-1}$ b)$10ms^{-1}$ c) $15ms^{-1}$ d)$20ms^{-1}$

382. Three particles start from the origin at the same time, one with a velocity v_1 along x-axis, the second along the y-axis with a velocity v_2 and the third along $x = y$ line. The velocity of the third so that the three may always lie on the same line is

a) $\frac{v_1 v_2}{v_1 + v_2}$ b)$\frac{\sqrt{2} v_1 v_2}{v_1 + v_2}$ c) $\frac{\sqrt{3} v_1 v_2}{v_1 + v_2}$ d)Zero

383. A ball is dropped downwards. After 1 second another ball is dropped downwards from the same point. What is the distance between them after 3 seconds

a) 25 m b)20 m c) 50 m d)9.8 m

384. The given graph shows the variation of velocity with displacement. Which one of the graph given below correctly represents the variation of acceleration with displacement

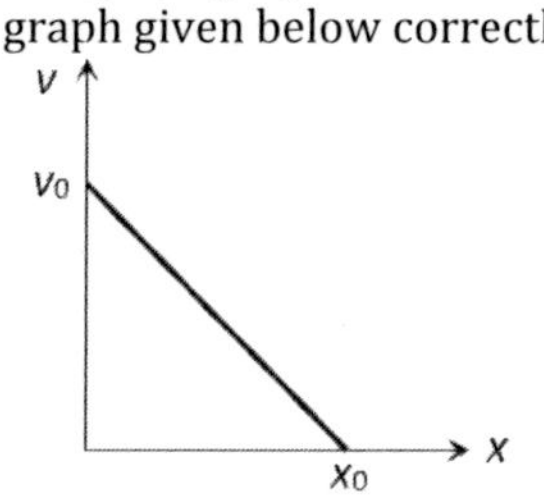

a)

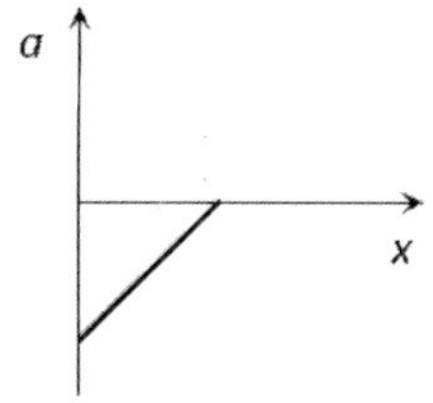

b) 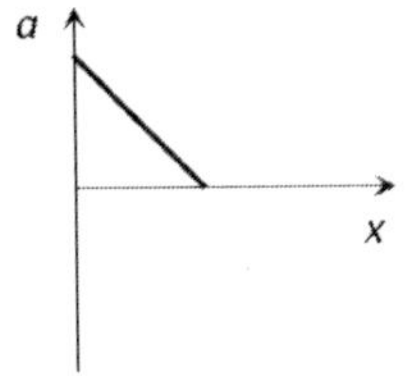

c) 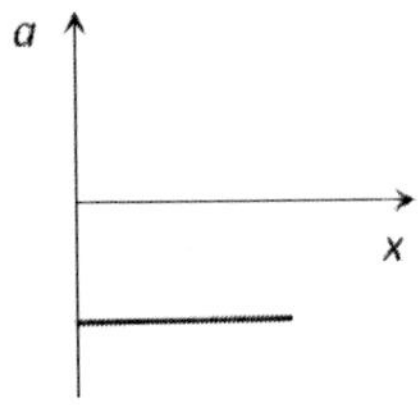

d) 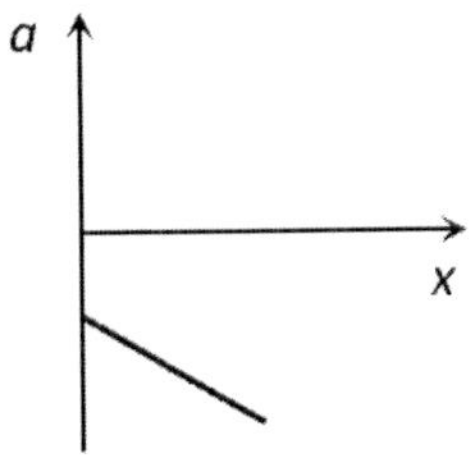

385. With what velocity a ball be projected vertically so that the distance covered by it in 5th second is twice the distance it covers in its 6th second ($g = 10\ m/s^2$)
a) 58.8 m/s b)49 m/s c) 65 m/s d)19.6 m/s

386. A body moving with a uniform acceleration crosses a distance of 65 m in the 5th second and 105 m in 9th second. How far will it go in 20 s?
a) 2040 m b)240 m c) 2400 m d)2004 m

387. A particle shows distance-time curve as given in this figure. The maximum instantaneous velocity of the particle is around the point

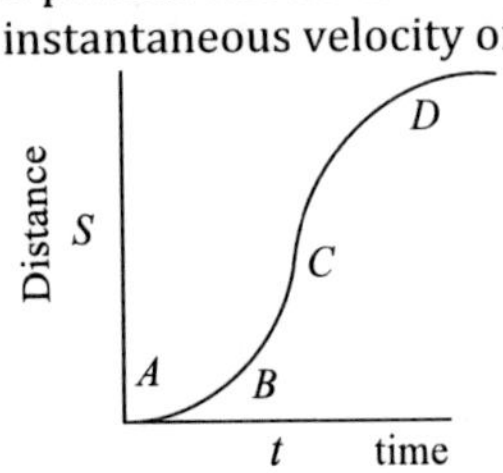

a) D b)A c) B d)C

388. A bus begins to move with an acceleration of 1 ms^{-2}. A man who is 48 m behind the bus starts running at 10 ms^{-1} to catch the bus. The man will be able to catch the bus after
a) 6 s b)5 s c) 3 s d)8 s

389. A Body moves 6 m north.8 m east and 10 m vertically upwards, what is its resultant displacement from initial position
a) $10\sqrt{2m}$ b)10 m c) $\frac{10}{\sqrt{2}}m$ d)$10 \times 2m$

390. The acceleration of a train between two stations 2 km apart is shown in the figure. The maximum speed of the train is

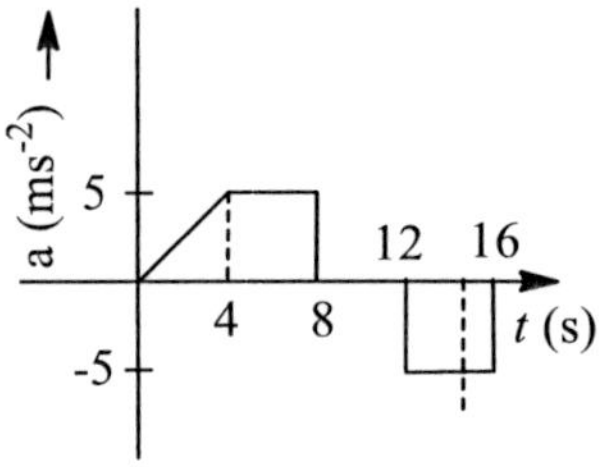

a) $60 ms^{-1}$ b) $30\ ms^{-1}$ c) $120\ ms^{-1}$ d) $90\ ms^{-1}$

391. A stone is allowed to fall from the top of a tower 100m high and at the same time another stone is projected vertically upwards from the ground with a velocity of $254 ms^{-1}$. The two stones will meet after

a) 4 s b) 0.4 s c) 0.04 s d) 40 s

392. A stone thrown vertically upwards attains a maximum height of 45m. In what time the velocity of stone become equal to one-half the velocity of throw? (Given $g = 10 ms^{-2}$)

a) 2 s b) 1.5 s c) 1 s d) 0.5 s

393. Look at the graph (a) to (d) carefully and indicate which of these possibly represents one dimensional motion of a particle?

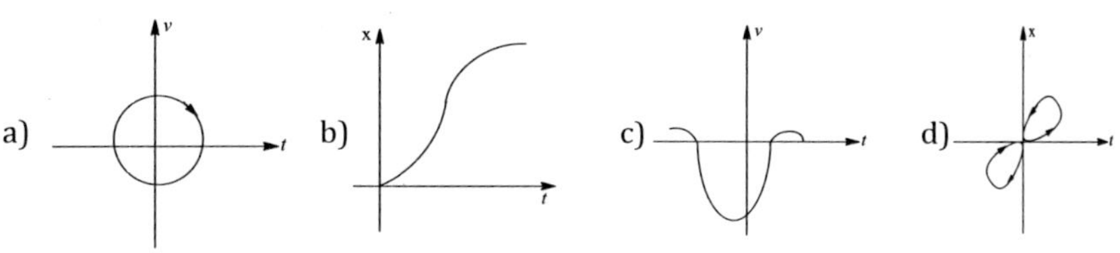

394. A boy standing at the top of a tower of 20 m height drops a stone. Assuming $g = 10\ ms^{-2}$, the velocity with which it hits the ground is

a) 5.0 m/s b) 10.0 m/s c) 20.0 m/s d) 40.0 m/s

395. Time taken by an object falling from rest to cover the height of h_1 and h_2 is respectively t_1 and t_2 then the ratio of t_1 to t_2 is

a) $h_1 : h_2$ b) $\sqrt{h_1} : \sqrt{h_2}$ c) $h_1 : 2h_2$ d) $2h_1 : h_2$

396. The relation between time t and distance x is $t = ax^2 + bx$, where a and $b$constants are. The acceleration is

a) $-2abv^2$ b) $2\ bv^3$ c) $-2av^3$ d) $2av^2$

397. A particle starts from rest at $t = 0$ and moves in a straight line with an acceleration as shown below. The velocity of the particle at $t = 3s$ is

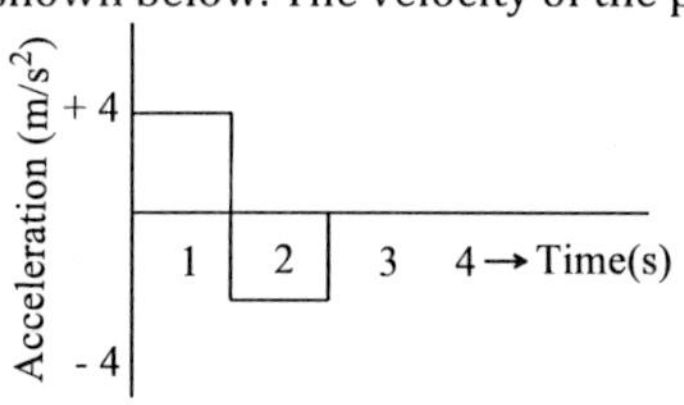

a) $2\ ms^{-1}$ b) $4\ ms^{-1}$ c) $6\ ms^{-1}$ d) $8\ ms^{-1}$

398. Two stones of equal masses are dropped from a rooftop of height h one after another. Their separation distance against time will

a) Remain the same b) Increase c) Decrease d) Be zero

399. A bullet moving with a velocity of 200 cm/s penetrates a wooden block and comes to rest after traversing 4cm inside it. What velocity is needed for travelling distance of 9cm in same block

a) 100 cm/s b) 136.2cm/s c) 300cm/s d) 250 cm/s

400. A body travelling with uniform acceleration crosses two points A and B with velocities $20ms^{-1}$ and d30 ms^{-1} respectively. The speed of the body at the mid-point of A and B is nearest to

a) $25.5\ ms^{-1}$ b) $25\ ms^{-1}$ c) $24ms^{-1}$ d) $10\sqrt{6}\ ms^{-1}$

401. A wheel of radius 1 m rolls forward half a revolution on a horizontal ground. The magnitude of the displacement of the point of wheel initially in contact with the ground is

a) 2π b) $\sqrt{2}\pi$ c) $\sqrt{\pi^2+4}$ d) π

402. A particle when thrown, moves such that it passes from same height at 2 and $10s$, the height is

a) g b) $2g$ c) $5g$ d) $10g$

403. The distance between two particles moving towards each other is decreasing at the rate of $6m/sec$. If these particles travel with same speeds and in the same direction, then the separation increase at the rate of $4m/sec$. The particles have speeds as

a) $5m/sec$: $1m/sec$ b) $4\ m/sec$: $1m/sec$ c) $4\ m/sec$: $2m/sec$ d) $5\ m/sec$: $2m/sec$

404. A cricket ball is thrown up with a speed of 19.6 ms^{-1}. The maximum height it can reach is

a) 9.8 m b) 19.6 m c) 29.4 m d) 39.2 m

405. For a particle moving in a straight line, the displacement of the particle at time t is given by $S = t^3 - 6t^2 + 3t + 7$. What is the velocity of the particle when its acceleration is zero?

a) $-9ms^{-1}$ b) $-12ms^{-1}$ c) $3\ ms^{-1}$ d) $42\ ms^{-1}$

406. A balloon rises from rest with a constant acceleration $g/8$. A stone is released from it when it has risen to height h. The time taken by the stone to reach the ground is

a) $4\sqrt{\frac{h}{g}}$ b) $2\sqrt{\frac{h}{g}}$ c) $\sqrt{\frac{2h}{g}}$ d) $\sqrt{\frac{g}{h}}$

407. Two balls are dropped to the ground from different heights. One ball is dropped 2 s after the other but they both strike the ground at the same time. If the first ball takes 5 s to reach the ground, then the difference in initial heights is ($g = 10\ ms^{-2}$)

a) 20 m b) 80 m c) 170 m d) 40 m

408. From a balloon rising vertically upwards as 5 ms^{-1} a stone is thrown up at 10 ms^{-1} relative to the balloon. Its velocity with respect to ground after 2 s is (assume $g = 10\ ms^{-2}$)

a) Zero b) $5ms^{-1}$ c) 10 ms^{-1} d) 20 ms^{-1}

409. A body dropped from top of a tower fall through 60 m during the last two second of its fall. The height of tower is ($g = 10\ ms^{-2}$)

a) 95 m b) 60 m c) 80 m d) 90 m

410. A particle is projected upwards. The times corresponding to height h while ascending and while descending are t_1 and t_2 respectively. The velocity of projection will be

a) gt_1 b) gt_2 c) $g(t_1+t_2)$ d) $\frac{g(t_1+t_2)}{2}$

411. From an elevated point A, a stone is projected vertically upwards. When the stone reaches a distance $h$below A, its velocity is double of what was at a height above A. The greatest height attained by the stone is

a) $\frac{h}{3}$ b) $\frac{2h}{2}$ c) $\frac{h}{2}$ d) $\frac{5h}{3}$

412. A body is thrown vertically upwards with a velocity u.Find the true statement from the following

a) Both velocity and acceleration are zero at its highest point

b) Velocity is maximum and acceleration is zero at the highest point

c) Velocity is maximum and acceleration is g downwards at its highest point
d) Velocity is zero at the highest point and maximum height reached is $u^2/2g$

413. A particle starting from rest falls from a certain height. Assuming that the value of acceleration due to gravity remains the same throughout motion, its displacement in three successive half second intervals are S_1, S_2, S_3.
Then,
a) $S_1:S_2:S_3:1:5:9$ b) $S_1:S_2:S_3:1:2:3$ c) $S_1:S_2:S_3:1:1:1$ d) $S_1:S_2:S_3:1:3:5$

414. A particle moves along a straight line such that its displacement at any time t is given by $S = t^3 - 6t^2 + 3t + 4$ *metres*
The velocity when the acceleration is zero is
a) $3ms^{-1}$ b) $-12ms^{-1}$ c) $42ms^{-1}$ d) $-9ms^{-1}$

415. A car moving with speed of 40 km/h can be stopped by applying brakes after atleast 2 m. If the same car is moving with a speed of 80 km/h, what is the minimum stopping distance
a) 8 m b) 2 m c) 4 m d) 6 m

416. A train has a speed of 60 km/h.for the first one hour and 40 km/h for the next half hour. Its average speed in km/h is
a) 50 b) 53.33 c) 48 d) 70

417. A 120 $m$long train is moving in a direction with speed 20 m/s. A train B moving with 30 m/s in the opposite direction and 130 m long crosses the first train in a time
a) 6 s b) 36 s c) 38 s d) None of these

418. From the top of tower a body A is projected vertically up, another body B is horizontally thrown and a third body C is thrown vertically down with same velocity. Then
a) B strikes the ground with more velocity
b) C strikes the ground with less velocity
c) A, B, C strike the ground with same velocity
d) A and Cstrike the ground with more velocity than$B$

419. Two trains are moving with equal speed in opposite directions along two parallel railway tracks. If the wind is blowing with speed u along the track so that the relative velocities of the trains with respect to the wind are in the ratio 1: 2, then the speed of each train must be
a) $3u$ b) $2u$ c) $5u$ d) $4u$

420. The wind appears to blow from the north to a man moving in the north-east direction. When he doubles his velocity the wind appears to move in the direction $\cot^{-1}(2)$ east of north. The actual direction of the wind is
a) $\sqrt{2}v$ towards east b) $\frac{v}{\sqrt{2}}$ towards west c) $\sqrt{2}\,v$ towards west d) $\frac{v}{\sqrt{2}}$ towards east

421. A ball is projected upwards from a height h above the surface of the earth with velocity v. The time at which the ball strikes the ground is
a) $\frac{v}{g}+\frac{2hg}{\sqrt{2}}$ b) $\frac{v}{g}\left[1-\sqrt{1+\frac{2h}{g}}\right]$ c) $\frac{v}{g}\left[1+\sqrt{1+\frac{2gh}{v^2}}\right]$ d) $\frac{v}{g}\left[1+\sqrt{v^2+\frac{2h}{g}}\right]$

422. A stone dropped from a balloon which is at a height h, reaches the ground after t second. From the same balloon, if two stones are thrown, one upwards and the other downwards, with the same velocity u and they reach the ground after $t_1$and t_2 second respectively, then
a) $t = t_1 - t_2$ b) $t = \frac{t_1 + t_2}{2}$ c) $t = \sqrt{t_1 t_2}$ d) $t = \sqrt{t_1^2 - t_2^2}$

423. A body falls freely from the top of a tower. It covers 36% of the total height in the last second before striking the ground level. The height of the tower is
a) 50 m b) 75 m c) 100m d) 125 m

424. Figure shows the graphical variation of displacement with time for the case of a particle moving along a straight line. The accelerations of the particle during the intervals OA, AB, BC and CD are respectively

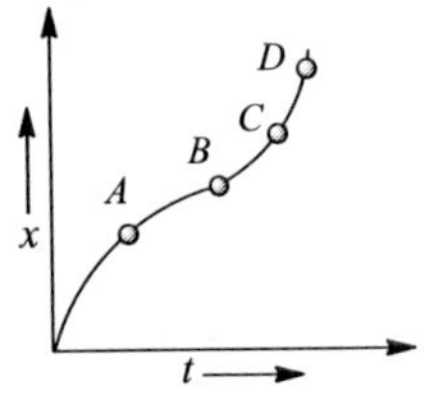

a) $OAABBCCD$ b) $-\ 0\ +\ 0$ c) $+\ 0\ +\ +$ d) $-\ 0\ -\ 0$

425. Two balls of same size but the density of one is greater than that of the other are dropped from the same height, then which ball will reach the earth first (air resistance is negligible)
a) Heavy ball b) Light ball
c) Both simultaneously d) Will depend upon the density of the balls

426. A body is moving according to the equation $x = at + bt^2 - ct^3$ where x = displacement and a, b and c are constants. The acceleration of the body is
a) $a + 2bt$ b) $2b + 6ct$ c) $2b - 6ct$ d) $3b - 6ct^2$

427. A particle moves along x-axis as
$x = 4(t-2) + a(t-2)^2$
Which of the following is true
a) The initial velocity of the particle is 4 b) The acceleration of particle is $2a$
c) The particle is at origin at $t = 0$ d) None of these

428. The position x of a particle varies with time t as $x = at^2 - bt^3$. The acceleration of the particle will be zero at time t equal to
a) $\frac{a}{b}$ b) $\frac{2a}{3b}$ c) $\frac{a}{3b}$ d) Zero

429. A stone is dropped from a height h. Simultaneously, another stone is thrown up from the ground which reaches a height $4\ h$. The two stones cross other after time
a) $\sqrt{\frac{h}{8g}}$ b) $\sqrt{8gh}$ c) $\sqrt{2gh}$ d) $\sqrt{\frac{h}{2g}}$

430. A particle moving with a uniform acceleration travels $24\ m$ and $64\ m$ in the first two consecutive intervals of 4 *sec* each.
Its initial velocity is
a) $1\ m/sec$ b) $10\ m/sec$ c) $5\ m/sec$ d) $2\ m/sec$

431. Consider the acceleration, velocity and displacement of a tennis ball as it falls to the ground and bounces back. Directions of which of these changes in the process
a) Velocity only b) Displacement and velocity
c) Acceleration, velocity and displacement d) Displacement and acceleration

432. What determines the nature of the path followed by the particle
a) Speed b) Velocity c) Acceleration d) Both (b) and (c)

433. A ball is thrown vertically upwards with a velocity of 25 ms^{-1} from the top of a tower of height 30 m. How long will it travel before it hits ground?
a) 6 s b) 5 s c) 4 s d) 12 s

434. A particle moves in a straight line with a constant acceleration. It changes its velocity from 10 ms^{-1} to 20 ms^{-1} while passing through a distance 135 m in t second. The value of t is
a) 12 b) 9 c) 10 d) 1.8

435. If the velocity of a car is increased by 20%, then the minimum distance in which it can be stopped increase by

a) 44% b)55% c) 66% d)88%

436. A cat moves from X to Y with a uniform speed $v_u$and returns toX with a uniform speed v_d. The average speed for this ground trip is

a) $-\frac{2v_d v_u}{v_d + v_u}$ b)$\sqrt{v_u v_d}$ c)$\frac{v_d v_u}{v_d + v_u}$ d)$\frac{v_u+v_d}{2}$

437. A particle is projected with velocity v_0 along $x - axis$.The deceleration on the particle is proportional to the square of the distance from the origin i.e., $a = -ax^2$.The distance at which the particle stops is

a) $\sqrt{\frac{3v_0}{2\alpha}}$ b)$\left(\frac{3v_0}{2\alpha}\right)^{\frac{1}{3}}$ c) $\sqrt{\frac{3v_0^2}{2\alpha}}$ d)$\left(\frac{3v_0^2}{2\alpha}\right)^{\frac{1}{3}}$

438. The displacement-time graphs of two moving particles make angles of 30° and 45° with the x −axis. The ratio of their velocities is

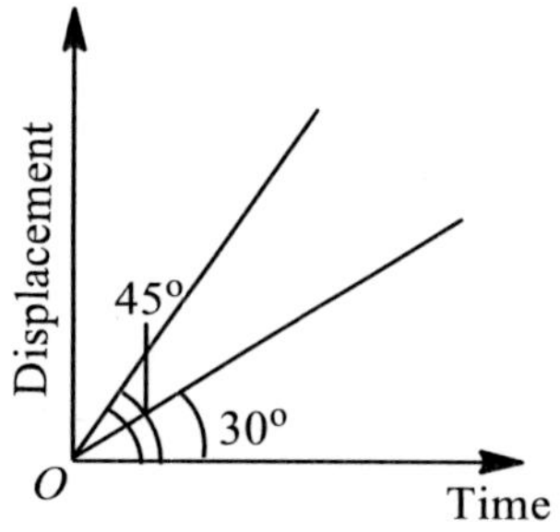

a) $1 : \sqrt{3}$ b)$1 : 2$ c) $1 : 1$ d)$\sqrt{3} : 2$

439. A particle located at $x = 0$ at time $t = 0$, starts moving along the positive x −direction with a velocity v that varies as $v = \alpha\sqrt{x}$. The displacement of the particle varies with time as

a) t^2 b)t c) $t^{1/2}$ d)t^3

440. Two boys are standing at the ends A and B of a ground where $AB = a$. The boy at B starts running in a direction perpendicular to AB with velocity v_1. The boy at A starts running simultaneously with velocity v and catches the other boy in a time t, where t is

a) $a/\sqrt{v^2 + v_1^2}$

b) $\sqrt{a^2/(v^2 - v_1^2)}$

c) $a/(v - v_1)$

d) $a/(v + v_1)$

441. A bullet emerges from a barrel of length 1.2 m with a speed of 640 ms^{-1}. Assuming constant acceleration, the approximate time that it spends in the barrel after the gun is fired is

a) 4 ms b)40 ms c) 400 μs d)1 s

442. Three balls A, B, C are thrown from a height h with equal speed upwards, downwards and horizontally respectively. What is the relation among speeds v_A, v_B, v_C with which they hit the ground?

a) $v_A = v_B = v_C$ b)$v_A > v_C > v_B$ c) $v_A = v_B > v_C$ d)$v_A < v_C < v_B$

443. The relation between time and distance is $t = \alpha x^2 + \beta x$, where α and β are constants. The retardation is

a) $2\alpha v^3$ b)$2\beta v^3$ c) $2\alpha\beta v^3$ d)$2\beta^2 v^3$

444. A 120 $m$long train is moving in a direction with speed 20 m/s. A train B moving with 30 m/s in the opposite direction and 130 m long crosses the first train in a time
a) 6 s b)36 s c) 38 s d)None of these

445. An object is projected upwards with a velocity of $100 m/s$. It will strike the ground after (approximately)
a) 10 sec b)20 sec c) 15 sec d)5 sec

446. The velocity of a particle is $v = v_0 + gt + ft^2$. If its position is $x = 0$ at $t = 0$, then its displacement after unit time $(t = 1) is$
a) $v_0 = 2g + 3f$ b)$v_0 + g/2 + f/3$ c) $v_0 + g + f$ d)$v_0 + g/2 + f$

447. The motion of a particle is described by the equation $x = a + bt^2$ where $a = 15\ cm$ and $b = 3\ cm/s^2$.Its instantaneous velocity at time 3 sec will be
a) 36 cm/sec b)18 cm/sec c) 16 cm/sec d)32 cm/sec

448. Two boys are standing at the ends A and B of a ground where $AB = a$. The boy at B starts running in a direction perpendicular to AB with velocity v_1. The boy at A starts running simultaneously with velocity v and catches the other boy in a time t, where t is
a) $a/\sqrt{v^2 + v_1^2}$ b) $\sqrt{a^2/(v^2 - v_1^2)}$ c) $a/(v - v_1)$ d)$a/(v + v_1)$

449. A particle moves along a parabolic path $y = 9x^2$ in such a way that the x-component of velocity remains constant and has a value $\frac{1}{3}$ms^{-1}. The acceleration of the particle is
a) $\frac{1}{3}\hat{j}$ms^{-2} b)$3\hat{j}$ms^{-2} c)$\frac{2}{3}\hat{j}$ms^{-2} d)$2\hat{j}$ms^{-2}

450. A particle is thrown upwards, then correct $v - t$ graph will be

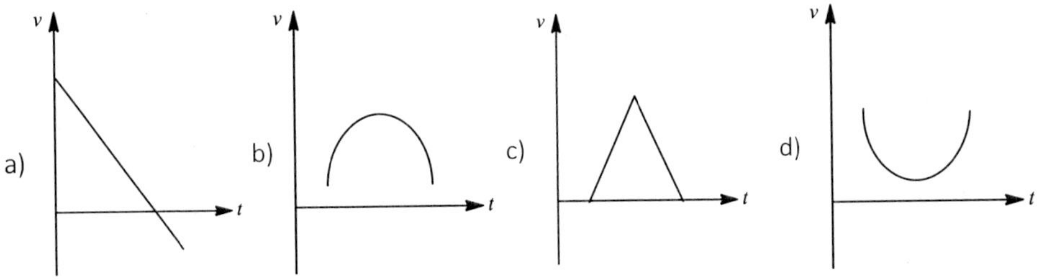

451. A ball is thrown vertically upwards from the top of a tower at 4.9 ms^{-1}.It strikes the pond near the base of the tower after 3 $seconds$. The height of the tower is
a) 73.5 m b)44.1 m c) 29.4 m d)None of these

452. A stone dropped from of the tower touches the ground in 4 sec. The height of the tower is aboout
a) 80 m b)40 m c) 20 m d)160 m

453. A body is thrown vertically upwards with velocity u. The distance travelled by it in the fifth and the sixth seconds are equal. The velocity u is given by ($g = 9.8\ m/s^2$)
a) 24.5 m/s b)49.0 m/s c) 73.5 m/s d)98.0 m/s

454. A man goes 10 m towards North, then 20 m towards east then displacement is
a) 22.5m b)25m c) 25.5m d)30m

455. Figure shows the acceleration-time graphs of a particle. Which of the following represents the corresponding velocity-time graphs?

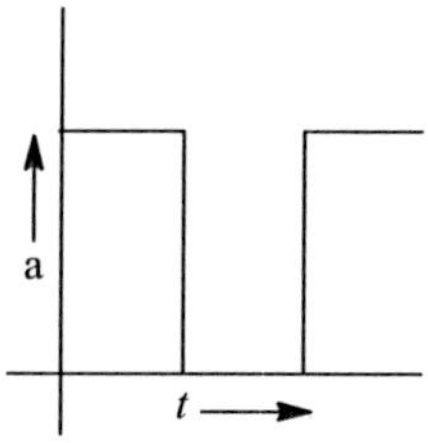

a) 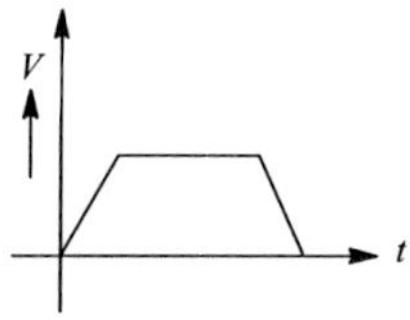

b) 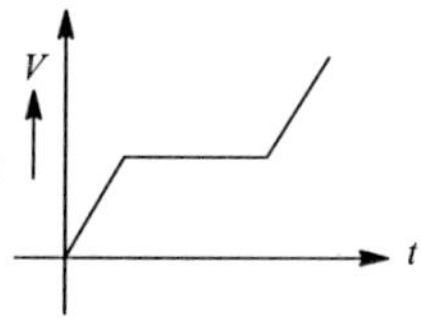

c) 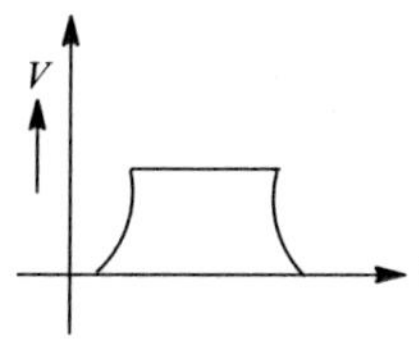

d) 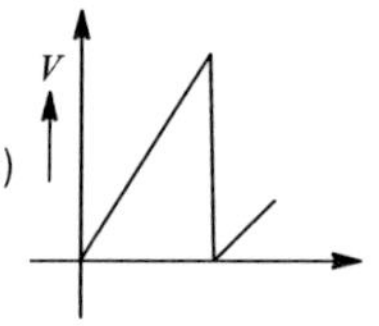

456. A body starting from rest moves with uniform acceleration. The distance covered by the body in time t is proportional to

a) $\sqrt{t}$ b) $t^{3/2}$ c) t^2 d) t^3

457. A body thrown vertically upwards with an initial velocity u reaches maximum height in 6 seconds. The ratio of the distances travelled by the body in the first second and the seventh second is

a) 1 : 1 b) 11 : 1 c) 1 : 2 d) 1 : 11

458. A particle is projected up with an initial velocity of 80 ft/sec. The ball will be at a height of 96 ft from the ground after

a) 2.0 and 3.0 sec b) Only at 3.0 sec c) Only at 2.0 sec d) After 1 and 2 sec

459. A ball is dropped from top of a tower of $100m$ height. Simultaneously another ball was thrown upward from bottom of the tower with a speed of 50 m/s ($g = 10\ m/s^2$). They will cross each other after

a) $1s$ b) $2s$ c) $3s$ d) $4s$

460. When a bullet is fired at a target, its velocity decreases by half after penetration 30 cm into it. The additional thickness it will penetrate before coming to rest is

a) 30 cm b) 40 cm c) 10 cm d) 50 cm

461. A particle located at $x = 0$ at time $t = 0$, starts moving along the positive x-direction with a velocity 'v' that varies as $v = a\sqrt{x}$. The displacement of the particle varies with time as

a) t b) $t^{1/2}$ c) t^3 d) t^2

462. A bus begins to move with an acceleration of $1ms^{-2}$. A man who is $48m$ behind the bus starts running at 10 ms^{-1} to catch the bus. The man will be able to catch the bus after

a) $6s$ b) $5s$ c) $3s$ d) $8s$

463. A man walks on a straight road from his home to a market 2.5 km away with a speed of 5 km/h. Finding the market closed, he instantly turns and walks back home with a speed of 7.5 km/h. The average speed of the man over the interval of time 0 to 40 min. Is equal to

a) 5 km/h b) $\frac{25}{4}$ km/h c) $\frac{30}{4}$ km/h d) $\frac{45}{8}$ km/h

464. The displacement of a particle undergoing rectilinear motion along the x-axis is given by $x = (2t^2 + 21t^2 + 60t + 6)$. The acceleration of the particle when its velocity is zero is

a) $36ms^{-2}$ b) $9ms^{-2}$ c) $-9ms^{-2}$ d) $-18ms^{-2}$

465. The $x-t$ graph shown in the figure represents

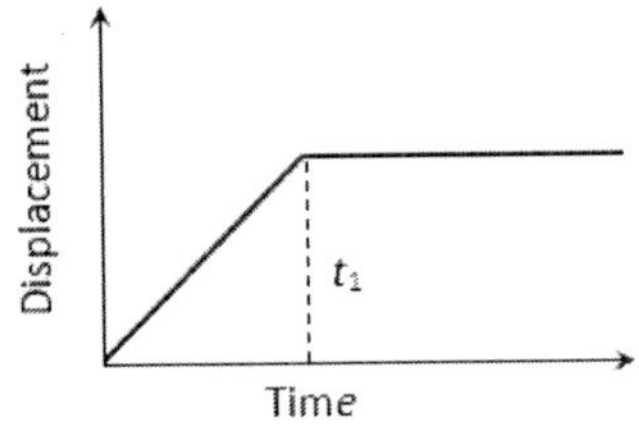

a) Constant velocity
b) Velocity of the body is continuously changing
c) Instantaneous velocity
d) The body travels with constant speed upto time t_1 and then stops

466. A particle moves for 20 *seconds* with velocity 3 m/s and then velocity 4 m/s for another 20 *seconds* and finally moves with velocity 5 m/s for next 20 *seconds*. What is the average velocity of the particle

a) 3 m/s b) 4 m/s c) 5 m/s d) Zero

467. A body moving with uniform acceleration, describes 40 m in the first 5 s and 65 m in next 5 s. its initial velocity will be

a) 4 ms^{-1} b) 2.5 ms^{-1} c) 3 ms^{-1} d) 11 ms^{-1}

468. A particle is moving with constant acceleration from A to B in a straight line AB. If u and v are the velocities at A and B respectively then its velocity at the midpoint C will be

a) $\left(\frac{u^2+v^2}{2u}\right)^2$ b) $\frac{u+v}{2}$ c) $\frac{v-u}{2}$ d) $\sqrt{\frac{u^2+v^2}{2}}$

469. A ball A is thrown up vertically with speed u and at the same instant another ball B is released from a height h. At time t, the speed of A relative to B is

a) u b) $2u$ c) $u-gt$ d) $\sqrt{(u^2-gt)}$

470. In the following graph, distance travelled by the body in metres is

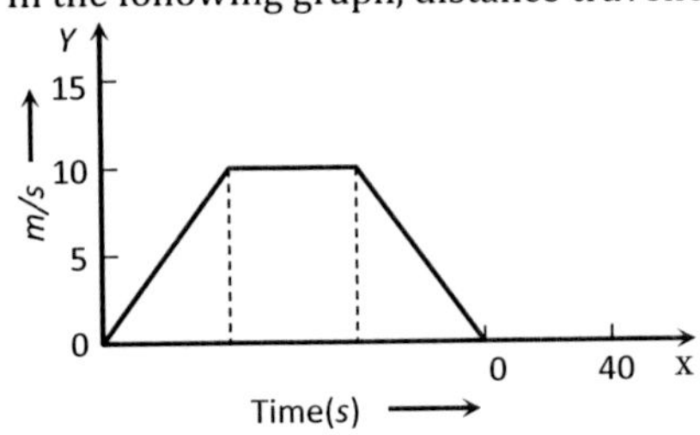

a) 200 b) 250 c) 300 d) 400

471. The ratio of the numerical values of the average velocity and average speed of a body is always

a) Unity b) Unity or less c) Unity or more d) Less than unity

472. Which of the following options is correct for the object having a straight line motion represented by the following graph?

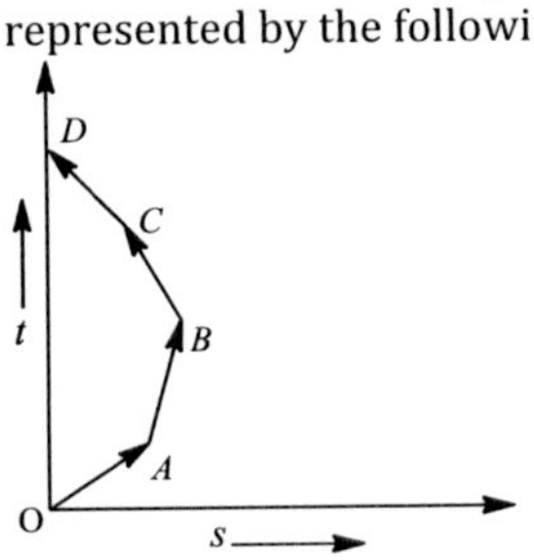

a) The object moves with constantly increasing velocity from O to A and then it moves with constant velocity.
b) Velocity of the object increases uniformly
c) Average velocity is zero
d) The graph shown is impossible

473. Velocity of a body on reaching the point from which it was projected upwards, is
a) $v = 0$ b) $v = 2u$ c) $v = 0.5u$ d) $v = u$

474. The acceleration a of a particle starting from rest varies with time according to relation $a = \alpha t + \beta$. The velocity of the particle after a time t will be
a) $\frac{\alpha t^2}{2} + \beta$ b) $\frac{\alpha t^2}{2} + \beta t$ c) $\alpha t^2 + \frac{1}{2}\beta t$ d) $\frac{(\alpha t^2 + \beta t)}{2}$

475. If the velocity of particle is given by $v = (180 - 16x)^{1/2}\ m/s$, then its acceleration will be
a) Zero b) $8\ m/s^2$ c) $-8\ m/s^2$ d) $4\ m/s^2$

476. A boat takes two hours to travel 8 km and back in still water. If the velocity of water 4 kmh^{-1}, the time taken for going ups tream 8km and coming back is
a) 2h b) 2 h 40 min
c) 1 h 20 min d) Cannot be estimated with the information given

477. A body freely falling from the rest has a velocity $'v'$ after it falls through a height 'h'. The distance it has to fall down for its velocity to become double, is
a) $2h$ b) $4h$ c) $6h$ d) $8h$

478. A ball is dropped on a floor and bounces back to a height somewhat less than the original height, which of the curves depicts its motion correctly?

a) 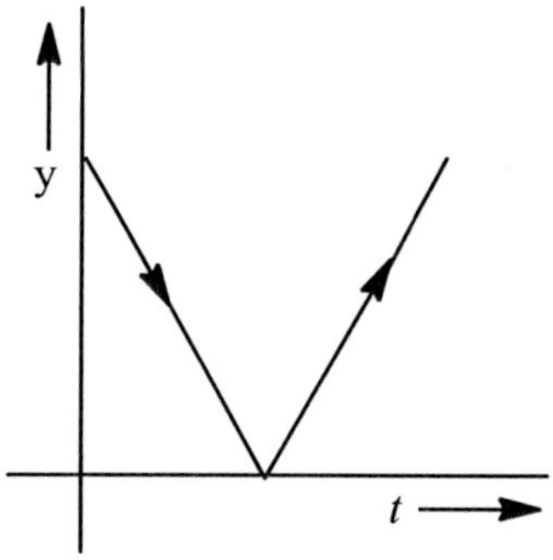

b) 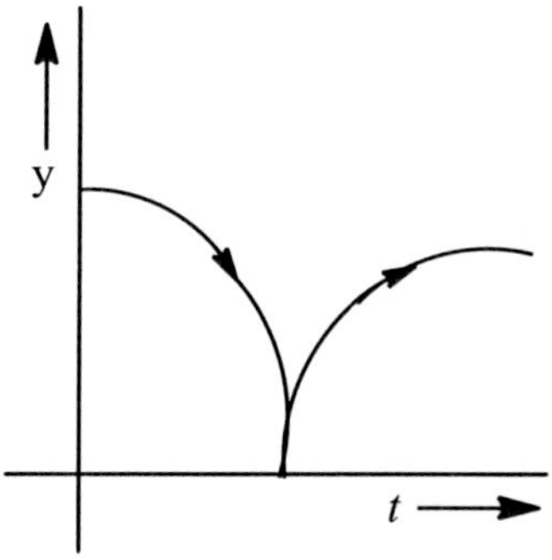

c) 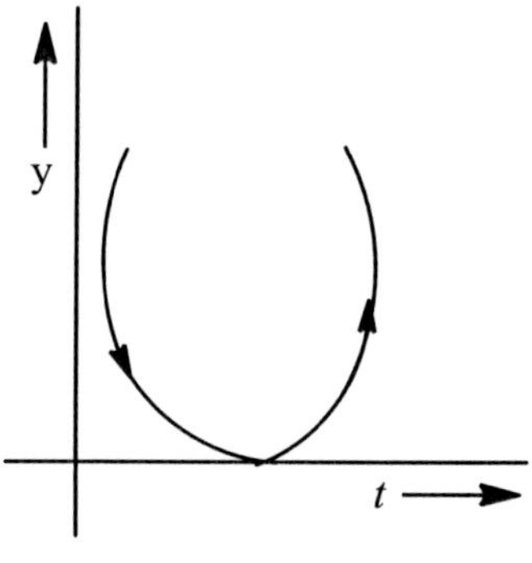

d) 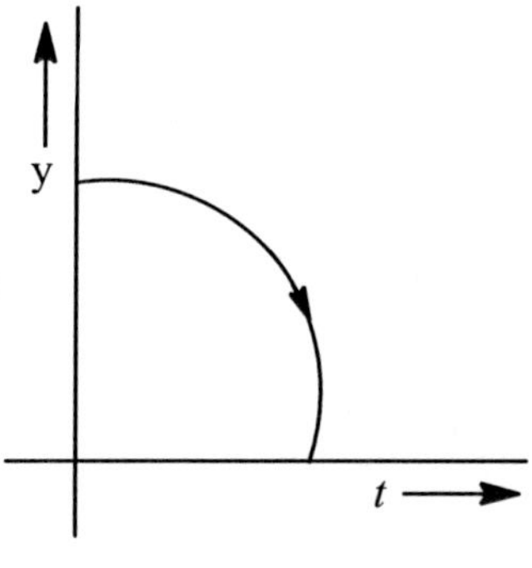

479. The velocity-time relation of an electron starting from rest is given by $v = kt$where $k = 2\text{ms}^{-1}$. The distance traversed in first 3 s is

a) 9 m b) 16 m c) 27 m d) 36 m

480. A particle moves for 20 s with velocity 3 ms^{-1} and then moves with velocity 4 ms^{-1} for another 20 s and finally moves with velocity 5 ms^{-1} for next 20 s. What is the average velocity of the particle?

a) 3 ms^{-1} b) 4 ms^{-1} c) 5 ms^{-1} d) Zero

481. The motion of a particle is described by the equation $x = a + bt^2$ where $a = 15\ cm$ and $b = 3\ cm/s^2$.Its instantaneous velocity at time 3 sec will be

a) $36\ cm/sec$ b) $18\ cm/sec$ c) $16\ cm/sec$ d) $32\ cm/sec$

482. From the top of a tower, a stone is thrown up and reaches the ground in time $t_1 = 9$s. A second stone is thrown down with the same speed and reaches the ground in time $t_2 =$ 4 s. A third stone is released from rest and reaches the ground in time t_3, which is equal to

a) 6.5 s b) 6.0 s c) $\frac{5}{36}$ s d) 65 s

483. A body is moving from rest under constant acceleration and let S_1 be the displacement in the first $(P - 1)$ sec and S_2 be the displacement in the first P sec. The displacement in $(P^2 - P + 1)^{th}$ see will be

a) $S_1 + S_2$ b) S_1S_2 c) $S_1 - S_2$ d) S_1 / S_2

484. For a moving body at any instant of time

a) If the body is not moving, the acceleration is necessarily zero
b) If the body is slowing, the retardation is negative
c) If the body is slowing, the distance is negative
d) If displacement, velocity and acceleration at that instant are known, we can find the displacement at any given time in future

485. A particle is projected up with an initial velocity of $80\ ft/sec$. The ball will be at a height of $96\ ft$ from the ground after

a) 2.0 and 3.0 sec b) Only at 3.0 sec c) Only at 2.0 sec d) After 1 and 2 sec

486. The $x - t$ graph shown in the figure represents

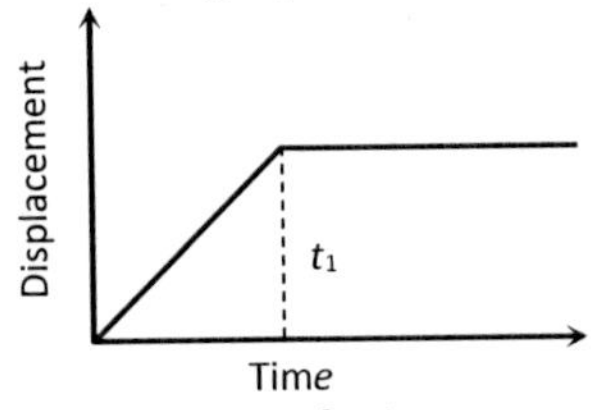

a) Constant velocity
b) Velocity of the body is continuously changing
c) Instantaneous velocity
d) The body travels with constant speed upto time t_1 and then stops

487. The distance x covered by a particle in one-dimensional motion varies with time $t$as $x^2 = at^2 + 2bt + c$. The accelerating of the particle varies as

a) x^{-3} b) $x^{3/2}$ c) x^2 d) $x^{-2/3}$

488. A body of 5 kg is moving with a velocity of $20m/s$. If a force of $100N$ is applied on it for $10s$in the same direction as its velocity, what will now be the velocity of the body

a) $200\ m/s$ b) $220\ m/s$ c) $240\ m/s$ d) $260\ m/s$

489. A body moves 6m north, 8m east and 10m vertically upwards, what is its resultant displacement from initial position?

a) $10\sqrt{2}$m b) 10m c) $\frac{10}{\sqrt{2}}$m d) 10×2 m

490. Two trains travelling on the same track are approaching each other with equal speeds of $40 m/s$. The drivers of the trains begin to decelerate simultaneously when they are just $2.0 km$ apart. Assuming the decelerations to be uniform and equal, the value of the deceleration to barely avoid collision should be

a) $11.8\ m/s^2$ b) $11.0\ m/s^2$ c) $2.1\ m/s^2$ d) $0.8\ m/s^2$

491. A car starts from rest and moves with uniform acceleration a on a straight road from time $t = 0$ to $t = T$. After that, a constant deceleration brings it to rest. In this process the average speed of the car is

a) $\frac{aT}{4}$ b) $\frac{3aT}{2}$ c) $\frac{aT}{2}$ d) aT

492. A stone falls freely from rest and the total distance covered by it in the last second of its motion equals the distance covered by it in the first three seconds of its motion. The stone remains in the air for

a) $6\ s$ b) $5\ s$ c) $7\ s$ d) $4\ s$

493. The numerical ratio of average velocity to average speed is

a) Always less than one b) Always equal to one
c) Always more than one d) Equal to or less than one

494. A body sliding on a smooth inclined plane required 4s to reach the bottom, starting from rest at the top. How much time does it takes to cover one-fourth the distance starting from rest at top?

a) 1 s b) 2 s c) 4 s d) 16 s

495. A man walks on a straight road from his home to a market $2.5\ km$ away with a speed of $5\ km/h$. Finding the market closed, he instantly turns and walks back home with a speed of $7.5\ km/h$. The average speed of the man over the interval of time 0 to $40\ min$. Is equal to

a) $5\ km/h$ b) $\frac{25}{4}\ km/h$ c) $\frac{30}{4}\ km/h$ d) $\frac{45}{8}\ km/h$

496. A stone is shot straight upward with a speed of $20\ m/sec$ from a tower $200\ m$ high. The speed with which it strikes the ground is approximately

a) $60\ m/sec$ b) $65\ m/sec$ c) $70\ m/sec$ d) $75\ m/sec$

497. A particle starts from rest. Its acceleration (a) versus time (t) is as shown in the figure. The maximum speed of the particle will be

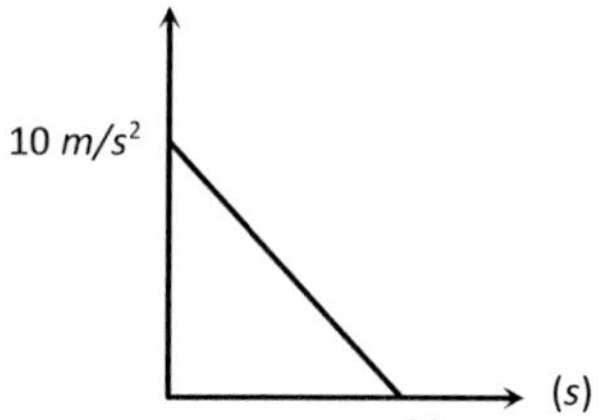

a) $110\ m/s$ b) $55\ m/s$ c) $550\ m/s$ d) $660\ m/s$

498. A bullet moving with a speed of $100\ ms^{-1}$ can just penetrate two planks of equal thickness. Then, the number of such planks penetrated by the same bullet when the speed is doubled will be

a) 6 b) 10 c) 4 d) 8

499. A particle starts its motion from rest under the action of a constant force. If the distance covered in first 10 *seconds* is S_1 and that covered in the first 20 *seconds* is S_2, then

a) $S_2 = 2S_1$ b) $S_2 = 3S_1$ c) $S_2 = 4S_1$ d) $S_2 = S_1$

500. Two balls A and B are thrown simultaneously from the top of a tower. A is thrown vertically up with a speed of $4 ms^{-1}$. B is thrown vertically down with a speed of $4\ ms^{-1}$. The ball A and B hit the ground with speed v_A and v_B respectively. Then,

a) $v_A < v_B$ b) $v_A > v_B$ c) $v_A \gtreqless v_B$ d) $v_A = v_B$

501. A frictionless wire AB is fixed on a sphere of radius R. A very small spherical ball slips on this wire. The time taken by this ball to slip from A to B is

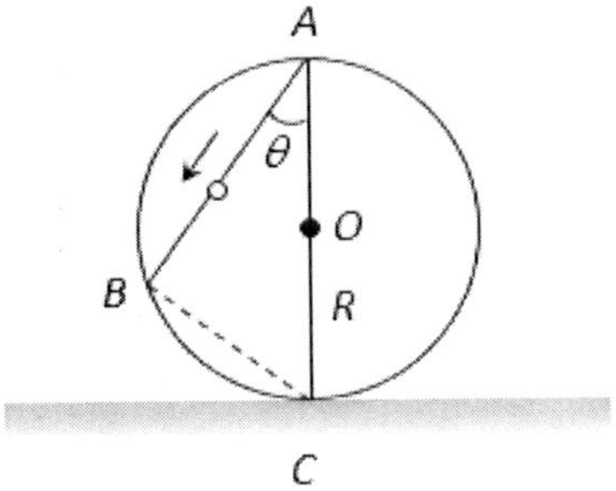

a) $\frac{2\sqrt{gR}}{g\cos\theta}$ b) $2\sqrt{gR}.\frac{\cos\theta}{g}$ c) $2\sqrt{\frac{R}{g}}$ d) $\frac{gR}{\sqrt{g\cos\theta}}$

502. A student is standing at a distance of 50 m from the bus. As soon as the bus begins its motion with an acceleration of 1 ms^{-2}, the students starts running towards the bus with a uniform velocity u. Assuming the motion to be along a straight road, the minimum value of u, so that the student is able to catch the bus is

a) 8 ms^{-1} b) 5 ms^{-1} c) 12 ms^{-1} d) 10 ms^{-1}

503. A stone thrown upward with a speed u from the top of the tower reaches the ground with a velocity $3u$. The height of the tower is

a) $3u^2/g$ b) $4u^2/g$ c) $6u^2/g$ d) $9u^2/g$

504. $v-t$ graph for a particle is as shown. The distance travelled in the first 4 s is

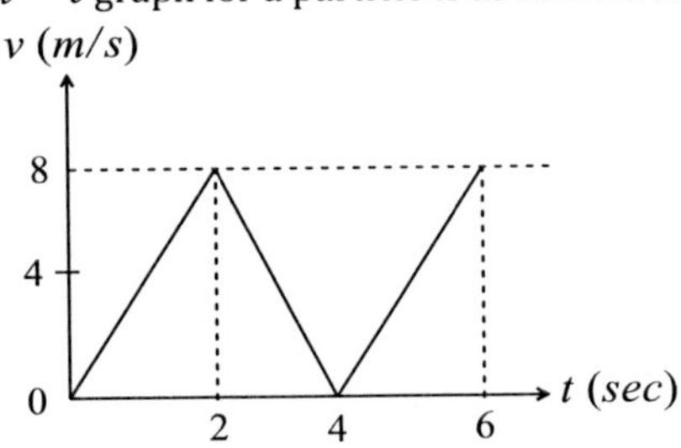

a) $12m$ b) $16m$ c) $20m$ d) $24m$

505. The graph between the displacement x and t for a particle moving in a straight line is shown in figure. During the interval OA, AB, BC and CD, the acceleration of the particle is

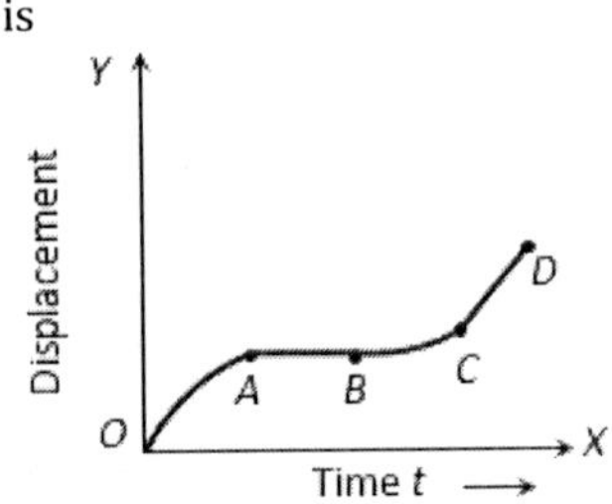

OA, AB, BC, CD

a) + 0 + +
b) - 0 + 0
c) + 0 - +
d) - 0 - 0

506. A particle starts from rest at $t = 0$ and moves in a straight line with an acceleration as shown below. The velocity of the particle at $t = 3s$ is

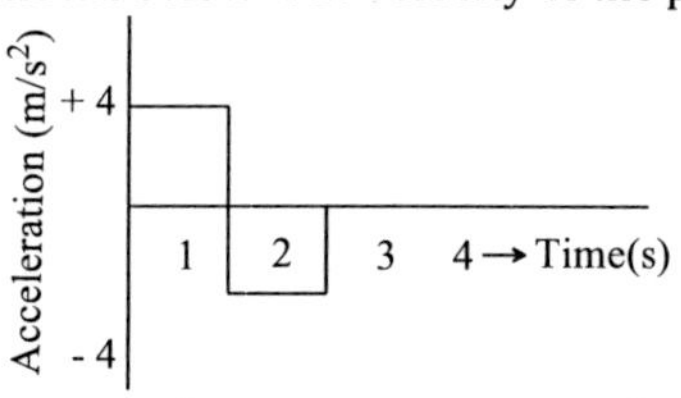

a) $2\ ms^{-1}$ b) $4\ ms^{-1}$ c) $6\ ms^{-1}$ d) $8\ ms^{-1}$

507. A particle is thrown vertically upwards. If it velocity at half of the maximum height is $10\ m/sec$, then maximum height attained by it is (Take $g = 10\ m/\sec^2$)

a) $8\ m$ b) $10\ m$ c) $12\ m$ d) $16\ m$

508. If a car covers $2/5^{th}$ of the total distance with v_1 speed and $3/5^{th}$ distance with v_2 then average speed is

a) $\frac{1}{2}\sqrt{v_1 v_2}$ b) $\frac{v_1 + v_2}{2}$ c) $\frac{2v_1 v_2}{v_1 + v_2}$ d) $\frac{5v_1 v_2}{3v_1 + 2v_2}$

509. In the given $v - t$ graph, the distance travelled by the body in 5 will be

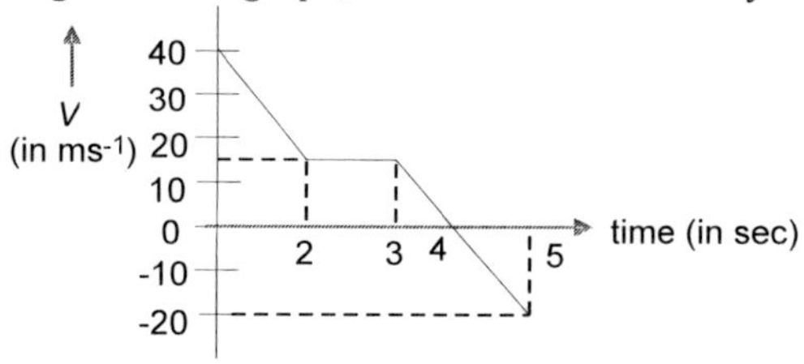

a) 20 m b) 40 m c) 80 m d) 100 m

510. A car starts from station and moves along the horizontal road by a machine delivering constant power. The distance covered by the car in time t is proportional to

a) t^2 b) $t^{3/2}$ c) $t^{2/3}$ d) t^3

511. The displacement-time graph of moving particle is shown below

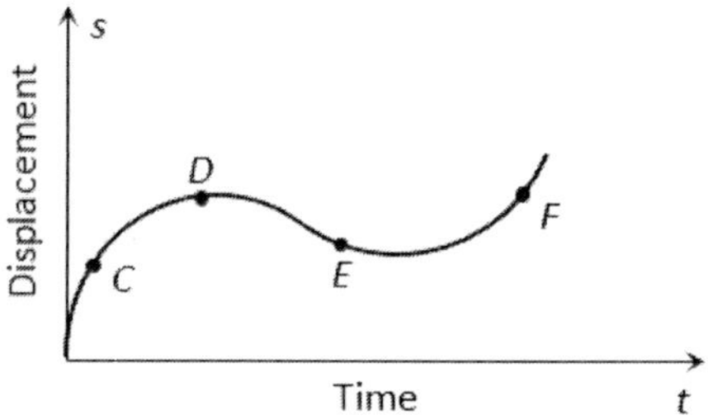

The instantaneous velocity of the particle is negative at the point

a) D b) F c) C d) E

512. A body starts from rest, with uniform acceleration a. The acceleration of a body as function of time t is given by the equation $a = pt$ where p is constant, then the displacement of the particle in the time interval $t = 0$ to $t = t_1$ will be

a) $\frac{1}{2}pt_1^3$ b) $\frac{1}{3}pt_1^2$ c) $\frac{1}{4}pt_1^2$ d) $\frac{1}{6}pt_1^3$

513. A particle has an initial velocity $3\hat{i} + 4\hat{j}$and an acceleration of $0.4\hat{i} + 0.3\hat{j}$. Its speed after 10 s is

a) 10 unit b) 7 unit c) $7\sqrt{2}$unit d) 8.5 unit

514. The distance-time graphs of a particle at time t makes angle 45° with the time axis. After two seconds, it makes an angle 60° with the time axis. What is the average acceleration of the particle?
a) 1/2 b) $\sqrt{3}/2$ c) $(\sqrt{3}-1)/2$ d) $(\sqrt{3}+1)/2$

515. What is the relation between displacement, time and acceleration in case of a body having uniform acceleration
a) $S = ut + \frac{1}{2}ft^2$ b) $S = (u+f)\,t$ c) $S = v^2 - 2fs$ d) None of these

516. A particle has an initial velocity of $3\hat{\imath} + 4\hat{\jmath}$ and an acceleration of $0.4\hat{\imath} + 0.3\hat{\jmath}$. Its speed after 10 s is
a) 10 units b) $7\sqrt{2}$ units c) 7 units d) 8.5 units

517. Two identical metal spheres are released from the top of a tower after t seconds of each other such that they fall along the same vertical line. If air resistance is neglected, then at any instant of time during their fall
a) The difference in their displacements remains the same
b) The difference between their speeds remains the same
c) The difference between their heights above ground is proportional to t^2
d) The difference between their displacements is proportional to t

518. A ball is dropped downwards. After 1 second another ball is dropped downwards from the same point. What is the distance between them after 3 seconds
a) 25 m b) 20 m c) 50 m d) 9.8 m

519. A ball released from the top of a tower travels $\frac{11}{36}$ of the height of the tower in the last second of its journey. The height of the tower is (Take $g = 10 = ms^{-2}$)
a) 11m b) 36m c) 47m d) 180m

520. A body moves from rest with a constant acceleration of 5 m/s^2. Its instantaneous speed (in m/s) at the end of 10 sec is
a) 50 b) 5 c) 2 d) 0.5

521. An athlete completes one round of a circular track of radius R in 40 sec. What will be his displacement at the end of 2 min. 20 sec
a) Zero b) $2R$ c) $2\pi R$ d) $7\pi R$

522. The graph of displacement v/s time is

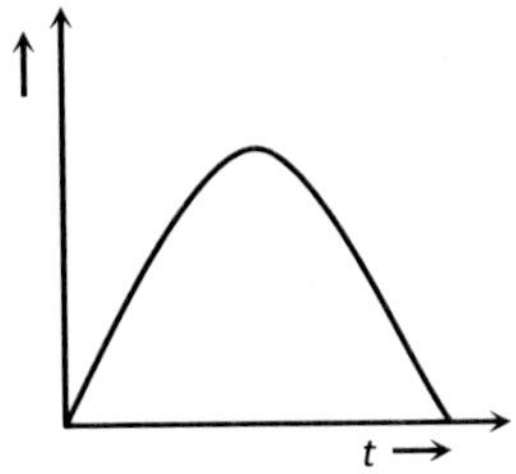

Its corresponding velocity-time graph will be

a)
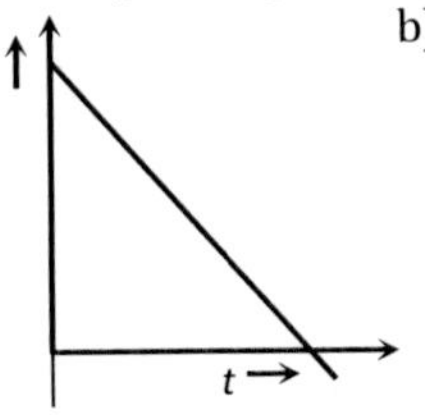

b)
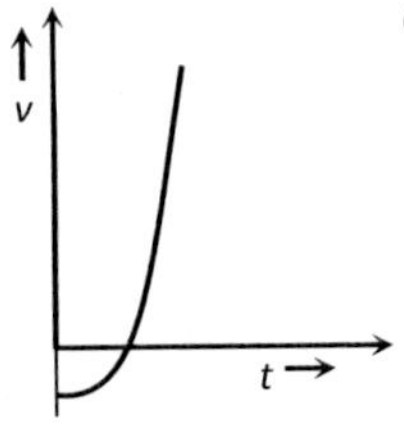

c)
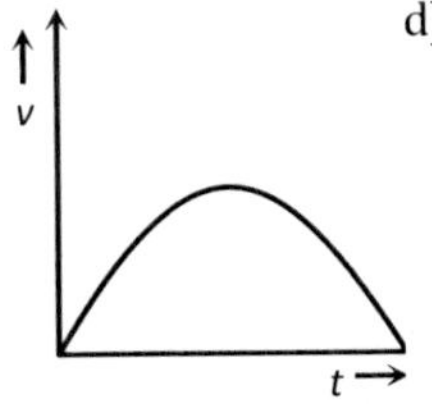

d)
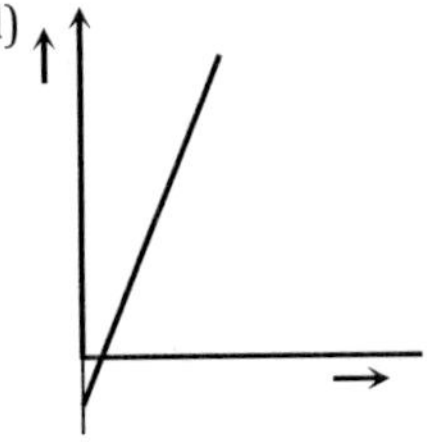

523. The displacement of a particle starting from rest ($at\ t = 0$) is given by $s = 6t^2 - t^3$. The time in seconds at which the particle will attain zero velocity again, is
a) 2 b) 4 c) 6 d) 8

524. Two balls A and B of same masses are thrown from the top of the building. A, thrown upward with velocity V and B, thrown downward with velocity V, then

a) Velocity of A is more than B at the ground
b) Velocity of B is more than A at the ground
c) Both $A\&B$ strike the ground with same velocity
d) None of these

525. A body travels for 15 sec starting from rest with constant acceleration. If it travels distances S_1, S_2 and S_3 in the first five seconds, second five secods and next five seconds respectively the relation between S_1, S_2 and S_3 is

a) $S_1 = S_2 = S_3$
b) $5S_1 = 3S_2 = S_3$
c) $S_1 = \frac{1}{3}S_2 = \frac{1}{5}S_3$
d) $S_1 = \frac{1}{5}S^2 = \frac{1}{3}S_3$

526. A balloon starts rising from the ground with an acceleration of 1.25 m/s^2 after 8s, a stone is released from the balloon. The stone will $(g - 10\ m/s^2)$

a) Reach the ground in 4 second
b) Begin to move down after being released
c) Have a displacement of 50 m
d) Cover a distance of 40 m in reaching the ground

527. A ball is thrown up under gravity $(g = 10\ m/sec^2)$. Find its velocity after 1.0 sec at a height of 10m

a) 5 $m/\sec^2$
b) 5 m/sec
c) 10 m/sec
d) 15 m/sec

528. The area under acceleration-time graph gives

a) Distance travelled
b) Change in acceleration
c) Force acting
d) Change in velocity

529. From the top of a tower two stones, whose masses are in the ratio 1: 2 are thrown one straight up with an initial speed u and the second straight down with the same speed u. Then, neglecting air resistance

a) The heavier stone hits the ground with a higher speed
b) The lighter stone hits the ground with a higher speed
c) Both the stones will have the same speed when they hit the ground
d) The speed can't be determined with the give data

530. An aircraft is flying at a height of 34000m above the ground. If the angle subtended at a ground observation point by the aircraft positions 10s apart is 30°, then the speed of the aircraft is

a) 19.63ms^{-1}
b) 1963 ms^{-1}
c) 108 ms^{-1}
d) 196.3 ms^{-1}

531. Two trains travelling on the same track are approaching each other with equal speeds of 40m/s. The drivers of the trains begin to decelerate simultaneously when they are just 2.0km apart. Assuming the decelerations to be uniform and equal, the value of the deceleration to barely avoid collision should be

a) 11.8 m/s^2
b) 11.0 m/s^2
c) 2.1 m/s^2
d) 0.8 m/s^2

532. A steam boat goes across a lake and comes back (i) on a quiet day when the water is still and (ii) on a rough day when there is a uniform current so as to help the journey onwards and to impede the journey back. If the speed of the launch on both days was same, the time required for complete journey on the rough day, as compared to the quiet day will be

a) More
b) Less
c) Same
d) None of these

533. The x and$y$ coordinates of a particles at any time t are given by $x = 7t + 4t^2$ and $y = 5t$, where x and y are in metre and t in second. The acceleration of particle at $t = 5$s is

a) Zero
b) 8ms^{-2}
c) 20 ms^{-2}
d) 40 ms^{-2}

534. A body projected vertically upwards with velocity u returns to the starting point in 4 seconds. If $g = 10\ m/\sec^2$, the value of u is

a) 5 m/sec
b) 10 m/sec
c) 15 m/sec
d) 20 m/sec

535. A ball is dropped vertically from a height d above the ground. It hits the ground and bounces up vertically to a height $d/2$. Neglecting subsequent motion and air resistance, its velocity v varies with the height h above the ground as

a) 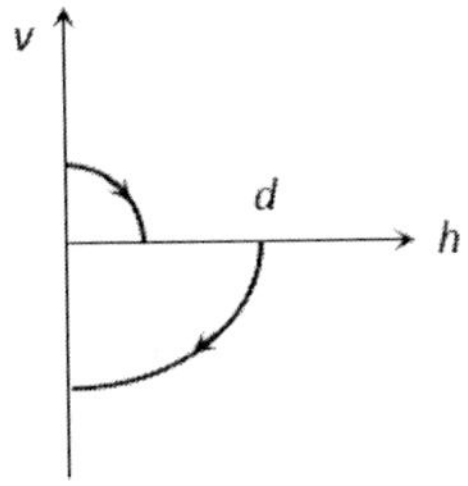

b) 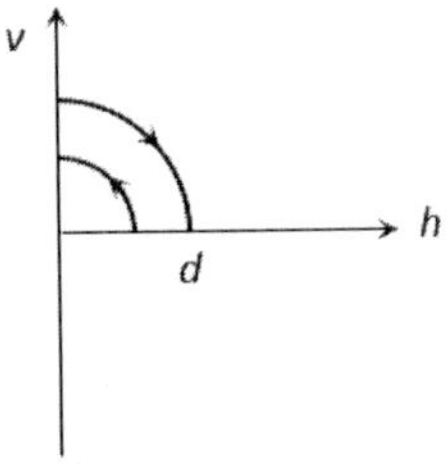

c) 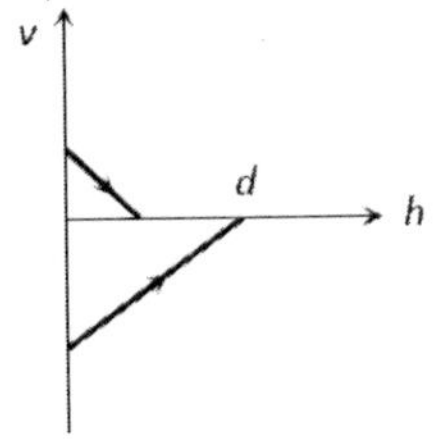

d) 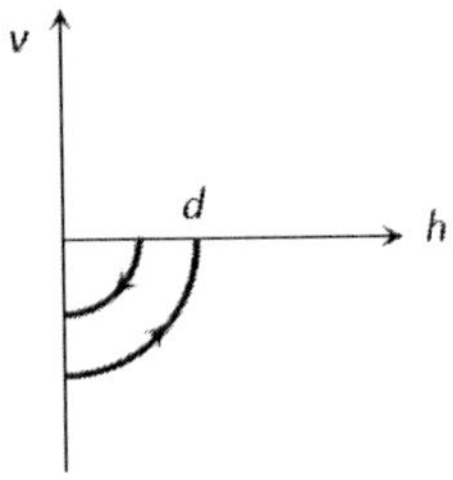

536. A body A is thrown up vertically from the ground with a velocity V_0 and another body B is simultaneously dropped from a height H. They meet at a height $\frac{H}{2}$ if V_0 is equal to

a) $\sqrt{2gH}$ b) $\sqrt{gH}$ c) $\frac{1}{2}\sqrt{gH}$ d) $\sqrt{\frac{2g}{H}}$

537. The displacement of a particle starting from rest $(at\ t = 0)$ is given by $s = 6t^2 - t^3$. The time in seconds at which the particle will attain zero velocity again, is

a) 2 b) 4 c) 6 d) 8

538. A particle covers 4m, 5m, 6m, and 7m, in 3rd, 4th, 5th and 6th second respectively. The particle starts

a) With an initial non-zero velocity and moves with uniform acceleration
b) From rest and moves with uniform velocity
c) With an initial velocity and moves with uniform velocity
d) From rest and moves with uniform acceleration

539. The velocity-time and acceleration-time graphs of a particle are given as

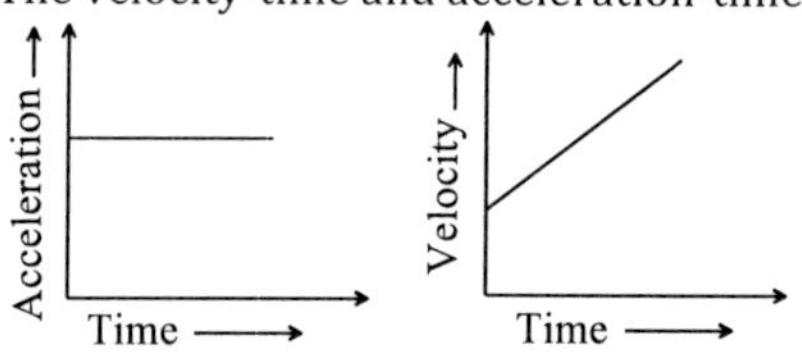

Its position-time graph may be given as

a) 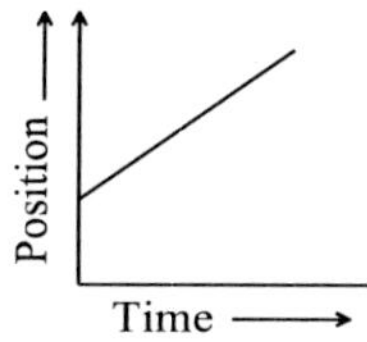

b) 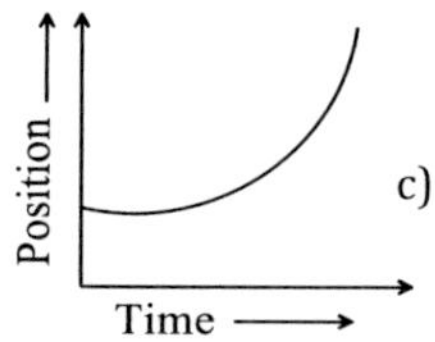

c) 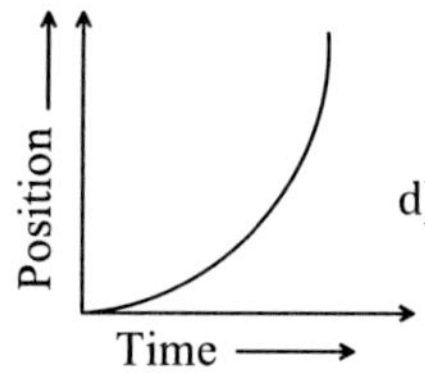

d) 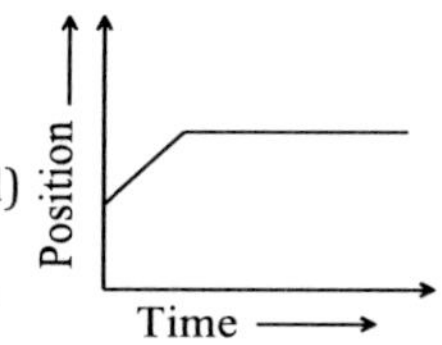

540. A boat crosses a river from port A to port B, which are just on the opposite side. The speed of the water is V_W and that of boat is V_B relative to still water. Assume $V_B = 2V_W$. What is the time taken by the boat, if It has to cross the river directly on the AB line

a) $\frac{2D}{V_B\sqrt{3}}$ b) $\frac{\sqrt{3}D}{2V_B}$ c) $\frac{D}{V_B\sqrt{2}}$ d) $\frac{D\sqrt{2}}{V_B}$

541. The displacement of the body along $x-$ axis depends on time as $\sqrt{x} = t + 1$. Then the velocity of body

a) Increases with time b) Decreases with time c) Independent of time d) None of these

542. An aeroplane files around a square field ABCD of each side 1000 km. Its speed along AB is 250 $km\ h^{-1}$, along DA 100 $km\ h^{-1}$. Its average speed (in $km\ h^{-1}$) over the entire trip is

a) 225.5 b) 175.5 c) 125.5 d) 190.5

543. A thief is running away on a straight road in a jeep moving with a speed of 9ms^{-1}. A police man chases him on a motor cycle moving at a speed of 10ms^{-1}. If the instantaneous. Separation of the jeep from the motor cycle is 100m, how long will it take for the police man to catch the theif.

a) 1 s b) 19 s c) 90 s d) 100 s

544. A body starting from rest moves with constant acceleration. The ratio of distance covered by the body during the 5th second to that covered in 5 s is

a) $\frac{9}{25}$ b) $\frac{3}{5}$ c) $\frac{25}{5}$ d) $\frac{1}{25}$

545. Two cars move in the same direction along parallel roads. One of them is a 100 m long travelling with a velocity of 7.5ms^{-1}. How long will it take for the first car to overtake the second car?

a) 24 s b) 40 s c) 60 s d) 80 s

546. Two bodies are thrown simultaneously from a tower with same initial velocity v_0: one vertically upwards, the other vertically downwards. The distance between the two bodies after time t is

a) $2v_0t + \frac{1}{2}gt^2$ b) $2v_0t$ c) $v_0t + \frac{1}{2}gt^2$ d) v_0t

547. A body falls from rest in the gravitational field of the earth. The distance travelled in the fifth second of its motion is $(g = 10\ m/s^2)$

a) $25m$ b) $45m$ c) $90m$ d) $125m$

548. The initial velocity of a body moving along a straight line is 7 m/s. It has a uniform acceleration of 4 m/s^2. The distance covered by the body in the 5^{th}second of its motion is

a) 25 m b)35 m c) 50 m d)85 m

549. A body is thrown vertically upwards with a velocity u.Find the true statement from the following

a) Both velocity and acceleration are zero at its highest point

b) Velocity is maximum and acceleration is zero at the highest point

c) Velocity is maximum and acceleration is g downwards at its highest point

d) Velocity is zero at the highest point and maximum height reached is $u^2/2g$

550. $v - t$ graph for a particle is as shown. The distance travelled in the first 4 s is

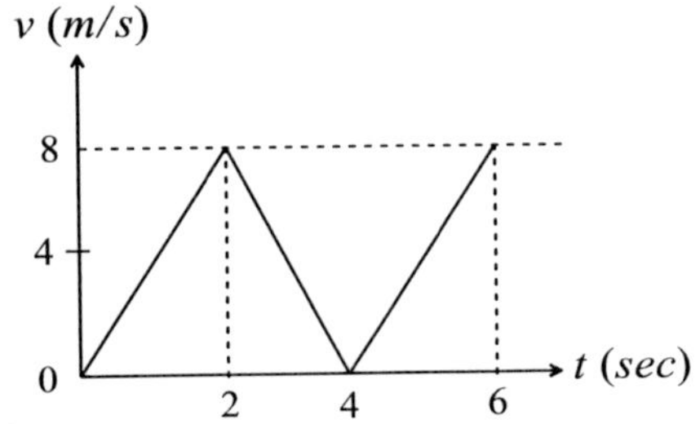

a) 12m b)16m c) 20m d)24m

551. A train is moving slowly on a straight track with a constant speed of 2 ms^{-1}. A passenger in that train starts walking at a steady aped of a 2 ms^{-1} to the back of the train in the opposite direction of the motion of the train. So to an observer standing on the platform directly in front of that passenger. The velocity of the passenger appears to be

a) 4 ms^{-1} b)2 ms^{-1}

c) 2 ms^{-1} in the opposite direction of the train d)Zero

552. A particle moves along a semicircle of radius 10$m$in 5 seconds. The average velocity of the particle is

a) 2π ms^{-1} b)4π ms^{-1} c) 2 ms^{-1} d)4 ms^{-1}

553. The effective acceleration of a body, when thrown upwards with acceleration a will be:

a) $\sqrt{a - g^2}$ b)$\sqrt{a^2 + g^2}$ c) $(a - g)$ d)$(a + g)$

554. Three different objects of masses m_1, m_2 and m_3 are allowed to fall from rest and from the same point 'O' along three different frictionless paths. The speeds of the three objects, on reaching the ground, will be in the ratio of

a) $m_1 : m_2 : m_3$ b)$m_1 : 2m_2 : 3m_3$ c) $1 : 1 : 1$ d)$\frac{1}{m_1} : \frac{1}{m_2} : \frac{1}{m_3}$

555. The velocity-time graph of a particle moving along a straight line is shown in figure. The displacement of the body in 5s is

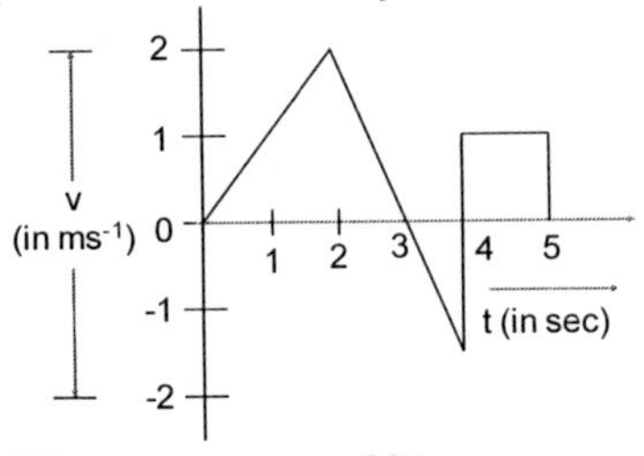

a) 0.5m b)1m c) 2m d)3m

556. Two trains travelling on the same track are approaching each other with equal speeds of $40 ms^{-1}$. The drivers of the trains begin to decelerate simultaneously when they are just 2 km apart. If the decelerations are both uniform and equal, then the value of deceleration to barely avoid collision should be

a) $0.8 ms^{-2}$ b) $2.1 ms^{-2}$ c) $11.0 ms^{-2}$ d) $13.2 ms^{-2}$

557. A particle starts from rest. Its acceleration (a) versus time (t) is as shown in the figure. The maximum speed of the particle will be

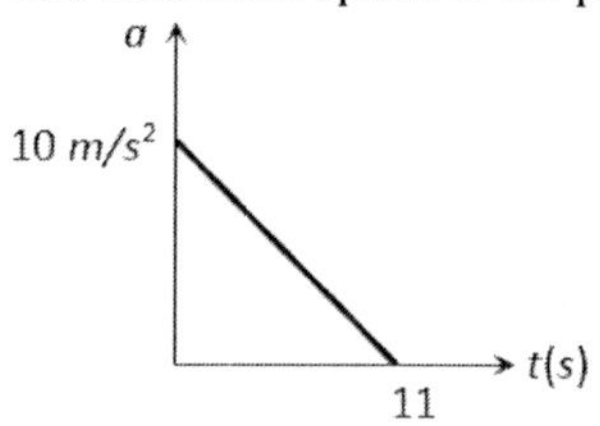

a) $110\ m/s$ b) $55\ m/s$ c) $550\ m/s$ d) $660\ m/s$

558. The velocity of a particle moving in a straight line varies with time in such a manner that v *versus* t graph is velocity is v_m and the total time of motion is t_0

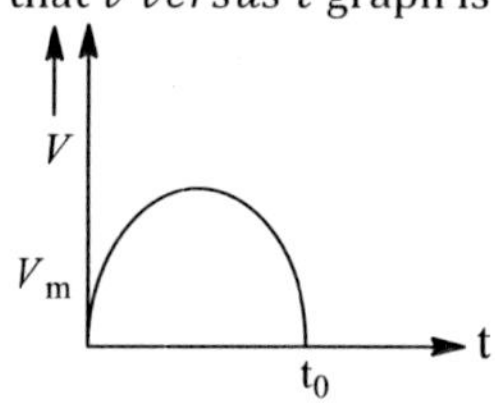

(i) Average velocity of the particle is $\frac{\pi}{4} v_m$

(ii)such motion cannot be realized in practical terms

a) Only (i) is correct b) Only (ii) is correct

c) Both (i)and (ii) are correct d) Both (i)and (ii) are wrong

559. Which of the following $v - t$ graphs represents the motion of a ball falling freely from rest under gravity and rebounding from a metallic surface?

a) 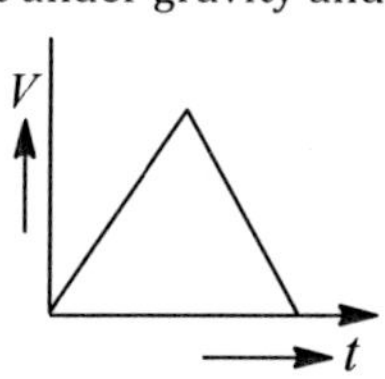

b) 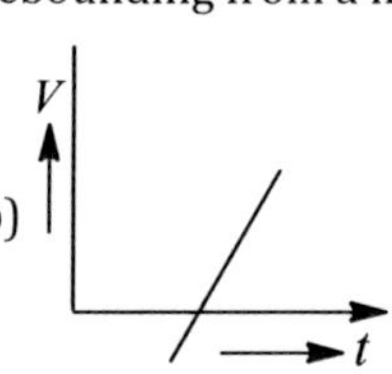

c) 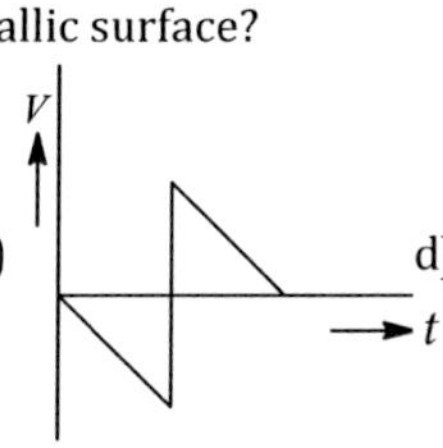

d) 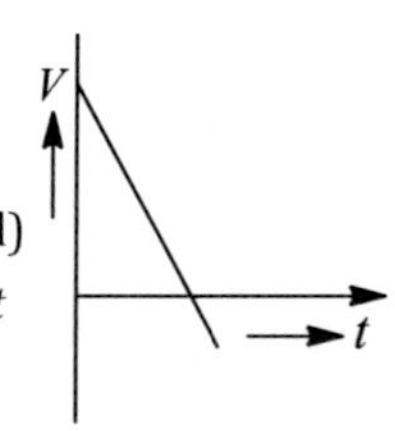

560. A train moves from one station to another 2 hours time. Its speed-time graph during this motion is shown in the figure. The maximum acceleration during the journey is

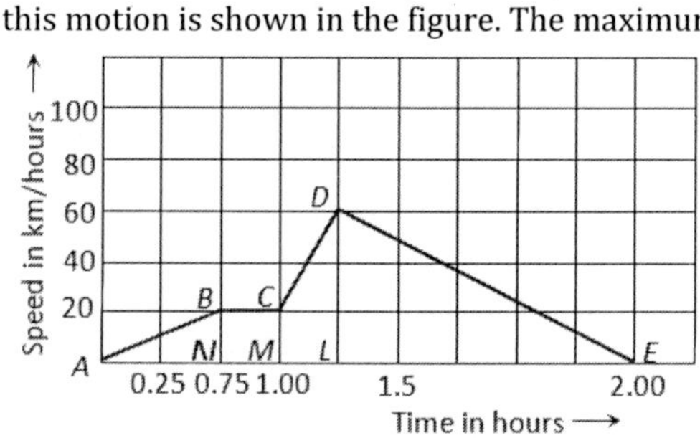

a) $140\ km\ h^{-2}$

b) $160\ km\ h^{-2}$

c) $100\ km\ h^{-2}$
d) $120\ km\ h^{-2}$

561. Which of the following is a one dimensional motion
a) Landing of an aircraft
b) Earth revolving around the sun
c) Motion of wheels of moving train
d) Train running on a straight track

562. The distance travelled by an object along a straight line in time t is given by $s = 3 - 4t + 5t^2$, the initial velocity of the object is
a) 3 unit
b) -3 unit
c) 4 unit
d) -4 unit

563. A stone dropped from of the tower touches the ground in 4 sec. The height of the tower is aboout
a) $80\ m$
b) $40\ m$
c) $20\ m$
d) $160\ m$

564. Which graph represents the uniform acceleration

a) 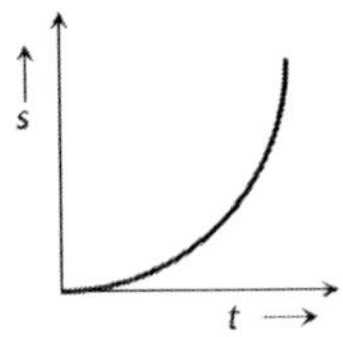

b) 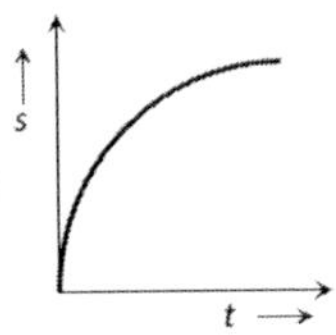

c) 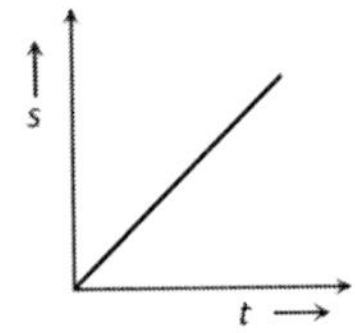

d) 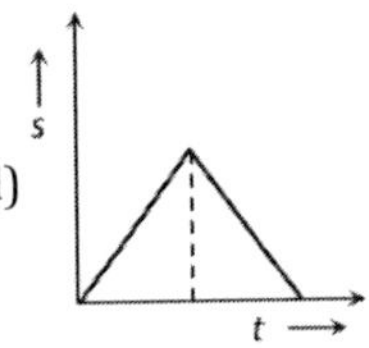

565. Equation of displacement for any particle is $s = 3t^3 + 7t^2 + 14t + 8m$. Its acceleration at time $t = 1$ sec is
a) $10\ m/s^2$
b) $16\ m/s^2$
c) $25\ m/s^2$
d) $32\ m/s^2$

566. Two particles held at different heights a and b above the ground are allowed to fall from rest. The ratio of their velocities on reaching the ground is
a) $a : b$
b) $\sqrt{a} : \sqrt{b}$
c) $a^2 : b^2$
d) $a^3 : b^3$

567. A body is moving according to the equation $x = at + bt^2 - ct^3$ where x = displacement and a, b and c are constants. The acceleration of the body is
a) $a + 2bt$
b) $2b + 6ct$
c) $2b - 6ct$
d) $3b - 6ct^2$

568. A particle moves for 20 $seconds$ with velocity 3 m/s and then velocity 4 m/s for another 20 $seconds$ and finally moves with velocity 5 m/s for next 20 $seconds$. What is the average velocity of the particle
a) $3\ m/s$
b) $4\ m/s$
c) $5\ m/s$
d) Zero

569. A body is thrown vertically up from the ground. It reaches a maximum height of $100m$ in $5sec$. After what time it will reach the ground from the maximum height position
a) 1.2 sec
b) 5 sec
c) 10 sec
d) 25 sec

570. An object is dropped from rest. Its v-t graph is

a) 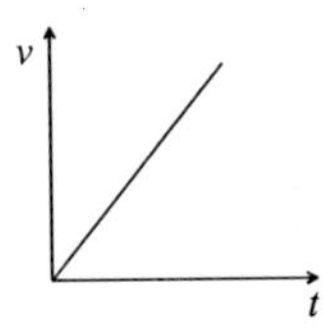

b) 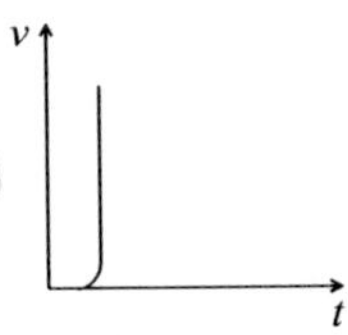

c) 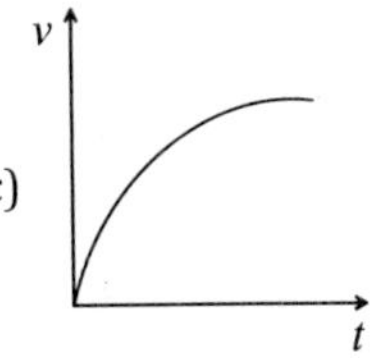

d) 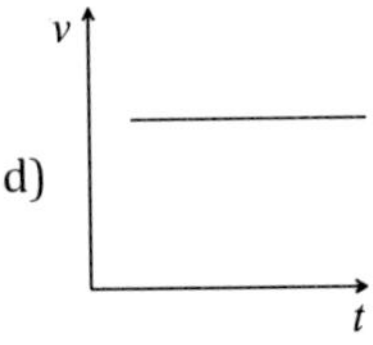

571. Which of the following graph represents uniform motion

a) 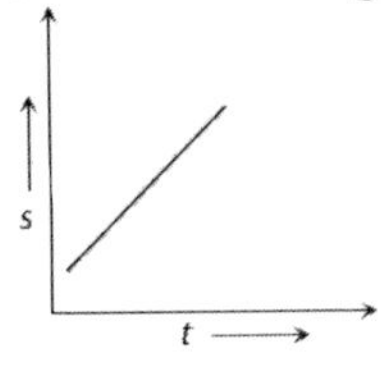

b) 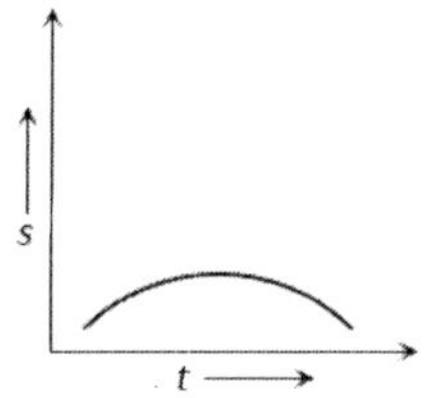

c) 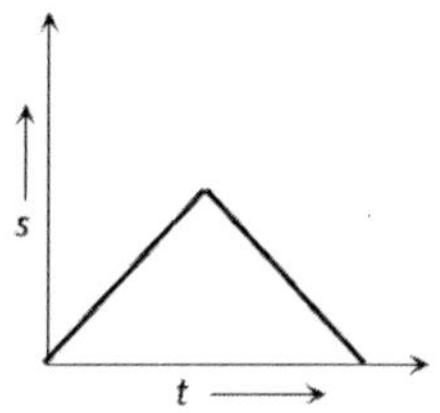

d) 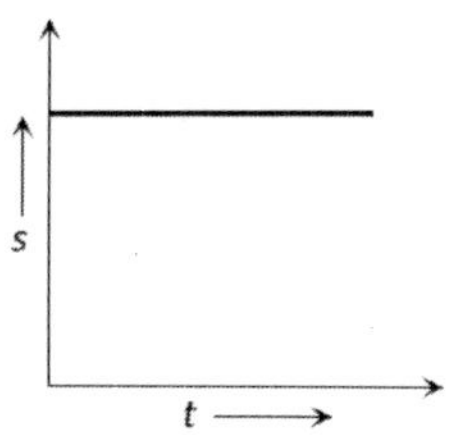

572. A particle is projected upwards. The times corresponding to height h while ascending and while descending are t_1 and t_2 respectively. The velocity of projection will be

a) gt_1 b) gt_2 c) $g(t_1 + t_2)$ d) $\frac{g(t_1 + t_2)}{2}$

573. For a moving body at any instant of time

a) If the body is not moving, the acceleration is necessarily zero
b) If the body is slowing, the retardation is negative
c) If the body is slowing, the distance is negative
d) If displacement, velocity and acceleration at that instant are known, we can find the displacement at any given time in future

574. A stone dropped from a building of height h and it reaches after t seconds on earth. From the same building if two stones are thrown (one upwards and other downwards) with the same velocity u and they reach the earth surface after t_1 and t_2 seconds respectively, then

a) $t = t_1 - t_2$ b) $t = \frac{t_1 + t_2}{2}$ c) $t = \sqrt{t_1 t_2}$ d) $t = t_1^2 t_2^2$

575. A train moves from one station to another 2 hours time. Its speed-time graph during this motion is shown in the figure. The maximum acceleration during the journey is

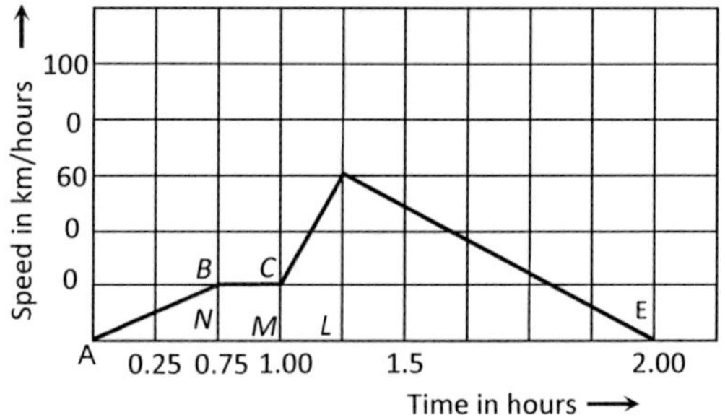

a) $140\ km\ h^{-2}$ b) $160\ km\ h^{-2}$ c) $100\ km\ h^{-2}$ d) $120\ km\ h^{-2}$

576. A bus moves over a straight level road with a constant acceleration a. A body in the bus drops a ball outside. The acceleration of the ball with respect to the bus and the earth are respectively
a) a and g b) $a + \text{g}$ and $\text{g} - a$ c) $\sqrt{a^2 + \text{g}^2}$ and g d) $\sqrt{a^2 + \text{g}^2}$ and a

577. A body is thrown vertically upwards with velocity u. The distance travelled by it in the fifth and the sixth seconds are equal. The velocity u is given by ($g = 9.8\ m/s^2$)
a) $24.5\ m/s$ b) $49.0\ m/s$ c) $73.5\ m/s$ d) $98.0\ m/s$

578. A particle moves a distance x in time t according to equation $x = (t + 5)^{-1}$. The acceleration of particle is proportional to
a) $(\text{Velocity})^{2/3}$ b) $(\text{Velocity})^{3/2}$ c) $(\text{distance})^2$ d) $(\text{distance})^{-2}$

579. Two cars A and Bat rest at same point initially. If A starts with uniform velocity of $40\ m/sec$ and B starts in the same direction with constant acceleration of $4 m/s^2$, then B will catch A after how much time
a) $10\ sec$ b) $20\ sec$ c) $30\ sec$ d) $35\ sec$

580. A body is at rest at $x = 0$. At $t = 0$, it starts moving in the positive x-direction with a constant acceleration. At the same instant another body passes through $x = 0$ moving in the positive x-direction with a constant speed. The position of the first body is given by $x_1(t)$ after time 't' and that of the second body by $x_2(t)$ after the same time interval. Which of the following graphs correctly describes $(x_1 - x_2)$ as a function of time 't'

a) 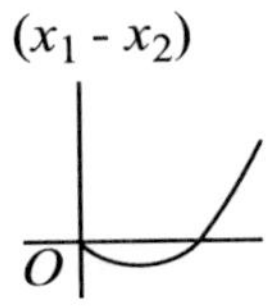

b) 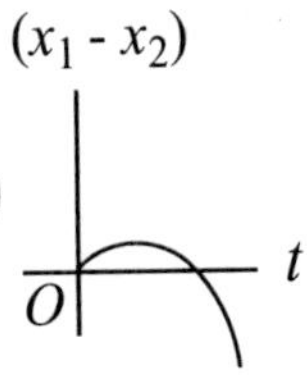

c) 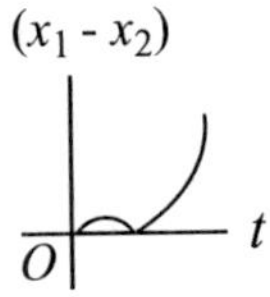

d) 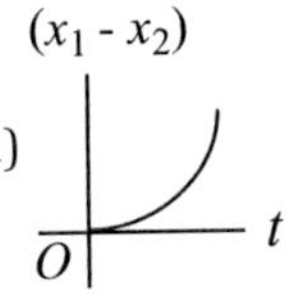

581. A ball is dropped from top of a tower of $100m$ height. Simultaneously another ball was thrown upward from bottom of the tower with a speed of $50\ m/s$ ($g = 10\ m/s^2$). They will cross each other after
a) $1s$ b) $2s$ c) $3s$ d) $4s$

582. A ball is projected upwards from a height h above the surface of the earth with velocity v. The time at which the ball strikes the ground is
a) $\frac{v}{g} + \frac{2hg}{\sqrt{2}}$ b) $\frac{v}{g}\left[1 - \sqrt{1 + \frac{2h}{g}}\right]$ c) $\frac{v}{g}\left[1 + \sqrt{1 + \frac{2gh}{v^2}}\right]$ d) $\frac{v}{g}\left[1 + \sqrt{v^2 + \frac{2h}{g}}\right]$

583. The correct statement from the following is
a) A body having zero velocity will not necessarily have zero acceleration
b) A body having zero velocity will necessarily have zero acceleration
c) A body having uniform speed can have only uniform acceleration
d) A body having non-uniform velocity will have zero acceleration

584. A body falling from a high Minaret travels $40\ m$ in the last 2 seconds of its fall to ground. Height of Minaret in meters is (take $g = 10\ m/s^{-2}$)
a) 60 b) 45 c) 80 d) 50

585. A man is $45\ m$ behind the bus when the bus start accelerating from rest with acceleration $2.5\ m/s^2$. With what minimum velocity should the man start running to catch the bus
a) $12\ m/s$ b) $14\ m/s$ c) $15\ m/s$ d) $16\ m/s$

586. A body has speed of V,2V and 3V in first 1/3 of distance S, seconds 1/3 of S and third 1/3 of S respectively. Its average speed will be
a) V b) $2V$ c) $\frac{18}{11}V$ d) $\frac{11}{18}V$

587. Two balls are dropped from height h and $2h$ respectively from the earth surface. The ratio of time of these balls to reach the earth is

a) $1 : \sqrt{2}$ b) $\sqrt{2} : 1$ c) $2 : 1$ d) $1 : 4$

588. Two balls A and B of same masses are thrown from the top of the building. A, thrown upward with velocity V and B, thrown downward with velocity V, then

a) Velocity of A is more than B at the ground
b) Velocity of B is more than A at the ground
c) Both A&B strike the ground with same velocity
d) None of these

589. If the velocity of a particle is $(10 + 2t^2) m/s$, then the average acceleration of the particle between $2s$ and $5s$ is

a) $2 m/s^2$ b) $4 m/s^2$ c) $12 m/s^2$ d) $14 m/s^2$

590. A body thrown vertically up to reach its maximum height in t second. The total time from the time of projection to reach a point at half of its maximum height while returning (in second) is

a) $\sqrt{2}\,t$ b) $\left(1 + \frac{1}{\sqrt{2}}\right) t$ c) $\frac{3t}{2}$ d) $\frac{t}{\sqrt{2}}$

591. The displacement of a particle is proportional to the cube of time elapsed. How does the acceleration of the particle depends on time obtained

a) $a \propto t^2$ b) $a \propto 2t$ c) $a \propto t^3$ d) $a \propto t$

592. If a ball is thrown vertically upwards with speed u, the distance covered during the last t seconds of its ascent is

a) $\frac{1}{2} g t^2$ b) $ut - \frac{1}{2} g t^2$ c) $(u - gt)t$ d) ut

593. Two cars A and B are moving with same speed of 45 km/hr along same direction. If a third car C coming from the opposite direction with a speed of 36 km/hr meets two cars in an interval of 5 minutes, the distance of separation of two cars A and B should be (in km)

a) 6.75 b) 7.25 c) 5.55 d) 8.35

594. A ball is thrown vertically upwards with a velocity of 25 ms^{-1} from the top of a tower of height 30 m. How long will it travel before it hits ground

a) $6s$ b) $5s$ c) $4s$ d) $12s$

595. The displacement of a body is given by $2s = gt^2$, where g is a constant. The velocity of the body at any time t is

a) gt b) $\frac{gt}{2}$ c) $\frac{gt^2}{2}$ d) $\frac{gt^3}{2}$

596. Consider the acceleration, velocity and displacement of a tennis ball as it falls to the ground and bounces back. Directions of which of these changes in the process

a) Velocity only b) Displacement and velocity
c) Acceleration, velocity and displacement d) Displacement and acceleration

597. The engine of motorcycle can produce a maximum acceleration 5 m/s^2. Its brakes can produce a maximum retardation 10 m/s^2. What is the minimum time in which it can over a distance of 1.5 km

a) 30 sec b) 15 sec c) 10 sec d) 5 sec

598. A particle moves along a straight line such that its displacement at any time t is given by $s = t^3 - 6t^2 + 3t + 4$. The velocity when its acceleration is zero is

a) 2 ms^{-1} b) 12 ms^{-1} c) -9 ms^{-1} d) 2 ms^{-1}

599. Water drops fall at regular intervals from a tap which is 5 m above the ground. The third drop is leaving the tap at the instant the first drop touches the ground. How far above the ground is the second drop at that instant

a) 2.50 m b) 3.75 m c) 4.00 m d) 1.25 m

600. A body is thrown vertically up with a velocity u. It passes three points A, B and C in its upward journey with velocities $\frac{u}{2}, \frac{u}{3}$ and $\frac{u}{4}$ respectively. The ratio of the separations between points A and B between B and C, ie, $\frac{AB}{BC}$ is

a) 1 b) 2 c) $\frac{10}{7}$ d) $\frac{20}{7}$

601. The velocity-time and acceleration-time graphs of a particle are given as

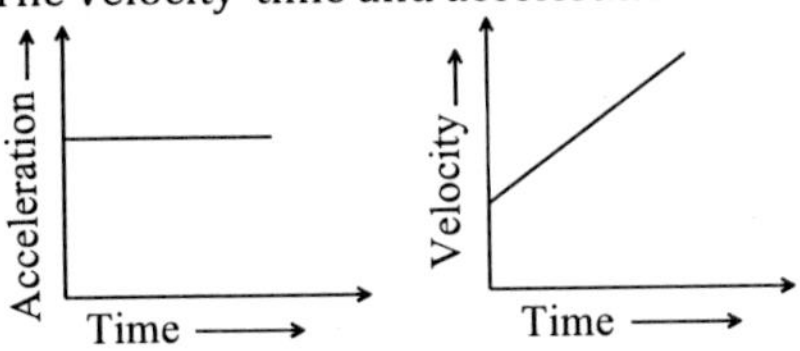

Its position-time graph may be given as

a) 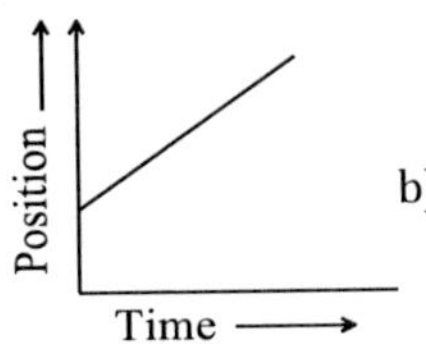

b) 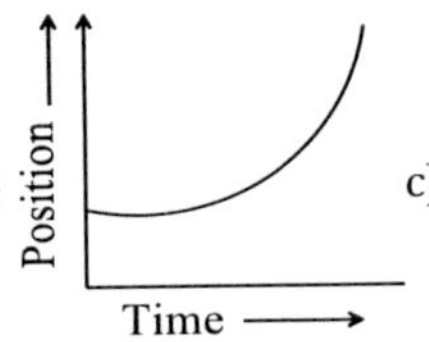

c) 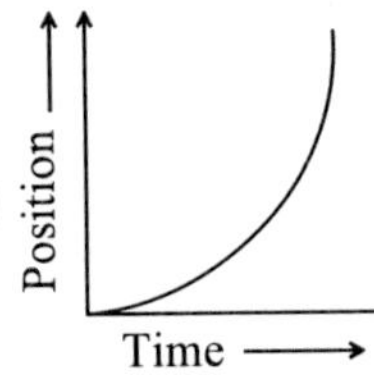

d) 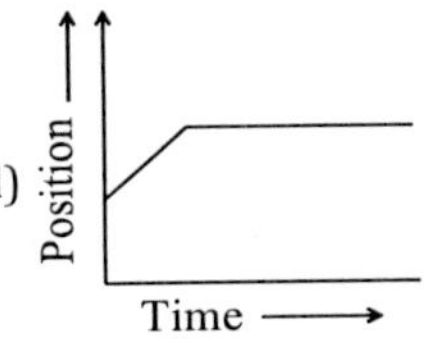

602. A body A moves with a uniform acceleration a and zero initial velocity. Another body B, starts from the same point moves in the same direction with a constant velocity v. The two bodies meet after a time$t$. The value of t is

a) $\frac{2v}{a}$ b) $\frac{v}{a}$ c) $\frac{v}{2a}$ d) $\sqrt{\frac{v}{2a}}$

603. If the velocity of a particle is $(10 + 2t^2) m/s$, then the average acceleration of the particle between $2s$ and $5s$ is

a) $2m/s^2$ b) $4m/s^2$ c) $12m/s^2$ d) $14m/s^2$

604. A body of mass 10 kg is moving with a constant velocity of 10 ms^{-1}. When a constant force acts for 4 s on it, it moves with velocity 2 ms^{-1} in the opposite direction. The acceleration produced in it is

a) 3 ms^{-2} b) -3 ms^{-2} c) 0.3 ms^{-2} d) -0.3 ms^{-2}

605. Acceleration of a particle changes when

a) Direction of velocity changes b) Magnitude of velocity changes
c) Both of above d) Speed changes

606. When a ball is thrown up vertically with velocity V_0, it reaches a maximum height of 'h'. If one wishes to triple the maximum height then the ball should be thrown with velocity

a) $\sqrt{3}V_0$ b) $3V_0$ c) $9V_0$ d) $3/2V_0$

607. The displacement-time graph of a moving particle is shown below. The instantaneous velocity of the particle is negative at the point

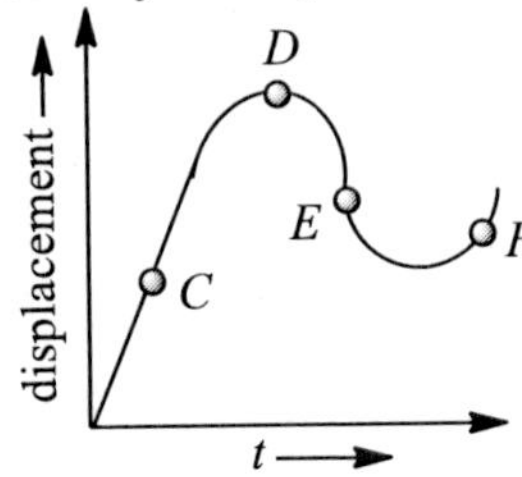

a) C b) D c) E d) F

608. A wheel of radius 1 m rolls forward half a revolution on a horizontal ground. The magnitude of the displacement of the point of wheel initially in contact with the ground is

a) 2π

b) $\sqrt{2}\pi$

c) $\sqrt{\pi^2+4}$

d) π

609. Velocity-time (v-t) graph for a moving object is shown in the figure. Total displacement of the object during the same interval when there is non-zero acceleration and retardation is

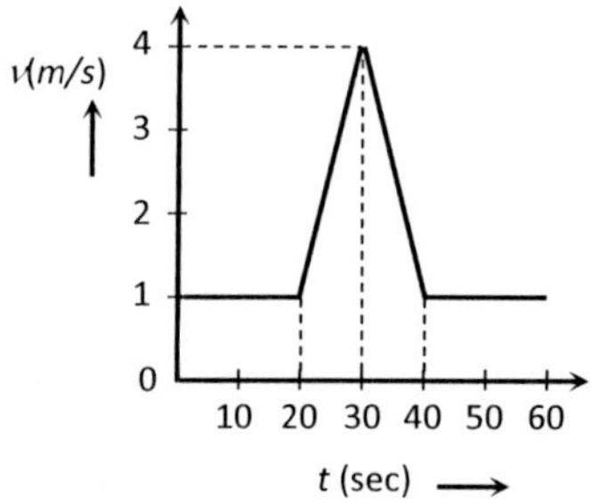

a) 60 m b) 50 m c) 30 m d) 40 m

610. The distance-time graph of a particle at time t makes angle 45° with time axis. After 1s, it makes angle 60° with time axis. What is the acceleration of the particle?

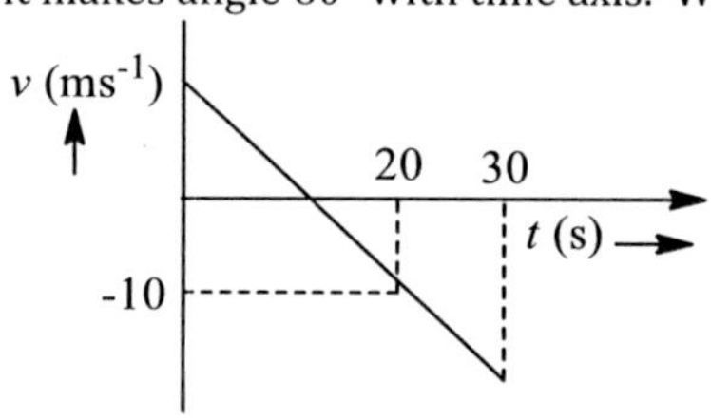

a) $\sqrt{3}-1$ unit b) $\sqrt{3}+1$ unit c) $\sqrt{3}$unit d) 1 unit

611. A body starting from rest moves with constant acceleration. The ratio of distance covered by the body during the 5th sec to that covered in 5 sec is

a) 9/15 b) 3/5 c) 25/9 d) 1/25

612. A body is at rest at $x=0$. At $t=0$, it starts moving in the positive x-direction with a constant acceleration. At the same instant another body passes through $x=0$ moving in the positive x-direction with a constant speed. The position of the first body is given by $x_1(t)$ after time 't' and that of the second body by $x_2(t)$ after the same time interval. Which of the following graphs correctly describes (x_1-x_2) as a function of time 't'

a) 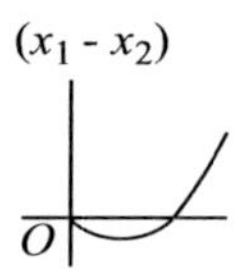

b) 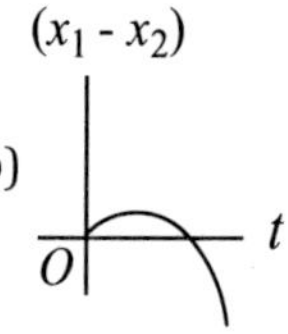

c) 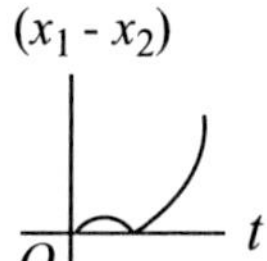

d) 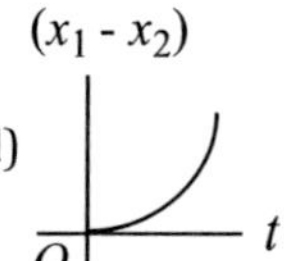

613. Select the incorrect statements from the following

S1 : Average velocity is path length divided by time interval

S2 : In general, speed is greater than the magnitude of the velocity

S3 : A particle moving in a given direction with a non-zero velocity can have zero speed

S4 : The magnitude of average velocity is the average speed

a) S2 and S3 b) S1 and S4 c) S1, S3 and S4 d) All four statements

614. Which of the following graph represents uniform motion

a) 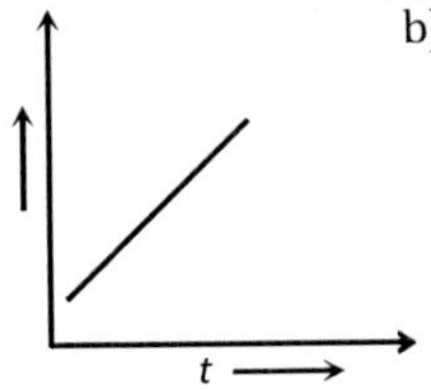

b) 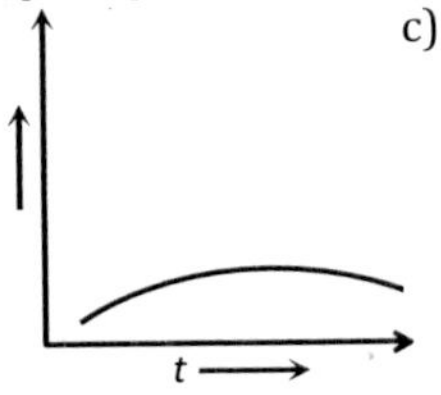

c) 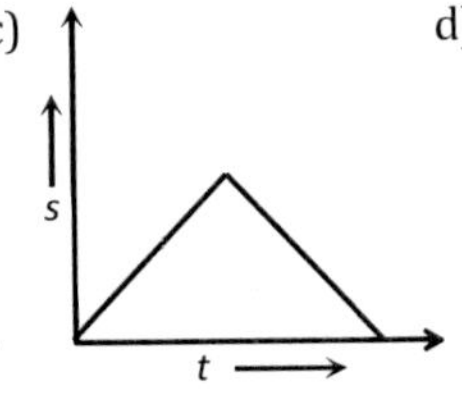

d) 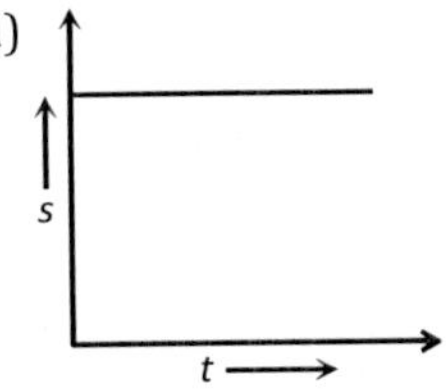

615. A body A starts from rest with an acceleration a_1. After 2 seconds, another body B starts from rest with an acceleration a_2.If they travel equal distances in the 5th second, after the start of A, then the ration $a_1 : a_2$ is equal to

a) $5:9$ b) $5:7$ c) $9:5$ d) $9:7$

616. A graph is drawn between velocity and time for the motion of a particle. The area under the curve between the time intervals t_1 and t_2 gives

a) Momentum of the particle b) Displacement of the particle
c) Acceleration of the particle d) Change in velocity of the particle

617. A particle is dropped vertically from rest a height. The time taken by it to fall through successive distances of 1 m each will then be

a) All equal, being equal to $\sqrt{2/g}$ second
b) In the ratio of the square roots of the integers 1, 2, 3.......
c) In the ratio of the difference in the square roots of the integers i.e. $\sqrt{1}, (\sqrt{2}-\sqrt{1}), (\sqrt{3}-\sqrt{2}), (\sqrt{4}-\sqrt{3})$......
d) In the ratio of the reciprocal of the square roots of the integers i.e.,. $\frac{1}{\sqrt{1}}, \frac{1}{\sqrt{2}}, \frac{1}{\sqrt{3}}, \frac{1}{\sqrt{4}}$

618. A body having uniform acceleration of $10 ms^{-2}$ has a velocity of $100 ms^{-1}$. In what time, the velocity will be doubled?

a) 8 s b) 10 s c) 12 s d) 14 s

619. A bird flies for 4 s with a velocity of $|t-2| m/s$ in a straight line, where t is time in seconds. It covers a distance of

a) $2\ m$ b) $4\ m$ c) $6\ m$ d) $8\ m$

620. A body of $mass$ 10 kg is moving with a constant velocity of 10 m/s.When a constant force acts for 4 $seconds$ on it, it moves with a velocity 2 m/sec in the opposite direction. The acceleration produced in it is

a) $3 m/sec^2$ b) $-3 m/sec^2$ c) $0.3 m/sec^2$ d) $-0.3 m/sec^2$

621. A car is moving along a straight road with uniform, acceleration. It passes through two points P and Q separated by a distance with velocities $30 kmh^{-1}$ and $40 kmh^{-1}$ respectively. The velocity of car midway between P and Q is

a) $33.3 km^{-1}$ b) $1\ km^{-1}$ c) $25\sqrt{2} km^{-1}$ d) $35.35\ km^{-1}$

622. A ball A is thrown up vertically with speed u and at the same instant another ball B is released from a height h. At time t, the speed of A relative to B is

a) u b) $2u$ c) $u-gt$ d) $\sqrt{(u^2-gt)}$

623. A particle starts from rest at $t=0$ and undergoes an acceleration a in ms^{-2} with time t in seconds which is as shown

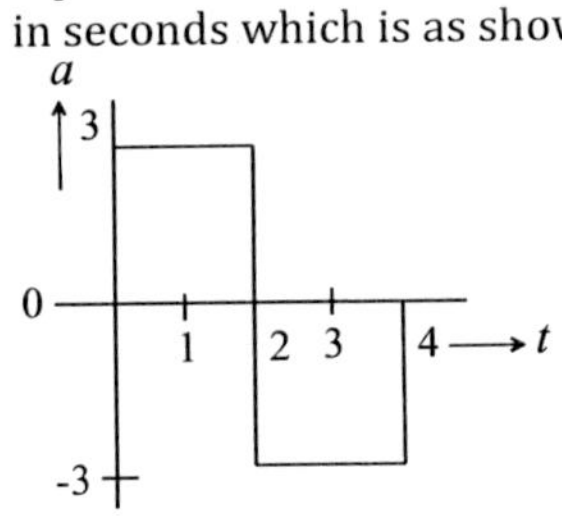

Which one of the following plot represents velocity V in ms^{-1} versus time t in seconds

a) 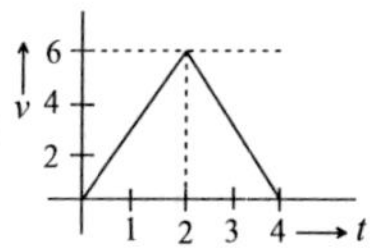

b) 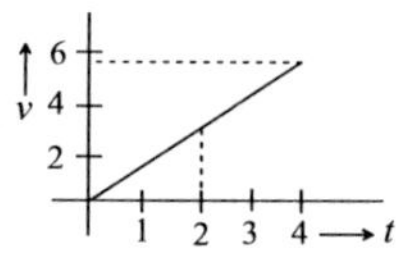

c) 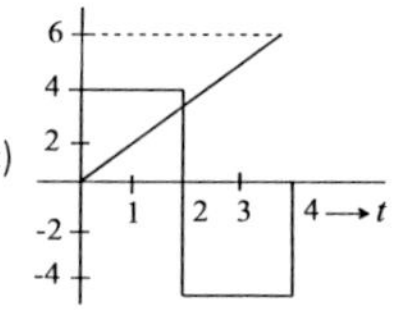

d) 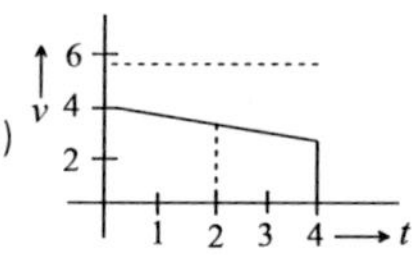

624. The motion of a body falling from rest in a resistive medium is described by the equation $\frac{dv}{dt} = a - bv$, where a and b are constants. The velocity at any time t is

a) $a(1 - b^{2t})$ b) $\frac{a}{b}(1 - e^{-bt})$ c) abe^{-t} d) $ab^2(1 - t)$

625. The velocity of a body depends on time according to the equation $v = 20 + 0.1t^2$. The body is undergoing

a) Uniform acceleration
b) Uniform retardation
c) Non-uniform acceleration
d) Zero acceleration

626. A body of 5 kg is moving with a velocity of $20 m/s$. If a force of $100N$ is applied on it for $10s$in the same direction as its velocity, what will now be the velocity of the body

a) $200\ m/s$ b) $220\ m/s$ c) $240\ m/s$ d) $260\ m/s$

627. A body starting from rest moves with uniform acceleration. The distance covered by the body in time t is proportional to

a) $\sqrt{t}$ b) $t^{3/2}$ c) $t^{2/3}$ d) t^2

628. The position of a particle x (in metre) at a time t second is given by the relation $r = (3t\hat{\mathbf{i}} - t^2\hat{\mathbf{j}} + 4\hat{\mathbf{k}})$. Calculate the magnitude of velocity of the particle after 5 s.

a) 3.55 m/s b) 5.03 m/s c) 8.75 m/s d) 10.44 m/s

629. For a body moving with relativistic speed, if the velocity is doubled, then

a) Its linear momentum is doubled
b) Its linear momentum will be less than double
c) Its linear momentum will be more than double
d) Its linear momentum remains unchanged

630. The engine of motorcycle can produce a maximum acceleration 5 m/s^2. Its brakes can produce a maximum retardation 10 m/s^2. What is the minimum time in which it can over a distance of 1.5 km

a) 30 sec b) 15 sec c) 10 sec d) 5 sec

631. The relation $3t = \sqrt{3x} + 6$ describes the displacement of a particle in one direction where x is in $metres$ and t in sec. The displacement, when velocity is zero, is

a) 24 $metres$ b) 12 $metres$ c) 5 $metres$ d) Zero

632. A body A starts from rest with an acceleration a_1. After 2 seconds, another body B starts from rest with an acceleration a_2.If they travel equal distances in the 5th second, after the start of A, then the ration $a_1 : a_2$ is equal to

a) 5 : 9 b) 5 : 7 c) 9 : 5 d) 9 : 7

633. Which graph represents a state of rest for an object

a) 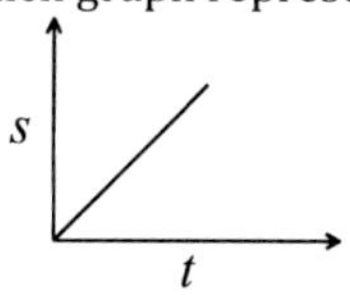

b) 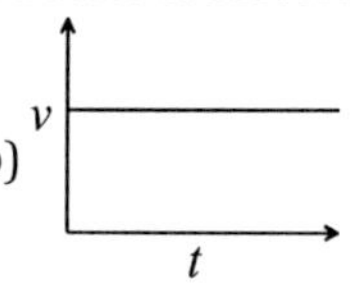

c) 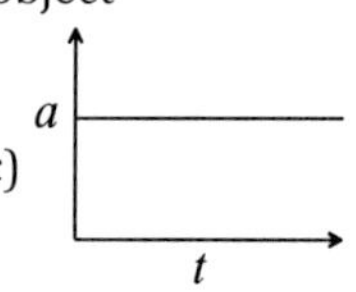

d) 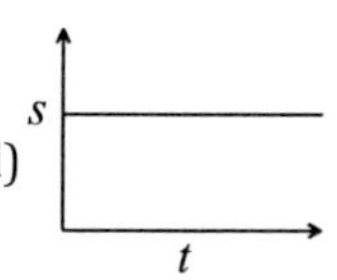

634. The displacement x of a particle varies with time t, $x = ae^{-\alpha t} + be^{\beta t}$, where a, b, α and β are positive constants. The velocity of the particle will

a) Go on decreasing with time
b) Be independent of α and β
c) Drop to zero when $\alpha = \beta$
d) Go on increasing with time

635. The distance travelled by a particle is proportional to the square of time, then the particle travels with

a) Uniform acceleration b) Uniform velocity
c) Increasing acceleration d) Decreasing velocity

636. An elevator is going up. The variation in the velocity of the elevator is as given in the graph. What is the height to which the elevator takes the passengers?

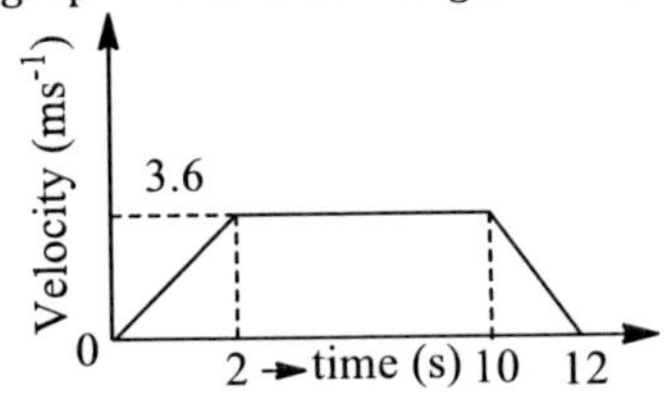

a) 3.6 m b) 28.8 m c) 36.0 m d) 72.0 m

637. A body moves for a total of nine second starting from rest with uniform acceleration and then with uniform retardation, which is twice the value of acceleration and then stops. The duration of uniform acceleration

a) $3\ s$ b) $4.5\ s$ c) $5\ s$ d) $6\ s$

638. A ball is thrown up under gravity ($g = 10\ m/sec^2$). Find its velocity after 1.0 sec at a height of $10m$

a) $5\ m/sec^2$ b) $5\ m/sec$ c) $10\ m/sec$ d) $15\ m/sec$

639. A particle travels $10m$ in first 5 sec and $10m$ in next 3 sec. Assuming constant acceleration what is the distance travelled in next 2 sec

a) $8.3\ m$ b) $9.3\ m$ c) $10.3\ m$ d) None of above

640. An express train is moving with a velocity v_1. Its driver finds another train is moving on the same track in the same direction with velocity v_2. To escape collision, driver applies a retardation a on the train. The minimum time of escaping collision will be

a) $t = \frac{v_1 - v_2}{a}$ b) $t = \frac{v_1^2 - v_2^2}{2}$ c) None d) Both

641. From the top of a tower of height 50m, a ball is thrown vertically upwards with a certain velocity. It hits the ground 10 s after it is thrown up. How much time does it take to cover a distance AB where A and B are two points 20m and 40m below the edge of the tower? ($g = 10ms^{-2}$)

a) 2.0 s b) 1.0 s c) 0.5 s d) 0.4 s

642. A train of 150 m length is going towards north direction at a speed of $10m/sec$. A parrot flies at the speed of 5 m/sec towards south direction parallel to the railway track. The time taken by the parrot to cross the train is

a) 12 sec b) 8 sec c) 15 sec d) 10 sec

643. A particle starts its motion from rest under the action of a constant force. If the distance covered in first 10 $seconds$ is S_1 and that covered in the first 20 $seconds$ is S_2, then

a) $S_2 = 2S_1$ b) $S_2 = 3S_1$ c) $S_2 = 4S_1$ d) $S_2 = S_1$

644. The three initial and final position of a man on the x –axis are given as (i) $(-8m, 7m)$ (ii) $(7m, -3m)$ and (iii) $(-7m, 3m)$
Which pair gives the negative displacement

a) (i) b) (ii) c) (iii) d) (i) and (iii)

645. The displacement- time graph for two particles A and B are straight lines inclined at angles of 30° and 60° with the time axis. The ratio of velocities of $V_A : V_B$ is

a) $1:2$ b) $1:\sqrt{3}$ c) $\sqrt{3}:1$ d) $1:3$

646. A body starts from rest with uniform acceleration. If its velocity after n second is v, then its displacement in the last 2 s is

a) $\frac{2v(n+1)}{n}$ b) $\frac{v(n+1)}{n}$ c) $\frac{v(n-1)}{n}$ d) $\frac{2v(n-1)}{n}$

647. The distance travelled by a particle is proportional to the square of time, then the particle travels with

a) Uniform acceleration
b) Uniform velocity
c) Increasing acceleration
d) Decreasing velocity

648. From the top of a tower, a particle is thrown vertically downwards with a velocity of 10 m/sec. The ratio of the distances, covered by it in the 3rd and 2nd seconds of the motion is (Take $g = 10\ m/s^2$)

a) 5 : 7
b) 7 : 5
c) 3 : 6
d) 6 : 3

649. A particle starts from rest at $t = 0$ and moves in a straight line with acceleration as shown in figure. The velocity of the particle at $t = 3$ s is

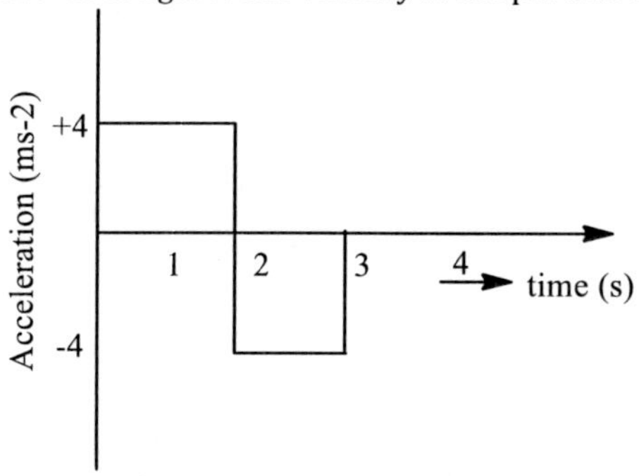

a) 2 ms^{-1}
b) 4 ms^{-1}
c) 6 ms^{-1}
d) 8 ms^{-1}

650. A body travels for 15 sec starting from rest with constant acceleration. If it travels distances S_1, S_2 and S_3 in the first five seconds, second five secods and next five seconds respectively the relation between S_1, S_2 and S_3 is

a) $S_1 = S_2 = S_3$
b) $5S_1 = 3S_2 = S_3$
c) $S_1 = \frac{1}{3}S_2 = \frac{1}{5}S_3$
d) $S_1 = \frac{1}{5}S^2 = \frac{1}{3}S_3$

651. A particle has an initial velocity of $3\hat{i} + 4\hat{j}$ and an acceleration of $0.4\hat{i} + 0.3\hat{j}$. Its speed after 10 s is

a) 10 units
b) $7\sqrt{2}$ units
c) 7 units
d) 8.5 units

652. An object is moving with a uniform acceleration which is parallel to its instantaneous direction of motion. The displacement (s) –velocity (v) graph of this object is

a)
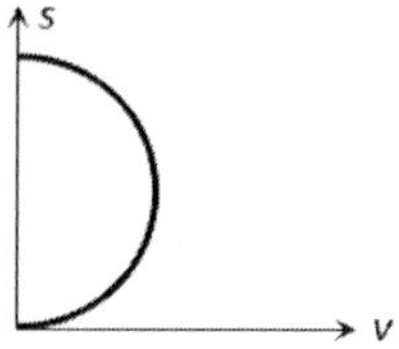

b)
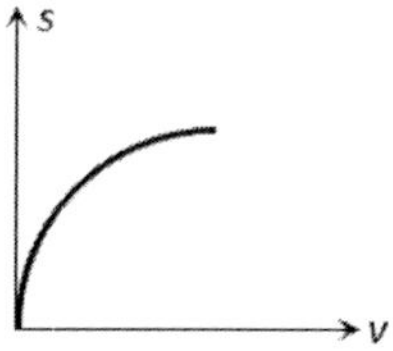

c)
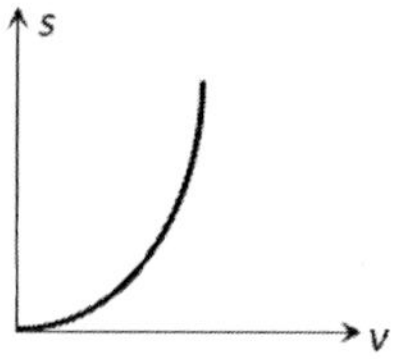

d) 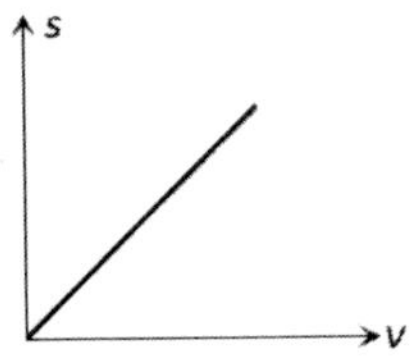

653. A particle is moving with constant acceleration from A to B in a straight line AB. If u and v are the velocities at A and B respectively, then its velocity at the midpoint c will be

a) $\left(\frac{u^2+v^2}{2u}\right)^2$ b) $\frac{u+v}{2}$ c) $\frac{v-u}{2}$ d) $\sqrt{\frac{u^2+v^2}{2}}$

654. A balloon is rising vertically up with a velocity of $29\ ms^{-1}$. A stone is dropped from it and it reaches the ground in 10 seconds. The height of the balloon when the stone was dropped from it is ($g = 9.8\ ms^{-2}$)

a) $100\ m$ b) $200\ m$ c) $400\ m$ d) $150\ m$

655. A cyclist starts from the centre O of a circular park of radius one kilometre, reaches the edge P of the park, then cycles along the circumference and returns to the centre along QO as shown in figure. If the round trip takes ten minutes, the net displacement and average speed of the cyclist (in metre and kilometre per hour) is

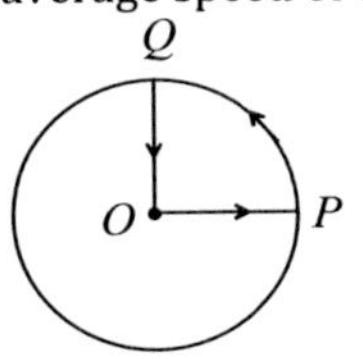

a) 0, 1 b) $\frac{\pi+4}{2}$, 0 c) 21.4, $\frac{\pi+4}{2}$ d) 0, 21.4

656. The motor of an electric train can give it an acceleration of 1 ms^{-2} and brakes can give a negative acceleration of 3 ms^{-2}. The shortest time in which the train can make a trip between the two stations 1215 m apart is

a) 113.6 s b) 56.9 s c) 60 s d) 55 s

657. Two stones of equal masses are dropped from a rooftop of height h one after another. Their separation distance against time will

a) Remain the same b) Increase c) Decrease d) Be zero

658. Two cars are moving in the same direction with a speed of 30kmh^{-1}. They are separated from each other by 5 km. Third car moving in the opposite direction meets the two cars after an interval of 4 min. What is the speed of the third car?

a) 30 kmh^{-1} b) 35 kmh^{-1} c) 40 kmh^{-1} d) 45 kmh^{-1}

659. A very large number of balls are thrown vertically upwards in quick succession in such a way that the next ball is thrown when the previous one is at the maximum height. If the maximum height is $5m$, the number of ball thrown per minute is (take $g = 10\ ms^{-2}$)

a) 120 b) 80 c) 60 d) 40

660. Velocity-time curve for a body projected vertically upwards is

a) Parabola b) Ellipse c) Hyperbola d) Straight line

661. If a freely falling body travels in the last second a distance equal to the distance travelled by it in the first three second, the time of the travel is

a) 6 sec b) 5 sec c) 4 sec d) 3 sec

662. The numerical ratio of displacement to the distance covered is always

a) Less than one b) Equal to one
c) Equal to or less than one d) Equal to or greater than one

663. A point initially at rest moves along x-axis. Its acceleration varies with time as $a - (6t + 5)$in ms^{-2}. If it starts from origin, the distance covered in 2s is
a) 20 m b)18 m c) 16 m d)25 m

664. Two balls are dropped to the ground from different heights. One ball is dropped $2s$ after the other but they both strike the ground at the same time. If the first ball takes $5s$ to reach the ground, then the difference in initial heights is $(g = 10\ ms^{-2})$
a) $20m$ b)$80m$ c) $170m$ d)$40m$

665. The path of a particle moving under the influence of a force fixed in magnitude and direction is
a) Straight line b)Circle c) Parabola d)Ellipse

666. A boat travels 50 km east, then 120 km north and finally it comes back to the starting point through the shortest distance. The total time of journey is 3 h. What is the average speed, in kmh^{-1}, over the entire trip?
a) Zero b)100 c) 17 d)33.33

667. If a car at rest accelerates uniformly to a speed of 144 km/h in 20 s. Then it covers a distance of
a) 20 m b)400 m c) 1440m d)2880 m

668. An aeroplane flies 400m north and 300m south and then flies 1200m upwards, then net displacement is
a) 1500m b)1400m c) 1300m d)1200m

669. A body is projected up with a speed 'u' and the time taken by it is T to reach the maximum height H.Pick out the correct statement
a) It reaches $H/2$ in $T/2\ sec$ b)It acquires velocity $u/2$ in $T/2sec$
c) Its velocity is $u/2$ at $H/2$ d)Same velocity at $2T$

670. A body falls from rest, its velocity at the end of first second is $(g = 32ft/sec)$
a) 16 ft/sec b)32 ft/sec c) 64 ft/sec d)24 ft/sec

671. A bullet moving with a velocity of 200 cm/s penetrates a wooden block and comes to rest after traversing 4cm inside it. What velocity is needed for travelling distance of 9cm in same block
a) 100 cm/s b)136.2cm/s c) 300cm/s d)250 cm/s

672. A constant force acts on a body of mass 0.9 kg at rest for 10s.If the body moves a distance of 250 m, the magnitude of the force is
a) 3N b)3.5N c) 4.0N d)4.5N

673. The displacement of the particle varies with time according to the relation $x = \frac{k}{b}[1 - e^{-bt}]$.Then the velocity of the particle is
a) $k(e^{-bt})$ b)$\frac{k}{b^2 e^{-bt}}$ c) $k\ b\ e^{-bt}$ d)None of these

674. A ball is dropped vertically downwards from a height d above the ground. It hits the ground and bounces up vertically to a height $d/2$. Neglecting subsequent motion and air resistance, its speed varies with the height h above the ground as

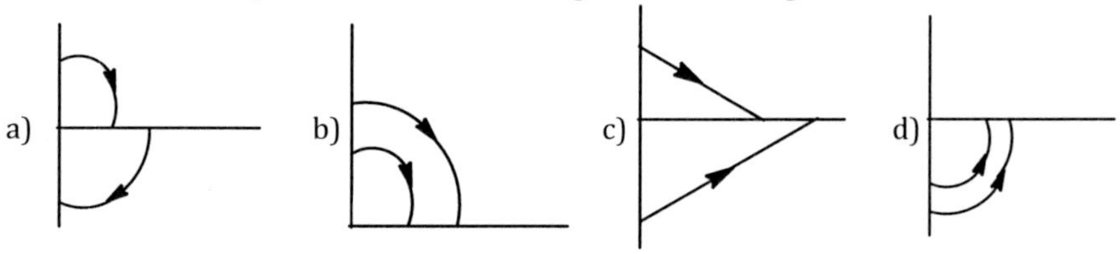

675. An object is projected upwards with a velocity of 100m/s. It will strike the ground after (approximately)
a) 10 sec b)20 sec c) 15 sec d)5 sec

676. Free fall of an object (in vacuum) is a case of motion with

a) Uniform velocity b) Uniform acceleration c) Variable acceleration d) Constant momentum

677. Which of the following statements is correct?

a) When air resistance is negligible, the time of ascent is less than the time of descent

b) When air resistance is not negligible, time of ascent is less than the time of descent

c) When air resistance is not negligible, the time ascent is greater than the time of descent

d) When air resistance is not negligible, the time of ascent is lesser than the time of descent

678. A body falls freely from the top of a tower. It covers 36% of the total height in the last second before striking the ground level. The height of the tower is

a) $50\ m$ b) $75\ m$ c) $100m$ d) $125\ m$

679. A particle starts from rest and traverses a distance $2x$ with uniform acceleration, then moves uniformly over a further distance $4x$ and finally comes to rest after moving a further distance $6x$ under uniform retardation. Assuming entire motion to be rectilinear motion, the ratio of average speed over the journey to the maximum speed on its way is

a) $4/5$ b) $3/5$ c) $2/5$ d) $1/5$

680. The acceleration 'a' in m/s^2 of a particle is given by $a = 3t^2 + 2t + 2$ where t is the time. If the particle starts out with a velocity $u = 2m/s$ at $t = 0$, then the velocity at the end of 2 *seconds* is

a) $12\ m/s$ b) $18\ m/s$ c) $27\ m/s$ d) $36\ m/s$

681. Displacement (x) of a particle is related to time (t) as

$$x = at + bt^2 - ct^3$$

Where a, b and c are constants of the motion. The velocity of the particle when its acceleration is zero is given by

a) $a + \frac{b^2}{c}$ b) $a + \frac{b^2}{2c}$ c) $a + \frac{b^2}{3c}$ d) $a + \frac{b^2}{4c}$

682. For the velocity-time graph shown in figure below the distance covered by the body in last two seconds of its motion is what fraction of the total distance covered by it in all the seven seconds

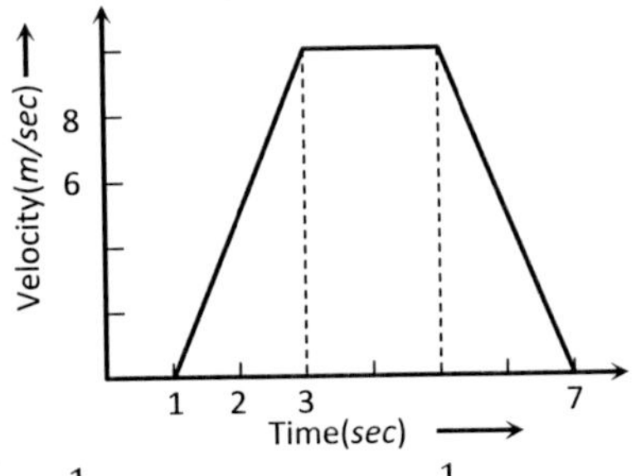

a) $\frac{1}{2}$ b) $\frac{1}{4}$ c) $\frac{1}{3}$ d) $\frac{2}{3}$

683. A man throws balls with the same speed vertically upwards one after the other at an interval of two seconds. What should be the speed of the throw so that more than two balls are in the sky at any time (Given $g = 9.8\ m/s^2$)

a) At least $0.8\ m/s$ b) Any speed less than $19.6\ m/s$

c) Only with speed $19.6\ m/s$ d) More than 19.6

684. Water drops fall from a tap on the floor 5m below at regular intervals of time, the first drop striking the floor when the fifth drop begins to fall. The height at which the third drop will be, from ground, at the instant when first drop strikes the ground, will be ($g = 10ms^{-2}$)

a) 1.25 m b) 2.15 m c) 2.73 m d) 3.75 m

685. From the top of a tower, a particle is thrown vertically downwards with a velocity of $10\ m/sec$. The ratio of the distances, covered by it in the 3rd and 2nd seconds of the motion is (Take $g = 10\ m/s^2$)

a) 5 : 7 b) 7 : 5 c) 3 : 6 d) 6 : 3

686. A car moving with a speed of 50 kmh^{-1}, can be stopped by brakes after at least 6 m. if the same car is moving at a speed of 100 kmh^{-1}, the minimum stopping distance is

a) 12 m b) 18 m c) 24 m d) 6 m

687. The numerical ratio of displacement to the distance covered is always

a) Less than one b) Equal to one

c) Equal to or less than one d) Equal to or greater than one

688. An aeroplane files 400 m north and 300 m south and then files 1200 m upwards then net displacement is

a) 1200 m b) 1300 m c) 1400 m d) 1500 m

689. A particle moving in a straight line covers half the distance with a speed of 3ms^{-1}. The other half of the distance is covered in two equal time intervals with speeds of 4.5ms^{-1} and 7.5ms^{-1}. The average speed of the particle during this motion is

a) 4ms^{-1} b) 5ms^{-1} c) 5.5 ms^{-1} d) 4.8 ms^{-1}

690. A particle starts from rest at $t = 0$ and undergoes an acceleration a in ms^{-2} with time t in second which is as shown

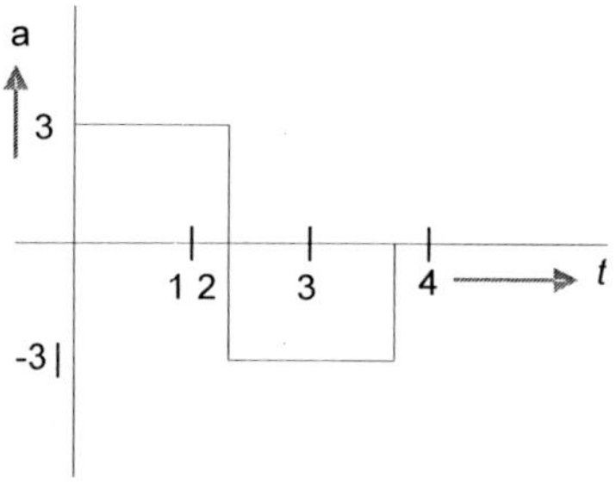

Which one of the following plot represents velocity v in ms^{-1} *versus* time t in second?

a)

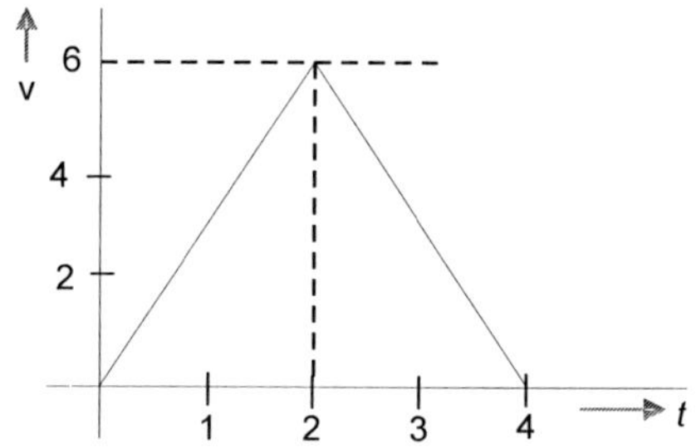

b)

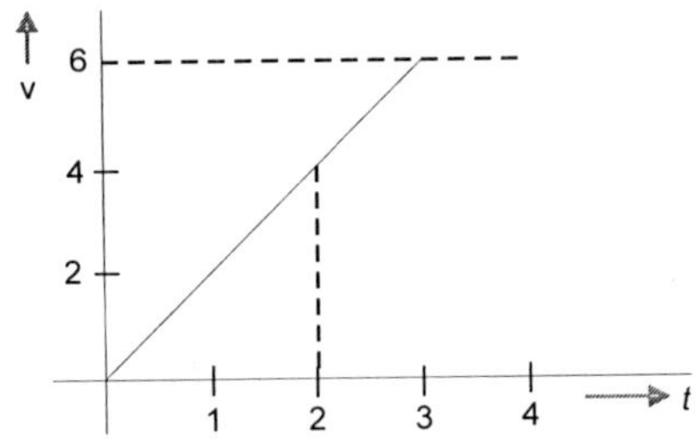

c)

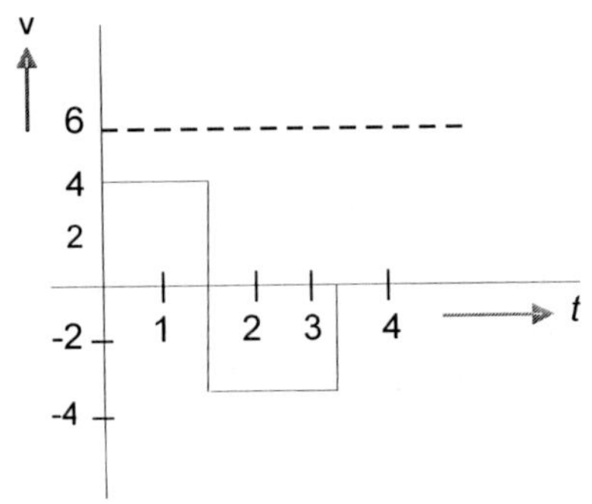

d)

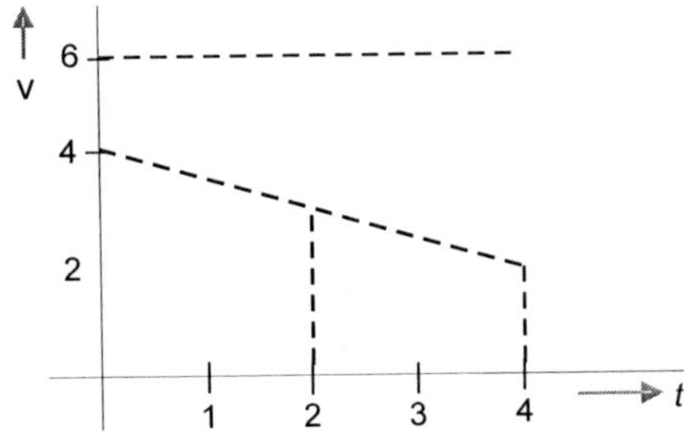

691. A car, starting from rest, accelerates at the rate f through a distance S, then continues at constant speed for time t and then decelerates at the rate $\frac{f}{2}$ to come to rest. If the total distance traversed is 15 S, then

a) $S=\frac{1}{2}ft^2$ b) $S=\frac{1}{4}ft^2$ c) $S=\frac{1}{72}ft^2$ d) $S=\frac{1}{6}ft^2$

692. The acceleration due to gravity on the planet A is 9 times the acceleration due to gravity on the planet B. A man jumps to a height of $2m$ on the surface of A. What is the height of jump by the same person on the planet B

a) $18\ m$ b) $6\ m$ c) $\frac{2}{3}m$ d) $\frac{2}{9}m$

693. A 2 m wide truck is moving with a uniform speed $v_0 = 8\text{ms}^{-1}$ along a straight horizontal road. A pedestrian starts to cross the road with a uniform speed v when the truck is 4m away from him. The minimum value of v, so that he can cross the raod safely is

a) 2.62ms^{-1} b) 4.6ms^{-1} c) 3.57ms^{-1} d) 1.414ms^{-1}

694. A particle covers half of its total distance with speed v_1 and the rest half distance with speed v_2. Its average speed during the complete journey is

a) $\frac{v_1^2 v_2^2}{v_1^2 + v_2^2}$ b) $\frac{v_1 + v_2}{2}$ c) $\frac{v_1 v_2}{v_1 + v_2}$ d) $\frac{2v_1 v_2}{v_1 + v_2}$

695. A point particle starting from rest has a velocity that increase linearly with time such that $v = pt$,where $p = 4\text{ms}^{-2}$. The distance covered in the first 2 s will be

a) 6 m b) 4 m c) 8 m d) 10 m

696. A ball which is at rest is dropped from a height h metre. As it bounces off the floor its speed is 80% of what it was just before touching the ground. The ball then rise to nearly a height

a) 0.94 h b) 0.80 h c) 0.75 h d) 0.64 h

697. Four marbles are dropped from the top of a tower one after the other with an interval of one second. The first one reaches the ground 4 seconds. When the first one reaches the ground the distances between the first and second, the second and third and the third and forth will be respectively

a) 35,25 and 15 m b) 30,20 and 10 m c) 20,10 and 5 m d) 40,30 and 20 m

698. Acceleration of a body when displacement equation is $3s = 9t + 5t^2$ is

a) 5/3 b) 14/3 c) 10/3 d) 19/3

699. If the velocity of a particle is given by $v = (180 - 16x)^{1/2}\text{ms}^{-1}$, then its acceleration will be

a) Zero b) 8 ms^{-2} c) -8 ms^{-2} d) 4 ms^{-2}

700. A lift in which a man is standing, is moving upward with a speed of 10ms^{-1}. The man drops a coin from a height of 4.9m and if $g = 9.8\text{ ms}^{-2}$,then the coin reaches the floor of the lift after a time

a) $\sqrt{2}$s b) 1 s c) $\frac{1}{2}$s d) $\frac{1}{\sqrt{3}}$s

701. A train started from rest from a station and accelerated at 2 ms^{-2} for 10 s. Then, it ran at constant speed for 30 s and thereafter it decelerated at 4 ms^{-2} until it stopped at the next station. The distance between two stations is

a) 650 m b) 700 m c) 750 m d) 800 m

702. A body is thrown vertically up with a velocity u. It passes three points A, B and C in its upward journey with velocities $\frac{u}{2}, \frac{u}{3}$ and $\frac{u}{4}$ respectively. The ratio of the separations between points A and B and between B and C *i.e.*, $\frac{AB}{BC}$ is

a) 1 b) 2 c) $\frac{10}{7}$ d) $\frac{20}{7}$

703. A boat moves with a speed of 5 km/h relative to water in a river flowing with a speed of 3 km/h and having a width of 1 km. The minimum time taken around a round trip is

a) 5 min b) 60 min c) 20 min d) 30 min

704. The relation $3t = \sqrt{3x} + 6$ describes the displacement of a particle in one direction where x is in *metres* and t in sec. The displacement, when velocity is zero, is

a) 24 *metres* b) 12 *metres* c) 5 *metres* d) Zero

705. A car moves a distance of 200 m. It covers first half of the distance at speed 60 kmh^{-1} and the second half at speed v. If the average speed is 40 kmh^{-1}, the value of v is

a) 30 kmh^{-1} b) 13 kmh^{-1} c) 60 kmh^{-1} d) 40 kmh^{-1}

706. A 210 m long train is moving due North at a speed of 25 m/s. A small bird is flying due South a little above the train with speed 5 m/s. The time taken by the bird to cross the train is

a) $6s$ b) $7s$ c) $9s$ d) $10s$

707. The displacement of a particle is given by $y = a + bt + ct^2 - dt^4$. The initial velocity and acceleration are respectively

a) $b, -4d$ b) $-b, 2c$ c) $b, 2c$ d) $2c, -4d$

708. A particle projected vertically upwards attains a maximum height H. If the ratio of the times to attain a height $h(h < H)$ is $\frac{1}{3}$. Then

a) $4h = 3H$ b) $3h = 4H$ c) $3h = H$ d) $4h = H$

709. Spotting a police car, you brake a parsche from a speed of 100kmh^{-1} to a speed of 80.0 kmh^{-1} during a displacement of 88.0m, at a constant acceleration. What is the acceleration?

a) -2.5ms^{-2} b) 1.58 ms^{-2} c) -1.58ms^{-2} d) 2.5 ms^{-2}

710. The ratios of the distances traversed, in successive intervals of time by a body, falling from rest are

a) $1:3:5:7:9:\ldots$ b) $2:4:6:8:10:\ldots$ c) $1:4:7:10:13:\ldots$ d) None of these

711. A body moves from rest with a constant acceleration of 5 m/s^2. Its instantaneous speed (in m/s) at the end of 10 sec is

a) 50 b) 5 c) 2 d) 0.5

712. A ball is thrown vertically upward. Ignoring the air resistance, which one of the following plot represents the velocity-time plot for the period ball remains in air?

a) b) c) d)

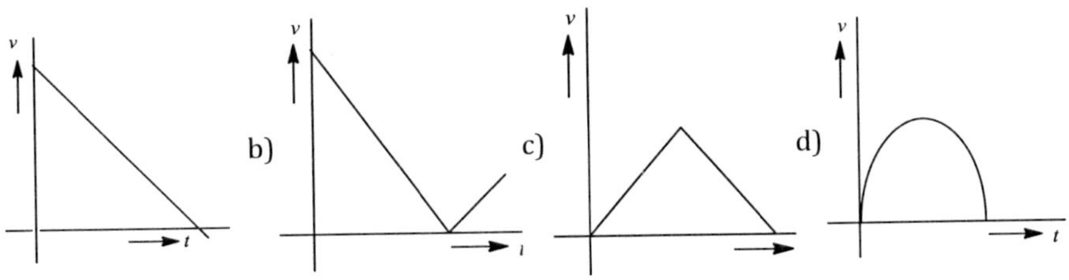

713. A cyclist starts from the centre O of a circular park of radius 1 km, reaches the edge P of the park, then cycles along the circumference and returns to the point O as shown in figure. If the round trip takes 10 min, the net displacement and average speed of the cyclist (in metre and kilometer per hour) are

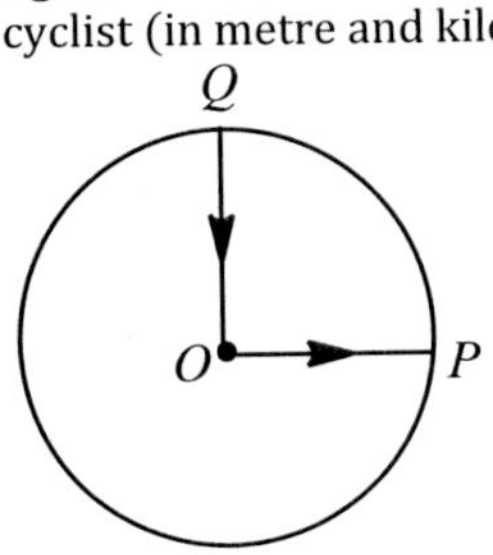

a) 0, 1 b) $\frac{\pi + 4}{2}, 0$ c) $214, \frac{\pi + 4}{2}$ d) 0, 21.4

714. A ball A is thrown up vertically with a speed u and at the same instant another ball B is released from a height h. At time t, the speed of A relative to B is

a) u b) $2u$ c) $u - \text{gt}$ d) $\sqrt{(u^2 - \text{gt})}$

715. An object, moving with a speed of 6.25 m/s, is decelerated at a rate given by $\frac{dv}{dt} = -2.5\sqrt{v}$ where v is the instantaneous speed. The time taken by the object, to come to rest, would be

a) 2 s b) 4 s c) 8 s d) 1 s

716. A point initially at rest moves along x-axis. Its acceleration varies with time as $a = (6t + 5)m/s^2$. If it starts from origin, the distance covered in 2 s is

a) 20 m b) 18 m c) 16 m d) 25 m

717. A ball kicked vertically up attains a height of 19.6m and returns to the point of throw. If the ball is in air for four second, then the value of acceleration due to gravity is

a) 4.9 ms^{-2} b) 9.8ms^{-2} c) 10 ms^{-2} d) $2 \times 9.8ms^{-2}$.

718. Two trains each 50 m long are travelling in opposite direction with velocity 10 m/s and 15 m/s. The time of crossing is

a) 2s b) 4s c) $2\sqrt{3}s$ d) $4\sqrt{3}\,s$

719. In a race for 100m dash, the first and the second runners have a gap of one metre at the mid way stage. Assuming the first runner goes steady, by what percentage should the second runner increases his speed just to win the race.

a) 2% b) 4% c) More than 4% d) Less than 4%

720. A rocket is fired upwards. Its engine explodes fully is 12s. The height reached by the rocket as calculated from its velocity-time graph is

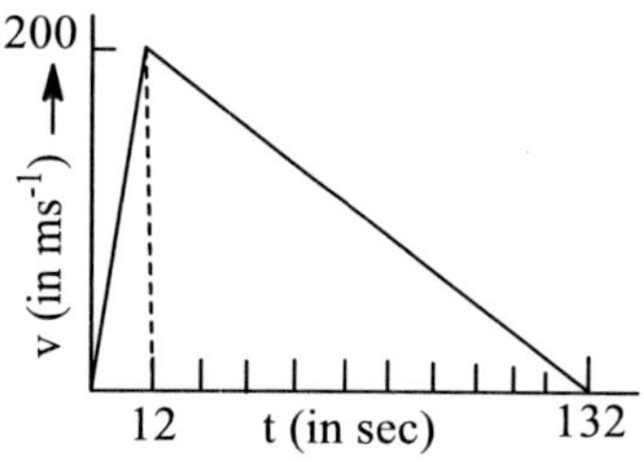

a) 1200×60m b) 1200×132m c) $\frac{1200}{12}$m d) 1200×12^2m

721. The position x of a particle with respect to time t along$x$-axis is given by $x = 9t^2 - t^3$ where x is in metres and t in second. What will be the position of this particle when it achieves maximum speed along the $+x$ direction

a) 32 m b) 54 m c) 81 m d) 24 m

722. A particle moving in a straight line with uniform acceleration is observed to be a distance a from a fixed point initially. It is at distances b, c, d from the same point after $n, 2n, 3n$ second. The acceleration of the particle is

a) $\frac{c - 2b + a}{n^2}$ b) $\frac{c + b + a}{9n^2}$ c) $\frac{c + 2b + a}{4n^2}$ d) $\frac{c - b + a}{n^2}$

723. A boat moves with a speed of 5 km/h relative to water in a river flowing with a speed of 3 km/h and having a width of 1 km. The minimum time taken around a round trip is

a) 5 min b) 60 min c) 20 min d) 30 min

724. A bullet emerges from a barrel of length 1.2 m with a speed of 640 ms^{-1}. Assuming constant acceleration, the approximate time that it spends in the barrel after the gun is fired is

a) 4 ms b) 40 ms c) 400 μs d) 1 s

725. A body of mass m is resting on a wedge of angle θ as shown in figure. The wedge is given at acceleration α. What is the value of a son that the mass m just falls freely?

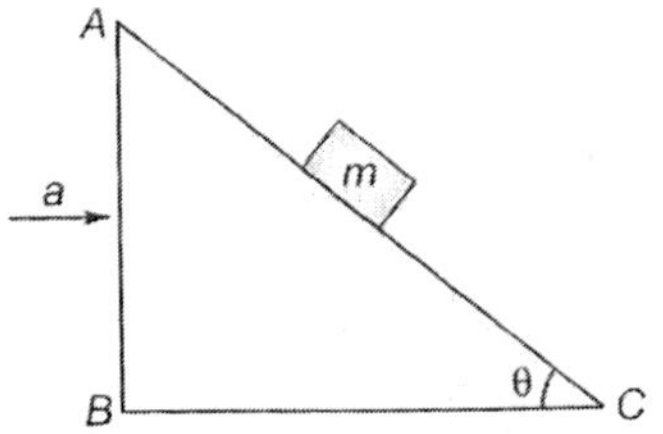

a) g b) g sin θ c) g tan θ d) g cot θ

726. Time taken by an object falling from rest to cover the height of h_1 and h_2 is respectively t_1 and t_2 then the ratio of t_1 to t_2 is

a) $h_1 : h_2$ b) $\sqrt{h_1} : \sqrt{h_2}$ c) $h_1 : 2h_2$ d) $2h_1 : h_2$

727. A body is released from the top of a tower of height h metre. It takes T second to reach the ground. Where is the ball at the time $\frac{T}{2}$ second?

a) At $\frac{h}{4}$ metre from the ground
b) At $\frac{h}{2}$ metre from the ground
c) At $\frac{3h}{4}$ metre from the ground
d) Depends upon the mass and volume of the ball

728. A body of mass m is thrown upwards at an angle θ with the horizontal with velocity v. While rising up the velocity of the mass after t second will be

a) $\sqrt{(v\cos\theta)^2 + (v\sin\theta)^2}$
b) $\sqrt{(v\cos\theta - v\sin\theta)^2 - gt}$
c) $\sqrt{v^2 + g^2t^2 - (2v\sin\theta)gt}$
d) $\sqrt{v^2 + g^2 + g^2 - (2v\cos\theta)gt}$

729. A body stars from rest and falls vertically from a height of 19.6m. If $g = 9.8\text{ms}^{-2}$, then the time taken by the body to fall through the last metre of its fall, is

a) 2.00 s b) 0.05 s c) 0.45 s d) 1.95 s

730. Acceleration of a particle changes when

a) Direction of velocity changes
b) Magnitude of velocity changes
c) Both of above
d) Speed changes

731. The initial and final position vectors for a particle are respectively $(-3.0\text{ m})\hat{\mathbf{i}} + (2.0\text{ m})\hat{\mathbf{j}} + (8.0\text{ m})\hat{\mathbf{k}}$ and $(9.0\text{ m})\hat{\mathbf{i}} + (2.0\text{ m})\hat{\mathbf{j}} + (-8.0\text{ m})\hat{\mathbf{k}}$. The displacement of the particle is

a) $(6.0\text{ m})\hat{\mathbf{i}} + (4.0\text{ m})\hat{\mathbf{j}} + (16.0\text{ m})\hat{\mathbf{k}}$
b) $(6.0\text{ m})\hat{\mathbf{i}}$
c) $(12.0\text{ m})\hat{\mathbf{i}} - (16.0\text{ m})\hat{\mathbf{j}}$
d) $(12.0\text{ m})\hat{\mathbf{i}}$

732. A car A is travelling on a straight level road with a uniform speed of 60 km/h. It is followed by another car B which is moving with a speed of $70km/h$. When the distance between them is 2.5 km, the car B is given a deceleration of 20 km/h^2. After how much time will B catch up with A.

a) 1 hr b) $1/2hr$ c) $1/4hr$ d) $1/8hr$

733. A body starts from rest. What is the ratio of the distance travelled by the body during the 4th and 3rd second

a) $\frac{7}{5}$ b) $\frac{5}{7}$ c) $\frac{7}{3}$ d) $\frac{3}{7}$

734. A body travelling with uniform acceleration crosses two points A and B with velocities $20ms^{-1}$ and d30 ms^{-1} respectively. The speed of the body at the mid-point of A and B is nearest to

a) 25.5 ms^{-1} b) 25 ms^{-1} c) $24ms^{-1}$ d) $10\sqrt{6}\ ms^{-1}$

735. An object start sliding on a frictionless inclined plane and from same height another object start falling freely

a) Both will reach with same speed
b) Both will reach with the same acceleration
c) Both will reach in same time
d) None of above

736. A graph is drawn between velocity and time for the motion of a particle. The area under the curve between the time intervals t_1 and t_2 gives

a) Momentum of the particle
b) Displacement of the particle
c) Acceleration of the particle
d) Change in velocity of the particle

737. The distance between two particles moving towards each other is decreasing at the rate of $6m/sec$. If these particles travel with same speeds and in the same direction, then the separation increase at the rate of $4m/sec$. The particles have speeds as

a) $5m/sec: 1m/sec$ b) $4\ m/sec: 1m/sec$ c) $4\ m/sec: 2m/sec$ d) $5\ m/sec: 2m/sec$

738. A particle moving in a straight line and passes through a point O with a velocity of $6ms^{-1}$. The particle moves with a constant retardation of $2ms^{-2}$ for 4 s and there after moves with a constant velocity. How long after leaving O does the particle return to O

a) $3s$ b) $8s$ c) $6\ m$ d) $8\ m$

739. Two trains one of 100 m and another of length 125m, are moving in mutually opposite directions along parallel lines, meet each other, each with speed 10m/s.
If their acceleration are $0.3m/s^2$ and $0.2m/s^2$ respectively, then the time taken to pass each other will be

a) 5 s b) 10 s c) 15 s d) 20 s

740. A body starts from rest and moves with uniform acceleration. Which of the following graphs represents it motion?

a) 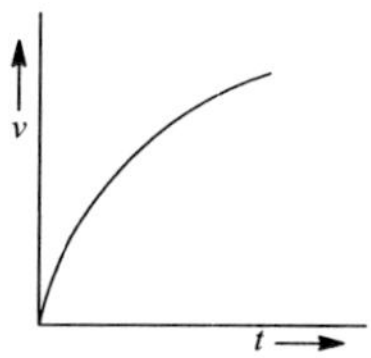

b) 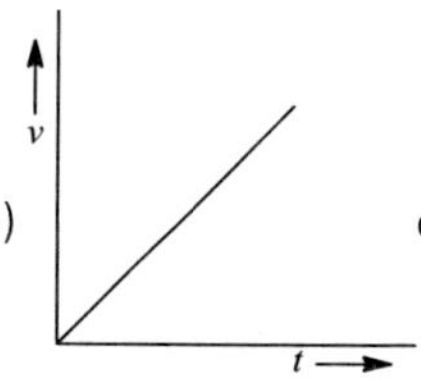

c) 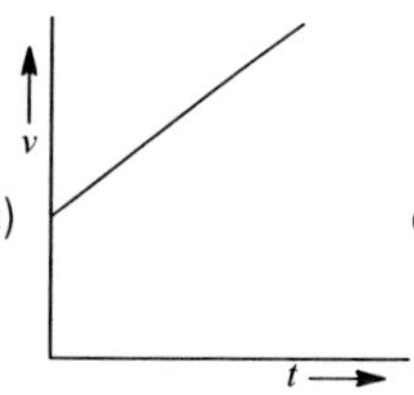

d) 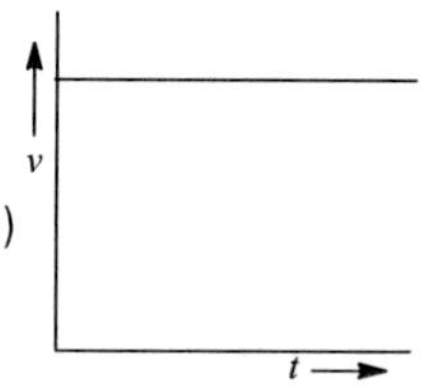

741. A point starts moving in a straight line with a certain acceleration. At a time t after beginning of motion the acceleration suddenly becomes retardation of the same value. The time in which the point returns to the initial point is

a) $\sqrt{2t}$
b) $(2+\sqrt{2})t$
c) $\frac{t}{\sqrt{2}}$
d) Cannot be predicted unless acceleration is given

742. The speed of body moving with uniform acceleration is u.This speed is doubled while covering distance S, its speed would be become

a) $\sqrt{3}u$ b) $\sqrt{5}u$ c) $\sqrt{11}u$ d) $\sqrt{7}u$

743. The position of a particle at any instant t is given by $x = a\cos\omega t$. The speed-time graph of the particle is

a) 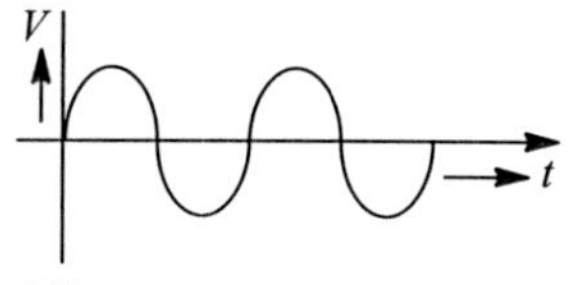

b) 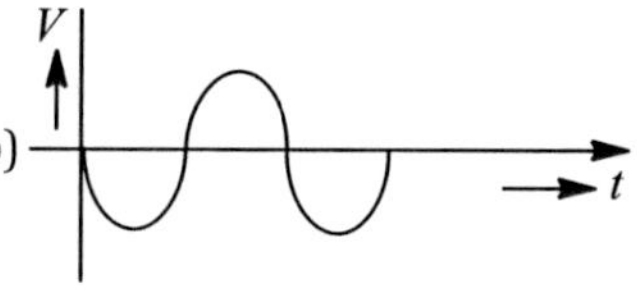

c) 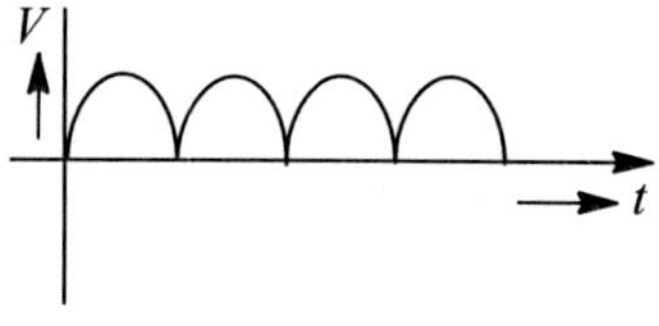

d) 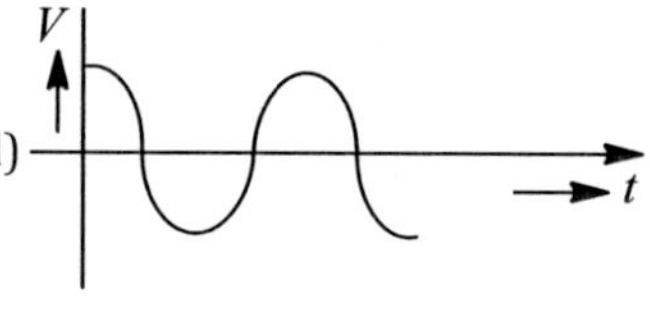

744. A police jeep is chasing with velocity of $45\ km/h$a thief in another jeep moving with velocity $153\ km/h$. Police fires a bullet with muzzle velocity of $180\ m/s$. The velocity with which it will strike the car of the thief is

a) $150\ m/s$ b)$27\ m/s$ c) $450\ m/s$ d)$250\ m/s$

745. A body is travelling in a straight line with a uniformly increasing speed. Which one of the plot represents the changes in distance (s) travelled with time (t)

a) 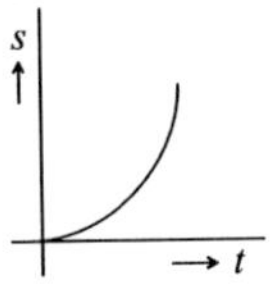

b) 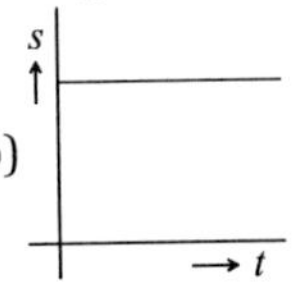

c) 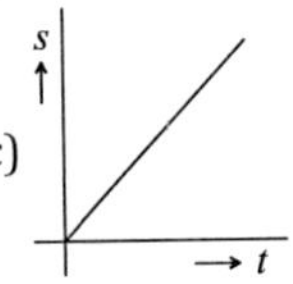

d) 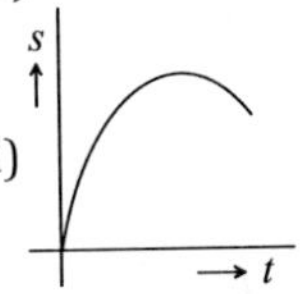

746. An athlete completes one round of a circular track of radius R in 40 sec. What will be his displacement at the end of 2 min. 20 sec

a) Zero b)$2R$ c) $2\pi R$ d)$7\pi R$

747. The variation of velocity of a particle moving along a straight line is shown in the figure. The distance travelled by the particle in 4s is

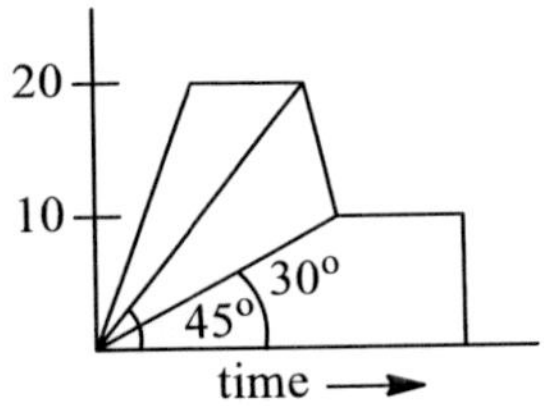

a) 25 m b)30 m c) 55 m d)60 m

748. The acceleration of a particle increasing linearly with time $t$is bt. The particle starts from the origin with an initial velocity v_0. The distance travelled by the particle in time t will be

a) $v_0 t+\frac{1}{6}bt^3$ b)$v_0 t+\frac{1}{3}bt^2$ c) $v_0 t+\frac{1}{3}bt^3$ d)$v_0 t+\frac{1}{3}bt^2$

749. A car starts from rest and accelerates uniformly to a speed of 180 kmh^{-1} in 10 $seconds$. The distance covered by the car in this time interval is

a) $500\ m$ b)$250\ m$ c) $100\ m$ d)$200\ m$

750. A ball is thrown vertically upwards with an initial velocity 1.4 ms^{-1} returns in 2 s. The total displacement of the ball will be

a) 22.4 m b)Zero c) 33.6 d)44.8 m

751. A particle moving in a straight line and passes through a point O with a velocity of $6ms^{-1}$.The particle moves with a constant retardation of $2ms^{-2}$ for 4 s and there after moves with a constant velocity. How long after leaving O does the particle return to O

a) $3s$ b)$8s$ c) $6\ m$ d)$8\ m$

752. A body of mass m thrown horizontally with velocity $v$from the top of tower of height h, touches the level ground a distance of 250 m from the foot of the tower. A body of mass, 2 m thrown horizontally with velocity $\frac{v}{2}$, from the top of tower of height 4 h will touch the level ground at a distance x from the foot of tower. The value of x is

a) 250 m b)500 m c) 125 m d)$250\sqrt{2}$m

753. A body projected vertically upwards crosses a point twice in its journey at a height h just after t_1 and t_2 second. Maximum height reached by the body is

a) $\frac{g}{4}(t_1+t_2)^2$ b)$g\left(\frac{t_1+t_2}{4}\right)^2$ c) $2g\left(\frac{t_1+t_2}{4}\right)^2$ d)$\frac{g}{4}(t_1 t_2)$

754. A body is falling freely under gravity. The distances covered by the body in first, second and third minute of its motion are in the ratio

a) 1 : 4 :9 b)1 : 2 : 3 c) 1 : 3 : 5 d)1 : 5 : 6

755. A man is 45 m behind the bus when the bus start accelerating from rest with acceleration 2.5 m/s^2. With what minimum velocity should the man start running to catch the bus

a) 12 m/s b) 14 m/s c) 15 m/s d) 16 m/s

756. The given graph shows the variation of velocity with displacement. Which one of the graph given below correctly represents the variation of acceleration with displacement?

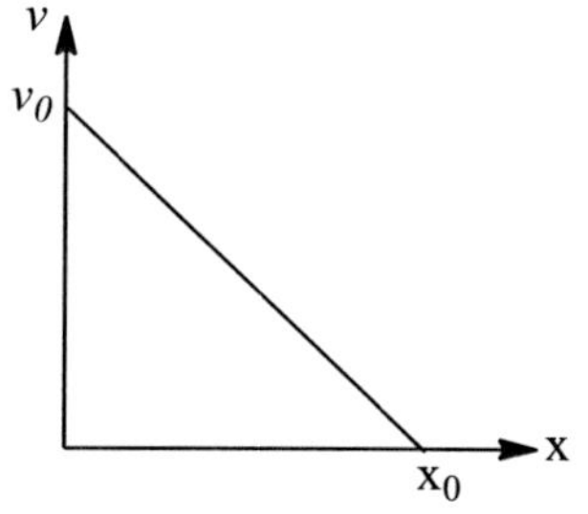

a)

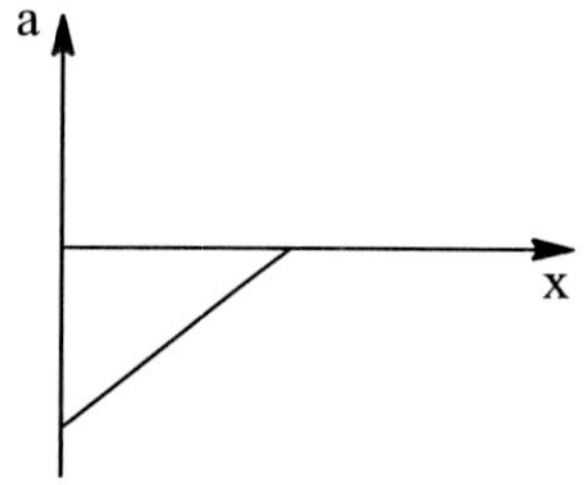

b)

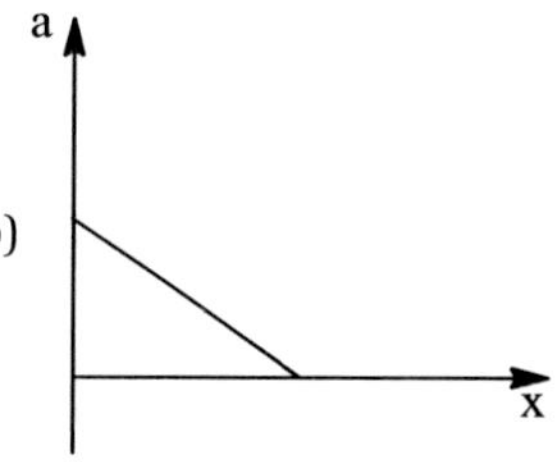

c)

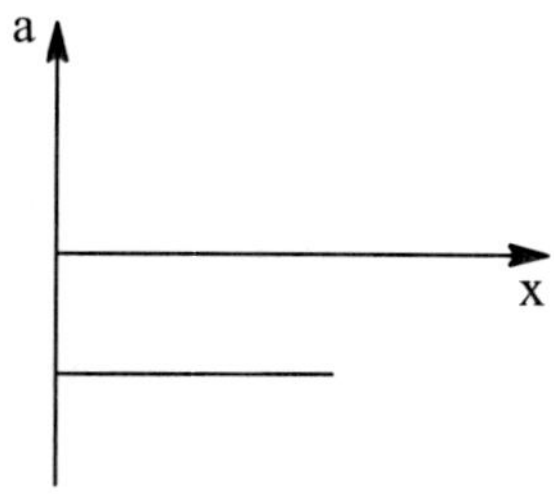

d)

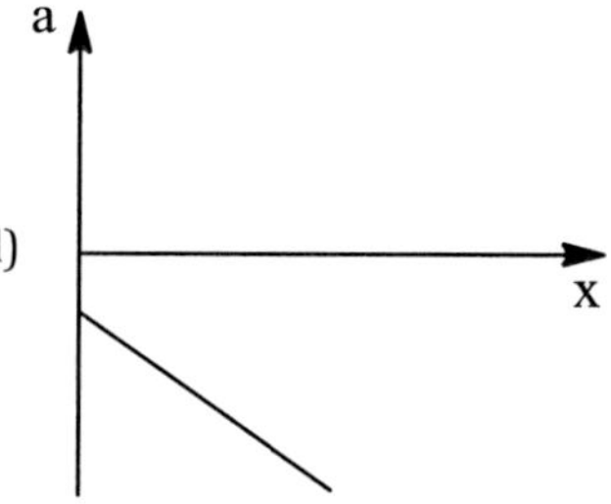

757. The acceleration of a particle increases linearly with time t as $6t$. If the initial velocity of the particle is zero and the particle starts from the origin, then the distance travelled by the particle in time t will be

a) t b) t^2 c) t^3 d) t^4

758. A body starts from origin and moves along x-axis such that at any instant velocity is $v_t = 4t^3 - 2t$ where t is in second and v_t in ms^{-1}. The acceleration of the particle when it is 2m from the origin is

a) $28ms^{-2}$ b) $22ms^{-2}$ c) $12ms^{-2}$ d) $10ms^{-2}$

759. A bullet fired into a fixed target loses half of its velocity after penetrating 3 cm. How much further it will penetrate before coming to rest assuming that if faces constant resistance to motion

a) 1.5 cm b) 1.0 cm c) 3.0 cm d) 2.0 cm

760. A particle is moving with uniform acceleration along a straight line. The average velocity of the particle from P to Q is $8 ms^{-1}$ and that from Q to S is $12 ms^{-1}$. If $QS - PQ$, then the average velocity from P to S is

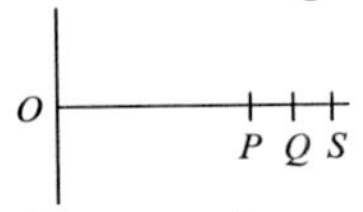

a) $9.6 ms^{-1}$ b) $12.87 ms^{-1}$ c) $64 ms^{-1}$ d) $327 ms^{-1}$

761. A stone thrown upward with a speed u from the top of the tower reaches the ground with a velocity $3u$. The height of the tower is

a) $3u^2/g$ b) $4u^2/g$ c) $6u^2/g$ d) $9u^2/g$

762. A stone is dropped into water from a bridge 44.1m above the water. Another stone is thrown vertically downward 1s later. Both strike the water simultaneously. What was the initial speed of the second stone?

a) $12.25 ms^{-1}$ b) $14.75 ms^{-1}$ c) $16.23 ms^{-1}$ d) $17.15 ms^{-1}$

763. A boggy of uniformly moving train is suddenly detached from train and stops after covering some distance. The distance covered by the boggy and distance covered by the train in the same time has relation

a) Both will be equal b) First will be half of second
c) First will be 1/4 of second d) No definite ratio

764. Velocity-time curve for a body projected vertically upwards is

a) Parabola b) Ellipse c) Hyperbola d) Straight line

765. If the figure below represents a parabola, identify the physical quantities representing Y and X for constant acceleration

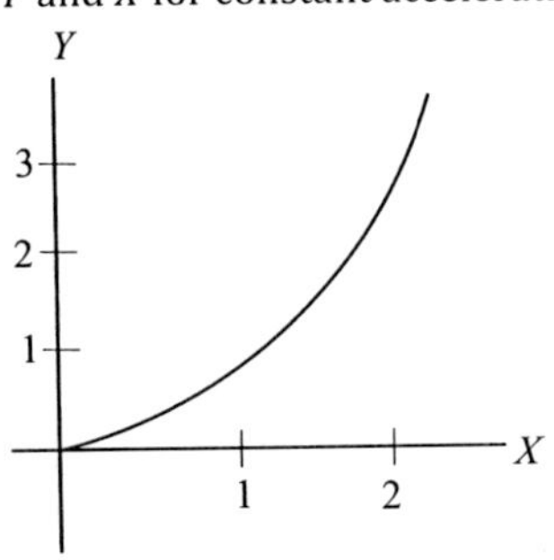

a) X = time, Y = velovity b) X = velocity, Y = time
c) X = time, Y = displacement d) X = time, Y = acceleration

766. A particle moving with a uniform acceleration along a straight line covers distance a and b in successive intervals of p and q second. The acceleration of the particle is

a) $\frac{pq(p+q)}{2(bp-aq)}$ b) $\frac{2(aq-bp)}{pq(p-q)}$ c) $\frac{bp-aq}{pq(p-q)}$ d) $\frac{2(bp-aq)}{pq(p-q)}$

767. A person moves 30 m north and then 20 m towards east and finally $30\sqrt{2}$m in south-west direction. The displacement of the person from the origin will be

a) 10 m along north b) 10 m along south c) 10 m along west d) Zero

768. The position of a particle moving along x-axis at certain times is given below:

$t(s)$	0	1	2	3
$x(m)$	-2	0	6	16

Which of the following describes the motion correctly

a) Uniform accelerated
b) Uniform decelerated
c) Non-uniform accelerated
d) There is not enough data for generalization

769. A body moves with initial velocity 10 ms^{-1}. If it covers a distance of 20m in 2s, then acceleration of the body is

a) Zero b) $10ms^{-2}$ c) $5ms^{-2}$ d) $2ms^{-2}$

770. A projectile is fired vertically upwards with an initial velocity u. After an interval of T seconds a second projectile is fired vertically upwards, also with initial velocity is

a) They meet at time $t=\frac{u}{g}$ and at a height $\frac{u^2}{2G}+\frac{gT^2}{8}$

b) They meet at time $t=\frac{u}{g}+\frac{T}{2}$ and at a height $\frac{u^2}{2g}+\frac{gT^2}{8}$

c) They meet at time $t=\frac{u}{g}+\frac{T}{2}$ and at a height $\frac{u^2}{2g}-\frac{gT^2}{8}$

d) They never meet

771. A very large number of balls are thrown vertically upwards in quick succession in such a way that the next ball is thrown when the previous one is at the maximum height. If the maximum height is 5m, the number of ball thrown per minute is (take $g=10\ ms^{-2}$)

a) 120 b) 80 c) 60 d) 40

772. A body starts from rest and moves with uniform acceleration. Which of the following graphs represent its motion

a) 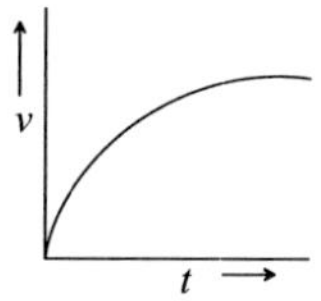

b) 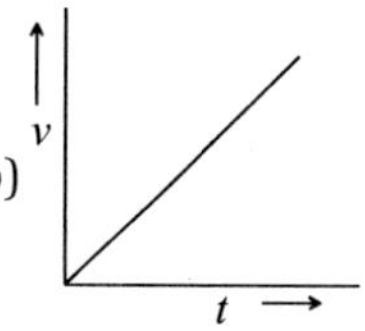

c) 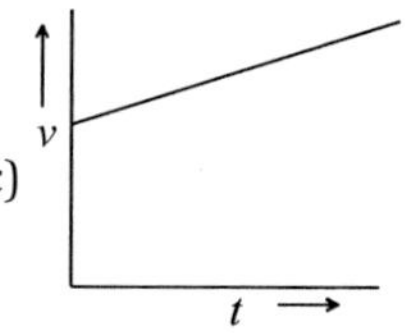

d) 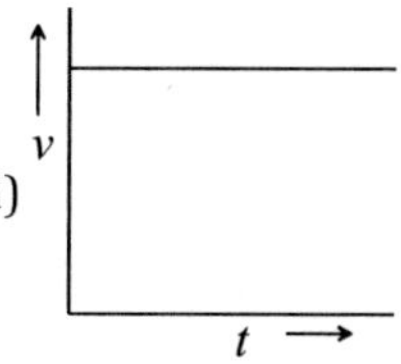

773. A person going towards east in a car with a velocity of 25kmh^{-1}, a train appears to move towards north with a velocity of $25\sqrt{3}\text{kmh}^{-1}$. The actual velocity of the train will be

a) 25 kmh^{-1} b) 50 kmh^{-1} c) 5 kmh^{-1} d) 35 kmh^{-1}

774. Two balls are dropped to the ground from different heights. One ball is dropped 2s after the other but they both strike the ground at the same time. If the first ball takes 5s to reach the ground, then the difference in initial heights is ($g=10\ ms^{-2}$)

a) 20m b) 80m c) 170m d) 40m

775. A body dropped from a height h with an initial speed zero, strikes the ground with a velocity 3 km/h. Another body of same mass is dropped from the same height h with an initial speed $-u' = 4km/h$. Find the final velocity of second body with which it strikes the ground

a) 3 km/h b) 4 km/h c) 5 km/h d) 12 km/h

776. The displacement-time graphs of two particles A and B are straight lines making angles of respectively 30° and 60° with the time axis. If the velocity of A is v_A and that of B is v_B, then the value of $\frac{v_A}{v_B}$ is

a) $\frac{1}{2}$ b) $\frac{1}{\sqrt{3}}$ c) $\sqrt{3}$ d) $\frac{1}{3}$

777. The velocity of a particle is $v=v_0+gt+ft^2$. If its position is $x=0$ at $t=0$, then its displacement after unit time $(t=1) is$

a) $v_0=2g+3f$ b) $v_0+g/2+f/3$ c) v_0+g+f d) $v_0+g/2+f$

778. A body starting from rest, accelerates at a constant rate a m/s^2 for some time after which it decelerates at a constant rate $b\ m/s^2$ to come to rest finally. If the total time elapsed is $t\ sec$, the maximum velocity attained by the body is given by

a) $\frac{ab}{a+b}t\ m/s$ b) $\frac{ab}{a-b}t\ m/s$ c) $\frac{2ab}{a+b}t\ m/s$ d) $\frac{2ab}{a-b}t\ m/s$

779. A stone is shot straight upward with a speed of 20 m/sec from a tower 200 m high. The speed with which it strikes the ground is approximately

a) 60 m/sec b) 65 m/sec c) 70 m/sec d) 75 m/sec

780. Two identical metal spheres are released from the top of a tower after t seconds of each other such that they fall along the same vertical line. If air resistance is neglected, then at any instant of time during their fall

a) The difference in their displacements remains the same
b) The difference between their speeds remains the same
c) The difference between their heights above ground is proportional to t^2
d) The difference between their displacements is proportional to t

781. A man throws a ball vertically upward and it rises through 20 m and returns to his hands. What was the initial velocity (u) of the ball and for how much time (T) it remained in the air [$g = 10m/s^2$]

a) $u = 10m/s, T = 2s$ b) $u = 10m/s, T = 4s$ c) $u = 20m/s, T = 2s$ d) $u = 20m/s, T = 4s$

782. A particle moving in a straight line covers half the distance with speed of 3 m/s. The other half of the distance is covered in two equal time intervals with speed of 4.5 m/s and 7.5 m/s respectively. The average speed of the particle during the motion is

a) 4.0 m/s b) 5.0 m/s c) 5.5 m/s d) 4.8 m/s

783. A small block sides without friction down an inclined plane starting from rest. Let S_n be the distance travelled from time $t = n - 1$ to $t = n$. Then $\frac{S_n}{S_{n+1}}$ is

a) $\frac{2n-1}{2n}$ b) $\frac{2n+1}{2n-1}$ c) $\frac{2n-1}{2n+1}$ d) $\frac{2n}{2n+1}$

784. The initial velocity of a particle is u (at $t = 0$) and the acceleration f is given by at. Which of the following relation is valid

a) $v = u + at^2$ b) $v = u + a\frac{t^2}{2}$ c) $v = u + at$ d) $v = u$

785. The area under acceleration-time graph gives

a) Distance travelled b) Change in acceleration
c) Force acting d) Change in velocity

786. The motion of a particle along a straight line is described by equation :

$x = 8 + 12t - t^3$

Where x is in metre and t in second. The retardation of the particle when its velocity becomes zero, is

a) $24ms^{-2}$ b) Zero c) $6ms^{-2}$ d) $12ms^{-2}$

787. A body is travelling in a straight line with a uniformly increasing speed. Which one of the plot represents the change in distance (s) travelled with time (t)?

a) 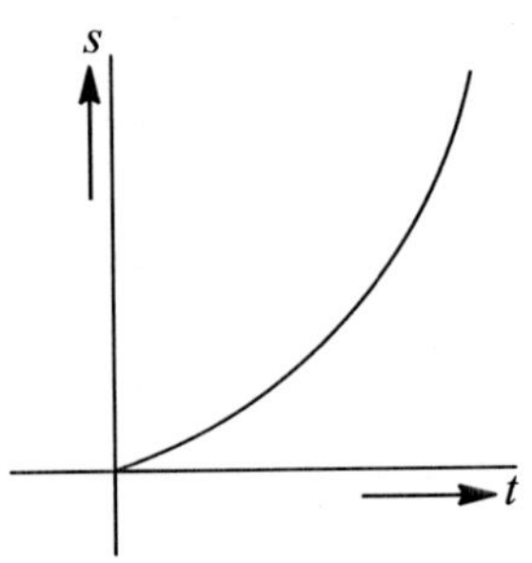

b) 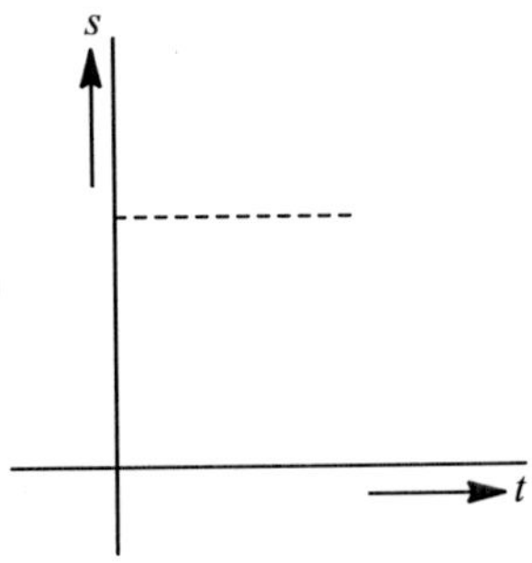

c) 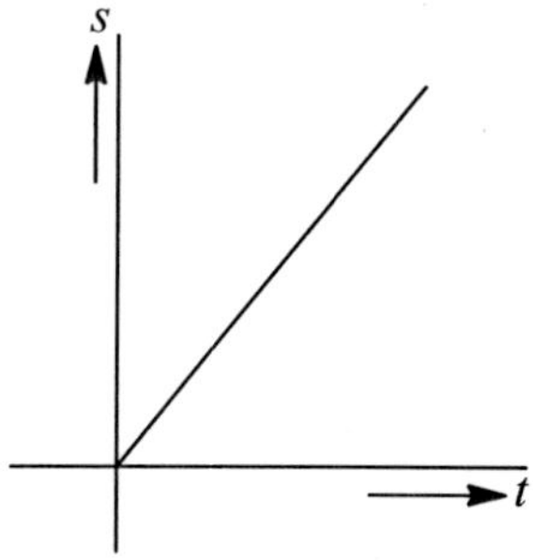 d) 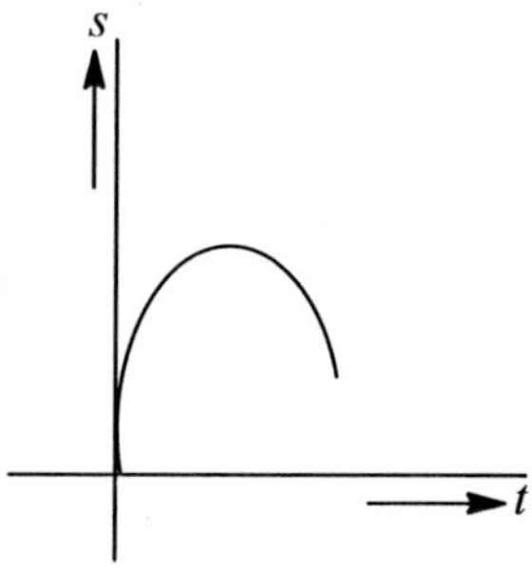

788. A particle crossing the origin of co-ordinates at time $t = 0$, moves in the xy −plane with a constant acceleration a in the y − direction. If its equation of motion is $y = bx^2$ (b is a constant), its velocity component in the x −direction is

a) $\sqrt{\frac{2b}{a}}$ b) $\sqrt{\frac{a}{2b}}$ c) $\sqrt{\frac{a}{b}}$ d) $\sqrt{\frac{b}{a}}$

789. A ball is thrown vertically upwards. Which of the following graph/graphs represent velocity-time graph of the ball during its flight (air resistance is neglected)

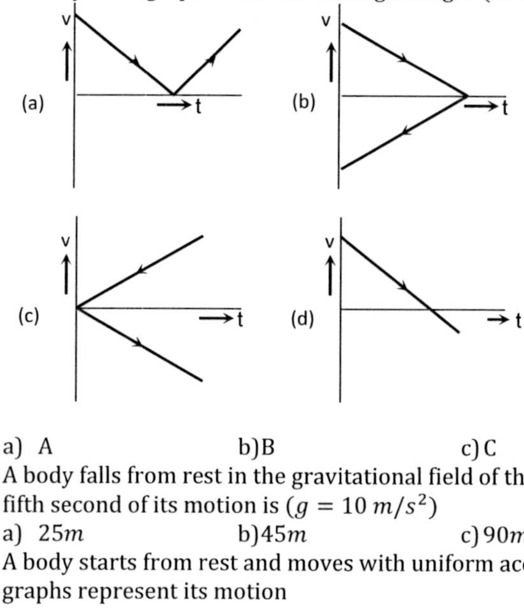

a) A b) B c) C d) D

790. A body falls from rest in the gravitational field of the earth. The distance travelled in the fifth second of its motion is ($g = 10\ m/s^2$)

a) $25m$ b) $45m$ c) $90m$ d) $125m$

791. A body starts from rest and moves with uniform acceleration. Which of the following graphs represent its motion

a) 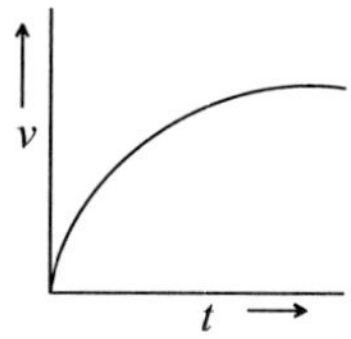 b) 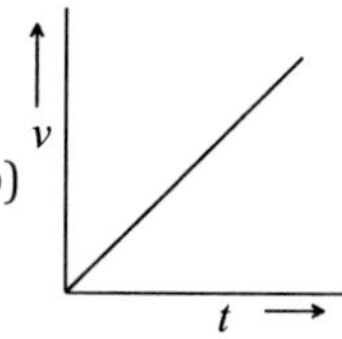 c) 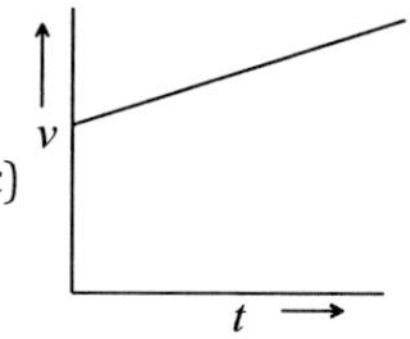 d) 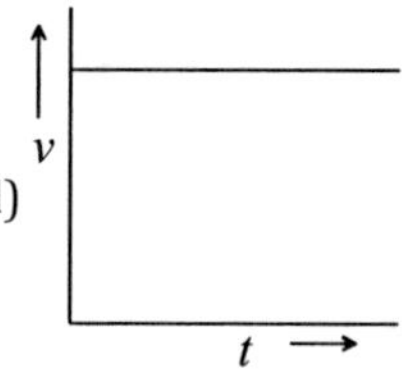

792. A car travels half the distance with constant velocity of 40 $kmph$ and the remaining half with a constant velocity of 60 $kmph$. The average velocity of the car in $kmph$ is

a) 40 b) 45 c) 48 d) 50

793. A body, thrown upwards with some velocity, reaches the maximum height of $20m$. Another body with double the mass thrown up, with double initial velocity will reach a maximum height of

a) $200\ m$ b) $16\ m$ c) $80\ m$ d) $40\ m$

794. The relation between time and distance is $t = \alpha x^2 + \beta x$, where α and β are constants. The retardation is

a) $2\alpha v^3$ b) $2\beta v^3$ c) $2\alpha\beta v^3$ d) $2\beta^2 v^3$

795. A goa can travel at a speed of 8 kmh^{-1} in still water on a lake. In the flowing water of a stream, it can move at 8kmh^{-1} relative to the water in the stream. If the stream speed is 3kmh^{-1}, how fast can the boat move past a tree on the shore in travelling (i) upstream (ii) downstream?

a) 5kmh^{-1} and 11 kmh^{-1} b) 11kmh^{-1} and 5 kmh^{-1}
c) 8kmh^{-1} and 8 kmh^{-1} d) 5$kmh^{-1}$and 5 kmh^{-1}.

796. An aeroplane files around a square field ABCD of each side 1000 km. Its speed along AB is 250 $km\ h^{-1}$, along DA 100 $km\ h^{-1}$. Its average speed (in $km\ h^{-1}$) over the entire trip is

a) 225.5 b) 175.5 c) 125.5 d) 190.5

797. A body starting from rest, accelerates at a constant rate a m/s^2 for some time after which it decelerates at a constant rate $b\ m/s^2$ to come to rest finally. If the total time elapsed is $t\ sec$, the maximum velocity attained by the body is given by

a) $\frac{ab}{a+b} t\ m/s$ b) $\frac{ab}{a-b} t\ m/s$ c) $\frac{2ab}{a+b} t\ m/s$ d) $\frac{2ab}{a-b} t\ m/s$

798. A body starts from rest, with uniform acceleration. If its velocity after n seconds is v,then its displacement in the last two seconds is

a) $\frac{2v(n+1)}{n}$ b) $\frac{v(n+1)}{n}$ c) $\frac{v(n-1)}{n}$ d) $\frac{2v(n-1)}{n}$

799. The displacement-time graph of moving particle is shown below

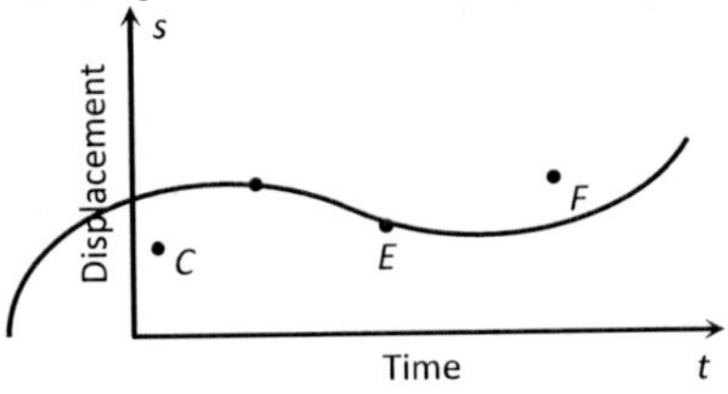

The instantaneous velocity of the particle is negative at the point

a) D b) F c) C d) E

800. The coordinates of a moving particle at any time t are given by $x = \alpha t^3$ and $y = \beta t^3$. The speed of the particle at time t is given by

a) $3t\sqrt{\alpha^2 + \beta^2}$ b) $3t^2\sqrt{\alpha^2 + \beta^2}$ c) $t^2\sqrt{\alpha^2 + \beta^2}$ d) $\sqrt{\alpha^2 + \beta^2}$

801. The velocity of particle is $v = v_0 + gt + ft^2$. If its position is $x = 0$ at $t = 0$, then its displacement after unit time ($t = 1$) is

a) $v_0 + 2g + 3f$ b) $v_0 + g/2 + f/3$ c) $v_0 + g + f$ d) $v_0 + g/2 + f$

802. The acceleration due to gravity on the planet A is 9 times the acceleration due to gravity on the planet B. A man jumps to a height of $2m$ on the surface of A. What is the height of jump by the same person on the planet B

a) $18\ m$ b) $6\ m$ c) $\frac{2}{3} m$ d) $\frac{2}{9} m$

803. A car accelerates from rest at a constant rate of 2$ms^{-2}$for sometime. Then, it retards at a constant rate of 4$ms^{-2}$and comes to rest. If the total time for which it remains in motion is 3s, what is the total distance travelled?

a) 2m b) 3m c) 4m d) 6m

804. The acceleration 'a' in m/s^2 of a particle is given by $a = 3t^2 + 2t + 2$ where t is the time. If the particle starts out with a velocity $u = 2m/s$ at $t = 0$, then the velocity at the end of 2 *seconds* is

a) 12 m/s b)18 m/s c) 27 m/s d)36 m/s

805. A particle is moving on a straight line path with constant acceleration directed along the direction of instantaneous velocity. Which of the following statements are false about the motion of particle?

a) Particle may reverse the direction of motion
b) Distance covered is not equal to magnitude of displacement
c) The magnitude of average velocity is less than average speed
d) All of the above

806. Two spheres of same size, one of mas 2 kg and another of mass 4 kg, are dropped simultaneously from the top of Qutub Minar (height $= 72m$). When they are 1 m above the ground, the two spheres have the same

a) Momentum b)Kinetic energy c) Potential energy d)Acceleration

807. A ball dropped from the 9th story of a multi-storeyed building reaches the ground in 3s. In the first second of its free fall, it passes through n stories, where n is equal to (Take $g = 10ms^{-2}$)

a) 1 b)2 c) 3 d)4

808. A stone is dropped from a height h. Simultaneously, another stone is thrown up from the ground which reaches a height 4 h. The two stones cross other after time

a) $\sqrt{\frac{h}{8g}}$ b)$\sqrt{8gh}$ c) $\sqrt{2gh}$ d) $\sqrt{\frac{h}{2g}}$

809. A body is moving with uniform acceleration describes 40 m in the first 5 sec and 65 m in next 5 sec. Its initial velocity will be

a) 4 m/s b)2.5 m/s c) 5.5 m/s d)11 m/s

810. The area under acceleration-time graph gives

a) Distance in travelled b)Change in acceleration
c) Force acting d)Change in velocity

811. An aeroplane files 400 m north and 300 m south and then files 1200 m upwards then net displacement is

a) 1200 m b)1300 m c) 1400 m d)1500 m

812. A bullet fired into a fixed target loses half of its velocity after penetrating 3 cm. How much further it will penetrate before coming to rest assuming that if faces constant resistance to motion

a) 1.5 cm b)1.0 cm c) 3.0 cm d)2.0 cm

813. Which of the following 4 statements is false

a) A body can have zero velocity and still be accelerated
b) A body can have a constant velocity and still have a varying speed
c) A body can have a constant speed and still have a varying velocity
d) The direction of the velocity of a body can change when its acceleration is constant

814. In the above question, the nearest distance between the two persons is

a) 10 m b)9 m c) 8 m d)7.2 m

815. A parachutist after bailing out falls 50 m without friction. When parachute opens, it decelerates at 2 m/s^2. He reaches the ground with a speed of 3 m/s. At what height, did he bail out

a) 293 m b)111 m c) 91 m d)182 m

816. A truck and a car are moving with equal velocity. On applying the brakes both will stop after certain distance, then

a) Truck will cover less distance before rest b)Car will cover less distance before rest
c) Both will cover equal distance d)None

817. A body is moving with uniform acceleration describes 40 m in the first 5 sec and 65 m in next 5 sec. Its initial velocity will be

a) 4 m/s b) 2.5 m/s c) 5.5 m/s d) 11 m/s

818. A balloon is at a height of 81 m and is ascending upwards with a velocity of 12 m/s. A body of $2kg$ weight is dropped from it. If $g = 10\ m/s^2$, the body will reach the surface of earth in

a) 1.5 s b) 4.025 s c) 5.4 s d) 6.75 s

819. A particle moving along a straight line has a velocity v ms^{-1}, when it cleared a distance of y metre. These two are connected by the relation $v = \sqrt{49 + y}$. When its velocity is 1ms^{-1}, its acceleration (in ms^{-2}) is

a) 1 b) 2 c) 7 d) 0.5

820. A particle is dropped vertically from rest a height. The time taken by it to fall through successive distances of 1 m each will then be

a) All equal, being equal to $\sqrt{2/g}$ second

b) In the ratio of the square roots of the integers 1, 2, 3.......

c) In the ratio of the difference in the square roots of the integers i.e. $\sqrt{1}, (\sqrt{2} - \sqrt{1}), (\sqrt{3} - \sqrt{2}), (\sqrt{4} - \sqrt{3})$......

d) In the ratio of the reciprocal of the square roots of the integers i.e.,. $\frac{1}{\sqrt{1}}, \frac{1}{\sqrt{2}}, \frac{1}{\sqrt{3}}, \frac{1}{\sqrt{4}}$

821. The acceleration experienced by a moving boat after is engine is cut-off, is given by $a = -kv^3$, where k is a constant. If v_0 is the magnitude of velocity at cut-off, then the magnitude of the velocity at time t after the cut off is

a) $\frac{v_0}{2ktv_0^2}$ b) $\frac{v_0}{1 + 2\ ktv_0^2}$ c) $\frac{v_0}{\sqrt{1 - 2\ k\ v_0^2}}$ d) $\frac{v_0}{\sqrt{1 + 2\ ktv_0^2}}$

822. A body begins to walk eastward along a street in front of his house and the graph of his position from home is shown in the following figure. His average speed for the whole time interval is equal to

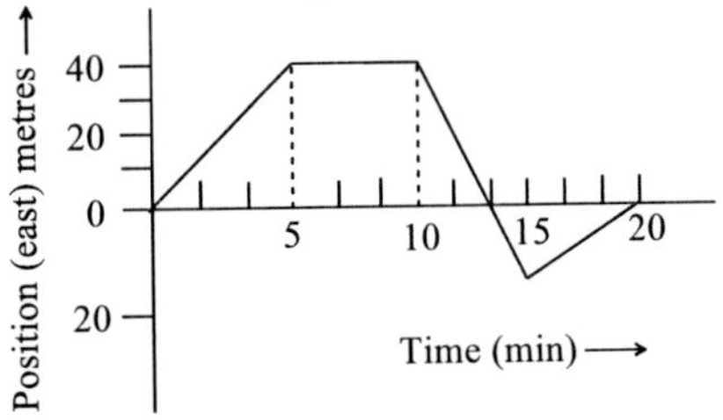

a) 8 m/min b) 6 m/min c) $\frac{8}{3}$ m/min d) 2 m/min

823. A particle is moving with constant acceleration from A to B in a straight line AB. If u and v are the velocities at A and B respectively then its velocity at the midpoint C will be

a) $\left(\frac{u^2 + v^2}{2u}\right)^2$ b) $\frac{u + v}{2}$ c) $\frac{v - u}{2}$ d) $\sqrt{\frac{u^2 + v^2}{2}}$

824. A particle moves along the sides AB, BC, CD of a square of side 25 m wuth a velocity of 15 ms^{-1}. Its average velocity is

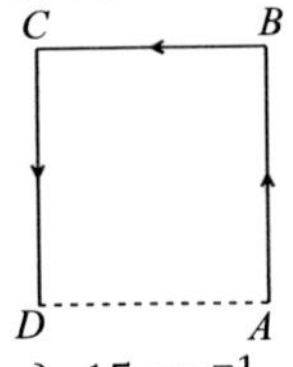

a) 15 ms^{-1} b) 10ms^{-1} c) 7.5ms^{-1} d) 5ms^{-1}

825. The retardation experienced by a moving motor boat, after its engine is cut off, is given by

$\frac{du}{dt} = -kv^3$,where$k$ is a constant.

If $v_0$is the magnitude of the velocity at cut-off, the magnitude of the velocity at time t after the cut-off is

a) v_0 b)$\frac{v_0}{2}$ c) $v_0 e^{-kt}$ d)$\frac{v_0}{\sqrt{2v_0^2kt+1}}$

826. If a body starts from rest and travels 120 cm in the 6th second, then what is the acceleration

a) 0.20 m/s^2 b)0.027 m/s^2 c) 0.218m/s^2 d)0.03m/s^2

827. The given graph shows the variation of velocity with displacement. Which one of the graph given below correctly represents the variation of acceleration with displacement

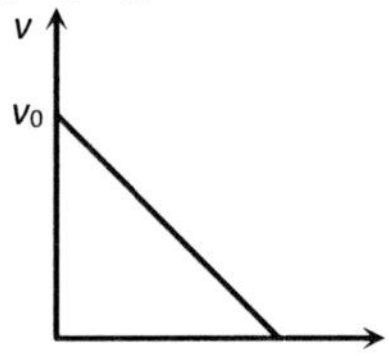

a) 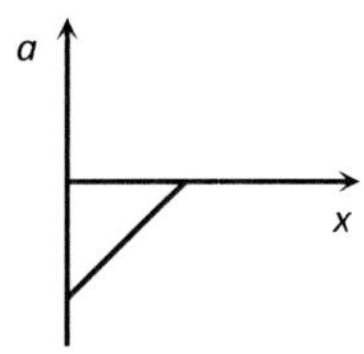

b) 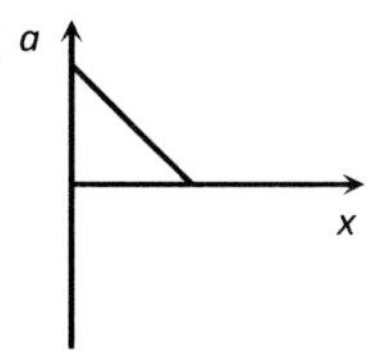

c) 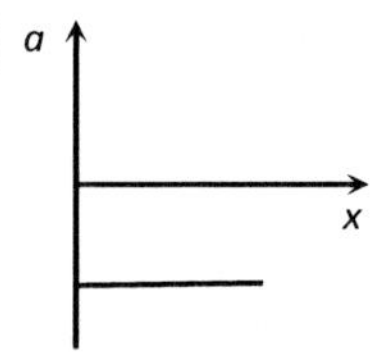

d) 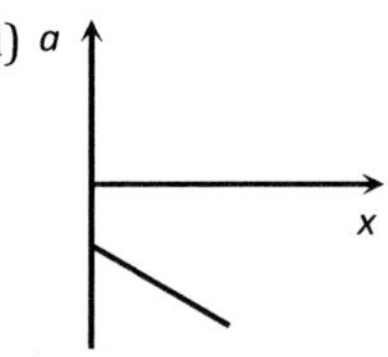

828. A metro train starts from rest and in five seconds achieves 108 kmh^{-1}. After that it moves with constant velocity and comes to rest after travelling 45 m with uniform retardation. If total distance travelled is 395 m, find total time of travelling

a) 12.2 s b)15.3 s c) 9 s d)17.2 s

829. From the top of a tower, a particle is thrown vertically downwards with a velocity of $10ms^{-1}$. The ratio of the distances covered by it in the 3rd and 2nd seconds of its motion is (Given, $g = 10ms^{-2}$)

a) 7:5 b)3:4 c) 4:3 d)6:5

830. An aircraft is flying at a height of 34000m above the ground. If the angle subtended at a ground observation point by the aircraft positions 10s apart is 30°, then the speed of the aircraft is

a) 19.63ms^{-1} b)1963 ms^{-1} c) 108 ms^{-1} d)196.3 ms^{-1}

831. A stone is thrown with an initial speed of 4.9 m/s from a bridge in vertically upward direction. It falls down in water after 2 sec. The height of the bridge is

a) 4.9 m b)9.8 m c) 19.8 m d)24.7 m

832. The velocity of a bullet is reduced from 200 m/s to 100 m/s while travelling through a wooden block of thickness 10 cm. The retardation, assuming it to be uniform, will be

a) $10 \times 10^4\ m/s^2$ b)$12 \times 10^4\ m/s^2$ c) $13.5 \times 10^4\ m/s^2$ d)$15 \times 10^4\ m/s^2$

833. A particle moves in a straight line so that its displacement x metre at time t second is given by

$t = \sqrt{x^2 - 1}$

Its acceleration in ms^{-2}at time t second is

a) $\frac{1}{x^3}$ b)$\frac{t^2}{x^3}$ c)$\frac{1}{x} - \frac{t^2}{x^3}$ d)$\frac{t^2}{x^3} - \frac{1}{x^2}$

834. From the top of a tower of two stones, whose masses are in the ratio 1 : 2 are thrown on straight up with an initial speed u and the second straight down with the same speed u. Then neglecting air resistance

a) The heavier stone hits the ground with a higher speed
b) The lighter stone hits the ground with a higher speed
c) Both the stones will have the same speed when they hit the ground
d) The speed can't be determined with the given data

835. A bus start from rest with an acceleration of 1 ms^{-2}. A man who is 48m behind the bus starts with a uniform velocity of $10ms^{-1}$. The minimum time after which the ma will catch the bus

a) 4.8 s b)8 s c) 10 s d)12 s

836. A body A is thrown up vertically from the ground with a velocity V_0 and another body B is simultaneously dropped from a height H. They meet at a height $\frac{H}{2}$ if V_0 is equal to

a) $\sqrt{2gH}$ b)$\sqrt{gH}$ c)$\frac{1}{2}\sqrt{gH}$ d) $\sqrt{\frac{2g}{H}}$

837. A body starts to fall freely under gravity. The distance covered by it in first, second and third *second* are in ratio

a) 1: 3: 5 b)1: 2: 3 c) 1: 4: 9 d)1: 5: 6

838. A car travels equal distances in the same direction with velocities 60kmh^{-1},20 km h^{-1} and 10 km h^{-1}respectively. The average velocity of the car over the whole journey of motion is

a) 8 ms^{-1} b)7 ms^{-1} c) 6 ms^{-1} d)5 ms^{-1}

839. Two bodies of different masses are dropped from heights of 16 m and 25 m respectively. The ratio of the time taken by them to reach the ground is

a) $\frac{25}{16}$ b)$\frac{5}{4}$ c)$\frac{4}{5}$ d)$\frac{16}{25}$

840. A car moving with a velocity of 10 m/s can be stopped by the application of a constant force F in a distance of 20 m. If the velocity of the car is 30 m/s. It can be stopped by this force in

a) $\frac{20}{3}m$ b)20 m c) 60 m d)180 m

3
MOTION IN A STRAIGHT LINE
ANSWER KEY

1)	b	2)	d	3)	c	4)	b
5)	b	6)	d	7)	d	8)	b
9)	d	10)	c	11)	b	12)	a
13)	d	14)	a	15)	b	16)	a
17)	b	18)	c	19)	c	20)	d
21)	d	22)	d	23)	d	24)	a
25)	c	26)	c	27)	b	28)	c
29)	c	30)	a	31)	c	32)	a
33)	c	34)	b	35)	c	36)	b
37)	b	38)	b	39)	c	40)	d
41)	a	42)	b	43)	d	44)	b
45)	c	46)	b	47)	a	48)	c
49)	a	50)	a	51)	b	52)	a
53)	c	54)	d	55)	b	56)	d
57)	d	58)	b	59)	c	60)	d
61)	c	62)	a	63)	c	64)	a
65)	d	66)	b	67)	a	68)	a
69)	c	70)	c	71)	d	72)	d
73)	c	74)	c	75)	b	76)	b
77)	b	78)	d	79)	a	80)	b
81)	c	82)	c	83)	d	84)	d
85)	d	86)	b	87)	c	88)	d
89)	c	90)	a	91)	b	92)	d
93)	b	94)	b	95)	a	96)	a
97)	b	98)	c	99)	d	100)	c
101)	c	102)	b	103)	c	104)	c
105)	d	106)	b	107)	b	108)	c
109)	c	110)	c	111)	b	112)	c
113)	c	114)	c	115)	c	116)	c
117)	b	118)	c	119)	b	120)	a
121)	d	122)	a	123)	b	124)	c

125)	c	126)	a	127)	c	128)	b
129)	d	130)	d	131)	a	132)	d
133)	c	134)	a	135)	c	136)	c
137)	a	138)	a	139)	d	140)	a
141)	c	142)	b	143)	b	144)	c
145)	d	146)	d	147)	b	148)	c
149)	d	150)	c	151)	b	152)	c
153)	b	154)	c	155)	a	156)	d
157)	b	158)	c	159)	c	160)	c
161)	a	162)	a	163)	b	164)	a
165)	b	166)	d	167)	b	168)	b
169)	b	170)	a	171)	a	172)	a
173)	a	174)	d	175)	b	176)	c
177)	d	178)	b	179)	c	180)	b
181)	a	182)	b	183)	d	184)	a
185)	c	186)	d	187)	c	188)	c
189)	b	190)	d	191)	a	192)	a
193)	b	194)	d	195)	a	196)	a
197)	c	198)	c	199)	c	200)	c
201)	a	202)	a	203)	c	204)	a
205)	a	206)	a	207)	a	208)	d
209)	b	210)	b	211)	c	212)	d
213)	c	214)	d	215)	b	216)	d
217)	d	218)	c	219)	c	220)	b
221)	b	222)	a	223)	b	224)	c
225)	b	226)	b	227)	a	228)	c
229)	a	230)	d	231)	d	232)	d
233)	c	234)	c	235)	c	236)	a
237)	c	238)	d	239)	d	240)	a
241)	d	242)	a	243)	b	244)	c
245)	b	246)	a	247)	b	248)	a
249)	c	250)	a	251)	c	252)	c
253)	b	254)	a	255)	b	256)	d
257)	b	258)	c	259)	d	260)	b
261)	c	262)	b	263)	a	264)	a
265)	b	266)	c	267)	c	268)	c

269)	a	270)	c	271)	b	272)	b
273)	d	274)	c	275)	c	276)	b
277)	a	278)	b	279)	c	280)	b
281)	a	282)	a	283)	c	284)	a
285)	c	286)	a	287)	a	288)	b
289)	a	290)	b	291)	d	292)	a
293)	d	294)	b	295)	d	296)	c
297)	b	298)	c	299)	b	300)	d
301)	b	302)	a	303)	a	304)	b
305)	b	306)	c	307)	d	308)	c
309)	a	310)	b	311)	b	312)	b
313)	b	314)	c	315)	b	316)	b
317)	d	318)	c	319)	a	320)	d
321)	b	322)	d	323)	b	324)	c
325)	c	326)	d	327)	c	328)	d
329)	b	330)	a	331)	d	332)	d
333)	b	334)	d	335)	a	336)	a
337)	d	338)	c	339)	a	340)	d
341)	d	342)	d	343)	c	344)	d
345)	b	346)	d	347)	d	348)	d
349)	a	350)	a	351)	a	352)	a
353)	a	354)	b	355)	a	356)	c
357)	b	358)	a	359)	a	360)	d
361)	a	362)	d	363)	d	364)	a
365)	d	366)	c	367)	a	368)	a
369)	a	370)	b	371)	d	372)	d
373)	d	374)	b	375)	b	376)	a
377)	a	378)	d	379)	c	380)	b
381)	a	382)	b	383)	a	384)	a
385)	c	386)	c	387)	d	388)	d
389)	a	390)	b	391)	a	392)	b
393)	b	394)	c	395)	b	396)	c
397)	b	398)	b	399)	c	400)	a
401)	c	402)	d	403)	a	404)	b
405)	a	406)	b	407)	b	408)	b
409)	c	410)	d	411)	d	412)	d

413)	d	414)	d	415)	a	416)	b
417)	d	418)	d	419)	a	420)	d
421)	c	422)	c	423)	d	424)	a
425)	c	426)	c	427)	b	428)	c
429)	a	430)	a	431)	b	432)	d
433)	a	434)	b	435)	a	436)	a
437)	d	438)	a	439)	a	440)	b
441)	a	442)	c	443)	a	444)	d
445)	b	446)	b	447)	b	448)	b
449)	d	450)	a	451)	c	452)	a
453)	b	454)	a	455)	b	456)	c
457)	b	458)	a	459)	b	460)	c
461)	d	462)	d	463)	d	464)	d
465)	d	466)	b	467)	c	468)	d
469)	a	470)	a	471)	b	472)	c
473)	d	474)	b	475)	c	476)	b
477)	b	478)	b	479)	a	480)	b
481)	b	482)	b	483)	a	484)	d
485)	a	486)	d	487)	a	488)	b
489)	a	490)	d	491)	c	492)	b
493)	d	494)	b	495)	d	496)	b
497)	b	498)	d	499)	c	500)	d
501)	c	502)	d	503)	b	504)	b
505)	b	506)	b	507)	b	508)	d
509)	d	510)	b	511)	d	512)	a
513)	c	514)	c	515)	a	516)	b
517)	b	518)	a	519)	d	520)	a
521)	b	522)	a	523)	b	524)	c
525)	c	526)	a	527)	b	528)	d
529)	c	530)	d	531)	d	532)	a
533)	b	534)	d	535)	a	536)	b
537)	b	538)	a	539)	b	540)	a
541)	a	542)	d	543)	d	544)	a
545)	a	546)	b	547)	b	548)	a
549)	d	550)	b	551)	d	552)	d
553)	c	554)	c	555)	d	556)	a

557)	b	558)	c	559)	c	560)	b
561)	d	562)	d	563)	a	564)	a
565)	d	566)	b	567)	c	568)	b
569)	b	570)	a	571)	a	572)	d
573)	d	574)	c	575)	b	576)	c
577)	b	578)	b	579)	b	580)	a
581)	b	582)	c	583)	a	584)	b
585)	c	586)	c	587)	a	588)	c
589)	d	590)	b	591)	d	592)	a
593)	a	594)	a	595)	a	596)	b
597)	a	598)	c	599)	b	600)	d
601)	b	602)	a	603)	d	604)	b
605)	c	606)	a	607)	c	608)	c
609)	b	610)	a	611)	a	612)	a
613)	c	614)	a	615)	a	616)	b
617)	c	618)	b	619)	b	620)	b
621)	d	622)	a	623)	a	624)	b
625)	c	626)	b	627)	d	628)	d
629)	c	630)	a	631)	d	632)	a
633)	d	634)	d	635)	a	636)	c
637)	d	638)	b	639)	a	640)	a
641)	d	642)	d	643)	c	644)	b
645)	d	646)	d	647)	a	648)	b
649)	b	650)	c	651)	b	652)	c
653)	d	654)	b	655)	d	656)	b
657)	b	658)	d	659)	c	660)	d
661)	b	662)	c	663)	b	664)	b
665)	a	666)	b	667)	b	668)	d
669)	b	670)	b	671)	c	672)	d
673)	a	674)	b	675)	b	676)	b
677)	d	678)	d	679)	b	680)	b
681)	c	682)	b	683)	d	684)	d
685)	b	686)	c	687)	c	688)	a
689)	a	690)	a	691)	c	692)	a
693)	c	694)	d	695)	c	696)	d
697)	a	698)	c	699)	c	700)	b

701)	d	702)	d	703)	d	704)	d
705)	a	706)	b	707)	c	708)	a
709)	c	710)	a	711)	a	712)	a
713)	d	714)	a	715)	a	716)	b
717)	b	718)	b	719)	c	720)	a
721)	b	722)	a	723)	d	724)	a
725)	d	726)	b	727)	c	728)	c
729)	b	730)	c	731)	c	732)	b
733)	a	734)	a	735)	a	736)	b
737)	a	738)	b	739)	b	740)	b
741)	b	742)	d	743)	c	744)	a
745)	a	746)	b	747)	c	748)	a
749)	b	750)	b	751)	b	752)	a
753)	c	754)	c	755)	c	756)	a
757)	c	758)	b	759)	b	760)	a
761)	b	762)	a	763)	b	764)	d
765)	c	766)	b	767)	c	768)	c
769)	a	770)	c	771)	c	772)	b
773)	b	774)	b	775)	c	776)	d
777)	b	778)	a	779)	b	780)	b
781)	d	782)	a	783)	c	784)	b
785)	d	786)	d	787)	a	788)	b
789)	d	790)	b	791)	b	792)	c
793)	c	794)	a	795)	a	796)	d
797)	a	798)	d	799)	d	800)	b
801)	b	802)	a	803)	d	804)	b
805)	d	806)	d	807)	a	808)	a
809)	c	810)	d	811)	a	812)	b
813)	b	814)	c	815)	a	816)	c
817)	c	818)	c	819)	d	820)	c
821)	d	822)	b	823)	d	824)	d
825)	d	826)	c	827)	a	828)	d
829)	a	830)	d	831)	b	832)	d
833)	a	834)	c	835)	b	836)	b
837)	a	838)	d	839)	c	840)	d

3
MOTION IN A STRAIGHT LINE
HINTS AND SOLUTIONS

1 **(b)**
$H_{max} \propto u^2$, It body projected with double velocity then maximum height will become four times $i.e. 200\ m$

2 **(d)**
Let the car accelerate at rate α for time t_1 then maximum velocity attained,
$v = 0 + \alpha t_1 = \alpha t_1$
Now, the car decelerates at a rate β for time $(t - t_1)$ and finally comes to rest. Then,
$0 = v - \beta(t - t_1) \Rightarrow 0 = \alpha t_1 - \beta t + \beta t_1$
$\Rightarrow t_1 = \frac{\beta}{\alpha + \beta} t$
$\therefore v = \frac{\alpha\beta}{\alpha + \beta} t$

3 **(c)**
For same direction relative velocity $= |v_1 - v_2|$

Distance covered, $d = \frac{(v_1 - v_2)^2}{2a}$

For no collision, $d > \frac{(v_1 - v_2)^2}{2a}$

4 **(b)**
$v = u + at \Rightarrow -2 = 10 + a \times 4 \Rightarrow a = -3m/sec^2$

5 **(b)**
Bullet will take $\frac{100}{1000} = 0.1\ sec$ to reach target.
During this period vertical distance (downward) travelled by the bullet $= \frac{1}{2}gt^2 = \frac{1}{2} \times 10 \times (0.1)^2 m = 5cm$
So the gun should be aimed 5 cm above the target

6 **(d)**
Distance is defined as length of the path between two points.

In this case

Disance = Area of $v - t$ graph in 4 s
= Area of trangle
$= \frac{1}{2}$base $\times$ height

$= \frac{1}{2} \times 4 \times 8 = 16$ m

7 **(d)**
$\because v = 0 + na \Rightarrow a = v/n$
Now, distance travelled in $n\ sec. \Rightarrow S_n = \frac{1}{2}an^2$ and
distance travelled in $(n - 2)sec \Rightarrow S_{n-2} = \frac{1}{2}a(n - 2)^2$
$\therefore$ Distance travelled in last 2 seconds,
$= S_n - S_{n-2} = \frac{1}{2}an^2 - \frac{1}{2}a(n - 2)^2$
$\frac{a}{2}[n^2 - (n - 2)^2] = \frac{a}{2}[n + (n - 2)][n - (n - 2)]$

$= a(2n - 2) = \frac{v}{n}(2n - 2) = \frac{2v(n - 1)}{n}$

8 **(b)**
$H_{max} \propto u^2$, It body projected with double velocity then maximum height will become four times $i.e. 200\ m$

9 **(d)**
Since $x = 1.2t^2$ which is in form $x = \frac{1}{2}at^2$
Thus the motion is uniformly accelerated

10 **(c)**
$v^2 = u^2 + 2as$, If $u = 0$, then $v^2 \propto S$
$i.e.$, Graph should be parabola symmetric to displacement axis

11 **(b)**
Time taken by first drop to reach the ground $t = \sqrt{\frac{2h}{g}} \Rightarrow t = \sqrt{\frac{2 \times 5}{10}} = 1$ sec
As the water drops fall at regular intervals from a tap therefore time difference between any two drops $= \frac{1}{2}$sec
In this given time, distance of second drop from the tap $= \frac{1}{2}g\left(\frac{1}{2}\right)^2 = \frac{10}{8} = 1.25\ m$

Its distance from the ground $= 5 - 1.25 = 3.75m$

12 **(a)**
Effective speed of bullet
= speed of bullet + speed of police jeep
$= 180\ m/s + 45\ km/h$
$= (180 + 12.5)\ m/s$
$= 192.5\ m/s$
Speed of thief's jeep $= 153\ km/h$
$= 42.5\ m/s$
Velocity of bullet $w.r.t.$ thief's car $= 192.5 - 42.5 = 150\ m/s$

13 **(d)**
Velocity of particle $= \frac{\text{Total displacement}}{\text{Total time}}$
$= \frac{\text{Diameter of circle}}{5} = \frac{2 \times 10}{5} = 4\ m/s$

14 **(a)**
$t = \sqrt{\frac{2h}{g}} \Rightarrow \frac{t_1}{t_2} = \sqrt{\frac{h_1}{h_2}} = \sqrt{\frac{1}{2}} = \frac{1}{\sqrt{2}}$

15 **(b)**
$$Average\ velocity = \frac{Total\ distance\ covered}{Time\ of\ flight}$$
$$= \frac{2H_{max}}{2u/g}$$
$$\Rightarrow v_{av} = \frac{2u^2/2g}{2u/g} \Rightarrow v_{av} = u/2$$
Velocity of projection $= v$ [Given]
$\therefore v_{av} = u/2$

16 **(a)**

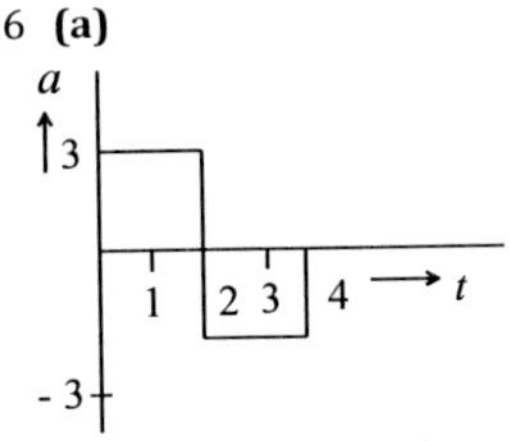

Takingthe motion from 0 to 2 s
$u = 0, a = 3ms^{-2}, t = 2s, v = ?$
$v = u + at = 0 + 3 \times 2 = 6ms^{-1}$
Taking the motion from 2 s to 4 s
$v = 6 + (-3)(2) = 0ms^{-1}$

17 **(b)**
Speed can never be negative. Hence (b) is correct.

18 **(c)**
Let initial velocity of body at point A is v, AB is 3 cm.

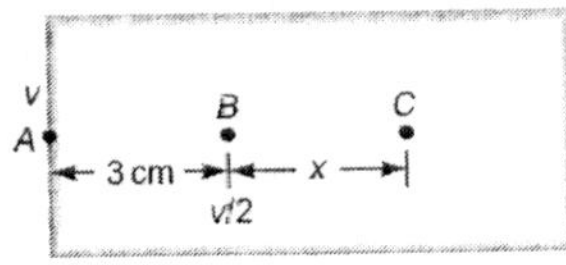

From $v^2 = u^2 - 2as$
$$\left(\frac{v}{2}\right)^2 = v^2 - 2a \times 3$$
$$a = \frac{v^2}{8}$$
Let on penetrating 3 cm in a wooden block, the body moves x distance form B to C.
So, for B to C
$$u = \frac{v}{2}, v = 0,$$
$$s = x, a = \frac{v^2}{8} \text{(deceleration)}$$
$$\therefore (0)^2 = \left(\frac{v}{2}\right)^2 - 2 . \frac{v^2}{8} . x$$
$$x = 1$$

19 **(c)**
For first case $v^2 - 0^2 = 2gh \Rightarrow (3)^2 = 2gh$
For second case $v^2 = (-u)^2 + 2gh = 4^2 + 3^2 \therefore v = 5km/h$

20 **(d)**
The velocity-time graph for the given situation can be drawn as below.
Magnitudes of slope of $OA = f$

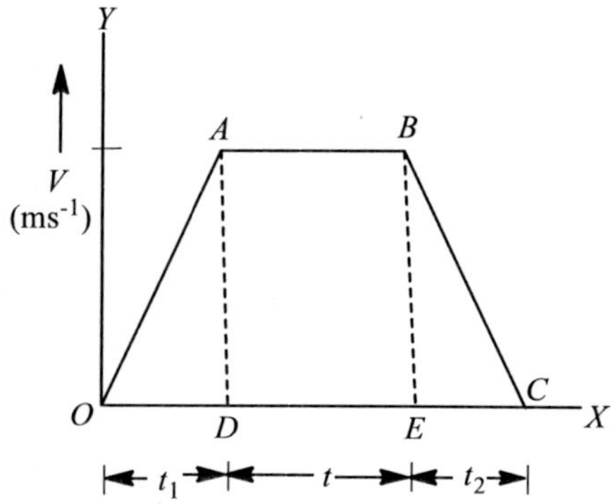

And slope of $BC = \frac{f}{2}$

$$v = f\,t_1 = \frac{f}{2}t_2$$

$$t_2 = 2t_1$$

In graph area of ΔOAD gives

Distance, $S = \frac{1}{2}\,f\,t_1^2$ ....(i)

Area of rectangle $ABED$ gives distance travelled in time t.

$$S_2 = (f\,t_1)t$$

Distance travelled in time $t_2 =$

$$S_3 = \frac{1}{2}\frac{f}{2}(2t_1)^2$$

Thus, $S_1 + S_2 + S_3 = 15\,S$

$$S + (ft_1)t + ft_1^2 = 15\,S$$

$$S + (ft_1)t + 2S = 15\,S \qquad \left(S = \frac{1}{2}\,ft_1^2\right)$$

$(ft_1)t = 12\,S$ ...(ii)

From Eqs. (i) and (ii), we have

$$\frac{12\,S}{S} = \frac{(ft_1)t}{\frac{1}{2}(ft_1)t_1}$$

$$\therefore t_1 = \frac{t}{6}$$

From Eq. (i), we get

$$\therefore S = \frac{1}{2}\,f(t_1)^2$$

$$\therefore S = \frac{1}{2}\,f\left(\frac{t}{6}\right)^2 = \frac{1}{72}\,ft^2$$

21 **(d)**
The nature of the path is decided by the direction of velocity, and the direction of acceleration. The trajectory can be a straight line, circle or a parabola depending on these factors

22 **(d)**

$$A \Rightarrow \frac{u^2}{4} - u^2 = -2gh_1$$

$$B \Rightarrow \frac{u^2}{9} - u^2 = -2gh_2$$

$$C \Rightarrow \frac{u^2}{16} - u^2 = -2gh_3$$

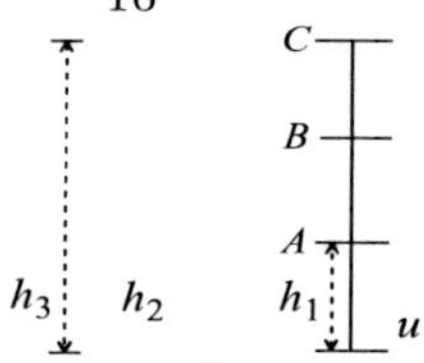

$$\therefore AB = \frac{u^2}{2g}\left\{\frac{8}{9} - \frac{3}{4}\right\} = \frac{u^2}{2g}\cdot\frac{5}{36}$$

$$BC = \frac{u^2}{2g}\left\{\frac{15}{16} - \frac{8}{9}\right\} = \frac{u^2}{2g}\cdot\frac{7}{144}$$

$$\therefore \frac{AB}{BC} = \frac{5}{36} \times \frac{144}{7} = \frac{20}{7}$$

23 **(d)**
Given, $x = 6t^2 - t^3$

$\frac{dx}{dt} = 12t - 3t^2$ ....(i)

$$\frac{dx}{dt} = 0 \Rightarrow t = 4\text{ s}$$

Now, again differentiating Eq. (i), we get

$$\frac{d^2x}{dt^2} = 12 - 6t = 12 - 6(4) = -12$$

Since, $\frac{d^2x}{dt^2}$ is negative, hence $t = 4$ s gives the maximum value for $x - t$ curve.

Moreover, acceleration $a = \frac{d^2x}{xt^2}$, at $t = 0, \frac{d^2x}{dt^2} = 12\ \text{ms}^{-2}$

24 **(a)**
Velocity required by body in 10 *sec*
$v = 0 + 2 \times 10 = 20\ m/s$
And distance travelled by it in 10 *sec*
$S_1 = \frac{1}{2} \times 2 \times (10)^2 = 100\ m$
Then it moves with constant velocity ($20\ m/s$) for 30 *sec*
$S_2 = 20 \times 30 = 600\ m$
After that due to retardation ($4\ m/s^2$) it stops
$S_3 = \frac{v^2}{2a} = \frac{(20)^2}{2 \times 4} = 50m$
Total distance travelled $S_1 + S_2 + S_3 = 750m$

25 **(c)**
Average velocity $= \frac{\text{Displacement}}{\text{Time interval}}$
A particle moving in a given direction with non-zero velocity cannot have zero speed.
In general, average speed is not equal to magnitude of average velocity. However, it can be so if the motion is along a straight line without change in direction

26 **(c)**
From acceleration time graph, acceleration is constant for first part of motion so, for this part velocity of body increases uniformly with time and as $a = 0$ then the velocity becomes constant. Then again increased because of acceleration

28 **(c)**
In a straight line-path with constant velocity distance travelled = displacement

ie, $s = D$

29 **(c)**
When packet is released from the balloon, it acquires the velocity of balloon of value $12\ m/s$. Hence velocity of packet after 2 *sec*, will be
$v = u + gt = 12 - 9.8 \times 2 = -76\ m/s$

30 **(a)**
Displacement $= (2 \times 4 - 2 \times 2 + 2 \times 4) = 12$m
$= 2 \times 4 + 2 \times 2 + 2 \times 4 = 20$m

31 **(c)**
Think of the lope of the given displacement-time graph at different points and you would arrive at the correct answer.
Alternatively, look at the self-illustrative figure.

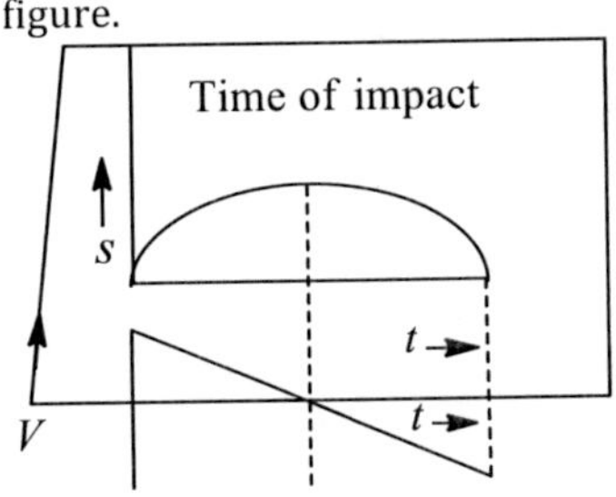

32 **(a)**
$$v = \frac{dx}{dt} = \frac{d}{dt}\left[\frac{k}{b}(1 - e^{-bt})\right]$$
$$= \frac{k}{b}[0 - (-b)e^{-bt}]$$
$$= ke^{-bt}$$

33 **(c)**
Let student catch the bus after t sec. So it will cover distance ut
Similarly distance travelled by the bus will be $\frac{1}{2}at^2$ for the given condition
$$ut = 50 + \frac{1}{2}at^2 = 50 + \frac{t^2}{2} \quad [a = 1\ m/s^2]$$
$$\Rightarrow u = \frac{50}{t} + \frac{t}{2}$$
To find the minimum value of u
$\frac{du}{dt} = 0$, so we get $t = 10\ sec$, then $u = 10\ m/s$

34 **(b)**
$u = 12\text{ms}^{-1}, \text{g} = 9.8\text{ms}^{-2}, t = 10s$
$S = \left(12 \times 10 + \frac{1}{2} \times 9.8 \times 10 \times 10\right)\text{m}$
$= (120 + 4.9 \times 100)\text{m} = 610\text{m}$

35 **(c)**

$$t=\sqrt{\frac{2h}{(g+a)}}=\sqrt{\frac{2\times 2.7}{(9.8+1.2)}}=\sqrt{\frac{5.4}{11}}$$

$$=\sqrt{0.49}=0.7 \text{ sec}$$

As $u=0$ and lift is moving upward with acceleration

36 **(b)**

Time of ascent $=\frac{u}{g}=6$ sec $\Rightarrow u=60m/s$

Distance in first second $h_{\text{first}}=60-\frac{g}{2}(2\times 1-1)=55m$

Distance in seventh second will be equal to the distance in first second of vertical downward motion

$$h_{\text{seventh}}=\frac{g}{2}(2\times 1-1)=5\ m$$

$$\Rightarrow h_{\text{first}}/h_{\text{seventh}}=11:1$$

37 **(b)**

$v=g\times t=32\times 1=32ft/sec$

38 **(b)**

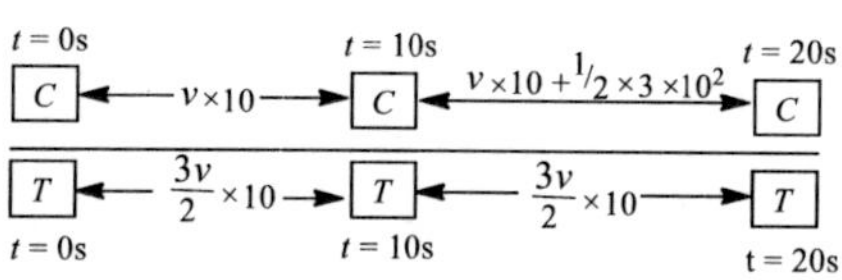

The diagram is showing the position of car and truck at various instants.

$$v\times 20+\frac{1}{2}\times 3\times 10^2=\frac{3v}{2}\times 20$$

$$\frac{3}{2}\times 100=\frac{v}{2}\times 20$$

$$v=15\text{ ms}^{-1}$$

39 **(c)**

$y=a+bt+ct^2-dt^4$

$\therefore v=\frac{dy}{dt}=b+2ct-4dt^3$ and $a=\frac{dv}{dt}=2c-12dt^2$

Hence, at $t=0$, $v_{\text{initial}}=b$ and $a_{\text{initial}}=2c$

40 **(d)**

$\vec{r}=3t\hat{\imath}-t^2\hat{\jmath}+4\hat{k}$

Velocity, $\vec{v}=\frac{d\vec{r}}{dt}=\frac{d}{dt}(3t\hat{\imath}-t^2\hat{\jmath}+4\hat{k})=3\hat{\imath}-2t\hat{\jmath}$

At $t=5s\Rightarrow\vec{v}=3\hat{\imath}-10\hat{\jmath}$

$|\vec{v}|=\sqrt{(3)^2+(-10)^2}=\sqrt{9+100}$

$$=\sqrt{109}=10.44\ ms^{-1}$$

42 **(b)**

The time of fall is independent of the mass

43 **(d)**

$$\text{Average speed}=\frac{\text{Total distance travelled}}{\text{Total time taken}}$$

$$=\frac{x}{\frac{2x/5}{v_1}+\frac{3x/5}{v_2}}=\frac{5v_1v_2}{3v_1+2v_2}$$

44 **(b)**

Constant velocity means constant speed as well as same direction throughout

46 **(b)**

$$x=\frac{1}{t+5}\Rightarrow v=\frac{dx}{dt}=-\frac{1}{(t+5)^2}$$

Acceleration, $a=\frac{dv}{dt}=\frac{2}{(t+5)^3}\Rightarrow a\propto(\text{velocity})^{3/2}$

47 **(a)**

Here, $u=10\ ms^{-1}, t=2s, S=20\ m$

Using $S=ut+\frac{1}{2}at^2$

$$\therefore\ 20=10\times 2+\frac{1}{2}\times a\times 2^2$$

$0=2a\Rightarrow a=0$

48 **(c)**

For upward motion

Effective acceleration $=-(g+a)$

And for downward motion

Effective acceleration $=(g-a)$

But both are constants. So the slope of speed-time graph will be constant

49 **(a)**

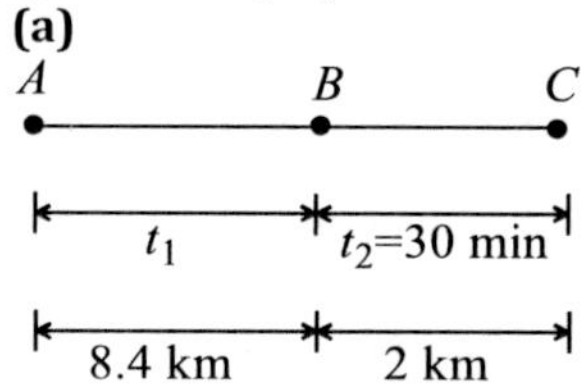

$$\text{Average speed } \bar{v}=\frac{\text{Total distance travelled}}{\text{Total time taken}}$$

$$=\frac{8.4km+2km}{t_1+t_2}=\frac{10.4\ km}{\left(\frac{8.4\ km}{70\ km/h}\right)+\frac{1}{2}h}$$

$$=\frac{10.4\ km}{0.12h+0.5h}=16.8\ km/h$$

50 **(a)**
Slope of velocity-time graph measures acceleration. For graph (a) slope is zero. Hence $a = 0 i.e.$ motion is uniform

51 **(b)**
Let 'a' be the retardation of boggy then distance covered by it be S. If u is the initial velocity of boggy after detaching from train ($i.e.$ uniform speed of train)
$v^2 = u^2 + 2as \Rightarrow 0 = u^2 - 2as \Rightarrow s_b = \frac{u^2}{2a}$
Time taken by boggy to stop
$v = u + at \Rightarrow 0 = u - at \Rightarrow t = \frac{u}{a}$
In this time t distance travelled by train $=$
$s_t = ut = \frac{u^2}{a}$
Hence ratio $\frac{s_b}{s_1} = \frac{1}{2}$

52 **(a)**
When the stone is released from the balloon. Its height
$h = \frac{1}{2}at^2 = \frac{1}{2} \times 1.25 \times (8)^2 = 40\ m$ and velocity
$v = at = 1.25 \times 8 = 10\ m/s$
Time taken by the stone to reach the ground

$$t = \frac{v}{g}\left[1 + \sqrt{1 + \frac{2gh}{v^2}}\right] = \frac{10}{10}\left[1 + \sqrt{1 + \frac{2 \times 10 \times 40}{(10)^2}}\right]$$

$= 4 sec$

53 **(c)**
When a particle is moving with uniform acceleration, let v be the velocity of particle at a distance s,
Then average velocity $= \frac{0+v}{2} = \frac{v}{2}$
Time taken, $t_1 = \frac{s}{(v/2)} = \frac{2s}{v}$
When particle moves with uniform velocity, time taken, $t_2 = \frac{2s}{v}$
When particle moves with uniform acceleration,
Time taken, $t_3 = \frac{3s}{(0+v)/2} = \frac{6s}{v}$
Total time$= t_1 + t_2 + t_3 = \frac{2s}{v} + \frac{2s}{v} + \frac{6s}{v} = \frac{10s}{v}$
$\therefore v_{av} = \frac{s+2s+3s}{10s/v} = \frac{6v}{10}$ or $\frac{v_{av}}{v} = \frac{6}{10} = \frac{3}{5}$

54 **(d)**
$x = ae^{-\alpha t} + be^{\beta t}$
Velocity $v = \frac{dx}{dt} = \frac{d}{dt}(ae^{-\alpha t} + be^{\beta t})$
$= a.e^{-\alpha t}(-\alpha) + be^{\beta t}(\beta)$
$= -a\alpha e^{-\alpha t} + b\beta e^{\beta t}$
Acceleration $= -a\alpha e^{-\alpha t}(-\alpha) + b\beta e^{bt}.\beta$
$= a\,\alpha^2 e^{-\alpha t} + b\beta^2 e^{\beta t}$
Acceleration is positive so velocity goes on increasing with time

55 **(b)**
Let the initial velocity of ball be u
Time of rise $t_1 = \frac{u}{g+a}$ and height reached $= \frac{u^2}{2(g+a)}$
Time of fall t_2 is given by
$\frac{1}{2}(g-a)t_2^2 = \frac{u^2}{2(g+a)}$
$\Rightarrow t_2 = \frac{u}{\sqrt{(g+a)(g-a)}} = \frac{u}{(g+a)}\sqrt{\frac{g+a}{g-a}}$
$\therefore t_2 > t_1$ because $\frac{1}{g+a} < \frac{1}{g-a}$

56 **(d)**
$u = at, x = \int u\,dt = \int at\,dt = \frac{at^2}{2}$
For $t = 4$ sec, $x = 8a$

57 **(d)**
In 's-t' graph (positive -time)
The straight line parallel with time axis represent state of rest

58 **(b)**
Speed of stone in a vertically upward direction is $4.9\ m/s$. So for vertical downward motion we will consider $u = -4.9\ m/s$
$h = ut + \frac{1}{2}gt^2 = -4.9 \times 2 + \frac{1}{2} \times 9.8 \times (2)^2 = 9.8\ m$

59 **(c)**
Parachute bails out at height H from ground. Velocity at A

$v = \sqrt{2\,gh}$

$= \sqrt{2 \times 9.8 \times 50} = \sqrt{980}\text{ms}^{-1}$

The velocity at ground $v_1 = 3\text{ ms}^{-1}$ (given)

Acceleration $= -2\text{ ms}^{-2}$ (given_

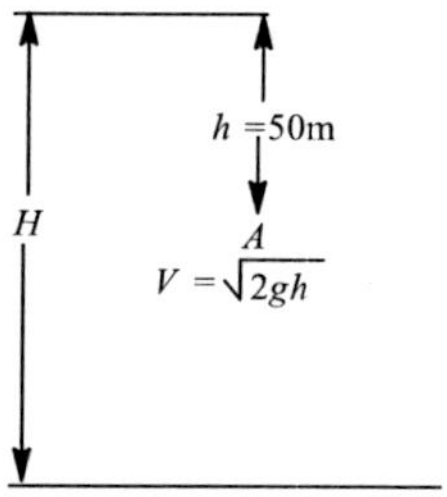

$\therefore H - h = \dfrac{v^2 - v_1^2}{2 \times 2}$

$= \dfrac{980 - 9}{4} = \dfrac{971}{4} = 242.75$

$\therefore \quad H = 242.75 + h$

$= 242.75 + 50 = 293\text{ m}$

60 **(d)**

$u = 72\,kmph = 20\,m/s, v = 0$

By using $v^2 = u^2 - 2as \Rightarrow a = \frac{u^2}{2s} = \frac{(20)^2}{2\times200} = 1\,m/s^2$

61 **(c)**

$(2v)^2 - v^2 = 2gh'$ or $4v^2 - v^2 = 2gh'$ or $3v^2 = 2gh'$ or $3 \times 2gh = 2gh'$ or $h' = 3h$

62 **(a)**

For first marble, $h_1 = \frac{1}{2}g \times 16 = 8g$

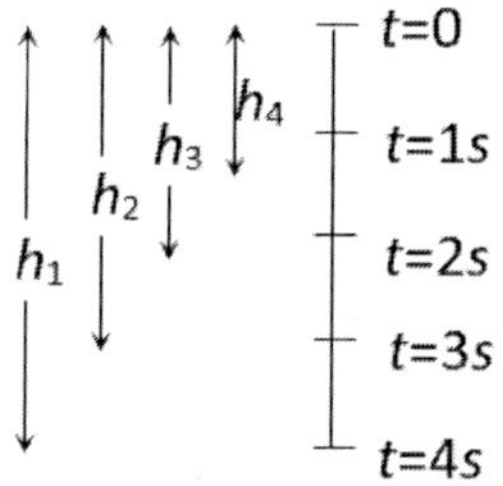

For second marble, $h_2 = \frac{1}{2}g \times 9 = 4.5g$

For third marble, $h_3 = \frac{1}{2}g \times 4 = 2g$

For fourth marble, $h_4 = \frac{1}{2}g \times 1 = 0.5g$

$\therefore h_1 - h_2 = 8g - 4.5g = 3.5g = 35m.$

$h_2 - h_3 = 4.5g - 2g = 2.5g = 25m$ and

$h_3 - h_4 = 2g - 0.5g = 1.5g = 15m$

63 **(c)**

$\dfrac{dx}{dt} = 4t^3 - 2t$

or $dx = 4t^3dt - 2t\,dt$

Integrating, $x = \frac{4t^4}{4} - \frac{2t^2}{2} = t^3 - t^2$

When $x = 2, t^4 - t^2 - 2 = 0,$

$t^2 = \dfrac{-(-1) \pm \sqrt{1+8}}{2}$

or $t^2 = \frac{1\pm3}{2} = 2$ (ignoring – ve sign)

Again, $\frac{d^2x}{dt^2} = 12t^2 - 2$

When $t^2 = 2$, acceleration$=12\times 2 - 2 = 22\text{ms}^{-2}$

64 **(a)**

Here, $u = 0, a = g$

Distance travelled in n^{th} second is given by

$D_n = u + \dfrac{a}{2}(2n-1) \;\therefore D_n \propto (2n-1)$

$\therefore D_1 : D_2 : D_3 : D_4 : D_5 \ldots = 1 : 3 : 5 : 7 : 9 : \ldots$

65 **(d)**

I is not possible because total distance covered by a particle increases with time

II is not possible because at a particular time, position cannot have two values

III is not possible because at a particular time, velocity cannot have two values

IV is not possible because speed can never be negative

66 **(b)**
Given, $x = 4(t-2) + a(t-2)^2$

$$v = \frac{dx}{dt} = 4 + 2a(t-2)$$

At $t = 0$, $v = 4(1-a)$

Acceleration $a = \frac{d^2x}{dt^2} = 2a$

67 **(a)**
Using
$V = u + at$
$V = gt$ ...(i)
Comparing with $y = mx + c$
Equation (i) represents a straight line passing through origin inclined x-axis (slope -g)

68 **(a)**
After balling out from point A parachutist falls freely under gravity. The velocity acquired by it will 'v'

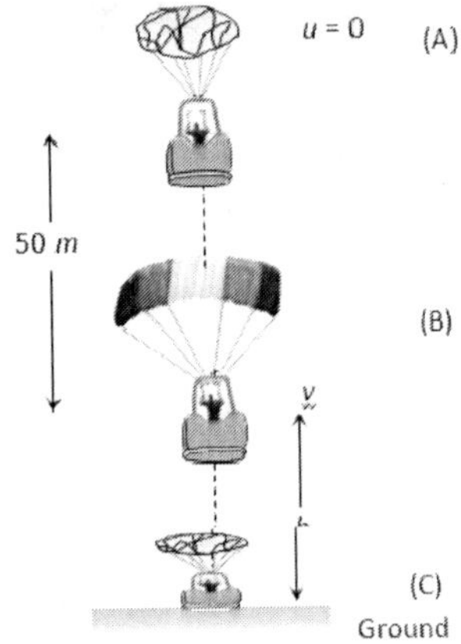

From $v^2 = u^2 + 2as = 0 + 2 \times 9.8 \times 50 = 980$
[As $u = 0, a = 9.8 m/s^2, s = 50\ m$]
At point B, parachute opens and it moves with retardation of 2 m/s^2 and reach at ground (point C) with velocity of 3 m/s
For the part 'BC' by applying the equation $v^2 = u^2 + 2as$
$v = 3m/s, u = \sqrt{980}\ m/s, a = -2m/s^2, s = h$

$$\Rightarrow (3)^2 = (\sqrt{980})^2 + 2 \times (-2) \times h \Rightarrow 9 = 980 - 4h$$

$$\Rightarrow h = \frac{980 - 9}{4} = \frac{971}{4} = 242.7 \cong 243\ m$$

So, the total height by which parachutist bail out $= 50 + 243 = 293\ m$

70 **(c)**

$$s = ut + \frac{1}{2}at^2$$

For Ist body

$u = 0$ and $a = g$ [freely falling body]

Distance covered in 2 s,

$$s_1 = 0 + \frac{1}{2}g(3)^2$$

For IInd body

Distance covered in 2 s,

$$s_2 = 0 + \frac{1}{2}g(2)^2$$

$$\therefore\ s_1 - s_2 = \frac{1}{2}g[(3)^2 - (2)^2]$$

$$= \frac{1}{2}g(9-4) = 25\text{ m}$$

71 **(d)**
The straight vertical line in the graph represents change in the direction of velocity.

72 **(d)**
Body reaches the point of projection with same velocity

73 **(c)**
Since, the object accelerates from rest, its initial velocity u is zero. $ie, u = 0$

From first equation of motion

$$v = u + at$$

$$\therefore\quad 27.5 = 0 + a \times 10$$

Or $\quad a = 2.75\ ms^{-2}$

Hence, distance covered in first 10 s

$$s_1 = ut + \frac{1}{2}at^2$$

$$= 0 + \frac{1}{2} \times 2.75 \times (10)^2$$
$$= 137.5\ m$$

Distance covered in next 10 s with uniform velocity of $27.5\ ms^{-1}$

$s_2 = 27.5 \times 10 = 275\ m$

Total distance covered

$s = 137.5 + 275 = 412.5\ m$

74 **(c)**
$S_n \propto (2n - 1)$. In equal time interval of 2 *seconds*
Ratio of distance = 1 : 3 : 5

75 **(b)**
Let two balls meet at depth h from platform
So $h = \frac{1}{2}g(18)^2 = v(12) + \frac{1}{2}g(12)^2 \Rightarrow$

$v = 75\ ms^{-1}$

76 **(b)**

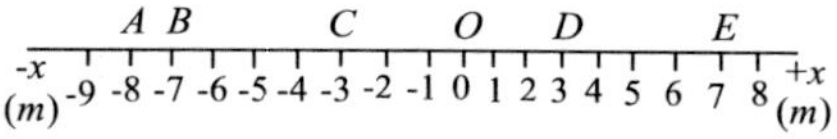

(i) The displacement of the main from A to E is $\Delta x = x_2 - x_1 = 7m - (-8m) = +15m$ directed in the positive x-direction
(ii) The displacement of the man from E to C is $\Delta x = -3m - (7m) = -10m$ directed in the negative x-direction
(iii) The displacement of the man from B to D is $\Delta x = 3m - (-7m) = +10m$ directed in the positive x-axis

77 **(b)**

$$\int_{6.25}^{0} \frac{dv}{\sqrt{v}} = -2.5 \int_0^t dt$$
$$\left|2\sqrt{v}\right|_{6.25}^{0} = -2.5t$$
$$2\sqrt{6.25} = 2.5t$$
$$t = 2sec$$

78 **(d)**

$$V_{av} = \frac{S + S}{\frac{S}{v_1} + \frac{S}{v_2}} = \frac{2v_1v_2}{v_1 + v_2}$$

79 **(a)**

$$t = \sqrt{\frac{2 \times 19.6}{9.8}} = 2s\ ;\ 1.9 = \sqrt{\frac{2 \times h}{9.8}}$$

or $h = \frac{1.9 \times 1.9 \times 9.8}{2}$m=17.689m
Required distance = 19.6 − 17.689 m = 1.9m.

80 **(b)**

$$\frac{(S)_{(last\ 2s)}}{(S)_{7s}}$$
$$= \frac{\frac{1}{2} \times 2 \times 10}{\frac{1}{2} \times 2 \times 10 + 2 \times 10 + \frac{1}{2} \times 2 \times 10} = \frac{1}{4}$$

81 **(c)**

Stopping distance= $\frac{\text{Kinetic energy}}{\text{Retarding force}} = \frac{\frac{1}{2}mu^2}{F}$
$= \frac{u^2}{2\mu g}[F = \mu mg]$
So both will cover equal distance

82 **(c)**
Let the stone falls through a height h in $t\ s$
Here, $u = 0, a = g$
Using $D_n = u + \frac{a}{2}(2n - 1)$
Distance travelled by the stone in the last second is
$\frac{9h}{25} = \frac{g}{2}(2t - 1)\quad [\because u = 0]$...(i)
Distance travelled by the stone in $t\ s$ is
$h = \frac{1}{2}gt^2 \qquad [\because u = 0]$ ...(ii)
Divide (i) by (ii), we get
$\frac{9}{25} = \frac{(2t - 1)}{t^2}$
$9t^2 = 50t - 25$
$9t^2 - 50t + 25 = 0$
On solving, we get
$t = 5s$ or $t = \frac{5}{9}s$
Substituting $t = 5s$ in (ii), we get
$h = \frac{1}{2} \times 9.8 \times (5)^2 = 122.5\ m$

83 **(d)**
Since $x = 1.2t^2$ which is in form $x = \frac{1}{2}at^2$
Thus the motion is uniformly accelerated

85 **(d)**
Given $a = 19.6m/s^2 = 2g$
Resultant velocity of the rocket after 5 *sec*
$v = 2g \times 5 = 10g\ m/s$

Height achieved after 5 $sec, h_1 = \frac{1}{2} \times 2g \times 25 = 245m$

On switching off the engine it goes up to height h_2 where its velocity becomes zero $0 = (10g)^2 - 2gh_2 \Rightarrow h_2 = 490m$

$\therefore$ Total height of rocket $= 245 + 490$
$= 735\ m$

86 **(b)**

Velocity at the time of striking the floor,
$u = \sqrt{2gh_1} = \sqrt{2 \times 9.8 \times 10} = 14m/s$

Velocity with which it rebounds
$v = \sqrt{2gh_2} = \sqrt{2 \times 9.8 \times 2.5} = 7\ m/s$

$\therefore$ Change in velocity $\Delta v = 7 - (-14)$
$= 21m/s$

$\therefore$ Acceleration $= \frac{\Delta v}{\Delta t} = \frac{21}{0.01}$
$= 2100m/s^2$(upwards)

87 **(c)**

Since, $v = (t-2)$,so $v \propto t$. On plotting a graph between v and t we get a straight line AB and BC as shown in figure. The distance covered in $4s$ is equal to the area under the velocity-time graph = Area of ΔOAB ∓Area of ΔBCD

$= \frac{2\times2}{2} + \frac{2\times2}{2} = 2 + 2 = 4$m

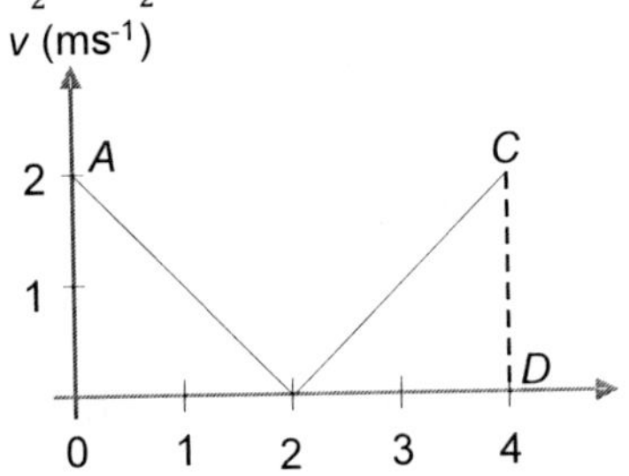

88 **(d)**

The separation between the two bodies, two seconds after the release of second body

$= \frac{1}{2} \times 9.8[(3)^2 - (2)^2] = 24.5\ m$

89 **(c)**

Total distance to be covered for crossing the bridge
= length of train + length of bridge
$= 150m + 850m = 1000m$

Time $= \frac{\text{Distance}}{\text{Velocity}} = \frac{1000}{45\times\frac{5}{18}} = 80\ sec$

90 **(a)**

$$h_{\max} = \frac{u^2}{2g} = \frac{(15)^2}{2 \times 10} = 11.25m$$

91 **(b)**

Let the total distance travelled by the body is $2S$. If t_1 is the time taken by the body to travel first half of the distance, then

$$t_1 = \frac{S}{2}$$

Let t_2 be the time taken by the body for each time interval for the remaining half journey.

$$\therefore \qquad S = 3t_2 + 5t_2 = 8t_2$$

So, average speed

$$= \frac{\text{Total distance travelled}}{\text{Total time taken}}$$

$$= \frac{2S}{t_1 + 2t_2} = \frac{2S}{\frac{S}{2} + \frac{S}{4}}$$

$$= \frac{8}{3}\text{ms}^{-1}$$

92 **(d)**

$s = 3t^3 + 7t^2 + 14t + 8m$

$a = \frac{d^2s}{dt^2} = 18t + 14$ at $t = 1$ sec $\Rightarrow a = 32\ m/s^2$

94 **(b)**

Given, $a = \alpha t + \beta$

$$\frac{dv}{dt} = \alpha t + \beta$$

$$\int_0^t dv = \int_0^t \alpha t\, dt + \int_0^t \beta\, dt$$

$$v = \frac{\alpha t^2}{2} + \beta t$$

95 **(a)**

As the clear from figure

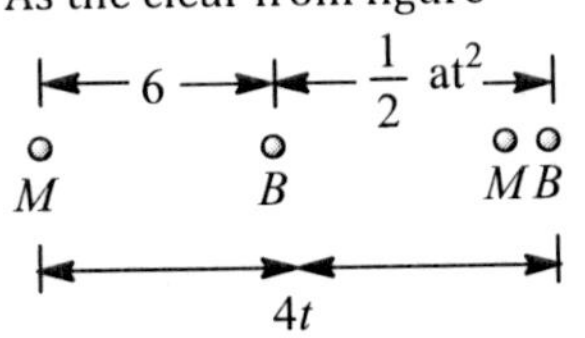

$4t = 6 + \frac{1}{2} \times 1.2 \times t^2$

or $4t = 6 + 0.6t^2$

96 **(a)**

We know that the velocity of body is given by the slope of displacement – time graph so it is clear that initially slope of the graph is positive and after some time it becomes zero (corresponding to the peak of graph) and it will become negative

97 **(b)**

For one dimensional motion along a plane

$$S = ut + \frac{1}{2}at^2 \Rightarrow 9.8 = 0 + \frac{1}{2}g\sin 30°t^2$$
$$\Rightarrow t = 2\text{sec}$$

98 **(c)**

$$v_{av} = \frac{2v_1v_2}{v_1 + v_2} = \frac{2 \times 40 \times 60}{100} = 48\ kmph$$

99 **(d)**

Initial velocity of balloon with respect to ground

$v = 10 + 5 = 15\ m/sec$ upward

After 2 seconds its velocity, $v = u - gt$ $= 15 - 10 \times 2 = -5\ m/sec = 5\ m/sec$ (downward)

100 **(c)**

$$S_n = u + \frac{a}{2}(2n - 1) \Rightarrow 1.2$$
$$= 0 + \frac{a}{2}(2 \times 6 - 1)$$
$$\Rightarrow a = \frac{1.2 \times 2}{11} = 0.218\ m/s^2$$

101 **(c)**

Let the ball be at height h at time t and time $(t + \Delta t)$.

Then,

$h = ut - \frac{1}{2}gt^2$ ....(i)

and $h = u(t + \Delta t) - \frac{1}{2}g(t - \Delta t)^2$ ...(ii)

Equating Eqs. (i) and (ii), we get

$t = \frac{2u - g\Delta t}{2g}$ ....(iii)

Substituting Eq. (iii) in Eq. (i), we get

$$h = \frac{4u^2 - g^2(\Delta t)^2}{8g}$$
$$\Rightarrow u = \frac{1}{2}\sqrt{8gh + g^2(\Delta t)^2}$$

102 **(b)**

Let v_w be velocity of water and v_b be the velocity of motor boat in still water. If x is the distance covered, then as per question

$x = (v_b + v_w) \times 6 = (v_b - v_w) \times 10$

On solving, $v_w = v_b/4$

$\therefore\ x = [v_b + v_b/4] \times 6 = 7.5v_b$

Time taken by motor boat to cross the same distance in still water is

$$t = \frac{x}{x_b} = \frac{7.5}{v_b} = 7.5\text{h}$$

103 **(c)**

From acceleration time graph, acceleration is constant for first part of motion so, for this part velocity of body increases uniformly with time and as $a = 0$ then the velocity becomes constant. Then again increased because of acceleration

104 **(c)**

Given : Initial velocity of a body $u = 0$ ... (i)

Let s be the distance covered by a body in time t

$\therefore s = ut + \frac{1}{2}at^2$ or $s = \frac{1}{2}at^2$ [Using (i)]

$\Rightarrow s \propto t^2$

105 **(d)**

$$\Delta t = \sqrt{\frac{2 \times 12}{10}} - \sqrt{\frac{2 \times 10}{10}}$$

$= 1.549\text{s} - 1.414\text{s} = 0.135\text{s}$

106 **(b)**

Let v be the speed of boatman is still water.

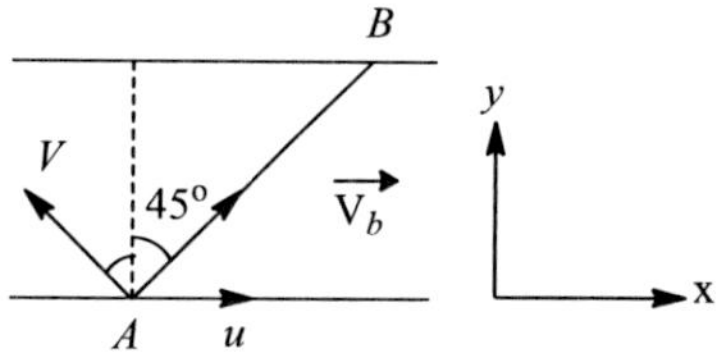

Resultant of v and u should be along AB.

Components of $\vec{v}_b$ (absolute velocity of boatman)along x and y direction are,

$v_x = u - v\sin\theta$

Further, $\tan 45° = \frac{v_y}{v_x}$

or $1 = \frac{v\cos\theta}{u - v\sin\theta}$

$$v = \frac{u}{\sin\theta + \cos\theta} = \frac{u}{\sqrt{2}\sin(\theta + 45°)}$$

v is minimum at,

$\theta + 45° = 90°$ or $\theta = 45°$

and $v_{\min} = \frac{u}{\sqrt{2}}$

107 **(b)**

For stone to be dropped from rising balloon of velocity 29 m/s

$u = -29\ m/s, t = 10sec$

$$\therefore h = -29 \times 10 + \frac{1}{2} \times 9.8 \times 100$$

$= -290 + 490 = 200\ m$

108 **(c)**

From acceleration-velocity graph, we have $a = k\ v$ where k is a constant which represents the slope of the given line.

As, $a = v\frac{dv}{ds}$, so $v = \frac{dv}{ds} = kv$

or $\frac{dv}{ds} = k =$ a constant

Thus, the slope of velocity-displacement graphs is same as that of acceleration velocity. Which is constant.

109 **(c)**

For first part,

$u = 0, t = T$ and acceleration $= a$

$\therefore v = 0 + aT = aT$ and $S_1 = 0 + \frac{1}{2}aT^2 = \frac{1}{2}aT^2$

For Second part,

$u = aT$, retardation $= a_1, v = 0$ and time taken $= T_1$ (let)

$\therefore 0 = u - a_1T_1 \Rightarrow aT = a_1T_1$

And from $v^2 = u^2 - 2aS_2 \Rightarrow S_2 = \frac{u^2}{2a_1} = \frac{1}{2}\frac{a^2T^2}{a_1}$

$S_2 = \frac{1}{2}aT \times T_1 \left(\text{As} a_1 = \frac{aT}{T_1}\right)$

$\therefore v_{av} = \frac{S_1 + S_2}{T + T_1} = \frac{\frac{1}{2}aT^2 + \frac{1}{2}aT \times T_1}{T + T_1}$

$= \frac{\frac{1}{2}aT(T + T_1)}{T + T_1} = \frac{1}{2}aT$

110 **(c)**

If a stone is dropped from height h then

$h = \frac{1}{2}gt^2$ ...(i)

if a stone is thrown upward with velocity u then

$h = -u\ t_1 + \frac{1}{2}gt_1^2$ ...(ii)

If a stone is thrown downward with velocity u then

$h = ut_2 + \frac{1}{2}gt_2^2$ ...(iii)

From (i) (ii) and (iii) we get

$-ut_1 + \frac{1}{2}gt_1^2 = \frac{1}{2}gt^2$ ...(iv)

$ut_2 + \frac{1}{2}gt_2^2 = \frac{1}{2}gt^2$ ...(v)

Dividing (iv) and (v) we get

$\therefore \frac{-ut_1}{ut_2} = \frac{\frac{1}{2}g(t^2 - t_1^2)}{\frac{1}{2}g(t^2 - t_2^2)}$

Or $-\frac{t_1}{t_2} = \frac{t^2 - t_1^2}{t^2 - t_2^2}$

By solving $t = \sqrt{t_1t_2}$

111 **(b)**

$\overrightarrow{v_{ct}} = \overrightarrow{v_c} - \overrightarrow{v_t}$

$\overrightarrow{v_{ct}} = \overrightarrow{v_c} + (-\overrightarrow{v_t})$

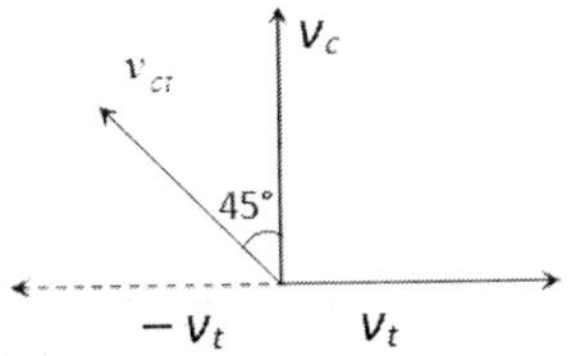

Velocity of car $w.r.t.$ train (v_{ct}) is towards West-North

112 **(c)**

$1 = \frac{v}{t_1}, 3 = \frac{v}{t_2}, 1200 = \frac{1}{2}(t_1 + t_2)v,$

$1200 = \frac{1}{2}\left(v + \frac{v}{3}\right)v = \frac{1}{2}\frac{4v^2}{3} = \frac{2v^2}{3}$

or $v^2 = 1800,$

$\therefore\ 1200 = \frac{1}{2}t \times \sqrt{1800}$

$t = \frac{2400}{\sqrt{1800}}s = \frac{2400}{42.43}s = 56.6$ s

113 **(c)**

$81 = -12t + \frac{1}{2} \times 10 \times t^2$

or $5t^2 - 12t - 81 = 0$

or $t = \frac{12 \pm \sqrt{144 + 4 \times 5 \times 81}}{2 \times 5}$

$= \frac{12 \pm \sqrt{1764}}{10} = \frac{12 \pm 42}{10} = \frac{54}{10} = 5.4$s

[negative sign has been ignored]

114 **(c)**

Given $\overrightarrow{AB}$ = Velocity of boat = km/hr

$\overrightarrow{AC}$ = Resultant velocity of boat = $10\ km/hr$

$\overrightarrow{BC}$ = Velocity of river

$= \sqrt{AC^2 - AB^2}$

$= \sqrt{(10)^2 - (8)^2} = 6\ km/hr$

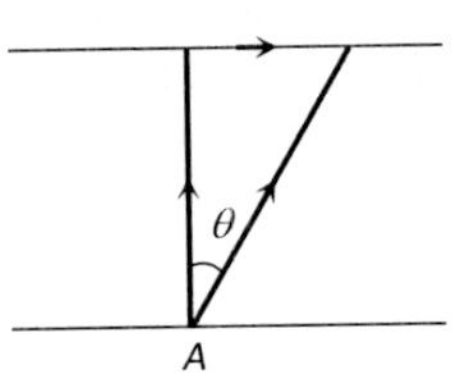

115 **(c)**

Total distance to be covered for crossing the bridge

= length of train + length of bridge

$= 150m + 850m = 1000m$

Time $= \frac{\text{Distance}}{\text{Velocity}} = \frac{1000}{45 \times \frac{5}{18}} = 80\ sec$

116 **(c)**

Distance travelled by motor bike at $t =$ 18s

$s_{\text{bike}} = s_1 = \frac{1}{2}(18)(60)$=540 m

Distance travelled by car at $t = 18$s

$s_{\text{car}} = s_2$=(18)(60)=720 m

Therefore, separation between them at $t = 18$s is 180m. Let, separation between them decreases to zero at time t beyond 18s.

Hence, $s_{\text{bike}} = 540 + 60t$ and $s_{\text{car}} = 720 + 40t$

$s_{\text{car}} - s_{\text{bike}} = 0$

$\Rightarrow \quad 720 + 40t = 540 + 60t$

$\Rightarrow \quad t = 9$s beyond 18s or

Hence, $t = (18 + 9)$s=27s from start and distant travelled by both is s_{bike}=s_{car} = 1080m

117 **(b)**

Distance = Area under $v - t$ graph =$A_1 + A_2 + A_3 + A_4$

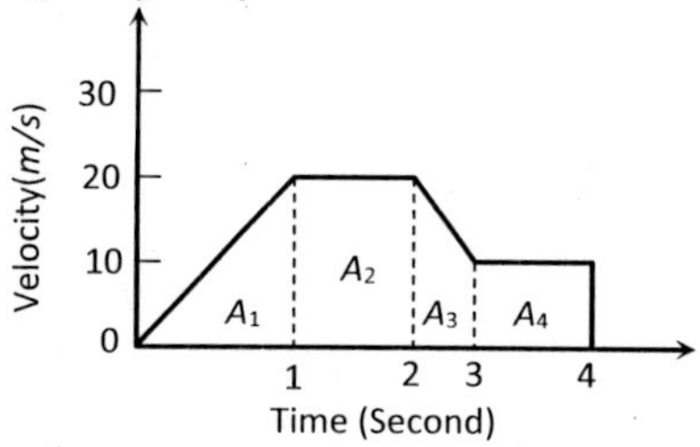

$$= \frac{1}{2} \times 1 \times 20 + (20 \times 1) + \frac{1}{2}(20 + 10) \times 1 + (10 \times 1)$$

$= 10 + 20 + 15 + 10 = 55m$

118 **(c)**

We know that gravity is a universal force with which all bodies are attracted towards the earth. Hence, g is same for both the balls. Also, if t is the time taken by the balls to reach the ground, then from equation of motion.

$$s = ut + \frac{1}{2}\text{g}t^2$$

$$\Rightarrow \quad t = \sqrt{\frac{2(s - ut)}{\text{g}}}$$

Since s, u and g are same for both, hence time taken by both the balls is same.

119 **(b)**

The velocity time graph for given problem is shown in the figure.

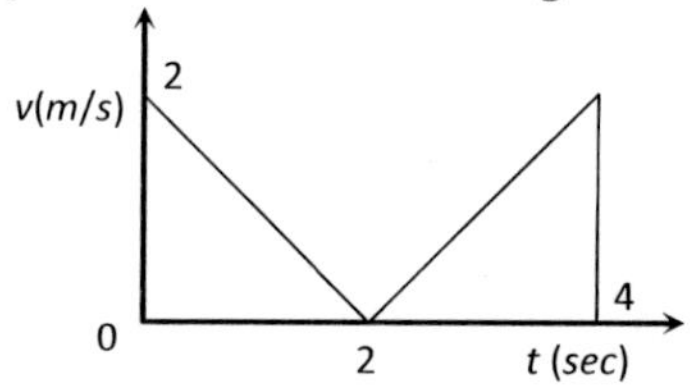

Distance travelled $S =$

Area under curve $= 2 + 2 = 4m$

120 **(a)**

When the body is projected vertically upward then at the highest point its velocity is zero but acceleration is not equal to zero (g = 9.8m/s^2)

121 **(d)**

If t_1 and t_2 are the time, when body is at the same height then,

$$h = \frac{1}{2}gt_1t_2 = \frac{1}{2} \times g \times 2 \times 10 = 10\ g$$

122 **(a)**

If u is the initial velocity then distance covered by it in 2 *sec*

$$S = ut + \frac{1}{2}at^2 = u \times 2 + \frac{1}{2} \times 10 \times 4$$

$$= 2u + 20 \quad \text{... (i)}$$

Now distance covered by it in 3rd*sec*

$$S_{3^{\text{rd}}} = u + \frac{1}{2}(2 \times 3 - 1)10$$

$$= u + 25 \quad \text{... (ii)}$$

From (i) and (ii), $2u + 20 = u + 25 \Rightarrow u = 5$

$\therefore S = 2 \times 5 + 20 = 30\ m$

123 **(b)**

Let two boys meet at point C after time t from the starting. Then,

$$AC = vt, BC = v_1t$$

$$(AC)^2 = (AB)^2 + (BC)^2$$

$$\Rightarrow \quad v^2t^2 = a^2 + v_1^2t^2$$

By showing, we get $t = \sqrt{\frac{a^2}{v^2 - v_1^2}}$

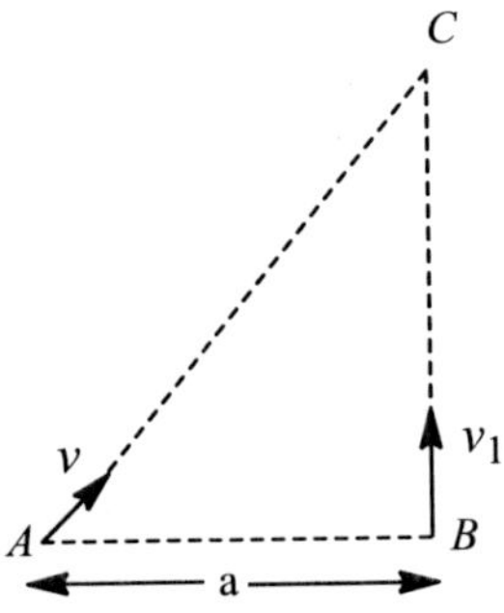

124 **(c)**

$h_{n^{th}} = u - \frac{g}{2}(2n-1)$

$h_{5^{th}} = u - \frac{10}{2}(2 \times 5 - 1) = u - 45$

$h_{6^{th}} = u - \frac{10}{2}(2 \times 6 - 1) = u - 55$

Given $h_{5^{th}} = 2 \times h_{6^{th}}$.By solving we get $u = 65\ m/s$

125 **(c)**

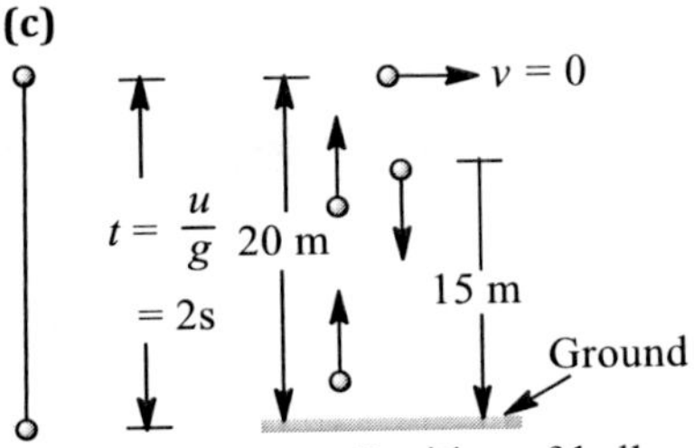

Position of balls

$h_1 = \frac{1}{2}gt^2 = \frac{1}{2} \times 10 \times 1^2 = 5\text{m}$

$h_2 = \frac{1}{2}gt^2 = \frac{1}{2} \times 10 \times 2^2 = 20\text{m}$

From ground, 5m, 20m, 15m (shown in figure)

127 **(c)**

For upward motion

Effective acceleration $= -(g+a)$

And for downward motion

Effective acceleration $= (g-a)$

But both are constants. So the slope of speed-time graph will be constant

128 **(b)**

Distance = Area under $v-t$ graph $= A_1 + A_2 + A_3 + A_4$

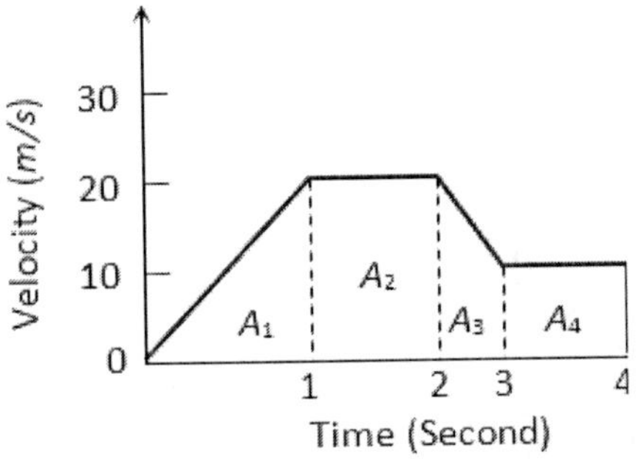

$= \frac{1}{2} \times 1 \times 20 + (20 \times 1) + \frac{1}{2}(20+10) \times 1 + (10 \times 1)$

$= 10 + 20 + 15 + 10 = 55m$

129 **(d)**

Since $c >> v$ (negligible)

130 **(d)**

Average acceleration $= \frac{\Delta v}{\Delta t}$

$= \frac{\sqrt{2gh'} - (-\sqrt{2gh})}{\Delta t} = \frac{\sqrt{2gh'} + \sqrt{2gh}}{\Delta t}$

$= \frac{\sqrt{2 \times 10 \times 2.5} + \sqrt{2 \times 10 \times 10}}{0.01}\text{ms}^{-2}$

$= \frac{\sqrt{15} + \sqrt{200}}{0.01}\text{ms}^{-2} = \frac{5\sqrt{2} + 10\sqrt{2}}{0.01}\text{ms}^{-2}$

$= \frac{15\sqrt{2}}{0.01}\text{ms}^{-2} = 1500\sqrt{2}\text{ms}^{-2}$

The upward velocity has been taken as positive. Since average acceleration is positive therefore its direction is vertically upward.

131 **(a)**

$h = -4.9 \times 3 + \frac{1}{2} \times 9.8 \times 3 \times 3 = 5.9 \times 3 \times 2\text{m} = 29.4\text{m}$

132 **(d)**

Distance, $x = b_0 + b_1 t + b_2 t^2$

Velocity $v = \left(\frac{dx}{dt}\right) = b_1 + 2b_2 t$

Acceleration, $a = \frac{d^2x}{dt^2} = 2b_2$

133 **(c)**

Net acceleration of a body when thrown upward = acceleration of body – acceleration due to gravity $= a - g$

134 **(a)**

Displacement = Summation of all the area with sign

$= (A_1) + (-A_2) + (A_3)$
$= (2 \times 4) + (-2 \times 2) + (2 \times 2)$

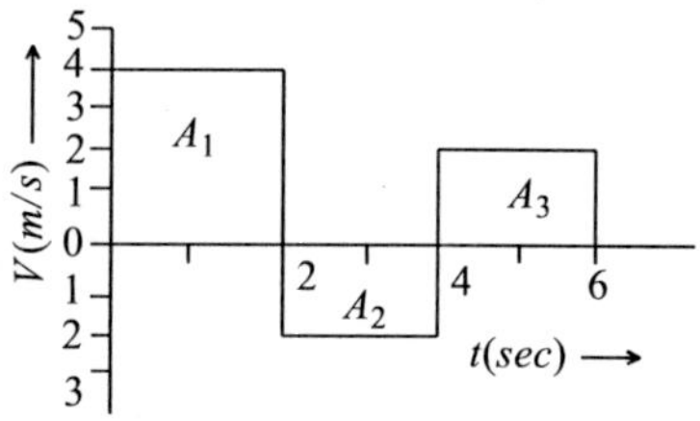

$\therefore$ Displacement $= 8\ m$
Distance = Summation of all the areas without sign
$= |A_1| + |-A_2| + |A_3| = |8| + |-4| + |4|$
$= 8 + 4 + 4$
$\therefore$ Distance $= 16\ m$

135 **(c)**
For first ball $h = \frac{1}{2}gt^2$ ...(i)
For second ball
$(h - 20) = \frac{1}{2}g(t-1)^2$ ...(ii)

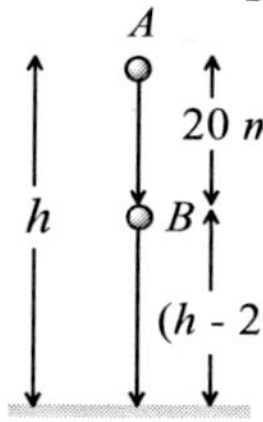

Subtract equation (ii) from (i)
$20 = -5 + 10t \Rightarrow \therefore\ t = 2.5$ sec
Hence, height $h = \frac{1}{2} \times 10 \times (2.5)^2 =$ 31.2 m

136 **(c)**
In first case: $s_1 = ut_1 + \frac{1}{2}at_1^2$
$200 = 2u + 2a$ $(\because t_1 = 2s)$
$u + a = 100$ ...(i)
In second case: $s_2 = ut_2 + \frac{1}{2}at_2^2$
$420 = 6u + 18a$ $(\because t_2 = 2 + 4 + 6s)$
$3a + u = 70$ ...(ii)
Solving Eqs. (i) and (ii), we get
$a = -15\text{ms}^{-2}$
and $u = 115\text{cms}^{-1}$
$\therefore\ v = u + at$
$= 115 - 15 \times 7 = 10\text{cms}^{-1}$

137 **(a)**
$S_1 = \frac{1}{2}ft^2, S_2 = -v_0t - \frac{1}{2}gt^2$, Clearly, $(S_1 - S_2) \propto t$

138 **(a)**
For the given condition initial height $h = d$ and velocity of the ball is zero. When the ball moves downward its velocity increases and it will be maximum when the ball hits the ground & just after the collision it becomes half and in opposite direction. As the ball moves upward its velocity again decreases and becomes zero at height $d/2$. This explanation match with graph (A)

139 **(d)**
For first ball

$$\frac{1}{2}gt^2 = 176.4$$

$$\Rightarrow \quad t = \sqrt{\frac{176.4 \times 2}{10}}$$

$$t = 5.9 \text{ s}$$

For second ball, $t = 3.9$ s

$$u(3.9) + \frac{1}{2}g(3.9)^2 = 176.4$$

$$\Rightarrow \quad 3.9u + \frac{10}{2}(3.9)^2 = 176.4$$

$$\Rightarrow \quad u = 25.7 \text{ ms}^{-1}$$

This value is approximated to 24.5 ms^{-1}.

140 **(a)**
$v = 18 \times \frac{5}{18}\text{ms}^{-1} = 5\text{ms}^{-1}$,
$h_{max} = \frac{5 \times 5}{2 \times 10}\text{m} = \frac{25}{20}\text{m} = 1.25\text{m}$

141 **(c)**
$t = \sqrt{\frac{2h}{(g+a)}} = \sqrt{\frac{2 \times 2.7}{(9.8 + 1.2)}} = \sqrt{\frac{5.4}{11}}$
$= \sqrt{0.49} = 0.7$ sec
As $u = 0$ and lift is moving upward with acceleration

142 **(b)**
Distance average speed $= \frac{2v_1v_2}{v_1+v_2} = \frac{2 \times 2.5 \times 4}{2.5+4}$
$= \frac{200}{65} = \frac{40}{13} km/hr$

143 **(b)**

Initial speed of car,

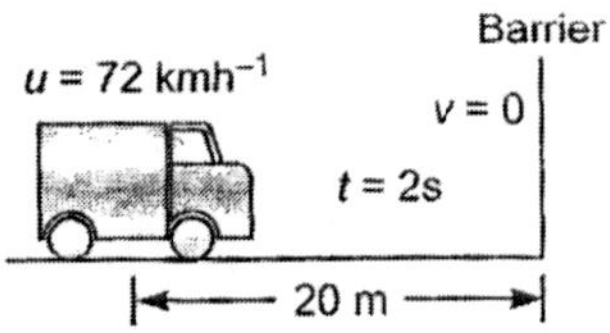

$u = 72 \text{ kmh}^{-1} = 72 \times \frac{5}{18} \text{ms}^{-1}$
$= 20 \text{ ms}^{-1}$

Distance from barrier $s = 20$ m

Time taken by car to hit the barrier, $t =$ 2 s

Using first equation of motion $v = u + at$

$a = \frac{v - u}{t}$

$\therefore a = \frac{0 - 20}{2} = -10 \text{ ms}^{-2}$

−vesign indicates that acceleration is retarding or it is deceleration which decreases the speed of car.

144 **(c)**

$\frac{dv}{dt} = bt \Rightarrow dv = bt\, dt \Rightarrow v = \frac{bt^3}{2} + K_1$

At $t = 0, v = v_0 \Rightarrow K_1 = v_0$

We get $v = \frac{1}{2}bt^2 + v_0$

Again $\frac{dx}{dt} = \frac{1}{2}bt^2 + v_0$

$\Rightarrow x = \frac{1}{2}\frac{bt^3}{3} + v_0 t + K_2$

At $t = 0, x = 0 \Rightarrow K_2 = 0$

$\therefore x = \frac{1}{6}bt^3 + v_0 t$

145 **(d)**

$\frac{\tan 30°}{\tan 45°} = \frac{1}{\sqrt{3}} + 1 = 1 : \sqrt{3}$

146 **(d)**

The separation between the two bodies, two seconds after the release of second body

$= \frac{1}{2} \times 9.8[(3)^2 - (2)^2] = 24.5\, m$

147 **(b)**

$\vec{v} = \vec{u} + \vec{a}t$

$v = (2\hat{\imath} + 3\hat{\jmath}) + (0.3\hat{\imath} + 0.2\hat{\jmath}) \times 10$

$= 5\hat{\imath} + 5\hat{\jmath}$

$|\vec{v}| = 5\sqrt{2}$

148 **(c)**

$h = \frac{1}{2}gt^2 \Rightarrow t = \sqrt{2h/g}$

$t_a = \sqrt{\frac{2a}{g}}$ and $t_b = \sqrt{\frac{2b}{g}} \Rightarrow \frac{t_a}{t_b} = \sqrt{\frac{a}{b}}$

149 **(d)**

$S \propto u^2 \Rightarrow \frac{S_1}{S_2} = \left(\frac{1}{4}\right)^2 = \frac{1}{16}$

150 **(c)**

A body is moving on a straight line with constant velocity. Between A and B, the straight line is the shortest distance. This is the distance travelled. The particle starts at A and reaches B along the straight line. Therefore displacement is also $AB, D = S$

A ———————— *B*

151 **(b)**

For vertically upward motion, $h_1 = v_0 t - \frac{1}{2}gt^2$ and for vertically downward motion, $h_2 = v_0 t + \frac{1}{2}g\,t^2$

$\therefore$ Total distance covered in t sec h

$= h_1 + h_2 = 2v_0 t$

152 **(c)**

Let car starts from point A from rest

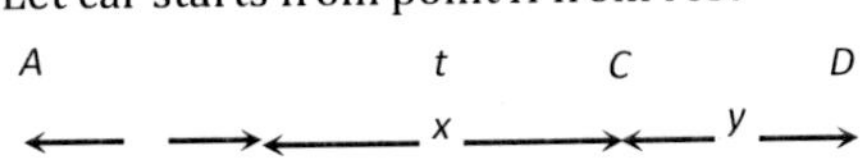

moves up to point B with acceleration f

Velocity of car at point B, $v = \sqrt{2fS}$

[As $v^2 = u^2 + 2as$]

Car moves distance BC with this constant velocity in time t

$x = \sqrt{2fS}.t$ [As $s = ut$] ... (i)

So the velocity of car at point C also will be $\sqrt{2fS}$ and finally car stops after covering distance y

Distance $CD \Rightarrow y = \frac{(\sqrt{2fS})^2}{2(f/2)} = \frac{2fS}{f} = 2S$ ...(ii)

So, the total distance $AD = AB + BC + CD = 15S$ [Given]

$\Rightarrow S + x + 2S = 15\,S \Rightarrow x = 12S$

Substituting the value of x in equation (i) we get

$x = \sqrt{2fS}.t \Rightarrow 12S = \sqrt{2fS}.t \Rightarrow 144S^2 = 2fS.t^2$

$\Rightarrow S = \frac{1}{72}ft^2$

153 **(b)**

Given, $a = \frac{dv}{dt} = 6t + 5$

Or $dv = (6t + 5)\,dt$

Integrating, we get

$\int_0^v dv = \int_0^t (6t + 5)\,dt$

Or $v = \left(\frac{6t^2}{2} + 5t\right)$

Again $v = \frac{ds}{dt}$

$\therefore \quad ds = \left(\frac{6t^2}{2} + 5t\right)dt$

Integrating again, we get

$\int_0^s ds = \int_0^t \left(\frac{6t^2}{2} + 5t\right)dt$

$\therefore \quad s = \frac{3t^3}{3} + \frac{5t^2}{2}$

When, $t = 2$ s, $s = 3 \times \frac{2^3}{3} + \frac{5 \times 2^2}{2} = 3 \times \frac{8}{3} + \frac{5 \times 4}{2}$

$= 8 + 10 = 18$ m

154 **(c)**

$\Delta x = \frac{1}{2}gt^2 - \frac{1}{2}g(t-1)^2$

$= \frac{1}{2}g[t^2 - (t-1)^2] = \frac{1}{2}g(2t-1)$

$= \frac{1}{2} \times 9.8 \times 5\text{m} = 24.5\text{m}$

155 **(a)**

$v = \sqrt{2 \times 9.8 \times 100} = \sqrt{1960}\text{ms}^{-1}$

$\frac{\sqrt{1960}+0}{2} = \frac{2}{t}$ or $t = \frac{4}{\sqrt{1960}}\text{s} = \frac{4}{44.27}\text{s} = 0.09\text{s}$

156 **(d)**

$h = ut - \frac{1}{2}gt^2$

$= 10 \times 1 - \frac{1}{2} \times 10 \times 1$

$= 10 - 5 = 5m$

157 **(b)**

$x = 4(t-2) + a(t-2)^2$

At $t = 0, x = -8 + 4a = 4a - 8$

$v = \frac{dx}{dt} = 4 + 2a(t-2)$

At $t = 0, v = 4 - 4a = 4(1-a)$

But acceleration, $a = \frac{d^2x}{dt^2} = 2a$

158 **(c)**

Displacement of the particle will be zero because it comes back to its starting point

Average speed $= \frac{\text{Total distance}}{\text{Total time}} = \frac{30m}{10sec} = 3\,m/s$

159 **(c)**

$v = \sqrt{2gh} = \sqrt{2 \times 10 \times 20} = 20m/s$

160 **(c)**

Instantaeneous velocity $= v = \frac{\Delta x}{\Delta t}$

By using the data from the table

$v_1 = \frac{0-(-2)}{1} = 2m/s, \quad v_2 = \frac{6-0}{1} = 6m/s$

$v_3 = \frac{16-6}{1} = 10m/s$

So, motion is non-uniform but accelerated

161 **(a)**

From first equation of motion

$v = u + at$

As object starts from rest, so $u = 0$

$\therefore \quad v = at$ or $v \propto t$

Therefore, $v - t$ graph is a straight line passing through O.

162 **(a)**

Slope of velocity-time graph measures acceleration. For graph (a) slope is zero. Hence $a = 0 i.e.$ motion is uniform

163 **(b)**

The velocity of balloon at height h, $v = \sqrt{2\left(\frac{g}{8}\right)h}$

When the stone released from this balloon, it will go upward with velocity, $= \frac{\sqrt{gh}}{2}$ (Same as that of balloon). In this condition time taken by stone to reach the ground

$$t = \frac{v}{g}\left[1 + \sqrt{1 + \frac{2gh}{v^2}}\right]$$

$$= \frac{\sqrt{gh}/2}{g}\left[1 + \frac{2gh}{gh/4}\right]$$

$$= \frac{2\sqrt{gh}}{g} = 2\sqrt{\frac{h}{g}}$$

164 **(a)**

$\vec{r} = x\hat{i} + y\hat{j} + z\hat{k} \therefore r = \sqrt{x^2 + y^2 + z^2}$

$r = \sqrt{6^2 + 8^2 + 10^2} = 10\sqrt{2}m$

165 **(b)**

Average speed is the ratio of distance to time taken

Distance travelled from 0 to $5s = 40\ m$

Distance travelled from 5 to $10s = 0\ m$

Distance travelled from 10 to $15s = 60\ m$

Distance travelled from 15to $20s = 20$

So, total distance $= 40 + 0 + 60 + 20 = 120\ m$

Total time taken $= 20\ minutes$

Hence, average speed

$$= \frac{\text{distance travelled } (m)}{\text{time (min)}} = \frac{120}{20}$$

$$= 6\ m/min$$

166 **(d)**

$$S \propto u^2 \Rightarrow \frac{S_1}{S_2} = \left(\frac{1}{4}\right)^2 = \frac{1}{16}$$

167 **(b)**

According to given relation acceleration $a = \alpha t + \beta$

As $a = \frac{dv}{dt} \Rightarrow \alpha t + \beta = \frac{dv}{dt}$

Since particle starts from rest, its initial velocity is zero

$i.e.$, At time $t = 0$, velocity $= 0$

$$\Rightarrow \int_0^v dv = \int_0^t (\alpha t + \beta)dt \Rightarrow v = \frac{\alpha t^2}{2} + \beta t$$

168 **(b)**

Bullet will take $\frac{100}{1000} = 0.1\ sec$ to reach target.

During this period vertical distance (downward) travelled by the bullet $= \frac{1}{2}gt^2 = \frac{1}{2} \times 10 \times (0.1)^2 m = 5cm$

So the gun should be aimed 5 cm above the target

169 **(b)**

Distance average speed $= \frac{2v_1 v_2}{v_1 + v_2} = \frac{2 \times 2.5 \times 4}{2.5 + 4}$

$$= \frac{200}{65} = \frac{40}{13} km/hr$$

170 **(a)**

Let t_1 and t_2 be times taken by the car to go from X to Y and then from Y to X respectively.

Then, $t_1 + t_2 = \frac{XY}{v_u} + \frac{XY}{v_d} = XY\left(\frac{v_u + u_d}{v_u v_d}\right)$

Total distance travelled

$$= XY + XY = 2\ XY$$

Therefore, average speed of the car for this round trip is

$$\text{Average speed} = \frac{\text{distance travelled}}{\text{time taken}}$$

$$v_{\text{av}} = \frac{2\ XY}{XY\left(\frac{v_u + v_d}{v_u v_d}\right)} \text{ or } v_{\text{av}} = \frac{2 v_u v_d}{v_u + v_d}$$

171 **(a)**

Average velocity $= \frac{2v_1 v_2}{v_1 + v_2}$

Given, $v_{av} = 40$ km/h, $v_1 = 60$ km/h and $v_2 = ?$

$$\therefore \quad 40 = \frac{2 \times 60 \times v_2}{60 + v_2}$$

$$80 v_2 = 2400$$

$$v_2 = 30 \text{ km/h}$$

172 **(a)**

$v = \sqrt{3x + 16}$ or $v^2 = 3x + 16$ or $v^2 - 16 = 3x$

Comparing with $v^2 - u^2 = 2aS$, we get,

$u = 4$ units , $2a = 3$

or $a = 1.5$ units.

173 **(a)**

$S \propto u^2 \therefore \frac{S_1}{S_2} = \left(\frac{u_1}{u_2}\right)^2 \Rightarrow \frac{2}{S_2} = \frac{1}{4} \Rightarrow S_2 = 8\,m$

174 **(d)**

In case of motion under gravity, time taken to go up is equal to the time taken to fall down through the same distance.

Time of desent(t_2) = time of ascent(t_1) $= \frac{u}{g}$

$\therefore$ Total time of flight $T = t_1 + t_2 = \frac{2u}{g}$

Or $u = \frac{g(t_1+t_2)}{2}$

175 **(b)**

$$\text{Time} = \frac{\text{Total length}}{\text{Relative velocity}} = \frac{50+50}{10+15} = \frac{100}{25} = 4\,sec$$

176 **(c)**

$h = ut + \frac{1}{2}gt^2, t = 3$ sec, $u = -4.9 m/s$

$\Rightarrow h = -4.9 \times 3 + 4.9 \times 9 = 29.4\,m$

177 **(d)**

Relative velocity

$= 10 + 5 = 15\,m/sec$

$\therefore t = \frac{150}{15} = 10\,sec$

178 **(b)**

$$H_{\max} = \frac{u^2}{2g} = \frac{19.6 \times 19.6}{2 \times 9.8} = 19.6\,m$$

179 **(c)**

Let both balls meet at point P after time t

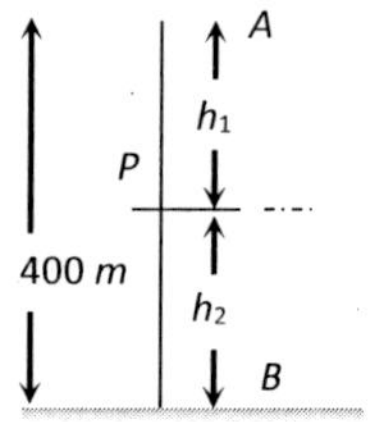

The distance travelled by ball A, $h_1 = \frac{1}{2}gt^2$

The distance travelled by ball B, $h_2 = ut - \frac{1}{2}gt^2$

$h_1 + h_2 = 400\,m \Rightarrow ut = 400, t = 400/50 = 8\,sec$

$\therefore h_1 = 320m$ and $h_2 = 80\,m$

180 **(b)**

$\overrightarrow{v_{ct}} = \overrightarrow{v_c} - \overrightarrow{v_t}$

$\overrightarrow{v_{ct}} = \overrightarrow{v_c} + (-\overrightarrow{v_t})$

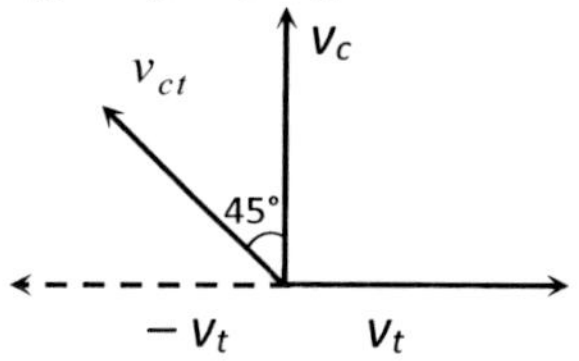

Velocity of car $w.r.t.$ train (v_{ct}) is towards West-North

181 **(a)**

Le the initial velocity $= u$

And acceleration $= a$

In Ist case $s_1 = ut_1 + \frac{1}{2}at_1^2$

$200 = 2u + 2a \quad (\because t_1 = 2\text{ s})$

Or $u + a = 100$ ... (i)

In IInd case

$s_2 = ut_2 + \frac{1}{2}at_2^2$

$420 = 6u + 18a \quad (\because t_2 = 2 + 4 = 6\text{ s})$

Or $\quad 3a + u = 70 \quad$ ...(ii)

Solving Eqs. (i) and (ii), we get

$$a = -15 \text{ ms}^{-2}$$

And $\quad u = 115 \text{ ms}^{-1}$

$$v = u + at$$

$$= 115 - 15 \times 7 = 10 \text{ ms}^{-1}$$

182 **(b)**

$h = vt - \frac{1}{2}gt^2$ or $\frac{1}{2}gt^2 - vt + h = 0$

or $gt^2 - 2vt + 2h = 0 \Rightarrow t_1t_2 = \frac{2h}{g}$

$1 \times 3 = \frac{2h}{10}$ or $2h = 30\text{m}$ or $h = 15\text{m}$

183 **(d)**

$$a = \frac{dv}{dt} = \frac{dv}{dx}\frac{dx}{dt} = v\frac{dv}{dx} = -\alpha x^2 \text{ [Given]}$$

$$\Rightarrow \int_{v_0}^{0} v dv = \alpha \int_{0}^{S} x^2\, dx \Rightarrow \left[\frac{v^2}{2}\right]_{v_0}^{0}$$

$$= -\alpha \left[\frac{x^3}{3}\right]_0^S$$

$$\Rightarrow \frac{v_0^2}{2} = \frac{\alpha S^3}{3} \Rightarrow S = \left(\frac{3v^2}{2\alpha}\right)^{\frac{1}{3}}$$

184 **(a)**

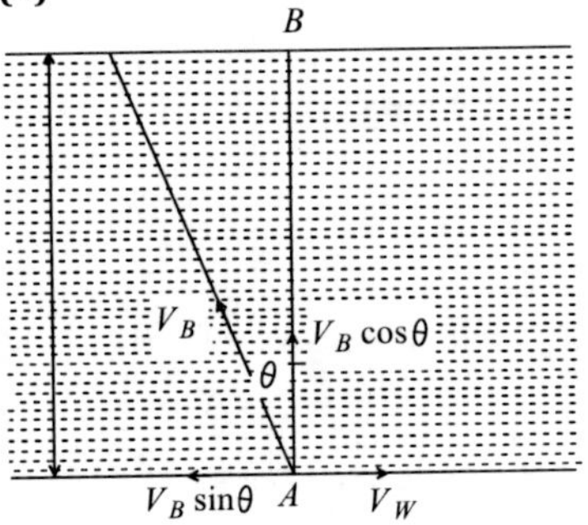

From figure, $V_B \sin\theta = V_W$

$\sin\theta = \frac{V_W}{V_B} = \frac{1}{2} \Rightarrow \theta = 30° \quad [\because V_B = 2V_W]$

Time taken to cross the river,

$$t = \frac{D}{V_B \cos\theta} = \frac{D}{V_B \cos 30°} = \frac{2D}{V_B\sqrt{3}}$$

185 **(c)**

Distance travelled by the particle is

$x = 40 + 12t - t^3$

We know that, speed is rate of change of distance i.e.

$v = \frac{dx}{dt}$

$\therefore v = \frac{d}{dt}(40 + 12t - t^3) = 0 + 12 - 3t^2$

But final velocity $v = 0$

$\therefore 12 - 3t^2 = 0$

$\Rightarrow t^2 = \frac{12}{3} = 4 \Rightarrow t = 2s$

Hence, distance travelled by the particle before coming to rest is given by

$x = 40 + 12(2) - (2)^3 = 40 + 24 - 8$

$= 64 - 8 = 56m$

186 **(d)**

Slope of line $= -\frac{2}{3}$

Equation of line is $(v - 20) = -\frac{2}{3}(s - 0)$

$\Rightarrow v = 20 - \frac{2}{3}s \quad$ ...(i)

Velocity at $s = 15m$ *ie*,

$v = \left.\frac{ds}{dt}\right|_{s=15\text{m}} = 20 - \frac{2}{3}(15) = 10\text{ms}^{-1}$

Differentiate Eq. (i) with respect to time,

acceleration $= \frac{dv}{dt} = \frac{2}{3}\frac{ds}{dt}$

$\therefore \left.\frac{dv}{dt}\right|_{s=15\text{m}} = -\frac{2}{3}\left.\frac{ds}{dt}\right|_{s=15\text{m}} = -\frac{20}{3}\text{ms}^{-2}$

187 **(c)**

(i) A body having constant speed can have varying velocity as direction may change.

(ii) Position-time graphs for two objects with zero relative velocity are parallel.

(iii) For a given time interval,

Distance $\geq$ | displacement |

$\therefore$ Average speed $\geq$ | average velocity |

Therefore, all the options are true.

188 **(c)**

$$\frac{dx}{dt}=2at-3bt^2 \Rightarrow \frac{d^2x}{dt^2}=2a-6bt=0$$
$$\Rightarrow t=\frac{a}{3b}$$

189 **(b)**

Let v be the velocity of the train after time t_1. Then $v=\alpha t_1=\beta t_2; x_1=\frac{1}{2}\alpha t_1^2$ and $x_2=\frac{1}{2}\beta t_2^2$

$\therefore \frac{\beta}{\alpha}=\frac{t_1}{t_2}$ and $\frac{x_1}{x_2}=\frac{\alpha t_1^2}{\beta t_2^2}=\frac{\alpha}{\beta}\times\frac{\beta^2}{\alpha^2}=\frac{\beta}{\alpha}$

$\therefore \frac{x_1}{x_2}=\frac{\beta}{\alpha}=\frac{t_1}{t_2}$

190 **(d)**

$S=\left(4\times 1.5+4\times 1.5+\frac{1}{2}\times 10\times 1.5\times 1.5\right)\text{m}$

$=(6+6+5\times 2.25)\text{m}=23.25\text{m}$

191 **(a)**

$\tan(90°-\theta)=\frac{20}{15}$

$\therefore \cot\theta=\frac{20}{15}=\frac{4}{3}$

$\Rightarrow \theta=37°$

$\therefore \theta=37°+23°$

$=60°$

192 **(a)**

$1900=\frac{1}{2}\times 0.4[(120)^2-(120-t^2)]$

or $\frac{1900}{0.2}=(240-t)t$ or $9500=240t-t^2$

or $t^2-240t+9500=0$

or $t^2-190t-50t+9500=0$

or $t(t-190)-50(t-190)=0$

or $(t-50)(t-190)=0 \Rightarrow t=50\text{s},190\text{s}$

Rejecting $t=190$ s, we get $t=50$ s

193 **(b)**

Let height of minaret is H and body take time T to fall from top to bottom

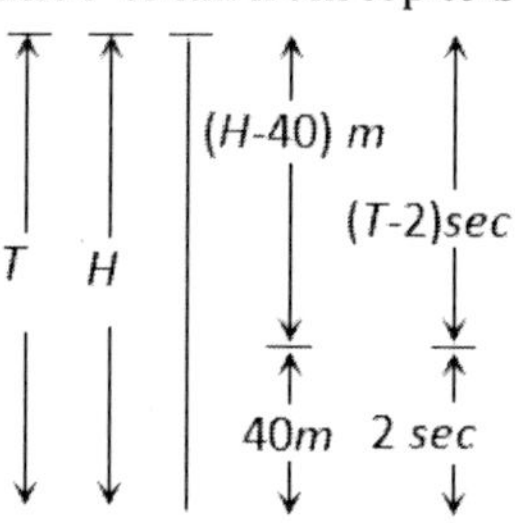

$H=\frac{1}{2}gT^2$ ...(i)

In last 2 sec body travels distance of 40 m so in $(T-2)$ sec distance travelled = $(H-40)m$

$(H-40)=\frac{1}{2}g(T-2)^2$ ...(ii)

By solving (i) and (ii), $T=3\,sec$ and $H=45m$

194 **(d)**

At the point A, the tangent to the curve is parallel to time axis. So, velocity at A is zero. But acceleration is not zero. Note that the displacement corresponding to the point A is not zero.

195 **(a)**

$$S_n=u+\frac{a}{2}(2n-1)\Rightarrow S_3$$
$$=0+\frac{4/3}{2}(2\times 3-1)$$
$$\Rightarrow S_3=\frac{10}{3}m$$

196 **(a)**

Since slope of graph remains constant for velocity-time graph

197 **(c)**

Relativistic momentum $=\frac{m_0v}{\sqrt{1-v^2/c^2}}$

If velocity is doubled then the relativistic mass also increases. Thus value of linear momentum will be more than double

198 **(c)**

For shortest possible path man should swim with an angle $(90+\theta)$ with downstream

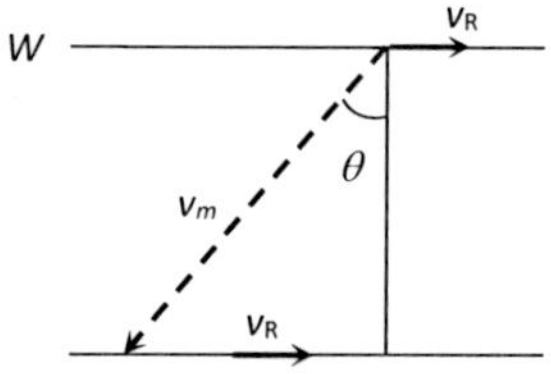

From the fig,

$\sin\theta=\frac{v_r}{v_m}=\frac{5}{10}=\frac{1}{2}$

$\Rightarrow\therefore \theta=30°$

So angle with downstream $=90°+30°=120°$

199 **(c)**

Given, $s=2$ m, $u=80\text{ ms}^{-1}, v=0$

From $v^2=u^2-2as$

$\therefore \quad (0)^2=(80)^2-2\times a\times 2$

Or $a = \frac{80 \times 80}{4} = 1600 \text{ ms}^{-2}$

200 **(c)**

Second law of motion gives

$$h = ut + \frac{1}{2}gT^2$$

or $h = 0 + \frac{1}{2}gT^2 (\because u = 0)$

$\therefore$ $T = \sqrt{\left(\frac{2h}{g}\right)}$

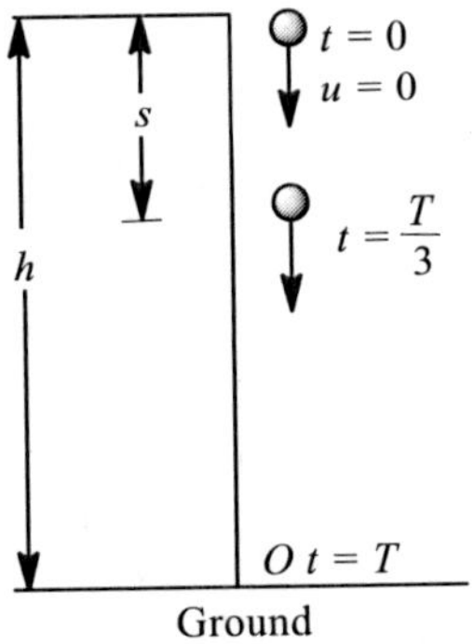

At $t = \frac{T}{3}$s,

$$s = 0 + \frac{1}{2}g\left(\frac{T}{3}\right)^2$$

Or $s = \frac{1}{2}g . \frac{T^2}{9}$

Or $s = \frac{g}{18} \times \frac{2h}{g}\left(\because T = \sqrt{\frac{2h}{g}}\right)$

$\therefore$ $s = \frac{h}{9}$m

Hence, the position of ball from the ground

$$= h - \frac{h}{9} = \frac{8h}{9}\text{m}.$$

201 **(a)**

The velocity of the particle is

$\frac{dx}{dt} = \frac{d}{dt}(2 - 5t + 6t^2) = (0 - 5 + 12t)$

For initial velocity $t = 0$, hence $v = -5\ m/s$

202 **(a)**

$S_n = u + \frac{g}{2}(2n - 1)$; when $u = 0, S_1 : S_2 : S_3 = 1 : 3 : 5$

203 **(c)**

Acceleration of the body along AB is $g \cos \theta$

Distance travelled in time $t\ sec = AB = \frac{1}{2}(g\ cos\ \theta)t^2$

From $\Delta\ ABC, AB = 2R \cos \theta$; $2R \cos \theta = \frac{1}{2}g \cos \theta t^2$

$\Rightarrow t^2 = \frac{4R}{g}$ or $t = 2\sqrt{\frac{R}{g}}$

204 **(a)**

The distance covered by the ball during the last t *seconds* of its upward motion = Distance covered by it in first t seconds of its downward motion

From $h == ut + \frac{1}{2}gt^2$

$h = \frac{1}{2}g\ t^2$ [As $u = 0$ for it downward motion]

206 **(a)**

Here, $s_n = \frac{1}{n}an^2$

s_{nth} = distance travelled in n second – distance travelled in $(n - 1)$ second

$= \left(\frac{2n - 1}{2}\right)a$

$\therefore \frac{s_{nth}}{s_n} = \frac{2n - 1}{n^2} = \frac{2}{n} - \frac{1}{n^2}$

207 **(a)**

Given, $s = 1.2$ m, $v = 640 \text{ ms}^{-1}$, $a = ?$, $u = 0; t = ?$

We have the third equation of motion

$$2as = v^2 - u^2$$

$$2a \times 1.2 = 640 \times 640$$

Or $a = \frac{8 \times 64 \times 10^3}{3}$

And by first equation of motion

$$v = u + at$$

Or $\qquad t = \frac{v}{a} = \frac{15}{4} \times 10^{-3}$

$$= 3.75 \times 10^{-3}\text{s} \approx 4 \text{ ms}$$

209 **(b)**

Free fall of an object (in vacuum) is a case of motion with uniform acceleration

211 **(c)**

$h = \frac{1}{2}gt^2$, (parabolic)

$v = -gt$and after the collision, $v = gt$ (straight line)

Collision is perfectly elastic then ball reaches to same height again and again with same velocity

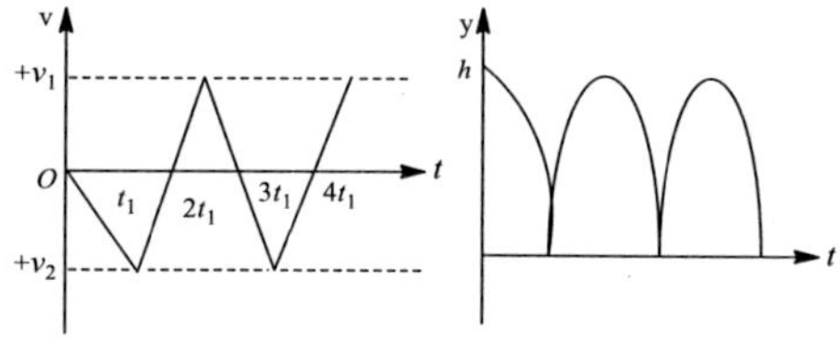

212 **(d)**

Since $c >> v$ (negligible)

213 **(c)**

When packet is released from the balloon, it acquires the velocity of balloon of value $12\ m/s$. Hence velocity of packet after $2\ sec$, will be

$v = u + gt = 12 - 9.8 \times 2 = -76\ m/s$

214 **(d)**

Interval of ball throw = $2sec$

If we want that minimum three (more than two) ball remain in air then time of flight of first ball must be greater than $4\ sec$

$T > 4\ sec$

$\frac{2u}{g} > 4 \text{ sec} \Rightarrow u > 19.6\ m/s$

For $u = 19.6$, first ball will just about to strike the ground (in air)

Second ball will be at highest point (in air)

Third ball will be at point of projection or at ground (not in air)

215 **(b)**

Between time interval 20 sec to 40 sec, there is non-zero acceleration and retardation. Hence distance travelled during this interval

= Area between time interval 20 sec to 40 sec

$= \frac{1}{2} \times 20 \times 3 + 20 \times 1 = 30 + 20 = 50\ m$

216 **(d)**

$$\frac{v_A}{v_B} = \frac{\tan\theta_A}{\tan\theta_B} = \frac{\tan 30°}{\tan 60°} = \frac{1/\sqrt{3}}{\sqrt{3}} = \frac{1}{3}$$

217 **(d)**

$v = \frac{ds}{dt} = 3t^2 - 12t + 3$ and $a = \frac{dv}{dt} = 6t - 12$

For $a = 0$, we have $t = 2$ and at $t = 2, v = -9\ ms^{-1}$

218 **(c)**

Using Newton's equation of motion

$$v^2 = u^2 + 2as$$

$\therefore \qquad (5)^2 = (20)^2 + 2(a)100$

Or $\qquad a = -\frac{400-25}{200} = -\frac{375}{200}\text{ms}^{-2}$

Therefore, fore F = mass m × acceleration a

$$= 20 \times \left(-\frac{375}{200}\right) = -37.5 \text{ N}$$

220 **(b)**

$\vec{v} = \vec{u} + \vec{a}t$

$v = (2\hat{\imath} + 3\hat{\jmath}) + (0.3\hat{\imath} + 0.2\hat{\jmath}) \times 10$

$= 5\hat{\imath} + 5\hat{\jmath}$

$|\vec{v}| = 5\sqrt{2}$

221 **(b)**

Let car B catches, car A after $'t'$sec, then

$60t + 2.5 = 70t - \frac{1}{2} \times 20 \times t^2$

$\Rightarrow 10t^2 - 10t + 2.5 = 0 \Rightarrow t^2 - t + 0.25 = 0$

$$\therefore t = \frac{\sqrt{1 - 4 \times (0.25)}}{2} = \frac{1}{2}hr$$

222 **(a)**

$$S_n = u + \frac{a}{2}(2n - 1) \Rightarrow S_3$$

$$= 0 + \frac{4/3}{2}(2 \times 3 - 1)$$

$$\Rightarrow S_3 = \frac{10}{3}m$$

223 **(b)**

$$\frac{|\text{Average velocity}|}{|\text{Average speed}|} = \frac{|\text{displacement}|}{|\text{distance}|} \leq 1$$

Because displacement will either be equal or less than distance. It can never be greater than distance

224 **(c)**

$h = -vt_1 + \frac{1}{2}gt_1^2$ or $\frac{h}{t_1} = -v + \frac{1}{2}gt_1$

...(i)

$h = vt_2 + \frac{1}{2}gt_1^2$ or $-\frac{h}{t_2} = -v + \frac{1}{2}gt_2$

...(ii)

$$\therefore \frac{h}{t_1} + \frac{h}{t_2} = \frac{1}{2}g(t_1 + t_2)$$

or $h = \frac{1}{2}gt_1t_2$

For falls under gravity from the top of the tower

$$h = \frac{1}{2}gt^2$$

$$\therefore \frac{1}{2}gt_1t_2 = \frac{1}{2}gt^2 \Rightarrow t = \sqrt{t_1t_2}$$

225 **(b)**

Let at point A initial velocity of body is equal to zero

for path AB: $v^2 = 0 + 2gh$ ... (i)

for path AC: $(2v)^2 = 0 + 2gx$

$4v^2 = 2gx$ ...(ii)

Solving (i) and (ii), $x = 4h$

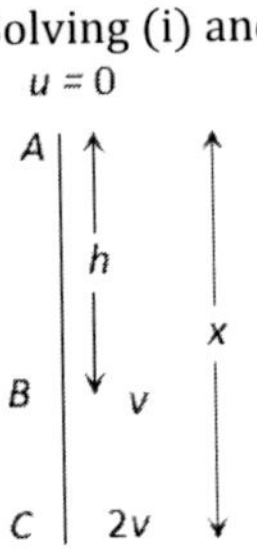

227 **(a)**

Displacement = Area of upper trapezium- Area of lower trapezium

$= \frac{1}{2}(2 + 4) \times 4 - \frac{1}{2}(2 + 4)2 = 12 - 6 =$ 6m

228 **(c)**

$$\frac{dv}{dt} = bt \Rightarrow dv = bt\,dt \Rightarrow v = \frac{bt^3}{2} + K_1$$

At $t = 0, v = v_0 \Rightarrow K_1 = v_0$

We get $v = \frac{1}{2}bt^2 + v_0$

Again $\frac{dx}{dt} = \frac{1}{2}bt^2 + v_0$

$$\Rightarrow x = \frac{1}{2}\frac{bt^3}{3} + v_0t + K_2$$

At $t = 0, x = 0 \Rightarrow K_2 = 0$

$$\therefore x = \frac{1}{6}bt^3 + v_0t$$

229 **(a)**

$$v_{rel} = 45 + 36 = 81 \text{ kmh}^{-1}$$

$$= 81 \times \frac{5}{18}\text{ms}^{-1}$$

$$s_{rel} = v_{rel} \times t = 81 \times \frac{5}{18} \times (5 \times 60)$$

$$= \frac{81 \times 5 \times 5 \times 60}{18}$$

$- 6750 \text{ m} = 6.75 \text{ km}$

230 **(d)**

Displacement form 0 to 5 s = 40 m

Displacement from 5 to 10 s = 40 m

Displacement from 0 to 15 s = − 20 m

And displacement from 15 to 20 s = 0 m

$\therefore$ Net displacement = 40 + 40 − 20 + 0 = 60 m

Total time taken = 5 + 5 + 15 + 5 = 30 min.

Hence, average speed

$$= \frac{\text{displacement (m)}}{\text{time (min)}}$$

$$= \frac{60}{30} = 2 \text{ m min}^{-1}.$$

231 **(d)**

Let acceleration is a and retardation is $-2a$

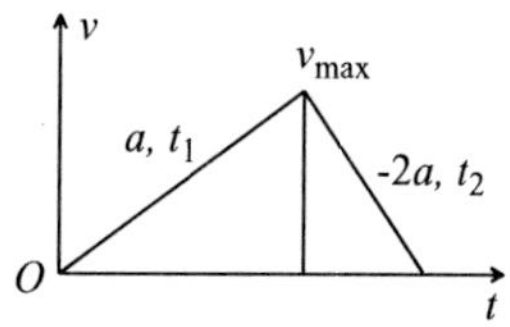

Then for acceleration motion

$t_1 = \frac{v}{a}$ ...(i)

For retarding motion

$t_2 = \frac{v}{2a}$ ...(ii)

Given $t_1 + t_2 = 9 \Rightarrow \frac{v}{a} + \frac{v}{2a} = 9 \Rightarrow \frac{3v}{2a} = 9 \Rightarrow \frac{v}{a} = 6$

Hence, duration of acceleration, $t_1 = \frac{v}{a} = 6\ sec$

232 **(d)**

Acceleration due to gravity is independent of mass of body

233 **(c)**

This person cannot walk straight more than 10 steps.

Distance covered in 10 steps

$= 10 \times 0.8 = 8$m

As the person has to turn after every 10 steps, the only way to have maximum displacement is walk as shown in the figure.

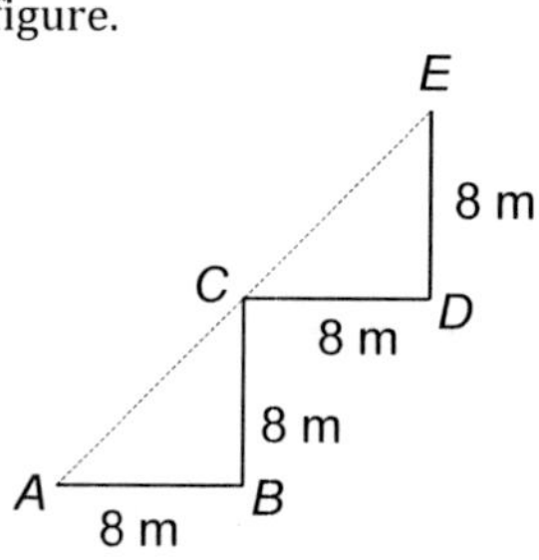

The maximum displacement $= AE = AC + CE$

$= 8\sqrt{2} + 8\sqrt{2} = 16\sqrt{2}$m

234 **(c)**

Vertical component of velocities of both the balls are same and equal to zero. So

$t = \sqrt{\frac{2h}{g}}$

235 **(c)**

$v = \frac{\text{Total distance}}{\text{Time taken}}$

$= \frac{x}{\frac{x/3}{v} + \frac{x/3}{3v} + \frac{x/3}{2v} + \frac{18}{11}v}$

236 **(a)**

If t_1 and t_2 are the time taken by particle to cover first and second half distance respectively.

$t_1 = \frac{x/2}{3} = \frac{x}{6}$

$x_1 = 4.5t_2$ and $x_2 = 7.5t_2$

So, $x_1 + x_2 = \frac{x}{2}$

$\Rightarrow 4.5t_2 + 7.5t_2 = \frac{x}{2}$

$t_2 = \frac{x}{24}$

Total time $t = t_1 + 2t_2 = \frac{x}{6} + \frac{x}{12} = \frac{x}{4}$

So, average speed = 4 m/s

237 **(c)**

Given $\overrightarrow{AB}$ = Velocity of boat = km/hr

$\overrightarrow{AC}$ = Resultant velocity of boat = $10\ km/hr$

$\overrightarrow{BC}$ = Velocity of river

$= \sqrt{AC^2 - AB^2}$

$= \sqrt{(10)^2 - (8)^2} = 6\ km/hr$

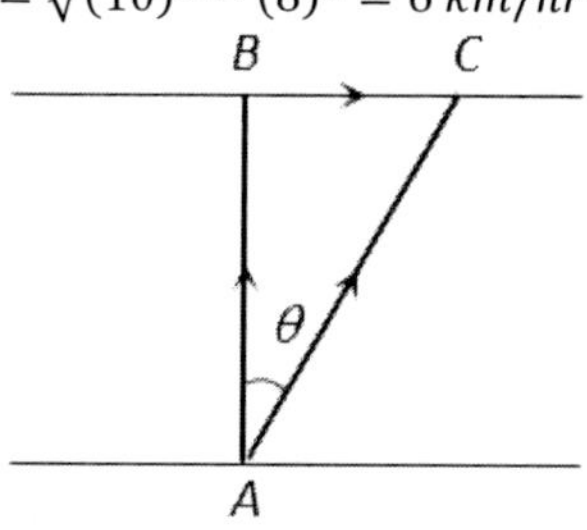

239 **(d)**

$u = 72\ kmph = 20\ m/s, v = 0$

By using $v^2 = u^2 - 2as \Rightarrow a = \frac{u^2}{2s} = \frac{(20)^2}{2\times200} = 1\ m/s^2$

240 **(a)**

Distance travelled in 4 sec

$24 = 4u + \frac{1}{2}\alpha \times 16$ ... (i)

Distance travelled in 8 sec

$88 = 8u + \frac{1}{2}a \times 64$ ... (ii)

After solving (i) and (ii), we get $u = 1 m/s$

241 **(d)**

Let acceleration is a and retardation is $-2a$. Then for accelerating motion

$t_1 = \frac{v}{a}$ ....(i)

For retarding motion, $t_2 = \frac{v}{2a}$ ....(ii)

Given,

$$t_1 + t_2 = 9 \Rightarrow \frac{v}{a} + \frac{v}{2a} = 9 \Rightarrow \frac{3v}{2a} = 9$$
$$\Rightarrow \frac{v}{a} = 6$$

Hence, duration of acceleration, $t_1 = 6$ s

242 **(a)**

$\vec{r} = 20\hat{i} + 10\hat{j} \therefore r = \sqrt{20^2 + 10^2}$
$= 22.5\ m$

243 **(b)**

$$\int_{6.25}^{0} \frac{dv}{\sqrt{v}} = -2.5 \int_0^t dt$$

$|2\sqrt{v}|_{6.25}^{0} = -2.5t$

$2\sqrt{6.25} = 2.5t$

$t = 2 sec$

244 **(c)**

For shortest possible path man should swim with an angle $(90 + \theta)$ with downstream

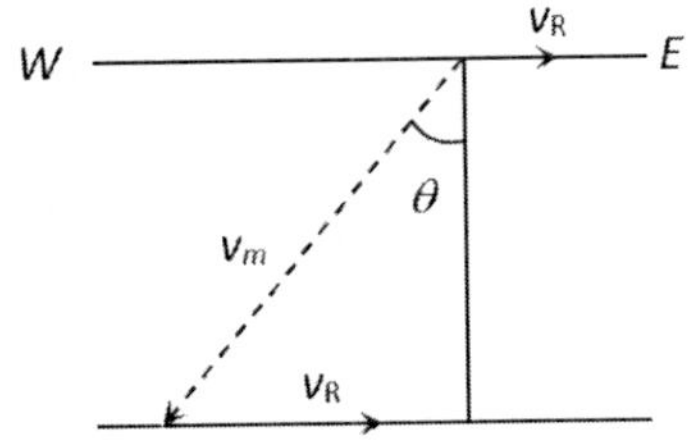

From the fig,

$\sin\theta = \frac{v_r}{v_m} = \frac{5}{10} = \frac{1}{2}$

$\Rightarrow \therefore\ \theta = 30°$

So angle with downstream $= 90° + 30° = 120°$

246 **(a)**

Let initial $(t = 0)$ velocity of particle$= u$

For first 5 sec of motion $s_5 = 10\ metre$

$s = ut + \frac{1}{2}at^2 \Rightarrow 10 = 5u + \frac{1}{2}a(5)^2$

$2u + 5a = 4$

For first 8 sec of motion $s_8 = 20\ metre$

$20 = 8u + \frac{1}{2}a(8)^2 \Rightarrow 2u + 8a = 5$

By solving $u = \frac{7}{6} m/s$ and $a = \frac{1}{3} m/s^2$

Now distance travelled by particle in Total 10 sec

$s_{10} = u \times 10 + \frac{1}{2}a(10)^2$

By substituting the value of u and a we will get $s_{10} = 28.3\ m$

so the distance in last 2 $sec = s_{10} - s_8$

$= 28.3 - 20 = 8.3m$

247 **(b)**

Let particle thrown with velocity u and its maximum height is H then $H = \frac{u^2}{2g}$

When particle is at height $H/2$, then its speed is 10 m/s

From equation $v^2 = u^2 - 2gh$

$(10)^2 = u^2 - 2g\left(\frac{H}{2}\right) = u^2 - 2g\frac{u^2}{4g} \Rightarrow u^2$
$= 200$

Maximum height $\Rightarrow H = \frac{u^2}{2g} = \frac{200}{2\times 10} =$ 10 m

248 **(a)**

$\sqrt{x} = t + 1$

Squaring both sides, we get

$x = (t + 1)^2 = t^2 + 2t + 1$

Differentiating it w.r.t time t, we get

$\frac{dx}{dt} = 2t + 2$

Velocity, $v = \frac{dx}{dt} = 2t + 2$

249 **(c)**

Let particle start from O and travels distance

$d_1(OA), d_2(AB), d_3(BC)$

From equation of motion, we have

$s = ut + \frac{1}{2}at^2$

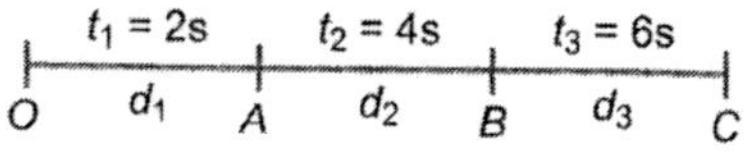

For OA : $t = 2\text{ s}, u = 0$

$$d_1 = \frac{1}{2}a(2)^2 = 2a$$

For OB : $t = 4\text{ s}, u = 0$

$$\therefore \quad s_2 = \frac{1}{2}a(4)^2 = 8a$$

$$d_2 = 8a - 2a = 6a$$

For OC : $t = 6\text{ s}, u = 0$

$$\therefore \quad S = \frac{1}{2}a(6)^2 = 18a$$

Distance in last 2 s $= 18a - 8a = 10a$

$$\therefore \quad d_1 : d_2 : d_3 = 2a : 6a : 10a$$

$$d_1 : d_2 : d_3 = 1 : 3 : 5$$

250 **(a)**

$v = u - gt$

At max height $\quad v^2 = u^2 - 2gh$

$t = \frac{u}{g} \quad h = \frac{u^2}{2g}$

$\frac{t_1}{t_2} = \frac{2}{3}\frac{h_1}{h_2} = \left(\frac{2}{3}\right)^2 = \frac{4}{9}$

251 **(c)**

Let student catch the bus after t sec. So it will cover distance ut

Similarly distance travelled by the bus will be $\frac{1}{2}at^2$ for the given condition

$ut = 50 + \frac{1}{2}at^2 = 50 + \frac{t^2}{2} \quad [a = 1\ m/s^2]$

$\Rightarrow u = \frac{50}{t} + \frac{t}{2}$

To find the minimum value of u

$\frac{du}{dt} = 0$, so we get $t = 10\ sec$, then $u = 10\ m/s$

(c)

252 Speed of the object at reaching the ground $v = \sqrt{2gh}$

If heights are equal then velocity will also be equal

253 **(b)**

Let two balls meet at depth h from platform

So $h = \frac{1}{2}g(18)^2 = v(12) + \frac{1}{2}g(12)^2 \Rightarrow$

$v = 75\ ms^{-1}$

254 **(a)**

$$h_{\max} = \frac{u^2}{2g} = \frac{(15)^2}{2 \times 10} = 11.25m$$

255 **(b)**

Given, $x = 2t^2 + 6t + 25$ and $y = t^2 + 2t + 1$

$$\therefore \frac{dx}{dt} = 4t + 6 \text{ and } \frac{dy}{dx} = 2 + 2$$

At $t = 10$ s

$$\frac{dx}{dt} = 4(10) + 6 = 46 \text{ and } \frac{dy}{dt} = 2\,(10) + 2 = 22$$

$$v = \sqrt{\left(\frac{dx}{dt}\right)^2 + \left(\frac{dy}{dt}\right)^2}$$

$$= \sqrt{(46)^2 + (22)^2} \simeq 51\text{m/s}$$

256 **(d)**

Time of flight $T = \frac{2u}{g} 4\text{ sec} \Rightarrow u = 20\ m/s$

257 **(b)**

There, $x_2 = vt$and $x_1 = \frac{at^2}{2}$

$$x_1 - x_2 = -\left(vt - \frac{at^2}{2}\right)$$

So, the graph would be like.

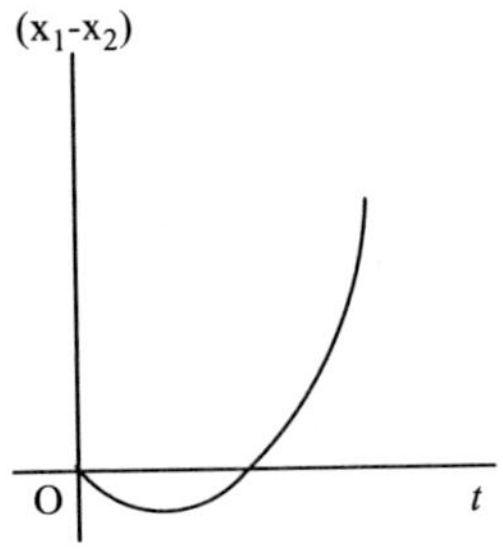

258 (c)

Let the particle touches the sphere t the point A.

Let $PA = 1$

$\therefore PB = \frac{l}{2}$

In $\Delta OPB, \cos \alpha = \frac{PB}{r}$

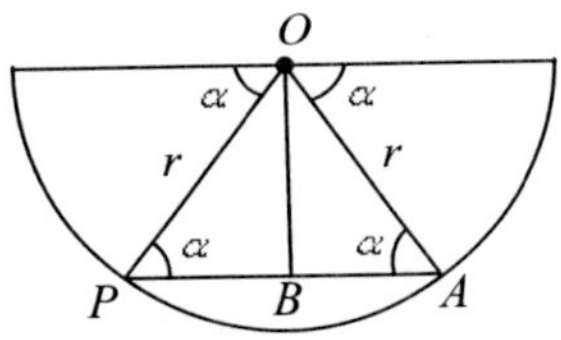

$\therefore PB = r \cos a$

or $\frac{l}{2} = r \cos a$

$\therefore l = 2r \cos \alpha$

But $l = \frac{1}{2} a_0 t^2$

$$\therefore t = \sqrt{\left(\frac{2l}{a_0}\right)} = \sqrt{\left(\frac{2 \times 2r \cos a}{a_0}\right)}$$

$$= \sqrt{\left(\frac{4 r \cos a}{a_0}\right)}$$

259 (d)

$h = ut - \frac{1}{2} gt^2$

$= 10 \times 1 - \frac{1}{2} \times 10 \times 1$

$= 10 - 5 = 5m$

260 (b)

For one dimensional motion along a plane

$S = ut + \frac{1}{2} at^2 \Rightarrow 9.8 = 0 + \frac{1}{2} g \sin\ 30° t^2$

$\Rightarrow t = 2\text{sec}$

261 (c)

$v = \frac{dx}{dt} = 0 + 12t - 3t^2 = 0$

$\Rightarrow \qquad t = 2\text{ s}$

Hence, distance travelled by the particle before coming to rest is given by

$$x = 40 + 12(2) - (2)^3$$

$$= 40 + 24 - 8 = 64 - 8$$

$$= 56\text{ m}$$

262 (b)

Region OA shows that graph bending toward time axis $i.e.$ acceleration is negative.

Region AB shows that graph is parallel to time axis$i.e.$ velocity is zero. Hence acceleration is zero.

Region BC shows that graph is bending towards displacement axis $i.e.$ acceleration is positive.

Region CD shows that graph having constant slope $i.e.$ velocity is constant. Hence acceleration is zero

264 (a)

$h = -vt + \frac{1}{2}gt^2$ or $ft^2 - 2vt - h=0$

$$t = \frac{-(-2v) \pm \sqrt{4v^2 + 4gh}}{2g}$$

$$= \frac{2v \pm 2\sqrt{v^2 + gh}}{2g}$$

$$= \frac{v}{g} \pm \frac{[v^2 + 2gh]^{1/2}}{g} = \frac{v}{g}\left[1 \pm \sqrt{1 + \frac{2gh}{v^2}}\right]$$

Now, retain only the positive sign.

265 (b)

Let $a_1$and $a_2$be the retardations offered to be bullet by wood and iron respectively.

For $A \rightarrow B \rightarrow C$,

$v_1^2 - u^2 = 2a_1(4)$, and $0^2 - v_1^2 = 2a_2(1)$

Adding, we get

$-u^2 = 2(4a_1 + a_2) \quad ...(i)$

For $A' \rightarrow B' \rightarrow C'$,

$v_2^2 - u^2 = 2a_2(2)$

and $0^2 - v_2^2 = 2a_1(2)$

Adding, we get

$-u^2 = 2(2a_1 + 2a_2) \quad ....(ii)$

Equating Eqs. (i) and (ii) and solving, we get

$4a_1 + a_2 = 2a_1 + 2a_2$

$\Rightarrow a_2 = 2a_1$

266 **(c)**

Mass does not affect maximum height

$H = \frac{u^2}{2g} \Rightarrow H \propto u^2$, So if velocity is doubled then height will become four times.i.e.$H = 20 \times 4 = 80m$

267 **(c)**

$\frac{1}{2}gt^2 - \frac{1}{2}g(t-2)^2 = 40$

or $\frac{1}{2} \times 10(2t-2)(2) = 40$ or $2t - 2 = 4$ or $t = 3s$

268 **(c)**

Let the stone falls through a height h in t s

Here, $u = 0, a = g$

Using $D_n = u + \frac{a}{2}(2n-1)$

Distance travelled by the stone in the last second is

$\frac{9h}{25} = \frac{g}{2}(2t-1)$ [$\because u = 0$]...(i)

Distance travelled by the stone in t s is

$h = \frac{1}{2}gt^2$ [$\because u = 0$] ...(ii)

Divide (i) by (ii), we get

$\frac{9}{25} = \frac{(2t-1)}{t^2}$

$9t^2 = 50t - 25$

$9t^2 - 50t + 25 = 0$

On solving, we get

$t = 5s$ or $t = \frac{5}{9}s$

Substituting $t = 5s$ in (ii), we get

$h = \frac{1}{2} \times 9.8 \times (5)^2 = 122.5\ m$

269 **(a)**

Horizontal velocity of dropped packet $= u$

Vertical velocity $= \sqrt{2gh}$

$\therefore$ Resultant velocity at earth

$= \sqrt{u^2 + 2gh}$

270 **(c)**

$S_n = \frac{1}{2}g\cos\theta\,(2n-1), S_{n+1} = \frac{1}{2}g\cos\theta\,\{2(n+1) - 1\}$

$\frac{S_n}{S_{n+1}} = \frac{2n-1}{2n+1}$

271 **(b)**

In this problem point starts moving with uniform acceleration a and after time t (Position B) the direction of acceleration get reversed i.e. the retardation of same value works on the point. Due to this velocity of points goes on decreasing and at position C its velocity becomes zero. Now the direction of motion of point reversed and it moves from C to A under the effect of acceleration a.

We have to calculate the total time in this motion.

Starting velocity at position A is equal to zero.

Velocity at position $B \Rightarrow v = at$ [As $u = 0$]

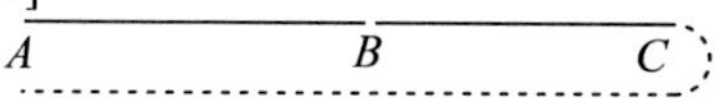

Distance between A and B, $S_{AB} = \frac{1}{2}at^2$

As same amount of retardation works on a point and it comes to rest therefore

$S_{BC} = S_{AB} = \frac{1}{2}a\,t^2$

$\therefore S_{AC} = S_{AB} + S_{BC} = a\,t^2$ and time required to cover this distance is also equal to t.

$\therefore$ Total time taken for motion between A and C $= 2t$

Now for the return journey from C to A ($S_{AC} = at^2$)

$S_{AC} = u\,t + \frac{1}{2}at^2 \Rightarrow at^2 = 0 + \frac{1}{2}at_1^2 \Rightarrow t_1 = \sqrt{2}t$

Hence total time in which point returns to initial point

$T = 2t + \sqrt{2}\,t = (2 + \sqrt{2})t$

272 **(b)**

If acceleration is variable (depends on time) then

$v = u + \int (f)dt = u + \int (a\,t)dt = u + \frac{a\,t^2}{2}$

273 **(d)**

Initial velocity of balloon with respect to ground

$v = 10 + 5 = 15\ m/sec$ upward

After 2 seconds its velocity, $v = u - gt = 15 - 10 \times 2 = -5\ m/sec = 5\ m/sec$ (downward)

274 **(c)**

$h = \frac{1}{2}gt^2 \Rightarrow t = \sqrt{2h/g}$

$t_a = \sqrt{\frac{2a}{g}}$ and $t_b = \sqrt{\frac{2b}{g}} \Rightarrow \frac{t_a}{t_b} = \sqrt{\frac{a}{b}}$

275 **(c)**

Displacement of the particle will be zero because it comes back to its starting point

Average speed $= \frac{\text{Total distance}}{\text{Total time}} = \frac{30m}{10sec}$
$= 3\ m/s$

276 **(b)**

$x = 9t^2 - t^3; v = \frac{dx}{dt} = 18t - 3t^2$, For maximum speed

$\frac{dv}{dt} = \frac{d}{dt}[18t - 3t^2] = 0 \Rightarrow 18 - 6t = 0$

$\therefore t = 3\ \text{sec}$

i.e., Particle achieve maximum speed at $t = 3$ sec. At this instant position of this particle, $x = 9t^2 - t^3$

$= 9(3)^2 - (3)^3 = 81 - 27 = 54\ m$

277 **(a)**

$v_1 - v_2 = at$

or $t = \frac{v_1 - v_2}{a}$.

278 **(b)**

Let the initial velocity of ball be u

Time of rise $t_1 = \frac{u}{g+a}$ and height reached

$= \frac{u^2}{2(g+a)}$

Time of fall t_2 is given by

$\frac{1}{2}(g-a)t_2^2 = \frac{u^2}{2(g+a)}$

$\Rightarrow t_2 = \frac{u}{\sqrt{(g+a)(g-a)}}$

$= \frac{u}{(g+a)}\sqrt{\frac{g+a}{g-a}}$

$\therefore t_2 > t_1$ because $\frac{1}{g+a} < \frac{1}{g-a}$

279 **(c)**

Distance travelled by the particle is

$x = 40 + 12t - t^3$

We know that, speed is rate of change of distance i.e.

$v = \frac{dx}{dt}$

$\therefore v = \frac{d}{dt}(40 + 12t - t^3) = 0 + 12 - 3t^2$

But final velocity $v = 0$

$\therefore 12 - 3t^2 = 0$

$\Rightarrow t^2 = \frac{12}{3} = 4 \Rightarrow t = 2s$

Hence, distance travelled by the particle before coming to rest is given by

$x = 40 + 12(2) - (2)^3 = 40 + 24 - 8$

$= 64 - 8 = 56m$

280 **(b)**

Area under acceleration-time graph gives the change in velocity. Hence,

$v_{max} = \frac{1}{2} \times 10 \times 11 = 55\ \text{ms}^{-1}$

Therefore, the correct option is (b).

281 **(a)**

$s = ut + \frac{1}{2}at^2$

$s = \frac{1}{2}at^2$ $\quad [\because u = 0]$

It is an equation of parabola

282 **(a)**

$\frac{1}{2}at^2 = vt \Rightarrow t = \frac{2v}{a}$

283 **(c)**

Acceleration $= a = \frac{dv}{dt} = 0.1 \times 2t = 0.2t$

Which is time dependent *i.e.* non-uniform acceleration

284 **(a)**

Average speed $= \frac{2v_d v_u}{v_d + v_u}$

285 **(c)**

From given figure, it is clear that the net displacement is zero. So average velocity will be zero

286 **(a)**

From equation of motion, we have

$h = ut + \frac{1}{2}gt^2$

taking upward direction as negative and downward direction as positive, we have

$h = 65$ m,

$u = -12\ \text{ms}^{-1}$ and $g = 10\ \text{ms}^{-2}$

$\therefore \quad 65 = -12t + \frac{1}{2} \times 10 \times t^2$

$\therefore \quad 5t^2 - 12t - 65 = 0$

$\Rightarrow \quad (t - 5)(5t + 13) = 0$

$\therefore \quad t = 5 \text{ s}$

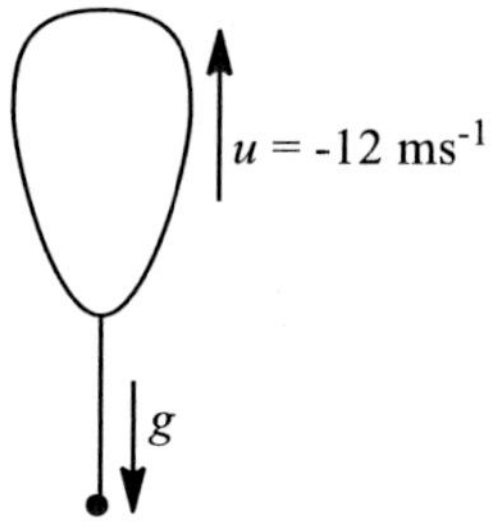

287 **(a)**

Displacement = Summation of all the area with sign

$= (A_1) + (-A_2) + (A_3)$

$= (2 \times 4) + (-2 \times 2) + (2 \times 2)$

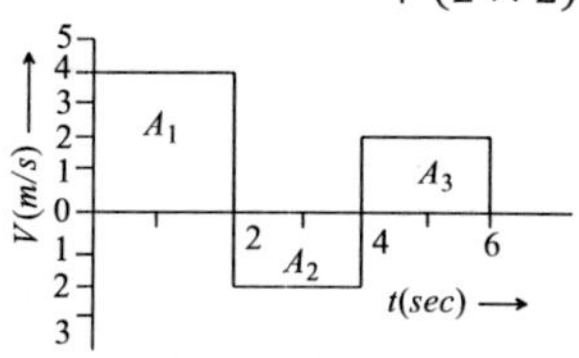

$\therefore$ Displacement $= 8\,m$

Distance = Summation of all the areas without sign

$= |A_1| + |-A_2| + |A_3| = |8| + |-4| + |4|$

$= 8 + 4 + 4$

$\therefore$ Distance $= 16\,m$

288 **(b)**

Distance travelled by train in first 1 *hour* is 60 *km* and distance in next 1/2 hour is 20 *km*.

So Average speed $= \frac{\text{Total distance}}{\text{Total time}} = \frac{60+20}{3/2}$

$= 53.33\ km/hour$

289 **(a)**

Distance b/w the cars A and B remains constant. Let the distance be 'x'

Velocity of C w.r.t. A and $BV = 45 + 36 = 81\ km/h$

Distance $= 81 \times \frac{5}{60} = 6.75\ Km$

290 **(b)**

The distance travelled in t sec in upward motion is

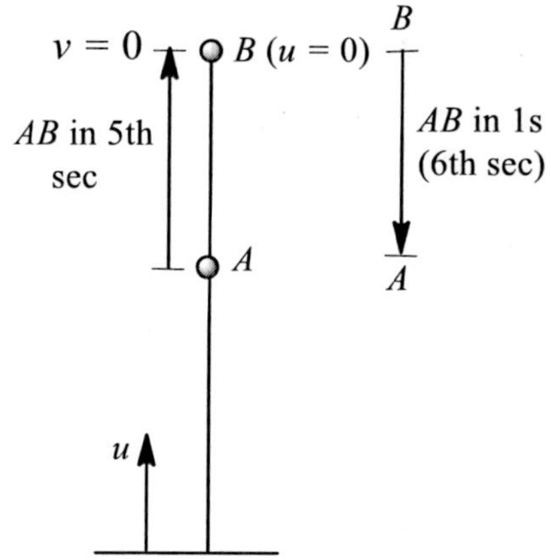

$$s = u - \frac{1}{2}\,g(2t - 1)$$

$$\therefore \quad AB = u - \frac{1}{2}g(2 \times 5 - 1)$$

$$AB = u - \frac{1}{2}\,9g$$

Distance travelled in 1 s in the downward direction is

$$BA = 0 + \frac{1}{2}\,g(1)^2$$

It is given that these distance are equal. Therefore,

$$u - \frac{9g}{2} = \frac{1}{2}g$$

$$\Rightarrow \quad u = 5 \times 9.8 = 49\ \text{ms}^{-1}$$

291 **(d)**

Let the body after time $t/2$ be at $x$from the top, then

$$x = \frac{1}{2}g\frac{t^2}{4} = \frac{gt^2}{8} \quad \text{... (i)}$$

$$h = \frac{1}{2}gt^2 \quad \text{... (ii)}$$

Eliminate t from (i) and (ii), we get $x = \frac{h}{4}$

$\therefore$Height of the body from the ground $= h - \frac{h}{4} = \frac{3h}{4}$

292 **(a)**

Time taken by each ball to reach highest point,

$t = \frac{1}{n}$s.

As the juggler throws the second ball, when the first ball is at its highest point, so $v = 0$

Using $v = u + at$, we have $0 = u + (-g)(1/n)$

or $u = (g/n)$

Also $v^2 = u^2 + 2as$

$\therefore 0 = (g/n)^2 + 2(-g)h$ or $h = \frac{g}{2n^2}$.

293 **(d)**

$\vec{r} = 3t\hat{\imath} - t^2\hat{\jmath} + 4\hat{k}$

Velocity, $\vec{v} = \frac{d\vec{r}}{dt} = \frac{d}{dt}(3t\hat{\imath} - t^2\hat{\jmath} + 4\hat{k}) = 3\hat{\imath} - 2t\hat{\jmath}$

At $t = 5s \Rightarrow \vec{v} = 3\hat{\imath} - 10\hat{\jmath}$

$|\vec{v}| = \sqrt{(3)^2 + (-10)^2} = \sqrt{9 + 100}$

$= \sqrt{109} = 10.44\ ms^{-1}$

294 **(b)**

Velocity at the time of striking the floor,

$u = \sqrt{2gh_1} = \sqrt{2 \times 9.8 \times 10} = 14m/s$

Velocity with which it rebounds

$v = \sqrt{2gh_2} = \sqrt{2 \times 9.8 \times 2.5} = 7\ m/s$

$\therefore$ Change in velocity $\Delta v = 7 - (-14)$

$= 21m/s$

$\therefore$ Acceleration $= \frac{\Delta v}{\Delta t} = \frac{21}{0.01}$

$= 2100m/s^2$(upwards)

295 **(d)**

$$u = at, x = \int u\,dt = \int at\,dt = \frac{at^2}{2}$$

For $t = 4$ sec, $x = 8a$

296 **(c)**

In order that the man catches the bus, let his minimum velocity be v, then from equation of motion.

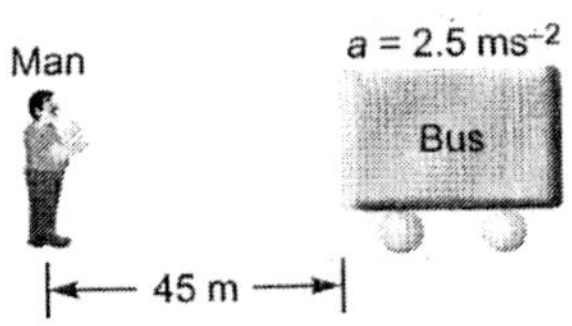

$v^2 = u^2 + 2as$

We have, $u = 0, a = 2.5\ \text{ms}^{-2}, s = 45$ m

$\therefore \quad v^2 = 2 \times 2.5 \times 45$

$\Rightarrow \quad v = \sqrt{225} = 15\ \text{ms}^{-1}$

297 **(b)**

At maximum height velocity $v = 0$

We know that $v = u + at$, hence

$0 = u - gT \Rightarrow u = gT$

When $v = \frac{u}{2}$, then

$$\frac{u}{2} = u - gt \Rightarrow gt = \frac{u}{2} \Rightarrow gt = \frac{gT}{2} \Rightarrow t = \frac{T}{2}$$

Hence at $t = \frac{T}{2}$, it acquires velocity $\frac{u}{2}$

298 **(c)**

Let both balls meet at point P after time t

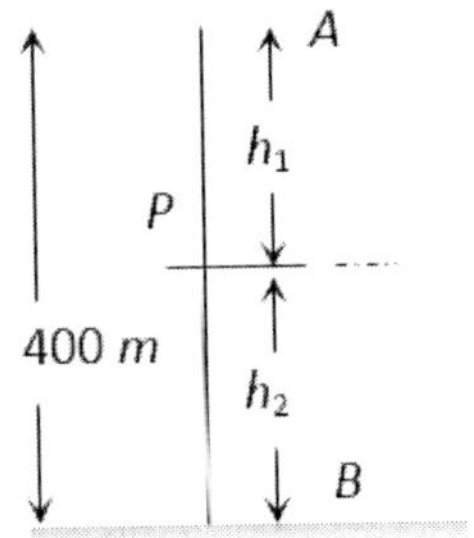

The distance travelled by ball A, $h_1 = \frac{1}{2}gt^2$

The distance travelled by ball B, $h_2 = ut - \frac{1}{2}gt^2$

$h_1 + h_2 = 400\ m \Rightarrow ut = 400, t = 400/50 = 8\ sec$

$\therefore h_1 = 320m$ and $h_2 = 80\ m$

299 **(b)**

The distance traveled in last second

$S_{\text{Last}} = u + \frac{g}{2}(2t - 1) = \frac{1}{2} \times 9.8(2t - 1)$

$= 4.9(2t - 1)$

and distance traveled in first three second,

$S_{\text{Three}} = 0 + \frac{1}{2} \times 9.8 \times 9 = 44.1\ m$

According to problem $S_{\text{Last}} = S_{\text{Three}}$

$\Rightarrow 4.9(2t - 1) = 44.1 \Rightarrow 2t - 1 = 9$

$\Rightarrow t = 5\ sec$

300 **(d)**

For a stone which is thrown downwards from a balloon rising upwards, the equation of motion is

$$h = -ut + \frac{1}{2}gt^2$$

$$= -29 \times 10 + \frac{1}{2} \times 9.8 \times (10)^2$$

$$= -290 + 490 = 200 \text{ m}$$

302 **(a)**

Horizontal velocity of dropped packet = u

Vertical velocity = $\sqrt{2gh}$

∴ Resultant velocity at earth

$$= \sqrt{u^2 + 2gh}$$

304 **(b)**

The displacement equation is given by

$$x = a_0 + \frac{a_1 t}{2} - \frac{a_2 t^2}{3}$$

Velocity = rate of change of displacement

ie, $v = \frac{dx}{dt}$

$$= \frac{d}{dt}\left(a_0 + \frac{a_1 t}{2} - \frac{a_2 t^2}{3}\right)$$

$$= 0 + \frac{a_1}{2} - \frac{2a_2 t}{3}$$

$$= \frac{a_1}{2} - \frac{2a_2 t}{3}$$

Acceleration = rate of change of velocity

ie, $a = \frac{dv}{dt}$

$$= \frac{d}{dt}\left(\frac{a_1}{2} - \frac{2a_2}{3}t\right)$$

$$= 0 - \frac{2a_2}{3}$$

$$= -\frac{2a_2}{3}$$

305 **(b)**

Relative velocity of one train $w.r.t.$ other $= 10 + 10 = 20m/s$

Relative acceleration$=0.3 + 0.2 = 0.5m/s^2$

If train crosses each other then from $s = ut + \frac{1}{2}at^2$

$As, s = s_1 + s_2 = 100 + 125 = 225$

$\Rightarrow 225 = 20t + \frac{1}{2} \times 0.5 \times 0.5 \times t^2$

$\Rightarrow 0.5t^2 + 40t - 450 = 0$

$$\Rightarrow t = \frac{-40 \pm \sqrt{1600 + 4.(005) \times 450}}{1} = -40 \pm 50$$

$\therefore t = 10sec$(Taking +ve value)

306 **(c)**

A body is moving on a straight line with constant velocity. Between A and B, the straight line is the shortest distance. This is the distance travelled. The particle starts at A and reaches B along the straight line. Therefore displacement is also $AB, D = S$

A ——————— *B*

307 **(d)**

Stopping distance $s \propto u^2$

$$\Rightarrow \quad \frac{s_2}{40} = \left(\frac{90 \times \frac{5}{18}}{50 \times \frac{5}{18}}\right)^2$$

$$\Rightarrow \quad s_2 = 129.6 \text{ m}$$

308 **(c)**

Vertical component of velocities of both the balls are same and equal to zero. So $t = \sqrt{\frac{2h}{g}}$

309 **(a)**

Distance = Area covered between velocity and time axis

$= \frac{1}{2}(30 + 10)10 = 200\, meter$

310 **(b)**

$u = 0, v = 180\ km\ h^{-1} = 50\ ms^{-1}$

Time taken $t = 10s$

$a = \frac{v-u}{t} = \frac{50}{10} = 5\ ms^{-2}$

$\therefore$ Distance covered $S = ut + \frac{1}{2}at^2$

$= 0 + \frac{1}{2} \times 5 \times (10)^2 = \frac{500}{2} = 250\ m$

311 **(b)**

(i) $\frac{\frac{1}{2}\times OA \times AC}{OA} + \frac{\frac{1}{2}\times AB \times AC}{AB} = 1$

(ii) $\tan 60° = \frac{CA}{OA}$ and $\tan 30° = \frac{CA}{AB}$

$OA \tan 60° = AB \tan 30°$

or $\frac{OA}{AB} = \frac{\tan 30°}{\tan 60°} = \frac{1}{\sqrt{3}} \times \frac{1}{\sqrt{3}} = \frac{1}{3}$

312 **(b)**

$\therefore\ v = |t-2|\text{ms}^{-1}$

$v = t - 2$, when $t > 2s$

$v = 2 - t$, when $t < 2s$

$\therefore\ a = \frac{dv}{dt} = 1\text{ms}^{-1}$ when $t > 2s$

$a = -1\text{ms}^{-1}$ when $t < 2s$

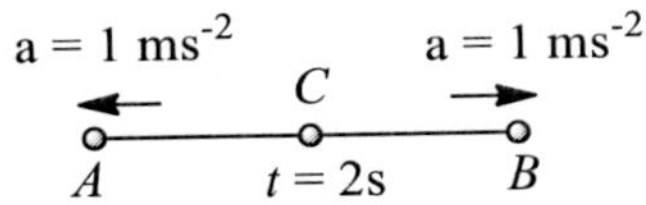

In the direction of motion from A to C, bee decelerates but for C to B, bee accelerates.

Let $AC = s_1, BC = s_2$

$u_A = 2\text{ms}^{-1}, t = 0$

$u_c = 0$ at $t = 4$s

$\therefore\ s_1 = \left(\frac{u_A + u_C}{2}\right)t_1$

$s_2 = \left(\frac{u_C + u_B}{2}\right)t_2$

$\therefore\ s = s_1 + s_2 = \left(\frac{2+0}{2}\right)2 + \left(\frac{0+2}{2}\right)2 = 4\text{m}$

313 **(b)**

Here $v = 144\ km/h = 40 m/s$

$v = u + at \Rightarrow 40 = 0 + 20 \times a \Rightarrow a = 2\ m/s^2$

$\therefore s = \frac{1}{2}at^2 = \frac{1}{2} \times 2 \times (20)^2 = 400\ m$

314 **(c)**

For first ball $h = \frac{1}{2}gt^2$ ...(i)

For second ball

$(h - 20) = \frac{1}{2}g(t-1)^2$ ...(ii)

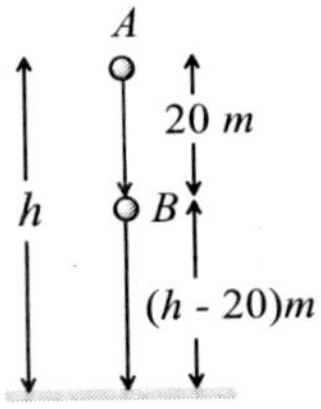

Subtract equation (ii) from (i)

$20 = -5 + 10t \Rightarrow \therefore\ t = 2.5$ sec

Hence, height $h = \frac{1}{2} \times 10 \times (2.5)^2 = 31.2\ m$

315 **(b)**

$v^2 = u^2 + 2as$

$a = \frac{v^2 - u^2}{2s} = \frac{(20)^2 - (10)^2}{2 \times 135} = \frac{300}{270}$

$= \frac{10}{9} m/s^2$

From first equation of motion

$v = u + at \Rightarrow t = \frac{v-u}{a} = \frac{20-10}{10/9}$

$= \frac{10}{10/9} = 9\ sec$

317 **(d)**

Let the man will be able to catch the bus after $t\ s$

Then

$10t = 48 + \frac{1}{2} \times 1 \times t^2$

$t^2 - 20t + 96 = 0$

$(t-12)(t-8) = 0$

$t = 8s$ and $t = 12s$

Thus the man will be able to catch the bus after $8s$

318 **(c)**

From equation of motion, we have

$$s = ut + \frac{1}{2}at^2$$

When $s = 24$ m, $t = 4$ s,

$\therefore \qquad 24 = 4u + \frac{1}{2}a(4)^2$

Or $\qquad 24 = 4u + 8a$

Or $\qquad 6 = u + 2a \qquad \ldots(i)$

When body travels a total distance of $24 + 64 = 88$ m in 8 s, we get

$$88 = 8u + \frac{1}{2}a\,(8)^2$$

Or $\quad 88 = 8u + 32a$

Or $\quad 11 = u + 4a \quad \ldots.(ii)$

Solving Eqs. (i) and (ii), we get

$$u = 1\ \text{ms}^{-1}$$

319 **(a)**

If t_0 is the reaction time, then the distance covered during decelerated motion is $10 - 10t_0$.

Now, in the first case,

$10^2 = 2a(10 - 10t_0)$ ...(i)

Similarly, in the second case,

$20^2 = 2a(30 - 20t_0)$ ...(ii)

Again, in the third case,

$15^2 = 2a(x - 5t_0)$ ...(iii)

Dividing Eq.(ii) by Eq. (i), $\frac{20^2}{10^2} = \frac{30-20t_0}{10-10t_0}$

or $40 - 40t_0 = 30 - 20t_0$

or $20t_0 = 10$ or $t_0 = \frac{1}{2}s$

Dividing Eq. (iii) by Eq. (i), we get

$\frac{225}{100} = \frac{x-15t_0}{10-10t_0}$ or $\frac{9}{4} = \frac{x-15\times\frac{1}{2}}{10-10\times\frac{1}{2}}$

$45 = 4x - 30$ or $4x = 75$

or $x = \frac{75}{4}\text{m} = 18.75\text{m}$

320 **(d)**

$$v = \alpha\sqrt{x} \Rightarrow \frac{dx}{dt} = \propto \sqrt{x} \Rightarrow \frac{dx}{\sqrt{x}} = \alpha dt$$

By integrating both sides $\int x^{-1/2}\,dx = \int \alpha\, dt$

$$\Rightarrow \frac{\sqrt{x}}{1/2} = \alpha t \Rightarrow \sqrt{x} = \frac{1}{2}\alpha t \Rightarrow x = \frac{1}{4}\alpha^2 t^2$$

$$\therefore x \propto t^2$$

321 **(b)**

Given acceleration $a = 6t + 5$

$$\therefore a = \frac{dv}{dt} = 6t + 5, dv = (6t + 5)dt$$

Integrating it, we have $\int_0^v dv = \int_0^t (6t + 5)dt$

$v = 3t^2 + 5t + C$, where C is a constant of integration

When $t = 0, v = 0$ so $C = 0$

$\therefore v = \frac{ds}{dt} = 3t^2 + 5t$ or, $ds = (3t^2 + 5t)dt$

Integrating it within the conditions of motion, *i.e.*, as t changes from 0 to $2s$, s changes from 0 to s, we have

$$\int_0^s ds = \int_0^2 (3t^2 + 5t)dt$$

$$\therefore s = \left[t^3 + \frac{5}{2}t^2\right]_0^2 = 8 + 10 = 18\ m$$

322 **(d)**

Given $a = 19.6 m/s^2 = 2g$

Resultant velocity of the rocket after 5 *sec*

$v = 2g \times 5 = 10g\ m/s$

Height achieved after 5 *sec*, $h_1 = \frac{1}{2} \times 2g \times 25 = 245m$

On switching off the engine it goes up to height h_2 where its velocity becomes zero

$0 = (10g)^2 - 2gh_2 \Rightarrow h_2 = 490m$

$\therefore$ Total height of rocket $= 245 + 490$

$= 735\ m$

323 **(b)**

Between time interval 20s the 4s, there is non-zero acceleration and retardation. Hence, distance travelled during this interval = Area between time interval 20 s to 40 s

$$= \frac{1}{2} \times 20 \times 3 + 20 \times 1$$

$$= 30 + 20 = 50\ \text{m}$$

324 **(c)**

$$h = ut + \frac{1}{2}gt^2 \Rightarrow 81$$

$$= -12t + \frac{1}{2} \times 10 \times t^2$$

$$\Rightarrow t = 5.4\ sec$$

325 **(c)**

For no collision, the speed of car A should be reduced to v_B before the cars meet, *ie*, final relative velocity of car A with respect to car B is zero *ie*, $v_r = 0$

Here initial relative velocity, $u_r = v_A - v_B$

Relative acceleration, $a_r = -a - 0 = -a$

Let relative displacement $= s_r$

The equation

$v_r^2 = u_r^2 + 2a_r s_r$

$(0)^2 = (v_A - v_B) - 2as,$

$$s_r = \frac{(v_A - v_B)^2}{2a}$$

For no collision, $s_r \leq s$

$ie, \dfrac{(v_A - v_B)}{2a} \leq s$

326 **(d)**

From equation of motion

$$s = ut + \frac{1}{2}at^2$$

$$x = 0 + \frac{1}{2} \times 10t^2$$
$$= 5t^2 \quad \text{...(i)}$$

$$x + 3 = 0 + \frac{1}{2} \times 10(t + 0.5)^2$$

$$x + 3 = 5\left(t^2 + \frac{1}{4} + t\right) \quad \text{...(ii)}$$

Subtract Eq. (i) from Eq. (ii)

$$3 = 5\left(\frac{1}{4} + t\right) = \frac{5}{4} + 5t$$

$$3 - \frac{5}{4} = 5t$$

$$\frac{7}{4} = 5t \Rightarrow t = \frac{7}{20} \text{ s}$$

Now, $\quad v = u + at$

$$v = 0 + 10 \times \frac{7}{20} = 3.5 \text{ ms}^{-1}$$

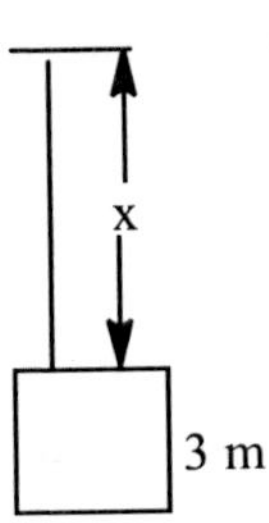

327 **(c)**

$v = (180 - 16x)^{1/2}$

As $a = \dfrac{dv}{dt} = \dfrac{dv}{dx} \cdot \dfrac{dx}{dt}$

$\therefore a = \dfrac{1}{2}(180 - 16x)^{-1/2} \times (-16)\left(\dfrac{dx}{dt}\right)$

$= -8(180 - 16x)^{-1/2} \times v$

$= -8(180 - 16x)^{-1/2} \times (180 - 16x)^{1/2}$

$= -8 \; m/s^2$

328 **(d)**

Instantaneous velocity is given by the slope of the curve at that instant $v = \dfrac{ds}{dt} = \tan\theta$ from the figure it is clear that slope of the curve is maximum at point 'C'

329 **(b)**

Let A and B will meet after time t sec. it means the distance travelled by both will be equal

$S_A = ut = 40t$ and $S_B = \frac{1}{2}at^2 = \frac{1}{2} \times 4 \times t^2$

$S_A = S_B \Rightarrow 40t = \dfrac{1}{2}4t^2 \Rightarrow t = 20 \; sec$

330 **(a)**

Velocity required by body in 10 sec

$v = 0 + 2 \times 10 = 20 \; m/s$

And distance travelled by it in 10 sec

$S_1 = \dfrac{1}{2} \times 2 \times (10)^2 = 100 \; m$

Then it moves with constant velocity ($20 \; m/s$) for 30 sec

$S_2 = 20 \times 30 = 600 \; m$

After that due to retardation ($4 \; m/s^2$) it stops

$S_3 = \dfrac{v^2}{2a} = \dfrac{(20)^2}{2 \times 4} = 50m$

Total distance travelled $S_1 + S_2 + S_3 = 750m$

331 **(d)**

$x = 8 + 12t + t^3$

$v = 0 + 12 - 3t^2 = 0$

$3t^2 = 12$

$t = 2 \; sec$

$a = \dfrac{dv}{dt} = 0 - 6t$

$a[t = 2] = -12 \; m/s^2$

Retardation $= 12 \; m/s^2$

332 **(d)**

Average velocity $= \dfrac{Total\ Displacement}{Time\ taken} = \dfrac{25}{75/15} = 5m/s$

333 **(b)**

Let the stone remains in air for t s. From $S = ut + \frac{1}{2}gt^2$

Here, $u = 0$, $S = \frac{1}{2}gt^2$

Total distance travelled by the stone in last second is

$$D = S_t - S_{t-1} = \frac{1}{2}gt^2 - \frac{1}{2}g(t-1)^2$$

Distance travelled by the stone in first three seconds is

$$S_3 = \frac{1}{2} \times g \times 3^2 = \frac{9}{2}g$$

According to given problem, $D = S_3$

$\therefore \frac{g}{2}(2t-1) = \frac{9}{2}g$ or $2t - 1 = 9 \Rightarrow t = 5s$

334 **(d)**

$u = 0, S = 250m, t = 10\text{sec}$

$$S = ut + \frac{1}{2}at^2 \Rightarrow 250 = \frac{1}{2}a[10]^2 \Rightarrow a = 5m/s^2$$

So, $F = ma = 0.9 \times 5 = 4.5N$

335 **(a)**

Time taken by the car to cover first half of the distance is

$$t_1 = \frac{100}{60}$$

Time taken by the car to cover speed half of the distance is

$$t_2 = \frac{100}{v}$$

Average speed , $v_{av} = \frac{\text{Total distance travelled}}{\text{TOtal time taken}}$

$$v_{av} = \frac{100 + 100}{t_1 + t_2} \Rightarrow 40 = \frac{200}{\frac{100}{60} + \frac{100}{v}}$$

$$\frac{1}{60} + \frac{1}{v} = \frac{1}{20} \Rightarrow \frac{1}{v} = \frac{1}{20} - \frac{1}{60}$$

$$\frac{1}{v} = \frac{2}{60} = \frac{1}{30}$$

$v = 30km\ h^{-1}$

336 **(a)**

$$\because a = \frac{dv}{dt} = 2(t-1) \Rightarrow dv = 2(t-1)dt$$

$$\Rightarrow v = \int_0^5 2(t-1)dt = 2\left[\frac{t^2}{2} - t\right]_0^5$$

$$= 2\left[\frac{25}{2} - 5\right]$$

$$= 15\ m/s$$

337 **(d)**

$$\text{Average speed} = -\frac{\text{Total distance}}{\text{Total time}} = \frac{x}{t_1+t_2}$$

$$= \frac{x}{\frac{x/3}{v_1} + \frac{2x/3}{v_2}} = \frac{1}{\frac{1}{3\times20} + \frac{2}{3\times60}} = 36\ km/hr$$

338 **(c)**

Distance covered by bus in 100 s

$$= 100 \times 10 = 1000 \text{ m}$$

Distance to be covered by scooterist

$$= 1000 + 1000$$
$$= 2000 \text{ m}$$

$$\therefore \text{Speed of scooterist} = \frac{2000}{100} = 20 \text{ ms}^{-1}$$

339 **(a)**

According to kinematics equation,

$v = v_0 - g\,t$

Upward direction is taken a positive and downward direction is taken as negative. Hence, $v - t$graphs is straight line having negative slope.

340 **(d)**

$x \propto t^3 \therefore x = Kt^3$

$$\Rightarrow v = \frac{dx}{dt} = 3Kt^2 \text{and } a = \frac{dv}{dt} = 6Kt$$

i. e. $a \propto t$

341 **(d)**

I is not possible because total distance covered by a particle increases with time

II is not possible because at a particular time, position cannot have two values

III is not possible because at a particular time, velocity cannot have two values

IV is not possible because speed can never be negative

342 **(d)**

$$\text{Average speed} = -\frac{\text{Total distance}}{\text{Total time}} = \frac{x}{t_1+t_2}$$

$$= \frac{x}{\frac{x/3}{v_1} + \frac{2x/3}{v_2}} = \frac{1}{\frac{1}{3\times20} + \frac{2}{3\times60}} = 36\ km/hr$$

343 **(c)**

$S_n \propto (2n-1)$. In equal time interval of $2\ seconds$

Ratio of distance = 1 : 3 : 5

344 **(d)**

As $v^2 = u^2 + 2as \Rightarrow (2y)^2 = u^2 + 2as \Rightarrow = 3u^2$

Now, after covering an additional distance s, if velocity becomes v,then,

$$v^2 = u^2 + 2a(2s) = u^2 + 4as = u^2 + 6u^2 = 7u^2$$

$\therefore v = \sqrt{7}u$

345 **(b)**

The time of fall is independent of the mass

346 **(d)**

$u = 200\ m/s, v = 100\ m/s, s = 0.1\ m$

$$a = \frac{u^2 - v^2}{2s} = \frac{(200)^2 - (100)^2}{2 \times 0.1}$$
$$= 15 \times 10^4\ m/s^2$$

347 **(d)**

In the positive region the velocity decreases linearly (during rise) and in the negative region velocity increases linearly (during fall) and the direction is opposite to each other during rise and fall, hence fall is shown in the negative region

348 **(d)**

Let the car accelerate at rate α for time t_1 then maximum velocity attained,

$v = 0 + at_1 = at_1$

Now, the car decelerates at a rate β for time $(t - t_1)$ and finally comes to rest. Then,

$0 = v - \beta(t - t_1) \Rightarrow 0 = \alpha t_1 - \beta t + \beta t_1$

$$\Rightarrow t_1 = \frac{\beta}{\alpha + \beta} t$$

$$\therefore v = \frac{\alpha\beta}{\alpha + \beta} t$$

349 **(a)**

$$\because a = \frac{dv}{dt} = 2(t - 1) \Rightarrow dv = 2(t - 1)dt$$

$$\Rightarrow v = \int_0^5 2(t - 1)dt = 2\left[\frac{t^2}{2} - t\right]_0^5$$
$$= 2\left[\frac{25}{2} - 5\right]$$
$$= 15\ m/s$$

350 **(a)**

The velocity of the particle is

$$\frac{dx}{dt} = \frac{d}{dt}(2 - 5t + 6t^2) = (0 - 5 + 12t)$$

For initial velocity $t = 0$, hence $v = -5\ m/s$

351 **(a)**

If u is the initial velocity then distance covered by it in 2 *sec*

$$S = ut + \frac{1}{2}at^2 = u \times 2 + \frac{1}{2} \times 10 \times 4$$
$$= 2u + 20 \quad \text{... (i)}$$

Now distance covered by it in 3rd *sec*

$$S_{3^{rd}} = u + \frac{1}{2}(2 \times 3 - 1)10$$
$$= u + 25 \quad \text{... (ii)}$$

From (i) and (ii), $2u + 20 = u + 25 \Rightarrow u = 5$

$\therefore S = 2 \times 5 + 20 = 30\ m$

(a)

352 Let the speed of trains be x

$$\therefore \frac{x - u}{x + u} = \frac{1}{2} \Rightarrow 2x - 2u = x + u \Rightarrow x = 3u$$

353 **(a)**

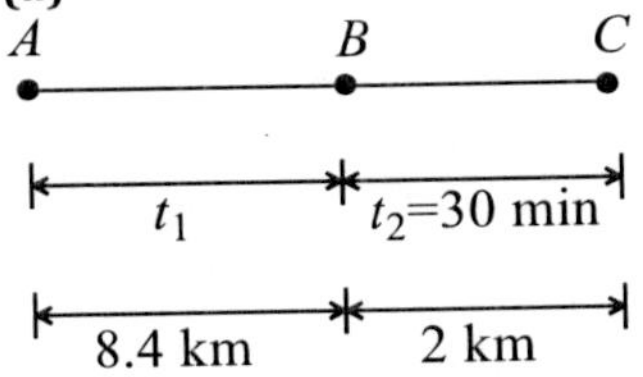

$$\text{Average speed } \bar{v} = \frac{\text{Total distance travelled}}{\text{Total time taken}}$$

$$= \frac{8.4km + 2km}{t_1 + t_2} = \frac{10.4\ km}{\left(\frac{8.4\ km}{70\ km/h}\right) + \frac{1}{2}h}$$

$$= \frac{10.4\ km}{0.12h + 0.5h} = 16.8\ km/h$$

354 **(b)**

Time of ascent = Time of descent=5*sec*

355 **(a)**

$S_n = u + \frac{a}{2}(2n - 1) = \frac{a}{2}(2n - 1)$ because $u = 0$

Hence $\frac{S_4}{S_3} = \frac{7}{5}$

356 **(c)**

From given figure, it is clear that the net displacement is zero. So average velocity will be zero

357 **(b)**

Average velocity

$$= \frac{\text{Total distance covered}}{\text{Time of flight}} = \frac{2H_{max}}{2u/g}$$

$$\Rightarrow v_{av} = \frac{2u^2/2g}{2u/g} \Rightarrow v_{av} = u/2$$

Velocity of projection = v [Given]

$\therefore v_{av} = u/2$

358 **(a)**

$H_{\max} \propto u^2 \therefore u \propto \sqrt{H_{\max}}$

i.e. to triple the maximum height, ball should be thrown with velocity $\sqrt{3}\,u$

359 **(a)**

$$S = ut + \frac{1}{2}gt^2$$

$30 = -25t + \frac{10}{2}t^2$ or $t^2 - 5t - 6 = 0$

Or $(t - 6)(t + 1) = 0$ Take positive root

$\therefore t = 6$ sec

360 **(d)**

$S \propto u^2$. If u becomes 3 times then S will become 9 times

$i.e. 9 \times 20 = 180m$

361 **(a)**

If t_1 and$2t_2$ are the time taken by particle to cover first and second half distance respectively

$t_1 = \frac{x/2}{3} = \frac{x}{6}$...(i)

$x_1 = 4.5\ t_2$ and $x_2 = 7.5\ t_2$

So, $x_1 + x_2 = \frac{x}{2} \Rightarrow 4.5\ t_2 + 7.5\ t_2 = \frac{x}{2}$

$t_2 = \frac{x}{24}$ ...(ii)

Total time $t = t_1 + 2t_2 = \frac{x}{6} + \frac{x}{12} = \frac{x}{4}$

So, average speed $= 4\ m/sec$

362 **(d)**

$$h = \frac{1}{2}gt^2 \Rightarrow t = \sqrt{\frac{2 \times 1960}{9.8}} = 20\text{ s}$$

$$\therefore \quad s = AB = ut$$

$$= 600 \times \frac{20}{60 \times 60}$$

$$= 3.33\text{ km}$$

363 **(d)**

The braking retardation will remain same and assumed to be constant, let it be a.

From the 3rd equation of motion, $v^2 = u^2 + 2as$

Ist case : $0 = \left(60 \times \frac{5}{18}\right)^2 - 2a \times s_1$

Or $s_1 = \frac{(60 \times 5/18)^2}{2a}$

IInd case : $0 = \left(120 \times \frac{5}{18}\right)^2 - 2a \times s_2$

Or $s_2 = \frac{(120 \times 5/18)^2}{2a}$

$\therefore \frac{s_1}{s_2} = \frac{1}{4} \Rightarrow s_2 = 4s_1 = 4 \times 20 = 80\text{ m}$

365 **(d)**

Net displacement $= 0$ and total distance $= OP + PQ + QO$

$= 1 + \frac{2\pi \times 1}{4} + 1 = \frac{14.28}{4}\ km$

Average speed $= \frac{14.28}{4 \times 10/60}$

$= \frac{6 \times 14.28}{4} = 21.42\ km/h$

366 **(c)**

For first projectile, $h_1 = ut - \frac{1}{2}gt^2$

For second projectile, $h_2 = u(t - T) - \frac{1}{2}g(t - T)^2$

When both meet i.e. $h_1 = h_2$

$$ut - \frac{1}{2}gt^2 = u(t - T) - \frac{1}{2}g(t - T)^2$$

$$\Rightarrow uT + \frac{1}{2}gT^2 = gtT \Rightarrow t = \frac{u}{g} + \frac{T}{2}$$

and $h_1 = u\left(\frac{u}{g} + \frac{T}{2}\right) - \frac{1}{2}g\left(\frac{u}{g} + \frac{T}{2}\right)^2$

$$= \frac{u^2}{2g} - \frac{gT^2}{8}$$

367 **(a)**

$$S_n = u + \frac{a}{2}[2n - 1]$$

$$S_{5^{th}} = 7 + \frac{4}{2}[2 \times 5 - 1] = 7 + 18 = 25m$$

368 **(a)**

(a) Take vertically upward direction as positive and vertically downward direction as negative.

369 **(a)**

As the train are moving in the same direction. So the initial relative speed $(v_1 - v_2)$ and by applying retardation final relative speed becomes zero

From $v = u - at \Rightarrow 0 = (v_1 - v_2) - at \Rightarrow t = \frac{v_1 - v_2}{a}$

370 **(b)**

If a in the relative acceleration, then

$3 = \frac{1}{2}a \times 5 \times 5$ or $a = \frac{6}{25}\text{ms}^{-2}$

Again, $S = \frac{1}{2} \times \frac{6}{25} \times 10 \times 10 = 12\text{m}$

371 **(d)**

Given, $x = 2 - 5t + 6t^2$ ... (i)

Differentiating with respect to t

$$\frac{dx}{dt} = -5 + 12t$$

At $t = 0$ s, $\frac{dx}{dt} = -5 \text{ ms}^{-1}$

Hence, initial velocity $v = -5 \text{ ms}^{-1}$

372 **(d)**

$u = \sqrt{2gh} = \sqrt{2 \times 10 \times 20} = 20 m/s$

and $T = \frac{2u}{g} = \frac{2\times20}{10} = 4\ sec$

373 **(d)**

$v_1 = \frac{u + v_1'}{2} = \frac{u + u + at_1}{2} = u + \frac{1}{2}at_1$

$v_2 = \frac{v_1' + v_2'}{2}$

$= \frac{(u + at_1) + u + a(t_1 + t_2)}{2}$

$= u + at_1 + \frac{1}{2}at_2$

$v_3 = \frac{v_2' + v_3'}{2}$

$= \frac{(u + at_1 + at_2) + u + a(t_1 + t_2 + t_3)}{2}$

$= u + at_1 + at_2 + \frac{1}{2}at_3$

Then, $v_1 - v_2 = -\frac{1}{2}a(t_1 + t_2)$

$v_2 - v_3 = -\frac{1}{2}a(t_2 + t_3)$

$\therefore \frac{v_1 - v_2}{v_2 - v_3} = \frac{t_1 + t_2}{t_2 + t_3}$

374 **(b)**

Let initial velocity of body a point A is v, AB is 40 cm.

From $v^2 = u^2 - 2as$

$\Rightarrow \left(\frac{v}{2}\right)^2 = v^2 - 2a \times 40$

Or $a = \frac{3v^2}{320}$

Let on penetrating 40 cm in a wooden block, the body moves x distance from B to C.

So, for B to C

$u = \frac{v}{2}, v = 0$

$s = x, a = \frac{3v^2}{320}$ (deceleration)

$\therefore (0)^2 = \left(\frac{v}{2}\right)^2 - 2 \times \frac{3v^2}{320} \times x$

Or $x = \frac{40}{3}$ cm

375 **(b)**

Relative velocity of bird $w.r.t.$ train = $25 + 5 = 30\ m/s$

Time taken by the bird to cross the train

$t = \frac{210}{30} = 7\ sec$

376 **(a)**

Distance covered in 5th second

$S_{5^{th}} = u + \frac{a}{2}(2n - 1) = 0 + \frac{a}{2}(2 \times 5 - 1)$

$= \frac{9a}{2}$

and distance covered in 5 second,

$S_5 = ut + \frac{1}{2}at^2 = 0 + \frac{1}{2} \times a \times 25 = \frac{25a}{2}$

$\therefore \frac{S_{5th}}{S_5} = \frac{9}{25}$

377 **(a)**

$v = u - gt$

At max height $v^2 = u^2 - 2gh$

$t = \frac{u}{g}$ $h = \frac{u^2}{2g}$

$\frac{t_1}{t_2} = \frac{2}{3}\frac{h_1}{h_2} = \left(\frac{2}{3}\right)^2 = \frac{4}{9}$

378 **(d)**

Let the body after time $t/2$ be at $x$from the top, then

$x = \frac{1}{2}g\frac{t^2}{4} = \frac{gt^2}{8}$ ... (i)

$h = \frac{1}{2}gt^2$ ... (ii)

Eliminate t from (i) and (ii), we get $x = \frac{h}{4}$

$\therefore$Height of the body from the ground = $h - \frac{h}{4} = \frac{3h}{4}$

379 **(c)**

Since acceleration due to gravity is independent of mass, hence time is also independent of mass (or density) of object

380 **(b)**

$a = \frac{dv}{dt} = 6t + 5$ or $dv = (6t + 5)dt$

Integrating it, we have

$v = \frac{6t^2}{2} + 5t + C = 3t^2 + 5t + C$

where C is constant of integration

When $v = 0$ so $c = 0$

$\therefore v = \frac{ds}{dt} = 3t^2 + 5t$ or $ds = (3t^2 + 5t)dt$

Integrating it within the conditions of motion *ie*, as t changes from 0 to 1 s changes from 0 to s, we have $s = t^3 + \frac{5t^2}{2} = 1 + \frac{5}{2} = 3.5\text{m}$.

382 **(b)**

Let time interval be chosen as 1 s

$\frac{PA}{PB} = \frac{OA}{OB} = \frac{v_x}{v_y}$

So, $P(x, y)$divides AB is the ratio $v_x : v_y$.

Using section formula,

$x = \frac{v_x \times 0 + v_y \times v_x}{v_x + v_y} = \frac{v_x v_y}{v_x + v_y}$

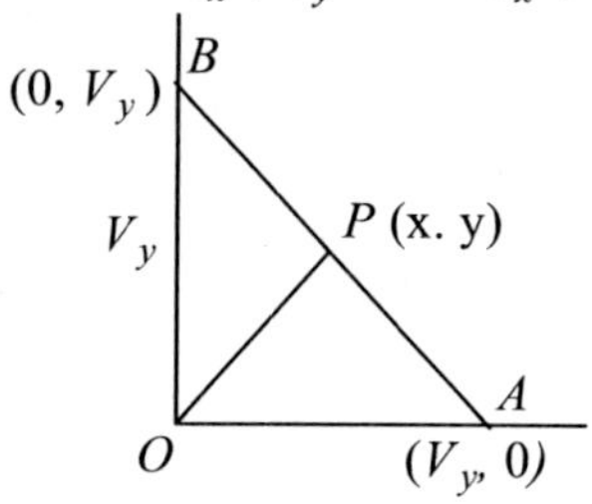

$y = \frac{v_x v_y + v_y \times 0}{v_x + v_y} = \frac{v_x v_y}{v_x + v_y}$

$v = \sqrt{x^2 + y^2} = \sqrt{2}\frac{v_x v_y}{v_x + v_y}$

Now, replace v_x by v_1 and $v_y$by v_2.

$v = \frac{\sqrt{2}v_1 v_2}{v_1 + v_2}$

383 **(a)**

Distance between the balls = Distance travelled by first ball in 3 seconds – Distance travelled by second ball in 2 seconds $= \frac{1}{2}g(3)^2 - \frac{1}{2}g(2)^2 = 45 - 20 = 25\ m$

384 **(a)**

Given line have positive intercept but negative slope. So its equation can be written as

$v = -mx + v_0$ ....(i) [where $m = \tan\theta = \frac{v_0}{x_0}$]

By differentiating with respect to time we get

$\frac{dv}{dt} = -m\frac{dx}{dt} = -mv$

Now substituting the value of v from eq. (i) we get

$\frac{dv}{dt} = -m[-mx + v_0] = m^2x - mv_0 \therefore a = m^2x - mv_0$

i.e. the graph between a and x should have positive slope but negative intercept on a-axis. So graph (a) is correct

385 **(c)**

$h_{n^{th}} = u - \frac{g}{2}(2n - 1)$

$h_{5^{th}} = u - \frac{10}{2}(2 \times 5 - 1) = u - 45$

$h_{6^{th}} = u - \frac{10}{2}(2 \times 6 - 1) = u - 55$

Given $h_{5^{th}} = 2 \times h_{6^{th}}$.By solving we get $u = 65\ m/s$

386 **(c)**

We know that,

$s_t = u + \frac{1}{2}a(2t - 1)$

$\therefore \quad 65 = u + \frac{1}{2}a[10 - 1]$

$\Rightarrow \quad 65 = u + \frac{9}{2}a \quad$ ... (i)

Again, $t = 9s$ and $s = 105$ m

$\therefore \quad 105 = u + \frac{1}{2}a[18 - 1]$

Or $\quad 105 = u + \frac{17}{2}a \quad$ ... (ii)

On substracting Eq. (i) from Eq. (ii)

$105 - 65 = \frac{17}{2}a - \frac{9}{2}a$

Or $\quad 40 = \frac{8}{2}a$

Or $\quad 40 = 4a$

Or $a = 10 \text{ ms}^{-2}$

On putting the value of a in Eq. (i),

Or $65 = u + \frac{9}{2} \times 10$

Or $65 = u + 45$

Or $u = 20 \text{ ms}^{-1}$

Now we know that from second equation of motion

$$s = ut + \frac{1}{2}at^2$$

$$\therefore \quad s = 20 \times 20 + \frac{1}{2} \times 10 \times (20)^2$$

Or $s = 400 + 5 \times 400$

Or $s = 2400 \text{ m}$

387 **(d)**

Instantaneous velocity is given by the slope of the curve at that instant $v = \frac{ds}{dt} = \tan\theta$ from the figure it is clear that slope of the curve is maximum at point 'C'

388 **(d)**

Given, $a = 1 \text{ m/s}, s = 48 \text{ m}$

By equation of motion

$$48 = 10t + \frac{1}{2}at^2$$

$$t = 8\,s$$

389 **(a)**

$\vec{r} = x\hat{i} + y\hat{j} + z\hat{k} \ \therefore r = \sqrt{x^2 + y^2 + z^2}$

$r = \sqrt{6^2 + 8^2 + 10^2} = 10\sqrt{2}m$

390 **(b)**

From the give graph. From 0s to 8s, particle is accelerated, then from 8 to 12 s, particle moves with constant acceleration. The form 12 to 16s, the particle is in the condition of deceleration. Hence, maximum velocity will be during 8s to 12s. During 0 to 4s, the acceleration will be function of time. The equation of straight lines is

$a = \frac{5}{4}t$

$\therefore \frac{dv}{dt} = \frac{5}{4}t$

$\therefore v = \int_0^t \frac{5}{4}t\,dt = \frac{5}{4}\frac{t^2}{2} = \frac{5}{8}t^2$

The velocity at $t = 4\text{s}$ is $u = 10\text{ms}^{-1}$

The distance travelled during 4 to 8s is

$s_2 = ut + \frac{1}{2}at^2$

$= 10 \times 4 + \frac{1}{2} \times 5 \times 4^2$

$= 40 + 40 = 80\text{m}$

The velocity at $t = 8\text{s}$ is

$v = 10 + 5 \times 4 = 30\text{ms}^{-1}$

This is the maximum velocity.

Tricky approach: The area of $a - t$ graph gives change in velocity. The area of the graph from 0 to 8s

$= v - u = \frac{1}{2} \times 4 \times 5 + 4 \times 5 = 30$

But $u = 0$

$v = 30\text{ms}^{-1}$

391 **(a)**

$x = \frac{1}{2}\text{g}t^2, 100 - x = 25x - \frac{1}{2}\text{g}t^2;$

Adding $25t = 100$ or $t = 4\text{ s}$

392 **(b)**

Let us solve the problem in terms initial velocity, relative acceleration and relative displacement of the coin with respect to floor of the lift.

$u = 10 - 10 = \text{ms}^{-1}, a = 9.8\text{ms}^{-2}, S = 4.9\text{m}, t = ?$

$4.9 = 0 \times t + \frac{1}{2} \times 9.8 \times t^2$ or $4.9t^2 = 4.9$

or $t = 1\text{s}$

$15 = 30 - 10t$ or $10t = 15$ or $t = 1.5\text{s}$

393 **(b)**

In figure (b), the particle is slowly accelerated first and reaches a constant velocity for the straight line portion. Then velocity decreases are finally it stops when it reaches the top straight line portion.

Hence, figure (b) represents one dimensional motion of a particle.

394 **(c)**

$v = \sqrt{2gh} = \sqrt{2 \times 10 \times 20} = 20 m/s$

395 **(b)**

$t = \sqrt{\frac{2h}{g}} \Rightarrow \frac{t_1}{t_2} = \sqrt{\frac{h_1}{h_2}}$

396 **(c)**

Given, $t = ax^2 + bx$

Differentiating w.r.t. t

$\frac{dx}{dt} = 2\,ax\frac{dx}{dt} + b\frac{dx}{dt}$

$v = \frac{dx}{dt} = \frac{1}{(2ax\ + b)}$

Again differentiating w.r.t. t

$\frac{d^2x}{dt^2} = \frac{-2a}{(2\,ax + b)^2}\frac{dx}{dt}$

$\therefore f = \frac{d^2x}{dt^2} = \frac{-1}{(2\,ax + b)^2} \cdot \frac{2a}{(2ax + b)}$

Or $f = \frac{-2a}{(2ax + b)^3}$

$\therefore f = -2av^3$

397 **(b)**

Velocity of graph = Area of a-t graph

$= (4 \times 1.5) - (2 \times 1) = 4 m/s$

398 **(b)**

$h = \frac{1}{2}gt^2$

$h' = \frac{1}{2}g(t - t_0)^2$

$h - h' = \frac{1}{2}g[t^2 - (t - t_0)^2]$

$= \frac{1}{2}g[t^2 - t^2 - t_0^2 + 2tt_0]$

$\Delta h = \frac{1}{2}gt_0(2t - t_0)$

Δh is increasing with time

399 **(c)**

As $v^2 = v^2 - 2as \Rightarrow u^2 = 2as$ $(\because v = 0)$

$\Rightarrow u^2 \propto s \Rightarrow \frac{u_2}{u_1} = \left(\frac{s_2}{s_1}\right)^{1/2}$

$\Rightarrow u_2 = \left(\frac{9}{4}\right)^{\frac{1}{2}} u_1 = \frac{3}{2}u_1 = 300 m/s$

400 **(a)**

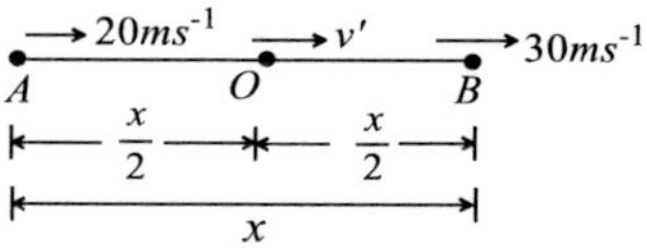

$\because v'' = \sqrt{\frac{(v_1)^2 + (v_2)^2}{2}} = \sqrt{\frac{900 + 400}{2}}$

$= \sqrt{650} = 25.5\ ms^{-1}$

401 **(c)**

Horizontal distance covered by the wheel in half revolution $= \pi R$

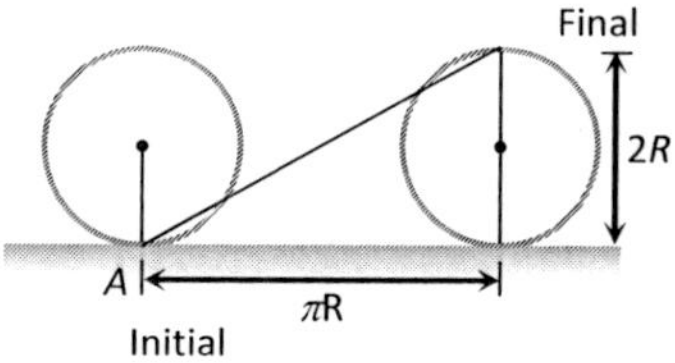

So the displacement of the point which was initially in contact with ground

$= AA' = \sqrt{(\pi R)^2 + (2R)^2}$

$= R\sqrt{\pi^2 + 4} = \sqrt{\pi^2 + 4}$ [As $R = 1m$]

402 **(d)**

If t_1 and t_2 are the time, when body is at the same height then,

$h = \frac{1}{2}gt_1t_2 = \frac{1}{2} \times g \times 2 \times 10 = 10\,g$

403 **(a)**

When two particles moves towards each other then

$v_1 - v_2 = 6$...(i)

When these particles moves in the same direction then

$v_1 - v_2 = 4$...(ii)

By solving $v_1 = 5$ and $v_2 = 1\ m/s$

404 **(b)**

$H_{\max} = \frac{u^2}{2g} = \frac{19.6 \times 19.6}{2 \times 9.8} = 19.6\ m$

405 **(a)**

$v = 3t^2 = 12t + 3; a = 6t - 12$

When $a = 0, 6t - 12 = 0$

or $6t = 12$ or $t = 2$ s When $t = 2$ s,

$v = 3 \times 2 \times 2 - 12 \times 2 + 3 = -9 \text{ms}^{-1}$

406 **(b)**

The velocity of balloon at height h, $v = \sqrt{2\left(\frac{g}{8}\right)h}$

When the stone released from this balloon, it will go upward with velocity, $= \frac{\sqrt{gh}}{2}$ (Same as that of balloon). In this condition time taken by stone to reach the ground

$$t = \frac{v}{g}\left[1 + \sqrt{1 + \frac{2gh}{v^2}}\right]$$

$$= \frac{\sqrt{gh}/2}{g}\left[1 + \frac{2gh}{gh/4}\right]$$

$$= \frac{2\sqrt{gh}}{g} = 2\sqrt{\frac{h}{g}}$$

407 **(b)**

$$h_1 = \frac{1}{2}gt_1^2 = \frac{10}{2} \times (5)^2 = 125 \text{ m}$$

$$h_2 = \frac{1}{2}gt_2^2 = \frac{10}{2} \times (3)^2 = 45 \text{ m}$$

$\therefore \quad h_1 - h_2 = 125 - 45 = 80 \text{ m}$

408 **(b)**

From equation of motion, we have

$v = u + gt$

Taking downward direction negative

$u = 10 + 5 = 15 \text{ms}^{-1}, g = 10 \text{ ms}^{-2}, t - 2 \text{ s}$

$\therefore v = 15 - 2 \times 10 = -2 \text{ ms}^{-1}$

409 **(c)**

Let body reaches the ground in t sec.

$\therefore$ Velocity of body after $(t - 2)$ sec from equation of motion.

$v = u + gt'$

And $t' = t - 2$

$\therefore v = g(t - 2)$

Distance covered in last two sec

$$h' = g(t - 2) \times 2 + \frac{1}{2}g(2)^2$$

$60 = 20(t - 2) + 20$

Or $t = 4$ s

Hence, height of tower is given buy

$$h = ut + \frac{1}{2}gt^2$$

$$h = \frac{1}{2}gt^2 [\because u = 0]$$

$$= \frac{1}{2} \times 10 \times (4)^2 = 80 \text{ m.}$$

410 **(d)**

If t_1 and t_2 are time of ascent and descent respectively then time of flight $T = t_1 + t_2 = \frac{2u}{g}$

$$\Rightarrow u = \frac{g(t_1 + t_2)}{2}$$

411 **(d)**

Let u be the velocity with which the stone is projected vertically upwards.

Given that, $v_{-h} = 2v_h$

$(v_{-h})^2 = 4v_h^2$

$\therefore u^2 - 2g(-h) = 4(u^2 - 2gh)$

$\therefore u^2 = \frac{10gh}{3}$

Now, $h_{max} = \frac{u^2}{2g} = \frac{5h}{3}$

412 **(d)**

At highest point $v = 0$ and $H_{max} = \frac{u^2}{2g}$

413 **(d)**

$(S' \propto t^2$. Now, $S_1' : S_2' : S_3' :: \frac{1}{4} : 1 : \frac{9}{4}$ or 1:4:9

For successive intervals,

$S_1 : S_2 : S_3 :: 1 : (4 - 1) : (9 - 4)$

or $S_1 : S_2 : S_3 :: 1 : 3 : 5$

414 **(d)**

$v = \frac{ds}{dt} = 3t^2 - 12t + 3$ and $a = \frac{dv}{dt} = 6t - 12$

For $a = 0$, we have $t = 2$ and at $t = 2, v = -9 \; ms^{-1}$

415 **(a)**

$S \propto u^2 \therefore \frac{S_1}{S_2} = \left(\frac{u_1}{u_2}\right)^2 \Rightarrow \frac{2}{S_2} = \frac{1}{4} \Rightarrow S_2 = 8\,m$

416 **(b)**

Distance travelled by train in first 1 *hour* is 60 *km* and distance in next 1/2 hour is 20 *km*.

So Average speed $= \frac{\text{Total distance}}{\text{Total time}} = \frac{60+20}{3/2}$

$= 53.33\ km/hour$

417 **(d)**

Total distance $= 130 + 120 = 250\ m$

Relative velocity $= 30 - (-20) = 50\ m/s$

Hence $t = 250/50 = 5\ s$

419 **(a)**

Let the speed of trains be x

$\therefore \frac{x-u}{x+u} = \frac{1}{2} \Rightarrow 2x - 2u = x + u \Rightarrow x = 3u$

420 **(d)**

$\vec{v}_{man} = \frac{v}{\sqrt{2}}\hat{\imath} + \frac{v}{\sqrt{2}}\hat{\jmath}$

Let $\vec{v}_{wind} = a\hat{\imath} + b\hat{\jmath}$

$\Rightarrow \vec{v}_{wind/man} = \left(a - \frac{v}{\sqrt{2}}\right)\hat{\imath} + \left(b - \frac{v}{\sqrt{2}}\right)\hat{\jmath}$

$\Rightarrow \tan\theta = \frac{b - \frac{v}{\sqrt{2}}}{a - \frac{v}{\sqrt{2}}} = \tan 270°$

$\Rightarrow a - \frac{v}{\sqrt{2}} = 0$

$\Rightarrow a = \frac{v}{\sqrt{2}} \Rightarrow \vec{v}_{wind} = \frac{v}{\sqrt{2}}\hat{\imath} + b\hat{\jmath}$

when the man doubles his speed

$\vec{v}'_{man} = 2\left(\frac{v}{\sqrt{2}}\hat{\imath} + \frac{v}{\sqrt{2}}\hat{\jmath}\right) = \sqrt{2}\,(v\hat{\imath} + v\hat{\jmath})$

$\Rightarrow \vec{v}'_{wind/man} = \left(\frac{v}{\sqrt{2}} - \sqrt{2}\,v\right)\hat{\imath} + (b - \sqrt{2}\,v)\hat{\jmath}$

$\Rightarrow \tan\theta' = \frac{b - \sqrt{2}}{\frac{v}{\sqrt{2}} - \sqrt{2}\,v} = \frac{2v - \sqrt{2}\,b}{v}$

But $\theta' = 270° - \cot^{-1}(2)$

$\Rightarrow \tan[270° - \cot^{-1}(2)] = \frac{2v - \sqrt{2}b}{v}$

$\Rightarrow \cot[\cot^{-1}(2)] = \frac{2v - \sqrt{2}b}{v}$

$\Rightarrow 2v = 2v - \sqrt{2}\,b = b = 0$

$\therefore \vec{v}_{wind} = \frac{v}{\sqrt{2}}\hat{\imath}$

421 **(c)**

Since direction of v is opposite to the direction of g and h so from equation of motion

$h = -vt + \frac{1}{2}gt^2 \Rightarrow gt^2 - 2vt - 2h = 0$

$\Rightarrow t = \frac{2v \pm \sqrt{4v^2 + 8gh}}{2g} \Rightarrow t = \frac{v}{g}\left[1 + \sqrt{1 + \frac{2gh}{v^2}}\right]$

422 **(c)**

From equation of motion, we have

$$s = ut + \frac{1}{2}gt^2$$

Where, u is initial velocity, g the acceleration due to gravity and t the time.

For upward motion

$$h = -ut_1 - \frac{1}{2}gt_1^2 \quad \text{... (i)}$$

for downward motion

$$h = -ut_2 + \frac{1}{2}gt_2^2 \quad \text{... (ii)}$$

multiplying Eq. (i) by t_2 and Eq. (ii) by t_1 and subtracting Eq. (ii) by Eq. (i), we get

$$h(t_2 - t_1) = \frac{1}{2}gt_1t_2(t_2 - t_1)$$

$$h = \frac{1}{2}gt_1t_2 \quad \text{... (iii)}$$

When stone is dropped from rest $u = 0$, reaches the ground in t second.

$$\therefore \quad h = \frac{1}{2}gt^2 \quad \text{... (iv)}$$

Equating Eqs. (iii) and (iv), we get

$$\frac{1}{2}gt^2 = \frac{1}{2}g\,t_1t_2$$

$$\Rightarrow \quad t^2 = t_1t_2 \Rightarrow t = \sqrt{t_1t_2}$$

423 **(d)**

Let height of tower is h and body takes t time to reach to ground when it fall freely

$\therefore h = \frac{1}{2}gt^2$

In last second i.e. t^{th} sec body travels $=0.36\,h$

It means in rest of the time i.e. in $(t-1)sec$ it travels

$= h - 0.36\,h = 0.64\,h$

Now applying equation of motion for $(t-1)sec$

$0.64\,h = \frac{1}{2}g(t-t)^2$

From (i) and (ii) we get, $t = 5$ sec and $h = 125m$

424 **(a)**

The portion OA of the graphs is convex upward. It represents negative acceleration. The portion AB represents negative acceleration. The portion AB represents that x is not changing with time. Clearly, it is a case of zero acceleration. The portion BC of the graph is concave upward. It represents positive acceleration. The portion CD is a straight line sloping upward to the right. It represents uniform velocity and hence acceleration is zero

425 **(c)**

Since acceleration due to gravity is independent of mass, hence time is also independent of mass (or density) of object

426 **(c)**

$x = at + bt^2 - ct^3, a = \frac{d^2x}{dt^2} = 2b - 6ct$

427 **(b)**

$x = 4(t-2) + a(t-2)^2$

At $t = 0, x = -8 + 4a = 4a - 8$

$v = \frac{dx}{dt} = 4 + 2a(t-2)$

At $t = 0, v = 4 - 4a = 4(1-a)$

But acceleration, $a = \frac{d^2x}{dt^2} = 2a$

428 **(c)**

$\frac{dx}{dt} = 2at - 3bt^2 \Rightarrow \frac{d^2x}{dt^2} = 2a - 6bt = 0$

$\Rightarrow t = \frac{a}{3b}$

429 **(a)**

For first stone $u = 0$ and

For second stone $\frac{u^2}{2g} 4h \Rightarrow u^2 = 8gh$

$\therefore u = \sqrt{8gh}$

Now, $h_1 = \frac{1}{2}gt^2$

$h_2 = \sqrt{8ght - \frac{1}{2}gt^2}$

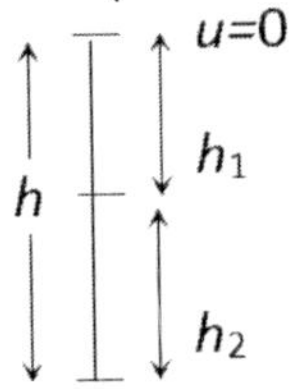

$u = \sqrt{8gh}$

Where, t =time cross each other

$\therefore h_1 + h_2 = h$

$\Rightarrow \frac{1}{2}gt^2 + \sqrt{8ght} - \frac{1}{2}gt^2 = h \Rightarrow t = \frac{h}{\sqrt{8gh}} = \sqrt{\frac{h}{8g}}$

430 **(a)**

Distance travelled in 4 sec

$24 = 4u + \frac{1}{2}\alpha \times 16$ ... (i)

Distance travelled in 8 sec

$88 = 8u + \frac{1}{2}\alpha \times 64$ ... (ii)

After solving (i) and (ii), we get $u = 1m/s$

431 **(b)**

Only directions of displacement and velocity gets changed, acceleration is always directed vertically downward

432 **(d)**

The nature of the path is decided by the direction of velocity, and the direction of acceleration. The trajectory can be a straight line, circle or a parabola depending on these factors

433 **(a)**

$h = ut + \frac{1}{2}gt^2$

$\therefore \quad 30 = -25t + \frac{10}{2}t^2$

Or $t^2 - 5t - 6 = 0$

Or $\quad (t-6)(t+1) = 0$

$\therefore \qquad t = 6\text{ s}$

434 **(b)**

$v^2 = u^2 + 2as$

$a = \frac{v^2 - u^2}{2s} = \frac{(20)^2 - (10)^2}{2 \times 135} = \frac{300}{270}$

$= \frac{10}{9} m/s^2$

From first equation of motion

$v = u + at \Rightarrow t = \frac{v-u}{a} = \frac{20-10}{10/9}$

$= \frac{10}{10/9} = 9\ sec$

435 **(a)**

$0^2 - v^2 = -2aS$ or $v^2 \propto S; \left(v + \frac{20v}{100}\right)^2 \propto S'$

or $\frac{S'}{S} = \left(1 + \frac{20}{100}\right)^2 = \frac{36}{25}$

$\therefore \left(\frac{S'}{S} - 1\right) \times 100 = \left(\frac{36}{25} - 1\right) \times 100$

$= 44\%$

436 **(a)**

Average speed $= \frac{2v_d v_u}{v_d + v_u}$

437 **(d)**

$a = \frac{dv}{dt} = \frac{dv}{dx}\frac{dx}{dt} = v\frac{dv}{dx} = -\alpha x^2$ [Given]

$\Rightarrow \int_{v_0}^{0} v dv = \alpha \int_0^S x^2\, dx \Rightarrow \left[\frac{v^2}{2}\right]_{v_0}^{0}$

$= -\alpha \left[\frac{x^3}{3}\right]_0^S$

$\Rightarrow \frac{v_0^2}{2} = \frac{\alpha S^3}{3} \Rightarrow S = \left(\frac{3v^2}{2\alpha}\right)^{\frac{1}{3}}$

438 **(a)**

Slope of displacement time-graph is velocity

$\frac{v_1}{v_2} = \frac{\tan\theta_1}{\tan\theta_2} = \frac{\tan 30°}{\tan 45°} = \frac{1}{\sqrt{3}}$

$v_1 : v_2 = 1 : \sqrt{3}$

439 **(a)**

Velocity $v = \alpha\sqrt{x}$

$\frac{dx}{dt} = \alpha\sqrt{x}\left(\because v = \frac{dx}{dt}\right)$

Or $\frac{dx}{\sqrt{x}} = \alpha\, dt$

Integrating

$\int_0^x \frac{dx}{\sqrt{x}} = \int_0^t \alpha\, dt$

[$\because$ at $t = 0, x$
$= 0$ ad let at any time t, paricle is at x]

Or $\frac{x^{1/2}}{1/2} = \alpha t$

Or $x^{1/2} = \frac{\alpha}{2}t$

Or $\quad x = \frac{\alpha^2}{4} \times t^2 \Rightarrow x \propto t^2$

440 **(b)**

Let two boys meet at point C after time 't' from the starting. Then $AC = vt, BC = v_1t$

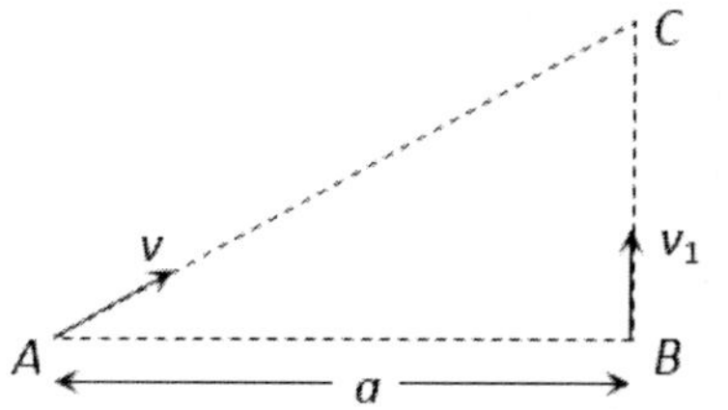

$(AC)^2 = (AB)^2 + (BC)^2 \Rightarrow v^2t^2$
$= a^2 + v_1^2t^2$

By solving we get $t = \sqrt{\frac{a^2}{v^2 - v_1^2}}$

441 **(a)**

$s = 1.2\ m$

$v = 640\ ms^{-1}$

$a = ?; u = 0;\ t = ?$

$2\ as = v^2 - u^2$

$\Rightarrow 2a \times 1.2 = 640 \times 640 \Rightarrow a$

$= \frac{8 \times 64 \times 10^3}{3}$

$v = u + at \Rightarrow t = \frac{v}{a} = \frac{15}{4} \times 10^{-3}$
$= 3.75 \times 10^{-3}\ s$
$\approx 4\ ms$

442 **(c)**
Let the initial velocity of balls A, B and C are equal and its magnitude is u. Since, the ball A is projected with velocity u in upward direction, so when it will come back to the projection point, its velocity remains same. So, the final velocity of ball A, when it hits the ground is given as

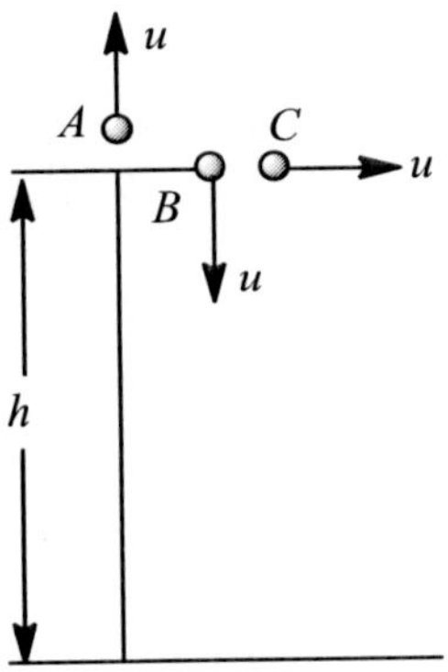

$v_A^2 = u^2 + 2gh$

Hence, $v_A = \sqrt{u^2 + 2gh}$ ... (i)

And the final velocity of ball B is

$v_B = \sqrt{u^2 + 2gh}$ ... (ii)

But the initial vertical velocity of the ball C is zero.

So, $v_C^2 = \sqrt{(0)^2 + 2gh}$

$\Rightarrow \quad v_C = \sqrt{2gh}$ ....(iii)

Hence, it is clear from Eqs. (i), (ii) and (iii), we get

$v_A = v_B > v_C$

443 **(a)**
$\frac{dt}{dx} = 2\alpha x + \beta \Rightarrow v = \frac{1}{2\alpha x + \beta}$
$\because a = \frac{dv}{dt} = \frac{dv}{dx}.\frac{dx}{dt}$
$a = v\frac{dv}{dx} = \frac{-v.2\alpha}{(2\alpha x + \beta)^2} = -2\alpha.v.v^2$
$= -2\alpha v^3$
$\therefore$ Retardation $= 2\alpha v^3$

444 **(d)**
Total distance $= 130 + 120 = 250\ m$
Relative velocity $= 30 - (-20) = 50\ m/s$
Hence $t = 250/50 = 5\ s$

445 **(b)**
Time of flight $= \frac{2u}{g} = \frac{2 \times 100}{10} = 20$ sec

446 **(b)**
$\int_0^x dx = \int_0^1 (v_0 + gt + ft^2)dt$
$x = v_0 + g\left(\frac{1}{2}\right) + f\left(\frac{1}{3}\right)$

447 **(b)**
$x = a + bt^2, v = \frac{dx}{dt} = 2bt$
Instantaneous velocity $v = 2 \times 3 \times 3 = 18\ cm/sec$

448 **(b)**
Let two boys meet at point C after time 't' from the starting. Then $AC = vt, BC = v_1t$

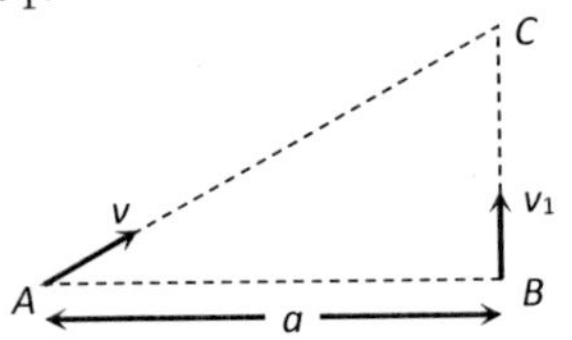

$(AC)^2 = (AB)^2 + (BC)^2 \Rightarrow v^2t^2$
$= a^2 + v_1^2t^2$
By solving we get $t = \sqrt{\frac{a^2}{v^2 - v_1^2}}$

449 **(d)**
$x = 30 + 180 = 120\text{m}$
$y = 9x^2$
$\Rightarrow \frac{dy}{dt} = v_y = 18x\frac{dx}{dt} = 18x\left(\frac{1}{3}\right)$
$\Rightarrow \frac{dy}{dt} = v_y = 6x$
$a_y = \frac{d^2y}{dt^2} = \frac{dv_y}{dt} = 6\left(\frac{dx}{dt}\right) = 6\left(\frac{1}{3}\right)$
$\Rightarrow a_y = 2\text{ms}^{-2}$along$y$-axis.
Hence, $a_y = (2\text{ms}^{-2})\hat{j}$

450 **(a)**

Taking initial position as origin and direction of motion (*ie*, vertically up) as positive. As the particle is thrown with initial velocity, at highest point its velocity is zero and then it returns back to its reference position. This situation is best depicted in figure of option (a).

In figure, AB part denotes upward motion and BC part denotes downward motion.

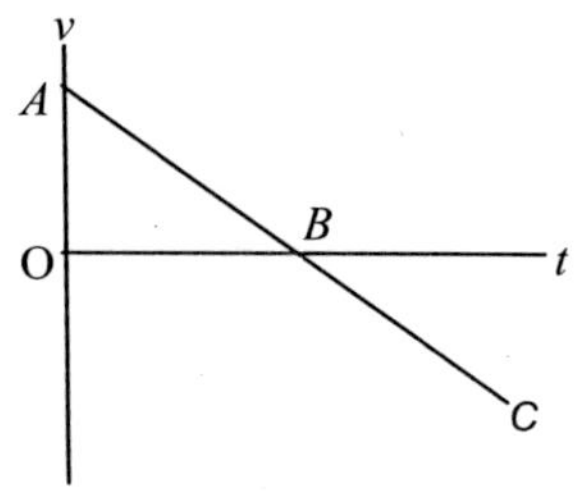

451 **(c)**

$h = ut + \frac{1}{2}gt^2, t = 3 \text{ sec}, u = -4.9 m/s$

$\Rightarrow h = -4.9 \times 3 + 4.9 \times 9 = 29.4\ m$

452 **(a)**

$h = \frac{1}{2}gt^2 = \frac{1}{2} \times 10 \times (4)^2 = 80\ m$

453 **(b)**

The given condition is possible only when body is at its highest position after 5 seconds

It means time of ascent = 5 *sec*

and time of flight $T = \frac{2u}{g} = 10$

$\Rightarrow u = 49\ m/s$

454 **(a)**

$\vec{r} = 20\hat{\imath} + 10\hat{\jmath}\ \therefore r = \sqrt{20^2 + 10^2}$
$= 22.5\ m$

455 **(b)**

Since acceleration is constant, therefore there is uniform increase in velocity. So, the $v - t$ graph is a straight line slopping upward to the right. When acceleration becomes zero, velocity is constant. So $v - t$ graph is a straight line parallel to the time-axis.

456 **(c)**

Given : Initial velocity of a body $u = 0$ ... (i)

Let s be the distance covered by a body in time t

$\therefore s = ut + \frac{1}{2}at^2$ or $s = \frac{1}{2}at^2$ [Using (i)]

$\Rightarrow s \propto t^2$

457 **(b)**

Time of ascent $= \frac{u}{g} = 6 \text{ sec} \Rightarrow u = 60 m/s$

Distance in first second $h_{\text{first}} = 60 - \frac{g}{2}(2 \times 1 - 1) = 55m$

Distance in seventh second will be equal to the distance in first second of vertical downward motion

$h_{\text{seventh}} = \frac{g}{2}(2 \times 1 - 1) = 5\ m$

$\Rightarrow h_{\text{first}}/h_{\text{seventh}} = 11:1$

458 **(a)**

$h = ut - \frac{1}{2}gt^2 \Rightarrow 96 = 80t - \frac{32}{2}t^2$

$\Rightarrow t^2 - 5t + 6 = 0 \Rightarrow t = 2sec$ or $3\ sec$

459 **(b)**

$h_1 = \frac{1}{2}gt^2, h_2 = 50t - \frac{1}{2}gt^2$

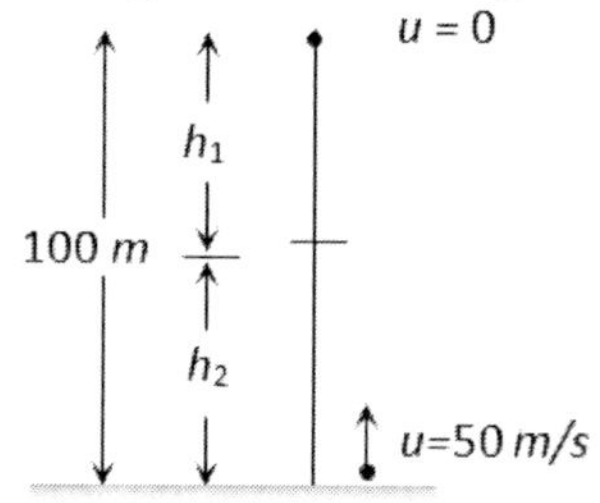

Given $h_1 + h_2 = 100\ m$

$\Rightarrow 50t = 100 \Rightarrow t = 2\ sec$

460 **(c)**

Let the velocity of the bullet when it strikes the target is v cm s^{-1}.

After penetrating 30 cm, velocity becomes half *ie*, $\frac{v}{2}$.

From equation $v^2 = u^2 + 2as$

$\therefore \quad \left(\frac{v}{2}\right)^2 = v^2 + 2a \times 30$

Or $\quad -60a = v^2 - \frac{v^2}{4}$

Or $-60a = \frac{3v^2}{4}$

$\therefore \quad a = -\frac{v^2}{80} \text{cm s}^{-2}$

Let the bullet further penetrates x cm before coming to rest, therefore

$v'^2 = u'^2 + 2\,as'$

$$0 = \left(\frac{v}{2}\right)^2 + 2\left(-\frac{v^2}{80}\right) x$$

$\frac{v^2 x}{40} = \frac{v^2}{4}$

$$x = 10 \text{ cm}$$

461 **(d)**

$v = \alpha\sqrt{x} \Rightarrow \frac{dx}{dt} = \propto \sqrt{x} \Rightarrow \frac{dx}{\sqrt{x}} = \alpha dt$

By integrating both sides $\int x^{-1/2}\, dx = \int \alpha\, dt$

$\Rightarrow \frac{\sqrt{x}}{1/2} = \alpha t \Rightarrow \sqrt{x} = \frac{1}{2}\alpha t \Rightarrow x = \frac{1}{4}\alpha^2 t^2$

$$\therefore x \propto t^2$$

462 **(d)**

Let the man will be able to catch the bus after t s

Then

$10t = 48 + \frac{1}{2} \times 1 \times t^2$

$t^2 - 20t + 96 = 0$

$(t - 12)(t - 8) = 0$

$t = 8s$ and $t = 12s$

Thus the man will be able to catch the bus after $8s$

463 **(d)**

Man walks from his home to market with a speed of 5 km/h. Distance= 2.5 km and time $= \frac{d}{v} = \frac{2.5}{5} = \frac{1}{2}$ hr and he returns back with speed of 7.5 km/h in rest of time of 10 $minutes$

Distance $= 7.5 \times \frac{10}{60} = 1.25$ km

So, Average speed $= \frac{\text{Total distance}}{\text{TOtal time}}$

$= \frac{(2.5 + 1.25)km}{(40/60)hr} = \frac{45}{8} km/hr$

464 **(d)**

$x = 2t^3 + 21t^2 + 60t + 6$

$\therefore v = \frac{dx}{dt} = 6t^2 + 42t + 60$

But , $v = 0$ (given)

$t^2 + 7t + 10 = 0$

$\Rightarrow \; t = -5\text{s}$

or $t = -2\text{s}$

$a = \frac{dv}{dt} = 12t + 42$

$a\,|_{t=5s} = -60 + 42 = -18\text{ms}^{-2}$

$a\,|_{t=-2s} = -24 + 42 = 18\text{ms}^{-2}$

465 **(d)**

Up to time t_1 slope of the graph is constant and after $t_1$slope is zero $i.e.$ the body travel with constant speed up to time t_1 and then stops

466 **(b)**

Time average velocity $= \frac{v_1+v_2+v_3}{3} = \frac{3+4+5}{3} = 4m/s$

467 **(c)**

Here, $s_1 = 40$ m, $s_2 = 65$ m,

$t_1 = 5$ s, $a = ?$

$$a = \frac{s_2 - s_1}{t^2} = \frac{(65 - 40) \times 2}{(5)^2}$$

$$= \frac{50}{25} = 2 \text{ ms}^{-2}$$

Now, $s_1 = ut + \frac{1}{2}\, at^2$

$$40 = 5u + \frac{1}{2} \times 2 \times 25$$

Or $5u = 15$ or $u = 3 \text{ ms}^{-1}$

468 **(d)**

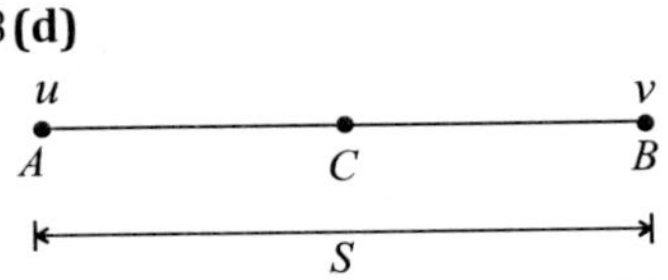

Let S be the distance between AB and a be constant acceleration of a particle.

Then

$v^2 - u^2 = 2aS$

Or $aS = \frac{v^2 - u^2}{2}$ ... (i)

Let v_c be velocity of a particle at midpoint C

$\therefore\ v_c^2 - u^2 = 2a\left(\frac{S}{2}\right)$

$v_c^2 = u^2 + aS = u^2 + \frac{v^2-u^2}{2}$[Using (i)]

$v_c = \sqrt{\frac{u^2 + v^2}{2}}$

469 **(a)**

At time t

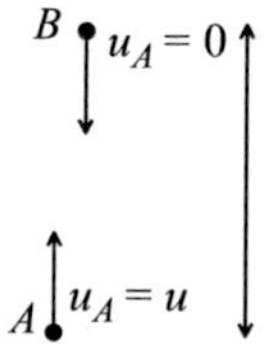

Velocity of $A, v_A = u - gt$ upward

Velocity of $B, v_B = gt$ downward

It we assume that height h is smaller than or equal to the maximum height reached by A, then at every instant v_A and v_B are in opposite directions

$\therefore V_{AB} = v_A + V_B$

$= u - gt + gt$ [Speeds in opposite directions get added]

$= u$

470 **(a)**

Distance = Area covered between velocity and time axis

$= \frac{1}{2}(30 + 10)10 = 200\ meter$

471 **(b)**

$\frac{|\text{Average velocity}|}{|\text{Average speed}|} = \frac{|\text{displacement}|}{|\text{distance}|} \leq 1$

Because displacement will either be equal or less than distance. It can never be greater than distance

472 **(c)**

Average velocity is defined as the displacement divided by time.

In the given graph, displacement is zero.

Hence, Average velocity $= \frac{\text{total displacement}}{\text{total time}} = \frac{0}{t} = 0$

473 **(d)**

Body reaches the point of projection with same velocity

474 **(b)**

According to given relation acceleration

$a = \alpha t + \beta$

As $a = \frac{dv}{dt} \Rightarrow \alpha t + \beta = \frac{dv}{dt}$

Since particle starts from rest, its initial velocity is zero

$i.e.$, At time $t = 0$, velocity = 0

$\Rightarrow \int_0^v dv = \int_0^t (\alpha t + \beta)dt \Rightarrow v = \frac{\alpha t^2}{2} + \beta t$

475 **(c)**

$v = (180 - 16x)^{1/2}$

As $a = \frac{dv}{dt} = \frac{dv}{dx}.\frac{dx}{dt}$

$\therefore a = \frac{1}{2}(180 - 16x)^{-1/2} \times (-16)\left(\frac{dx}{dt}\right)$

$= -8(180 - 16x)^{-1/2} \times v$

$= -8(180 - 16x)^{-1/2} \times (180 - 16x)^{1/2}$

$= -8\ m/s^2$

476 **(b)**

Boat covers distance of 16 km in a still water in hours

$ie\ \ v_B = \frac{16}{2}$ =8kmh^{-1}

Now, velocity of water

$v_W = 4\text{kmh}^{-1}$

Time taken for going upstream

$t_1 = \frac{8}{v_B - v_w} = \frac{8}{8-4} = 2\text{h}$

(As water current oppose the motion of boat) Time taken for going downstream

$t_2 = \frac{8}{v_B + v_w} = \frac{8}{8+4} = \frac{8}{12}\text{h}$

(As water current helps the motion of boat)

$\therefore$ Total time $= t_1 + t_2$

$= \left(2 + \frac{8}{12}\right)$h=2 h 40 min

477 **(b)**

Let at point A initial velocity of body is equal to zero

for path AB: $v^2 = 0 + 2gh$ ... (i)

for path AC: $(2v)^2 = 0 + 2gx$

$4v^2 = 2gx$...(ii)

Solving (i) and (ii), $x = 4h$

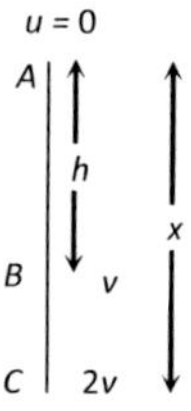

478 **(b)**

When a ball is dropped on a floor then

$$y = \frac{1}{2} gt^2 \qquad ... (i)$$

So, the graph between $y$and t is a parabola. Here y decrease as time decrease. When the ball is bounces back then

$$y = ut - \frac{1}{2} gt^2 \qquad ... (ii)$$

Eq. (ii) is also the form of a general equation of parabola so the graph between $y$and t will be a parabola. Here$y$ increases when the time increases. Hence, the required graph between $y$and $t$is shown in the figure.

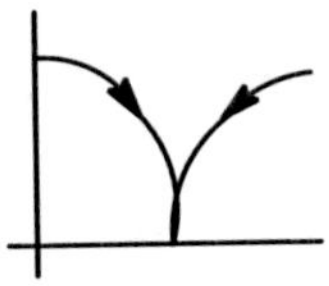

479 **(a)**

Given that $u = 0$ (the electron starts from rest), At any time $v = kt = 2t$

$a = \frac{du}{dt} = 2\text{ms}^{-1}$(Constant acceleration)

Now $s = ut + \frac{1}{2} at^2$

$= 0 \times 3 + \frac{1}{2} \times 2 \times (3)^2$

$= 9\text{m}$

480 **(b)**

Average velocity is that uniform velocity with which the object will cover the same displacement in same interval of time as it does with its actual variable velocity during that time interval.

Here, total distance covered

$$= (3 \text{ ms}^{-1} \times 20 \text{ s}) + (4 \text{ ms}^{-1} \times 20 \text{ s}) + (5 \; ms^{-1} \times 20 \text{ s})$$

$$= (60 + 80 + 100) = 240 \text{ m}$$

Total time taken $= 20 + 20 + 20 = 60$ s

$\therefore$ Average velocity $= \frac{240}{60} = 4 \text{ ms}^{-1}$

481 **(b)**

$x = a + bt^2, v = \frac{dx}{dt} = 2bt$

Instantaneous velocity $v = 2 \times 3 \times 3 =$ 18 cm/sec

482 **(b)**

Let u be the initial upward velocity of the ball from A and let h be the height of the tower.

Taking the downward motion of the first stone from A to the ground, we have

$h = -ut_1 + \frac{1}{2} gt_1^2$ ...(i)

Taking the downward motion of the second stone from A to the ground, we have

$h = ut_2 + \frac{1}{2} gt_2^2$ ...(ii)

Multiplying Eq.(i) t_2 and Eq. (ii) by t_1 and adding we get

$h(t_1 + t_2) = \frac{1}{2} gt_1 t_2 (t_1 + t_2)$

So, $h = \frac{1}{2} gt_1 t_2$ ...(iii)

For falls under gravity from the top of the tower

$h = \frac{1}{2} gt_3^2$ ...(iv)

From Eqs. (iii) and (iv), $t_3^2 = t_1 t_2$ or $t_3 = \sqrt{t_1 t_2}$

$= \sqrt{6 \times 4} = 6$ s

483 **(a)**

From the equation of motion

$$S = ut + \frac{1}{2} at^2$$

$$S_1 = 0 + \frac{1}{2} a(P-1)^2$$

And $S_2 = 0 + \frac{1}{2} aP^2$

From $S_n = u + \frac{a}{2}(2n - 1)$

$$S_{(P^2-P+1)^{th}} = \frac{a}{2}[2(P^2 - P + 1) - 1]$$

$$= \frac{a}{2}[2P^2 - 2P + 1] = S_1 + S_2$$

485 **(a)**

$$h = ut - \frac{1}{2}gt^2 \Rightarrow 96 = 80t - \frac{32}{2}t^2$$

$\Rightarrow t^2 - 5t + 6 = 0 \Rightarrow t = 2sec$ or $3\ sec$

486 **(d)**

Up to time t_1 slope of the graph is constant and after $t_1$slope is zero *i.e.* the body travel with constant speed up to time t_1 and then stops

487 **(a)**

$x^2 = at^2 + 2bt + c$

Differentiating w.r.t. time, we get

$2x\frac{dx}{dt} = 2at + 2b$ or $xv = at + b$ or $v = \frac{at+b}{x}$

Again differentiating w.r.t. time, we get

$x\ \frac{dx}{dt} + v\ \frac{dx}{dt} = a$

or $xA + v^2 = a$ or $xA = a - v^2$

or $xA = a - \left(\frac{at+b}{x}\right)^2$ or $xA = \frac{ax^2 - a^2t^2 - b^2 - 2abt}{x^2}$

or $xA = \frac{a^2t^2 + 2abt + ac - a^2t^2 - b^2 - 2abt}{x^2}$

or $A = \frac{ac - b^2}{x^3}$ or $A \propto x^{-3}$

488 **(b)**

$$v = u + at = u + \left(\frac{F}{m}\right)t$$

$$= 20 + \left(\frac{100}{5}\right) \times 10$$

$$= 220\ m/s$$

489 **(a)**

$\mathbf{r} = x\hat{\mathbf{i}} + y\hat{\mathbf{j}} + z\hat{\mathbf{k}}$

$\therefore \mathbf{r} = \sqrt{x^2 + y^2 + z^2}$

$\mathbf{r} = \sqrt{6^2 + 8^2 + 10^2} = 10\sqrt{2}$m

490 **(d)**

Both trains will travel a distance of 1 km before to come in rest. In this case by using $v^2 = u^2 + 2as$

$\Rightarrow 0 = (40)^2 + 2a \times 1000 \Rightarrow a = -0.8\ m/s^2$

491 **(c)**

For first part,

$u = 0, t = T$ and acceleration $= a$

$\therefore v = 0 + aT = aT$ and $S_1 = 0 + \frac{1}{2}aT^2 = \frac{1}{2}aT^2$

For Second part,

$u = aT$, retardation $= a_1, v = 0$ and time taken $= T_1$ (let)

$\therefore 0 = u - a_1T_1 \Rightarrow aT = a_1T_1$

And from $v^2 = u^2 - 2aS_2 \Rightarrow S_2 = \frac{u^2}{2a_1} = \frac{1}{2}\frac{a^2T^2}{a_1}$

$S_2 = \frac{1}{2}aT \times T_1 \left(\text{As} a_1 = \frac{aT}{T_1}\right)$

$\therefore v_{av} = \frac{S_1 + S_2}{T + T_1} = \frac{\frac{1}{2}aT^2 + \frac{1}{2}aT \times T_1}{T + T_1}$

$= \frac{\frac{1}{2}aT(T + T_1)}{T + T_1} = \frac{1}{2}aT$

492 **(b)**

Let the stone remains in air for t s. From $S = ut + \frac{1}{2}gt^2$

Here, $u = 0,\ S = \frac{1}{2}gt^2$

Total distance travelled by the stone in last second is

$D = S_t - S_{t-1} = \frac{1}{2}gt^2 - \frac{1}{2}g(t-1)^2$

Distance travelled by the stone in first three seconds is

$S_3 = \frac{1}{2} \times g \times 3^2 = \frac{9}{2}g$

According to given problem, $D = S_3$

$\therefore \frac{g}{2}(2t - 1) = \frac{9}{2}g$ or $2t - 1 = 9 \Rightarrow t = 5s$

493 **(d)**

The average speed

$$v_{av} = \frac{\text{length of path } ACB}{\text{time interval } (t_2 - t_1)} \quad \text{... (i)}$$

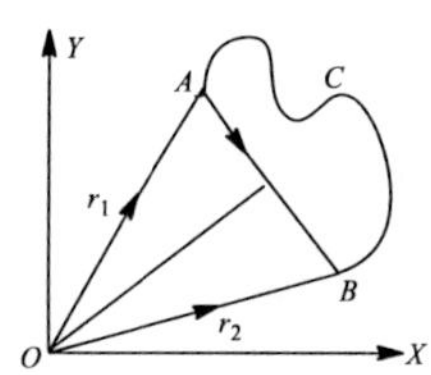

And average velocity,

$$\mathbf{v}_{av} = \frac{\text{displacement}}{\text{time interval}} = \frac{\mathbf{r}_2 - \mathbf{r}_1}{t_2 - t_1} \quad \text{... (ii)}$$

But we know that distance is always be greater than or equal to magnitude of displacement. So the average speed will always be greater than or equal to the magnitude of average velocity.

From Eqs. (i) and (ii)

$$\frac{\mathbf{v}_{av}}{v_{av}} = \frac{\text{displacement}}{\text{length of path (distance)}} \leq 1$$

494 **(b)**

Using $t = \sqrt{\frac{2h}{a}}, t \propto \sqrt{l}, \frac{t_2}{t_1} = \sqrt{\frac{l/4}{l}} = \frac{1}{2}$

or $t_2 = \frac{t_1}{2} = \frac{4}{2} = 2\text{s}$

495 **(d)**

Man walks from his home to market with a speed of 5 km/h. Distance= 2.5 km and time $= \frac{d}{v} = \frac{2.5}{5} = \frac{1}{2}$ hr and he returns back with speed of 7.5 km/h in rest of time of 10 *minutes*

Distance $= 7.5 \times \frac{10}{60} = 1.25\ km$

So, Average speed $= \frac{\text{Total distance}}{\text{TOtal time}}$

$= \frac{(2.5 + 1.25)km}{(40/60)hr} = \frac{45}{8} km/hr$

496 **(b)**

Speed of stone in a vertically upward direction is $20m/s$. So for vertical downward motion we will consider $u = -20\ m/s$

$v^2 = u^2 + 2gh = (-20)^2 + 2 \times 9.8 \times 200$
$= 4320\ m/s$

$\therefore v \simeq 65m/s$

497 **(b)**

The area under acceleration time graph gives change in velocity. As acceleration is zero at the end of 11 *sec*

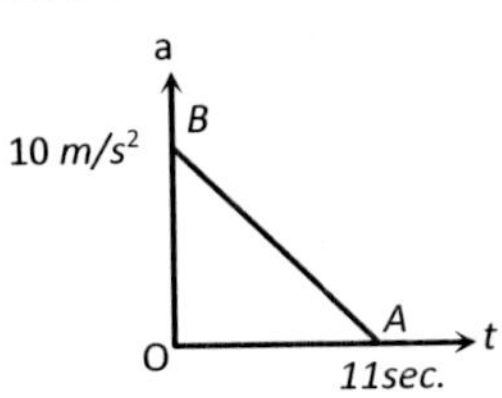

i.e. v_{max} =Area of ΔOAB

$= \frac{1}{2} \times 11 \times 10 = 55\ m/s$

498 **(d)**

Let the thickness of each plank is d.

From equation of motion

$$v^2 = u^2 + 2as \quad \text{... (i)}$$

Ist case

$$s = 2d, u = 100\ \text{ms}^{-1}, v = 0$$

$\therefore \quad 0 = (100)^2 + 2a \times 2d$

Or $\quad 4ad = -100 \times 100$

Or $\quad a = -\frac{100 \times 100}{4d}$

$\therefore \quad a = -\frac{2500}{d} \quad \text{... (ii)}$

IInd case Let the bullet with double the previous speed will penetrate n planks of equal thickness$d$.

Now, $v = 0, u = 200\ \text{ms}^{-1}, a = -\frac{2500}{d}, s = nd$

$$0 = (200)^2 - 2 \times \frac{2500}{d} \times nd$$

Or $\quad n = \frac{200 \times 200}{2 \times 2500} = 8$

499 **(c)**

$S \propto t^2 \Rightarrow \frac{S_1}{S_2} = \left(\frac{10}{20}\right)^2 \Rightarrow S_2 = 4S_1$

500 **(d)**

When A returns to the 'level' of top of tower, its downward velocity is 4ms^{-1}. This velocity is the same a that of B. So, both A and B hit the ground with the same velocity.

501 **(c)**

Acceleration of the body along AB is $g\cos\theta$

Distance travelled in time $t\ sec = AB = \frac{1}{2}(g\cos\theta)t^2$

From $\Delta ABC, AB = 2R\cos\theta$; $2R\cos\theta = \frac{1}{2}g\cos\theta t^2$

$\Rightarrow t^2 = \frac{4R}{g}$ or $t = 2\sqrt{\frac{R}{g}}$

502 **(d)**

The student is able to catch the bus if in time t the distance travelled by him is equal to the distance travelled by bus in time t

ie, $s_1 = s_2$ ... (i)

From Eq. (i)

$$0 + \frac{1}{2}at^2 = ut - d$$

Or $at^2 - 2ut + 2d = 0$

It is quadratic equation

So, $t = \frac{+2u \pm \sqrt{4u^2 - 8ad}}{2} = \frac{+2u \pm 2\sqrt{u^2 - 2ad}}{2}$

For t to be real

$$u \geq \sqrt{2ad} \geq \sqrt{2 \times 1 \times 50} = 10\ \text{ms}^{-1}$$

503 **(b)**

$v^2 = u^2 + 2gh \Rightarrow (3u)^2 = (-u)^2 + 2gh \Rightarrow h = \frac{4u^2}{g}$

504 **(b)**

Distance covered =Area enclosed by $v - t$ graph

= Area of triangle = $\frac{1}{2} \times 4 \times 8 = 16\ m$

505 **(b)**

Region OA shows that graph bending toward time axis $i.e.$ acceleration is negative.

Region AB shows that graph is parallel to time axis $i.e.$ velocity is zero. Hence acceleration is zero.

Region BC shows that graph is bending towards displacement axis $i.e.$ acceleration is positive.

Region CD shows that graph having constant slope $i.e.$ velocity is constant. Hence acceleration is zero

506 **(b)**

Velocity of graph = Area of a-t graph
$= (4 \times 1.5) - (2 \times 1) = 4m/s$

507 **(b)**

Let particle thrown with velocity u and its maximum height is H then $H = \frac{u^2}{2g}$

When particle is at height $H/2$, then its speed is 10 m/s

From equation $v^2 = u^2 - 2gh$

$(10)^2 = u^2 - 2g\left(\frac{H}{2}\right) = u^2 - 2g\frac{u^2}{4g} \Rightarrow u^2 = 200$

Maximum height $\Rightarrow H = \frac{u^2}{2g} = \frac{200}{2 \times 10} = 10\ m$

508 **(d)**

$$\text{Average speed} = \frac{\text{Total distance travelled}}{\text{Total time taken}}$$

$$= \frac{x}{\frac{2x/5}{v_1} + \frac{3x/5}{v_2}} = \frac{5v_1v_2}{3v_1 + 2v_2}$$

509 **(d)**

Area between $v - t$ graph and time-axis

$$= \frac{1}{2} \times 2 \times 20 + 3 \times 20 + \frac{1}{2} \times 1 \times 20 + \frac{1}{2} \times 1 \times 20$$

$= 100$ m.

510 **(b)**

Power, $P = \frac{W}{t}$

$P = \frac{Fs}{t}, P = \frac{mas}{t} (\because F = ma)$

$P = \frac{mv\,s}{t^2}, \left(\because a = \frac{v}{t}\right)$

$P = \frac{ms \cdot s}{t^3} \left(\because v = \frac{s}{t}\right)$

$Pt^2 = ms^3$

$\therefore s \propto t^{3/2}$

511 **(d)**

Slope of displacement time graph is negative only at point time E

512 **(a)**

Given, $\frac{dv}{dt} = pt$

$dv = pt\ dt$

$$\int_0^v dv = P\int_0^{t_1} t\ dt$$

$v = \frac{1}{2}pt_1^2$

Again, $x = \frac{1}{2}p\int_0^{t_1} t_1^2 dt = \frac{1}{2}pt_1^3$

513 **(c)**

$\mathbf{u} = 3\hat{\mathbf{i}} + 4\hat{\mathbf{j}}, \mathbf{a} = 0.4\hat{\mathbf{i}} + 0.3\hat{\mathbf{j}}$

Speed $\mathbf{v} = \mathbf{u} + \mathbf{a}t$

$= 3\hat{\mathbf{i}} + 4\hat{\mathbf{j}} + (0.4\hat{\mathbf{i}} + 0.3\hat{\mathbf{j}})10$

$= 3\hat{\mathbf{i}} + 4\hat{\mathbf{j}} + 4\hat{\mathbf{i}} + 3\hat{\mathbf{j}} = 7\hat{\mathbf{i}} + 7\hat{\mathbf{j}}$

$v = \sqrt{7^2 + 7^2} = 7\sqrt{2}$ unit

514 **(c)**

Initial velocity $u = \tan 45° = 1$

Velocity after 2s, $v = \tan 60° = \sqrt{3}$

$\therefore$ Average acceleration, $a_{av} = \frac{v-u}{t} = \frac{\sqrt{3}-1}{2}$.

516 **(b)**

$\vec{v} = \vec{u} + \vec{a}t \Rightarrow \vec{v} = 3\hat{\imath} + 4\hat{\jmath} + (0.4\hat{\imath} + 0.3\hat{\jmath}) \times 10$

$v = 3\hat{\imath} + 4\hat{\jmath} + 4\hat{\imath} + 3\hat{\jmath} = 7\hat{\imath} + 7\hat{\jmath}$

$\Rightarrow |\vec{v}| = 7\sqrt{2}$ units

518 **(a)**

Distance between the balls = Distance travelled by first ball in 3 seconds – Distance travelled by second ball in 2 seconds $= \frac{1}{2}g(3)^2 - \frac{1}{2}g(2)^2 = 45 - 20 = 25\ m$

519 **(d)**

$\frac{11h}{36} = \frac{9.8}{2}(2n-1)$

or $\frac{11}{36} \times \frac{1}{2} \times 9.8n^2 = \frac{9.8}{2}(2n-1)$

or $2n - 1 = \frac{11}{36}n^2$ or $11n^2 = 72n - 36$

or $11n^2 - 72n + 36 = 0$

or $11n^2 - 66n - 6n + 36 = 0$

or $11n(n-6) - 6(n-6) = 0$

or $(11n - 6)(n - 6) = 0$

$\Rightarrow n = 6$ (Rejecting fractional value)

$h = \frac{1}{2} \times 10 \times 6 \times 6\text{m} = 180\text{m}$

520 **(a)**

$v = u + at \Rightarrow v = 0 + 5 \times 10 = 50\ m/s$

521 **(b)**

Total time of motion is 2 *min* 20 *sec* = 140 *sec*

As time period of circular motion is 40 *sec* so in 140 *sec*. Athlete will complete 3.5 revolution *i.e.*,He will be at diametrically opposite point

i.e.,Displacement = $2R$

522 **(a)**

We know that the velocity of body is given by the slope of displacement – time graph so it is clear that initially slope of the graph is positive and after some time it becomes zero (corresponding to the peak of graph) and it will become negative

523 **(b)**

$v = \frac{ds}{dt} = 12t - 3t^2$

Velocity is zero for $t = 0$ and $t = 4$ sec

524 **(c)**

$v^2 = u^2 + 2gh \Rightarrow v = \sqrt{u^2 + 2gh}$

So for both the cases velocity will be equal

525 **(c)**

If the body starts from rest and moves with constant acceleration then the ratio of distances in consecutive equal time interval $S_1 : S_2 : S_3 = 1 : 3 : 5$

526 **(a)**

When the stone is released from the balloon. Its height

$h = \frac{1}{2}at^2 = \frac{1}{2} \times 1.25 \times (8)^2 = 40\ m$ and velocity

$v = at = 1.25 \times 8 = 10\ m/s$

Time taken by the stone to reach the ground

$$t = \frac{v}{g}\left[1 + \sqrt{1 + \frac{2gh}{v^2}}\right]$$

$$= \frac{10}{10}\left[1 + \sqrt{1 + \frac{2 \times 10 \times 40}{(10)^2}}\right]$$

$= 4 sec$

530 **(d)**

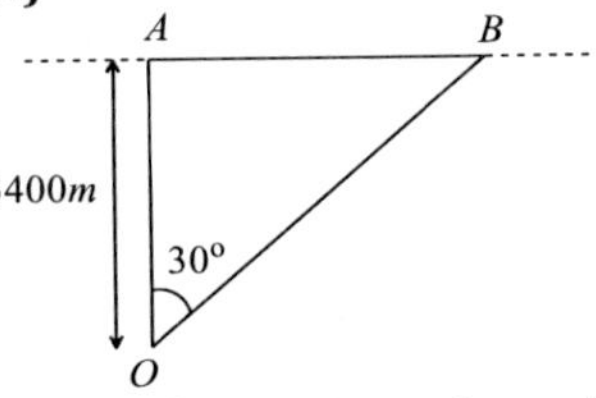

O is the observation point at the ground. A and B are the positions of aircraft for

which $\angle AOB = 30°$. Time taken by aircraft from A to B is $10s$
ΔAOB

$$\tan 30° = \frac{AB}{3400}$$

$$AB = 3400 \tan 30° = \frac{3400}{\sqrt{3}} m$$

$\therefore$ Speed of aircraft,

$$v = \frac{AB}{10} = \frac{3400}{10\sqrt{3}} = 196.3\ ms^{-1}$$

531 **(d)**

Both trains will travel a distance of $1\ km$ before to come in rest. In this case by using $v^2 = u^2 + 2as$
$\Rightarrow 0 = (40)^2 + 2a \times 1000 \Rightarrow a = -0.8\ m/s^2$

532 **(a)**

Here, $t_1 = \frac{x}{v} + \frac{x}{v} = \frac{2x}{v}, t_2 = \frac{x}{v+\omega} = \frac{x}{v-\omega} + \frac{2xv}{v^2-\omega^2}$

or $t_2 = \frac{2xv}{v^2\left(1-\frac{\omega^2}{v^2}\right)} = \frac{2x}{v\ \left(1-\frac{\omega^2}{v^2}\right)}$

or $t_2 = \frac{t_1}{1-\frac{\omega^2}{v^2}}$

or $t_2 > t_1$

533 **(b)**

$$a = \sqrt{a_x^2 + a_y^2}$$

$$= \left[\left(\frac{d^2x}{dy}\right)^2 + \left(\frac{d^2y}{dt^2}\right)^2\right]^{\frac{1}{2}}$$

Here, $\frac{d^2y}{dx^2} = 0$

Hence, $a = \frac{d^2x}{dt^2} = 8\text{ms}^{-2}$

534 **(d)**

Time of flight $T = \frac{2u}{g} 4 \text{ sec} \Rightarrow u = 20\ m/s$

535 **(a)**

For the given condition initial height $h = d$ and velocity of the ball is zero. When the ball moves downward its velocity increases and it will be maximum when the ball hits the ground & just after the collision it becomes half and in opposite direction. As the ball moves upward its velocity again decreases and becomes zero at height $d/2$. This explanation match with graph (A)

536 **(b)**

Let the two bodies A and B respectively meet at a time t, at a height $\frac{H}{2}$ from the ground

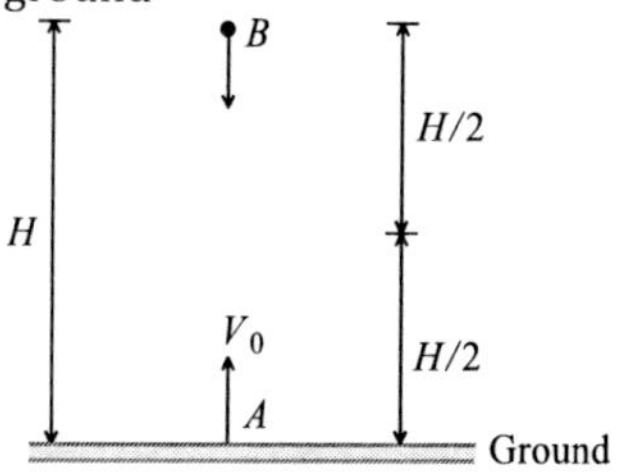

Using $S = ut + \frac{1}{2}at^2$

For a body $A, u = V_0, a = -g, S = \frac{H}{2}$

$\therefore \frac{H}{2} = V_0t - \frac{1}{2}gt^2$ ...(i)

For body $B, u = 0, a = +g, S = \frac{H}{2}$

$\therefore \frac{H}{2} = \frac{1}{2}gt^2$ ...(ii)

Equating equations (i) and (ii), we get

$V_0t - \frac{1}{2}gt^2 = \frac{1}{2}gt^2 \Rightarrow V_0t = gt^2$ or $t = \frac{V_0}{g}$

Substituting the value of t in equation (i), we get

$$\frac{H}{2} = V_0 \times \left(\frac{V_0}{g}\right) - \frac{1}{2}g\left(\frac{V_0}{g}\right)^2 = \frac{V_0^2}{g} - \frac{1}{2}\frac{V_0^2}{g}$$

$\frac{H}{2} = \frac{1}{2}\frac{V_0^2}{g}$ or $V_0^2 = gH \Rightarrow V_0 = \sqrt{gH}$

537 **(b)**

$$v = \frac{ds}{dt} = 12t - 3t^2$$

Velocity is zero for $t = 0$ and $t = 4$ sec

538 **(a)**

$4 = u + \frac{a}{2}(2 \times 3 - 1)$or $4 = u + \frac{5a}{2}$,

$5 = u + \frac{a}{2}(2 \times 4 - 1)$ or $5 = u + \frac{7a}{2}$

Subtracting, $1 = \frac{7a}{2} - \frac{5a}{2} = \frac{2a}{2} = a$

Again, $4 = u + \frac{5}{2}$ or $u = 4 - \frac{5}{2} = 1.5\text{ms}^{-1}$

So, the initial velocity is non-zero and acceleration is uniform.

540 **(a)**

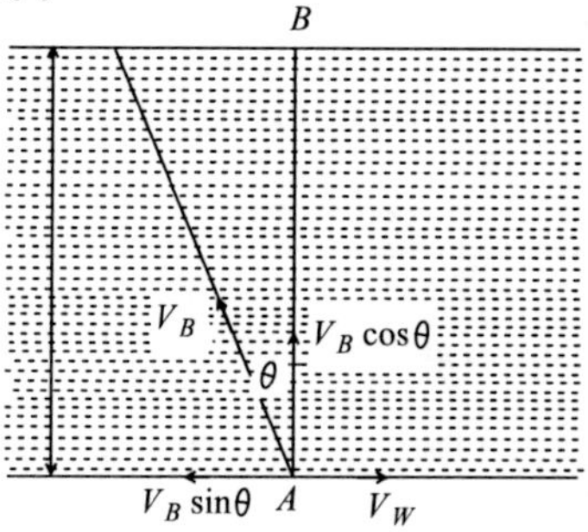

From figure, $V_B \sin\theta = V_W$

$\sin\theta = \frac{V_W}{V_B} = \frac{1}{2} \Rightarrow \theta = 30°$ $[\because V_B = 2V_W]$

Time taken to cross the river,

$t = \frac{D}{V_B \cos\theta} = \frac{D}{V_B \cos 30°} = \frac{2D}{V_B\sqrt{3}}$

541 **(a)**

$\sqrt{x} = t + 1$

Squaring both sides, we get

$x = (t+1)^2 = t^2 + 2t + 1$

Differentiating it w.r.t time t, we get

$\frac{dx}{dt} = 2t + 2$

Velocity, $v = \frac{dx}{dt} = 2t + 2$

542 **(d)**

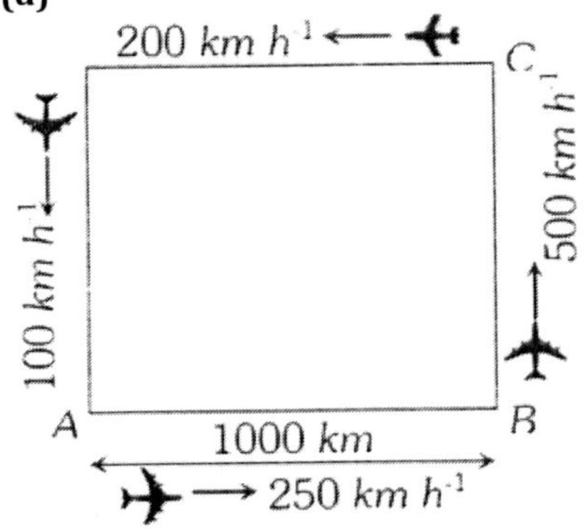

Let t_{AB}, t_{BC}, t_{CD} and t_{DA} be the time taken by the aeroplane to go from A to B, B to C, C to D and D to A respectively

$\therefore t_{AB} = \frac{1000\ km}{250\ km\ h^{-1}} = 4h$

$\therefore t_{BC} = \frac{1000\ km}{500\ km\ h^{-1}} = 2h$

$\therefore t_{CD} = \frac{1000\ km}{200\ km\ h^{-1}} = 5h$

$\therefore t_{DA} = \frac{1000\ km}{100\ km\ h^{-1}} = 10h$

Average speed= $\frac{\text{Total distance covered}}{\text{Total time taken}}$

$= \frac{1000km + 1000km + 1000km + 1000km}{t_{AB} + t_{BC} + t_{CD} + t_{DA}}$

$= 190.5\ km\ h^{-1}$

543 **(d)**

Relative velocity of police man w.r.t. the time$10 - 9 = 1\text{ms}^{-1}$. Since the relative separation between them is 100 m, hence, the time taken will be = relative separation/relative velocity =100/1=100s

544 **(a)**

Distance covered in 5 s

$s_1 = \frac{1}{2}\ at^2$

$= \frac{1}{2}a(5)^2 = \frac{25a}{2}$

Distance covered in 5th second

$s_2 = \frac{1}{2} \times a(2 \times 5 - 1) = \frac{9}{2}\ a$

$\therefore \quad \frac{s_2}{s_1} = \frac{9}{25}$

545 **(a)**

From the figure, the relative displacement is

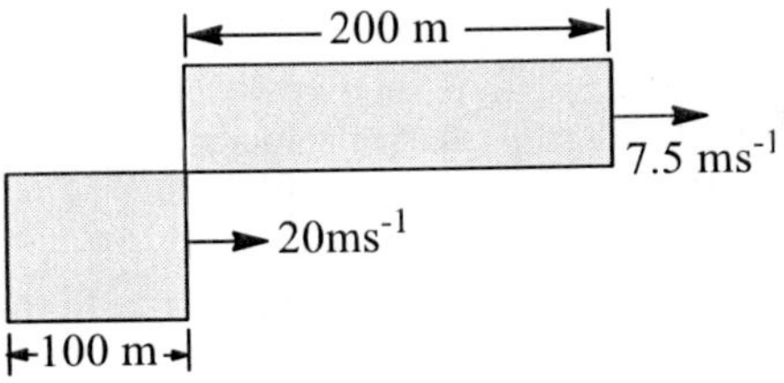

$s_{\text{rel}} = (200 + 100)\text{m} = 300\text{m}$

$v_{\text{rel}} = v_1 - v_2 = (20 - 7.5)\text{ms}^{-1}$

$= 12.5\text{ms}^{-1}$

$\therefore t = \frac{s_{\text{rel}}}{v_{\text{rel}}} = \frac{300}{12.5} = 24\text{s}$

546 **(b)**

For vertically upward motion, $h_1 = v_0 t - \frac{1}{2}gt^2$ and for vertically downward motion, $h_2 = v_0 t + \frac{1}{2}g\ t^2$

$\therefore$ Total distance covered in t sec h

$= h_1 + h_2 = 2v_0 t$

547 **(b)**

$h_n = \frac{g}{2}(2n - 1) \Rightarrow h_{5\text{th}} = \frac{10}{2}(2 \times 5 - 1)$

$= 45\ m$

548 **(a)**

$S_n = u + \frac{a}{2}[2n - 1]$

$S_{5^{th}} = 7 + \frac{4}{2}[2 \times 5 - 1] = 7 + 18 = 25m$

549 **(d)**

At highest point $v = 0$ and $H_{\max} = \frac{u^2}{2g}$

550 **(b)**

Distance covered =Area enclosed by $v-t$ graph

$=$ Area of triangle $=\frac{1}{2}\times 4\times 8=16\ m$

552 **(d)**

Velocity of particle $=\frac{\text{Total displacement}}{\text{Total time}}$

$=\frac{\text{Diameter of circle}}{5}=\frac{2\times 10}{5}=4\ m/s$

553 **(c)**

Net acceleration of a body when thrown upward = acceleration of body – acceleration due to gravity $=a-g$

554 **(c)**

Speed of the object at reaching the ground $v=\sqrt{2gh}$

If heights are equal then velocity will also be equal

555 **(d)**

Displacement (in magnitude)

$=\frac{1}{2}\left(3\times 2-\frac{1}{2}\times 1\times 2+1\times 1\right)$m=3m

556 **(a)**

Let us calculate relative deceleration by considering relative velocity

Using, $v^2-u^2=2aS, 0^2-80^2=2\times a\times 2000$

or $a=-\frac{80\times 80}{4000}=-\frac{64}{40}\text{ms}^{-2}=-1.6\text{ms}^{-2}$

Deceleration of each train is

$\frac{1.6}{2}\text{ms}^{-2} ie, 0.8\text{ ms}^{-2}$

557 **(b)**

The area under acceleration time graph gives change in velocity. As acceleration is zero at the end of 11 *sec*

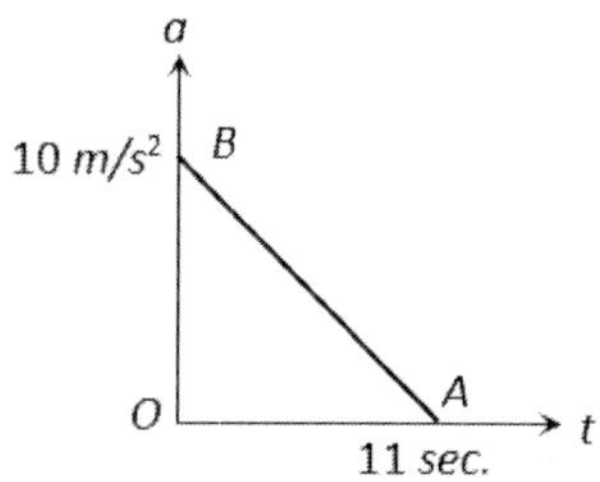

i.e. $v_{\max}$ =Area of ΔOAB

$=\frac{1}{2}\times 11\times 10=55\ m/s$

558 **(c)**

The displacement of the particle is determined by the area bounded by the curve. This area is

$s=\frac{\pi}{4}v_m t_0$

The average velocity is

$<v>=\frac{s}{t_0}=\frac{\pi}{4}v_m$

Such motion cannot be realized in practical terms since at the initial and final moment, the acceleration (which is slope of $v-t$) is infinitely large. Hence, both (i) and (ii) are correct.

559 **(c)**

On rebound, there is an instantaneous change in the direction of velocity. This settle the answer.

560 **(b)**

Maximum acceleration will be represented by CD part of the graph

Acceleration $=\frac{dv}{dt}=\frac{(60-20)}{0.25}=160\ km/h^2$

562 **(d)**

$S=3-4t+5t^2$

Velocity $\frac{ds}{dt}=-4+10\,t$

Hence, initial velocity will be

$\left|u=\frac{ds}{dt}\right|_{t=0}=-4$ unit

563 **(a)**

$h=\frac{1}{2}gt^2=\frac{1}{2}\times 10\times(4)^2=80\ m$

564 **(a)**

Since slope of graph remains constant for velocity-time graph

565 **(d)**

$s=3t^3+7t^2+14t+8m$

$a=\frac{d^2s}{dt^2}=18t+14$ at $t=1$ sec $\Rightarrow a=32\ m/s^2$

566 **(b)**

$v\propto\sqrt{h}\therefore\frac{v_1}{v_2}=\sqrt{\frac{a}{b}}$

So, (b) is the correct choice.

The velocity acquired by a body in falling freely from rest through height h is $\sqrt{2gh}$.

$[u=0, v=?,' a'=g, 'S'=h, v^2-u^2=2aS]$

567 **(c)**

$x=at+bt^2-ct^3, a=\frac{d^2x}{dt^2}=2b-6ct$

568 **(b)**

Time average velocity $=\frac{v_1+v_2+v_3}{3}=\frac{3+4+5}{3}=4m/s$

569 **(b)**

Time of ascent = Time of descent=5sec

570 **(a)**

Using

$V = u + at$

$V = gt$ ...(i)

Comparing with $y = mx + c$

Equation (i) represents a straight line passing through origin inclined x-axis (slope -g)

571 **(a)**

This graph shows uniform motion because line having a constant slope

572 **(d)**

If t_1 and t_2 are time of ascent and descent respectively then time of flight $T = t_1 + t_2 = \frac{2u}{g}$

$\Rightarrow u = \frac{g(t_1 + t_2)}{2}$

574 **(c)**

If a stone is dropped from height h then

$h = \frac{1}{2}gt^2$...(i)

if a stone is thrown upward with velocity u then

$h = -u\, t_1 + \frac{1}{2}gt_1^2$ ...(ii)

If a stone is thrown downward with velocity u then

$h = ut_2 + \frac{1}{2}gt_2^2$...(iii)

From (i) (ii) and (iii) we get

$-ut_1 + \frac{1}{2}gt_1^2 = \frac{1}{2}gt^2$ ...(iv)

$ut_2 + \frac{1}{2}gt_2^2 = \frac{1}{2}gt^2$ ...(v)

Dividing (iv) and (v) we get

$\therefore \frac{-ut_1}{ut_2} = \frac{\frac{1}{2}g(t^2 - t_1^2)}{\frac{1}{2}g(t^2 - t_2^2)}$

Or $-\frac{t_1}{t_2} = \frac{t^2 - t_1^2}{t^2 - t_2^2}$

By solving $t = \sqrt{t_1 t_2}$

575 **(b)**

Maximum acceleration will be represented by CD part of the graph

Acceleration $= \frac{dv}{dt} = \frac{(60-20)}{0.25} = 160\ km/h^2$

576 **(c)**

Let $\vec{a}_{rel}$ = acceleration of ball with respect to ground – acceleration of bus with respect to ground

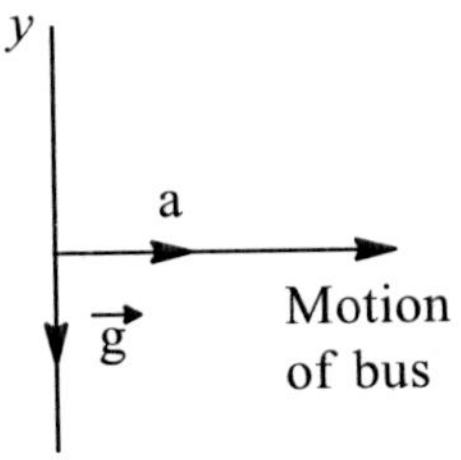

$= -g\hat{j} - a\hat{j}$

$|\vec{a}_{rel}| = \sqrt{g^2 + a^2}$

Hence, (c) is correct.

577 **(b)**

The given condition is possible only when body is at its highest position after 5 seconds

It means time of ascent = 5 sec

and time of flight $T = \frac{2u}{g} = 10$

$\Rightarrow u = 49\ m/s$

578 **(b)**

$x = \frac{1}{t+5} \Rightarrow v = \frac{dx}{dt} = -\frac{1}{(t+5)^2}$

Acceleration, $a = \frac{dv}{dt} = \frac{2}{(t+5)^3} \Rightarrow a \propto$ (velocity)$^{3/2}$

579 **(b)**

Let A and B will meet after time t sec. it means the distance travelled by both will be equal

$S_A = ut = 40t$ and $S_B = \frac{1}{2}at^2 = \frac{1}{2} \times 4 \times t^2$

$S_A = S_B \Rightarrow 40t = \frac{1}{2}4t^2 \Rightarrow t = 20\ sec$

580 **(a)**

$x_1 = \frac{1}{2}at^2$ and $x_2 = ut \therefore x_1 - x_2 = \frac{1}{2}\ at^2 - ut$

$y = \frac{1}{2}at^2 - ut$.This equation is of parabola

$\frac{dy}{dt} = at - u$ and $\frac{d^2y}{dt^2} = a$

As $\frac{d^2y}{dt^2} > 0 i.e.,$ graph shows possess minima at $t = \frac{u}{a}$

581 **(b)**

$h_1 = \frac{1}{2}gt^2, h_2 = 50t - \frac{1}{2}gt^2$

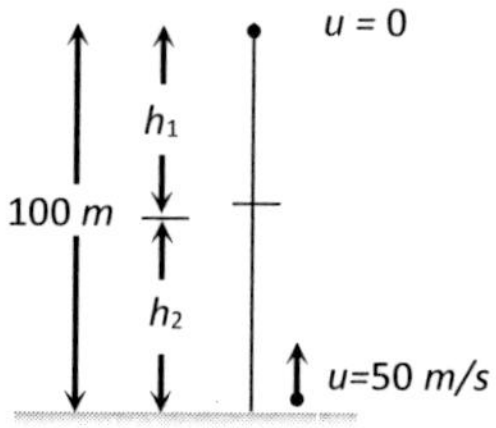

Given $h_1 + h_2 = 100\ m$
$\Rightarrow 50t = 100 \Rightarrow t = 2\ sec$
(c)

582 Since direction of v is opposite to the direction of g and h so from equation of motion

$$h = -vt + \frac{1}{2}gt^2 \Rightarrow gt^2 - 2vt - 2h = 0$$

$$\Rightarrow t = \frac{2v \pm \sqrt{4v^2 + 8gh}}{2g} \Rightarrow t$$

$$= \frac{v}{g}\left[1 + \sqrt{1 + \frac{2gh}{v^2}}\right]$$

583 **(a)**
When the body is projected vertically upward then at the highest point its velocity is zero but acceleration is not equal to zero ($g = 9.8 m/s^2$)

584 **(b)**
Let height of minaret is H and body take time T to fall from top to bottom

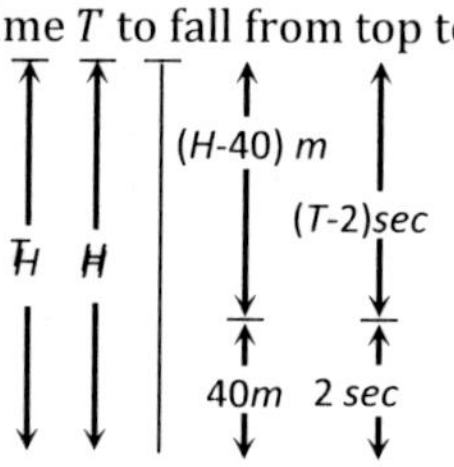

$H = \frac{1}{2}gT^2$ ...(i)
In last 2 sec body travels distance of 40 m so in $(T-2)$ sec distance travelled = $(H-40)m$
$(H-40) = \frac{1}{2}g(T-2)^2$ ...(ii)
By solving (i) and (ii), $T = 3\ sec$ and $H = 45m$

585 **(c)**
Let man will catch the bus after 't'sec. So he will cover distance ut
Similarly distance travelled by the bus will be $\frac{1}{2}at^2$
For the given condition

$$ut = 45 + \frac{1}{2}at^2 = 45 + 1.25t^2 \quad [\text{As } a = 2.5\ m/s^2]$$

$$\Rightarrow u = \frac{45}{t} + 1.25t$$

To find the minimum value of u
$\frac{du}{dt} = 0$ sp we get $t = 6\ sec$ then,

$$u = \frac{45}{6} + 1.25 \times 6 = 7.5 + 7.5 = 15\ m/s$$

586 **(c)**

$$v = \frac{\text{Total distance}}{\text{Time taken}}$$

$$= \frac{x}{\frac{x/3}{v} + \frac{x/3}{3v} + \frac{x/3}{2v} + \frac{18}{11}v}$$

587 **(a)**

$$t = \sqrt{\frac{2h}{g}} \Rightarrow \frac{t_1}{t_2} = \sqrt{\frac{h_1}{h_2}} = \sqrt{\frac{1}{2}} = \frac{1}{\sqrt{2}}$$

588 **(c)**
$v^2 = u^2 + 2gh \Rightarrow v = \sqrt{u^2 + 2gh}$
So for both the cases velocity will be equal

589 **(d)**
Average acceleration $= \frac{\text{Change in velocity}}{\text{Time taken}} = \frac{v_2 - v_1}{t_2 - t_1}$

$$= \frac{[10 + 2(5)^2] - [10 + 2(2)^2]}{3}$$

$$= \frac{60 - 18}{3} 14 m/s^2$$

590 **(b)**
The ball is thrown vertically upwards, then according to equation of motion.

$(0)^2 - u^2 = -2gh$ ... (i)

And $0 = u - gt$ ... (ii)

From Eqs. (i) and (ii),

$$h = \frac{gt^2}{2}$$

When the ball is falling downwards after reaching the maximum height

$$s = ut' + \frac{1}{2}g(t')^2$$

$$\frac{h}{2} = (0)t' + \frac{1}{2}g(t')^2$$

$$\Rightarrow \qquad t' = \sqrt{\frac{h}{g}}$$

$$t' = \frac{t}{\sqrt{2}}$$

Hence, the total time from the time of projection of reach a point at half of its maximum height while returning $= t + t'$

$$= t + \frac{t}{\sqrt{2}} = \left(1 + \frac{1}{\sqrt{2}}\right)t$$

591 **(d)**

$x \propto t^3 \therefore x = Kt^3$

$\Rightarrow v = \frac{dx}{dt} = 3Kt^2$ and $a = \frac{dv}{dt} = 6Kt$

i.e. $a \propto t$

592 **(a)**

The distance covered by the ball during the last t *seconds* of its upward motion = Distance covered by it in first t seconds of its downward motion

From $h == ut + \frac{1}{2}gt^2$

$h = \frac{1}{2}g\,t^2$ [As $u = 0$ for it downward motion]

593 **(a)**

Distance b/w the cars A and B remains constant. Let the distance be 'x'

Velocity of C w.r.t. A and $BV = 45 + 36 = 81\ km/h$

Distance $= 81 \times \frac{5}{60} = 6.75\ Km$

594 **(a)**

$$S = ut + \frac{1}{2}gt^2$$

$30 = -25t + \frac{10}{2}t^2$ or $t^2 - 5t - 6 = 0$

Or $(t - 6)(t + 1) = 0$ Take positive root

$\therefore t = 6$ sec

595 **(a)**

$$s = \frac{1}{2}gt^2, v = \frac{1}{2}g \times 2t = gt$$

596 **(b)**

Only directions of displacement and velocity gets changed, acceleration is always directed vertically downward

597 **(a)**

If a body starts from rest with acceleration α and then retards with retardation β and comes to rest. The total time taken for this journey is t and distance covered is S

Then $S = \frac{1}{2}\frac{\alpha\beta t^2}{(\alpha + \beta)} = \frac{1}{2}\frac{5 \times 10}{(5 + 10)} \times t^2$

$\Rightarrow 1500 = \frac{1}{2}\frac{5 \times 10}{(5 + 10)} \times t^2 \Rightarrow t = 30\ sec$

598 **(c)**

Given, $s = t^3 - 6t^2 + 3t + 4$

Velocity $v = \frac{ds}{dt} = 3t^2 - 12t + 3$ ... (i)

Acceleration $a = \frac{dv}{dt} = 6t - 12$ ... (ii)

Since, acceleration is zero, so, $6t - 12 = 0$, or $t = 2$ s

So, velocity $v$at

$t = 2$ s, is $= 3 \times 2^2 - 12 \times 2 + 3$
$= -9\ \text{ms}^{-1}$

599 **(b)**

Time taken by first drop to reach the ground $t = \sqrt{\frac{2h}{g}} \Rightarrow t = \sqrt{\frac{2\times5}{10}} = 1$ sec

As the water drops fall at regular intervals from a tap therefore time difference between any two drops $= \frac{1}{2}$ sec

In this given time, distance of second drop from the tap $= \frac{1}{2}g\left(\frac{1}{2}\right)^2 = \frac{10}{8} = 1.25\ m$

Its distance from the ground $= 5 - 1.25 = 3.75m$

600 **(d)**

The equation of motion

$$\left(\frac{u}{2}\right)^2 = u^2 - 2g\,(AO)$$

$$2g \times AO = u^2 - \frac{u^2}{4} = \frac{3u^2}{4}$$

$$AO = \frac{3u^2}{8g}$$

When particle will reach at point B

$$\left(\frac{u}{3}\right)^2 = u^2 - 2g(OB)$$

$$OB = \frac{8u^2}{18g}$$

When particle will reach at point C

$$\left(\frac{u}{4}\right)^2 = u^2 - 2g\,(OC)$$

$$OC = \frac{15u^2}{32g}$$

$$AB = OB - OA = \frac{u^2}{g}\left[\frac{8}{18} - \frac{3}{8}\right] = \frac{5u^2}{72g}$$

$$BC = OC - OB = \frac{u^2}{g}\left[\frac{15}{32} - \frac{8}{18}\right]$$

The ratio, $\frac{AB}{BC} = \frac{20}{7}$

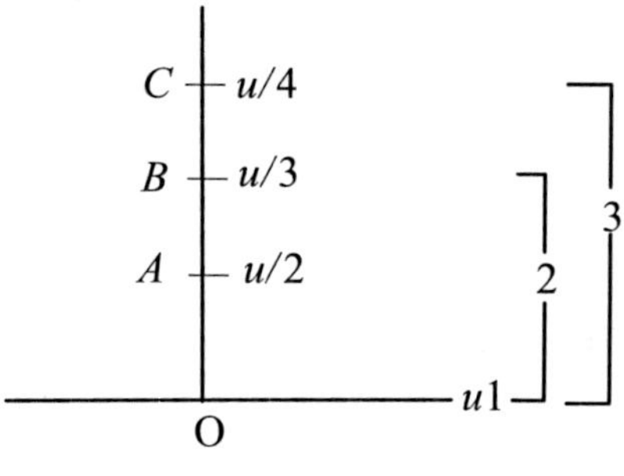

602 **(a)**

$$\frac{1}{2}at^2 = vt \Rightarrow t = \frac{2v}{a}$$

603 **(d)**

Average acceleration $= \frac{\text{Change in velocity}}{\text{Time taken}} = \frac{v_2 - v_1}{t_2 - t_1}$

$$= \frac{[10 + 2(5)^2] - [10 + 2(2)^2]}{3}$$

$$= \frac{60 - 18}{3} 14 m/s^2$$

604 **(b)**

$$v = u + at$$

$$\Rightarrow -2 = 10 + a \times 4$$

$$\therefore \quad a = -3\ \text{ms}^{-2}$$

605 **(c)**

Because acceleration is a vector quantity

606 **(a)**

$H_{\text{max}} \propto u^2 \therefore u \propto \sqrt{H_{\text{max}}}$

i.e. to triple the maximum height, ball should be thrown with velocity $\sqrt{3}\,u$

607 **(c)**

Slope is negative at the point E.

608 **(c)**

Horizontal distance covered by the wheel in half revolution $= \pi R$

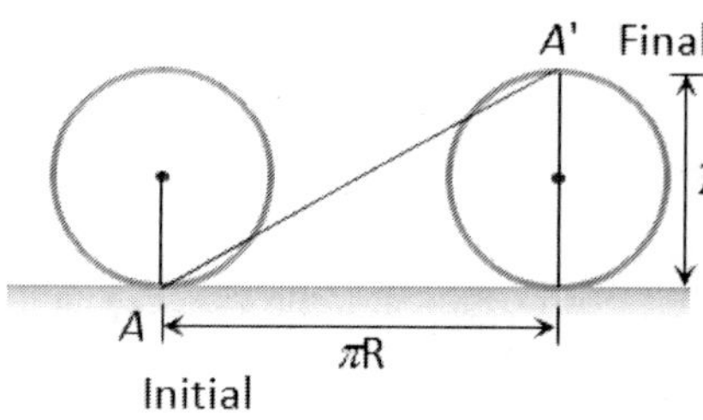

So the displacement of the point which was initially in contact with ground

$= AA' = \sqrt{(\pi R)^2 + (2R)^2}$

$= R\sqrt{\pi^2 + 4} = \sqrt{\pi^2 + 4}$ [As $R = 1m$]

609 **(b)**

Between time interval 20 *sec* to 40 *sec*, there is non-zero acceleration and retardation. Hence distance travelled during this interval

= Area between time interval 20 *sec* to 40 *sec*

$= \frac{1}{2} \times 20 \times 3 + 20 \times 1 = 30 + 20 = 50\ m$

610 **(a)**

The slope of distance-time graphs speed.
The change in velocity in 1 s

$= \tan 60° - \tan 45° = \sqrt{3} - 1$

$\therefore$ Acceleration $= \frac{\Delta v}{\Delta t} = \frac{\sqrt{3}-1}{1} = (\sqrt{3} - 1)$unit

611 **(a)**

Distance covered in 5th second

$S_{5^{th}} = u + \frac{a}{2}(2n-1) = 0 + \frac{a}{2}(2\times 5 - 1) = \frac{9a}{2}$

and distance covered in 5 second,

$S_5 = ut + \frac{1}{2}at^2 = 0 + \frac{1}{2}\times a \times 25 = \frac{25a}{2}$

$\therefore \frac{S_{5^{th}}}{S_5} = \frac{9}{25}$

612 **(a)**

$x_1 = \frac{1}{2}at^2$ and $x_2 = ut$ $\therefore x_1 - x_2 = \frac{1}{2}at^2 - ut$

$y = \frac{1}{2}at^2 - ut$.This equation is of parabola

$\frac{dy}{dt} = at - u$ and $\frac{d^2y}{dt^2} = a$

As $\frac{d^2y}{dt^2} > 0$ *i.e.*, graph shows possess minima at $t = \frac{u}{a}$

613 **(c)**

Average velocity $= \frac{\text{Displacement}}{\text{Time interval}}$

A particle moving in a given direction with non-zero velocity cannot have zero speed.
In general, average speed is not equal to magnitude of average velocity. However, it can be so if the motion is along a straight line without change in direction

614 **(a)**

This graph shows uniform motion because line having a constant slope

615 **(a)**

According to problem
Distance travelled by body A in 5th *sec* and distance travelled by body B in3^{rd} *sec*. of its motion are equal.

$0 + \frac{a_1}{2}(2\times 5 - 1) = 0 + \frac{a_2}{2}[2\times 3 - 1]$

$9a_1 = 5a_2 \Rightarrow \frac{a_1}{a_2} = \frac{5}{9}$

616 **(b)**

Area under the velocity–time curve over a given time interval gives the displacement of the particle

617 **(c)**

$h = ut + \frac{1}{2}gt^2 \Rightarrow 1 = 0 \times t_1 + \frac{1}{2}gt_1^2 \Rightarrow t_1 = \sqrt{2/g}$

Velocity after travelling 1 m distance

$v^2 = u^2 + 2gh \Rightarrow v^2 = (0)^2 + 2g \times 1$

$\Rightarrow v = \sqrt{2g}$

For second 1 m distance

$1 = \sqrt{2g} \times t_2 + \frac{1}{2}gt_2^2$

$\Rightarrow gt_2^2 + 2\sqrt{2g}t_2 - 2 = 0$

$t_2 = \frac{-2\sqrt{2g} \pm \sqrt{8g + 8g}}{2g} = \frac{-\sqrt{2} \pm 2}{\sqrt{g}}$

Taking +ve sign $t_2 = (2 - \sqrt{2})/\sqrt{g}$

$\therefore \frac{t_1}{t_2} = \frac{\sqrt{2/g}}{(2-\sqrt{2})/\sqrt{g}} = \frac{1}{\sqrt{2}-1}$ and so on

618 **(b)**

$v = u + a\,t$

$2\times 100 = 100 + 10t$ or $t = 10$ s

619 **(b)**

The velocity time graph for given problem is shown in the figure.

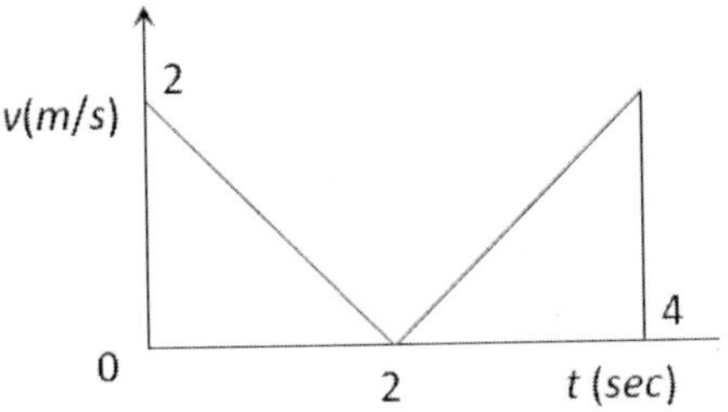

Distance travelled $S =$
Area under curve $= 2 + 2 = 4m$

620 **(b)**

$v = u + at \Rightarrow -2 = 10 + a \times 4 \Rightarrow a = -3m/sec^2$

621 **(d)**

$40^2 - 30^2 = 2aS, v^2 - 30^2 = 2a\frac{S}{2}$

or $2(v^2 - 30^2) = 2aS$

Comparing, $2(v^2 - 900) = 1600 - 900 = 700$

or $v^2 = 900 + 350 = 1250$ or $v = 35.35\ \text{kmh}^{-1}$

(a)

At time t

622 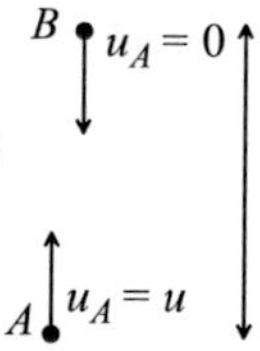

Velocity of $A, v_A = u - gt$ upward
Velocity of $B, v_B = gt$ downward
It we assume that height h is smaller than or equal to the maximum height reached by A, then at every instant v_A and v_B are in opposite directions
$\therefore V_{AB} = v_A + V_B$
$= u - gt + gt$ [Speeds in opposite directions get added]
$= u$

623 **(a)**

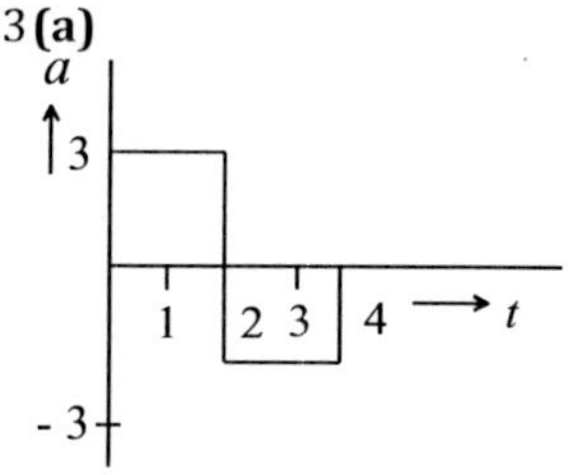

Takingthe motion from 0 to 2 s
$u = 0, a = 3ms^{-2}, t = 2s, v = ?$
$v = u + at = 0 + 3 \times 2 = 6ms^{-1}$
Taking the motion from 2 s to 4 s
$v = 6 + (-3)(2) = 0ms^{-1}$

624 **(b)**

$\frac{dv}{dt} = a - bv$

or $\int_0^v \frac{dv}{a-bv} = \int_0^t dt$

or $-\frac{1}{b}[\log(a - bv)]_0^v = t$

or $\left(\frac{a-bv}{a}\right) = e^{-bt}$

or $a - \frac{b}{a}v = e^{-bt}$

$v = \frac{a}{b}(1 - e^{-bt})$

625 **(c)**

Acceleration $= a = \frac{dv}{dt} = 0.1 \times 2t = 0.2t$
Which is time dependent $i.e.$ non-uniform acceleration

626 **(b)**

$$v = u + at = u + \left(\frac{F}{m}\right)t$$

$$= 20 + \left(\frac{100}{5}\right) \times 10$$

$$= 220\ m/s$$

627 **(d)**

The distance covered by a body moving with uniform acceleration is given by

$$s = ut + \frac{1}{2}at^2$$

As body starts from rest, therefore initial velocity $u = 0$

$\therefore$ Distance covered by the body

$$s = \frac{1}{2}at^2$$

Or $\quad s \propto t^2$

628 **(d)**

Given, $\mathbf{r} = 3t\hat{\mathbf{i}} - t^2\hat{\mathbf{j}} + 4\hat{\mathbf{k}}$

$$\therefore \quad \frac{d\mathbf{r}}{dt} = \mathbf{v} = 3\hat{\mathbf{i}} - 10\hat{\mathbf{j}}$$

At $t = 5$ s

$$\mathbf{v} = 3\hat{\mathbf{i}} - 10\hat{\mathbf{j}}$$

$$v = \sqrt{(3)^2 + (10)^2} = \sqrt{109} = 10.44\ \text{m/s}$$

629 **(c)**

Relativistic momentum $= \frac{m_0 v}{\sqrt{1-v^2/c^2}}$
If velocity is doubled then the relativistic mass also increases. Thus value of linear momentum will be more than double

630 **(a)**

If a body starts from rest with acceleration α and then retards with retardation β and comes to rest. The total time taken for this journey is t and distance covered is S

Then $S = \frac{1}{2}\frac{\alpha\beta t^2}{(\alpha + \beta)} = \frac{1}{2}\frac{5 \times 10}{(5 + 10)} \times t^2$

$\Rightarrow 1500 = \frac{1}{2}\frac{5 \times 10}{(5 + 10)} \times t^2 \Rightarrow t = 30\ sec$

631 **(d)**

$3t = \sqrt{3x} + 6 \Rightarrow 3x = (3t - 6)^2$
$\Rightarrow x = 3t^2 - 12t + 12$

$v = \frac{dx}{dt} = 6t - 12$, for $v = 0, t = 2$ sec
$x = 3(2)^2 - 12 \times 2 + 12 = 0$

632 **(a)**
According to problem
Distance travelled by body A in 5th *sec* and distance travelled by body B in3^{rd} *sec*. of its motion are equal.
$0 + \frac{a_1}{2}(2 \times 5 - 1) = 0 + \frac{a_2}{2}[2 \times 3 - 1]$
$9a_1 = 5a_2 \Rightarrow \frac{a_1}{a_2} = \frac{5}{9}$

633 **(d)**
In 's-t' graph (positive -time)
The straight line parallel with time axis represent state of rest

634 **(d)**
$x = ae^{-\alpha t} + be^{\beta t}$
Velocity $v = \frac{dx}{dt} = \frac{d}{dt}(ae^{-\alpha t} + be^{\beta t})$
$= a.e^{-\alpha t}(-\alpha) + be^{\beta t}(\beta)$
$= -a\alpha e^{-\alpha t} + b\beta e^{\beta t}$
Acceleration $= -a\alpha e^{-\alpha t}(-\alpha) + b\beta e^{bt}.\beta$
$= a\,\alpha^2 e^{-\alpha t} + b\beta^2 e^{\beta t}$
Acceleration is positive so velocity goes on increasing with time

635 **(a)**
$s \propto t^2$[Given] $\therefore s = Kt^2$
Acceleration $a = \frac{d^2s}{dt^2} = 2k$ [constant]
It means the particle travels with uniform acceleration

636 **(c)**
Height $= \frac{1}{2}(12 + 8)3.6$m$=36$m

637 **(d)**
Let acceleration is a and retardation is $-2a$

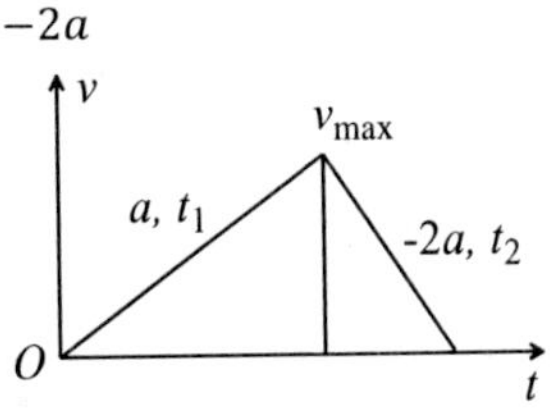

Then for acceleration motion
$t_1 = \frac{v}{a}$ ...(i)
For retarding motion
$t_2 = \frac{v}{2a}$ ...(ii)
Given $t_1 + t_2 = 9 \Rightarrow \frac{v}{a} + \frac{v}{2a} = 9 \Rightarrow \frac{3v}{2a} = 9 \Rightarrow \frac{v}{a} = 6$
Hence, duration of acceleration, $t_1 = \frac{v}{a} = 6$ *sec*

639 **(a)**
Let initial ($t = 0$) velocity of particle$= u$
For first 5 *sec* of motion $s_5 = 10$ *metre*
$s = ut + \frac{1}{2}at^2 \Rightarrow 10 = 5u + \frac{1}{2}a(5)^2$
$2u + 5a = 4$
For first 8 sec of motion $s_8 = 20$ *metre*
$20 = 8u + \frac{1}{2}a(8)^2 \Rightarrow 2u + 8a = 5$
By solving $u = \frac{7}{6}m/s$and $a = \frac{1}{3}m/s^2$
Now distance travelled by particle in Total 10 *sec*
$s_{10} = u \times 10 + \frac{1}{2}a(10)^2$
By substituting the value of u and a we will get $s_{10} = 28.3\ m$
so the distance in last 2 *sec* $= s_{10} - s_8$
$= 28.3 - 20 = 8.3m$

640 **(a)**
As the train are moving in the same direction. So the initial relative speed $(v_1 - v_2)$ and by applying retardation final relative speed becomes zero
From $v = u - at \Rightarrow 0 = (v_1 - v_2) - at \Rightarrow t = \frac{v_1 - v_2}{a}$

641 **(d)**
Let the body be projected upwards with velocity u from top of tower. Taking vertical downward motion of boy form top of tower to ground, we have
$u = -u, a = \text{g} = 10\text{ms}^{-2}, s = 50\text{m}, t = 10\text{s}$
As $s = ut + \frac{1}{2}at^2$,
So, $50 = -u \times 10 + \frac{1}{2} \times 10 \times 10^2$
On solving $u = 45\text{ms}^{-1}$
If t_1 and t_2 are the timings taken by the ball to reach points A and B respectively, then
$20 = 45t_1 + \frac{1}{2} \times 10 \times t_1^2$
and $40 = -45t_2 + \frac{1}{2} \times 10 \times t_2^2$
On solving, we get $t_1 = 9.4$ s and $t_2 = 9.8$s
Time taken to cover the distance AB
$= (t_2 - t_1) = 9.8 = 9.4 = 0.4$s.

642 **(d)**
Relative velocity
$= 10 + 5 = 15\ m/sec$
$\therefore t = \frac{150}{15} = 10\ sec$

643 **(c)**

$$S \propto t^2 \Rightarrow \frac{S_1}{S_2} = \left(\frac{10}{20}\right)^2 \Rightarrow S_2 = 4S_1$$

644 **(b)**

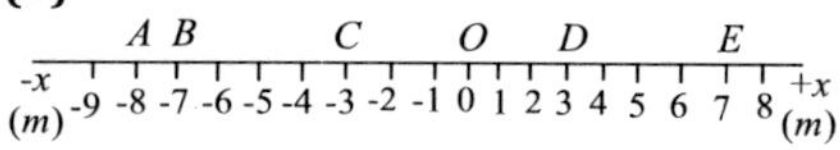

(i) The displacement of the main from A to E is $\Delta x = x_2 - x_1 = 7m - (-8m) = +15m$ directed in the positive x-direction
(ii) The displacement of the man from E to C is $\Delta x = -3m - (7m) = -10m$ directed in the negative x-direction
(iii) The displacement of the man from B to D is $\Delta x = 3m - (-7m) = +10m$ directed in the positive x-axis

645 **(d)**

$$\frac{v_A}{v_B} = \frac{\tan\theta_A}{\tan\theta_B} = \frac{\tan 30°}{\tan 60°} = \frac{1/\sqrt{3}}{\sqrt{3}} = \frac{1}{3}$$

646 **(d)**

As $v = 0 + na$

$$\Rightarrow \quad a = \frac{v}{n}$$

Now, distance travelled in n sec

$$\Rightarrow \quad S_n = \frac{1}{2}an^2$$

Distance travelled in $(n-2)$ sec

$$\Rightarrow \quad S_{n-2} = \frac{1}{2}a(n-2)^2$$

Distance travelled in last 2 s,

$$S_n - S_{n-2} = \frac{1}{2}an^2 - \frac{1}{2}a(n-2)^2$$
$$= \frac{a}{2}[n^2 - (n-2)^2]$$
$$= \frac{a}{2}[n + (n-2)][n - (n-2)]$$
$$= a(2n-2) = \frac{v}{n}(2n-2)$$
$$= \frac{2v(n-1)}{n}$$

647 **(a)**

$s \propto t^2$[Given] $\therefore s = Kt^2$
Acceleration $a = \frac{d^2s}{dt^2} = 2k$ [constant]
It means the particle travels with uniform acceleration

648 **(b)**

$$S_{3^{rd}} = 10 + \frac{10}{2}(2 \times 3 - 1) = 35\ m$$
$$S_{2^{nd}} = 10 + \frac{10}{2}(2 \times 2 - 1) = 25\ m$$
$$\Rightarrow \frac{S_{3^{rd}}}{S_{2^{nd}}} = \frac{7}{5}$$

649 **(b)**

Velocity at 3s = total algebraic sum of area under the curve

$$v = 4 \times 2 - 4 \times 1 = 4\ \text{m/s}$$

650 **(c)**

If the body starts from rest and moves with constant acceleration then the ratio of distances in consecutive equal time interval $S_1 : S_2 : S_3 = 1:3:5$

651 **(b)**

$\vec{v} = \vec{u} + \vec{a}t \Rightarrow \vec{v} = 3\hat{i} + 4\hat{j} + (0.4\hat{i} + 0.3\hat{j}) \times 10$
$v = 3\hat{i} + 4\hat{j} + 4\hat{i} + 3\hat{j} = 7\hat{i} + 7\hat{j}$
$\Rightarrow |\vec{v}| = 7\sqrt{2}$ units

652 **(c)**

$v^2 = u^2 + 2as$, If $u = 0$, then $v^2 \propto S$
i.e., Graph should be parabola symmetric to displacement axis

653 **(d)**

We have standard result for this type of velocity

$$v = \sqrt{\frac{u^2 + v^2}{2}}$$

654 **(b)**

For stone to be dropped from rising balloon of velocity 29 m/s
$u = -29\ m/s, t = 10 sec$
$\therefore h = -29 \times 10 + \frac{1}{2} \times 9.8 \times 100$
$= -290 + 490 = 200\ m$

655 **(d)**

Net displacement = 0 and total distance $= OP + PQ + QO$

$$= 1 + \frac{2\pi \times 1}{4} + 1 = \frac{14.28}{4}\ km$$

Average speed $= \frac{14.28}{4 \times 10/60}$

$$= \frac{6 \times 14.28}{4} = 21.42\ km/h$$

656 **(b)**

Let s_1 be the distance travelled by train with acceleration 1ms^{-2} for time t_1 and s_2 be the distance travelled by train with retardation $3\ \text{ms}^{-2}$ for time t_2. If v is the velocity of train after time t_1, then

$$v = 1 \times t_1 \quad \text{... (i)}$$

And $s_1 = \frac{1}{2} \times 1 \times t_1^2 = \frac{t_1^2}{2}$ ... (ii)

$$v = 3t_2 \quad \text{... (iii)}$$

And $s_2 = vt_2 - \frac{1}{2} \times 3 \times t_2^2$

From Eqs. (i) and (iii),

$$t_1 = 3t_2 \text{or } t_2 = \frac{t_1}{3}$$

$$\therefore s_1 + s_2 = \frac{t_1^2}{2} + t_1 \times \frac{t_1}{3} - \frac{3}{2} \times \frac{t_1^2}{9}$$

$$= \frac{2}{3} t_1^2$$

$$1215 = \frac{2}{3} t_1^2$$

$$t_1 = \sqrt{\frac{3 \times 1215}{2}} = 42.69\ \text{s}$$

Total time $= t_1 + t_2 = t_1 + \frac{t_1}{3} = 56.9$ s

657 **(b)**

$$h = \frac{1}{2} g t^2$$

$$h' = \frac{1}{2} g (t - t_0)^2$$

$$h - h' = \frac{1}{2} g [t^2 - (t - t_0)^2]$$

$$= \frac{1}{2} g [t^2 - t^2 - t_0^2 + 2tt_0]$$

$$\Delta h = \frac{1}{2} g t_0 (2t - t_0)$$

Δh is increasing with time

658 **(d)**

This problem can be solved by using the concept of relative velocity

$v + 30 = 75$

or $v = 45\text{kmh}^{-1}$

659 **(c)**

Maximum height of ball $= 5\ m$

So velocity of projection $\Rightarrow u = \sqrt{2gh} = 10\ m/s$

Time interval between two balls (time of ascent)

$$= \frac{u}{g} = 1\ sec = \frac{1}{60}\ min$$

So number of ball thrown per min = 60

660 **(d)**

Because acceleration due to gravity is constant so the slope of line will be constant *i.e.*, velocity time curve for a body projected vertically upwards is straight line

661 **(b)**

The distance traveled in last second

$$S_{\text{Last}} = u + \frac{g}{2}(2t - 1) = \frac{1}{2} \times 9.8(2t - 1)$$
$$= 4.9(2t - 1)$$

and distance traveled in first three second,

$$S_{\text{Three}} = 0 + \frac{1}{2} \times 9.8 \times 9 = 44.1\ m$$

According to problem $S_{\text{Last}} = S_{\text{Three}}$

$\Rightarrow 4.9(2t - 1) = 44.1 \Rightarrow 2t - 1 = 9$

$\Rightarrow t = 5\ sec$

662 **(c)**

Since displacement is always less than or equal to distance, but never greater than distance. Hence numerical ratio of displacement to the distance covered is always equal to or less than one

663 **(b)**

Given, $a = \frac{dv}{dt} = 6t + 5$

Integrating it $\int_0^v dv = \int_0^t (6t + 5) dt$

$$v = \frac{6t^2}{2} + 5t$$

As $v = \frac{ds}{dt}$, so $ds = \left(\frac{6t^2}{2} + 5t\right) dt$

$$s = 3\frac{t^3}{3} + \frac{5t^2}{2}$$

Where $t = 2\text{s}, s = 3 \times \frac{2^3}{3} + \frac{5 \times 2^2}{2} = 18\text{m}$

664 **(b)**

$$S_2 = \frac{1}{2} g t_2^2 = \frac{10}{2} \times (3)^2 = 45\ m$$

$$S_1 = \frac{1}{2}gt_1^2 = \frac{10}{2} \times (5)^2 = 125\ m$$

$\therefore S_1 - S_2 = 125 - 45 = 80\ m$

666 **(b)**

$$\text{Time} = \frac{\text{distance}}{\text{average speed}}$$

$\Rightarrow$ Average speed

$$= \frac{\text{distance}}{\text{time}}$$

$$v = \frac{130 + 120 + 50}{3}$$

$$= \frac{300}{3} = 100\ \text{kmh}^{-1}$$

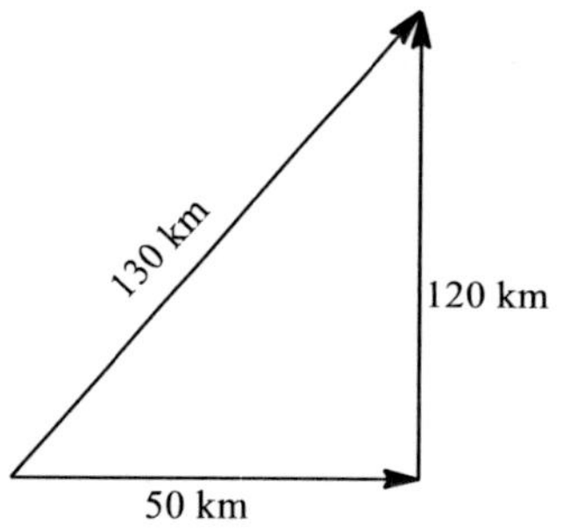

667 **(b)**

Here $v = 144\ km/h = 40m/s$

$v = u + at \Rightarrow 40 = 0 + 20 \times a \Rightarrow a = 2\ m/s^2$

$\therefore s = \frac{1}{2}at^2 = \frac{1}{2} \times 2 \times (20)^2 = 400\ m$

668 **(d)**

An aeroplane flies 400m north and 300m south so the net displacement is 100m towards north.

Then it flies 1200m upward so $r = \sqrt{(100)^2 + (1200)^2}$

$= 1204\ \text{m} \simeq 1200\ \text{m}$

669 **(b)**

At maximum height velocity $v = 0$

We know that $v = u + at$, hence

$0 = u - gT \Rightarrow u = gT$

When $v = \frac{u}{2}$, then

$$\frac{u}{2} = u - gt \Rightarrow gt = \frac{u}{2} \Rightarrow gt = \frac{gT}{2} \Rightarrow t = \frac{T}{2}$$

Hence at $t = \frac{T}{2}$, it acquires velocity $\frac{u}{2}$

670 **(b)**

$v = g \times t = 32 \times 1 = 32ft/sec$

671 **(c)**

As $v^2 = v^2 - 2as \Rightarrow u^2 = 2as\ \ (\because v = 0)$

$$\Rightarrow u^2 \propto s \Rightarrow \frac{u_2}{u_1} = \left(\frac{s_2}{s_1}\right)^{1/2}$$

$$\Rightarrow u_2 = \left(\frac{9}{4}\right)^{\frac{1}{2}} u_1 = \frac{3}{2}u_1 = 300m/s$$

672 **(d)**

$u = 0, S = 250m, t = 10\text{sec}$

$$S = ut + \frac{1}{2}at^2 \Rightarrow 250 = \frac{1}{2}a[10]^2 \Rightarrow a = 5m/s^2$$

So, $F = ma = 0.9 \times 5 = 4.5N$

673 **(a)**

$$v = \frac{dx}{dt} = \frac{d}{dt}\left[\frac{k}{b}(1 - e^{-bt})\right] = \frac{k}{b}[0 - (-b)e^{-bt}] = ke^{-bt}$$

674 **(b)**

For downward motion.

$v^2 = u^2 + 2gh$

For upward motion,

$v^2 = u^2 - 2gh$

The graph will be parabolic in both case. But velocities will be opposite. But speed never be negative.

Hence, (b) is correct.

675 **(b)**

$$\text{Time of flight} = \frac{2\text{u}}{g} = \frac{2 \times 100}{10} = 20\ \text{sec}$$

676 **(b)**

Free fall of an object (in vacuum) is a case of motion with uniform acceleration

678 **(d)**

Let height of tower is h and body takes t time to reach to ground when it fall freely

$$\therefore h = \frac{1}{2}gt^2$$

In last second i.e. $t^{th}\ sec$ body travels $= 0.36\ h$

It means in rest of the time i.e. in $(t - 1)sec$ it travels

$= h - 0.36\ h = 0.64\ h$

Now applying equation of motion for $(t-1)sec$

$0.64\,h = \frac{1}{2}g(t-t)^2$

From (i) and (ii) we get, $t = 5$ sec and $h = 125m$

679 **(b)**

Let t_1, t_2 and t_3 be the time taken by the particles to cover the distance $2x$, $4x$ and $6x$ respectively. Let v be the velocity of the particle at Bie, maximum velocity. The particle moves with uniform acceleration from A to B.

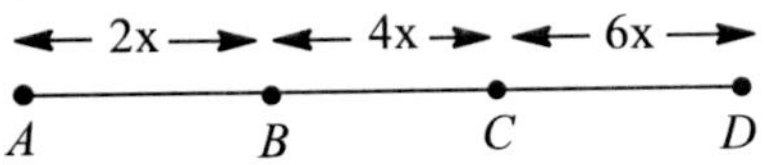

Acceleration a_y uniform motion Retardation a_y

For motion for A to B.

Average velocity $= \frac{0+v}{2} = \frac{v}{2}$

Time taken, $t_1 = \frac{2x}{v/2} = \frac{4x}{v}$

Particle moves with uniform retardation from C to D.

Time taken , $t_3 = \frac{6x}{(0+v)/2} = \frac{12x}{v}$

Total time $= t_1 + t_2 + t_3$

$= \frac{4x}{v} + \frac{4x}{v} + \frac{12x}{v} = \frac{20x}{v}$

$v_{av} = \frac{2x+4x+6x}{20x/v} = \frac{12v}{20}$

or $\frac{v}{v} = \frac{12}{20} = \frac{3}{5}$.

680 **(b)**

$v = u + \int a dt = u + \int (3t^2 + 2t + 2)dt$

$= u + \frac{3t^3}{3} + \frac{2t^2}{2} + 2t = u + t^3 + t^2 + 2t$

$= 2 + 8 + 4 + 4 = 18\,m/s$ (As $t = 2\,sec$)

681 **(c)**

Displacement $x = at + bt^2 - ct^3$

Velocity, $v = \frac{dx}{dt} = a + 2bt - 3ct^2$ ... (i)

And acceleration, $a = \frac{a^2x}{dt^2} = 2b - 6ct$ ... (ii)

When acceleration $a = 0, t = \frac{b}{3c}$

Substituting the value of t in Eq. (i), we get

$$v = a + \frac{b^2}{3c}$$

682 **(b)**

$$\frac{(S)_{(last\ 2s)}}{(S)_{7s}} = \frac{\frac{1}{2} \times 2 \times 10}{\frac{1}{2} \times 2 \times 10 + 2 \times 10 + \frac{1}{2} \times 2 \times 10}$$

$$= \frac{1}{4}$$

683 **(d)**

Interval of ball throw $= 2sec$

If we want that minimum three (more than two) ball remain in air then time of flight of first ball must be greater than $4\,sec$

$T > 4\,sec$

$\frac{2u}{g} > 4\text{ sec} \Rightarrow u > 19.6\,m/s$

For $u = 19.6$, first ball will just about to strike the ground (in air)

Second ball will be at highest point (in air)

Third ball will be at point of projection or at ground (not in air)

684 **(d)**

By the time 5th water drop starts falling, the first water drop reaches the ground.

As $u = 0, h = \frac{1}{2}gt^2 = \frac{1}{2} \times 10 \times t^2$

or $5 = \frac{1}{2} \times 10 \times t^2$ or $t = 1$ s

Hence, the interval of each water drop $= \frac{1s}{4} = 0.25$s.

When the 5th drop starts its journey towards ground, the third drop travels in air for

$t_1 = 0.25 + 0.25 = 0.5$s

$\therefore$ Height (distance) covered by 3rd drop in air is

$h_1 = \frac{1}{2}gt_1^2 = \frac{1}{2} \times 10 \times (0.5)^2$

$= 5 \times 0.25 = 1.25$m

So, third water drop will be at a height of $= 5 - 1.25 = 3.75$m

685 **(b)**

$$S_{3^{\text{rd}}} = 10 + \frac{10}{2}(2 \times 3 - 1) = 35\, m$$

$$S_{2^{\text{nd}}} = 10 + \frac{10}{2}(2 \times 2 - 1) = 25\, m$$

$$\Rightarrow \frac{S_{3^{\text{rd}}}}{S_{2^{\text{nd}}}} = \frac{7}{5}$$

686 **(c)**

$$v^2 = u^2 + 2as$$

$$0 = \left(50 \times \frac{5}{18}\right)^2 + 2a \times 6$$

$$a = -16\ \text{ms}^{-2}(a = \text{retardation})$$

Again, $v^2 = u^2 + 2as$

$$0 = \left(100 \times \frac{5^2}{18}\right) - 16 \times 2 \times s$$

$$s = \frac{(100\ \times 5)^2}{18 \times 18 \times 32} = 24.1 \approx 24\ \text{m}$$

687 **(c)**

Since displacement is always less than or equal to distance, but never greater than distance. Hence numerical ratio of displacement to the distance covered is always equal to or less than one

688 **(a)**

An aeroplane files 400 m north and 300 m south so the net displacement is 100 m towards north

Then it files 1200 m upwards so $r = \sqrt{(100)^2 + (1200)^2}$

$= 1204\, m \simeq 1200\, m$

The option should be 1204 m, because this value mislead one into thinking that net displacement is in upward direction only

689 **(a)**

Average speed for other half of distance

$$= \frac{4.5 + 7.5}{2}\ \text{ms}^{-1} = 6\text{ms}^{-1}$$

Average speed during whole motion

$$= \frac{2 \times 3 \times 6}{3 + 6}\ \text{ms}^{-1} = 4\text{ms}^{-1}$$

690 **(a)**

A particle starts from rest at $t = 0$

The equation of motion

$$v = u + at = 0 + 3 \times 2 = 6\ \text{ms}^{-1}$$

The velocity for next 2 s

$$v' = v + at$$

$$= 6 - 3 \times 2 = 0$$

$$v' = 0$$

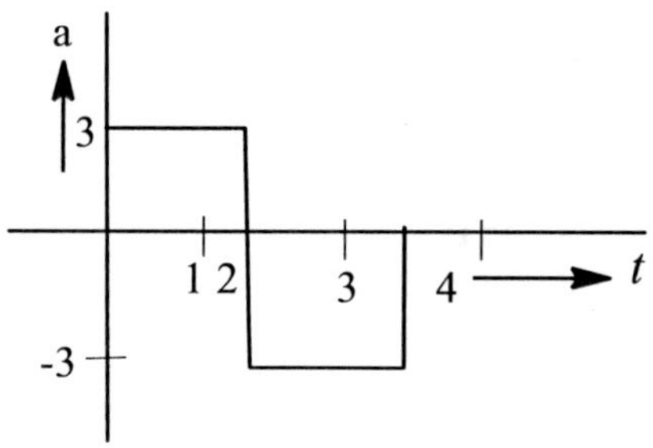

Hence, $v - t$ graph will be as shown.

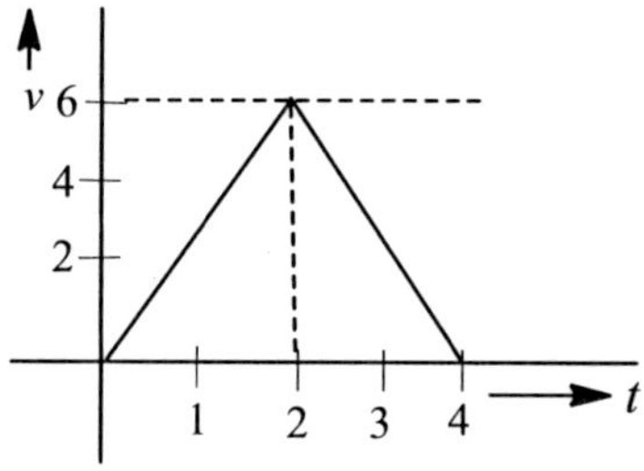

691 **(c)**

Let car starts from point A from rest moves up to point B with acceleration f

Velocity of car at point B, $v = \sqrt{2fS}$

$[\text{As}\, v^2 = u^2 + 2as]$

Car moves distance BC with this constant velocity in time t

$x = \sqrt{2fS}.t$ [As $s = ut$] ... (i)

So the velocity of car at point C also will be $\sqrt{2fS}$ and finally car stops after covering distance y

Distance $CD \Rightarrow y = \frac{(\sqrt{2fS})^2}{2(f/2)} = \frac{2fS}{f} = 2S$ ...(ii)

So, the total distance $AD = AB + BC + CD = 15S$ [Given]

$\Rightarrow S + x + 2S = 15\,S \Rightarrow x = 12S$

Substituting the value of x in equation (i) we get

$x = \sqrt{2fS}.t \Rightarrow 12S = \sqrt{2fS}.t \Rightarrow 144S^2 = 2fS.t^2$

$\Rightarrow S = \frac{1}{72} ft^2$

692 **(a)**

$H_{max} = \frac{u^2}{2g} \Rightarrow H_{max} \propto \frac{1}{g}$

On planet B value of g is 1/9 times to that of A. So value of H_{max} will become 9 times $i.e.\, 2 \times 9 = 18\, metre$

693 **(c)**

Let the man starts crossing the road at an angle θ with the roadside. For safe crossing, the condition is that the man must cross the road by the time truck describes the distance(4+2 cot θ),

So, $\frac{4+2\cot\theta}{8} = \frac{2l\sin\theta}{v}$

or $v = \frac{8}{2\sin\theta=2\cos\theta}$

For minimum $v = \frac{dv}{d\theta} = 0$

or $\frac{-8(2\cos\theta-\sin\theta)}{(2\sin\theta+\cos\theta)^2} = 0$

or $2\cos\theta - \sin\theta = 0$

or $\tan\theta = 2$, so, $\sin\theta = \frac{2}{\sqrt{5}}, \cos\theta = \frac{1}{\sqrt{5}}$

$\therefore\ v_{min} = \frac{6}{2\left(\frac{2}{\sqrt{5}}\right) + \frac{1}{\sqrt{5}}}$

$= \frac{8}{\sqrt{5}} = 3.57\text{ms}^{-1}$

694 **(d)**

$V_{av} = \frac{S+S}{\frac{S}{v_1} + \frac{S}{v_2}} = \frac{2v_1v_2}{v_1+v_2}$

695 **(c)**

Given, $v = pt$

$\int_0^X dx = p\int_0^2 t\,dt$

$= \frac{pt^2}{2} = \frac{4\times4}{2} = 8\text{m}$

696 **(d)**

$v = \sqrt{2\,gh}$ ... (i)

After rebounce, $v^2 = u^2 - 2gh$

Or $u^2 = v^2 + 2\,gh'$

$\therefore\quad u^2 = 2\,gh'$ ... (ii)

$\therefore\quad \frac{v^2}{u^2} = \frac{2gh}{2gh'}$

Or $h' = h \times \frac{u^2}{v^2}$

$= h \times \left(\frac{80}{100}\right)^2$

$= 0.64\,h$

697 **(a)**

For first marble, $h_1 = \frac{1}{2}g \times 16 = 8g$

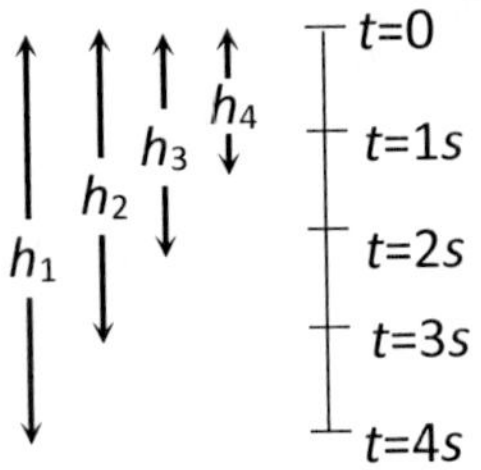

For second marble, $h_2 = \frac{1}{2}g \times 9 = 4.5g$

For third marble, $h_3 = \frac{1}{2}g \times 4 = 2g$

For fourth marble, $h_4 = \frac{1}{2}g \times 1 = 0.5g$

$\therefore h_1 - h_2 = 8g - 4.5g = 3.5g = 35m$.

$h_2 - h_3 = 4.5g - 2g = 2.5g = 25m$ and

$h_3 - h_4 = 2g - 0.5g = 1.5g = 15m$

698 **(c)**

Given, $3s = 9t + 5t^2$ or $s = \frac{1}{3}(9t + 5t^2)$

Velocity $v = \frac{ds}{dt} = \frac{1}{3}(9 + 10t)$

Acceleration $a = \frac{d}{dt}\left(\frac{ds}{dt}\right) = \frac{d^2s}{dt^2} = \frac{10}{3}$ units.

699 **(c)**

Given, $v = (180 - 16x)^{1/2}$

Or $v^2 = 180 - 16x$

Differentiating with respect to t, we get

$$2v\frac{dv}{dt} = 0 - 16\frac{dx}{dt}$$

$$2v\frac{dv}{dt} = -16v$$

$$\Rightarrow \quad \frac{dv}{dt} = -8$$

Hence, particle decelerates at the rate of 8 ms^{-2}.

700 **(b)**

We will solve the problem is terms of relative initial velocity, relative acceleration and relative displacement of the coin with respect to the floor of the lift.

$u = 10 - 10 = 0\text{ms}^{-1}, a = 9.8\text{ms}^{-2}, s = 4.9\text{m}, t$=?

$4.9 = 0 \times t + \frac{1}{2} \times 9.8 \times t^2$

or $4.9t^2 = 4.9$ or $t = 1\text{s}$

701 **(d)**

From first equation of motion, we have

$$v = u + at$$

Given, $u = 0, a_1 = 2\text{ ms}^{-2}$

$$t = 10\text{ s},$$

$\therefore \quad v_1 = 2 \times 10 = 20\text{ ms}^{-1}$

In the next 30 s, the constant velocity becomes

$v_2 = v_1 + a_2 t_2$

Given, $v_1 = 20\text{ ms}^{-1}, a_2 = 2\text{ ms}^{-2}, t_2 = 30\text{ s}$

$$\therefore \quad v_2 = 20 + 2 \times 30 = 80\text{ ms}^{-1}.$$

When it decelerates, then

$v_3^2 = u^2 - 2a_3 s$

Here, $v_3 = 0$ (train stops), $v_2 = 80\text{ ms}^{-1}$,

$a_3 = 4\text{ ms}^{-2}$

$$0 = (80)^2 - 2 \times 4 \times s$$

Or $\quad s = \frac{80 \times 80}{8} = 800\text{ m}.$

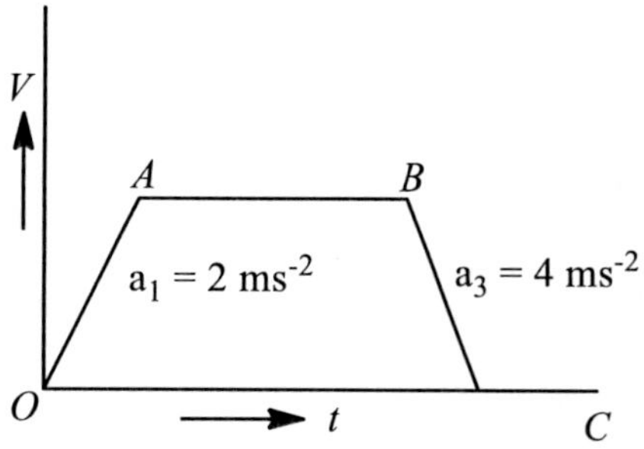

702 **(d)**

$A \Rightarrow \frac{u^2}{4} - u^2 = -2gh_1$

$B \Rightarrow \frac{u^2}{9} - u^2 = -2gh_2$

$C \Rightarrow \frac{u^2}{16} - u^2 = -2gh_3$

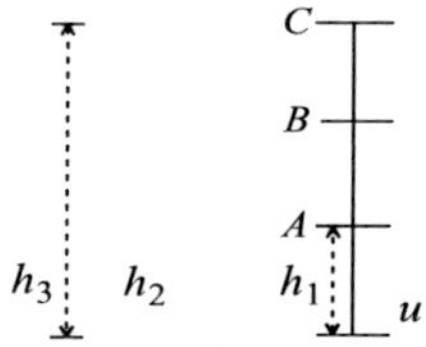

$\therefore AB = \frac{u^2}{2g}\left\{\frac{8}{9} - \frac{3}{4}\right\} = \frac{u^2}{2g} \cdot \frac{5}{36}$

$BC = \frac{u^2}{2g}\left\{\frac{15}{16} - \frac{8}{9}\right\} = \frac{u^2}{2g} \cdot \frac{7}{144}$

$\therefore \frac{AB}{BC} = \frac{5}{36} \times \frac{144}{7} = \frac{20}{7}$

703 **(d)**

For the round trip he should cross perpendicular to the river

$\therefore$ Time for trip to that side $= \frac{1km}{4km/hr} = 0.25hr$

To come back, again he take 0.25 hr to cross the river. Total time is 30 min, he goes to the other bank and come back at the same point

704 **(d)**

$3t = \sqrt{3x} + 6 \Rightarrow 3x = (3t - 6)^2$

$\Rightarrow x = 3t^2 - 12t + 12$

$v = \frac{dx}{dt} = 6t - 12$, for $v = 0, t = 2$ sec

$x = 3(2)^2 - 12 \times 2 + 12 = 0$

705 **(a)**

Time taken by the car to cover first half of the distance is

$t_1 = \frac{100}{60}$

Time taken by the car to cover speed half of the distance is

$t_2 = \frac{100}{v}$

Average speed , $v_{av} = \frac{\text{Total distance travelled}}{\text{TOtal time taken}}$

$v_{av} = \frac{100 + 100}{t_1 + t_2} \Rightarrow 40 = \frac{200}{\frac{100}{60} + \frac{100}{v}}$

$\frac{1}{60} + \frac{1}{v} = \frac{1}{20} \Rightarrow \frac{1}{v} = \frac{1}{20} - \frac{1}{60}$

$\frac{1}{v} = \frac{2}{60} = \frac{1}{30}$

$v = 30km\ h^{-1}$

706 **(b)**

Relative velocity of bird $w.r.t.$ train $=$ $25 + 5 = 30\ m/s$

Time taken by the bird to cross the train

$t = \frac{210}{30} = 7\ sec$

707 **(c)**

$y = a + bt + ct^2 - dt^4$

$\therefore v = \frac{dy}{dt} = b + 2ct - 4dt^3$ and $a = \frac{dv}{dt} = 2c - 12dt^2$

Hence, at $t = 0, v_{\text{initial}} = b$ and $a_{\text{initial}} = 2c$

709 **(c)**

Assume that the motion is along the positive direction of x-axis. For simplicity, let us take the beginning of the braking to be a t time $t = 0$, at position x_0

Therefore, $x - x_0 = v_0\left(\frac{v - v_0}{a}\right) + \frac{1}{2}a\left(\frac{v - v_0}{a}\right)^2$

Solving for a and substituting known data then yield

$a = \frac{v^2 - v_0}{2(x - x_0)}$

Here, $v_0 = 100\text{kmh}^{-1} = 27.78\text{ms}^{-1}, x - x_0 = 88.0$

And $v = 80\text{kmh}^{-1} = 22.22\text{m}^{-1}$

$\therefore\ a = \frac{(22.22)^2 - (27.78)^2}{2(88.0)}$

$= -1.58\text{ms}^{-2}$

710 **(a)**

Here, $u = 0, a = g$

Distance travelled in n^{th} second is given by

$D_n = u + \frac{a}{2}(2n - 1)\ \therefore D_n \propto (2n - 1)$

$\therefore D_1 : D_2 : D_3 : D_4 : D_5 \ldots = 1 : 3 : 5 : 7 : 9 : \ldots$

711 **(a)**

$v = u + at \Rightarrow v = 0 + 5 \times 10 = 50\ m/s$

712 **(a)**

During upward motion the velocity is decreasing while during downward motion the velocity is increasing in downward direction.

The graph plot is as shown.

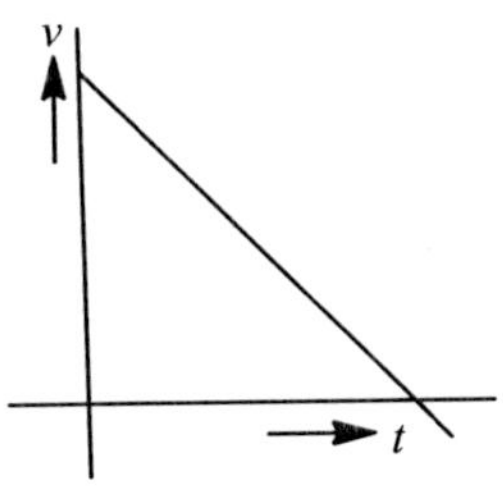

713 **(d)**

Since, the initial position of cyclist coincides with final position, so his net displacement is zero.

Average speed $= \frac{\text{total distance travelled}}{\text{total time taken}}$

$= \frac{OP + PQ + QO}{10}\text{km min}^{-1}$

$$= \frac{1 + \frac{\pi}{2} \times 1 + 1}{10} \text{km min}^{-1}$$

$$= \frac{\pi + 4}{20} \times 60 \text{ kmh}^{-1}$$
$$= 21.4 \text{ kmh}^{-1}$$

714 **(a)**

From equation of motion, we have

$$v = u + gt$$

Where u is initial velocity and t is time.

For ball $A, v_A = u - gt$

For ball $B, u_B = 0 + gt$

$\therefore$ Relative speed, $\Delta\mathbf{v} = \mathbf{v}_A - \mathbf{v}_B$

$$= (u - gt)\hat{\mathbf{j}} - (-gt\hat{\mathbf{j}}) = u\hat{\mathbf{j}}$$

715 **(a)**

$$\frac{dv}{dt} = -2.5\sqrt{v}$$

$$\Rightarrow \frac{dv}{\sqrt{v}} = -2.5dt$$

$$\Rightarrow \int_{6.25}^{0} v^{-1/2}\, dv = -2.5 \int_0^t dt$$

$$\Rightarrow -2.5[t]_0^t = \left[2v^{1/2}\right]_{6.25}^{0}$$

$$\Rightarrow t = 2 \text{ s}$$

716 **(b)**

Given acceleration $a = 6t + 5$

$\therefore a = \frac{dv}{dt} = 6t + 5, dv = (6t + 5)dt$

Integrating it, we have $\int_0^v dv = \int_0^t (6t + 5)dt$

$v = 3t^2 + 5t + C$, where C is a constant of integration

When $t = 0, v = 0$ so $C = 0$

$\therefore v = \frac{ds}{dt} = 3t^2 + 5t$ or, $ds = (3t^2 + 5t)dt$

Integrating it within the conditions of motion, *i. e.*, as t changes from 0 to $2s, s$ changes from 0 to s, we have

$$\int_0^s ds = \int_0^2 (3t^2 + 5t)dt$$

$$\therefore s = \left[t^3 + \frac{5}{2}t^2\right]_0^2 = 8 + 10 = 18\, m$$

717 **(b)**

$$2 = \sqrt{\frac{2\times19.6}{g}} \text{ or } g = \frac{2\times19.6}{4} = 9.8\text{ms}^{-2}$$

718 **(b)**

$$\text{Time} = \frac{\text{Total length}}{\text{Relative velocity}} = \frac{50 + 50}{10 + 15}$$
$$= \frac{100}{25} = 4\, sec$$

719 **(c)**

Let v_1, v_2 be the initial speeds of first and second runners. Let t be time by them when the first runner has completed 50m. During this time, the second runner has covered a distance = 50 − 1 = 49m.

So, $t = \frac{50}{v_1} = \frac{49}{v_2}$ ...(i)

Suppose, the second runner increases his speed to v_3 so that he covers the remaining distance (= 51m) in times t. So

$$t = \frac{51}{v_3} = \frac{49}{v_2}$$

or $v_3 = \frac{51}{49}v_2$

or $v_3 = \left(1 + \frac{2}{49}\right)v_2$ or $\frac{v_3}{v_2} - 1 = \frac{2}{49}$

or $\frac{v_3 - v_2}{v_2} = \frac{2}{49}$

or % increase $= \frac{2}{49} \times 100 = 4.1\%$

720 **(a)**

Height reached $= \frac{1}{2} \times 132 \times 1200\text{m} = 66 \times 1200\text{m}$

721 **(b)**

$x = 9t^2 - t^3; v = \frac{dx}{dt} = 18t - 3t^2$, For maximum speed

$$\frac{dv}{dt} = \frac{d}{dt}[18t - 3t^2] = 0 \Rightarrow 18 - 6t = 0$$
$$\therefore t = 3 \text{ sec}$$

i.e., Particle achieve maximum speed at $t = 3$ sec. At this instant position of this particle, $x = 9t^2 - t^3$

$= 9(3)^2 - (3)^3 = 81 - 27 = 54\, m$

722 **(a)**

$$b - a = un + \frac{1}{2}An^2$$
$$2b - 2a = 2un + An^2$$

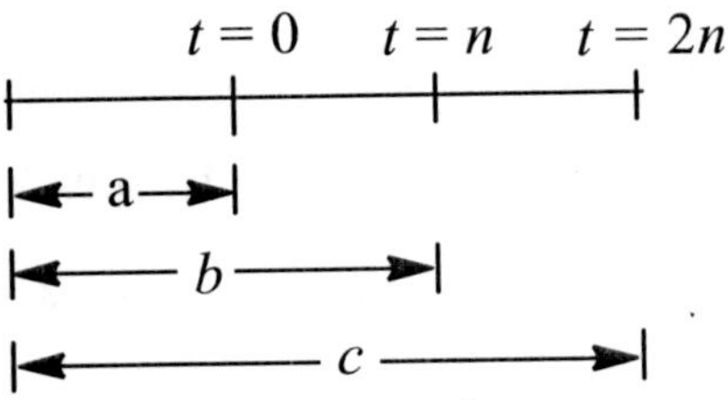

Again, $c - a = u(2n) + \frac{1}{2}A(2n)^2$

Subtracting, $c - a - 2b + 2a = An^2$

$A = \dfrac{c - 2b + a}{n^2}$

723 **(d)**

For the round trip he should cross perpendicular to the river

$\therefore$ Time for trip to that side $= \frac{1km}{4km/hr} =$ $0.25hr$

To come back, again he take 0.25 hr to cross the river. Total time is 30 min, he goes to the other bank and come back at the same point

724 **(a)**

$s = 1.2\ m$

$v = 640\ ms^{-1}$

$a = ?; u = 0;\ t = ?$

$2\ as = v^2 - u^2$

$\Rightarrow 2a \times 1.2 = 640 \times 640 \Rightarrow a$

$= \dfrac{8 \times 64 \times 10^3}{3}$

$v = u + at\ \Rightarrow t = \dfrac{v}{a} = \dfrac{15}{4} \times 10^{-3}$

$= 3.75 \times 10^{-3}\ s$

$\approx 4\ ms$

725 **(d)**

The horizontal acceleration a of the wedge should be such that in time the wedge moves the horizontal distance BC. The body must fall through a vertical distance AB under gravity. Hence,

$BC = \frac{1}{2}at^2$ and $AB = \frac{1}{2}gt^2$

$\tan\theta = \frac{AB}{BC} = \frac{g}{a}$or $a = \frac{g}{\tan\theta} = g\cot\theta$

726 **(b)**

$$t = \sqrt{\frac{2h}{g}} \Rightarrow \frac{t_1}{t_2} = \sqrt{\frac{h_1}{h_2}}$$

727 **(c)**

$h = \frac{1}{2}gT^2, h' = \frac{1}{2}g\left(\frac{T}{2}\right)^2\ h' = \frac{h}{4},$

Height above the ground $= h - \frac{h}{4} = \frac{3h}{4}$

728 **(c)**

Instantaneous velocity of running mass after t sec will be

$$v_t = \sqrt{v_x^2 + v_y^2}$$

Where, $v_x = v\sin\theta - gt$ = vertical component of velocity,

$v_y = v\cos\theta$ =horizontal component of velocity

$$v_t = \sqrt{(v\cos\theta)^2 + (v\sin\theta - gt)^2}$$

$$v_t = \sqrt{v^2 + g^2t^2 - 2v\sin\theta\, gt}$$

729 **(b)**

$\Delta t = \sqrt{\frac{2\times19.6}{9.8}}s - \sqrt{\frac{2\times18.6}{9.8}}s = 2 - 1.95 =$ $0.05s$

730 **(c)**

Because acceleration is a vector quantity

731 **(c)**

$\mathbf{r}_i = (-3.0\ \text{m})\hat{\mathbf{i}} + (2.0\ \text{m})\hat{\mathbf{j}} + (8.0\ \text{m})\hat{\mathbf{k}}$

$r_f = (9.0\ \text{m})\hat{\mathbf{i}} + (2.0\ \text{m})\hat{\mathbf{j}} + (-8.0\ \text{m})\hat{\mathbf{k}}$

$\therefore$Displacement $= \boldsymbol{r}_f - \boldsymbol{r}_i$

$= \left[(9.0\ \text{m})\hat{\mathbf{i}} + (2.0\ \text{m})\hat{\mathbf{j}} + (-8.0\ \text{m})\hat{\mathbf{k}}\right]$

$-[(-3.0\text{m})\hat{\mathbf{i}} + (2.0\text{m})\hat{\mathbf{j}} + (8.0\text{m})\hat{\mathbf{k}}]$

$= [(12.0\text{m})\hat{\mathbf{i}} - (16.0\text{m})\hat{\mathbf{j}}]$

732 **(b)**

Let car B catches, car A after $'t'$sec, then

$60t + 2.5 = 70t - \dfrac{1}{2} \times 20 \times t^2$

$\Rightarrow 10t^2 - 10t + 2.5 = 0 \Rightarrow t^2 - t + 0.25 = 0$

$\therefore t = \dfrac{\sqrt{1 - 4 \times (0.25)}}{2} = \dfrac{1}{2}hr$

733 **(a)**

$S_n = u + \frac{a}{2}(2n - 1) = \frac{a}{2}(2n - 1)$ because $u = 0$

Hence $\frac{S_4}{S_3} = \frac{7}{5}$

734 **(a)**

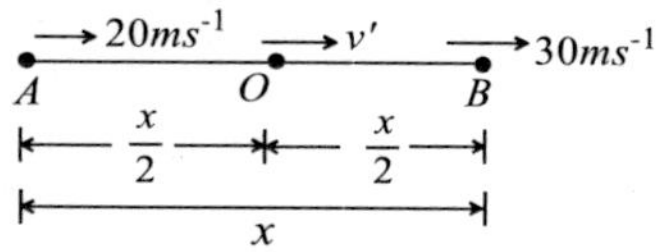

$\because v'' = \sqrt{\frac{(v_1)^2 + (v_2)^2}{2}} = \sqrt{\frac{900 + 400}{2}}$

$= \sqrt{650} = 25.5\ ms^{-1}$

736 **(b)**

Area under the velocity–time curve over a given time interval gives the displacement of the particle

737 **(a)**

When two particles moves towards each other then

$v_1 - v_2 = 6$...(i)

When these particles moves in the same direction then

$v_1 - v_2 = 4$...(ii)

By solving $v_1 = 5$ and $v_2 = 1\ m/s$

738 **(b)**

Let the particle moves toward right with velocity $6\ m/s$. Due to retardation after time $t_1$its velocity becomes zero

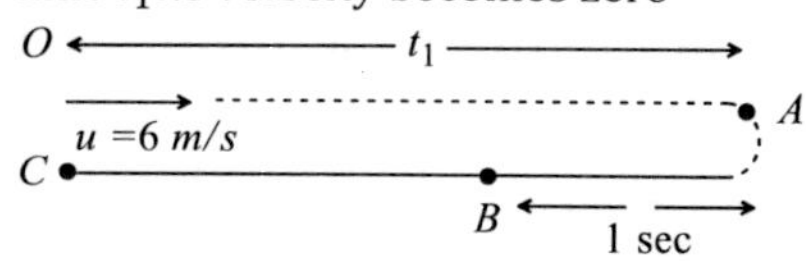

From $v = u - at \Rightarrow 0 = 6 - 2t_1 \Rightarrow t_1 = 3sec$

But retardation work on it for 4 sec.It means after reaching point A direction of motion get reversed and acceleration works on the particle for next one second.

$S_{OA} = ut_1 - \frac{1}{2}at_1^2 = 6 \times 3 - \frac{1}{2}(2)(3)^2 = 18 - 9 = 9m$

$S_{AB} = \frac{1}{2} \times 2 \times (1)^2 = 1m$

$\therefore S_{BC} = s_{oa} - s_{aB} = 9 - 1 = 8m$

Now velocity of the particle at pint B in return journey $v = 0 + 2 \times 1 = 2m/s$

In return journey from B to C, particle moves with constant velocity $2m/s$ to cover the distance $8m$.

Time taken $= \frac{\text{Distance}}{\text{Velocity}} = \frac{8}{2} = 4\ sec$

Total time taken by particle to return at point $O\ is$

$\Rightarrow T = t_{OA} + t_{AB} + t_{BC} = 3 + 1 + 4$

$= 8\ sec$

739 **(b)**

Relative velocity of one train $w.r.t.$ other $= 10 + 10 = 20m/s$

Relative acceleration$=0.3 + 0.2 = 0.5m/s^2$

If train crosses each other then from $s = ut + \frac{1}{2}at^2$

$As, s = s_1 + s_2 = 100 + 125 = 225$

$\Rightarrow 225 = 20t + \frac{1}{2} \times 0.5 \times 0.5 \times t^2$

$\Rightarrow 0.5t^2 + 40t - 450 = 0$

$\Rightarrow t = \frac{-40 \pm \sqrt{1600 + 4.(005) \times 450}}{1}$

$= -40 \pm 50$

$\therefore t = 10sec$(Taking +ve value)

740 **(b)**

Acceleration, $a = \frac{dv}{dt}$

As $a =$ constant

Then $\frac{dv}{dt} =$ constant or $v = kt$

Hence, the correct graph is (b).

741 **(b)**

In this problem point starts moving with uniform acceleration a and after time t (Position B) the direction of acceleration get reversed i.e. the retardation of same value works on the point. Due to this velocity of points goes on decreasing and at position C its velocity becomes zero. Now the direction of motion of point reversed and it moves from C to A under the effect of acceleration a.

We have to calculate the total time in this motion.

Starting velocity at position A is equal to zero.

Velocity at position $B \Rightarrow v = at$ [As $u = 0$]

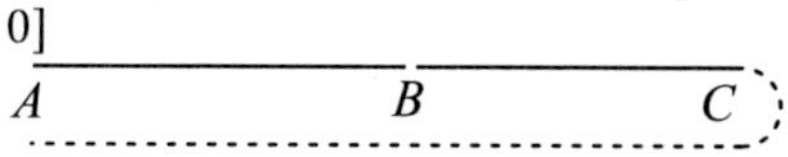

Distance between A and $B, S_{AB} = \frac{1}{2}at^2$

As same amount of retardation works on a point and it comes to rest therefore

$S_{BC} = S_{AB} = \frac{1}{2}a\,t^2$

$\therefore S_{AC} = S_{AB} + S_{BC} = a\,t^2$ and time required to cover this distance is also equal to t.

$\therefore$ Total time taken for motion between A and $C = 2t$

Now for the return journey from C to A ($S_{AC} = at^2$)

$$S_{AC} = u\,t + \frac{1}{2}at^2 \Rightarrow at^2 = 0 + \frac{1}{2}at_1^2 \Rightarrow t_1 = \sqrt{2}t$$

Hence total time in which point returns to initial point

$T = 2t + \sqrt{2}\,t = (2 + \sqrt{2})t$

742 **(d)**

As $v^2 = u^2 + 2as \Rightarrow (2y)^2 = u^2 + 2as \Rightarrow = 3u^2$

Now, after covering an additional distance s, if velocity becomes v, then,

$$v^2 = u^2 + 2a(2s) = u^2 + 4as = u^2 + 6u^2 = 7u^2$$

$\therefore v = \sqrt{7}u$

743 **(c)**

$\therefore\ x = a\cos\omega t$

$$\therefore v = \frac{dx}{dt} = -a\omega\sin\omega t$$

The instantaneous speed is given by modulus of instantaneous velocity.

$\therefore$ speed$= |u| = |-a\omega\sin\omega t|$

Hence, (c) is correct.

744 **(a)**

Effective speed of bullet

= speed of bullet + speed of police jeep

$= 180\ m/s + 45\ km/h$

$= (180 + 12.5)\ m/s$

$= 192.5\ m/s$

Speed of thief's jeep $= 153\ km/h$

$= 42.5\ m/s$

Velocity of bullet $w.r.t.$ thief's car $= 192.5 - 42.5 = 150\ m/s$

745 **(a)**

$$s = ut + \frac{1}{2}at^2$$

$s = \frac{1}{2}at^2$ $\qquad [\because u = 0]$

It is an equation of parabola

746 **(b)**

Total time of motion is $2\ min\ 20\ sec = 140\ sec$

As time period of circular motion is $40\ sec$ so in $140\ sec$. Athlete will complete 3.5 revolution $i.e.$,He will be at diametrically opposite point

$i.e.$,Displacement $= 2R$

747 **(c)**

$$S = \frac{1}{2}\times 1\times 20 + 1\times 20 + \frac{1}{2}\times(20 + 10)\times 1 + 1\times 10$$

$= 55$m

748 **(a)**

$\frac{dv}{dt} = bt$ or $dv = bt\,dt$

$\int_{v_0}^{v} dv = \int_0^t bt\,dt$or $\ v - v_0 = \frac{bt^2}{2}$

or $\ v = v_0 = +\frac{bt^2}{2}$

or $\ dx = v_0dt + \frac{bt^2}{2}dt$

$$\int_0^x dx = \int_0^t v_0dt + \frac{b}{2}\int_0^t t^2dt$$

or $x = v_0t + \frac{1}{2}\frac{bt^3}{3} = v_0t + \frac{bt^3}{6}$

749 **(b)**

$u = 0, v = 180\ km\,h^{-1} = 50\ ms^{-1}$

Time taken $t = 10s$

$$a = \frac{v - u}{t} = \frac{50}{10} = 5\ ms^{-2}$$

$\therefore$ Distance covered $S = ut + \frac{1}{2}at^2$

$$= 0 + \frac{1}{2}\times 5\times(10)^2 = \frac{500}{2} = 250\ m$$

751 **(b)**

Let the particle moves toward right with velocity $6\ m/s$. Due to retardation after time $t_1$its velocity becomes zero

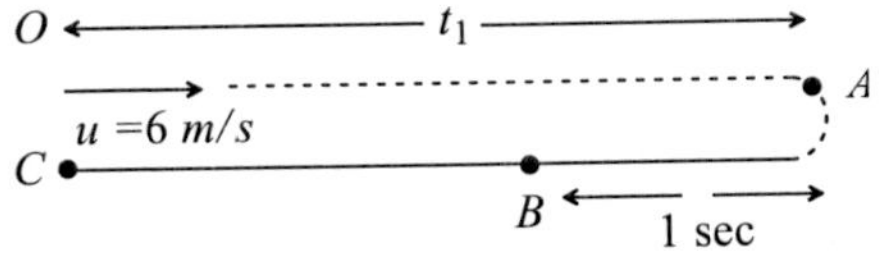

From $v = u - at \Rightarrow 0 = 6 - 2t_1 \Rightarrow t_1 = 3sec$

But retardation work on it for 4 sec.It means after reaching point A direction of motion get reversed and acceleration works on the particle for next one second.

$S_{OA} = ut_1 - \frac{1}{2}at_1^2$$=6\times 3 - \frac{1}{2}(2)(3)^2 = 18 - 9 = 9m$

$$S_{AB} = \frac{1}{2}\times 2\times(1)^2 = 1m$$

$\therefore S_{BC} = s_{oa} - s_{aB} = 9 - 1 = 8m$

Now velocity of the particle at pint B in return journey $v = 0 + 2\times 1 = 2m/s$

In return journey from B to C, particle moves with constant velocity $2m/s$ to cover the distance $8m$.

$$\text{Time taken} = \frac{\text{Distance}}{\text{Velocity}} = \frac{8}{2} = 4\ sec$$

Total time taken by particle to return at point O *is*

$\Rightarrow T = t_{OA} + t_{AB} + t_{BC} = 3 + 1 + 4$
$= 8\ sec$

752 **(a)**

Time, $t = \sqrt{\frac{2h}{g}}$

Distance from the foot of the tower

$$d = vt = v\sqrt{\frac{2h}{g}}$$

$$= 250 \text{ m}$$

When velocity $= \frac{v}{2}$

And height of tower $= 4h$

Then, distance $x = \frac{v}{2}\sqrt{\frac{2(4h)}{g}}$

$$x = v\sqrt{\frac{2h}{g}} = 250 \text{ m}$$

753 **(c)**

Time taken by the body to reach the point A is t_1 (During upward journey).

The body crosses this point again (during downward journey) after t_2, *ie*, the body takes the time $(t_2 - t_1)$ to come again at point A.

So, the time taken by the body to reach at point B (a maximum height).

$$t = t_1\left(\frac{t_2 - t_1}{2}\right)$$

[$\because$ Time pf ascending
= Time of descending]

$$t = \frac{t_1 + t_2}{2}$$

So, maximum height $H = \frac{1}{2}\text{ g}t^2$

$$= \frac{1}{2}\text{g}\left(\frac{t_1 + t_2}{2}\right)^2$$

$$= 2\text{g}\left(\frac{t_1 + t_2}{4}\right)^2$$

754 **(c)**

Distance covered in a particular time is

$$s_n = u + \frac{1}{2}\text{g}(2\text{n} - 1)$$

$$s_1 = 0 + \frac{1}{g}(2 \times 1 - 1) = \frac{g}{2}$$

$$s_2 = 0 + \frac{1}{2}\text{g}(2 \times 2 - 1) = \frac{3}{2}\text{g}$$

And $s_3 = 0 + \frac{1}{2}\text{ g}(2 \times 3 - 1) = \frac{5}{2}\text{ g}$

Hence, the required ration is

$$s_1 : s_2 : s_3 = \frac{g}{2} : \frac{3}{2}\text{g} : \frac{5}{2}\text{g}$$

$$= 1 : 3 : 5$$

755 **(c)**

Let man will catch the bus after 't'*sec*. So he will cover distance ut

Similarly distance travelled by the bus will be $\frac{1}{2}at^2$

For the given condition

$u\,t = 45 + \frac{1}{2}a\,t^2 = 45 + 1.25t^2$ [As $a = 2.5\ m/s^2$]

$\Rightarrow u = \frac{45}{t} + 1.25t$

To find the minimum value of u

$\frac{du}{dt} = 0$ sp we get $t = 6\ sec$ then,

$u = \frac{45}{6} + 1.25 \times 6 = 7.5 + 7.5 = 15\ m/s$

756 (a)

The $v-x$ equation from the given graph can be written as,

$$v=\left(-\frac{v_0}{x_0}\right)x + v_0 \quad \text{...(i)}$$

$$\therefore \quad a=\frac{dv}{dt}=\left(-\frac{v_0}{x_0}\right)\frac{dx}{dt}=\left(-\frac{v_0}{x_0}\right)v$$

Substituting v from Eq. (i), we get

$$a=\left(-\frac{v_0}{x_0}\right)\left[\left(-\frac{v_0}{x_0}\right)x+v_0\right]$$

$$a=\left(\frac{v_0}{x_0}\right)^2 x-\frac{v_0^2}{x_0}$$

Thus, $a-x$ graph is a straight line with positive slope and negative intercept.

757 (c)

$\frac{dv}{dt}=6t$ or $dv=6t, mv=\frac{6t^2}{2}=3t^2$,

$dx=3t^2dt \Rightarrow x=3\frac{t^3}{2}=t^3$

758 (b)

$v_t=4t^3-2t$

$\Rightarrow \frac{dx_t}{dt}=4t^3-2t$

$\Rightarrow \int dx_t=\int 4t^3dt-\int 2tdt$

$\Rightarrow x_t=t^4-t^2$

Since, $x_t=2\text{m}$

$t=\sqrt{2}\text{s}$ (rejecting negative time)

Now acceleration,

$a_t=\frac{dv_t}{dt}=12t^2-2=12(2)-2$

$=22\text{ms}^{-2}$

759 (b)

Let initial velocity of the bullet $=u$

After penetrating 3 cm its velocity becomes $\frac{u}{2}$

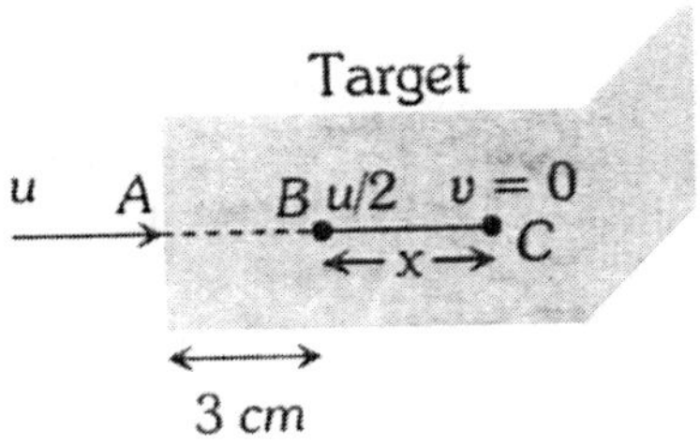

From $v^2=u^2-2as$

$\left(\frac{u}{2}\right)^2=u^2-2a(3)$

$\Rightarrow 6a=\frac{3u^2}{4}\Rightarrow a=\frac{u^2}{8}$

Let further it will penetrate through distance x and stops at point C

For distance $BC, v=0, u=u/2, s=x, a=u^2/8$

For $v^2=u^2-2as \Rightarrow 0=\left(\frac{u}{2}\right)^2-2\left(\frac{u^2}{8}\right).x \Rightarrow x=1\ cm$

760 (a)

Average velocity $=\frac{2\times8\times12}{8+12}\text{ms}^{-1}=9.6\text{ms}^{-1}$

761 (b)

$v^2=u^2+2gh \Rightarrow (3u)^2=(-u)^2+2gh$

$\Rightarrow h=\frac{4u^2}{g}$

762 (a)

$t=\sqrt{\frac{2\times44.1}{9.8}}\text{s}=\sqrt{9}\text{s}=3\text{s}$,

$44.1=v\times2+\frac{1}{2}\times9.8\times2\times2$

or $2v=44.1-4.9\times4=24.5$

or $v=\frac{24.5}{2}\text{ms}^{-1}=12.25\text{ms}^{-1}$

763 (b)

Let 'a' be the retardation of boggy then distance covered by it be S. If u is the initial velocity of boggy after detaching from train ($i.e.$ uniform speed of train)

$v^2=u^2+2as \Rightarrow 0=u^2-2as \Rightarrow s_b=\frac{u^2}{2a}$

Time taken by boggy to stop

$v=u+at \Rightarrow 0=u-at \Rightarrow t=\frac{u}{a}$

In this time t distance travelled by train $=s_t=ut=\frac{u^2}{a}$

Hence ratio $\frac{s_b}{s_1}=\frac{1}{2}$

764 **(d)**

Because acceleration due to gravity is constant so the slope of line will be constant *i.e.*, velocity time curve for a body projected vertically upwards is straight line

765 **(c)**

Since, body has uniform acceleration. So, velocity of particle is increasing with time. Hence, this is displacement-time graph with X as time axis and Y as displacement axis.

766 **(b)**

According to problem, when

$s = a, t = p$

$\because\ s = ut + \frac{1}{2}ft^2$ (here, f =acceleration)

$\therefore\ a = up + \frac{fp^2}{2}$ (i)

For $s = b, t = q$

$b = uq + \frac{fq^2}{2}$ (ii)

After solving Eqs. (i) and (ii),

$$f = \frac{2(aq - bp)}{pq(p - q)}$$

767 **(c)**

Taking the starting point as O, we have 30 m north OA, 20 m east AB, and finally $30\sqrt{2}$ m$(S - W)BD$.

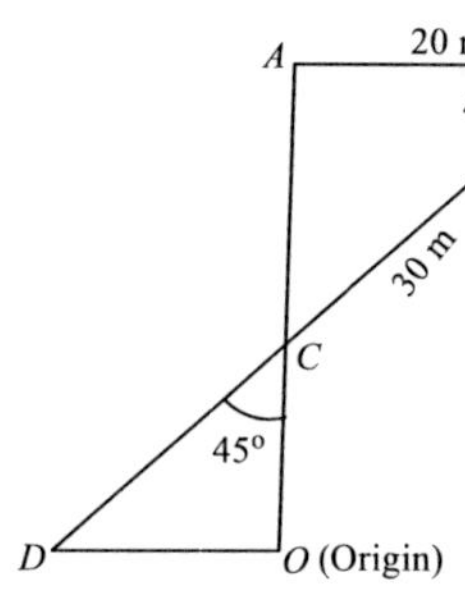

From $\Delta\ CAB$,

$$AC = 20 \text{ m}, OC = 10 \text{ m}$$

In $\Delta\ OCD$,

$$OD = OC, OD = 10 \text{ m}$$

Hence, final displacement from origin is 10 m.

768 **(c)**

Instantaeneous velocity $= v = \frac{\Delta x}{\Delta t}$

By using the data from the table

$v_1 = \frac{0 - (-2)}{1} = 2m/s,\quad v_2 = \frac{6 - 0}{1} = 6m/s$

$v_3 = \frac{16 - 6}{1} = 10m/s$

So, motion is non-uniform but accelerated

769 **(a)**

Here, $u = 10\ ms^{-1}, t = 2s, S = 20\ m$

Using $S = ut + \frac{1}{2}at^2$

$\therefore\ 20 = 10 \times 2 + \frac{1}{2} \times a \times 2^2$

$0 = 2a \Rightarrow a = 0$

770 **(c)**

For first projectile, $h_1 = ut - \frac{1}{2}gt^2$

For second projectile, $h_2 = u(t - T) - \frac{1}{2}g(t - T)^2$

When both meet i.e. $h_1 = h_2$

$ut - \frac{1}{2}gt^2 = u(t - T) - \frac{1}{2}g(t - T)^2$

$\Rightarrow uT + \frac{1}{2}gT^2 = gtT \Rightarrow t = \frac{u}{g} + \frac{T}{2}$

and $h_1 = u\left(\frac{u}{g} + \frac{T}{2}\right) - \frac{1}{2}g\left(\frac{u}{g} + \frac{T}{2}\right)^2$

$= \frac{u^2}{2g} - \frac{gT^2}{8}$

771 **(c)**

Maximum height of ball $= 5\ m$

So velocity of projection $\Rightarrow u = \sqrt{2gh} = 10\ m/s$

Time interval between two balls (time of ascent)

$= \frac{u}{g} = 1\ sec = \frac{1}{60}\ min$

So number of ball thrown per min $= 60$

772 **(b)**

$v = u + at$ As $u = 0, v = at$

The graph (b) is correct as $v = 0$ at $t = 0$, and in the straight line graph $y = mx, y = v, m = a$ and $t = x$

773 **(b)**

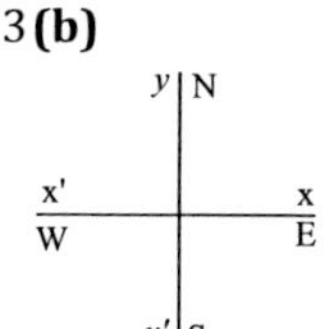

Let $\vec{v}_t$= actual velocity of train

$\vec{v}_t$= actual velocity of car

$\therefore$ $\vec{v}_{tc}$= relative velocity of train with respect to car

$\therefore$ $\vec{v}_c$=25î

and $\vec{v}_{tc}$=25$\sqrt{3}$ĵ

$\therefore$ $\vec{v}_{tc} = \vec{v}_t - \vec{v}_c$

$\therefore$ $\vec{v}_t = \vec{v}_{tc} + \vec{v}_c$

= 25$\sqrt{3}$ĵ+25î

$$v_t = \sqrt{(25\sqrt{3})^2 + (25)^2}$$

= 25$\sqrt{4}$= 50kmh^{-1}

774 **(b)**

$$S_2 = \frac{1}{2}gt_2^2 = \frac{10}{2} \times (3)^2 = 45\,m$$

$$S_1 = \frac{1}{2}gt_1^2 = \frac{10}{2} \times (5)^2 = 125\,m$$

$\therefore S_1 - S_2 = 125 - 45 = 80\,m$

775 **(c)**

For first case $v^2 - 0^2 = 2gh \Rightarrow (3)^2 = 2gh$

For second case $v^2 = (-u)^2 + 2gh = 4^2 + 3^2 \therefore v = 5km/h$

776 **(d)**

$$\frac{v_A}{v_B} = \frac{\tan 30°}{\tan 60°} = \frac{1}{\sqrt{3}} \times \frac{1}{\sqrt{3}} = \frac{1}{3}$$

777 **(b)**

$$\int_0^x dx = \int_0^1 (v_0 + gt + ft^2)dt$$

$$x = v_0 + g\left(\frac{1}{2}\right) + f\left(\frac{1}{3}\right)$$

778 **(a)**

Total time of motion = t

Duration of acceleration = t'

Duration of deceleration = $t - t'$

Given $u = 0, a$ = constant acceleration and

b = constant deceleration

$v = 0 + at'$

Also $0 = v - b(t - t')$

$\therefore v = at'$

From (ii), $-v = -bt + bt'$

$\Rightarrow -at' = -bt + bt'$

$$\Rightarrow (a+b)t' = bt \Rightarrow t' = \frac{b}{(a+b)}t$$

But $v = at'$

$\therefore$ Maximum velocity attained = at'

$$\Rightarrow v = \frac{ab}{(a+b)}t\,m/s$$

779 **(b)**

Speed of stone in a vertically upward direction is $20m/s$. So for vertical downward motion we will consider $u = -20\,m/s$

$v^2 = u^2 + 2gh = (-20)^2 + 2 \times 9.8 \times 200$
$= 4320\,m/s$

$\therefore v \simeq 65m/s$

781 **(d)**

$u = \sqrt{2gh} = \sqrt{2 \times 10 \times 20} = 20m/s$

and $T = \frac{2u}{g} = \frac{2\times 20}{10} = 4\,sec$

782 **(a)**

If t_1 and$2t_2$ are the time taken by particle to cover first and second half distance respectively

$t_1 = \frac{x/2}{3} = \frac{x}{6}$...(i)

$x_1 = 4.5\,t_2$ and $x_2 = 7.5\,t_2$

So, $x_1 + x_2 = \frac{x}{2} \Rightarrow 4.5\,t_2 + 7.5\,t_2 = \frac{x}{2}$

$t_2 = \frac{x}{24}$ ...(ii)

Total time $t = t_1 + 2t_2 = \frac{x}{6} + \frac{x}{12} = \frac{x}{4}$

So, average speed = $4\,m/sec$

783 **(c)**

$$S_n = \frac{1}{2}g\cos\theta\,(2n-1), S_{n+1} = \frac{1}{2}g\cos\theta\,\{2(n+1) - 1\}$$

$$\frac{S_n}{S_{n+1}} = \frac{2n-1}{2n+1}$$

784 **(b)**

If acceleration is variable (depends on time) then

$$v = u + \int (f)dt = u + \int (a\,t)dt = u + \frac{a\,t^2}{2}$$

786 **(d)**

$x = 8 + 12t + t^3$

$v = 0 + 12 - 3t^2 = 0$

$3t^2 = 12$

$t = 2\,sec$

$$a = \frac{dv}{dt} = 0 - 6t$$

$a[t = 2] = -12\,m/s^2$

Retardation = $12\,m/s^2$

787 **(a)**

The equation of motion

$$s = ut + \frac{1}{2}at^2$$
$$= 0 + \frac{1}{2}at^2 = \frac{1}{2}at^2$$

The graph plot is as shown.

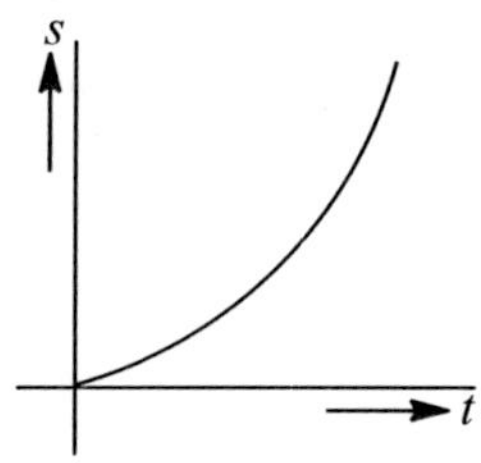

788 **(b)**

Given, $y = bx^2$

$$\frac{dy}{dt} = 2bx\frac{dx}{dt} \quad \text{... (i)}$$
$$\frac{dy}{dt} = at \quad (\because v_y = u_y + a_y t)$$
$$at = 2bx\frac{dx}{at}$$
$$at\,dt = 2bx\,dx$$

Take integration of both sides

$$\int at\,dt = \int 2bx\,dx$$
$$\frac{at^2}{2} = bx^2 + c \quad \text{... (ii)}$$

At $t = 0, x = 0, c = 0$

Then $\frac{at^2}{2} = bx^2$

$$x = \sqrt{\frac{at^2}{2b}} = \sqrt{\frac{a}{2b}}t$$

$$\therefore \quad v_x = \frac{dx}{dt} = \sqrt{\frac{a}{2b}}$$

789 **(d)**

In the positive region the velocity decreases linearly (during rise) and in the negative region velocity increases linearly (during fall) and the direction is opposite to each other during rise and fall, hence fall is shown in the negative region

790 **(b)**

$$h_n = \frac{g}{2}(2n-1) \Rightarrow h_{5^{\text{th}}} = \frac{10}{2}(2\times 5 - 1)$$
$$= 45\,m$$

791 **(b)**

$v = u + at$ As $u = 0, v = at$

The graph (b) is correct as $v = 0$ at $t = 0$, and in the straight line graph $y = mx, y = v, m = a$ and $t = x$

792 **(c)**

$$v_{av} = \frac{2v_1v_2}{v_1 + v_2} = \frac{2\times 40\times 60}{100} = 48\,kmph$$

793 **(c)**

Mass does not affect maximum height

$H = \frac{u^2}{2g} \Rightarrow H \propto u^2$, So if velocity is doubled then height will become four times.i.e.$H = 20 \times 4 = 80m$

794 **(a)**

$$\frac{dt}{dx} = 2\alpha x + \beta \Rightarrow v = \frac{1}{2\alpha x + \beta}$$
$$\because a = \frac{dv}{dt} = \frac{dv}{dx}.\frac{dx}{dt}$$
$$a = v\frac{dv}{dx} = \frac{-v.2\alpha}{(2\alpha x + \beta)^2} = -2\alpha.v.v^2$$
$$= -2\alpha v^3$$

$\therefore$ Retardation $= 2\alpha v^3$

795 **(a)**

For upstream motion, kmh^{-1}=5kmh^{-1}

For downstream motion,

$v = (8+3)\text{kmh}^{-1} = 11\text{kmh}^{-1}$

796 **(d)**

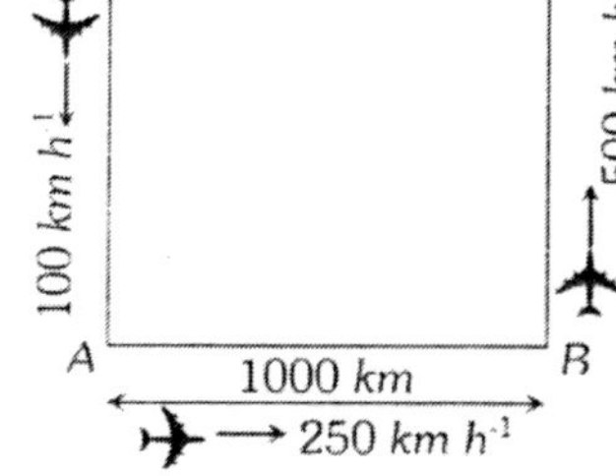

Let t_{AB}, t_{BC}, t_{CD} and t_{DA} be the time taken by the aeroplane to go from A to B, B to C, C to D and D to A respectively

$\therefore t_{AB} = \dfrac{1000\ km}{250\ km\ h^{-1}} = 4h$

$\therefore t_{BC} = \dfrac{1000\ km}{500\ km\ h^{-1}} = 2h$

$\therefore t_{CD} = \dfrac{1000\ km}{200\ km\ h^{-1}} = 5h$

$\therefore t_{DA} = \dfrac{1000\ km}{100\ km\ h^{-1}} = 10h$

Average speed$= \dfrac{\text{Total distance covered}}{\text{Total time taken}}$

$= \dfrac{1000km + 1000km + 1000km + 1000k}{t_{AB} + t_{BC} + t_{CD} + t_{DA}}$

$= 190.5\ km\ h^{-1}$

797 **(a)**

Total time of motion $= t$
Duration of acceleration $= t'$
Duration of deceleration $= t - t'$
Given $u = 0, a =$ constant acceleration and
$b =$ constant deceleration
$v = 0 + at'$
Also $0 = v - b(t - t')$
$\therefore v = at'$
From (ii), $-v = -bt + bt'$
$\Rightarrow -at' = -bt + bt'$

$\Rightarrow (a + b)t' = bt \Rightarrow t' = \dfrac{b}{(a + b)}t$

But $v = at'$
$\therefore$ Maximum velocity attained $= at'$

$\Rightarrow v = \dfrac{ab}{(a + b)}t\ m/s$

798 **(d)**

$\because v = 0 + na \Rightarrow a = v/n$
Now, distance travelled in $n\ sec. \Rightarrow S_n = \frac{1}{2}an^2$ and
distance travelled in $(n - 2)sec \Rightarrow S_{n-2} = \frac{1}{2}a(n - 2)^2$
$\therefore$ Distance travelled in last 2 seconds,
$= S_n - S_{n-2} = \frac{1}{2}an^2 - \frac{1}{2}a(n - 2)^2$

$\dfrac{a}{2}[n^2 - (n - 2)^2] = \dfrac{a}{2}[n + (n - 2)][n - (n - 2)]$

$= a(2n - 2) = \dfrac{v}{n}(2n - 2) = \dfrac{2v(n - 1)}{n}$

799 **(d)**

Slope of displacement time graph is negative only at point time E

800 **(b)**

$x = \alpha t^3, y = \beta t^3$

$v_x = \dfrac{dx}{dt} = 3\alpha t^2$

$v_y = \dfrac{dy}{dt} = 3\ \beta t^2$

Resultant velocity, $v = \sqrt{v_x^2 + v_y^2}$

$= \sqrt{9\alpha^2 t^4 + 9\beta^2 t^4}$
$= 3t^2\sqrt{\alpha^2 + \beta^2}$

801 **(b)**

$v = v_0 + gt + ft^2$

Or $\frac{dx}{dt} = v_0 + gt + ft^2$

Or $dx = (v_0 + gt + ft^2)\ dt$

So, $\int_0^x dx = \int_0^1 (v_0 + gt + ft^2)\ dt$

Or $x = v_0 + \frac{g}{2} + \frac{f}{3}$

802 **(a)**

$H_{\max} = \dfrac{u^2}{2g} \Rightarrow H_{\max} \propto \dfrac{1}{g}$

On planet B value of g is $1/9$ times to that of A. So value of $H_{\max}$ will become 9 times i.e. $2 \times 9 = 18\ metre$

803 **(d)**

Using, $v = u + at$ or $v - u = at$, we find that if $|\vec{a}|$is, it t is the time for acceleration, then $\frac{t}{2}$is the time for retardation
Now, $t + \frac{t}{2} = 3$ or $\frac{3t}{2} = 3$ or $t = 2s$
$S = \frac{1}{2} \times 2 \times 2 \times 2 + \frac{1}{2} \times 4 \times 1 \times 1 = (4 + 2)$m=6m

804 **(b)**

$v = u + \int a dt = u + \int (3t^2 + 2t + 2)dt$

$= u + \dfrac{3t^3}{3} + \dfrac{2t^2}{2} + 2t = u + t^3 + t^2 + 2t$

$= 2 + 8 + 4 + 4 = 18\ m/s$ (As $t = 2\ sec$)

805 **(d)**

Since, acceleration is in the direction of instantaneous velocity, so particle always moves in forward direction.
Hence, (d) is correct.

806 **(d)**

Acceleration due to gravity is independent of mass of body

807 **(a)**

$9y = \frac{1}{2} \times 10 \times 3 \times 3$ or $y = 5\text{m}$

Again, $n \times 5 = \frac{1}{2} \times 10 \times 1 \times 1 = 5$ or $n = 1$

808 **(a)**

For first stone $u = 0$ and

For second stone $\frac{u^2}{2g} 4h \Rightarrow u^2 = 8gh$

$\therefore u = \sqrt{8gh}$

Now, $h_1 = \frac{1}{2}gt^2$

$h_2 = \sqrt{8ght - \frac{1}{2}gt^2}$

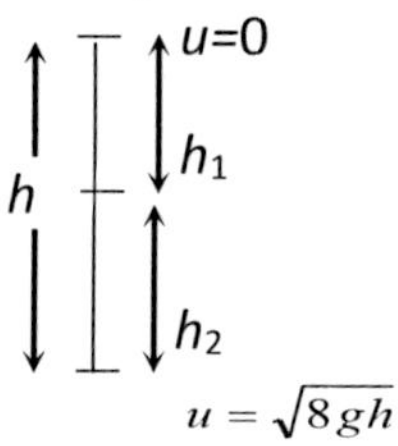

$u = \sqrt{8gh}$

Where, t = time cross each other

$\therefore h_1 + h_2 = h$

$\Rightarrow \frac{1}{2}gt^2 + \sqrt{8ght} - \frac{1}{2}gt^2 = h \Rightarrow t = \frac{h}{\sqrt{8gh}} = \sqrt{\frac{h}{8g}}$

809 **(c)**

$\because S_1 = ut + \frac{1}{2}at^2$ ... (i)

and velocity after first t *sec*

$v = u + at$

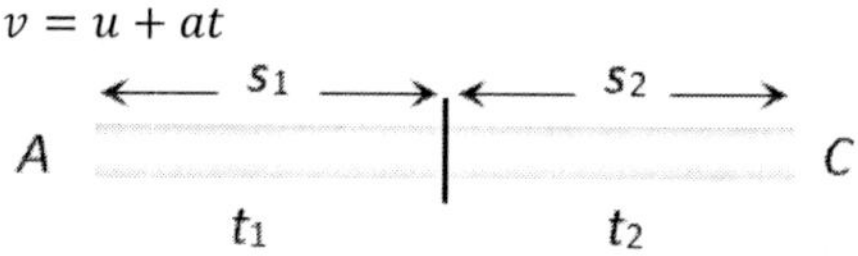

$t_1 = t_2 = t$ (given)

Now, $S_2 = vt + \frac{1}{2}at^2$

$= (u + at)t + \frac{1}{2}at^2$ ... (ii)

Equation (ii) – (i) $\Rightarrow S_2 - S_1 = at^2$

$\Rightarrow a = \frac{S_2 - S_1}{t^2} = \frac{65 - 40}{(5)^2} = 1 m/s^2$

From equation (i), we get,

$S_1 = ut + \frac{1}{2}at^2 \Rightarrow 40 = 5u + \frac{1}{2} \times 1 \times 25$

$\Rightarrow 5u = 27.5 \therefore u = 5.5\ m/s$

810 **(d)**

The area under acceleration-time graph gives change in velocity.

811 **(a)**

An aeroplane files 400 m north and 300 m south so the net displacement is 100 m towards north

Then it files 1200 m upwards so $r = \sqrt{(100)^2 + (1200)^2}$

$= 1204\ m \simeq 1200\ m$

The option should be 1204 m, because this value mislead one into thinking that net displacement is in upward direction only

812 **(b)**

Let initial velocity of the bullet $= u$

After penetrating 3 cm its velocity becomes $\frac{u}{2}$

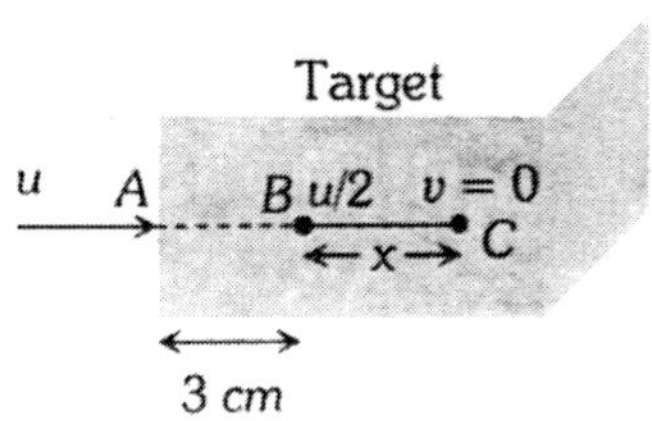

From $v^2 = u^2 - 2as$

$\left(\frac{u}{2}\right)^2 = u^2 - 2a(3)$

$\Rightarrow 6a = \frac{3u^2}{4} \Rightarrow a = \frac{u^2}{8}$

Let further it will penetrate through distance x and stops at point C

For distance $BC, v = 0, u = u/2, s = x, a = u^2/8$

For $v^2 = u^2 - 2as \Rightarrow 0 = \left(\frac{u}{2}\right)^2 - 2\left(\frac{u^2}{8}\right).x \Rightarrow x = 1\ cm$

813 **(b)**

Constant velocity means constant speed as well as same direction throughout

814 **(c)**

At D then the distance CD is least.
Therefore, $OC = (10 - 3t)$m

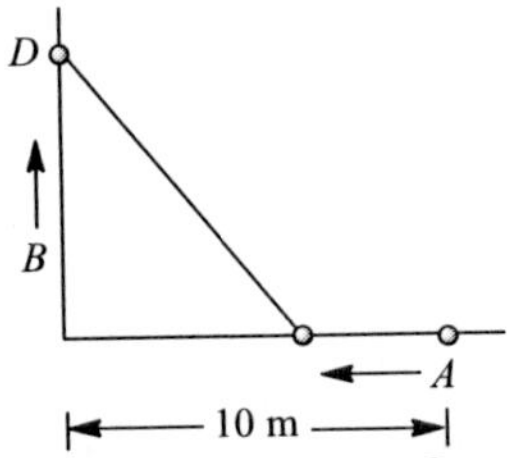

$\therefore CD^2 = OC^2 + OD^2$
$= (10 - 3t)^2 + (4t)^2$
$= 25t^2 - 60t + 100$
or $CD = [(5t - 6)^2 + 64]^{1/2}$
CD is least if $(5t - 6)^2 = 0$ or $5t - 6 = 0$
or $5t = 6$ or $6 = t/5 = 1.2$s
So, $AC = 3 \times 1.2 = 3.6$m: $CD = (64)^{1/2}$=8m.

815 **(a)**

After balling out from point A parachutist falls freely under gravity. The velocity acquired by it will 'v'

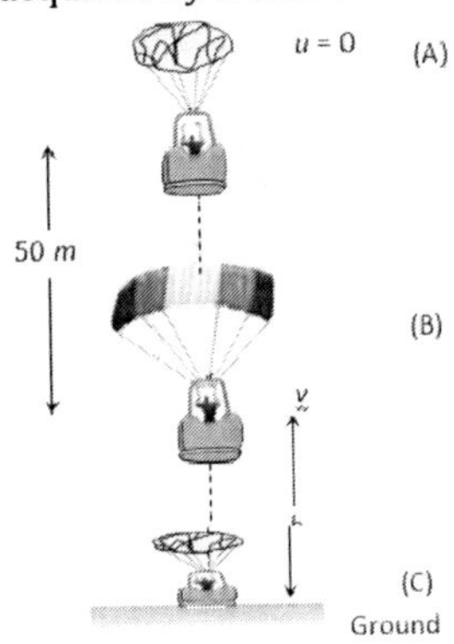

From $v^2 = u^2 + 2as = 0 + 2 \times 9.8 \times 50 = 980$
[As $u = 0, a = 9.8m/s^2, s = 50\ m$]
At point B, parachute opens and it moves with retardation of 2 m/s^2 and reach at ground (point C) with velocity of 3 m/s
For the part 'BC' by applying the equation
$v^2 = u^2 + 2as$
$v = 3m/s, u = \sqrt{980}\ m/s, a = -2m/s^2, s = h$
$\Rightarrow (3)^2 = \left(\sqrt{980}\right)^2 + 2 \times (-2) \times h \Rightarrow 9 = 980 - 4h$
$\Rightarrow h = \frac{980 - 9}{4} = \frac{971}{4} = 242.7 \cong 243\ m$
So, the total height by which parachutist bail out $= 50 + 243 = 293\ m$

816 **(c)**

Stopping distance= $\frac{\text{Kinetic energy}}{\text{Retarding force}} = \frac{\frac{1}{2}mu^2}{F}$
$= \frac{u^2}{2\mu g}[F = \mu mg]$
So both will cover equal distance

817 **(c)**

$\because S_1 = ut + \frac{1}{2}at^2$ ... (i)
and velocity after first t sec
$v = u + at$

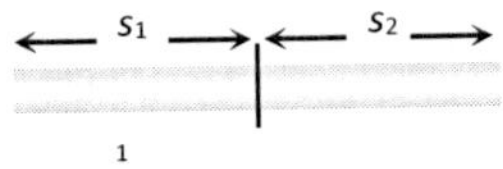

= $t_2 = t$ (given)

Now, $S_2 = vt + \frac{1}{2}at^2$
$= (u + at)t + \frac{1}{2}at^2$ ... (ii)
Equation (ii) – (i) $\Rightarrow S_2 - S_1 = at^2$
$\Rightarrow a = \frac{S_2 - S_1}{t^2} = \frac{65 - 40}{(5)^2} = 1m/s^2$
From equation (i), we get,
$S_1 = ut + \frac{1}{2}at^2 \Rightarrow 40 = 5u + \frac{1}{2} \times 1 \times 25$
$\Rightarrow 5u = 27.5 \therefore u = 5.5\ m/s$

818 **(c)**

$h = ut + \frac{1}{2}gt^2 \Rightarrow 81$
$= -12t + \frac{1}{2} \times 10 \times t^2$
$\Rightarrow t = 5.4\ sec$

819 **(d)**

$v = \sqrt{49 + y}, a = \frac{dv}{dy} \cdot \frac{dy}{dt} = v\frac{dv}{dy}$
$= (49 + y)^{1/2} \times \frac{1}{2}(49 + y)^{1/2-1} = \frac{1}{2}\text{ms}^{-2}$

820 **(c)**

$h = ut + \frac{1}{2}gt^2 \Rightarrow 1 = 0 \times t_1 + \frac{1}{2}gt_1^2 \Rightarrow t_1 = \sqrt{2/g}$
Velocity after travelling 1 m distance
$v^2 = u^2 + 2gh \Rightarrow v^2 = (0)^2 + 2g \times 1$
$\Rightarrow v = \sqrt{2g}$
For second 1 m distance
$1 = \sqrt{2g} \times t_2 + \frac{1}{2}gt_2^2$
$\Rightarrow gt_2^2 + 2\sqrt{2g}t_2 - 2 = 0$
$t_2 = \frac{-2\sqrt{2g} \pm \sqrt{8g + 8g}}{2g} = \frac{-\sqrt{2} \pm 2}{\sqrt{g}}$
Taking +ve sign $t_2 = (2 - \sqrt{2})/\sqrt{g}$

$\therefore \frac{t_1}{t_2} = \frac{\sqrt{2/g}}{(2-\sqrt{2})/\sqrt{g}} = \frac{1}{\sqrt{2}-1}$ and so on

821 **(d)**

Given, acceleration $a = -kv^3$

Initial velocity at cut-off, $v_1 = v_0$

Initial time of cut-off, $t = 0$

And final time after cut-off, $t_2 = t$

Again, $a = \frac{dv}{dt} = -kv^3$

Or $\frac{dv}{v^3} = -kdt$

Integrating both sides, with in the condition of motion.

$$\int_{v_0}^{v} \frac{dv}{v^3} = -\int_0^t k\,dt$$

Or $\left[-\frac{1}{2v^2}\right]_{v_0}^{v} = -[kt]_0^t$

Or $\frac{1}{2v^2} - \frac{1}{2v_0^2} = kt$

Or $v = \frac{v_0}{\sqrt{1+2\,kt\,v_0^2}}$

822 **(b)**

Average speed is the ratio of distance to time taken

Distance travelled from 0 to $5s = 40\ m$

Distance travelled from 5 to $10s = 0\ m$

Distance travelled from 10 to $15s = 60\ m$

Distance travelled from 15to $20s = 20$

So, total distance $= 40 + 0 + 60 + 20 = 120\ m$

Total time taken $= 20\ minutes$

Hence, average speed

$= \frac{\text{distance travelled } (m)}{\text{time (min }\ )} = \frac{120}{20}$

$= 6\ m/min$

823 **(d)**

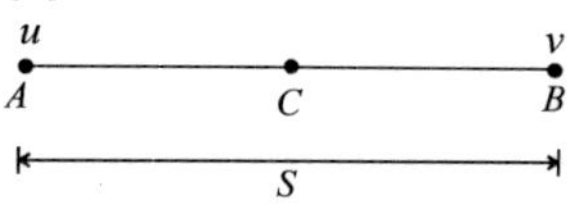

Let S be the distance between AB and a be constant acceleration of a particle. Then

$v^2 - u^2 = 2aS$

Or $aS = \frac{v^2 - u^2}{2}$ ... (i)

Let v_c be velocity of a particle at midpoint C

$\therefore v_c^2 - u^2 = 2a\left(\frac{S}{2}\right)$

$v_c^2 = u^2 + aS = u^2 + \frac{v^2-u^2}{2}$[Using (i)]

$v_c = \sqrt{\frac{u^2 + v^2}{2}}$

824 **(d)**

Average velocity $= \frac{Total\ Displacement}{Time\ taken} = \frac{25}{75/15} = 5m/s$

825 **(d)**

$$\int_{v_0}^{v} \frac{dv}{kv^3} = \int_0^t dt$$

$-\frac{1}{k}\int_{v_0}^{v} v^{-3}dv = t$ or $-\frac{1}{k}\left|\frac{v^{-3+1}}{-3+1}\right|_{v_0}^{v} = t$

or $\frac{1}{2k}\left[\frac{1}{v^2} - \frac{1}{v_0^2}\right] = t$ or $\frac{1}{v^2} - \frac{1}{v_0^2} = 2kt$

or $\frac{1}{v^2} = \frac{1}{v_0^2} + 2kt$ or $\frac{1}{v^2} = \frac{1+2v_0^2kt}{v_0^2}$

or $v = \frac{v_0}{\sqrt{2v_0^2kt+1}}$

826 **(c)**

$S_n = u + \frac{a}{2}(2n-1) \Rightarrow 1.2$

$= 0 + \frac{a}{2}(2 \times 6 - 1)$

$\Rightarrow a = \frac{1.2 \times 2}{11} = 0.218\ m/s^2$

827 **(a)**

Given line have positive intercept but negative slope. So its equation can be written as

$v = -mx + v_0$ ....(i) [where $m = \tan\theta = \frac{v_0}{x_0}$]

By differentiating with respect to time we get

$\frac{dv}{dt} = -m\frac{dx}{dt} = -mv$

Now substituting the value of v from eq. (i) we get

$\frac{dv}{dt} = -m[-mx + v_0] = m^2x - mv_0 \therefore a = m^2x - mv_0$

i.e. the graph between a and x should have positive slope but negative intercept on a-axis. So graph (a) is correct

828 **(d)**

Given, $v = 10 \text{ g kmh}^{-1} = 30 \text{ mh}^{-1}$

From first equation of motion.

$$v = u + at$$

$\therefore \quad 30 = 0 + a \times 5 \quad (\because u = 0)$

Or $\quad a = 6 \text{ ms}^{-2}$

So, distance travelled by metro train in 5 s

$$s_1 = \frac{1}{2}at^2 = \frac{1}{2} \times (6) \times (5)^2 = 75 \text{ m}$$

Distance travelled before coming to rest

$$= 45 \text{ m}$$

So, from third equation of motion

$$0^2 = (30)^2 - 2a' \times 45$$

Or $a' = \frac{30 \times 30}{2 \times 45} = 10 \text{ ms}^{-2}$

Time taken in travelling 45 m is

$$t_3 = \frac{30}{10} = 3 \text{ s}$$

Now, total distance = 395 m

ie, $\quad 75 + s' + 45 = 395 \text{ m}$

Or $s' = 395 - (75 + 45) = 275 \text{ m}$

$\therefore \quad t_2 = \frac{275}{30} = 9.2 \text{ s}$

Hence, total time taken in whole journey

$$= t_1 + t_2 + t_3$$

$$= 5 + 9.2 + 3 = 17.2 \text{ s}$$

829 **(a)**

$S_{3rd} = 10 + \frac{10}{2}(2 \times 3 - 1) = 35$

$S_{2nd} = 10 + \frac{10}{2}(2 \times 2 - 1) = 25$

$\therefore \frac{S_{3rd}}{S_{2nd}} = \frac{35}{25} \; ie = \frac{7}{5}$

830 **(d)**

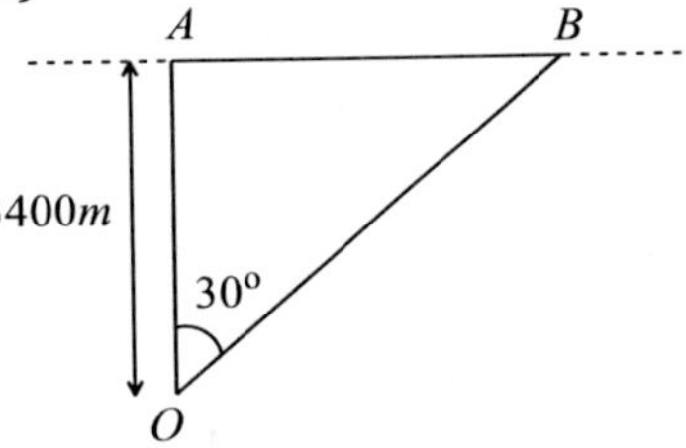

O is the observation point at the ground. A and B are the positions of aircraft for which $\angle AOB = 30°$. Time taken by aircraft from A to B is $10s$

ΔAOB

$\tan 30° = \frac{AB}{3400}$

$AB = 3400 \tan 30° = \frac{3400}{\sqrt{3}} m$

$\therefore$ Speed of aircraft,

$v = \frac{AB}{10} = \frac{3400}{10\sqrt{3}} = 196.3 \; ms^{-1}$

831 **(b)**

Speed of stone in a vertically upward direction is $4.9 \; m/s$. So for vertical downward motion we will consider $u = -4.9 \; m/s$

$h = ut + \frac{1}{2}gt^2 = -4.9 \times 2 + \frac{1}{2} \times 9.8 \times (2)^2 = 9.8 \; m$

832 **(d)**

$u = 200 \; m/s, v = 100 \; m/s, s = 0.1 \; m$

$a = \frac{u^2 - v^2}{2s} = \frac{(200)^2 - (100)^2}{2 \times 0.1} = 15 \times 10^4 \; m/s^2$

833 **(a)**

$t^2 = x^2 - 1$ or $x^2 = t^2 + 1$

Or $2x\frac{dx}{dt} = 2t$ or $xv = t$

Or $2x\frac{dx}{dt} + v\frac{dx}{dt} = 1$

Or $x\frac{dv}{dt} = 1 - v^2$

Or $\frac{dv}{dt} = \frac{1-v^2}{x} = \frac{1-\frac{t^2}{x^2}}{x} = \frac{x^2-t^2}{x^3}$

But $x^2 - t^2 = 1$

$\therefore \frac{dv}{dt} = \frac{1}{x^3}$

834 **(c)**

Both the stones will have the same speed when they hit the ground.

835 **(b)**

$10t = 48 + \frac{1}{2} \times 1 \times t^2$ or $t^2 - 20t + 96 = 0$

or $t^2 - 8t - 12t + 96 = 0$ or $t(t - 8) - 12(t - 8) = 0$

or $(t - 12)(t - 8) = 0$ or $t = 8$ s or 12 s

But we are interested in minimum time.

836 **(b)**

Let the two bodies A and B respectively meet at a time t, at a height $\frac{H}{2}$ from the ground

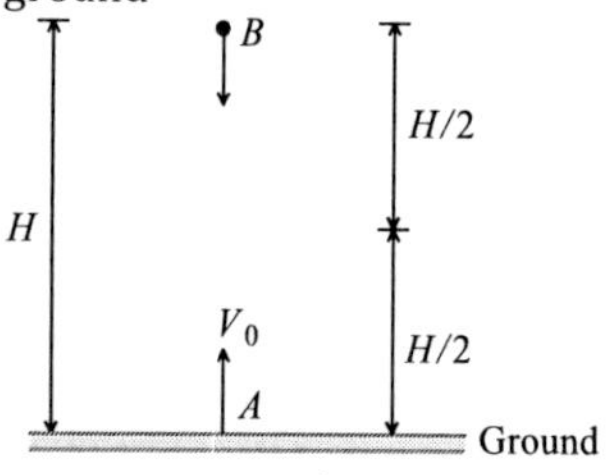

Using $S = ut + \frac{1}{2}at^2$

For a body $A, u = V_0, a = -g, S = \frac{H}{2}$

$\therefore \frac{H}{2} = V_0 t - \frac{1}{2}gt^2$...(i)

For body $B, u = 0, a = +g, S = \frac{H}{2}$

$\therefore \frac{H}{2} = \frac{1}{2}gt^2$ ...(ii)

Equating equations (i) and (ii), we get

$V_0 t - \frac{1}{2}gt^2 = \frac{1}{2}gt^2 \Rightarrow V_0 t = gt^2$ or $t = \frac{V_0}{g}$

Substituting the value of t in equation (i), we get

$\frac{H}{2} = V_0 \times \left(\frac{V_0}{g}\right) - \frac{1}{2}g\left(\frac{V_0}{g}\right)^2 = \frac{V_0^2}{g} - \frac{1}{2}\frac{V_0^2}{g}$

$\frac{H}{2} = \frac{1}{2}\frac{V_0^2}{g}$ or $V_0^2 = gH \Rightarrow V_0 = \sqrt{gH}$

837 **(a)**

$S_n = u + \frac{g}{2}(2n - 1)$; when $u = 0, S_1 : S_2 : S_3 = 1 : 3 : 5$

838 **(d)**

Average velocity$= \frac{3x}{\frac{x}{60} + \frac{x}{20} + \frac{x}{10}} = \frac{3x}{\frac{x+3x+6x}{60}}$

$= \frac{3x \times 60}{10x} = 18\text{kmh}^{-1}$

$= \frac{18 \times 5}{18}\text{ms}^{-1} = 5\text{ms}^{-1}$

839 **(c)**

$h = 0 + \frac{1}{2}\text{g}t^2 \Rightarrow t^2 \propto h$

$\therefore \frac{t_1}{t_2} = \sqrt{\frac{h_1}{h_2}} = \sqrt{\frac{16}{25}} = \frac{4}{5}$

840 **(d)**

$S \propto u^2$. If u becomes 3 times then S will become 9 times

$i.e. 9 \times 20 = 180m$

4
MOTION IN A PLANE

SINGLE CORRECT ANSWER TYPE

1. If a person can throw a stone to maximum height of h metre vertically, then the maximum distance through which it can be thrown horizontally by the same person is

a) $\frac{h}{2}$ b) h c) $2h$ d) $3h$

2. A pendulum bob on a 2 m string is displaced 60° from the vertical and then released. What is the speed of the bob as it passes through the lowest point in its path

a) $\sqrt{2}\, m/s$ b) $\sqrt{9.8}\, m/s$ c) $4.43\, m/s$ d) $1/\sqrt{2} m/s$

3. A wheel is subjected to uniform angular acceleration about its axis. Initially its angular velocity is zero. In the first 2 sec, it rotates through an angle θ_1. In the next 2 sec, it rotates through an additional angle θ_2. The ratio of θ_2 / θ_1 is

a) 1 b) 2 c) 3 d) 5

4. A car is moving in a circular horizontal track of radius 10m with a constant speed of 10 ms^{-1}. The angle made by the rod with track is

a) Zero b) 30° c) 45° d) 60°

5. A cyclist is travelling on a circular section of highway of radius 2500 ft at the speed of 60 mile h^{-1}. The cyclist suddenly applies the brakes causing the bicycle to slown sown at constant rate. Knowing that after 8 s the speed has been reduced to 45 mileh^{-1}. The acceleration of the bicycle immediately after the brakes have been applied is

a) 2 ft/s^2 b) 4.14 ft/s^2 c) 3.10 ft/s^2 d) 2.75 ft/s^2

6. Two stones are projected from the same speed but making different angels with the horizontal. Their horizontal ranges are equal. The angle of projection of one is $\pi/3$ and the maximum height reached by it is 102 m. Then maximum height reached by the other in metre is

a) 336 b) 224 c) 56 d) 34

7. A particle moves in circle of radius 25 cm at the rate of two revolutions per second. The acceleration of particle is

a) $2\pi^2 \text{ms}^{-2}$ b) $4\pi^2 \text{ms}^{-2}$ c) $8\pi^2 \text{ms}^{-2}$ d) $\pi^2 \text{ms}^{-2}$

8. A small particle of mass m is projected at an angle θ with the x-axis with an initial velocity v_0 in the x-y plane as shown in the figure. At a time $t < \frac{v_0 \sin\theta}{g}$, the angular momentum of the particle is

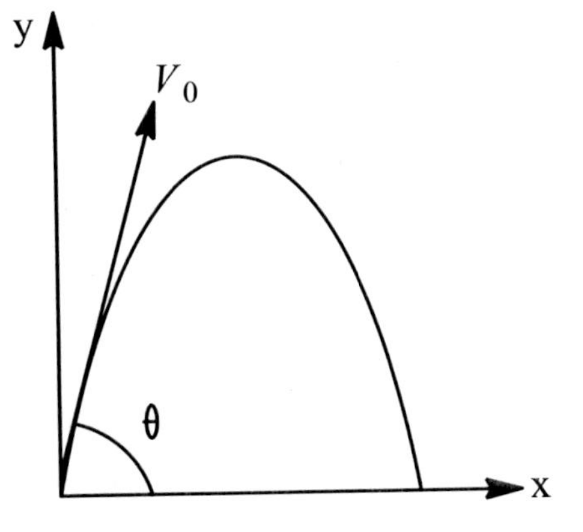

a) $-mgv_0t^2 \cos\theta\, \hat{\mathbf{j}}$ b) $mgv_0t \cos\theta\, \hat{\mathbf{k}}$ c) $-\frac{1}{2}mgv_0t^2 \cos\theta\, \hat{\mathbf{k}}$ d) $\frac{1}{2}mgv_0t^2 \cos\theta\, \hat{\mathbf{i}}$

9. When a body moves in a circular path, no work is done by the force since
a) force and displacement are perpendicular other
b) the force is always away from the center
c) there is no displacement
d) there is no net force

10. Consider a vector $\vec{F} = 4\hat{i} - 3\hat{j}$. Another vector that is perpendicular to $\vec{F}$ is
a) $4\hat{i} + 3\hat{j}$
b) $6\hat{j}$
c) $7\hat{j}$
d) $3\hat{i} - 4\hat{j}$

11. A particle is projected with a velocity 200 ms^{-1} at an angle of 60°. At the highest point, it explodes into three particles of equal masses. One goes vertically upwards with a velocity 100 ms^{-1}, the second particle goes vertically downwards. What is the velocity of third particle?
a) 120 ms^{-1} making 60° angle with horizontal
b) 200 ms^{-1} making 60° angle with horizontal
c) 300ms^{-1}
d) 200 ms^{-1}

12. The maximum height attained by a projectile is increased by 10% by increasing its speed of projection, without changing the angle of projection. The percentage increase in the horizontal range will be
a) 5%
b) 10%
c) 15%
d) 20%

13. A motorcycle is going on an overbridge of radius R. The driver maintains a constant speed. As the motorcycle is ascending on the overbridge, the normal force on it
a) Increases
b) Decreases
c) Remains the same
d) Fluctuates

14. A car is moving on a circular level road of radius of curvature 300 m. If the coefficient of friction is 0.3 and acceleration due to gravity 10 ms^{-2}, the maximum speed the car can have is (in km h^{-1})
a) 30
b) 81
c) 108
d) 162

15. A body of mass 5 kg is moving in a circle of radius 1 m with an angular velocity of 2 $radian$/sec. The centripetal force is
a) 10 N
b) 20 N
c) 30 N
d) 40 N

16. If the sum of the two unit vectors is also a unit vector, then magnitude of their difference is
a) $\sqrt{2}$
b) $\sqrt{3}$
c) $\sqrt{4}$
d) $\sqrt{7}$

17. A force $\vec{F} = -K(y\hat{i} + x\hat{j})$ (where K is a positive constant) acts on a particle moving in the $x - y$ plane. Starting from the origin, the particle is taken along the positive x-axis to the point $(a, 0)$ and then parallel to the y-axis to the (a, a). The total work done by the force $\vec{F}$ on the particle is
a) $-2Ka^2$
b) $2Ka^2$
c) $-Ka^2$
d) Ka^2

18. A projectile can have the same range R for two angles of projection. If T_1 and T_2 be the times of flights in the two cases, then the product of the two times of flights is directly proportional to
a) $\frac{1}{R^2}$
b) $\frac{1}{R}$
c) R
d) R^2

19. A projectile shot into air at some angle with the horizontal has a range of 200 m. If the time of flight is 5 s, then the horizontal component of the velocity of the projectile at the highest point of trajectory is
a) 40 ms^{-1}
b) 0 ms^{-1}
c) 9.8 ms^{-1}
d) Equal to the velocity of projection of the projectile

20. If $\vec{A}=2\hat{i} + 3\hat{j} + 4\hat{k}$ and $\vec{B} = 4\hat{i} + 3\hat{j} + 2\hat{k}$, then angle between $\vec{A}$ and $\vec{B}$ is
a) $\sin^{-1}(25/29)$
b) $\sin^{-1}(29/25)$
c) $\cos^{-1}(25/29)$
d) $\cos^{-1}(29/25)$

21. A body slides down a frictionless track which ends in a circular loop of diameter D. Then the minimum height h of the body in terms of D so that it may just complete the loop, is

a) $h = \frac{5}{2}D$ b) $h = \frac{3}{2}D$ c) $h = \frac{5}{4}D$ d) $h = 2D$

22. A particle moves in a circular path with decreasing speed. Choose the correct statement
a) Angular momentum remains constant
b) Acceleration ($\vec{a}$) is towards the centre
c) Particle moves in a spiral path with decreasing radius
d) The direction of angular momentum remains constant

23. A stone of mass 1 kg tied to a light inextensible string of length $L = \frac{10}{3} m$ is whirling in a circular path of radius L in a vertical plane. If the ratio of the maximum tension in the string to the minimum tension in the string is 4 and if g is taken to be $10m/\sec^2$, the speed of the stone at the highest point of the circle is
a) $20m/sec$ b) $10\sqrt{3}m/sec$ c) $5\sqrt{2}m/sec$ d) $10m/sec$

24. A body moves along a circular path of radius 5 m. The coefficient of friction between the surface of path and the body is 0.5. The angular velocity, in radians/sec, with which the body should move so that it does not leave the path is ($g = 10ms^{-2}$)
a) 4 b) 3 c) 2 d) 1

25. Two particles of equal mass are connected to a rope AB of negligible mass such that one is at end A and other dividing the length of rope in the ratio 1 : 2from B. The rope is rotated about end B in a horizontal plane. Ratio of tensions in the smaller part to the other is (ignore effect of gravity)
a) 4 : 3 b) 1 : 4 c) 1 : 2 d) 1 : 3

26. The velocity of projection of an oblique projectile is $\vec{\mathbf{v}} = 3\hat{\mathbf{i}} + 2\hat{\mathbf{j}}$ (in ms^{-1}). The speed of the projectile at the highest point of the trajectory is
a) 3 ms^{-1} b) 2 ms^{-1} c) 1 ms^{-1} d) Zero

27. The centripetal acceleration of a body moving in a circle of radius 100 m with a time period of 2 s will be
a) 98.5 ms^{-2} b) 198.5 ms^{-2} c) 49.29 ms^{-2} d) 985.9 ms^{-2}

28. A particle of mass m is moving in a horizontal circle of radius r, under a centripetal force $= \frac{k}{r^{2\prime}}$ where k is a constant.
a) The potential energy of the particle is zero
b) The potential energy of the particle is $\frac{k}{r}$
c) The total energy of the particle is $-\frac{k}{2r}$
d) The Kinetic energy of the particle is $-\frac{k}{r}$

29. A plumb line is suspended from a ceiling of a car moving with horizontal acceleration of$a$. What will be the angle of inclination with vertical?
a) $\tan^{-1}\left(\frac{a}{g}\right)$ b) $\tan^{-1}\left(\frac{g}{a}\right)$ c) $cas^{-1}\left(\frac{a}{g}\right)$ d) $cas^{-1}\left(\frac{g}{a}\right)$

30. A monkey can jump a maximum horizontal distance of 20 m. Then the velocity of the monkey is
a) 10 ms^{-1} b) 14 ms^{-1} c) 20 ms^{-1} d) 24 ms^{-1}

31. A ball is projected with kinetic energy K at an angle of 45° to the horizontal. At the highest point during its flight, its kinetic energy will be
a) K b) $K/\sqrt{2}$ c) $K/2$ d) Zero

32. A bullet fired at an angle of 30° with the horizontal hits the ground 3 km away. By adjusting its angle of projection, one can hope to hit a target 5 km away. Assume the muzzle speed to be same and the air resistance is negligible
a) possible to hit a target 5 km away b) not possible to hit a target 5 km away

c) prediction is not possible
d) None of the above

33. A particle undergoes uniform circular motion. About which point on the plane of the circle, will the angular momentum of the particle remain conserved?
a) center of the circle
b) on the circumference of the circle
c) inside the circle
d) outside the circle

34. When a ceiling fan is switched on, it makes 10 rotations in the first 4 s. How many rotations will it make in the next 4 s? (Assuming uniform angular acceleration)
a) 10
b) 20
c) 40
d) 30

35. The adjacent sides of a parallelogram are represented by co-initial vectors $2\hat{\imath} + 3\hat{\jmath}$and $\hat{\imath} + 4\hat{\jmath}$. The area of the parallelogram is
a) 5 units along z-axis
b) 5 units in $x - y$ plane
c) 3 units in $x - z$ plane
d) 3 units in $y - z$ plane

36. A body of mass m thrown horizontally with velocity v, from the top of tower of height h touches the level ground at distance of 250 m from the foot of the tower. A body of mass $2m$ thrown horizontally with velocity $\frac{v}{2}$, from the top of tower of height $4h$ will touch the level ground at a distance x from the foot of tower. The value of x is
a) 250 m
b) 500 m
c) 125 m
d) $250\sqrt{2}$ m

37. For an object thrown at 45° to horizontal, the maximum height (H) and horizontal range (R) are related as
a) $R = 16H$
b) $R = 8H$
c) $R = 4H$
d) $R = 2H$

38. A stone is just released from the window of a train moving along a horizontal straight track. The stone will hit the ground following
a) Straight path
b) Circular path
c) Parabolic path
d) Hyperbolic path

39. For a particle in uniform circular motion the acceleration **a** at a point $P(R, \theta)$ on the circle of the radius R is (here θ is measured from the x –axis)
a) $-\frac{v^2}{R}\cos\theta\,\hat{\mathbf{i}} + \frac{v^2}{R}\sin\theta\,\hat{\mathbf{j}}$
b) $-\frac{v^2}{R}\sin\theta\,\hat{\mathbf{i}} + \frac{v^2}{R}\cos\theta\,\hat{\mathbf{j}}$
c) $-\frac{v^2}{R}\cos\theta\,\hat{\mathbf{i}} - \frac{v^2}{R}\sin\theta\,\hat{\mathbf{j}}$
d) $-\frac{v^2}{R}\hat{\mathbf{i}} + \frac{v^2}{R}\hat{\mathbf{j}}$

40. A body is thrown with a velocity of 9.8 m/s making an angle of 30° with the horizontal. It will hit the ground after a time
a) 1.5 s
b) 1 s
c) 3 s
d) 2 s

41. A particle rests on the top of a hemisphere of radius R. Find the smallest horizontal velocity that must be imparted to the particle if it is to leave the hemisphere without sliding down it
a) $\sqrt{gR}$
b) $\sqrt{2gR}$
c) $\sqrt{3gR}$
d) $\sqrt{5gR}$

42. A small object placed on a rotating horizontal turn table just slips when it is placed at a distance of 4 cm from the axis of rotation, if the angular velocity of the turn table is doubled the object slips when its distance from the axis of rotation is
a) 1 cm
b) 2 cm
c) 4 cm
d) 8 cm

43. A particle is projected with velocity $2\sqrt{gh}$ so that it just clears two walls of equal height h, which are at a distance of $2h$ from each other. What is the time interval of passing between the two walls?
a) $\frac{2h}{g}$
b) $\sqrt{\frac{2h}{g}}$
c) $\sqrt{\frac{h}{g}}$
d) $2\sqrt{\frac{h}{g}}$

44. A can filled with water is revolved in a vertical of radius 4 m and the water does not fall down. The time period for a revaluation is about
a) 2 s
b) 4 s
c) 8 s
d) 10 s

45. The angular velocity of a wheel is 70 rad/sec. If the radius of the wheel is 0.5 m, then linear velocity of the wheel is
a) 70 m/s
b) 35 m/s
c) 30 m/s
d) 20 m/s

46. A stone is projected from the ground with velocity 50 m/s at an angle of 30°. It crosses a wall after 3 sec. How far beyond the wall the stone will strike the ground ($g = 10\ m/\sec^2$)

a) 90.2 m b) 89.6 m c) 86.6 m d) 70.2 m

47. The co-ordinates of a moving particle at time t are given by $x = ct^2$ and $y = bt^2$. The instantaneous speed of the particle is

a) $2t(b+c)$ b) $2t(b+c)^{1/2}$ c) $2t(c^2 - b^2)$ d) $2t(c^2 + b^2)^{1/2}$

48. A particle is acted upon by a force of constant magnitude which is always perpendicular to the velocity of the particle. The motion of the particle takes place in a plane it follows that

a) Its velocity is constant b) Its acceleration is constant
c) Its kinetic energy is constant d) It moves in a straight line

49. A circular road of radius 1000 m has banking angle 45°. The maximum safe speed of a car having mass 2000 kg will be, if the coefficient of friction between tyre and road is 0.5

a) 172 m/s b) 124 m/s c) 99 m/s d) 86 m/s

50. A small sphere is hung by a string fixed to a wall. The sphere is pushed away from the wall by a stick. The force acting on the sphere are shown in figure. Which of the following statements is wrong?

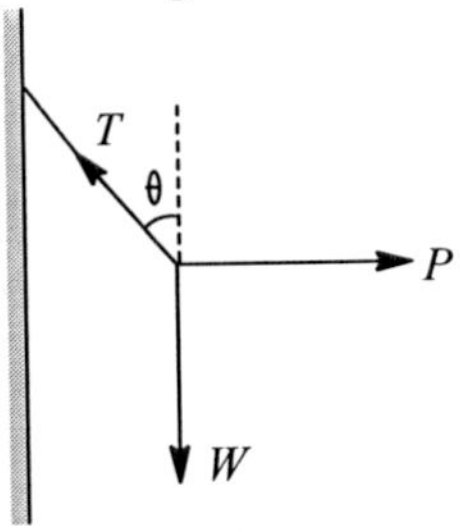

a) $P = W\tan\theta$ b) $\vec{T} + \vec{P} + \vec{W} = 0$ c) $T^2 = P^2 + W^2$ d) $T = P + W$

51. An aircraft executes a horizontal loop with a speed of 150 m/s with its wings banked at an angle of 12°. The radius of the loop is ($g = 10\ m/s^2$, $\tan 12° = 0.2126$)

a) 10.6 km b) 9.6 km c) 7.4 km d) 5.8 km

52. Two forces, each equal to F, act as shown in figure. Their resultant is

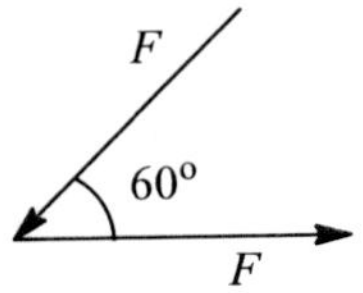

a) $\frac{F}{2}$ b) F c) $\sqrt{3}\,F$ d) $\sqrt{5}\,F$

53. Radius of the curved road on national highway is R. Width of the road is b. The outer edge of the road is raised by h with respect to inner edge so that a car with velocity v can pass safe over it. The value of h is

a) $\frac{v^2b}{Rg}$ b) $\frac{v}{Rgb}$ c) $\frac{v^2R}{g}$ d) $\frac{v^2b}{R}$

54. A particle is moving along a circular path with a uniform speed. How does its angular velocity change when it completes half of the circular path?

a) No change b) Increases c) Decreases d) Cannot say

55. A body moving along a circular path of radius R with velocity v, has centripetal acceleration a. If its velocity is made equal to $2v$, then its centripetal acceleration is

a) $4a$ b) $2a$ c) $\frac{a}{4}$ d) $\frac{a}{2}$

56. A body of mass 1 kg is rotating in a vertical circle of radius 1m. What will be the difference in its kinetic energy at the top and bottom of the circle? (Take $g = 10\ ms^{-2}$)

a) 10 J b) 20 J c) 30 J d) 50 J

57. In the case of an oblique projectile, the velocity is perpendicular to acceleration

a) Once only b) Twice c) Thrice d) Four times

58. A body of mass $m\ kg$ is rotating in a vertical circle at the end of a string of length r metre. The difference in the kinetic energy at the top and bottom of the circle is

a) $\frac{mg}{r}$ b) $\frac{2mg}{r}$ c) $2mgr$ d) mgr

59. A body of mass 1 kg thrown with a velocity of $10\ ms^{-1}$ at an angle of $60°$ with the horizontal. Its momentum at the highest point is

a) $2\ kg\ ms^{-1}$ b) $3\ kg\ ms^{-1}$ c) $4\ kg\ ms^{-1}$ d) $5\ kg\ ms^{-1}$

60. A stone is projected from the ground with velocity $50\ ms^{-1}$ and angle of $30°$. It crosses a wall after 3 s. How far beyond the wall the stone will strike the ground?

a) 80.5 m b) 85.6 m c) 86.6 m d) 75.2 m

61. A bob of mass M is suspended by a massless string of length L. The horizontal velocity V at position A is just sufficient to make it reach the point B. The angle θ at which the speed of the bob is half of that at A, satisfies

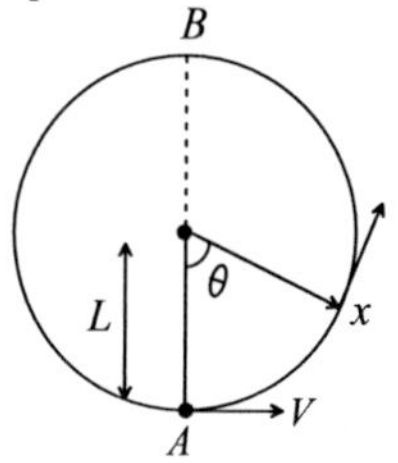

a) $\theta = \frac{\pi}{4}$ b) $\frac{\pi}{4} < \theta < \frac{\pi}{4}$ c) $\frac{\pi}{2} < \theta < \frac{3\pi}{4}$ d) $\frac{3\pi}{4} < \theta < \pi$

62. For a particle in uniform circular motion, thee acceleration $\vec{a}$ at a point $P(R, \theta)$ on the circle of radius R is (Here θ is measured from the x-axis)

a) $\frac{v^2}{R}\hat{\imath} + \frac{v^2}{R}\hat{\jmath}$ b) $-\frac{v^2}{R}\cos\theta\,\hat{\imath} + \frac{v^2}{R}\sin\theta\,\hat{\jmath}$

c) $-\frac{v^2}{R}\sin\theta\,\hat{\imath} + \frac{v^2}{R}\cos\theta\,\hat{\jmath}$ d) $-\frac{v^2}{R}\cos\theta\,\hat{\imath} - \frac{v^2}{R}\sin\theta\,\hat{\jmath}$

63. A man can thrown a stone 100 m away. The maximum height to which he can throw vertically is

a) 200 m b) 100 m c) 50 m d) 25 m

64. Roads are banked on curves so that

a) The speeding vehicles may not fall outwards
b) The frictional force between the road and vehicle may be decreased
c) The wear and tear of tyres may be avoided
d) The weight of the vehicle may be decreased

65. A piece of wire is bent in the shape of a parabola $y = kx^2$ (y-axis vertical) with a bead of mass m on it. The bead can side on the wire without friction. It stays at the lowest point of the parabola when the wire is at rest. The wire is now accelerated parallel to the x-axis with a constant acceleration a. The distance of the new equilibrium position of the bead, where the bead can stay at rest with respect to the wire, from the y-axis is

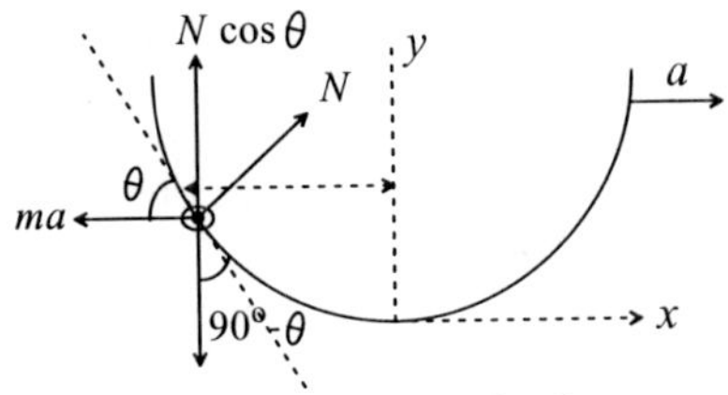

a) a/gk b) $a/2gk$ c) $2a/gk$ d) $a/4gk$

66. 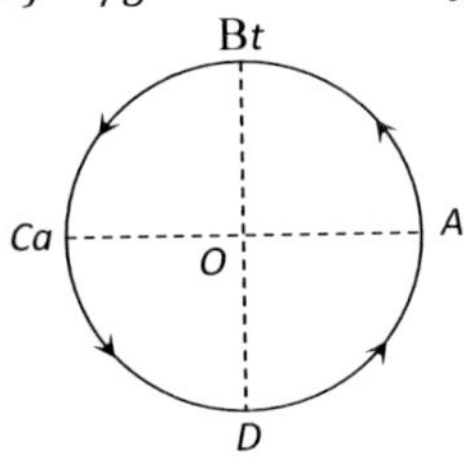

Figure shows a body of mass m moving with a uniform speed v along a circle of radius r. The change in velocity in going from A to B is

a) $v\sqrt{2}$ b) $v/\sqrt{2}$ c) v d) zero

67. A particle starts from the origin of coordinates at time $t = 0$ and moves in the $x - y$ plane with a constant acceleration α in the y-diretion. Its equation of motion is $y = \beta x^2$. Its velocity component in the x-direction is

a) Variable b) $\sqrt{\frac{2\alpha}{\beta}}$ c) $\frac{\alpha}{2\beta}$ d) $\sqrt{\frac{\alpha}{2\beta}}$

68. A body is revolving with a uniform speed v in a circle of radius r. The tangential acceleration is

a) $\frac{v}{r}$ b) $\frac{v^2}{r}$ c) Zero d) $\frac{v}{r^2}$

69. A project is projected with a velocity of 20 m/s making an angle of 45° with horizontal. The equation for the trajectory is $h = Ax - Bx^2$ where h is height, x is horizontal distance, A and B are constants. The ratio $A : B$ is ($g = 10\ ms^{-2}$)

a) 1 : 5 b) 5 : 1 c) 1 : 40 d) 40 : 1

70. If the equation for the displacement of a particle moving on a circular path is given by $(\theta) = 2t^3 + 0.5$, where θ is in radians and t in seconds, then the angular velocity of the particle after 2 sec from its start is

a) 8 rad/sec b) 12 rad/sec c) 24 rad/sec d) 36 rad/sec

71. Two masses M and m are attached to a vertical axis by weightless threads of combined length l. They are set in rotational motion in a horizontal plane about this axis with constant angular velocity ω . If the tensions in the threads are the same during motion, the distance of M from the axis is

a) $\frac{Ml}{M+m}$ b) $\frac{ml}{M+m}$ c) $\frac{M+m}{M}l$ d) $\frac{M+m}{m}l$

72. A bomber plane moves horizontally with a speed of 500 m/s and a bomb released from it, strikes the ground in 10 sec. Angle at which it strikes the ground will be ($g = 10\ m/s^2$)

a) $\tan^{-1}\left(\frac{1}{5}\right)$ b) $\tan^{-1}\left(\frac{1}{2}\right)$ c) $\tan^{-1}(1)$ d) $\tan^{-1}(5)$

73. An electric fan has blades of length 30 cm as measured from the axis of rotation. If the fan is rotating at 1200 $r.p.m.$ The acceleration of a point on the tip of the blade is about

a) 1600 $m/\sec^2$ b) 4740 $m/\sec^2$ c) 2370 $m/\sec^2$ d) 5055 $m/\sec^2$

74. A particle moves in a plane with constant acceleration in a direction different from the initial velocity. The path of the particle will be
a) A straight line b) An arc of a circle c) A parabola d) An ellipse

75. A force is inclined at 60° to the horizontal. If its rectangular component in the horizontal direction is 50 N, then magnitude of the force in the vertical direction is
a) 25 N b) 75 N c) 87 N d) 100 N

76. The maximum speed with which a car is driven round a curve of radius 18 m without skidding (where, $g = 10\text{ms}^{-2}$ and the coefficient of friction between rubber tyres and the roadway is 0.2) is
a) 36.0 km h^{-1} b) 18.0 km h^{-1} c) 21.6 km h^{-1} d) 14.4 km h^{-1}

77. The equation of trajectory of a projectile is $y = 10x - \left(\frac{5}{9}\right)x^2$. if we assume $g = 10\text{ms}^{-2}$, the range of projectile (in metre) is
a) 36 b) 24 c) 18 d) 9

78. The minimum velocity at the lowest point, so that the string just slack at the highest point in a vertical circle of radius l
a) $\sqrt{gl}$ b) $\sqrt{3gl}$ c) $\sqrt{5gl}$ d) $\sqrt{7gl}$

79. The maximum height attained by a projectile when thrown at an angle θ with the horizontal is found to be half the horizontal range. Then θ is equal to
a) $\tan^{-2}(2)$ b) $\frac{\pi}{6}$ c) $\frac{\pi}{4}$ d) $\tan^{-1}\left(\frac{1}{2}\right)$

80. The magnitude of the centripetal force acting on a body of *mass m* executing uniform motion in a circle of radius r with speed v is
a) mvr b) mv^2/r c) v/r^2m d) v/rm

81. The kinetic energy K of a particle moving along a circle or radius R depends on the distance covered $s$as $K = a\,s^2$. The force acting on the particle is
a) $2\,a\,s\,R$ b) $2\,a\,s[1 + s^2/R^2]^{1/2}$ c) $2\,a\,s$ d) $2\,as^2/R$

82. A body is projected at an angle θ to the horizontal with kinetic energy E_k. The potential energy at the highest point of the trajectory is
a) E_k b) $E_k\cos^2\theta$ c) $E_k\sin^2\theta$ d) $E_k\tan^2\theta$

83. In a loop-the-loop, a body starts at a height $h = 2R$. The minimum speed with which the body must be pushed down initially in order that it may be able to complete the vertical circle is
a) $\sqrt{2gR}$ b) $\sqrt{gR}$ c) $\sqrt{3gR}$ d) $2\sqrt{gR}$

84. A stone is swinging in a horizontal circle 0.8 m in diameter, at 30rev/min. A distant light causes a shadow of the stone to be formed on a nearby wall. What is the amplitude of the motion of the shadow? What is the frequency?
a) 0.4 m, 1.5 Hz b) 0.4 m, 0.5 Hz c) 0.8 m, 0.5 Hz d) 0.2 m, 0.5 Hz

85. A particle is projected at an angle of 60° above the horizontal with a speed of 10 ms^{-1}. After some time the direction of its velocity makes an angle of 30° above the horizontal. The speed of the particle at this instant is
a) $\frac{5}{\sqrt{3}}\text{ms}^{-1}$ b) $5\sqrt{3}\text{ ms}^{-1}$ c) 5 ms^{-1} d) $\frac{10}{\sqrt{3}}\text{ms}^{-1}$

86. A particle of mass m attracted with a string of length l is just revolving on the vertical circle without slacking of the string. If v_A, v_B and v_D are speed at position A, B and D then

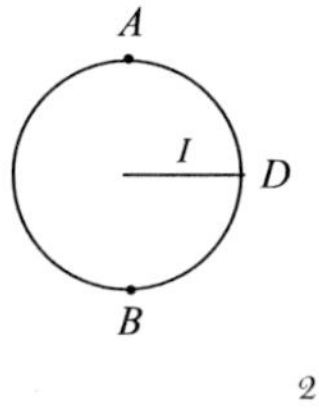

a) $v_B > v_D > v_A$ b) Tension in string at $D = 3\,mg$

c) $v_D = \sqrt{3gl}$ d) All of the above

87. A car is travelling with linear velocity v on a circular road of radius r. If it is increasing its speed at the rate of $'a'm/s^2$, then the resultant acceleration will be

a) $\sqrt{\left\{\frac{v^2}{r^2} - a^2\right\}}$ b) $\sqrt{\left\{\frac{v^4}{r^2} + a^2\right\}}$ c) $\sqrt{\left\{\frac{v^4}{r^2} - a^2\right\}}$ d) $\sqrt{\left\{\frac{v^2}{r^2} + a^2\right\}}$

88. The acceleration of a vehicle travelling with speed of 400ms^{-1} as it goes round a curve of radius 160 m, is

a) 1 kms^{-2} b) 100 ms^{-2} c) 10 ms^{-2} d) 1 ms^{-2}

89. A 500 kg car takes a round turn of radius 50 m with a velocity of 36 km h^{-1}. The centripetal force, is

a) 250 N b) 750 N c) 1000 N d) 1200 N

90. If a cyclist moving with a speed of $4.9\,m/s$ on a level road can take a sharp circular turn of radius $4\,m$, then coefficient of friction between the cycle tyres and road is

a) 0.41 b) 0.51 c) 0.61 d) 0.71

91. If $\vec{A}\cdot\vec{B} = AB$, then the angle between $\vec{A}$ and $\vec{B}$ is

a) 0° b) 45° c) 90° d) 180°

92. The angular amplitude of a simple pendulum is θ_0. The maximum tension in its string will be

a) $mg(1-\theta_0)$ b) $mg(1+\theta_0)$ c) $mg(1-\theta_0^2)$ d) $mg(1+\theta_0^2)$

93. A body is projected up a smooth inclined plane with a velocity v_0 from the point A as shown in figure. The angle of inclination is 45° and top B of the plane is connected to a well of diameter 40 m. If the body just manages to cross the well, what is the value of v_0? Length of the inclined plane is $20\sqrt{2}\text{m}$, and $\text{g} = 10\text{ms}^{-2}$

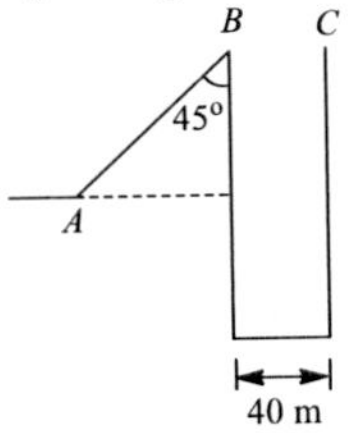

a) 20 ms^{-1} b) $20\sqrt{2}\text{ms}^{-1}$ c) 40 ms^{-1} d) $40\sqrt{2}\text{ ms}^{-1}$

94. A sphere of mass m is tied to end of a string of length l and rotated through the other end along a horizontal circular path with speed v. The work done in full horizontal circle is

a) 0 b) $\left(\frac{mv^2}{l}\right).2\pi l$ c) $mg.2\pi$ d) $\left(\frac{mv^2}{l}\right).(l)$

95. A bucket filled with water is tied to a rope of length 0.5 m and is rotated in a circular path in vertical pane. The least velocity it should have at the lowest point of circle so that water dose not spill is, $(g = 10\text{ ms}^{-2})$

a) $\sqrt{5}\text{ms}^{-1}$ b) $\sqrt{10}\text{ms}^{-1}$ c) 5 ms^{-1} d) $2\sqrt{5}\text{ms}^{-1}$

96. An object is projected at an angle of 45° with the horizontal. The horizontal range and the maximum height reached will be in the ratio

a) 1 : 2 b) 2 : 1 c) 1 : 4 d) 4 : 1

97. Three identical particles are joined together by a thread as shown in figure. All the three particles are moving in a horizontal plane. If the velocity of the outermost particle is v_0, then the ratio of tensions in the three sections of the string is

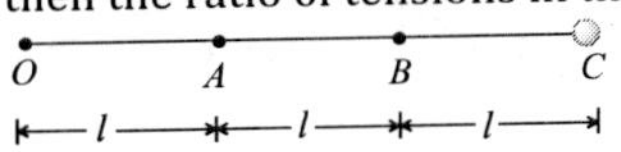

a) 3 : 5 : 7 b) 3 : 4 : 5 c) 7 : 11 : 6 d) 3 : 5 : 6

98. The horizontal range and the maximum height of a projectile are equal. The angle of projection of the projectile is

a) $\theta = \tan^{-1}\left(\frac{1}{4}\right)$ b) $\theta = \tan^{-1}(4)$ c) $\theta = \tan^{-1}(2)$ d) $\theta = 45°$

99. A man can throw a stone to a maximum distance of 80 m. The maximum height to which it will rise in metre, is

a) 30 m b) 20 m c) 10 m d) 40 m

100. A bob of mass *M* is suspended by a massless string of length *L*. The horizontal velocity v at position *A* is just sufficient to make it reach the point *B*. The angle θ at which the speed of the bob is half of that at *A*, satisfies

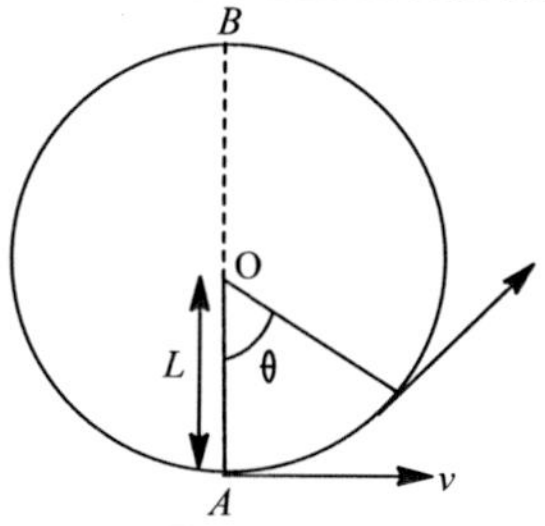

a) $\theta = \frac{\pi}{4}$ b) $\frac{\pi}{4} < \theta < \frac{\pi}{2}$ c) $\frac{\pi}{2} < \theta < \frac{3\pi}{4}$ d) $\frac{3\pi}{4} < \theta < \pi$

101. Given that centripetal force $F = -k/r^2$. The total energy is

a) $-k/r^2$ b) k/r c) $-k/2r^2$ d) $-k/2r$

102. The wheel of toy car rotates about axis. It slows down from 400 rps to 200 rps in 2s. Then its angular retardation in rads^{-2} is

a) 200 π b) 100 c) 400 π d) None of these

103. The magnitude of resultant of three vectors of magnitude 1, 2 and 3 whose directions are those of the sides of an equilateral triangle taken in order is

a) zero b) $2\sqrt{2}$ unit c) $4\sqrt{3}$ unit d) $\sqrt{3}$ unit

104. The horizontal range of a projectile is $4\sqrt{3}$ times its maximum height. Its angle of projection will be

a) 45° b) 60° c) 90° d) 30°

105. When a body moves in a circular path, no work is done by the force since,

a) There is no displacement
b) There is no net force
c) Force and displacement are perpendicular to each other
d) The force is always away from the centre

106. A particle is moving on a circular path with constant speed, then its acceleration willbe

a) Zero b) External radial acceleration
c) Internal radial acceleration d) Constant acceleration

107. When the road is dry and coefficient of friction is μ, the maximum speed of a car in a circular path is 10 ms^{-1}. If the road becomes wet and $\mu' = \mu/2$, what is the maximum speed permitted?

a) 5ms^{-1} b) 10 ms^{-1} c) $10\sqrt{2}$ ms^{-1} d) $5\sqrt{2}$ ms^{-1}

108. A particle is moving with a constant speed v in a circle. What is the magnitude of average after half rotation?

a) $2v$ b) $2\frac{v}{\pi}$ c) $\frac{v}{2}$ d) $\frac{v}{2\pi}$

109. A projectile is fired with a velocity v at an angle θ with the horizontal. The speed of the projectile when its direction of motion makes an angle β with the horizontal is

a) $v \cos\theta$ b) $v \cos\theta \cos\beta$ c) $v \cos\theta \sec\beta$ d) $v \cos\theta \tan\beta$

110. The radio of angular speeds of minute hand and hour hand of a watch is

a) 1 : 12 b) 12 : 1 c) 6 : 1 d) 1 : 6

111. Three vectors $\vec{A}, \vec{B}$ and $\vec{C}$ satisfy the relation $\vec{A} \cdot \vec{B} = 0$ and $\vec{A} \cdot \vec{C} = 0$. If $\vec{B}$ and $\vec{C}$ are not lying in the same plane then $\vec{A}$ is parallel to

a) $\vec{B}$ b) $\vec{C}$ c) $\vec{B} \times \vec{C}$ d) $\vec{B} \cdot \vec{C}$

112. A car of mass 2000 kg is moving with a speed of 10 ms^{-1} on a circular path of radius 20 m on a level road. What must be the frictional force between the car and the road so that the car does not slip?

a) 10^4N b) 10^3N c) 10^5N d) 10^2N

113. The vectors $\vec{a}$ and $\vec{b}$ are such that $|\vec{a} + \vec{b}| = |\vec{a} - \vec{b}|$. What is the angle between $\vec{a}$ and $\vec{b}$?

a) 0° b) 90° c) 60° d) 180°

114. The maximum horizontal range of a projectile is 400 m. The maximum value of the height attained by it will be

a) 100 m b) 200 m c) 400 m d) 800 m

115. The second's hand of a watch has length 6 cm. Speed of end point and magnitude of difference of velocities at two perpendicular positions will be

a) 6.28 and 0 mm/s b) 8.88 and 4.44 mm/s c) 8.88 and 6.28 mm/s d) 6.28 and 8.88 mm/s

116. A cane filled with water is revolved in a vertical circle of radius 4 m and the water just does not fall down. The time period of revolution will be

a) 1 sec b) 10 sec c) 8 sec d) 4 sec

117. A heavy small sized sphere is suspended by a string of length l. The sphere is rotated uniformly in a horizontal circle with the string making an angle θ with the vertical. The time period of this conical pendulum is

a) $2\pi\sqrt{\frac{l \tan\theta}{g}}$ b) $2\pi\sqrt{l \sin\theta/g}$ c) $2\pi\sqrt{\frac{l}{g}}$ d) $2\pi\sqrt{\frac{l \cos\theta}{g}}$

118. The angular speed of seconds needle in a mechanical watch is

a) $\frac{\pi}{30}\ rad/s$ b) $2\pi\ rad/s$ c) $\pi\ rad/s$ d) $\frac{60}{\pi}\ rad/s$

119. The minimum speed for a particle at the lowest point of a vertical circle of radius r, to describe the circle is v. If the radius of the circle is reduced to one-fourth its value, the corresponding minimum speed will be

a) $v/4$ b) $v/2$ c) $2v$ d) $4v$

120. A particle of mass m is projected with a velocity v making an angle of 30° with the horizontal. The magnitude of angular momentum of the projectile about the point of projection when the particle is at its maximum height h is

a) $\frac{\sqrt{3}}{2}\frac{mv^2}{g}$ b) Zero c) $\frac{mv^3}{\sqrt{2g}}$ d) $\frac{\sqrt{3}}{16}\frac{mv^3}{g}$

121. A cyclist moves in such a way that he track 60° turn after 100 m. What is the displacement when to takes seventh turn?

a) 100 m b) 200 m c) $100\sqrt{3}$ m d) $100\sqrt{3}$ m

122. A body is moving in a circular path with acceleration a. If its velocity gets doubled, find the ratio of acceleration after and before the change

a) 1 : 4 b) $\frac{1}{4}$: 2 c) 2 : 1 d) 4 : 1

123. A weightless thread can bear tension upto 37 N. A. stone of mass 500 g is tied to it and revolved in a circular path of radius 4 m in a vertical plane. If $g = 10\ ms^{-2}$, then the maximum angular velocity of the stone will be

a) 2 rad s^{-1} b) 4 rad s^{-1} c) 8 rad s^{-1} d) 16 rad s^{-1}

124. A ball is projected with velocity V_o at an angle of elevation 30° . Mark the correct statement
a) Kinetic energy will be zero at the highest point of the trajectory
b) Vertical component of momentum will be conserved
c) Horizontal component of momentum will be conserved
d) Gravitational potential energy will be minimum at the highest point of the trajectory

125. What is the unit vector along $\hat{i}+\hat{j}$?
a) $\frac{\hat{i}+\hat{j}}{\sqrt{2}}$ b) $\sqrt{2}(\hat{i}+\hat{j})$ c) $\hat{i}+\hat{j}$ d) $\hat{k}$

126. A child travelling in a train throws a ball outside with a speed V. According to a child who is standing on the ground, the speed of the ball is
a) Same as V b) Greater than V c) Less than V d) None of these

127. A weightless thread can bear tension upto3.7 kg wt. A stone of mass 500 gms is tied to it and revolved in a circular path of radius 4 m in a vertical plane. If $g = 10\ ms^{-2}$, then the maximum angular velocity of the stone will be
a) 4 $radians/sec$ b) 16 $radians/sec$ c) $\sqrt{21}$ $radians/sec$ d) 2 $radians/sec$

128. A car of mass 800 kg moves on a circular track of radius 40 m. If the coefficient of friction is 0.5, then maximum velocity with which the car can move is
a) 7 m/s b) 14 m/s c) 8 m/s d) 12 m/s

129. Which of the following statements is false for a particle moving in a circle with a constant angular speed?
a) The velocity vector is tangent to the circle
b) The acceleration vector is tangent to the circle
c) The acceleration vector point to the center of the circle
d) The velocity and acceleration vectors are perpendicular to each other

130. A particle moves in a circular orbit under the action of a central attractive force inversely proportional to the distance $'r'$. The speed of the particle is
a) Proportional to r^2 b) Independent of r c) Proportional to r d) Proportional to $1/r$

131. A particle A is projected from the ground with an initial velocity of 10 ms^{-1}at an angle of 60° with horizontal. From what height h should an another particle B be projected horizontally with velocity 5ms^{-1} so that both the particles collide in ground at point C if both are projected simultaneously? (g = 10 ms^{-2})

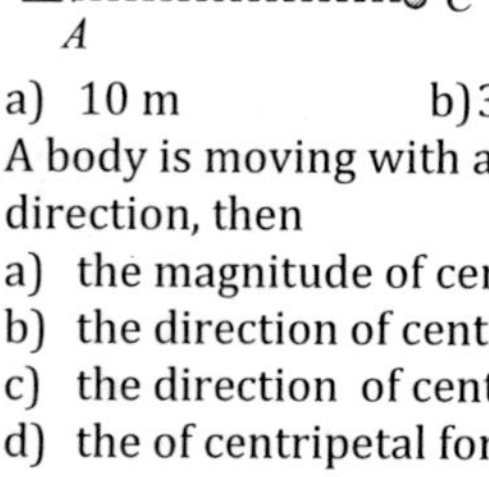

a) 10 m b) 30 m c) 15 m d) 25 m

132. A body is moving with a certain velocity in a circular path. Now, the body reverses its direction, then
a) the magnitude of centripetal force remains same
b) the direction of centripetal force remains same
c) the direction of centripetal acceleration remains same
d) the of centripetal force does not change

133. An aeroplane is flying horizontally with a velocity of 216 kmh^{-1} and at a height of 1960 m. When it is vertically above a point A on the ground, a bomb is released from it. The bomb strikes the ground at point B. The distance AB is (ignoring air resistance)
a) 1200 m b) 0.33 km c) 3.33 km d) 33 km

134. A body 'A' moves with constant velocity on a straight line path tangential to the earth's surface. Another body 'B' is thrown vertically upwards, it goes to a height and falls back on earth. A third body 'C' is projected to an angle and follows a parabolic path as shown in figure

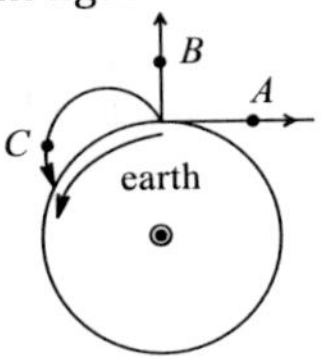

The bodies whose angular momentum relative to the center of the earth is conserved are

a) B only b) B and C c) A, B, C d) None of the above

135. A cart is moving horizontally along a straight line with constant speed 30 m/s. A projectile is to be fired from the moving cart in such a way that it will return to the cart after the cart has moved 80 m. At what speed (relative to the cart) must the projectile be fired (Take $g = 10\ m/s^2$)

a) 10 m/s b) $10\sqrt{8}\ m/s$ c) $\frac{40}{3}\ m/s$ d) None of these

136. If $\vec{A}, \vec{B}$ and $\vec{C}$ are the unit vectors along the incident ray, reflected ray and outward normal to the reflecting surface, then

a) $\vec{B} = \vec{A} - \vec{C}$ b) $\vec{B} = \vec{A} + (\vec{A} \cdot \vec{C})\vec{C}$ c) $\vec{B} = 2\vec{A} - \vec{C}$ d) $\vec{B} = \vec{A} - 2(\vec{A} \cdot \vec{C})\vec{C}$

137. The string of a pendulum of length l is displaced through 90° from the vertical and released. Then the minimum strength of the string in order to withstand the tension as the pendulum passes through the mean position is

a) mg b) 6 mg c) 3 mg d) 5 mg

138. A particle reaches its highest point when it has covered exactly one half of its horizontal range. The corresponding point on the displacement time graph is characterised by

a) Negative slope and zero curvature b) Zero slope and negative curvature
c) Zero slope and positive curvature d) positive slope and zero curvature

139. If the vector $\vec{A} = 2\hat{i} + 4\hat{j}$ and $\vec{B} = 5\hat{i} + p\hat{j}$ are parallel to each other, the magnitude of $\vec{B}$ is

a) $5\sqrt{5}$ b) 10 c) 15 d) $2\sqrt{5}$

140. An arrow is shot into air. Its range is 200 m and its time of flight is 5s. If $g = 10m/s^2$, then the horizontal component of velocity of the arrow is

a) 12.5 m/s b) 25 m/s c) 31.25 m/s d) 40 m/s

141. A stone of mass 2kg is tied to a string of length 0.5 m. If the breaking tension of the string is 900 N, then the maximum angular velocity, the stone can have in uniform circular motion is

a) 30 rads^{-1} b) 20 rads^{-1} c) 10 rads^{-1} d) 25 rads^{-1}

142. Four bodies P, Q, R and S are projected with equal velocities having angles of projection 15°, 30°, 45° and 60° with the horizontal respectively. The body having shortest range is

a) P b) Q c) R d) S

143. A plane surface is inclined making an angle θ with the horizontal. Form the bottom of this inclined plane, a bullet is fired with velocity v. The maximum possible range of the bullet on the inclined plane is

a) $\frac{v^2}{g}$ b) $\frac{v^2}{g(1+\sin\theta)}$ c) $\frac{v^2}{g(1-\sin\theta)}$ d) $\frac{v^2}{g(1+\sin\theta)^2}$

144. A body of mass 0.4 kg is whirled in a vertical circle making2 rev/sec. If the radius of the circle is 2 m, then tension in the string when the body is at the top of the circle, is

a) 41.56 N b) 89.86 N c) 109.86 N d) 115.86 N

145. For a particle in non-uniform accelerated circular motion
   a) Velocity is radial and acceleration is transverse only
   b) Velocity is transverse and acceleration is radial only
   c) Velocity is radial and acceleration has both radial and transverse components
   d) Velocity is transverse and acceleration has both radial and transverse components
146. An aeroplane is flying horizontally with a constant velocity of 100 kmh^{-1} at a height of 1 km from the ground level. At $t = 0$, it starts dropping packets at constant time intervals of T_0. If R represents the separation between two consecutive points of impact on the ground, then for the first three packets, R_1/R_2 is
   a) 1 b) >1
   c) <1 d) Sufficient data is not given
147. The angle which the bicycle and its rider must make with the vertical when going round a curve of 7 m radius at 5 ms^{-1} is
   a) 20° b) 15° c) 10° d) 5°
148. Average torque on a projectile of mass m, initial speed u and angles of projection θ, between initial and final position P and Q as shown in figure about the point of projection is

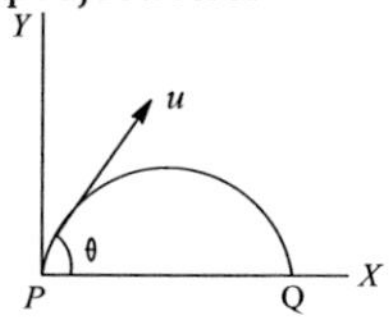

   a) $mu^2 \sin\theta$ b) $mu^2 \cos\theta$ c) $\frac{1}{2} mu^2 \sin 2\theta$ d) $\frac{1}{2} mu^2 \cos 2\theta$
149. Two bullets are fired simultaneously, horizontally and with different speeds from the same place. Which bullet will hit the ground first?
   a) The faster bullet b) The slower bullet
   c) Both will hit simultaneously d) Depends on the masses
150. A stone of mass m is tied to a string and is moved in a vertical circle of radius r making n revolutions per *minute*. The total tension in the string when the stone is at its lowest point is
   a) mg b) $m(g + \pi n r^2)$
   c) $m(g + \pi n r)$ d) $m\{g + (\pi^2 n^2 r)/900\}$
151. A projectile is thrown with velocity v making an angle θ with the horizontal. It just crosses the tops of two poles, each of height h, after 1s and 3s respectively. The time of flight of the projectile is
   a) 1 s b) 3 s c) 4 s d) 7.8 s
152. A coin placed on a rotating turn table just slips if it is placed at a distance of 8 cm from the centre. If angular velocity of the turn table is doubled, it will just slip at a distance of
   a) 1 cm b) 2 cm c) 4 cm d) 8 cm
153. A bomb is dropped from an aeroplane moving horizontally at constant speed. When air resistance is taken into consideration, the bomb
   a) Falls to earth exactly below the aeroplane b) Fall to earth behind the aeroplane
   c) Falls to earth ahead of the aeroplane d) Flies with the aeroplane
154. One end of a string of length l is connected to a particle of mass m and the other to a small peg on a smooth horizontal table. If the particle moves in a circle with speed v, the net force on the particle (directed towards the centre) is
   a) T b) $T - \frac{mv^2}{l}$ c) $T + \frac{mv^2}{l}$ d) Zero

155. Two bodies are projected from the same point with equal speeds in such directions that they both strike the same point on a plane whose inclination is β. If α be the angle of projection of the first body with the horizontal the ratio of their times of flight is

a) $\frac{\cos\alpha}{\sin(\alpha+\beta)}$ b) $\frac{\sin(\alpha+\beta)}{\cos\alpha}$ c) $\frac{\cos\alpha}{\sin(\alpha-\beta)}$ d) $\frac{\sin(\alpha-\beta)}{\cos\alpha}$

156. A car is circulating on the path of radius r and at any time its velocity is v and rate of increases of velocity is a. The resultant acceleration of the car will be

a) $\sqrt{\frac{v^2}{a^2}+r^2}$ b) $\sqrt{\frac{v^2}{r}+a}$ c) $\sqrt{\frac{v^4}{r^2}+a^2}$ d) $\left(\frac{v^2}{r}+a\right)$

157. Two bodies are projected from the same point with equal speeds in such directions that they both strike the same point on a plane whose inclination isβ. If ∝ be the angle of projection of the first body with the horizontal the ratio of their times of flight is

a) $\frac{\cos\propto}{\sin(\propto+\beta)}$ b) $\frac{\sin(\propto+\beta)}{\cos\propto}$ c) $\frac{\cos\propto}{\sin(\propto-\beta)}$ d) $\frac{\sin(\propto-\beta)}{\cos\propto}$

158. An athlete completes one round of a circular track of radius 10 m in 40 sec. The distance covered by him in 2 min 20 sec is

a) 70 m b) 140 m c) 110 m d) 220 m

159. A long horizontal rod has a bead which can slide along its length, and initially placed at a distance L from one end A of the rod. The rod is set in angular motion about A with constant angular acceleration α. If the coefficient of friction between the rod and the bead is μ, and gravity is neglected, then the time after which the bead starts slipping is

a) $\sqrt{\frac{\mu}{\alpha}}$ b) $\frac{\mu}{\sqrt{\alpha}}$ c) $\frac{1}{\sqrt{\mu\alpha}}$ d) Infinitesimal

160. A car runs at a constant speed on a circular track of radius 100 m, taking 62.8 s for every circular lap. The average velocity and average speed for each circular lap respectively is

a) 0,0 b) 0, 10 ms^{-1} c) 10 ms^{-1}, 10 ms^{-1} d) 10 ms^{-1}, 0

161. A 500 kg crane takes a turn of radius 50 m with velocity of 36 km/hr. The centripetal force is

a) 1200 N b) 1000 N c) 750 N d) 250 N

162. The relation between the time of flight of a projectile T_f and the time to reach the maximum height t_m is

a) $T_f = 2t_m$ b) $T_f = t_m$ c) $T_f = \frac{t_m}{2}$ d) $T_f = \sqrt{2}(t_m)$

163. A particle P is moving in a circle of radius r with a uniform speed v. C is the centre of the circle and AB is the diameter. The angular velocity of P about A and C is in ratio

a) 1 : 1 b) 1 : 2 c) 2 : 1 d) 4 : 1

164. Two cars of masses m_1 and m_2 are moving in circles of radii r_1 and r_2 respectively. Their speeds are such that they make complete circles in the same time t. The ratio of their centripetal acceleration is

a) $m_1r_1 : m_2r_2$ b) $m_1 : m_2$ c) $r_1 : r_2$ d) 1 : 1

165. In a circus stuntman rides a motorbike in a circular track of radius R in the vertical plane. The minimum speed at highest point of track will be

a) $\sqrt{2gR}$ b) $2gR$ c) $\sqrt{3gR}$ d) $\sqrt{gR}$

166. A particle describes a horizontal circle in a conical funnel whose inner surface is smooth with speed of 0.5 m/s. What is the height of the plane of circle from vertex of the funnel

a) 0.25 cm b) 2 cm c) 4 cm d) 2.5 cm

167. A road of 10 m width has radius of curvature 50 m. Its outer edge is raised above the inner edge by a distance of 1.5 m. The road is most suited for vehicles moving with velocity of

a) 8.5 ms^{-1} b) 6.5 ms^{-1} c) 5.5 ms^{-1} d) None of these

168. Consider the given velocity-time graph

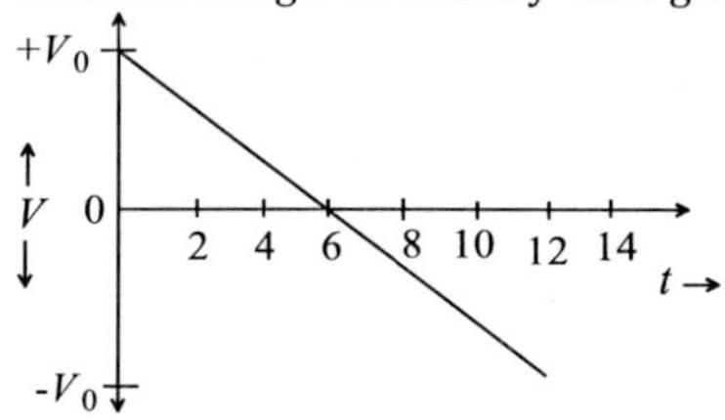

It represents the motion of
a) A projectile projected vertically upward, from a point
b) An electron in the hydrogen atom
c) A car with constant acceleration along a straight road
d) A bullet fired horizontally from the top of a tower

169. The maximum and minimum tensions in the string whirling in a circle of radius 2.5 m are in the ratio 5:3, then its velocity is
a) $\sqrt{98}\text{ms}^{-1}$ b) 7 ms^{-1} c) $\sqrt{490}\text{ms}^{-1}$ d) $\sqrt{4.9}\text{ms}^{-1}$

170. A particle of mass m is circulating on a circle of radius r having angular momentum L, then the centripetal force will be
a) L^2/mr b) L^2m/r c) L^2/mr^3 d) L^2/mr^2

171. Two forces, each of magnitude F, have a resultant of the same magnitude F. The angle between the two forces is
a) 45° b) 120° c) 150° d) 180°

172. Given $\vec{c} = \vec{a} \times \vec{b}$. The angle which $\vec{a}$ makes with $\vec{c}$ is
a) 0° b) 45° c) 90° d) 180°

173. A helicopter is flying horizontally at an altitude of 2 km with a speed of 100 ms^{-1}. A packet is dropped from it. The horizontal distance between the point where the packet is dropped and the point where it hits the ground is ($g = 10 \text{ ms}^{-2}$)
a) 2 km b) 0.2 km c) 20 km d) 4 km

174. A ball is projected with kinetic energy E at an angle of 45° to the horizontal. At the highest point during its flight, its kinetic energy will be
a) Zero b) $E/2$ c) $E/\sqrt{2}$ d) E

175. A body is acted upon by a constant force directed towards a fixed point. The magnitude of the force varies inversely as the square of the distance from the fixed point. What is the nature of the path?
a) Straight line b) Parabola c) Circle d) Hyperbola

176. A ball of mass 0.25 kg attached to the end of a string of length 1.96 m is moving in a horizontal circle. The string will break if the tension is more than 25N. What is the maximum speed with which the ball can be moved?
a) 14 ms^{-1} b) 3 ms^{-1} c) 3.92 ms^{-1} d) 5 ms^{-1}

177. In a projectile motion, velocity at maximum height is
a) $\dfrac{u\cos\theta}{2}$ b) $u\cos\theta$ c) $\dfrac{u\sin\theta}{2}$ d) None of these

178. A particle of mass m is projected with a velocity v making an angle of 45° with the horizontal. The magnitude of angular momentum of projectile about the point of projection when the particle is at its maximum height h is
a) Zero b) $\dfrac{mvh}{\sqrt{2}}$ c) $\dfrac{mvh^2}{\sqrt{2}}$ d) None of these

179. An object is moving in a circle of radius 100 m with a constant speed of 31.4 ms^{-1}. What is its average speed for one complete revolution?
a) Zero b) 31.4 ms^{-1} c) 3.14 ms^{-1} d) $\sqrt{2} \times 31.4 \text{ ms}^{-1}$

180. A missile is fired for maximum range with an initial velocity of $20\ m/s$. If $g = 10\ m/s^2$, the range of the missile is

a) $20\ m$ b) $40\ m$ c) $50\ m$ d) $60\ m$

181. What is the angular velocity of earth?

a) $\frac{2\pi}{86400}$ rad s^{-1} b) $\frac{2\pi}{3600}$ rad s^{-1} c) $\frac{2\pi}{24}$ rad s^{-1}l d) $\frac{2\pi}{6400}$ rad s^{-1}

182. A stone of mass 1 kg is tied at one end of string of length 1 m. It is whirled in a vertical circle at constant speed of 4 ms^{-1}. The tension in the string is 6 N when the stone is at ($g = 10ms^{-2}$)

a) Top of the circle b) Bottom of the circle c) Half way down d) None of these

183. What should be the coefficient of friction between the tyres and the road, when a car travelling at 60 km h^{-1} makes a level turn of radius 40 m?

a) 0.5 b) 0.66 c) 0.71 d) 0.80

184. A body of mass 2 kg attached to a string is whirled in a vertical circle of radius 5 m. The minimum speed of the body at lowest point so that the cord does not slacken even at the highest point is

a) 15.65 ms^{-1} b) 6.75 ms^{-1} c) 20.87 ms^{-1} d) 45.83 ms^{-1}

185. A car runs at a constant speed on a circular track of radius $100\ m$, taking 62.8 *seconds* for every circular loop. The average velocity and average speed for each circular loop respectively is

a) $10m/s, 10m/s$ b) $10\ m/s$,0 c) 0, 0 d) 0,10 m/s

186. A particle originally at rest at the highest point of a smooth vertical circle is slightly displaced. It will leave the circle at a vertical distance h below the highest point such that

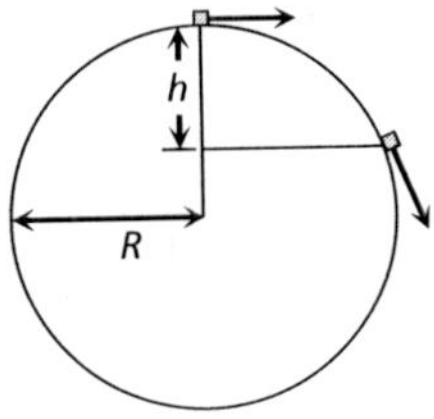

a) $h = R$ b) $h = \frac{R}{3}$ c) $h = \frac{R}{2}$ d) $h = \frac{2R}{3}$

187. The bob of a pendulum of mass m and length L is displaced, 90° from the vertical and gently released. In order that the string may not break upon passing through the lowest point, its minimum strength must be

a) mg b) $2\ mg$ c) $3\ mg$ d) $4\ mg$

188. Projection of $\vec{P}$ on $\vec{Q}$is

a) $\vec{P} \cdot \hat{Q}$ b) $\hat{p} \cdot \vec{Q}$ c) $\vec{P} \times \hat{Q}$ d) $\vec{P} \times \vec{Q}$

189. When a body moves with a constant speed along a circle

a) No work is done on it b) No acceleration is produced in the body

c) No force acts on the body d) Its velocity remains constant

190. A cyclist moves in such a way that he track 60° turn after 100 m. What is the displacement when to takes seventh turn?

a) 100 m b) 200 m c) $100\sqrt{3}$ m d) $100\sqrt{3}$ m

191. A ball of mass 0.1 kg. Is whirled in a horizontal circle of radius 1 m. By means of a string at an initial speed of 10 $R.P.M$. Keeping the radius constant, the tension in the string is reduced to one quarter of its initial value. The new speed is

a) $5\ r.p.m.$ b) $10\ r.p.m.$ c) $20\ r.p.m.$ d) $14\ r.p.m.$

192. A fan is making 600 revolutions per *minute*. If after some time it makes 1200 revolutions per *minute*, then increase in its angular velocity is

a) $10\pi\ rad/sec$ b) $20\pi\ rad/sec$ c) $40\pi\ rad/sec$ d) $60\pi\ rad/sec$

193. A particle moves in a circular path with decreasing speed. Choose the correct statement.

a) Angular momentum remains constant
b) Acceleration **(a)** is towards the center
c) Particle moves in a spiral path with decreasing radius
d) The direction of angular momentum remains constant

194. At what point of a projectile motion acceleration and velocity and velocity are perpendicular to each other

a) At the point of projection b) At the point of drop
c) At the topmost point d) Any where in between the point of projection and topmost point

195. The centripetal acceleration of a particle of mass m moving with a velocity v in a circular orbit of radius r is

a) v^2/r along the radius, towards the center
b) v^2/r along the radius, away from the center
c) mv^2/r along the redius, away from the center
d) mv^2/r along the radius, towards the center

196. Two projectiles A and B thrown with speed in the ratio $1:\sqrt{2}$ acquired the same heights. If A is thrown at an angle of 45° with the horizontal, the angle of projection of B will be

a) 0° b) 60° c) 30° d) 45°

197. A boy playing on the roof of a 10 m high building throws a ball with a speed of 10 ms^{-1} at an angle of 30° with the horizontal. How far from the throwing point will the ball be at the height of 10 m from the ground?
($g = 10\ ms^{-2}, \sin 30° = 1/2, \cos 30° = \sqrt{3}/2$

a) 5.20 m b) 4.33 m c) 2.60 m d) 8.66 m

198. A stone is tied to one end of a string and rotated in a horizontal circle with a uniform angular velocity. Let T be the tension in the string. If the length of the string is halved and its angular velocity is doubled, tension in the string will be

a) $T/4$ b) $T/2$ c) $2T$ d) $4T$

199. A stone thrown at an angle θ to the horizontal a projectile makes an angle $\pi/4$ with the horizontal, then its initial velocity and angle of projection are, respectively

a) $\frac{\sqrt{2h\sin\theta}}{g}$ b) $\frac{2\sqrt{2h\sin\theta}}{g}$ c) $2\sqrt{\frac{2h}{g}}$ d) $\sqrt{\frac{2h}{g}}$

200. A sphere is suspended by a thread of length l. What minimum horizontal velocity has to be imparted the ball for it to reach the height of the suspension

a) gl b) $2gl$ c) $\sqrt{gl}$ d) $\sqrt{2gl}$

201. If retardation produced by air resistance of projectile is one-tenth of acceleration due to gravity, the time to reach maximum height

a) Decreases by 11 percent b) Increases by 11 percent
c) Decreases by 9 percent d) Increases by 9 percent

202. A particle is projected from the ground with an initial speed of v at an angle θ with horizontal. The average velocity of the particle between its point of projection and highest point of trajectory is

a) $\frac{v}{2}\sqrt{1+2\cos^2\theta}$ b) $\frac{v}{2}\sqrt{1+\cos^2\theta}$ c) $\frac{v}{2}\sqrt{1+3\cos^2\theta}$ d) $v\cos\theta$

203. There are two forces each of magnitude 10 units. One inclined at an angle of 30° and the other at an angle of 135° to the positive direction of x −axis. The x and y components of the resultant are respectively.

a) $1.59\hat{i}$ and $12.07\hat{j}$ b) $10\hat{i}$ and $10\hat{j}$ c) $1.59\hat{i}$ d) $15.9\hat{i}$ and $12.07\hat{j}$

204. A particle is moving in a circle of radius R in such a way that at any instant the normal and tangential components of its acceleration are equal. If its speed at $t = 0$ is v_0, the time taken to complete the first revolution is

a) $\frac{R}{v_0}$ b) $\frac{R}{v_0}(1 - e^{-2\pi})$ c) $\frac{R}{v_0}e^{-2\pi}$ d) $\frac{2\pi R}{v_0}$

205. The tension in the string revolving in a vertical circle with a mass m at the end which is the lowest position

a) $\frac{mv^2}{r}$ b) $\frac{mv^2}{r} - mg$ c) $\frac{mv^2}{r} + mg$ d) mg

206. The ratio of the angular speed of minutes hand and hour hand of a watch is

a) 6 : 1 b) 12 : 1 c) 1 : 6 d) 1 : 12

207. A car runs at a constant speed on a circular track of radius 100 m, taking 62.8 s for every circular lap. The average velocity and average speed for each circular lap is

a) 0, 0 b) 0,10ms^{-1} c) 10 ms^{-1}, 10ms^{-1} d) 10 ms^{-1}, 0

208. An aeroplane moving horizontally with a speed of 720 km/h drops a food pocket, while flying at a height of 396.9 m. The time taken by a food pocket to reach the ground and its horizontal range is (Take $g = 9.8\ m/sec^2$)

a) 3 sec and 2000 m b) 5 sec and 500 m c) 8 sec and 1500 m d) 9 sec and 1800 m

209. A cyclist turns around a curve at 15 miles/hour. If the turns at double the speed, the tendency to overturn is

a) Doubled b) Quadrupled c) Halved d) Unchanged

210. A point mass m is suspended from a light thread of length l, fixed at O, is whirled in a horizontal circle at constant speed as shown. From your point of view, stationary with respect to the mass, the forces on the mass are

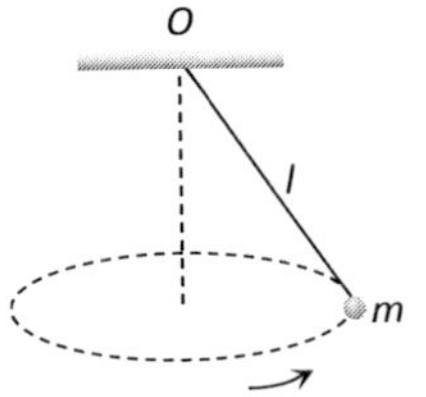

a) 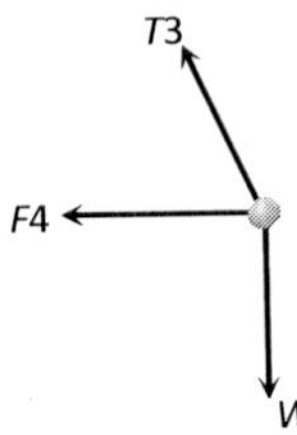

b) 

c) 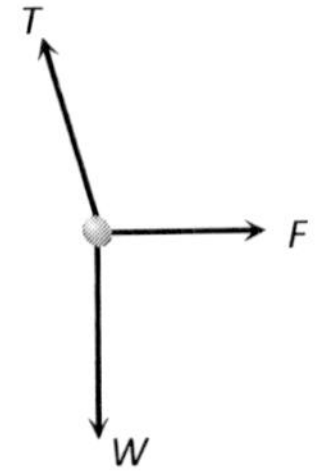

d) 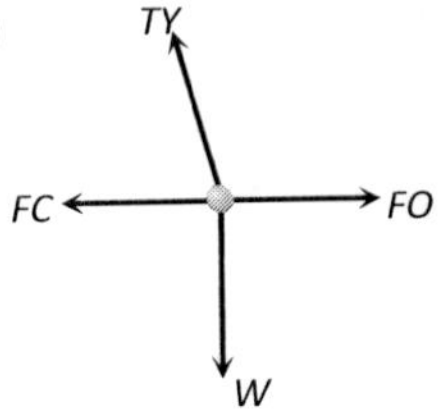

211. Two tall buildings are 40 m apart. With what speed must a ball be thrown horizontally from a window 145 m above the ground in one building, so that it will enter a window 22.5 m from the ground in the other?

a) 5 ms^{-1} b) 8 ms^{-1} c) 10 ms^{-1} d) 16 ms^{-1}

212. A wheel rotates with a constant angular velocity of 300 rpm. The angle through which the wheel rotates in one second is

a) π rad b) 5 π rad c) 10 π rad d) 20 π rad

213. A tube of length L is filled completely with an incompressible liquid of mass M and closed at both the ends. The tube is then rotated in a horizontal plane about one of its ends with a uniform angular velocity ω. The force exerted by the liquid at the other end is

a) $\frac{ML\omega^2}{2}$ b) $ML\omega^2$ c) $\frac{ML\omega^2}{4}$ d) $\frac{ML^2\omega^2}{2}$

214. A stone is thrown with a velocity v making an angle θ with the horizontal. At some instant, its velocity V is perpendicular to the initial velocity v. Then V is

a) $v \sin\theta$ b) $v \cos\theta$ c) $v \tan\theta$ d) $v \cot\theta$

215. If the length of the second's hand in a stop-clock is 3 cm, the angular velocity and linear velocity of the tip is

a) 0.2047 rads^{-1}, 0.0314 ms^{-1} b) 0.2547 rads^{-1}, 0.0314 ms^{-1}

c) 0.1472 rads^{-1}, 0.06314 ms^{-1} d) 0.1047 rads^{-1}, 0.00314 ms^{-1}

216. A cylindrical vessel partially filled with water is rotated about its vertical central axis. It's surface will

a) Rise equally b) Rise from the sides c) Rise from the middle d) Lowered quality

217. A man standing on a hill top projects a stone horizontally with speed v_0 as shown in figure. Taking the coordinate system as given in the figure. The coordinates of the point where the stone will hit the hill surface

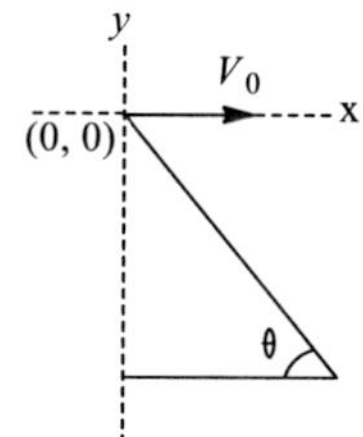

a) $\left(\frac{2v_0^2\tan\theta}{g}, \frac{-2v_0^2\tan^2\theta}{g}\right)$ b) $\left(\frac{2v_0^2}{g}, \frac{2v_0^2\tan^2\theta}{g}\right)$

c) $\left(\frac{2v_0^2\tan\theta}{g}, \frac{2v_0^2}{g}\right)$ d) $\left(\frac{2v_0^2\tan^2\theta}{g}, \frac{2v_0^2\tan\theta}{g}\right)$

218. A string of length L is fixed at one end and carries a mass M at the other end. The string makes $2/\pi$ revolutions per *second* around the vertical axis through the fixed end as shown in figure , then tension in the string is

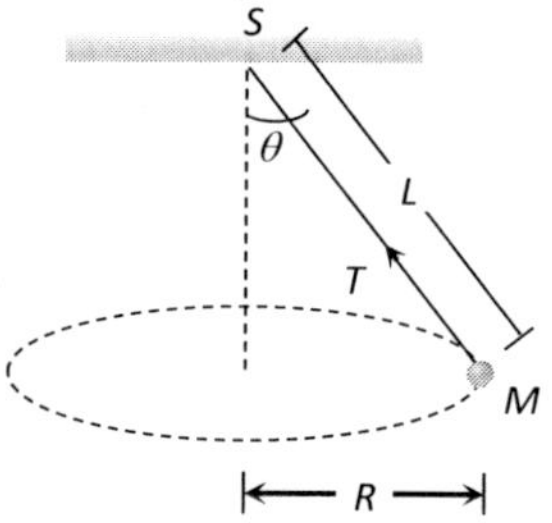

a) ML b) $2\,ML$ c) $4\,ML$ d) $16\,ML$

219. A particle P is sliding down a frictionless hemispherical bowl. It passes the point Aat $t = 0$. At this instant of time, the horizontal component of its velocity v. A bead Q of the same mass as P is ejected from A to $t = 0$ along the horizontal string AB (see figure) with the speed v. Friction between the bead and the string may be neglected. Let t_p and t_Q be the respective time taken by P and Q to reach the point B. Then

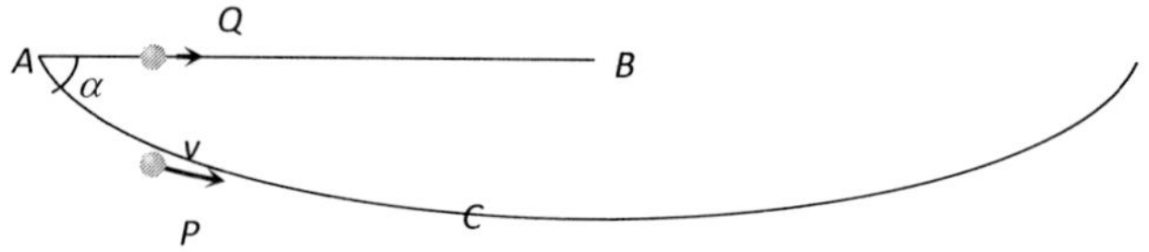

a) $t_p < t_Q$ b) $t_p = t_Q$ c) $t_p > t_Q$ d) All of these

220. The string of a pendulum of length l is displaced through 90° from the vertical and released. Then the minimum strength of the string in order to withstand the tension as the pendulum passes through the mean position is
a) mg b) $6\ mg$ c) $3\ mg$ d) $5\ mg$

221. Velocity vector and acceleration vector in a uniform circular motion are related as
a) Both in the same direction
b) Perpendicular to each other
c) Both in opposite direction
d) No related to each other

222. Two projectiles thrown from the same point at angles 60° and 30° with the horizontal attain the same height. The ratio of their initial velocities is
a) 1 b) 2 c) $\sqrt{3}$ d) $\frac{1}{\sqrt{3}}$

223. An aeroplane flying at a velocity of $900 kmh^{-1}$ loops the loop. If the maximum force pressing the pilot against the seat is five times its weight, the loop radius should be
a) 1594 m b) 1402 m c) 1315 m d) 1167 m

224. If KE of the particle of mass m performing UCM in a circle of radius r is E. Find the acceleration of the particale
a) $\frac{2E}{mr}$ b) $\left(\frac{2E}{mr}\right)^2$ c) $2\ Emr$ d) $\frac{4E}{mr}$

225. A particle of mass 100 g tied to a string is rotated along circle of radius 0.5 m. The breaking tension of string is 10 N. The maximum speed with which particle can rotated without breaking the string is
a) $10\ ms^{-2}$ b) $9.8\ ms^{-2}$ c) $7.7\ ms^{-2}$ d) $7.07\ ms^{-2}$

226. A boy is hanging from a horizontal branch of a tree. The tension in the arms will be maximum when the angle between the arms is
a) 0° b) 60° c) 90° d) 120°

227. A particle P is at the origin starts with velocity $\vec{\mathbf{v}} = (2\hat{\mathbf{i}} - 4\hat{\mathbf{j}})ms^{-1}$ with constant acceleration $(3\hat{\mathbf{i}} - 5\hat{\mathbf{j}})ms^{-2}$. After travelling for 2 s, its distance from the origin is
a) 10 m b) 10.2 m c) 9.8 m d) 11.7 m

228. An object of mass 5 kg is whirled round in a vertical circle of radius 2 m with a constant speed of $6\ ms^{-1}$. The maximum tension in the string is
a) 152 N b) 139 N c) 121 N d) 103 N

229. For a projectile the ratio of maximum height reached to the square of time of flight is $(g = 10\ ms^{-1})$
a) 5 : 1 b) 5 : 2 c) 5 : 4 d) 1 : 1

230. If the body is moving in a circle of radius r with a constant speed v, its angular velocity is
a) v^2/r b) vr c) v/r d) r/v

231. One end of a string of length l is connected to a particle of mass m and other to a small peg on a smooth horizontal table. If the particle moves in a circle with speed v, the net force on the particle (directed towards the centre) is
a) T b) $T - \frac{mv^2}{l}$ c) $T + \frac{mv^2}{l}$ d) zero

232. A body of mass m is projected at an angle of 45° with the horizontal. If air resistance is negligible, then total change in momentum when it strikes the ground is
a) $2mv$ b) $\sqrt{2}\ mv$ c) mv d) $mv/\sqrt{2}$

233. At the point of a projectile motion, acceleration and velocity are perpendicular to each other?
a) At the point of projection
b) At the point of drop
c) At the top most point
d) Anywhere in between the point of projection and two most point

234. A ball rolls of the top of stair-way with a horizontal velocity of magnitude 1.8 ms^{-1}. The steps are 0.20 m high and 0.20 m wide. Which step will the ball hit first?
a) First b) Second c) Third d) Fourth

235. A particle moves along a circle of radius $\left(\frac{20}{\pi}\right)$ m with constant tangential acceleration. If the velocity of the particle is 80 ms^{-1}, at the end of seconds revolution after motion has begun, the tangential acceleration is
a) 40 ms^{-2} b) 640 π ms^{-2} c) 1609 π ms^{-2} d) 40 π ms^{-2}

236. Tom and Dick are running forward with the same speed. They are throwing a rubber ball to each other at a constant speed v as seen by the thrower. According to Sam who is standing on the ground the speed of the ball is
a) Same as v b) Greater than v c) Less than v d) None of these

237. Two paper screens A and B are separated by a distance of 200 m. A bullet pierces A and then B. The hole in B is 40 cm below the hole in A. If the bullet is travelling horizontally at the time of hitting A, then the velocity of the bullet at A is
a) 200 ms^{-1} b) 400 ms^{-1} c) 600 ms^{-1} d) 700 ms^{-1}

238. A 2 kg stone tied at the end of a string 1 m long is whirled along a vertical circle at a constant speed of 4 ms^{-1}. The tension in the string has a value of 52 N when the stone is
a) At the top of the circle b) Half way down
c) At the bottom of the circle d) None of the above

239. A stone of mass m is tied to a string of length l and rotated in a circle with a constant speed v. If the string is released, the stone flies
a) Radially outwards b) Radially inwards
c) Tangentially outwards d) With an acceleration mv^2/l

240. If a cycle wheel of radius 4 m completes one revolution in two seconds. Then acceleration of a point on the cycle wheel will be
a) $\pi^2 m/s^2$ b) $2\pi^2 m/s^2$ c) $4\pi^2 m/s^2$ d) $8\pi m/s^2$

241. A projectile is thrown at angle βwith vertical. It reaches a maximum height H. The time taken to reach the highest point of its path is
a) $\sqrt{\frac{H}{g}}$ b) $\sqrt{\frac{2H}{g}}$ c) $\sqrt{\frac{H}{2g}}$ d) $\sqrt{\frac{2H}{g\cos\beta}}$

242. The radius vector and linear momentum are respectively given by vector $2\hat{i}+2\hat{j}+\hat{k}$ and $2\hat{i}-2\hat{j}+\hat{k}$. Their angular momentum is
a) $2\hat{i}-4\hat{j}$ b) $4\hat{i}-8\hat{k}$ c) $2\hat{i}-4\hat{j}+2\hat{k}$ d) $4\hat{i}-8\hat{j}$

243. A fighter plane is moving in a vertical circle of radius $'r'$. Its minimum velocity at the highest point of the circle will be
a) $\sqrt{3gr}$ b) $\sqrt{2gr}$ c) $\sqrt{gr}$ d) $\sqrt{gr/2}$

244. A particle moves in a circle of radius 30cm. Its liner speed is given by $v = 2t$, where t is in second and $v$in ms^{-1}. Find out its, radial and tangential acceleration at $t = 3$s, respectively,
a) 220 ms^{-2}, 50 ms^{-2} b) 100 ms^{-2}, 5 ms^{-2} c) 120 ms^{-2}, 2 ms^{-2} d) 110 ms^{-2}, 10 ms^{-2}

245. When a simple pendulum is rotated in a vertical plane with constant angular velocity, centripetal force is
a) Maximum at highest point b) Maximum at lowest point
c) Same at all lower point d) Zero

246. A particle is projected with velocity V_0 along x-axis. The deceleration on the particle is proportional to the square of the distance from the origin i.e. $a = \alpha x^2$, the distance at which the particle stops is
a) $\sqrt{\frac{3V_0}{2\alpha}}$ b) $\left(\frac{3V_0}{2\alpha}\right)^{\frac{1}{3}}$ c) $\sqrt{\frac{2V_0^2}{3\alpha}}$ d) $\left(\frac{3V_0^2}{2\alpha}\right)^{\frac{1}{3}}$

247. A particle leaves the origin with an initial velocity $\vec{\mathbf{v}} = (3.00\hat{\mathbf{i}})\text{ms}^{-1}$ and a constant acceleration $\vec{\mathbf{a}} = (-1.00\hat{\mathbf{i}} - 0.50\hat{\mathbf{j}})\ \text{ms}^{-2}$. When the particle reaches it maximum x-coordinate, what is its y-component a velocity?

a) $-2.0\ \text{ms}^{-1}$ b) $-1.0\ \text{ms}^{-1}$ c) $-1.5\ \text{ms}^{-1}$ d) $1.0\ \text{ms}^{-1}$

248. What is the angular velocity of earth

a) $\frac{2\pi}{86400}\ rad/sec$ b) $\frac{2\pi}{3600}\ rad/sec$ c) $\frac{2\pi}{24}\ rad/sec$ d) $\frac{2\pi}{6400}\ rad/sec$

249. The area o the parallelogram represented by the vectors. $\vec{A} = 4\hat{\imath} + 3\hat{\jmath}$ and $\vec{B} = 2\hat{\imath} + 4\hat{\jmath}$ is

a) 14 units b) 7.5 units c) 10 units d) 5 units

250. A cricket ball is hit at 30° with the horizontal with kinetric energy E_k. What is the kinetic energy at the highest point?

a) $E_k/2$ b) $3E_k/4$ c) $E_k/4$ d) Zero

251. A projectile A is thrown at an angle of 30° to the horizontal from point P. At the same time, another projectile B is thrown with velocity v^2 upwards from the point Q vertically below the highest point. For B to collide with A, $\frac{v_2}{v_1}$ should be

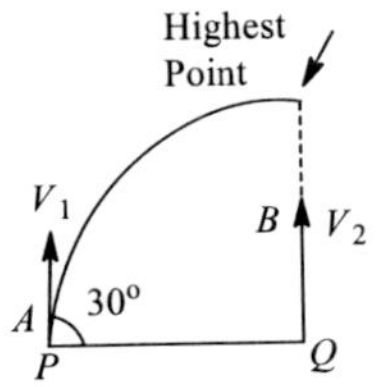

a) 1 b) 2 c) $\frac{1}{2}$ d) 4

252. A particle of mass m is moving in a circular path of constant radius r such that its centripetal acceleration a_c is varying with time as $a_c = k^2rt^4$, where k is a constant. The power delivered to the particle by the forces acting on its is

a) Zero b) $mk^2r^2t^2$ c) $\frac{1}{3}mk^2r^2t^2$ d) $2mk^2r^2t^3$

253. The angle of projection at which the horizontal range and maximum height of projectile are equal is

a) 45° b) $\theta = \tan^{-1}(0.25)$

c) $\theta = \tan^{-1} 4$or $(\theta = 76°)$ d) 60°

254. For motion in a plane with constant acceleration $\vec{a}$, initial velocity$\vec{v}_0$ and final velocity $\vec{v}$ after time t, we have

a) $\vec{v}.(\vec{v} - \vec{a}t) = \vec{v}_0.(\vec{v}_0 + \vec{a}t)$ b) $\vec{v}.\vec{v}_0 = at^2$

c) $\vec{v}.\vec{v}_0 = \vec{v}.\vec{v}_0 t$ d) $\vec{v}_0.\vec{v}_0 = \vec{a}.\vec{v}_0 t$

255. A small sphere is attached to a cord and rotates in a vertical circle about a point O. If the average speed of the sphere is increased, the cord is most likely to break at the orientation when the mass is at

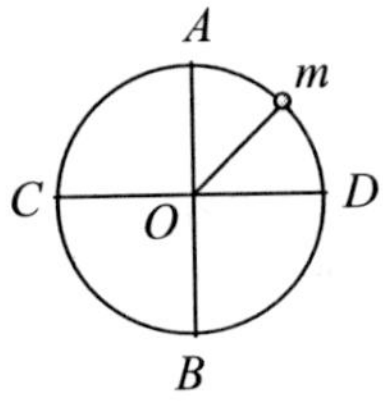

a) Bottom point B b) Top point A c) The point D d) The point C

256. A projectile is thrown in the upward direction making an angle of 60° with the horizontal direction with a velocity of147ms^{-1}. Then the time after which its inclination with the horizontal is 45°, is

a) 15 s b) 10.98 s c) 5.49 s d) 2.745 s

257. If the length of the second's hand in a stop clock is 3 cm the angular velocity and linear velocity of the tip is

a) $0.2047\ rad/sec., 0.0314\ m/sec$
b) $0.2547\ rad/sec., 0.314\ m/sec$
c) $0.1472\ rad/sec., 0.06314\ m/sec$
d) $0.1047\ rad/sec., 0.00314\ m/sec$

258. From an inclined plane two particles are projected with same speed at same angle θ, one up and other down the plane as shown in figure. Which of the following statements (s) is/are correct?

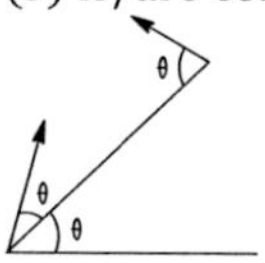

a) The time of flight of each particle is the same.
b) The particles will collide the plane with same speed
c) Both the particles strike the plane perpendicularly
d) The particles will collide in mid air if projected simultaneously and time of flight of each particle is less than the time of collision

259. The time taken by the projectile to reach from A to B is t, then the distance AB is equal to

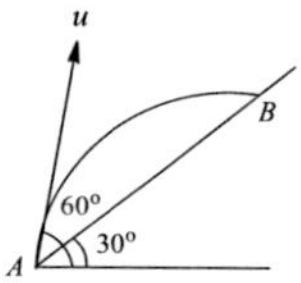

a) $2\,ut$
b) $\sqrt{3}\,ut$
c) $\frac{\sqrt{3}}{2}ut$
d) $\frac{ut}{\sqrt{3}}$

260. An arrow is projected into air. Its time of flight is 8 s and range 200 m. What is the maximum height reached by it? (Take $g = 10\ ms^{-2}$)

a) 31.25 m
b) 24.5 m
c) 18.25 m
d) 46.75 m

261. A particle of mass = 5 is moving with a uniform speed $v = 3\sqrt{2}$ in the XOY plane along the line $Y = X + 4$. The magnitude of the angular momentum of the particle about the origin is

a) 60 units
b) $40\sqrt{2}$ units
c) 7.5 units
d) zero

262. A shell is fired from a cannon with a velocity v at angle θwith horizontal. At the highest point, it explodes into two pieces of equal mass. One of the pieces retraces its path to the cannon. The speed of the other piece just after explosion is

a) $3\,v\cos\theta$
b) $2\,v\cos\theta$
c) $\frac{3}{2}v\cos\theta$
d) $\frac{\sqrt{3}}{2}v\cos\theta$

263. Given $\vec{A} = \hat{i} + 2\hat{j} - 3\hat{k}$. When a vector $\vec{B}$ is added to $\vec{A}$, We get a unit vector along X=axis. Then, $\vec{B}$ is

a) $-2\hat{j} + 3\hat{k}$
b) $-\hat{i} - 2\hat{j}$
c) $-\hat{i} + 3\hat{k}$
d) $2\hat{j} - 3\hat{k}$

264. A hollow sphere has radius 6.4 m. Minimum velocity required by a motor cyclist at bottom to complete the circle will be

a) 17.7 m/s
b) 10.2 m/s
c) 12.4 m/s
d) 16.0 m/s

265. $\vec{A} = 3\hat{i} - \hat{j} + 7\hat{k}$ and $\vec{B} = 5\hat{i} - \hat{j} + 9\hat{k}$ the direction cosine m of the vector $\vec{A} + \vec{B}$ is

a) Zero
b) $\frac{3}{\sqrt{31}}$
c) $\frac{9}{\sqrt{107}}$
d) 5

266. A car sometimes overturns while taking a turn. When it overturns, it is

a) The inner wheel which leaves the ground first
b) The outer wheel which leaves the ground first
c) Both the wheels leave the ground simultaneously
d) Either wheel leaves the ground first

267. A car of mass 1000 kg moves on a circular track of radius 20 m. If the coefficient of friction is 0.64, then the maximum velocity with which the car can move is

a) 22.4 ms^{-1} b) 5.6 ms^{-1} c) 11.2 ms^{-1} d) None of these

268. A projectile is thrown with a speed u at an angle θ to the horizontal. The radius of curvature of its trajectory when the velocity vector of the projectile makes an angle α with the horizontal is

a) $\frac{u^2\cos^2\alpha}{g\cos^2\theta}$ b) $\frac{2u^2\cos^2\alpha}{g\cos^2\theta}$ c) $\frac{u^2\cos^2\theta}{g\cos^3\alpha}$ d) $\frac{u^2\cos^2\theta}{g\cos^2\alpha}$

269. A stone of mass 1 kg is tied to a string 4 m long and is rotated at constant speed of 40 ms^{-1} in a vertical circle. The ratio of the tension at the top and the bottom is

a) 11 : 12 b) 39 : 41 c) 41 : 39 d) 12 : 11

270. The magnitudes of the two vectors $\vec{a}$ and $\vec{b}$ are a and b respectively. The vector product of $\vec{a}$ and $\vec{b}$ cannot be

a) equal to zero b) less than ab c) equal to ab d) greater than ab

271. A projectile is thrown at angle β with vertical. It reaches a maximum height H. The time taken to reach the highest point of its path is

a) $\sqrt{\frac{H}{g}}$ b) $\sqrt{\frac{2H}{g}}$ c) $\sqrt{\frac{H}{2g}}$ d) $\sqrt{\frac{H}{g\cos\beta}}$

272. A boy on a cycle pedals around a circle of 20 $metres$ radius at a speed of 20 $metres/sec$. The combined mass of the boy and the cycle is 90 kg. The angle that the cycle makes with the vertical so that it may not fall is ($g = 9.8\ m/sec^2$)

a) 60.25° b) 63.90° c) 26.12° d) 30.00°

273. A stone is thrown at an angle θ to the horizontal reaches a maximum height H. Then the time of flight of stone will be

a) $\sqrt{\frac{2H}{g}}$ b) $2\sqrt{\frac{2H}{g}}$ c) $\frac{2\sqrt{2H\sin\theta}}{g}$ d) $\frac{\sqrt{2H\sin\theta}}{g}$

274. A body crosses the topmost point of a vertical circle with critical speed. Its centripetal acceleration, when the string is horizontal will be

a) 6 g b) 3 g c) 2 g d) g

275. A particle revolves around a circular path. The acceleration of the particle is

a) Along the circumference of the circle b) Along the tangent
c) Along the radius d) Zero

276. If a particle covers half the circle of radius R with constant speed then

a) Change in momentum is mvr b) Change in $K.E.$ is $1/2\ mv^2$
c) Change in $K.E.$ is mv^2 d) Change in $K.E.$ is zero

277. If $\vec{A} + \vec{B} = \vec{C}$ and $A = \sqrt{3}$, $B = \sqrt{3}$ and $C = 3$, then the angle between $\vec{A}$ *and* $\vec{B}$ is

a) 0° b) 30° c) 60° d) 90°

278. The angle turned by a body undergoing circular motion depends on time as $\theta = \theta_0 + \theta_1 t + \theta_2 t^2$. Then the angular acceleration of the body is

a) θ_1 b) θ_2 c) $2\theta_1$ d) $2\theta_2$

279. Given $\vec{A} = 4\hat{i} + 6\hat{j}$ and $\vec{B} = 2\hat{i} + 3\hat{j}$. Which of the following is correct?

a) $\vec{A} \times \vec{B} = \vec{0}$ b) $\vec{A} \cdot \vec{B} = 24$
c) $\frac{|\vec{A}|}{|\vec{B}|} = \frac{1}{2}$ d) $\vec{A}$ and $\vec{B}$ are anti-parallel

280. A bullet is fired with a velocity u making an angle of 60° with the horizontal plane. The horizontal component of the velocity of the bullet when it reaches the maximum height is

a) u b) 0 c) $\frac{\sqrt{3}u}{2}$ d) $u/2$

281. A cart is moving horizontally along a straight line with constant speed30 ms^{-1}. A projectile is to be fired from the moving cart in such a way that it will return to the cart has moved 80 m. At what speed (relative to the cart) must the projectile be fired? (Take $g = 10\ ms^{-2}$)

a) $10\ ms^{-1}$ b) $10\sqrt{8} ms^{-1}$ c) $\frac{40}{3} ms^{-1}$ d) None of the above

282. A car is moving with high velocity when it has a turn. A force acts on it outwardly because of

a) Centripetal force b) Centrifugal force c) Gravitational force d) All the above

283. The area of parallelogram formed from the vectors $\vec{A} = \hat{i} - 2\hat{j} + 3\hat{k}$ and $\vec{B} = 3\hat{i} - 2\hat{j} + \hat{k}$ as adjacent sides is

a) $8\sqrt{3}$ units b) 64 units c) 32 units d) $4\sqrt{6}$ units

284. The maximum velocity (in ms^{-1}) with which a car driver must traverse a flat curve of radius 150 m and coefficient of friction 0.6 to avoid skidding is

a) 60 b) 30 c) 15 d) 25

285. The maximum and minimum tension in the string whirling in a circle of radius 2.5 m with constant velocity are in the ratio 5: 3 the the velocity is

a) $\sqrt{98}\ m/s$ b) $7\ m/s$ c) $\sqrt{490}\ m/s$ d) $\sqrt{4.9}$

286. A body is projected from the earth at angle 30° with the horizontal with some initial velocity. If its range is 20 m, the maximum height reached by its is (in metre)

a) $5\sqrt{3}$ b) $\frac{5}{\sqrt{3}}$ c) $\frac{10}{\sqrt{3}}$ d) $10\sqrt{3}$

287. Neglecting the air resistance, the time of flight of a projectile is determined by

a) $U_{vertical}$ b) $U_{horizontal}$

c) $U = U^2_{vertical} + U^2_{horizontal}$ d) $U = U(U^2_{vertical} + U^2_{horizontal})^{1/2}$

288. If the magnitudes of scalar and vector products of two vectors and are 6 and $6\sqrt{3}$ respectively, then the angle between two vectors is

a) 15° b) 30° c) 60° d) 75°

289. Three balls are dropped from the top of a building with equal speed at different angles. When the balls strike ground their velocities are v_1, v_2 and v_3 respectively, then

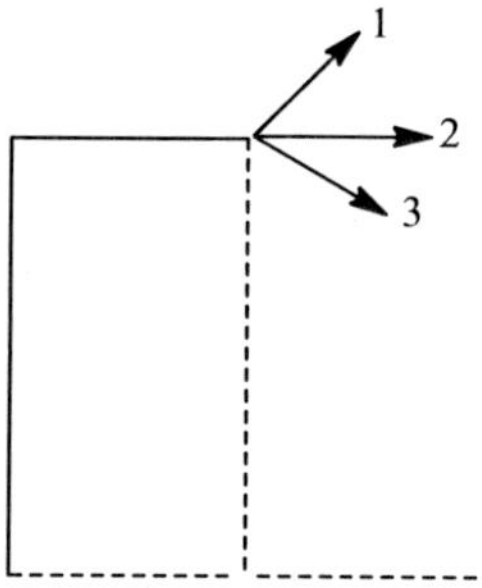

a) $v_1 > v_2 > v_3$ b) $v_3 > v_2 > v_1$ c) $v_1 = v_2 = v_3$ d) $v_1 < v_2 < v_3$

290. An object is projected at an angle of 45° with the horizontal. The horizontal range and maximum height reached will be in the ratio

a) 1:2 b) 2:1 c) 1:4 d) 4:1

291. A coin, placed on a rotating turn-table slips, when it is placed at a distance of 9 cm from the centre. If the angular velocity of the turn-table is tripled, it will just slip, if its distance from the centre is

a) 27 cm b) 9 cm c) 3 cm d) 1 cm

292. A particle is moving on a circular path of radius r with uniform velocity v. The change in velocity when the particle moves from P to Q is ($\angle POQ = 40°$)

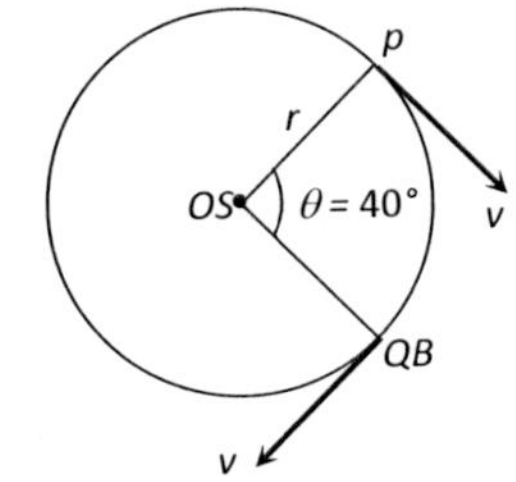

a) $2v\cos 40°$ b) $2v\sin 40°$ c) $2v\sin 20°$ d) $2v\cos 20°$

293. A coin is placed on a gramophone record rotating at a speed of 45 rpm. It flies away when the rotational speed is 50 rpm. If two such coins are placed over the other on the same record, both of them will fly away when rotational speed is
a) 100 rpm b) 25 rpm c) 12.5 rpm d) 50 rpm

294. A body is whirled in a horizontal circle of radius 20 cm. It has angular velocity of 10 rad/s. What is its linear velocity at any point on circular path
a) 10 m/s b) 2 m/s c) 20 m/s d) $\sqrt{2}\ m/s$

295. For a given velocity, a projectile has the same range R for two angles of projection if t_1 and t_2 are the times of flight in the two cases then
a) $t_1t_2 \propto R^2$ b) $t_1t_2 \propto R$ c) $t_1t_2 \propto \frac{1}{R}$ d) $t_1t_2 \propto \frac{1}{R^2}$

296. A projectile is fired with a velocity v at right angle to the slope which is inclined at an angle θ with the horizontal. What is the time of flight?
a) $\frac{2v^2}{g}\tan\theta$ b) $\frac{v^2}{g}\tan\theta$ c) $\frac{2v^2}{g}\sec\theta$ d) $\frac{2v^2}{g}\tan\theta\sec\theta$

297. The horizontal range of a projectile is $4\sqrt{3}$ times its maximum height. Its angle of projection will be
a) 45° b) 60° c) 90° d) 30°

298. A boy whirl is a stone in a horizontal circle of radius 1.5 m and at height 2.0 m above level ground. The string breaks and the stone flies off tangentially and strikes the ground after traveling a horizontal distance of 10 m. What is the magnitude of the centripetal acceleration of the stone while in circular motion?
a) 163 ms^{-2} b) 64 ms^{-2} c) 15.63 ms^{-2} d) 125 ms^{-2}

299. Two vectors $\vec{A}$ and $\vec{B}$ are inclined to each other at an angle θ. Which of the following is the unit vector perpendicular to both $\vec{A}$and $\vec{B}$?
a) $\frac{\vec{A}\times\vec{B}}{\vec{A}\cdot\vec{B}}$ b) $\frac{\hat{A}\cdot\hat{B}}{\sin\theta}$ c) $\frac{\vec{A}\times\vec{B}}{AB\sin\theta}$ d) $\frac{\hat{A}\times\hat{B}}{AB\cos\theta}$

300. It was calculated that a shell when fired from a gun with a certain velocity and at an angle of elevation $\frac{5\pi}{36}$ rad should strike a given target. In actual practice, it was found that a hill just prevented the trajectory. At what angle of elevation should the gun be to hit the target?
a) $\frac{5\pi}{36}$rad b) $\frac{11\pi}{36}$rad c) $\frac{7\pi}{36}$rad d) $\frac{13\pi}{36}$rad

301. A particle does uniform circular motion in a horizontal plane. The radius of the circle is 20 cm. The centripetal force acting on the particle is 10 N. It's kinetic energy is
a) 0.1 J b) 0.2 J c) 2.0 J d) 1.0 J

302. The condition of apparent weightlessness can be created momentarily when a plane flies over the top of a vertical circle. At a speed of 900 kmh^{-1}, the radius of the vertical circle that the pilot must use is
a) 10.6 km b) 8.5 kg c) 6.4 km d) 4.0 km

303. A particle is projected from a point A with velocity $u\sqrt{2}$ at an angle of 45° with horizontal as shown in figure. It strikes the plane BC at right angles. The velocity of the particle at the time of collision is

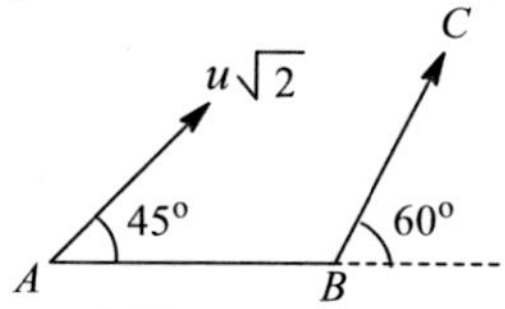

a) $\frac{\sqrt{3}u}{2}$ b) $\frac{u}{2}$ c) $\frac{2u}{\sqrt{3}}$ d) u

304. A fighter plane enters inside the enemy territory, at time $t = 0$ with velocity $v_0 = 250\ \text{ms}^{-1}$ and moves horizontally with constant acceleration $a = 20\text{ms}^{-2}$ (see figure). An enemy tank at the border, spot the plane and fire shots at an angle $\theta = 60°$ with the horizontal and with velocity $u = 600\ \text{ms}^{-1}$. At what altitude H of the plane it can be hit by the shot?

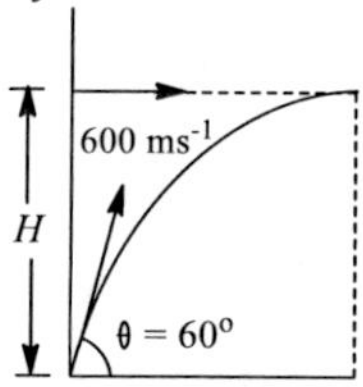

a) $1500\sqrt{3}\text{m}$ b) 125 m c) 1400 m d) 2473 m

305. At the height 80 m, an aeroplane is moving with 150 m/s. A bomb is dropped from it so as to hit a target. At what distance from the target should be bomb be dropped (Given $g = 10\ m/s^2$)

a) 605.3 m b) 600 m c) 80 m d) 230 m

306. An object moves at a constant speed along a circular path in a horizontal XY plane, with the centre at the origin. When the object is at $x = -2m$, its velocity is $-(4m/s)\hat{j}$. What is the object's acceleration when it is $y = 2m$

a) $-(8m/s^2)\hat{j}$ b) $-(8m/s^2)\hat{i}$ c) $-(4m/s^2)\hat{j}$ d) $(4m/s^2)\hat{i}$

307. The momentum of a particle is $\vec{P}$=2cost$\hat{i}$ + 2sint$\hat{j}$. What is the angle between the force $\vec{F}$ acting on the particle and the momentum $\vec{P}$

a) 65° b) 90° c) 150° d) 180°

308. If the angle of projection of a projectile is 30°, then how many times the horizontal range is larger than the maximum height?

a) 2 b) 3 c) $3\sqrt{4}$ d) $4\sqrt{3}$

309. A small roller coaster starts at point A with a speed u on a curved track as shown in figure

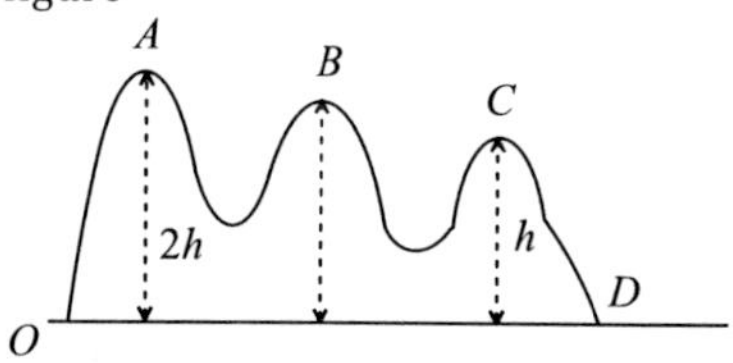

The friction between the roller coaster and the track is negligible and it always remains in contact with the track. The speed of the roller coaster at point D on the track will be

a) $(u^2 + gh)^{\frac{1}{2}}$ b) $(u^2 + 2gh)^{\frac{1}{2}}$ c) $(u^2 + 4gh)^{\frac{1}{2}}$ d) u

310. The vectors $2\hat{i} + 3\hat{j} - 2\hat{k}$, $5\hat{i} + a\hat{j} + \hat{k}$ and $-\hat{i} + 2\hat{j} + 3\hat{k}$ are coplanar when a is

a) −9 b) 9 c) −18 d) 18

311. Centripetal acceleration is

a) A constant vector
b) A constant scalar
c) A magnitude changing vector
d) Not a constant vector

312. The trajectory of a projectile in vertical plane in $y = ax - bx^2$, where a and b are constant and x and y are respectively horizontal and vertical distances of the projectile from the point of projection. The maximum height attained by the particle and the angle of projection from the horizontal are

a) $\frac{b^2}{4b}, \tan^{-1}(b)$
b) $\frac{a^2}{b}, \tan^{-1}(2b)$
c) $\frac{a^2}{4b}, \tan^{-1}(a)$
d) $\frac{2a^2}{b}, \tan^{-1}(a)$

313. A body of mass 100 g is rotating in a circular path of radius r with constant velocity. The work done in one complete revolution is

a) $100\ rJ$
b) $(r/100)J$
c) $(100/r)J$
d) Zero

314. A point P moves in counter-clockwise direction on a circular path as shown in the figure. The movement of P is such that it sweeps out length $s = t^3 + 5$, where s is in metre and t is in second. The radius of the path is 20 m. The acceleration of P when t=2s is nearly

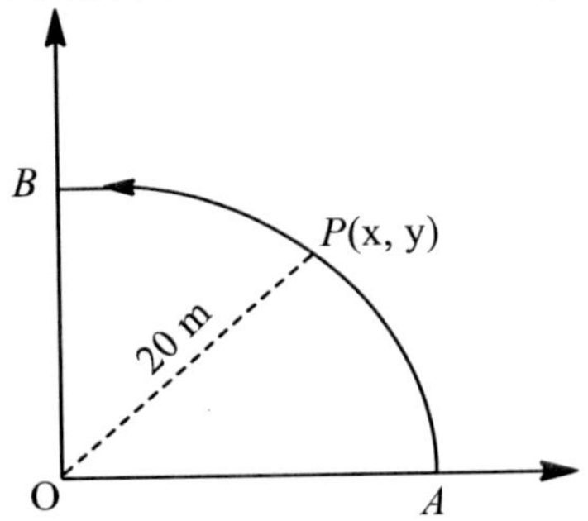

a) 13 ms^{-2}
b) 12 ms^{-2}
c) 7.2 ms^{-2}
d) 14 ms^{-2}

315. A projectile is thrown in the upward direction making an angle of 60° with the horizontal direction with a velocity of 147 ms^{-1}. Then the time after which its inclination with the horizontal is 45°, is

a) 15 s
b) 10.98 s
c) 5.49 s
d) 2.745 s

316. The kinetic energy k of a particle moving along a circle of radius R depends on the distance covered s as $k = as^2$ where a is a constant

a) $2a\frac{s^2}{R}$
b) $2as\left(1 + \frac{s^2}{R^2}\right)^{1/2}$
c) $2as$
d) $2a\frac{R^2}{s}$

317. A heavy mass is attached to a thin wire and is whirled in a vertical circle. The wire is most likely to break

a) When the mass is at the highest point of the circle
b) When the mass is the lowest point of the circle
c) When the wire is horizontal
d) At an angle of $\cos^{-1}(1/3)$ from the upward vertical

318. Three particles A, B and C are projected from the same point with the same initial speeds making angles 30°, 45° and 60° respectively with the horizontally. Which of the following statements is correct?

a) A,B and C have unequal ranges
b) Range of A and C are less than that of B
c) Range of A and C are equal and greater than that of B
d) A,B and C have equal ranges

319. If $\vec{A} = \vec{B}$, then which of the following is not correct

a) $\hat{A} = \hat{B}$
b) $\hat{A} \cdot \hat{B} = AB$
c) $|\vec{A}| = |\vec{B}|$
d) $A\hat{B} \mid \quad \mid B\hat{A}$

320. A bomb is dropped on an enemy post by an aeroplane flying horizontally with a velocity of 60 kmh^{-1} and at a height of 490 m. At the time of dropping the bomb, how far the aeroplane should be from the enemy post so that the bomb may directly hit the target?

a) $\frac{400}{3}$ m b) $\frac{500}{3}$ m c) $\frac{1700}{3}$ m d) 498 m

321. Two stones thrown at different angles have same initial velocity and same range. If H is the maximum height attained by one stone thrown at an angle of 30°, then the maximum height attained by the other stone is

a) $\frac{H}{2}$ b) H c) $2H$ d) $3H$

322. A ball of mass m is thrown vertically upwards. Another ball of mass $2m$ is thrown at an angle θ with the vertical. Both of them stay in air for same period of time. The heights attained by the two balls are in the ratio of

a) 2: 1 b) $1: \cos\theta$ c) 1: 1 d) $\cos\theta : 1$

323. A projectile of mass m is thrown with a velocity v making an angle of 45° with the horizontal. The change in momentum from departure to arrival along vertical direction, is

a) $2mv$ b) $\sqrt{2}\, mv$ c) mv d) $\frac{mv}{2}$

324. The resultant of two forces at right angle is 5N. When the angle between them is 120°, the resultant is $\sqrt{13}$. Then the force are

a) $\sqrt{12}$N, $\sqrt{13}$N b) $\sqrt{20}$N, $\sqrt{5}$N c) 3 N, 4 N d) $\sqrt{40}$N, $\sqrt{15}$N

325. A bridge is in the form of a semi-circle of radius $40m$. The greatest speed with which a motor cycle can cross the bridge without leaving the ground at the highest point is ($g = 10\ ms^{-2}$)
(frictional force is negligibly small)

a) 40 ms^{-1} b) 20 ms^{-1} c) 30 ms^{-1} d) 15 ms^{-1}

326. A particle crossing the origin of co-ordinates at time $t = 0$, moves in the xy-plane with a constant acceleration a in the y-direction. If its equation of motion is $y = bx^2$ (b is a constant), its velocity component in the x-direction is

a) $\sqrt{\frac{2b}{a}}$ b) $\sqrt{\frac{a}{2b}}$ c) $\sqrt{\frac{a}{b}}$ d) $\sqrt{\frac{b}{a}}$

327. A 500 kg car takes a round turn of radius 50 m with a velocity of 36 km/hr. The centripetal force is

a) 250 N b) 750 N c) 1000 N d) 1200 N

328. A body is thrown with a velocity of 10 ms^{-1} at an angle of 60° with the horizontal. Its velocity at the highest point is

a) 7 ms^{-1} b) 9 ms^{-1} c) 18.7 ms^{-1} d) 5 ms^{-1}

329. Two particles 1 and 2 are projected with same speed v as shown in figure. Particle 2 is on the ground and particle 1 is at a height h from the ground and at a horizontal distance s from particle 2. If a graph is plotted between v and s for the condition of collision of the two then (v on y-axis and s on x-axis)

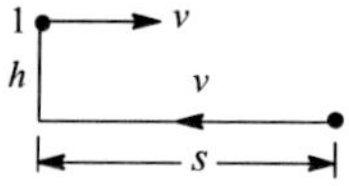

a) It will be a parabola passing through the origin

b) It will be straight line passing through the origin and having a slope of $\sqrt{\frac{g}{8h}}$

c) It will be a straight line passing through the origin and having a slope of $\sqrt{\frac{g}{4h}}$

d) It will be a straight line not passing through the origin

330. A stone tied to one end of rope and rotated in a circular motion. If the string suddenly breaks, then the stone travels
a) in perpendicular direction
b) in direction of centrifugal force
c) towards centripetal force
d) in tangential direction

331. A particle moves in a circle of radius 30 cm. Its linear speed is given by $v = 2t$ where t in second and $v$in m/s. Find out its radial and tangential acceleration at $t = 3\ sec$ respectively
a) $220\ m/\sec^2$, $50\ m/sec^2$ b) $100\ m/\sec^2$, $5\ m/sec^2$
c) $120\ m/\sec^2$, $2\ m/sec^2$ d) $110\ m/\sec^2$, $10\ m/sec^2$

332. A particle moves with constant speed v along a circular path of radius r and completes the circle in time T. The acceleration of the particle is
a) $2\pi v/T$ b) $2\pi r/T$ c) $2\pi r^2/T$ d) $2\pi v^2/T$

333. A particle is projected up an inclined plane with initial speed $v = 20\text{ms}^{-1}$ at an angle $\theta = 30°$ with plane. The component of its velocity perpendicular to plane when it strikes the plane is
a) $10\sqrt{3}\ \text{ms}^{-1}$ b) $10\ \text{ms}^{-1}$
c) $5\sqrt{3}\ \text{ms}^{-1}$ d) Data is insufficient

334. An aircraft is flying at a height of 3400 m above the ground. If the angle subtended at a ground observation point by the aircraft position 10 s apart is 30°, then the speed of the aircraft is
a) $19.63\ \text{ms}^{-1}$ b) $1963\ \text{ms}^{-1}$ c) $108\ \text{ms}^{-1}$ d) $196.3\ \text{ms}^{-1}$

335. A body constrained to move in Y direction, is subjected to a force given by $\vec{F} = (-2\hat{i} + 15\hat{j} + 6\hat{k})$N done by this force in moving the body through a distance of 10m along Y axis?
a) 190 J b) 160 J c) 150 J d) 20 J

336. A body moves with constant angular velocity on a circle. Magnitude of angular acceleration
a) $r\omega^2$ b) Constant c) Zero d) None of the above

337. The sum of two vectors $\vec{A}$ and $\vec{B}$ is at right angles to their difference. Then
a) $A = B$
b) $A = 2B$
c) $B = 2A$
d) $\vec{A}$ and $\vec{B}$ have the same direction

338. A projectile is projected with kinetic energy K. If it has the maximum possible horizontal range, then its kinetic energy at the highest point will be
a) 0.25 K b) 0.5 K c) 0.75 K d) 1.0 K

339. Two particles of equal masses are revolving in circular paths of radii r_1 and r_2 respectively with the same speed. The ratio of their centripetal forces is
a) $\frac{r_2}{r_1}$ b) $\sqrt{\frac{r_2}{r_1}}$ c) $\left(\frac{r_1}{r_2}\right)^2$ d) $\left(\frac{r_2}{r_1}\right)^2$

340. A scooter is going round a circular road of radius 100 m at a speed of 10 m/s. The angular speed of the scooter will be
a) 0.01 rad/s b) 0.1 rad/s c) 1 rad/s d) 10 rad/s

341. A car rounds an unbanked curve of radius 92 m without skidding at a speed of 26 ms^{-1}. The smallest possible coefficient of static friction between the tyres and the road is
a) 0.75 b) 0.60 c) 0.45 d) 0.30

342. A body of mass 5 kg is whirled in a vertical circle by a string 1 m long. Calculate velocity at the top of the circle for just looping the vertical loop
a) 3.1 ms^{-1} b) 7 ms^{-1} c) 9 ms^{-1} d) 7.3 ms^{-1}

343. A cricketer can throw a ball to a maximum horizontal distance of 100 m. The speed with which he throws the ball is (to the nearest integer)

a) 30 ms^{-1} b) 42 ms^{-1} c) 32 ms^{-1} d) 35 ms^{-1}

344. A cyclist riding the bicycle at a speed of $14\sqrt{3}\ ms^{-1}$ takes a turn around a circular road of radius $20\sqrt{3}\ m$ without skidding. Given $g = 9.8\ ms^{-2}$, what is his inclination to the vertical

a) 30° b) 90° c) 45° d) 60°

345. Two projectile are thrown with the same initial velocity at angles α and (90° − α) with the horizontal. The maximum heights attained by them are h_1 and h_2 respectively. Then $\frac{h_1}{h_2}$ is equal to

a) $\sin^2 \alpha$ b) $\cos^2 \alpha$ c) $\tan^2 \alpha$ d) 1

346. The average acceleration vector for a particle having a uniform circular motion is

a) A constant vector of magnitude v^2/r

b) A vector of magnitude v^2/r directed normal to the plane of the given uniform circular motion

c) Equal to the instantaneous acceleration vector at the start of the motion

d) A null vector

347. Consider a disc rotating in the horizontal plane with a constant angular speed ω about its centre O. The disc has a shaded region on one side of the diameter and an unshanded region on the other side as shown in the figure. When the disc is in the orientation as shown, two pebbles P and Q are simultaneously projected at an angle towards R. The velocity of projection is in the $y - z$ plane and is same for both pebbles with respect to the disc. Assume that (i) they land back on the disc before the disc has completed $\frac{1}{8}$ rotation. (ii) their range is less than half the disc radius, and (iii) ω remains constant throughout. Then

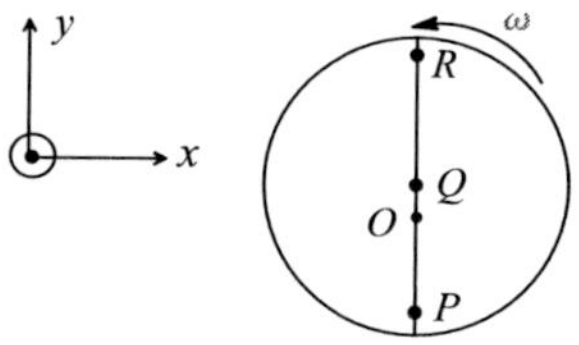

a) P lands in the shaded region and Q in the unshaded region

b) P lands in the unshaded region and Q in the shaded region

c) Both P and Q land in the unshaded region

d) Both P and Q land in the shaded region

348. A ball is moving to and fro about the lowest point A of a smooth hemispherical bowl. If it is able to rise up to a height of 20 cm on either side of A, its speed at A must be ($Take\ g = 10\ m/s^2$, massof the body 5 g)

a) 0.2 m/s b) 2 m/s c) 4 m/s d) 4.5 ms^{-1}

349. What is the angle between $(\hat{i} + 2\hat{j} + 2\hat{k})$ and $\hat{i}$

a) 0° b) $\pi/6$ c) $\pi/3$ d) None of these

350. A force of $(10\hat{i} - 3\hat{j} + 6\hat{k})$ N acts on a body of mass 100 g and displaces it from $(6\hat{i} + 5\hat{j} - 3\hat{k})$ m to $(10\hat{i} - 2\hat{j} + 7\hat{k})$ m. The work done is

a) 21 J b) 121 J c) 361 J d) 1000 J

351. A bridge is in the form of a semi-circle of radius 40 m. The greatest speed with which a motor cycle can cross the bridge without leaving the ground at the highest point is($g = 10\text{ms}^{-2}$)(frictional force is negligibly small)

a) 40 ms^{-1} b) 20 ms^{-1} c) 30 ms^{-1} d) 15 ms^{-1}

352. A body is projected at such angle that the horizontal range is three times the greatest height. The angel of projection is

a) 42°8′ b) 53°7′ c) 33°7′ d) 25°8′

353. The maximum and minimum tension in the string whirling in a circle of radius 2.5 m with constant velocity are in the ratio 5 : 3 then its velocity is

a) $\sqrt{98}\ ms^{-1}$ b) $7\ ms^{-1}$ c) $\sqrt{490}\ ms^{-1}$ d) $\sqrt{4.9}\ ms^{-1}$

354. A body of mass $\sqrt{3}$ kg is suspended by a string from a rigid support. The body is pulled horizontally by a force F until the string makes an angle of 30° with the vertical. The value o F and tension in the string are

a) 19.6 N; 19.6 N b) 9.8 N; 9.8 N c) 9.8 N, 19.6 N d) 19.6 N, 9.8 N

355. A sphere of mass 0.2 kg is attached to an inextensible string of length 0.5 m whose upper end is fixed to the ceiling. The sphere is made to describe a horizontal circle of radius 0.3 m. The speed of the sphere will be

a) $1.5\ m\ s^{-1}$ b) $2.5\ m\ s^{-1}$ c) $3.2\ m\ s^{-1}$ d) $4.7\ m\ s^{-1}$

356. A bullet is fired from a cannon with velocity 500 m/s . If the angle of projection is 15°and $g = 10\ m/s^2$. Then the range is

a) $25 \times 10^3\ m$ b) $12.5 \times 10^3\ m$ c) $50 \times 10^2\ m$ d) $25 \times 10^2\ m$

357. Find the maximum speed at which a car can turn round a curve of 30 m radius on a level road if the coefficient of friction between the tyres and the road is 0.4 (Acceleration due to gravity = $10\ ms^{-2}$)

a) $12\ ms^{-2}$ b) $10\ ms^{-2}$ c) $11\ ms^{-2}$ d) $15\ ms^{-2}$

358. A simple pendulum oscillates in a vertical plane. When it passes through the mean position, the tension in the string is 3 times the weight of the pendulum bob. What is the maximum displacement of the pendulum with respect to the vertical

a) 30° b) 45° c) 60° d) 90°

359. A particle is moving in a circle of radius R with constant speed$v$. If radius is doubled, then its centripetal force to keep the same speed gets

a) twice as great as before b) half

c) one-fourth d) remains constant

360. A ball is projected with kinetic energy E at an angle of 45° to the horizontal. At the highest point during its flight, its kinetic energy will be

a) Zero b) $E/2$ c) $E/\sqrt{2}$ d) E

361. Given that A and B are greater than 1. The magnitude of $(\vec{A} \times \vec{B})$ can not be

a) equal to AB b) less than AB c) more than AB d) equal to A/B

362. The string of pendulum of length l is displaced through 90° from the vertical and released. Then the minimum strength of the string in order to withstand the tension, as the pendulum passes through the mean position is

a) mg b) $3mg$ c) $5\ mg$ d) $6\ mg$

363. A car is moving on a circular road of diameter 50 m with a speed of $5\ ms^{-1}$. It is suddenly accelerated at rate $1\ ms^{-2}$. If the mass is 500 kg, find the net force acting on the car

a) 5 N b) 1000 N c) $500\sqrt{2}$N d) $500/\sqrt{2}$ N

364. A bomb is dropped from an aeroplane flying horizontally with a velocity 469 ms^{-1} at an altitude of 980 m. The bomb will hit the ground after a time

a) 2 s b) $\sqrt{2}$ s c) $5\sqrt{2}$ s d) $10\sqrt{2}$ s

365. Following forces start acting on a particle at rest at the origin of the co-ordinate system simultaneously $\vec{F}_1 = 5\hat{i} - 5\hat{j} + 5\hat{k}$, $F_2 = 2\hat{i} + 8\hat{j} + 6\hat{k}$, $F_3 = -6\hat{i} + 4\hat{j} - 7\hat{k}$, $F_4 = -\hat{i} - 3\hat{j} - 2\hat{k}$. The particle will move

a) in $x - y$ plane b) in $y - z$ plane c) in $x - z$ plane d) along x-axis

366. If $0.5\hat{i} + 0.8\hat{j} + c\hat{k}$ is a unit vector, then the value of c is

a) $\sqrt{0.11}$ b) $\sqrt{0.22}$ c) $\sqrt{0.33}$ d) $\sqrt{0.89}$

367. A cyclist goes round a circular path of circumference 34.3 m in $\sqrt{22}$s, the angle made by him with the vertical will be

a) 45° b) 40° c) 42° d) 48°

368. The simple sum of two co-initial vectors is 10 units. Their vector sum is 8 units. The resultant of the vectors is perpendicular to the smaller vector. The magnitudes of the two vectors are
a) 2 units and 14 units
b) 4 units and 12 units
c) 6 units and 10 units
d) 8 units and 8 units

369. A particle is projected with a velocity v such that its range on the horizontal plane is twice the greatest height attained by it. The range of the projectile is (where g is acceleration due to gravity)
a) $\frac{4v^2}{5g}$ b) $\frac{4g}{5v^2}$ c) $\frac{v^2}{g}$ d) $\frac{4v^2}{\sqrt{5}g}$

370. A mass of 2 kg is whirle'd in a horizontal circle by means of a string at an initial speed of 5 revolutions per minute. Keeping the radius constant, the tension in the string is double. The new speed is nearly
a) 2.25 rpm b) 7 rpm c) 10 rpm d) 14 rpm

371. A point of application of a force $\vec{F} = 5\hat{i} - 4\hat{j} + 2\hat{k}$ is moved from $\vec{r_1} = 2\hat{i} + 7\hat{j} + 4\hat{k}$ to $\vec{r_2} = 5\hat{i} + 2\hat{j} + 3\hat{k}$ the work done is
a) 22 units b) −22 units c) 33 units d) −33 units

372. A particle performing uniform circular motion has
a) Radial velocity and radial acceleration
b) A radial velocity and transverse acceleration
c) Transverse velocity and radial acceleration
d) Transverse velocity and transverse acceleration

373. The velocity of projection of an oblique projectile is $(6\hat{\mathbf{i}} + 8\hat{\mathbf{j}})\text{ms}^{-1}$. The horizontal range of the projectile is
a) 4.9 m b) 9.6 m c) 19.6 m d) 14 m

374. One o the rectangular components of a velocity of 60kmh^{-1} is 30 kmh^{-1}. The other rectangular component is
a) 30 kmh^{-1} b) $30\sqrt{3}\text{ kmh}^{-1}$ c) $30\sqrt{2}\text{ kmh}^{-1}$ d) Zero

375. The sum of the magnitudes of two forces acting at a point is 16 N. the resultant of these forces is perpendicular to the smaller force has a magnitude of 8 N. If the smaller force is magnitude x, then the value of x is
a) 2 N b) 4 N c) 6 N d) 7 N

376. Trajectories of two projectiles are shown in figure. Let T_1 and T_2 be the time periods and u_1 and u_2 their speeds of projection. Then

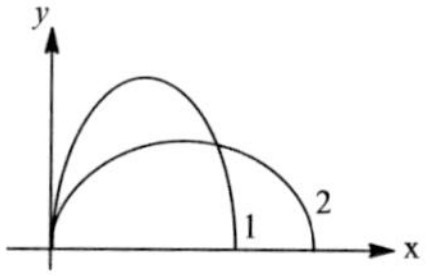

a) $T_2 > T_1$ b) $T_1 = T_2$ c) $u_1 > u_2$ d) $u_1 < u_2$

377. A car is moving in a circular horizontal track of radius 10 m with a constant speed of 10 m/sec. A plumb bob is suspended from the roof of the car by a light rigid rod of length 1.00 m. The angle made by the rod with track is
a) Zero b) 30° c) 45° d) 60°

378. The horizontal and vertical displacement x and y of a projectile at a given time t are given by $x = 6t$ metre and $y = 8t - 5t^2$ metre. The range of the projectile in metre is
a) 9.6 b) 10.6 c) 19.2 d) 38.4

379. A projectile is projected with a speed u making an angle 2θ with the horizontal. What is the speed when its direction of motion makes an angle θ with the horizontal
a) $(u\cos 2\theta)/2$ b) $u\cos\theta$ c) $u(2\cos\theta - \sec\theta)$ d) $u(\cos\theta - \sec\theta)$

380. Two stones are projected with the same velocity in magnitude but making different angles with the horizontal. Their ranges are equal. If the angel of projection of one is $\pi/3$ and its maximum height is y_1, the maximum height of the other will be

a) $3y_1$ b) $2y_1$ c) $\frac{y_1}{2}$ d) $\frac{y_1}{3}$

381. A glass marble projected horizontally from the top of a table falls at a distance x from the edge of the table. If h is the height of the table, then the velocity of projection is

a) $h\sqrt{\frac{g}{2x}}$ b) $x\sqrt{\frac{g}{2h}}$ c) gxh d) $gx + h$

382. A bucket full of water is revolved in vertical circle of radius $2m$. What should be the maximum time-period of revolution so that the water doesn't fall of the bucket

a) $1\ sec$ b) $2\ sec$ c) $3\ sec$ d) $4\ sec$

383. Which of the following sets of factors will affect the horizontal distance covered by an athlete in a long-jump event

a) Speed before he jumps and his weight b) The direction in which he leaps and the initial speed

c) The force with which he pushes the ground and his speed d) None of these

384. Two inclined planes are located as shown in figure. A particle is projected from the foot of one frictionless plane along its line with a velocity just sufficient to carry it to top after which the particle slides down the other frictionless inclined plane. The total time it will take to reach the point C is

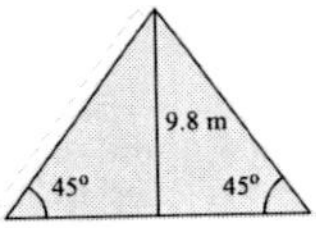

a) 2 s b) 3 s c) $2\sqrt{2}$ s d) 4 s

385. A ball rolls of the top of a stair way with a horizontal velocity u $\mathrm{ms^{-1}}$. If the steps are h metre and b metre wide, the ball hits the edge of nth step, the time taken by the ball is

a) $\frac{hu}{gb}$ b) $\frac{2hu}{gb}$ c) $\frac{2hu^2}{gb}$ d) $\frac{hu^2}{2gb}$

386. An airplane, diving at an angle of 53.0° with the vertical releases a projectile at an altitude of 730 m. The projectile hits the ground 5.00 s after being released. What is the speed of the aircraft?

a) 282 $\mathrm{ms^{-1}}$ b) 202 $\mathrm{ms^{-1}}$ c) 182 $\mathrm{ms^{-1}}$ d) 102 $\mathrm{ms^{-1}}$

387. A body of mass m hangs at one end of a string of length l, the other end of which is fixed. It is given a horizontal velocity so that the string would just reach where it makes an angle of 60° with the vertical. The tension in the string at mean position is

a) $2\ mg$ b) mg c) $3mg$ d) $\sqrt{3}mg$

388. A cyclist goes round a circular path of circumference 34.3 m in $\sqrt{22}\ sec$. the angle made by him, with the vertical, will be

a) 45° b) 40° c) 42° d) 48°

389. A1 kg stone at the end of 1 m long string is whirled in a vertical circle at constant speed of 4 m/sec. The tension in the string is 6 N, when the stone is at ($g = 10\ m/\sec^2$)

a) Top of the circle b) Bottom of the circle c) Half way down d) None of the above

390. A curved road of diameter 1.8 km is banked so that no friction is required at a speed of 30 $\mathrm{ms^{-1}}$. What is the banking angle?

a) 6° b) 16° c) 26° d) 0.6°

391. A stone tied to the end of a string $1m$ long is whirled in a horizontal circle with a constant speed. If the stone makes 22 revolution in 44 seconds, what is the magnitude and direction of acceleration of the stone

a) $\pi^2/4\ ms^{-2}$ and direction along the radius towards the centre

b) $\pi^2 ms^{-2}$ and direction along the radius away from the centre
c) $\pi^2 ms^{-2}$ and direction along the radius towards the centre
d) $\pi^2 ms^{-2}$ and direction along the tangent to the circle

392. A fly wheel rotates about a fixed axis and slows down from 300 rpm to 100 rpm in 2 min. Then its angular retardation in rad/min^2 is

a) $\frac{100}{\pi}$ b) 100 c) 100 π d) 200 π

393. A small mass m is attached to a massless string whose other end is fixed at P as shown in the figure. The mass is undergoing circular motion in the $x-y$ plane with centre at O and constant angular speed ω. If the angular momentum of the system, calculated about O and P are denoted by $\vec{L}_O$ and $\vec{L}_P$ respectively, then

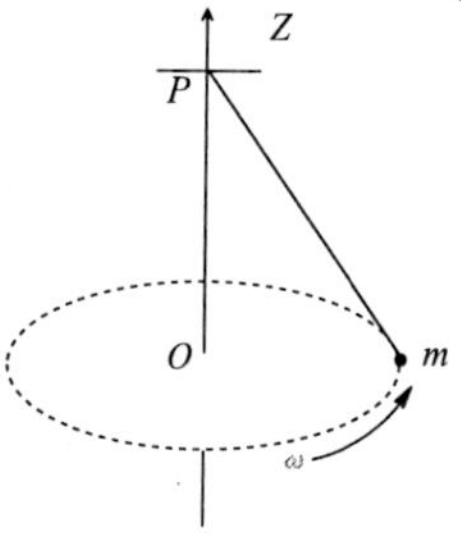

a) $\vec{L}_O$ and $\vec{L}_P$ do not vary with time
b) $\vec{L}_O$ varies with time while $\vec{L}_P$ remains constant
c) $\vec{L}_O$ remains constant while $\vec{L}_P$ varies with time
d) $\vec{L}_O$ and $\vec{L}_P$ both vary with time

394. A particle of mass m is moving in a circular path of constant radius r such that its centripetal acceleration a_c is varying with time t as, $a_c = k^2rt^2$, The power delivered to the particle by the forces acting on it is

a) $2\pi mk^2r^2t$ b) mk^2r^2t c) $\frac{mk^4r^2t^5}{3}$ d) Zero

395. A man projects a coin upwards from the gate of a uniformly moving train. The path of coin for the man will be

a) Parabolic
b) Inclined straight line
c) Vertical straight line
d) Horizontal straight line

396. A piece of marble is projected from earth's surface with velocity of 50ms^{-1}. 2 s later, it just clears a wall 5 m high. What is the angle of projection?

a) 45° b) 30° c) 60° d) None of these

397. A ball rolls off the top of a stairway with horizontal velocity $v_0 ms^{-1}$. If the steps are h metre high and w metre wide, the ball will hit the edge of nth step if

a) $n = \frac{2hv_0}{gw^2}$ b) $n = \frac{2hv_0^2}{gw}$ c) $n = \frac{hv_0^2}{gw^2}$ d) $n = \frac{2hv_0^2}{gw^2}$

398. A particle is projected from the ground at an angle of 60° with horizontal with speed $u = 20$ ms^{-1}. The radius of curvature of the path of the particle, when its velocity makes an angle of 30° with horizontal is (g $= 10$ ms^{-2})

a) 10.6 m b) 12.8 m c) 15.4 m d) 24.2 m

399. A particle moves along the parabolic path $y = ax^2$ in such a way that the x-component of the velocity remains constant, say c. The acceleration of the particle is

a) $ac\hat{\mathbf{k}}$ b) $2ac^2\hat{\mathbf{j}}$ c) $ac^2\hat{\mathbf{k}}$ d) $a^2c\hat{\mathbf{j}}$

400. An aeroplane is flying horizontally with a velocity of 600 km/h at a height of 1960 m. When it is vertically at a point Aon the ground, a bomb is released from it. The bomb strikes the ground at point B. The distance AB is

a) 1200 m b) 0.33 km c) 3.33 km d) 33 km

401. Given that $\vec{A}+\vec{B}+\vec{C}=0$. Out of three vectors, two are equal in magnitude and the magnitude of third vector is $\sqrt{2}$ times that of either of the two having equal magnitude. Then the angles between vectors are given by

a) $45°, 45°, 90°$ b) $90°, 135°, 135°$ c) $30°, 60°, 90°$ d) $45°, 60°, 90°$

402. The magnitude of the X and Y components of $\vec{A}$ are 7 and 6. Also the magnitudes of X and Y components of $\vec{A}+\vec{B}$ are 11 and 9 respectively. What is the magnitude of $\vec{B}$?

a) 5 b) 6 c) 8 d) 9

403. For what value of a, $\vec{A}=2\hat{i}+a\hat{j}+\hat{k}$ will be perpendicular to $\vec{B}=4\hat{i}-2\hat{j}-\hat{k}$

a) 4 b) zero c) 3 d) 1

404. A particle is thrown with a speed u at an angle θ with the horizontal. When the particle makes an angle α with the horizontal, its speed becomes v, whose values is

a) $u\cos\theta$ b) $u\cos\theta\cos\alpha$ c) $u\cos\theta\sec\alpha$ d) $u\sec\theta\cos\alpha$

405. A cyclist is moving on a circular track of radius 80 m with a velocity $v=36\text{kmh}^{-1}$. He has to lean from the vertical approximately through an angle (take $g=10\text{ms}^{-2}$

a) $\tan^{-1}(4)$ b) $\tan^{-1}\left(\frac{1}{3}\right)$ c) $\tan^{-1}\left(\frac{1}{4}\right)$ d) $\tan^{-1}\left(\frac{1}{8}\right)$

406. A man 80 kg is supported by two cables as shown in the figure. Then the ratio of tensions T_1 and T_2 is

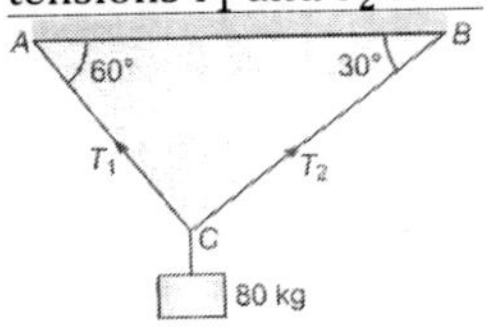

a) 1:1 b) $1:\sqrt{3}$ c) $\sqrt{3}:1$ d) 1:3

407. The resultant of two vectors $\vec{A}$ and $\vec{B}$ is perpendicular to the vector $\vec{A}$ and its magnitude is equal to half of the magnitude of vector $\vec{B}$. Then the angle between $\vec{A}$ and $\vec{B}$ is

a) 30° b) 45° c) 150° d) 120°

408. The horizontal range is four times the maximum height attained by a projectile. The angle of projection is

a) 90° b) 60° c) 45° d) 30°

409. An object is projected so that its horizontal range R is maximum. If the maximum height attained by the object is H, then the ratio of R/H is

a) 4 b) $\frac{1}{4}$ c) 2 d) $\frac{1}{2}$

410. An electric fan has blades of length 30 cm as measured from the axis of rotation. If the fan is rotating at 1200 rpm, the acceleration of a point on the tip of the blade is about

a) 1600 ms^{-2} b) 4740 ms^{-2} c) 2370 ms^{-2} d) 5055 ms^{-2}

411. The x and y components of a force are 2 N and −3N. The force is

a) $2\hat{i}-3\hat{j}$ b) $2\hat{i}+3\hat{j}$ c) $-2\hat{i}-3\hat{j}$ d) $3\hat{i}+2\hat{j}$

412. A projectile fired with initial velocity u at some angle θ has a range R. If the initial velocity be doubled at the same angle off projection, then the range will be

a) $2R$ b) $R/2$ c) R d) $4R$

413. A particle of mass m is executing uniform circular motion on a path of radius r. If p is the magnitude of its linear momentum. The radial force acting on the particle is

a) pmr b) $\frac{rm}{p}$ c) $\frac{mp^2}{r}$ d) $\frac{p^2}{rm}$

414. Two forces, each equal to $\frac{P}{2}$, act at right angles. Their effect may be neutralized by a third force acting along their bisector in the opposite direction with a magnitude of

a) P b) $\frac{P}{2}$ c) $\frac{P}{\sqrt{2}}$ d) $\sqrt{2}\,P$

415. Two projectiles A and B thrown with speeds in the ratio $1:\sqrt{2}$ acquired the same heights. If A is thrown at an angle of $45°$ with the horizontal, the angle of projection of B will be

a) $0°$ b) $60°$ c) $30°$ d) $45°$

416. Two particles are projected simultaneously in the same vertical plane, from the same point, both with different speeds and at different angles with horizontal. The path followed by one, as seed by the other, is

a) A vertical line
b) A parabola
c) A hyperbola
d) A straight line making a constant angle ($\neq 90°$) with the horizontal

417. A vector $\vec{A}$ when added to the vector $\vec{B} = 3\hat{i} + 4\hat{j}$ yields a resultant vector that is in the positive y direction and has a magnitude equal to that of $\vec{B}$. Find the magnitude of

a) $\sqrt{10}$ b) 10 c) 5 d) $\sqrt{15}$

418. A boy throws a ball upwards with velocity $u = 15 \text{ms}^{-1}$. The wind imparts a horizontal acceleration of 3ms^{-1} to the left. The angel θ at which the ball must be thrown so that the ball returns to the boy's hand is (use $g = 10 \text{ ms}^{-2}$)

a) $\tan^{-1}(0.4)$ b) $\tan^{-1}(0.2)$ c) $\tan^{-1}(0.3)$ d) $\tan^{-1}(0.15)$

419. A proton of velocity $(3\hat{i} + 2\hat{j}) \times 10^5 \text{ms}^{-1}$ enters a magnetic field $(2\hat{i} + 3\hat{k})$T. If the specific charge is $9.6 \times 10^7 \text{C kg}^{-1}$. The acceleration of the proton in ms^{-2} is

a) $(6\hat{i} - 9\hat{j} + 4\hat{k}) \times 9.6 \times 10^{12}$ b) $(6\hat{i} + 9\hat{j} + 4\hat{k}) \times 9.6 \times 10^{12}$
c) $(6\hat{i} - 9\hat{j} - 4\hat{k}) \times 9.6 \times 10^{12}$ d) $(6\hat{i} + 9\hat{j} - 4\hat{k}) \times 9.6 \times 10^{12}$

420. The path of a projectile in the absence of air drag is shown in the figure by dotted line. If the air resistance is not ignored then which one of the path is shown in the figure is appropriate for the projectile

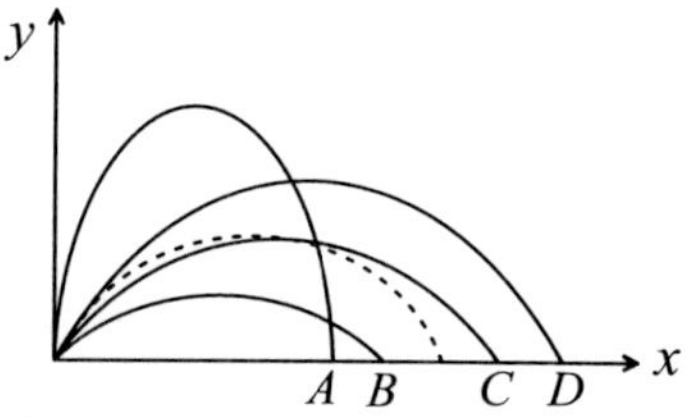

a) B b) A c) D d) C

421. A cannon on a level plane is aimed at an angle θ above the horizontal and a shell is fired with a muzzle velocity v_0 towards a vertical cliff a distance D away. Then the height from the bottom at which the shell strikes the side walls of the cliff is

a) $D\sin\theta - \frac{gD^2}{2v_0^2\sin^2\theta}$ b) $D\cos\theta - \frac{gD^2}{2v_0^2\cos^2\theta}$ c) $D\tan\theta - \frac{gD^2}{2v_0^2\cos^2\theta}$ d) $D\tan\theta - \frac{gD^2}{2v_0^2\sin^2\theta}$

422. Which of the following statements is false for a particle moving in a circle with a constant angular speed

a) The velocity vector is tangent to the circle
b) The acceleration vector is tangent to the circle
c) The acceleration vector points to the centre of the circle
d) The velocity and acceleration vectors are perpendicular to each other

423. A mass of 100 gm is tied to one end of a string 2 m long. The body is revolving in a horizontal circle making a maximum of 200 revolutions per min. The other end of the string is fixed at the centre of the circle of revolution. The maximum tension that the string can bear is (approximately)

a) 8.76 N b) 8.94 N c) 89.42 N d) 87.64 N

424. A gramophone disc is set revolving in a horizontal plane and reaches a steady state of motion of two revolutions per second. It is found that a small coin placed on the disc will remain there if its centre is not more than 5 cm from the axis of rotation; the coefficient of friction between the coin and the disc is

a) 0.2 b) 0.4 c) 0.6 d) 0.8

425. A body is projected with a speed u m/s at an angle β with the horizontal. The kinetic energy at the highest point is $\frac{3}{4}$ th of the initial energy. The values of β is

a) 30° b) 45° c) 60° d) 120°

426. Four persons K, L, M and N are initially at the corners of a square of side of length d. If every person starts moving, such that K is always headed towards L, L towards M, M is headed directly towards N and N towards K, then the four persons will meet after

a) $\frac{d}{v}$ sec b) $\frac{\sqrt{2d}}{v}$ sec c) $\frac{d}{\sqrt{2v}}$ sec d) $\frac{d}{2v}$ sec

427. A body of mass m is moving with a uniform speed v along a circle of radius r, what is the average acceleration in going from A to B?

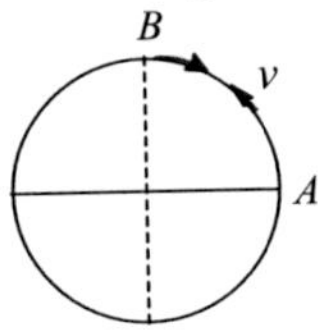

a) $2v^2/\pi r$ b) $2\sqrt{2}v^2/\pi r$ c) $v^2/\pi r$ d) None of these

428. A small block is shot into each of the four tracks as shown below. Each of the tracks rises to the same height. The speed with which the block enters the track is the same in all cases. At the highest point of the track, the normal reaction is maximum in

a) 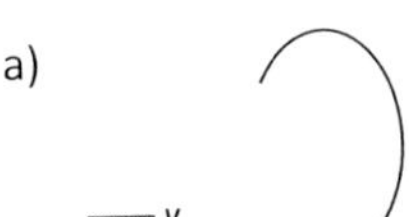

b) 

c) 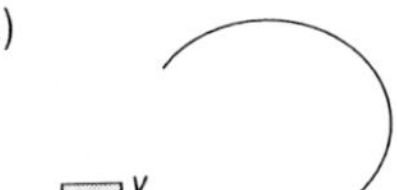

d) 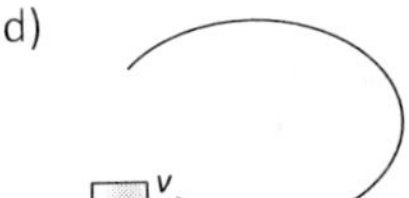

429. An unbanked curve has a radius of $60m$. The maximum speed at which car can make a turn if the coefficient of static friction is 0.75, is

a) 2.1 m/s b) 14 m/s c) 21 m/s d) 7 m/s

430. A particle of mass m is rotating in a horizontal circle of radius R and is attached to a hanging mass M as shown in the figure. The speed of rotation required by the mass m keep M steady is

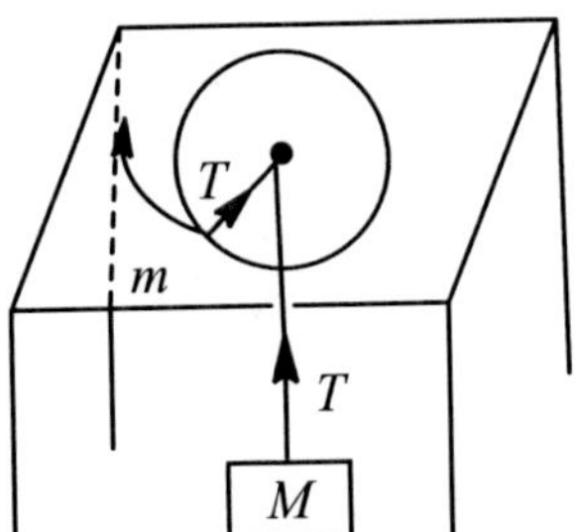

a) $\sqrt{\frac{mgR}{M}}$ b) $\sqrt{\frac{mgR}{m}}$ c) $\sqrt{\frac{mg}{MR}}$ d) $\sqrt{\frac{mR}{Mg}}$

431. Two bodies are projected from ground with equal speed 20 ms^{-1} from the same position in the same vertical plane to have equal range but at different angles above the horizontal. If one of the angle is 30° the sum of their maximum heights is (assume $g = 10\ ms^{-2}$)

a) 400 m b) 20 m c) 30 m d) 40 m

432. A particle is projected with velocity v_0 along x-axis. The deceleration on the particle is proportional to the square of the distance from the origin *ie* $a = \propto x^2$ the distance at which the particle stops is

a) $\sqrt{\frac{3v_0}{2\propto}}$ b) $\left(\frac{3v_0}{2\propto}\right)^{\frac{1}{3}}$ c) $\sqrt{\frac{2v_0^2}{3\propto}}$ d) $\left(\frac{3v_0^2}{2\propto}\right)^{\frac{1}{3}}$

433. A wheel completes 2000 revolutions to cover the 9.5 km distance, then the diameter of the wheel is

a) 1.5 m b) 1.5 cm c) 7.5 cm d) 7.5 m

434. A body can throw a stone up to a maximum height of $10m$. The maximum horizontal distance that the boy can throw the same stone up to will be

a) $20\sqrt{2}m$ b) $10m$ c) $10\sqrt{2}m$ d) $20m$

435. If the resultant of $\vec{A}$ and $\vec{B}$ makes angle α with $\vec{A}$ and β with $\vec{B}$ then

a) $\alpha < \beta$ always b) $\alpha < \beta$, if $A < B$ c) $\alpha < \beta$, if $A > B$ d) $\alpha < \beta$, if $A = B$

436. A body is projected at an angle θ with respect to horizontal direction with velocity u. The maximum range of the body is

a) $R = \frac{u^2 \sin 2\theta}{g}$ b) $R = \frac{u^2 \sin^2\theta}{2g}$ c) $R = \frac{u^2}{g}$ d) $R = u^2 \sin\theta$

437. A curved road of 50 m radius is banked at correct angle for a given speed. If the speed is to be doubled keeping the same banking angle, the radius of curvature of the road should be changed to

a) 25 m b) 100 m c) 150 m d) 200 m

438. A ball of mass 0.1 kg is suspended by a string. It is displaced through an angle of 60° and left. When the ball passes through the mean position, the tension in the string is

a) 19.6 N b) 1.96 N c) 9.8 N d) Zero

439. Galileo writes that for angles of projection of a projectile at angles $(45 + \theta)$ and $(45 - \theta)$, the horizontal ranges described by the projectile are in the ratio of (if $\theta \leq 45$)

a) 2 : 1 b) 1 : 2 c) 1 : 1 d) 2 : 3

440. A tennis ball rolls off the top of a stair case way with a horizontal velocity u ms^{-1}. If the steps are b metre wide and h metre high, the ball will hit the edge of the nth step, if

a) $n = \frac{2hu}{gb^2}$ b) $n = \frac{2hu^2}{gb^2}$ c) $n = \frac{2hu^2}{gb}$ d) $n = \frac{hu^2}{gb^2}$

441. If a body A of mass M is thrown with velocity V at an angle of 30° to the horizontal and another body B of the same mass is thrown with the same speed at an angle of 60° to the horizontal. The ratio of horizontal range of A to B will be

a) 1 : 3 b) 1 : 1 c) $1 : \sqrt{3}$ d) $\sqrt{3} : 1$

442. A pallet of mass 1 g is moving with an angular velocity of 1 $rads^{-1}$ along a circle of radius 1 m the centrifugal force is

a) 0.1 dyne b) 12 dyne c) 10 dyne d) 100 dyne

443. What will be the maximum speed of a car on a road-turn of radius 30m if the coefficient of friction between the tyres and the road is 0.4?

a) 10.84 ms^{-1} b) 9.84 ms^{-1} c) 8.84 ms^{-1} d) 6.84 ms^{-1}

444. Two particles A and B are projected with same speed so that ratio of their maximum heights reached is 3:1. If the speed of A is doubled without altering other parameters, the ratio of horizontal ranges attained by A and B is

a) 1:1 b) 2:1 c) 4:1 d) 3:2

445. When a ceiling fan is switched off its angular velocity reduces to 50% while it makes 36 rotations. How many more rotation will it make before coming to rest (Assume uniform angular retardation)

a) 18 b) 12 c) 36 d) 48

446. An object is tied to a string and rotated in a vertical circle of radius r. Constant speed is maintained along the trajectory. If $T_{\max}/T_{\min} = 2$, then v^2/rg is

a) 1 b) 2 c) 3 d) 4

447. What is the angle between $\hat{\imath} + \hat{\jmath} + \hat{k}$ and $\hat{\imath}$

a) $0°$ b) $\pi/6$ c) $\pi/3$ d) None of these

448. Which of the following is the graph between the height (h) of a projectile and time (t), when it is projected from the ground

a) 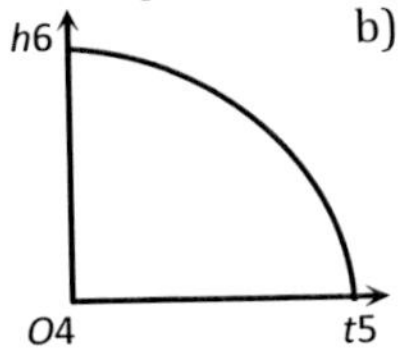

b) 

c) 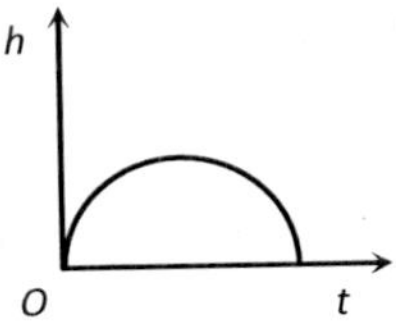

d) 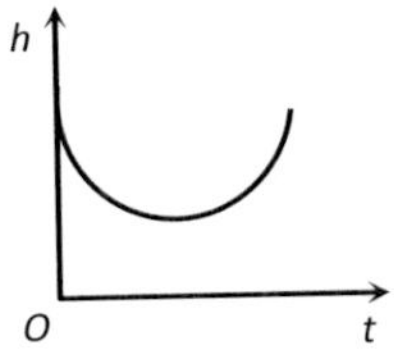

449. In a vertical circle of radius r, at what point in its path a particle has tension equal to zero if it is just able to complete the vertical circle

a) Highest point b) Lowest point
c) Any point d) At a point horizontally from the centre of circle of radius r

450. In the above question, if the angular velocity is kept same but the radius of the path is halved, the new force will be

a) $2F$ b) F^2 c) $F/2$ d) $F/4$

451. An unbanked curve has a radius of 60 m. The maximum speed at which the car make a turn is (Take $\mu = 0.75$)

a) 7 ms^{-1} b) 14 ms^{-1} c) 21 ms^{-1} d) 2.1 ms^{-1}

452. A ball of mass(m)0.5 kg is attached to the end of a string having length (L) 0.5 m. The ball is rotated on a horizontal circular path about vertical axis. The maximum tension that the string can bear is 324 N. The maximum possible value of angular velocity of ball (in rad/s) is

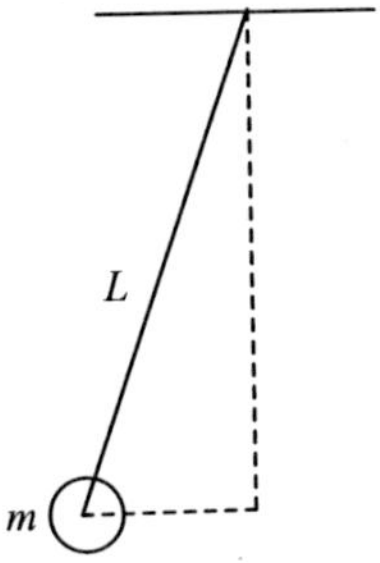

a) 9 b) 18 c) 27 d) 36

453. In resultant of which of the following sets of forces can not be zero

a) 10,20 and 40 b) 10,10 and 20 c) 10,20 and 20 d) 10,10 and 10

454. A particle of mass m is tied to one end of a string of length l and rotated through the other end along a horizontal circular path with speed v. The work done in half horizontal circle is

a) Zero b) $\left(\frac{mv^2}{l}\right)2\pi l$ c) $\left(\frac{mv^2}{l}\right)\pi l$ d) $\left(\frac{mv^2}{l}\right)l$

455. A particle moves with constant angular velocity in circular path of certain radius and is acted upon by a certain centripetal force F. If the angular velocity is doubled keeping radius the same, the new force will be

a) $2F$ b) F^2 c) $4F$ d) $F/2$

456. A projectile is fired at an angle of 30° to the horizontal such that the vertical component of its initial velocity is 80 ms^{-1}. Its time of flight is T. Its velocity at $t = \frac{T}{4}$ has a magnitude of nearly

a) 200 ms^{-1} b) 300 ms^{-1} c) 140 ms^{-1} d) 100 ms^{-1}

457. If a particle of mass m is moving in a horizontal circle of radius r with a centripetal force $(-K/r^2)$ the total energy is

a) $-\frac{K}{2r}$ b) $-\frac{K}{r}$ c) $-\frac{2K}{r}$ d) $-\frac{4K}{r}$

458. Two particles A and B are projected with same speed so that the ratio of their maximum heights reached is 3 : 1. If the speed of A is doubled without altering other parameters, the ratio of the horizontal ranges attained by A and B is

a) 1 : 1 b) 2 : 1 c) 4 : 1 d) 3 : 2

459. A gun is aimed at a target in a line of its barrel. The target is released and allowed to fall under gravity at the same instant the gun is fired. The bullet will

a) Pass above the target b) Pass below the target
c) Hit the target d) Certainly miss the target

460. A grass hopper finds that he can jump a maximum horizontal distance of 1 m. With what speed can be travel along the path if he speeds a negligible time on the ground

a) 9.8 ms^{-1} b) 4.42 ms^{-1} c) 2.21 ms^{-1} d) 3.13 ms^{-1}

461. A body of mass m is moving in a circle of radius r with a constant speed v. The force on the body is $\frac{mv^2}{r}$ and is directed towards the centre. What is the work done by this force in moving the body over half the circumference of the circle

a) $\frac{mv^2}{r} \times \pi r$ b) Zero c) $\frac{mv^2}{r^2}$ d) $\frac{\pi r^2}{mv^2}$

462. Given $\vec{P} \cdot \vec{Q} = 0$, then $|\vec{P} \times \vec{Q}|$ is

a) $|\vec{P}||\vec{Q}|$ b) Zero c) 1 d) $\sqrt{PQ}$

463. A car wheel is rotated to uniform angular acceleration about its axis. Initially its angular velocity is zero. It rotates through an angle θ_1 in the first 2 s , in the next 2 s, it rotates through an additional angle θ_2, the ratio of $\frac{\theta_2}{\theta_1}$ is

a) 1 b) 2 c) 3 d) 5

464. A mass of 2 kg is whirled in a horizontal circle by means of a string at an initial speed of 5 *revolutions per minute*. Keeping the radius constant the tension in the string is doubled. The new speed is nearly

a) 14 *rpm* b) 10 *rpm* c) 2.25 *rpm* d) 7 *rpm*

465. Which of the following is the altitude-time graph for a projectile thrown horizontally from the top of the tower

a) 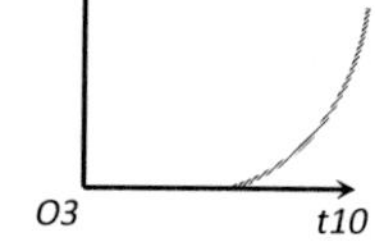

b) 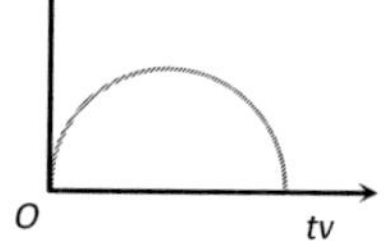

c) 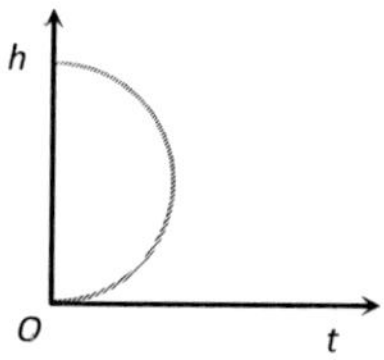

d) 

466. A car takes a turn around a circular curve. If it turns at double the speed, the tendency to overturn is

a) Halved b) Doubled c) Quadrupled d) Unchanged

467. A body is projected with speed v ms^{-1} at angle θ. The kinetic energy at the highest point is half of the initial kinetic energy. The value of θ is
a) 30° b) 45° c) 60° d) 90°
468. A ball of mass 0.25 kg attached to the end of string of length 1.96 m moving in a horizontal circle. The string will break if the tension is more than 25 N. What is the maximum speed with which the ball can be moved
a) 14 m/s b) 3 m/s c) 3.92 m/s d) 5 m/s
469. An aeroplane moving horizontally at a speed of 200 ms^{-1} and at a height of 8×10^3m is to drop a bomb on a target. At what horizontal distance from the target should the bomb be released? (Take g = 10 ms^{-2})
a) 9124 m b) 8714 m c) 8000 m d) 7234 m
470. The horizontal range of an oblique projectile is equal to the distance through which a projectile has to fall freely from rest to acquire a velocity equal to the velocity of projection in magnitude. The angle of projection is
a) 75° b) 60° c) 45° d) 30°
471. A body of mass m tied to a string is moved in a vertical circle of radius r. The difference in tensions at the lowest point and the highest point is
a) 2 mg b) 6 mg c) 4 mg d) 8 mg
472. The speed of projection of a projectile is increased by 10%, without changing the angel of projection. The percentage increase in the range will be
a) 10% b) 20% c) 15% d) 5%
473. If $\vec{A} = 2\hat{i} + 3\hat{j} + 6\hat{k}$ and $\vec{B} = 3\hat{i} - 6\hat{j} + 2\hat{k}$, then vector perpendicular to both $\vec{A}$ and $\vec{B}$ has magnitude k times that $(6\hat{i} + 2\hat{j} - 3\hat{k})$. That k is equal to
a) 1 b) 4 c) 7 d) 9
474. For a body moving in a circular path, a condition for no skidding if μ is the coefficient of friction, is
a) $\frac{mv^2}{r} \leq \mu mg$ b) $\frac{mv^2}{r} \geq \mu mg$ c) $\frac{v}{r} = \mu g$ d) $\frac{mv^2}{r} = \mu mg$
475. Given, $\vec{P} = \vec{A} + \vec{B}$ and $P = A + B$. The angle between $\vec{A}$ and $\vec{B}$ is
a) 0° b) $\frac{\pi}{4}$ c) $\frac{\pi}{2}$ d) π
476. A particle covers 50 m distance when projected with an initial speed. On the same surface it will cover a distance, when projected with double the initial speed
a) 100 m b) 150 m c) 200 m d) 250 m
477. A particle is moving in a vertical circle. The tensions in the string when passing through two positions at angles 30° and 60° from vertical (lowest position) are T_1 and T_2 respectively. then
a) $T_1 = T_2$ b) $T_2 > T_1$
c) $T_1 > T_2$ d) Tension in the string always remains the same
478. A ball is projected upwards from the top of tower with a velocity 50 ms^{-1} making an angle 30° with the horizontal. The height of tower is 70m. After how many seconds from the instant of throwing will the ball reach the ground?
a) 2 s b) 5 s c) 7 s d) 9 s
479. In the above question, the percentage increase in the time of flight of the projectile will be
a) 5% b) 10% c) 15% d) 20%
480. A particle originally at a rest at the highest point of a smooth circle in a vertical plane, is gently pushed and starts sliding along the circle. It will leave the circle at a vertical distance h below the highest point such that

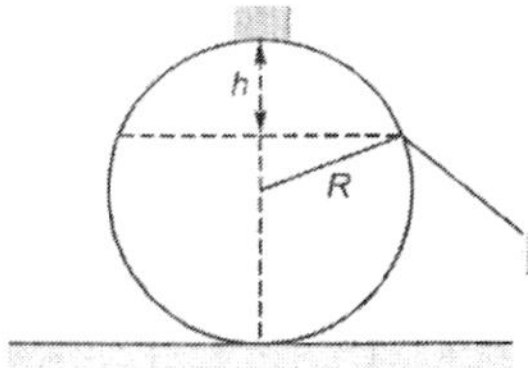

a) $h = 2R$ b) $h = \frac{R}{2}$ c) $h = R$ d) $h = \frac{R}{3}$

481. A projectile is fired at an angle of 45° with the horizontal. Elevation angle of the projectile at its highest point as seen from the point of projection, is:

a) $\tan^{-1}\left(\frac{\sqrt{3}}{2}\right)$ b) 45° c) 60° d) $\tan^{-1}\frac{1}{2}$

482. A roller coaster is designed such that riders experience 'weightlessness' as they go round the top of a hill whose radius of curvature is 20 m. The speed of the car at the top the hill is between

a) 14 ms^{-1} and 15 ms^{-1} b) 15 ms^{-1} and 16 ms^{-1} c) 16 ms^{-1} and 17 ms^{-1} d) 13 ms^{-1} and 14 mS^{-1}

483. A wheel of radius 1m rolls forward half a revolution on a horizontal ground. The magnitude to the displacement of the point of the wheel initially in contact with the ground is

a) 2π b) $\sqrt{2\pi}$ c) $\sqrt{\pi^2 + 4}$ d) π

484. A motor cyclist moving with a velocity of 72 $km/hour$ on a flat road takes a turn on the road at a point where the radius of curvature of the road is 20 m. The acceleration due to gravity is 10 $m/\sec^2$. In order to avoid skidding, he must not bend with respect to the vertical plane by an angle greater than

a) $\theta = \tan^{-1} 6$ b) $\theta = \tan^{-1} 2$ c) $\theta = \tan^{-1} 25.92$ d) $\theta = \tan^{-1} 4$

485. A body of mass 0.4 kg is whirled in a vertical circle making 2rev/s. If the radius of the circle is 2m, then tension in the string when the body is at the top of the circle is

a) 41.56 N b) 89.86 N c) 109.86 N d) 115.86 N

486. A particle is kept at rest at the top of a sphere of diameter 42 m. When disturbed slightly, it slides down. At what height $'h'$ from the bottom, the particle will leave the sphere

a) 14 m b) 28 m c) 35 m d) 7 m

487. In case of uniform circular motion which of the following physical quantity do not remain constant

a) Speed b) Momentum c) Kinetic energy d) Mass

488. The magnitude of the vectors product o two vectors is $\sqrt{3}$ times their scalar product. The angle between the two vectors is

a) 90° b) 60° c) 45° d) 30°

489. A projectile is thrown in the upward direction making an angle of 60° with the horizontal direction with velocity of147 ms^{-1}. Then the time after which its inclination with the horizontal is 45°, is

a) 25 s b) 10.98 s c) 5.49 s d) 2.745 s

490. If α is angular acceleration, ω is angular velocity and a is the centripetal acceleration then, which of the following is true?

a) $\alpha = \frac{\omega a}{v}$ b) $\alpha = \frac{v}{\omega a}$ c) $\alpha = \frac{va}{\omega}$ d) $\alpha = \frac{a}{\omega v}$

491. A car of mass 1000 kg negotiates a banked curve of radius 90 m on a frictionless road. If the banking angle is 45°, the speed of the car is

a) $20ms^{-1}$ b) $30ms^{-1}$ c) $5ms^{-1}$ d) $10ms^{-1}$

492. In uniform circular motion of a particle

a) Velocity is constant but acceleration is variable

b) Velocity is variable but acceleration in constant

c) Both speed and acceleration are constant
d) Speed is constant but acceleration is variable

493. The resultant of a system of forces shown in figure is a force of 10 N parallel to given forces through R, where PR equals

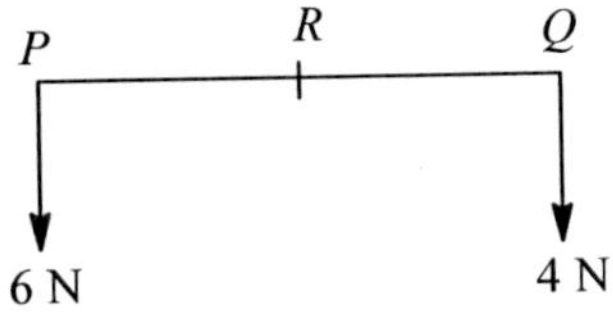

a) $(2/5)R\ Q$ b) $(3/5)R\ Q$ c) $(2/3)R\ Q$ d) $(1/2)R\ Q$

494. The angle of banking is independent of
a) speed of vehicle b) radius of curvature of road
c) height of inclination d) None of the above

495. In an atom for the electron to revolve around the nucleus, the necessary centripetal force is obtained from the following force exerted by the nucleus on the electron
a) Nuclear force b) Gravitational force c) Magnetic force d) Electrostatics force

496. The resultant of two forces, each P, acting at an angle θ is
a) $2P \sin\frac{\theta}{2}$ b) $2P \cos\frac{\theta}{2}$ c) $2P \cos\theta$ d) $P\sqrt{2}$

497. The resultant of two vectors of magnitudes $2A$ and $\sqrt{2}A$ acting at an angle $\theta$is $\sqrt{10}A$. The correct value of θ is
a) 30° b) 45° c) 60° d) 90°

498. A ball is projected upwards from the top of a tower with a velocity of 50 ms^{-1} making an angle of 60° with the vertical. If the height of the tower is 70 m, then the ball will reach the ground in (Take g = 10 ms^{-2})
a) 2 s b) 3 s c) 5 s d) 7 s

499. A projectile is given an initial velocity of $\hat{\mathbf{i}} + 2\hat{\mathbf{j}}$. The cartesian equation of its path is (g = 10 ms^{-2})
a) $y = x - 5x^2$ b) $y = 2x - 5x^2$ c) $y = 2x - 15x^2$ d) $y = 2x - 25\,x^2$

500. A ball thrown by a boy is caught by another after 2 *sec* some distance away in the same level. If the angle of projection is 30°, the velocity of projection is
a) 19.6 m/s b) 9.8 m/s c) 14.7 m/s d) None of these

501. On an unbanked road, a cyclist negotiating a bend of radius r at velocity v leans inwards by an angle
a) $\tan^{-1}\left(\frac{v^2}{2gr}\right)$ b) $\tan^{-1}\left(\frac{v^2}{gr}\right)$ c) $\tan^{-1}\left(\frac{rg}{v^2}\right)$ d) $\tan^{-1}\left(\frac{v}{gr}\right)$

502. A particle is moving in a circle of radius R with constant speed v, if radius is double then its centripetal force to keep the same speed should be
a) Doubled b) Halved c) Quadrupled d) Unchanged

503. The coordinates of a moving particle at any time $'t'$ are given by $x = \alpha t^3$ and $y = \beta t^3$. The speed of the particle at time $'t'$ is given by
a) $\sqrt{\alpha^2 + \beta^2}$ b) $3t\sqrt{\alpha^2 + \beta^2}$ c) $3t^2\sqrt{\alpha^2 + \beta^2}$ d) $t^2\sqrt{\alpha^2 + \beta^2}$

504. A string of length l is fixed at one end and caries a mass m at the other end. The string makes $2/\pi$ rps around a vertical axis through the fixed end. What is the tension in string?
a) $m\,l$ b) $16\,m\,l$ c) $4\,m\,l$ d) $2\,m\,l$

505. A particle of mass m moves with constant speed along a circular path of radius r under the action of a force F. Its speed is
a) $\sqrt{\frac{rF}{m}}$ b) $\sqrt{\frac{F}{r}}$ c) $\sqrt{Fmr}$ d) $\sqrt{\frac{F}{mr}}$

506. A stone is thrown at an angle θ to the horizontal reaches a maximum heights H. then the time of flight of stone will be

a) $\sqrt{\frac{2H}{g}}$ b) $2\sqrt{\frac{2H}{g}}$ c) $\frac{2\sqrt{2H\sin\theta}}{g}$ d) $\frac{\sqrt{2H\sin\theta}}{g}$

507. What should be the angular velocity of earth so that a body on its equator is weightless?

a) $\frac{1}{8000}$ rad s^{-1} b) $\frac{1}{8}$ rad s^{-1} c) $\frac{1}{800}$ rad s^{-1} d) $\frac{1}{80}$ rad s^{-1}

508. $(\vec{P}+\vec{Q})$ is a unit vector along X-axis. If $\vec{P}=\hat{\imath}-\hat{\jmath}+\hat{k}$, then what value is $\vec{Q}$?

a) $\hat{\imath}+\hat{\jmath}-\hat{k}$ b) $\hat{\jmath}-\hat{k}$ c) $\hat{\imath}+\hat{\jmath}+\hat{k}$ d) $\hat{\jmath}+\hat{k}$

509. A ball is projected from the ground at a speed of 10ms^{-1} making an angle of 30° with the horizontal. Another ball is simultaneously released from a point on the vertical line along the maximum height of the projectile. The initial height of the second ball is ($g=$ 10ms^{-2})

a) 6.25 m b) 2.5 m c) 3.75 m d) 5 m

510. A particle is moving in a circle with uniform speed$v$. In moving from a point to another diametrically opposite point

a) the momentum changes by mv b) the momentum changes by 2 mv

c) the kinetic energy changes by (1/2) mv^2 d) the kinetic energy changes by mv^2

511. If the range of a gun which fires a shell with muzzle speed V is R, then the angle of elevation of the gun is

a) $\cos^{-1}\left(\frac{V^2}{Rg}\right)$ b) $\cos^{-1}\left(\frac{gR}{V^2}\right)$ c) $\frac{1}{2}\left(\frac{V^2}{Rg}\right)$ d) $\frac{1}{2}\sin^{-1}\left(\frac{gR}{V^2}\right)$

512. On the centre of a frictionless table a small hole is made, through which a weightless string of length $2l$ is inserted. On the two ends of the string two balls of the same mass m are attached. Arrangement is made in such a way that half of the string is on the table top and half is hanging below. The ball on the table top is made to move in a circular path with a constant speed v. What is the centripetal acceleration of the moving ball

a) mvl b) g c) Zero d) $2mvl$

513. A body of mass 1 kg is moving in a vertical circular path of radius 1 m. The difference between the kinetic energies at its highest and lowest point is

a) 20 J b) 10 J c) $4\sqrt{5}$ J d) $10\sqrt{5}$ J

514. The angle of projection of a projectile for which the horizontal range and maximum height are equal is

a) $\tan^{-1}(2)$ b) $\tan^{-1}(4)$ c) $\cot^{-1}(2)$ d) 60°

515. A stone of mass 1 kg tied to the end of a string of length 1 m, is whirled in a horizontal circle with a uniform angular velocity 2 rads^{-1}. The tension of the string is (in newton)

a) 2 b) $\frac{1}{3}$ c) 4 d) $\frac{1}{4}$

516. A tachometer is a device to measure

a) Gravitational pull b) Speed of rotation c) Surface tension d) Tension in a spring

517. A body is thrown horizontally from the top of a tower of height 5 m. It touches the ground at a distance of 10 m from the foot of the tower. The initial velocity of the body is ($g=10\ ms^{-2}$)

a) 2.5 ms^{-1} b) 5 ms^{-1} c) 10 ms^{-1} d) 20 ms^{-1}

518. A point P moves in counter-clockwise direction on a circular path as shown in the figure. The movement of $''P''$ is such that it sweeps out a length $s=t^3+5$, where s is in metres and t is in seconds. The radius of the path is 20 m. The acceleration of $''P''$ when $t=2s$ is nearly

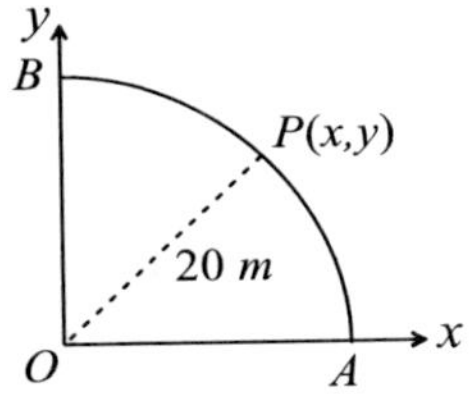

a) $14\ m/s^2$ b) $13\ m/s^2$ c) $12\ m/s^2$ d) $7.2\ m/s^2$

519. In figure, a particle is placed at the highest point A of a smooth sphere of radius r. It is given slight push, and it leaves the sphere at B, at a depth h vertically below A such that h is equal to

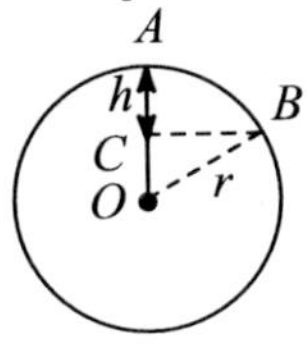

a) $\frac{r}{6}$ b) $\frac{1}{4}r$ c) $\frac{1}{3}r$ d) $\frac{1}{2}r$

520. A large number of bullets are fired in all directions with same speed v. What is the maximum area on the ground on which these bullets will spread

a) $\pi\frac{v^2}{g}$ b) $\pi\frac{v^4}{g^2}$ c) $\pi^2\frac{v^4}{g^2}$ d) $\pi^2\frac{v^2}{g^2}$

521. For a projectile thrown into space with a speed v, the horizontal range is $\frac{\sqrt{3}v^2}{2g}$. The vertical range is $\frac{v^2}{8g}$. The angle which the projectile makes with the horizontal initially is

a) 15° b) 30° c) 45° d) 60°

522. A string breaks if its tension exceeds 10 *newtons*. A stone of mass 250 *gm* tied to this string of length 10 *cm* is rotated in a horizontal circle. The maximum angular velocity of rotation can be

a) 20 *rad/s* b) 40 *rad/s* c) 100 *rad/s* d) 200 *rad/s*

523. Which one of the following statements is not correct in uniform circular motion

a) The speed of the particle remains constant
b) The acceleration always points towards the centre
c) The angular speed remains constant
d) The velocity remains constant

524. A stone is tied to one end of a string 50 *cm* long is whirled in a horizontal circle with a constant speed. If the stone makes 10 revolutions in 20 *s*, what is the magnitude of acceleration of the stone

a) $493\ cm/s^2$ b) $720\ cm/s^2$ c) $860\ cm/s^2$ d) $990\ cm/s^2$

525. An aeroplane flying horizontally with a speed of 360 kmh^{-1} releases a bomb at a height of 490 m from the ground. When will the bomb strike the ground?

a) 8 s b) 6 s c) 7 s d) 10 s

526. The horizontal range of a projectile $4\sqrt{3}$ times the maximum height achieved by it, then the angle of projection is

a) 30° b) 45° c) 60° d) 90°

527. With what minimum speed a particle be projected from the origin so that it is able to pass through a given point (30 m, 40 m)?

a) 30 ms^{-1} b) 40 ms^{-1} c) 50 ms^{-1} d) 60 ms^{-1}

528. What is the value of linear velocity, if $\vec{\omega} = 3\hat{\imath} - 4\hat{\jmath} + \hat{k}$ and $\vec{r} = 5\hat{\imath} - 6\hat{\jmath} + 6\hat{k}$

a) $6\hat{\imath} + 2\hat{\jmath} - 3\hat{k}$ b) $-18\hat{\imath} - 13\hat{\jmath} + 2\hat{k}$ c) $4\hat{\imath} - 13\hat{\jmath} + 6\hat{k}$ d) $6\hat{\imath} - 2\hat{\jmath} + 2\hat{k}$

529. A particle is moving with velocity $\mathbf{v} = k(y\hat{\mathbf{i}} + x\,\hat{\mathbf{j}})$,where k is a constant. The general equation for its path is

a) $y = x^2$ + constant b) $y^2 = x$ + constant c) xy = constant d) $y^2 = x^2$constant

530. The angular speed of a car increases from 600 rpm to 1200 rpm in 10 s. What is the angular acceleration of the car?

a) 600 rad s^{-1} b) 60 rad s^{-1} c) 60 πrad m s^{-1} d) 2 π rad s^{-1}

531. What is the numerical value of the vector $3\hat{\imath} + 4\hat{\jmath} + 5\hat{k}$?

a) $3\sqrt{2}$ b) $5\sqrt{2}$ c) $7\sqrt{2}$ d) $9\sqrt{2}$

532. A 500 kg car takes a round of radius 50 m with a velocity of 36 kmh^{-1}. The centripetal force is

a) 250 N b) 750 N c) 1000 N d) 1200 N

533. As per given future to complete the circular loop what should be the radius if initial height is 5 m

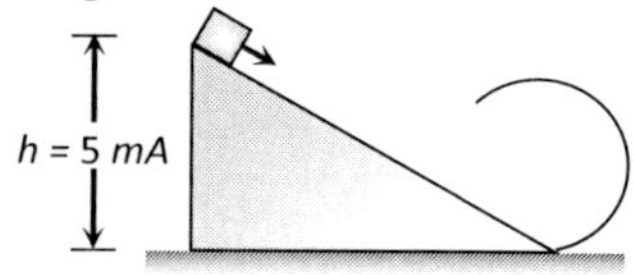

a) 4 m b) 3 m c) 2.5 m d) 2 m

534. A projectile moves from the ground such that its horizontal displacement is $x = Kt$ and vertical displacement is $y = Kt(1 - \alpha t)$, where K and α are constants and t is time. Find out total time of flight (T) and maximum height attained (Y_{max}) its

a) $T = \alpha, Y_{max} = \frac{K}{2\alpha}$ b) $T = \frac{1}{\alpha}, Y_{max} = \frac{2K}{\alpha}$ c) $T = \frac{1}{\alpha}, Y_{max} = \frac{K}{6\alpha}$ d) $T = \frac{1}{\alpha}, Y_{max} = \frac{K}{4\alpha}$

535. Two bodies of mass 10 kg and 5 kg moving in concentric orbits of radii R and r such that their periods are the same. Then the ratio between their centripetal acceleration is

a) R/r b) r/R c) R^2/r^2 d) r^2/R^2

536. A particle is projected with a speed v at 45° with the horizontal. The magnitude of angular momentum of the projectile about the point of projection when the particle is at its maximum height h is

a) Zero b) $\frac{mvh^2}{\sqrt{2}}$ c) $\frac{mvh}{\sqrt{2}}$ d) $\frac{mvh^3}{\sqrt{2}}$

537. If $\vec{P} = 2\hat{\imath} - 3\hat{\jmath} + \hat{k}$ and $\vec{Q} = 3\hat{\imath} - 2\hat{\jmath}$, then $\vec{P} \cdot \vec{Q}$ is

a) Zero b) 6 c) 12 d) 15

538. A body of mass m moves in a circular path with uniform angular velocity. The motion of the body has constant

a) Acceleration b) Velocity c) Momentum d) Kinetic energy

539. A ball is projected up an incline of 30° with a velocity of 30 ms^{-1} at an angle of 30° with reference to the inclined plane from the bottom of the inclined plane. If g = 10ms^{-2}, then the range on the inclined plane is

a) 12 m b) 60 m c) 120 m d) 600 m

540. Two bodies are thrown up at angles of 45°and 60°, respectively with the horizontal. If both bodies attain same vertical height , then the ratio of velocities with which these are thrown is

a) $\sqrt{\frac{2}{3}}$ b) $\frac{2}{\sqrt{3}}$ c) $\sqrt{\frac{3}{2}}$ d) $\frac{\sqrt{3}}{2}$

541. A proton in a cyclotron changes its velocity from 30kms^{-1} north to 40kms^{-1} east in 20 s. what is the average acceleration during this time

a) 2.5 kms^{-2}at 37° E of S b) 2.5kms^{-2}at 37° N of E

c) $2.5 kms^{-2}$ at 37° N of S   d) $2.5 kms^{-2}$ at 37° E of N

542. The minimum velocity (in ms^{-1}) with which a car driver must traverse a flat curve of radius of 150 m and coefficient of friction 0.6 to avoid skidding is

a) 60   b) 30   c) 15   d) 25

543. A bullet is to be fired with a speed of 20 ms^{-1} to hit a target 200m away on a level ground. If $g = 10\ ms^{-2}$, the gun should be aimed

a) directly at the target   b) 5 cm below the target
c) 5 cm above the target   d) 2 cm above the target

544. A projectile is projected with velocity kv_e in vertically upward direction from the ground into the space (v_e is the escape velocity and $k < 1$). If air resistance is considered to be negligible then the maximum height from the center of earth to which it can go will be ($R =$ radus of earth)

a) $\frac{R}{k^2+1}$ b) $\frac{R}{k^2-1}$ c) $\frac{R}{1-k^2}$ d) $\frac{R}{k+1}$

545. A ball thrown by one player reaches the other in 2 s. The maximum height attained by the ball above the point of projection will be ($g = 10\ ms^{-2}$)

a) 2.5 m   b) 5 m   c) 7.5 m   d) 10 m

546. A stone tied to a string of length L is whirled in a vertical circle, with the other end of the string at the centre. At a certain instant of time, the stone is at its lowest position, and has a speed u. The magnitude of change in its velocity as it reaches a position, where the string is horizontal is

a) $\sqrt{u^2 - 2gL}$ b) $\sqrt{2gL}$ c) $\sqrt{u^2 - gL}$ d) $\sqrt{2(u^2 - gL)}$

547. A cycle wheel of radius 0.4 m completes one revolution in one *second* then the acceleration of a point on the cycle wheel will be

a) $0.8\ m/s^2$ b) $0.4\ m/s^2$ c) $1.6\ \pi^2 m/s^2$ d) $0.4\ \pi^2 m/s^2$

548. Two projectiles A and B are thrown with velocities v and $\frac{v}{2}$ respectively. They have the same range. If B is thrown at an angle of 15° to the horizontal, a must have been thrown at an angle

a) $\sin^{-1}\left(\frac{1}{16}\right)$ b) $\sin^{-1}\left(\frac{1}{4}\right)$ c) $2\sin^{-1}\left(\frac{1}{4}\right)$ d) $\frac{1}{2}\sin^{-1}\left(\frac{1}{8}\right)$

549. A ball is projected from a certain point on the surface of a planet at a certain angel with the horizontal surface. The horizontal and vertical displacement x and y vary with time t in second as

$x = 10\sqrt{3}t$ and $y = 10t - t^2$

The maximum height attained by the ball is

a) 100 m   b) 75 m   c) 50 m   d) 25 m

550. The angular speed of a fly wheel making 120 *revolutions/minute* is

a) $2\pi\ rad/s$ b) $4\pi^2\ rad/s$ c) $\pi\ rad/s$ d) $4\pi\ rad/s$

551. An object moves along a straight line path from P to Q under the action of a force $(4\hat{i} - 3\hat{j} + 3\hat{k})$N. If the coordinates of P and Q, in metres, are (3,3,−1) and (2,−1,4) respectively, then the work done by the force is

a) +23 J   b) −23 J   c) 1015 J   d) $\sqrt{35}(4\hat{i} - 3\hat{j} + 2\hat{k})$ J

552. A particle has velocity $\sqrt{3rg}$ at the highest pint in vertical circle. Find the ratio of tensions at the highest and lowest point

a) 1 : 6   b) 1 : 4   c) 1 : 3   d) 1 : 2

553. A small body of mass m slides down from the top of a hemisphere of radius r. The surface of block and hemisphere are frictionless. The height at which the body lose contact with the surface of the sphere is

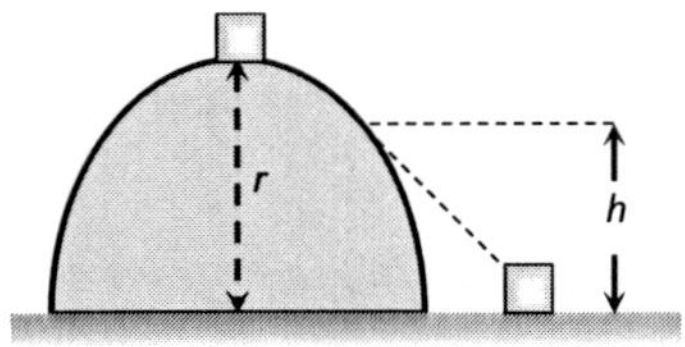

a) $\frac{3}{2}r$ b) $\frac{2}{3}r$ c) $\frac{1}{2}gt^2$ d) $\frac{v^2}{2g}$

554. A body is thrown upward from the earth surface with velocity 5 m/s and from a planet surface with velocity 3 m/s. Both follow the same path. What is the projectile acceleration due to gravity on the planet

a) 2 m/s^2 b) 3.5 m/s^2 c) 4 m/s^2 d) 5 m/s^2

555. The maximum range of a gun on horizontal terrain is 16 km. If $g = 10\ m/s^2$. What must be the muzzle velocity of the shell

a) 200 m/s b) 400 m/s c) 100 m/s d) 50 m/s

556. Ratio between maximum range and square of time of flight in projectile motion is

a) 10 : 49 b) 49 : 10 c) 98 : 10 d) 10 : 98

557. A body of mass m is suspended from a string of length l. What is minimum horizontal velocity that should be given to the body in its lowest position so that it may complete one full revolution in the vertical plane with the point of suspension as the centre of the circle

a) $v = \sqrt{2lg}$ b) $v = \sqrt{3lg}$ c) $v = \sqrt{4lg}$ d) $v = \sqrt{5lg}$

558. A tube of length L is filled completely with an incompressible liquid of mass M and closed at both the ends. The tube is then rotated in a horizontal plane about one of its ends with a uniform angular velocity ω. The force exerted by the liquid at the other end is

a) $\frac{ML\omega^2}{2}$ b) $ML\omega^2$ c) $\frac{ML\omega^2}{4}$ d) $\frac{ML^2\omega^2}{2}$

559. An object is being weighed on a spring balance moving around a curve of radius 100 m at a speed 7 ms^{-1}. The object has a weight of 60 kg-wt. The reading registered on the spring balance would be

a) 60.075 kg-wt b) 60.125 kg-wt c) 60.175 kg-wt d) 60.225 kg-wt

560. A vector $\overrightarrow{F_1}$ is along the positive Y-axis. If its vector product with another vector $\overrightarrow{F_1}$ is zero, then $\overrightarrow{F_2}$ could be

a) $4\hat{j}$ b) $\hat{j} + \hat{k}$ c) $\hat{j} - \hat{k}$ d) $-4\hat{i}$

561. Two stones are projected with same velocity v at an angle $\theta$and $(90° - \theta)$. If H and H_1 are greatest heights in the two paths, what is the relation between R, $H$and H_1?

a) $R = 4\sqrt{HH_1}$ b) $R = \sqrt{HH_1}$ c) $R = 4HH_1$ d) None of these

562. A sphere is suspended by a thread of length l. The minimum horizontal velocity which has to be imparted to the sphere for it to reach the height of suspension is

a) $2\sqrt{gR}$ b) $\sqrt{2gl}$ c) 2 g l d) g l

563. After one second the velocity of a projectile makes an angle of 45° with the horizontal. After another one second it is travelling horizontally. The magnitude of its initial velocity and angle of projection are (g = 10 ms^{-2})

a) 14.02 ms^{-1}, tan⁻b) 22.36 ms^{-1}, $\tan^{-1}$(2c) 14.62 ms^{-1}, 60° d) 22.36 ms^{-1}, 60°

564. A particle (A) is dropped from a height and another particle(B) is thrown in horizontal direction with speed of 5 m/sec from the same height. The correctstatement is

a) Both particles will reach at ground simultaneously
b) Both particles will reach at ground with same speed
c) Particle (A) will reach at ground first with respect to particle (B)
d) Particle (B) will reach at ground first with respect to particle (A)

565. A car is travelling at a velocity of 10 kmh^{-1} on a straight road. The driver of the car throws a parcel with a velocity of $10\sqrt{2}$ kmh^{-1} when the car is passing by a man standing on the side of the road. If the parcel is to reach the man, the direction of throw makes the following angle with direction of the car

a) 135° b) 45° c) $\tan^{-1}(\sqrt{2})60°$ d) $\tan\left(\frac{1}{\sqrt{2}}\right)$

566. Angle between $\vec{A}$ and $\vec{B}$is θ. What is the value of $\vec{A}.(\vec{B}\times\vec{A})$?

a) $A^2B\cos\theta$ b) $A^2B\sin\theta\cos\theta$ c) $A^2B\sin\theta$ d) zero

567. A car is moving with speed 30 m/sec on a circular path of radius 500 m. Its speed is increasing at the rate of $2m/\sec^2$, What is the acceleration of the car

a) $2m/\sec^2$ b) $2.7m/\sec^2$ c) $1.8m/\sec^2$ d) $9.8m/\sec^2$

568. The $x-y$ plane is the boundary between two transparent media. A medium I has a refractive index $\mu_1=\sqrt{2}$ and medium II has a refractive index $\mu_2=\sqrt{3}$. A ray of light in medium I, given by vector, $\vec{A}=\sqrt{3}\hat{\imath}-\hat{k}$is incident on the plane of separation. The unit vector in the direction of the refracted ray in medium II is

a) $\frac{1}{\sqrt{2}}(\hat{\imath}+\hat{k})$ b) $\frac{1}{\sqrt{2}}(\hat{\imath}+\hat{\jmath})$ c) $\frac{1}{\sqrt{2}}(\hat{k}-\hat{\imath})$ d) $\frac{1}{\sqrt{2}}(\hat{\imath}-\hat{k})$

569. The equation of motion of a projectile are given by $x=36\,t\,metre$and $2y=96\,t-9.8\,t^2\,metre$. The angle of projection is

a) $\sin^{-1}\left(\frac{4}{5}\right)$ b) $\sin^{-1}\left(\frac{3}{5}\right)$ c) $\sin^{-1}\left(\frac{4}{3}\right)$ d) $\sin^{-1}\left(\frac{3}{4}\right)$

570. What happens to the centripetal acceleration of a particle, when its speed is doubled and angular velocity is halved?

a) Doubled b) Halved
c) Remains unchanged d) Becomes 4 times

571. A boy playing on the roof of a 10 m high building throws a ball with a speed of 10ms^{-1} at an angle of 30° with the horizontal. How far from the throwing point will the ball be at the height of 10 m from the ground? [$g=10ms^{-2}$, $\sin 30°=1/2$, $\cos 60°=\sqrt{3}/2$]

a) 5.20 m b) 4.33 m c) 2.60 m d) 8.66 m

572. A body is fired vertically upward. At half the maximum height, the velocity of the body is 10 m/s. The maximum height raised by the body is

a) 0 m b) 10 m c) 15 m d) 20 m

573. A stone tied with string, is rotated in a vertical circle. The minimum speed with which the string has to be rotated

a) Is independent of the mass of the stone b) Is independent of the length of the string
c) Decreases with increasing mass of the stone d) Decreases with increasing length of the string

574. A body of mass 0.5 kg is projected under gravity with a speed of 98 m/s at an angle of 30° with the horizontal. The change in momentum (in magnitude) of the body is

a) 24.5 $N-s$ b) 49.0 $N-s$ c) 98.0 $N-s$ d) 50.0 $N-s$

575. At the height 80 m, an aeroplane is movingwith 150 ms^{-1}. A bomb is dropped from it, so as to hit a target. At what distancefrom the target should bomb be dropped? ($g=10\ ms^{-2}$)

a) 605.3 m b) 600m c) 80 m d) 230m

576. A ball is projected horizontally with a velocity of 5 m/s from the top of a building 19.6 m high. How long will the ball take to hit the ground

a) $\sqrt{2}\ s$ b) 2 s c) $\sqrt{3}\ s$ d) 3 s

577. The maximum speed of a car on a road-turn of radius 30 m, if the coefficient of friction between the tyres and the road is 0.4, will be

a) 10.84 m/sec b) 9.84 m/sec c) 8.84 m/sec d) 6.84 m/sec

578. A stone projected with a velocity u at an angle θ with the horizontal reaches maximum height H_1. When it is projected with velocity u at an angle $\left(\frac{\pi}{2}-\theta\right)$ with the horizontal, it reaches maximum height H_2. The relation between the horizontal range R of the projectile, H_1 and H_2 is

a) $R = 4\sqrt{H_1H_2}$ b) $R = 4(H_1 - H_2)$ c) $R = 4(H_1 + H_2)$ d) $R = \frac{H_1^2}{H_2^2}$

579. A body of mass m is thrown upwards at an angle θ with the horizontal with velocity v. While rising up the velocity of the mass after t seconds will be

a) $\sqrt{(v\cos\theta)^2 + (v\sin\theta)^2}$ b) $\sqrt{(v\cos\theta - v\sin\theta)^2 - gt}$
c) $\sqrt{v^2 + g^2t^2 - (2v\sin\theta)gt}$ d) $\sqrt{v^2 + g^2t^2 - (2v\cos\theta)gt}$

580. A car round an unbanked curve of radius 92 m without skidding at a speed of 26 ms^{-1}. The smallest possible coefficient of static friction between the tyres and the road is

a) 0.75 b) 0.60 c) 0.45 d) 0.30

581. The coefficient of friction between the tyres and the road is 0.25. The maximum speed with which a car can be driven round a curve a radius 40 m without skidding is (assume $g = 10\ ms^{-2}$)

a) 40 ms^{-1} b) 20 ms^{-1} c) 15 ms^{-1} d) 10 ms^{-1}

582. A body executing uniform circular motion has at any instant its velocity vector and acceleration vector

a) along the same direction b) in opposite direction
c) normal to each other d) not related to each other

583. A man throws a ball vertically upwards and it rises through 20 m and returns to his hands. What was the initial velocity (u) of the ball and for how much time (T) it remained in the air? (g = 10 ms^{-2})

a) $u = 10\ ms^{-1}; T = 2s$ b) $u = 10\ ms^{-1}; T = 4\ s$ c) $u = 20\ ms^{-1}; T = 2\ s$ d) $u = 20\ ms^{-1}; T = 4\ s$

584. An object is moving in a circle of radius 100 m with a constant speed of 31.4 m/s. What is its average speed for one complete revolution

a) Zero b) 31.4 m/s c) 3.14 m/s d) $\sqrt{2} \times 31.4\ m/s$

585. A force of $(7\hat{i} + 6\hat{k})$ N makes a body move on a rough plane with a velocity of $(3\hat{j} + 4\hat{k})ms^{-1}$. Calculate the power in watt

a) 24 b) 34 c) 21 d) 45

586. A 2 kg stone at the end of a string 1 m long is whirled in a vertical circle at a constant speed. The speed of the stone is 4 m/sec. The tension in the string will be 52 N, when the stone is

a) At the top of the circle b) At the bottom of the circle
c) Halfway down d) None of the above

587. If $\vec{A}\cdot\vec{B} = 0$ and $\vec{A}\times\vec{B} = 1$, then $\vec{A}$ and $\vec{B}$ are

a) Perpendicular unit vectors b) Parallel unit vectors
c) Parallel d) Perpendicular.

588. A boy throws a cricket ball from the boundary to the wicket-keeper. If the frictional force due to air cannot be ignored, the forces acting on the ball at the position X are respected by

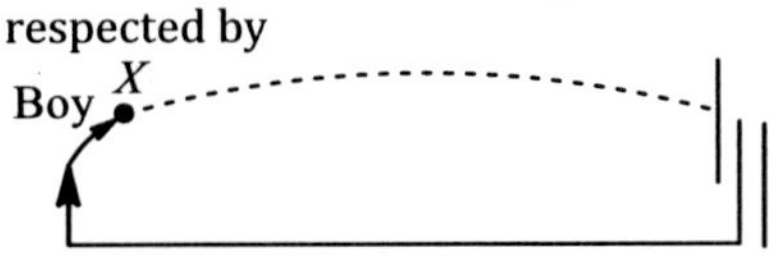

a)  b)  c) 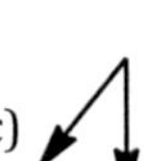 d) 

589. In a two dimensional motion of a particle, the particle moves from point A, position vector $\vec{r}_1$. If the magnitudes of these vectors are respectively, r_1=3 and $r_2 = 4$ and the angles they make with the x-axis are $\theta_1 = 75°$ and 15°, respectively, then find the magnitude of the displacement vector

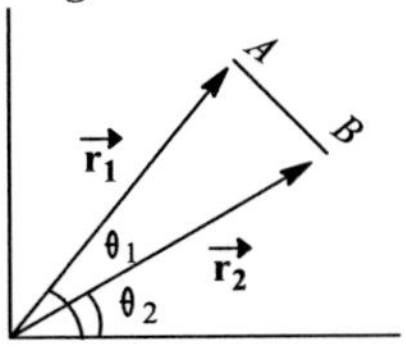

a) 15 b) $\sqrt{13}$ c) 17 d) $\sqrt{15}$

590. A body moves along a circular path of radius 10 m and the coefficient of friction is 0.5. What should be its angular speed in rad s^{-1}, if it is not to slip from the surface? (g = 9.8 ms^{-2})

a) 5 b) 10 c) 0.1 d) 0.7

591. The trajectory of a particle moving in vast maidan is as shown in the figure. The coordinates of a position A are (0,2). The coordinates of another point at which the instantaneous velocity is same as the average velocity between the points are

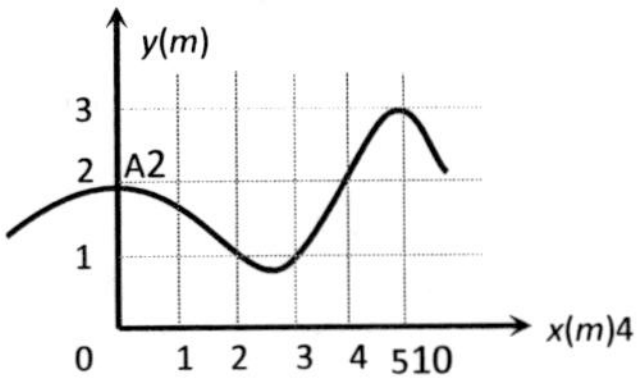

a) (1,4) b) (5,3) c) (3,4) d) (4,1)

592. A particle is moving with velocity $\vec{v} = K(y\hat{i} + x\hat{j})$, where K is a constant. The general equation for its path is

a) $y^2 = x^2 +$ constant b) $y = x^2 +$ constant c) $y^2 = x +$ constant d) $xy =$ constant

593. A bucket tied at the end a 1.6 m long string is whirled in a vertical circle with constant speed. What should be the minimum speed so that the water from the bucket does not spill, when the bucket is at the highest position (Take $g = 10m/\sec^2$)

a) 4 m/sec b) 6.25 m/sec c) 16 m/sec d) None of the above

594. A particle moves in a circle of radius 5 cm with constant speed and time period 0.2 πs. The acceleration of the particle is

a) 5 m/s^2 b) 15 m/s^2 c) 25 m/s^2 d) 36 m/s^2

595. A stone tied to a string of length L is whirled in a vertical circle with the other end of the string at the centre. At a certain instant of time, the stone is at its lowest position and has speed u. The magnitude of the change in its velocity as it reaches a position where the string is horizontal is

a) $\sqrt{u^2 - 2gL}$ b) $\sqrt{2gL}$ c) $\sqrt{u^2 - gl}$ d) $\sqrt{2(u^2 - gL)}$

596. If time of flight of a projectile is 10 seconds. Range is 500 $meters$. The maximum height attained by it will be

a) 125 m b) 50 m c) 100 m d) 150 m

597. A force $\vec{F} = 2\hat{i} + 2\hat{j}$ N displaces a particle through $\vec{S} = 2\hat{i} + 2\hat{k}$ m in 16 s. The power developed by $\vec{F}$ is

a) 0.25 J s^{-1} b) 25 J s^{-1} c) 225 J s^{-1} d) 450 J s^{-1}

598. A vector $\vec{A}$ points vertically upwards and $\vec{B}$ points upwards North. The vector product $\vec{A} \times \vec{B}$ is

a) Zero b) along East

c) along West d) vertically downwards

599. A body is projected horizontally with speed 20 ms^{-1}. The approximate displacement of the body after 5 s is

a) 80 m b) 120 m c) 160 m d) 320 m

600. A particle moves along a parabolic path $y = 9x^2$ in such a way that the x-componentes of velocity remains constant and has a value $\frac{1}{3}$ ms^{-1}. The acceleration of the projectile is

a) $\frac{1}{2}\hat{j}ms^{-2}$ b) $3\hat{j}ms^{-2}$ c) $\frac{2}{3}\hat{j}ms^{-2}$ d) $2\hat{j}ms^{-2}$

601. The range of particle when launched at an angle 15° with the horizontal is 1.5 km. What is the range of projectile when launched at an angle of 45° to the horizontal?

a) 3.0 km b) 1.5 km c) 6.0 km d) 0.75 km

602. What vector must be added to the sum of two vectors $2\hat{i} - \hat{j} + 3\hat{k}$ and $3\hat{i} - 2\hat{j} - 2\hat{k}$ so that the resultant is a unit vector along Z-axis.

a) $5\hat{i} + \hat{k}$ b) $-5\hat{i} + 3\hat{j}$ c) $3\hat{j} + 5\hat{k}$ d) $-3\hat{j} + 2\hat{k}$

603. A mass of 100 g is tied to one end of string 2 m long. The body is revolving in a horizontal circle making a maximum of 200 revolutions/min. The other end of the string is fixed at the centre of the circle of revolution. The maximum tension that the string can bear is approximately

a) 8.76 N b) 8.94 N c) 89.42 N d) 87.64 N

604. A bucket tied at the end of 11.6 m long string is whirled in a vertical circle with a constant speed. The minimum speed at which water from the bucket does not spill when it is a the highest position is

a) 4 ms^{-1} b) 6.25 ms^{-1} c) 2 ms^{-1} d) 16 ms^{-1}

605. A block of massm at the end of a string is whirled round in a vertical circle of radius R. The critical speed of the block at the top of its swing below which the string would slacken before the block reaches the top is

a) Rg b) $(Rg)^2$ c) R/g d) $\sqrt{Rg}$

606. A body of mass m is projected with a speed u making an angle α with the horizontal. The change in momentum suffered by the body along he y-axis between the starting point and the highest point of its path will be

a) $mu \cos\alpha$ b) $mu \sin\alpha$ c) $3\, mu \sin\alpha$ d) mu

607. A bob of mass 10 kg is attached to wire 0.3 m long. Its breaking stress is $4.8 \times 10^7 N/m^2$. The area of cross section of the wire is $10^{-6}m^2$. The maximum angular velocity with which it can be rotated in a horizontal circle

a) 8 rad/sec b) 4 rad/sec c) 2 rad/sec d) 1 rad/sec

608. A ball is projected upwards from the top of tower with a velocity 50 ms^{-1} making an angle 30° with the horizontal. The height of tower is 70 m. After how many seconds from the instant of throwing will the ball reach the ground

a) 2 s b) 5 s c) 7 s d) 9 s

609. A railway carriage has its centre of gravity at a height of 1 m above the rails, which are 1.5 m apart. The maximum safe speed at which it could travel round an unbanked curve of radius 100 m is

a) 12 ms^{-1} b) 18 ms^{-1} c) 22 ms^{-1} d) 27 ms^{-1}

610. Four concurrent coplanar forces in newton are acting at a point and keep it in equilibrium figure. Then values of F and θ are

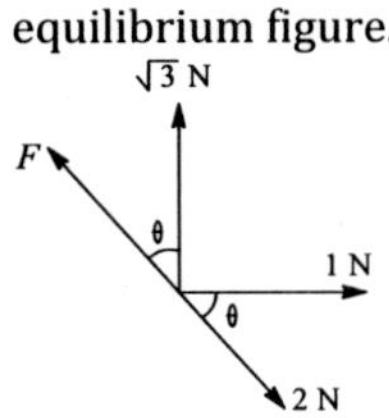

a) 1 N, 60° b) 2 N, 60° c) $\sqrt{2}$ N, 90° d) 2 N, 90°

611. Two bodies are projected from ground with equal speeds 20 m/sec from the same position in same vertical plane to have equal range but at different angle above the horizontal. If one of the angle is 30° the sum of their maximum heights is (assume $g = 10 m/s^2$)

a) 400 m b) 20 m c) 30 m d) 40 m

612. The length of second's hand in a watch is 1 cm. The change in velocity of its tip in 15 seconds is

a) Zero b) $\frac{\pi}{30\sqrt{2}}$ cm/sec c) $\frac{\pi}{30}$ cm/sec d) $\frac{\pi\sqrt{2}}{30}$ cm/sec

613. A particle A is projected from the ground with an initial velocity of 10 ms^{-1} at an angle 60° with horizontal. From what height should an another particle B be projected horizontally with velocity 5 ms^{-1} so that both the particles collide in ground at point C if both are projected simultaneously $g = 10 ms^{-2}$

a) 10 m b) 15 m c) 20 m d) 30 m

614. The X and Y components of vector $\vec{A}$ have numerical values 6 and 6 respectively and that of $(\vec{A} + \vec{B})$ have numerical values 10 and 9. What is the numerical value of $\vec{B}$?

a) 2 b) 3 c) 4 d) 5

615. A particle is projected up from a point at an angle θ with the horizontal direction. At any time t'. If p is the linear momentum, y is the vertical displacement, x is horizontal displacement, the graph among the following which does not represent the variation of kinetic energy KE of the particle

a) 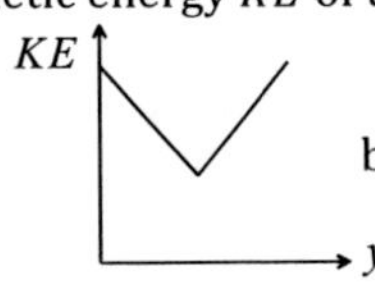

b) 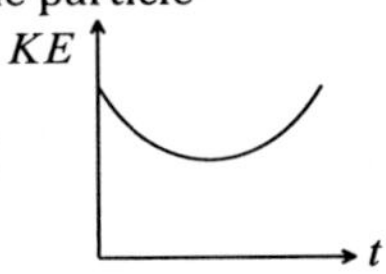

c) 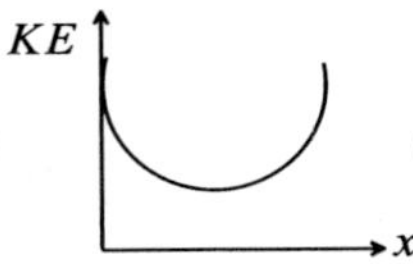

d) 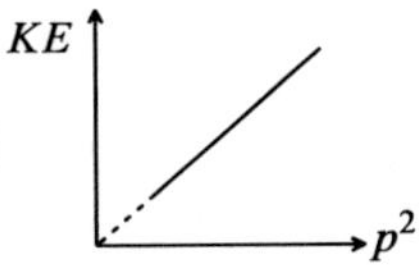

616. A 0.098 kg block slides down a frictionless track as shown. The vertical component of the velocity of block at A is

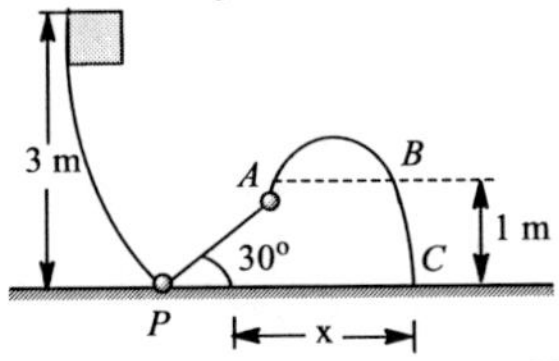

a) $\sqrt{g}$ b) $2\sqrt{g}$ c) $3\sqrt{g}$ d) $4\sqrt{g}$

617. What is the smallest radius of a circle at which a cyclist can travel if its speed is 36 kmh^{-1}, angle of inclination 45° and $g = 10 ms^{-2}$?

a) 20 m b) 10 m c) 30 m d) 40 m

618. Three forces of magnitudes 6N, 6N and $\sqrt{72}$ N at a corner of a cube along three sides as shown in figure. Resultant of these forces is

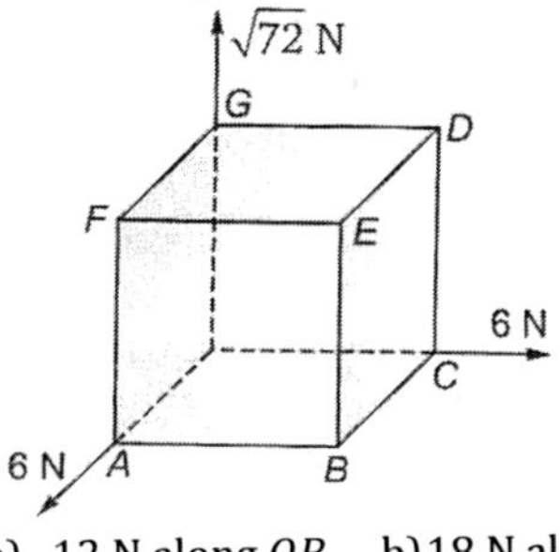

a) 12 N along OB b) 18 N along OA c) 18 N along OC d) 12 N along OE

619. A mass m is attached to the end of a rod of length l. The mass goes along a vertical circular path with the other end hinged at its centre. What should be the minimum velocity of the mass at the bottom of the circle so that the mass completes the circle?
a) $\sqrt{5gl}$ b) $\sqrt{2gl}$ c) $\sqrt{3gl}$ d) $\sqrt{4gl}$

620. The equation of motion of a projectile is $y = 12x - \frac{3}{4}x^2$. The horizontal component of velocity is 3ms^{-1}. What is the range of the projectile?
a) 18 m b) 16 m c) 12 m d) 21.6 m

621. A cannon of a level plane is aimed at an angle θ above the horizontal and a shell is fired muzzle velocity v_0 towards a cliff D distance away. The height at which the canon strikes the cliff is given by
a) $D \sin\theta - \frac{1}{2}\frac{gD^2}{v_0^2 \sin^2\theta}$ b) $D \cos\theta - \frac{1}{2}\frac{gD^2}{v_0^2 \sin^2\theta}$
c) $D \tan\theta - \frac{1}{2}\frac{gD^2}{v_0^2 \cos^2\theta}$ d) $D \tan\theta - \frac{1}{2}\frac{gD^2}{v_0^2 \sin^2\theta}$

622. At the top of the trajectory of a projectile, the direction of its velocity and acceleration are
a) perpendicular to each other b) parallel to each other
c) inclined to each other at angle of 45° d) antiparallel to each other

623. An aeroplane is flying horizontally with a velocity of 600 km/h and at a height of 1960 m. When it is vertically above a point A on the ground a bomb is released from it. The bomb strikes the ground at point B. The distance AB is
a) 1200 m b) 0.33 km c) 333.3 km d) 3.33 km

624. If the resultant of two forces $(A + B)$ and $(A - B)$ is $\sqrt{A^2 + B^2}$, then the angle between these forces is
a) $\cos^{-1}\left[-\frac{(A^2 - B^2)}{A^2 + B^2}\right]$ b) $\cos^{-1}\left[-\frac{(A^2 + B^2)}{(A^2 - B^2)}\right]$
c) $\cos^{-1}\left[-\frac{A^2 + B^2}{2(A^2 - B^2)}\right]$ d) $\cos^{-1}\left[-\frac{2(A^2 + B^2)}{A^2 - B^2}\right]$

625. A body of mass 1 kg tied to one end of string is revolved in a horizontal circle of radius 0.1 m with a speed of 3 $revolution/sec$, assuming the effect of gravity is negligible, then linear velocity, acceleration and tension in the string will be
a) 1.88 m/s, 35.5m/s^2, 35.5 N b) 2.88 m/s, 45.5m/s^2, 45.5 N
c) 3.88 m/s, 55.5m/s^2, 55.5 N d) None of these

626. A body is projected vertically upwards at time $t = 0$ and it is seen at a height H at instants t_1 and t_2 seconds during its flight. The maximum height attained is (g is acceleration due to gravity)
a) $\frac{g(t_2 - t_1)^2}{8}$ b) $\frac{g(t_1 + t_2)^2}{4}$ c) $\frac{g(t_1 + t_2)^2}{8}$ d) $\frac{g(t_2 - t_1)^2}{4}$

627. A particle describes a horizontal circle in a conical funnel whose inner surface is smooth with speed of 0.5 ms^{-1}. What is the height of the plane of circle from vertex of the funnel?
a) 0.25 cm b) 2 cm c) 4 cm d) 2.5 cm

628. A 100 kg car is moving with a maximum velocity of 9 m/s across a circular track of radius 30 m. The maximum force of friction between the road and the car is
a) 1000 N b) 706 N c) 270 N d) 200 N

629. If the radius of curvature of the path of two particles of same masses are in the ratio 1 : 2, then in order to have constant centripetal force, their velocity, should be in the ratio of
a) 1 : 4 b) 4 : 1 c) $\sqrt{2}$: 1 d) 1 : $\sqrt{2}$

630. A bomber plane moves horizontally with a speed of $500 ms^{-1}$ and a bomb released from it, strikes the ground in 10 s. Angel at which it strikes the ground will be ($g = 10 ms^{-2}$)

a) $\tan^{-1}\left(\frac{1}{5}\right)$ b) $\tan\left(\frac{1}{5}\right)$ c) $\tan^{-1}(1)$ d) $\tan^{-1}(5)$

631. A particle moves along a circle of radius $\left[\frac{20}{\pi}\right]$ m with constant tangential acceleration. If the velocity of the particle is 80 m/s at the end of the second revolution after motion has begun, the tangential acceleration is

a) $40\ ms^{-2}$ b) $640\ \pi ms^{-2}$ c) $160\ \pi ms^{-2}$ d) $40\ \pi ms^{-2}$

632. A bullet is to be fired with a speed of $2000\ ms^{-1}$ to hit a target 200 m away on a level ground. If $g = 10\ ms^{-2}$, the gun should be aimed

a) Directly at the target b) 5 cm below the target

c) 5 cm above the target d) 2 cm above the target

633. A wheel making 20 revolution per second is in a horizontal circle with a uniform angular velocity. Let T be the tension in the string. If the length of the string is halved and its angular velocity is doubled, tension in the string will be

a) π rad s^{-2} b) 2π rad s^{-2} c) 4π rad s^{-2} d) 8π rad s^{-2}

634. When a body is thrown with a velocity u making an angle θ with the horizontal plane, the maximum distance covered by it in horizontal direction is

a) $\frac{u^2 \sin\theta}{g}$ b) $\frac{u^2 \sin 2\theta}{2g}$ c) $\frac{u^2 \sin 2\theta}{g}$ d) $\frac{u^2 \cos 2\theta}{g}$

635. A body moving with constant speed in a circular path has

a) angular momentum b) constant acceleration c) constant velocity d) no work done

636. A body is tied to one end of the string and whirled in a vertical circle, the physical quantity which remains constant is

a) Momentum b) Speed c) Kinetic energy d) Total energy

637. The speed limit of a car over a roadways bridge in the form of a vertical arc is $9.8\ ms^{-1}$. The diameter of the are is

a) 19.6 m b) 9.8 m c) 39.2 m d) 4.9 m

638. Toy cart tied to the end of an unstretched string of length a, when revolved moves in a horizontal circle of radius $2a$ with a time period T. Now the toy cart is speeded up until it moves in a horizontal circle of radius $3a$ with a period T'. If Hook's law holds then

a) $T' = \sqrt{\frac{3}{2}}\,T$ b) $T' = \left(\frac{\sqrt{3}}{2}\right)T$ c) $T' = \left(\frac{3}{2}\right)T$ d) $T' = T$

639. A coastguard ship locates a pirate ship at a distance 560 m. It fires a cannon ball with an initial speed 82 m/s. At what angle from horizontal the ball must be fired so that it hits the pirate ship

a) 54° b) 125° c) 27° d) 18°

640. A body of mass 5 kg is moving in a circle of radius 1 m with an angular velocity of 2 rad s^{-1}. The centripetal force, is

a) 10 N b) 20 N c) 30 N d) 40 N

641. In hydrogen atom, the electron is moving round the nucleus with velocity $2.18 \times 10^6 ms^{-1}$ in an orbit of radius 0.528 Å. The acceleration of the electron is

a) $9 \times 10^{18}\ ms^{-2}$ b) $9 \times 10^{22}\ ms^{-2}$ c) $9 \times 10^{-22}\ ms^{-2}$ d) $9 \times 10^{12}\ ms^{-2}$

642. The force required to keep a body in uniform circular motion is

a) Centripetal force b) Centrifugal force c) Resistance d) None of the above

643. Given θ is the angle between $\vec{A}$ and $\vec{B}$. Then $|\hat{A} \times \hat{B}|$ is equal to

a) Sin θ b) Cos θ c) Tan θ d) Cot θ

644. A particle of mass m is projected with a velocity v making an angle of 45° with the horizontal. The magnitude of the angular momentum of the particle about the point of projection when the particle is at its maximum height, is

a) $m\sqrt{2gh^3}$ b) $\frac{mv^3}{\sqrt{2}g}$ c) $\frac{mv^3}{4\sqrt{2}g}$ d) Zero

645. If $\vec{A}_1$ and $\vec{A}_2$ are two non-collinear unit vectors and if $|\vec{A}_1 + \vec{A}_2| = \sqrt{3}$, then the value of $(\vec{A}_1 - \vec{A}_2) \cdot (2\vec{A}_1 + \vec{A}_2)$ is

a) 1 b) 1/2 c) 3/2 d) 2

646. A ball is projected with velocity u at an angle α with horizontal plane. Its speed when it makes an angle β with the horizontal is

a) $u \cos \alpha$ b) $\frac{u}{\cos \beta}$ c) $u \cos \alpha \cos \beta$ d) $\frac{u \cos \alpha}{\cos \beta}$

647. Find the maximum velocity for skidding for a car moved on a circular track of radius 100 m. The coefficient of friction between the road and tyre is 0.2

a) 0.14 m/s b) 140 m/s c) 1.4 km/s d) 14 m/s

648. A stone of mass 1 kg tied to a light inextensible string of length $L = \frac{10}{3}$ is whirling in a circular path of radius L in vertical plane. If the ratio of the maximum tension to the minimum tension in the string is 4. What is the speed of stone at the highest point of the circle? (Taking $g = 10 \text{ms}^{-2}$)

a) 10 ms^{-1} b) $5\sqrt{2}\text{ms}^{-1}$ c) $10\sqrt{3}\text{ms}^{-1}$ d) 20 ms^{-1}

649. A man is supported on a frictionless horizontal surface. It is attached to a string and rotates about a fixed centre at an angular velocity ω. The tension in the string if F. If the length of string and angular velocity are doubled, the tension is string is now

a) F b) $F/2$ c) 4 F d) 8 F

650. A very broad elevator is going up vertically with a constant acceleration of 2 ms^{-2}. At the instant when its velocity is 4 ms^{-1} a ball is projected from the floor of the list with a speed of 4 ms^{-1} relative to the floor at an elevation of 30°. The time taken by the ball to return the floor is ($g = 10 \text{ ms}^{-2}$)

a) 1/2 s b) 1/3 s c) 1/4 s d) 1 s

651. A particle is moving in a circular path with a constant speed v. If θ is the angular displacement, then starting from $\theta = 0°$, the maximum and maximum changes in the momentum will occur, when value of θ is respectively

a) 45° and 90° b) 90° and 180° c) 180° and 360° d) 90° and 270°

652. The angle between $\vec{A}$and $\vec{B}$ is θ, the value of the triple product $\vec{A} \cdot \vec{B} \times \vec{A}$ is

a) A^2B b) Zero c) $A^2B \sin\theta$ d) $A^2B \cos\theta$

653. A particle is projected with velocity $\sqrt{2gh}$,such that it just crosses two walls of height h and separated by h Find the angle of projection.

a) 15° b) 75° c) 60° d) 30°

654. A weightless thread can support tension upto30 N. A stone of mass 0.5 kg is tied to it and is revolved in a circular path of radius 2 m in a vertical plane. If $g = 10 \ m/s^2$, then the maximum angular velocity of the stone will be

a) 5 rad/s b) $\sqrt{30}$ rad/s c) $\sqrt{60}$ rad/s d) 10 rad/s

655. A body is tided with a string and is given a circular motion with velocity v in radius r. The magnitude of the acceleration

a) $\frac{v}{r}$ b) $\frac{v^2}{r}$ c) $\frac{v}{r^2}$ d) $\frac{v^2}{r^2}$

656. A body starts from rest from the origin with an acceleration of 6 m/s^2 along the x-axis and 8 m/s^2 along the y-axis. Its distance from the origin after 4 seconds will be

a) 56 m b) 64 m c) 80 m d) 128 m

657. The angle between the z-axis and the vector $\hat{\imath} + \hat{\jmath} + \sqrt{2}\hat{k}$ is

a) 30° b) 45° c) 60° d) 90°

658. A proton of mass $1.6 \times 10^{-27} kg$ goes round in a circular orbit of radius $0.10\ m$ under a centripetal force of $4 \times 10^{-13}\ N$. then the frequency of revolution of the proton is about

a) 0.08×10^{8} *cycles per sec* b) 4×10^{8} *cycles per sec*
c) 8×10^{8} *cycles per sec* d) 12×10^{8} *cycles per sec*

659. For an object thrown at 45° to horizontal, the maximum height (H) and horizontal range (R) are related as

a) $R = 16H$ b) $R = 8H$ c) $R = 4H$ d) $R = 2H$

660. The torque of a force $\vec{F} = -3\hat{i} + \hat{j} + 5\hat{k}$ acting at a point is $\vec{\tau}$. If the position vector of the point is $7\hat{i} + \hat{j} + \hat{k}$, then $\vec{\tau}$ is

a) $7\hat{i} - 8\hat{j} + 9\hat{k}$ b) $14\hat{i} - \hat{j} + 3\hat{k}$ c) $2\hat{i} - 3\hat{j} + 8\hat{k}$ d) $14\hat{i} - 38\hat{j} + 16\hat{k}$

661. The figure shows a circular path of a moving particle. If the velocity of the particle at same instant is $\mathbf{v} = -3\hat{\mathbf{i}} - 4\hat{\mathbf{j}}$, through which quadrant is the particle moving when clockwise and anti-clockwise respectively

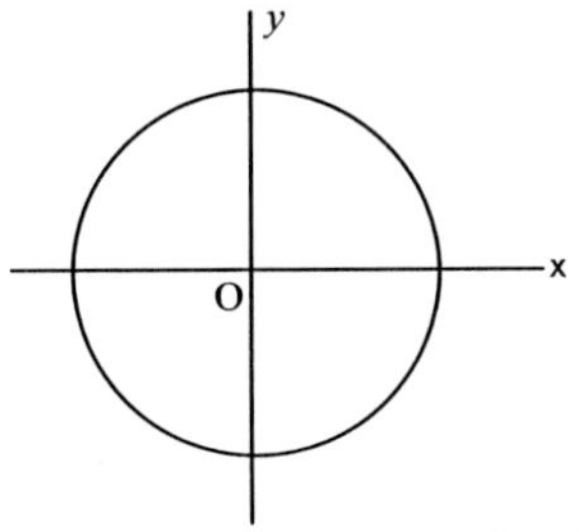

a) 1st and 4 th b) 2nd and 4th c) 2 nd and 3rd d) 3 rd and 4 th

662. A particle is thrown above, the correct $v - t$ graph will be

a) 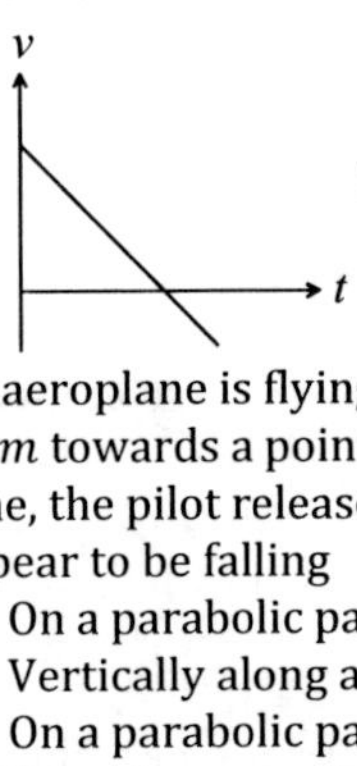
b) 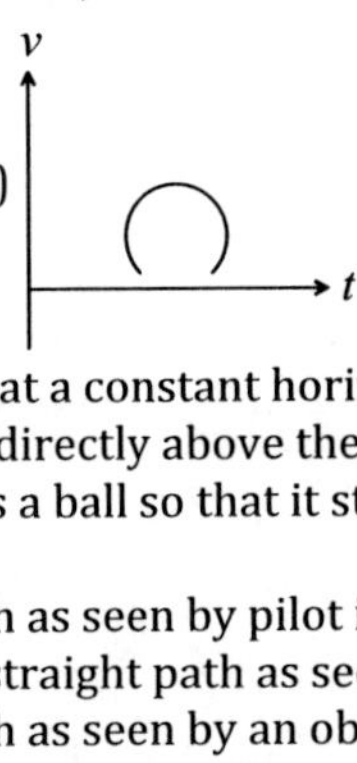
c) 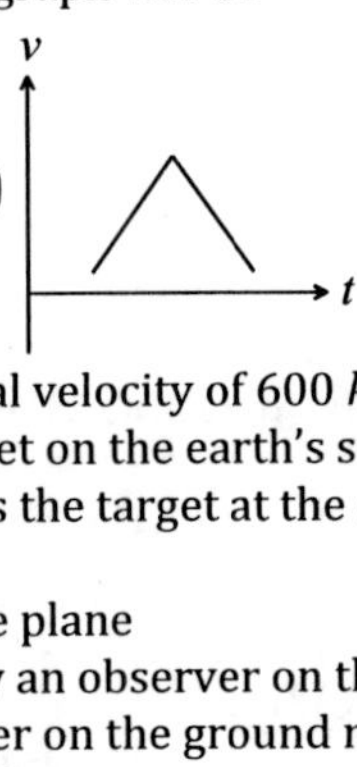
d) 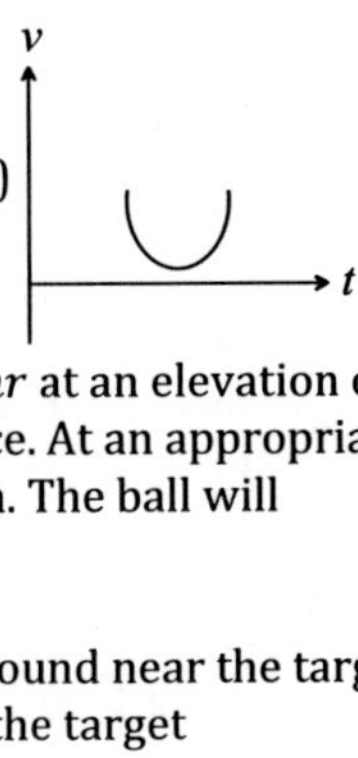

663. An aeroplane is flying at a constant horizontal velocity of 600 km/hr at an elevation of 6 km towards a point directly above the target on the earth's surface. At an appropriate time, the pilot releases a ball so that it strikes the target at the earth. The ball will appear to be falling

a) On a parabolic path as seen by pilot in the plane
b) Vertically along a straight path as seen by an observer on the ground near the target
c) On a parabolic path as seen by an observer on the ground near the target
d) On a zig-zag path as seen by pilot in the plane

664. A wheel completes 2000 revolutions to cover the 9.5 km. distance. Then the diameter of the wheel is

a) 1.5 m b) 1.5 cm c) 7.5 cm d) 7.5 m

665. Figure shows four paths for a kicked football. Ignoring the effects of air on the flight, rank the paths according to initial horizontal velocity component, highest first

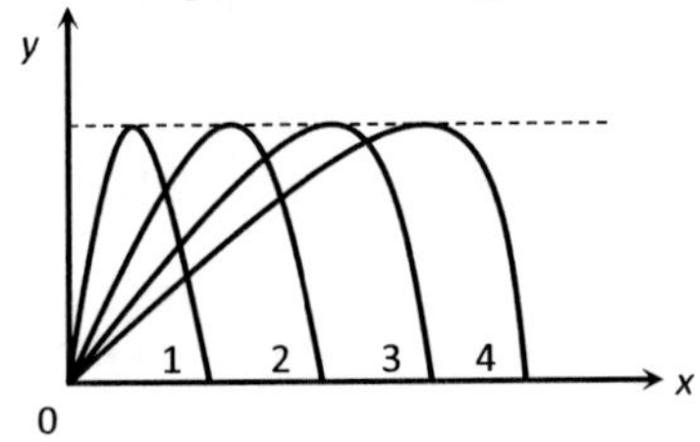

a) 1, 2, 3, 4 b) 2, 3, 4, 1 c) 3, 4, 1, 2 d) 4, 3, 2, 1

666. If a stone s to hit at a point which is at a distance d away and at a height h above the point from where the stone starts, then what is the value of initial sped u, if the stone is launched at an angle Q?

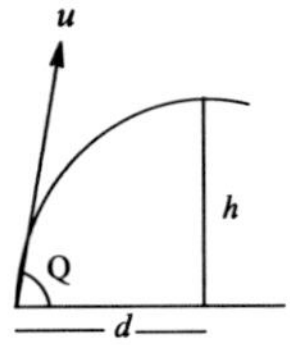

a) $\frac{g}{\cos\theta}\sqrt{\frac{d}{2(d\tan\theta - h)}}$ b) $\frac{d}{\cos\theta}\sqrt{\frac{g}{2(d\tan\theta - h)}}$ c) $\sqrt{\frac{gd^2}{h\cos^2\theta}}$ d) $\sqrt{\frac{gd^2}{(d-h)}}$

667. A small particle of mass m is projected at an angle θ with the x-axis with an initial velocity v_0 in the x-y plane as shown in the figure. At a time $t < \frac{v_0 \sin\theta}{g}$, the angular momentum of the particle is

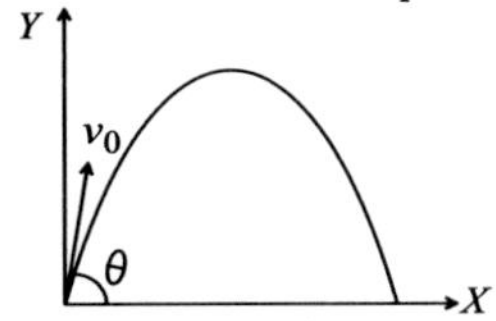

Where $\hat{\imath}, \hat{\jmath}$ and $\hat{k}$ are unit vectors along x, y and z-axis respectively.

a) $\frac{1}{2}mgv_0t^2\cos\theta\,\hat{\imath}$ b) $-mgv_0t^2\cos\theta\,\hat{\jmath}$ c) $mgv_0t\cos\quad\hat{k}$ d) $-\frac{1}{2}mgv_0t^2\cos\theta\,\hat{k}$

668. If the magnitude of the sum of the two vectors is equal to the difference of their magnitudes, then the angle between vectors is

a) 0° b) 45° c) 90° d) 180°

669. A particle of a mass m is projected with velocity v making an angle of 45° with the horizontal. The magnitude of the angular momentum of the particle about the point of projection when the particle is at its maximum height is (where g = acceleration due to gravity)

a) Zero b) $mv^3/(4\sqrt{2}g)$ c) $mv^3/(\sqrt{2}g)$ d) $mv^2/2g$

670. A toy cyclist completes one round of a square track of side 2 m in 40 s. What will be the displacement at the end of 3 min?

a) 52 m b) Zero c) 16 m d) $2\sqrt{2}$m

671. The angular velocity of a particle rotating in a circular orbit 100 times per minute is

a) 1.66 rad s^{-1} b) 10.47 rad s^{-1} c) 10.47 deg s^{-1} d) 60 rad s^{-1}

672. Two forces $\overrightarrow{F_1}$ and $\overrightarrow{F_2}$ are acting at right angles to each other. Then their resultant is

a) $F_1 + F_2$ b) $\sqrt{F_1^2 + F_2^2}$ c) $\sqrt{F_1^2 - F_2^2}$ d) $\frac{F_1 + F_2}{2}$

673. Two bodies are projected with the same velocity. If one is projected at an angle of 30° and the other at an angle of 60° to the horizontal, the ratio of the maximum heights reached is

a) 3 : 1 b) 1 : 3 c) 1 : 2 d) 2 : 1

674. If a_r and a_t represent radial and tangential accelerations, the motion of a particle will be uniformly circular if

a) $a_r = 0, a_t = 0$ b) $a_r \neq 0, a_t \neq 0$ c) $a_r \neq 0, a_t = 0$ d) $a_r = 0, a_t \neq 0$

675. The kinetic energy of a projectile at the highest point is half of the initial kinetic energy. What is the angle of projection with the horizontal?

a) 30° b) 45° c) 60° d) 90°

676. The distance r from the origin of a particle moving in $x - y$ plane varies with time as $r = 2t$ and the angle made by the radius vector with positive x-axis is $\theta = 4t$. Here, t is in second, r in metre and θ in radian. The speed of the particle at $t = 1$ s is

a) $10\ ms^{-1}$ b) $16\ ms^{-1}$ c) $10\ ms^{-1}$ d) $12\ ms^{-1}$

677. In uniform circular motion

a) Both the angular velocity and the angular momentum vary
b) The angular velocity varies but the angular momentum remains constant
c) Both the angular velocity and the angular momentum stay constant
d) The angular momentum varies but the angular velocity remains constant

678. A cyclist is travelling with velocity v on a banked curved road of radius R. The angle θ through which the cyclist leans inwards is given by

a) $\tan\theta = \frac{Rg}{v^2}$ b) $\tan\theta = v^2\ Rg$ c) $\tan\theta = \frac{v^2 g}{R}$ d) $\tan\theta = \frac{v^2}{Rg}$

679. A small cone filled with water is revolved in a vertical circle of radius 4 m and the water does not fall down. What must be the maximum period of revolution?

a) 4 s b) 2 s c) 1 s d) 6 s

680. In uniform circular motion, the velocity vector and acceleration vector are

a) Perpendicular to each other b) Same direction
c) Opposite direction d) Not related to each other

681. In a bicycle the radius of rear wheel is twice the radius of front wheel. If r_f and r_r are the radius, v_f and v_r are the speed of top most points of wheel, then

a) $v_r = 2v_f$ b) $v_f = 2v_r$ c) $v_f = v_r$ d) $v_f > v_r$

682. An object of mass 10 kg is whirled round a horizontal circle of radius 4 m by a revolving string inclined 30° to the vertical. If the uniform speed of the object is $5\ ms^{-1}$, the tension in the string (approximately) is

a) 720 N b) 960 N c) 114 N d) 125 N

683. A bullet is fired horizontally with a velocity of $80\ ms^{-1}$. During the first second,

a) It falls 9.8 m b) It falls $\frac{80}{9.8}$ m c) It does not fall at all d) It falls 4.9 m

684. A mass is supported on a frictionless horizontal surface. It is attached to a string and rotates about a fixed centre at an angular velocity ω_0. If the length of the string and angular velocity both are doubled, the tension in the string which was initially T_0 is now

a) T_0 b) $T_0/2$ c) $4T_0$ d) $8T_0$

685. An aeroplane is flying in a horizontal direction with a velocity $600 kmh^{-1}$ at a height of 1960 m. when it is vertically above the point A on the ground, a body is dropped from it. The body strikes the ground at point B. Calculate the distance AB.

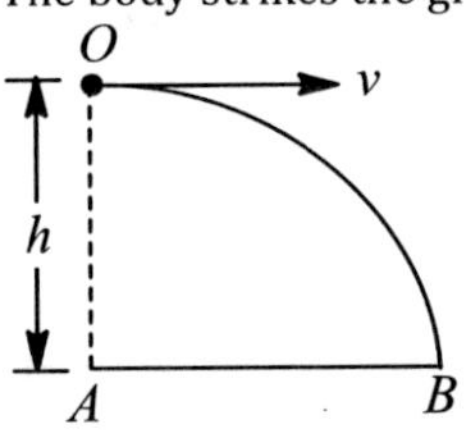

a) 3.33 km b) 333 km c) 33.3 km d) 3330 km

686. A cylinder full. Of water, is rotating about its own axis with uniform angular velocity ω. The shape of free surface of water will be

a) Parabola b) Elliptical c) Circular d) Spherical

687. A cricketer hits a ball with a velocity 25 m/s at 60° above the horizontal. How far above the ground it passes over a fielder 50 m from the bat (assume the ball is struck very close th the ground)

a) 8.2 m b) 9.0 m c) 11.6 m d) 12.7 m

688. Two identical discs of same radius R are rotating about their axes in opposite directions with the same constant angular speed ω. The discs are in the same horizontal plane. At time $t = 0$, the points P and Q are facing each other as shown in figure. The relative speed between the two points P and Q is V_r as function of times best represented by

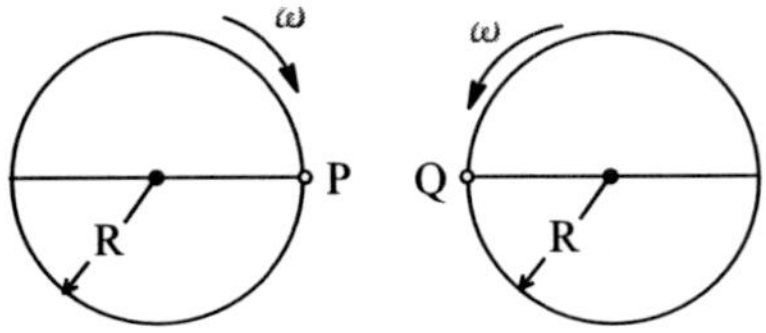

a) 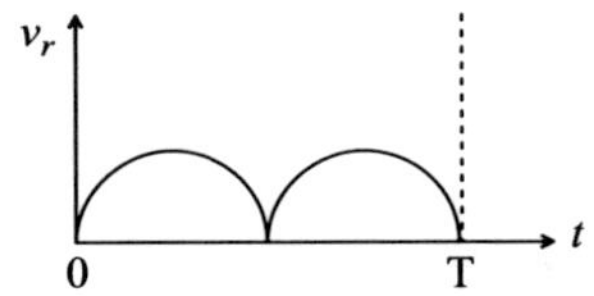

b) 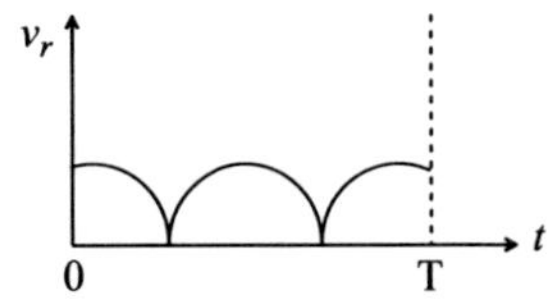

c) 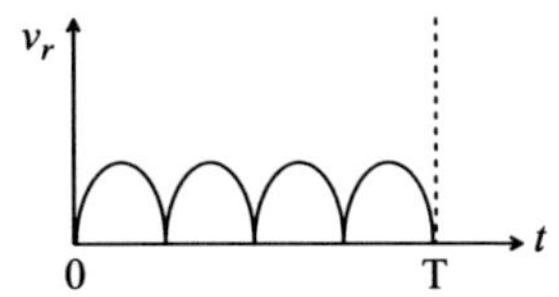

d) 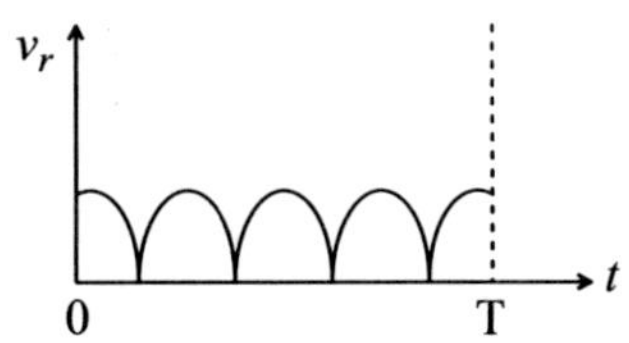

689. A car is moving rectilinearly on a horizontal path with acceleration a_0. A person sitting inside the car observes that an insect S is crawling up the screen with an acceleration a. If θ is the inclination of the screen with the horizontal the acceleration of the insect

a) Parallel to screen is $a + a_0 \cos\theta$ b) Along the horizontal is $a_0 - a\cos\theta$
c) Perpendicular to screen is $a_0 \sin\theta$ d) Perpendicular to screen is $a_0 \tan\theta$

690. A particle of mass M is moving in a horizontal circle of radius R with uniform speed V. When it moves from one point to a diametrically opposite point, its

a) Kinetic energy changes by $MV^2/4$ b) Momentum does not change
c) Momentum changes by $2MV$ d) Kinetic energy changes by MV^2

691. A ball is projected from the ground at a speed of $10 ms^{-1}$ making an angle of $30°$ with the horizontal. Another ball is simultaneously released from a point on the vertical line along the maximum height of the projectile. Both the balls collide at the maximum height of first ball. The initial height of the second ball is ($g = 10 ms^{-2}$)

a) $6.25m$ b) $2.5m$ c) $3.75m$ d) $5m$

692. A ball is thrown up at an angle with the horizontal. Then the total change of momentum by the instant it returns to ground is

a) Acceleration due to gravity × total time of flight
b) Weight of the ball × half the time of flight
c) Weight of the ball × total time of flight
d) Weight of the ball × horizontal range

693. A man, using a 70 kg garden roller on a level surface exerts a force of 200 N at 45° to the ground. What is the vertical force of the roller on the ground, if he pushed the roller? ($g = 10ms^{-2}$)

a) 70 N b) 200 N c) 560 N d) 840 N

694. An aeroplane moving horizontally at a speed of 200 m/s and at a height of 8.0×10^3 m is to drop a bomb on a target. At what horizontal distance from the target should the bomb be released

a) 7.234 km b) 8.081 km c) 8.714 km d) 9.124 km

695. A particle is projected with certain velocity at two different angels of projections with respect to horizontal plane so as to have same range R on a horizontal plane. If t_1 and t_2 are the time taken for the two paths, the which one of the following relations is correct?

a) $t_1t_2 = \frac{2R}{g}$ b) $t_1t_2 = \frac{R}{g}$ c) $t_1t_2 = \frac{R}{2g}$ d) $t_1t_2 = \frac{4R}{g}$

696. A particle moves in a circle of radius 25 cm at two revolutions per second. The acceleration of the particle in m/s^2 is

a) π^2 b) $8\pi^2$ c) $4\pi^2$ d) $2\pi^2$

697. The vector which can give unit vector along x-axis with $\vec{A} = 2\hat{i} - 4\hat{j} + 7\hat{k}$, $\vec{B} = 7\hat{i} + 2\hat{j} - 5\hat{k}$ and $\vec{C} = -4\hat{i} + 7\hat{j} + 3\hat{k}$ is

a) $4\hat{i} + 5\hat{j} + 5\hat{k}$ b) $-5\hat{i} - 5\hat{j} + 5\hat{k}$ c) $-4\hat{i} - 5\hat{j} - 5\hat{k}$ d) $4\hat{i} - 5\hat{j} + 5\hat{k}$

698. The acceleration of a train travelling with speed of 400 m/s as it goes round a curve of radius 160 m, is

a) 1 km/s^2 b) 100 m/s^2 c) 10 m/s^2 d) 1 m/s^2

699. A man standing on the roof of a house of height h throws one particle vertically downwards and another particle horizontally with the same velocity u. The ratio of their velocities when they reach the earth's surface will be

a) $\sqrt{2gh + u^2} : u$ b) 1: 2 c) 1: 1 d) $\sqrt{2gh + u^2} : \sqrt{2gh}$

700. An artillery piece which consistently shoots its shells with the same muzzle speed has a maximum range R. To hit a target which is $\frac{R}{2}$ from the gun and on the same level, the elevation angle of the gun should be

a) 15° b) 45° c) 30° d) 60°

701. A body is projected with initial velocity of $(8\hat{i} + 6\hat{j})ms^{-1}$. The horizontal range is

a) 9.6 m b) 14 m c) 50 m d) None of these

702. A stone is tied at one end of a 5m long string and whirled in a vertical circle. The minimum speed required to just cross the topmost position is

a) 5 ms^{-1} b) 7 ms^{-1} c) 57 ms^{-1} d) 75 ms^{-1}

703. A particle of mass m moves with constant speed along a circular path of radius r under the action of force F. Its speed is

a) $\sqrt{Fr/m}$ b) $\sqrt{F/r}$ c) $\sqrt{F\,m\,r}$ d) $\sqrt{F/mr}$

704. For a projection, $(\text{range})^2$ is 48 times of $(\text{maximum height})^2$ obtained. Find angle projection.

a) 60° b) 30° c) 45° d) 75°

705. A particle is projected up from a point at an angle with the horizontal. At any time t if p =linear momentum, y =vertical displacement, x =horizontal displacement, then the kinetic energy (K) of the particle plotted against these parameters can be

a) 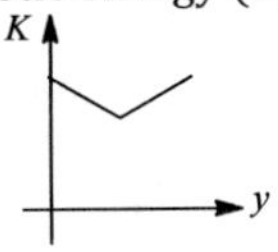

b) 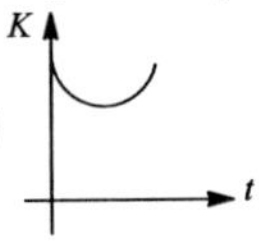

c) 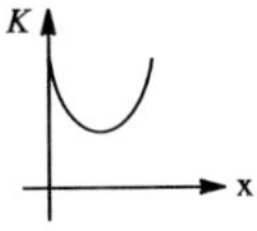

d) 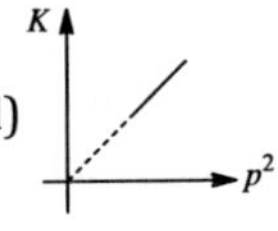

706. The equation of a projectile is $y = \sqrt{3}x - \frac{gx^2}{2}$. The angle of projection is given by

a) $\tan\theta = \frac{1}{\sqrt{3}}$ b) $\tan\theta = \sqrt{3}$ c) $\frac{\pi}{2}$ d) Zero

707. If a body is projected with an angle θ to the horizontal then

a) its velocity is always perpendicular to its acceleration
b) its velocity becomes zero as its maximum height
c) its velocity makes zero angle with the horizontal at its maximum height
d) the body just before hitting the ground, the direction of velocity coincides with the acceleration

708. The speed of revolution of a particle going around a circle is doubled and its angular speed is halved. What happens to the centripetal acceleration?

a) Becomes four times b) Double c) Halved d) Remains unchanged

709. Given $\vec{R} = \vec{A} + \vec{B}$ and $R = A = B$. The angle between $\vec{A}$ and $\vec{B}$ is

a) 60° b) 90° c) 120° d) 180°

710. A stone is projected with a velocity $20\sqrt{2}\text{ms}^{-1}$ at an angle of 45° to the horizontal. The average velocity of stone during its motion from starting point to its maximum height is $(g = 10\text{ms}^{-2})$

a) $5\sqrt{5}\text{ ms}^{-1}$ b) $10\sqrt{5}\text{ ms}^{-1}$ c) 20 ms^{-1} d) $20\sqrt{5}\text{ ms}^{-1}$

711. A body moving along a circular path of radius R with velocity v, has centripetal acceleration a. If its velocity is made equal to $2v$, then its centripetal acceleration is

a) $4a$ b) $2a$ c) $\frac{a}{4}$ d) $\frac{a}{2}$

712. The speed of a projectile at its maximum height is half of its initial speed. The angle of projection is

a) 60° b) 15° c) 30° d) 45°

713. If a_r and a_t represent radial and tangential accelerations, the motion of a particle will be uniformly circular if

a) $a_r = 0$ and $a_t = 0$ b) $a_r = 0$ but $a_t \neq 0$ c) $a_r \neq 0$ but $a_t = 0$ d) $a_r \neq 0$ and $a_t \neq 0$

714. In above question, if the centripetal force F is kept constant but the angular velocity is doubled, the new radius of the path (original radius R) will be

a) $2R$ b) $R/2$ c) $R/4$ d) $4R$

715. Two wires AC and BC are tied at C of small sphere of mass 5 kg, which revolves at a constant speed v in the horizontal circle of radius 1.6 m. The minimum value of v is

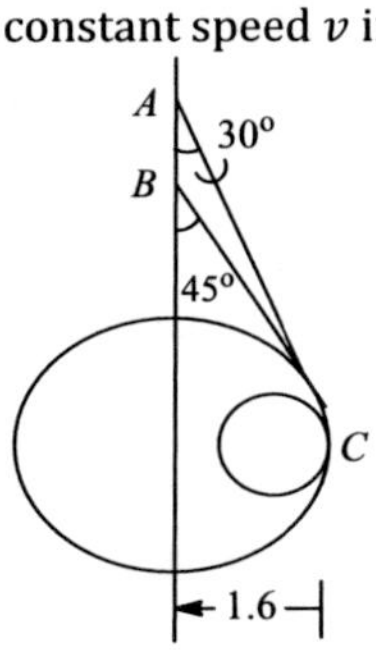

a) 3.01 ms^{-1} b) 4.01 ms^{-1} c) 8.2 ms^{-1} d) 3.96 ms^{-1}

716. If the resultant of the vectors $(\hat{i} + 2\hat{j} - \hat{k})$, $(\hat{i} - \hat{j} + 2\hat{k})$ and $\vec{C}$ is a unit vector along the y-direction, then $\vec{C}$ is

a) $-2\hat{i} - \hat{k}$ b) $-2\hat{i} + \hat{k}$ c) $2\hat{i} - \hat{k}$ d) $2\hat{i} + \hat{k}$

717. A small disc is on the top of a hemisphere of radius R. What is the smallest horizontal velocity v that should be given to the disc for it to leave the hemisphere and not slide down it? [There is no friction]

a) $v = \sqrt{2gR}$ b) $v = \sqrt{gR}$ c) $v = \frac{g}{R}$ d) $v = \sqrt{g^2R}$

718. What is the angle between $\vec{P}$ and $\vec{Q}$. The resultant of $(\vec{P} + \vec{Q})$ and $(\vec{P} - \vec{Q})$?

a) Zero b) $\tan^{-1}(P/Q)$

c) $\tan^{-1}(Q/P)$ d) $\tan^{-1}(P - Q)/(P + Q)$

719. A long horizontal rod has a bead, which can slide along its length and initially placed at a distance L from one end A of the rod. The rod is set in angular acceleration α. If the coefficient of friction, between the rod and the bead is μ and gravity is neglected, then the time after which the bead starts slipping is

a) $\sqrt{\mu/\alpha}$ b) $\mu/\sqrt{\alpha}$ c) $1/\sqrt{\mu\alpha}$ d) Infinitesimal

720. The earth moves round the sun in a near circular orbit of radius 1.5×10^{11} m. Its centripetal acceleration is

a) $1.5 \times 10^{-3}\, m/s^2$ b) $3 \times 10^{-3}\, m/s^2$ c) $6 \times 10^{-3}\, m/s^2$ d) $12 \times 10^{-3}\, m/s^2$

721. A particle is projected with speed v at an angle $\theta \left(0 < \theta < \frac{\pi}{2}\right)$ above the horizontal from a height H above the ground. If $v =$ speed with which particle hits the ground and $t =$ time taken by particle to reach ground, then

a) As θ increases, v decreases and t increases
b) As θ increases, v increases and t increases
c) As θ increases, v remains same and t increases
d) As θ increases, v remains same and t decreases

722. A simple pendulum is oscillating without damping. When the displacement of the bob is less than maximum, its acceleration vector $\vec{a}$ is correctly shown in

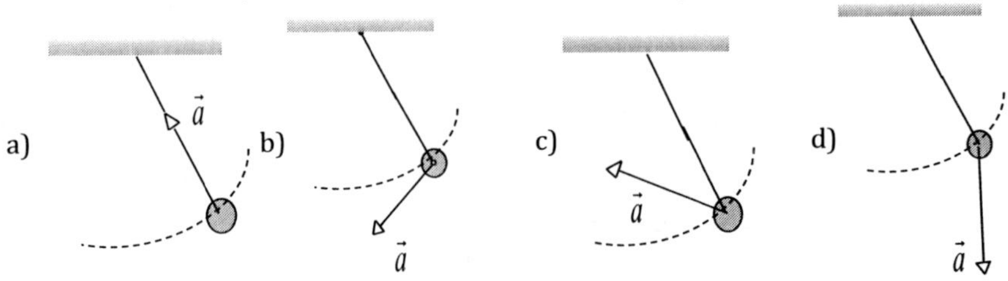

723. For a projectile, the ratio of maximum height reached to the square of flight time is ($g = 10\, ms^{-2}$)

a) 5 : 4 b) 5 : 2 c) 5 : 1 d) 10 : 1

724. From an inclined plane two particles are projected with same speed at same angle θ, one up and other down the plane as shown in figure, which of the following statements is/are correct?

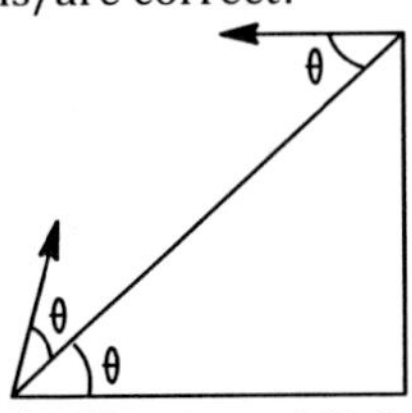

a) The time of flight of each particle is the same
b) The particles will collide the plane with same speed
c) Both the particles strike the plane perpendicularly
d) The particles will collide in mid air if projected simultaneously and time of flight of each particle is less than of collision

725. Two racing cars of masses m_1 and m_2 are moving in circles of radii r_1 and r_2 respectively. Their speeds are such that each makes a complete circle in the same duration of time t. The ratio of the angular speed of the first to the second car is

a) $m_1 : m_2$ b) $r_1 : r_2$ c) 1 : 1 d) $m_1 : r_1 : m_2 r_2$

726. A car moving with the speed of 10 m/s takes a circular turn of radius 20 m. The magnitude of the acceleration of the car is

a) 5.0 ms $^{-2}$ b) 50.0 ms $^{-2}$ c) 0.25 ms $^{-2}$ d) 0.5 ms $^{-2}$

727. Two vectors $\vec{a}$ and $\vec{b}$ are at an angle of 60° with each other. Their resultant makes an angle of 45° with $\vec{a}$. If $|\vec{b}| = 2$ units, then $|\vec{a}|$ is

a) $\sqrt{3}$ b) $\sqrt{3}-1$ c) $\sqrt{3}+1$ d) $\sqrt{3}/2$

728. A body crosses the topmost point of a vertical circle with critical speed. What will be its acceleration when the string is horizontal?

a) g b) 2g c) 3g d) 6 g

729. A particle is moving in a horizontal circle with constant speed. It has constant

a) Velocity b) Acceleration c) Kinetic energy d) Displacement

730. A bend in a level road has a radius of 80 m. Find the maximum speed which a car turning the bend may have without skidding, if $\mu = 0.25$

a) 24 ms^{-1} b) 4 ms^{-1} c) 14 ms^{-1} d) 9.8 ms^{-1}

731. An object of mass 2 m is projected with a speed of 100 ms^{-1} at angle $\theta = \sin^{-1}\left(\frac{3}{5}\right)$ to the horizontal. At the highest point, the object breaks into pieces of same mass m and the first one comes to rest. The distance between the point of projection and the point of landing of the bigger piece (in metre) is (given, $g = 10$ ms^{-2})

a) 3840 b) 1280 c) 1440 d) 960

732. Given $\vec{P} = 3\hat{j} + 4\hat{k}$ and $\vec{Q} = 2\hat{i} + 5\hat{k}$. The magnitude of the scalar product of these vectors is

a) 20 b) 23 c) 26 d) $5\sqrt{33}$

733. A cricketer can throw a ball to a maximum horizontal distance of 100 m. With the same effort, he throws the ball vertically upwards. The maximum height attained by the ball is

a) 100 m b) 80 m c) 60 m d) 50 m

734. Neglecting the air resistance, the time of flight of a projectile is determined by

a) U_{vertical} b) $U_{\text{horizontal}}$

c) $U = U^2_{\text{vertical}} + U^2_{\text{horizontal}}$ d) $U = \left(U^2_{\text{vertical}} + U^2_{\text{horizontal}}\right)^{1/2}$

735. The height y and the distance x along the horizontal plane of a projectile on a certain planet (with no surrounding atmosphere) are given by $y = 8t - 5t^2$ metre and $x = 6t$ metre, where t is in second. The velocity with which the projectile is projected, is

a) 14 ms^{-1} b) 10 ms^{-1} c) 8 ms^{-1} d) 6 ms^{-1}

736. A car when passes through a convex bridge exerts a force on it which is equal to

a) $Mg + \frac{Mv^2}{r}$ b) $\frac{Mv^2}{r}$ c) Mg d) None of these

737. A batsman hits a sixer and the ball touches the ground outside the cricket ground. Which of the following graph describes the variation of the cricket ball's vertical velocity v with time between the time t_1 as it hits the bat and time t_2 when it touches the ground

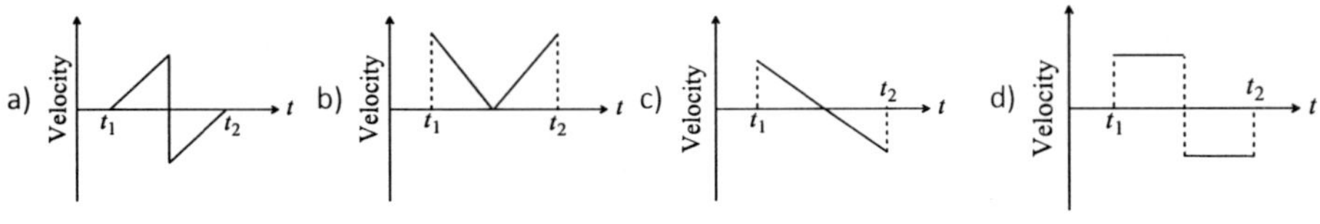

738. The time period of the second's hand of a watch is
a) 1 h b) 1 s c) 12 h d) 1 min

739. A particle is tied to 20cm long string. It performs circular motion in vertical plane. What is the angular velocity of the string when the tension in the string at the top is zero
a) 5 rad/sec b) 2 rad/sec c) 7.5 rad/sec d) 7 rad/sec

740. A particle moves in the $x-y$ plane with velocity $v_x = 8t - 2$and $v_y = 2$. If it passes through the pointy $x = 14$ and $y = 4$ at $t = 2s$, find the equation ($x - y$ relation) of the path
a) $x = y^2 - y + 2$ b) $x = 2y^2 + 2y - 3$
c) $x = 3y^2 + 5$ d) Can not be found from above data

741. A body just being revolved in a vertical circle of radius R with a uniform speed. The string breaks when the body is at the highest point. The horizontal distance covered by the body after the string breaks is
a) 2 R b) R c) $R\sqrt{2}$ d) 4 R

742. If $\vec{A}$ and $\vec{B}$ denote the sides of a parallelogram and its area is $\frac{1}{2}AB$(A and B are the magnitude of $\vec{A}$ and $\vec{B}$ respectively), the angle between $\vec{A}$ and $\vec{B}$ is
a) 30° b) 60° c) 45° d) 120°

743. Given $\vec{r} = 4\hat{j}$and $\vec{p} = 2\hat{i} + 3\hat{j} + \hat{k}$. The angular momentum is
a) $4\hat{i} - 8\hat{k}$ b) $8\hat{i} - 4\hat{k}$ c) $8\hat{j}$ d) $9\hat{k}$

744. A car is moving along a straight horizontal road with a speedv_0. If the coefficient of friction between tyres and the road is μ, the shortest distance in which the car can be stopped is
a) $\frac{v_0^2}{2\mu g}$ b) $\frac{v_0}{\mu g}$ c) $\left(\frac{v_0}{\mu g}\right)^2$ d) $\frac{v_0}{\mu}$

745. Work done when a force, $\vec{F} = (\hat{i} + 2\hat{j} + 3\hat{k})$N acting on a particle takes it from the point$\vec{r}_1 = (\hat{i} + \hat{j} + \hat{k})$m to the point $\vec{r}_2 = (\hat{i} + \hat{j} + 2\hat{k})$
a) −3 J b) −1 J c) zero d) 2 J

746. An aeroplane is flying with a uniform speed of 100 m/s along a circular path of radius 100 m. the angular speed of the aeroplane will be
a) 1 rad/sec b) 2 rad/sec c) 3 rad/sec d) 4 rad/sec

747. A particle comes round a circle of radius 1 m once. The time taken by it is 10 sec. The average velocity of motion is
a) 0.2 $\pi m/s$ b) 2 $\pi m/s$ c) 2 m/s d) Zero

748. A chain of 125 links is 1.25 m long and has mass of 2 kg with the ends fastened together. It is set for rotating at 50 revs^{-1}, centripetal force on each links is
a) 3.14 N b) 0.314 N c) 314 N d) None of these

749. A particle of mass m is fixed to one end of a light spring of force constant k and unstretched length l. The system is rotated about the other end of the spring with an angular velocity ω, in gravity free space. As shown in figure the increase in length of the spring will be
a) $\frac{m\,\omega^2 l}{k}$ b) $\frac{m\,\omega^2 l}{k - m\,\omega^2}$ c) $\frac{m\,\omega^2 l}{k + m\,\omega^2}$ d) None of these

750. The position of a particle moving in the xy- plane at any time t is given by $x = (3t^2 - 6t)metres, y = (t^2 - 2t)metres$. Select the correct statement about the moving particle from the following
a) The acceleration of the particle is zero at $t = 0$ second
b) The velocity of the particle is zero at $t = 0$ second
c) The velocity of the particle is zero at $t = 1$ second
d) The velocity and acceleration of the particle are never zero

751. A car is moving on a circular path and takes a turn. If R_1 and R_2 be the reactions on the inner and outer wheels respectively, then

a) $R_1 = R_2$ b) $R_1 < R_2$ c) $R_1 > R_2$ d) $R_1 \geq R_2$

752. Which one is Angular resolution fundamental quantity

a) Length b) Time c) Radian d) Angle

753. A particle is projected from horizontal making an angle 60° with initial velocity $40 ms^{-1}$. The time taken by the particle to make angel 45° from horizontal, is

a) 15 s b) 2.0 s c) 20 s d) 1.5 s

754. If a car is to travel with a speed v along the frictionless, banked circular track of radius r, the required angle of banking so that the car does skid is

a) $\theta = \tan^{-1}\left(\frac{v^2}{rg}\right)$ b) $\theta = \tan^{-1}\left(\frac{v}{rg}\right)$ c) $\theta = \tan^{-1}\left(\frac{r^2}{vg}\right)$ d) $\theta < \tan^{-1}\left(\frac{v^2}{rg}\right)$

755. A frictionless track $ABCDE$ ends in a circular loop of radius R, figure. A body slides down the track from point A which is at a height $h = 5$ cm. Maximum value of R for the body to successfully complete the loop is

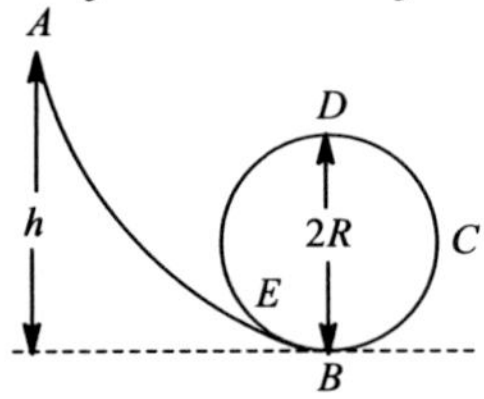

a) 5 cm b) 15/4 cm c) 10/3 cm d) 2 cm

756. A stone of mass 16 kg is attached to a string 144 m long and is whirled in a horizontal circle. The maximum tension the string can withstand is 16 *Newton*. The maximum velocity of revolution that can be given to the stone without breaking it, will be

a) $20\ ms^{-1}$ b) $16\ ms^{-1}$ c) $14\ ms^{-1}$ d) $12\ ms^{-1}$

757. When a projectile is projected at a certain angle with the horizontal, its horizontal range is R and time of flight is T_1. When the same projectile is throwing with the same speed at some other angle with the horizontal, its horizontal range is R and time of flight is T_2. The product of T_1 and T_2 is

a) $\frac{R}{g}$ b) $\frac{2R}{g}$ c) $\frac{3R}{g}$ d) $\frac{4R}{g}$

758. If $\vec{P} = 4\hat{i} - 2\hat{j} + 6\hat{k}$ and $\vec{Q} = \hat{i} - 2\hat{j} - 3\hat{k}$, then the angle which $\vec{P} + \vec{Q}$ makes with x-axis is

a) $\cos^{-1}\left(\frac{3}{\sqrt{50}}\right)$ b) $\cos^{-1}\left(\frac{4}{\sqrt{50}}\right)$ c) $\cos^{-1}\left(\frac{5}{\sqrt{50}}\right)$ d) $\cos^{-1}\left(\frac{12}{\sqrt{50}}\right)$

759. A small block is shot into each of the four tracks as shown below. Each of the frictionless track rises to the same height. The speed, which the block enters the tracks, is same in all cases. At the highest point of the track, normal reaction is maximum in

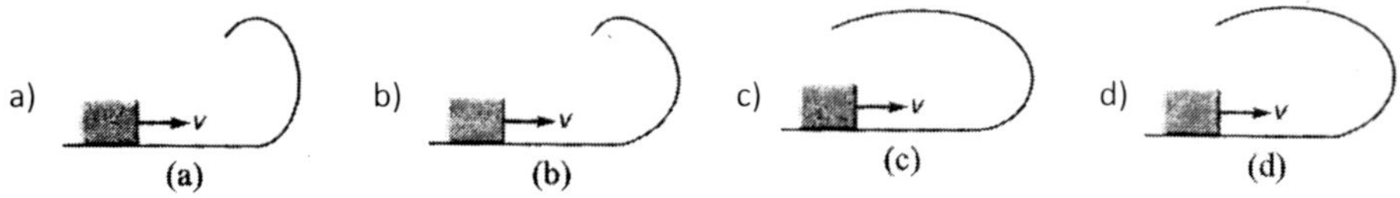

760. A particle moves in a circle with a uniform speed. When it goes from a point A to a diametrically opposite point B, the momentum of the particle changes by $\vec{\mathbf{p}}_A - \vec{\mathbf{p}}_B = 2\ \text{kg ms}^{-1}(\hat{\mathbf{j}})$ and the centripetal fore acting on it changes by $\vec{\mathbf{F}}_A - \vec{\mathbf{F}}_B = 8\ \text{N}(\hat{\mathbf{i}})$ where $\hat{\mathbf{i}}$ and $\hat{\mathbf{j}}$ are unit vectors along X and Y axes respectively. The angular velocity of the particle is

a) Dependent of its mass b) $4\ \text{rad s}^{-1}$ c) $\frac{2}{\pi}\ \text{rad s}^{-1}$ d) $16\ \mu\ \text{rad s}^{-1}$

761. A roller coaster is designed such that riders experience "weightlessness" as they go round the top of a hill whose radius of curvature is 20 m. The speed of the car at the top of the hill is between

a) 16 m/s and 17 m/s b) 13 m/s and 14 m/s c) 14 m/s and 15 m/s d) 15 m/s and 16 m/s

762. If a particle of mass m is moving in a horizontal circle of radius r with a centripetal force $(-k/r^2)$, the total energy is

a) $-\frac{k}{2r}$ b) $-\frac{k}{r}$ c) $-\frac{2k}{r}$ d) $-\frac{4k}{r}$

763. Two vectors $\vec{A}$ and $\vec{B}$ are inclined at an angle θ. Now if the vectors are interchanged then the resultant turns through an angle β. Which of the following relation is true

a) $\tan\frac{\alpha}{2} = \left(\frac{A-B}{A+B}\right)^2 \tan\frac{\theta}{2}$ b) $\tan\frac{\alpha}{2} = \left(\frac{A-B}{A+B}\right) \tan\frac{\theta}{2}$

c) $\tan\frac{\alpha}{2} = \left(\frac{A-B}{A+B}\right)^2 \cot\frac{\theta}{2}$ d) $\tan\frac{\alpha}{2} = \left(\frac{A-B}{A+B}\right) \cot\frac{\theta}{2}$

764. A string of length L is fixed at one end and the string makes $\frac{2}{\pi}$ rev/s around the vertical axis through, the fixed and as shown in the figure, then tension in the string is

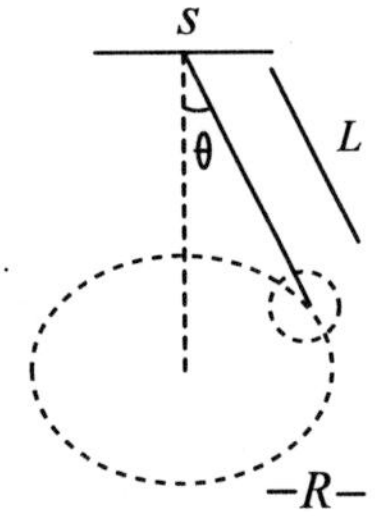

a) ML b) $2\ ML$ c) $4\ ML$ d) $16\ ML$

765. An electric fan has blades of length 30 cm measured from the axis of rotation. If the fan is rotating at 120 rpm, the acceleration of a point on the tip of the blade is

a) 1600 ms^{-2} b) 47.4 ms^{-2} c) 23.7 ms^{-2} d) 50.55 ms^{-2}

4
MOTION IN A PLANE
ANSWER KEY

1)	c	2)	c	3)	c	4)	c
5)	b	6)	d	7)	b	8)	c
9)	a	10)	b	11)	c	12)	b
13)	a	14)	c	15)	b	16)	b
17)	c	18)	c	19)	a	20)	c
21)	c	22)	d	23)	d	24)	d
25)	a	26)	a	27)	d	28)	a
29)	a	30)	b	31)	c	32)	b
33)	a	34)	d	35)	a	36)	a
37)	c	38)	c	39)	c	40)	b
41)	a	42)	a	43)	d	44)	b
45)	b	46)	c	47)	d	48)	c
49)	a	50)	d	51)	a	52)	b
53)	a	54)	a	55)	a	56)	b
57)	a	58)	c	59)	d	60)	c
61)	d	62)	d	63)	c	64)	a
65)	b	66)	a	67)	d	68)	c
69)	d	70)	c	71)	b	72)	a
73)	b	74)	c	75)	c	76)	c
77)	c	78)	c	79)	a	80)	b
81)	d	82)	c	83)	b	84)	b
85)	d	86)	d	87)	b	88)	a
89)	c	90)	c	91)	a	92)	d
93)	b	94)	a	95)	c	96)	d
97)	d	98)	b	99)	b	100)	d
101)	d	102)	a	103)	d	104)	d
105)	c	106)	c	107)	d	108)	b
109)	c	110)	b	111)	c	112)	a
113)	b	114)	a	115)	d	116)	d
117)	d	118)	a	119)	b	120)	d
121)	a	122)	d	123)	b	124)	c
125)	a	126)	d	127)	a	128)	b

129)	b	130)	b	131)	c	132)	a
133)	a	134)	d	135)	c	136)	d
137)	c	138)	b	139)	a	140)	d
141)	a	142)	a	143)	b	144)	d
145)	d	146)	a	147)	a	148)	c
149)	c	150)	d	151)	c	152)	b
153)	b	154)	a	155)	d	156)	c
157)	b	158)	d	159)	a	160)	b
161)	b	162)	a	163)	b	164)	c
165)	d	166)	d	167)	a	168)	a
169)	a	170)	c	171)	b	172)	c
173)	a	174)	b	175)	c	176)	a
177)	b	178)	b	179)	b	180)	b
181)	a	182)	d	183)	c	184)	a
185)	d	186)	b	187)	c	188)	b
189)	a	190)	a	191)	a	192)	b
193)	d	194)	c	195)	a	196)	c
197)	d	198)	c	199)	c	200)	d
201)	c	202)	c	203)	a	204)	b
205)	c	206)	b	207)	b	208)	d
209)	b	210)	c	211)	b	212)	c
213)	a	214)	d	215)	d	216)	b
217)	a	218)	d	219)	a	220)	c
221)	b	222)	d	223)	a	224)	a
225)	d	226)	d	227)	b	228)	b
229)	c	230)	c	231)	a	232)	b
233)	c	234)	d	235)	a	236)	b
237)	d	238)	c	239)	c	240)	c
241)	b	242)	b	243)	c	244)	c
245)	c	246)	d	247)	c	248)	a
249)	c	250)	b	251)	c	252)	d
253)	c	254)	a	255)	a	256)	d
257)	d	258)	a	259)	d	260)	a
261)	a	262)	a	263)	a	264)	a
265)	a	266)	a	267)	c	268)	c
269)	b	270)	d	271)	b	272)	b

273)	b	274)	b	275)	c	276)	d
277)	c	278)	d	279)	a	280)	d
281)	c	282)	b	283)	d	284)	b
285)	a	286)	b	287)	a	288)	c
289)	c	290)	d	291)	d	292)	c
293)	d	294)	b	295)	b	296)	a
297)	d	298)	a	299)	b	300)	d
301)	d	302)	c	303)	c	304)	a
305)	a	306)	a	307)	b	308)	d
309)	d	310)	d	311)	d	312)	c
313)	d	314)	d	315)	c	316)	b
317)	b	318)	b	319)	b	320)	b
321)	d	322)	c	323)	b	324)	c
325)	b	326)	b	327)	c	328)	d
329)	b	330)	d	331)	c	332)	a
333)	b	334)	d	335)	c	336)	c
337)	a	338)	b	339)	a	340)	b
341)	a	342)	a	343)	c	344)	d
345)	c	346)	d	347)	a	348)	b
349)	d	350)	b	351)	b	352)	b
353)	a	354)	d	355)	a	356)	b
357)	c	358)	d	359)	b	360)	b
361)	c	362)	b	363)	c	364)	d
365)	b	366)	a	367)	a	368)	c
369)	a	370)	b	371)	a	372)	c
373)	b	374)	b	375)	c	376)	d
377)	c	378)	a	379)	c	380)	d
381)	b	382)	c	383)	b	384)	d
385)	b	386)	b	387)	a	388)	a
389)	a	390)	a	391)	c	392)	d
393)	c	394)	b	395)	c	396)	c
397)	d	398)	c	399)	b	400)	c
401)	b	402)	a	403)	c	404)	c
405)	d	406)	c	407)	c	408)	c
409)	a	410)	b	411)	a	412)	d
413)	d	414)	c	415)	c	416)	d

417)	a	418)	c	419)	c	420)	a
421)	c	422)	b	423)	d	424)	d
425)	a	426)	a	427)	b	428)	a
429)	c	430)	b	431)	d	432)	d
433)	a	434)	d	435)	c	436)	c
437)	d	438)	b	439)	c	440)	b
441)	b	442)	d	443)	a	444)	c
445)	b	446)	c	447)	d	448)	c
449)	a	450)	c	451)	c	452)	d
453)	a	454)	a	455)	c	456)	c
457)	a	458)	c	459)	c	460)	d
461)	b	462)	a	463)	c	464)	d
465)	d	466)	c	467)	b	468)	a
469)	c	470)	a	471)	b	472)	b
473)	c	474)	a	475)	a	476)	c
477)	c	478)	c	479)	a	480)	d
481)	d	482)	a	483)	c	484)	b
485)	d	486)	c	487)	b	488)	b
489)	c	490)	a	491)	b	492)	d
493)	c	494)	d	495)	d	496)	b
497)	b	498)	d	499)	b	500)	a
501)	b	502)	b	503)	c	504)	b
505)	a	506)	b	507)	c	508)	b
509)	b	510)	b	511)	d	512)	b
513)	a	514)	b	515)	c	516)	b
517)	c	518)	a	519)	c	520)	b
521)	b	522)	a	523)	d	524)	a
525)	d	526)	a	527)	a	528)	b
529)	d	530)	d	531)	b	532)	c
533)	d	534)	d	535)	a	536)	c
537)	a	538)	d	539)	b	540)	c
541)	a	542)	b	543)	c	544)	c
545)	b	546)	d	547)	c	548)	d
549)	d	550)	d	551)	a	552)	b
553)	b	554)	b	555)	b	556)	b
557)	d	558)	a	559)	a	560)	a

561)	a	562)	b	563)	b	564)	a
565)	b	566)	d	567)	b	568)	d
569)	a	570)	c	571)	d	572)	b
573)	a	574)	b	575)	b	576)	b
577)	a	578)	a	579)	c	580)	a
581)	d	582)	c	583)	d	584)	b
585)	a	586)	b	587)	a	588)	c
589)	c	590)	d	591)	b	592)	a
593)	a	594)	a	595)	d	596)	a
597)	a	598)	c	599)	c	600)	d
601)	a	602)	b	603)	d	604)	a
605)	d	606)	b	607)	b	608)	c
609)	d	610)	d	611)	b	612)	d
613)	b	614)	d	615)	a	616)	a
617)	b	618)	d	619)	d	620)	b
621)	c	622)	a	623)	d	624)	c
625)	a	626)	c	627)	d	628)	c
629)	d	630)	a	631)	a	632)	c
633)	b	634)	c	635)	d	636)	d
637)	a	638)	a	639)	c	640)	b
641)	b	642)	a	643)	a	644)	c
645)	b	646)	d	647)	d	648)	a
649)	d	650)	b	651)	c	652)	b
653)	d	654)	a	655)	b	656)	c
657)	b	658)	a	659)	c	660)	d
661)	b	662)	a	663)	c	664)	a
665)	d	666)	b	667)	d	668)	d
669)	b	670)	d	671)	b	672)	b
673)	b	674)	c	675)	b	676)	b
677)	c	678)	d	679)	a	680)	a
681)	c	682)	d	683)	d	684)	d
685)	a	686)	a	687)	a	688)	a
689)	c	690)	c	691)	b	692)	c
693)	d	694)	b	695)	a	696)	c
697)	c	698)	a	699)	c	700)	a
701)	a	702)	b	703)	a	704)	b

705)	b	706)	b	707)	c	708)	d
709)	c	710)	b	711)	a	712)	a
713)	c	714)	c	715)	d	716)	a
717)	b	718)	a	719)	a	720)	c
721)	c	722)	c	723)	a	724)	a
725)	c	726)	a	727)	b	728)	c
729)	c	730)	c	731)	c	732)	c
733)	d	734)	a	735)	b	736)	d
737)	c	738)	d	739)	d	740)	a
741)	a	742)	a	743)	a	744)	a
745)	b	746)	a	747)	d	748)	c
749)	b	750)	c	751)	b	752)	d
753)	d	754)	d	755)	d	756)	d
757)	b	758)	c	759)	a	760)	b
761)	c	762)	a	763)	b	764)	d
765)	b						

4
MOTION IN A PLANE
HINTS AND SOLUTIONS

1 **(c)**

$h = \dfrac{u^2}{2g} \Rightarrow u^2 = 2gh$

Maximum horizontal distance

$R_{\max} = \dfrac{u^2}{g}$

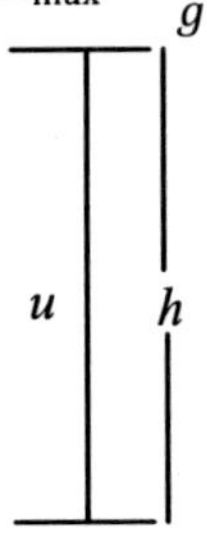

$R_{\max} = 2h$

2 **(c)**

$v = \sqrt{2gl(1 - \cos\theta)}$

$= \sqrt{2 \times 9.8 \times 2(1 - \cos 60°)} = 4.43\ m/s$

3 **(c)**

Using relation $\theta = \omega_0 t + \frac{1}{2}at^2$

$\theta_1 = \frac{1}{2}(\alpha)(2)^2 = 2\alpha$ ...(i)

As $\omega_0 = 0, t = 2\ sec$

Now using same equation for $t = 4\ sec, \omega_0 = 0$

$\theta_1 + \theta_2 = \frac{1}{2}\alpha(4)^2 = 8\alpha$ ...(ii)

From (i) and (ii), $\theta_1 = 2\alpha$ and $\theta_2 = 6\alpha$ $\therefore$ $\frac{\theta_2}{\theta_1} = 3$

4 **(c)**

Centrifugal force on rod,$F = \frac{mv^2}{r}$ along BF. Let θ be the angle which the rod makes with the vertical. Forces acting on the rod are shown in figure

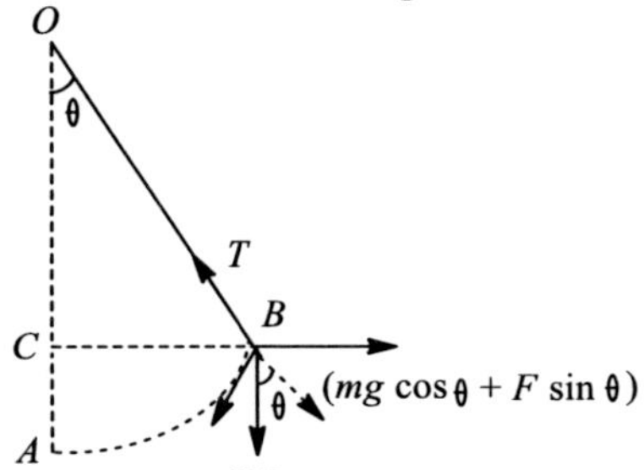

Resolving mg and F into two rectangular components, we have,

Forces parallel to rod,

$mg\cos\theta + \dfrac{mv^2}{r}\sin\theta = T$

Force perpendicular rod

$= mg\sin\theta - \dfrac{mv^2}{r}\cos\theta$

The rod will be balanced if

$mg\sin\theta = \dfrac{mv^2}{r}\cos\theta = 0$

or $mg\sin\theta = \frac{mv^2}{r}\cos\theta$

or $\tan\theta = \frac{v^2}{rg} = \frac{(10)^2}{10\times10} = 1 = \tan 45°$or

$\theta = 45°$

5 **(b)**

Tangential acceleration, $a_t = \dfrac{v_f - v_i}{t}$

$= \left(\dfrac{45-60}{8}\right)\dfrac{22}{15}\text{fts}^{-1}$

$= -\dfrac{11}{4}\text{fts}^{-2}$

Radial acceleration, $a_r = \dfrac{v^2}{r} = \dfrac{\left(60 \times \frac{22}{15}\right)^2}{2500}$

$= 3.1\ \text{fts}^{-2}$

Then, $a = \sqrt{a_r^2 + a_t^2} = 4.14\ \text{fts}^{-2}$

6 **(d)**

We know that if two stones have same horizontal range, then this implies that both are projected at θ and 90° − θ.

Given, $\theta = \dfrac{\pi}{3} = 60°$

$\therefore\ 90° - \theta = 90° - 60° = 30°$

For first stone,

Maximum height $= 102 = \dfrac{u^2\sin^2 60°}{2g}$

For second stone,

Maximum height, $h = \dfrac{u^2\sin^2 30°}{2g}$

$\therefore\ \dfrac{h}{102} = \dfrac{\sin^2 30°}{\sin^2 60°} = \dfrac{(1/2)^2}{(\sqrt{3}/2)^2}$

or $h = 102 \times \dfrac{1/4}{3/4} = 34$ m

7 **(b)**

Acceleration of the particle is

$a = r\omega^2 = r(2\pi n)^2$

$= 0.25 \times (2\pi \times 2)^2$

$= 16\pi^2 \times 0.25$

$= 4\pi^2 \text{ms}^{-2}$

8 **(c)**

$\mathbf{L} = m(\mathbf{r} \times \mathbf{v})$

$\mathbf{L} = m\left[v_0\cos\theta\, t\hat{\mathbf{i}} + \left(v_0\sin\theta\, t - \frac{1}{2}gt^2\right)\hat{\mathbf{j}}\right]$

$\times \left[v_0\cos\theta\, \hat{\mathbf{i}} + (v_0\sin\theta - gt)\hat{\mathbf{j}}\right]$

$= mv_0\cos\theta\, t\left[-\frac{1}{2}gt\right]\hat{\mathbf{k}}$

$= -\frac{1}{2}mgv_0t^2\cos\theta\hat{\mathbf{k}}$

9 **(a)**

When a body moves on a circular path then force and distance are perpendicular to each other. Therefore, work done by the force is

$W = F \cdot d\cos\theta$

$= F \cdot d\cos 90^\circ \;(\because \theta = 90^\circ)$

$= 0 \quad (\because \cos 90^\circ = 0)$

10 **(b)**

Since $\vec{F} = 4\hat{\imath} - 3\hat{\jmath}$ is lying in $X - Y$ plane, hence the vector perpendicular to $\vec{F}$ must be lying perpendicular to$X - Y$ plane *ie*, along Z-axis.

11 **(c)**

If a particle is projected with velocity u at an angle θ with the horizontal, the velocity of the particle at the highest point is

$v = u\cos\theta = 200\cos 60^\circ = 100 \text{ ms}^{-1}$

If m is the mass of the particle, then its initial momentum at highest point in the horizontal direction $= mv = m \times 100$. It means at the highest point, initially the particle has no momentum vertically upwards or downwards. Therefore, after explosion, the final momentum of the particles going upwards and downwards must be zero. Hence, the final momentum after explosion is the momentum of the third particle, in the horizontal direction. If the third particle moves with velocity v', then its momentum $= \frac{mv'}{3}$,

According to law of conservation of linear momentum,

We have $\frac{mv'}{3} = m \times 100$ or $v' =$ 300 ms^{-1}

12 **(b)**

Maximum height, $H = \frac{u^2\sin^2\theta}{2g}$

$\frac{dH}{du} = \frac{2u^2\sin^2\theta}{2g}$ or $dH = \frac{u\ \sin^2\theta}{g}du$

$\therefore \frac{dH}{H} = \frac{2du}{u}$

Since H is increased by 10%

So, $\frac{dH}{H} = \frac{10}{100} = \frac{1}{10} = \frac{2du}{u}$

Now horizontal range, $dR = \frac{2u}{g}\sin 2\theta\, du$

or $\frac{dR}{R} = \frac{2du}{u} = \frac{1}{10}$

$\therefore$ % increase in $R = \frac{dR}{R} \times 100$

$= \frac{1}{10} \times 100 = 10\ \%$

13 **(a)**

$R = mg\cos\theta - \frac{mv^2}{r}$

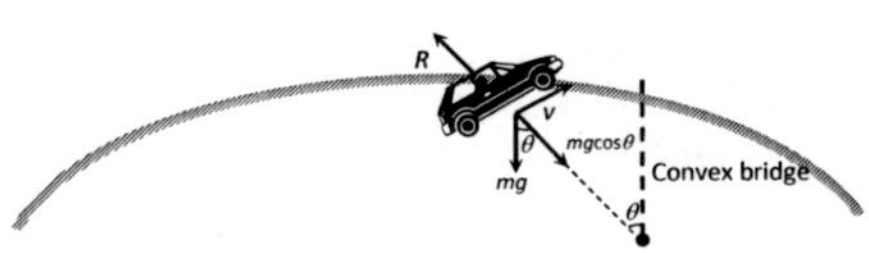

When θ decreases $\cos\theta$ increases $i.e.$, R increases

14 **(c)**

Here, $r = 300$ m, $\mu = 0.3$, $g = 10 \text{ ms}^{-2}$

$v_{\max} = \sqrt{\mu rg} = \sqrt{0.3 \times 300 \times 10}$

$= 30 \text{ ms}^{-1}$

$= 30 \times \frac{18}{5} \text{km h}^{-1} = 108 \text{ km h}^{-1}$

15 **(b)**

Centripetal force $= mr\omega^2 = 5 \times 1 \times (2)^2 = 20\ N$

16 **(b)**

Let $\hat{A} + \hat{B} = \hat{R}$ then using parallelogram law of vectors we have

$1 = (1^2 + 1^2 + 2.1.1\cos\theta)^{1/2}$

or $\quad 1 = 2(1 + \cos\theta)$

or $\quad \frac{1}{2} - 1 = \cos\theta$

or $\quad \cos\theta = -\frac{1}{2} = \cos 120^\circ$

or $\quad \theta = 120^\circ$

$\therefore |\hat{A} - \hat{B}| = |\hat{A} + (-\hat{B})|$.

Now the angle between $\hat{A}$ and $-\hat{B}$ is 60°

The resultant of $|\hat{A} + (-\hat{B})|$

$(1^2 + 1^2 + 2 \times 1 \times 1 \times \cos 60^\circ)^{1/2}$

$= \sqrt{3}$

17 **(c)**

Displacement, $\vec{r} = (a\hat{i} + a\hat{j}) - (a\hat{i}) = a\hat{j}$

$\vec{F} = -K(y\hat{i} + x\hat{j}) = -K(a\hat{i} + a\hat{j})$

Workdone, $W = \vec{F}.\vec{r}$

$= -K(a\hat{i} + a\hat{j}).a\hat{j} = -Ka^2$

18 **(c)**

We know that range of projectile is same for complementary angles *ie*, for θ and $(90° - \theta)$.

$\therefore T_1 = \frac{2u\sin\theta}{g}$

$T_2 = \frac{2u\sin(90° - \theta)}{g} = \frac{2u\cos\theta}{g}$

and $R = \frac{u^2\sin\theta}{g}$

Therefore, $T_1T_2 = \frac{2u\sin\theta}{g} \times \frac{2u\cos\theta}{g}$

$= \frac{2u^2(2\sin\theta\cos\theta)}{g^2}$

$= \frac{2u^2(2\sin\theta)}{g^2} = \frac{2R}{g}$

$\therefore T_1T_2 \propto R$

19 **(a)**

$R = \frac{v^2\sin 2\theta}{g} = 200, T = \frac{2v\sin\theta}{g} = 5$

Dividing, $\frac{v^2 \times 2\sin\theta\cos\theta}{g} \times \frac{g}{2v\sin\theta} = \frac{200}{5} =$ 40

or $v\cos\theta = 40\text{ms}^{-1}$

It may be noted here that the horizontal component of the velocity of projection remains the same during the flight of the projectile

20 **(c)**

$\cos\theta = \vec{A}\cdot\vec{B}/AB$

21 **(c)**

For looping the loop, the velocity at the lowest point of loop should be

$v = \sqrt{5gr} = \sqrt{5gD/2} = \sqrt{2gh}$ or $h = 5\,D/4$

22 **(d)**

Angular momentum is an axial vector. It is directed always in a fix direction (perpendicular to the plane of rotation either outward or inward), if the sense of rotation remain same

23 **(d)**

Since the maximum tension T_B in the string moving in the vertical circle is at the bottom and minimum tension T_T is at the top

$\therefore T_B = \frac{mv_B^2}{L} + mg$ and $T_T = \frac{mv_T^2}{L} - mg$

$\therefore \frac{T_B}{T_T} = \frac{\frac{mv_B^2}{L} + mg}{\frac{mv_T^2}{L} - mg} = \frac{4}{1}$ or $\frac{v_B^2 + gL}{v_T^2 - gL} = \frac{4}{1}$

Or $v_B^2 + gL = 4v_T^2 - 4gL$ but $v_B^2 = v_T^2 + 4gL$

$\therefore v_T^2 + 4gL + gL = 4v_T^2 - 4gL \Rightarrow 3v_T^2 = 9gL$

$\therefore v_T^2 = 3 \times g \times L = 3 \times 10 \times \frac{10}{3}$ or $v_T = 10\,m/\text{sec}$

24 **(d)**

Here, $r = 5$ m, $\mu = 0.5, \omega = ?, \text{g} = 10\text{ms}^{-2}$

$mr\omega^2 = F = \mu R = \mu\, mg$

$\omega = \sqrt{\frac{\mu g}{r}} = \sqrt{\frac{0.5 \times 10}{5}} = 1 \text{ rad s}^{-1}$

25 **(a)**

Tensions in the respective parts are shown in figure

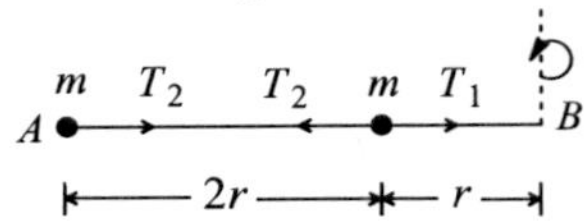

Let ω be angular velocity, then

$T_1 - T_2 = m\omega^2 \times r$ ...(i)

and $T_2 = m\omega^2(r + 2r)$

$T_2 = 3m\omega^2 r$ ...(ii)

From equation (i) and (ii)

$T_1 = 4m\omega^2 r \Rightarrow \frac{T_1}{T_2} = \frac{4}{3}$

26 **(a)**

At the highest point, velocity is horizontal

27 **(d)**

Centripetal acceleration, $a = \omega^2 r$

$= 4\pi^2 n^2 r = 4\pi^2\left(\frac{1}{2}\right)^2 \times 100$

$\pi^2 \times 100\text{ms}^{-2} = 985.9\text{ms}^{-2}$

28 **(a)**

For horizontal planes potential energy remains constant equal to zero, if we assumes surface to be the zero level.

29 **(a)**

Let the angle from the vertical be θ. The diagram showing the different forces is given

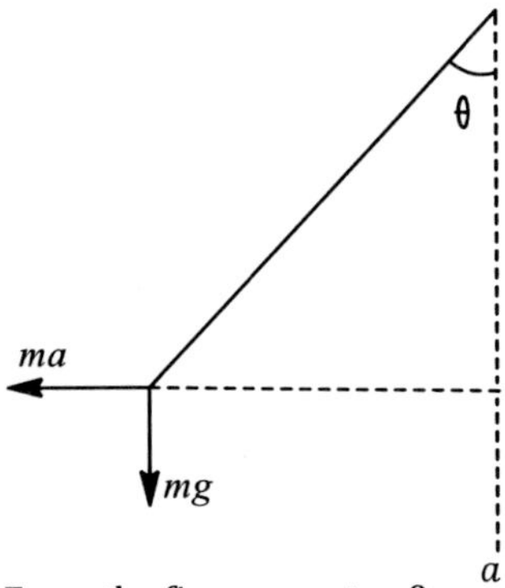

Form the figure, $\tan\theta = \frac{a}{g}$

$\theta = \tan^{-1}\frac{a}{g}$

30 **(b)**
$\frac{v}{g} = 20$ or$v^2 = 20g = 20 \times 9.8 = 196, v$
$= 14\ ms^{-1}$

31 **(c)**
Kinetic energy at highest point=
$K\cos^2 45° = \frac{K}{2}$

32 **(b)**
The body covers a horizontal distance *AB* during its flight. This horizontal range is given by

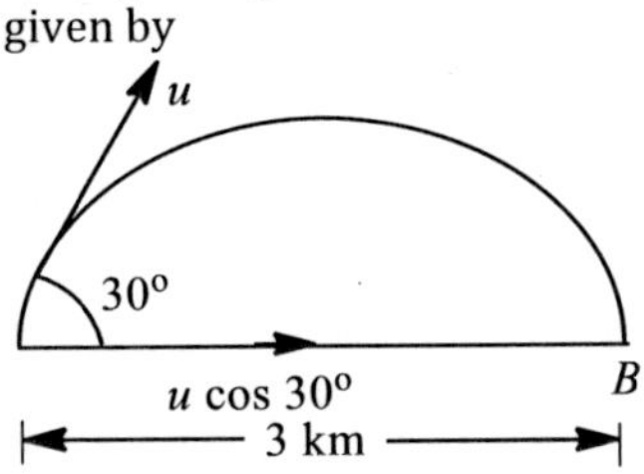

$R = \frac{u^2 \sin 2\theta}{g}$ ....(i)

For maximum horizontal range, $\sin 2\theta = 1$

$\therefore R_{max} = \frac{u^2}{g}$ ...(ii)

Given, $R = 3$km, $\theta = 30°$
∴ From Eq. (i)

$\frac{u^2}{g} = \frac{R}{\sin 2\theta} = \frac{3}{\sin 60°} = \frac{3\times 2}{\sqrt{3}} = \sqrt{3}\times 2$

$\frac{u^2}{g} = 3.464$m or $R_{max} = 3.46$cm

Hence, maximum range with velocity of projection u cannot be more than 3.464 m, Hence, it is not possible to hit a target 5 km away.

33 **(a)**
In uniform circular motion the only force acting on the particle is centripetal (towards centre). Torque of this force about the centre is zero. Hence angular momentum about centre remains conserved.

34 **(d)**
$\theta = 2\pi n = \omega_0 t + \frac{1}{2}\alpha t^2$

$2\pi \times 10 = \frac{1}{2}\alpha 4^2$ or $\alpha = \frac{40\pi}{16}$

Let it make N rotations in the first 8 s

Then, $2\pi N = \frac{1}{2}\alpha\, 8^2$

or $N = \frac{1}{2\pi}\times\frac{64}{2}\times\frac{40\pi}{16} = 40$

∴ The required number of rotations
$= 40 - 10 = 30$

35 **(a)**
$\vec{A}\times\vec{B} = (2\hat{i} + 3\hat{j})\times(\hat{i} + 4\hat{j})$
$= 8(\hat{i}\times\hat{j}) + 3(\hat{j}\times\hat{i}) = 8\hat{k} - 3\hat{k} = 5\hat{k}$

36 **(a)**
$t = \sqrt{\frac{2h}{g}}$

Distance from the foot of the tower

$d = vt = v\sqrt{\frac{2h}{g}} = 250$m

When velocity $= \frac{v}{2}$
and height of tower= $4h$
Then distance $x = \frac{v}{2}\sqrt{\frac{2(4h)}{g}}$

$x = v\sqrt{\frac{2h}{g}} = 250$ m

37 **(c)**
The range of particle is

$R = \frac{u^2 \sin 2\theta}{g}$

$R = \frac{u^2 \sin 2\times 45°}{g}$ $(\therefore \theta = 45°)$

or $R = \frac{u^2 \sin 90°}{g}$

or $R = \frac{u^2}{g}$ ....(i)

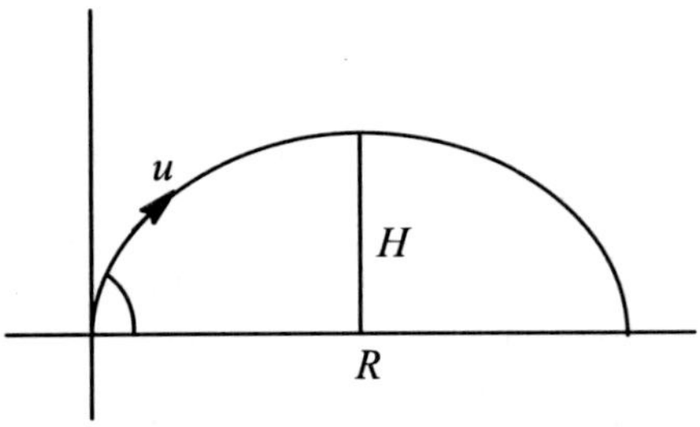

Now, the maximum height of the particle is

$$H = \frac{u^2 \sin^2\theta}{2g}$$

$$= \frac{u^2 \sin^2 45°}{2g} = \frac{u^2\left(\frac{1}{\sqrt{2}}\right)^2}{2g}$$

$$= \frac{u^2}{4g} \;...(ii)$$

Dividing Eqs. (i) by Eq. (ii),

$$\frac{R}{H} = \frac{u^2/g}{u^2/4g}$$

or $R = 4H$

38 **(c)**

Due to constant velocity along horizontal and vertical downward force of gravity stone will hit the ground following parabolic path

39 **(c)**

For a particle in uniform circular motion

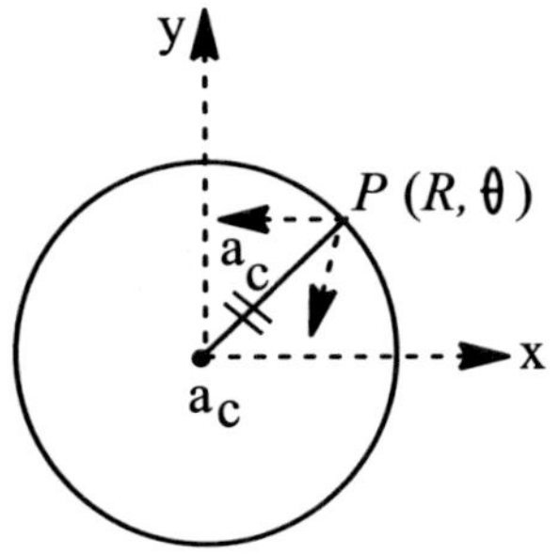

$$\mathbf{a} = \frac{v^2}{R} = \text{towards center of circle}$$

$$\mathbf{a} = \frac{v^2}{R}(-\cos\theta\,\hat{\mathbf{i}} - \sin\theta\,\hat{\mathbf{j}})$$

$$\text{or}\,\mathbf{a} = \frac{v^2}{R}\cos\theta\,\hat{\mathbf{i}} - \frac{v^2}{R}\sin\theta\,\hat{\mathbf{j}}$$

40 **(b)**

$$T = \frac{2u\sin\theta}{g} = \frac{2 \times 9.8 \times \sin 30}{9.8} = 1\,s$$

41 **(a)**

The velocity should be such that the centripetal acceleration is equal to the acceleration due to gravity $ie, v^2/R = g$

or $v = \sqrt{gR}$

42 **(a)**

The object will slip if centripetal force acting on it is more than friction force.

So, $mr\omega^2 > \mu mg$

$r\omega^2 \geq \mu g$

$r\omega^2 = \text{constant}$

$$\frac{r_1}{r_2} = \left(\frac{\omega_2}{\omega_1}\right)^2$$

$$\frac{4}{r_2} = \left(\frac{2\omega}{\omega}\right)^2$$

$r_2 = 1$ cm

43 **(d)**

Let Δt be the time interval. Then,

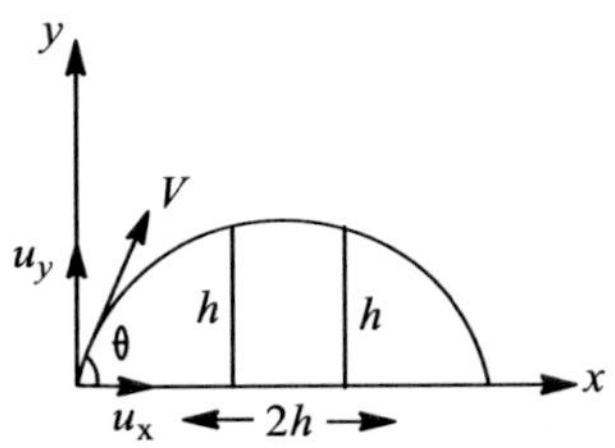

$2h = (u_x)(\Delta t)$

or $u_x = \frac{2h}{\Delta t}$ ...(i)

Further, $h = u_y t - \frac{1}{2}gt^2$

or $gt^2 - 2u_y t + 2h = 0$

$$\therefore\ t_1 = \frac{2u_y + \sqrt{4u_y^2 - 8gh}}{2g}$$

$$\text{and}\ t_2 = \frac{2u_y - \sqrt{4u_y^2 - 8gh}}{2g}$$

$$\Delta t = t_1 - t_2 = \frac{\sqrt{4u_y^2 - 8gh}}{g}$$

or $u_y^2 = \frac{g^2(\Delta t)^2}{4} + 2gh$

Given, $u_x^2 + u_y^2 = \left(2\sqrt{gh}\right)^2$

$$\therefore\ \frac{4h^2}{(\Delta t)^2} + \frac{g^2(\Delta t)^2}{4} + 2gh = 4gh$$

$$\frac{g^2}{4}(\Delta t)^4 - 2gh(\Delta t)^2 + 4h^2 = 0$$

$$(\Delta t)^2 = \frac{2gh \pm \sqrt{4g^2h^2 - 4g^2h^2}}{g^2/2} = \frac{4h}{g}$$

or $\Delta t = 2\sqrt{\frac{h}{g}}$

44 **(b)**
When a body is revolving in circular motion it is acted upon by a centripetal force directed towards the center. Water will not fall if weight is balanced by centripetal force. Therefore

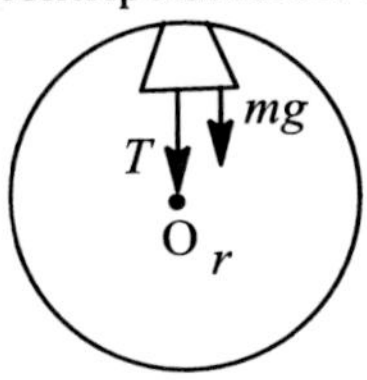

$mg = \frac{mv^2}{r}$

$\Rightarrow v^2 = rg$ ... (i)

Circumference of a circle is $2\pi r$.

Time of revoluation $= \frac{2\pi r}{v}$

Putting the value of v from Eq. (i), we get

$T = \frac{2\pi r}{\sqrt{gr}} = 2\pi\sqrt{\frac{r}{g}}$

Given, $r = 4$ m, $g = 9.8\ \frac{\text{m}^2}{\text{s}}$

$\therefore T = 2\pi\sqrt{\frac{4}{9.8}}$

$\Rightarrow T = \frac{4\pi}{\sqrt{9.8}} = 4\text{s}$

45 **(b)**
$v = r\omega = 0.5 \times 70 = 35 m/s$

46 **(c)**
Total time of flight $= \frac{2u\sin\theta}{g} = \frac{2\times50\times1}{2\times10} =$ $5s$

Time to cross the wall $= 3\ sec$ (Given)

Time in air after crossing the wall $=$ $(5-3) = 2\ sec$

$\therefore$ Distance travelled beyond the wall $=$ $(u\cos\theta)t$

$= 50 \times \frac{\sqrt{3}}{2} \times 2 = 86.6\ m$

47 **(d)**
$v_x = \frac{dx}{dt} = 2ct$ and $v_y = \frac{dy}{dt} = 2bt$

$\therefore v = \sqrt{v_x^2 + v_y^2} = 2t(c^2+b^2)^{1/2}$

48 **(c)**
When a force of constant magnitude acts on velocity of particle perpendicularly, then there is no change in the kinetic energy of particle.
Hence, kinetic energy remains constant.

49 **(a)**
The maximum velocity for a banked road with friction,

$v^2 = gr\left(\frac{\mu + \tan\theta}{1 - \mu\tan\theta}\right)$

$\Rightarrow v^2 = 9.8 \times 1000 \times \left(\frac{0.5+1}{1-0.5\times1}\right) \Rightarrow v$

$= 172 m/s$

50 **(d)**
Here $W = T(\cos\theta + \sin\theta) < T$

so $P + Q = T(\cos\theta + \sin\theta) < T$

Where as (a), (b) and (c) are correct and (d) is wrong.

51 **(a)**
The angle of banking, $\tan\theta = \frac{v^2}{rg}$

$\Rightarrow \tan 12^\circ = \frac{(150)^2}{r \times 10} \Rightarrow r = 10.6 \times 10^3 m$

$= 10.6\ km$

52 **(b)**
Note that the angle between two forces is 120° and not 60°.

$R^2 = F^2 + F^2 + 2F^2\cos120^\circ$

or $R^2 = 2F^2 + 2F^2\left(-\frac{1}{2}\right) = F^2$

or $R = F$

53 **(a)**
We know that $\tan\theta = \frac{v^2}{Rg}$ and $\tan\theta = \frac{h}{b}$

Hence $\frac{h}{b} = \frac{v^2}{Rg} \Rightarrow h = \frac{v^2 b}{Rg}$

54 **(a)**
There is no change in the angular velocity, when speed is constant

55 **(a)**
$\frac{v^2}{r} = a$, the centripetal acceleration [Given]

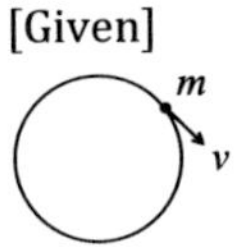

If v is doubled, $a'' = \frac{4v^2}{r} = 4a$

56 **(b)**
Difference in KE $= \frac{1}{2}m\left[\left(\sqrt{5gr}\right)^2 - \sqrt{gr}\right]^2$

$= 2mgr = 2 \times 1 \times 10 \times 1 = 20$ J

57 **(a)**
At the highest point of trajectory, the velocity becomes horizontal. So, it is perpendicular to acceleration (which is directed vertically downwards)

58 **(c)**
Difference in K.E. = Difference in P.E. = $2mgr$

59 **(d)**
$p = mv\cos\theta$
$= 1 \times 10 \times \cos 60° = 10\left(\frac{1}{2}\right) \text{kg ms}^{-1}$
$= 5 \text{ kg ms}^{-1}$

60 **(c)**
$T = \frac{2 \times 50 \times \frac{1}{2}}{10} = 5 \text{ s}$
Horizontal distance travelled in last 2 s
$= 50 \times \cos 30° \times 2 \text{ m}$
$= 100 \times \frac{2}{\sqrt{3}} \text{m} = 50\sqrt{3}\text{m} = 86.6 \text{ m}$

61 **(d)**
$V^2 = U^2 - 2g(L - L\cos\theta)$
$\frac{5gL}{4} = 5gL - 2gL(1 - \cos\theta)$

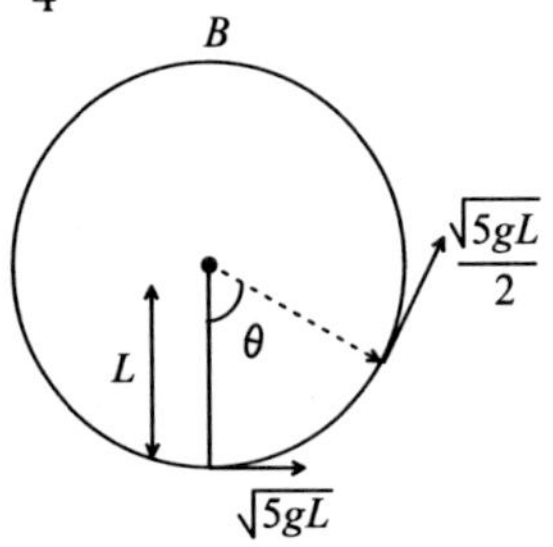

$5 = 20 - 8 + 8\cos\theta$
$\cos\theta = -\frac{7}{8}$
$\frac{3\pi}{4} < \theta < \pi$

62 **(d)**

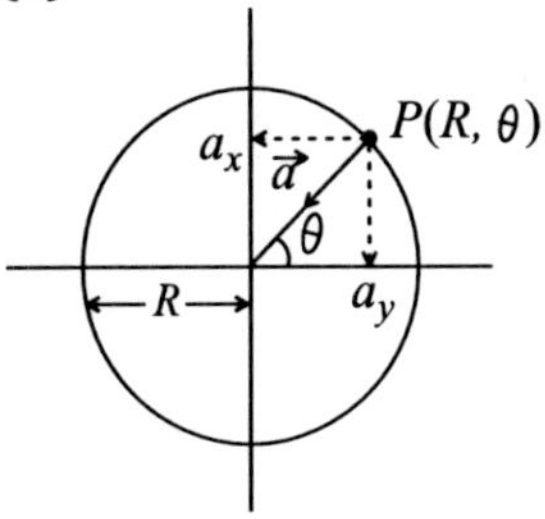

$\vec{a} = -\frac{v^2}{R}\cos\theta\hat{i} - \frac{v^2}{R}\sin\theta\hat{j}$

63 **(c)**
$\frac{v^2}{g} = 100 \text{ or} v^2 = 100 \text{ g}$
$h_{\text{max.}} = \frac{v^2}{2g} = \frac{100g}{2g} = 50\text{m}$

64 **(a)**
By doing so component of weight of vehicle provides centripetal force

65 **(b)**
$ma\cos\theta = mg\cos(90 - \theta)$
$\Rightarrow \frac{a}{g} = \tan\theta \Rightarrow \frac{a}{g} = \frac{dy}{dx}$
$\Rightarrow \frac{d}{dx}(kx)^2 = \frac{a}{g} \Rightarrow x = \frac{a}{2gk}$

66 **(a)**
$|\overrightarrow{\Delta v}| = 2v\sin(\theta/2) = 2v\sin\left(\frac{90}{2}\right)$
$= 2v\sin 45 = v\sqrt{2}$

67 **(d)**
$\frac{d^2y}{dt^2} = \alpha$ and $\frac{d^2x}{dt^2} = 0$
$\frac{dy}{dt} = 2\beta x.\frac{dx}{dt}$
$\frac{d^2y}{dt^2} = 2\beta\left[x.\frac{d^2x}{dt^2} + \left(\frac{dx}{dt}\right)^2\right]$
$\because \alpha = 2\beta v_x^2$
$\therefore v_x = \sqrt{\frac{\alpha}{2\beta}}$

68 **(c)**
In uniform circular motion only centripetal acceleration works

69 **(d)**
Standard equation of projectile motion
$y = x\tan\theta - \frac{gx^2}{2u^2\cos^2\theta}$
Comparing with given equation
$A = \tan\theta$ and $B = \frac{g}{2u^2\cos^2\theta}$
So $\frac{A}{B} = \frac{\tan\theta \times 2u^2\cos^2\theta}{g} = 40$
[As $\theta = 45°, u = 20 \ m/s, g = 10m/s^2$]

70 **(c)**
$\omega = \frac{d\theta}{dt} = \frac{d}{dt}(2t^3 + 0.5) = 6t^2$
At $t = 2 \ s, \omega = 6 \times (2)^2 = 24 \ rad/s$

71 **(b)**
If the both mass are revolving about the axis yy' and tension in both the threads are equal then

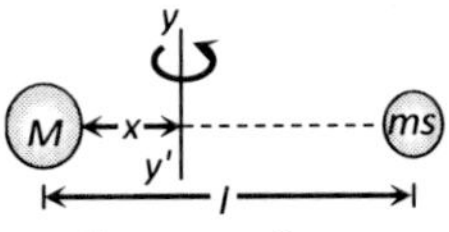

$M\omega^2 x = m\omega^2(l - x)$
$\Rightarrow Mx = m(l - x)$
$\Rightarrow x = \frac{ml}{M + m}$

72 **(a)**
Horizontal component of velocity $v_x = 500\ m/s$and vertical components of velocity while striking the ground

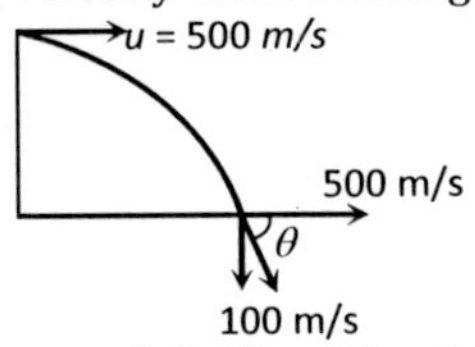

$v_y = 0 + 10 \times 10 = 100\ m/s$
∴ Angle with which it strikes the ground
$\theta = \tan^{-1}\left(\frac{v_y}{v_x}\right) = \tan^{-1}\left(\frac{100}{500}\right)$
$= \tan^{-1}\left(\frac{1}{5}\right)$

73 **(b)**
$\omega^2 R = 4\pi^2 n^2 r = 4\pi^2\left(\frac{1200}{60}\right)^2 \times 30$
$\times 10^{-2} = 4732\ m/s^2$

75 **(c)**
$A_x = 50, \theta = 60°$
Then $\tan\theta = A_y/A_x$ or $A_y = A_x \tan\theta$
Or $A_y = 50 \tan 60° = 50 \times \sqrt{3} = 87$ N

76 **(c)**
Here, $v_{max} = ?, r = 18$ m, g $= 10$ ms^{-2}
$\mu = 0.2$
$\frac{mv_{max}^2}{r} = F = \mu R = \mu mg$
$v_{max} = \sqrt{\mu rg} = \sqrt{0.2 \times 18 \times 10} = 6$ ms^{-1}
$= 6 \times \frac{18}{5}$ km h^{-1} $= 21.6$ km h^{-1}

77 **(c)**
Equation of projectile
$y = x - \left(\frac{5}{9}\right)x^2$
Standard equation
$y = x\tan\theta - \frac{g}{2u^2\cos^2\theta} \cdot x^2$
On comparing, we get
$\tan\theta = 10$
and $\frac{g}{2u^2\cos^2\theta} = \frac{5}{9}$
or $10u^2\cos^2\theta = 9g$
$g = 10$ ms^{-2}
∴ $u^2\cos^2\theta = 9$
range of projecticle $R = \frac{2u^2 \tan\theta \cdot \cos\theta}{g}$
$= \frac{2u^2 \tan\theta \cdot \cos\theta}{g}$
$(\because \sin\theta = \tan\theta \cdot \cos\theta)$
$\frac{2(u^2\cos^2\theta) \cdot \tan\theta}{g}$
$= \frac{2 \times 9 \times 10}{10} = 18$ m

79 **(a)**
Maximum height, $H = \frac{u^2\sin^2\theta}{2g}$
Range, $R = \frac{u^2\sin 2\theta}{g}$
Given, $H = \frac{R}{2}$
∴ $\frac{u^2\sin^2\theta}{2g} = \frac{u^2 2\sin\theta\cos\theta}{2g}$
or $\sin\theta = 2\cos\theta$
or $\tan\theta = 2$
or $\theta = \tan^{-1}(2)$

81 **(d)**
Kinetic energy $= \frac{1}{2}mv^2 = K = as^2$
or $mv^2 = 2as^2$
Centripetal force $= \frac{mv^2}{R} = \frac{2as^2}{R}$

82 **(c)**
Let v be the velocity of projection and $\theta$the angle of projection
Kinetic energy at highest point
$= \frac{1}{2}mv^2\cos^2\theta$ or $E_k\cos^2\theta$
Potential energy at highest point
$= E_k - E_k\cos^2\theta = E_k(1 - \cos^2\theta)$
$= E_k\sin^2\theta$

83 **(b)**
In going from C to A, potential energy lost = potential energy gained in going from A to B
For looping the loop, minimum velocity required at B is $\sqrt{gR}$. This must be the velocity of push down initially from C

84 **(b)**
The amplitude is the radius of the circle
$R = \frac{0.8}{2} = 0.4$ m
The frequency of the shadow is the same as that of the circular motion, so
$\omega = 30$ rev/min
$= 0.5$rev/s $= \pi$ rads^{-1}
and $v = \frac{\omega}{2\pi} = \frac{\pi}{2\pi} = 0.5$ Hz.

85 **(d)**
Let v the velocity of projectile at this instants. Horizontal component of velocity remains unchanged. Therefore,
$v\cos 30° = 10\cos 60°$

or $v\frac{\sqrt{3}}{2}=\frac{10}{2}$ or $v=\frac{10}{\sqrt{3}}$ ms^{-1}

86 **(d)**

At $A, v_A=\sqrt{gl}$

At $B, v_B=\sqrt{5gl}$

and at $D, v_D=\sqrt{3gl}$

Thus, $v_B > v_D > v_A$

Also, $T = 3\,mg(1+\cos\theta)$

So, $D, \theta = 90°$

$\therefore\ T = 3\,mg(1+\theta) = 3\,mg$

87 **(b)**

$$a_{resultant}=\sqrt{a_{radial}^2+a_{tangential}^2}$$

$$=\sqrt{\frac{v^4}{r^2}+a^2}$$

88 **(a)**

Given, $v = 400\ ms^{-1}, r = 160$ m, a =?

Centripetal force, $F=\frac{mv^2}{r}$

$ma=\frac{mv^2}{r}$

or $a=\frac{v^2}{r}$

So, $a=\frac{(400)^2}{160}=\frac{16\times10^4}{160}$

$=10^3 ms^{-2} = 1\ kms^{-2}$

89 **(c)**

$F=\frac{mv^2}{r}=\frac{500\times(10)^2}{50}=1000N$

90 **(c)**

$\mu=\frac{v^2}{rg}=\frac{(4.9)^2}{4\times9.8}=0.61$

91 **(a)**

$AB\cos\theta = AB$ or $\cos\theta = 1$ or $\theta = 0°$

92 **(d)**

Maximum tension in the string is

$T_{max}=mg+\frac{mv_B^2}{l}$

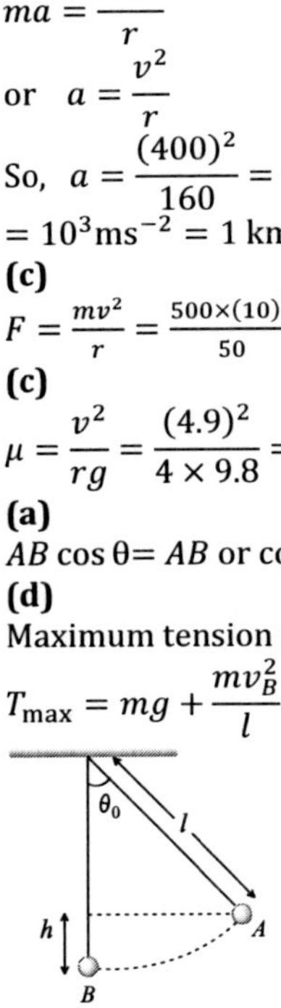

$=mg+\frac{2mgl}{l}(1-\cos\theta_0)$

$=mg+\frac{2mgl}{l}.2\sin^2\frac{\theta_0}{2}$

$\because\left(1-\cos\theta_0=2\sin^2\frac{\theta_0}{2}\right)$

[Since θ_0 is small]

$=mg(1+\theta_0^2)$

93 **(b)**

Let v be the velocity acquired by the body at B which will be moving making an angle 45° with the horizontal direction.

As the body just crosses the well so $\frac{v^2}{g}=$ 40

or $v^2 = 40g = 40\times10 = 400$

or $v = 20\ ms^{-1}$

Taking motion of the body from A to B along the inclined plane we have

$u=v_0, a=-g\sin45°=-\frac{10}{\sqrt{2}}ms^{-2}$

$s = 20m, v = 20ms^{-1}$

As $v^2 = u^2 + 2as$

$\therefore\ 400=v_0^2+2\left(-\frac{10}{\sqrt{2}}\right)\times20\sqrt{2}$

or $v_0^2 = 400 + 400 = 800$ or $v = 20\sqrt{2}ms^{-1}$

94 **(a)**

Work done by centripetal force in uniform circular motion is always equal to zero

95 **(c)**

For water not to spill out of the bucket,

$v_{min}=\sqrt{5gR}$ (at the lowest point)

$=\sqrt{5\times10\times0.5=5\ ms^{-1}}$

96 **(d)**

$R = 4H\cot\theta$, if $\theta = 45°$ then $R = 4H \Rightarrow \frac{R}{H}=\frac{4}{1}$

97 **(d)**

Let ω is the angular speed of revolution

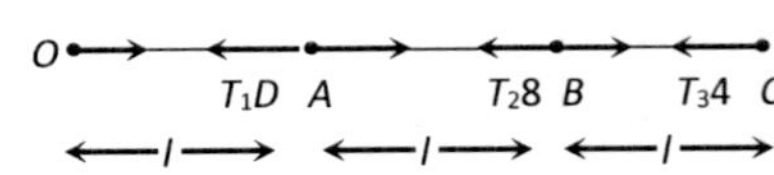

$T_3 = m\omega^2 3l$

$T_2 - T_3 = m\omega^2 2l \Rightarrow T_2 = m\omega^2 5l$

$T_1 - T_2 = m\omega^2 l \Rightarrow T_1 = m\omega^2 6l$

$T_3 : T_2 : T_1 = 3 : 5 : 6$

98 **(b)**

Horizontal range

$R=\frac{u^2\sin2\theta}{g}$ ...(i)

Maximum height

$H=\frac{u^2\sin^2\theta}{2g}$ ...(ii)

Here (i)=(ii)

$\frac{u^2\sin2\theta}{g}=\frac{u^2\sin^2\theta}{2g}$

$$2\cos\theta = \frac{\sin\theta}{2}$$
$$\theta = \tan^{-1}(4)$$

99 **(b)**

Maximum horizontal range = 80 m

$\because$ $\theta = 45°$ m

$\therefore$ $\frac{u^2}{g} = 80$ m

Maximum height, $h = \frac{u^2\sin^2\theta}{2g}$

$= \frac{80}{2}(\sin^2 45°) = 20$ m

100 **(d)**

Velocity of the bob at the point A

$v = \sqrt{5gL}$ .....(i)

$\left(\frac{v}{2}\right)^2 = v^2 - 2gh$ .... (ii)

$h = L(1 - \cos\theta)$ ... (iii)

Solving Eqs. (i), (ii)and (iii), we get

$\cos\theta = -\frac{7}{8}$

or $\theta = \cos^{-1}\left(-\frac{7}{8}\right) = 151°$

101 **(d)**

Centripetal force $F = -\frac{k}{r^2}$

$\frac{mv^2}{r} = \frac{k}{r^2} \Rightarrow mv^2 = \frac{k}{r}$

Kinetic energy $= \frac{1}{2}mv^2 = \frac{k}{2r}$

Since the centripetal force is a conservative force, and for a conservative force,

$F = \frac{dU}{dr} \Rightarrow U = -\int F \cdot dr$

$U = \int \frac{k}{r^2}dr = -\frac{k}{r}$

Toatal energy $= K + U = \frac{k}{2r} - \frac{k}{r} = -\frac{k}{2r}$

102 **(a)**

Given, $\omega_1 = 2\pi \times 400$ rad s^{-1}

$\omega_2 = 2\pi \times 200$ rad s^{-1}

$\therefore \alpha = \frac{2\pi(400-200)}{2} = 200\pi$ rad s^{-2}

103 **(d)**

The three vectors $\vec{A}, \vec{B}$ and $\vec{C}$ are represented as shown in figure (a) where $A = 1, B = 2$and $C = 3$. Here the sides of the equilateral triangle represent only the directions and not the magnitudes of the vectors.

In figure (b), these vector are drawn from a common point $O$and they are lying in $X - Y$ plane. Resolving these vectors into two rectangular components along X-axis and Y-axis, we have, the X-component of resultant vector as

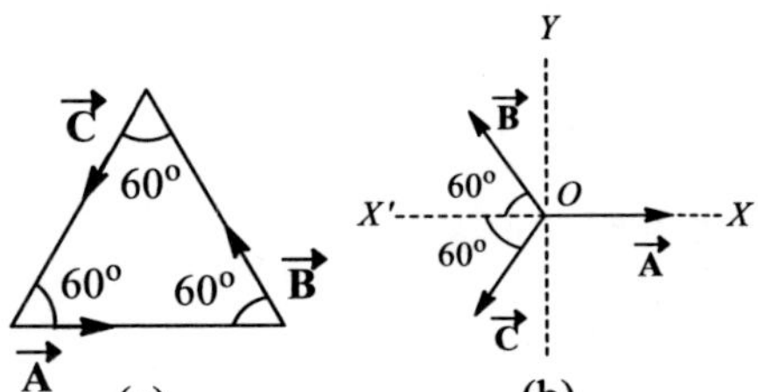

(a) (b)

$R_X = |\vec{A}| + |\vec{B}|\cos(180° - 60°) +$
$|\vec{C}|\cos(180° + 60°)$

$= -1 - 2\cos 60° - 3\cos 60°$

$= -1 - 2 \times \frac{1}{2} - 3 \times \frac{1}{2} = -\frac{3}{2}$

Y-component of resultant vector is

$R_Y = 0 + |B|\sin(180° - 60°)$
$+ |\vec{C}|\sin(180° + 60°)$

$= 0 + 2\sin 60° - 3\sin 60° = -\sin 60°$
$= \sqrt{3}/2$

Magnitude of resultant vector,

$R = \sqrt{R_X^2 + R_Y^2} = \sqrt{\left(-\frac{3}{2}\right)^2 + \left(-\frac{\sqrt{3}}{2}\right)^2}$

$= \sqrt{3}$ units

104 **(d)**

Let u be initial velocity of projection at angle θ with the horizontal. Then, horizontal range,

$R = \frac{u^2\sin 2\theta}{g}$

and maximum height $H = \frac{u^2\sin 2\theta}{2g}$

Given, $R = 4\sqrt{3}H$

$\therefore$ $\frac{u^2\sin 2\theta}{g} = 4\sqrt{3} \cdot \frac{u^2\sin^2\theta}{2g}$

$\therefore$ $2\sin\theta\cos\theta = 2\sqrt{3}\sin^2\theta$

or $\frac{\cos\theta}{\sin\theta} = \sqrt{3}$

or $\cot\theta = \sqrt{3} = \cot 30°$

105 **(c)**

$\because W = FS\cos\theta$ $\therefore \theta = 90°$

106 **(c)**

In uniform circular motion, acceleration causes due to change in direction and is directed radially towards centre

107 **(d)**

The maximum speed without skidding is

$v = \sqrt{\mu rg}$

$\therefore \frac{v_2}{v_1} = \sqrt{\frac{\mu_2}{\mu_1}} = \sqrt{\frac{\mu/2}{\mu}} = \frac{1}{\sqrt{2}}$

$v_2 = \frac{v_1}{\sqrt{2}} = 5\sqrt{2}\text{ms}^{-1}$

108 **(b)**

Time $T = \frac{2\pi r}{v}$

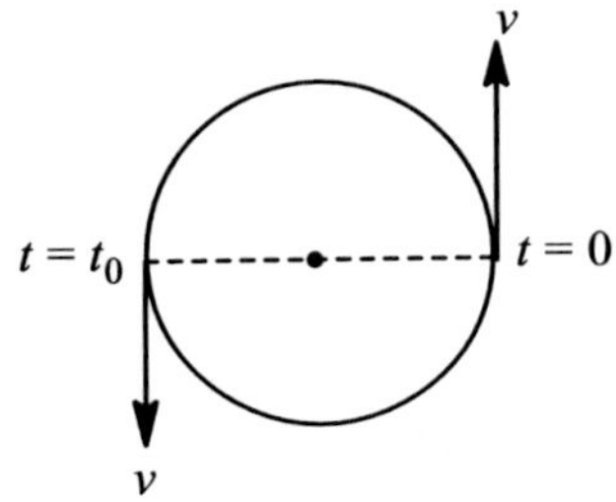

and $t_0 = \frac{T}{2} = \frac{\pi r}{\text{v}}$

$\therefore \text{v}_{av} = \frac{2r}{\pi r/\text{v}} = \frac{2\text{v}}{\pi}$

109 **(c)**

$V \cos\beta = v\cos\theta$

or $V = v\cos\theta\sec\beta$

110 **(b)**

$\frac{\omega_1}{\omega_2} = \frac{T_2}{T_1} = \frac{12\text{h}}{1\text{h}} = 12 : 1$

111 **(c)**

As, $\vec{A}\cdot\vec{B} = 0$ so $\vec{A}$ is perpendicular to $\vec{B}$. Also $\vec{A}.\vec{C} = 0$ means $\vec{A}$ is perpendicular to $\vec{C}$. Since $\vec{B}\times\vec{C}$ is perpendicular to $\vec{B}$ and $\vec{C}$, so $\vec{A}$ parallel to $\vec{B}\times\vec{C}$.

112 **(a)**

To avoid slipping friction force

$F = \frac{mv^2}{r}$

$F = \frac{2000\times10\times10}{20} = 10^4\text{N}$

113 **(b)**

$a^2 + b^2 + 2ab\cos\theta$

$= -a^2 + b^2 - 2ab\cos\theta$

or $4ab\cos\theta = 0$

But $4ab \neq 0 \therefore \cos\theta = 0$ or $\theta = 90°$

Again

$|\vec{a}+\vec{b}|$ and $|\vec{a}-\vec{b}|$ are the diagonals of parallelogram whose adjacent sides are $\vec{a}$ and $\vec{b}$.

Since $|\vec{a}+\vec{b}| = |\vec{a}-\vec{b}|$, therefore, the two diagonals of a parallelogram are equal. So, think of square. This leads to $\theta = 90°$.

114 **(a)**

$R_{\max} = \frac{u^2}{g} = 4H$ [For $\theta = 45°$]

$4H = 400 \Rightarrow H = 100\, m$

115 **(d)**

$v = r\omega = \frac{r\times2\pi}{T} = \frac{0.06\times2\pi}{60}$

$= 6.28 mm/s$

Magnitude of change in velocity $= |\vec{v_2} - \vec{v_1}|$

$= \sqrt{v_1^2 + v_2^2} = 8.88\, mm/s$ [As $v_1 = v_2$ $= 6.28 mm/s$]

116 **(d)**

For critical condition at the highest point

$\omega = \sqrt{g/R}$

$\Rightarrow T = \frac{2\pi}{\omega} = 2\pi\sqrt{R/g} = 2\times3.14\sqrt{4/9.8}$

$= 4$ sec

117 **(d)**

As is clear from figure

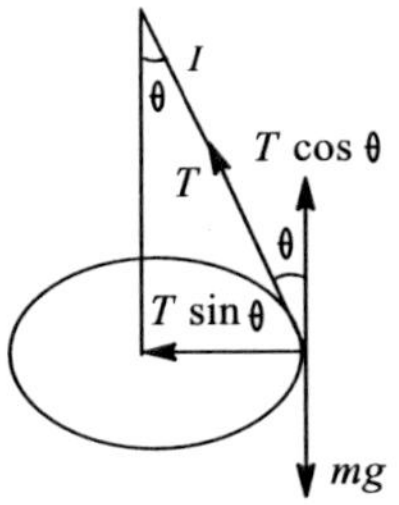

$T\sin\theta = \frac{mv^2}{r}, T\cos\theta = mg$

Dividing, we get

$\tan\theta = \frac{v^2}{rg} = \frac{r}{g}\left(\frac{2\pi}{T}\right)^2$

$\frac{2\pi}{T} = \sqrt{\frac{g\tan\theta}{r}} = \sqrt{\frac{g\tan\theta}{l\sin\theta}} = \sqrt{\frac{g}{l\cos\theta}}$

or $T = 2\pi\sqrt{\frac{l\cos\theta}{g}}$

118 **(a)**

$\omega = \frac{2\pi}{T} = \frac{2\pi}{60} = \frac{\pi}{30}\, rad/s$

119 **(b)**

$v = \sqrt{5gR}$

When $R' = \frac{R}{4}$

$v' = \sqrt{5gR'} = \sqrt{5gR/4} = \frac{1}{2}\sqrt{5gR} = \frac{1}{2}v$

120 **(d)**

Angular momentum of the projectile

$L = mv_h r_\perp$

$= m(v\cos\theta)h$

(where h is the maximum height)

$\Rightarrow = m(v\cos\theta)\left(\frac{v^2\sin^2\theta}{2g}\right)$

$L = \frac{mv^3\sin^2\theta\cos\theta}{2g} = \frac{\sqrt{3}mv^3}{16g}$

121 **(a)**

In 6 turns each of 60°, the cyclist traversed a regular hexagon path having each side 100 m. So, at 7th turn, he will be again at

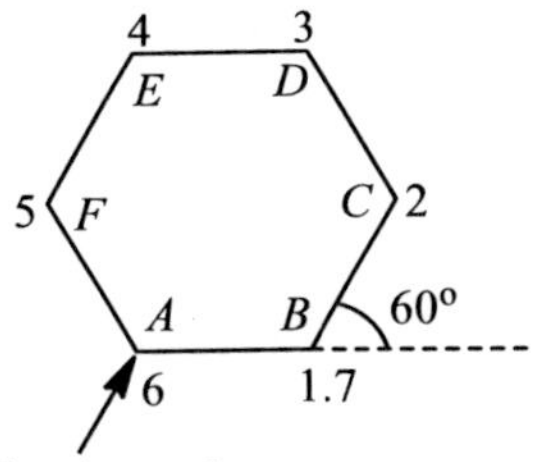

Starting point

Point B (as shown) which is at distance 100 m from starting point A. Hence, net displacement of cyclist is 100 m.

122 **(d)**

In a circular motion

$a = \frac{v^2}{r} \Rightarrow \frac{a_2}{a_1} = \left(\frac{v_2}{v_1}\right)^2 = \left(\frac{2v_1}{v_1}\right)^2 = 4$

123 **(b)**

Maximum tension in the thread is given by

$T_{max} = mg + \frac{mv^2}{r}$

or $T_{max} = mg + mrw^2 (\because v = r\omega)$

or $\omega^2 = \frac{T_{max} - mg}{mr}$

Given, $T_{max} = 37$ N, m = 500g

$= 0.5$ kg, $g = $ mg^{-2},

$r = 4$m

$\therefore \quad \omega^2 = \frac{37 - 0.5 \times 10}{0.5 \times 4} = \frac{37 - 5}{2}$

or $\omega^2 = 16$

or $\omega = 4$ rad s^{-1}

124 **(c)**

Since horizontal component of velocity is constant, hence momentum is constant

125 **(a)**

$\vec{A} = A\hat{A}$ or $\hat{A} = \frac{\vec{A}}{A}$

$\therefore$ required unit vector is $\frac{\hat{i}+\hat{j}}{|\hat{i}+\hat{j}|} = \frac{\hat{i}+\hat{j}}{\sqrt{2}}$

127 **(a)**

Max. tension that string can bear =

$3.7\ kgwt = 37\ N$

Tension at lowest point of vertical loop =

$mg + m\omega^2 r$

$= 0.5 \times 10 + 0.5 \times \omega^2 \times 4 = 5 + 2\omega^2$

$\therefore 37 = 5 + 2\omega^2 \Rightarrow \omega = 4$ rad/s

128 **(b)**

$v_{max} = \sqrt{\mu r g} = \sqrt{0.5 \times 40 \times 9.8}$

$= 14\ m/s$

129 **(b)**

For a particle moving in a circle with constant angular speed, velocity vector is always tangent to the circle and the acceleration vector always points towards the center of circle or is always along radius of the circle. Since, tangential vector is perpendicular to the acceleration vector. But in no case acceleration vector is tangent to the circle.

130 **(b)**

$\frac{mv^2}{r} \propto \frac{K}{r} \Rightarrow v \propto r^o$

i.e. speed of the particle is independent of r

131 **(c)**

Horizontal component of velocity of A is 10 cos 60° or 5 ms^{-1} which is equal to the velocity of B in horizontal direction. They will collide at C if time of flight of the particles are equal or $t_A = t_B$

$\frac{2u\sin\theta}{g} = \sqrt{\frac{2h}{g}}\left(\because h = \frac{1}{2}gt_B^2\right)$

or $h = \frac{2u^2\sin^2\theta}{g}$

$= \frac{2(10)^2\left(\frac{\sqrt{3}}{2}\right)^2}{10} = 15$ m

133 **(a)**

Horizontal velocity of aeroplane,

$u = \frac{216 \times 1000}{60 \times 100} = 60$ ms^{-1}

Time of flight, $T = \sqrt{\frac{2s}{g}} = \sqrt{\frac{2\times1960}{9.8}} = 20s$

Horizontal range, $AB = uT$

$= 60 \times 20 = 1200m$

135 **(c)**

As seen from the cart, the projectile moves vertically upward and comes back

The time taken by cart to cover 80 m

$\frac{s}{v} = \frac{80}{30} = \frac{8}{3}\ s$

For a projectile going upward , $a = -g = -10\ m/s^2, v = 0$

And $t = \frac{8/3}{2} = \frac{4}{3}\ s$

$\therefore v = u + a\,t \Rightarrow 0 = u - 10 \times \frac{4}{3} \Rightarrow u$

$= \frac{40}{3}\ ms^{-1}$

136 **(d)**

Let $\vec{A}, \vec{B}$ and $\vec{C}$ be as shown in figure. Let θ be the angle of incidence, which is also equal to the angle of reflection. Resolving these vectors in rectangular components, we have

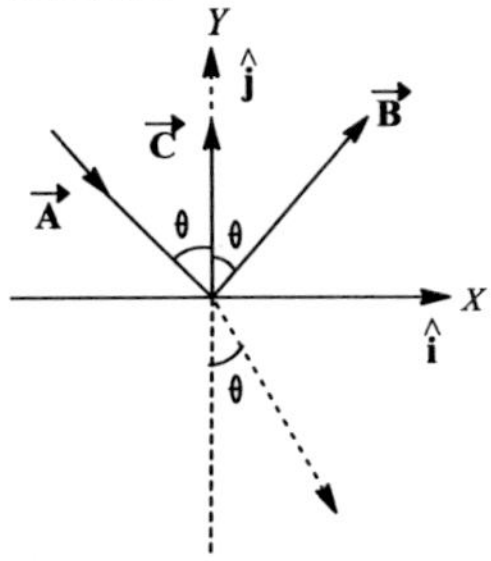

$\vec{A} = \sin\theta\hat{i} - \cos\theta\hat{j}$

$\vec{B} = \sin\theta\hat{i} + \cos\theta\hat{j}$

$\vec{B} - \vec{A} = 2\cos\theta\,\hat{j}$

or $\vec{B} = \vec{A} + 2\cos\theta\,\hat{j}$

Now $\vec{A}.\vec{C} = 2\cos\theta\hat{j}$ or $\vec{B} = \vec{A}\cos\theta\,\hat{j}$

$\therefore\ \vec{B} = \vec{A} - 2(\vec{A}.\vec{C})\hat{j}$ or $\vec{B} = \vec{A} - 2(\vec{A}.\vec{C})\vec{C}$

(as $\hat{j} = \vec{C}$)

137 **(c)**

Velocity at the lowest point, $v = \sqrt{2gl}$

At the lowest point, the tension in the string

$T = mg + \frac{mv^2}{l} = mg + \frac{m}{l}(2gl) = 3mg$

139 **(a)**

$\vec{A} = 2\hat{i} + 4\hat{j}, \vec{B} = 5\hat{i} + p\,\hat{j}$

$A = \sqrt{2^2 + 4^2} = \sqrt{20}$

$B = \sqrt{5^2 + p^2}$

$\vec{A}\cdot\vec{B} = 10 - 4p$

If $\vec{A} \parallel \vec{B}$ then

$\vec{A}\cdot\vec{B} = AB\cos 0° = AB$

$10 - 4p = \sqrt{20}\sqrt{25 + p^2}$

Square $100 + 16p^2 - 80p$

$= 20(25 + p^2) = 500 = 20p^2$

or $20p^2 - 16p^2 + 80p + 400 = 0$

or $p^2 + 20p + 100 = 0$

or $(p + 10)^2 = 0$

$\therefore\ p = -10$

$\therefore\ \vec{B} = 5\hat{i} + 10\hat{j}$

$B = \sqrt{5^2 + (10)^2} = \sqrt{125} = 5\sqrt{5}$

140 **(d)**

Range $= \frac{u^2 \sin 2\theta}{g} = 200\ m$

$\Rightarrow \frac{u^2(2\sin\theta\cos\theta)}{g} = 200\ m$ ...(i)

Time of flight $= \frac{2u\sin\theta}{g} = 5s$ ...(ii)

From equations (i) and (ii)

$u\cos\theta = 40\ m/s$

141 **(a)**

Here, Mass of a stone, $m = 2kg$

Length of a string, $r = 0.5\ m$

Breaking tension, $T = 900\ N$

As $T = mr\omega^2$ or $\omega^2 = \frac{T}{mr} = \frac{900}{2\times0.5} = 900$

$\omega = 30\ \text{rads}^{-1}$

142 **(a)**

When the angle of projection is very far from 45° then range will be minimum

143 **(b)**

$v_x = v\cos(\alpha - \theta)\,; v_y = v\sin(\alpha - \theta)$

$a_x = -g\sin\theta\,; a_y = -g\cos\theta$

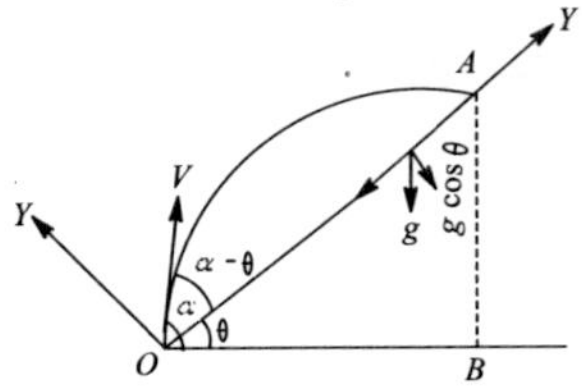

If T is the time of flight, then

$0 = v\sin(\alpha - \theta).T - \frac{1}{2}g\cos\theta\,.T^2$

or $T = \frac{2v\sin(\alpha-\theta)}{g\cos\theta}$

$OB = v\cos\alpha \times T$

Now, $\cos\theta = \frac{OB}{OA}$ or $OA = \frac{OB}{\cos\theta}$

or $OA = \frac{v\sin\alpha T}{\cos\theta}$

or $OA = v\cos\alpha \times \frac{2v\sin(\alpha-\theta)}{g\cos\theta} \times \frac{1}{\cos\theta}$

or $OA = \frac{v^2}{g\cos^2\theta}[\sin(2\alpha - \theta)\cos\alpha]$

or $OA = \frac{v^2}{g\cos^2\theta}[\sin(2\alpha - \theta) + \sin(-\theta)]$

or $OA = \frac{v^2}{g\cos^2\theta}[\sin(2\alpha - \theta) - \sin\theta]$

Clearly, the range $R(= OA)$ will be maximum when $\sin(2a - \theta)$ is maximum, *ie*, 1.

This would mean

$2\alpha - \theta = \frac{\pi}{2}$ or $\alpha \frac{\theta}{2} + \frac{\pi}{4}$

Maximum range up the inclined plane,

$$R_{\max} = \frac{v^2}{g\cos^2\theta}(1 - \sin\theta)$$

$$= \frac{v^2(1 - \sin\theta)}{g(1 - \sin^2\theta)}$$

$$= \frac{v^2(1 - \sin\theta)}{g(1 - \sin\theta)(1 + 1 - \sin\theta)}$$

$$= \frac{v^2}{g(1 + \sin\theta)}$$

144 **(d)**

Tension at the top of the circle, $T = m\omega^2 r - mg$

$T = 0.4 \times 4\pi^2 n^2 \times 2 - 0.4 \times 9.8$

$= 122.2N \approx 115.86N$

145 **(d)**

In non-uniform circular motion particle possess both centripetal as well as tangential acceleration

147 **(a)**

Here, $r = 7$ m, $v = 5$ ms^{-1}, $\theta = ?$

$$\tan\theta = \frac{v^2}{rg} = \frac{5 \times 5}{7 \times 9.8} = 0.364$$

$\theta = \tan^{-1}(0.364) = 20°$

148 **(c)**

Time of flight. $T = \frac{2u\sin\theta}{g}$

Horizontal range, $R = \frac{u^2\sin 2\theta}{g}$

Change in angular momentum,

$|d\vec{\mathbf{L}}| = |\vec{\mathbf{L}}_f - \vec{\mathbf{L}}_i|$ about point of projection

$$= (mu\sin\theta) \times \frac{u^2\sin 2\theta}{g}$$

$$= \frac{mu^3\sin\theta\sin 2\theta}{g}$$

Torque $|\vec{\tau}|$

$$= \frac{\text{change in angular momentum}}{\text{time of flight}}$$

$$= \left|\frac{d\vec{\mathbf{L}}}{T}\right|$$

149 **(c)**

When two bullets are fired simultaneously, horizontally with different speeds, then they cover different horizontal distance because there is no acceleration in this direction.

Since, horizontal distance(R)

$= \text{velocity} \times \text{time}$.

But there is a vertical acceleration towards the earth (g), so the vertical distance covered by both bullet are given by

$y = \frac{1}{2}gt^2$, which is independent of initial velocity.

So, both the bullets will hit the ground simultaneously .

150 **(d)**

$$T = mg + m\omega^2 r = m\{g + 4\pi^2 n^2 r\}$$

$$= m\left\{g + \left(4\pi^2\left(\frac{n}{60}\right)^2 r\right)\right\}$$

$$= m\left\{g + \left(\frac{\pi^2 n^2 r}{900}\right)\right\}$$

151 **(c)**

$$h = v\sin\theta\, t - \frac{1}{2}gt^2$$

or $\frac{1}{2}gt^2 - v\sin\theta\, t + h = 0$

$$t_1 + t_2 = -\frac{-v\sin\theta}{\frac{1}{2}g} \text{ or } t_1 + t_2 = \frac{2v\sin\theta}{g}$$

$= T$

or $T = (1 + 3)\text{s} = 4\text{s}$

152 **(b)**

We know, $F = mr\omega^2$

$r\omega^2 =$constant

$$\omega^2 \propto \frac{1}{r}$$

$$\left(\frac{\omega_2}{\omega_1}\right)^2 = \frac{r_1}{r_2}$$

$\frac{4\omega_1^2}{\omega_1^2} = \frac{8}{r_2}$ $\therefore r_2 = 2$ cm

153 **(b)**

Due to air resistance, it's horizontal velocity will decrease so it will fall behind the aeroplane

154 **(a)**

When particle moves in a circle, then the resultant force must act towards the centre and its magnitude F must satisfy,

$F = \frac{mv^2}{l}$

This resultant force is directed towards the centre and it is called centripetal force. This force originates from the tension T

Hence, $F = \frac{mv^2}{l} = T$

155 **(d)**

Let α'' be the angle of projection of the second body

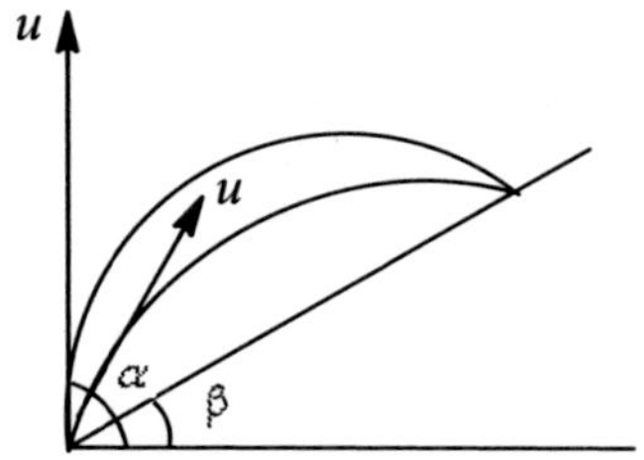

$$R = \frac{u^2}{g\cos\beta}[\sin(2\alpha - \beta)]$$

Range of both the bodies is same. Therefore,

$\sin(2\alpha - \beta) = \sin(2\alpha'' - \beta)$

or $2\alpha'' - \beta = \pi - (2\alpha - \beta)$

$\alpha'' = \frac{\pi}{2} - (\alpha - \beta)$

Now, $T = \frac{2u\sin(\alpha-\beta)}{g\cos\beta}$ and $T'' = \frac{2u\sin(\alpha''-\beta)}{g\cos\beta}$

$$\therefore \frac{T}{T''} = \frac{\sin(\alpha - \beta)}{\sin(\alpha'' - \beta)}$$

$$= \frac{\sin(\alpha - \beta)}{\sin\left\{\frac{\pi}{2} - (\alpha - \beta) - \beta\right\}}$$

$$= \frac{\sin(\alpha - \beta)}{\sin\left(\frac{\pi}{2} - \alpha\right)} = \frac{\sin(\alpha - \beta)}{\cos\alpha}$$

156 **(c)**

Resultant acceleration

$$= \sqrt{\left(\begin{matrix}\text{tangential}\\ \text{acceleration}\end{matrix}\right)^2 + \left(\begin{matrix}\text{centripetal}\\ \text{acceleration}\end{matrix}\right)^2}$$

$$= \sqrt{a^2 + \left(\frac{v^2}{r}\right)^2} = \sqrt{\frac{v^4}{r^2} + a^2}$$

157 **(b)**

Let α' be the angle of projection of the second body

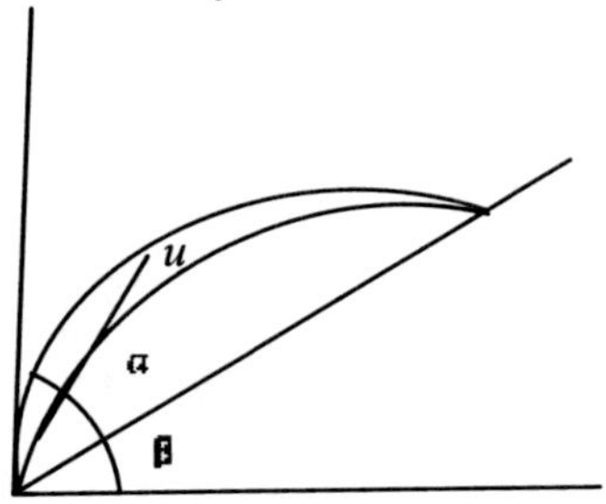

$$R = \frac{u^2}{g\cos^2\beta}[\sin(2\alpha - \beta) - \sin\beta]$$

Range of both the body is same. Therefore

$\sin(2\alpha - \beta) = \sin(2\alpha' - \beta)$

or $2\alpha' - \beta = \pi - (2\alpha - \beta)$

$\alpha' = \frac{\pi}{2} - (\alpha - \beta)$

Now, $T = \frac{2u\sin(\alpha - \beta)}{g\cos\beta}$ and T'

$$= \frac{2u\sin(\alpha' - \beta)}{g\cos\beta}$$

$$\therefore \frac{T}{T'} = \frac{\sin(\alpha - \beta)}{\sin(\alpha' - \beta)}$$

$$= \frac{\sin(\alpha - \beta)}{\sin\left\{\frac{\pi}{2} - (\alpha - \beta) - \beta\right\}}$$

$$\frac{\sin(\alpha - \beta)}{\sin\left(\frac{\pi}{2} - \alpha\right)} = \frac{\sin(\alpha - \beta)}{\cos\alpha}$$

158 **(d)**

Time period = 40 *sec*

No. of revolution $= \frac{\text{Total time}}{\text{Time period}} = \frac{140 \text{ sec}}{40 \text{ sec}} =$ 3.5 *Rev*.

So, distance $= 3.5 \times 2\pi R = 3.5 \times 2\pi \times 10 = 220m$.

159 **(a)**

Let the bead starts slipping after time t

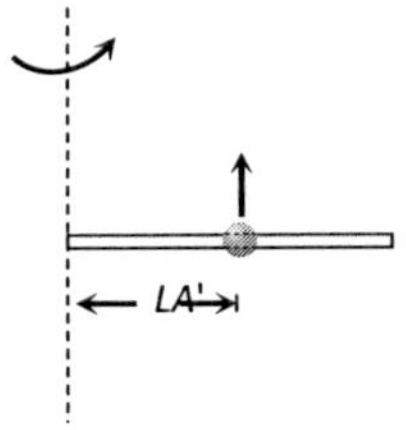

For critical condition Frictional force provides the centripetal force

$m\omega^2 L = \mu R = \mu m \times a_t = \mu L m\alpha$

$\Rightarrow m(\alpha t)^2 L = \mu m L\alpha$

$\Rightarrow t = \sqrt{\frac{\mu}{\alpha}}$ [As $\omega = \alpha t$]

160 **(b)**

On a circular path in completing one turn, the distance traveled is $2\pi r$ while displacement is zero.

Hence, average velocity $= \frac{\text{displacement}}{\text{time interval}}$

$$= \frac{0}{t} = 0$$

Average speed $= \frac{\text{distance}}{\text{time interval}}$

$$= \frac{2\pi r}{t} = \frac{2 \times 3.14 \times 100}{62.8} = 10 \text{ ms}^{-1}$$

161 **(b)**

$$F = \frac{mv^2}{r} = \frac{500 \times 100}{50} = 10^3\ N$$

162 **(a)**

Time to reach max. height $= t_m$

Time to reach back to ground $= t_m$

Total time of flight $T_f = t_m + t_m$

$T_f = 2t_m$

163 **(b)**

Angular velocity about $A, \omega_1 = v/2r$

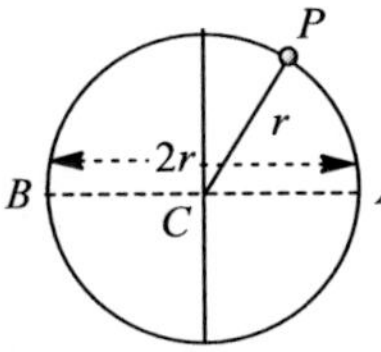

Angular velocity about, $\omega_1 = v/2r$

$\therefore\ \omega_1/\omega_2 = (v/2r)/(v/r) = 1/2$

164 **(c)**

They have same ω

Centripetal acceleration $= \omega^2 r$

$$\frac{a_1}{a_2} = \frac{\omega^2 r_1}{\omega^2 r_2} = \frac{r_1}{r_2}$$

165 **(d)**

Minimum speed at the highest point of vertical circular path $v = \sqrt{gR}$

166 **(d)**

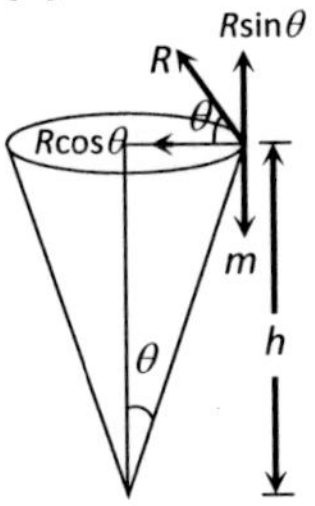

The particle is moving in circular path

From the figure, $mg = R\sin\theta$ ... (i)

$$\frac{mv^2}{r} = R\cos\theta \quad \text{... (ii)}$$

From equation (i) and (ii) we get

$\tan\theta = \frac{rg}{v^2}$ but $\tan\theta = \frac{r}{h}$

$$\therefore h = \frac{v^2}{g} = \frac{(0.5)^2}{10} = 0.025m = 2.5\ cm$$

167 **(a)**

$$\tan\theta = \frac{h}{b} = \frac{v^2}{rg},$$

$$v = \sqrt{\frac{hrh}{b}} = \sqrt{\frac{1.5 \times 50 \times 10}{10}} = 8.5\ \text{ms}^{-1}$$

169 **(a)**

Minimum tension $T_1 = \frac{mv^2}{r} - mg$

Maximum tension $T_2 = \frac{mv^2}{r} + mg$

Let $\frac{mv^2}{r} = x$

So, $T_1 = x - mg$ ...(i)

$T_2 = x + mg$ ...(ii)

Diving Eq. (i) by Eq. (ii)

$$\frac{T_1}{T_2} = \frac{x - mg}{x + mg}\left(\because \text{Given} \frac{T_1}{T_2} = \frac{3}{5}\right)$$

$$\therefore \frac{3}{5} = \frac{x - mg}{x + mg}$$

or $3x + 3mg = 5x - 5mg$

or $x = 4\ mg$

$ie, \frac{mv^2}{r} = 4mg$

$\therefore\ v^2 = 4\ rg$

or $v = \sqrt{4rg}$

or $v = \sqrt{4 \times 2.5 \times 9.8}$

$v = \sqrt{98}\text{ms}^{-1}$

170 **(c)**

Angular momentum $L = r \times p = r \times m \times v$

$$v = \frac{L}{mr} \quad \text{.... (i)}$$

Now, as centripetal force, F_c

$$= \frac{mv^2}{r} \quad \text{.... (ii)}$$

Substituting the value of v from Eq. (i) in Eq. (ii), we get

$$F_c = \frac{m}{r}\left[\frac{L}{mr}\right]^2 = \frac{L^2}{mr^3}$$

171 **(b)**

$F^2 = F^2 + F^2 + 2F^2\cos\theta$

or $F^2 = 2F^2(1 + \cos\theta)$

or $1 + \cos\theta = \frac{1}{2}$

or $\cos\theta = -\frac{1}{2}$ or $\theta = 120°$

$\therefore \cos 120° = -\frac{1}{2}$

172 **(c)**

The result follows from the definition of cross product.

173 **(a)**

$$t = \sqrt{\frac{2 \times 2000}{10}} = \sqrt{400} = 20\ \text{s}$$

$x = 100\text{ms}^{-1} \times 20\text{s} = 2000\ \text{m} = 2\text{km}$

174 **(b)**

New kinetic energy $E' = E = \cos^2\theta$

$= E\cos^2(45°)$

$\frac{E}{2}$

175 **(c)**

When the force acting on a body is directed towards a fixed point, then it changes only the direction of motion of the body without changing its speed. So, the particle will describe a circular motion

176 **(a)**

$25 = 0.25 \times v^2/1.96$

or $v = (25 \times 1.96/0.25)^{1/2} = 5 \times \frac{14}{5} =$

$14 ms^{-1}$

177 **(b)**

Only horizontal component of velocity $(u\cos\theta)$

178 **(b)**

The angular momentum of a particle is given by

$\vec{\mathbf{L}} = \vec{\mathbf{r}} \times m\vec{\mathbf{v}}$

$L = mvr\sin\theta$

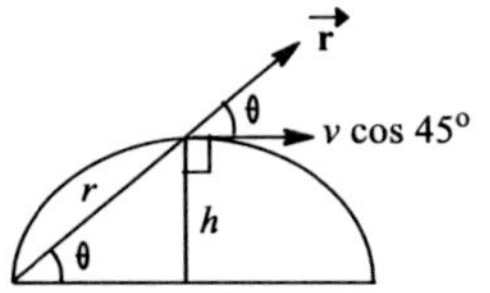

From figure,

$L = rm(v\cos 45°)\sin\theta$

$= \frac{mv}{\sqrt{2}}(r\sin\theta)$

$= \frac{mvh}{\sqrt{2}}\left(\because \sin\theta = \frac{h}{r}\right)$

179 **(b)**

The time taken by the particle for one complete revolution.

$t = \frac{2\pi r}{\text{speed}}$

$= \frac{2 \times 3.14 \times 100}{31.4} = 20s$

Hence, averge speed is

$v_{av} = \frac{2 \times 3.14 \times 100}{20} = 31.4\ ms^{-1}$

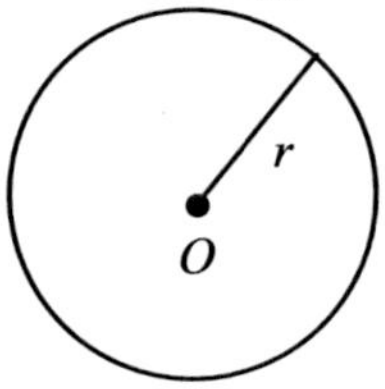

180 **(b)**

$R_{max} = \frac{u^2}{g} = \frac{(20)^2}{10} = 40\ m$

181 **(a)**

Time period of earth on its own axis

$T = 24$ h

$= 24 \times 60 \times 60$ s

$\therefore$ Angular velocity $\omega = \frac{2\pi}{T}$

$= \frac{2\pi}{24 \times 60 \times 60}$

$= \frac{2\pi}{86400}$ rads^{-1}

182 **(d)**

When the string makes an angle θ with the vertical, then

$T - mg\cos\theta = \frac{mv^2}{r}$

Substituting the values, we obtain

$6 - (1)(10)\cos\theta = \frac{1 \times (4)^2}{1}$

or $6 - 10\cos\theta = 16$

or $\cos\theta = -1 = \cos 180°$

$\therefore = 180°$

183 **(c)**

$\mu = \frac{v^2}{rg} = \frac{(60 \times 5/18)^2}{40 \times 9.8} = 0.71$

184 **(a)**

Minimum speed at the lowest point

$= \sqrt{5rg} = \sqrt{5 \times 5 \times 9.8} = 15.65\ ms^{-1}$

185 **(d)**

Net displacement in one loop = 0

Average velocity $= \frac{\text{net displacement}}{\text{time}} = \frac{0}{t} = 0$

Distance travelled in one rotation (loop) $= 2\pi r$

$\therefore$ Average speed $= \frac{\text{distance}}{\text{time}} = \frac{2\pi r}{t}$

$= \frac{2 \times 3.14 \times 100}{62.8} = 10 m/s.$

187 **(c)**

If v is velocity of the bob on reaching the lowest point, then $\frac{1}{2}mv^2 = mgL$

To void breaking, strength of the string

$T_L = \frac{mv^2}{L} + mg = \frac{2mg\,L}{L} + mg = 3\ mg$

188 **(b)**

Projection of $\vec{P}$ on $\vec{Q}$ is $P\cos\theta$

$P\cos\theta = \frac{PQ\cos\theta}{Q} = \frac{\vec{P}\cdot\vec{Q}}{Q} = \vec{P}\cdot\vec{Q}$

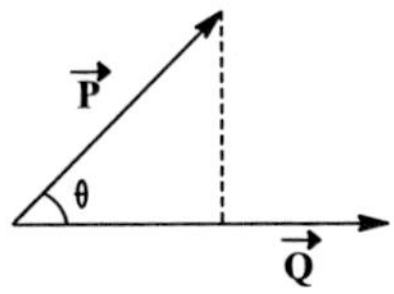

189 **(a)**
When speed is constant in circular motion, it means work done by centripetal force is zero

190 **(a)**
In 6 turns each of 60°, the cyclist traversed a regular hexagon path having each side 100 m. So, at 7th turn, he will be again at

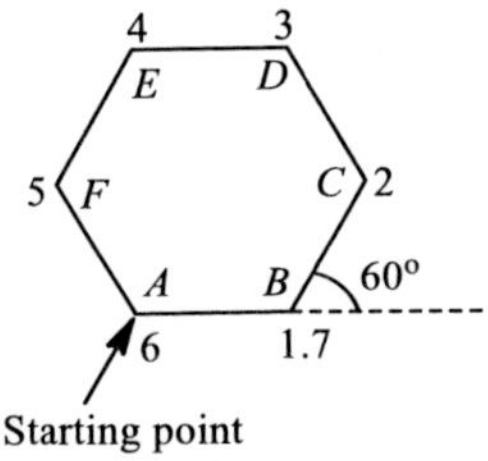

Point B (as shown) which is at distance 100 m from starting point A. Hence, net displacement of cyclist is 100 m.

191 **(a)**

$$T = m\omega^2 r \Rightarrow \omega \propto \sqrt{T} \therefore \frac{\omega_2}{\omega_1} = \sqrt{\frac{1}{4}} \Rightarrow \omega_2$$

$$= \frac{\omega_1}{2} = 5\ rpm$$

192 **(b)**
Increment in angular velocity $\omega = 2\pi(n_2 - n_1)$

$$\omega = 2\pi(1200 - 600)\frac{rad}{min}$$

$$= \frac{2\pi \times 600}{60}\frac{rad}{s}$$

$$= 20\pi\frac{rad}{s}$$

193 **(d)**
$\mathbf{L} = m(\mathbf{r} \times \mathbf{v})$
Direction of $(\mathbf{r} \times \mathbf{v})$, hence the direction of angular momentum remains the same.

196 **(c)**
Given condition $h_1 = h_2$

$$u_1^2 \sin^2 45° = u_2^2 \sin^2 \theta$$

$$\sin^2\theta = \frac{u_1^2}{u_2^2}\sin^2 45°$$

$$= \frac{1}{2} \cdot \frac{1}{2} = \frac{1}{4}$$

$$\sin\theta = \frac{1}{2} \Rightarrow \theta = 30°$$

197 **(d)**
The ball will be at point P when it is at height of 10 m from the ground. So, we have to find distance OP, whichcan be calculated by directly considering it as a projectile on a level (OX).

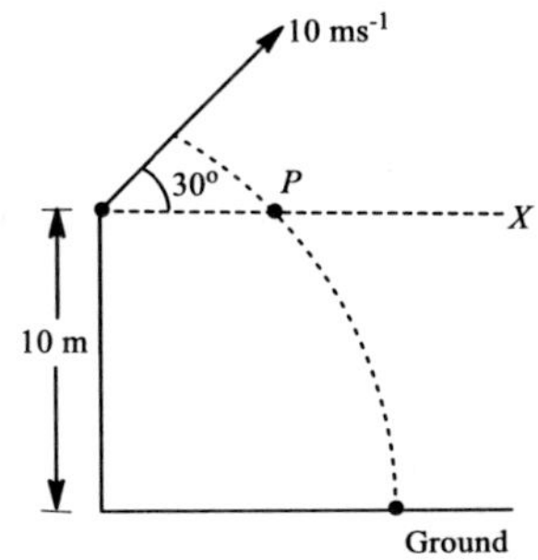

$$OP = R = \frac{u^2 \sin 2\theta}{g}$$

$$= \frac{10^2 \sin(2 \times 30°)}{10}$$

$$= \frac{10\sqrt{3}}{2} = 5\sqrt{3}$$

$$= 8.66\text{ m}$$

198 **(c)**
Tension in the string
$T = mv^2/r = (r\omega)^2 = mr\omega^2$
If $r_1 = r/2$ and $\omega_1 = 2\omega$, then
$T_1 = m(r/2)(2\omega)^2 = 2mr\omega^2 = 2T$

199 **(c)**

Here, $h = \frac{u^2 \sin^2\theta}{2g}$ or $\sqrt{\frac{2h}{g}} = \frac{u\sin\theta}{g}$

Time of flight, $T = \frac{2u\sin\theta}{g} = 2\sqrt{\frac{2h}{g}}$

200 **(d)**
Kinetic energy given to a sphere at lowest point = potential energy at the height of suspension

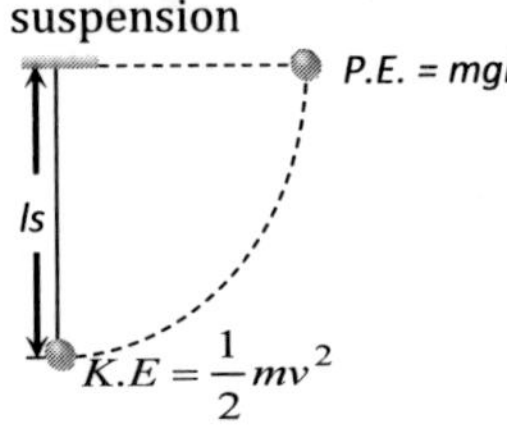

$$\Rightarrow \frac{1}{2}mv^2 = mgl$$

$$\therefore v = \sqrt{2gl}$$

201 **(c)**

Suppose t_0 be the time to reach maximum height in the absence of air resistance, then from the relation

$t_0 = \frac{u \sin\alpha}{g}$ ...(i)

When a is retardation caused by air resistance, then total retardation will be $g+a$

$t_1 = \frac{u\sin\alpha}{g+a} = \frac{u\sin\alpha}{g+\left(\frac{1}{10}\right)g} = \frac{10u\sin\alpha}{11g}$...(ii)

Now from equations (i) and (ii), we have

$\therefore t_1 = \frac{10}{11}t_0 \Rightarrow t_0 - t_1 = t_0 - \frac{10}{11}t_0$

$= \frac{1}{11}t_0 = 0.09\ t_0$

∴ Time will decrease by 9%

202 **(c)**

Average velocity $= \frac{\text{displacement}}{\text{time}}$

$V_{av} = \frac{\sqrt{H^2+\frac{R^2}{4}}}{T/2}$ ....(i)

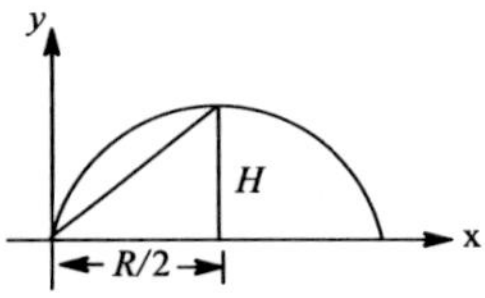

Here H = maximum height $= \frac{v^2\sin^2\theta}{2g}$

R = range $= \frac{v^2\sin 2\theta}{g}$

and T = time of flight $= \frac{2v\sin\theta}{g}$

Substituting in Eq. (i), we get

$v_{av} = \frac{v}{2}\sqrt{1+3\cos^2\theta}$

203 **(a)**

Here $\vec{A} - \overrightarrow{OP} = 10$ units along OP

$\vec{B} - (\overrightarrow{OQ}) = 10$ units along OQ

∴ $\angle XOP = 30°$ and $\angle XOQ = 135°$

∴ $\angle QOX' = 180° - 135° = 45°$

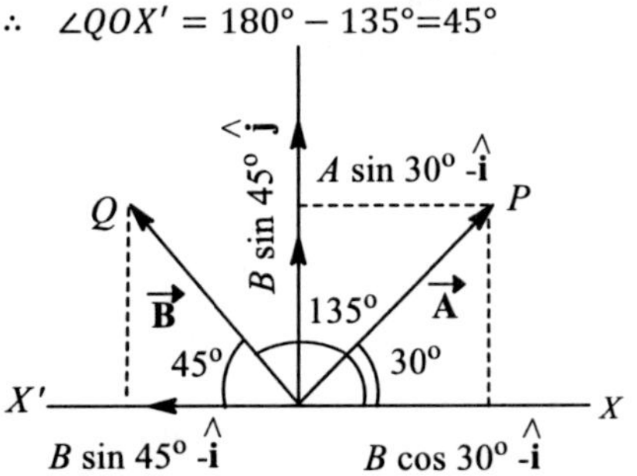

Resolving $\vec{A}$ and $\vec{B}$ into two rectangular components we have $A\cos 30°$ along OX and $A\sin 30°$ along OY. $B\cos 45°$ along OX' and $B\sin 45°$ along OY'.

Resultant component force along X-axis.

$(A\cos 30° - B\sin 45)\hat{i}$

$= \left(10\times\sqrt{3}/2 - 10\times 1/\sqrt{2}\right)\hat{i} = 1.59\hat{i}$

Resultant component force along Y-axis

$= (A\sin 30° + B\sin 45°)\hat{j}$

$= \left(10\times 1/2 + 10\times 1/\sqrt{2}\right)\hat{j} = 12.07\hat{j}$

204 **(b)**

When a particle moves in a circular motion, it is acted upon by centripetal force directed towards the centre. Hence, centripetal acceleration is

$a_N = \frac{dv}{dt} = \frac{v^2}{R}$

or $\int_0^t \frac{dt}{R} = \int_{v_0}^{v}\frac{dv}{v^2}$

or $t = -R\left[\frac{1}{v}\right]_{v_0}^{v}$

$v = \frac{v_0 R}{R - v_0 t}$

Also $\frac{dr}{dt} = \frac{v_0 R}{(R - v_0 t)}$

$\int_0^{2\pi R} dr = v_0 R\int_0^T \frac{dt}{R - v_0 t}$

$\Rightarrow\ T = \frac{R}{v_0}(1 - 1 - e^{-2\pi})$

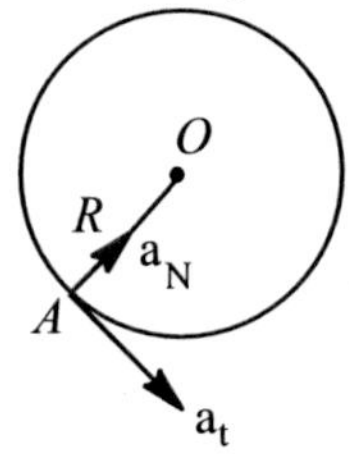

205 **(c)**

Tension = Centrifugal force + weight = $\frac{mv^2}{r} + mg$

206 **(b)**

Angular speed of minute hand,

$\omega_m = \frac{2\pi}{60\times 60}$ rad s^{-1}

Angular speed of hour hand,

$\omega_h = \frac{2\pi}{12\times 60\times 60}$ rad s^{-1}

$\therefore \frac{\omega_m}{\omega_h} = 12$

207 **(b)**

Here, $r = 100$ m, $t = 62.8$ s

In one circular loop, displacement = 0

∴ Velocity=0

Distance traveled $=2\pi r$

$\therefore$ Speed $= \frac{2\pi r}{t} = \frac{2.14\times100}{62.8} = 10\ \text{ms}^{-1}$

208 **(d)**

$t = \sqrt{\frac{2h}{g}} = \sqrt{\frac{2\times396.9}{9.8}} \simeq 9\ \text{sec}$ and $u =$ $720 km/hr = 200\ m/s$

$\therefore R = u \times t = 200 \times 9 = 1800\ m$

209 **(b)**

$F = \frac{mv^2}{r} \Rightarrow F \propto v^2$. If v becomes double then F(tendency to overturn) will become four times

210 **(c)**

$T =$ tension, $W =$ weight and F $=$ centrifugal force

211 **(b)**

$h = 145 - 22.5 = 122.5$ m

Now, $40 = v\sqrt{\frac{2\times122.5}{9.8}}$

or $40 = v \times 5$ or $v = 8\ \text{ms}^{-1}$

212 **(c)**

Frequency of wheel, $\nu = \frac{300}{60} = 5$ rps.

Angle described by wheel in one rotation $= 2\pi$ rad. Therefore, angle described by wheel in 1 s $= 2\pi \times 5$ rad $= 10\pi$ rad

213 **(a)**

$dM = \left(\frac{M}{L}\right) dx$

Force on $'dM'$ mass is

$dF = (dM)\omega^2 x$

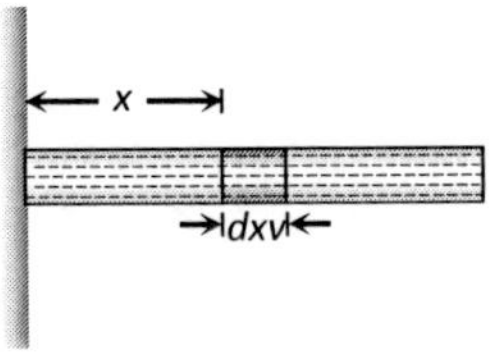

By integration we can get the force exerted by whole liquid

$\Rightarrow F = \int_0^L \frac{M}{L}\omega^2 x\, dx = \frac{1}{2}M\omega^2 L$

214 **(d)**

Equating horizontal components, we get

$V\cos(90° - \theta) = v\cos\theta$

or $V\sin\theta = v\cos\theta$ or $V = v\cot\theta$

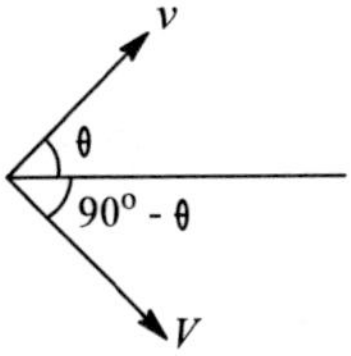

215 **(d)**

Angular velocity of second's hand

$= \frac{2\pi}{60} = \frac{\pi}{30} = \frac{3.14}{30}$

$= 0.1047\ \text{rad s}^{-1}$

Linear velocity, $v = r\omega$

$= 3 \times 10^{-2} \times 0.1047 = 0.00314\ \text{ms}^{-1}$

216 **(b)**

Due to centrifugal force

217 **(a)**

Range of the projectile on an inclined plane (down the plane) is,

$R = \frac{u^2}{g\cos^2\beta}[\sin(2\alpha + \beta) + \sin\beta]$

Here, $u = v_0, \alpha = 0$ and $\beta = \theta$

$\therefore R = \frac{2v_0^2\sin\theta}{g\cos^2\theta}$

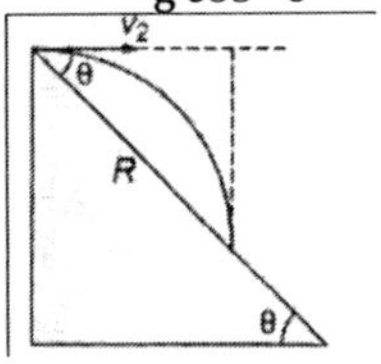

Now $x = R\cos\theta = \frac{2v_0^2\tan\theta}{g}$

and $y = -R\sin\theta = -\frac{2v_0^2\tan^2\theta}{g}$

218 **(d)**

$T\sin\theta = M\omega^2 R$...(i)

$T\sin\theta = M\omega^2 L\sin\theta$ ...(ii)

From (i) and (ii)

$T = M\omega^2 L$

$= M\ 4\pi^2 n^2 L$

$= M\ 4\pi^2\left(\frac{2}{\pi}\right)^2 L = 16\ ML$

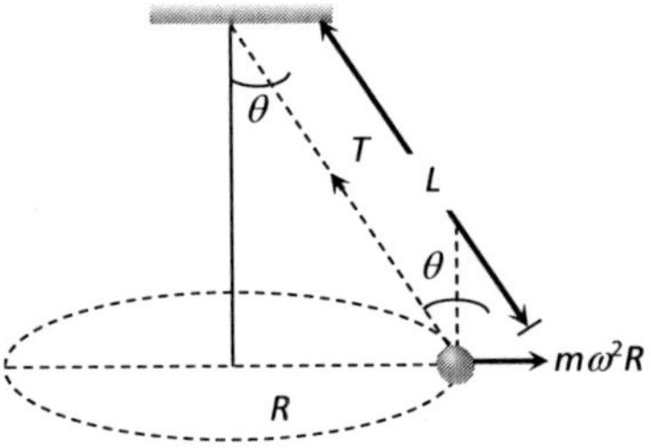

219 **(a)**

For particle P, motion between A and C will be an accelerated one while between C and B a retarded one. But in any case horizontal component of it's velocity will be greater than or equal to v on the other hand in case of particle Q, it is always equal to v. Horizontal displacement of both the particles are equal, so $t_P < t_Q$

220 **(c)**

Velocity at the lowest point

$v=\sqrt{2gl}$

At the lowest point, the tension in the string

$$T=mg+\frac{mv^2}{l}$$

$$=mg+\frac{m}{l}(2gl)=3\,mg$$

221 **(b)**

One circular motion, the force acts along the radius and displacement at a location is perpendicular to the radius *ie*, $\theta=90°$

As work done $=\vec{\mathbf{F}}\cdot\vec{\mathbf{s}}=Fs\cos 90°=0$

222 **(d)**

$$h_{\max.}=\frac{v^2\sin^2\theta}{2g}$$

In the given problem, $h_{\max.}$ is same in both the cases

$\therefore\ v_1^2\sin^2 60°=v_2^2\sin^2 30°$

or $\frac{v_1}{v_2}=\frac{\sin 30°}{\sin 60°}=\frac{1}{2}\times\frac{2}{\sqrt{3}}=\frac{1}{\sqrt{3}}$

223 **(a)**

Here, $v=900\text{ km h}^{-1}$

$$=\frac{900\times 1000}{60\times 60}\text{ms}^{-1}=250\text{ ms}^{-1}$$

Minimum force is at the bottom of the vertical circle

$$F_{\max}=\frac{mv^2}{r}+mg=5\,mg$$

$\therefore\ v^2=4\,gr$

or $r=\frac{v^2}{4g}=\frac{250\times250}{4\times980}=1594$ m

224 **(a)**

Kinetic energy

$$E=\frac{1}{2}mv^2$$

$$\text{or}\ \frac{1}{2}mr\frac{v^2}{r}=E$$

$$\text{or}\ \frac{1}{2}mra=E$$

$$\text{or}\ a=\frac{2E}{mr}$$

225 **(d)**

Centripetal force, $F=\frac{mv^2}{r}$

$$v=\sqrt{\left(\frac{rF}{m}\right)}=\sqrt{\frac{0.5\times10\times1000}{100}}$$

$=\sqrt{50}\text{ms}^{-1}=7.07\text{ms}^{-1}$

226 **(d)**

As maximum value of $T=mg$ from $2T\cos\theta=mg$.

$2\cos\theta=1,\cos\theta=\frac{1}{2},\theta=60°$

Angle between two arms $=2\theta=120°$

227 **(b)**

Since, acceleration is constant

$$\therefore\ \vec{\mathbf{s}}=\vec{\mathbf{u}}+\frac{1}{2}\vec{\mathbf{a}}t^2$$

$$=(2\hat{\imath}-4\hat{\jmath})t+\frac{1}{2}(3\hat{\imath}+5\hat{\jmath})t^2$$

$$=(2\hat{\imath}-4\hat{\jmath})2+\frac{1}{2}(3\hat{\imath}+5\hat{\jmath})2^2$$

$=10\hat{\imath}+2\hat{\jmath}$

$|\vec{\mathbf{s}}|=\sqrt{10^2+2^2}=\sqrt{104}=10.2$m

228 **(b)**

Here, $m=5$ kg, $r=2$m, $v=6\text{ ms}^{-1}$

The tension is maximum at the lowest point

$$T_{\max}=mg+\frac{mv^2}{r}$$

$$=5\times9.8+\frac{5\times6\times6}{2}$$

$=139$N

229 **(c)**

Maximum height, $H=\frac{u^2\sin^2\theta}{2g}$

Time of flight, $T=\frac{2u\sin\ \theta}{g}$

$$\therefore\ \frac{H}{T^2}=\frac{u^2\sin^2\theta/2g}{4u^2\sin^2\theta/g^2}=\frac{g}{8}=\frac{10}{8}=\frac{5}{4}$$

230 **(c)**

$v=r\omega\Rightarrow\omega=\frac{v}{r}=$ constant [As v and r are constant]

231 **(a)**

When particle moves in circle, then the resultant force must act towards the center and its magnitude F must satisfy

$$F=\frac{mv^2}{l}$$

This resultant force is directed towards the center and it is called centripetal force. This force originates form tension T.

$$\therefore F=\frac{mv^2}{l}=T$$

232 **(b)**

Change in momentum $=2mv\sin\theta=2mv\sin\frac{\pi}{4}=\sqrt{2}mv$

233 **(c)**

At the topmost point of the projectile there is only horizontal component of velocity and acceleration due to gravity is vertically downward, so velocity and acceleration are perpendicular to each other.

234 **(d)**

Given, $x = 0.20$ m, $y = 0.20$ m, $u = 1.8\text{ms}^{-1}$

Let the ball strike the nth step of stairs,

Vertical distance travelled

$= ny = n \times 0.20 = \frac{1}{2}gt^2$

Horizontal distance travelled, $nx = ut$

or $t = \frac{nx}{u}$

$\therefore ny = \frac{1}{2}g \times \frac{n^2x^2}{u^2}$

or $n = \frac{2u^2}{g}\frac{y}{x^2} = \frac{2 \times (1.8)^2 \times 0.20}{9.8 \times (0.20)^2}$

$= 3.3 = 4$

235 **(a)**

Initial angular velocity $\omega_0 = 0$. Final angular velocity $\omega = \frac{v}{r} = \frac{80}{(20/\pi)} = 4\pi$ rad s^{-1}

angle described, $\theta = 4\pi$ rad

$\therefore$ Angular acceleration, $\alpha = \frac{\omega^2 - \omega_0^2}{2\theta}$

$= \frac{(4\pi)^2 - 0}{2 \times 4\pi} = 2\pi$ rad s^{-2}

Linear acceleration, $a = \alpha r$

$= 2\pi \times \frac{20}{\pi} = 40$ ms^{-2}

237 **(d)**

$200 = ut$

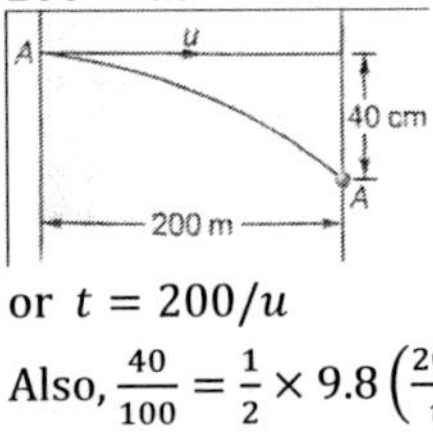

or $t = 200/u$

Also, $\frac{40}{100} = \frac{1}{2} \times 9.8\left(\frac{200}{u}\right)^2$

On solving $u = 700$ ms^{-1}

238 **(c)**

Here, $m = 2$ kg, $r = 1$ m, $v = 4$ ms^{-1}

Tension at the bottom of the circle,

$T_L = mg + \frac{mv^2}{r}$

$= 2 \times 10 + \frac{2 \times 4^2}{1} = 52\text{N}$

239 **(c)**

When a stone tied at the end of string is rotated in a circle, the velocity of the stone at an instant acts tangentially outwards the circle. When the string is released, the stone files off tangentially outwards *ie*, in the direction of velocity

240 **(c)**

Centripetal acceleration $= 4\pi^2n^2r = 4\pi^2\left(\frac{1}{2}\right)^2 \times 4 = 4\pi^2$

241 **(b)**

$H = \frac{v^2\cos^2\beta}{2\text{g}}$ or $v\cos\beta = \sqrt{2\text{g}H}$

$t = \frac{v\cos\beta}{\text{g}} = \frac{\sqrt{2\text{g}H}}{\text{g}}$ or $t = \sqrt{\frac{2H}{\text{g}}}$

242 **(b)**

Angular momentum $\vec{\text{L}} = \vec{\text{r}} \times \vec{\text{p}}$

$= (2\hat{\text{i}} + 2\hat{\text{j}} + \hat{\text{k}}) \times (2\hat{\text{i}} - 2\hat{\text{j}} + \hat{\text{k}}) = 4\hat{\text{i}} - 8\hat{\text{k}}$

244 **(c)**

Given that, radius of circle, $r = 30\text{cm} = 0.3\text{m}$

linear speed $v = 2t$

Now, radial acceleration $a_\text{R} = \frac{v^2}{r} = \frac{(2t)^2}{0.3}$

at $t = 3s$

$a_\text{R} = \frac{(2 \times 3)^2}{0.3}$

$\frac{36}{0.3} = 120\text{ms}^{-2}$

or $a_\text{R} = 120\text{ms}^{-2}$

and tangential acceleration $a_T = \frac{dv}{dt}$

$= \frac{d}{dt}(2t) = 2\text{ms}^{-2}$

245 **(c)**

In a vertical circular motion, centripetal force remains same at all points on circular path and always directed towards the centre of circular path

246 **(d)**

Acceleration $a = \alpha x^2 \Rightarrow \frac{dV}{dt} = \alpha x^2$

$\Rightarrow dV = \alpha x^2 dt \Rightarrow dV = \alpha x^2 dx\frac{dt}{dx}$

$\Rightarrow \int_0^{v_0} VdV = \int_0^x \alpha x^2 dx \Rightarrow \frac{V_0^2}{2} = \frac{\alpha . x^3}{3}$

$\Rightarrow x = \left[\frac{3V_0^2}{2\alpha}\right]^{-1/3}$

247 **(c)**

The velocity of the particle at any time t

$\vec{\mathbf{v}} = \vec{\mathbf{v}}0 + \vec{\mathbf{a}}t$

The x-component is

$v_x = v_{ax} + a_x t$

The y-component is

$v_y = v_{ay} + a_x = (-0.5t)\text{ms}^{-1}$

When the particle reaches its maximum x-coordinates, $v_x = 0$. That is

$3 - t = 0$

$\Rightarrow t = 3s$

The y-component of the velocity of this time is

$v_y = -0.5 \times 3 = -1.5 \text{ ms}^{-1}$

248 **(a)**

Angular velocity $= \frac{2\pi}{T} = \frac{2\pi}{24}\ rad/hr = \frac{2\pi}{86400}\ rad/s$

249 **(c)**

Area $= |\vec{A}| \times |\vec{B}| = |(4\hat{i} + 3\hat{j}) \times (2\hat{i} + 4\hat{j})| = |10\hat{k}| = 10$units

250 **(b)**

$E_k' = E_k \cos^2 30° = \frac{3E_k}{4}$

251 **(c)**

Equating velocities along the vertical,

$v_2 = v_1 \sin 30°$ or $\frac{v_2}{v_1} = \frac{1}{2}$

252 **(d)**

$a_c = k^2rt^4 = \frac{v^2}{r}$ or $v = krt^2$

The tangential acceleration is $a_T = \frac{dv}{dt} = 2krt$

The tangential force on the particle, $F_T = ma_T = 2mkrt$

Power delivered to the particle

$= F_T = ma_T = 2mkrt = F_T v$

$= (2mkrt)(krt)^2$

$= 2mk^2r^2t^3$

253 **(c)**

$R = 4H \cot\theta$

When $R = H$ then $\cot\theta = 1/4 \Rightarrow \theta = \tan^{-1}(4)$

254 **(a)**

Since $v^2 - v_0^2 = 2\vec{a}.\vec{s} = 2\vec{a}.\left(\frac{\vec{v}+\vec{v}_0}{2}\right)t$

or $\vec{v}.\vec{v} - \vec{v}_0.\vec{v}_0 = (\vec{v} + \vec{v}_0).\vec{a}t$

or $\vec{v}.(\vec{v} - \vec{a}t) = \vec{v}_0.(\vec{v}_0 + \vec{a}t)$

255 **(a)**

The cord is most likely to break at the orientation, when mass is at B as tension in the string at this point is maximum

256 **(d)**

At the two point of the trajectory during projectile motion, the horizontal component of the velocity is same. Then,

$u\cos 60° = v\cos 45°$

$147 \times \frac{1}{2} = v \times \frac{1}{\sqrt{2}}$ or $v = \frac{147}{\sqrt{2}}\frac{\text{m}}{\text{s}}$

Initially, $u_y = u\sin 60° = \frac{147\sqrt{3}}{2}$ m/s

Finally, $v_y = v\sin 45° = \frac{147}{\sqrt{2}} \times \frac{1}{\sqrt{2}}$

$= \frac{147}{\sqrt{2}}$ m/s

But $v_y = u_y + a_y t$ or $\frac{147}{2}$

$= \frac{147\sqrt{3}}{2} - 9.8\,t$

$9.8\,t = \frac{147}{2}(\sqrt{3} - 1)$ or $t = 5.49\ s$

257 **(d)**

$\omega = \frac{2\pi}{T} = \frac{2\pi}{60} = 0.1047\ rad/s$

And $v = \omega r = 0.1047 \times 3 \times 10^{-2} = 0.00314\ m/s$

258 **(a)**

Here, $\alpha = 2\theta, \beta = \theta$

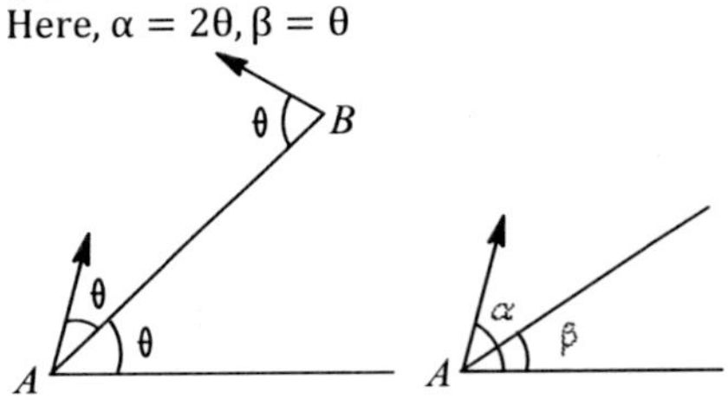

(a) (b)

Time of flight of A is, $T_1 = \frac{2u\sin(\alpha-\beta)}{g\cos\beta}$

$= \frac{2u\sin(2\theta - \theta)}{g\cos\theta} = \frac{2u}{g}\tan\theta$

Time of flight of B is, $T_2 = \frac{2u\sin\theta}{g\cos\theta} = \frac{2u}{g}\tan\theta$

So, $T_1 = T_2$. The acceleration of both the particles is g downwards. Therefore, relative acceleration between the two is zero or relative motion between the two is uniform. The relative velocity of A w.r.t. B is towards AB , therefore collision will take place between the two in mid air.

259 **(d)**

Horizontal component of velocity at A

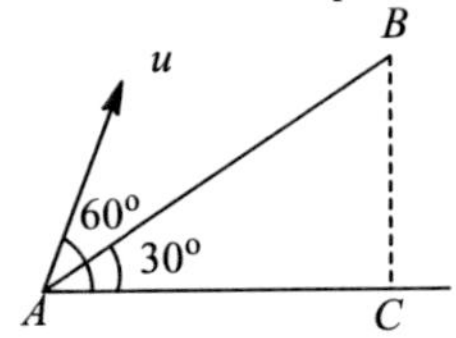

$v_H = u\cos 60° = \frac{u}{2}\ \therefore AC = u_H \times t = \frac{ut}{2}$

$$AB = AC\sec 30° = \frac{ut}{2} \times \frac{2}{\sqrt{3}} = \frac{ut}{2}$$

260 **(a)**

Given, $5 = \frac{2u\sin\theta}{2g}$ or $\frac{u\sin\theta}{g} = \frac{5}{2}$

Maximum height $= \frac{u^2\sin\theta}{2g} = \frac{g}{2}\left(\frac{u^2\sin^2\theta}{g^2}\right)$

$= \frac{g}{2} \times \left(\frac{5}{2}\right)^2 = \frac{10}{2} \times \frac{25}{4} = 31.25$ m

261 **(a)**

Motion is along the time ; $Y = X + 4$;
Differentiating it wrt time, we have

$\frac{dY}{dt} = \frac{dX}{dt}$ *ie*, $v_Y = v_X$

As, $v = (v_X^2 + v_Y^2)^{1/2} = 3\sqrt{2}$ and $v_X = v_Y$, therefore,

$(v_X^2 + v_X^2)^{1/2} = 3\sqrt{2}$ or $v_X = 3 = v_Y$

When $X = 0$, from the given equation,
$Y = 0 + 4 = 4$

Magnitude o angular momentum of particle

$= m\,v\,r = m\,v\,y \quad (\because y = r)$

$= 5 \times 3 \times 4 = 60$ units

262 **(a)**

According to law of conservation of linear momentum at the highest point.

$mv\cos\theta = \frac{m}{2}(-v\cos\theta) + \frac{m}{2}v_1$

or $v_1 = 3v\cos\theta$

263 **(a)**

$\vec{B} + (\hat{i} + 2\hat{j} - 3\hat{k}) = \hat{i}$

or $\quad \vec{B} = -2\hat{j} + 3\hat{k}$

264 **(a)**

$v_{min} = \sqrt{5\,gr} = 17.7\ m/sec$

265 **(a)**

$\vec{A} + \vec{B} = 8\hat{i} - 2\hat{j} + 16\hat{k}$

$m = \frac{0}{|\vec{A} + \vec{B}|} = 0$

266 **(a)**

Because the reaction on inner wheel decreases and becomes zero. So it leaves the ground first

267 **(c)**

$v = \sqrt{\mu\,rg} = \sqrt{0.64 \times 20 \times 10}$
$= 11.2\ \text{ms}^{-1}$

268 **(c)**

Refer figure when the velocity vector makes an angle α with the horizontal, the component of acceleration, perpendicular to velocity, *ie*, the centripetal acceleration is g cos α. As horizontal component of velocity remains unchanged in angular projection of projectile, hence

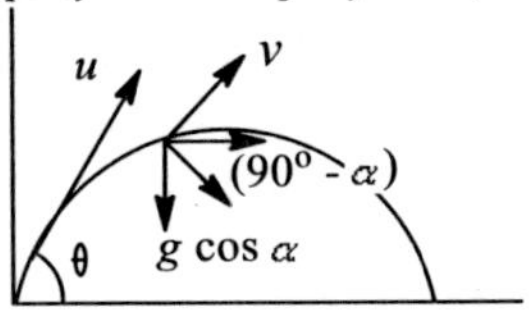

$v\cos\alpha = u\cos\theta$ or $v = \frac{u\cos\theta}{\cos\alpha}$

As, g cos α provides centripetal acceleration, hence

$g\cos\alpha = \frac{v^2}{r}$ or $\frac{v^2}{g\cos\alpha} = \frac{u^2\cos^2\theta}{g\cos^3\alpha}$

269 **(b)**

$T_{top} = \frac{mv^2}{r} - mg \quad$ ...(i)

$T_{bottom} = \frac{mv^2}{r} + mg \quad$ ...(ii)

$\frac{T_{top}}{T_{bottom}} = \frac{\frac{v^2}{r} - g}{\frac{v^2}{r} + g} = \frac{\frac{40\times 40}{4} - 10}{\frac{40\times 40}{4} + 10}$

$= \frac{400 - 10}{400 + 10} = \frac{390}{410} = \frac{39}{41}$

270 **(d)**

$|\vec{a} \times \vec{b}| = ab\ \sin\theta$

sinθcannot be greater than 1.

$\therefore |\vec{a} \times \vec{b}|$cannot be greater than ab.

271 **(b)**

Maximum height $H = \frac{v^2\cos^2\beta}{2g}$

or $v\cos\beta = \sqrt{2gH}$

$t = \frac{v\cos\beta}{g} = \frac{\sqrt{2gH}}{g}$

$t = \sqrt{\frac{2H}{g}}$

272 **(b)**

$\tan\theta = \frac{v^2}{rg} = \frac{400}{20 \times 9.8} \Rightarrow \theta = 63.9°$

273 **(b)**

$H = \frac{u^2\sin^2\theta}{2g}$ and $T = \frac{2u\sin\theta}{g} \Rightarrow T^2 = \frac{4\,u^2\sin^2\theta}{g^2}$

$\therefore \frac{T^2}{H} = \frac{8}{g} \Rightarrow T = \sqrt{\frac{8H}{g}} = 2\sqrt{\frac{2H}{g}}$

274 **(b)**

$v = \sqrt{3gr}$ and $a = \frac{v^2}{r} = \frac{3gr}{r} = 3g$

276 **(d)**

As momentum is vector quantity

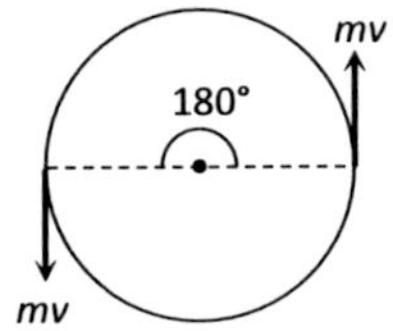

∴change in momentum

$\Delta P = 2mv\sin(\theta/2)$

$= 2mv\sin(90) = 2mv$

But kinetic energy remains always constant so change in kinetic energy is zero

277 **(c)**

$\vec{A} + \vec{B} = \vec{C}$ (given)

So, it is given that $\vec{C}$ is the resultant of $\vec{A}$ and $\vec{B}$

$\therefore \quad C^2 = A^2 + B^2 + 2AB\cos\theta$

$3^2 = 3 + 3 + 2 \times 3 \times \cos\theta$

3=6 cosθ or $\cos\theta = \frac{1}{2} \quad \Rightarrow \theta\ 60°$

278 **(d)**

Angular acceleration $= \frac{d^2\theta}{dt^2} = 2\theta_2$

279 **(a)**

$\vec{A} \times \vec{B} = (4\hat{i} + 6\hat{j}) \times (2\hat{i} + 3\hat{j})$

$= 12(\hat{i} \times \hat{j}) + 12(\hat{j} \times \hat{i})$

$= 12(\hat{i} \times \hat{j}) - 12(\hat{i} \times \hat{j})$

$= 0$

Again, $\vec{A} \cdot \vec{B} = (4\hat{i} + 6\hat{j}) \cdot (2\hat{j} + 3\hat{i}) = 8 + 18 = 26$

Again, $\frac{|\vec{A}|}{|\vec{B}|} = \frac{\sqrt{16+36}}{\sqrt{4+9}} \neq \frac{1}{2}$

Again, $\vec{B} = \frac{1}{2}\vec{A}$

280 **(d)**

At maximum height H, the horizontal component of the velocity of the bullet $= u\cos\theta = u\cos 60° = u/2$

281 **(c)**

As seen from the cart the projectile moves vertically upward and comes back.

The time taken by cart to cover 80 m

$= \frac{s}{v} = \frac{80}{30} = \frac{8}{3}$s

Given, $u = ?, v = 0, a = -g = 10\text{ms}^{-2}$

(for a projectile going upward)

and $t = \frac{8/3}{2} = \frac{4}{3}$s

From first equation of motion

$v = u + at$

$0 = u - 10 \times \frac{4}{3}$

$= \frac{40}{3}\text{ms}^{-1}$

283 **(d)**

$\vec{A} \times \vec{B} = (\hat{i} - 2\hat{j} + 3\hat{k}) \times (3\hat{i} - 2\hat{j} + \hat{k})$

$= -2\hat{k} - \hat{j} - 6(-\hat{k}) - 2\hat{i} + 9\hat{j} - 6(-\hat{i})$

$= 4\hat{i} + 8\hat{j} + 4\hat{k}$

Modulus is $\sqrt{4^2 + 8^2 + 4^2} = \sqrt{32 + 64}$

$= \sqrt{96} = 4\sqrt{6}$units.

284 **(b)**

$v = \sqrt{\mu r g} = \sqrt{0.6 \times 150 \times 10} = 30m/s$

285 **(a)**

In this problem it is assumed that particle although moving in a vertical loop but its speed remain constant

Tension at lowest point $T_{\max} = \frac{mv^2}{r} + mg$

Tension at highest point $T_{\min} = \frac{mv^2}{r} - mg$

$\frac{T_{\max}}{T_{\min}} = \frac{\frac{mv^2}{r} + mg}{\frac{mv^2}{r} - mg} = \frac{5}{3}$

By solving we get, $v = \sqrt{4gr} = \sqrt{4 \times 9.8 \times 2.5} = \sqrt{98}\, m/s$

286 **(b)**

Range, $R = \frac{u^2\sin 2\theta}{g}$

$\therefore 20 = \frac{u^2\sin(2 \times 30°)}{g}$

$\Rightarrow \frac{u^2}{g} = \frac{20}{\sin 60°} = \frac{20}{\sqrt{3}} \times 2 = \frac{40}{\sqrt{3}}$

Now, $H = \frac{u^2\sin^2\theta}{2g}$

$= \frac{40}{\sqrt{3}} \times \frac{\sin^2 30°}{2}$

$= \frac{40}{\sqrt{3}} \times \frac{\left(\frac{1}{2}\right)^2}{2} = \frac{5}{\sqrt{3}}$m

287 **(a)**

Time of flight $= \frac{2u\sin\theta}{g} = \frac{2u_y}{g} = \frac{2 \times u_{\text{vertical}}}{g}$

288 **(c)**

$\vec{A} \cdot \vec{B} = AB\cos\theta = 6$

and $|\vec{A} \times \vec{B}| = AB\sin\theta = 6\sqrt{3}$

$\therefore \frac{AB\sin\theta}{AB\cos\theta} = \frac{6\sqrt{3}}{6} = \sqrt{3}$

or $\tan\theta = \sqrt{3}$

and $\theta = 60°$

289 **(c)**

All the balls are projected from the same height, therefore their velocities will be equal.

So, $v_1 = v_2 = v_3$

290 **(d)**

Horizontal range, $R = \frac{u^2 \sin^2 45°}{2g}$

$= \frac{u^2}{g}$

Maximum height, $H = \frac{u^2 \sin^2 45°}{g} = \frac{u^2}{4g}$

$\therefore \frac{R}{H} = \frac{4}{1}$

291 **(d)**

In the given condition friction provides the required centripetal force and that is constant. $i.e. m\omega^2 r =$ constant

$\Rightarrow r \propto \frac{1}{\omega^2} \therefore r_2 = r_1 \left(\frac{\omega_1}{\omega_2}\right)^2 = 9\left(\frac{1}{3}\right)^2$

$= 1\ cm$

292 **(c)**

Change in velocity $= 2v \sin(\theta/2) = 2v \sin 20°$

293 **(d)**

A coin files off when centrifugal force just exceeds the force of friction *ie*,

$mr\omega^2 \geq \mu mg$

or $\omega \geq \frac{\sqrt{\mu g}}{r}$

Thus ω does not depend upon mass and will remain the same

294 **(b)**

$v = r\omega = 20 \times 10 cm/s = 2m/s$

295 **(b)**

For same range angle of projection should be θ and $90 - \theta$

So, time of flights $t_1 = \frac{2u \sin\theta}{g}$ and

$t_2 = \frac{2u \sin(90 - \theta)}{g} = \frac{2u \cos\theta}{g}$

By multiplying $= t_1 t_2 = \frac{4u^2 \sin\theta \cos\theta}{g^2}$

$t_1 t_2 = \frac{2}{g}\frac{(u^2 \sin 2\theta)}{g} = \frac{2R}{g} \Rightarrow t_1 t_2 \propto R$

296 **(a)**

We know that the range of projectile projected with velocity u, making an angle θ with the horizontal direction up the inclined plane, whose inclination with the horizontal direction is θ_0, is

$R = \frac{u^2}{g\cos^2\theta_0}[\sin(2\theta - \theta_0) - \sin\theta_0]$

Here, $u = v, \theta = (90° + \theta), \theta_0 = \theta$

$\therefore R = \frac{v^2}{g\cos^2\theta_0}\{\sin[2(90° + \theta)] - \theta\} - \sin\theta\}$

$= \frac{v^2}{g\cos^2\theta_0}[\sin(180° + \theta) - \sin\theta]$

$= -\frac{v^2}{g\cos^2\theta_0} 2\sin\theta = -\frac{2u^2}{g}\tan\theta \sec\theta$

$= \frac{2v^2}{g}\tan\theta$ (in magnitude)

297 **(d)**

$R = 4H\cot\theta$, if $R = 4\sqrt{3}H$ then $\cot\theta = \sqrt{3} \Rightarrow \theta = 30°$

298 **(a)**

$t = \sqrt{\frac{2h}{g}} = \sqrt{\frac{2\times2}{9.8}} = 0.64$ s

$v = \frac{10}{t} = 15.62\ ms^{-1}$

$\therefore a = \frac{v^2}{R} = 163\ ms^{-2}$

299 **(b)**

$\hat{n} = \frac{\vec{A} \times \vec{B}}{AB\sin\theta} = \frac{A\hat{A} \times B\hat{B}}{AB\sin\theta} = \frac{\hat{A} \times \hat{B}}{\sin\theta}$

300 **(d)**

Required angle $= \frac{\pi}{2} - \frac{5\pi}{36} = \frac{18\pi - 5\pi}{36} = \frac{13\pi}{36}$ rad

301 **(d)**

$\frac{mv^2}{r} = 10 \Rightarrow \frac{1}{2}mv^2 = 10 \times \frac{r}{2} = 1\,J$

302 **(c)**

Here, $v = 900$ km h^{-1} $= \frac{900\times1000}{60\times60} = 250\ ms^{-1}$

$g = 9.8\ ms^{-2}$

For apparent weightlessness, $\frac{mv^2}{r} = mg$

$r = \frac{v^2}{g} = \frac{250 \times 250}{9.8}$

$= 6377.5$ m $= 6.4$ km

303 **(c)**

Let v be the velocity at the time of collision

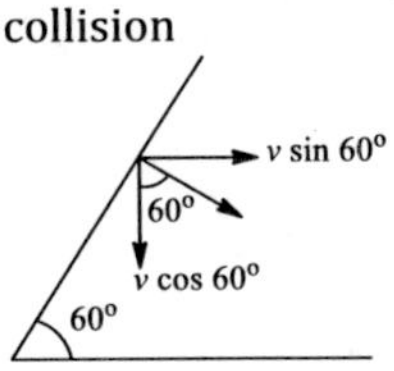

Then, $u\sqrt{2}\cos 45° = v \sin 60°$

$$(u\sqrt{2})\left(\frac{1}{\sqrt{2}}\right)=\frac{\sqrt{3}v}{2} \quad \therefore v=\frac{2}{\sqrt{3}}u$$

304 **(a)**

If it is being hit then

$d=v_0t+\frac{1}{2}at^2=(u\cos\theta)t$

or $t=\frac{u\cos\theta-v_0}{a/2}$

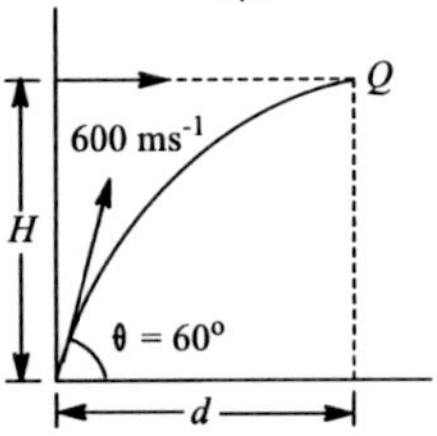

$\therefore\ t=\frac{600\times\frac{1}{2}-250}{10}=5\ \text{s}$

$H=(u\sin\theta)t-\frac{1}{2}\times gt^2$

$=600\times\frac{\sqrt{3}}{2}\times 5-\frac{1}{2}\times 10\times 25$

$H=2473$ m

305 **(a)**

The horizontal distance covered by the bomb,

$$BC=v_H\times\sqrt{\frac{2h}{g}}=150\sqrt{\frac{2\times 80}{10}}=600\ m$$

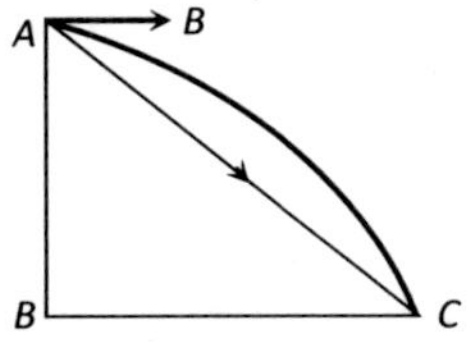

∴ The distance of target from dropping point of bomb,

$AC=\sqrt{AB^2+BC^2}=\sqrt{(80)^2+(600)^2}$
$=605.3\ m$

306 **(a)**

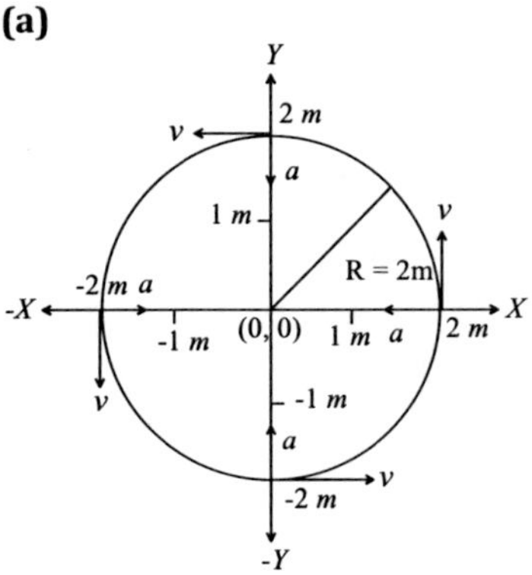

The radius of circular path is $2m$ and the speed of the object is $4m/s$

The magnitude of acceleration is

$a=\frac{v^2}{R}=\frac{16}{2}=8m/s^2$

The acceleration is directed towards the centre

Therefore, when an object is at $y=2m$, its acceleration is $-8\hat{j}\ m/s^2$

307 **(b)**

$\overrightarrow{\mathrm{F}}=\frac{d\overrightarrow{\mathrm{p}}}{dt}=(-2\sin t)\hat{\imath}+(2\cos t)\hat{\jmath}$

$\cos\hat{\theta}=\frac{\overrightarrow{\mathrm{F}}\cdot\overrightarrow{\mathrm{P}}}{Fp}=0$

$\therefore\ \theta=90°$

308 **(d)**

$\theta=30°$

$\frac{R}{H}=\frac{v^2(2\sin\theta\cos\theta)}{g}\times\frac{2g}{v^2\sin^2\theta}=\frac{4\cos\theta}{\sin\theta}$

or $R=4\cot 30°\times H=4\sqrt{3}$

309 **(d)**

There is no loss of energy. Therefore the final velocity is the same as the initial velocity

310 **(d)**

If the three vectors are coplanar then their scalar triple product is zero. So $(\vec{A}\times\vec{C}).\vec{B}=0$

or $\left[(2\hat{\imath}+3\hat{\jmath}-2\hat{k})\times(-\hat{\imath}+2\hat{\jmath}+3\hat{k})\right].\left[5\hat{\imath}+a\hat{\jmath}+\hat{k}\right]=0$

or $\left[13\hat{\imath}-4\hat{\jmath}+7\hat{k}\right].\left[5\hat{\imath}+a\hat{\jmath}+5\hat{k}\right]=0$

or $65-4a+7=0$ or $a=18$

311 **(d)**

Centripetal acceleration, $a_c=\frac{v^2}{R}$

Where v is the speed of an object and R is the radius of the circle

It is always directed towards the centre of the circle. Since v and R are constants for a given uniform circular motion, therefore the magnitude of centripetal acceleration is also constant. However, the direction of centripetal acceleration changes continuously. Therefore, a centripetal acceleration is not a constant vector

312 **(c)**

$y=ax-bx^2$

For height or y to be maximum

$\frac{dy}{dx}=0$ or $a-2bx=0$ or $x=\frac{a}{2b}$

$\therefore\ y_{max} = a\left(\frac{a}{2b}\right) - b\left(\frac{a}{2b}\right)^2 = \frac{a^2}{4b}$

and $\left(\frac{dy}{dx}\right)_{x=0} = a = \tan\theta$

where θ = angle of projection

$\therefore\ \theta \tan^{-1}(a)$

313 **(d)**

Work done in circular motion is always zero

314 **(d)**

Given, $s = t^3 + 5$

Speed, $v = \frac{ds}{dt} = 3t^2$

and rate of change of speed, $a_t = \frac{dv}{dt} = 6t$

$\therefore$ Tangential acceleration at $t = 2\ s$,

$a_t = 6 \times 2 = 12\ ms^{-2}$

and at $t = 2s, v = 3(2)^2 = 12 ms^{-1}$

$\therefore$ Centripetal acceleration, $a_c = \frac{v^2}{R}$

$= \frac{144}{20} ms^{-2}$

$\therefore$ Net acceleration $= a_t^2 + a_i^2 \approx 14 ms^{-2}$

315 **(c)**

At the two points of the trajectory during projection, the horizontal component of the velocity is the same

$\Rightarrow u \cos 60° = v \cos 45°$

$\Rightarrow 147 \times \frac{1}{2} = v \times \frac{1}{\sqrt{2}} \Rightarrow v = \frac{147}{\sqrt{2}}\ m/s$

Vertical component of $u = u \sin 60° = \frac{147\sqrt{3}}{2}\ m$

Vertical component of $v = v \sin 45° = \frac{147}{\sqrt{2}} \times \frac{1}{\sqrt{2}}$

$= \frac{147}{2}\ m$

but $v_y = u_y + a_y^t \Rightarrow \frac{147}{2} = \frac{147\sqrt{3}}{2} - 9.8t$

$\Rightarrow 9.8t = \frac{147}{2}(\sqrt{3} - 1) \Rightarrow t = 5.49\ s$

316 **(b)**

According to given problem $\frac{1}{2}mv^2 = as^2$

$\Rightarrow v = s\sqrt{\frac{2a}{m}}$

So $a_R = \frac{v^2}{R} = \frac{2as^2}{mR}$ ...(i)

Further more as

$a_t = \frac{dv}{dt} = \frac{dv}{ds}.\frac{ds}{dt} = v\frac{dv}{ds}$...(ii)

[By chain rule]

Which is light of equation (i) i.e. $v = s\sqrt{\frac{2a}{m}}$ yields

$a_t = \left[s\sqrt{\frac{2a}{m}}\right]\left[\sqrt{\frac{2a}{m}}\right] = \frac{2as}{m}$...(iii)

So that $a = \sqrt{a_R^2 + a_t^2} = \sqrt{\left[\frac{2as^2}{mR}\right]^2 \left[\frac{2as}{m}\right]^2}$

Hence $a = \frac{2as}{m}\sqrt{1 + [s/R]^2}$

$\therefore F = ma = 2as\sqrt{1 + [s/R]^2}$

317 **(b)**

Because here tension is maximum

318 **(b)**

When a body is projected at an angle θ with the horizontal with initial velocity u, then the horizontal range R of projectile is

$R = \frac{u^2 \sin 2\theta}{g}$

Clearly, for maximum horizontal range $\sin 2\theta = 1$ or $2\theta = 90°$ or $\theta = 45°$. Hence, in order to achieve maximum range, the body should be projected at 45°.

In this case

$R_{max} = \frac{u^2}{g}$

Hence, range of A and C are less than that of B.

319 **(b)**

$\hat{A} \cdot \hat{B} = (1)(1)\cos 0° = 1 \neq AB$.

320 **(b)**

$t = \sqrt{\frac{2h}{g}} = \sqrt{\frac{2 \times 490}{9.8}} = \sqrt{100} = 10\ s$

$x = vt = \left(60 \times \frac{5}{18}\right) ms^{-1} \times 10\ s = \frac{500}{3} m$

321 **(d)**

Since range is given to be the same therefore the other angle is $(90° - 30°)$, *ie*, 60°

$H = \frac{v^2 \sin^2 30°}{2g} = \frac{1}{4}\left[\frac{v^2}{2g}\right]$

$H' = \frac{v^2 \sin^2 60°}{2g} = \frac{3}{4}\left[\frac{v^2}{2g}\right]$

$\frac{H'}{H} = \frac{3}{4} \times \frac{4}{1} = 3$ or $H' = 3H$

322 **(c)**

The vertical component of velocity of the balls will be same if they stay in air for the same period of time. Hence vertical height attained will be same

323 **(b)**

Change in momentum is the product of force and time

$$\Delta p = mg \times \frac{2\sin\theta}{g}$$

$$= 2mv\sin\theta = 2mv\sin 45° = \frac{2mv}{\sqrt{2}}$$

$$= \sqrt{2}\,mv$$

324 **(c)**

Let A and B be the two forces. As per question

$\sqrt{A^2 + B^2} = 5$

or $A^2 + B^2 = 25$ (i)

and $A^2 + B^2 + 2AB\cos 120° = 13$

or $25 + 2AB \times (-1/2) = 13$

or $AB = 25 - 13 = 12$

or $2AB = 24$ (ii)

Solving (i) and (ii), we get

$A = 3\text{N}$

and $B = 4\text{ N}$

325 **(b)**

$v = \sqrt{gr} = \sqrt{10 \times 40} = 20ms^{-1}$

326 **(b)**

$y = bx^2$

Differentiating w.r.t. t, on both sides, we get

$$\frac{dy}{dt} = b2x\frac{dx}{dt} \Rightarrow v_y = 2bxv_x$$

Again, differentiating w.r.t. t, on both sides, we get

$$\frac{dv_y}{dt} = 2bv_x\frac{dx}{dt} + 2bx\frac{dv_x}{dt} = 2bv_x^2 + 0$$

$\left[\frac{dv_x}{dt} = 0\right.$, because the particle has constant acceleration along y-direction]

As per question

$$\frac{dv_y}{dt} = a = 2bv_x^2;\ v_x^2 = \frac{a}{2b} \Rightarrow v_x = \sqrt{\frac{a}{2b}}$$

327 **(c)**

$$v = 36\frac{km}{h} = 10\frac{m}{s} \therefore F = \frac{mv^2}{r}$$

$$= \frac{500 \times 100}{50}$$

$$= 1000\,N$$

328 **(d)**

$v\cos\theta = 10\cos 60° = 5\text{ ms}^{-1}$

329 **(b)**

Assuming particle 2 to be at rest, substituting in

$$y = x\tan\theta - \frac{gx^2}{2u^2\cos^2\theta} \quad (\theta = 0°)$$

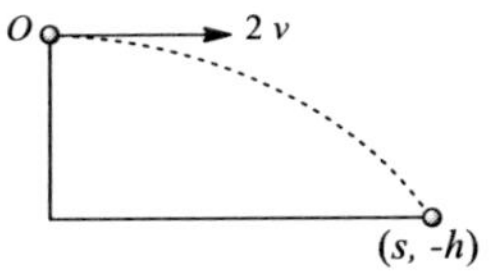

We have $-h = \frac{-g}{2(4v^2)}$

or $v = \sqrt{\frac{g}{8h}}$

Which is a straight line passing through origin with slope $\sqrt{\frac{g}{8h}}$

330 **(d)**

If the string suddenly breaks, the centripetal force will be zero only tangential force will be present, then the stone travels in tangential direction.

331 **(c)**

$$a_T = \frac{dv}{dt} = \frac{d}{dt}(2t) = 2m/s^2$$

$$a_c = \frac{V^2}{r} = \frac{(2 \times 3)^2}{30 \times 10^{-2}} = 120\,m/s^2$$

332 **(a)**

Acceleration $= \omega^2 r = \frac{v^2}{r} = \omega v = \frac{2\pi}{T}v$

333 **(b)**

Component of velocity perpendicular to plane remains the same (in opposite direction)

ie, $u\sin\theta = 20\sin 30° = 10\text{ ms}^{-1}$

334 **(d)**

$$\tan\theta = \frac{L}{A}$$

$$\tan 30° = \frac{10v}{3400}$$

$$v = \frac{340}{\sqrt{3}} = 196.3\text{ m/s}$$

335 **(c)**

Since displacement is long the Y-direction, hence displacement $\vec{s} = 10\hat{j}$.

Work done $= \vec{F}\cdot\vec{s} = (-2\hat{i} + 15\hat{j} + 6\hat{k}).\,10\hat{j} = 150\text{ J}$

336 **(c)**

$\alpha = \frac{d\omega}{dt} = 0$ [As ω = constant]

337 **(a)**

$(\vec{A} + \vec{B})\cdot(\vec{A} - \vec{B}) = 0$

$A^2 - B^2 = 0$ or $A = B$.

338 **(b)**

Since range is max, therefore $\theta = 45°$

Hence, $V_x = V\cos\theta = V\cos 45° = \frac{V}{\sqrt{2}}$

At the highest point, the net velocity of the projectile is

$V_x = V\cos 45°$

$\therefore \text{K.E.} = \frac{1}{2}mV_x^2 = \frac{1}{2}m\frac{V^2}{2} = 0.5\,K$

339 **(a)**

$F = \frac{mv^2}{r}$. If m and v are constants then

$F \propto \frac{1}{r}$

$\therefore \frac{F_1}{F_2} = \left(\frac{r_2}{r_1}\right)$

340 **(b)**

$\omega = \frac{v}{r} = \frac{10}{100} = 0.1 rad/s$

341 **(a)**

Here, $r = 92$ m, $v = 26$ ms^{-1}, $\mu = ?$

As $\frac{mv^2}{r} = F = \mu R = \mu\, mg$

$\mu = \frac{v^2}{rg} = \frac{26\times 26}{92\times 9.8} = 0.75$

342 **(a)**

$v_H = \sqrt{rg} = \sqrt{1\times 9.8} = 3.1$ ms^{-1}

343 **(c)**

$R_{\max} = \frac{u^2}{g} = 100 \Rightarrow u = 10\sqrt{10}$

$= 32\,m/s$

344 **(d)**

$\theta = \tan^{-1}\left(\frac{v^2}{rg}\right) = \tan^{-1}\left[\frac{(14\sqrt{3})^2}{20\sqrt{3}\times 9.8}\right]$

$= \tan^{-1}[\sqrt{3}] = 60°$

345 **(c)**

$h_1 = \frac{v^2\sin^2\alpha}{2g}, h_2 = \frac{v^2\cos^2\alpha}{2g}, \frac{h_1}{h_2}$

$= \tan^2\alpha$

346 **(d)**

In complete revolution change in velocity becomes zero so average acceleration will be zero

347 **(a)**

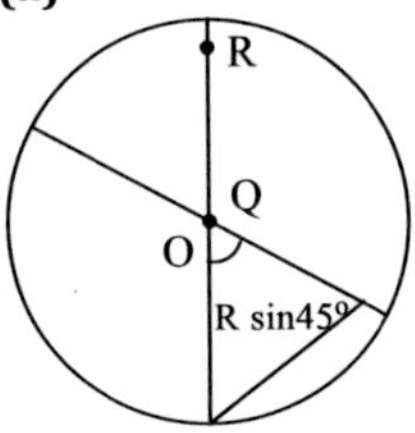

To reach the unshaded portion particle P needs to travel horizontal range greater than $R\sin 45°$ or $(0.7\,R)$ but its range is less than $\frac{R}{2}$. So it will fall on shaded portion

Q is near to origin, its velocity will be nearly along QR so its will fall in unshaded portion

348 **(b)**

$v = \sqrt{2gh} = \sqrt{2\times 10\times 0.2} = 2\,m/s$

349 **(d)**

$\cos\theta = \frac{(\hat{\imath}+2\hat{\jmath}+2\hat{k}).\hat{\imath}}{(1^2+2^2+2^2)^{1/2}} = \frac{1}{\sqrt{3}} = \frac{1}{3}$

$= 0.4472 = \cos 63°12'$

350 **(b)**

$\vec{S} = (10\hat{\imath} - 2\hat{\jmath} + 7\hat{k}) - (6\hat{\imath} + 5\hat{\jmath} - 3\hat{k})$

$= 4\hat{\imath} - 7\hat{\jmath} + 10\hat{k}$

$\vec{W} = \vec{F}\cdot\vec{S}$

$= (10\hat{\imath} - 3\hat{\jmath} + 6\hat{k})\cdot(4\hat{\imath} - 7\hat{\jmath} + 10\hat{k})$

$= (40 + 21 + 60)$J$=121$ J

351 **(b)**

Given, $r = 40$ m and g $= 10$m/s^2

we have $v = \sqrt{gr}$

$= 10\times 40 = \sqrt{400}$

$= 20$ m/s

352 **(b)**

Let a body be projected at a velocity u at an angle θ with the horizontal. Then horizontal range covered is given by

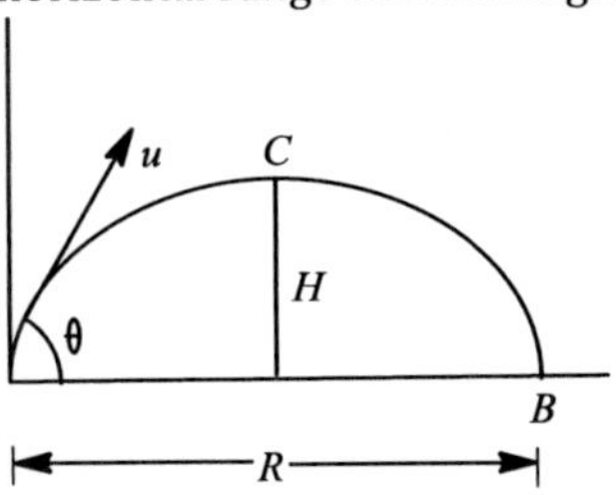

$R = \frac{u^2\sin 2\theta}{g}$ ... (i)

and height H is

$H = \frac{u^2\sin^2\theta}{2g}$ ... (ii)

Given, $R = 3H$

$\frac{u^2\sin 2\theta}{g} = 3\times\frac{u^2\sin^2\theta}{2g}$

Also, $\sin 2\theta = 2\sin\theta\cos\theta$

$\therefore \frac{u^2 2\sin\theta\cos\theta}{g} = 3\times\frac{u^2\sin^2\theta}{2g}$

or $2\cos\theta = 1.5\sin\theta$

or $\tan\theta = \frac{2}{1.5} = 1.33$

or $\theta = 53°7''$

Hence, angle of projection is $53°7'$

353 **(a)**

In this problem it is assumed that particle although moving in a vertical loop but it speed remain constant

Tension at lowest point $T_{\max} = \frac{mv^2}{r} + mg$

Tension at highest point $T_{\min} = \frac{mv^2}{r} - mg$

$$\frac{T_{\max}}{T_{\min}} = \frac{\frac{mv^2}{r} + mg}{\frac{mv^2}{r} - mg} = \frac{5}{3}$$

$\therefore v = \sqrt{4gr} = \sqrt{4 \times 9.8 \times 2.5}$

$= \sqrt{98}\text{ms}^{-1}$

354 **(d)**

$T\cos 30° = mg$

or $T = \frac{mg}{\cos 30°} = \frac{\sqrt{3} \times 9.8}{\sqrt{3}/2} = 19.6\text{N}$

$F = T\sin 30° = 19.6 \times \frac{1}{2} = 9.8\text{N}$

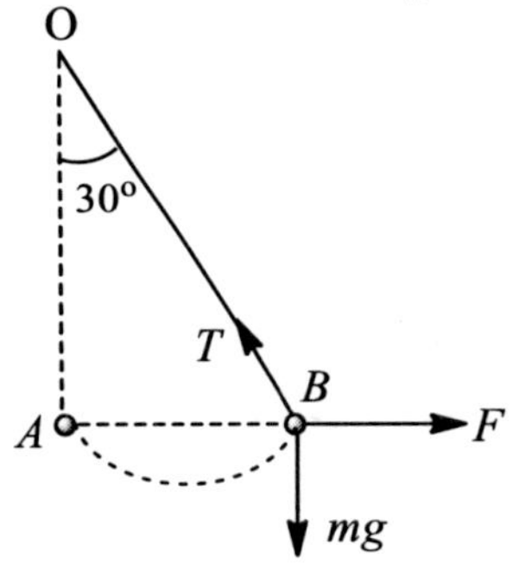

355 **(a)**

In figure, $T\sin\theta = \frac{mv^2}{r}; T\cos\theta = mg;$

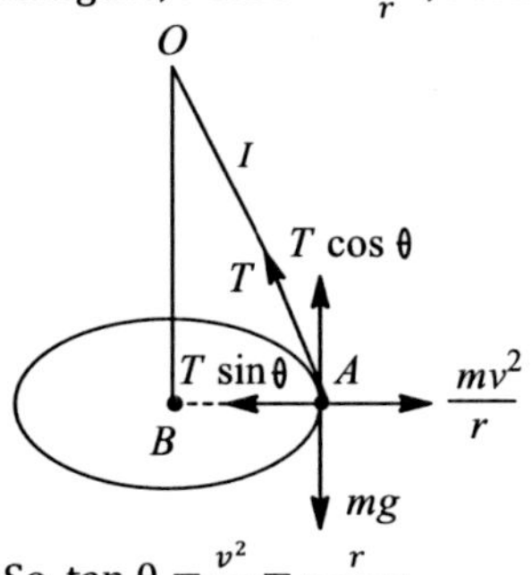

So, $\tan\theta = \frac{v^2}{rg} = \frac{r}{\sqrt{l^2 - r^2}}$

$$v = \left[\frac{r^2 g}{(l^2 - r^2)^{1/2}}\right]^{1/2}$$

$$= \left[\frac{0.09 \times 10}{(0.25 - 0.09)^{1/2}}\right]^{1/2}$$

$= 1.5\text{ ms}^{-1}$

356 **(b)**

$$R = \frac{u^2 \sin 2\theta}{g} = \frac{(500)^2 \times \sin 30°}{10}$$

$= 12.5 \times 10^3\ m$

357 **(c)**

Centripetal force is provided by friction, so

$\frac{mv^2}{r} < f_L$ ie, $\frac{mv^2}{r} < \mu mg$

i.e, $v < \sqrt{\mu gr}$ so that, $v_{\max} = \sqrt{\mu gr}$

Here, $\mu = 0.4, r = 30\text{m}$ and $g = 10\text{ms}^{-2}$

$\therefore v_{\max} = \sqrt{0.4 \times 30 \times 10} = 11\text{m/s}$

358 **(d)**

Tension at mean position, $mg + \frac{mv^2}{l} = 3mg$

$v = \sqrt{2gl}$

And if the body displaces by angle θ with the vertical

Then $v = \sqrt{2gl(1 - \cos\theta)}$

Comparing (i) and (ii), $\cos\theta = 0 \Rightarrow \theta = 90°$

359 **(b)**

Centripetal force is given by

$$F = \frac{mv^2}{R}$$

$$\Rightarrow F \propto \frac{1}{R}$$

or $\frac{F_2}{F_1} = \frac{R_1}{R_2}$

Given, $r_2 = 2r_1$

$\therefore \frac{F_2}{F_1} = \frac{R_1}{2R_1} = \frac{1}{2}$

or $F_2 = \frac{F_1}{2}$

therefore, centripetal force will become half.

360 **(b)**

$$E' = E\cos^2\theta = E\cos^2(45°) = \frac{E}{2}$$

361 **(c)**

$|\vec{A} \times \vec{B}| = AB\sin\theta$. As $\sin\theta \leq 1$, therefore $AB\sin\theta$ can not be more than AB.

362 **(b)**

$$T = mg + \frac{mv^2}{l} = mg + 2mg = 3mg$$

Where $v = \sqrt{2gl}$ from $\frac{1}{2}mv^2 = mgl$

363 **(c)**

Here, $r = 25\text{m}, v = 5\text{ ms}^{-1}, \text{m} = 500\text{ kg}$

$a_t = 1\text{ ms}^{-2}$,

$$a_r = \frac{v^2}{r} = \frac{5 \times 5}{25} = 1\text{ ms}^{-2}$$

$$a_{net} = \sqrt{a_r^2 + a_t^2} = \sqrt{1^2 + 1^2} = \sqrt{2}\text{ms}^{-2}$$

$F = ma_{net} = 500\sqrt{2}\text{N}$

364 **(d)**

$$t = \sqrt{\frac{2h}{g}} = \sqrt{\frac{2 \times 980}{9.8}} = 10\sqrt{2}\text{ s}$$

365 **(b)**

$\vec{F}_1 + \vec{F}_2 + \vec{F}_3 + \vec{F}_4 = (5\hat{i} - 5\hat{j} + 5\hat{k}) + (2\hat{i} + 8\hat{j} + 6\hat{k})$

$+(-6\hat{i} + 4\hat{j} - 7\hat{k}) + (-\hat{i} - 3\hat{j} - 2\hat{k})$

$= 4\hat{i} + 2\hat{k}$

This fore is in $y - z$ plane. Therefore, particle will move in $y - z$ plane.

366 **(a)**

$(0.5)^2 + (0.8)^2 + c^2 = 1$

$0.25 + 0.64 + c^2 = 1$

or $c^2 = 1 - 0.25 - 0.64 = 0.11$

or $c = \sqrt{0.11}$

367 **(a)**

Here, $2\pi r = 34.3 \Rightarrow r = \frac{3403}{2\pi}$ and $v = \frac{2\pi r}{T} = \frac{2\pi r}{\sqrt{22}}$

Angle of banking $\theta = \tan^{-1}\left(\frac{v^2}{rg}\right) = 45°$

368 **(c)**

$P + Q = 16$ (i)

$P^2 + Q^2 + 2PQ\cos\theta = 64$ (ii)

$\tan 90° = \frac{Q\sin\theta}{P + Q\cos\theta}$

$\infty = \frac{Q\sin\theta}{P + Q\cos\theta}$

$\Rightarrow P + Q\cos\theta = 0$ or $Q\cos\theta = -P$

From Eq. (ii)

$P^2 + Q^2 + 2P(-P) = 64$ or $Q^2 - P^2 = 64$

or $(Q - P)(Q + P) = 64$

or $Q - P = \frac{64}{16} = 4$ (iii)

Adding Eq. (i) and (iii), we get

$2Q = 20$ or $Q = 10$ units

From (i), $P + 10 = 16$ or $P = 6$ units

369 **(a)**

$R = 2H$ Given

We know $R = 4H \cot\theta \Rightarrow \cot\theta = \frac{1}{2}$

From triangle we can say that $\sin\theta = \frac{2}{\sqrt{5}}, \cos\theta = \frac{1}{\sqrt{5}}$

$\therefore$Range of projectile $R = \frac{2v^2 \sin\theta \cos\theta}{g}$

$= \frac{2v^2}{g} \times \frac{2}{\sqrt{5}} \times \frac{1}{\sqrt{5}} = \frac{4v^2}{5g}$

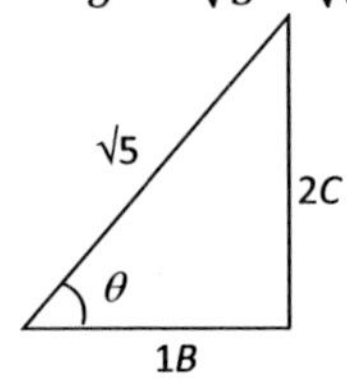

370 **(b)**

$F = mr\omega^2 = mr(2\pi v)^2$ *ie*, $F \propto v^2$

$\frac{2F}{F} = \left(\frac{v'}{v}\right)^2$ or $v' = v\sqrt{2} = 5\sqrt{2} = 7\text{rpm}$

371 **(a)**

Displacement, $\vec{r} = (\vec{r}_2 - \vec{r}_1)$ and

workdone $= \vec{F}.\vec{r}$

372 **(c)**

A particle performing a uniform circular motion has a transverse velocity and radial acceleration

373 **(b)**

$\vec{v} = 6\hat{i} - 8\hat{j}$

Comparing with

$\vec{v} = v_x\hat{i} = v_y = \hat{j}$, we get

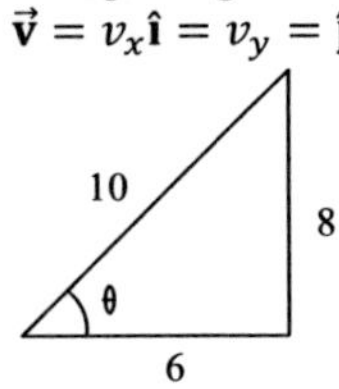

and $u_x = 6 \text{ ms}^{-2}$

Also, $u^2 = v_x^2 + v_y^2$

36+64=100

or $v = 10\text{ms}^{-1}$

$\sin\theta = \frac{8}{10}$ and $\cos\theta = \frac{6}{10}$

$R = \frac{v^2 \sin 2\theta}{g} = \frac{2v^2 \sin\theta \cos\theta}{g}$

$R = 2 \times 10 \times 10 \times \frac{8}{10} \times \frac{6}{10} \times \frac{1}{10}\text{m}$
$= 9.6 \text{ m}$

374 **(b)**

$60^2 = 30^2 + v^2$ or $v^2 = 90 \times 30$

or$v = 30\sqrt{3}\text{ km h}^{-1}$.

375 **(c)**

$x + y = 16$, Also $y^2 = 8^2 + x^2$

or $y^2 = 64 + (16 - y)^2$

$[\because x = 16 - y]$

or $y^2 = 64 + 256 + y^2 - 32y$

or $32y = 320$ or $y = 10\text{N}$

$\therefore x + 10 = 16$ or $x = 6\text{N}$

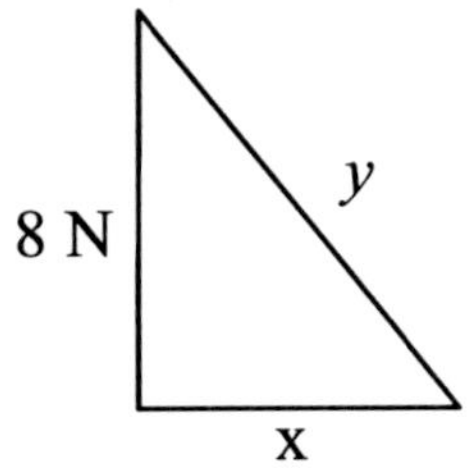

376 **(d)**

Maximum height and time of flight depend on the vertical component of initial velocity

$H_1 = H_2 \Rightarrow u_{y_1} = u_{y_2}$

Here $T_1 = T_2$

Range $R = \frac{u^2 \sin 2\theta}{g} = \frac{2(u \sin\theta)(u\cos\theta)}{g}$

$= \frac{2u_x u_y}{g}$

$R_2 > R_1 \therefore u_{x_2} > u_{x_1}$ or $u_2 > u_1$

377 **(c)**

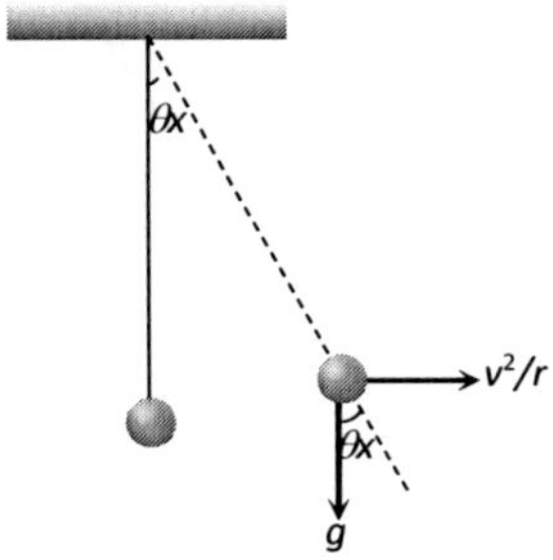

$\tan\theta = \frac{v^2}{rg}$

$\therefore \theta = \tan^{-1}\left(\frac{v^2}{rg}\right) = \tan^{-1}\left(\frac{10 \times 10}{10 \times 10}\right)$

$\therefore \theta = \tan^{-1}(1) = 45°$

378 **(a)**

$x = (u\cos\theta)t = 6t$

$y = (u\sin\theta)\,t - \frac{1}{2}gt^2 = 8t - 5t^2$

Therefore, $u\sin\theta = 8$

$u\cos\theta = 6$

Range. $R = \frac{u^2 2\sin 2\theta}{g}$

$= \frac{u^2 \times 2\sin\theta\cos\theta}{g}$

$= \frac{2(u\sin\theta)\,(\cos\theta)}{g}$

$= \frac{2(8)(6)}{10} = 9.6$ m

379 **(c)**

In projectile motion, horizontal component of velocity remains constant

$\therefore v\cos\theta = u\cos 2\theta$

$\Rightarrow v = \frac{u\cos 2\theta}{\cos\theta} = \frac{u(2\cos^2\theta - 1)}{\cos\theta}$

$= u(2\cos\theta - \sec\theta)$

380 **(d)**

Given $\theta_1 = \pi/3 = 30°$

Horizontal range is same if $\theta_1 + \theta_2 = 90°$

$\therefore \quad \theta_2 = 90° - 30° = 60°$

$y_1 = \frac{u^2 \sin^2 30°}{2g}$ and $y_2 \frac{u^2 \sin^2 60°}{2g}$

$\therefore \frac{y_2}{y_1} = \frac{\sin^2 30°}{\sin^2 60°} = \left(\frac{1/4}{\sqrt{3}/4}\right)^2 = \frac{1}{2}$ or y_2

$= \frac{y_1}{3}$

381 **(b)**

$t = \sqrt{\frac{2h}{g}}, x = v\sqrt{\frac{2h}{g}}$ or $v = x\sqrt{\frac{g}{2h}}$

382 **(c)**

Minimum angular velocity $\omega_{\min} = \sqrt{g/R}$

$\therefore T_{\max} = \frac{2\pi}{\omega_{\min}} = 2\pi\sqrt{\frac{R}{g}} = 2\pi\sqrt{\frac{2}{10}} \cong 3s$

383 **(b)**

Range $= \frac{u^2 \sin 2\theta}{g}$. It is clear that range is proportional to the direction (angle) and the initial speed.

384 **(d)**

The time of ascent = time of descent = t_0

T = total time of flight = $2t_0$

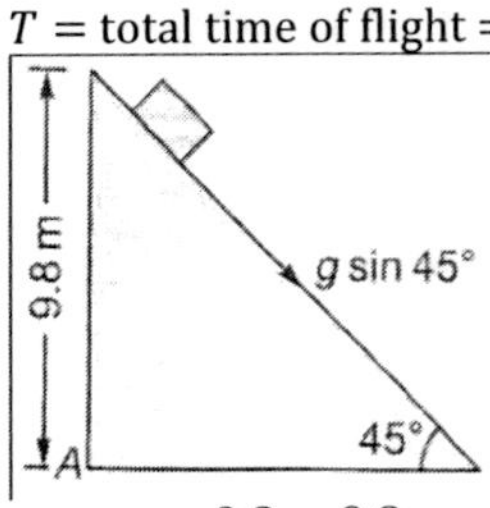

$\sin 45° = \frac{9.8}{BC} = \frac{9.8}{s}$

$\therefore \quad s = 9.8\sqrt{2}$

$\therefore \quad s = ut + \frac{1}{2}at^2$

$s = 0 \times t + \frac{1}{2}(g\sin 45°)t_0^2$

or $9.8\sqrt{2} = \frac{9.8}{2\sqrt{2}}t_0^2$

$\therefore \; t_0^2 = 4$

$\therefore \quad t_0 = 2s$

$\therefore\ T = 2t_0 - 4s$

385 **(b)**

The ball reaches nth step in time t, then
$bn = ut$
or $t = bn/u$,
$nh = \frac{1}{2}gt^2 = \frac{1}{2}g \times \frac{b^2n^2}{u^2}$; so $n = \frac{2u^2h}{gb^2}$
Time taken to travel vertical distance nh is

$$t = \sqrt{\frac{2nh}{g}} = \sqrt{\frac{2h}{g} \times \frac{2u^2h}{gb^2}} = \frac{2uh}{gb}$$

386 **(b)**

Since the projectile is released its initial velocity is the same as the velocity of the plane at the time of release
Take the origin at the point of release
Let x and $y(= -730\text{m})$ be the coordinates of the point on the ground where the projectile hits and let t be the time when it hits. Then

$$y = -v_0 t \cos\theta - \frac{1}{2}gt^2$$

where $\theta = 53.0°$
This equation gives

$$v_0 = -\frac{y + \frac{1}{2}gt^2}{t\cos\theta}$$

$$= -\frac{-730 + \frac{1}{2}(9.8)(5)^2}{5\cos 53°} = 202\ \text{ms}^{-1}$$

387 **(a)**

When body is released from the position p (inclined at angle θ from vertical) then velocity at mean position
$v = \sqrt{2gl(1-\cos\theta)}$
$\therefore$ Tension at the lowest point $= mg + \frac{mv^2}{l}$

$$= mg + \frac{m}{l}[2gl(1-\cos 60)] = mg + mg$$
$$= 2mg$$

388 **(a)**

$2\pi r = 34.3 \Rightarrow r = \frac{34.3}{2\pi}$ and $v = \frac{2\pi r}{T} = \frac{2\pi r}{\sqrt{22}}$

Angle of banking $\theta = \tan^{-1}\left(\frac{v^2}{rg}\right) = 45°$

389 **(a)**

$$mg = 1 \times 10 = 10N, \frac{mv^2}{r} = \frac{1 \times (4)^2}{1}$$
$$= 16$$

Tension at the top of circle $= \frac{mv^2}{r} - mg = 6N$

Tension at the bottom of circle $= \frac{mv^2}{r} + mg = 26N$

390 **(a)**

$\tan\theta = \frac{v^2}{rg} = \frac{30\times30}{900\times9.8} = 0.102$
$\therefore \theta = 6°$

391 **(c)**

$$a = \frac{v^2}{r} = \omega^2 r = 4\pi^2 n^2 r = 4\pi^2\left(\frac{22}{44}\right)^2 \times 1$$
$$= \pi^2 m/s^2$$

and its direction is always along the radius and towards the centre

392 **(d)**

$\omega_1 = 2\pi \times 300$ rad/min
$\omega_2 = 2\pi \times 100$ rad/min
Angular retardation

$$\alpha = \frac{\omega_1 - \omega_2}{2}$$
$$= \frac{2\pi \times 300 - 2\pi \times 100}{2}$$
$$= 2\pi \times 100\ \text{rad/min}^2$$
$$= 200\pi\ \text{rad/min}^2$$

394 **(b)**

Here the tangential acceleration also exists which requires power
Given that $a_C = k^2rt^2$ and $a_C = \frac{v^2}{r} \therefore \frac{v^2}{r} = k^2rt^2$
Or $v^2 = k^2r^2t^2$ or $v = krt$
Tangential acceleration $a = \frac{dv}{dt} = kr$
Now force $F = m \times a = mkr$
So power $P = F \times v = mkr \times krt = mk^2r^2t$

395 **(c)**

Because horizontal velocity is same for coin and the observer. So relative horizontal displacement will be zero

396 **(c)**

Horizontal component $= u\cos\theta$
Vertical component $= u\sin\theta$
$g = -10\text{ms}^{-2}, u = 50\text{ms}^{-1}, h = 5\text{m}, t = 2\text{ s}$

$$h = u_y t + \frac{1}{2}gt^2$$

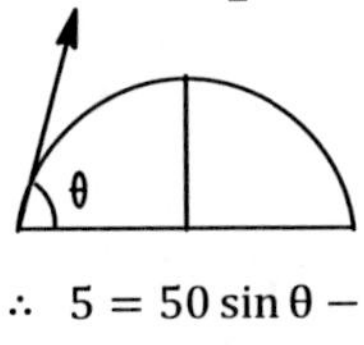

$$\therefore\ 5 = 50\sin\theta - \frac{1}{2} \times 10 \times 4$$
or $5 = 50\sin\theta - 20$

or $\sin\theta = \frac{25}{50} = \frac{1}{2}$

$\therefore\ \theta = 30°$

397 **(d)**

For horizontal motion,

$nw = v_0 t$ or $t = \dfrac{nw}{v_0}$

For vertical motion, $nh = \frac{1}{2}gt^2$

or $\frac{1}{2}g\left(\frac{n^2w^2}{v_0^2}\right) = nh$ or $n = \frac{2hv_0^2}{gw^2}$

398 **(c)**

Let v be the velocity of particle when it makes 30° with horizontal. Then

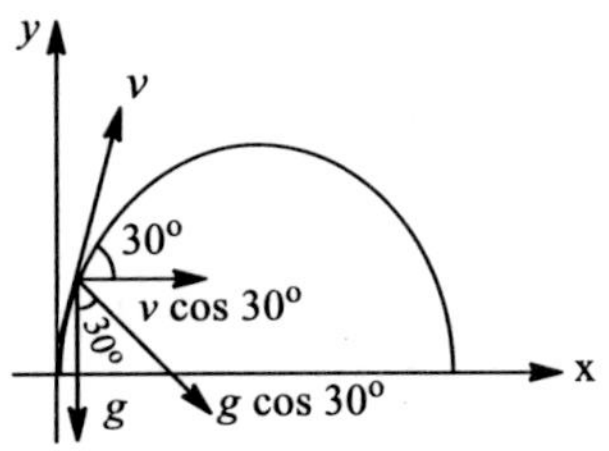

$v\cos 30° = u\cos 60°$

or $v = \frac{u\cos 60°}{\cos 30°} = \frac{20\left(\frac{1}{2}\right)}{\left(\frac{\sqrt{3}}{2}\right)} = \frac{20}{\sqrt{3}}\,\text{ms}^{-1}$

Now, $g\cos 30° = \frac{v^2}{R}$

or $R = \frac{v^2}{g\cos 30°} = \frac{\left(\frac{20}{\sqrt{3}}\right)^2}{(10)\frac{\sqrt{3}}{2}}$

$= 15.4$ m

400 **(c)**

Horizontal displacement of the bomb

AB = Horizontal velocity × time available

$$AB = u \times \sqrt{\frac{2h}{g}} = 600 \times \frac{5}{18} \times \sqrt{\frac{2 \times 1960}{9.8}}$$

$$= 3.33\ km$$

401 **(b)**

If $|\vec{A}| = |\vec{B}| = x$, then $|\vec{C}| = \sqrt{2}x$

Now, $\vec{A} + \vec{B} = -\vec{C}$

or $(\vec{A}+\vec{B})\cdot(\vec{A}+\vec{B}) = (-\vec{C})\cdot(-\vec{C})$

or $\cos\theta = 0$ or $\theta = 90°$

or $\vec{A}\cdot\vec{A} + \vec{C}\cdot\vec{C} + 2\vec{A}\cdot\vec{C} = B^2$

or $x^2 + 2x^2 + 2x^2\sqrt{2}\cos\theta = x^2$

or $\cos\theta = -\frac{1}{\sqrt{2}}$

$\therefore$ or $\theta = 135°$

Again, $\vec{B} + \vec{C} = -\vec{A}$

or $(\vec{B}+\vec{C})\cdot(\vec{B}+\vec{C}) = -(-\vec{A})\cdot(-\vec{A})$

or $x^2 + 2x^2 + 2x^2\sqrt{2}\cos\theta = x^2$

or $\cos\theta = -\frac{2x^2}{2x^2\sqrt{2}\cos\theta} = -\frac{1}{2}$ or $\theta =$ 135°

402 **(a)**

Let $\vec{A} + \vec{B} = \vec{R}$. Given $A_x = 7$ and $A_y = 6$

Also $R_x = 11$ and $R_y = 9$. Therefore,

$B_x = R_x - A_x = 11 - 7 = 4$

and $B_y = R_y - A_y = 9 - 6 = 3$

Hence, $B = \sqrt{B_x^2 + B_y^2} = \sqrt{4^2 + 3^2} = 5$

403 **(c)**

$\vec{A} \perp \vec{B}$, if $\vec{A}.\vec{B} = AB\cos 90° = 0$

$(2\hat{\imath} + a\hat{\jmath} + \hat{k})\cdot(4\hat{\imath} - 2\hat{\jmath} - \hat{k}) = 0$

or $8 - 2a - 2 = 0$ or $a = 3$.

404 **(c)**

In projectile motion given angular projection, the horizontal component velocity remains unchanged. Hence

$v\cos\alpha = u\cos\theta$ or $v = u\cos\theta\sec\alpha$

405 **(d)**

Angle made by the cyclist with vertical is given by

$\tan\theta = \dfrac{v^2}{rg}$

$\therefore\ \theta = \tan^{-1}\left(\dfrac{10 \times 10}{80 \times 10}\right)\left(\because v = 36 \times \dfrac{5}{18} = 10\ \text{ms}^{-1}\right)$

$= \tan^{-1}\left(\dfrac{1}{8}\right)$

406 **(c)**

From figure in equilibrium position

$T_1\sin 30° = T_2\sin 60°$

or $T_1 \times \frac{1}{2} = T_2 \times \frac{\sqrt{3}}{2}$

or $\frac{T_1}{T_2} = \sqrt{3}$

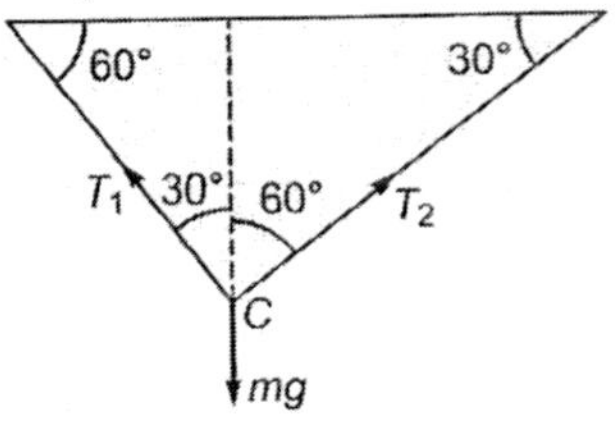

407 **(c)**

$\tan 90° = \frac{B\sin\theta}{A + B\cos\theta}$ or $A + B\cos\theta = 0$

or $\cos\theta = -A/B$ ....(i)

$R = \dfrac{B}{2} = [A^2 + B^2 + 2AB\cos\theta]^{1/2}$

or $\frac{B^2}{4} = A^2 + B^2 + 2AB(-A/B) = B^2 - A^2$

or $\frac{A^2}{B^2} = \frac{3}{4}$ or $\frac{A}{B} = \frac{\sqrt{3}}{2}$

From (i), $\cos\theta = -\frac{\sqrt{3}}{2} = \cos 150°$

408 **(c)**

$R = 4H \cot\theta$, if $R = 4H$ then $\cot\theta = 1 \Rightarrow \theta = 45°$

410 **(b)**

Given, $f = 1200$ rpm,

$r = 30 \text{ cm} = \frac{30}{100}\text{m}$

Acceleration of a point at the tip of the blade

= centripetal acceleration

$= \omega^2 r = (2\pi f)^2 r$

$= \left(2 \times \frac{22}{7} \times \frac{1200}{60}\right)^2 \times \frac{30}{100}$

$= 4740 \text{ ms}^{-2}$

411 **(a)**

$\vec{F} = F_x\hat{i} + F_y\hat{j}$ or $\vec{F} = 2\hat{i} - 3\hat{j}$.

412 **(d)**

$R = \frac{u^2 \sin 2\theta}{g}$ $\therefore R \propto u^2$. If initial velocity be doubled then range will become four times

413 **(d)**

Radial force $= \frac{mv^2}{r} = \frac{m}{r}\left(\frac{p}{m}\right)^2 = \frac{p^2}{mr}$[As $p = mv$]

414 **(c)**

$F = \sqrt{\left(\frac{P}{2}\right)^2 + \left(\frac{P}{2}\right)^2} = \sqrt{2}\frac{P}{2} = \frac{P}{\sqrt{2}}$

415 **(c)**

For projectile A

Maximum height, $H_A = \frac{u_A^2 \sin^2 45°}{2g}$

For projectile B

Maximum height $H_B = \frac{u_B^2 \sin^2 \theta}{2g}$

As per equation

$H_A = H_B$

$\frac{u_A^2 \sin^2 45°}{2g} = \frac{u_B^2 \sin^2 \theta}{2g}$

$\Rightarrow \frac{\sin^2\theta}{\sin^2 45°} = \frac{u_A^2}{u_B^2}$

$\Rightarrow \sin^2\theta = \left(\frac{u_A}{u_B}\right)^2 \sin^2 45°$

$\Rightarrow \sin^2\theta = \left(\frac{1}{\sqrt{2}}\right)^2 \left(\frac{1}{\sqrt{2}}\right)^2 \left[\because \frac{u_A}{u_B} = \frac{1}{\sqrt{2}} \text{ (Given)}\right]$

$\Rightarrow \sin\theta = \frac{1}{2} \Rightarrow \theta \sin^{-1}\left(\frac{1}{2}\right) = 30°$

416 **(d)**

Let $\vec{\mathbf{u}}_1$ and $\vec{\mathbf{u}}_2$ be the initial velocities of the two particles and θ_1 andθ_2 be their angles of projection with the horizontal The velocities of the two particles after time $t$are,

$\vec{\mathbf{v}}_1 = (u_1\cos\theta_1)\hat{\mathbf{i}} + (u_1 \sin\theta_1 - gt)\hat{\mathbf{j}}$and

$\vec{\mathbf{v}}_1 = (u_1\cos\theta_1)\hat{\mathbf{i}} + (u_2 \sin\theta_2 - gt)\hat{\mathbf{j}}$

Their relative velocity is $\vec{\mathbf{v}}_{12} = \vec{\mathbf{v}}_1 - \vec{\mathbf{v}}_2$

$= (u_1\cos\theta_1 - u_2\cos\theta_1)\,\hat{\mathbf{i}} + (u_1\sin\theta_1 - u_2\sin\theta_2)\hat{\mathbf{j}}$

Which is a constant. So the path followed by one, as seen by the other is a straight line, making a constant angle with the horizontal

417 **(a)**

Given $\vec{C} = |\vec{B}|\hat{j} \Rightarrow \vec{C} = 5\hat{j}$

Let $\vec{C} = \vec{A} + \vec{B} = A + 3\hat{i} + 4\hat{j}$

$5\hat{j} = A + 3\hat{i} + 4\hat{j}$

$\Rightarrow \vec{A} = -3\hat{i} + \hat{j}$

$|\vec{A}| = \sqrt{(3)^2 + (1)^2}$

$= \sqrt{10}$

418 **(c)**

Here $u_y = u\cos\theta = 15\cos\theta$

$u_x = u\sin\theta = 15\sin\theta$

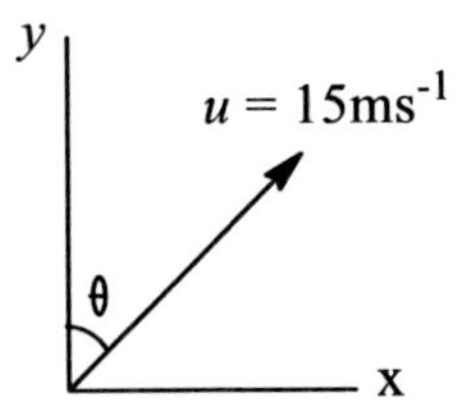

Time of flight of the ball is

$T = \frac{2u_y}{g} = \frac{2\times 15\cos\theta}{10} = 3\cos\theta$ ...(i)

The boy will catch the ball in time T, displacement of the ball in horizontal direction should also be zero, so

$0 = u_x T - \frac{1}{2}a_x T^2$

or $T = \frac{2u_x}{a_x} = \frac{2(15\sin\theta)}{3} = 10\sin\theta$ ...(ii)

From Eqs. (i) and (ii), $3\cos\theta = 10\sin\theta$

or $\tan\theta = \frac{3}{10} = 0.3$ or $\theta = \tan^{-1}(0.3)$

419 **(c)**

$q(\vec{v} \times \vec{B}) = \frac{mv^2}{r}$

or acceleration $= \frac{v^2}{r} = \frac{q}{m}(\vec{v} \times \vec{B})$

$= 9.6 \times 10^7 \times [(3\hat{i} + 2\hat{j})10^5 \times (2\hat{i} + 3\hat{k})]$

$= 9.6 \times 10^{12} \times [6\hat{i} - 9\hat{j} - 4\hat{k}]\text{ms}^{-2}$

420 **(a)**

If air resistance is taken into consideration then range and maximum height, both will decrease

421 **(c)**

Equation of trajectory for oblique projectile motion

$$y = x\tan\theta - \frac{gx^2}{2u^2\cos^2\theta}$$

Substituting $x = D$ and $u = v_0$

$$h = D\tan\theta - \frac{gD^2}{2u_0^2\cos^2\theta}$$

423 **(d)**

Maximum tension $= m\omega^2 r = m \times 4\pi^2 \times n^2 \times r$

By substituting the values we get $T_{\max} = 87.64\ N$

424 **(d)**

Here, $n = 2$ rps, $r = 5$ cm $= 0.05$ m, $\mu = ?$

As $\frac{mv^2}{r} = F = \mu R = \mu mg$

$$\mu = \frac{v^2}{rg} = \frac{(r2\pi n)^2}{rg} = \frac{4\pi^2 n^2 r}{g}$$

$$= 4 \times \frac{22}{7} \times \frac{22}{7} \times \frac{2^2 \times 0.05}{9.8} = 0.8$$

425 **(a)**

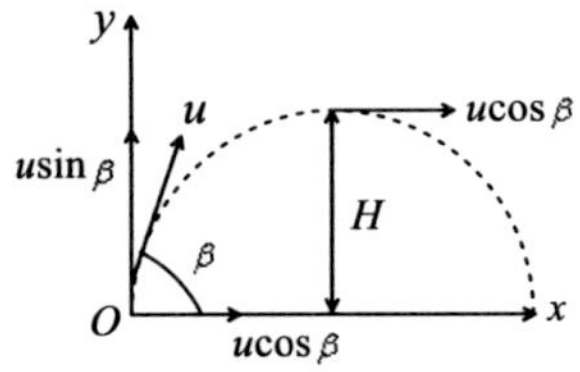

Initial kinetic energy at the point of projection O is

$$K = \frac{1}{2}mu^2$$

Where, m = mass of the body

u = initial velocity of the projection

At the highest point (i.e. at maximum height H)

Velocity is $v = u\cos\beta$,

Where β is the angle of projection

$\therefore$ Kinetic energy at the highest point is

$$K'' = \frac{1}{2}mv^2 \Rightarrow K'' = \frac{1}{2}m(u\cos\beta)^2$$
$$= \frac{1}{2}mu^2\cos^2\beta$$

According to given problem, $K'' = \frac{3}{4}K$

$$\frac{1}{2}mu^2\cos^2\beta = \frac{3}{4}\left(\frac{1}{2}mu^2\right) \Rightarrow \cos^2\beta = \frac{3}{4}$$

$\cos\beta = \frac{\sqrt{3}}{2}$ or $\beta = \cos^{-1}\left(\frac{\sqrt{3}}{2}\right) = 30°$

426 **(a)**

It is obvious from considerations of symmetry that at any moment of time all of time all of the persons will be at the corners of square whose side gradually decreases (see fig.) and so they will finally meet at the centre of the square O

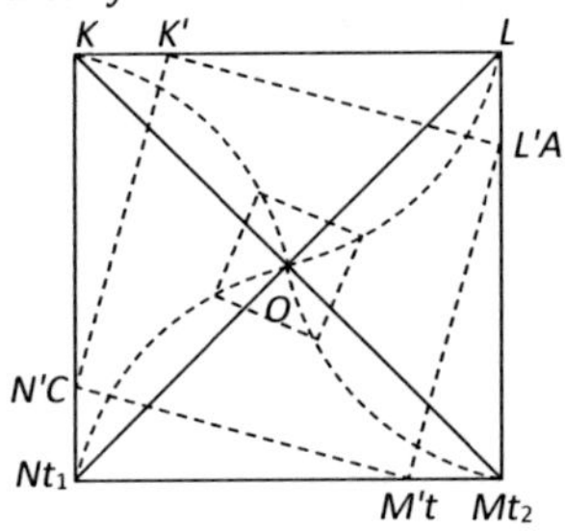

The speed of each person along the line joining his initial position and O will be

$v\cos 45 = v/\sqrt{2}$

As each person has displacement

$d\cos 45 = d/\sqrt{2}$ to reach the centre, the four persons will meet at the centre of the square O after time

$$\therefore t = \frac{d/\sqrt{2}}{v/\sqrt{2}} = \frac{d}{v}$$

427 **(b)**

Here, $T = \frac{2\pi r}{4v} = \frac{\pi r}{2v}$

Change in velocity is going from A to $B = v\sqrt{2}$

Average acceleration $= \frac{v\sqrt{2}}{\pi r/2v} = \frac{2\sqrt{2}v^2}{\pi r}$

428 **(a)**

Normal reaction at the highest point

$$R = \frac{nv^2}{r} - mg$$

Reaction is inversely proportional to the radius of the curvature of path and radius is minimum for path depicted in (a)

429 **(c)**

$$V_{\max} = \sqrt{\mu rg} = \sqrt{0.75 \times 60 \times 9.8}$$
$$= 21 m/s$$

430 **(b)**

To keep the mass M steady, let T is the tension in the string joining the two. Then for particle m,

$T = \frac{mv^2}{R}$ ... (i)

For mass M,

$T = Mg$ ... (ii)

From Eqs. (i) and (ii)

$$\frac{mv^2}{R} = Mg \Rightarrow v = \sqrt{\frac{MgR}{m}}$$

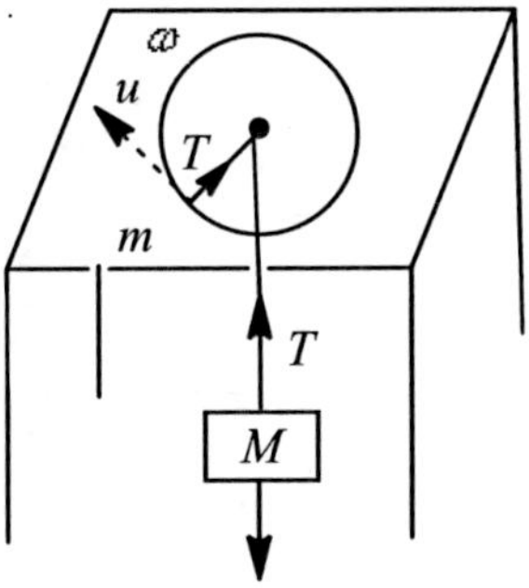

431 **(d)**

Maximum height attained is given by

$$h_{\max} = \frac{u^2}{2g}$$

Given, $u = 20\text{ms}^{-1}$

$$h_{\max} = \frac{(20)^2}{2 \times 10} = 20\text{m}$$

For the second body also $h_{\max} = 20\text{m}$

$\therefore$ Sum of maximum height =
20m+20m=40m

432 **(d)**

Given, initial velocity $= v_0$

final velocity $= 0$

deceleration $a = -\alpha x^2$ ....(i)

Let the distance travelled by the particle be s,

Now, we know that

$$a = \frac{dv}{dt} = \frac{dv}{dt} \times \frac{dx}{dx} = \frac{vdv}{dx}$$

or $a = v\dfrac{dv}{dx}$ ...(ii)

From Eqs. (i)and (ii)

$$v\frac{dv}{dx} = -\alpha x^2$$

or $vdv = -\alpha x^2 dx$

On integrating with limit $v_0 \to 0$ and $0 \to s$

$$\therefore \int_{v_0}^{0} vdv = \int_0^s -\alpha x^2 dx$$

or $\left[\dfrac{v^2}{2}\right]_{v_0}^{0} = -\alpha\left[\dfrac{x^3}{3}\right]_0^s$

$$-\frac{v_0^2}{2} = -\frac{\alpha(s)^3}{3}$$

or $\dfrac{v_0^2}{2} = \dfrac{\alpha s^3}{3}$

or $\dfrac{3v_0^2}{2\alpha} = s^3$

or $s = \left[\dfrac{3v_0^2}{2\alpha}\right]^{1/3}$

433 **(a)**

Distance covered in n revoulation=n$2\pi r = n\pi D$

$\Rightarrow 2000\pi D = 9500$

$$\Rightarrow D = \frac{9500}{2000 \times \pi} = 1.5\text{m}$$

434 **(d)**

$$h_{\max} = \frac{u^2}{2g} = 10 \quad [\because \theta = 90^\circ]$$

$u^2 = 200$

$$R_{\max} = \frac{u^2}{g} = 20m$$

435 **(c)**

$\alpha < \beta$ if $\vec{B} < \vec{A}$ or $B < A$

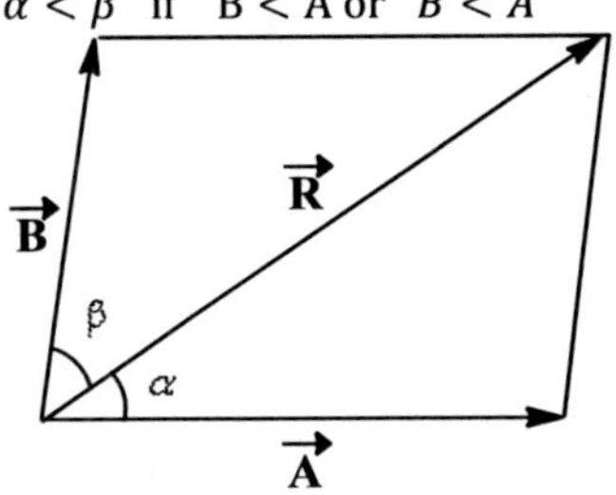

436 **(c)**

$R = \dfrac{u^2 \sin 2\theta}{g}$, At $\theta = 45^\circ, R = \max$.

$$\therefore R_{\max} = \frac{u^2}{g}$$

437 **(d)**

Here, $r = 50$ m

As $\tan\theta = \frac{v^2}{rg}$, therefore, when speed v is doubled; r must be made 4 times, if θ remains the same

$\therefore$ New radius of curvature,

$r' = 4r = 4 \times 50 \text{ m} = 200 \text{ m}$

438 **(b)**

$$T = mg + \frac{mv^2}{l} = mg + \frac{m}{l}[2gl(1 - \cos\theta)]$$

$$= mg + 2mg(1 - \cos 60^\circ) = 2mg = 2 \times 0.1 \times 9.8 = 1.96N$$

439 **(c)**

For angle $(45° - \theta), R = \frac{u^2 \sin(90°-2\theta)}{g} = \frac{u^2 \cos 2\theta}{g}$

For angle$(45° + \theta), R = \frac{u^2 \sin(90°+2\theta)}{g} = \frac{u^2 \cos 2\theta}{g}$

440 **(b)**

$nh = \frac{1}{2}gt^2 \Rightarrow t = \sqrt{\left(\frac{2nh}{g}\right)}$ ...(i)

Horizontal distance travelled by ball

$nb = ut, nb = u\sqrt{\left(\frac{2nh}{g}\right)}$ ....(ii)

Squaring Eq. (ii), we get

$n^2b^2 = \frac{u^2 2nh}{g}$

$\therefore n = \frac{2u^2h}{gb^2}$

441 **(b)**

For complementary angles range will be equal

442 **(d)**

Given, $m = 1 \times 10^{-3}$kg, ω

$= 1$ rad $s^{-1}$and $r = 1$m

Hence, centrifugal force $= m\omega^2 r = 10^{-3}$N

$1\text{N} = 10^5$dyne

$\therefore$ Centrifugal force $= 100$ dyne

443 **(a)**

Maximum speed $v = \sqrt{\mu rg}$

$= \sqrt{0.4 \times 30 \times 9.8}$

$= 10.84\text{ms}^{-1}$

444 **(c)**

$H = \frac{u^2\sin^2\theta}{2g} \Rightarrow \frac{H_1}{H_2} = \frac{u^2\sin^2\theta_1}{u^2\sin^2\theta_2}$

$\frac{3}{1} = \frac{\sin^2\theta_1}{\sin^2\theta_2} \Rightarrow \frac{\sin\theta_1}{\sin\theta_2} = \frac{\sqrt{3}}{1}$

Logically, we can conclude that

$\theta_1 = 60°, \quad \theta_2 = 30°$

Again $R = \frac{u^2\sin 2\theta}{g}$

$\therefore \frac{R_1}{R_2} = \frac{4\, u^2\sin 2\theta_1}{u^2\sin 2\theta_2}$

$\frac{R_1}{R_2} = \frac{4\sin 2(60°)}{\sin 2(30°)} = \frac{4\sin 120°}{\sin 60°}$

or $\frac{R_1}{R_2} = \frac{4 \times \frac{\sqrt{3}}{2}}{\frac{\sqrt{3}}{2}} = 4$

445 **(b)**

By using equation $\omega^2 = \omega_0^2 - 2\alpha\theta$

$\left(\frac{\omega_0}{2}\right)^2 = \omega_0^2 - 2\alpha(2\pi n) \Rightarrow \alpha = \frac{3}{4}\frac{\omega_0^2}{4\pi \times 36}, (n = 36)$ ..(i)

Now let fan completes total n' revolution from the starting to come to rest

$0 = \omega_0^2 - 2\alpha(2\pi n') \Rightarrow n' = \frac{\omega_0^2}{4\alpha\pi}$

Substituting the value of α from equation (i)

$n' = \frac{\omega_0^2}{4\pi}\frac{4 \times 4\pi \times 36}{3\omega_0^2} = 48$ revolution

Number of rotation $= 48 - 36 = 12$

446 **(c)**

At the lowest point, $\frac{mv^2}{r} = T_L - mg$ ...(i)

At the highest point, $\frac{mv^2}{r} = T_H + mg$ ...(ii)

As $\frac{T_{\max}}{T_{\min}} = \frac{T_L}{T_H} = 2$

$\therefore T_L = 2T_H$

From Eqs. (i) and (ii),

$2T_H - mg = T_H + mg$

$T_H = 2mg$

From Eq. (ii), $\frac{mv^2}{r} = 3mg$ or $\frac{v^2}{rg} = 3$

447 **(d)**

$\vec{A} = \hat{i} + \hat{j} + \hat{k}; A = \sqrt{1^2 + 1^2 + 1^2} = \sqrt{3}$

$\cos\theta = \frac{\vec{A}.\hat{i}}{\sqrt{3}} = \frac{1}{\sqrt{3}} = \frac{\sqrt{3}}{3}$

$= \frac{1.732}{3} = 0.5773 = \cos 54°44'$

$\theta = 54°44'$

450 **(c)**

$F = m\omega^2 R \therefore F \propto R$ (m and ω are constant)

If radius of the path is halved, then force will also become half

451 **(c)**

From $\frac{mv^2}{r} = F = \mu\, mg$

$\therefore v = \sqrt{\mu rg} = \sqrt{0.75 \times 60 \times 10} = \sqrt{450}$

$= 21\text{ms}^{-1}$

452 **(d)**

$T\cos\theta$ component will cancel mg.

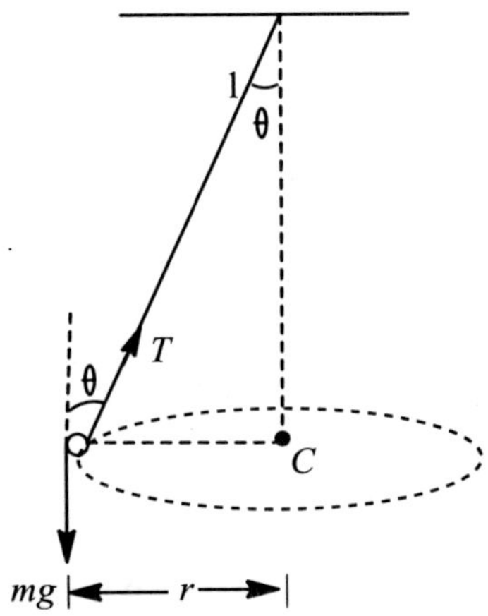

$T \sin \theta$ Component will provide necessary centripetal force the ball towards center C.

$\therefore T \sin \theta = mr\omega^2 = m(l \sin \theta)\omega^2$

or $T = ml\omega^2 \Rightarrow \omega = \sqrt{\frac{T}{\text{ml}}}$ rad/s

or ω_{max}

$= \sqrt{\frac{T_{\text{max}}}{ml}} = \sqrt{\frac{324}{0.5 \times 0.5}} = 36$ rad/s

453 **(a)**

One force must lie in between sum and difference of two other forces.

454 **(a)**

Here, centripetal force,

$F = \frac{mv^2}{l}$

But the angle between force and displacement is 90° because the direction of centripetal force is always towards the center and direction of displacement is always tangential.

Then work done

$W = \mathbf{F} \cdot \mathbf{s} = Fs \cos 90°$

$\Rightarrow W = 0$

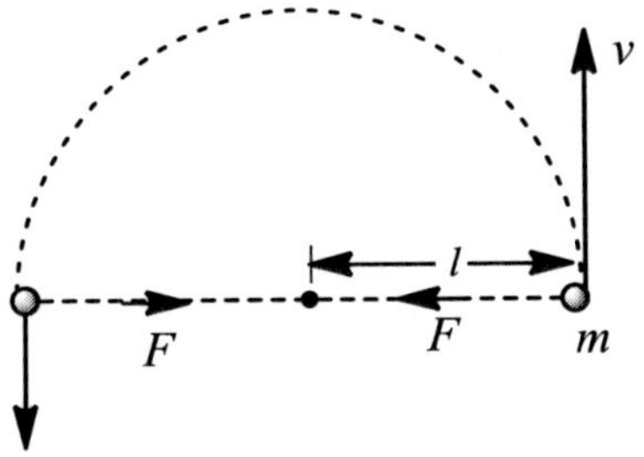

455 **(c)**

$F = m\omega^2 R \;\therefore F \propto \omega^2$ (m and R are constant)

If angular velocity is doubled force will becomes four times

456 **(c)**

$\frac{u_x}{u_y} = \cot 30° = \sqrt{3} \therefore u_x = 80\sqrt{3}$ ms^{-1}

$T = \frac{2u_y}{\text{g}} = \frac{2 \times 80}{10} = 16$ s

At $t = \frac{T}{4} = 4\, s, v_x = 80\sqrt{3}$ ms^{-1}

$v_y = 80 - 10 \times 4 = 40$ ms^{-1}

$\therefore 0\, v = \sqrt{(80\sqrt{3})^2 + (40)^2} = 140$ ms^{-1}

457 **(a)**

From $F = -\frac{dU}{dr}, dU = -F\, dr$

$U = \int -F\, dr = \int \frac{K}{r^2} dr = -\frac{K}{r}$

$\text{KE} = \frac{1}{2} \text{PE} = \frac{K}{2r}$

Total energy = KE + PE

$= \frac{K}{2r} - \frac{K}{r} = -\frac{K}{2r}$

458 **(c)**

$H = \frac{u^2 \sin^2 \theta}{2g} \Rightarrow \frac{H_1}{H_2} = \frac{\sin^2 \theta_1}{\sin^2 \theta_2}$

$\frac{\sqrt{3}}{1} = \frac{\sin \theta_1}{\sin \theta_2}$ So, $\frac{\cos \theta_1}{\cos \theta_2} = \frac{1}{\sqrt{3}}$

$\frac{R_1}{R_2} = \frac{(2u)^2 \sin 2\theta_1}{u^2 \sin 2\theta_2} = \frac{4. \sin \theta_1 \cos \theta_1}{\sin \theta_2 \cos \theta_2} = \frac{4}{1}$

459 **(c)**

Because vertical downward displacement of both (target and bullet) will be equal

460 **(d)**

It spends negligible time on earth *ie*, it performs projectile motion

Here maximum range $R_{\text{max}} = 1$ m

$\frac{u^2}{\text{g}} = 1$

$u^2 = 1 \times 9.8$

$u = \sqrt{9.8} = 3.13$ ms^{-1}

461 **(b)**

Work done by centripetal force is always zero

462 **(a)**

$\vec{\text{P}}.\vec{\text{Q}} = 0 \Rightarrow \vec{\text{P}} \perp \vec{\text{Q}}$ or $\theta = 90°$

$|\vec{\text{P}} \times \vec{\text{Q}}| = PQ \sin 90° = PQ$ or $|\vec{\text{P}}||\vec{\text{Q}}|$

463 **(c)**

$\alpha = \frac{\omega}{t}$ and $\omega = \frac{\theta}{t}$

$\therefore\ \alpha = \frac{\theta}{t^2}$

But α =constant

So, $\frac{\theta_1}{\theta_1+\theta_2} = \frac{(2)^2}{(2+2)^2}$

or $\frac{\theta_1}{\theta_1+\theta_2} = \frac{1}{4}$

or $\frac{\theta_1+\theta_2}{\theta_1} = \frac{4}{1}$

or $1 + \frac{\theta_2}{\theta_1} = \frac{4}{1}$

$\therefore\ \frac{\theta_2}{\theta_1} = 3$

464 **(d)**

Tension in the string $T = m\omega^2 r = 4\pi^2 n^2 mr$

$\therefore T \propto n^2 \Rightarrow \frac{n_2}{n_1} = \sqrt{\frac{T_2}{T_1}} \Rightarrow n_2 = 5\sqrt{\frac{2T}{T}}$
$= 7\ rpm$

466 **(c)**

$F = \frac{mv^2}{r}$. When v is doubled, F becomes 4 times

$\therefore$ Tendency to overturn is quadrupted

467 **(b)**

Given $(KE)_{\text{highest}} = \frac{1}{2}(\text{KE})$

$\frac{1}{2}mv^2\cos^2\theta = \frac{1}{2}.\frac{1}{2}mv^2$

$\cos^2\theta = \frac{1}{2} \Rightarrow \cos\theta = \sqrt{\frac{1}{2}}$

$\Rightarrow\ \theta = 45°$

468 **(a)**

$T = \frac{mv^2}{r} \Rightarrow 25 = \frac{0.25 \times v^2}{1.96} \Rightarrow v$
$= 14\ m/s$

469 **(c)**

$t = \sqrt{\frac{2 \times 8 \times 10^3}{10}} = 40\text{ s}$

$x = vt = 200 \times 40 = 8000\text{m}$

470 **(a)**

Using $v^2 - u^2 = 2as$, we get

$s = \frac{v^2}{2g}$

Now, $\frac{v^2\sin 2\theta}{g} = \frac{v^2}{2g}$ or $\sin 2\theta = \frac{1}{2}$

or $\sin 2\theta = \sin 30°$ or $\theta = 15°$

The other possible angle of projection is $(90° - 15°)$,

$ie, 75°$

471 **(b)**

$T_L - T_H = 6\ mg$

472 **(b)**

$H = \frac{u^2\sin^2\theta}{2g}$

$dH = \frac{2u\ \ \sin^2\theta}{2g}du$

$\therefore\ \frac{dH}{H} = \frac{2du}{u} = 2 \times \frac{1}{10}$

$\therefore$% increase in $H = \frac{dH}{H} \times 100$

$= \frac{2}{10} \times 100 = 20\%$

473 **(c)**

Let $\vec{C}$ be a vector perpendicular to $\vec{A}$ and $\vec{B}$

Then as per question $k\vec{C} = \vec{A} \times \vec{B}$

or $k = \frac{(\vec{A} \times \vec{B})}{\vec{C}}$

$= \frac{(2\hat{i} + 3\hat{j} + 6\hat{k}) \times (3\hat{i} - 6\hat{j} + 2\hat{k})}{(6\hat{k} + 2\hat{j} - 3\hat{k})}$

$= \frac{(42\hat{i} + 14\hat{j} - 21\hat{k})}{(6\hat{i} + 2\hat{j} - 3\hat{k})} = 7$

474 **(a)**

The value of frictional force should be equal or more than required centripetal force. i.e. $\mu mg \geq \frac{mv^2}{r}$

475 **(a)**

$|\vec{P}| = A + B \Rightarrow |\vec{P}|^2 = (A+B)^2$

$|\vec{A} + \vec{B}|^2 = (A+B)^2$

or $A^2 + B^2 + 2AB\cos\theta = A^2 + B^2 + 2AB$

or $\cos\theta = 1$ or $\theta = 0°$

476 **(c)**

$R = \frac{u^2\sin 2\theta}{g} = R \propto u^2$. So if the speed of projection doubled, the range will becomes four times,

$i.e., 4 \times 50 = 200\ m$

477 **(c)**

Tension, $T = \frac{mv^2}{r} + mg\cos\theta$

For, $\theta = 30°, T_1 = \frac{mv^2}{r} + mg\cos 30°$

$\theta = 60°, T_2 = \frac{mv^2}{r} + mg\cos 60°\ \therefore T_1 > T_2$

478 **(c)**

Taking vertical downward motion of projectile from point of projection to ground, we have

$u = -50\sin 30° = -25\text{ ms}^{-1}$

$a = +10\text{ ms}^{-2}, s = 70\text{ m}, t = ?$

$\therefore\ s = ut + \frac{1}{2}at^2;$

So, $70 = -25 \times t + \frac{1}{2} \times 10 \times t^2$

or $5t^2 - 25t - 70 = 0$ or $t^2 - 5t - 14 = 0$

On solving $t = 7$ s

479 **(a)**

Time of flight, $T = \frac{2u\sin\theta}{g}$

$\therefore\ dT = \frac{2du\sin\theta}{g}$

Now, $\therefore\ \frac{dT}{T} = \frac{du}{u} = \frac{1}{20}$

$\therefore$ % increase in $T = \frac{dT}{T} \times 100$

$= \frac{1}{20} \times 100 = 5\%$

480 **(d)**

From law of conservation of energy, potential energy of fall gets converted to kinetic energy.

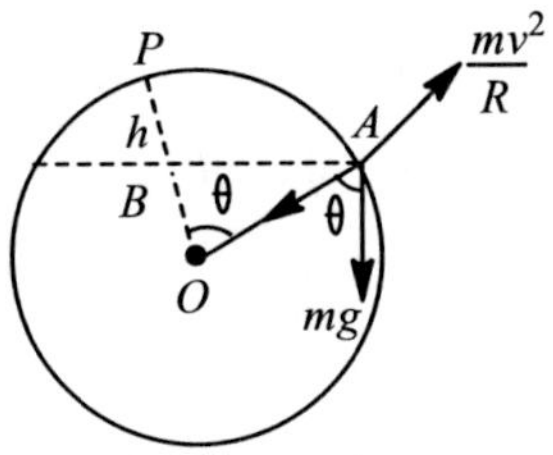

$\therefore$ PE = KE

$mgh = \frac{1}{2}mv^2$

or $v = \sqrt{2gh}$ .... (i)

Also, the horizontal component of force is equal centrifugal force.

$\therefore\ mg\cos\theta = \frac{mv^2}{R}$ ... (ii)

From Eq. (i)

$\therefore\ mg\cos\theta = \frac{2mgh}{R}$ ... (iii)

From ΔAOB,

$\cos\theta = \frac{(R-h}{R}$

$\Rightarrow mg\left(\frac{(R-h)}{R}\right) = \frac{2mgh}{R}$

$\Rightarrow\ 3h = R$

$\Rightarrow h = \frac{R}{3}$

481 **(d)**

$H = \frac{u^2\sin^2 45°}{2g} = \frac{u^2}{4g}$ ...(i)

$R = \frac{u^2\sin 90°}{g} = \frac{u^2}{g}$

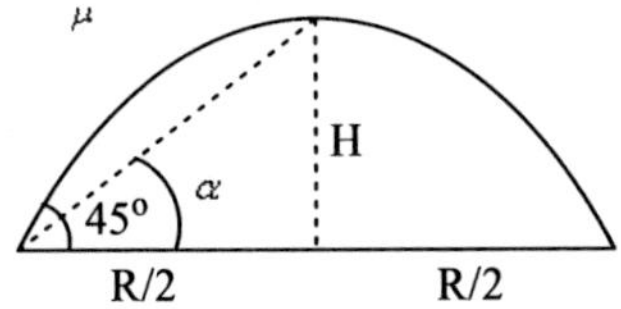

$\therefore \frac{R}{2} = \frac{u^2}{2g}$ ...(ii)

$\therefore \tan\alpha = \frac{H}{R/2}$

$= \frac{\frac{u^2}{4g}}{\frac{u^2}{2g}} = \frac{1}{2}\ \therefore\ \alpha = \tan^{-1}\left(\frac{1}{2}\right)$

482 **(a)**

Balancing the force, we get

$Mg - N = M\frac{v^2}{R}$

For weightlessness, $N = 0$

$\therefore\ \frac{Mv^2}{R} = Mg$

or $v = \sqrt{Rg}$

Putting the values, $R = 20$ m, $g = 10.0$ ms^{-2}

So, $v = \sqrt{20 \times 10.0} = 14.14$ ms^{-1}

Thus, the speed of the car at the top of the hill is between 14 ms^{-1} and 10 ms^{-1}

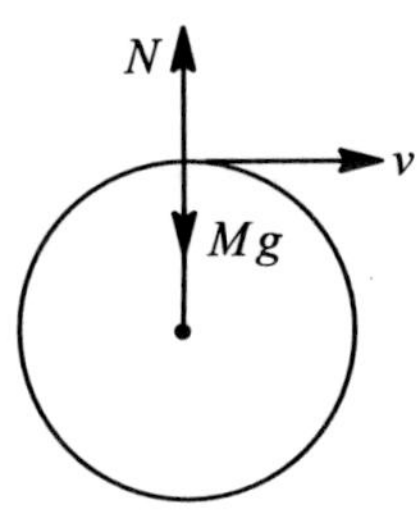

483 **(c)**

When wheel rolls half a revolution, the point (P) of the wheel which is in contact with the ground initially, moves at the top of the wheel as shown.

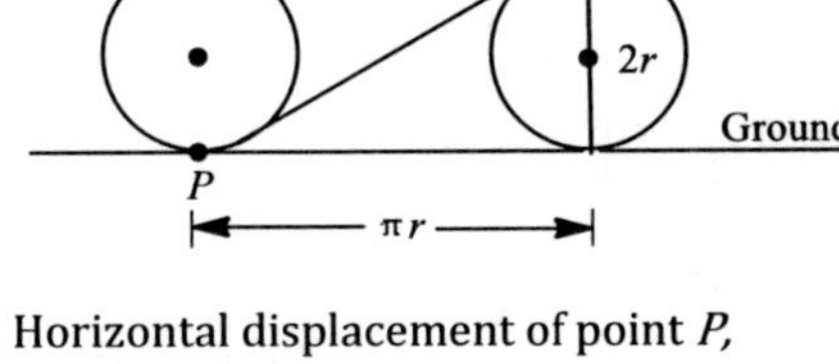

Horizontal displacement of point P, $y = 2r$

Net displacement $= \sqrt{x^2 + y^2}$

$= \sqrt{(\pi r)^2 + (2r)^2}$
$= r\sqrt{\pi^2 + 4}$
$= \sqrt{\pi^2 + 4}$

484 **(b)**

$v = 72km/hour = 20m/sec$

$\theta = \tan^{-1}\left(\frac{v^2}{rg}\right) = \tan^{-1}\left(\frac{20 \times 20}{20 \times 20}\right)$
$= \tan^{-1}(2)$

485 **(d)**

Tension at the top of the circle
$T = m\omega^2 r - mg$
$T = 0.4 \times 4\pi^2 n^2 \times 2 - 0.4 \times 9.8$
$= 115.86$ N

486 **(c)**

As we know for hemisphere the particle will leave the sphere at height $h = 2r/3$

$h = \frac{2}{3} \times 21 = 14m$

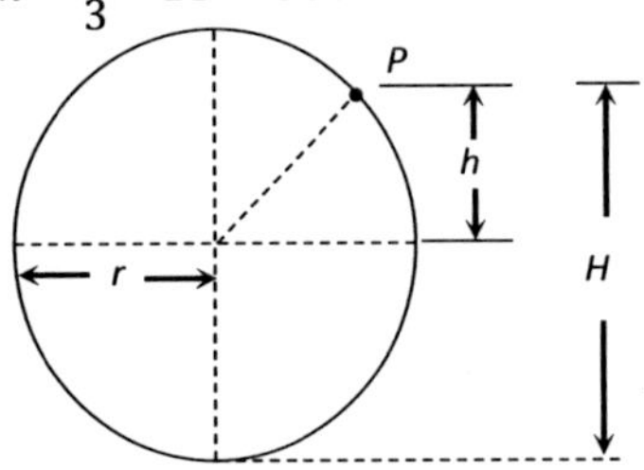

But from the bottom
$H = h + r = 14 + 21 = 35\ metre$

487 **(b)**

It is a vector quantity

488 **(b)**

Given , $|\vec{A} \times \vec{B}| = \sqrt{3}\vec{A} \cdot \vec{B}$
$AB \sin\theta = \sqrt{3}\ AB\cos\theta$
or $\tan\theta = \sqrt{3}$
$\theta = 60°$

489 **(c)**

Horizontal component of velocity at angle 60°
= Horizontal component of velocity at angle 45°
ie, $\cos 60° = u\cos 45°$

$147 \times \frac{1}{2} = v \times \frac{1}{\sqrt{2}}$

or $v = \frac{147}{\sqrt{2}}\text{ms}^{-1}$

vertical component of velocity at angle 60°

$u = u\sin 60° = \frac{147\sqrt{3}}{2}$m

vertical component of velocity at angle 45° $= v\sin 45°$

$= \frac{147}{\sqrt{2}} \times \frac{1}{\sqrt{2}} = \frac{147}{2}$m

But $v_y = u_y + at$

$\therefore \frac{147}{2} = \frac{147\sqrt{3}}{2} - 9.8t$

or $9.8t = \frac{147}{2}(\sqrt{3} - 1)$

$\therefore t = 5.49$ s

490 **(a)**

Centripetal acceleration
$a = \frac{v}{t}$

Angular acceleration $\alpha = \frac{\omega}{t} = \frac{\omega v}{vt}$

$\therefore \alpha = \frac{\omega a}{v}$

491 **(b)**

For banking $\tan\theta = \frac{V^2}{Rg}$

$\tan 45 = \frac{V^2}{90 \times 10} = 1$
$V = 30\ m/s$

493 **(c)**

Equating the moments about R
$6 \times PR = 4 \times RQ$
$PR = \frac{4}{6}RQ = \frac{2}{3}RQ$

495 **(d)**

Electrostatic force provides necessary centripetal force for circular motion of electron.

496 **(b)**

$R^2 = P^2 + P^2 + 2P^2\cos\theta$ or $R^2 = 2P^2 + 2P^2\cos\theta$
or $R^2 = 2P^2(1 + \cos\theta)$
or $R^2 = 2P^2\left(1 + 2\cos^2\frac{\theta}{2} - 1\right)$
or $R^2 = 4P^2\cos^2\frac{\theta}{2}$
or $R = 2P\cos\frac{\theta}{2}$

497 **(b)**

$10A^2 = 4A^2 + 2A^2 + 2 \times 2A \times \sqrt{2}A \times \cos\theta$
or $4A^2 = 4\sqrt{2}A\cos\theta$
or $\cos\theta = \frac{1}{\sqrt{2}} \Rightarrow \theta = 45°$

498 **(d)**

$u = -50\cos 60°\ \text{ms}^{-1} = -25\text{ms}^{-1}$
$a = 10\ \text{ms}^{-2}, t = ?, s = 70$ m
Using $s = ut + \frac{1}{2}at^2$, we get
$70 = -25 + 5t^2$ or $5t^2 - 25t - 70 = 0$
or $t^2 - 5t - 14 = 0$ or $t^2 - 7t + 2t - 14 = 0$

or $t(t-7)+2(t-7)=0$ or $(t-7)(t+2)=0$

or $t = 7s$ [Rejecting –ve value of t]

499 **(b)**

Given, $\vec{\mathbf{u}} = \hat{\mathbf{i}} + 2\hat{\mathbf{j}} = u_x\hat{\mathbf{i}} + u_y\hat{\mathbf{j}}$

Then $u_x = u\cos\theta$

and $u_y = 2 = u\cos\theta$

$\therefore \tan\theta = \dfrac{u\sin\theta}{u\cos\theta} = \dfrac{2}{1} = 2$

The equation of trajectory of a projectile motion is

$$y = x\tan\theta - \frac{gx^2}{2u^2\cos^2\theta}$$

$$= x\tan\theta - \frac{gx^2}{2(u\cos\theta)^2}$$

$$= x \times 2 - \frac{10 \times x^2}{2(1)^2} = 2x - 5x^2$$

500 **(a)**

$$T = \frac{2u\sin\theta}{g} \Rightarrow u = \frac{T \times g}{2\sin\theta} = \frac{2 \times 9.8}{2 \times \sin 30}$$
$$= 19.6\ m/s$$

501 **(b)**

$\tan\theta = \frac{v^2}{rg}$ or $\theta = \tan^{-1}\left(\frac{v^2}{rg}\right)$

502 **(b)**

$F = \frac{mv^2}{r}$. For same mass and same speed if radius is doubled then force should be halved

503 **(c)**

$x = \alpha t^3$ and $y = \beta t^3$ [Given]

$v_x = \frac{dx}{dt} = 3\alpha t^2$ and $v_y = \frac{dy}{dt} = 3\beta t^2$

Resultant velocity $= v = \sqrt{v_x^2 + v_y^2} = 3t^2\sqrt{\alpha^2 + \beta^2}$

504 **(b)**

$T\sin\theta = mr\omega^2 = m(l\sin\theta)\omega^2$

or $T = ml\omega^2 = ml\left(2\pi \times \frac{2}{\pi}\right)^2 = 16ml$

505 **(a)**

$$F = \frac{mv^2}{r} \Rightarrow v = \sqrt{\frac{rF}{m}}$$

506 **(b)**

Maximum height $H = \dfrac{u^2\sin^2\theta}{2g}$

and times of flight $T = \dfrac{2u\sin\theta}{g}$

or $T^2 = \dfrac{4u^2\sin^2\theta}{g^2}$

$\therefore \dfrac{T^2}{H} = \dfrac{8}{g}$

or $T = \sqrt{\dfrac{8H}{g}} = 2\sqrt{\dfrac{2H}{g}}$

507 **(c)**

For weightlessness state of a body on equator

$mg = mR\omega^2$

or $\omega = \sqrt{\frac{g}{R}} = \sqrt{\frac{10}{6400 \times 100}} = \frac{1}{800}$ rads^{-1}

508 **(b)**

$\vec{P} + \vec{Q} = \hat{\imath}$

$\vec{Q} = \hat{\imath} - \hat{\imath} + \hat{\jmath} - \hat{k}$

$= \hat{\jmath} - \hat{k}$

509 **(b)**

Maximum height of projectile

$$H = \frac{u^2\sin^2\theta}{2g}$$

$$\therefore H = \frac{(10)^2 \times \sin^2(30°)}{2 \times 10}$$

$$= \frac{5}{4} = 1.25 \text{ m}$$

Time to reach maximum height

$$t = \frac{u\sin\theta}{g}$$

$$\therefore t = \frac{10 \times \sin 30°}{10} = 0.5\text{s}$$

So, distance of vertical fall in 0.5s

$$h = \frac{1}{2}gt^2$$

or $h = \frac{1}{2} \times 10 \times (0.5)^2 = 1.25$ m

$\therefore$ Height of second ball $= 1.25 + 1.25$
$= 2.5$ m

510 **(b)**

Initial velocity

$v_1 = v$

Final velocity $v_2 = -v$

Initial momentum $p_1 = mv$

Final momentum $p_1 = m(-v) = -mv$

Change in momentum $\Delta p = p_1 - p_2$

$= mv - (-mv)$

$= 2mvj$

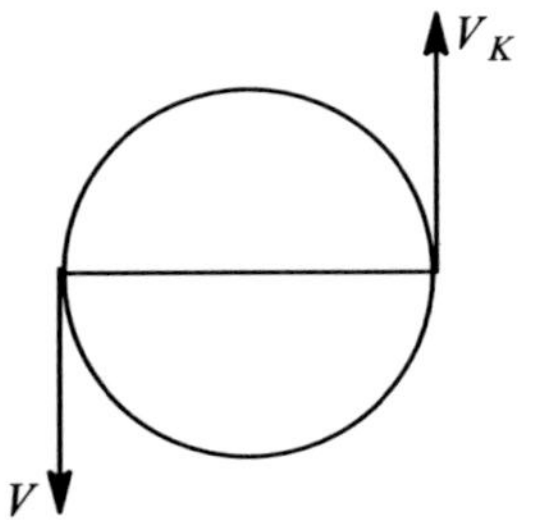

511 **(d)**

$$R = \frac{v^2 \sin 2\theta}{g} \Rightarrow \theta = \frac{1}{2}\sin^{-1}\left(\frac{gR}{v^2}\right)$$

512 **(b)**

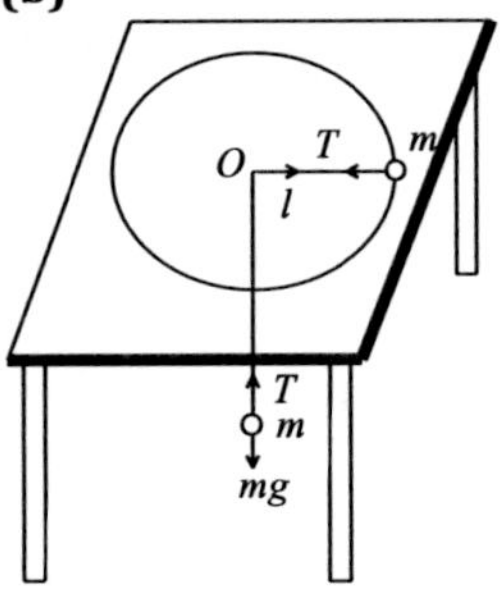

Tension T in the string will provide centripetal force

$\Rightarrow \frac{mv^2}{l} = T$ ...(i)

Also, tension T is provided by the hanging ball of mass m,

$\Rightarrow T = mg$ ...(ii)

$$mg = \frac{mv^2}{l} \Rightarrow g = \frac{v^2}{l}$$

It is the centripetal acceleration of a moving ball

513 **(a)**

$$(\text{KE})_L - (\text{KE})_H = \frac{1}{2}m(v_L^2 - v_H^2)$$

$$= \frac{1}{2}m(5\,gr - gr) = 2mgr$$

$= 2 \times 1 \times 10 \times 1 = 20$ J

514 **(b)**

Given, $R = H$

$$\frac{u^2 \sin 2\alpha}{g} = \frac{u^2 \sin^2 \alpha}{2g}$$

or $2 \sin\alpha \cos\alpha = \frac{\sin^2 \alpha}{2}$

or $\frac{\sin\alpha}{\cos\alpha} = 4$ or $\tan\alpha = 4$

$\therefore\ a = \tan^{-1}(4)$

515 **(c)**

The tension of the string,

$T = mr\omega^2$

$= 1 \times 1 \times (2)^2 = 4\text{N}$

517 **(c)**

$$S = u \times \sqrt{\frac{2h}{g}} \Rightarrow 10 = u\sqrt{2 \times \frac{5}{10}} \Rightarrow u$$

$$= 10\,m/s$$

518 **(a)**

As $S = t^3 + 5$

$$\frac{ds}{dt} = 3t^2 = v$$

$$\therefore a_t = \frac{dv}{dt} = 6t$$

at $t = 2\,sec$

$$|\vec{a}| = \sqrt{a_c^2 + a_t^2}$$

$$= \sqrt{\left(\frac{v^2}{R}\right)^2 + a_t^2} = \sqrt{\left(\frac{4t^4}{R}\right)^2 + \left(\frac{dv}{dt}\right)^2}$$

$$= \sqrt{(7.2)^2 + 144}$$

$$|\vec{a}| = 14m/s^2$$

519 **(c)**

If v velocity acquired at B, then

$v^2 = 2gh$

The particle will leave the sphere at B, when

$$\frac{mv^2}{r} \geq mg\cos\theta$$

$$\frac{2gh}{r} = \frac{g(r-h)}{r},$$

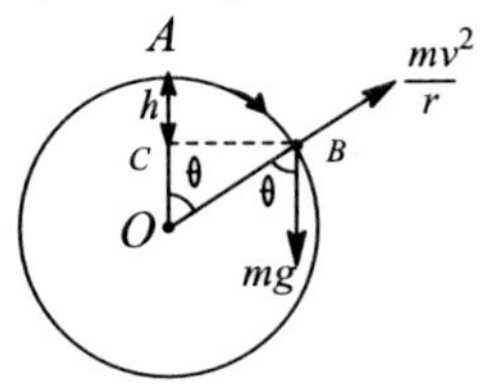

Which gives $h = \frac{r}{3}$

520 **(b)**

Area in which bullet will spread $= \pi r^2$

For maximum area, $r = R_{\max} = \frac{v^2}{g}$ [When $\theta = 45°$]

Maximum area $\pi R_{\max}^2 = \pi\left(\frac{v^2}{g}\right)^2 = \frac{\pi v^4}{g^2}$

521 **(b)**

$$\frac{v^2 \sin 2\theta}{g} = \frac{\sqrt{3}v^2}{2g}$$

or $\sin 2\theta = \frac{\sqrt{3}}{2}$ or $2\theta = 60°$ or $\theta = 30°$

Let us cross-check with the help of data for vertical range

$$\frac{v^2 \sin^2\theta}{2g} = \frac{v^2}{8g} \quad \text{or} \quad \sin^2\theta = \frac{1}{4}$$

or $\sin\theta = \frac{1}{2}$ or $\theta = 30°$

522 **(a)**

$T = m\omega^2 r \Rightarrow 10 = 0.25 \times \omega^2 \times 0.1 \Rightarrow \omega = 20\, rad/s$

524 **(a)**

$a = 4\pi^2 n^2 r = 4\pi^2 \left(\frac{1}{2}\right)^2 \times 50 = 493\, cm/s^2$

525 **(d)**

$t = \sqrt{\frac{2 \times 490}{9.8}} = \sqrt{\frac{2 \times 49 \times 100}{98}} = \sqrt{100}\text{ s} = 10\text{ s}$

526 **(a)**

$\frac{u^2 \sin 2\theta}{g} = 4\sqrt{3} \times \frac{u^2 \sin\theta}{2g}$

or $\frac{u^2}{g} 2\sin\theta\cos\theta = 2\sqrt{3}\frac{u^2}{g}\sin^2\theta$

527 **(a)**

Using the equation for projectile motion,

$y = x\tan\theta - \frac{gx^2}{2u^2}(1 + \tan^2\theta)$, we have

$40 = 30\tan\theta - \frac{g(30)^2}{2u}(1 + \tan^2\theta)$

or $900\tan^2\theta - 6u^2\tan\theta + (900 + 8u^2) = 0$

For real value of θ

$(6u^2)^2 \geq 4 \times 900(900 + 8u^2)$

or $(u^4 - 800u^2) \geq 90000$

or $(u^2 - 400)^2 \geq 200000$

or $u^2 \geq 900$ or $u \geq 30\text{ms}^{-1}$

528 **(b)**

$\vec{v} = \vec{\omega} \times \vec{r} = \begin{vmatrix} \hat{\imath} & \hat{\jmath} & \hat{k} \\ 3 & -4 & 1 \\ 5 & -6 & 6 \end{vmatrix} = -18\hat{\imath} - 13\hat{\jmath} + 2\hat{k}$

529 **(d)**

$\text{v} = ky\hat{\mathbf{i}} + kx\,\hat{\mathbf{j}}$

$\frac{dx}{dt} = ky, \frac{dy}{dt} = kx$

$\frac{dy}{dx} = \frac{dy}{dt} \times \frac{dt}{dx} = \frac{kx}{ky}$

$ydy = xdx$

$y^2 = x^2 + \text{c}$

530 **(d)**

$\alpha = \frac{\omega_2 - \omega_1}{t} = \frac{2\pi v_2 - 2\pi v_1}{t}$

$= \frac{2\pi\left(\frac{1200}{60} - \frac{600}{60}\right)}{10} = 2\pi\text{ rads}^{-2}$

531 **(b)**

Required numerical value is

$\sqrt{3^2 + 4^2 + 5^2}$, *ie*, $\sqrt{50}$ or $5\sqrt{2}$

532 **(c)**

Given, $m = 500$ kg,

$v = 36\text{ kmh}^{-1} = 36 \times \frac{5}{18} = 10\text{ ms}^{-1}$ and $r = 50$ m

Centripetal force $F = \frac{mv^2}{r}$

Hence, $F = \frac{500 \times (10)^2}{50} = 1000$ N

533 **(d)**

$h = \frac{5}{2}r \Rightarrow r = \frac{2}{5} \times h = \frac{2}{5} \times 5 = 2\,metre$

534 **(d)**

During time of flight vertical displacement becomes zero

i.e., $y = 0$

$KT(1 - \alpha T) = 0 \Rightarrow T = \frac{1}{\alpha}$

$\therefore$ Time taken by particle to attain max, height $t = T/2 = \frac{1}{2\alpha}$; $\therefore Y_{\max} = K.\frac{1}{2\alpha}\left(1 - \alpha.\frac{1}{2\alpha}\right) \Rightarrow = \frac{K}{4\alpha}$

535 **(a)**

$\frac{a_R}{a_r} = \frac{\omega_{R\times R}^2}{\omega_r^2 \times r} = \frac{T_r^2}{T_R^2} \times \frac{R}{r} = \frac{R}{r}\,[As\ T_r = T_R]$

536 **(c)**

Velocity of particle at maximum height h is $v' = v\cos\theta$ where $v =$ initial velocity of particle at which it is projected, $\theta =$ angle of projection

Angular momentum, $L = mv'h = mv\cos\theta\, h$

$= mvh\cos 45° = \frac{mvh}{\sqrt{2}}$

537 **(a)**

$(2\hat{\imath} - 3\hat{\jmath} + \hat{k}).(3\hat{\imath} + 3\hat{\jmath}) = 6(\hat{\imath}.\hat{\imath}) - 6(\hat{\jmath}.\hat{\jmath}) = 0$

539 **(b)**

$R = \frac{2 \times 30 \times 30 \sin 30° \cos 60°}{10\cos^2 30°}$

$= 180 \times \frac{1}{2} \times \frac{1}{2} \times \frac{2 \times 2}{3}\text{m} = 60\text{ m}$

540 **(c)**

$H_{\max} = \frac{u^2\sin^2\theta}{2g}$

According to problem $\frac{u_1^2\sin^2 45°}{2g} = \frac{u_2^2\sin^2 60°}{2g}$

$\Rightarrow \frac{u_1^2}{u_2^2} = \frac{\sin^2 60°}{\sin^2 30°} \Rightarrow \frac{u_1}{u_2} = \frac{\sqrt{3}/2}{1/\sqrt{2}} = \sqrt{\frac{3}{2}}$

541 **(a)**

Here, $\overrightarrow{v_1}$=30km h^{-1} due north = $\overrightarrow{OA}$

$\overrightarrow{v_2}$ = 40kmh^{-1} due east = $\overrightarrow{OB}$

Change in velocity in 20 s

$\Delta\vec{v} = \overrightarrow{v_2} - \overrightarrow{v_1} = \overrightarrow{v_2} + (\overrightarrow{-v_1})$

$= \overrightarrow{OB} + \overrightarrow{OC} = \overrightarrow{OD}$

$|\Delta\vec{v}| = \sqrt{v_2^2 + v_1^2} = \sqrt{40^2 + 30^2}$

$= 50$ kmh^{-1}

Acceleration , $\vec{a} = \frac{|\Delta \vec{v}|}{\Delta t}$

$= \frac{50}{20} = 2.5$ kmh^{-2}

Tan $\beta = \frac{v_1}{v_2} = \frac{30}{40}$

$= 0.75 = \tan 37°$

$\therefore\ \beta = 37°$ north of east

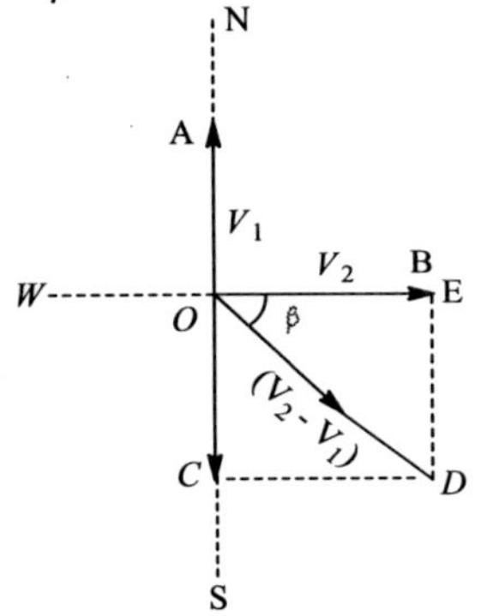

542 **(b)**

Using the relation

$\frac{mv^2}{r} = \mu R, \qquad R = mg$

$\frac{mv^2}{r} = \mu mg \Rightarrow v^2 = \mu rg$

$v^2 = 0.6 \times 150 \times 10$

$v = 30$ms^{-1}

543 **(c)**

Horizontal distance of target is 200 m.

Speed of bullet = 2000 ms^{-1}

Time taken by bullet to cover the horizontal distance

$t = \frac{200}{2000} = \frac{1}{10}$s

During$\frac{1}{10}$ s, the bullet will fall down vertically due to gravitational acceleration.

Therefore, height above the target, so that the bullet hits the target is

$h = u + \frac{1}{2}gt^2 = \left(0 \times \frac{1}{10}\right) + \frac{1}{2} \times 10 \times (0.1)^2$

$= 0.05$m $= 5$cm

544 **(c)**

Kinetic energy = potential energy

$\frac{1}{2}m(kv_e)^2 = \frac{mgh}{1 + \frac{h}{R}}$

$\Rightarrow \frac{1}{2}mk^2 2gR = \frac{mgh}{1 + \frac{h}{R}} \Rightarrow h = \frac{Rk^2}{1 - k^2}$

Height of projectile from the earth's surfac $= h$

Height from centre $r = R + h$

$= R + \frac{Rk^2}{1 - k^2}$

By solving $r = \frac{R}{1 - k^2}$

545 **(b)**

Since the ball reaches from one player to another in 2 s, so the time period of the flite, $T = 2$s

or $\frac{2u\sin\theta}{g} = 2$

or $u\sin\theta = g$ ..(i)

Now, we know that the maximum height of the projection.

$H = \frac{u^2\sin^2\theta}{2g}$

or $H = \frac{(u\sin\theta)^2}{2g}$

On putting the value of$u\sin\theta$ from Eq.(i), we have

$H = \frac{g^2}{2g} = \frac{g}{2}$

or $H = \frac{g}{2} = \frac{10}{2}$ m or $H = 5$m

546 **(d)**

The velocity at B is v, where

$v^2 = u^2 - 2g\,L$, in vertically upward direction. As is clear from figure change in velocity

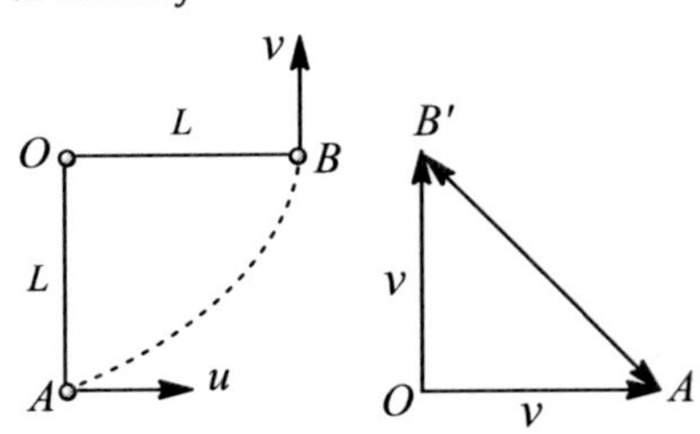

$\overrightarrow{\mathbf{OB'}} = \overrightarrow{\mathbf{OA}} = \overrightarrow{\mathbf{A'B'}}$

$= \sqrt{u^2 + v^2} = \sqrt{u^2 + (u^2 - 2gL)}$

$= \sqrt{2(u^2 - gL)}$

547 **(c)**

Centripetal acceleration= $4\pi^2 n^2 r = 4\pi^2 \times (1)^2 \times 0.4 = 1.6\pi^2$

548 **(d)**

$$R = \frac{v^2 \sin\ 2\theta}{g}$$

In the given problem v^2 sin2θ = constant

$$v^2 \sin 2\theta = \left(\frac{v}{2}\right)^2 \sin 30° = \frac{v^2}{8}$$

or $\sin 2\theta = \frac{1}{8}$ or $2\theta = \sin^{-1}\left[\frac{1}{8}\right]$ or $\theta = \frac{1}{2}\sin^{-1}\left[\frac{1}{8}\right]$

549 **(d)**

$$v_y = \frac{d}{dt}(y) = \frac{d}{dt}(10t) - \frac{d}{dt}(t^2) = 10 - 2t$$

At maximum height, $v_y = 0$

$\therefore$ $10 - 2t = 0$ or $2t = 10$ or $t = 5$ s

$\therefore$ $y = (10 \times 5 - 5 \times 5)\text{m} = 25$ m

550 **(d)**

$$120\ rev/\text{min} = 120 \times \frac{2\pi}{60}\ rad/\text{sec} = 4\pi\ rad/sec$$

551 **(a)**

$$(\overrightarrow{\text{PO}}) = \vec{r} = (2-3)\hat{\imath} + (-1-3)\hat{\jmath} + (4+1)\hat{k}$$

$= -1\hat{\imath} - 4\hat{\jmath} + 5\hat{k}$

Work done $= \vec{F} \cdot \vec{r}$

$= (4\hat{\imath} - 3\hat{\jmath} + 3\hat{k}) \cdot (-1\hat{\imath} - 4\hat{\jmath} + 5\hat{k})$

$= -4 + 12 + 15 = 23$ J

552 **(b)**

$$T_{\text{top}} = \frac{mv^2}{r} - mg = 2mg$$

$$\frac{T_{\text{top}}}{T_{\text{bottom}}} = \frac{2mg}{2mg + 6mg} = \frac{1}{4}$$

554 **(b)**

$v^2 = u^2 + 2as$

At max. height $v = 0$ and for upward direction $a = -g$

$\therefore u^2 = 2gs \Rightarrow s = \frac{u^2}{2g}$; $\because s_e = s_p$

$$\left(\frac{u_e}{u_p}\right)^2 = \left(\frac{g_e}{g_p}\right) \Rightarrow \left(\frac{5}{3}\right)^2 = \frac{9.8}{g_p} \Rightarrow g_p = 3.5\ m/s^2$$

555 **(b)**

$$R_{\text{max}} = \frac{u^2}{g} = 16 \times 10^3$$

$\Rightarrow u = 400 m/s$

556 **(b)**

For maximum range $\theta = 45°$

$$\frac{R_{\text{max}}}{T^2} = \frac{u^2 \sin 2\theta}{g} \Big/ \frac{4u^2 \sin^2\theta}{g^2}$$

$$\Rightarrow \frac{R_{\text{max}}}{T^2} = \frac{\sin 90° \times g}{4 \times \sin^2 45°} = \frac{49}{10}$$

557 **(d)**

For looping the loop minimum velocity at the lowest point should be $\sqrt{5gl}$

558 **(a)**

The centre of gravity of other tube will be at length $L/2$

So radius $r = L/2$

Centripetal force, $= M\, r\, \omega^2 = M(L/2)\omega^2 = ML\omega^2/2$

559 **(a)**

Here, $r = 100$m, $v = 7\text{ms}^{-1}$, $m = 60$kg

Reading registered= resultant force =?

Two force are acting, weight mg and centripetal force $\frac{mv^2}{r}$ at 90° to each other

$\therefore$ Resultant force $= \sqrt{(mg)^2 + \left(\frac{mv^2}{r}\right)^2} = mg\left[1 + \left(\frac{v^2}{rg}\right)^2\right]^{1/2}$

$$= 60 \times 9.8\left[1 + \left(\frac{7 \times 7}{100 \times 9.8}\right)^2\right]^{1/2}$$

$= 60.075 \times 9.8$ N $= 60.075$ kg-wt

560 **(a)**

$\vec{F_1} = F_1\hat{\jmath}$; $\vec{F_1} \times \vec{F_2}$ is equal to zero only I angle between $\vec{F_1}$ and $\vec{F_2}$ is either 0° or 180°. So $\vec{F_2}$ will be $4\hat{\jmath}$.

561 **(a)**

Range of projectile,

$R = \frac{2u^2 \sin^2\theta}{g}$ ...(i)

Height, $H = \frac{u^2 \sin^2\theta}{2g}$ ...(ii)

$H_1 = \frac{u^2 \sin^2(90° - \theta)}{2g} = \frac{u^2 \cos^2\theta}{g}$ ...(iii)

Then $HH_1 = \frac{u^2 \sin^2\theta \times u^2 \cos^2\theta}{2g \times 2g}$ ...(iv)

From Eq. (i) we get,

$$R^2 = \frac{4u^2 \sin^2\theta \times u^2 \cos^2\theta \times 4}{2g \times 2g}$$

$R = \sqrt{16HH_1}$ [from Eq. (iv)]

$\therefore R = 4\sqrt{HH_1}$

562 **(b)**

To reach the height of suspension, $h = l$

$v = \sqrt{2gh} = \sqrt{2gl}$

563 **(b)**

Time of flight of this particle, $T = 4$ s. if u is its initial speed and θ is the angle of projection, then

$T = 4 = \frac{2u\sin\theta}{g}$ or $u\sin\theta = 2g$ ...(i)

After 1 s, the velocity vector of particle makes an angle of 45° with horizontal, so

$v_x = v_y$ $ie, u\cos\theta = (u\sin\theta) - gt$

or $u\cos\theta = 2g - g$ $(\because t = 1s)$

or $u\cos\theta = g$ ...(ii)

Squaring and adding Eqs. (i) and (ii), we have

$u^2 = 5g^2 = 5(10)^2 = 500$

or $u = \sqrt{500} = 22.36\ ms^{-1}$

Dividing Eq. (i) by Eq. (ii), we have

$\tan\theta = 2$ or $\theta = \tan^{-1}(2)$

564 **(a)**

For both cases $t = \sqrt{\frac{2h}{g}}$ = constant

Because vertical downward component of velocity will be zero for both the particles

565 **(b)**

Let v_1 be the velocity of the car and v_2 be the velocity of the parcel. The parcel is thrown at an angle $\theta$from O. It reaches the mass at M.

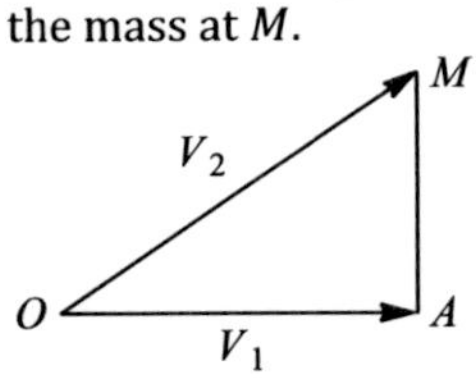

$\therefore \cos\theta = \frac{v_1}{v_2} = \frac{10}{10\sqrt{2}} = \frac{1}{\sqrt{2}} = \cos 45°$

So $\theta = 45°$

566 **(d)**

$\vec{A}\cdot(\vec{B}\times\vec{A}) = (\vec{A}\times\vec{A})\cdot\vec{B} = (\vec{0})\cdot\vec{B} = 0$

567 **(b)**

Net acceleration in nonuniform circular motion,

$$a = \sqrt{a_t^2 + a_c^2} = \sqrt{(2)^2 + \left(\frac{900}{500}\right)^2}$$
$$= 2.7\ m/s^2$$

a_t = tangential acceleration

a_c = centripetal acceleration $= \frac{v^2}{r}$

568 **(d)**

From figure here $\vec{A} = \sqrt{3}\hat{\imath} - \hat{k}$

so $\tan i_1 = \frac{\sqrt{3}}{1} = \sqrt{3} = \tan 60°$

$i_1 = 60°$

Using snell's law $\mu_1 \sin i_1 = \mu_2 \sin r$

or $\sin r = \frac{\mu_1}{\mu_2}\sin i_1 = \frac{\sqrt{2}}{\sqrt{3}}\sin 60° = \frac{1}{\sqrt{2}} = \sin 45°$

or $r = 45°$

The unit vector in the direction of the refracted ray will be

$\hat{n} = 1\sin 45°\hat{\imath} - 1\cos 45°\hat{k} = \frac{1}{\sqrt{2}}(\hat{\imath} - \hat{k})$

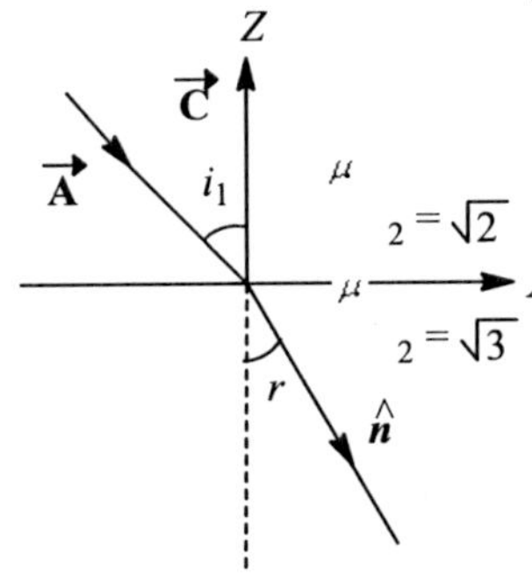

569 **(a)**

$x = 36t \therefore v_x = \frac{dx}{dt} = 36\ m/s$

$y = 48t - 4.9t^2 \therefore v_y = 48 - 9.8t$

at $t = 0$ $v_x = 36$ and $v_y = 48\ m/s$

So, angle of projection $\theta = \tan^{-1}\left(\frac{v_y}{v_x}\right) = \tan^{-1}\left(\frac{4}{3}\right)$

Or $\theta = \sin^{-1}(4/5)$

570 **(c)**

From $v = r\omega$, when v is doubled and ω halved, r must be 4 times. Therefore, centripetal acceleration

$= \frac{v^2}{r} = r\omega^2$ will remain unchanged

571 **(d)**

We know that $R = \frac{u^2 \sin 2\theta}{g}$

$= \frac{10 \times 10 \times \sin 60°}{10} = 10 \times \frac{\sqrt{3}}{2}$

$= 5 \times 1.732 = 8.66$ m

572 **(b)**

$v^2 = u^2 + 2as$ ...(i)

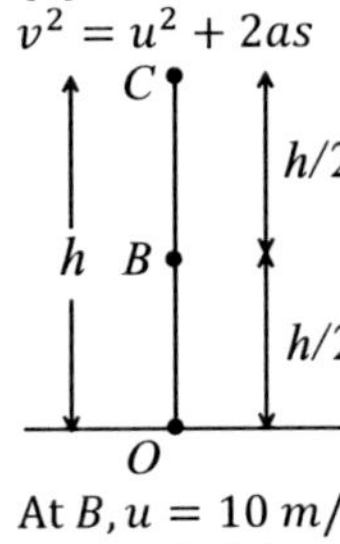

At $B, u = 10\ m/s$

at max. height, $v = 0$

$a = -10\ m/s^2; s = h/2$

From equation (i)

$0 = (10)^2 + 2(-10)h/2 \Rightarrow h = 10\ m$

574 **(b)**

Change in momentum $= 2mu \sin\theta$

$= 2 \times 0.5 \times 98 \times \sin 30 = 49\ N\text{-}s$

575 **(b)**

Time taken by the bomb to reach the ground is given by

$h_{OA} = \frac{1}{2} g t_{OB}^2$

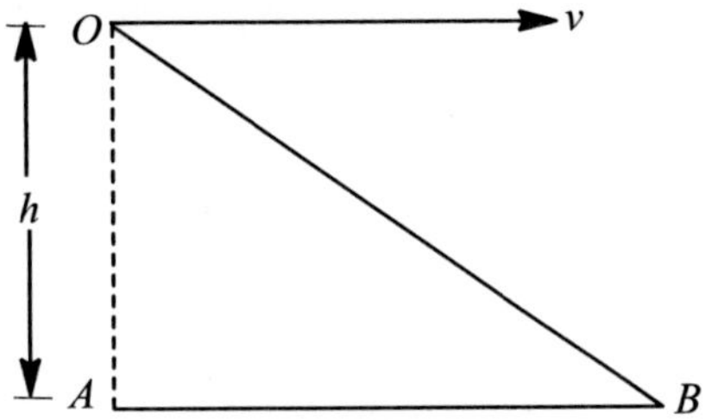

we have $t_{OB} = \sqrt{\frac{2h_{OA}}{g}}$

given, $h_{OA} = 80\text{m}$, $g = 10\ \text{ms}^{-2}$

$\therefore\ t_{OB} = \sqrt{\frac{2 \times 80}{10}}\ 4\text{s}$

Horizontal velocity of bomb

$v = 150\ \text{ms}^{-1}$

Horizontal distance covered by the bomb

$AB = vt_{OB}$

$= 150 \times 4$

$= 600\ \text{m}$

Hence, the bomb should be dropped 600 m before the target.

576 **(b)**

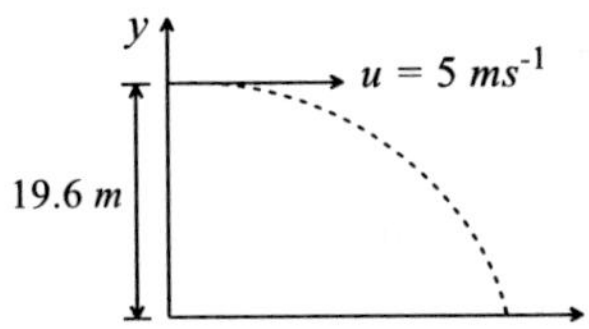

Let $t\ s$ be time taken by the ball to hit the ground

$\therefore H = \frac{1}{2} gt^2 \Rightarrow t = \sqrt{\frac{2H}{g}} = \sqrt{\frac{2 \times 19.6 m}{9.8\ ms^{-2}}}$

$= 2s$

577 **(a)**

$v = \sqrt{\mu r g} = \sqrt{0.4 \times 30 \times 9.8}$

$= 10.84\ m/s$

578 **(a)**

$H_1 = \frac{u^2 \sin^2\theta}{2g}$ and $H_2 = \frac{u^2 \sin^2(90-\theta)}{2g} = \frac{u^2\cos^2\theta}{2g}$

$H_1 H_2 = \frac{u^2 \sin^2\theta}{2g} \times \frac{u^2 \cos^2\theta}{2g}$

$= \frac{(u^2 \sin 2\theta)^2}{16g^2} = \frac{R^2}{16}$

$\therefore R = 4\sqrt{H_1 H_2}$

579 **(c)**

Instantaneous velocity of rising mass after t sec will be

$v_t = \sqrt{v_x^2 + v_y^2}$

Where $v_x = v\cos\theta$ =Horizontal component of velocity

$v_y = v \sin\theta - gt$

= Vertical component of velocity

$v_t = \sqrt{(v\cos\theta)^2 + (v\sin\theta - gt)^2}$

$v_t = \sqrt{v^2 + g^2t^2 - 2v\sin\theta\ gt}$

580 **(a)**

Here, $r = 92$ m, $v = 26\ \text{ms}^{-1}$, $\mu = ?$

As $\frac{mv^2}{r} = F = \mu R = \mu\, mg$

$\mu = \frac{v^2}{rg} = \frac{26 \times 26}{92 \times 9.8} = 0.75$

581 **(d)**

$v = \sqrt{\mu r g} = \sqrt{0.25 \times 40 \times 10} = 10\ m/s$

582 **(c)**

An object moving in uniform circular motion is moving around the perimeter of the circle with a constant speed. While the speed of object is constant, its velocity is changing. Velocity being a vector quantity has a constant magnitude but a changing direction. The direction is always directed tangent line is always pointing in a new direction. Also when it is moving in circular motion towards the centre, hence acceleration is perpendicular to velocity.

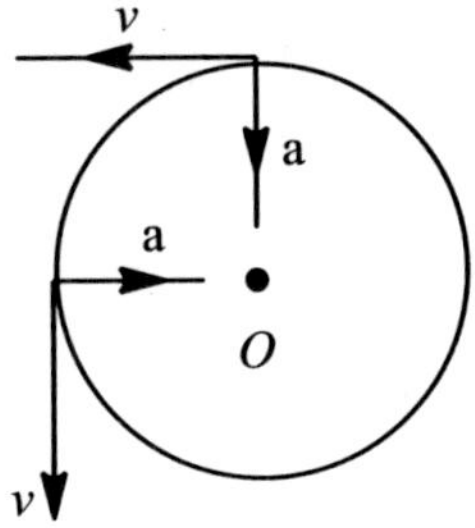

583 **(d)**

$u=\sqrt{2gh}=\sqrt{2\times10\times20}=20\text{ ms}^{-1}$

Time of ascent= time of descent

$=\sqrt{\frac{2h}{g}}=\sqrt{\frac{2\times20}{10}}=2\text{s}$

$\therefore$ time of flight = 2 +2 = 4s

584 **(b)**

As the speed is constant throughout the circular motion therefore its average speed is equal to instantaneous speed

585 **(a)**

Power, $P=\vec{F}.\vec{v}=(7\hat{\imath}+6\hat{k}).(3\hat{\jmath}+4\hat{k})=24$

586 **(b)**

$mg=20N$ and $\frac{mv^2}{r}=\frac{2\times(4)^2}{1}=32N$

It is clear that 52 N tension will be at the bottom of the circle. Because we know that $T_{\text{Bottom}}=mg+\frac{mv^2}{r}$

587 **(a)**

$\vec{A}\cdot\vec{B}=0\Rightarrow\vec{A}\perp\vec{B}$

Now, $\vec{A}\times\vec{B}=\vec{1}$ or $AB\sin\theta=1$

$AB\sin90°=1$ or $AB=1\Rightarrow A=1$ and $B=1$

So, $\vec{A}$ and $\vec{B}$ are perpendicular unit vectors.

588 **(c)**

The forces acting on the ball will be (i) in the direction opposite to its motion *ie*, frictional force and(ii) weight mg.

589 **(c)**

Displacement = AB

angle between $\vec{r_1}$ and $\vec{r_2}$

$\theta=75°-15°=60°$

From figure

$AB^2=r_1^2+r_2^2-2r_1r_2\cos\theta$

$=3^2+4^2-2\times3\times4\cos60°$

$=13$

$AB=\sqrt{13}$

590 **(d)**

For body to move on circular path. Frictional force provides the necessary centripetal force,

ie, frictional force = centripetal force

or $\mu mg=\frac{mv_0^2}{r}=mr\omega^2$

or $\mu g=r\omega^2$

$\therefore$ $0.5\times9.8=10\,\omega^2$

or $\omega=0.7\text{ rad s}^{-1}$

592 **(a)**

$v=K(y\hat{\imath}+x\hat{\jmath})$

$v_x=Ky$

$\frac{dx}{dt}=Ky$

Similarly, $\frac{dy}{dt}=Kx$

Hence $\frac{dy}{dx}=\frac{x}{y}$

$\Rightarrow y\,dy=x\,dx$, by integrating

$y^2=x^2+c$

593 **(a)**

Centripetal velocity at highest point =

$\sqrt{gR}=\sqrt{10\times1.6}=4m/s$

594 **(a)**

$a=\omega^2R=\left(\frac{2\pi}{0.2\pi}\right)^2(5\times10^{-2})=5\,m/s^2$

595 **(d)**

$\frac{1}{2}mu^2-\frac{1}{2}mv^2=mgL$

$\Rightarrow v=\sqrt{u^2-2gL}$

$|\vec{v}-\vec{u}|=\sqrt{u^2+v^2}=\sqrt{u^2+u^2-2gL}$

$=\sqrt{2(u^2-gL)}$

596 **(a)**

$T=\frac{2u\sin\theta}{g}=10\text{ Sec}\Rightarrow u\sin\theta$

$=50\,m/s$

$\therefore H=\frac{u^2\sin^2\theta}{2g}=\frac{(u\sin\theta)^2}{2g}=\frac{50\times50}{2\times10}$

$=125\,m$

597 **(a)**

$\vec{P}=\frac{\vec{F}\cdot\vec{S}}{t}$

$=\frac{(2\hat{\imath}+2\hat{\jmath})\cdot(2\hat{\imath}+2\hat{k})}{16}\text{Js}^{-1}=\frac{4}{16}\text{Js}^{-1}$

$=0.25\text{Js}^{-1}$

598 **(c)**

Apply right handed screw rule for the direction of $(\vec{A}\times\vec{B})$

599 **(c)**

$x=20\times5=100\text{m}$

$y=\frac{1}{2}\times10\times5\times5=125\text{m}$

$r=\sqrt{100^2+125^2}=160\text{m}$

600 **(d)**

Given, equation is

$y=9x^2$ ....(i)

Since, x-component of velocity remains constant, we have

$\frac{dx}{dt}=\frac{1}{3}\text{ms}^{-1}$ ...(ii)

From Eq. (i), we have y-component of velocity.

$\frac{dy}{dt} = 18x \cdot \left(\frac{dx}{dt}\right)^2$

$\frac{dy}{dt} = 18\left(\frac{dx}{dt}\right)^2 = 18 \times \left(\frac{1}{3}\right)^2 = 2\text{ms}^{-2}$

$\therefore \mathbf{a}_y = 2\hat{\mathbf{j}}\text{ms}^{-2}$

601 **(a)**

The horizontal range $R_x = \frac{u^2 \sin 2\theta}{g}$

When projected at angle of 15°

$R_{x1} = \frac{u^2}{g}\sin(2 \times 15) = \frac{u^2}{2g} = 1.5\text{km}$

When projected at angle of 45°

$R_{x1} = \frac{u^2}{g}\sin(2 \times 45°)\frac{u^2}{g}$

$= \frac{2u^2}{2g} = 2 \times 1.5 = 3.0$ km

602 **(b)**

$\vec{A} = 2\hat{\imath} - \hat{\jmath} + 3\hat{k}$; $\vec{B} = 3\hat{\imath} - 2\hat{\jmath} - 2\hat{k}$; $\vec{C} = ?$

$\vec{R} = \hat{k} = \vec{A} + \vec{B} + \vec{C}$

$\hat{k} = (2\hat{\imath} - \hat{\jmath} + 3\hat{k}) + (3\hat{\imath} - 2\hat{\jmath} - 2\hat{k}) + \vec{C}$

$= 5\hat{\imath} - 3\hat{\jmath} + \hat{k} + \vec{C}$

$\therefore \vec{C} = -5\hat{\imath} + 3\hat{\jmath}$

603 **(d)**

Maximum tension, $F = mr\omega^2 = mr4\pi^2v^2$
$= 0.1 \times 2 \times 4 \times \pi^2 \times (200/60)^2 =$
87.64 N

604 **(a)**

For water not to spil, $\frac{mv^2}{r} = mg$

or $v = \sqrt{r\,g} = \sqrt{1.6 \times 10} = 4\text{ ms}^{-1}$

605 **(d)**

At highest point $\frac{mv^2}{R} = mg \Rightarrow v = \sqrt{gR}$

606 **(b)**

At the highest point, velocity along y-axis is zero. Therefore, change in linear momentum
$= m(u \sin\alpha - 0) = mu\sin\alpha$

607 **(b)**

Centripetal force = breaking force
$\Rightarrow m\omega^2 r$ = breaking stress × cross sectional area

$\Rightarrow m\omega^2 r = p \times A \Rightarrow \omega = \sqrt{\frac{p \times A}{mr}}$

$= \sqrt{\frac{4.8 \times 10^7 \times 10^{-6}}{10 \times 0.3}}$

$\therefore \omega = 4\,rad/sec$

608 **(c)**

The vertical component of velocity of projection
$= -50\sin 30° = -25\,m/s$

If t be the time taken to reach the ground,

$h = ut + \frac{1}{2}gt^2 \Rightarrow 70 = -25t + \frac{1}{2} \times 10t^2$

$\Rightarrow 70 = -25t + 5t^2 \Rightarrow t^2 - 5t - 14 = 0 \Rightarrow t = -2s$ and $7s$

Since, $t = -2\ s$ is not valid $\therefore t = 7s$

609 **(d)**

Here, $h = 1$ m, $r = 100$ m, $2x = 1.5$ m

For no skidding

$\frac{mv^2}{r} \times h = mg\,x$

$v = \sqrt{\frac{g\,r\,x}{h}} = \sqrt{\frac{9.8 \times 100 \times 0.75}{1}}$

$v = 27.1\text{ ms}^{-1}$

610 **(d)**

In equilibrium position along y-direction

$2\sin 60° = \sqrt{3} + F\cos\theta$

or $2 \times \frac{\sqrt{3}}{2} = \sqrt{3} + F\cos\theta$ or $F\cos\theta = 0$

As $F \neq 0$

$\therefore \cos\theta = 0$ or $\theta = 90°$

Along x-direction, $F\sin 90° = 1 + 2\cos 60°$

$= 1 + 2 \times \frac{1}{2}$

$F = 2$N

611 **(b)**

$H_1 + H_2 = \frac{u^2}{2g}(\sin^2 30° + \sin^2 60°)$

$= \frac{20^2}{2 \times 10}\left(\frac{1}{4} + \frac{3}{4}\right) = 20\ m$

612 **(d)**

In 15 second's hand rotate through 90°

Change in velocity $|\overrightarrow{\Delta v}| = 2v\sin(\theta/2)$

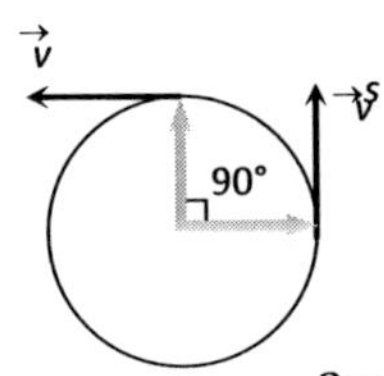

$= 2(r\omega)\sin(90°/2) = 2 \times 1 \times \frac{2\pi}{T} \times \frac{1}{\sqrt{2}}$

$= \frac{4\pi}{60\sqrt{2}} = \frac{\pi\sqrt{2}}{30}\frac{cm}{sec}$ [As $T = 60$ sec]

613 **(b)**

Both the particles will meet at C, if their time of flight is the same. The time of flight of A is

$$T = \frac{2u \sin\theta}{g} = \frac{2 \times 10 \times \sin 60°}{10} = \sqrt{3}\text{s}$$

For vertical downward motion of particle B from B to C

We have

$$h = \frac{1}{2}gT^2 = \frac{1}{2} \times 10 \times (\sqrt{3})^2 = 15 \text{ m}$$

614 **(d)**

$B_x = 10 - 6 = 4$ and $B_y = 9 - 6 = 3$

so, $B = (B_x^2 + B_y^2)^{1/2} = \sqrt{4^2 + 3^2} = \sqrt{16 + 9}$

$= \sqrt{25} = 5$

616 **(a)**

$\frac{1}{2}mv^2 = mg(3 - 1) = 2mg$ or $v = \sqrt{4}g = 2\sqrt{g}$

Vertical component at $A = 2\sqrt{g} \sin 30° = \sqrt{g}$

617 **(b)**

From, $\tan\theta = \frac{v^2}{rg}$

$$r = \frac{v^2}{g\tan\theta} = \frac{10 \times 10}{10 \times \tan 45°} = 10 \text{ m}$$

618 **(d)**

The resultant of 5 N along OC and 5 N along OA is

$R = \sqrt{6^2 + 6^2}$

$= \sqrt{72}$N along OB

The resultant of $\sqrt{72}$ N along OB and $\sqrt{72}$ N along OG is

$R' = \sqrt{72 + 72} = 12$N along OE.

619 **(d)**

As the mass is attached to the end of a rod. Which does not stacken, therefore, taking $v = 0$ at the highest point, from

$v^2 - u^2 = 2a\,s$

$0 - u^2 = 2(-g)2l$

$\therefore u = \sqrt{4gl}$

620 **(b)**

Given, $y = 12x - \frac{3}{4}x^2$

$u_x = 3 \text{ ms}^{-1}$

$$v_y = \frac{dy}{dt} = 12\frac{dx}{dt} - \frac{3}{2}x\frac{dx}{dt}$$

At $x = 0, v_y = u_y = 12\frac{dx}{dt} = 12u_x = 12 \times 3 = 36 \text{ ms}^{-1}$

$$a_y = \frac{d}{dt}\left(\frac{dy}{dt}\right) = 12\frac{d^2x}{dt^2} - \frac{3}{2}\left(\frac{dx}{dt} + x\frac{d^2x}{dt^2}\right)$$

But $\frac{d^2x}{dt^2} = a_x = 0$, hence

$$a_y = -\frac{3}{2}\frac{dx}{dt} = -\frac{3}{2}u_x = -\frac{3}{2} \times 3 - \frac{9}{2}\text{ms}^{-2}$$

Range $R = \frac{2u_xu_y}{a_y} = \frac{2\times3\times12}{9/2} = 16$m

621 **(c)**

Time taken to cover horizontal distance D with constant horizontal velocity, $t = D/v_0\cos\theta$. Taking vertical motion for time t, we have

$$h = v_0 \sin\theta \times t - \frac{1}{2}gt^2$$

$$= v_0 \sin\theta \times \frac{D}{v_0\cos\theta} - \frac{1}{2}g\left(\frac{D}{v_0\cos\theta}\right)^2$$

$$= D\tan\theta - \frac{1}{2}\frac{gD^2}{v_0^2\cos^2\theta}$$

622 **(a)**

Direction of velocity is always tangent to path, so at the top of trajectory it is in horizontal direction and acceleration due to gravity is always in vertically downward direction. Hence, **v** and **g** are perpendicular to each other.

623 **(d)**

$$\text{Height, } h = \frac{1}{2}gt^2 \Rightarrow t = \sqrt{\frac{2 \times 1960}{9.8}}$$

$$= 20\text{s}$$

$$s = AB = ut = 600 \times \frac{20}{60 \times 60} = 3.33 \text{ km}$$

624 **(c)**

Here, $P = (A + B),\ Q = (A - B)$,

$R = \sqrt{A^2 + B^2}$;

$\cos\theta = \frac{R^2 - P^2 - Q^2}{2PQ}$

$$= \frac{(A^2 + B^2) - (A + B)^2 - (A - B)^2}{2(A + B)(A - B)}$$

$$= -\left[\frac{A^2 + B^2}{2(A^2 - B^2)}\right]$$

$$\therefore \quad \theta = \cos^{-1}\left[-\frac{A^2 + B^2}{2(A^2 - B^2)}\right]$$

625 **(a)**

Linear velocity,

$v = \omega r = 2\pi nr = 2 \times 3.14 \times 3 \times 0.1$

$= 1.88\ m/s$

Acceleration, $a = \omega^2 r = (6\pi)^2 \times 0.1 = 35.5\ m/s^2$

Tension in string, $T = m\omega^2 r = 1 \times (6\pi)^2 \times 0.1 = 35.5\ N$

626 **(c)**

Let time taken by the body to fall from point C to B be t'' Then, $t_1 + 2t'' = t_2$

$t'' = \left(\frac{t_2 - t_1}{2}\right)$ ...(i)

Total time taken, to reach point C

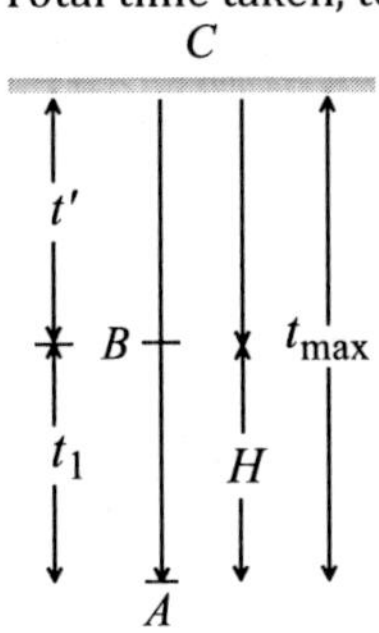

$T = t_1 + t''$

$= t_1 + \frac{t_2 - t_1}{2}$

$= \frac{2t_1 + t_2 - t_1}{2} = \left(\frac{t_1 + t_2}{2}\right)$

Then maximum height attained

$H_{max} = \frac{1}{2} g(T)^2 = \frac{1}{2} g\left(\frac{T_1 + t_2}{2}\right)^2$

$= \frac{1}{2} g . \frac{(t_1 + t_2)^2}{4}$

$\Rightarrow H_{max} = \frac{1}{8} g . (t_1 + t_2)^2 m$

627 **(d)**

The various forces acting on the particle, are its weight mg acting vertically downwards, normal reaction N. Equating the vertical forces, we have

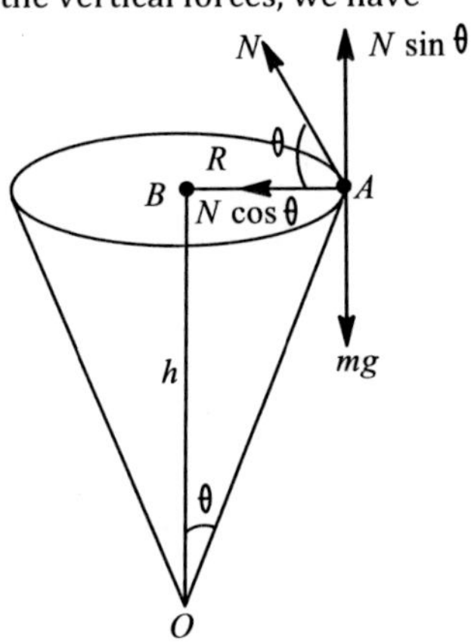

$N \sin\theta = mg$ ... (i)

Also, centripetal force,

$\frac{mv^2}{R} = N\cos\theta$ .... (ii)

From Eqs. (i) and (ii), we get

$\tan\theta = \frac{Rg}{v^2}$ .... (iii)

Also, from triangle OAB,

$\tan\theta = \frac{R}{h}$ ..... (iv)

Equating Eqs. (iii) and (iv), we get

$h = \frac{v^2}{g}$

Given, $v = 0.5 ms^{-1}$ and $g = 10\ ms^{-2}$

$\therefore\ h \frac{(0.5)^2}{10} = 0.025\ m = 2.5\ cm$

628 **(c)**

Maximum force of friction = centripetal force

$\frac{mv^2}{r} = \frac{100 \times (9)^2}{30} = 270\ N$

629 **(d)**

The centripetal force, $F = \frac{mv^2}{r} \Rightarrow r = \frac{mv^2}{F}$

$\therefore r \propto v^2$ or $v \propto \sqrt{r}$ [If m and F are constant]

$\Rightarrow \frac{v_1}{v_2} = \sqrt{\frac{r_1}{r_2}} = \sqrt{\frac{1}{2}}$

630 **(a)**

Horizontal component of velocity $v_x = 500 ms^{-1}$ and vertical components of velocity while striking the ground

$v_y = 0 + 10 \times 10 = 100\ ms^{-1}$

$\therefore$ Angle with which it strikes the ground

$\theta = \tan^{-1}\left(\frac{v_y}{v_x}\right) = \tan^{-1}\left(\frac{100}{500}\right)$

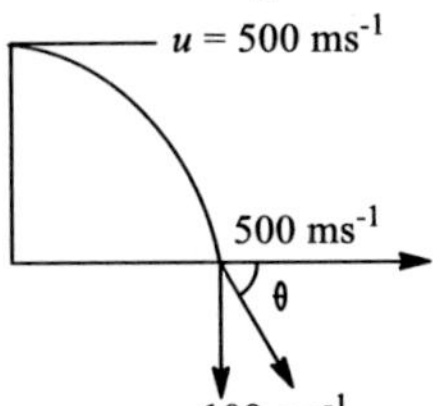

$\theta = \tan^{-1}\left(\frac{1}{5}\right)$

631 **(a)**

Given, $r = (20/\pi) m$

$v = 80\ m/s$

$\theta = 2$ rev $= 4\pi rad$

$\omega_0 = 0$

From the equation
$\omega^2 = \omega_0^2 + 2\alpha\theta$, we have
$\omega^2 = 2\alpha\theta$
or $\frac{v^2}{r^2} = 2 \cdot \frac{a}{r}\theta$
or $a = \frac{v^2}{2r\theta} = \frac{(80)^2}{2\times(20/\pi)\times4\pi}$
$= 40 \text{ ms}^{-2}$

632 **(c)**
Let t be time taken by the bullet to hit the target
$\therefore 200\, m = 2000\, ms^{-1} t$
$\Rightarrow t = \frac{200m}{2000ms^{-1}} = \frac{1}{10} s$
For vertical motion,
Here $u = 0$
$\therefore h = \frac{1}{2} g t^2$
$h = \frac{1}{2} \times 10 \times \left(\frac{1}{10}\right)^2 = \frac{1}{20} m = 5\, cm$
$\therefore$ Gun should be aimed 5 cm above the target

633 **(b)**
$\omega_1 = 2\pi r = 2\pi \times 20, \omega_2 = 0, t = 20a, \alpha = ?$
As $\omega_2 = \omega_1 + \alpha t$
$\therefore \alpha = \frac{\omega_2 - \omega_1}{t} = \frac{40\pi - 0}{20} = 2\pi \text{ rad s}^{-2}$

636 **(d)**
Momentum, speed and kinetic energy change continuously in a vertical circular motion. The physical quantity which remains constant is the total energy.

637 **(a)**
$v = \sqrt{rg}$
$2r = \frac{2v^2}{g} = \frac{2\times9.8\times9.8}{9.8} = 19.6 \text{ m}$

638 **(a)**
Extension is first case
$\Delta l_1 = 2a - a = a$
From Hooke's law
$F = k\Delta l_1$
$mr\omega^2 = ka$
$m(2a)\omega^2 = ka$
$2ma\left(\frac{2\pi}{T}\right)^2 = ka$
In second case,
$m(3a)\left(\frac{2\pi}{T'}\right)^2 = k(3a - 2a)$
On dividing Eq. (ii) by Eq. (i)
$T' = \sqrt{\frac{3}{2}} T$

639 **(c)**
Horizontal range $R = \frac{u^2 \sin 2\theta}{g}$
Substituting the given values we get
$560 = \frac{82 \times 82 \times \sin 2\theta}{9.8}$
$\Rightarrow \sin 2\theta = \frac{560 \times 9.8}{82 \times 82} = \frac{5488}{6724}$
$\Rightarrow \sin 2\theta = 0.82 \Rightarrow 2\theta = \sin^{-1}(0.82)$
$\Rightarrow 2\theta = 55.1° \Rightarrow \theta \approx 27°$

640 **(b)**
Centripetal force $= mr\omega^2 = 5 \times 1 \times 4 =$ 20N

641 **(b)**
Acceleration of electron
$= \frac{v^2}{r} = \frac{(2.18 \times 10^6)^2}{0.528 \times 10^{-10}} = 9 \times 10^{22} \text{ ms}^{-2}$

643 **(a)**
$|\hat{A} \times \hat{B}| = (1)(1)\sin\theta = \sin\theta$.

644 **(c)**
The speed of projectile is v and angle of projection is45°.
$v_x = v \cos 45° = \frac{v}{\sqrt{2}}$
$v_y = v \sin 45° - gt = \frac{v}{\sqrt{2}} - gt$
At highest point $v_y = 0, v_x = \frac{v}{\sqrt{2}}$
Maximum height achieved,
$H = \frac{v^2 \sin^2 45°}{2g} = \frac{v^2}{4g}$
Now, angular momentum about O
$= \frac{mv}{\sqrt{2}} \cdot \frac{v^2}{4g} = \frac{mv^3}{4\sqrt{2}g}$

645 **(b)**
Here, $A_1 = A_2 = 1$
and $A_1^2 + A_2^2 + 2A_1A_2\cos\theta = (\sqrt{3})^2 = 3$
or $1+1+2\times 1 \times 1 \times\cos\theta = 3$ or$\cos\theta = \frac{1}{2}$
Now, $(\vec{A}_1 - \vec{A}_2) \cdot (2\vec{A}_1 + \vec{A}_2)$
$= 2A_1^2 - A_2^2 - A_1A_2\cos\theta$
$= 2 \times 1^2 - 1^2 - 1 \times 1 \times \frac{1}{2} = \frac{1}{2}$

646 **(d)**
$v \cos \beta = u \cos \alpha$
$v = \frac{u \cos \alpha}{\cos \beta}$

647 **(d)**
$v_{\max} = \sqrt{\mu r g} = \sqrt{0.2 \times 100 \times 9.8}$
$= 14 m/s$

648 **(a)**

$$\frac{T_{max}}{T_{min}} = \frac{\frac{mv^2}{L}+mg}{\frac{mv^2}{L}-mg} = 2 \quad ...(i)$$

Simplifying Eq. (i), we get,

$$v_H = \sqrt{3gL} = \sqrt{\frac{3\times 10\times 10}{3}} = 10\text{ ms}^{-1}$$

649 **(d)**

Tension is string = centrifugal force

In first case, $F = m\,r\,\omega^2$

In second case, $F' = m(2r)(2\omega)^2 = 8mr\,\omega^2 = 8\,F$

650 **(b)**

Components of velocity of ball relative to lift are

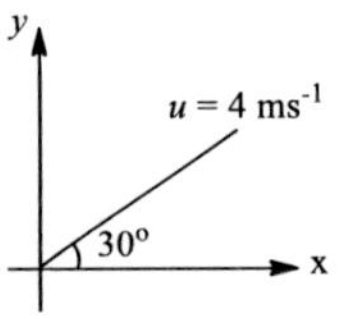

$u_x = 4\cos 30° = 2\sqrt{3}\text{ms}^{-1}$

and $u_y = 4\sin 30° = 2\text{ms}^{-1}$

and acceleration of ball relative to lift is 12 ms^{-2} in negative y-direction or vertically downwards. Hence, time of flight

$$T = \frac{2u_y}{12} = \frac{u_y}{6} = \frac{2}{6} = \frac{1}{3}\text{s}$$

651 **(c)**

When $\theta = 180°$, the particle will be at diametrically opposite point, where its velocity is opposite to the initial directions of motion. The change in momentum $= mv - (-mv) = 2mv$ (maximum). When $\theta = 360°$, the particle is at the initial position with momentum m. Change in momentum $mv - mv = 0$ (minimum)

652 **(b)**

If any two vectors are parallel or equal, then the scalar triple product is zero.

653 **(d)**

For vertically upward motion of a projectile,

$$y = (u\sin\alpha)t - \frac{1}{2}gt^2$$

or $h = (u\sin\alpha)t - \frac{1}{2}gt^2$

or $gt^2 - (2u\sin\alpha)t + 2h = 0$

$$\therefore t = \frac{2u\sin\alpha \pm \sqrt{(4u^2\sin^2\alpha) - 8gh}}{2g}$$

If two roots of quadratic Eq.(i) are t_1, t_2 then

$$t_1 = \frac{2u\sin\alpha + \sqrt{4u^2\sin^2\alpha - 8gh}}{2g}$$

$$t_2 = \frac{2u\sin\alpha - \sqrt{4u^2\sin^2\alpha - 8gh}}{2g}$$

If particle crosses the wall at times t_1 and t_2 respectively, then time of flight t is

$t = \sqrt{t_1t_2}$

or $t^2 = t_1t_2$

$$\therefore \left(\frac{2u\sin\alpha}{g}\right)^2 = \frac{(2u\sin\alpha)^2 - (4u^2\sin^2\alpha - 8gh)}{4g^2}$$

or $\frac{4u^2\sin^2\alpha}{g^2} = \frac{8gh}{4g^2}$

or $2u^2\sin^2\alpha = gh$

Given, $u = \sqrt{2gh}$

$\therefore\ 2(2gh)\sin^2\alpha = gh$

or $\sin^2\alpha = \frac{1}{4}$

or $\sin\alpha = \frac{1}{2}$

$\therefore\ \alpha = 30°$

654 **(a)**

$$T_{max} = m\omega_{max}^2 r + mg \Rightarrow \frac{T_{max}}{m} = \omega^2 r + g$$

$$\Rightarrow \frac{30}{0.5} - 10 = \omega_{max}^2 r \Rightarrow \omega_{max} = \sqrt{\frac{50}{r}}$$

$$= \sqrt{\frac{50}{2}} = 5\,rad/s$$

655 **(b)**

Centripetal acceleration $a_c = v^2/r$

It acts along the radius and directed towards the centre of the circular path

656 **(c)**

$$S_x = u_xt + \frac{1}{2}a_xt^2 \Rightarrow S_x = \frac{1}{2}\times 6\times 16 = 48\,m$$

$$S_y = u_yt + \frac{1}{2}a_yt^2 \Rightarrow S_y = \frac{1}{2}\times 8\times 16 = 64\,m$$

$$S = \sqrt{S_x^2 + S_y^2} = 80\,m$$

657 **(b)**

$$\cos\theta = \frac{(\hat{k}).(\hat{i}+\hat{j}+\sqrt{2}\hat{k})}{1\sqrt{1^2+1^2(\sqrt{2})^2}}$$

or $\cos\theta = \frac{\sqrt{2}}{2} = \frac{1}{\sqrt{2}}$ or $\theta = 45°$

658 **(a)**

$m4\pi^2 n^2 r = 4 \times 10^{-13} \Rightarrow n$
$= 0.08 \times 10^8$ *cycles/sec.*

659 **(c)**

For $\theta = 45°$

$$H_{\max} = \frac{u^2 \sin^2 45°}{2g} = \frac{u^2}{4g}\left[\because \sin 45° = \frac{1}{\sqrt{2}}\right]$$

$$R = \frac{u^2 \sin 90°}{g} = \frac{u^2}{g}; \therefore \frac{R}{H} = \frac{u^2}{g} \times \frac{4g}{u^2} = 4$$

$$\Rightarrow R = 4H$$

660 **(d)**

$$\vec{\tau} = \vec{r} \times \vec{F} = \begin{vmatrix} \hat{\imath} & \hat{\jmath} & \hat{k} \\ 7 & 2 & 1 \\ -3 & 1 & 5 \end{vmatrix}$$

$= \hat{\imath}[15 - 1] + \hat{\jmath}[-3 - 35] + k[7 + 9]$
$= 14\hat{\imath} - 38\hat{\jmath} + 16\hat{k}$

661 **(b)**

The figure shows a circular path of moving particle. At any instant velocity of particle.

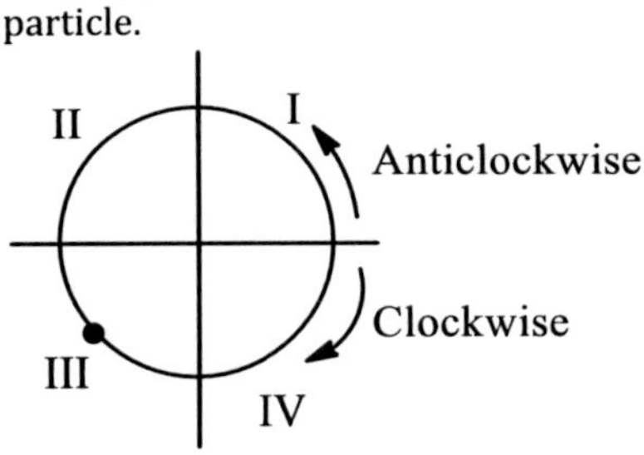

$v = -3\hat{\mathbf{i}} - 4\hat{\mathbf{j}} = (-3, -4)$ (in coordinate from).

The coordinates of velocity show that particle is in 3rd quadrant at that instant. While moving clockwise particle will enter into 4th quadrant and these into 3rd and while moving anticlockwise particle will enter into 2nd quadrant and then into 3rd quadrant.

$\therefore$ 4th and 2nd quadrants.

662 **(a)**

Taking initial position as origin and direction of motion (*i.e.*, vertically up) as positive. As the particle is thrown with initial velocity, at highest point its velocity is zero and then it returns back to its reference position. This situation is best depicted in figure of option (a)

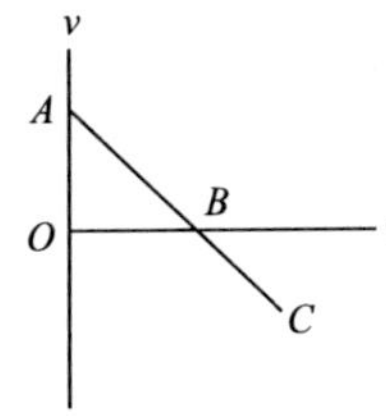

In figure, AB part denotes upward motion and BC part denotes downward motion

663 **(c)**

The pilot will see the ball falling in straight line because the reference frame is moving with the same horizontal velocity but the observer at rest will see the ball falling in parabolic path

664 **(a)**

Distance covered in $'n'$ revolution $= n\, 2\pi r = n\pi D$

$\Rightarrow 2000\pi D = 9500$[As $n = 200$, distance $= 9500\ m$]

$$\Rightarrow D = \frac{9500}{2000 \times \pi} = 1.5\ m$$

665 **(d)**

$$R = \frac{u^2 \sin 2\theta}{g} = \frac{2u_x v_y}{g}$$

$\therefore$ Range $\propto$ horizontal initial velocity (u_x)

In path 4 range is maximum so football possess maximum horizontal velocity in the path

666 **(b)**

$$h = (u \sin\theta)t - \frac{1}{2}gt^2$$

$$d = (u\cos\theta)t \text{ or } t = \frac{d}{u\cos\theta}$$

$$h = u\sin\theta \,.\, \frac{d}{u\cos\theta} - \frac{1}{2}g.\frac{d^2}{u^2\cos^2\theta}$$

$$u = \frac{d}{\cos\theta}\sqrt{\frac{g}{2(d\tan\theta - h)}}$$

667 **(d)**

$\because$ Angular momentum $\vec{L} = \vec{r} \times \vec{p}$

Where

$$\vec{r} = v_0\cos\theta\, t\hat{\imath} + \left(v_0\sin\theta t - \frac{1}{2}gt^2\right)\hat{\jmath}$$

$$\vec{p} = m[v_0\cos\theta\,\hat{\imath} + (v_0\sin\theta - gt)\hat{\jmath}]$$

$$\vec{L} = \vec{r} \times \vec{p} = -\frac{1}{2}mgv_0 t^2\cos\theta\,\hat{k}$$

668 **(d)**

$\sqrt{P^2 + Q^2 + 2PQ\cos\theta} = (P - Q)$

$\Rightarrow P^2 + Q^2 + 2PQ\cos\theta = P^2 + Q^2 - 2PQ$

$\Rightarrow 2PQ(1+\cos\theta)=0$

but $2PQ \neq 0$
$\therefore\ 1+\cos\theta=0$ or $\cos\theta=-1$
or $\theta=180°$

669 **(b)**

$\vec{L}=\vec{r}\times m\vec{v}=H\,mv\cos\theta$

$$=\frac{v\sin^2\theta}{2g}mv\cos\theta$$

$$=\frac{mv^3}{4\sqrt{2}g}$$

670 **(d)**

Displacement is distance from initial to final position In 40s cyclist completes =1 round

$\therefore$ In 3 min(180 s) cyclist will complete

$=4\frac{1}{2}$ round Displacement for 4 round is z

Displacement for $\frac{l}{2}$ round

$=$ length of diagonal

$=2\sqrt{2}$m.

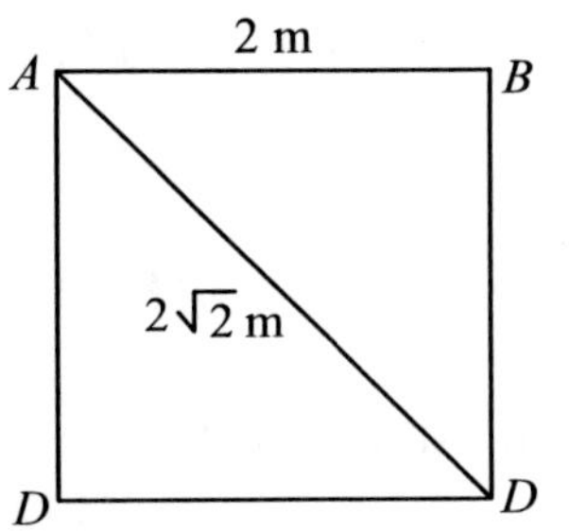

671 **(b)**

$\omega=2\pi n/t=2\pi\times100/60=10.47$ rad s^{-1}

672 **(b)**

$F^2=F_1^2+F_2^2+2F_1F_2\cos90°$

or $F^2=F_1^2+F_2^2\Rightarrow F=\sqrt{F_1^2+F_2^2}$

673 **(b)**

As $H=\frac{u^2\sin^2\theta}{2g}\ \therefore\frac{H_1}{H_2}=\frac{\sin^2\theta_1}{\sin\theta_2}=\frac{\sin^2 30°}{\sin^2 60}=\frac{1/4}{3/4}=\frac{1}{3}$

674 **(c)**

For uniform circular motion $a_t=0$

$a_r=\frac{v^2}{r}\neq0$

675 **(b)**

$(\text{KE})_H=\frac{1}{2}(\text{KE})_i$

$\frac{1}{2}mv^2\cos^2\theta=\frac{1}{2}\left(\frac{1}{2}mv^2\right)=\frac{1}{4}mv^2$

or $\cos^2\theta=\frac{1}{2}$ or $\cos\theta=\frac{1}{\sqrt{2}}$ or $\theta=45°$

676 **(b)**

Here, $r=2t,\theta=4t$

$l=r\theta=(2t)(4t)=8t^2$

$v=\frac{dl}{dt}=\frac{d}{dt}(8t)^2=16t$

$=16\times1=16\text{ ms}^{-1}$

677 **(c)**

$L=I\omega$. In U.C.M. $\omega=$ constant $\therefore L$ $=$ constant.

678 **(d)**

$R\cos\theta=mg$ ... (i)

$R\sin\theta=\frac{mv^2}{R}$ ... (ii)

From, Eqs. (i) and (ii) we get

$\therefore\ \tan\theta=\frac{v^2}{Rg}$

679 **(a)**

Water will not fall down, if weight, $mg=$ centrifugal force

$=mr\omega^2=mr\left(\frac{2\pi}{T}\right)^2$

680 **(a)**

Because velocity is always tangential and centripetal acceleration is radial.

681 **(c)**

Speeds at the top point of each wheel will equal and is equal to the speed of centre of mass.

682 **(d)**

In figure, $\sin30°=\frac{AB}{OA}$

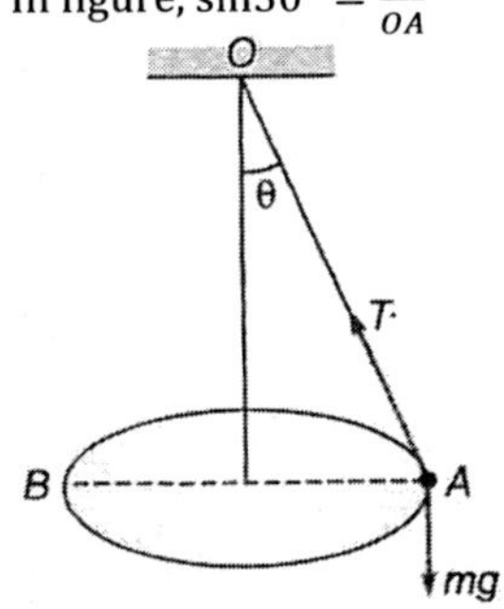

or $OA=\frac{AB}{\sin30°}=\frac{4}{1/2}=8$m

$\frac{T}{AO}=\frac{F}{AB}=\frac{mg}{OB}$

$F=\frac{AO}{AB}\times F=\frac{AO}{AB}\frac{mv^2}{r}=\frac{8}{4}\times10\times\frac{5^2}{4}$

≈125 N

683 **(d)**

$s=0\times1+\frac{1}{2}\times9.8\times1\times1=4.9$ m

684 **(d)**

Tension in the string $T_0 = mR\omega_0^2$

In the second case $T = m(2R)(4\omega_0^2) = 8mR\omega_0^2 = 8T_0$

685 **(a)**

From $h = \frac{1}{2}gt^2$

We have $t_{OB} = \sqrt{\frac{2h_{OA}}{g}}$

$= \sqrt{\frac{2 \times 1960}{9.8}} = 20s$

Horizontal distance $AB = vt_{OB}$

$= \left(600 \times \frac{5}{18}\right)(20)$

$= 3333.33\text{ m} = 3.33\text{ km}$

686 **(a)**

The shape of free surface of water is parabolic, because of difference in centrifugal force ($F = mr\omega^2$, which is proportional to r)

687 **(a)**

Horizontal component of velocity

$v_x = 25\cos 60° = 12.5\ m/s$

Vertical component of velocity

$v_y = 25\sin 60° = 12.5\sqrt{3}\ m/s$

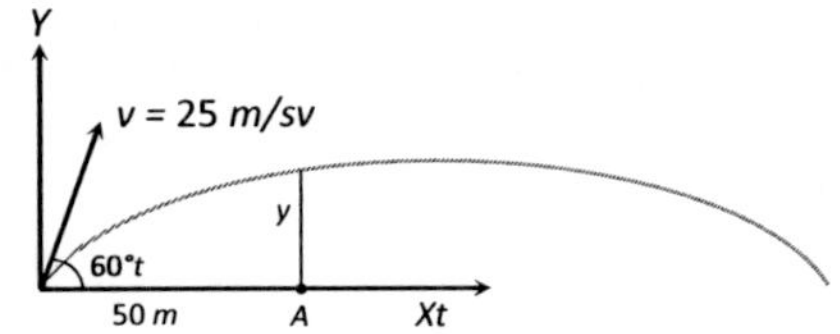

Tim to over 50 m distance $t = \frac{50}{12.5} = 4\ sec$

The vertical height y is given by

$$y = v_y t - \frac{1}{2}gt^2 = 12.5\sqrt{3} \times 4 - \frac{1}{2} \times 9.8 \times 16 = 8.2\ m$$

688 **(a)**

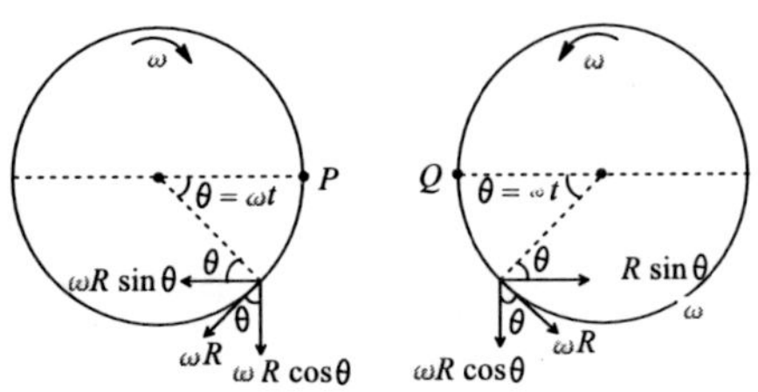

So, $V_r = 2\omega R\sin(\omega t)$

At $t = T/2, V_r = 0$

So two half cycles will take place

689 **(c)**

Acceleration of insect with respect to car $\vec{\mathbf{a}}_{sc}$ is a in the direction shown in figure.

Absolute acceleration of insect is

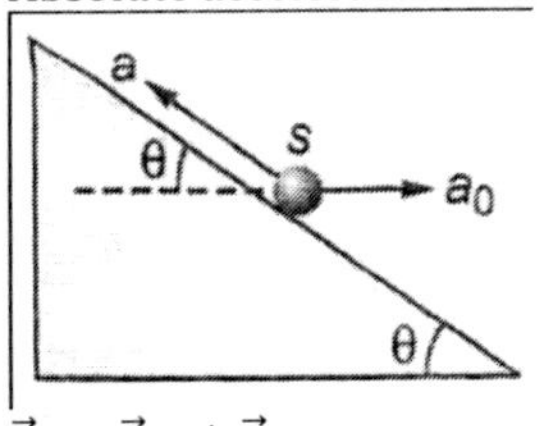

$\vec{\mathbf{a}}_s = \vec{\mathbf{a}}_{sc} + \vec{\mathbf{a}}_c$

Component of $\vec{\mathbf{a}}_s$ along horizontal is $a_0 - a\cos\theta$ and perpendicular to screen is $a_0 \sin\theta$

691 **(b)**

Maximum height of projectile, $h_0 = \frac{u^2\sin^2\theta}{2g}$

$\therefore h_0 = \frac{(10)^2 \times \sin^2 30°}{2 \times 10} = \frac{5}{4} = 1.25\ m$

Time for attaining maximum height , $t = \frac{u\sin\theta}{g}$

$\therefore t = \frac{10 \times \sin 30°}{10} = 0.5$ sec

$\therefore$ Distance of vertical fall in 0.5 sec, $S = \frac{1}{2}gt^2$

$\Rightarrow S = \frac{1}{2} \times 10 \times (0.5)^2 = 1.25m$

$\therefore$ Height of second ball $= 1.25 + 1.25 = 2.50m$

692 **(c)**

Change in momentum of the ball $= mv\sin\theta - (-mv\sin\theta)$

$= 2mv\sin\theta = 2mgv\frac{\sin\theta}{g}$

$= mg \times \frac{2v\sin\theta}{g}$

= weight of the ball × total time of flight

693 **(d)**

Vertical force on the roller = weight of roller + component of force in vertical downward direction = $(70 \times 10 + 200\cos 45°)$N

694 **(b)**

Horizontal distance travelled by the bomb

$S = u \times t$

$= 200 \times \sqrt{\frac{2h}{g}} = 200 \times \sqrt{\frac{2 \times 8 \times 10^3}{9.8}}$

$= 8.081\ km$

695 **(a)**

If the horizontal range is the same then the angle of projection of an object is θ or $(90° - \theta)$ with the horizontal direction.

So, the angle of projection of first particle is θ and the other particle is $(90° - \theta)$

$$t_1 = \frac{2u\sin\theta}{g}$$

$$t_2 = \frac{2u\sin\theta(90° - \theta)}{g}$$

$$t_2 = \frac{2u\cos\theta}{g}$$

$$\because \; t_1 t_2 = \frac{2u\sin\theta}{g} \cdot \frac{2u\cos\theta}{g}$$

$$t_1 t_2 = \frac{2u^2\sin2\theta}{g^2}$$

$$\text{or } \; t_1 t_2 = \frac{2R}{g}\left(\because R = \frac{u^2\sin2\theta}{g}\right)$$

696 **(c)**

Since, $n = 2, \omega = 2\pi \times 2 = 4\pi \; rad/s^2$

So acceleration $= \omega^2 r = (4\pi)^2 \times \frac{25}{100} \; m/s^2 = 4\pi^2$

697 **(c)**

The vector is $\hat{\imath} - [\vec{A} + \vec{B} + \vec{C}]$

$= \hat{\imath} - [(2\hat{\imath} - 4\hat{\jmath} + 7\hat{k}) + (7\hat{\imath} + 2\hat{\jmath} - 5\hat{k}) + (-4\hat{\imath} + 7\hat{\jmath} + 3\hat{k})]$

$= -4\hat{\imath} - 5\hat{\jmath} - 5\hat{k}$

698 **(a)**

$$a = \frac{v^2}{r} = \frac{(400)^2}{160} = 10^3 m/s^2 = 1 \; km/s^2$$

699 **(c)**

When a particle is thrown in vertical downward direction with velocity u then final velocity at the ground level

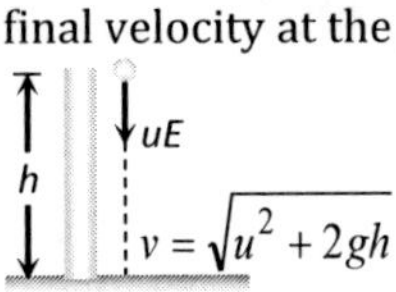

$v^2 = u^2 + 2gh \; \therefore v = \sqrt{u^2 + 2gh}$

Another particle is thrown horizontally with same vertically then at the surface of earth

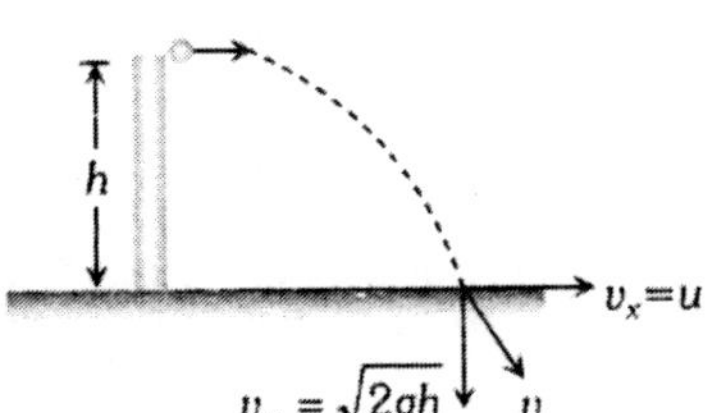

Horizontal component of velocity $v_x = u$

$\therefore$ Resultant velocity, $v = \sqrt{u^2 + 2gh}$

For both the particle final velocities when they reach the earth's surface are equal

700 **(a)**

$$R_{\max} = R = \frac{u^2}{g}$$

$$\Rightarrow u^2 = Rg$$

Now, as range $= \dfrac{u^2\sin2\theta}{g}$

then $\dfrac{R}{2} = \dfrac{Rg\sin2\theta}{g}$

$$\Rightarrow \sin2\theta = \frac{1}{2} = \sin30° \Rightarrow \theta = 15°$$

701 **(a)**

Here,

$v = \sqrt{(8)^2 + (6)^2} = 10$ and $\tan\theta = \frac{6}{8}$

$\therefore$ Hypotenuse, $h = 10\text{m}$

$$\therefore \; \sin\theta = \frac{6}{10}, \cos\theta = \frac{8}{10}$$

$$R = \frac{2u^2\sin\theta\cos\theta}{g}$$

$$R = \frac{2 \times 10 \times 10 \times \frac{6}{10} \times \frac{8}{10}}{10}$$

$$R = \frac{96}{10} = 9.6\text{m}$$

702 **(b)**

$$v = \sqrt{gr} = \sqrt{9.8 \times 5} = 7 \text{ ms}^{-1}$$

703 **(a)**

$$F = \frac{mv^2}{r} \text{ or } \; v = \sqrt{\frac{Fr}{m}}$$

704 **(b)**

Let the angle of projection be α.

$$\therefore \text{Range}, R = \frac{u^2\sin2\alpha}{g}$$

and maximum height $H = \dfrac{u^2\sin^2\alpha}{2g}$

Now, it is given that,

$(\text{Range})^2 = 48(\text{maximum height})^2$

$$\therefore \; \left(\frac{u^2\sin2\alpha}{g}\right)^2 = 48\left(\frac{u^2\sin^2\alpha}{2g}\right)^2$$

$$\text{or } \frac{u^2\sin2\alpha}{g} = 4\sqrt{3}\left(\frac{u^2\sin^2\alpha}{2g}\right)$$

$$\text{or } \frac{2\sin\alpha\cos\alpha}{4\sqrt{3}} = \frac{\sin^2\alpha}{2}$$

$$\text{or } \tan\alpha = \frac{4}{4\sqrt{3}} = \frac{1}{\sqrt{3}}$$

$$\therefore \; \alpha = 30°$$

705 **(b)**

From conservation of mechanical energy

$K = K_i - mgy$ ....(i)

(Here K_i= initial kinetic energy = constant)

$ie, K - y$ graph is straight line. It first decreases linearly becomes minimum at highest point and then becomes equal to K_i in the similar manner. Therefore, $K - y$ graph should be as shown below

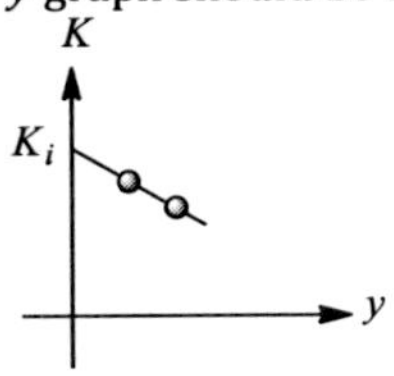

Eq. (i) we can written as

$$K = K_i - mg\left(u_y t - \frac{1}{2}gt^2\right)$$

$ie, K - y$graph is a parabola. Kinetic energy first decreases and then increases.

Eq. (i) can also written as

$$K = K_i - mg\left(x\tan\theta - \frac{gx^2}{2u_x^2}\right)$$

Again $K - x$ graph is a parabola

Further, $p^2 = 2Km$ $ie, p^2 = 2Km$ $ie, p^2 = K$ or K *versus* p^2

Graph is a straight line passing through origin

706 **(b)**

Computing the given equation with

$y = x\tan\theta - \frac{gx^2}{2v^2\cos^2\theta}$, we get

$\tan\theta = \sqrt{3}$

708 **(d)**

$r_1 = v/\omega; r_2 = 2v/(\omega/2) = 4v/\omega = 4r_1$

$a_1 = v^2/r_1; a_2 = (2v)^2/r_2 = 4v^2/r_1$

$= v^2/r_1 = a_1$

709 **(c)**

$R^2 = R^2 + R^2 + 2R^2\cos\theta$ or $R^2 = 2R^2 + 2R^2\cos\theta$

$\frac{1}{2} = 1 + \cos\theta$ or $\cos\theta = -\frac{1}{2}$ or $\theta = 120°$

710 **(b)**

Refer figure are when projectile is at A, then

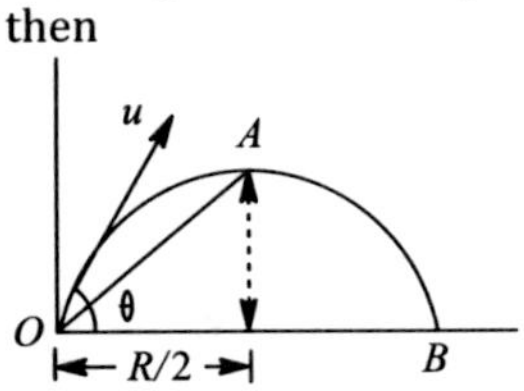

$$OC = \frac{R}{2} = \frac{1}{2}\frac{u^2}{g}\sin 2\theta$$

$$= \frac{1}{2} \times \frac{(20\sqrt{2})^2}{10}\sin 2 \times 45°$$

$$= 40 \text{ m}$$

$$AC = H = \frac{u^2\sin^2\theta}{2g} = \frac{(20\sqrt{2})^2}{2 \times 10}\sin^2 45°$$

$\therefore$ Displacement, $OA = \sqrt{OC^2 + CA^2} = \sqrt{40^2 + 20^2}$

Time of projectile from O to A

$$= \frac{1}{2}\left(\frac{2u\sin\theta}{g}\right) = \frac{u\sin\theta}{2g}$$

$$= \frac{(20\sqrt{2})\ \sin 45°}{10}$$

$$= 2\text{s}$$

$$\therefore \text{ Average velocity } = \frac{\text{displacement}}{\text{time}}$$

$$= \frac{\sqrt{40^2 + 20^2}}{2} = 10\sqrt{5}\text{ ms}^{-1}$$

711 **(a)**

Centripetal force

$$\frac{mv^2}{R} = ma$$

or $a = \frac{v^2}{R}$

$\therefore \frac{a_1}{a_2} = \frac{v_1^2}{v_2^2}$

Here, $v_1 = v,\ v_2 = 2v,\qquad a_1 = a$

$\therefore \frac{a}{a_2} = \frac{v^2}{(2v)^2} = \frac{1}{4}$

or $a_2 = 4a$

712 **(a)**

$v' = v_0\cos\theta$

$\frac{v_0}{2} = v_0\cos\theta$

$\cos\theta = \frac{1}{2}$

$\theta = 60°$

713 **(c)**

In uniform circular motion tangential acceleration remains zero but magnitude of radial acceleration remains constant.

714 **(c)**

$F = m\omega^2 R$ $\therefore R \propto \frac{1}{\omega^2}$ (m and F are constant)

If ω is doubled then radius will become $1/4$ times i.e. $R/4$

715 **(d)**

From force diagram shown in figure

$T_1 \cos 30° + T_2 \cos 45° = mg$ ...(i)

$T_1 \sin 30° + T_2 \sin 45° = \frac{mv^2}{r}$ ...(ii)

After solving Eq. (i) and eq. (ii), we get

$$T_1 = \frac{mg - \frac{mv^2}{r}}{\left(\frac{\sqrt{3}-1}{2}\right)}$$

But $T_1 > 0$

$$\therefore \frac{mg - \frac{mv^2}{r}}{\frac{\sqrt{3}-1}{2}} > 0$$

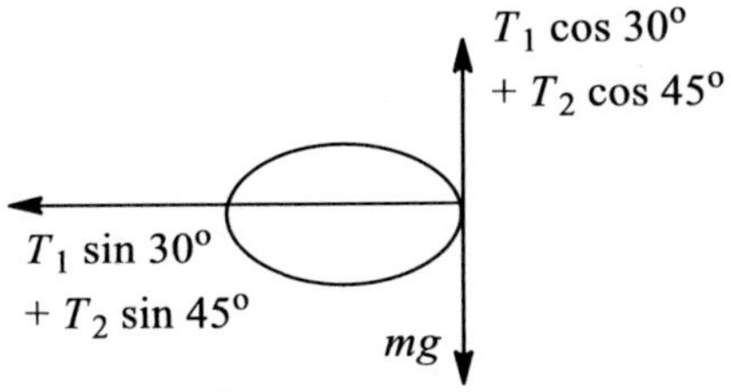

or $mg > \frac{mv^2}{r}$

or $v < \sqrt{rg}$

$\therefore v_{max} = \sqrt{rg} = \sqrt{1.6 \times 9.8} = 3.96 \text{ ms}^{-1}$

716 **(a)**

Given $(\hat{i} + 2\hat{j} - \hat{k}) + (\hat{i} - \hat{j} + 2\hat{k}) + \vec{C} = \hat{j}$

$\therefore \vec{C} = \hat{j} - (\hat{i} - 2\hat{j} - \hat{k}) - (\hat{i} + \hat{j} + 2\hat{k})$

$= -2\hat{i} - \hat{k}$.

718 **(a)**

Resultant $\vec{R} = (\vec{P} + \vec{Q}) + (\vec{P} - \vec{Q}) = 2\vec{P}$.

Thus angle between $\vec{R}$ and $\vec{P}$ is 0°.

719 **(a)**

Tangential acceleration $a = L\alpha$

$\therefore$ Normal relation $N = Ma = ML\alpha$

$\therefore$ Frictional force $F = mN = \mu ML\alpha$

For no sliding along the length frictional force ≥ centripetal force

ie, $\mu ML\alpha \geq ML\omega^2$

As $\omega = \omega_0 + \alpha t = \alpha t$

$\therefore \mu ML\alpha \geq ML(\alpha t)^2 \Rightarrow t = \sqrt{\frac{\mu}{\alpha}}$

720 **(c)**

Centripetal acceleration, $a_c = w^2 R$

$$= \frac{(2\pi)^2 \times 1.5 \times 10^{11}}{(365 \times 86400)^2} = 6 \times 10^{-3} ms^{-2}$$

721 **(c)**

From figure,

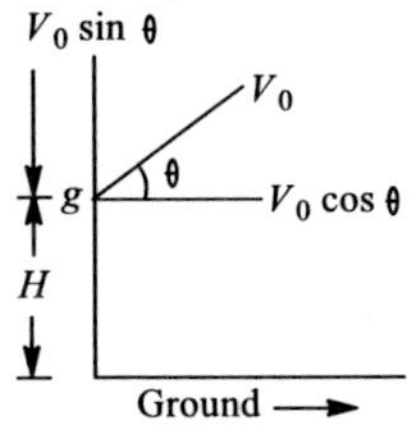

$$H = (-v_0 \sin\theta)t + \frac{1}{2}gt^2$$

$v_x = v_0 \cos\theta$

$v_y^2 = (v_0 \sin\theta)^2 + 2gH$

$v = \sqrt{v_x^2 + v_y^2}$ at ground

$v = \sqrt{v_0^2 + 2gH}$

It means speed is independent of angle of projection

Also, $\frac{1}{2}gt^2 = H + t\, v_0 \sin\theta$

From this, where θ increases, t increases

Hence, (c) is correct

722 **(c)**

a_c = centripetal acceleration

a_t = tangential acceleration

a_N = net acceleration = Resultant of a_c and a_t

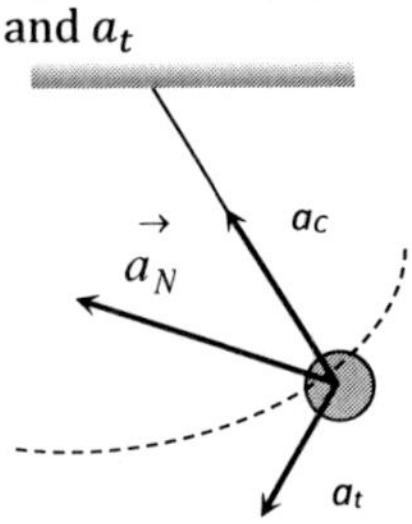

723 **(a)**

$H = \frac{u^2 \sin^2\theta}{2g}$ and $T = \frac{2u \sin\theta}{g}$

So $\frac{H}{T^2} = \frac{u^2 \sin^2\theta/2g}{4u^2 \sin^2\theta/g^2} = \frac{g}{8} = \frac{5}{4}$

724 **(a)**

Here, $\alpha = 2\theta, \beta = \theta$

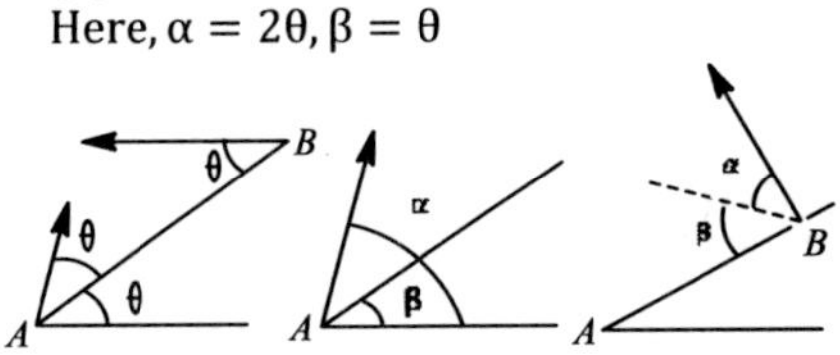

Time of flight of A is,

$T_1 = \frac{2u \sin(\alpha - \beta)}{g \cos \beta}$

$= \frac{2u \sin(2\theta - \theta)}{g \cos \theta}$

$= \frac{2u}{g} \tan \theta$

Time of flight of B is, $T_2 = \frac{2u \sin \theta}{g \cos \theta}$

$= \frac{2u}{g} \tan \theta$

So, $T_1 = T_2$. The acceleration of both the particles is g downwards. Therefore, relative acceleration between the two I zero or relative motion between the two is uniform. The relative velocity of A w.r.t. B is towardsAB, therefore collision will take place between the two in mid air.

725 **(c)**

As time periods are equal therefore ration of angular speeds will be one. $\omega = \frac{2\pi}{T}$

726 **(a)**

Centripetal acceleration $= \frac{v^2}{r} = \frac{(10)^2}{20}$

$= 5\text{m/s}^2$

727 **(b)**

$\text{Tan } 45° = \frac{2\sin 60°}{a + 2\cos 60°} = \frac{\sqrt{3}}{a+1}$

$1 = \frac{\sqrt{3}}{a+1}$

or $a + 1 = \sqrt{3}$

$a = \sqrt{3} - 1$

728 **(c)**

The body crosses the top most position of a vertical circle with critical velocity, so the velocity at the lowest point of vertical circle $u = \sqrt{5gr}$

Velocity of the body when string is horizontal is

$v^2 = u^2 - 2gr = 5gr - 2gr = 3gr$

$\therefore$ Centripetal acceleration $= \frac{v^2}{r} = \frac{3gr}{r} =$ 3g

729 **(c)**

$\text{K.E.} = \frac{1}{2} mv^2$.Which is scalar, so it remains constant

730 **(c)**

$v = \sqrt{\mu rg} = \sqrt{0.25 \times 80 \times 9.8} = 14 \text{ ms}^{-1}$

731 **(c)**

Horizontal range of the object fired,

$R = \frac{u^2 \sin 2\theta}{g}$

At the highest point, when object is exploded into two equal masses, then

$2mu \cos \theta = m(0) + mv$

or $v = 2u \cos \theta$

It means, the horizontal velocity becomes double at the highest point, hence it will cover double the distance during the remaining flight.

$\therefore$ Total horizontal range of the other part

$= \frac{R}{2} + R = \frac{3R}{2}$

$= \frac{3}{2} \frac{u^2 \sin 2\theta}{g}$

$= \frac{3}{2} \times \frac{(100)^2 \times 2 \sin \theta \cos \theta}{g}$

$= \frac{3}{2} \times \frac{(100)^2 \times 2 \times \frac{3}{5} \times \frac{4}{5}}{10} = 1440 \text{ m}$

732 **(c)**

$\vec{P}.\vec{Q} = (3\hat{j} + 4\hat{k}).(2\hat{i} + 5\hat{k}) = 6 + 20 = 26$

733 **(d)**

Maximum range $= \frac{u^2}{g} = 100\ m$

Maximum height $= \frac{u^2}{2g} = \frac{100}{2} = 50\ m$

734 **(a)**

Time of flight (T) is $2t$.

$\therefore \quad T = t2 = \frac{2\ u \sin \theta}{g}$

$= \frac{2}{8} \times U_{\text{vertical}}$

735 **(b)**

$x = (u\cos\theta)t = 6$

$u\cos\theta = \frac{x}{t} = 6$

$y = (u\sin\theta)t = -\frac{1}{2}gt^2$

$y = 8t - 5t^2 \Rightarrow u\sin\theta = 8$

$u = 10\text{m/s}$

736 **(d)**

$F = mg - \frac{mv^2}{r}$

737 **(c)**

At time t_1 the velocity of ball will be maximum and it goes on decreasing with respect to time

At the highest point of path its velocity becomes zero, then it increases but direction is reversed

This explanation matches with graph (c)

738 **(d)**

Second's hand of a watch completes its one rotation in 1 min. So, its time period is 1 min.

739 **(d)**

$$\omega = \sqrt{\frac{g}{r}} = \sqrt{\frac{9.8}{0.2}} = 7 rad/s$$

740 **(a)**

$v_x = 8t - 2$

or $\frac{dx}{dt} = 8t - 2$

or $\int_{14}^{x} dx = \int_{2}^{t} (8t - 2)dt$

or $x - 14 = [4t^2 - 2t]_2^t = 4t^2 - 2t - 12$

or $x = 4t^2 - 2t + 2$ ...(i)

Further, $v_y = 2$

or $\frac{dy}{dt} = 2$

$$\therefore \int_4^y dy = \int_2^t 2dt$$

or $y - 4 = [2t]_2^t = 2t - 4$

or $y = 2t$

or $t = \frac{y}{2}$ ...(ii)

Substituting value of t from Eq. (ii) in Eq. (i), we have

$x = y^2 - y + 2$

741 **(a)**

As the body just completes the circular path, hence critical speed at the highest point.

$v_H = \sqrt{gR}$

which is totally horizontal.

As the string breaks at the highest point, hence form here onwards the body will follow parabolic path. Time taken by the body to reach the ground

$$t = \sqrt{\frac{2h}{g}} = \sqrt{\frac{2 \times 2R}{g}}$$

Hence, horizontal distance covered by the body

$= v_H \times t$

$$= \sqrt{gR} \times \sqrt{\frac{4R}{g}} = 2R$$

742 **(a)**

Area of parallelogram $= |A \times B|$

$$AB \sin\theta = \frac{1}{2} AB$$

$$\therefore \sin\theta = \frac{1}{2}, \theta = 30°$$

743 **(a)**

$$\vec{L} = \vec{r} \times \vec{p} = \begin{vmatrix} \hat{i} & \hat{j} & \hat{k} \\ 0 & 4 & 0 \\ 2 & 3 & 1 \end{vmatrix}$$

$= \hat{i}[4 - 0] + \hat{j}[0 - 0] + \hat{k}[0 - 8] = 4\hat{i} - 8\hat{k}$

744 **(a)**

Retarding force $F = ma = \mu R = \mu\, mg$

$a = \mu g$

Now, from equation of motion, $v^2 = u^2 - 2as$

$\therefore 0 = u^2 - 2as$

$$\therefore s = \frac{u^2}{2a} = \frac{u^2}{2\mu g} = \frac{v_0^2}{2\mu g}$$

745 **(b)**

Displacement , $\vec{r} = \vec{r}_2 - \vec{r}_1$

$= (\hat{i} - \hat{j} + 2\hat{k}) - (\hat{i} - \hat{j} + 2\hat{k}) = -2\hat{i} + \hat{k}$

$\therefore$workdone, $W = \vec{F} \cdot \vec{r} = (\hat{i} + 2\hat{j} + 3\hat{k}) \cdot (-2\hat{j} + \hat{k}) = -1$ J

746 **(a)**

$$\omega = \frac{v}{r} = \frac{100}{100} = 1 rad/s$$

747 **(d)**

In complete revolution total displacement is zero so average velocity is zero

748 **(c)**

$F = mr\omega^2$

$$= \frac{2}{125} \times \frac{1.25}{2\pi} \times (100\pi)^2 \quad (\because \omega = 2\pi f = 2 \times 50 \times \pi)$$

$= 100\pi N = 314N$

749 **(b)**

Let x be increase in length of the spring. The particle would move in a circular path of radius $(l + x)$. Centripetal force = force due to the spring

$m(l + x)\omega^2 = kx$

$$\therefore x = \frac{m\omega^2 l}{k - m\omega^2}$$

750 **(c)**

$v_x = \frac{dx}{dt} = \frac{d}{dt}(3t^2 - 6t) = 6t - 6$. At $t = 1, v_x = 0$

$v_y = \frac{dy}{dt} = \frac{d}{dt}(t^2 - 2t) = 2t - 2$. At $t = 1, v_y = 0$

Hence $v = \sqrt{v_x^2 + v_y^2} = 0$

751 **(b)**

Reaction on inner wheel $R_1 = \frac{1}{2} M \left[g - \frac{v^2 h}{ra}\right]$

Reaction on outer wheel $R_2 = M\left[g + \frac{v^2h}{ra}\right]$

where, r = radius of circular path, $2a$ =distance between two wheels and h = height of centre of gravity of car

753 **(d)**

At 45°, $v_x = v_y$

or $u_x = u_y - gt$

$\therefore \; t = \frac{u_y - u_x}{g}$

$= \frac{40(\sin 60° - \sin 30°)}{9.8} = 1.5$ s

755 **(d)**

For successfully completing the loop,

$h = \frac{5}{4}R \Rightarrow R = \frac{2h}{5} = \frac{2 \times 5}{5} = 2$cm

756 **(d)**

Maximum tension $= \frac{mv^2}{r} = 16N$

$\Rightarrow \frac{16 \times v^2}{144} = 16 \Rightarrow v = 12\, m/s$

757 **(b)**

The two angles of projection are clearly θ and (90° − θ)

$T_1 = \frac{2v \sin\theta}{g}$ and $T_2 = \frac{2v\sin(90° - \theta)}{g}$

$T_1T_2 = \frac{2(v)^2(2\sin\theta\cos\theta)}{g \times g} = \frac{2R}{g}$

758 **(c)**

$\vec{P} + \vec{Q} = 5\hat{i} - 4\hat{j} + 3\hat{k}$

$\cos\alpha = \frac{5}{\sqrt{5^2+4^2+3^2}} = \frac{5}{\sqrt{50}}$

or $\alpha = \cos^{-1}\left(\frac{5}{\sqrt{50}}\right)$

759 **(a)**

Since the block rises to the same heights in all the four cases, from conversation of energy, speed of the block at highest point will be same in all four cases. Say it is v_0. Equation of motion will be

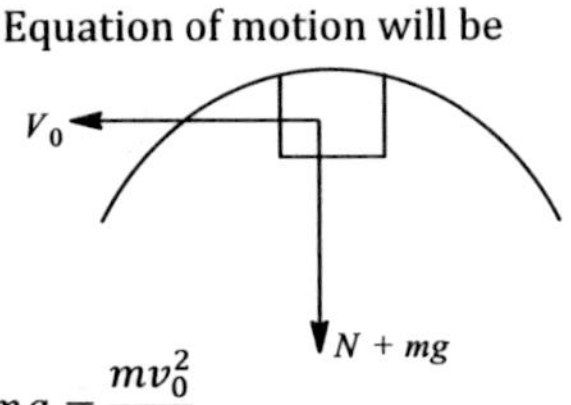

$N + mg = \frac{mv_0^2}{R}$

or $N = \frac{mv_0^2}{R} - mg$

R (the radius of curvature) in first case minimum. Therefore, normal reaction N will be maximum in first case.

760 **(b)**

$\vec{\mathbf{p}}_B - \vec{\mathbf{p}}_A = m(\vec{\mathbf{v}}_B - \vec{\mathbf{v}}_A) = mv(\hat{\mathbf{j}} + \hat{\mathbf{i}})$

$= 2mv\hat{\mathbf{j}} = 2\text{kg ms}^{-1}$ ...(i)

$\vec{\mathbf{F}}_B - \vec{\mathbf{F}}_A = \frac{mv^2}{R}(-\hat{\mathbf{i}}) - \frac{mv^2}{R}(+\hat{\mathbf{i}})$

$= \frac{2mv^2}{R}(-\hat{\mathbf{i}}) = 8\text{N}$ ...(ii)

Dividing Eq. (ii) by Eq. (i), $\frac{v}{R} = \omega = 4$ rad s^{-1}

762 **(a)**

$\frac{mv^2}{r} = \frac{k}{r^2} \Rightarrow mv^2 = \frac{k}{r} \therefore \text{K.E.} = \frac{1}{2}mv^2$

$= \frac{k}{2r}$

P.E. $= \int F\,dr = \int \frac{k}{r^2}\,dr = -\frac{k}{r}$

$\therefore$ Total energy = K.E. + P.E. $= \frac{k}{2r} - \frac{k}{r} = -\frac{k}{2r}$

763 **(b)**

$\vec{A} = A\hat{A} = B\hat{B}$. let θ be the angle between $\vec{A}$ and $\vec{B}$.

As per question,

$\cos\alpha = \frac{(A\hat{A} + B\hat{B}).(A\hat{B} + B\hat{A})}{|A\hat{A} + B\hat{B}||A\hat{B} + B\hat{A}|}$

or $\cos\alpha = \frac{2AB + (A^2+B^2)\cos\theta}{(\sqrt{A^2+B^2+2AB}\cos\theta)^2}$

or $2AB + (A^2 + B^2)\cos\theta = (A^2 + B^2)\cos\alpha + 2AB\cos\theta\cos\alpha$

or $2AB(1 - \cos\alpha\cos\theta)$

$= (A^2 + B^2)(\cos\alpha - \cos\theta)$

or $\frac{2AB}{A^2+B^2} = \frac{\cos\alpha - \cos\theta}{1 - \cos\alpha\cos\theta}$

or $\frac{2AB}{(A^2+B^2)} = \frac{\cos\alpha - \cos\theta}{1 - \cos\alpha\cos\theta}$

or $\frac{2AB + (A^2+B^2)}{(A^2+B^2) - AB}$

$= \frac{(\cos\alpha\cos\theta) + (1 - \cos\alpha\cos\theta)}{(1 - \cos\alpha\cos\theta) + (\cos\alpha\cos\theta)}$

or $\frac{(A+B)^2}{(A-B)^2} = \frac{(1+\cos\alpha)(1-\cos\theta)}{(1+\cos\theta)(1-\cos\alpha)}$

$= \frac{\tan^2\theta/2}{\tan^2\theta/2}$

or $\tan\frac{\alpha}{2} = \left(\frac{A-B}{A+B}\right)\tan\frac{\theta}{2}$

764 **(d)**

$T\sin\theta = M\omega^2R$ ...(i)

$T\sin\theta = M\omega^2L\sin\theta$ ...(ii)

$T = M\omega^2L$

$= M \cdot 4\pi^2n^2L$

$= M \cdot 4\pi^2\left(\frac{2}{\pi}\right)^2 L$

$= 16\,ML$

765 **(b)**

Centripetal acceleration $= r4\pi^2v^2$

5
LAWS OF MOTION
SINGLE CORRECT ANSWER TYPE

1. An ice cart of mass 60 kg rests on a horizontal snow patch with coefficient of static friction $\frac{1}{3}$. Assuming that there is no vertical acceleration, find the magnitude of the maximum horizontal force required to move the ice cart ($g = 9.8\ \mathrm{ms^{-2}}$)
   a) 100 N b) 110 N c) 209 N d) 196 N
2. A monkey of mass 20 kg is holding a vertical rope.The rope will not break when a mass of 25 kg is suspended from it but will break if the mass exceeds 25 kg. What is the maximum acceleration with which the monkey can climb up along the rope ($g = 10\ m/s^2$)
   a) $10\ m/s^2$ b) $25\ m/s^2$ c) $2.5\ m/s^2$ d) $5\ m/s^2$
3. A player kicks a football of mass 0.5 kg and the football begins to move with a velocity of 10 m/s. If the contact between the leg and the football lasts for $\frac{1}{50}$ s, then the force on the ball should be
   a) 2500 N b) 1250 N c) 250 N d) 625 N
4. An 80 kg person is parachuting and is experiencing a downward acceleration of $2.8\mathrm{ms^{-2}}$. The mass of the parachute is 5 kg. The upward force on the open parachute is (Take $g = 9.8\mathrm{ms^{-2}}$)
   a) 595 N b) 675 N c) 456 N d) 925 N
5. A block of mass 15 kg is resting on a rough inclined plane as shown in figure. The block is tied by a horizontal string which has a tension of 50 N. The coefficient of friction between the surfaces of contact is ($g = 10\ \mathrm{ms^{-2}}$)

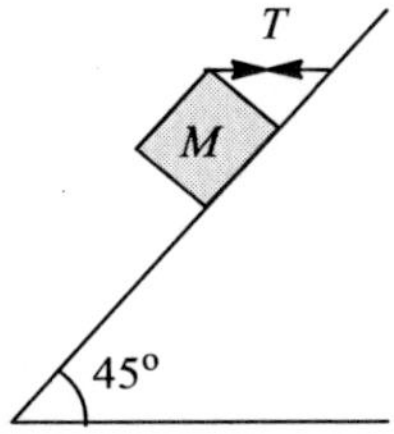

   a) 1/2 b) 3/4 c) 2/3 d) 1/4
6. The rate of the mass of the gas emitted from rear of a rocket is initially $0.1\mathrm{kgs^{-1}}$. If the speed of the gas relative to the rocket is 50 $\mathrm{ms^{-1}}$ and mass of the rocket is 2 kg, then the acceleration of the rocket (in $\mathrm{ms^{-2}}$) is
   a) 5 b) 5.2 c) 2.5 d) 25
7. A 100 kg block is suspended with the help of three string A, B and C. The tension in the string C is

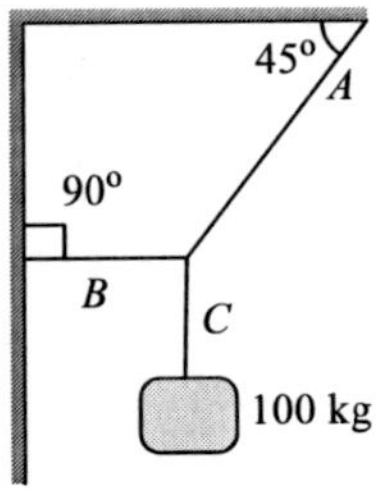

a) 50 gN b) 100 gN c) 20 gN d) 20 gN

8. If the resultant of all the external forces acting on a system of particles is zero, then form an inertial frame, one can surely say that
a) Linear momentum of the system does not change in time
b) Kinetic energy of the system does not change in time
c) Angular momentum of the system does not change in time
d) Potential energy of the system does not change in time

9. Which of the following quantities measured from different inertial reference frames are same
a) Force b) Velocity c) Displacement d) Kinetic energy

10. Maximum value of static friction is called
a) Limiting friction b) Rolling friction c) Normal reaction d) Coefficient of friction

11. Two bodies of mass 4 kg and 6 kg are attached to the ends of a string passing over a pulley. The 4 kg mass is attached to the table by another string. The tension in this string T_1 is

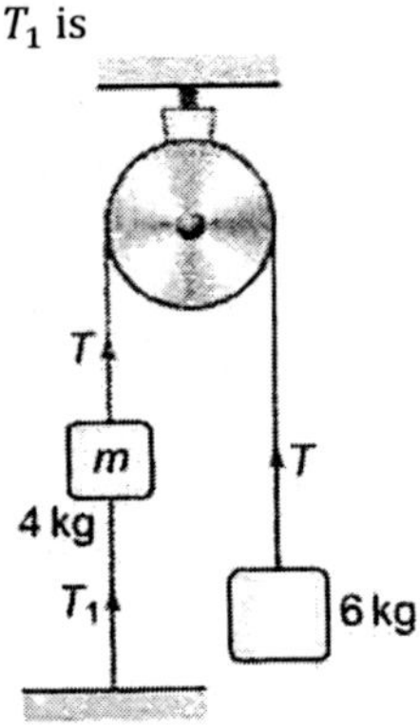

a) 19.6 N b) 25 N c) 10.6 N d) 10 N

12. A vessel containing water is given a constant acceleration a towards the right, along a straight horizontal path. Which of the following diagram represents the surface of the liquid

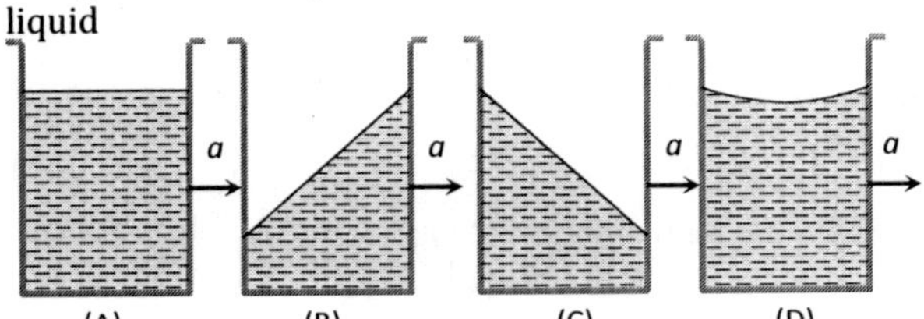

a) A b) B c) C d) D

13. A marble block of mass 2 kglying on ice when given a velocity of 6 m/s is stopped by friction in 10s. Then the coefficient of friction is
a) 0.01 b) 0.02 c) 0.03 d) 0.06

14. Which of the following is the correct order of forces
a) Weak <gravitational forces < strong forces (nuclear) < electrostatic
b) Gravitational < Weak < (electrostatic) < strong force
c) Gravitational <electrostatic <Weak <strong force
d) Weak < gravitational < electrostatic < strong forces

15. A block of mass m is in contact with the cart C as shown in the figure

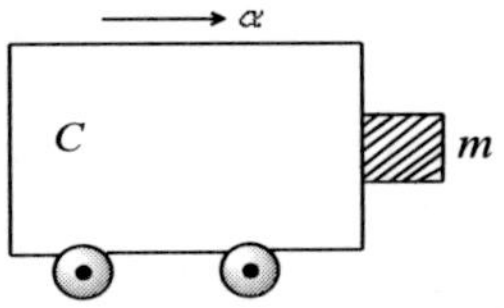

The coefficient of static friction between the block and the cart is μ. The acceleration α of the cart that will prevent the block from falling satisfies

a) $\alpha < \frac{g}{\mu}$ b) $\alpha > \frac{mg}{\mu}$ c) $\alpha > \frac{g}{\mu m}$ d) $\alpha \geq \frac{g}{\mu}$

16. A block of mass m_1 reacts on a horizontal table. A string tied to the block is passed on a frictionless pulley fixed at the end of the table and to the other end of string is hung another block of mass m_2. The acceleration of the system is

a) $\frac{m_2 g}{(m_1 + m_2)}$ b) $\frac{m_1 g}{(m_1 + m_2)}$ c) g d) $\frac{m_2 g}{m_1}$

17. If a street light of mass M is suspended from the end of a uniform rod of length L in different possible patterns as shown in figure, then

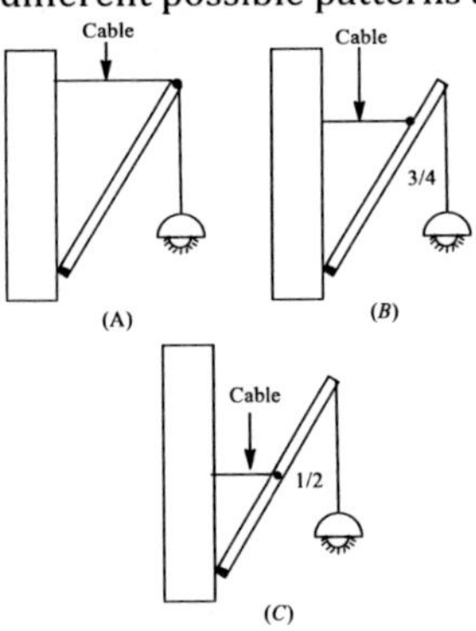

a) Pattern A is more sturdy b) Pattern B is more sturdy
c) Pattern C is more sturdy d) All will have same sturdiness

18. If a ladder weighing 250 N is placed against a smooth vertical wall having coefficient of friction between it and floor is 0.3, then what is the maximum force of friction available at the point of contact between the ladder and the floor

a) 75 N b) 50 N c) 35 N d) 25 N

19. A particle of mass m moving with velocity u makes an elastic one dimensional collision with a stationary particle of mass m. They are in contact for a very short time T. Their force of interaction increases from zero to F_0 linearly in time $T/2$, and decreases linearly to zero in further time $T/2$. The magnitude of F_0 is

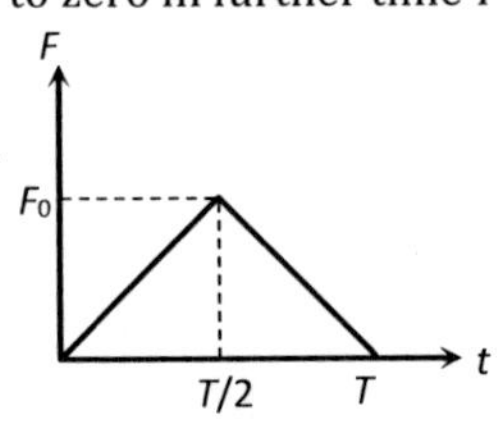

a) mu/T b) $2mu/T$ c) $mu/2T$ d) None of these

20. Pulling force making an angle θ to the horizontal is applied on a block of weight W placed on a horizontal table. If the angle of friction is α, then the magnitude of force required to move the body is equal to

a) $\frac{W \sin \alpha}{g \tan(\theta - \alpha)}$ b) $\frac{W \cos \alpha}{\cos(\theta - \alpha)}$ c) $\frac{W \sin \alpha}{\cos(\theta - \alpha)}$ d) $\frac{W \tan \alpha}{\sin(\theta - \alpha)}$

21. Two blocks of masses $m_1 = 4$ kg and $m_2 = 2$ kg are connected to the ends of a string which passes over a massless, frictionless pulley. The total downwards thrust on the pulley is nearly

a) 27 N b) 54 N c) 0.8 N d) Zero

22. Consider the following statement. When jumping from some height, you should bend your knees as you come to rest instead of keeping your legs stiff. Which of the following relations can be useful in explaining the statement?

a) $\Delta \mathbf{p}_1 = -\Delta \mathbf{p}_2$ b) $\Delta E = -\Delta(\text{PE} + \text{KE}) = 0$

c) $\mathbf{F}\, \Delta t = m \Delta \mathbf{v}$ d) $\Delta \mathbf{x} \propto \Delta \mathbf{F}$
Where symbols have their usual meaning

23. At a certain instant of time the mass of rocket going up vertically is 100 kg. If it is ejecting 5 kg of gas per second at a speed of 400 m/s, the acceleration of the rocket would be (Taking $g = 10$ m/s^2)

a) 20 m/s^2 b) 10 m/s^2 c) 2 m/s^2 d) 1 m/s^2

24. Two masses A and B of 10 kg and 5 kg respectively are connected with a string passing over a frictionless pulley fixed at the corner of a table as shown. The coefficient of static friction of A with table is 0.2. The minimum mass of C that may be placed on A to prevent it from moving is

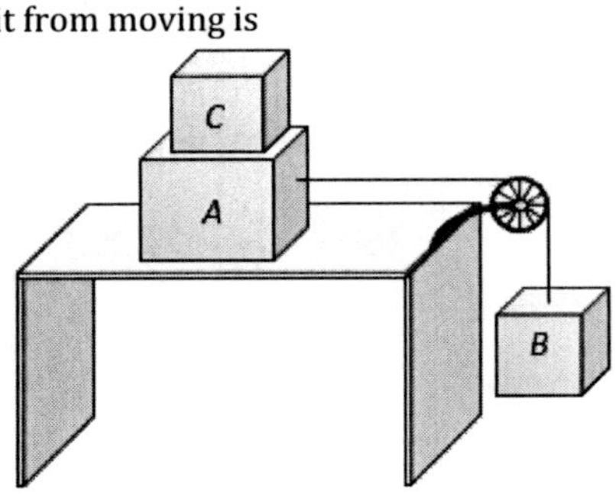

a) 15 kg b) 10 kg c) 5 kg d) 12 kg

25. A bullet of mass 5 g is shot from a gun of mass 5 kg. The muzzle velocity of the bullet is 5000 ms^{-1}. The recoil velocity of the gun is

a) 0.5 ms^{-1} b) 0.25 ms^{-1} c) 1 ms^{-1} d) Data is insufficient

26. The masses of 10 kg and 20 kg respectively are connected by a massless spring as shown in figure. A force of 200 N acts on the 20 kg mass. At the instant shown, the 10 kg mass has acceleration 12 m/sec^2. What is the acceleration of 20 kg mass

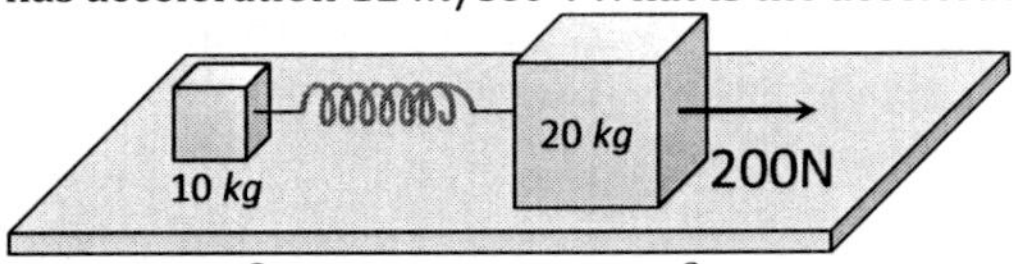

a) 12 m/sec^2 b) 4 m/sec^2 c) 10 m/sec^2 d) Zero

27. A steel wire can withstand a load up to 2940 N. A load of 150 kg is suspended from a rigid support. The maximum angle through which the wire can be displaced from the mean position, so that the wire does not break when the load passes through the position of equilibrium, is

a) 30° b) 60° c) 80° d) 85°

28. The upper half of an inclined plane of inclination θ is perfectly smooth while the lower half is rough. A body starting from the rest at top comes back to rest at the bottom if the coefficient of friction for the lower half is given by

a) $\mu = \sin\theta$ b) $\mu = \cot\theta$ c) $\mu = 2\cos\theta$ d) $\mu = 2\tan\theta$

29. Which of the four arrangements in the figure correctly shows the vector addition of two forces $\vec{F_1}$ and $\vec{F_2}$ to yield the third force $\vec{F_3}$

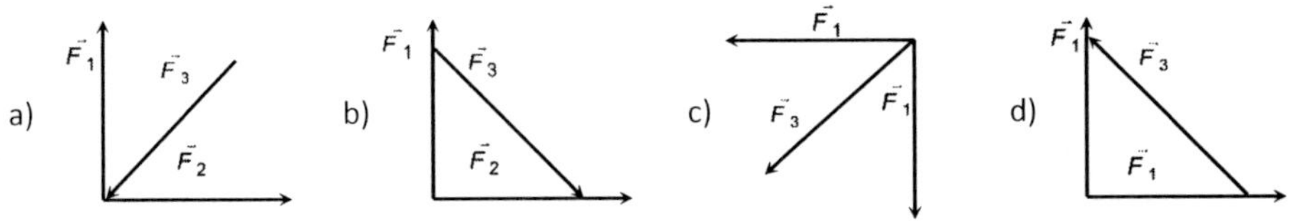

30. A particle moves in the $x - y$ plane under the influence of a force such that its linear momentum is

$\vec{p}(t) = A[\hat{\imath}\cos(kt) - \hat{\jmath}\sin(kt)]$

Where A and k are constants. The angle between the force and momentum is

a) $0°$ b) $30°$ c) $45°$ d) $90°$

31. The spring balance inside a lift suspends an object. As the lift begins to ascent, the reading indicated by the spring balance will

a) Increase b) Decrease

c) Remain unchanged d) Depend on the speed of ascend

32. Two blocks of equal masses m are released from the top of a smooth fixed wedge as shown in the figure. The acceleration of the centre of mass of the two blocks is

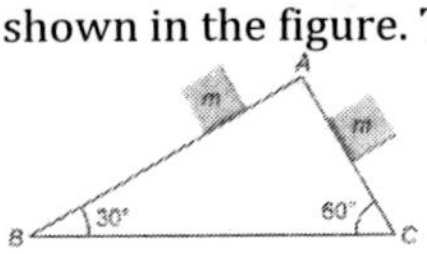

a) g b) $\frac{g}{2}$ c) $\frac{3g}{4}$ d) $\frac{g}{\sqrt{2}}$

33. A brick of mass 2 kg begins to slide down on a plane inclined at an angle of 45° with the horizontal. The force of friction will be

a) $19.6\sin 45°$ b) $19.6\cos 45°$ c) $9.8\sin 45°$ d) $9.8\cos 45°$

34. System shown in figure is in equilibrium and at rest. The spring and string are massless, now the string is cut. The acceleration of mass 2 m and m just after the string is cut will

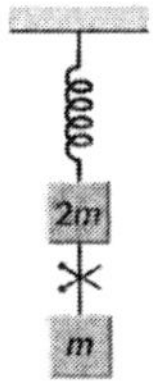

be

a) $g/2$ upward, g downward b) g upward, $g/2$ downward

c) g upward, $2g$ downward d) $2g$ upward, g downward

35. A gramophone record is revolving with an angular velocity ω. A coin isplaced at a distance r from the centre of the record. The static coefficient of friction is μ. The coin will revolve with the record if

a) $r \geq \frac{\mu g}{\omega^2}$ b) $r = \mu g \omega^2$ c) $r < \frac{\omega^2}{\mu g}$ d) $r \leq \frac{\mu g}{\omega^2}$

36. A body sitting on the topmost berth in the compartment of a train which is just going to stop on a railway station, drops an apple aiming at the open hand of his brother sitting vertically below his hands at a distance of about 2 m. The apple will fall

a) Precisely on the hand of his brother

b) Slightly away from the hand of his brother in the direction of motion of the train

c) Slightly away from the hand of his brother in the direction opposite to the direction of motion of the train

d) None of the above

37. Two masses $m_1 = 1\text{kg}$ and $m_2 = 2\text{kg}$ are connected by a light inextensible string and suspended by means of a weightless pulley as shown in figure.

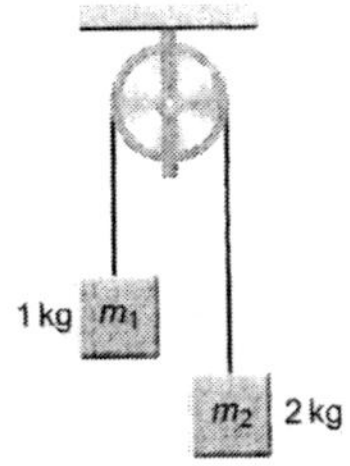

Assuming that both the masses start from rest, the distance travelled by the centre of mass in 2 s is (take $g = 10\text{m/s}^2$)

a) $\frac{20}{9}$ m b) $\frac{40}{9}$ m c) $\frac{2}{3}$ m d) $\frac{1}{3}$ m

38. For ordinary terrestrial experiments, the observer is an inertial frame in the following cases is

a) A child revolving in a giant wheel

b) A driver in a sports car moving with a constant high speed of 200 kmh^{-1} on a straight rod

c) The pilot of an aeroplane which is taking off

d) A cyclist negotiating a sharp curve

39. A large force is acting on a body for a short time. The impulse imparted is equal to the change in

a) Acceleration b) Momentum c) Energy d) Velocity

40. A block is kept on an inclined plane of inclination θ of length l. The velocity of particle at the bottom of inclined plane is (the coefficient of friction is μ)

a) $\sqrt{2gl(\mu\cos\theta - \sin\theta)}$ b) $\sqrt{2gl(\sin\theta - \mu\cos\theta)}$ c) $\sqrt{2gl(\sin\theta + \mu\cos\theta)}$ d) $\sqrt{2gl(\cos\theta + \mu\sin\theta)}$

41. A ball of mass 1 kg hangs in equilibrium from two strings OA and OB as shown in figure. What are the tensions in strings OA and OB? (Take $\text{g} = 10\ \text{ms}^{-2}$)

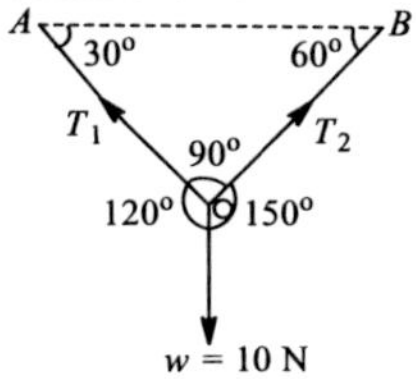

a) 5 N, zero b) Zero, N c) 5 N, $5\sqrt{3}$ N d) $5\sqrt{3}$ N, 5 N

42. A solid sphere of mass 2 kg is resting inside a cube as shown in the figure. The cube is moving with a velocity $v = (5t\hat{i} + 2t\hat{j})m/s$. Here t is the time in second. All surface are smooth. The sphere is at rest with respect to the cube. What is the total force exerted by the sphere on the cube. (Take $g = 10\ m/s^2$)

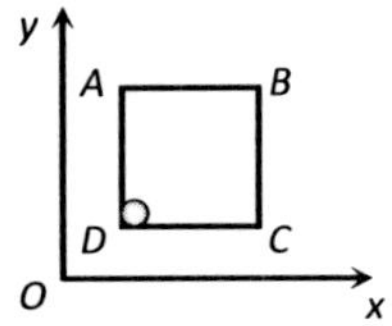

a) $\sqrt{29}N$ b) 29 N c) 26 N d) $\sqrt{89}$ N

43. In the given arrangement, $n$number of equal masses are connected by strings of negligible masses. The tension in the string connected to nth mass is

a) $\frac{mMg}{nm+M}$ b) $\frac{mMg}{nmM}$ c) mg d) mng

44. A block of mass m is placed on a smooth wedge of inclination θ. The whole system is accelerated horizontally so that the block does not slip on the wedge. The force exerted by the wedge on the block (g is acceleration due to gravity) will be

a) $mg\cos\theta$ b) $mg\sin\theta$ c) mg d) $mg/\cos\theta$

45. A body is imparted motion from rest to move in a straight line. If it is then obstructed by an opposite force, then

a) The body may necessarily change direction
b) The body is sure to slow down
c) The body will necessarily continue to move in the same direction at the same speed
d) None of these

46. A person used force (F), shown in figure to move a load with constant velocity on given surface

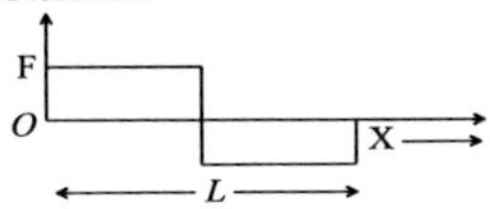

Identify the correct surface profile

a) 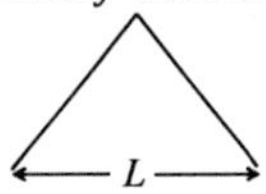 b) 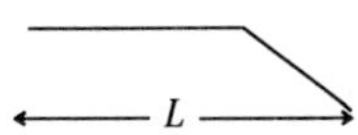 c) 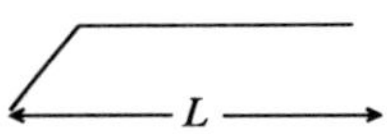 d) 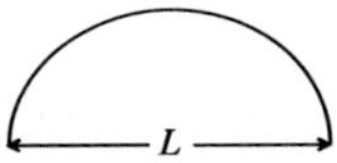

47. The acceleration of system of two bodies over the wedge as shown in figure is

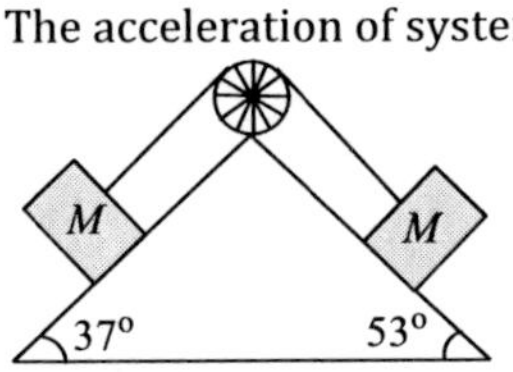

a) $1ms^{-1}$ b) $2ms^{-2}$ c) $0.52ms^{-2}$ d) $10ms^{-2}$

48. A body of mass 6 kg moves in a straight line according to the equation
$x = t^3 - 75t$,
Where x denotes the distance in metre and t the time in second. The force on the body at $t = 4$ s is

a) 64 N b) 72 N c) 144 N d) 36 N

49. An object is subjected to a force in the north-east direction. To balance this force, a second force should be applied in the direction

a) North-East b) South c) South-west d) West

50. A body of mass 0.05 kg is observed to fall with an acceleration of 9.5 ms^{-2}. The opposing force of air on the body is ($g = 9.8\ ms^{-2}$)

a) 0.015 N b) 0.15 N c) 0.030 N d) Zero

51. Diwali rockets are ejecting 50 g of gases per second at a velocity of 400 ms^{-1}. The accelerating force on the rocket will be

a) 22 dyne b) 20 N c) 20 dyne d) 100 N

52. Which of the following statements is not true

a) The coefficient of friction between two surfaces increases as the surface in contact are made rough
b) The force of friction acts in a direction opposite to the applied force
c) Rolling friction is greater than sliding friction
d) The coefficient of friction between wood and wood is less than 1

53. A truck is moving on a frictionless surface with uniform velocity of 10 ms^{-1}. A leak occurs in the water tank of the truck at the rate of 2 kgs^{-1}. What is the speed of truck after 50 s if the mass of truck is 100 kg and mass of water in truck initially was 100 kg?

a) 20 ms^{-1} b) 10 ms^{-1} c) 5 ms^{-1} d) None of these

54. A body of mass 1000 kg is moving horizontally with a velocity 50 ms^{-1}. A mass of 250 kg is added. Find the final velocity

a) 40 ms^{-1} b) 23 ms^{-1} c) 12 ms^{-1} d) 32.5 ms^{-1}

55. In the figure shown, $m_1 = 10$ kg, $m_2 = 6$ kg, $m_3 = 4$ kg. If $T_3 = 40$ N, $T_2 = ?$

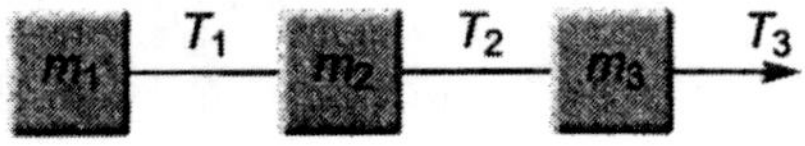

a) 13 N b) 32 N c) 25 N d) 35 N

56. A body takes time t to reach the bottom of an inclined plane of angle θ with the horizontal. If the plane is made rough, time taken now is $2t$. The coefficient of the friction of the rough surface is

a) $\frac{3}{4}\tan\theta$ b) $\frac{2}{3}\tan\theta$ c) $\frac{1}{4}\tan\theta$ d) $\frac{1}{2}\tan\theta$

57. A 5000 kg rocket is set for vertical firing. The exhaust speed is800 ms^{-1}. To give an initial upward acceleration of 20 ms^{-2}, the amount of gas ejected per second to supply the needed thrust will be ($g = 10\ ms^{-2}$)

a) 127.5 kgs^{-1} b) 187.5 kgs^{-1} c) 185.5 kgs^{-1} d) 137.5 kgs^{-1}

58. A block of mass M is pulled along a horizontal frictionless surface by a rope of mass m. Force$P$ is applied at one end of the rope. The force which the rope exerts on the block is

a) $\frac{P}{M(m+M)}$ b) $\frac{P}{M-m}$ c) $\frac{Pm}{M-m}$ d) $\frac{PM}{m+M}$

59. A body of mass M is kept on a rough horizontal surface (friction coefficient μ). A person is trying to pull the body by applying a horizontal force but the body is not moving. The force by the surface on the body is$F$, where

a) $F = Mg$ b) $F = \mu Mgf$

c) $Mg \le F \le Mg\sqrt{1+\mu^2}$ d) $Mg \ge F \ge Mg\sqrt{1+\mu^2}$

60. Three concurrent co-planar forces 1 N, 2 N and 3 N acting along different directions on a body

a) Can keep the body in equilibrium if 2 N and 3 N act at right angle

b) Can keep the body inequilibrium if 1 N and 2 N act at right angle

c) Cannot keep the body in equilibrium

d) Can keep the body in equilibrium if 1 N and 3 N act at an acute angle

61. A car turns a corner on a slippery road at a constant speed of 10 m/s. If the coefficient of friction is 0.5, the minimum radius of the arc in meter in which the car turns is

a) 20 b) 10 c) 5 d) 4

62. In the arrangement shown, the pulleys are fixed and ideal, the strings are light, $m_1 > m_2$ and S is a spring balance which is itself massless. The reading of S (in unit of mass) is

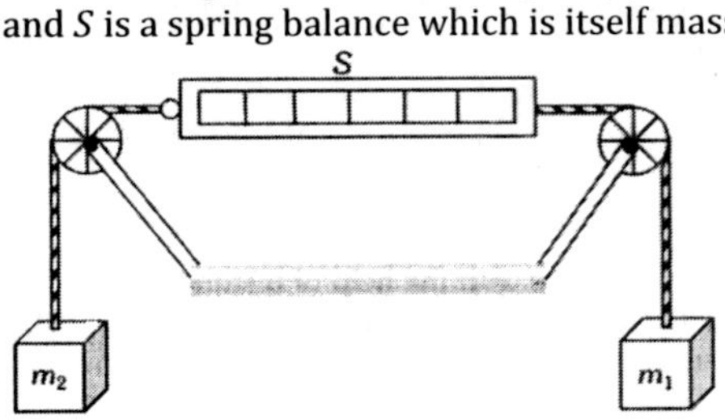

a) $m_1 - m_2$ b) $\frac{1}{2}(m_1 + m_2)$ c) $\frac{m_1 m_2}{m_1 + m_2}$ d) $\frac{2m_1 m_2}{m_1 + m_2}$

63. Which activity is not based upon friction

a) Writing b) Speaking c) Hearing d) Walking

64. A rope of mass0.1 kg is connected at the same height of two opposite walls. It is allowed to hang under its own weight. At the contact point between the rope and the wall, the rope makes an angle $\theta = 10°$ with respect to the horizontal. The tension in the rope at its midpoint between the wall is

a) 2.78 N b) 2.56 N c) 2.82 N d) 2.71 N

65. Two masses M and m are connected by a weightless string. They are pulled by a force F on a frictionless horizontal surface. The tension in the string will be

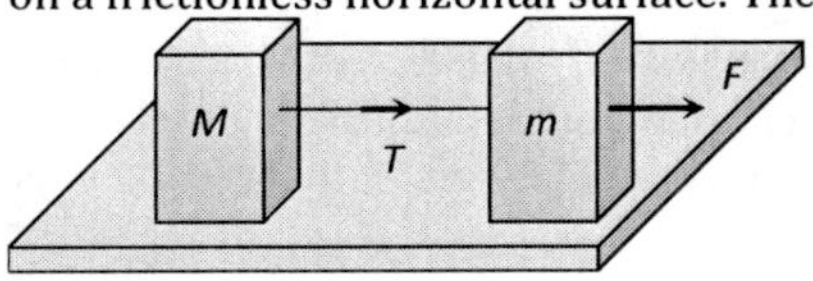

a) $\frac{FM}{m+M}$ b) $\frac{F}{M+m}$ c) $\frac{FM}{m}$ d) $\frac{Fm}{M+m}$

66. The variation of momentum with time of one of the body in a two body collision is shown in fig. The instantaneous force is maximum corresponding to point

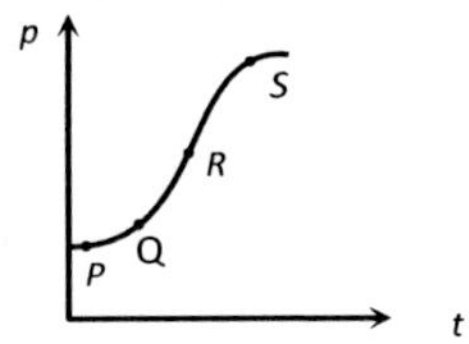

a) P b) Q c) R d) S

67. A motorcyclist of mass m is to negotiate a curve of radius r with a speed v. The minimum value of the coefficient of friction so that this negotiation may take place safely, is

a) v^2rg b) $\frac{v^2}{gr}$ c) $\frac{gr}{v^2}$ d) $\frac{g}{v^2r}$

68. If a body of mass m is carried by a lift moving with an upward acceleration a, then the forces acting on the body are (i) the reaction R on the floor of the lift upwards (ii) the weight mg of the body acting vertically downwards. The equation of motion will be given by

a) $R = mg - ma$ b) $R = mg + ma$ c) $R = ma - mg$ d) $R = mg \times ma$

69. A packet of weight w is dropped with the help of a parachute and on striking the ground comes to rest with retardation equal to twice the acceleration due to gravity. What is the force exerted on the ground?

a) w b) 2 w c) 3 w d) 4 w

70. Two forces are such that the sum of their magnitudes is 18 N and their resultant is perpendicular to the smaller force and magnitude of resultant is 12 N. Then the magnitudes of the forces are

a) 12 N, 6 N b) 13 N, 5 N c) 10 N, 8 N d) 16 N, 2 N

71. A horizontal force of 10 N is necessary to just hold a block stationary against a wall. The coefficient of friction between the block and the wall is 0.2. The weight of the block is

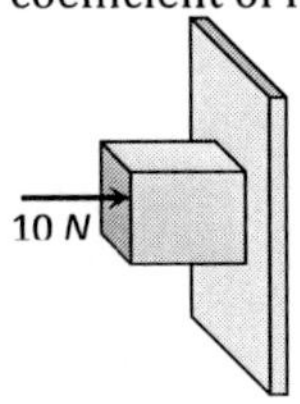

a) 2 N b) 20 N c) 50 N d) 100 N

72. A car is moving with uniform velocity on a rough horizontal road. Therefore, according to Newton's first law of motion
a) No force is being applied by its engine
b) A force is surely being applied by its engine
c) An acceleration is being produced in the car
d) The kinetic energy of the car is increasing

73. A 2 kg block is lying on a smooth table which is connected by a body of mass 1 kg by a string which passes through a pulley. The 1 kg mass is hanging vertically. The acceleration of block and tension in the string will be
a) $3.27 m/s^2, 6.54 N$ b) $4.38 m/s^2, 6.54 N$ c) $3.27 m/s^2, 9.86 N$ d) $4.38 m/s^2, 9.86 N$

74. A machine gun mounted on a 2000 kg car on a horizontal frictionless surface fires 10 bullets per second. If 10 g be the mass of each bullet and 500 ms^{-1}, the velocity of each bullet, then the acceleration of the car will be
a) $\frac{1}{10}$ ms^{-2} b) $\frac{1}{20}$ ms^{-2} c) $\frac{1}{40}$ ms^{-2} d) $\frac{1}{60}$ ms^{-2}

75. A rocket of mass 100 kg burns 0.1 kg of fuel per second. If velocity of exhaust gas is 1 kms^{-1}, then it lifts with an acceleration of
a) 1000 ms^{-2} b) 100 ms^{-2} c) 10 ms^{-2} d) 1 ms^{-2}

76. A boy of mass 0.25 kg is projected with muzzle velocity 100 ms^{-1} from a tank of mass 100 kg. What is the recoil velocity of the tank
a) 5 ms^{-1} b) 25 ms^{-1} c) 0.5 ms^{-1} d) 0.25 ms^{-1}

77. A force vector applied on a mass is represented as $\vec{F} = 6\hat{i} - 8\hat{j} + 10\hat{k}$ and accelerates with 1 m/s^2. What will be the mass of the body
a) $10\sqrt{2}\ kg$ b) $2\sqrt{10}\ kg$ c) $10\ kg$ d) $20\ kg$

78. A rope of length L is pulled by a constant force F. What is the tension in the rope at distance x from the end when the force is applied?
a) $\frac{F(L-x)}{L}$ b) $\frac{FL}{L-x}$ c) $\frac{FL}{x}$ d) $\frac{Fx}{L-x}$

79. A person is sitting in a left accelerating upwards. Measured weight of person will be
a) Less than actual weight b) Equal to actual weight
c) More than actual weight d) None of the above

80. A block of mass m is pushed with a velocity u towards a movable wedge of mass nm and height h, figure. All the surfaces are smooth. The minimum value of u for which the block will reach the top of wedge is
a) $\sqrt{2gh\left(1-\frac{1}{n}\right)}$ b) $\sqrt{2gh\left(1+\frac{1}{n}\right)}$ c) $\sqrt{3gh}$ d) $\sqrt{2gh}$

81. A block of mass 10 kg is placed on an inclined plane. When the angle of inclination is 30°, the block just begin to slide down the plane. The force of static friction is
a) 10 $kg\ wt$ b) 89 $kg\ wt$ c) 49 $kg\ wt$ d) 5 $kg\ wt$

82. The block A in figure weight 100 N. The coefficient of static friction between the block and the table is 0.25. The weight of the block B is maximum for the system to be in equilibrium. The value of T_1 is

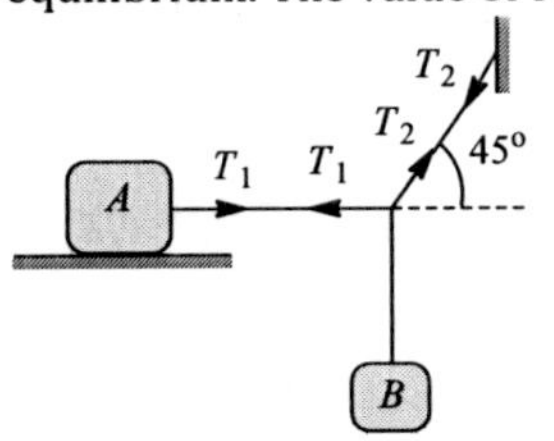

a) 0.25 N b) 25 N c) 100 N d) 100.25 N

83. A body of 10 kg is acted by a force of 129.4 N if $g = 9.8\ m/s^2$. The acceleration of the block is 10 m/s^2. What is the coefficient of kinetic friction

a) 0.03 b) 0.01 c) 0.30 d) 0.25

84. A lift of mass 1000 kg is moving with an acceleration of 1 m/s^2 in upward direction. Tension developed in the string, which is connected to the lift is ($g = 9.8\ m/s^2$)

a) 9,800 N b) 10,000 N c) 10, 800 N d) 11, 000 N

85. A block is kept on a frictionless inclined surface with angle of inclinationα. The incline is given an acceleration a to keep the block stationary. Then a is equal to

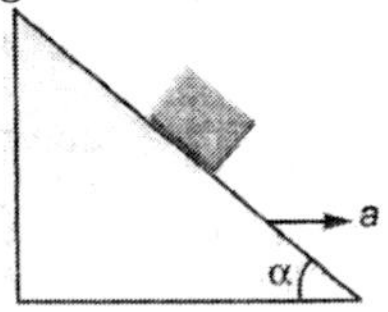

a) $g/\tan\alpha$ b) g cosec α c) g d) $g\tan\alpha$

86. A coin of mass 10 g is placed over a book of length 50 cm. the coin is on the verge of sliding when one end of the book is lifted to 10 cm up. The coefficient of static friction between the book and the coin is

a) 1.0 b) 0.4 c) 0.3 d) 0.2

87. A uniform rope of length l lies on a table. If the coefficient of friction is μ, then the maximum length l_1 of the part of this rope which can overhang from the edge of the table without sliding down is

a) $\frac{1}{\mu}$ b) $\frac{1}{\mu + l}$ c) $\frac{\mu l}{1+\mu}$ d) $\frac{\mu l}{\mu - 1}$

88. A boy of mass 100 g is sliding from an inclined plane of inclination 30°. What is the frictional force experienced if $\mu = 1.7$

a) $1.7 \times \sqrt{2} \times \frac{1}{\sqrt{3}} N$ b) $1.7 \times \sqrt{3} \times \frac{1}{2} N$ c) $1.7 \times \sqrt{3} N$ d) $1.7 \times \sqrt{2} \times \frac{1}{3} N$

89. The figure shows the position – time $(x - t)$ graph of one-dimensional motion of a body of mass 0.4 kg . The magnitude of each impulse is

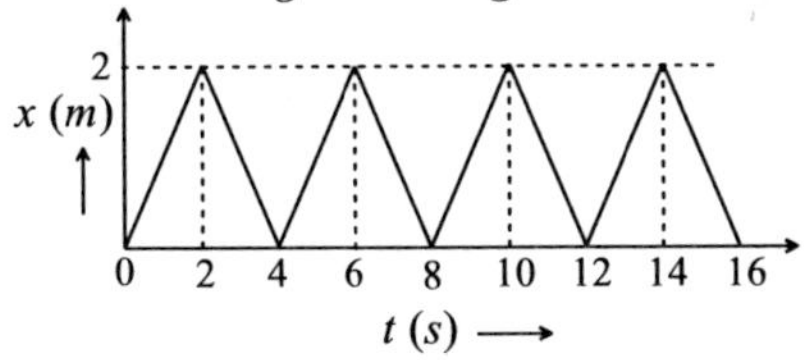

a) 0.2 Ns b) 0.4 Ns c) 0.8 Ns d) 1.6 Ns

90. A person is measuring his weight by standing on a weighing machine inside a lift. When the lift is at rest, the machine shows his weight to be 55 kg. In between the floor when the lift is moving up with a constant speed of 10 km/hr, he again measures his weight, which is

a) 55 kg b) 65 kg c) 50 kg d) 45 kg

91. A cricket ball of mass 150 g collides straight with a bat with a velocity of $10ms^{-1}$. Batsman hits it straight back with a velocity of $20ms^{-1}$. If ball remains in contact with bat for 0. 1s, then average force exerted by the bat on the ball is

a) 15 N b) 45 N c) 150 N d) 4.5 N

92. A satellite in force-free space sweeps stationary interplanetary dust at rate

$$\frac{dM}{dt} = \alpha\ v,$$

where M is the mass, v is the velocity of satellite and α is a constant

What is the deceleration of the satellite?

a) $\frac{-2\alpha v^2}{M}$ b) $-\alpha v^2/M$ c) $-\alpha v^2$ d) $\frac{\alpha\ v^2}{M}$

93. A rocket with a lift-off mass 3.5×10^4 kg is blasted upward with an initial acceleration of 10 ms^{-2}. Then the initial thrust of the blast is

a) 3.5×10^5N b) 7.0×10^5 N c) 14.0×10^5 N d) 1.75×10^5N

94. The tension in the spring is

a) Zero b) 2.5 N c) 5 N d) 10 N

95. A motor cycle and a car are moving on a horizontal road with the same velocity. If they are brought to rest by the application of brakes, which provided equal retardation, then

a) Motor cycle will stop at shorter distance b) Car will stop at a shorter distance
c) Both will stop at the same distance d) Nothing can be predicted

96. A light string passes over a frictionless pulley. To one of its ends a mass of 6 kg is attached. To its other end a mass of 10 kg is attached. The tension in the thread will be

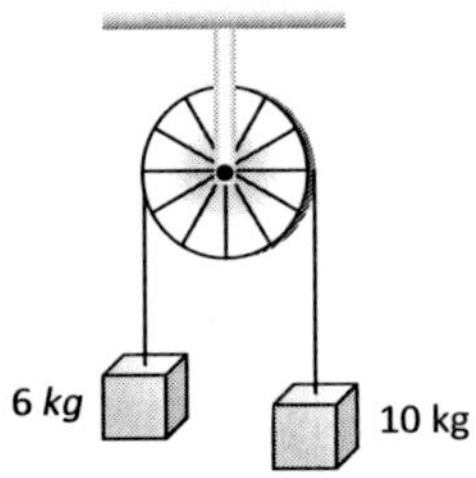

a) 24.5 N b) 2.45 N c) 79 N d) 73.5 N

97. A flat plate moves normally with a speed v_1 towards a horizontal jet of water of uniform area of cross-section. The jet discharges water at the rate of volume V per second at a speed of v_2. The density of water is ρ. Assume that water splashes along the surface of the plate at right angles to the original motion. The magnitude of the force acting on the plate due to the jet of water is

a) $\rho V v_1$ b) $\rho V(v_1 + v_2)$ c) $\frac{\rho V}{v_1 + v_2} v_1^2$ d) $\rho \left[\frac{V}{v_2}\right] (v_1 + v_2)^2$

98. Three equal weights A, B and C of mass 2 kg each are hanging on a string passing over a fixed frictionless pulley as shown in the figure. The tension in the string connecting weights B and C is

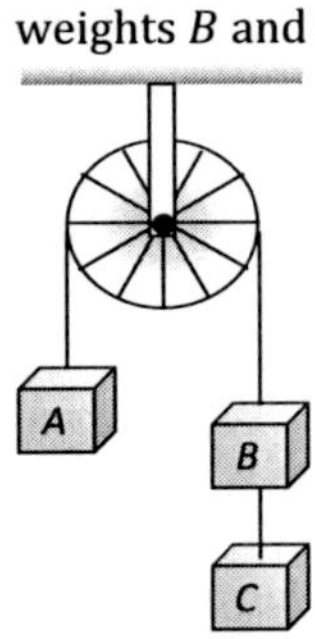

a) Zero b) 13 N c) 3.3 N d) 19.6 N

99. A student attempts to pull himself up by tugging on his hair. He will not succeed

a) As the force exerted in small
b) The frictional force while gripping, is small
c) Newton's law of inertia is not applicable to living beings
d) As the force applied is internal to the system

100. The force required just to move a body up an inclined plane is double the force required just to prevent the body sliding down. If the coefficient of friction is 0.25, the angle of inclination of the plane is

a) 36.8° b) 45° c) 30° d) 42.6°

101. An aircraft is moving with velocity of 300 ms^{-1}. If all the forces acting on it are balanced, then

a) It still moves with the same velocity
b) It will be just floating at the same point in space
c) It will fall down instantaneously
d) It will lose its velocity gradually

102. A block rests on a rough inclined plane making an angle of 30° with the horizontal. The coefficient of static friction between the block and the plane is 0.8. If the frictional force on the block is 10 N, the mass of the block (in kg) is (take $g = 10\ m/s^2$)

a) 2.0
b) 4.0
c) 1.6
d) 2.5

103. Three weight A, B and C are connected by string as shown in the figure. The system moves over a frictionless pulley. The tension in the string connecting A and B is (where g is acceleration due to gravity)

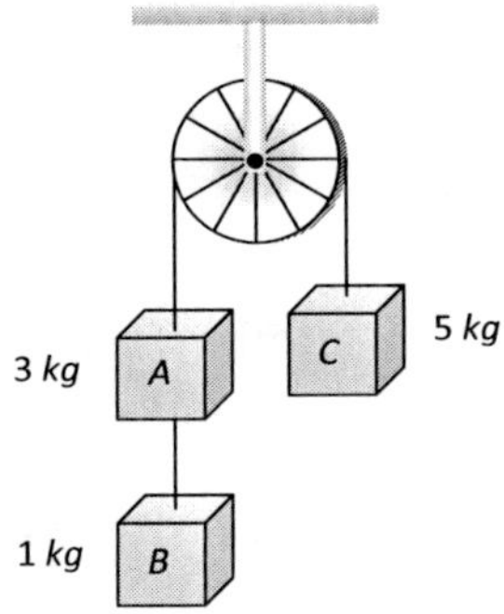

a) g
b) $\frac{g}{9}$
c) $\frac{8g}{9}$
d) $\frac{10g}{9}$

104. What is the maximum value of the force F such that the block shown in the arrangement, does not move?

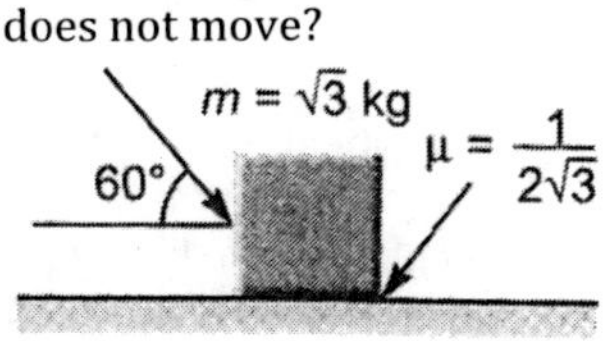

a) 20 N
b) 10 N
c) 12 N
d) 15 N

105. A frictionless inclined plane of length l having inclination θ is placed inside a lift which is accelerating downward with an acceleration $a(< g)$. If a block is allowed to move, down the inclined plane, from rest, then the time taken by the block to slide from top of the inclined plane to the bottom of the inclined plane is

a) $\sqrt{\frac{2l}{g}}$
b) $\sqrt{\frac{2l}{g-a}}$
c) $\sqrt{\frac{2l}{g+a}}$
d) $\sqrt{\frac{2l}{(g-a)\sin\theta}}$

106. A spacecraft of mss M moving with velocity v in free space explodes and breaks into two pieces. After the explosion, a mass m of the space craft is left stationary. The velocity of the other part is

a) $\frac{mv}{M-m}$
b) $\frac{Mv}{M-m}$
c) $\frac{M+m}{M}$
d) $\frac{Mv}{M}$

107. In the arrangement shown in figure, the strings are light and inextensible. The surface over which blocks are placed is smooth. What is the acceleration of each block?

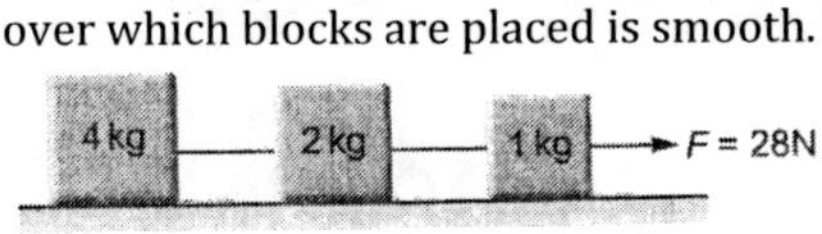

a) 8 ms^{-2}
b) 4 ms^{-2}
c) 2 ms^{-2}
d) 14 ms^{-2}

108. A car having a mass of 1000 kg is moving at a speed of 30 $metres/sec$. Brakes are applied to bring the car to rest. If the frictional force between the tyres and the road surface is 5000 $newtons$, the car will come to rest in

a) 5 $seconds$ b) 10 $seconds$ c) 12 $seconds$ d) 6 $seconds$

109. A particle of mass 2 kg is initially at rest. A force acts on it whose magnitude changes with time. The force time graph is shown below

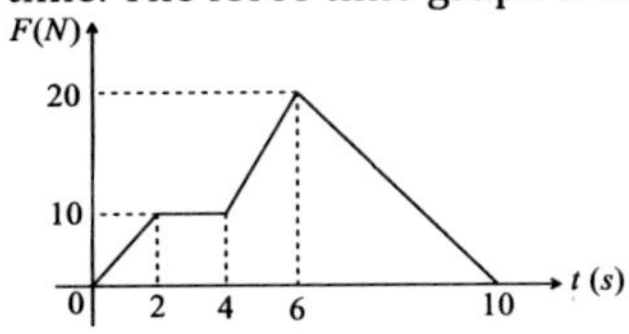

The velocity of the particle after 10 s is

a) 20 ms^{-1} b) 10 ms^{-1} c) 75 ms^{-1} d) 50 ms^{-1}

110. A student unable to answer a question on Newton's law of motion attempts to pull himself up by tugging on his hair. He will not succeed

a) As the force exerted is small

b) The frictional force while gripping, is small

c) Newton's law of inertia is not applicable to living beings

d) As the force applied is internal to the system

111. A block moves down a smooth inclined plane of inclination θ. Its velocity on reaching the bottom is v. If it slides down a rough inclined plane of same inclination, its velocity on reaching the bottom is v/n, where n is a number greater than 1. The coefficient of friction is given by

a) $\mu = \tan\theta\left(1-\frac{1}{n^2}\right)$ b) $\mu = \cot\theta\left(1-\frac{1}{n^2}\right)$

c) $\mu = \tan\theta\left(1-\frac{1}{n^2}\right)^{1/2}$ d) $\mu = \cot\theta\left(1-\frac{1}{n^2}\right)^{1/2}$

112. A pendulum bob of mass 50 gm is suspended from the ceiling of an elevator. The tension in the string if the elevator goes up with uniform velocity is approximately

a) 0.30 N b) 0.40 N c) 0.42 N d) 0.50 N

113. A uniform metal chain is placed on a rough table such that one end of chain hangs down over the edge of the table. When one-third of its length hangs over the edge, the chain starts sliding. Then, the coefficient of static friction is

a) $\frac{3}{4}$ b) $\frac{1}{4}$ c) $\frac{2}{3}$ d) $\frac{1}{2}$

114. Two blocks of masses m and $2m$ are connected by a light string passing over a frictionless pulley. As shown in the figure, the mass m is placed on a smooth inclined plane of inclination 30° and $2m$ hangs vertically. If the system is released, the blocks move with an acceleration equal to

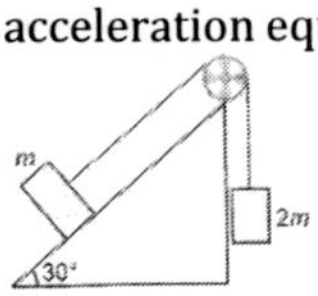

a) $\frac{g}{4}$ b) $\frac{g}{3}$ c) $\frac{g}{2}$ d) g

115. A horizontal force of 10 N is necessary to just hold a block stationary against a wall. The coefficient of friction between the block and the wall is 0.2. The weight of the block is

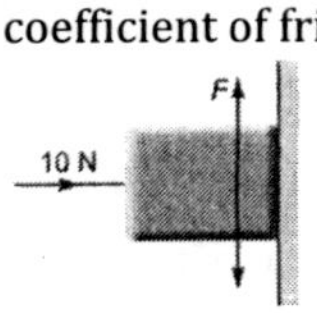

a) 20 N b) 50 N c) 100 N d) 2 N

116. A Bullet of mass 10 gm is fired from a gun of mass 1 kg. If the recoil velocity is 5 m/s, the velocity of the muzzle is

a) 0.05 m/s b) 5 m/s c) 50 m/s d) 500 m/s

117. A heavy uniform chain lies on horizontal table top. If the coefficient of friction between the chain and the table surface is 0.25, then the maximum fraction of the length of the chain that can hang over one edge of the table is

a) 20% b) 25% c) 35% d) 15%

118. The maximum speed of a car on a road turn of radius 30 m; if the coefficient of friction between the tyres and the road is 0.4; will be

a) 9.84 m/s b) 10.84 m/s c) 7.84 m/s d) 5.84 m/s

119. The sum of the magnitudes of two forces acting at a point is 18 N and the magnitude of their resultant is 12 N. If the resultant is at 90° with the smaller force, the magnitude of the forces in N are

a) 6,12 b) 11,7 c) 5,13 d) 14,4

120. A block of mass M placed on a frictionless horizontal table is pulled by an other block of mass m hanging vertically by a massless string passing over a frictionless pulley. The tension in the string is

a) $\frac{m}{M+m}g$ b) $\frac{M}{M+m}g$ c) $\frac{M+m}{Mm}g$ d) $\frac{Mm}{M+m}g$

121. The engine of a jet aircraft applies a thrust force of $10^5 N$ during take off and causes the plane to attain a velocity of 1 km/sec in 10 sec. The mass of the plane is

a) 10^2 kg b) 10^3 kg c) $10^4 kg$ d) $10^5 kg$

122. Three forces starts acting simultaneously on a particle moving with velocity $\vec{v}$. These forces are represented in magnitude and direction by the three sides of a triangle ABC (as shown). The particle will now move with velocity

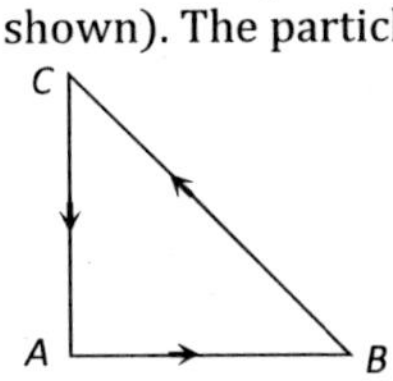

a) $\vec{v}$ remaining unchanged b) Less than $\vec{v}$

c) Greater than $\vec{v}$ d) $\vec{v}$ in the direction of the largest force BC

123. A wagon weighing 1000 kg is moving with a velocity 50 km/h on smooth horizontal rails. A mass of 250 kg is dropped into it. The velocity with which it moves now is

a) 2.5 $km/hour$ b) 20 $km/hour$ c) 40 $km/hour$ d) 50 $km/hour$

124. The mass of man when standing on lift is 60 kg. What is the weight when he is standing on lift which is moving upwards with acceleration 4.9 ms^{-2} ?

a) 882 kg b) 600 N c) 306 N d) Zero

125. A cricket ball of mass 250 g collides with a bat with velocity 10 m/s and returns with the same velocity within 0.01 second . The force acted on bat is

a) 25 N b) 50 N c) 250 N d) 500 N

126. A block of mass 5 kg is on a rough horizontal surface and is at rest. Now a force of 24 N is imparted to it with negligible impulse. If the coefficient of kinetic friction is 0.4 and g = 9.8 m/s^2, then the acceleration of the block is

a) 0.26 m/s^2 b) 0.39 m/s^2 c) 0.69 m/s^2 d) 0.88 m/s^2

127. A smooth block is released at rest on a45° incline and then slides a distance d. The time taken to slide is n times as much to slide on rough incline than on a smooth incline. The coefficient of friction is

a) $\mu_k = 1 - \frac{1}{n^2}$ b) $\mu_k = \sqrt{1 - \frac{1}{n^2}}$ c) $\mu_s = 1 - \frac{1}{n^2}$ d) $\mu_s = \sqrt{1 - \frac{1}{n^2}}$

128. A plate of mass M is placed on a horizontal frictionless surface (see figure), and a body of mass m is placed on this plate. The coefficient of dynamics friction between this body and the plate is μ. If a force $2\mu mg$ is applied to the body of mass m along the horizontal, the acceleration of the plate will be

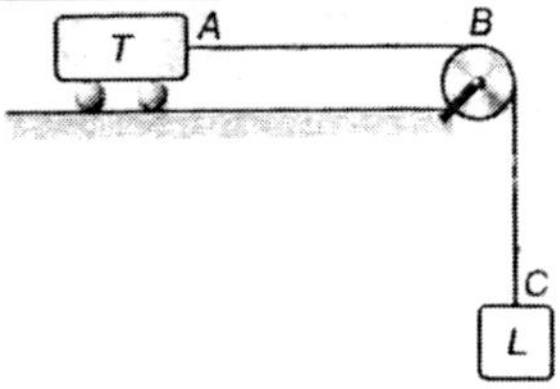

a) $\frac{\mu m}{M}g$ b) $\frac{\mu m}{(M+m)}g$ c) $\frac{2\mu m}{M}g$ d) $\frac{2\mu m}{(M+m)}g$

129. A 60 kg man stands on a spring scale in a lift. At some instant he finds that the scale reading has changed from 60 kg to 50 kg for a while and then comes again to 60 kg mark. What should he conclude?

a) The lift was in constant motion upwards b) The lift was in constant motion downwards

c) The lift while in motion suddenly stopped d) The lift while in motion upwards suddenly stopped

130. Two blocks of masses M and m are connected by a string passing over a pulley as shown in

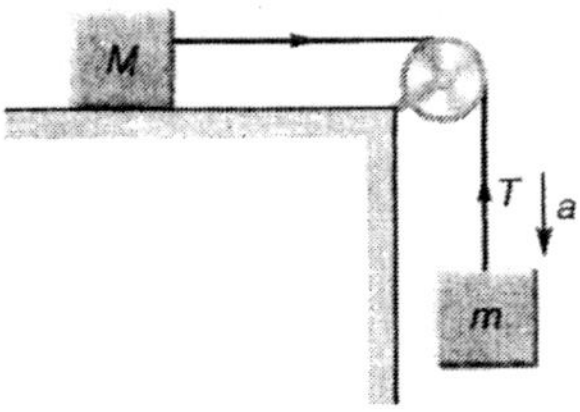

the figure. The downward acceleration of the block with mass m is

a) $M/(m+M)g$ b) $mg/(m+M)$ c) $(m+M)/mg$ d) $(n+M)/Mg$

131. Two blocks A and B of masses $2m$ and m, respectively, are connected by a massless and inextensible string. The whole system is suspended by a massless spring as shown in the figure. The magnitudes of acceleration of A and B , immediately after the spring is cut, are respectively

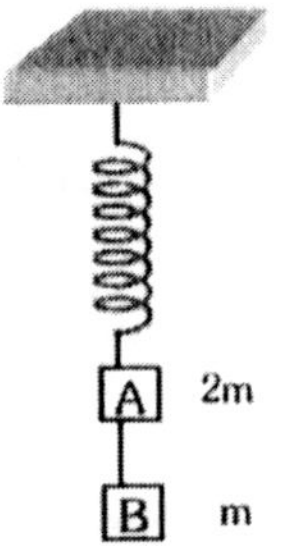

a) $g, g/2$ b) $g/2, g$ c) g, g d) $g/2, g/2$

132. A false balance has equal arms. An object weigh X when placed in one pan and Y when placed in other pan, then the weight W of the object is equal to

a) $\sqrt{XY}$ b) $\frac{X+Y}{2}$ c) $\frac{X^2+Y^2}{2}$ d) $\frac{2}{\sqrt{X^2+Y^2}}$

133. A 5000 kg rocket is set for vertical firing. The exhaust speed is 800 ms^{-1}. To give an initial upward acceleration of 20 ms^{-2} the amount of gas ejected per second to supply the needed thrust will be ($g = 10\ ms^{-2}$)

a) 127.5 $kg\ s^{-1}$ b) 187.5 $kg\ s^{-1}$ c) 185.5 $kg\ s^{-1}$ d) 137.5 $kg\ s^{-1}$

134. An object at rest in space suddenly explodes into three parts of same mass. The momentum of the two parts are $2p\hat{\mathbf{i}}$and $p\hat{\mathbf{j}}$. The momentum of the third part

a) Will have a magnitude $p\sqrt{3}$ b) Will have a magnitude $p\sqrt{5}$

c) Will have a magnitude p d) Will have a magnitude $2p$

135. A rifle of 20 kg mass can fire 4 bullets per second. The mass of each bullet is 35×10^{-3} kg and its final velocity 400 ms^{-1}. Then what force must be applied on the rifle so that it does not move backwards while firing the bullets?

a) 80 N b) 28 N c) −112 N d) −56 N

136. A body of mass 5 kg is suspended by a spring balance on an inclined plane as shown in figure. The spring balance measure

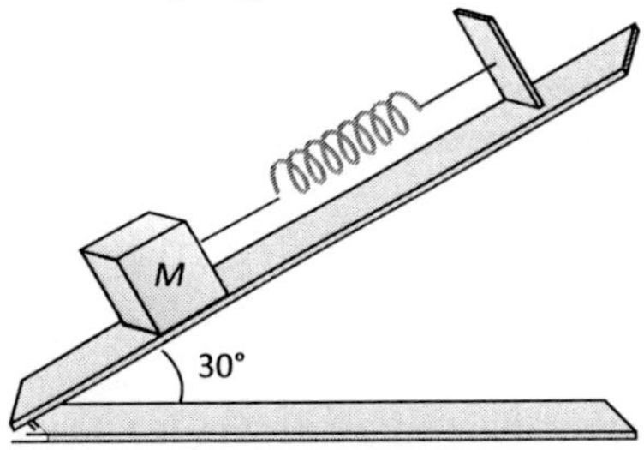

a) 50 N b) 25 N c) 500 N d) 10 N

137. A vehicle of mass 120 kg is moving with a uniform velocity of 108kmh^{-1}. The force required to step the vehicle in 10 s is

a) 90 N b) 180 N c) 360 N d) 720 N

138. A man is standing at the centre of frictionless pond of ice. How can he get himself to the shore

a) By throwing his shirt in vertically upward direction b) By spitting horizontally

c) He will wait for the ice to melt in pond d) Unable to get at the shore

139. A block of mass M is attached to the lower end of a vertical rope of mass. An upward force P acts on the upper end of the rope. The system is free to move. The force exerted by the rope on the block is $\frac{PM}{M+m}$

a) In all cases b) Only if the rope is uniform

c) In gravity-free space only d) Only if $P > (M+m)g$

140. A block B is placed on block A. The mass of block Bis less that the mass of block A. Friction exists between the blocks, whereas the ground on which the block A is placed is taken to be smooth. A horizontal force F, increasing linearly with time begins to act on B. The acceleration a_A and a_B of blocks A and B respectively are plotted against t. The correctly plotted graph is

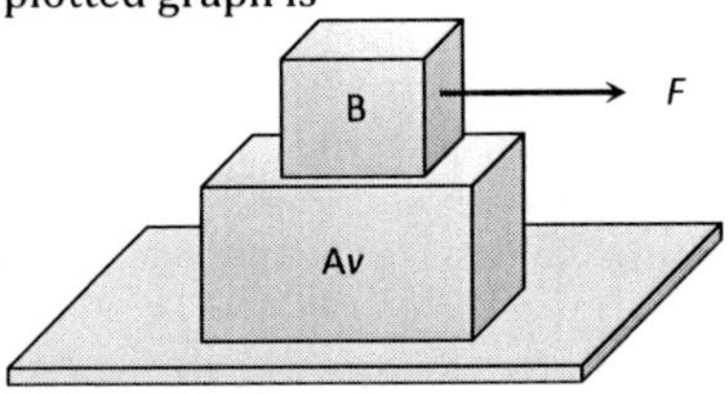

a) 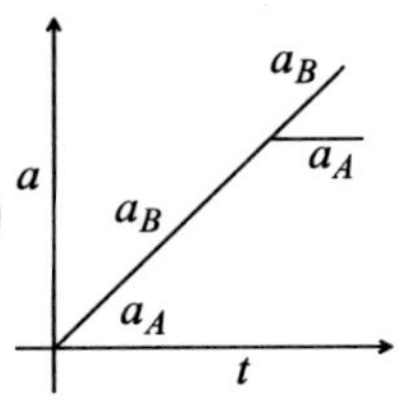

b) 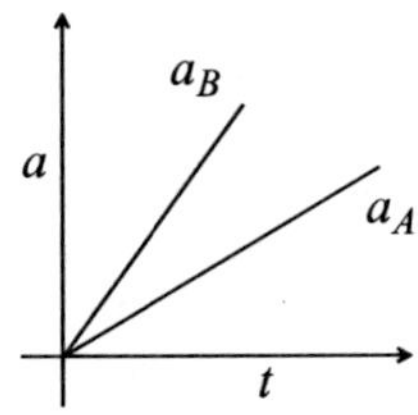

c) 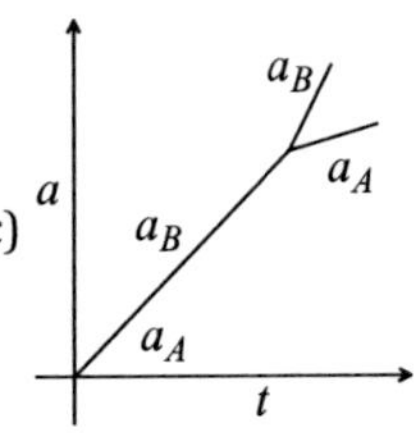

d) 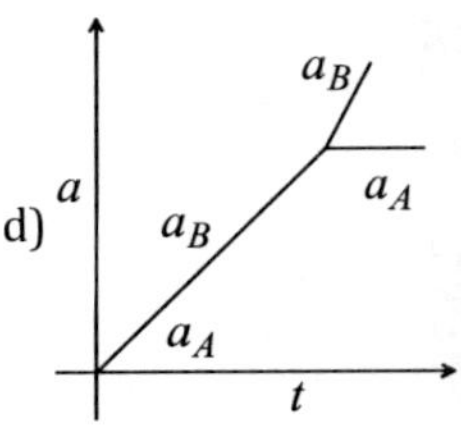

141. A plumb line is suspended from a ceiling of a car moving with horizontal acceleration of a. What will be the angle of inclination with vertical

a) $\tan^{-1}(a/g)$ b) $\tan^{-1}(g/a)$ c) $\cos^{-1}(a/g)$ d) $\cos^{-1}(g/a)$

142. A machine gun fires a bullet of mass 40 g with a velocity $1200 ms^{-1}$. The man holding it can exert a maximum force of 144 N on the gun. How many bullets can be fired per second at the most?

a) Only one b) Three
c) Can fire any number of bullets d) 144×48

143. A 10 kg stone is suspended with a rope of breaking strength 30 kg wt. The minimum time in which the stone can be raised through a height 10 m starting from rest is (taking $g = 10\ N/kg$)

a) 0.5 $seconds$ b) 1.0 $seconds$ c) $\sqrt{\frac{2}{3}}$ $seconds$ d) 2.0 $seconds$

144. The masses M_1, M_2 and M_3 are 5,2 and 3 kg respectively. These have been joined using massless, inextensible pieces of strings as shown in figure. If whole system is going upward with an acceleration of 2 ms^{-2}, then the value of tensions T_1, T_2 and T_3 respectively are

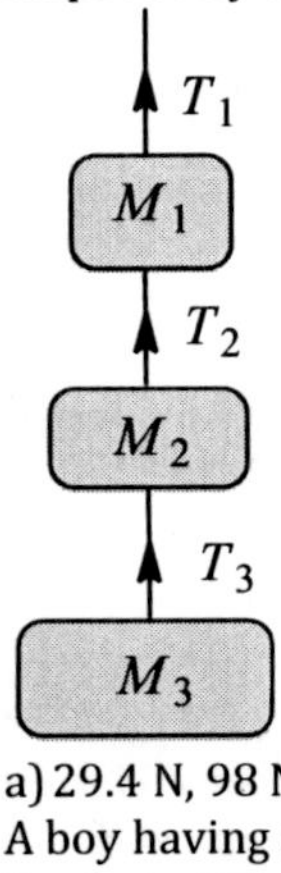

a) 29.4 N, 98 N, 98 N b) 98 N, 49 N, 29.4 N c) 118 N, 59 N, 35.4 N d) 35.4 N, 118 N, 59 N

145. A boy having a mass equal to 40 $kilograms$ is standing in an elevator. The force felt by the feet of the boy will be greatest when the elevator
($g = 9.8\ metres/\sec^2$)

a) Stands still
b) Moves downward at a constant velocity of 4 $metres/sec$
c) Accelerates downward with an acceleration equal to 4 $metres/\sec^2$
d) Accelerates upward with an acceleration equal to 4 $metres/\sec^2$

146. A person is standing in an elevator. In which situation he finds his weight less than actual weight

a) The elevator moves upward with constant acceleration
b) The elevator moves downward with constant acceleration
c) The elevator moves upward with uniform velocity
d) The elevator moves downward with uniform velocity

147. A man slides down on a telegraphic pole with acceleration equal to one-fourth of acceleration due to gravity. The frictional force between man and pole is equal to in terms of man's weight w

a) $\frac{w}{4}$ b) $\frac{w}{2}$ c) $\frac{3w}{4}$ d) w

148. A rod of length L and weight W is kept horizontally. A small weight w is hung at one end. If the system balances on a fulcrum placed at T then

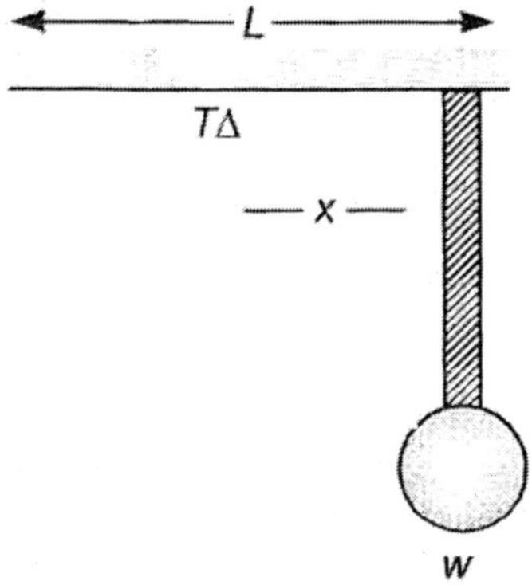

a) $x=\frac{L}{2}$ b) $x=\frac{wL}{2(W+w)}$ c) $x=\frac{wL}{W}$ d) None of these

149. A block of mass m on a rough horizontal surface is acted upon by two forces as shown in figure. For equilibrium of block, the coefficient of friction between block and surface is

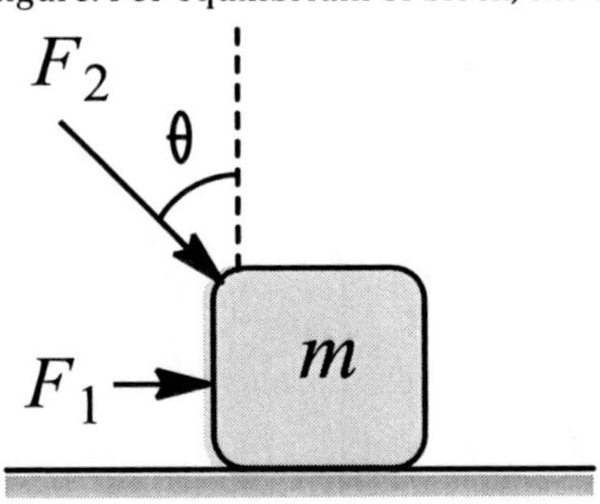

a) $\frac{F_1+F_2\sin\theta}{mg+F_2\cos\theta}$ b) $\frac{F_1\sin\theta+F_2}{mg+F_2\sin\theta}$ c) $\frac{F_1+F_2\cos\theta}{mg+F_2\sin\theta}$ d) $\frac{F_1\sin\theta-F_2}{mg-F_2\cos\theta}$

150. A body of mass 10 kg is lying on a rough plane inclined at an angle of 30° to the horizontal and the coefficient of friction is 0.5. the minimum force required to pull the body up the plane is

a) 914 N b) 91.4 N c) 9.14 N d) 0.914 N

151. A block of mass 10 kg is placed on a rough horizontal surface having coefficient of friction $\mu=0.5$. If a horizontal force of 100 N is acting on it, then acceleration of the block will be

a) 0.5 m/s^2 b) 5 m/s^2 c) 10 m/s^2 d) 15 m/s^2

152. A block of mass M is attached to the lower end of a vertical spring. The spring is hung from a ceiling and has force constant value k. The mass is released from rest with the spring initially unstretched. The maximum extension produced in the length of the spring will be

a) 1 Mg/k b) $2Mg/k$ c) 4 Mg/k d) $Mg/2k$

153. A bullet of mass 10 g moving with 300 ms^{-1} hits a block of ice of mass 5 kg and drops dead. The velocity of ice is

a) 50 cm/s b) 60 cm/s c) 40 cm/s d) 200 cm/s

154. A lift accelerated downward with acceleration 'a'. A man in the lift throws a ball upward with acceleration $a_0 (a_0 < a)$. Then acceleration of ball observed by observer, which is on earth, is

a) $(a + a_0)$ upward b) $(a - a_0)$ upward c) $(a + a_0)$ downward d) $(a - a_0)$ downward

155. The pulley and strings shown in figure are smooth and of negligible mass. For the system to remain in equilibrium, the angle θ should be

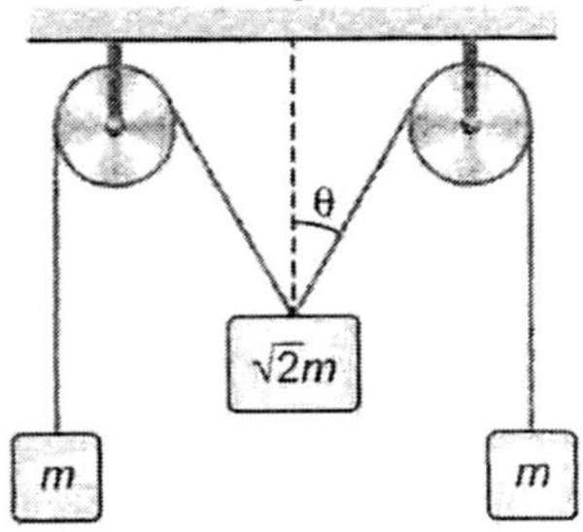

a) 0° b) 30° c) 45° d) 60°

156. Two blocks of masses m and $2m$ are connected by a light string passing over a frictionless pulley. As shown in the figure, the mass m is placed on a smooth inclined plane of inclination 30° and $2m$ hangs vertically. If the system is released, the blocks move with an acceleration equal to

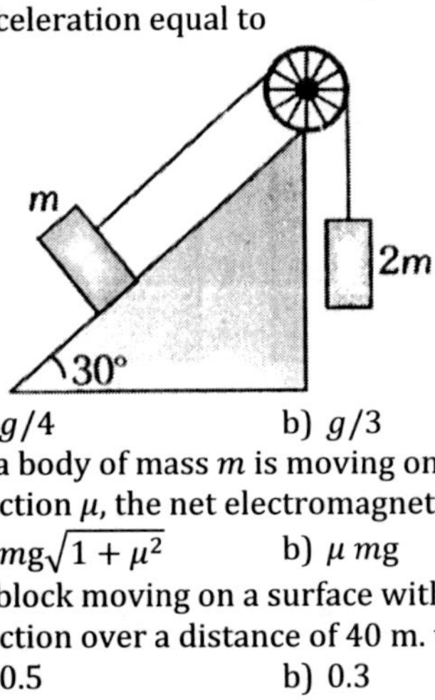

a) $g/4$ b) $g/3$ c) $g/2$ d) g

157. If a body of mass m is moving on a rough horizontal surface of coefficient of kinetic friction μ, the net electromagnetic force exerted by surface on the body is

a) $mg\sqrt{1+\mu^2}$ b) μmg c) mg d) $mg\sqrt{1-\mu^2}$

158. A block moving on a surface with velocity 20 ms^{-1} comes to rest because of surface friction over a distance of 40 m. taking $g = 10\ ms^{-2}$, the coefficient of dynamic friction is

a) 0.5 b) 0.3 c) 0.2 d) 0.1

159. A man weighing 60 kg is standing on a trolley weighing 240 kg. The trolley is resting on frictionless horizontal rails. If the man starts walking on the trolley with a velocity of 1 ms^{-1}, then after 4 s, his displacement relative to the ground is

a) 6 m b) 4.8 m c) 3.2 m d) 2.4 m

160. When the speed of a moving body is doubled

a) Its acceleration is doubled b) Its momentum is doubled
c) Its kinetic energy is doubled d) Its potential energy is doubled

161. A boy B lies on a smooth horizontal table and another body A is placed on B. The coefficient of friction between A and B is μ. What acceleration given to B will cause slipping to occur between A and B

a) μg b) g/μ c) μ/g d) $\sqrt{\mu g}$

162. Impulse is

a) A scalar
b) Equal to change in the momentum of a body
c) Equal to rate of change of momentum of a body
d) A force

163. A body of mass 40 kg resting on a rough horizontal surface is subjected to a force P which is just enough to start the motion of the body. If $\mu_s = 0.5, \mu_k = 0.4, g = 10 \text{ ms}^{-2}$ and the force P is continuously applied on the body, then the acceleration of the body is

a) Zero b) 1 ms^{-2} c) 2 ms^{-2} d) 2.4 ms^{-2}

164. The coefficient of limiting friction μ is defined as

a) $\mu = \frac{R}{F}$ b) $\mu = \sqrt{\frac{F}{R}}$ c) $\mu = \frac{F}{R}$ d) $\mu = \sqrt{\frac{R}{F}}$

165. A block B is pushed momentarily along a horizontal surface with an initial velocity v. If μ is the coefficient of sliding friction between B and the surface, block B will come to rest after a time

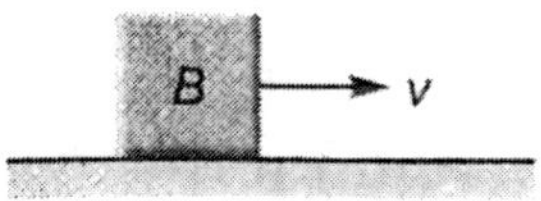

a) $\frac{v}{g\mu}$ b) $\frac{g\mu}{v}$ c) $\frac{g}{v}$ d) $\frac{v}{g}$

166. A body of mass 10 kg slides along a rough horizontal surface. The coefficient of friction is $1/\sqrt{3}$. Taking $g = 10\ m/s^2$, the least force which acts an angle of 30° to the horizontal is

a) 25 N b) 100 N c) 50 N d) $\frac{50}{\sqrt{3}}\ N$

167. Two masses of 4 kg and 5 kg are connected by a string passing through a frictionless pulley and are kept on a frictionless table as soon as shown in the figure. The acceleration of 5 kg mass is

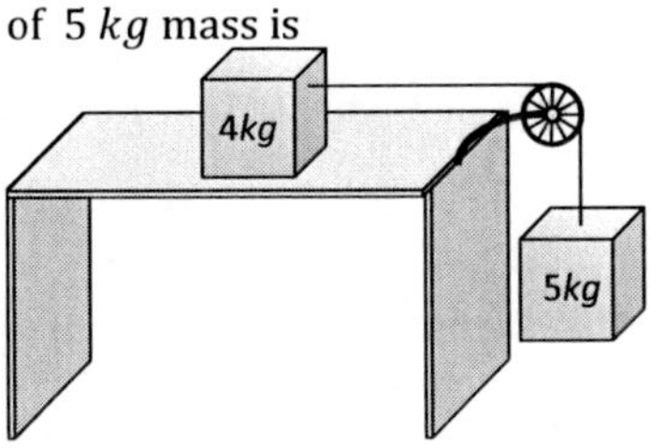

a) 49 m/s^2 b) 5.44 m/s^2 c) 19.5 m/s^2 d) 2.72 m/s^2

168. The motion of a rocket is based on the principle of conservation of

a) Mass b) Kinetic energy c) Linear momentum d) Angular momentum

169. 10000 small balls, each weighing 1 g, strike one square centimeter of area per second with a velocity 100 ms^{-1} in a normal direction and rebound with the same velocity. The value of pressure on the surface will be

a) $2 \times 10^3 \text{Nm}^{-2}$ b) $2 \times 10^5 \text{Nm}^{-2}$ c) 10^7Nm^{-2} d) $2 \times 10^7 \text{Nm}^{-2}$

170. A block is kept on a frictionless inclined surface with angle of inclination 'α'. The incline is given an acceleration 'a' to keep the block stationary. Then a is equal to

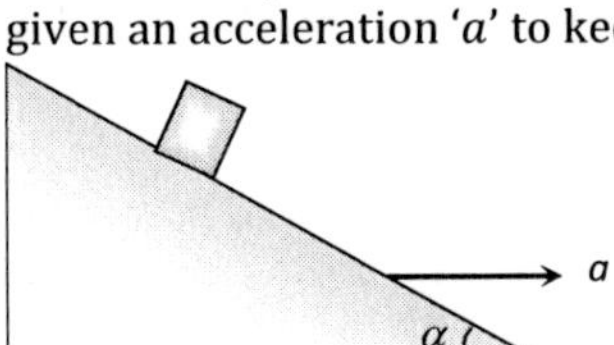

a) g b) $g \tan \alpha$ c) $g/\tan \alpha$ d) g cosec α

171. In the figure shown, a block of weight 10 N is resting on a horizontal surface. The coefficient of static friction between the block and the surface $\mu_s = 0.4$. A force of 3.5 N will keep the block in uniform motion, once it has been set in motion. A horizontal force of 3N is applied to the block then the block will

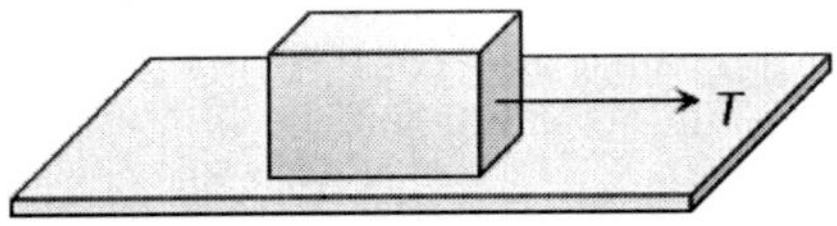

a) Move over the surface with constant velocity
b) Move having accelerated motion over the surface
c) Not move
d) First it will move with a constant velocity for some time and then will have accelerated motion

172. A rough vertical board has an acceleration $'a'$ so that a 2 kg block pressing against it does not fall. The coefficient of friction between the block and the board should be

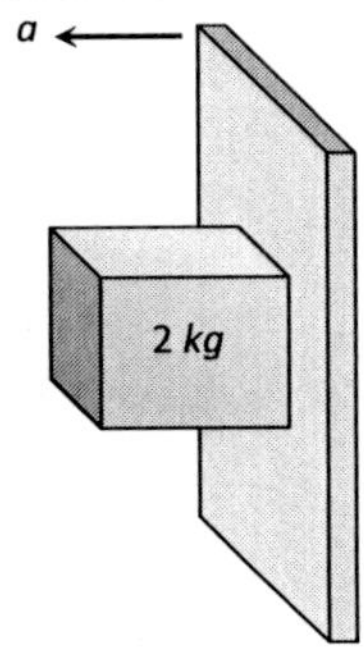

a) $> g/a$ b) $< g/a$ c) $= g/a$ d) $> a/g$

173. A light spring balance hangs from the hook of the other light spring balance and a block of mass $M\ kg$ hangs from the former one. Then the true statement about the scale reading is
a) Both the scales read $M/2\ kg$ each
b) Both the scales read $M\ Kg$ each
c) The scale of the lower one reads $M\ kg$ and of the upper one zero
d) The reading of the two scales can be anything but the sum of the reading will be $M\ kg$

174. While waiting in a car at a stoplight, and 80 kg man and his car are suddenly accelerated to a speed of 5 ms^{-1} as a result or rear end colllison. If the time of impact is 0.4 s, find the average force on the man
a) 100 N b) 200 N c) 500 N d) 1000 N

175. A body of weight 2 kg is suspended as shown in figure. The tension T_1 in the horizontal string (in kg-wt) is

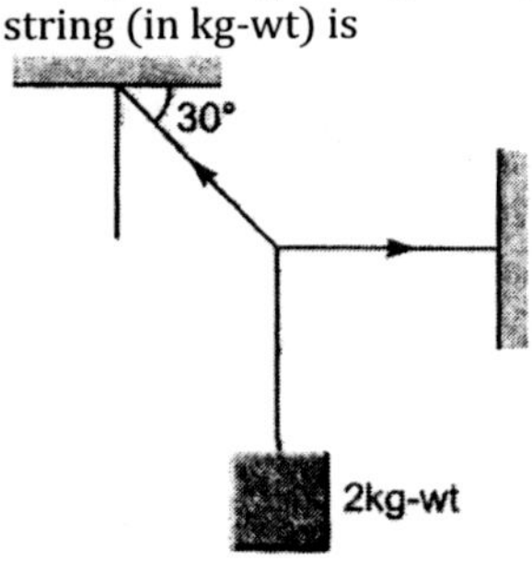

a) $2/\sqrt{3}$ b) $\sqrt{3}/2$ c) $2\sqrt{3}$ d) 2

176. Rocket engines lift a rocket from the earth surface because hot gas with high velocity
a) Push against the earth b) Push against the air
c) React against the rocket and push it up d) Heat up the air which lifts the rocket

177. A force of 98 N is required to just start moving a body of mass 100 kg over ice. The coefficient of static friction is
a) 0.6 b) 0.4 c) 0.2 d) 0.1

178. Consider the following statements about the blocks shown in the diagram that are being pushed by a constant force on a frictionless table

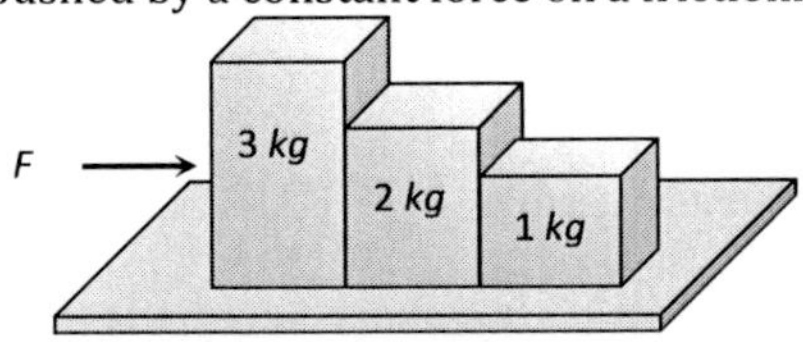

A. All blocks move with the same acceleration
B. The net force on each block is the same
Which of these statements are/is correct
a) A only b) B only c) Both A and B d) Neither A nor B

179. Two masses $m_1 = 5\ kg$ and $m_2 = 4.8\ kg$ tied to a string are hanging over a light frictionless pulley. What is the acceleration of the masses when they are free to move $(g = 9.8\ m/s^2)$

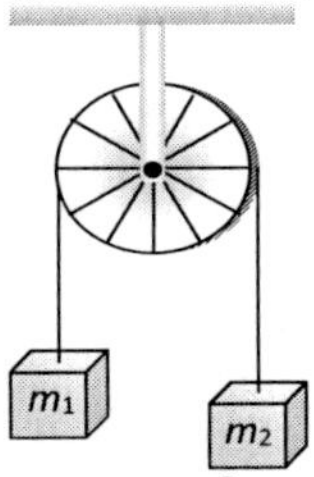

a) $0.2\ m/s^2$ b) $9.8\ m/s^2$ c) $5\ m/s^2$ d) $4.8\ m/s^2$

180. A dynamometer D is attached to two blocks of masses 6 kg and 4 kg. Forces of 20 N and 10 N are applied on the blocks as shown in figure. The dynamometer reads

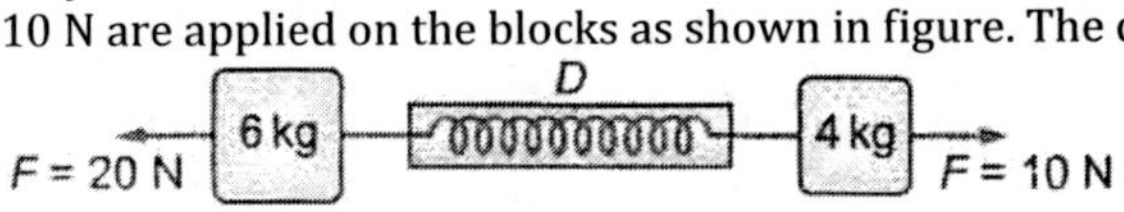

a) 10 N b) 20 N c) 6 N d) 14 N

181. A block A of mass 7 kg is placed on a frictionless table. A thread tied to it passes over a frictionless pulley and carries a body B of mass 3 kg at the other end, as in figure. The acceleration of the system is (given g = 10 ms^{-2})

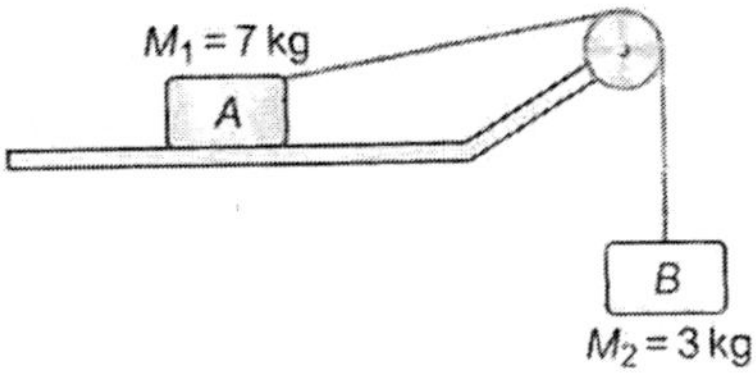

a) 100 ms^{-2} b) 3 ms^{-2} c) 10 ms^{-2} d) 30 ms^{-2}

182. Which one of the following is not a force
a) Impulse b) Tension c) Thrust d) Air resistance

183. A second's pendulum is mounted in a rocket. Its period of oscillation decreases when the rocket
a) Comes down with uniform acceleration
b) Moves round the earth in a geostationary orbit
c) Moves up with a uniform velocity
d) Moves up with uniform acceleration

184.

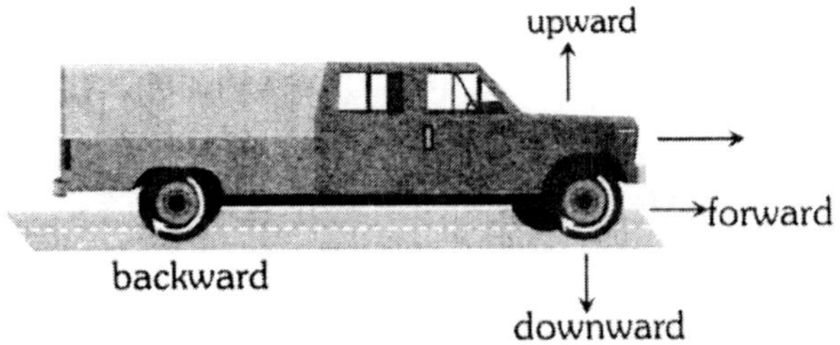

Direction of frictional force between wheel of the car and road is

a) Upward b) Forward c) Backward d) Downward

185. A man drags a block through $10m$ on rough surface ($\mu = 0.5$). A force of $\sqrt{3}kN$ acting at $30°$ to the horizontal. The work done by applied force is

a) Zero b) 15 kJ c) 5 kJ d) 10 kJ

186. A body of mass 3 kg is acted on by a force which varies as shown in the graph below. The momentum acquired is given by

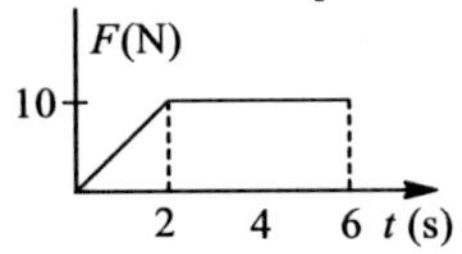

a) Zero b) 5 N-s c) 30 N-s d) 50 N-s

187. A motor car has a width 1.1 m between wheels. Its centre of gravity is 0.62 m above the ground and the coefficient of friction between the wheels and the road is 0.8. What is the maximum possible speed, if the centre of gravity inscribes a circle of radius 15 m? (Road surface is horizontal)

a) 7.64 m/s b) 6.28 m/s c) 10.84 m/s d) 11.23 m/s

188. **Statement I** A cloth covers a table. Some dishes are kept on it. The cloth can be pulled out without dislodging the dishes from the table.

Statement II For every action there is an equal and opposite reaction.

a) Statement I is true, statement II is true; statement II is a correct explanation for statement I

b) Statement I is true, statement II is true; statement II is not a correct explanation for statement I

c) Statement I is true, statement II is false

d) Statement I is false, statement II is true

189. A mass m hangs with a help of a string wrapped around a pulley on a frictionless bearing. The pulley has mass m and radius R. Assuming pulley to be a perfect uniform circular disc, the acceleration of the mass m, if the string does not slip on the pulley, is

a) $\frac{3}{2}g$ b) g c) $\frac{2}{3}g$ d) $\frac{g}{3}$

190. A force of 50 $dynes$ is acted on a body of mass 5 g which is at rest for an interval of 3 seconds, then impulse is

a) $0.15 \times 10^{-3}\ N-s$ b) $0.98 \times 10^{-3}\ N-s$ c) $1.5 \times 10^{-3}\ N-s$ d) $2.5 \times 10^{-3}\ N-s$

191. A ball of mass 0.5 kg moving with a velocity of 2 m/sec strikes a wall normally and bounces back with the same speed. If the time of contact between the ball and the wall is one millisecond, the average force exerted by the wall on the ball is

a) 2000 N b) 1000 N c) 5000 N d) 125 N

192. A hockey player receives a corner shot at a speed of 15 ms^{-1} at an angle of $30°$ with the y-axis and then shoots the ball of mass 100 g along the negative x-axis with a speed of 30 ms^{-1}. If it remains in contact with the hockey stick for 0.01 s, the force imparted to the ball in the x-direction is

a) 281.25 N b) 187.5 N c) 562.5 N d) 375 N

93. The mass of a body measured by a physical balance in a lift at rest is found to be m. If the lift is going up with an acceleration a, its mass will be measured as

a) $m\left(1-\frac{a}{g}\right)$ b) $m\left(1+\frac{a}{g}\right)$ c) m d) Zero

94. Same force acts on two bodies of different masses 3 kg and 5 kg initially at rest. The ratio of times required to acquire same final velocity is

a) 5: 3 b) 25: 9 c) 9: 25 d) 3: 5

95. A body of mass 3kg is acted on by a force which varies as shown in the graph below. The momentum acquired is given by

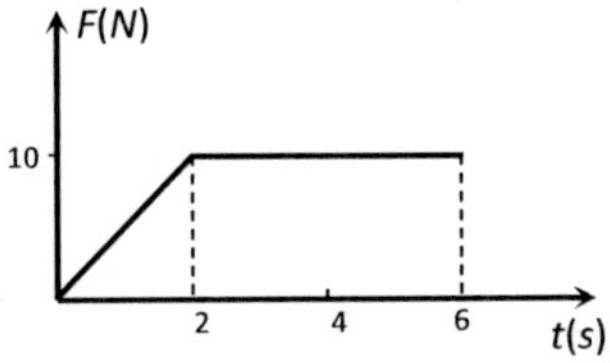

a) Zero b) 5 N-s c) 30 N-s d) 50 N-s

96. In the above Question, if the string C is stretched slowly, then

a) The portion AB of the string will break b) The portion BC of the string will break
c) None of the strings will break d) None of the above

97. Which one of the following is not a contact force

a) Viscous force b) Air resistance c) Friction d) Magnetic force

98. Figure represents a painter in a crate which hangs along the side of a building. When the painter of mass 100 kg pulls the rope, the force exerted by him on the floor of the crate is 450 N. If the crate weighs 125 kg, then the acceleration in the rope is (Given $g = 10\ ms^{-2}$)

a) 1 ms^{-2} b) 2 ms^{-2} c) 3 ms^{-2} d) 4 ms^{-2}

99. A rod length AB is moving with ends remaining in contact with frictionless wall and floor. If at the instant shown, the velocity of end B is 3 ms^{-1} towards negative x-direction, then magnitude of velocity of end A will be

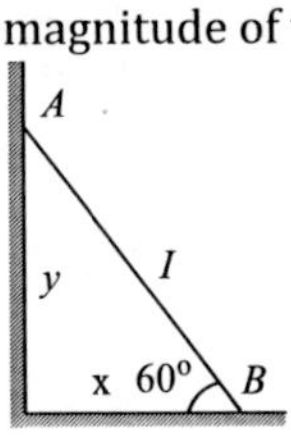

a) 3 ms^{-1} b) $\sqrt{3}\ ms^{-1}$ c) 1.5 ms^{-1} d) 2 ms^{-1}

00. An iron block of mass 5 kg is kept on a trolley. If the trolley is being pushed with an acceleration of 5 m/s^2, What will be the force of friction between the block and the trolley surface? (Take the coefficient of static friction between the block and the surface to be 0.8)

a) Zero b) 5 N c) 4 N d) 25 N

01. For the system shown in figure, the pulleys are light and frictionless. The tension in the string will be

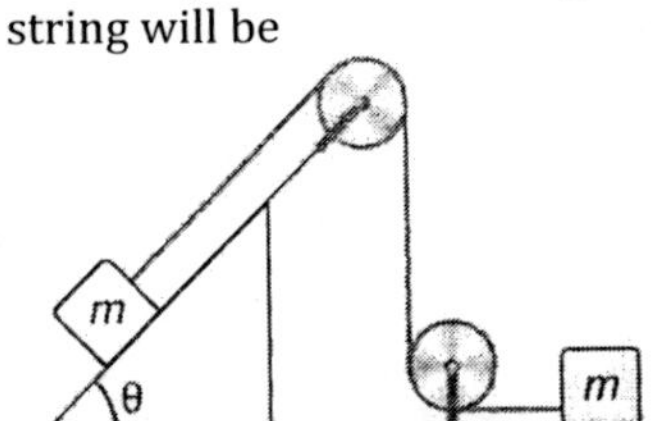

a) $\frac{2}{3}mg\sin\theta$ b) $\frac{3}{2}mg\sin\theta$ c) $\frac{1}{2}mg\sin\theta$ d) $2mg\sin\theta$

202. A body of mass 4 kg is accelerated upon by a constant force, travels a distance of 5 m in the first second and a distance of 2 m in the third second. The force acting on the body is

a) $6N$ b) $8\ N$ c) $2\ N$ d) $4\ N$

203. 10,000 small balls, each weighing 1 gm, strike one square cm of area per second with a velocity 100 m/sin a normal direction and rebound with the same velocity. The value of pressure on the surface will be

a) $2 \times 10^3\ N/m^2$ b) $2 \times 10^5\ N/m^2$ c) $10^7\ N/m^2$ d) $2 \times 10^7\ N/m^2$

204. A bob of mass 0.450 kg hangs from the massless string of a long simple pendulum. A bullet of mass 0.50 kg is fired vertically from below into the bob. The bullet gets embedded into the bob and the combination rises vertically through a height of 1.8 m. If $g = 10\ ms^{-2}$, then the velocity of the bullet is

a) $6\ ms^{-1}$ b) $60 ms^{-1}$ c) $600\ ms^{-1}$ d) $6000 ms^{-1}$

205. Force of 4 N is applied on a body of mass 20 kg. The work done in 3rd second is

a) 2 J b) 4 J c) 16 J d) 1.2 J

206. The figure shows the position-time $(x - t)$ graph of one-dimensional motion of a body of mass 0.4 kg. The magnitude of each impulse is

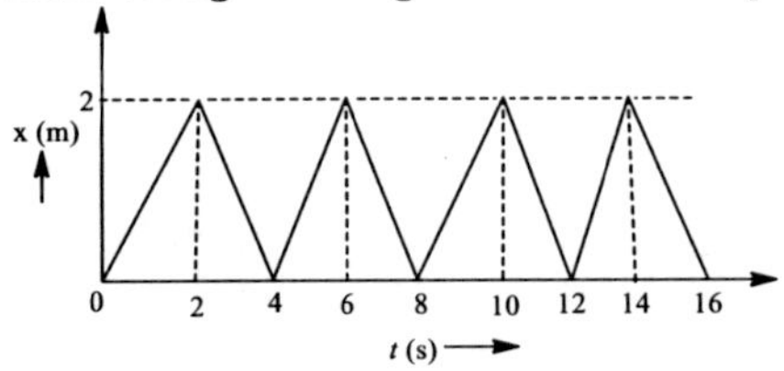

a) 0.4 Ns b) 0.8 Ns c) 1.6 Ns d) 0.2 Ns

207. A block of mass M is pulled along a horizontal frictionless surface by a rope of mass$m$. If a force P is applied at the free end of the rope, the force exerted by the rope on the block is

a) $\frac{Pm}{M+m}$ b) $\frac{Pm}{M-m}$ c) P d) $\frac{PM}{M+m}$

208. An army vehicle of mass 1000 kg is moving with a velocity of 10 m/s and is acted upon by a forward force of 1000 N due to the engine and a retarding force of 500 N due to friction. What will be its velocity after 10 s

a) $5\ m/s$ b) $10\ m/s$ c) $15\ m/s$ d) $20\ m/s$

209. A train is moving with velocity 20 m/s on this dust is falling at the rate 50kg/min. The extra force requested to move this train with a constant velocity will be

a) 16.66 N b) 1200 N c) 1000 N d) 166.6 N

210. The resultant of two forces, one double the other in magnitude, is perpendicular to the smaller of the two forces. The angle between the two forces is

a) 60° b) 120° c) 150° d) 90°

211. A piece of wire is bent in the shape of a parabola $y = kx^2$ (y-axis vertical) with a bead of mass m on it. The bead can slide on the wire without friction. It stays at the lowest point of the parabola when the wire is at rest. The wire is now accelerated parallel to the x-axis with a constant acceleration a. The distance of the new equilibrium position of the bead, where the bead can stays at rest with respect to the wire, from the y-axis is

a) $\frac{a}{gk}$ b) $\frac{a}{2gk}$ c) $\frac{2a}{gk}$ d) $\frac{a}{4gk}$

212. Two bodies of masses m_1 and m_2 are connected by a light, inextensible string which passes over a frictionless pulley. If the pulley is moving upward with uniform acceleration g, then the tension in the string is

a) $\frac{4m_1m_2}{m_1+m_2}g$ b) $\frac{m_1m_2}{4m_1m_2}g$ c) $\frac{m_1m_2}{m_1+m_2}g$ d) $\frac{m_1-m_2}{m_1+m_2}g^2$

213. A block of mass m is resting on a smooth horizontal surface. One end of a uniform rope of mass $\left(\frac{m}{3}\right)$ is fixed to the block, which is pulled in the horizontal direction by applying force F at the other end. The tension in the middle of the rope is

a) $\frac{8}{6}F$ b) $\frac{1}{7}F$ c) $\frac{1}{8}F$ d) $\frac{7}{8}F$

214. A ball of mass 0.2 kg is thrown vertically upwards by applying a force by hand. If the hand moves 0.2 m while applying the force and the ball goes upto 2 m height further, find the magnitude of the force. Consider $g = 10\ m/s^2$

a) 16 N b) 20 N c) 22 N d) 4 N

215. Three blocks of masses m_1, m_2 and m_3 are connected by massless strings as shown on a frictionless table. They are pulled with a force $T_3 = 40\ N$. If $m_1 = 10\ kg, m_2 = 6\ kg$ and $m_3 = 4\ kg$, the tension T_2 will be

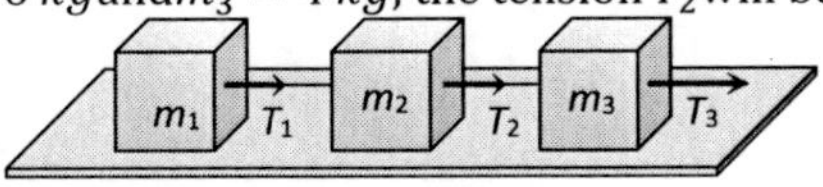

a) 20 N b) 40 N c) 10 N d) 32 N

216. A horizontal force F is applied on a block mass m placed on a rough inclined plane of inclination θ. The normal reaction N is

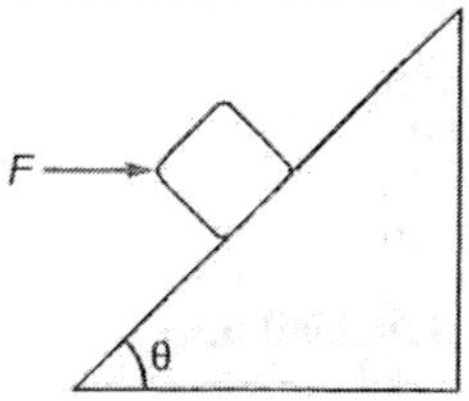

a) $mg\cos\theta$ b) $mg\sin\theta$ c) $mg\cos\theta - F\cos\theta$ d) $mg\cos\theta + F\sin\theta$

217. Block A of mass m and block B of mass $2m$ are placed on a fixed triangular wedge by means of a massless, inextensible string and a frictionless pulley as shown in figure. The wedge is inclined at 45° to the horizontal on both the sides. The coefficient of friction between the block A and the wedge is 2/3 and that between the block B and the wedge is 1/3 and both the blocks A and B are released from rest, the acceleration of A will be

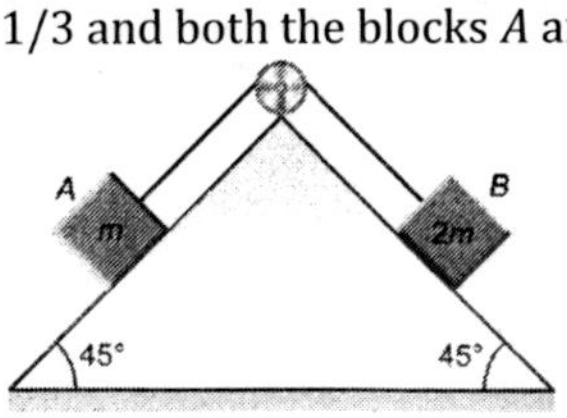

a) −1 b) 1.2 c) 0.2 d) Zero

218. When a bullet is fired at a target, its velocity decreases by half after penetrating 30 cm into it. The additional thickness it will penetrate before coming to rest is

a) 30 cm b) 40 cm c) 10 cm d) 50 cm

219. A mass of 100 g strikes the wall with speed 5 ms^{-1} at an angle as shown in figure and it rebounds with the same speed. If the contact time is 2×10^{-3} s, what is the force applied?

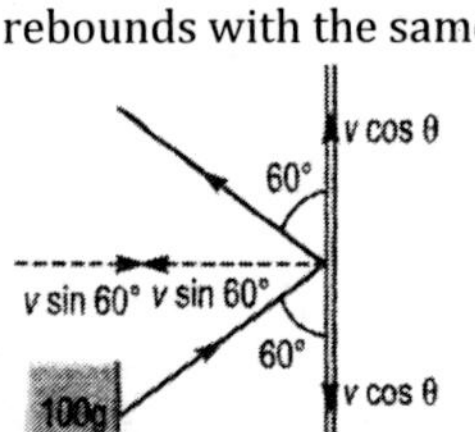

a) $250\sqrt{3}$ N to right b) 250 N to right c) $250\sqrt{3}$ N to left d) 250 N to left

220. The linear momentum ρ of a body moving in one dimension varies with time according to the equation $\rho = a + bt^2$ where a and b are positive constants. The net force acting on the body is

a) A constant
b) Proportional to t^2
c) Inversely proportional to t
d) Proportional to t

221. Two fixed frictionless in inclined plane making an angle 30° and 60° with the vertical are shown in the figure. Two blocks A and B are placed on the two planes. What is the relative vertical acceleration of A with respect to B?

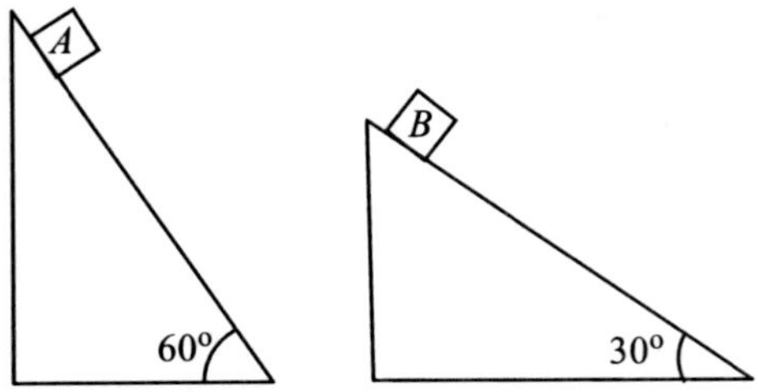

a) 4.9 ms^{-2} in horizontal direction
b) 9.8 ms^{-2} in vertical direction
c) Zero
d) 4.9 ms^{-2} in vertical direction

222. A rocket is propelled by a gas which is initially at a temperature of 4000 K. The temperature of the gas falls to 1000 K as it leaves the exhaust nozzle. The gas which will acquire the largest momentum while leaving the nozzle, is

a) Hydrogen
b) Helium
c) Nitrogen
d) Argon

223. A block of mass 200 kg is being pulled up by men on an inclined plane at angle of 45° as shown. The coefficient of static friction is 0.5. Each man can only apply a maximum force of 500 N. Calculate the number of men required for the block to just start moving up the plane

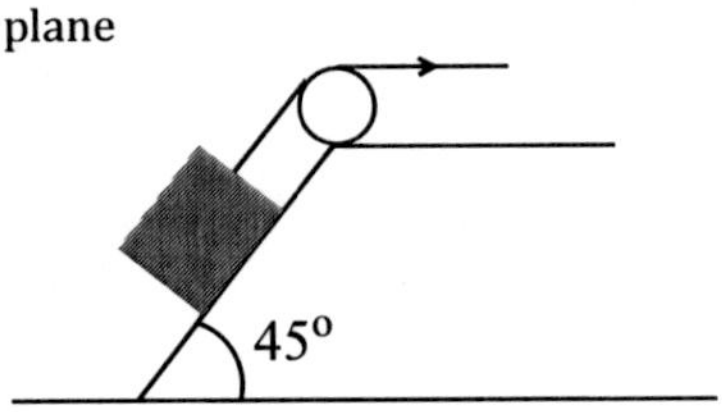

a) 10
b) 15
c) 5
d) 3

224. A man of 50 kg is standing at one end on a boat of length 25 m and mass 200 kg. If he starts running and when he reaches the other end, he has a velocity 2 ms^{-1} with respect to the boat. The final velocity of the boat is (in ms^{-1})

a) $\frac{2}{5}$
b) $\frac{2}{3}$
c) $\frac{8}{5}$
d) $\frac{8}{3}$

225. A body floats in a liquid contained in a beaker. If the whole system as shown in figure falls freely under gravity, then the upthrust on the body due to liquid is

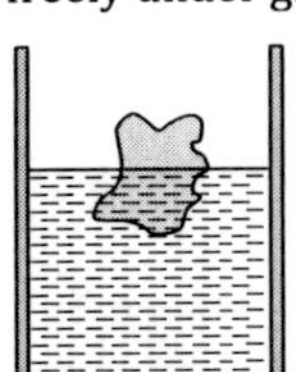

a) Zero
b) Equal to the weight of liquid displaced
c) Equal to the weight of the body in air
d) None of these

226. A block of mass 2 kg is kept on the floor. The coefficient of static friction is 0.4. If a force F of 2.5 $Newtons$ is applied on the block as shown in the figure, the frictional force between the block and the floor will be

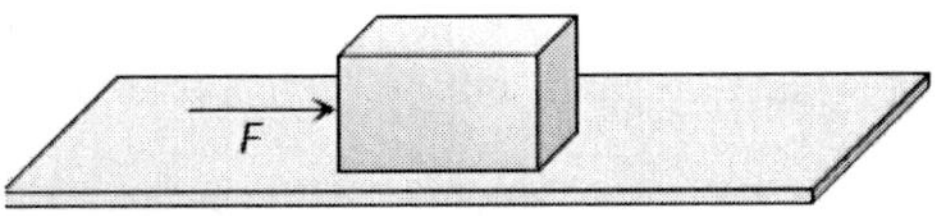

a) 2.5 N b) 5 N c) 7.84 N d) 10 N

227. A light string passing over a smooth light pulley connects two blocks of masses m_1 and m_2 (vertically). If the acceleration of the system is $g/8$then the ratio of the masses is

a) 8 : 1 b) 9 : 7 c) 4 : 3 d) 5 : 3

228. A particle of mass m, initially at rest, is acted upon by a variable force F for a brief interval of time T. It begins to move with a velocity u after the force stops acting. F is shown in the graph as a function of time. The curve is semicircle

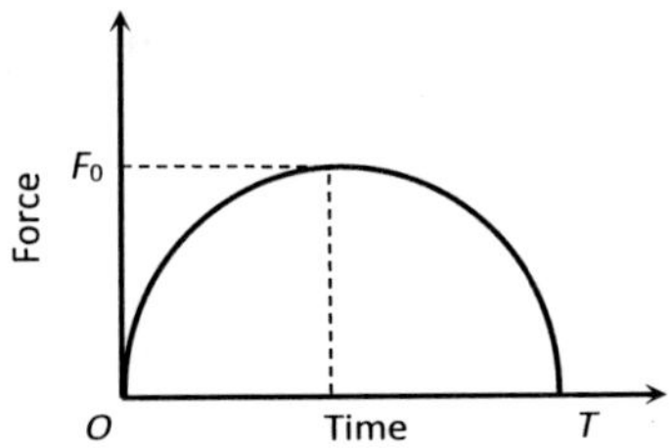

a) $u = \frac{\pi F_0^2}{2m}$ b) $u = \frac{\pi T^2}{8m}$ c) $u = \frac{\pi F_0 T}{4m}$ d) $u = \frac{F_0 T}{2m}$

229. In a gravitational force field a particle is taken from A to B along different paths as shown in figure. Then

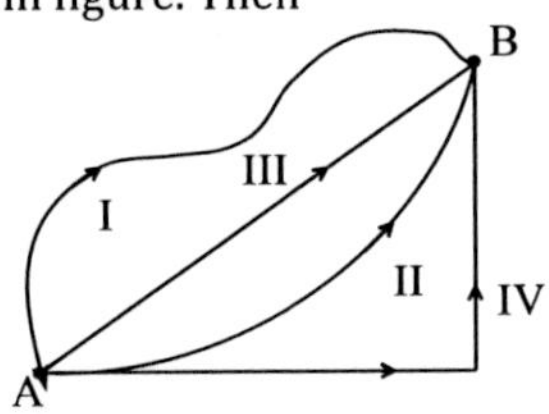

a) Work done along path I will be maximum

b) Work done along path III will be minimum

c) Work done along path IV will be minimum

d) Work done along all the paths will be the same

230. A 100 g iron ball having velocity 10 m/s collides with a wall at an angle 30° and rebounds with the same angle. If the period of contact between the ball and wall is 0.1 *second* , then the force experienced by the wall is

a) 10 N b) 100 N c) 1.0 N d) 0.1 N

231. A force of 100 dyne acts a mass of 5 gram for 10 s. The velocity produced is

a) 2000 cms^{-1} b) 200 cms^{-1} c) 20 cms^{-1} d) 2 cms^{-1}

232. An astronaut of weight Mg is in a rocket accelerating upward with an acceleration of 4g. The apparent weight of the astronaut will be

a) 5 Mg b) 4 Mg c) Mg d) Zero

233. A block mass 10 kg is kept on a horizontal surface. A force F is acted on the block as shown in figure. For what minimum value of F, the block will be lifter up?

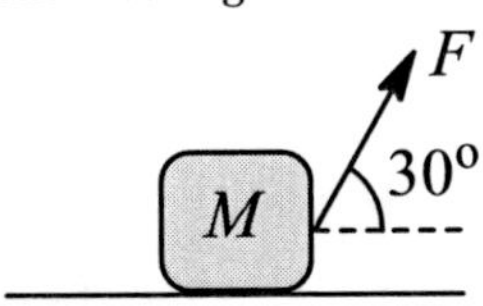

a) 98 N b) 49 N c) 200 N d) None of these

234. A machine gun fires a bullet of mass 40 g with a velocity 1200 ms^{-1} . The man holding it can exert a maximum force of 144 N on the gun. How many bullets can he fire per second at the most

a) One b) Four c) Two d) Three

235. The velocity of a body of mass 20 kg decreases from 20 $m/s$to 5 m/s in a distance of 100 m. Force on the body is

a) $-27.5\ N$ b) $-47.5\ N$ c) $-37.5\ N$ d) $-67.5\ N$

236. Two masses 8 kg and 12 kg are connected at the two ends of a string that goes over a frictionless pulley. Calculate the acceleration of the masses and the tension in the string. (Take $g = 10\ \mathrm{m/s^2}$)

a) $8\ \mathrm{m/s^2}$, 144 N b) $4\ \mathrm{m/s^2}$, 112 N c) $6\ \mathrm{m/s^2}$, 128 N d) $2\ \mathrm{m/s^2}$, 96 N

237. A body of mass 2 kg is being dragged with uniform velocity of 2 m/s on a rough horizontal plane. The coefficient of friction between the body and the surface is 0.20. The amount of heart generated in 5 sec is

($J = 4.2\ joule/cal$and $g = 9.8\ m/s^2$)

a) 9.33 cal b) 10.21 cal c) 12.67 cal d) 13.34 cal

238. The minimum velocity (in $\mathrm{ms^{-1}}$) with which a car driver must traverse a flat curve of radius 150 m and coefficient of friction 0.6 to avoid skidding is

a) 60 $\mathrm{ms^{-1}}$ b) 30 $\mathrm{ms^{-1}}$ c) 15 $\mathrm{ms^{-1}}$ d) 25 $\mathrm{ms^{-1}}$

239. A pendulum bob of mass 50 g is suspended from the ceiling of an elevator. The tension in the string if the elevator goes up with uniform velocity is approximately

a) 0.30 N b) 0.40 N c) 0.42 N d) 0.50 N

240. A block of mass 4 kg is kept on a rough horizontal surface. The coefficient of static friction is 0.8. If a force of 19 N is applied on the block parallel to the floor, then the force of friction between the block and floor is

a) 32 N b) 18 N c) 19 N d) 9.8 N

241. A ball falls from 20 m height on floor and rebounds to 5 m. Time of contact is 0.02 s. Find acceleration during impact.

a) 1200 $\mathrm{ms^{-2}}$ b) 1000 $\mathrm{ms^{-2}}$ c) 2000 $\mathrm{ms^{-2}}$ d) 1500 $\mathrm{ms^{-2}}$

242. In an elevator moving vertically up with an acceleration g, the force exerted on the floor by a passenger of mass M is

a) Mg b) $\frac{1}{2}Mg$ c) Zero d) $2\ Mg$

243. In the given figure the pulley is assumed massless and frictionless. If the friction force on the object of mass m is f, then its acceleration in terms of the force F will be equal to

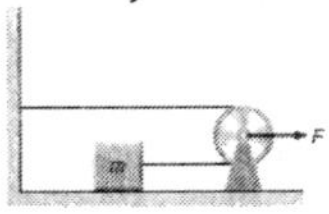

a) $(F-f)/m$ b) $\left(\frac{F}{2}-f\right)/m$ c) F/m d) None of these

244. The two particles of mass m each are tied at the ends of a light string of length $2a$. The whole system is kept on a frictionless horizontal surface with the string held tight so that each mass is at a distance 'a' from the center P (as shown in the figure). Now, the mid-point of the string is pulled vertically upwards with a small but constant force F. As a result, the particles move towards each other on the surface. The magnitude of acceleration, when the separation between them becomes $2x$, is

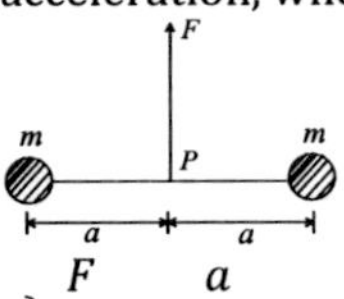

a) $\frac{F}{2m}\frac{a}{\sqrt{a^2-x^2}}$ b) $\frac{F}{2m}\frac{x}{\sqrt{a^2-x^2}}$ c) $\frac{F}{2m}\frac{x}{a}$ d) $\frac{F}{2m}\frac{\sqrt{a^2-x^2}}{x}$

245. A body weight 8 g when placed in one pan and 18 g when placed on the other pan of a false balance. If the beam is horizontal when both the pans are empty. The true weight of the body is

a) 13 g b) 12 g c) 15.5 g d) 15 g

246. Block A of mass 2 kg is placed over a block B of mass 8 kg. The combination is placed on a rough horizontal surface. If $g = 10 ms^{-2}$, coefficient of friction between B and floor $=0.5$, coefficient of friction between A and $B = 0.4$ and a horizontal force of 10 N is applied on 8 kg block, then the force of friction between A and B is

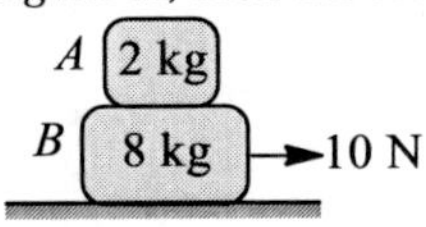

a) 100 N b) 50 N c) 40 N d) None of these

247. A man of mass 60 kg and a boy mass 30 kg are standing together on frictionless ice surface. If they push each other apart, man moves away with a speed of $0.4 ms^1$ relative to ice after 5 s. They will be away from each other at a distance of

a) 9.0 m b) 3.0 m c) 6.0 m d) 30 m

248. A player kicks a football of mass 0.5 kg and the football begins to move with a velocity of 10 m/s. If the contact between the leg and the football lasts for $\frac{1}{50}$ sec, then the force acted on the football should be

a) 2500 N b) 1250 N c) 250 N d) 625 N

249. Two blocks masses 1 kg and 2 kg rest on a smooth horizontal table. When the 2 kg block is pulled by a certain force F, the tension T in the string is 1.5 N. The value of F is

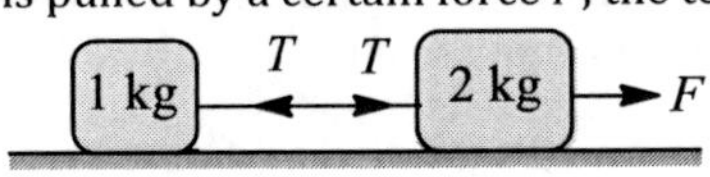

a) 1.5 N b) 2.5 N c) 3.5 N d) 4.5 N

250. A block rests on a rough inclined plane making an angle of 30° with the horizontal. The coefficient of static friction between the block and the plane is 0.8. If the frictional force on the block is 10 N, the mass of the block (in kg) is (Take $g = 10 ms^{-2}$)

a) 2.0 b) 4.0 c) 1.6 d) 2.5

251. A bag of sand of mass m is suspended by a rope. A bullet of mass $\frac{m}{20}$ is fired at it with a velocity v and gets embedded into it. The velocity of the bag finally is

a) $\frac{v}{20} \times 21$ b) $\frac{20v}{21}$ c) $\frac{v}{20}$ d) $\frac{v}{21}$

252. Two carts of masses 200 kg and 300 kg on horizontal rails are pushed apart. Suppose the coefficient of friction between the carts and the rails are same. If the 200 kg cart travels a distance of 36 m and stops, then the distance travelled by the cart weighing 300 kg is

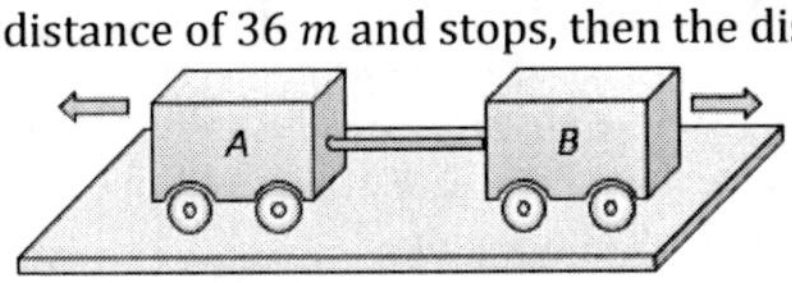

a) 32 m b) 24 m c) 16 m d) 12 m

253. If a block slides down a plane inclined at 30° with horizontal, then the coefficient of friction between the block and inclined plane is

a) 0.5 b) $\frac{1}{3}$ c) $\frac{1}{\sqrt{2}}$ d) $\frac{1}{\sqrt{3}}$

254. A cylinder roll up an inclined plane, reaches some height and then rolls down (without slipping throughout these motions). The directions of frictional force acting on the cylinder are

a) Up the inclined while ascending and down the incline while descending

b) Up the incline while ascending as well as descending

c) Down the incline while ascending and up the incline while descending
d) Down the incline while ascending as well as descending

255. A machine gun fires n bullets per second, each of mass m. If the speed of each bullet is u, then the force of recoil is

a) mng b) mnv c) $mnvg$ d) $\frac{mnv}{g}$

256. In the arrangement shown in figure, if a force 2 mg is applied at the free end of the rope, the mass m will ascend with an acceleration of

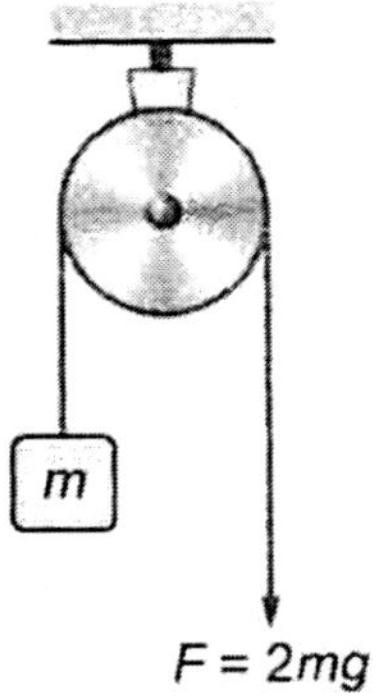

a) $\frac{g}{3}$ b) $\frac{g}{2}$ c) g d) 2 g

257. A partly hanging uniform chain of length chain of length L is resting on a rough horizontal table. $l$is the maximum possible length that can hang in equilibrium. The coefficient of friction between the chain and table is

a) $\frac{l}{L-l}$ b) $\frac{L}{l}$ c) $\frac{l}{L}$ d) $\frac{lL}{L+l}$

258. Two blocks, 4 kg and 2 kg are sliding down an incline plane as shown in figure. The acceleration of 2 kg block is

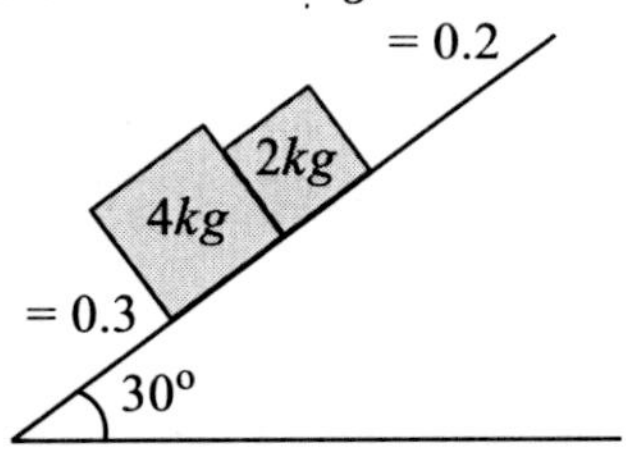

a) 1.66 m/s^2 b) 2.66 m/s^2 c) 3.66 m/s^2 d) 4.66 m/s^2

259. A block of mass m is connected to another block of mass M by a spring (massless) of spring constant$k$. The blocks are kept on a smooth horizontal plane. Initially the blocks are at rest and the spring is unstretched. Then a constant force F starts acting on the block of mass M to pull it. Find the force on the block of mass m

a) $\frac{mF}{M}$ b) $\frac{(M+m)F}{m}$ c) $\frac{mF}{(m+M)}$ d) $\frac{MF}{(m+M)}$

260. Which of the following quantities measured from different inertial reference frames are same?

a) Force b) Velocity c) Displacement d) Kinetic energy

261. A solid disc of mass M is just held in air horizontally by throwing 40 stones per sec vertically upwards to strike the disc each with a velocity 6 ms^{-1}. If the mass of each stone is 0.05 kg what is the mass of the disc ($g = 10ms^{-2}$)

a) 1.2 kg b) 0.5 kg c) 20 kg d) 3 kg

262. If a person with a spring balance and a body hanging from it goes up and up in an aeroplane, then the reading of the weight of the body as indicated by the spring balance will

a) Go on increasing
b) Go on decreasing
c) First increase and then decrease
d) Remain the same

263. A vehicle of mass m is moving on a rough horizontal road with momentum P. If the coefficient of friction between the tyres and the road be μ, then the stopping distance is

a) $\frac{P}{2\mu mg}$
b) $\frac{P^2}{2\mu mg}$
c) $\frac{P}{2\mu m^2 g}$
d) $\frac{P^2}{2\mu m^2 g}$

264. Consider a car moving along a straight horizontal road with a speed of 72 km/h. If the coefficient kinetic friction between the tyres and the road is 0.5, the shortest distance in which the car can be stopped is $[g = 10\ ms^{-2}]$

a) 30 m
b) 40 m
c) 72 m
d) 20 m

265. The velocity of a body at time $t = 0$ is $10\sqrt{2}\ m/s$ in the north-east direction and it is moving with an acceleration of $2\ m/s^2$ directed towards the south. The magnitude and direction of the velocity of the body after 5 sec will be

a) 10 m/s, towards east
b) 10 m/s, towards north
c) 10 m/s, towards south
d) 10 m/s, towards north-east

266. A block of mass 5 kg, resting on a horizontal surface, is connected by a cord, passing over a light frictionless pulley to a hanging block of mass 5 kg. The coefficient of kinetic friction between the block and the surface is 0.5. Tension in the cord is [Take $g = 9.8\ ms^{-2}$]

a) 49 N
b) 36 N
c) 36.75 N
d) 2.45 N

267. The backside of a truck is open and a box of 40 kg is placed 5m away from the rear end. The coefficient of friction of the box with the surface of the truck is 0.15. The truck starts from rest with 2 m/s^2acceleration. Calculate the distance covered by the truck when the box falls off

a) 20 m
b) 30 m
c) 40 m
d) 50 m

268. An object of mass 5 kg is attached to the hook of a spring balance and the balance is suspended vertically from the roof of a lift. The reading on the spring balance when the lift is going up with an acceleration of $0.25 ms^{-2}$ is $(g = 10 ms^{-2})$

a) 51.25 N
b) 48.75 N
c) 52.75 N
d) 47.25 N

269. A body of weight 50 N placed on a horizontal surface is just moved by a force of 28.2 N. The frictional force and the normal reaction are

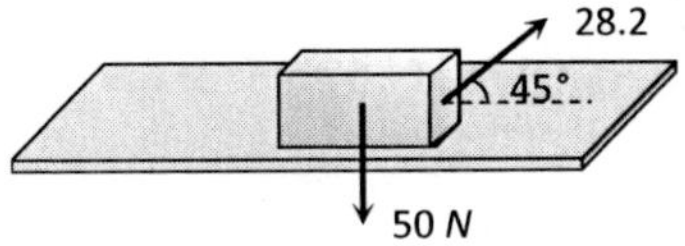

a) 10 N, 15 N
b) 20 N, 30 N
c) 2 N, 3 N
d) 5 N, 6 N

270. The tension in the string in the pulley system shown in the figure is

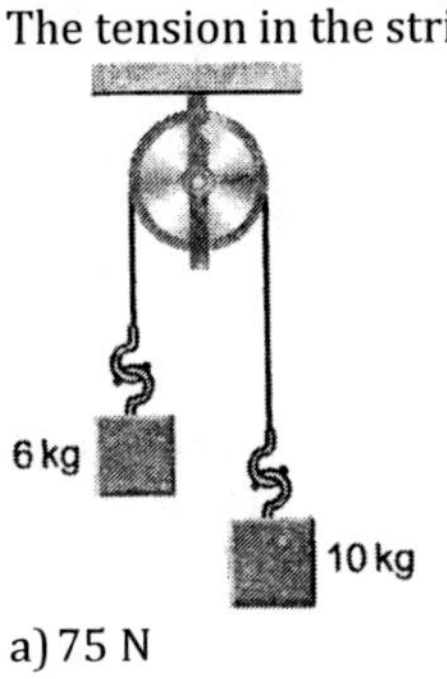

a) 75 N
b) 80 N
c) 7.5 N
d) 30 N

271. A block of mass 50 kg slides over a horizontal distance of $1m$. If the coefficient of friction between their surface is 0.2, then work done against friction is

a) 98 J b) 72 J c) 56 J d) 34 J

272. If in a stationary lift, a man is standing with a bucket full of water, having a hole at its bottom. The rate of flow of water through this hole is R_0. If the lift starts to move up and down with same acceleration and then that rates of flow of water are R_u and R_d, then

a) $R_0 > R_u > R_d$ b) $R_u > R_0 > R_d$ c) $R_d > R_0 > R_u$ d) $R_u > R_d > R_0$

273. For ordinary terrestrial experiments, the observer in an inertial frame in the following cases is

a) A child revolving in a giant wheel

b) A driver in a sport car moving with a constant high speed of 200km h^{-1} on a straight rod

c) The pilot of an aeroplane which is taking off

d) A cyclist negotiating a sharp curve

274. A spring balance is attached to the ceiling of a lift. A man hangs his bag on the spring and spring reads 49 N, when the lift is stationary. If the lift moves downward with an acceleration of 5 ms^{-2}, the reading of the spring balance will be

a) 24 N b) 74 N c) 15 N d) 49 N

275. A particle of mass m is at rest at the origin at time $t = 0$. It is subjected to a force $F(t) = F_0 e^{-bt}$ in the x direction. Its speed $v(t)$ is depicted by which of the following curves

a) 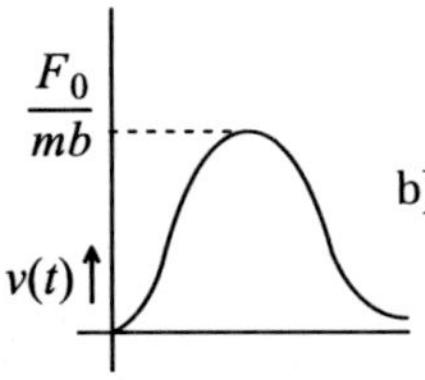

b) 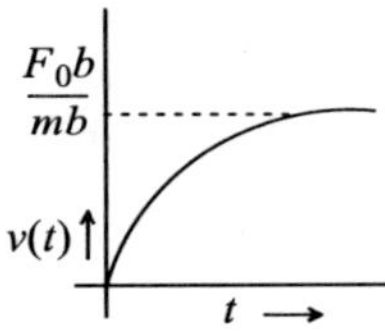

c) 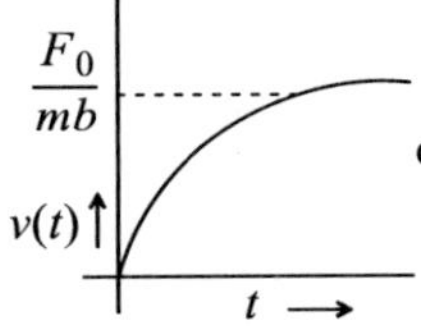

d) 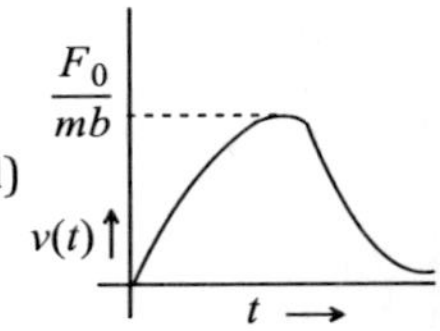

276. A stationary body of mass 3 kg explodes into three equal pieces. Two of the pieces fly off in two mutually perpendicular directions, one with velocity of $3\,\hat{i}ms^{-1}$ and the other with a velocity of $4\,\hat{j}ms^{-1}$. If the explosion occurs in 10^{-4}s, the force acting on the third piece in newtons is

a) $(3\,\hat{i} + 4\,\hat{j}) \times 10^{-4}$ b) $(3\,\hat{i} - 4\,\hat{j}) \times 10^{-4}$ c) $(3\,\hat{i} + 4\,\hat{j}) \times 10^{4}$ d) $-(3\,\hat{i} + 4\,\hat{j}) \times 10^{4}$

277. A body of mass 5 kg starts from the origin with an initial velocity $\mathbf{u} = (30\hat{i} + 40\hat{j})\ ms^{-1}$. If a constant force $(-6\hat{i} - 5\hat{j})$N acts on the body, the time in which the y component of the velocity becomes zero is

a) 5 s b) 20 s c) 40 s d) 80 s

278. A player caught a cricket ball of mass 150 g moving at the rate of 20 ms^{-1}. If the catching process be completed in 0.1 s, the force of blow exerted by the ball on the hands of the player is

a) 0.3 N b) 30 N c) 300 N d) 3000 N

279. In an air collision between an aeroplane and a bird, the force experienced by the bird as compared to that of the aeroplane is

a) Very high b) Equal c) Less d) Zero

280. A mass of 5 kg is suspended by a rope of length 2 m from a ceiling. A force of 50 N in the horizontal direction is applied at the mid-point of the rope. The angle made by rope with the vertical, in equilibrium is

a) 50° b) 60° c) 30° d) 45°

281. Two masses A and B of 15 kg and 10 kg are connected with a string passing over a frictionless pulley fixed at the corner of a table (as shown in figure). The coefficient of friction between the table and block is 0.4. The minimum mass of C, that may be placed on A to prevent it from moving is

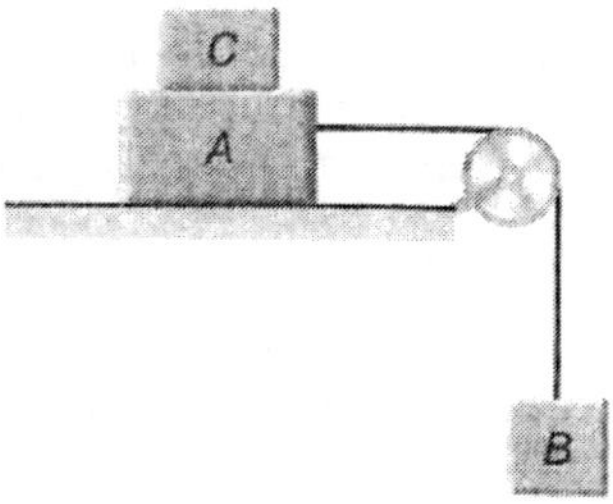

a) 10 kg b) 5 kg c) Zero d) 15 kg

282. A body of mass 8 kg is suspended through two light springs X and Y connected in series as shown in figure. The reading is X and Y respectively are

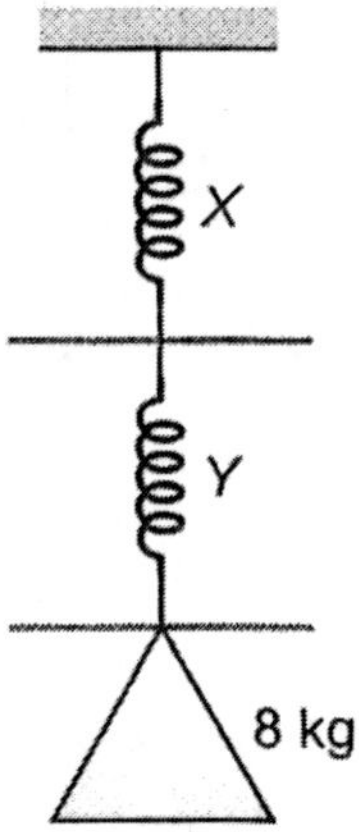

a) 8 kg, zero b) Zero, 8 kg c) 6 kg, 2 kg d) 8 kg, 8 kg

283. A trolley T (mass 5 kg) on a horizontal smooth surface is pulled by a load L (2 kg) through a uniform rope ABC of length 2 m and mass 1kg. As the load falls from $BC = 0$ to $BC = 2$ m, its acceleration (in ms^{-2}) changes from (Take $g = 10ms^{-2}$)

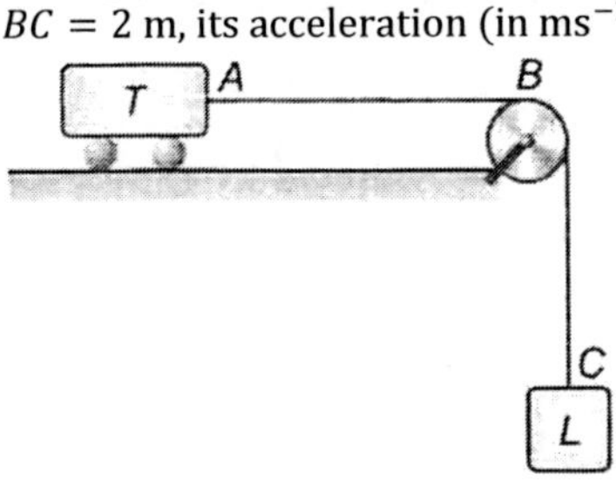

a) 20/6 to 20/5 b) 20/8 to 30/8 c) 20/5 to 30/6 d) None of these

284. A uniform metal chain is placed on a rough table such that one end of it hangs down over the edge of the table. When one-third of its length hangs over the edge, the chain starts sliding. Then, the coefficient of static friction is

a) 3/4 b) 1/4 c) 2/3 d) 1/2

285. In which of the following cases forces may not be required to keep the

a) Particle going in a circle b) Particle going along a straight line

c) The momentum of the particle constant d) Acceleration of the particle constant

286. A stationary bomb explodes into three pieces. One piece of 2 kg mass moves with a velocity of 8 ms^{-1} at right angles to the other pieces of mass 1 kg moving with a velocity of 12 ms^{-1}. If the mass of the third piece is 0.5 kg, then its velocity is

a) 10 ms^{-1} b) 20 ms^{-1} c) 30 ms^{-1} d) 40 ms^{-1}

287. A 40 kg slab rests on a frictionless floor as shown in the figure. A 10 Kg block rests on the top of the slab. The static coefficient of friction between the block and slab is 0.60 while the kinetic friction is 0.40. The 10 kg block is acted upon by a horizontal force 100 N. If $g = 9.8\ m/s^2$, the resultant acceleration of the slab will be

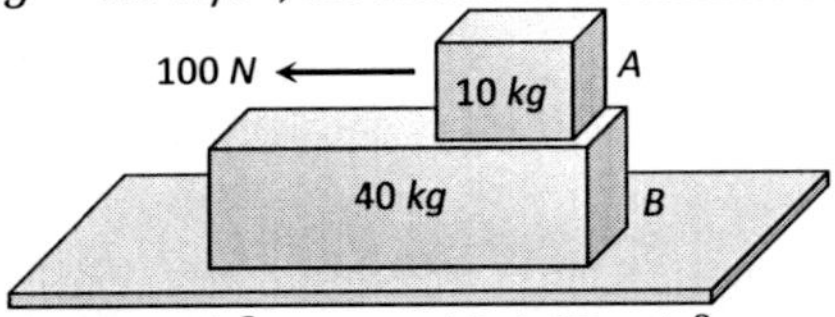

a) $0.98\ m/s^2$ b) $1.47\ m/s^2$ c) $1.52\ m/s^2$ d) $6.1\ m/s^2$

288. The normal reaction on a body placed in a lift moving up with constant acceleration 2 ms^{-1} is 120 N. Mass of body is (Take $g = 10\ ms^{-2}$)

a) 10 kg b) 15 kg c) 12 kg d) 5 kg

289. A rope of length 5 m is kept on frictionless surface and a force of 5N is applied to one of its end. Find tension in the rope at 1 m from this end

a) 1 N b) 3 N c) 4 N d) 5 N

290. A balloon has 5 g of air. A small hole is pierced into it. The air escapes at a uniform rate with a velocity of 4 cms^{-1}. If the balloon shrinks completely in 2.5 s the average force acting on the balloon is

a) 2 dyne b) 50 dyne c) 8 dyne d) 8 N

291. A car starts from rest to cover a distance s. The coefficient of friction between the road and the tyresis μ. The minimum time in which the car can cover the distance is proportional to

a) μ b) $\sqrt{\mu}$ c) $1/\mu$ d) $1/\sqrt{\mu}$

292. One day on a spacecraft corresponds to 2 days on the earth. The speed of the spacecraft relative to the earth is

a) $1.5 \times 10^8\ ms^{-1}$ b) $2.1 \times 10^8\ ms^{-1}$ c) $2.6 \times 10^8\ ms^{-1}$ d) $5.2 \times 10^8\ ms^{-1}$

293. 100 g of an iron ball having velocity 10 ms^{-1} collides with wall at an angle 30° and rebounds with the same angle. If the period of contact between the ball and wall is 0.1 s, then the average force experienced by the wall is

a) 10 N b) 100 N c) 1.0 N d) 0.1 N

294. A body, whose momentum is constant, must have constant

a) Force b) Velocity c) Acceleration d) All of these

295. A force of 50 $dynes$ is acted on a body of mass 5 g which is at rest for an interval of 3 seconds, then impulse is

a) $0.15 \times 10^{-3}\ Ns$ b) $0.98 \times 10^{-3}\ Ns$ c) $1.5 \times 10^{-3} Ns$ d) $2.5 \times 10^{-3}\ Ns$

296. Two bodies A and B of masses 10 kg and 15 kg respectively kept on a smooth, horizontal surface are tied to the ends of a light string. If T represents the tension in the spring when a horizontal force $F = 500\ N$ is applied to A (as shown in figure 1) and T″ be the tension when it is applied to B (figure 2), then which of the following is true

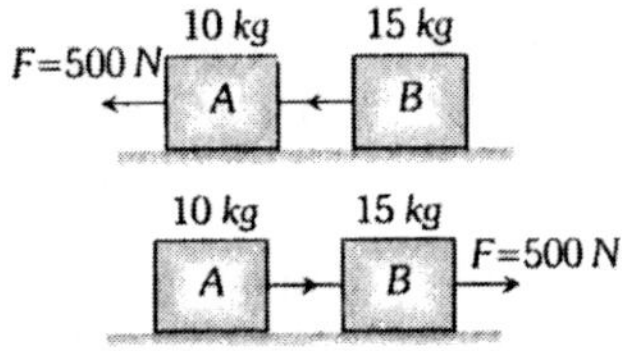

a) $T = T' = 500\ N$ b) $T = T' = 250\ N$

c) $T = 200\ N, T' = 300\ N$ d) $T = 300N, T' = 200\ N$

297. A block of mass m is placed on a smooth inclined plane of inclination θ with the horizontal. The force exerted by the plane on the block has a magnitude

a) mg b) $\frac{mg}{\cos\theta}$ c) $mg\cos\theta$ d) $mg\tan\theta$

298. A force-time graph for a linear motion of a body is shown in the figure. The change in linear momentum between 0 and 7 s is

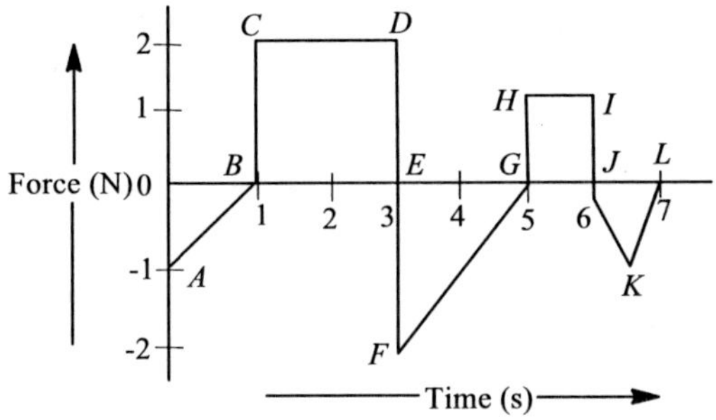

a) 2 Ns b) 3 Ns c) 4 Ns d) 5 NS

299. A stone is accelerated upwards by a cord whose breaking strength is three times the weight of the stone. The maximum acceleration with which the stone can be moved up without breaking the cord is

a) g b) 2g c) 3g d) 4g

300. An ice cube is kept on an inclined plane of angle 30°. Coefficient of kinetic friction between block and incline plane is $1/\sqrt{3}$. What is acceleration of block

a) Zero b) $2\ m/s^2$ c) $1.5\ m/s^2$ d) $5\ m/s^2$

301. A body of mass 1000 kg is moving horizontally with a velocity 50 m/s. A mass of 250 kg is added. Find the final velocity

a) 40 m/s b) 20 m/s c) $30\sqrt{2}\ m/s$ d) 50 m/s

302. The acceleration of the 500 g block in figure is

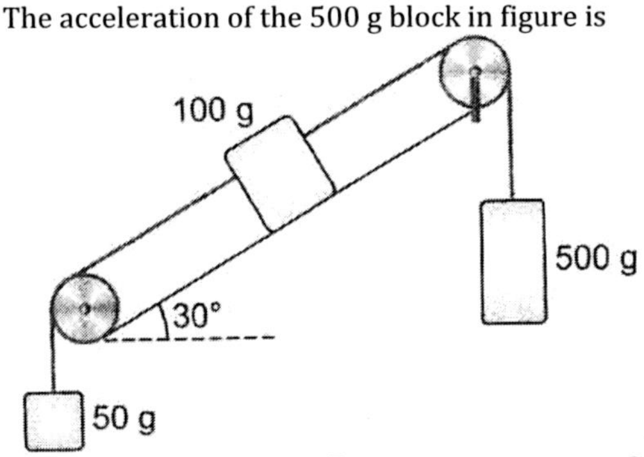

a) $\frac{6g}{13}$ downwards b) $\frac{7g}{13}$ downwards c) $\frac{8g}{13}$ downwards d) $\frac{9g}{13}$ upwards

303. A force of 20N is applied on a body of mass 5 kg resting on a horizontal plane. The body gains a kinetic energy of 10 *joule* after it moves a distance 2 m. The frictional force is

a) 10 N b) 15 N c) 20 N d) 30 N

304. An object is kept on a smooth inclined plane of 1 in l. The horizontal acceleration to be imparted to the inclined plane so that the object is stationary relative to the inclined is

a) $g\sqrt{l^2-1}$ b) $g(l^2-1)$ c) $\frac{g}{\sqrt{l^2-1}}$ d) $\frac{g}{l^2-1}$

305. A block at rest slides down a smooth inclined plane which makes an angle 60° with the vertical and it reaches the ground in t_1 second. Another block is dropped vertically from the same point and reaches the ground in t_2 second. Then the ratio of $t_1 : t_2$ is

a) 1:2 b) 2:1 c) 1:3 d) $1:\sqrt{2}$

306. A body of mass m rests on horizontal surface. The coefficient of friction between the body and the surface is μ. If the mass is pulled by a force P as shown in the figure, the limiting friction between body and surface will be

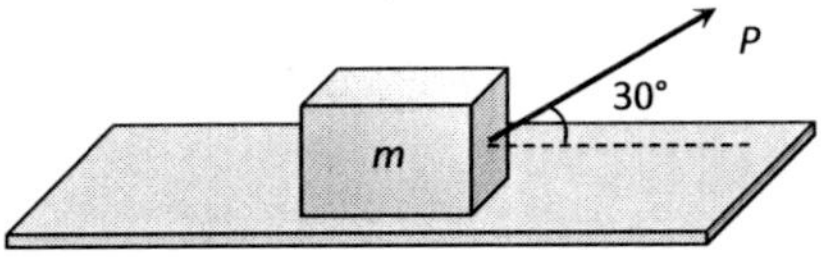

a) μmg b) $\mu\left[mg+\left(\frac{P}{2}\right)\right]$ c) $\mu\left[mg-\left(\frac{P}{2}\right)\right]$ d) $\mu\left[mg-\left(\frac{\sqrt{3}P}{2}\right)\right]$

307. When force F_1, F_2, F_3 are acting on a particle of mass m such that F_2 and F_3 are mutually perpendicular, then the particle remains stationary. If the force F_1 is now removed, then the acceleration of the particle is

a) F_1/m b) F_2F_3/mF_1 c) $(F_2 - F_3)/m$ d) F_2/m

308. A ball of mass 0.2 kg is thrown vertically upward by applying a force by hand. If the hand moves 0.2 m while applying the force and the ball goes up to 2 m height further, find the magnitude of the force. Consider $g = 10\text{ms}^{-2}$

a) 4 N b) 16 N c) 20 N d) 22 N

309. A block of mass 1 kg slides down on a rough inclined plane of inclination 60° starting from its top. If the coefficient of kinetic friction is 0.5 and length of the plane is 1 m, then work done against friction is (Take $g = 9.8\ m/s^2$)

a) 9.82 J b) 4.94 J c) 2.45 J d) 1.96 J

310. Two blocks are in contact on a frictionless table. One has mass m and other $2m$. A force f is applied on $2m$ as shown in figure. Next the same force F is applied from the right on m. In the two cases respectively, the force of contact between the two blocks will be

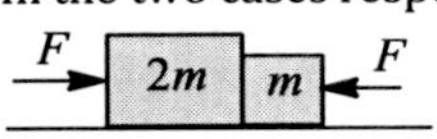

a) 2 : 1 b) 1 : 3 c) 1 : 2 d) 3 : 1

311. A body of mass $m = 3.513\ kg$ is moving along the x-axis with a speed of 5.00 ms^{-1}. The magnitude of its momentum is recorded as

a) 17.565 $kg\ ms^{-1}$ b) 17.56 $kg\ ms^{-1}$ c) 17.57 $kg\ ms^{-1}$ d) 17.6 $kg\ ms^{-1}$

312. A block of mass M placed on a frictionless horizontal table is pulled by another block of mass m hanging vertically by a massless string passing over a frictionless pully. The tension in the string is

a) $\frac{m}{M+m}g$ b) $\frac{M}{M+m}g$ c) $\frac{M+m}{Mm}g$ d) $\frac{Mm}{M+m}g$

313. In the motion of a racket, physical quantity which is conserved is

a) Angular momentum b) Linear momentum c) Force d) Work

314. A body of mass 2 kg is moving on the ground comes to rest after some time. The coefficient of kinetic friction between the body and the ground is 0.2. The retardation in the body is

a) 9.8 m/s^2 b) 4.73 m/s^2 c) 2.16 m/s^2 d) 1.96 m/s^2

315. A body is coming with a velocity of 72 kmh^{-1} on a rough horizontal surface of coefficient of friction 0.5. If the acceleration due to gravity is 10 ms^{-2}, find the minimum distance it can be stopped

a) 400 m b) 40 m c) 0.40 m d) 4 m

316. A particle moves in the X-Yplane under the influence of a force such that its linear momentum is $\mathbf{p}(t) = A[\hat{\mathbf{i}}\cos(kt) - \hat{\mathbf{j}}\sin(kt)]$ where A and$k$ are constant. The angle between the force and the momentum is

a) 0° b) 30° c) 45° d) 90°

317. A bullet of mass 0.05 kg moving with a speed of 80 ms^{-1} enters a wooden block and is stopped after a distance of 0.40 m. The average resistive force exerted by the block on the bullet is

a) 300 N b) 20 N c) 400 N d) 40 N

318. Two rectangular blocks A and B of masses 2 kgand 3 kg respectively are connected by a spring of spring constant 10.8 Nm^{-1} and are placed on a frictionless horizontal surface. The block A was given an initial velocity of 0.15 ms^{-1} in the direction shown in the figure. The maximum compression of the spring during the motion is

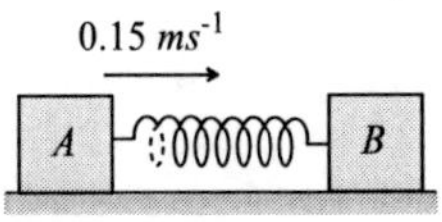

a) 0.01 m b) 0.02 m c) 0.05 m d) 0.03 m

319. The fixed frictionless inclined planes making an angle 30° and 60° with the vertical are shown in the figure. Two blocks Aand B are placed on the two planes. What is the relative vertical acceleration of A with respect to B

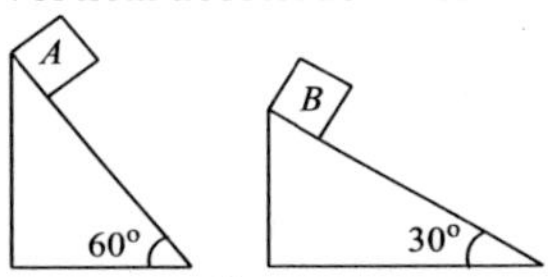

a) 4.9 ms^{-2} in vertical direction b) 4.9 ms^{-2} in horizontal direction
c) 9.8 ms^{-2} in vertical direction d) Zero

320. Newton's third law of motion leads to the law of conservation of

a) Angular momentum b) Energy c) Mass d) Momentum

321. A 500 kg horse pulls a cart of mass 1500 kg along a level road with an acceleration of 1 ms^{-2}. If the coefficient of sliding friction is 0.2, then the force exerted by the horse in forward direction is

a) 3000 N b) 4000 N c) 5000 N d) 6000 N

322. Three weights W, $2W$ and $3W$ are connected to identical springs suspended from a rigid horizontal rod. The assembly of the rod and the weights fall freely. The positions of the weights from the rod are such that

a) $3W$ will be farthest b) W will be farthest
c) All will be at the same distance d) $2W$ will be farthest

323. A shell is fired from cannon with a velocity v ms^{-1} at an angle of θ with the horizontal direction. At the highest point in its path, it explodes into 2 pieces of equal masses. One of the pieces retraces its path to the cannon. The speed of the other piece immediately after the explosion is

a) $\sqrt{3/2}\, v \cos\theta$ b) $v \cos\theta$ c) $2v \cos\theta$ d) $3v \cos\theta$

324. When a horse pulls a wagon, the force that causes the horse to move forward is the force

a) The ground exerts on it b) It exerts on the ground
c) The wagon exerts on it d) It exerts on the wagon

325. To avoid slipping while walking on ice, one should take smaller steps because of the

a) Friction of ice is large b) Larger normal reaction
c) Friction of ice is small d) Smaller normal reaction

326. Three guns are aimed at the centre of a circle. They are mounted on the circle, 120° apart. They fire in a timed sequence, such that the three bullets collide at the centre and mash into a stationary lump. Two of the bullets have identical masses of 4.5 g and speed of v_1 and v_2. The third bullet has a mass of 2.50 g and a speed of 575 ms^{-1}
Find the unknown speeds.

a) 200 ms^{-1} each b) 145 $ms^{-1}$and 256 ms^{-1}
c) 536 ms^{-1} and 320 ms^{-1} d) None of the above

327. A child weighing 25 kg slides down a rope hanging from the branch of a tall tree. If the force of friction acting against him is 2 N, what is the acceleration of the child (Take $g = 9.8\ m/s^2$)

a) $22.5\ m/s^2$ b) $8\ m/s^2$ c) $5\ m/s^2$ d) $9.72\ m/s^2$

328. A 2 kg mass starts from rest on an inclined smooth surface with inclination 30° and length 2 m. How much will it travel before coming to rest on a frictional surface with frictional coefficient of 0.25

a) $4\ m$ b) $6\ m$ c) $8\ m$ d) $2\ m$

329. Starting from rest, the time taken by a body sliding down on a rough inclined plane at 45° with the horizontal is twice the time taken to travel on a smooth plane of same inclination and same distance. Then the coefficient of kinetic friction is

a) 0.25 b) 0.33 c) 0.50 d) 0.75

330. A jet engine works on the principle of

a) Conservation of mass b) Conservation of energy
c) Conservation of linear momentum d) Conservation of angular momentum

331. A box of mass m kg is placed on the rear side of an open truck accelerating at t ms^{-2}. The coefficient of friction between the box and the surface below it is 0.4. The net acceleration of the box with respect to the truck is zero. The value of m is [Given g = 10 ms^{-2}]

a) 4 kg b) 8 kg c) 9.78 kg d) It could be any value

332. Two balls of masses m_1 and m_2 are separated from each other by a powder charge placed between them. The whole system is at rest on the ground. Suddenly the powder charge explodes and masses are pushed apart. The mass m_1 travels a distance s_1 and stops. If the coefficients of friction between the balls and ground are same, the mass m_2 stops after traveling the distance

a) $s_2 = \frac{m_1}{m_2} s_1$ b) $s_2 = \frac{m_2}{m_1} s_1$ c) $s_2 = \frac{m_1^2}{m_2^2} s_1$ d) $s_2 = \frac{m_2^2}{m_1^2} s_1$

333. In the above question, the acceleration of mass m is

a) $\frac{F}{m}$ b) $\frac{F-T}{m}$ c) $\frac{F+T}{m}$ d) $\frac{F}{M}$

334. A block is at rest on an inclined plane making an angle α with the horizontal. As the angle α of the incline is increased, the block starts slipping when the angle of inclination becomes θ. The coefficient of static friction between the block and the surface of the incline plane is or A body starts sliding down at an angle θ to horizontal. Then coefficient of friction equal to

a) $\sin\theta$ b) $\cos\theta$ c) $\tan\theta$ d) Independent of θ

335. A given object takes n times more time to slide down a 45° rough inclined plane as it takes so slide down a perfectly smooth 45° incline. The coefficient of kinetic friction between the object and the incline is

a) $\frac{1}{1-n^2}$ b) $1-\frac{1}{n^2}$ c) $\sqrt{1-\frac{1}{n^2}}$ d) $\sqrt{\frac{1}{1-n^2}}$

336. An iron nail is dropped from a height h from the level of a sand bed. If it penetrates through a distance x in the sand before coming to rest, then average force exerted by the sand on nail is

a) $mg\left(\frac{h}{x}+1\right)$ b) $mg\left(\frac{x}{h}+1\right)$ c) $mg\left(\frac{h}{x}-1\right)$ d) $mg\left(\frac{x}{h}-1\right)$

337. A block compartment containing gas is moving with some acceleration in horizontal direction. Neglect effect of gravity. Then the pressure in the compartment is

a) Same everywhere b) Lower in front side c) Lower in rear side d) Lower in upper side

338. The average resisting force that must act on a 5 kg mass to reduce its speed from 65 cm/s to 15 cm/s in 0.2s is

a) 12.5 N b) 25 N c) 50 N d) 100 N

339. A body of 2 kg has an initial speed 5 ms^{-1}. A force acts on it for some time in the direction of motion. The force time graph is shown in figure. The final speed of the body

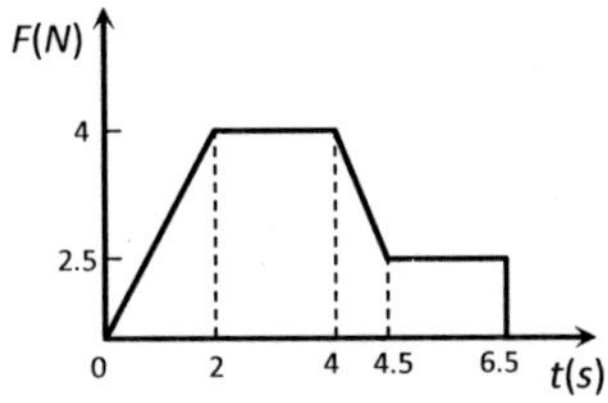

a) 9.25 ms^{-1} b) 5 ms^{-1} c) 14.25 ms^{-1} d) 4.25 ms^{-1}

340. A block of mass m is on an inclined plane of angle θ. The coefficient of friction between the block and the plane is μ and $\tan\theta > \mu$. The block is held stationary by applying a force E parallel to the plane. The direction of force pointing up the plane is taken to be positive. As P is varied from $P_1 = mg(\sin\theta - \mu\cos\theta)$ to $P_2 = mg(\sin\theta + \mu\cos\theta)$, the frictional force f $versus$ P graph will look like

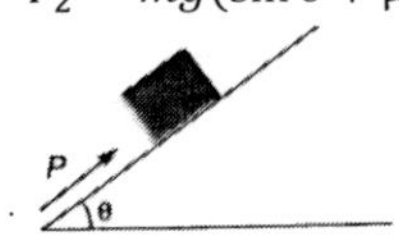

a)
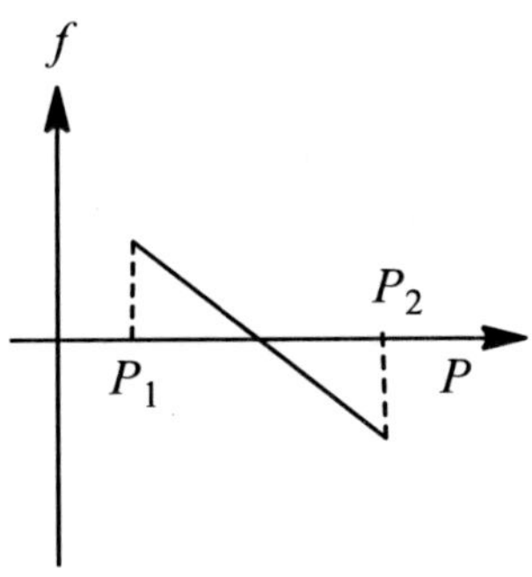

b)
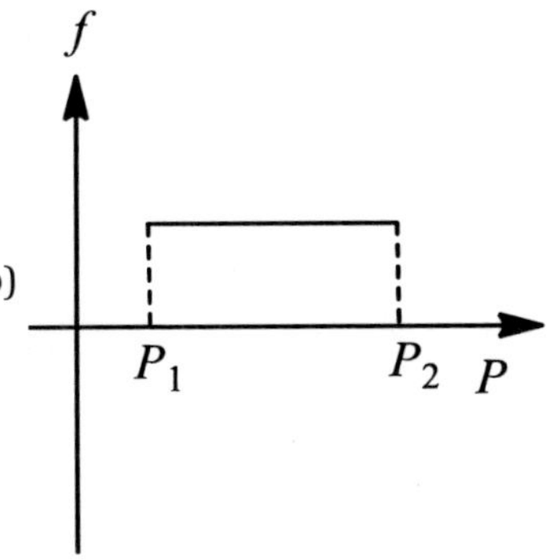

c)
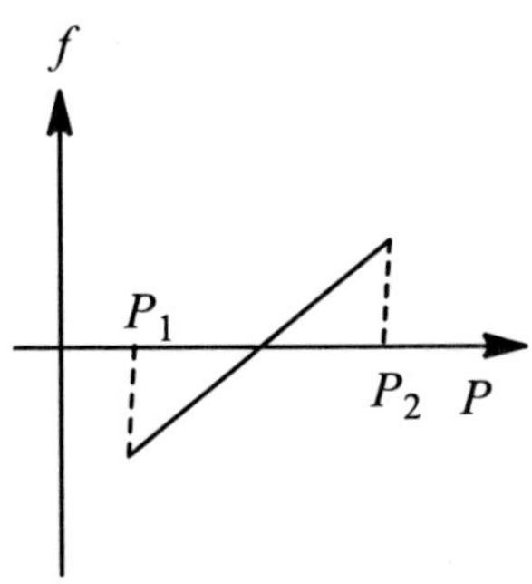

d)
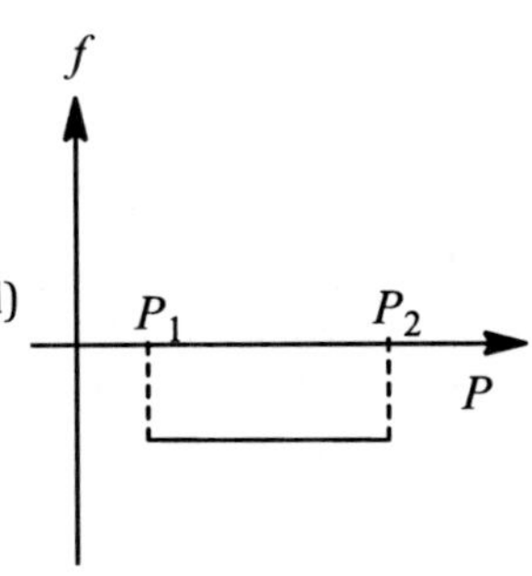

341. A body is moving along a rough horizontal surface with an initial velocity 6 m/s. If the body comes to rest after travelling 9m, then the coefficient of sliding friction will be

a) 0.4 b) 0.2 c) 0.6 d) 0.8

342. In the above ques., the height to which the lift takes the passenger is

a) 3.6 $meters$ b) 8 $meters$ c) 1.8 $meters$ d) 36 $meters$

343. Sand is being dropped on a conveyor belt at the rate of M kg/s. The force necessary to keep the belt moving with a constant velocity of v m/s will be

a) $\frac{Mv}{2}$ newton b) Zero c) Mv newton d) 2 Mv newton

344. 300 *joule* of work is done in sliding up a 2 *kg* block on an inclined plane to a height of 10 *metres*. Taking value of acceleration due to gravity 'g' to be 10 m/s^2, work done against friction is

a) 100 J b) 200 J c) 300 J d) Zero

345. A mass of 10 kg is suspended from a string balance. It is pulled aside by a horizontal string so that it makes an angle of 60° with the vertical. The new reading of the balance is

a) 20 kg-wt b) 10 kg-wt c) $10\sqrt{3}$ kg-wt d) $20\sqrt{3}$ kg-wt

346. A block of mass $\sqrt{3}$ kg resting on a horizontal surface. A force F is applied on the block as shown in figure. If coefficient of friction between the block be $\frac{1}{2\sqrt{3}}$, what can be the maximum value of force F so that block does not start moving? (Take $g = 10\ ms^{-2}$)

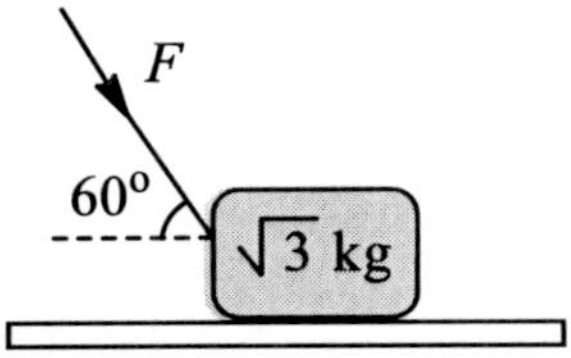

a) 20 N b) 10 N c) 12N d) 15 N

347. A force of 1200 N acts on a 0.5 kg steel ball as result of collision lasting 25 ms. If the force is in a direction opposite to the initial velocity of 14 ms^{-1} then the final speed of the steel ball would be

a) 24 ms^{-1} b) 35 ms^{-1} c) 12 ms^{-1} d) 46 ms^{-1}

348. A block of weight 5N is pushed against a vertical wall by a force 12N. The coefficient of friction between the wall and block is 0.6. The magnitude of the force exerted by the wall on the block is

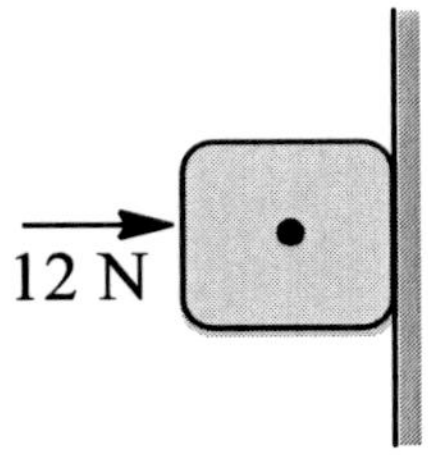

a) 12 N b) 5 N c) 7.2 N d) 13 N

349. The one –rupee coins are put on top of each other on a table. Each coin has a mass m. Which of the following statements is not true

a) The force on the 6th (counted from the bottom) due to all the coins on its top is equal to 4 mg (downwards)

b) The force on the 6th coin due to 7th coin is 4mg (downwards)

c) The reaction of the 6th coin on the 7th coin is 4 mg (upwards)

d) The total force on the 10th coin is 9 mg (downwards)

350. A stationary bomb explodes into three pieces. One piece of 2 kg mass moves with a velocity of 8 ms^{-1} at right angles to the other piece of mass 1 kg moving with a velocity of 12 ms^{-1}. If the mass of the third piece is 0.5 kg, then its velocity is

a) 10 ms^{-1} b) 20 ms^{-1} c) 30 ms^{-1} d) 40 ms^{-1}

351. Newton's second and third laws of motion lead to the conservation of

a) Linear momentum b) Angular momentum c) Potential energy d) Kinetic energy

352. A chain lies on a rough horizontal table. It starts sliding when one-fourth of its length hangs over the edge of the table. The coefficient of static friction between the chain and the surface of the table is

a) $\frac{1}{2}$ b) $\frac{1}{3}$ c) $\frac{1}{4}$ d) $\frac{1}{5}$

353. The minimum force required to start pushing a body up a rough (frictional coefficient μ) inclined plane is F_1 while the minimum force needed to prevent it from sliding down is F_2. If the inclined plane makes an angle θ from the horizontal such that $\tan\theta = 2\mu$, then the ratio $\frac{F_1}{F_2}$ is

a) 4 b) 1 c) 2 d) 3

354. A body of mass 2 kg is kept by pressing to a vertical wall by a force of 100 N. The coefficient of friction between wall and body is 0.3. Then the frictional force is equal to

a) 6 N b) 20 N c) 600 N d) 700 N

355. A circular disc with a groove along its diameter is placed horizontally. A block of mass 1kg is placed as shown. The co-efficient of friction between the block and all surfaces of groove in contact is $\mu = 2/5$. The disc has an acceleration of 25 m/s^2. Find the acceleration of the block with respect to disc

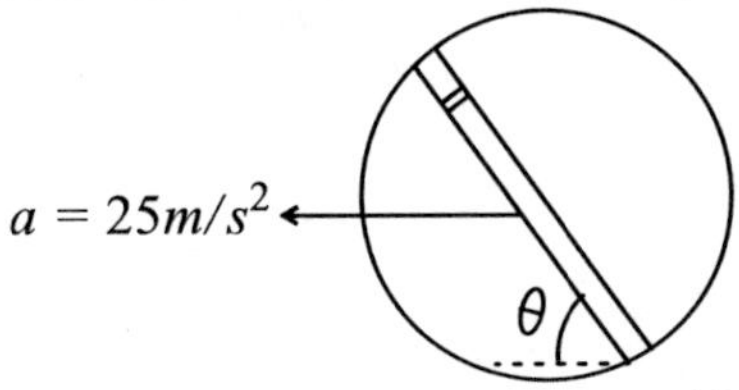

$\cos\theta = 4/5$ $\sin\theta = 3/5$

a) 10 m/s^2 b) 5 m/s^2 c) 20 m/s^2 d) 1 m/s^2

356. A solid disc of mass M is just held in air horizontal by throwing 40 stones per sec vertically upwards to strike the disc each with a velocity 6 ms^{-1}. If the mass of each stone is 0.05 kg. What is the mass of the disc ($g = 10$ ms^{-2})

a) 1.2 kg b) 0.5 kg c) 20 kg d) 3 kg

357. A block of mass $\sqrt{3}$ kg rests on a horizontal frictionless XY plane. What would be the acceleration of the block if it is subjected to two forces as shown in figure?

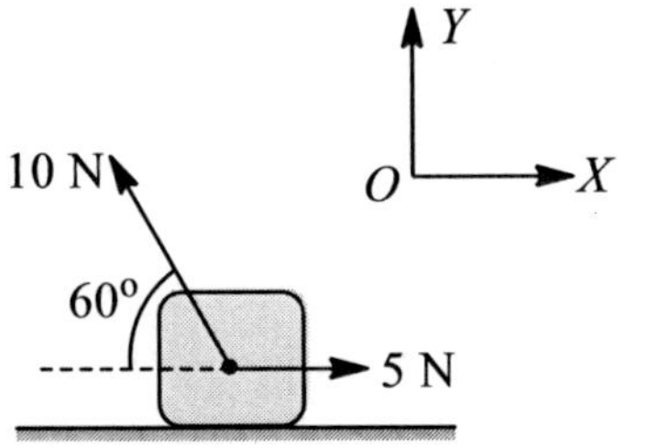

a) 2.5 ms^{-2} b) 5 ms^{-2} along Y-axis
c) 10 ms^{-2} along X-axis d) 15 ms^{-2} along Y-axis

358. Two forces of magnitude F have a resultant of the same magnitude F. The angle between the two forces is

a) 45° b) 120° c) 150° d) 60°

359. A block A of mass 7 kg is placed on a frictionless table. A thread tied to it passes over a frictionless pulley and carries a body B of mass 3 kg at the other end. The acceleration of the system is (given $g = 10$ ms^{-2})

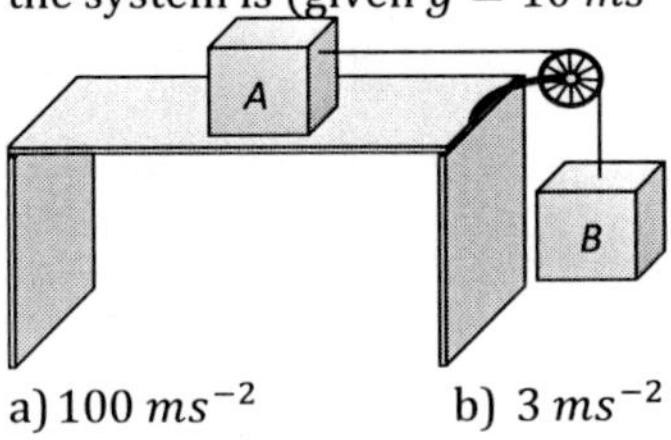

a) 100 ms^{-2} b) 3 ms^{-2} c) 10 ms^{-2} d) 30 ms^{-2}

360. Two blocks of masses 1 kg and 2 kg are connected by a metal wire going over a smooth pulley as shown in figure. The breaking stress of the metal is $2 \times 10^9 Nm^{-2}$. What should be the minimum radius of the wire used if it is not to break? Take $g = 10\ ms^{-2}$

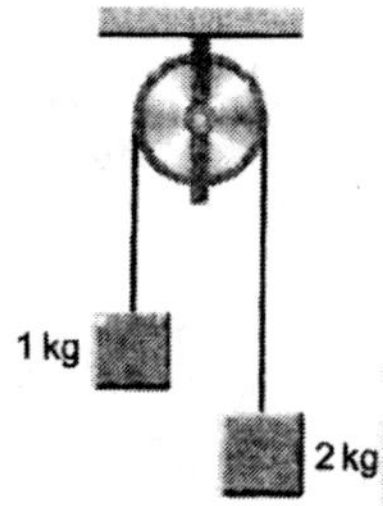

a) 4.6×10^{-5}m b) 4.6×10^{-6} m c) 2.5×10^{-6}m d) 2.5×10^{-5}m

361. When forces F_1, F_2, F_3 are acting on a particle of mass m such that F_2 and F_3 are naturally perpendicular, then the particle remains stationary. If the force F_1 is now removed then the acceleration of the particle is

a) F_1/m b) F_2F_3/mF_1 c) $(F_2 - F_3)/m$ d) F_2/m

362. The minimum force required to move a body up an inclined plane is three times the minimum force required to prevent it from sliding down the plane. If the coefficient of friction between the body and the inclined plane is $\frac{1}{2\sqrt{3}}$, the angle of the inclined plane is

a) 60° b) 45° c) 30° d) 15°

363. Two block of masses 7 kg and 5 kg are placed in contact with each other on a smooth surface. If a force of 6 N is applied on the heavier mass, the force on the lighter mass is

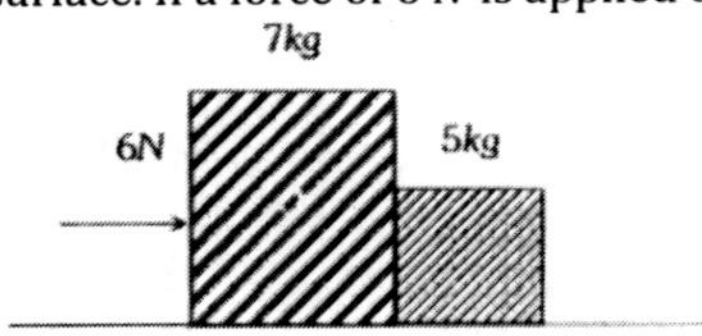

a) 3.5 N b) 2.5 N c) 7 N d) 5 N

364. A particle moves in a circular path with decreasing speed. Choose the correct statement

a) Angular momentum remains constant

b) Acceleration $\vec{a}$ is towards the centre

c) Particle moves in a spiral path with decreasing radius

d) The direction of angular momentum remains constant

365. Figures I, II, III and IV depict variation of force with time

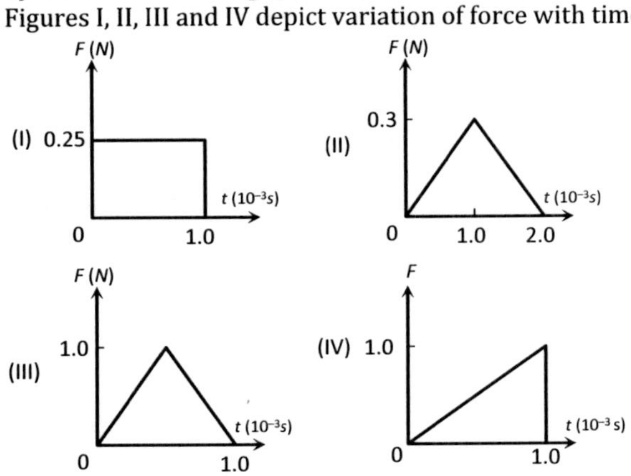

The impulse is highest in the case of situations depicted. Figure

a) I and II b) III and I c) III and IV d) IV only

366. A body of weight 2 $kg$is suspended as shown in the figure. The tension T_1 in the horizontal string (in $kg\ wt$) is

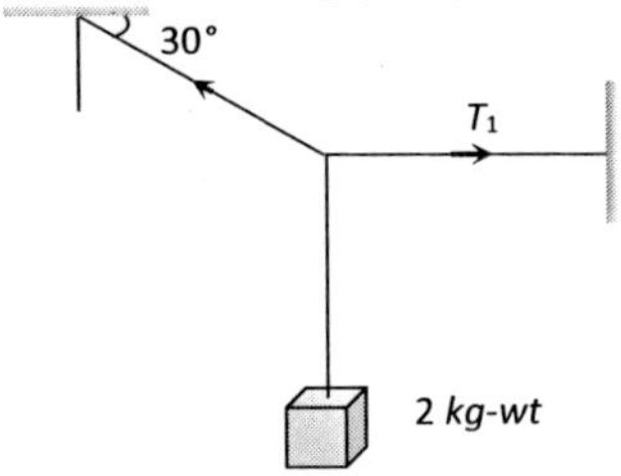

a) $2/\sqrt{3}$ b) $\sqrt{3}/2$ c) $2\sqrt{3}$ d) 2

367. A block is dragged on a smooth horizontal plane with the help of a light rope which moves with a velocity v as shown in figure. The horizontal velocity of the block is

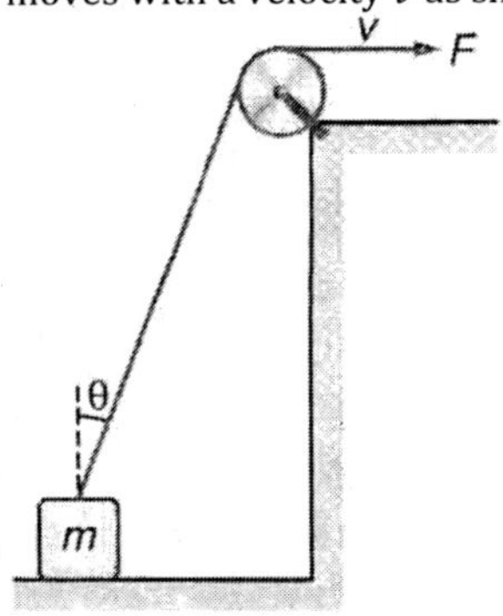

a) v b) $v \sin\theta$ c) $\frac{v}{\sin\theta}$ d) $\frac{v}{\cos\theta}$

368. A block of mass 4 kg is suspended through two light spring balances A and B. Then A and B, Then A and B will read respectively

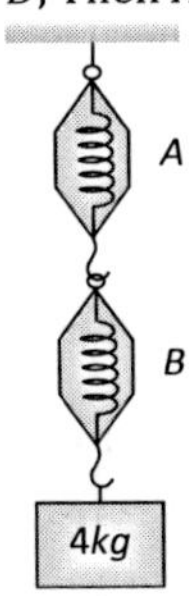

a) 4 kg and zero kg b) Zero kg and 4 kg c) 4 kg and 4 kg d) 2 kg and 2 kg

369. A cold soft drink is kept on the balance. When the cap is open, then the weight

a) Increases b) Decreases

c) First increases then decreases d) Remains same

370. An ice tube is kept on an inclined plane of angle 30°. Coefficient of kinetic friction between block and inclined plane is $\frac{1}{\sqrt{3}}$. What is the acceleration of block?

a) Zero b) 2 ms^{-2} c) 1.5 ms^{-2} d) 5 ms^{-2}

371. A man weighs 80 kg. He stands on a weighing scale in a lift which is moving upwards with a uniform acceleration of $5 m/s^2$. What would be the reading on the scale. ($g = 10\ m/s^2$)

a) 400 N b) 800 N c) 1200 N d) Zero

372. A 20 kg block is initially at rest on a rough horizontal surface. A horizontal force of 75 N is required set the block in motion. After it is in motion, a horizontal force of 60 N is required to keep the block moving with constant speed.The coefficient of static friction is

a) 0.38 b) 0.44 c) 0.52 d) 0.60

373. A block of mass m lying on a rough horizontal plane is acted upon by a horizontal force P and another force Q inclined at an angle θ to the vertical. The block will remain in equilibrium if the coefficient of friction between it and the surface is

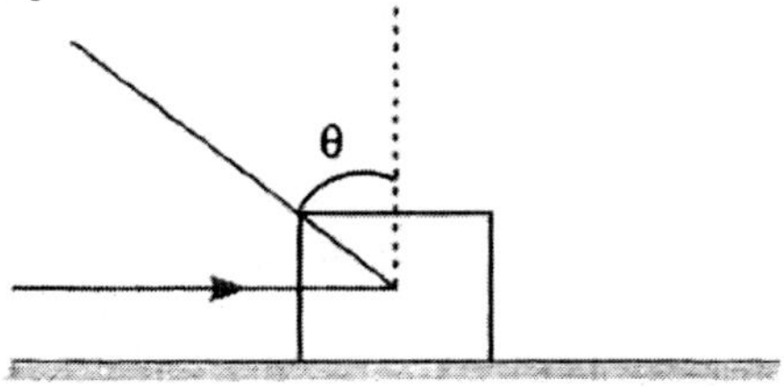

a) $\frac{P + Q\sin\theta}{mg + Q\cos\theta}$ b) $\frac{P\cos\theta + Q}{mg - Q\sin\theta}$ c) $\frac{P + Q\cos\theta}{mg + Q\sin\theta}$ d) $\frac{P\sin\theta - Q}{mg - Q\cos\theta}$

374. The mass of ship is 2×10^7 kg. On applying a force of 25×10^5 N , it is displaced through 25 m. After the displacement, the velocity acquired by the ship will be

a) 12.5 m/s b) 5 m/s c) 3.7 m/s d) 2.5 m/s

375. When a force F acts on a body of mass m, the acceleration produced in the body is a. If three equal forces $F_1 = F_2 = F_3 = F$ act on the same body as shown in figure, the acceleration produced is

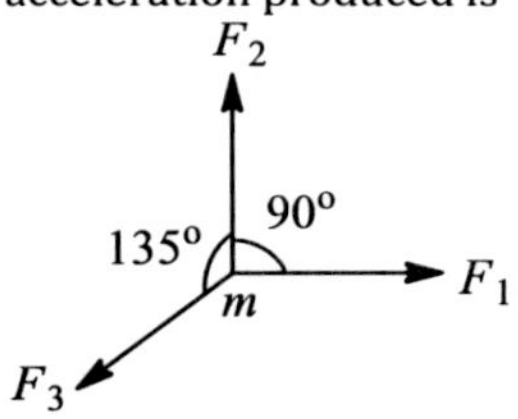

a) $(\sqrt{2} - 1)a$ b) $(\sqrt{2} + 1)a$ c) $\sqrt{2}a$ d) a

376. A block of 1 kg is stopped against a wall by applying a force F perpendicular to the wall. If $\mu = 0.2$ then minimum value of F will be

a) 980 N b) 49 N c) 98 N d) 490 N

377. A train is moving with velocity 20 m/sec. On this dust is falling at the rate of 50 $kg/minute$. The extra force required to move this train with constant velocity will be

a) 16.66 N b) 1000 N c) 166.6 N d) 1200 N

378. Two masses $m_1 = 5$ kg and $m_2 = 4.8$ kg tried to a string are hanging over a light frictionless pulley. What is the acceleration of the masses when lift is free to move? ($g = 9.8$ ms^{-2})

a) 0.2 ms^{-2} b) 9.8 ms^{-2} c) 5 ms^{-2} d) 4.8 ms^{-2}

379. In the figure, the ball A is released from rest when the spring is at its natural length. For the block B of mass M to leave contact with the ground at same stage, the minimum mass of A must be

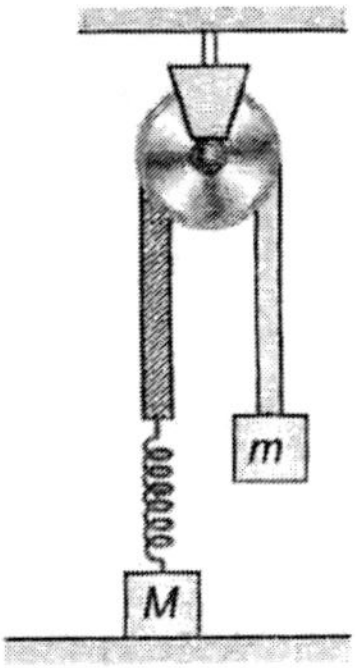

a) $2M$
b) M
c) $\frac{M}{2}$
d) A function of M and the force constant of the spring

380. A 1000 kg lift is supported by a cable that can support 2000 kg. The shortest distance in which the lift can be stopped when it is descending with a speed of 2.5 ms^{-1} is [Take g = 10 ms^{-2}]
a) 1 m b) 2 m c) $\frac{5}{32}$m d) $\frac{5}{16}$m

381. A box is placed on an inclined plane and has to be pushed down. The angle of inclination is
a) Equal to angle of friction b) More than angle of friction
c) Equal to angle of repose d) Less than angle of repose

382. A rocket with a lift-off mass 10^5 kg is blasted upward with an initial acceleration of 5 ms^{-2}. If g = 10 ms^{-2}, then the initial thrust of the blast is
a) 1.5×10^2N b) 1.5×10^3N c) 1.5×10^5N d) 1.5×10^6N

383. The maximum speed that can be achieved without skidding by a car on a circular unbanked road of radius R and coefficient of static friction μ, is
a) μRg b) $Rg\sqrt{\mu}$ c) $\mu\sqrt{Rg}$ d) $\sqrt{\mu Rg}$

384. A force of 5 N acts on a body of weight 9.8 N. What is the acceleration produced in m/sec^2
a) 49.00 b) 5.00 c) 1.46 d) 0.51

385. A player caught a cricket ball of mass 150 gmmoving at the rate of 20 m/sec. If the catching process be completed in 0.1 sec the force of the blow exerted by the ball on the hands of player is
a) 0.3 N b) 30 N c) 300 N d) 3000 N

386. A disc of mass 10 g is kept floating horizontally in air by firing bullets, each of mass 5g, with the same velocity at the same rate of 10 bullets per second. The bullets rebound with the same speed in positive direction. The velocity of each bullet at the time of impact is
a) 196 cms^{-1} b) 98 cms^{-1} c) 49 cms^{-1} d) 392 cms^{-1}

387. A bullet of mass 10 g moving with 300 m/s hits a block of ice of mass 5 kg and drops dead. The velocity of ice is
a) 50 cm/s b) 60 cm/s c) 40 cm/s d) 30 cm/s

388. A particle is moving with a constant speed along a straight line path. A force is not required to
a) Increase its speed b) Decrease the momentum
c) Change in direction d) Keep it moving with uniform velocity

389. The coefficient of static friction, μ_s, between block A of mass 2 kg and the table as shown in the figure is 0.2. What would be the maximum mass value of block B so that the two blocks do not move? The string and the pulley are assumed to be smooth and massless $(g = 10\ m/s^2)$

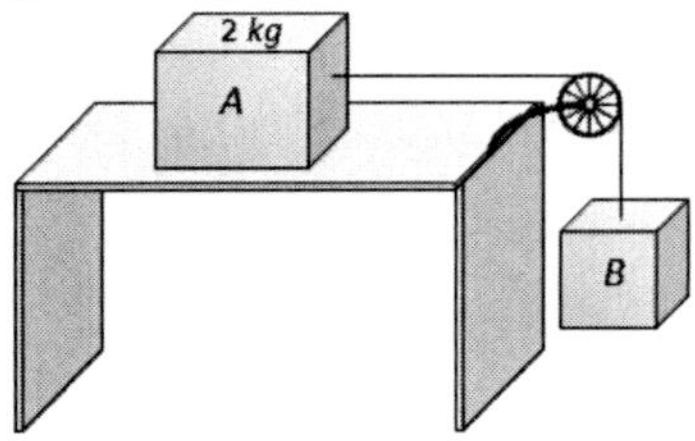

a) 2.0 kg b) 4.0 kg c) 0.2 kg d) 0.4 kg

390. A coin is placed on a horizontal platform which undergoes vertical simple harmonic motion of angular frequency ω. The amplitude of oscillation is gradually increased. The coin will leave contact with the platform for the first time

a) At the mean position of the platform b) For an amplitude of g/ω^2
c) For an amplitude of g^2/ω^2 d) At the highest position of the platform

391. The coefficient of static friction μ_s between block A of mass 2 kg and the table as shown in the figure is 0.2. What would be the maximum mass value of block B so that the two blocks do not move? The string and the pully are assumed to be smooth and massless $(\text{g} = 10\ \text{ms}^{-2})$

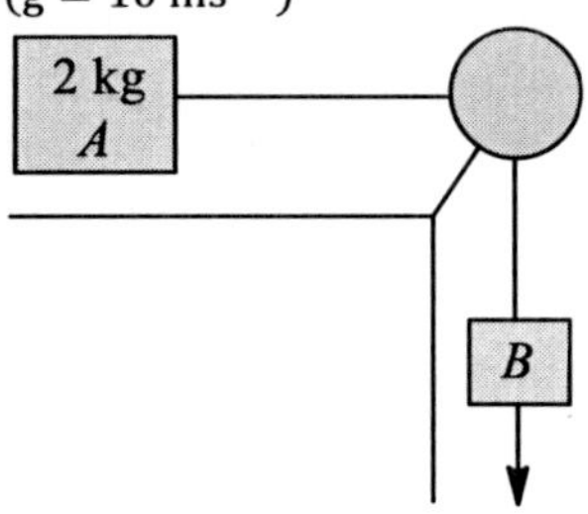

a) 2.0 kg b) 4.0 kg c) 0.2 kg d) 0.4 kg

392. The adjacent figure is the part of a horizontally stretched net. Section AB is stretched with a force of 10 N. The tension in the section BC and BF are

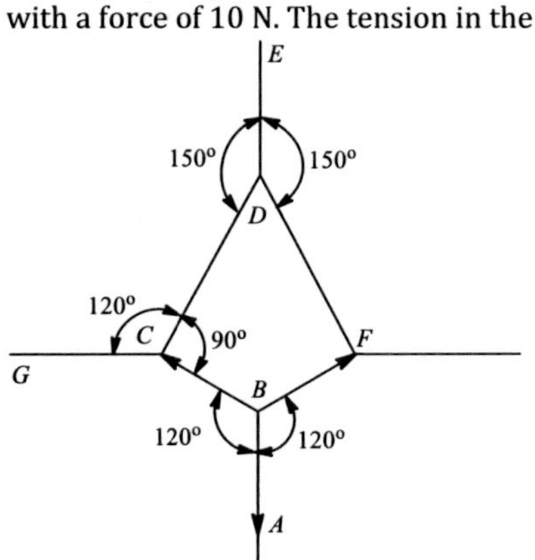

a) 10 N, 11N b) 10 N, 6 N
c) 10 N, 10 N d) Can't be calculated due to insufficient data

393. A stick of 1 m is moving with velocity of $2.7 \times 10^8\ ms^{-1}$. What is the apparent length of the stick ($c = 3 \times 10^8\ ms^{-1}$)

a) 10 m b) 0.22 m c) 0.44 m d) 2.4 m

394. A block of mass m lying on a rough horizontal plane is acted upon by a horizontal force P and another force Q inclined at an angle θ to the vertical. The block will remain in equilibrium, if the coefficient of friction between it and the surface is

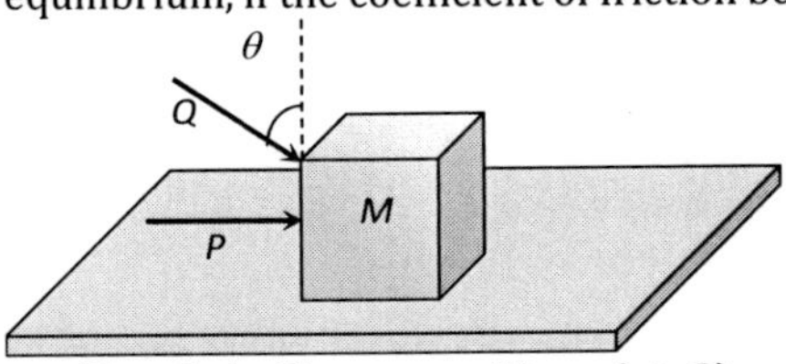

a) $\frac{(P + Q \sin \theta)}{(mg + Q \cos \theta)}$ b) $\frac{(P \cos \theta + Q)}{(mg - Q \sin \theta)}$ c) $\frac{(P + Q \cos \theta)}{(mg + Q \sin \theta)}$ d) $\frac{(P \sin \theta - Q)}{(mg - Q \cos \theta)}$

395. The time in which a force of 2 N produces a change in momentum of $0.4\ kg - ms^{-1}$ in the body is

a) 0.2 s b) 0.02 s c) 0.5 s d) 0.05 s

396. A ball of mass 0.2 kg is thrown normally against a wall with a speed of 15 ms^{-1}. The impulse of the force by the ball on the wall is

a) 0.35 Ns b) 0.7 Ns c) 7 Ns d) 70 Ns

397. The linear momentum of a particle varies with time t as $p = a + bt + ct^2$ Which of the following statements is correct?

a) Force varies with time in a quadratic manner

b) Force is time-dependent

c) The velocity of the particle is proportional to time

d) The displacement of the particle is proportional to t

398. As shown in the figure, two equalmasses each of 2 kg are suspended from a spring balance. The reading of the spring balance will be

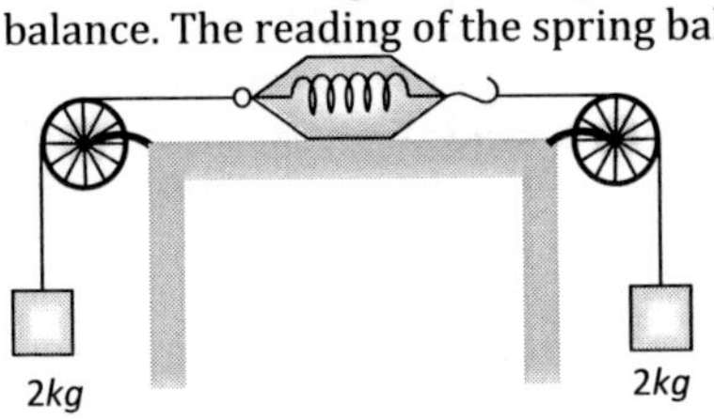

a) Zero b) 2 kg

c) 4 kg d) Between zero and 2 kg

399. Physical independence of force is consequence of

a) Third law of motion b) Second law of motion c) First law of motion d) All of the above

400. A conveyor belt is moving at a constant speed of 2 m/s. A box is gently dropped on it. The coefficient of friction between them is $\mu = 0.5$. The distance that the box will move relative to belt before coming to rest on it, taking $g = 10\ ms^{-2}$, is

a) Zero b) 0.4 m c) 1.2 m d) 0.6 m

401. Steady rain, giving 5 mm an hour, turns suddenly into a downpour giving 20 mm an hour and the speed of the rain drops falling vertically on to a flat roof simultaneously doubles. The pressure exerted by the falling rain on the roof is raised by a factor of

a) 4 b) 8 c) 16 d) 32

402. A body of mass 4 kg is accelerated upon by a constant force, travel a distance of 5 m in the first second and a distance of 2 m in the third second. The force acting on the body is

a) 2 N b) 4 N c) 6 N d) 8 N

403. Two particle of equal mass are connected to a rope AB of negligible mass, such that one is at end A and the other dividing the length of the rope in the ratio 1:2 from$B$. The rope is rotated about end B in a horizontal plane. Ratio of the tension in the smaller part to the other is (ignore effect of gravity)

a) 4:3 b) 1:4 c) 1:2 d) 1:3

404. An open carriage in a goods train is moving with a uniform velocity of $10ms^{-1}$. If the rain adds water with zero velocity at the rate of $5\ kgs^{-1}$, then the additional force applied by the engine to maintain the same velocity of the train is

a) 0.5 N b) 2.0 N c) 50 N d) 25 N

405. A body of mass m is suspended by two strings making angle α and β with the horizontal as shown in figure. Tensions in the two strings are

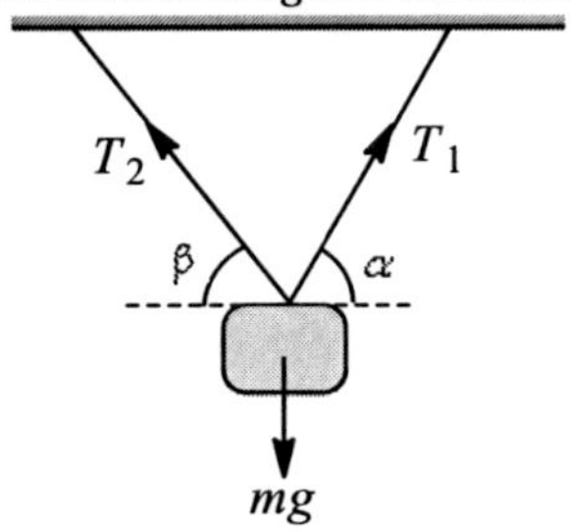

a) $T_1 = \frac{mg\cos\beta}{\sin(\alpha+\beta)} = T_2$ b) $T_1 = \frac{mg\sin\beta}{\sin(\alpha+\beta)} = T_2$

c) $T_1 = \frac{mg\cos\beta}{\sin(\alpha+\beta)}; T_2 = \frac{mg\cos\alpha}{\sin(\alpha+\beta)}$ d) None of the above

406. A person of mass 60 kg is inside a lift of mass 940 kg and presses the button one control panel. The lift starts moving upwards with an acceleration $1.0\ m/s^2$. If $g = 10\ ms^{-2}$, the tension in the supporting cable is

a) 1200 N b) 8600 N c) 9680 N d) 11000 N

407. A stationary body of mass 3 kg explodes into three equal pieces. Two of the pieces fly off in two mutually perpendicular directions, one with a velocity of $3\hat{i}\ ms^{-1}$ and the other with a velocity of $4\hat{j}\ ms^{-1}$. If the explosion occurs in 10^{-4} s, the average force acting on the third piece in newton is

a) $(3\hat{i}+4\hat{j})\times10^{-4}$ b) $(3\hat{i}-4\hat{j})\times10^{-4}$ c) $(3\hat{i}+4\hat{j})\times10^{4}$ d) $-(3\hat{i}+4\hat{j})\times10^{4}$

408. Three solids of masses m_1, m_2 and m_3 are connected with weightless string in succession and are placed on a frictionless table. If the mass m_3 is dragged with a force T, the tension in the string between m_2 and m_3 is

a) $\frac{m_2}{m_1+m_2+m_3}T$ b) $\frac{m_3}{m_1+m_2+m_3}T$ c) $\frac{m_1+m_2}{m_1+m_2+m_3}T$ d) $\frac{m_2+m_3}{m_1+m_2+m_3}T$

409. If the surface is smooth, the acceleration of the block m_2 will be

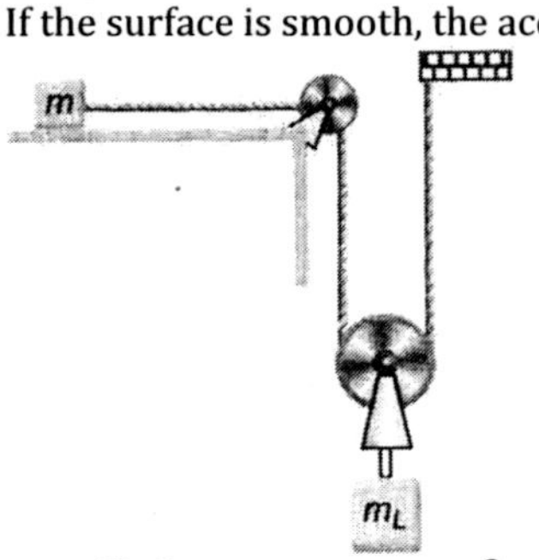

a) $\frac{m_2 g}{4m_1+m_2}$ b) $\frac{2m_2 g}{4m_1+m_2}$ c) $\frac{2m_1 g}{m_1+4m_2}$ d) $\frac{2m_1 g}{m_1+m_2}$

410. A block of mass 2 kg is placed on the floor. The coefficient of static friction is 0.4. If a force of 2.8 N is applied on the block parallel to floor, the force of friction between the block and floor (Taking $g = 10\ ms^{-2}$) is

a) 2.8 N b) 8 N c) 2 N d) Zero

411. A point mass m is moving along inclined plane with acceleration a with respect to smooth triangular block. The triangular block is moving horizontally with acceleration a_0. The value of a is

a) $g \sin\theta + a_0 \cos\theta$ b) $g \sin\theta - a_0 \cos\theta$ c) $g \cos\theta - a_0 \sin\theta$ d) None of these

412. If a street light of mass M is suspended from the end of a uniform rod of length L in different possible patterns as shown in figure, then

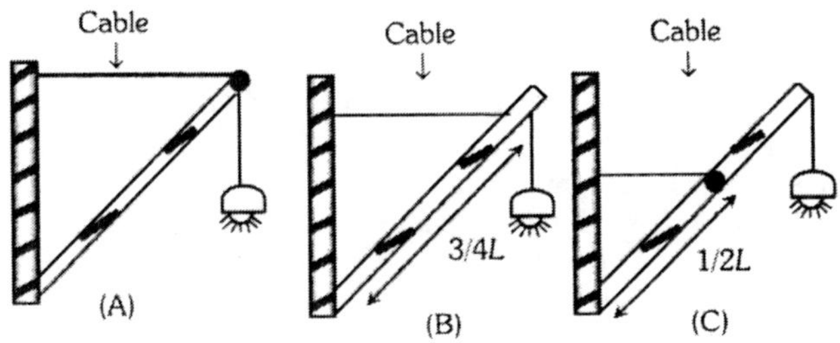

a) Pattern A is sturdier b) Pattern B is sturdier

c) Pattern C is sturdier d) All will have same sturdiness

413. A body of mass $5kg$ rests on a rough horizontal surface of coefficient of friction 0.2. The body is pulled through a distance of $10m$ by a horizontal force of 25 N. The kinetic energy acquired by it is ($g = 10\ ms^2$)

a) 330 J b) 150 J c) 100 J d) 50 J

414. A ball of mass 0.2 kg rests on a vertical post of height 5 m. A bullet of mass 0.01 kg, travelling with a velocity v m/s in a horizontal direction, hits the centre of the ball. After the collision, the ball and bullet travel independently. The ball hits the ground at a distance of 20 m and the bullet at ball hits the ground at a distance of 100 m from the foot of the post. The initial velocity v of the bullet is

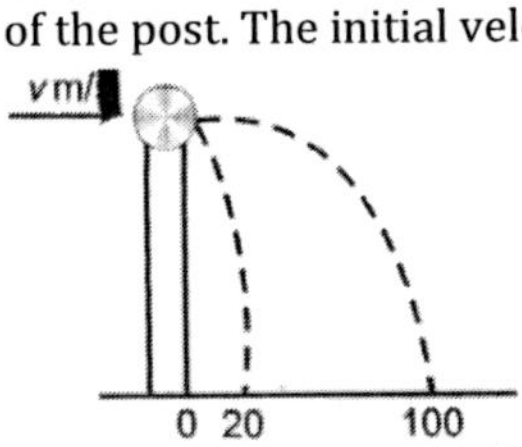

a) 250 m/s b) $250\sqrt{2}$m/s c) 400 m/s d) 500 m/s

415. A block of mass 5 kg is moving horizontally at a speed of 1.5 m/s. A perpendicular force of $5N$ acts on it for 4 sec. What will be the distance of the block from the point where the force started acting

a) 10 m b) 8 m c) 6 m d) 2 m

416. A spring balance and a physical balance are kept in a lift. In these balances equal masses are placed. If now the lift starts moving upwards with constant acceleration, then

a) The reading of spring balance will increase and the equilibrium position of the physical balance will disturb

b) The reading of spring balance will remain unchanged and physical balance will remain in equilibrium

c) The reading of spring balance will decrease and physical balance will remain in equilibrium

d) The reading of spring balance will increase and the physical balance will remain in equilibrium

417. A wooden box of mass 8 kg slides down an inclined plane of inclination 30° to the horizontal with a constant acceleration of 0.4 ms^{-2}. What is the force of friction between the box and inclined plane? [$g = 10\ ms^{-2}$]

a) 36.8 N b) 76.8 N c) 65.6 N d) 97.8 N

418. A 60 kg weight is dragged on a horizontal surface by a rope upto 2 $metres$. If coefficient of friction is $\mu = 0.5$, the angle of rope with the surface is 60° and $g = 9.8 m/sec^2$, then work done is

a) 294 $joules$ b) 315 $joules$ c) 588 $joules$ d) 197 $joules$

419. A block of base 10 cm × 10 cm and height 15 cm is kept on an inclined plane. The coefficient of friction between them is $\sqrt{3}$. The inclination θ of this inclined plane from the horizontal plane is gradually increased from 0°. Then

a) At $\theta = 30°$, the block will start sliding down the plane

b) The block will remain at rest on the plane up to certain θ and then it will topple

c) At $\theta = 60°$, the block will start sliding down the plane and continue to do so at higher angles

d) At $\theta = 60°$, the block will start sliding down the plane and on further increasing θ, it will topple at certain θ

420. A monkey climbs up and another monkey climbs down a rope hanging from a tree with same uniform acceleration separately. If the respective masses of monkeys are in the ratio 2 : 3, the common acceleration must be

a) $g/5$ b) $6g$ c) $g/2$ d) g

421. The resultant of two forces acting at an angle of 120° is 10 kg-wt and is perpendicular to one of the forces. That force is

a) $\frac{10}{\sqrt{3}}$ kg-wt b) 10 kg-wt c) $20\sqrt{3}$kg-wt d) $10\sqrt{3}$kg-wt

422. A shell of rest at the origin explodes into three fragments of masses 1 kg, 2 kg and m kg. The 1 kg and 2 kg pieces fly off with speed of 5 ms^{-1} along x-axis and 6 ms^{-1} along y-axis respectively. If the m kg piece flies off with speed of 6.5 ms^{-1}, the total mass of the shell must be

a) 4 kg b) 5 kg c) 3.5 kg d) 4.5 kg

423. Refer to the system shown in figure. The acceleration of the masses is

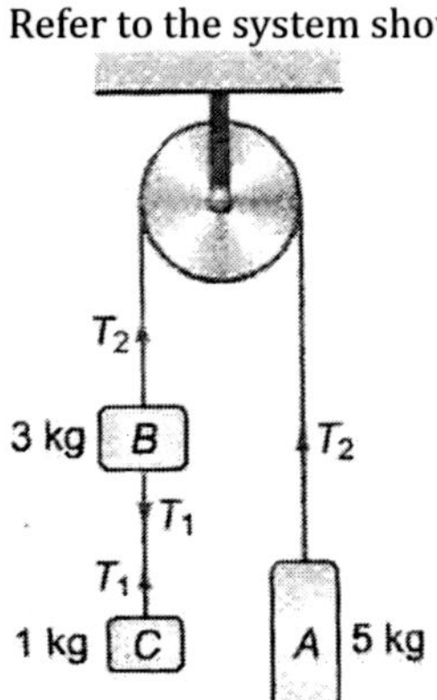

a) $\frac{g}{3}$ b) $\frac{g}{6}$ c) $\frac{g}{9}$ d) $\frac{g}{12}$

424. Mass of 3 kg descending vertically downward supports a mass of 2 kg by means the end of 5 s, the string breaks. How much higher the 2 kg mass will go further?

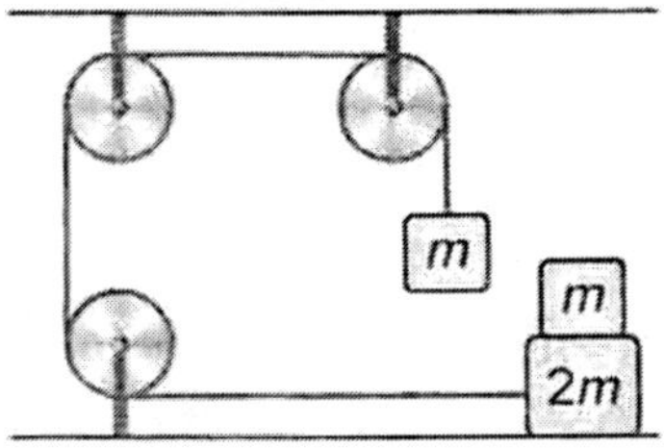

a) 4.9 m b) 9.8 m c) 19.6 m d) 2.45 m

425. A lift is going up. The total mass of the lift and the passenger is 1500 kg the variation in the speed of the lift is as given in the graph. The tension in the rope pulling the lift at $t = 11^{\text{th}} sec$ will be

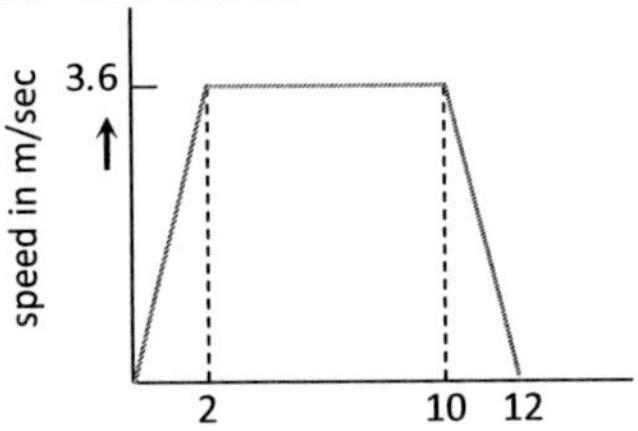

a) 17400 N b) 14700 N c) 12000 N d) Zero

426. A wooden block of mass 1 kg is attached to the hook of a spring balance. The spring balance is then raised with an acceleration of 9.8 ms^{-2}. The apparent weight of the body is

a) 1 kg-wt b) 2 kg-wt c) 3 kg-wt d) 4 kg-wt

427. The engine of a car produces an acceleration of $6ms^{-2}$ in the car. If this car pulls another car of the same mass, then the acceleration would be

a) $6ms^{-2}$ b) $12ms^{-2}$ c) $3ms^{-2}$ d) $1.5ms^{-2}$

428. A wooden wedge of mass M and inclination angle α rests on a smooth floor. A block of mass m is kept on wedge. A force $\vec{\mathbf{P}}$ is applied on the wedge as shown in figure, such that a block remains stationary with respect to wedge. The magnitude of force $\vec{\mathbf{P}}$ is

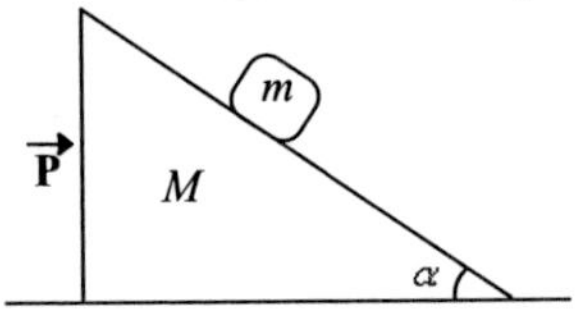

a) $(M + m)g \tan \alpha$ b) $g \tan \alpha$ c) $mg \cos \alpha$ d) $(M + m)g \operatorname{cosec} \alpha$

429. A bird is sitting in a large closed cage which is placed on a spring balance. It records a weight of 25 N. The bird (mass $m = 0.5\ kg$) files upward in the cage with an acceleration of 2 m/s^2. The spring balance will now record a weight of

a) 24 N b) 25 N c) 26 N d) 27 N

430. The ratio of the weight of a man in a stationary lift and when it is moving downward with uniform acceleration 'a' is 3: 2,. The value of 'a' is (g- Acceleration due to gravity of the earth)

a) $\frac{3}{2}g$ b) $\frac{g}{3}$ c) $\frac{2}{3}g$ d) g

431. On the horizontal surface of a truck ($\mu = 0.6$), a block of mass 1 kg is placed. If the truck is accelerating at the rate of $5m/sec^2$ then frictional force on the block will be

a) 5 N b) 6 N c) 5.88 N d) 8 N

432. A body of mass M at rest explodes into three pieces, two of which of mass $M/4$ each are thrown off in perpendicular directions with velocities of 3 m/s and 4 m/srespectively. The third piece will be thrown off with a velocity of

a) 1.5 m/s b) 2.0 m/s c) 2.5 m/s d) 3.0 m/s

433. A light string passing over a smooth light pulley connects two blocks of masses m_1 and m_2 (vertically). If the acceleration of the system is (g/8), then the ratio of masses is

a) 8 : 1 b) 9 : 7 c) 4 : 3 d) 5 : 3

434. A gun fires N bullets per second, each of mass m with velocity v. The force exerted by the bullets on the gun is

a) vNm b) $\frac{mv}{N}$ c) mvN^2 d) $\frac{mv^2}{N}$

435. Two masses of 5 kg and 10 kg are connected to a pulley as shown. What will be the acceleration of the system (g = acceleration due to gravity)

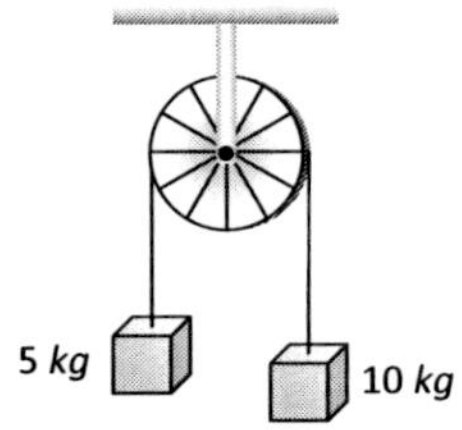

a) g b) $\frac{g}{2}$ c) $\frac{g}{3}$ d) $\frac{g}{4}$

436. A block B is pushed momentarily along a horizontal surface with an initial velocity V. If μ is the coefficient of sliding friction between B and the surface, block B will come to rest after a time

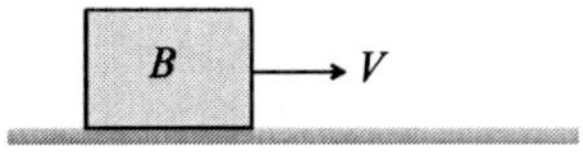

a) $V/(g\mu)$ b) $g\mu/V$ c) g/V d) V/g

437. A vehicle of 100 kg is moving with a velocity of 5 m/sec. To stop it in $\frac{1}{10}$ sec, the required force in opposite direction is

a) 5000 N b) 500 N c) 50 N d) 1000 N

438. The pulley and strings shown in the figure are smooth and of negligible mass. For the system to remain in equilibrium, the angle θ should be

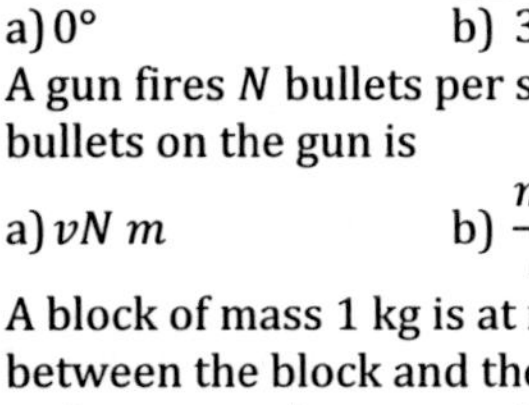

a) 0° b) 30° c) 45° d) 60°

439. A gun fires N bullets per second, each of mass $m$with velocity v. The force exerted by the bullets on the gun is

a) $vN\ m$ b) $\frac{mv}{N}$ c) mvN^2 d) $\frac{mv^2}{N}$

440. A block of mass 1 kg is at rest on a horizontal table. The coefficient of static friction between the block and the table is 0.5. If $g = 10ms^{-2}$, then the magnitude of the force acting upwards at an angle of 60° from the horizontal that will just start the block moving is

a) 5 N b) 5.36 N c) 74.6 N d) 10 N

441. A mass 1 kg is suspended by a thread. It is (i) lifted up with an acceleration 4.9 m/s^2 (ii) lowered with an acceleration 4.9 m/s^2
The ratio of the tension is
a) 3: 1 b) 1: 3 c) 1: 2 d) 2: 1
442. A satellite in force-free space sweeps stationary interplanetary dust at a rate $dM/dt = \alpha v$ where M is the mass, v is the velocity of the satellite and α is a constant. What is the deacceleration of the satellite
a) $-2\alpha v^2/M$ b) $-\alpha v^2/M$ c) $+\alpha v^2/M$ d) $-\alpha v^2$
443. A nucleus disintegrates into two nuclear parts which have their velocities in the ratio 2 : 1. The ratio of their nuclear sizes will be
a) $2^{1/3}: 1$ b) $1: 3^{1/2}$ c) $3^{1/2}: 1$ d) $1: 2^{1/3}$
444. A body moves along a circular path of radius 10m and the coefficient of friction is 0.5. What should be its angular speed in rad/s if it is not to slip from the surface ($g = 9.8\ m/s^2$)
a) 5 b) 10 c) 0.1 d) 0.7
445. A sphere is accelerated upwards by a cord whose braking strength is four times its weight. The maximum acceleration with which the sphere can move up without breaking the cord is
a) g b) $3g$ c) $2g$ d) $4g$
446. Two masses $m_1 = 1\ kg$ and $m_2 = 2kg$ are connected by a light inextensible string and suspended by means of a weightless pulley as shown in the figure. Assuming that both the masses start from rest, the distance travelled by the centre of mass in two seconds is (Take $g = 10ms^{-2}$)

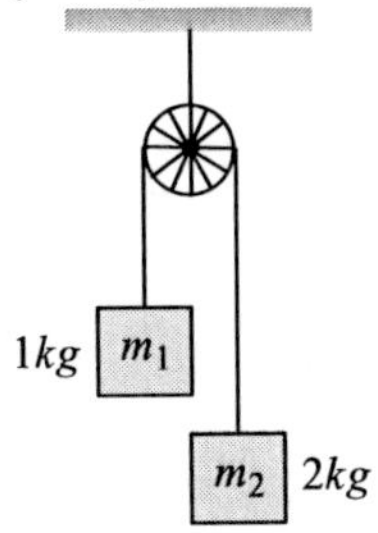

a) $\frac{20}{9}\ m$ b) $\frac{40}{9}\ m$ c) $\frac{2}{3}\ m$ d) $\frac{1}{3}\ m$
447. A shell of mass 10 kg is moving with a velocity of 10 ms^{-1} when it blasts and forms two parts of mass 9 kg and 1 kg respectively. If the 1st mass is stationary, the velocity of the 2nd is
a) 1 ms^{-1} b) 10 ms^{-1} c) 100 ms^{-1} d) 1000 ms^{-1}
448. The mass of a lift is 2000 kg. When the tension in the supporting cable is 28000 N, then its acceleration is
a) 30 ms^{-2} downwards b) 4 ms^{-2} upwards c) 4 ms^{-2} downwards d) 14 ms^{-2} upwards
449. When a bus suddenly takes a turn, the passengers are thrown outwards because of
a) Inertia of direction b) Acceleration of motion
c) Speed of motion d) Both (b) and (c)
450. A shell initially at rest explodes into two pieces of equal mass, the two pieces will
a) Move with different velocities in different directions
b) Move with the same velocity in opposite directions
c) Move with the same velocity in the same directions
d) Be at rest

451. Three forces start acting simultaneously on a particle moving with velocity **v**. These forces are represented in magnitude and direction by the three sides of a triangle ABC (as shown). The particle will now move with velocity

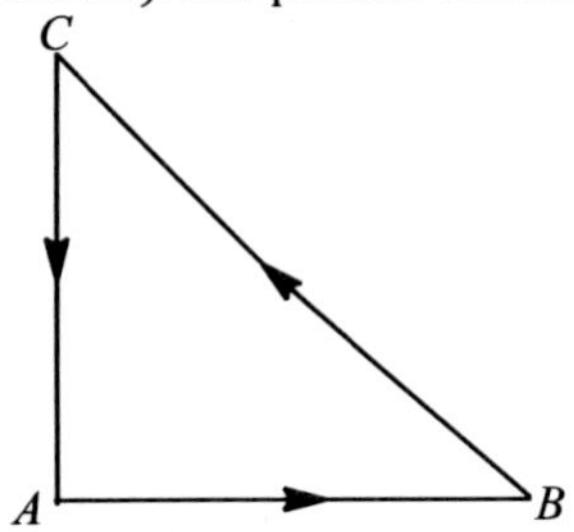

a) Less than **v**
b) Greater than **v**
c) v in the direction of largest force BC
d) **v** remaining unchanged

452. A man wants to slide down a rope. The breaking load for the rope $\frac{2}{3}$rd of the weight of the man. With what minimum acceleration should fireman slide down?

a) $\frac{g}{4}$
b) $\frac{g}{3}$
c) $\frac{2g}{3}$
d) $\frac{g}{6}$

453. A blumb bob is hung from the ceiling of a train compartment. The train moves on an inclined track of inclination 30° with horizontal. Acceleration of train up the plane is $a = 9/2$. The angle which the string supporting the bob makes with normal to the ceiling in equilibrium is

a) 30°
b) $\tan^{-1}\left(\frac{2}{\sqrt{3}}\right)$
c) $\tan^{-1}\left(\frac{\sqrt{3}}{2}\right)$
d) $\tan^{-1}(2)$

454. A lift is moving upwards with a uniform velocity v in which a block of mass m is lying. The frictional force offered by the block, when coefficient of the frictional is μ, will be

a) Zero
b) mg
c) μmg
d) $2\mu mg$

455. A shell of mass 10 kg is moving with a velocity of 10 ms^{-1} when it blasts and forms two parts of mass 9 kg and 1 kg respectively. If the 1st mass is stationary, the velocity of the 2nd is

a) 1 m/s
b) 10 m/s
c) 100 m/s
d) 1000 m/s

456. The rate of mass of the gas emitted from rear of a rocket is initially 0.1 kg/sec. If the speed of the gas relative to the rocket is 50 m/sec and mass of the rocket is 2 kg, then the acceleration of the rocket in m/sec^2 is

a) 5
b) 5.2
c) 2.5
d) 25

457. A machine gun fires 20 bullets per second into a target. Each bullet weighs 150 gms and has a speed of 800 m/sec. Find the force necessary to hold the gun in position

a) 800 N
b) 1000 N
c) 1200 N
d) 2400 N

458. Force required to move a mass of 1 kg at rest on a horizontal rough plane ($\mu = 0.1$ and $g = 9.8\ \text{ms}^{-2}$) is

a) 0.98 N
b) 0.49 N
c) 9.8 N
d) 4.9 N

459. An explosion blows a rock into three parts. Two parts go off at right angles to each other. These two are, 1 kg first part moving with a velocity of 12 ms^{-1} and 2 kg second part moving with a velocity of 8 ms^{-1}. If the third part files off with a velocity of 4 ms^{-1}, its mass would be

a) 5 kg
b) 7 kg
c) 17 kg
d) 3 kg

460. A block of base 10 cm × 10 cm and height 15 cm is kept on an inclined plane. The coefficient of friction between them is $\sqrt{3}$. The inclination θ of this inclined plane from the horizontal plane is gradually increased from 0°. Then,

a) at $0 = 30°$, the block will start sliding down the plane
b) The block will remain at the rest on the plane up to certain θ and then it will topple

c) At $\theta = 60°$, the block will start sliding down the plane and continue to do so at higher angles

d) At $\theta = 60°$, the block will start sliding down the plane and on further increasing θ, it will topple at certain θ

461. A block of mass m is resting on a smooth horizontal surface. One end of a uniform rope of mass $(m/3)$ is fixed to the block, which is pulled in the horizontal direction by applying a force F at the other end. The tension in the middle of the rope is

a) $\frac{8}{7} F$ b) $\frac{1}{7} F$ c) $\frac{1}{8} F$ d) $\frac{7}{8} F$

462. Two masses m_1 and m_2 $(m_1 > m_2)$ are connected by massless flexible and inextensible string passed over massless and frictionless pulley. The acceleration of centre of mass is

a) $\left(\frac{m_1 - m_2}{m_1 + m_2}\right)^2 g$ b) $\frac{m_1 - m_2}{m_1 + m_2} g$ c) $\frac{m_1 + m_2}{m_1 - m_2} g$ d) Zero

463. The resultant of two forces $3P$ and $2P$ is R. If the first force is doubled then the resultant is also doubled. The angle between the two forces is

a) 60° b) 120° c) 70° d) 180°

464. A rocket standing vertically on a launch pad has to start moving up with practically zero velocity. If the initial mass of the rocket is 5×10^5 kg, then the rate of burning of the fuel should be [Take g = 10 ms^{-2} and velocity of exhaust gases = 10kms^{-1}]

a) 10 kgs^{-1} b) 50 kgs^{-1} c) 500 kgs^{-1} d) 5000 kgs^{-1}

465. The momentum is most closely related to

a) Force b) Impulse c) Power d) K.E.

466. A block weighing W is held against a vertical wall by applying a horizontal force F. The minimum value of F needed to hold the block is

a) Less than W b) Equal to W c) Greater than W d) Data is insufficient

467. A block is lying static on the floor. The maximum value of static frictional force on the block is $10 N$. If a horizontal force of 8 N is applied to the block, what will be the frictional force on the block

a) 2 N b) 18 N c) 8 N d) 10 N

468. Force required to move a mass of 1 kg at rest on a horizontal rough plane ($\mu = 0.1$ and $g = 9.8\, m/s^2$) is

a) 0.98 N b) 0.49 N c) 9.8 N d) 4.9 N

469. A block of mass 200 kg is being pulled up by men on an inclined plane at angle of 45° as shown in the figure. The coefficient of static friction is 0.5. Each man can only apply a maximum force of 500 N. Calculate the number of men required for the block to just start moving up the plane.

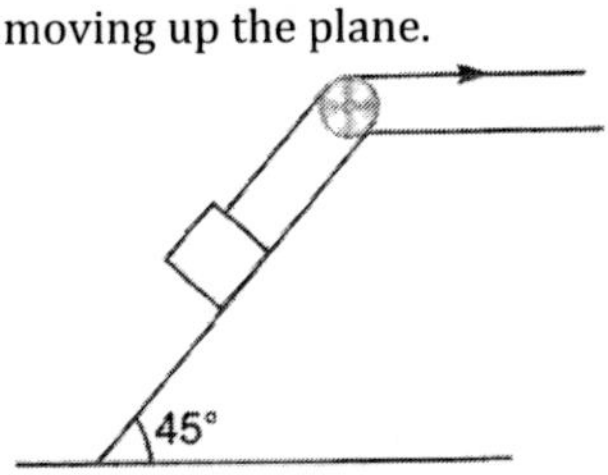

a) 10 b) 15 c) 5 d) 3

470. Two blocks of masses 3 m and 2 m are in contact on a smooth table. A force P is first applied horizontally on block of mass 3 m and then on mass 2 m. The contact forces between the two blocks in the two cases are in the ratio

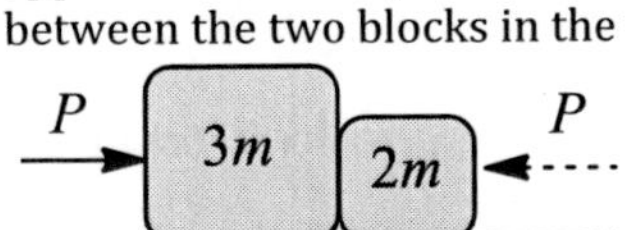

a) 1 : 2 b) 2 : 3 c) 3 : 2 d) 5 : 3

471. If a force of 250 N act on body, the momentum acquired is 125 kg-m/s. What is the period for which force acts on the body
a) 0.5 sec b) 0.2 sec c) 0.4 sec d) 0.25 sec

472. The average force necessary to stop a bullet of mass 20 g moving with a speed of 250 m/s, as it penetrates into the wood for a distance of 12 cm is
a) $2.2 \times 10^3\ N$ b) $3.2 \times 10^3\ N$ c) $4.2 \times 10^3\ N$ d) $5.2 \times 10^3\ N$

473. In the figure given below, the position –time graph of a particle of mass 0.1 kg is shown. The impulse at $t = 2\ sec$ is

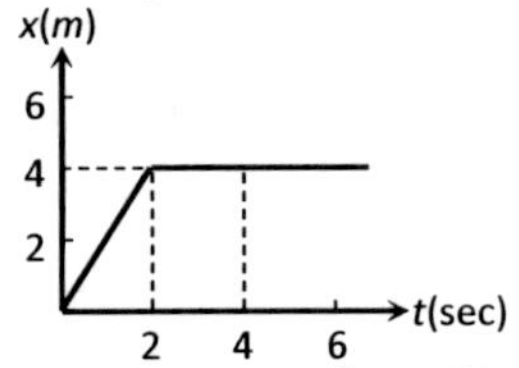

a) $0.2\ kg\ m\ \text{sec}^{-1}$ b) $-0.2\ kg\ m\ \text{sec}^{-1}$ c) $0.1\ kg\ m\ \text{sec}^{-1}$ d) $-0.4\ kg\ m\ \text{sec}^{-1}$

474. Two weights w_1 and w_2 are suspended from the ends of a light string over a smooth fixed pulley. If the pulley is pulled up with acceleration g, the tension in the string will be
a) $\frac{4w_1w_2}{w_1 + w_2}$ b) $\frac{2w_1w_2}{w_1 + w_2}$ c) $\frac{w_1 - w_2}{w_1 + w_2}$ d) $\frac{w_1w_2}{2(w_1 + w_2)}$

475. The engine of a car produces acceleration 4 m/s^2 in the car. If this car pulls another car of same mass, what will be the acceleration produced
a) 8 m/s^2 b) 2 m/s^2 c) 4 m/s^2 d) 1/2 m/s^2

476. Two small balls of same size and masses m_1 and $m_2 (m_1 > m_2)$ are tied by a thin weightless thread and dropped from a certain height. Training upward buoyancy force F into account, the tension T of the thread during the flight after the motion of the ball becomes uniform will be
a) $(m_1 - m_2)g$ b) $(m_1 - m_2)g/2$ c) $(m_1 + m_2)g$ d) $(m_1 + m_2)g/2$

477. A coin of mass 10 kg is placed over a book of length 50 cm. The coin is on the verge of sliding when one end of the book is lifted 10 cm up. The coefficient of static friction between the book and the coin is
a) 1.0 b) 0.4 c) 0.3 d) 0.2

478. If coefficient of friction between an inset and bowl is μ and radius of the bowl is r, the maximum height to which the insect can crawl in the bowl is
a) $r\left[1 - \frac{1}{\sqrt{1 + \mu^2}}\right]$ b) $\frac{r}{\sqrt{1 + \mu^2}}$ c) $r\sqrt{1 + \mu^2}$ d) $r[\sqrt{1 + \mu^2} - 1]$

479. Consider the following two statements :
I. Linear momentum of a system of particles is zero.
II. Kinetic energy of a system of particles is zero. Then
a) I does not imply II and II does not imply I b) I implies II but II does not imply I
c) I does not imply II but II implies I d) I implies II and II implies I

480. Two elastic blocks P and Q of equal masses m and connected by a massless spring rest on a smooth horizontal surface, as in figure. A third block R of the same mass M strikes the block P. After the collision, P and Q will

a) Always move in same direction
b) Sometimes move in same direction and sometime move in opposite directions
c) Always move in opposite directions
d) Be at rest with respect to each other

481. A man is standing on a balance and his weight is measured. If he takes a step in the left side, then weight

a) Will decrease
b) Will increase
c) Remains same
d) First decreases then increases

482. The coefficient of kinetic friction between a 20 kg box and the floor is 0.40. How much work does a pulling force do on the box in pulling it 8.0 m across the floor at constant speed? The pulling force is directed 37° above the horizontal

a) 343 J
b) 482 J
c) 14.4 J
d) None of these

483. Work done by a frictional force is

a) Negative
b) Positive
c) Zero
d) All of the above

484. Two blocks are connected by a string as shown in the diagram. The upper block is hung by another string. A force F applied on the upper string produces an acceleration of $2\ m/s^2$ in the upward direction in both the blocks. If T and T' be the tensions in the two parts of the string, then ($g = 9.8\ m/s^2$)

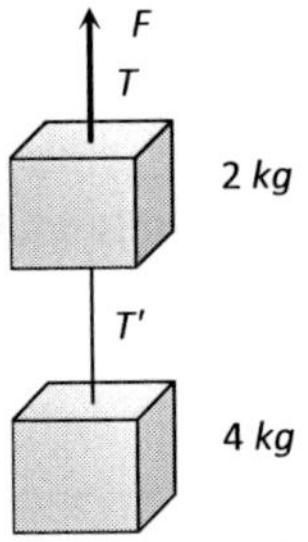

a) $T = 70.8\ N$ and $T' = 47.2\ N$
b) $T = 58.8\ N$ and $T' = 47.2\ N$
c) $T = 70.8\ N$ and $T' = 58.8\ N$
d) $T = 70.8\ N$ and $T' = 0$

485. A body shown in figure is accelerating downward with acceleration $2\ m/s^2$. The tension in the string is

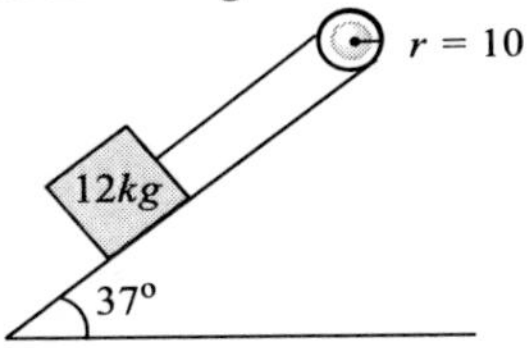

a) 48 N
b) 50 N
c) 30 N
d) 42 N

486. The pulleys and the strings shown in figure are smooth and of negligible mass. For the system to remain in equilibrium, the angle θ should be

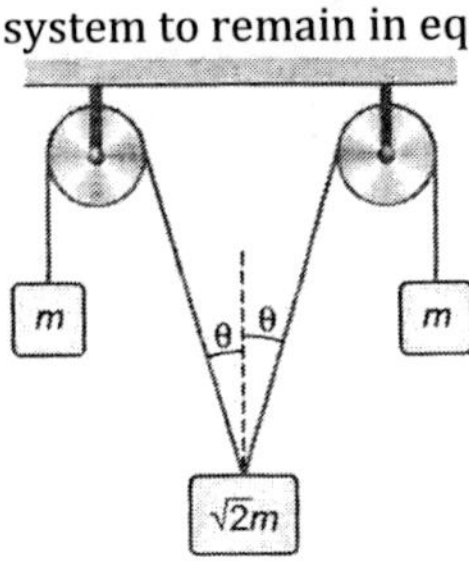

a) 0°
b) 30°
c) 45°
d) 60°

487. A car is moving along a straight horizontal road with a speed v_0. If the coefficient of friction between the tyres and the road is μ, the shortest distance in which the car can be stopped is

a) $\frac{v_0^2}{2\mu g}$
b) $\frac{v_0}{\mu g}$
c) $\left(\frac{v_0}{\mu g}\right)^2$
d) $\frac{v_0}{\mu}$

488. An elevator and its load have a total mass of 800 kg. The elevator is originally moving downwards at 10 ms^{-1}, it slows down to stop with constant acceleration in a distance of 25 m. Find the tension T in the supporting cable while the elevator is being brought to rest. (Take g = 10 ms^{-2})

a) 8000 N　　b) 1600 N　　c) 9600 N　　d) 6400 N

489. Refer to the system shown in figure. The ratio of tension T_1 and T_2 is

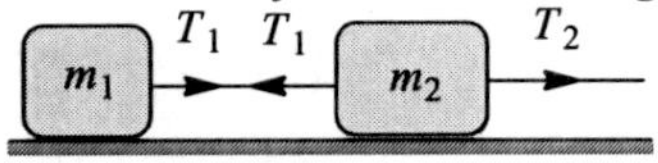

a) $\frac{m_1}{m_1+m_2}$　　b) $\frac{m_2}{m_1+m_2}$　　c) $\frac{m_1}{m_2}$　　d) $\frac{m_2}{m_1}$

490. A force of 750 N is applied to a block of mass 102 kg to prevent it from sliding on a plane with an inclination angle 30° with the horizontal. If the coefficients of static friction and kinetic friction between the block and the plane are 0.4 and 0.3 respectively, then the frictional force acting on the block is

a) 750 N　　b) 500 N　　c) 345 N　　d) 250 N

491. In figure a block of weight 10 N is shown resting on a horizontal surface. The coefficient of static friction between the block and surface is $\mu_s = 0.4$. A force of 3.5 N will keep the block in uniform motion, once it has been in motion. A horizontal force of 3 N is applied to the block. The block will there

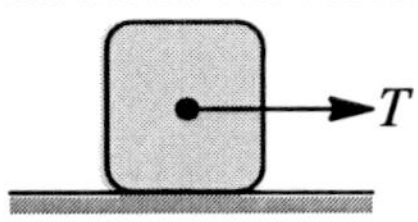

a) Move over the surface with a constant velocity
b) Moves having accelerated motion over the surface
c) Not move
d) First move with a constant velocity for sometime and then will have accelerated motion

492. Two weights ω_1 and ω_2 are suspended from the ends of a light string passing over a smooth fixed pulley. If the pulley is pulled up at an acceleration g, the tension in the string will be

a) $\frac{4\omega_1\omega_2}{\omega_1+\omega_2}$　　b) $\frac{2\omega_1\omega_2}{\omega_1+\omega_2}$　　c) $\frac{\omega_1\omega_2}{\omega_1+\omega_2}$　　d) $\frac{\omega_1\omega_2}{2(\omega_1+\omega_2)}$

493. A smooth inclined plane of length L having inclination θ with the horizontal is inside a lift which is moving down with retardation a. The time taken by α body to slide down the inclined plane, from rest, will be

a) $\sqrt{\frac{2L}{(g+a)\sin\theta}}$　　b) $\sqrt{\frac{2L}{(g-a)\sin\theta}}$　　c) $\sqrt{\frac{2L}{g\sin\theta}}$　　d) $\sqrt{\frac{2L}{\alpha\sin\theta}}$

494. The backside of a truck is open and a box of 40 kg is placed 5 m away from the rear end. The coefficient of friction of the box with the surface of the truck is 0.15. The truck starts from rest with 2 m/s^2 acceleration. Calculate the distance covered by the truck when the box falls off.

a) 20 m　　b) 30 m　　c) 40 m　　d) 50 m

495. A block of mass 2 kg rests on horizontal surface. If a horizontal force of 5 N is applied on the block, the frictional force on, it is ($\mu_k = 0.4, \mu_s = 0.5$)

a) 5 N　　b) 10 N　　c) 8 N　　d) Zero

496. A block of mass 2 kg is at rest on a floor. The coefficient of static friction between block and the floor is 0.54. A horizontal force of 2.8 N is applied to the block. What should be the frictional force between the block and the floor? (Take $g = 10$ m/s^2)

a) 8.8 N　　b) 5.8 N　　c) 2.8 N　　d) 10.8 N

497. A small block slides without friction down an inclined plane starting from rest. Let s_n be the distance travelled from time $t = n - 1$ to $t = n$. Then

$\frac{s_n}{s_n + 1}$ is

a) $\frac{2n-1}{2n}$ b) $\frac{2n+1}{2n-1}$ c) $\frac{2n-1}{2n+1}$ d) $\frac{2n}{2n+1}$

498. A person is sitting in a travelling train and facing the engine. He tosses up a coin and the coin falls behind him. It can be concluded that the train is

a) Moving forward and gaining speed b) Moving forward and losing speed
c) Moving forward with uniform speed d) Moving backward with uniform speed

499. A block is gently placed on a conveyor belt moving horizontally with constant speed. After 4s the velocity of the block becomes equal to the velocity of belt. If the coefficient of friction between the block and the belt is 0.2, then velocity of the conveyor belt is

a) 2 ms^{-1} b) 4 ms^{-1} c) 6 ms^{-1} d) 8 ms^{-1}

500. Two bodies of mass 3 kg and 4 kg are suspended at the ends of massless string passing over a frictionless pulley. The acceleration of the system is ($g = 9.8\ m/s^2$)

a) 4.9 m/s^2 b) 2.45 m/s^2 c) 1.4 m/s^2 d) 9.5 m/s^2

501. A force of 100 dynes acts on a mass of 5 $gram$for10 sec. The velocity produces is

a) 2000 cm/sec b) 200 cm/sec c) 20 cm/sec d) 2 cm/sec

502. A ball of mass m moves with speed v and it strikes normally with a wall and reflected back normally, if its time of contact with wall is t then find force exerted by ball on wall

a) $\frac{2mv}{t}$ b) $\frac{mv}{l}$ c) mvt d) $\frac{mv}{2t}$

503. Three blocks of masses m_1, m_2 and m_3 are placed on a horizontal frictionless surface. A force of 40 N pulls the system then calculate the value of T, if $m_1 = 10$ kg, $m_2 = 6$ kg, $m_3 = 4$ kg

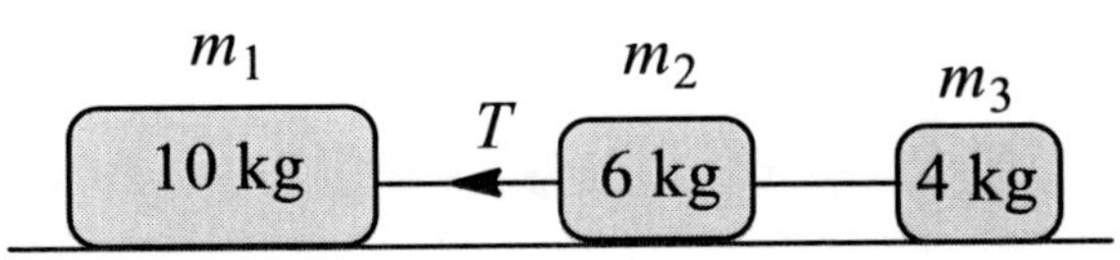

a) 40 N b) 20 N c) 10 N d) 5 N

504. Consider car moving on a straight road with a speed of 100 m/s. The distance at which car can be stopped is [$\mu_k = 0.5$]

a) 100 m b) 400 m c) 800 m d) 1000 m

505. A spring balance is attached to the ceiling of a lift. A man hangs his bag on the spring and the spring reads 49 N, when the lift is stationary. If the lift moves downward with an acceleration of 5 m/s^2, the reading of the spring balance will be

a) 49 N b) 24 N c) 74 N d) 15 N

506. The force-time ($F - t$) curve of a particle executing linear motion is as shown in the figure. The momentum acquired by the particle in time interval from zero to 8 $second$ will be

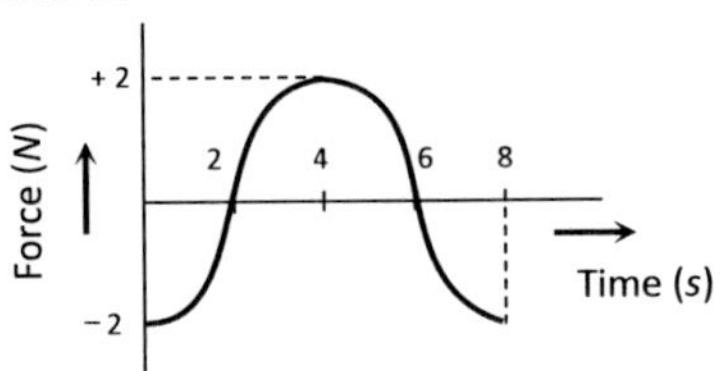

a) −2 N-s b) +4 N-s c) 6 N-s d) Zero

507. A body of mass 2 kg is kept by pressing to a vertical wall by a force of 100 N. The friction between wall and body is 0.3. Then the frictional force is equal to

a) 6 N b) 20 N c) 600 N d) 700 N

508. Newton's first law of motion describes the following

a) Energy b) Work c) Inertia d) Moment of inertia

509. If two forces of 5 N each are acting along X and Y axes, then the magnitude and direction of resultant is

a) $5\sqrt{2}, \pi/3$ b) $5\sqrt{2}, \pi/4$ c) $-5\sqrt{2}, \pi/3$ d) $-5\sqrt{2}, \pi/4$

510. A block is kept on an inclined plane of inclination θ and length l. The velocity of particle at the bottom of incline is (the coefficient of friction is μ)

a) $\sqrt{2gl(\mu\cos\theta - \sin\theta)}$ b) $\sqrt{2gl(\sin\theta - \mu\cos\theta)}$

c) $\sqrt{2gl(\sin\theta + \mu\cos\theta)}$ d) $\sqrt{2gl(\cos\theta - \mu\sin\theta)}$

511. Newton's Second law gives the measure of

a) Acceleration b) Force c) Momentum d) Angular momentum

512. The upper half of an inclined plane with inclination ϕ is perfectly smooth, while the lower half is rough. A body starting from rest at the top will again come to rest at the bottom if coefficient of friction for the lower half is given by

a) $2\sin\phi$ b) $2\cos\phi$ c) $2\tan\phi$ d) $\tan\phi$

513. A gun fires bullet each of mass 1 g with velocity of 10 ms^{-1} by exerting a constant force of 5 g weight. Then the number of bullets fired per second is
(Take $g = 10\ ms^{-2}$)

a) 50 b) 5 c) 10 d) 25

514. If force on a rocket having exhaust velocity of 300 m/sec is 2010 N, then rate of combustion of the fuel is

a) 0.7 kg/s b) 1.4 kg/s c) 0.07 kg/s d) 10.7 kg/s

515. A force of 19.6 N when applied parallel to the surface just moves a body of mass 10 kg kept on a horizontal surface. If a 5 kg mass is kept on a horizontal surface. If a 5 kg mass is kept on the first mass, the force applied parallel to the surface to just move the combined body is

a) 29.4 N b) 39.2 N c) 18.6 N d) 42.6 N

516. The adjacent figure is the part of a horizontally stretched net. Section AB is stretched with a force of 10 N. The tensions in the section BC and BF are

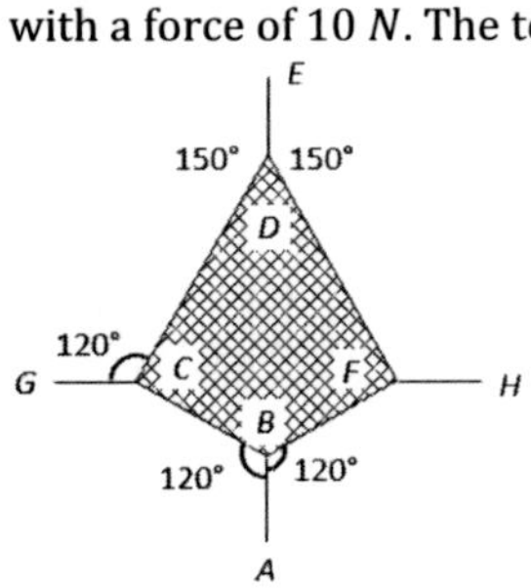

a) 10 N, 11 N

b) 10 N, 6 N

c) 10 N, 10 N

d) Can't calculate due to insufficient data

517. A block at rest slides down a smooth inclined plane which makes an angle 60° with the vertical and it reaches the ground in t_1 seconds. Another block is dropped vertically from the same point and reaches the ground in t_2 seconds.
Then the ratio of $t_1 : t_2$ is

a) 1: 2 b) 2: 1 c) 1: 3 d) $1:\sqrt{2}$

518. A block of mass 'm' is connected to another block of mass 'M' by a spring (massless) of spring constant 'K'. The blocks are kept on a smooth horizontal plane. Initially the blocks

are at rest and the spring is unstretched. Then a constant force 'F' starts acting on the block of mass 'M' to pull it. Find the force on the block of mass 'm'

a) $\frac{mF}{M}$ b) $\frac{(M+mF)}{m}$ c) $\frac{mF}{(m+M)}$ d) $\frac{MF}{(m+M)}$

519. A 40 kg slab rests on a frictionless floor. A 10 kg block rests on top of the slab. The static coefficient of friction between the block and the slab is 0.60 while the kinetic coefficient of friction is 0.40. The 10 kg block is acted upon by a horizontal force of 100 N. If $g = 9.8\ ms^{-2}$, the resulting acceleration of the slab will be

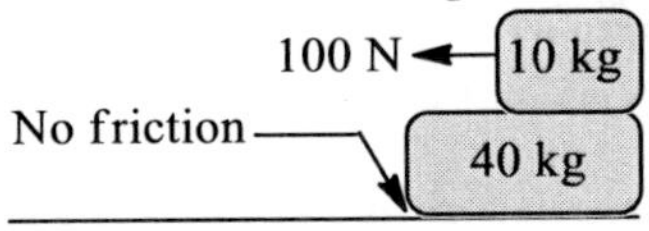

a) $1.47\ ms^{-2}$ b) $1.69\ ms^{-2}$ c) $9.8\ ms^{-2}$ d) $0.98\ ms^{-2}$

520. A light spring balance hangs from the hook of the other light spring balance and a block of mass M kg hangs from the former one. Then the true statement about the scale reading is
a) Both the scales read M kg each
b) The scale of the lower one reads M kg and of the upper one zero
c) The reading of the two scales can be anything but the sum of reading will be M kg
d) Both the scales read $M/2$ kg

521. A man getting down a running bus falls forward because
a) Due to inertia of rest, road is left behind and man reaches forward
b) Due to inertia of motion upper part of body continues to be in motion in forward direction while feet come to rest as soon as they touch the road
c) He leans forward as a matter of habit
d) Of the combined effect of all the three factors stated in (a), (b) and (c)

522. Two masses of 3 kg and 5 kg are suspended from the ends of an unstreachablemassless cord passing over a frictionless pulley. When the masses are released, the pressure on the pulley is

a) 2 kgf b) 7.5 kgf c) 8 kgf d) 15 kgf

523. A block can slide on a smooth inclined plane of inclination θ kept on the floor of a lift. When the lift is descending with a retardation a, the acceleration of the block relative to the incline is

a) $(g+a)\sin\theta$ b) $(g-a)$ c) $g\sin\theta$ d) $(g-a)\sin\theta$

524. A lift is moving down with acceleration a. A man in the lift drops a ball inside the lift. The acceleration of the ball as observed by the man in the lift and a man standing stationary on the ground are respectively

a) g, g b) g − a, g − a c) g − a, g d) a, g

525. An object is kept on a smooth inclined plane of 1 in l. The horizontal acceleration to be imparted to the inclined plane so that the object is stationary relative to the inclined is

a) $g\sqrt{l^2-1}$ b) $g(l^2-1)$ c) $\frac{g}{\sqrt{l^2-1}}$ d) $\frac{g}{l^2-1}$

526. If the coefficient of static friction between the tyres and road is 0.5, what is the shortest distance in which an automobile can be stopped when travelling at $72\ kmh^{-1}$?

a) 50 m b) 60 m c) 40.8 m d) 80.16 m

527. A man is standing at a spring platform. Reading of spring balance is 60 $kg\ wt$. If man jumps outside platform, then reading of spring balance
a) First increases then decreases to zero
b) Decreases
c) Increases
d) Remains same

528. A particle moves in x-y plane under the action of force **F** and the value of its linear momentum **p** at a given time $t$is $p_x = 2\cos t$, $p_y = 2\sin t$. Then the angle θ between **F** and **p** at a given time t is

a) θ = 30° b) θ = 180° c) θ = 0° d) θ = 90°

529. If the normal force is doubled, the coefficient of friction is
a) Not changed b) Halved c) Doubled d) Tripled

530. When two surfaces are coated with a lubricant, then they
a) Stick to each other b) Slide upon each other c) Roll upon each other d) None of these

531. A rope of length L is pulled by a constant force F. What is the tension in the rope at a distance x from the end where the force is applied
a) $\frac{FL}{x}$ b) $\frac{F(L-x)}{L}$ c) $\frac{FL}{L-x}$ d) $\frac{Fx}{L-x}$

532. Observer O_1 is in a lift going upwards and O_2 is on the ground. Both apply Newton's law, and measure normal reaction on the body
a) Both measure the same value b) Both measure zero
c) Both measure different value d) No sufficient data

533. A ball of mass $150g$ starts moving with an acceleration of $20\ m/s^2$. When hit by a force, which acts on it for $0.1\ sec$ the impulsive force is
a) $0.5\ N\text{-}s$ b) $0.1\ N\text{-}s$ c) $0.3\ N\text{-}s$ d) $1.2\ N\text{-}s$

534. If a bullet of mass $5\ gm$ moving with velocity $100\ m/sec$, penetrates the wooden block upto $6\ cm$. Then the average force imposed by the bullet on the block is
a) $8300\ N$ b) $417\ N$ c) $830\ N$ d) Zero

535. In the first second of its flight, rocket ejects $1/60$ of its mass with a velocity of $2400\ ms^{-1}$. The acceleration of the rocket is
a) $19.6\ ms^{-2}$ b) $30.2\ ms^{-2}$ c) $40\ ms^{-2}$ d) $49.8\ ms^{-2}$

536. An insect crawls up a hemispherical surface very slowly, figure. The coefficient of friction between the insect and the surface is $1/3$. If the line joining the centre of the hemispherical surface to the insect makes an angle α with the vertical, the maximum possible value of α is given by

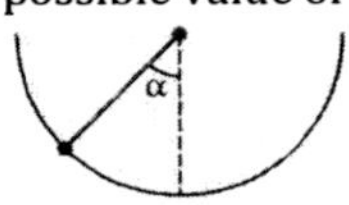

a) $\cot\alpha = 3$ b) $\sec\alpha = 3$ c) $\text{cosec}\ \alpha = 3$ d) None of these

537. Block A of mass 2 kg is placed over block B of mass 8 kg. The combination is placed over a rough horizontal surface. Coefficient of friction between B and the floor is 0.5. Coefficient of friction between A and B is 0.4. A horizontal force of 10 N is applied on block B. The force of friction between A and B is

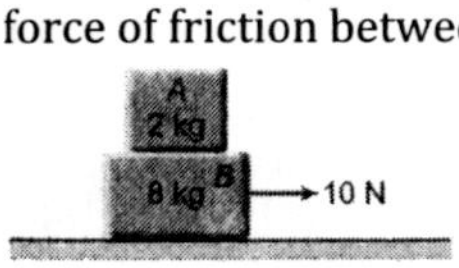

(b) 50 N
(d) 100 N

a) Zero b) 50 N c) 40 N d) 100 N

538. A body is under the action of two mutually perpendicular forces of 3 N and 4 N. The resultant force acting on the body is
a) 7 N b) 1 N c) 5 N d) Zero

539. Which of the following graph depicts spring constant k versus length l of the spring correctly

a) 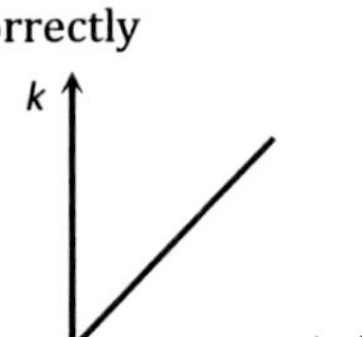

b) 

c) 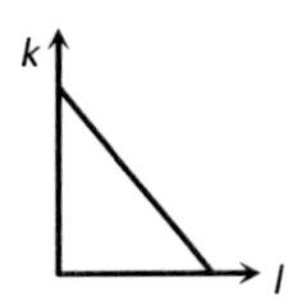

d) 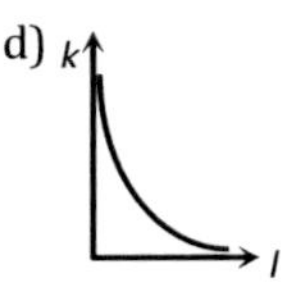

540. A fireman of mass 60 kg slides down a pole. He is pressing the pole with a force of 600 N. The coefficient of friction between the hands and the pole is 0.5, with what acceleration will the fireman slide down ($g = 10\ m/s^2$)

a) $1\ m/s^2$ b) $2.5\ m/s^2$ c) $10\ m/s^2$ d) $5\ m/s^2$

541. Which of the four arrangements in the figure correctly shows the vector addition of two forces F_1 and F_2 to yield the third force F_3?

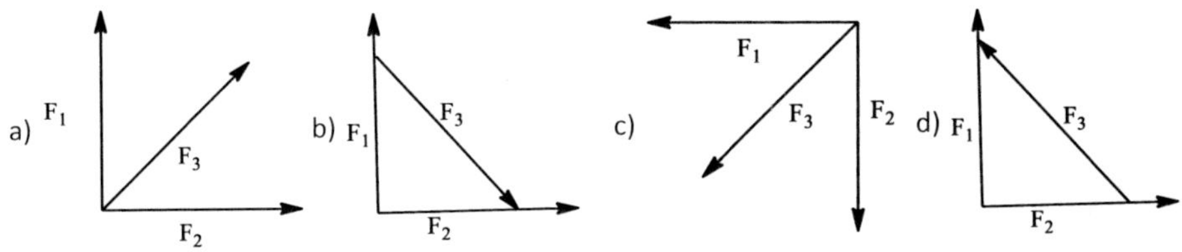

542. A book is lying on the table. What is the angle between the action of the book on the table and the reaction of the table on the book

a) 0° b) 30° c) 45° d) 180°

543. A shell at rest at the origin explodes into three fragments of masses 1 kg, 2kg and m kg. The 1 kg and 2 kg pieces fly off with speeds off $5ms^{-1}$ along x-axis and $6ms^{-1}$ along y-axis respectively. If the m kg piece files off with a speed of 6.5 ms^{-1}, the total mass of the shell must be

a) 4 kg b) 5 kg c) 3.5 kg d) 4.5 kg

544. Three blocks of masses m_1, m_2 and m_3 are connected by massless strings as shown on a frictionless table in figure. They are pulled with a force $T_3 = 40$ N. If $m_1 = 10$ kg, $m_2 = 6$ kg and $m_3 = 4$ kg, the tension T_2 will be

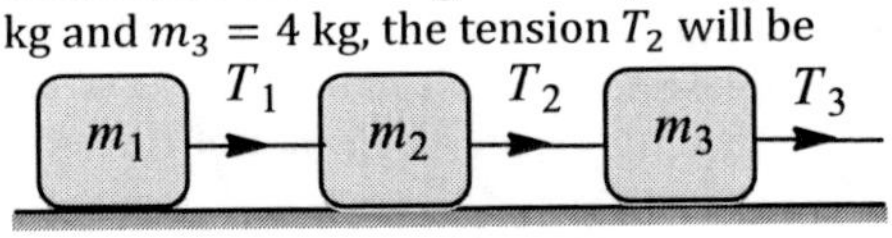

a) 20 N b) 40 N c) 10 N d) 32 N

545. A particle of mass 0.3 kg is subjected to a force $F = -kx$ with $k = 15\ N/m$. What will be its initial acceleration if it is released from a point 20 cm away from the origin

a) $5\ m/s^2$ b) $10\ m/s^2$ c) $3\ m/s^2$ d) $15\ m/s^2$

546. A person sitting in an open car moving at constant velocity throws a ball vertically up into air. The ball falls

a) Outside the car
b) In the car ahead of the person
c) In the car to the side of the person
d) Exactly in the hand which threw it up

547. If the mass of $A = 10$ kg, coefficient of static friction = 0.22, coefficient of kinetic friction = 0.2, then minimum mass of B to start motion is

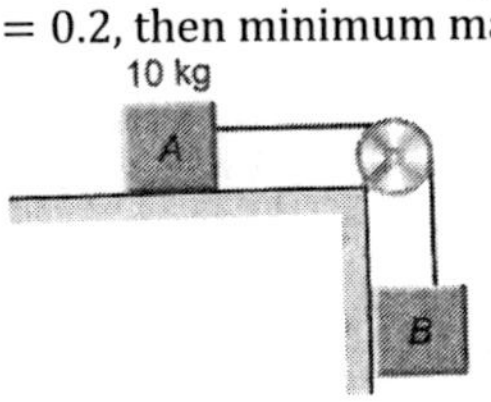

a) 2 kg b) 2.2 kg c) 4.8 kg d) 3.4 kg

548. Which one of the following is not used to reduce friction

a) Oil b) Ball bearings c) Sand d) Graphite

549. Rocket propulsion is associated with

a) The conservation of angular momentum
b) The conservation of mass
c) The conservation of mechanical energy
d) Newton's III law of motion

550. Consider the following statement: When jumping from some height, you should bend your knees as you come to rest, instead of keeping your legs stiff. Which of the following relations can be useful in explaining the statement

a) $\Delta\overrightarrow{P_1} = -\Delta\overrightarrow{P_2}$ b) $\Delta E = -\Delta(PE + KE) = 0$

c) $\vec{F}\Delta t = m\Delta\vec{v}$ d) $\Delta\vec{x} \propto \Delta\vec{F}$

551. A body presses a book against the front wall such that the book does not move. The force of friction between the wall and the book is

a) Towards right b) Towards left c) Downwards d) Upwards

552. A person used force (F), shown in figure move a load with constant velocity on give surface.

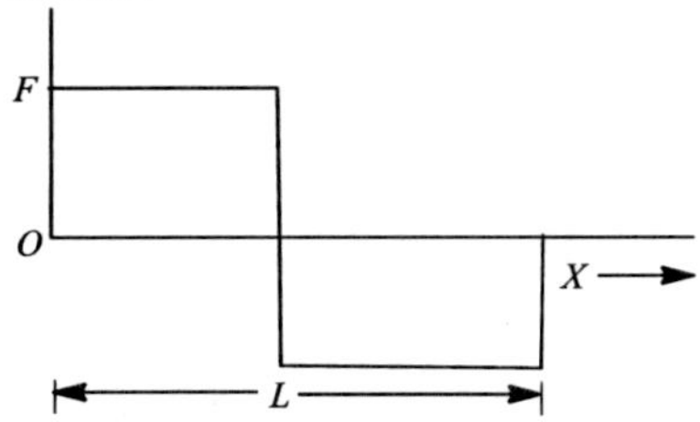

Identify the correct surface profile

a) 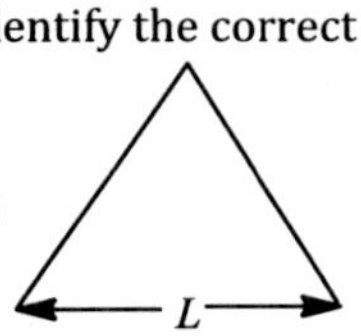

b) 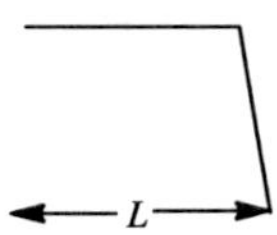

c) 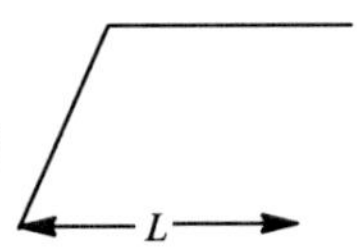

d) 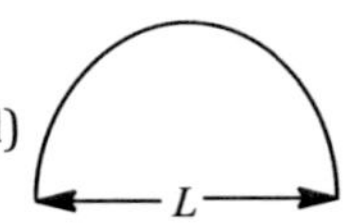

553. In the figure a smooth pulley of negligible weight is suspended by a spring balance. Weights of 1 kg and 5 kg are attached to the opposite ends of a string passing over the pulley and move with acceleration because of gravity. During their motion, the spring balance reads a weight of

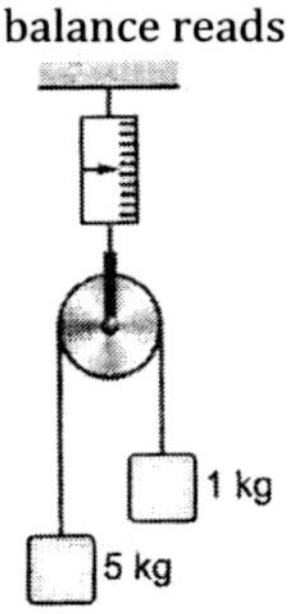

a) 6 kg b) Less than 6 kg

c) More than 6 kg d) May be more or less than 6 kg

554. A unidirectional force F varying with time t as shown in the figure acts on a body initially at rest for a short duration $2T$. Then the velocity acquired by the body is

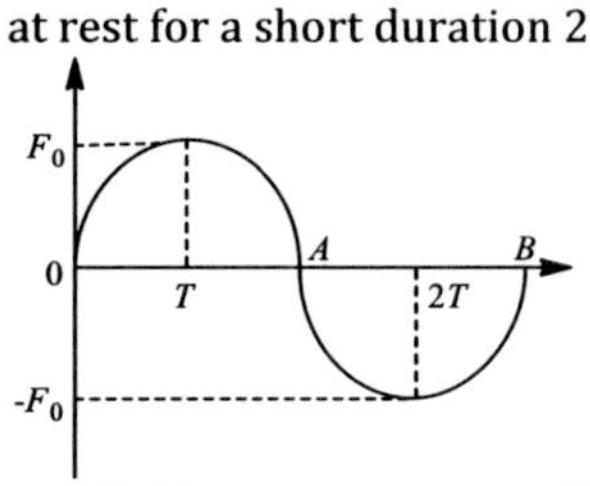

a) $\frac{\pi F_0 T}{4m}$ b) $\frac{\pi F_0 T}{2m}$ c) $\frac{F_0 T}{4m}$ d) Zero

555. A diwali rocket is ejecting 0.05 kg of gases per second at a velocity of 400 m/sec. The accelerating force on the rocket is

a) 20 $dynes$ b) 20 N c) 22 $dynes$ d) 1000 N

556. In relativity which is constant between two frames of reference

a) Acceleration b) Conservation of mass c) Space interval d) Velocity

557. A body of mass 5 kg is suspended by a spring balance on an inclined plane as shown in figure. The spring balance, measure

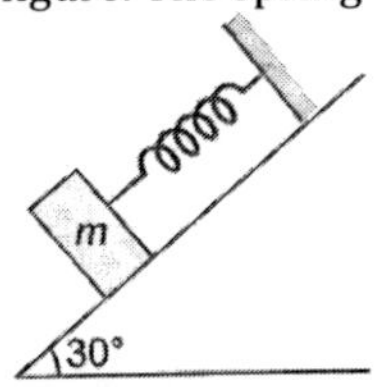

a) 50 N b) 25 N c) 500 N d) 10 N

558. Impulse is

a) A scalar b) Equal to change in the momentum of a body c) Equal to rate of change of momentum of a body d) A force

559. A block of mass $M = 5\ kg$ is resting on a rough horizontal surface for which the coefficient of friction is 0.2. When a force $F = 40\ N$ is applied, the acceleration of the block will be($g = 10\ ms^2$)

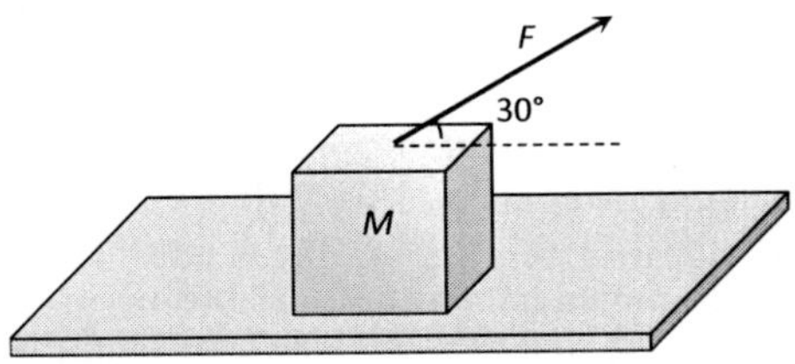

a) 5.73 m/sec^2 b) 8.0 m/sec^2 c) 3.17 m/sec^2 d) 10.0 m/sec^2

560. A block P of mass m is placed on a frictionless horizontal surface. Another block Q of same mass is kept on $P$and connected to the wall with the help of a spring of spring constant k as shown in the figure. μ_s is the coefficient of friction between P and Q. The blocks move together performing SHM of amplitude A. The maximum value of the friction force between$P$ and Q is

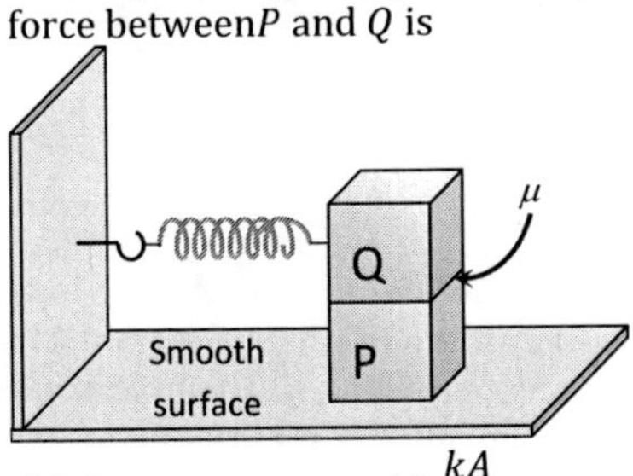

a) kA b) $\frac{kA}{2}$ c) Zero d) $\mu_s mg$

561. A pulley fixed to the ceiling carries a string with blocks of mass m and $3m$ attached to its ends. The masses of string and pulley are negligible. When the system is released, its centre of mass moves with what acceleration

a) 0 b) $g/4$ c) $g/2$ d) $-g/2$

562. The monkey B shown in figure is holding on to the tail of the monkey A which is climbing up a rope. The masses of the monkeys A and B are 5 kg and 2 kg respectively. If A can tolerate a tension of 30 N in its tail, what force should it apply on the rope in order to carry the monkey B with it? (Take g = 10 ms^{-2})

a) 105 N b) 108 N c) 10.5 N d) 100 N

563. A body of mass 5 kg starts from the origin an initial velocity $\vec{u} = 30\hat{\imath} + 40\hat{\jmath}\ ms^{-1}$. If a constant force $\vec{F} = -(\hat{\imath} + 5\hat{\jmath})N$ acts on the body, the time in which the $y-$ component of the velocity becomes zero is

a) 5 *seconds* b) 20 *seconds* c) 40 *seconds* d) 80 *seconds*

564. A mass of 6 kg is suspended by a rope of length 2 m from a ceiling. A force of 50 N is applied in the horizontal direction at the mid-point of the rope. The angel made by the rope, with the vertical, in equilibrium position will be (take g = 10 ms^{-2}, neglect the mass of the rope)

a) 90° b) 60° c) 50° d) 40°

565. A bird weighs 2 kg and is inside a closed cage of 1 kg. If it starts flying, then what is the weight of the bird and cage assembly

a) 1.5 kg b) 2.5 kg c) 3 kg d) 4 kg

566. A closed compartment containing gas is moving with some acceleration in horizontal direction. Neglect effect of gravity. Then the pressure in the compartment is

a) same everywhere b) lower in front side c) lower in rear side d) lower in upper side

567. A 24 kg block resting on a floor has a rope tied to its top. The maximum tension, the rope can withstand without breaking is 310 N. The minimum time in which the block can be lifted a vertical distance of 4.6 m by pulling on the rope is

a) 1.2 s b) 1.3 s c) 1.7 s d) 2.3 s

568. A bullet is fired from a gun. The force on the bullet is given by $F = 600 - 2 \times 10^5 t$, where F is in *newtons* and t in seconds. The force on the bullet becomes zero as soon as it leaves the barrel. What is the average impulse imparted to the bullet

a) 9 Ns b) Zero c) 0.9 Ns d) 1.8 Ns

569. A given object taken n times as much time to slide down a 45° rough incline as it takes to slide down a perfectly smooth 45° incline. The coefficient of kinetic friction between the object and the incline is given by

a) $\left(1 - \frac{1}{n^2}\right)$ b) $\frac{1}{1-n^2}$ c) $\sqrt{\left(1 - \frac{1}{n^2}\right)}$ d) $\sqrt{\frac{1}{1-n^2}}$

570. A block P of mass m is placed on a horizontal surface. Another block Q of same mass is kept on P and connected to the wall with the help of a spring of spring constant k as shown in the figure. μ_s is the coefficient of friction between P and Q. The blocks move together performing SHM of amplitude A. The maximum value of the friction force between P and Q is

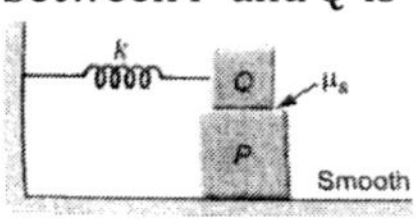

a) kA b) $\frac{kA}{2}$ c) Zero d) $\mu_s mg$

571. A cricket ball of mass 0.25 kg with speed 10 m/s collides with a bat and returns with same speed with in 0.01 s. The force acted on bat is

a) 25 N b) 50 N c) 250 N d) 500 N

572. A rocket of mass 1000 kg is exhaust gases at a rate of 4 kgs^{-1} with a velocity 3000 ms^{-1}. The thrust developed on the rocket is

a) 12000 N b) 120 N c) 800 N d) 200 N

573. A 1.5 kg ball drops vertically on a floor hitting with a speed of 25ms^{-1} . It rebounds with an initial speed of 15 ms^{-1}. If the ball was in contact for only 0.03 seconds, the force exerted on the floor by the ball is

a) 2000 N b) 3000 N c) 3500 N d) 4000 N

574. Starting from rest, a body slides down a 45° inclined plane in twice the time it takes to slide down the same distance in the absence of friction. The coefficient of friction between the body and the inclined plane is

a) 0.33 b) 0.25 c) 0.75 d) 0.80

575. In figure, the blocks A, B and C each of mass m have acceleration a_1, a_2 and a_3 respectively. F_1 and F_2 are external forces of magnitude 2 mg and mg respectively. Then

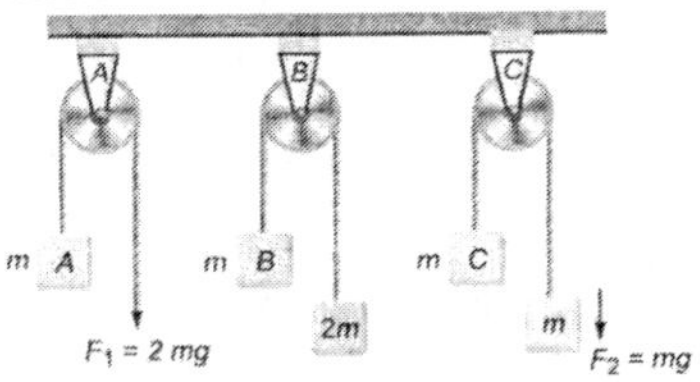

a) $a_1 = a_2 = a_3$ b) $a_1 > a_3 > a_2$ c) $a_1 = a_2, a_2 = a_3$ d) $a_1 = a_2, a_1 = a_3$

576. Human heart is pumping blood with constant velocity v m s^{-1} at the rate of M kg s^{-1}. The force required for this is (in N)

a) M b) Mv c) $\frac{M}{v}$ d) $v\frac{dM}{dt}$

577. An automobile travelling with a speed of 60 km/h, can brake to stop within a distance of 20 m. If the car is going twice at fast, $i.e.$ 120 km/h, the stopping distance will be

a) 20 m b) 40 m c) 60 m d) 80 m

578. A body of mass 1.0 kg is falling with an acceleration of 10 m/sec^2. Its apparent weight will be ($g = 10\ m/sec^2$)

a) 1.0 kg wt b) 2.0 kg wt c) 0.5 kg/wt d) Zero

579. A passenger is travelling in a train moving at 72 kmh^{-1}. His suitcase is kept on the berth. The driver of the train applies brakes such that the speed of the train decreases at a constant rate of 36 kmh^{-1} in 5 s. What should be the minimum coefficient of friction between the suitcase and the berth if the suitcase is not the slide during retardation of the train?

a) $\frac{10}{49}$ b) $\frac{10}{98}$ c) $\frac{28}{49}$ d) $\frac{30}{98}$

580. A man fires a bullet of mass 200 $g$at a speed of 5 m/s . The gun is of one kg mass. By what velocity the gun rebounds backwards

a) 0.1 m/s b) 10 m/s c) 1 m/s d) 0.01 m/s

581. When the speed of a moving body is doubled

a) Its acceleration is doubled b) Its momentum is doubled

c) Its kinetic energy is doubled d) Its potential energy is doubled

582. A marble block of mass 2 kg lying on ice when given a velocity of 6 ms^{-1} is stopped by friction in 10 s. Then the coefficient of friction is

a) 0.02 b) 0.03 c) 0.06 d) 0.01

583. A 5 kg stationary bomb is exploded in three parts having mass 1: 1: 3 respectively. Parts having same mass move in perpendicular directions with velocity 39ms^{-1}, then the velocity of bigger part will be

a) $10\sqrt{2}\text{ ms}^{-1}$ b) $\frac{10}{\sqrt{2}}\text{ms}^{-1}$ c) $13\sqrt{2}\text{ ms}^{-1}$ d) $\frac{15}{\sqrt{2}}\text{ms}^{-1}$

584. Which of the following groups of forces could be in equilibrium

a) $3\ N, 4\ N, 5N$ b) $4\ N, 5\ N, 10\ N$ c) $30\ N, 40\ N, 80\ N$ d) $1\ N, 3\ N, 5\ N$

585. Formula for true force is

a) $F = ma$ b) $F = \frac{mdv}{dt}$ c) $F = \frac{dmv}{dt}$ d) $F = \frac{md^2x}{dt^2}$

586. If rope of lift breaks suddenly, the tension exerted by the surface of lift (a =acceleration of lift)

a) mg b) $m(g + a)$ c) $m(g - a)$ d) 0

587. A body of weight 64 N is pushed with just enough force to start it moving across a horizontal floor and the same force continues to act afterwards. If the coefficients of static and dynamic friction are 0.6 and 0.4 respectively, the acceleration of the body will be (Acceleration due to gravity = g)

a) $\frac{g}{6.4}$ b) $0.64\ g$ c) $\frac{g}{32}$ d) $0.2\ g$

588. In the above question, if the lift is moving upwards with a uniform velocity, then the frictional resistance offered by the body is

a) Mg b) μMg c) $2\mu Mg$ d) Zero

589. A block of mass 0.1 kg is held against a wall by applying a horizontal force of 5 N on the block. If the coefficient of friction between the block and the wall is 0.5, the magnitude of the frictional force acting on the block is

a) 2.5 N b) 0.98 N c) 4.9 N d) 0.49 N

590. If the resultant of all the external forces acting on a system of particles is zero, then from an inertial frame, one can surely say that

a) Liner momentum of the system does not change in time

b) Kinetic energy of the system does not change in time

c) Angular momentum of the system does not change in time

d) Potential energy of the system does not change in time

591. A stone weighing 1 kg and sliding on ice with a velocity of 2 m/s is stopped by friction in 10 sec.The force of friction (assuming it to be constant) will be

a) $-20\ N$ b) $-0.2\ N$ c) $0.2\ N$ d) $20\ N$

592. With what minimum acceleration can a fireman slide down a rope while breaking strength of the rope is $\frac{2}{3}$ of the weight?

a) $\frac{2}{3}g$ b) g c) $\frac{1}{3}g$ d) Zero

593. A lift is moving downwards with an acceleration equal to acceleration due to gravity. A body of mass m kept on the floor of the lift is pulled horizontally. If the coefficient of friction is μ, then the frictional resistance offered by the body is

a) mg b) μmg c) $2\mu mg$ d) Zero

594. The resultant force of 5 N and 10 N can not be

a) 12 N b) 8 N c) 4 N d) 5 N

595. A boy of mass 40 kg is climbing a vertical pole at a constant speed. If the coefficient of friction between his palms and the pole is 0.8 and $g = 10\ m/s^2$, the horizontal force that he is applying on the pole is

a) 300 N b) 400 N c) 500 N d) 600 N

596. When a body is moving on a surface, the force of friction is called

a) Static friction b) Dynamic friction c) Limiting friction d) Rolling friction

597. It is easier to roll a barrel than pull it along the road. This statement is

a) False b) True c) Uncertain d) Not possible

598. A body of mass m collides against a wall with a velocity v and rebounds with the same speed. Its change of momentum is

a) $2\ mv$ b) mv c) $-mv$ d) Zero

599. Two iron blocks of equal mass but with double area slide down an inclined plane with friction coefficient μ. If the first block with surface area A experience a frictional force f, then the second block with surface area $2A$ will experience a frictional force

a) $f/2$ b) f c) $2f$ d) $4f$

600. A particle of mass 2 kg is initially at rest. A force acts on it whose magnitude changes with time. The force- time graph is shown below

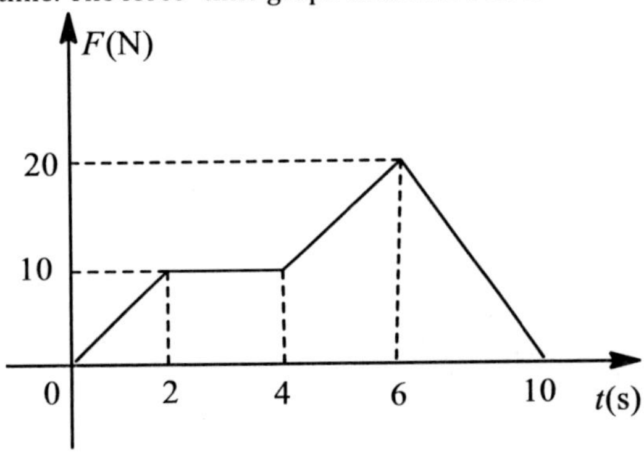

The velocity of the particle after 10 s is

a) $20\ ms^{-1}$ b) $10\ ms^{-1}$ c) $75\ ms^{-1}$ d) $50\ ms^{-1}$

601. There is a simple pendulum hanging from the ceiling of a lift. When the lift is stand still, the time period of the pendulum is T. If the resultant acceleration becomes $g/4$, then the new time period of the pendulum is

a) $0.8\ T$ b) $0.25\ T$ c) $2\ T$ d) $4\ T$

602. One end of a massless rope, which passes over a massless and frictionless pulley P is tied to a hook C while the other end is free. Maximum tension that the rope can bear is 360 N. with what value of minimum safe acceleration (in ms^{-2}) can a monkey of 60 kg move down on the rope

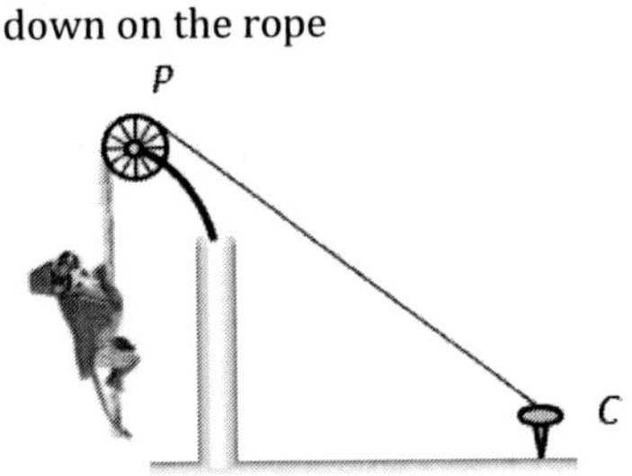

a) 16 b) 6 c) 4 d) 8

603. A spring balance, A reads 2 kg with a block m suspended from it. A balance B reads 5 kg when a breaker filled with liquid is put on the pan of the balance. The two balances are now so arranged that the hanging mass is inside the liquid as shown in figure. In this situation

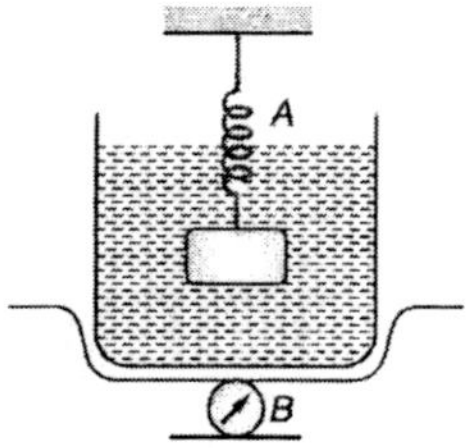

a) The balance A will read more than 2 kg
b) The balance B will read more than 5 kg
c) The balance A will read less than 2 kg and B will read more than 5 kg
d) The balance A and B will read 2 kg and 5 kg

604. A body of mass 10 kg is acted upon by two forces each of magnitude 10 N making an angle of 60° with each other. Find the net acceleration of the body
a) $2\sqrt{3}ms^{-2}$ b) $\sqrt{3}ms^{-2}$ c) $3\sqrt{3}ms^{-2}$ d) $4\sqrt{3}ms^{-2}$

605. A motorcycle is travelling on a curved track of radius 500 m. If the coefficient of between road and tyres is 0.5, the speed avoiding skidding will be
a) 50 m/s b) 75 m/s c) 25 m/s d) 35 m/s

606. A passenger train is running on a railways track with a speed v_1. While driving, the driver discovers that another goods train is travelling in front of the passenger train with a speed $v_2 (v_1 > v_2)$. Its retardation after applying brakes is a. The least distance the passenger train must travel to avoid collision with goods train is
a) $\frac{v_1^2 - v_2^2}{2a}$ b) $\frac{v_2 - v_1}{a}$ c) $\frac{v_2 + v_1}{2a}$ d) $\frac{v_2^2 + v_1^2}{2a}$

607. The acceleration of the system shown in figure is

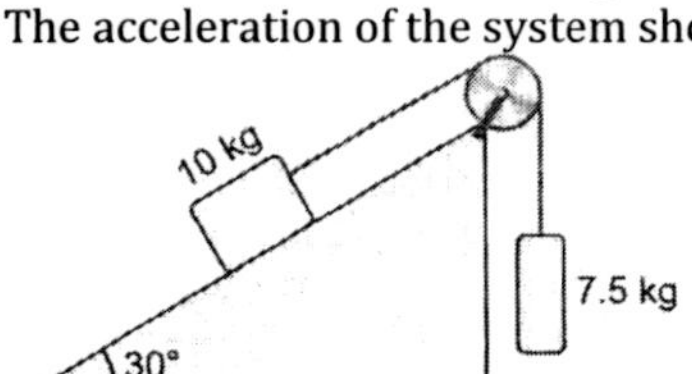

a) $\frac{3.5}{17.5}g$ b) $\frac{7.5}{17.5}g$ c) $\frac{14.5}{17.5}g$ d) $\frac{g}{7}$

608. If μ_s, μ_k and μ_r are coefficients of static friction, sliding friction and rolling friction, then
a) $\mu_s < \mu_k < \mu_r$ b) $\mu_k < \mu_r < \mu_s$ c) $\mu_r < \mu_k < \mu_s$ d) $\mu_r < \mu_k < \mu_s$

609. What is the maximum value of the force F such that the block shown in the arrangement, does not move

a) 20 N b) 10 N c) 12 N d) 15 N

610. A mass of 1 kg is suspended by a string A. Another string C is connected to its lower end (see figure). If a sudden jerk is given to C, then

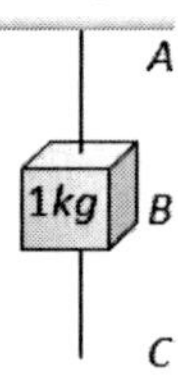

a) The portion AB of the string will break
b) The portion BC of the string will break
c) None of the strings will break
d) The mass will start rotating

3

611. The engine of a car produces an acceleration of 6 ms^{-2} in the car. If this car pulls another car of the same mass, then the acceleration would be

a) 6 ms^{-2} b) 12 ms^{-2} c) 3 ms^{-2} d) 1.5 ms^{-2}

612. A block A with mass 100 kg is resting on another block B of mass 200 kg. As shown in figure a horizontal rope tied to a wall holds it. The coefficient of friction between A and B is 0.2 while coefficient of friction between B and the ground is 0.3. The minimum required force F to start moving B will be

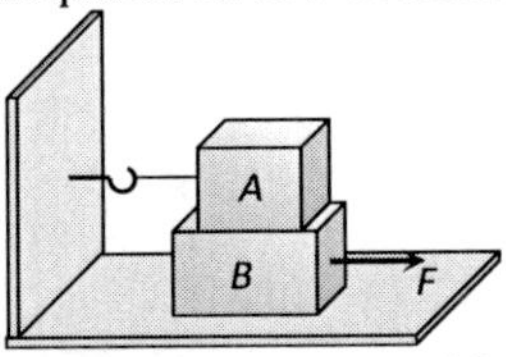

a) 900 N b) 100 N c) 1100 N d) 1200 N

613. Swimming is possible on account of

a) First law of motion b) Second law of motion
c) Third law of motion d) Newton's law of gravitation

614. A block is moving up an inclined plane of inclination 60° with velocity of 20 ms^{-1} and stops after 2.00 s. If $g = 10 ms^{-2}$ then the approximate value of coefficient of friction is

a) 3 b) 3.3 c) 0.27 d) 0.33

615. Three blocks of masses 2 kg, 3 kg and 5 kg are connected to each other with light string and are then placed on a frictionless surface as shown in the figure. The system is pulled by a force $F = 10\ N$, then tension $T_1 =$

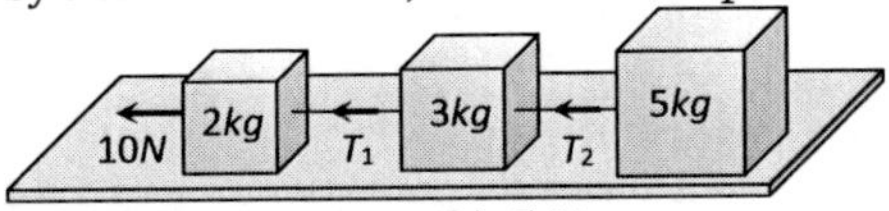

a) 1 N b) 5 N c) 8 N d) 10 N

616. Two persons are holding a rope of negligible weight tightly at its ends so that it is horizontal. A 15 kg weight is attached to rope at the mid-point which now no more remains horizontal. The minimum tension required to completely straighten the rope is

a) 15 kg b) 15/2 kg c) 5 kg d) Infinitely large

617. A block of mass 50 kg can slide on a rough horizontal surface. The coefficient of friction between the block and the surface is 0.6. The least force of pull acting at an angle of 30° to the upward drawn vertical which causes the block to just slide is

a) 29.43 N b) 219.6 N c) 21.96 N d) 294.3 N

618. A particle moves in the xy- plane under the action of a force $\boldsymbol{F}$ such that the components of its linear momentum $\boldsymbol{p}$ at any time t are $p_x = 2\cos t$, $p_y = 2\sin t$. The angle between $\boldsymbol{F}$ and $\boldsymbol{p}$ at time t is

a) 90° b) 0° c) 180° d) 30°

619. A body of 5 kg weight kept on a rough inclined plane of angle 30° starts sliding with a constant velocity. Then the coefficient of friction is (assume $g = 10\ m/s^2$)

a) $1/\sqrt{3}$ b) $2/\sqrt{3}$ c) $\sqrt{3}$ d) $2\sqrt{3}$

620. A force of 10 Newton acts on a body of mass 20 kg for 10 seconds. Change in its momentum is

a) 5 $kg\ m/s$ b) 100 $kg\ m/s$ c) 200 $kg\ m/s$ d) 1000 $kg\ m/s$

621. In a rocket of mass 1000 kg fuel is consumed at a rate of 40 kg/s. The velocity of the gases ejected from the rocket is $5 \times 10^4\ m/s$. The thrust on the rocket is

a) $2 \times 10^3\ N$ b) $5 \times 10^4 N$ c) $2 \times 10^6\ N$ d) $2 \times 10^9\ N$

622. The time period of a simple pendulum measured inside a stationary lift is found to be T. If the lift starts accelerating upwards with an acceleration $g/3$, the time period is

a) $T\sqrt{3}$ b) $T\sqrt{3}/2$ c) $T/\sqrt{3}$ d) $T/3$

623. A rocket with a lift-off mass $3.5 \times 10^4\ kg$ is blasted upwards with an initial acceleration of $10\ m/s^2$. Then the initial thrust of the blast is

a) $1.75 \times 10^5\ N$ b) $3.5 \times 10^5 N$ c) $7.0 \times 10^5\ N$ d) $14.0 \times 10^5\ N$

624. A block A weighing 100 kg rests on a block B and is tied with a horizontal string to the wall at C. Block B weighs 200kg. The coefficient of friction between A and B is 0.25 and between B and the surface is $\frac{1}{3}$. The horizontal force P necessary to move the block B should be $\left(g = 10 \frac{\text{m}}{\text{s}^2}\right)$

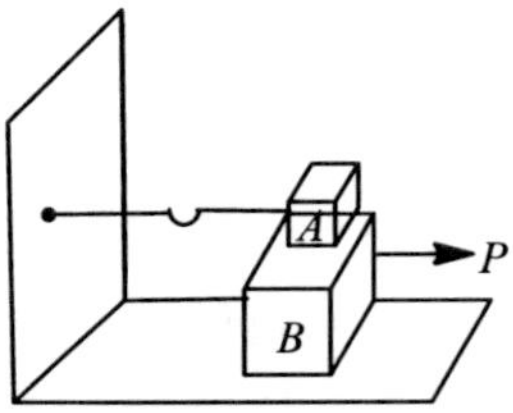

a) 1150 N b) 1250 N c) 1300 N d) 1420 N

625. A ball of mass 400 gm is dropped from a height of $5m$. A boy on the ground hits the ball vertically upwards with a bat with an average force of 100 *newton* so that it attains a vertical height of 20 m. The time for which the ball remains in contact with the bat is [$g = 10\ m/s$]

a) $0.12\ s$ b) $0.08\ s$ c) $0.04\ s$ d) $12\ s$

626. When a train stops suddenly, passengers in the running train feel an instant jerk in the forward direction because

a) The back of seat suddenly pushes the passengers forward
b) Inertia of rest stops the train and takes the body forward
c) Upper part of the body continues to be in the state of motion whereas the lower part of the body in contact with seat remains at rest
d) Nothing can be said due to insufficient data

627. Which of the following is correct, when a person walks on a rough surface

a) The frictional force exerted by the surface keeps him moving
b) The force which the man exerts on the floor keeps him moving
c) The reaction of the force which the man exerts on floor keeps him moving
d) None of the above

628. The spring balance inside a lift suspends an object. As the lift begins to ascend, the reading indicated by the spring balance will

a) Increase b) Decrease
c) Remain unchanged d) Depend on the speed of ascend

629. A cork is submerged in water by a spring attached to the bottom of a pail. When the pail is kept in a elevator moving with an acceleration downwards, the spring length

a) Increases b) Decreases c) Remains unchanged d) Data insufficient

630. A ball of mass m moves with speed v and it strikes normally with a wall and reflected back normally. If its time of contact with wall is t, then find force exerted by ball on the wall

a) $\frac{2mv}{t}$ b) $\frac{mv}{t}$ c) mvt d) $\frac{mv}{2t}$

631. A 60 kg man stands on a spring scale in the lift. At some instant he finds, scale reading has changed from 60 kg to 50 kg for a while and then comes back to the original mark. What should we conclude

a) The lift was in constant motion upwards
b) The lift was in constant motion downwards

c) The lift while in constant motion upwards, is stopped suddenly
d) The lift while in constant motion downwards, is suddenly stopped

632. A body is sliding down an inclined plane having coefficient of friction 0.5. If the normal reaction is twice that of the resultant downward force along the incline, the angle between the inclined plane and the horizontal is
a) $15°$ b) $30°$ c) $45°$ d) $60°$

633. Two masses of M and $4M$ are moving with equal kinetic energy. The ratio of their linear momentum is
a) 1:8 b) 1:4 c) 1:2 d) 4:1

634. Two masses M and $M/2$ are joined together by means of light inextensible string passed over a frictionless pulley as shown in the figure. When the bigger mass is released, the small one will ascend with an acceleration of

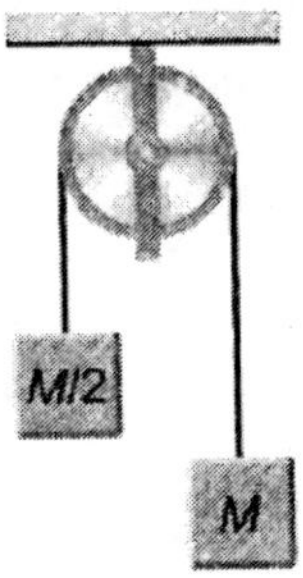

a) $\frac{g}{3}$ b) $\frac{3g}{2}$ c) $\frac{g}{2}$ d) g

635. When a body is stationary
a) There is no force acting on it
b) The force acting on it is not in contact with it
c) The combination of forces acting on it balances each other
d) The body is in vacuum

636. Three forces F_1, F_2 and F_3 together keep a body in equilibrium. If $F_1 = 3\ N$ along the positive x-axis, $F_2 = 4N$ along the positive y-axis, then the third force F_3 is
a) $5\ N$ making an angle $\theta = \tan^{-1}\left(\frac{3}{4}\right)$ with negative y-axis
b) $5\ N$ making an angle $\theta = \tan^{-1}\left(\frac{4}{3}\right)$ with negative yaxis
c) $7\ N$ making an angle $\theta = \tan^{-1}\left(\frac{3}{4}\right)$ with negative yaxis
d) $7\ N$ making an angle $\theta = \tan^{-1}\left(\frac{4}{3}\right)$ with negative yaxis

637. A rocket of mass 100 kg burns 0.1 kg of fuel per sec. If velocity of exhaust gas is 1 km/sec, then it lifts with an acceleration of
a) $1000\ ms^{-2}$ b) $100\ ms^{-2}$ c) $10\ ms^{-2}$ d) $1\ ms^{-2}$

638. If two particles collide at constant temperature then which of the following is conserved?
a) Kinetic energy b) Momentum c) Temperature d) Velocity

639. A box is lying on an inclined plane what is the coefficient if static friction if the box starts sliding when an angle of inclination is $60°$
a) 1.173 b) 1.732 c) 2.732 d) 1.677

640. A man of weight mg is moving up in a rocket with acceleration 4g. the apparent weight of the man in the rocket is
a) Zero b) $4\ mg$ c) $5\ mg$ d) mg

641. Two masses A and B of 10 kg and 5 kg respectively, are connected with a string passing over a frictionless pulley fixed at the corner of a table as shown in figure. The coefficient of friction of A with the table is 0.2. The minimum mass of C that may be placed on A to prevent it front moving is

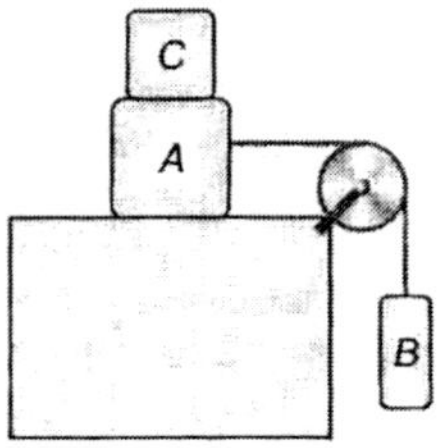

a) 15 kg b) 5 kg c) 10 kg d) Zero

642. The coefficient of friction between a body and the surface of an inclined plane at 45° is 0.5. If $g = 9.8\ m/s^2$, the acceleration of the body downwards in m/s^2 is

a) $\frac{4.9}{\sqrt{2}}$ b) $4.9\sqrt{2}$ c) $19.6\sqrt{2}$ d) 4.9

643. The mass of a body measured by a physical balance in the lift at rest is found to be m. If the lift is going up with an acceleration a, its mass will be measured as

a) $m\left(1 - \frac{a}{g}\right)$ b) $m\left(1 + \frac{a}{g}\right)$ c) m d) Zero

644. An object placed on an inclined plane starts sliding when the angle of incline becomes 30°. The coefficient of static friction between the object and the plane is

a) $\frac{1}{\sqrt{3}}$ b) $\sqrt{3}$ c) $\frac{1}{2}$ d) $\frac{\sqrt{3}}{2}$

645. A rider on horse back falls when horse starts running all of a sudden because

a) Rider is taken back

b) Rider is suddenly afraid of falling

c) Inertia of rest keeps the upper part of body at rest whereas lower part of the body moves forward with the horse

d) None of the above

646. A gardner waters the plants by a pipe of diameter 1mm. The water comes out at the rate or 10 cm^3/sec. The reactionary force exerted on the hand of the gardner is

a) Zero b) $1.27 \times 10^{-2}\ N$ c) $1.27 \times 10^{-4}\ N$ d) $0.127\ N$

647. A bullet is fired from a gun. The force on the bullet is given by $F = 600 - 2 \times 10^5 t$ where F is in newton and t in second. The force on the bullet becomes zero as soon as it leaves the barrel. What is the average in pulse imparated to the bullet?

a) 9 Ns b) 1.8 ns c) 0.9 Ns d) 0.3 Ns

648. A gun of mass 10 kg fires 4 bullets per second. The mass of each bullet is 20 g and the velocity of the bullet when it leaves the gun is 300 ms^{-1}. The force required to hold the gum while firing is

a) 6 N b) 8 N c) 24 N d) 240 N

649. A shell is fired from a cannon with velocity v ms^{-1} at an angle θ with the horizontal direction. At the highest point in its path it explodes into two pieces of equal mass. One of the pieces retraces its path to the cannon and the speed in m/s of the piece immediately after the explosion is

a) $3v\cos\theta$ b) $2v\cos\theta$ c) $\frac{3v}{2}\cos\theta$ d) $\frac{\sqrt{3}v\cos\theta}{2}$

5
LAWS OF MOTION
ANSWER KEY

1)	d	2)	c	3)	c	4)	a
5)	a	6)	c	7)	b	8)	a
9)	a	10)	a	11)	a	12)	c
13)	d	14)	b	15)	d	16)	a
17)	a	18)	a	19)	b	20)	c
21)	b	22)	c	23)	b	24)	a
25)	a	26)	b	27)	b	28)	d
29)	c	30)	d	31)	a	32)	b
33)	a	34)	a	35)	d	36)	b
37)	a	38)	b	39)	b	40)	b
41)	c	42)	c	43)	a	44)	d
45)	b	46)	a	47)	a	48)	c
49)	c	50)	a	51)	b	52)	c
53)	d	54)	a	55)	b	56)	a
57)	b	58)	d	59)	c	60)	c
61)	a	62)	d	63)	c	64)	c
65)	a	66)	c	67)	b	68)	b
69)	c	70)	b	71)	a	72)	b
73)	a	74)	c	75)	d	76)	d
77)	a	78)	a	79)	c	80)	b
81)	d	82)	b	83)	c	84)	c
85)	d	86)	d	87)	c	88)	b
89)	c	90)	a	91)	b	92)	d
93)	a	94)	c	95)	c	96)	d
97)	d	98)	b	99)	d	100)	a
101)	a	102)	a	103)	d	104)	a
105)	d	106)	b	107)	b	108)	d
109)	d	110)	d	111)	a	112)	d
113)	d	114)	c	115)	d	116)	d
117)	a	118)	b	119)	c	120)	d
121)	b	122)	a	123)	c	124)	a

125)	d	126)	d	127)	a	128)	a
129)	d	130)	a	131)	b	132)	b
133)	b	134)	b	135)	d	136)	b
137)	c	138)	b	139)	c	140)	d
141)	a	142)	b	143)	b	144)	c
145)	d	146)	b	147)	c	148)	b
149)	a	150)	b	151)	b	152)	b
153)	b	154)	d	155)	c	156)	c
157)	a	158)	a	159)	c	160)	b
161)	a	162)	b	163)	b	164)	c
165)	a	166)	c	167)	b	168)	c
169)	d	170)	b	171)	c	172)	a
173)	b	174)	d	175)	c	176)	c
177)	d	178)	a	179)	a	180)	d
181)	b	182)	a	183)	d	184)	b
185)	b	186)	d	187)	c	188)	b
189)	c	190)	c	191)	a	192)	d
193)	c	194)	d	195)	d	196)	a
197)	d	198)	b	199)	b	200)	d
201)	c	202)	d	203)	d	204)	b
205)	a	206)	b	207)	d	208)	c
209)	a	210)	b	211)	d	212)	a
213)	d	214)	c	215)	d	216)	d
217)	d	218)	c	219)	a	220)	d
221)	d	222)	d	223)	c	224)	a
225)	a	226)	a	227)	b	228)	c
229)	d	230)	a	231)	b	232)	a
233)	c	234)	d	235)	c	236)	d
237)	a	238)	b	239)	d	240)	a
241)	d	242)	d	243)	d	244)	b
245)	b	246)	b	247)	c	248)	c
249)	d	250)	a	251)	d	252)	c
253)	d	254)	a	255)	b	256)	d
257)	a	258)	b	259)	c	260)	d
261)	a	262)	c	263)	d	264)	b
265)	a	266)	c	267)	a	268)	a

269)	b	270)	a	271)	a	272)	b
273)	b	274)	a	275)	c	276)	d
277)	c	278)	b	279)	b	280)	d
281)	a	282)	d	283)	b	284)	d
285)	c	286)	d	287)	a	288)	a
289)	c	290)	c	291)	d	292)	c
293)	a	294)	b	295)	c	296)	d
297)	c	298)	a	299)	b	300)	a
301)	a	302)	c	303)	b	304)	c
305)	b	306)	c	307)	a	308)	d
309)	c	310)	c	311)	d	312)	d
313)	b	314)	d	315)	b	316)	d
317)	c	318)	c	319)	a	320)	d
321)	d	322)	c	323)	d	324)	a
325)	c	326)	d	327)	d	328)	a
329)	d	330)	c	331)	d	332)	c
333)	b	334)	c	335)	b	336)	a
337)	b	338)	a	339)	c	340)	a
341)	b	342)	d	343)	c	344)	a
345)	a	346)	a	347)	d	348)	d
349)	d	350)	d	351)	a	352)	b
353)	d	354)	b	355)	a	356)	a
357)	b	358)	b	359)	b	360)	a
361)	a	362)	c	363)	b	364)	d
365)	c	366)	c	367)	c	368)	c
369)	c	370)	a	371)	c	372)	a
373)	a	374)	d	375)	a	376)	b
377)	a	378)	a	379)	c	380)	d
381)	d	382)	d	383)	d	384)	b
385)	b	386)	b	387)	b	388)	d
389)	d	390)	b	391)	d	392)	c
393)	c	394)	a	395)	a	396)	c
397)	b	398)	b	399)	c	400)	b
401)	b	402)	c	403)	d	404)	c
405)	c	406)	d	407)	d	408)	c
409)	a	410)	a	411)	b	412)	a

413)	b	414)	d	415)	a	416)	d
417)	a	418)	b	419)	b	420)	d
421)	a	422)	b	423)	c	424)	a
425)	c	426)	b	427)	c	428)	a
429)	c	430)	b	431)	a	432)	c
433)	b	434)	a	435)	c	436)	a
437)	a	438)	c	439)	a	440)	b
441)	a	442)	c	443)	d	444)	d
445)	b	446)	a	447)	c	448)	b
449)	a	450)	b	451)	d	452)	b
453)	b	454)	a	455)	c	456)	c
457)	d	458)	a	459)	a	460)	d
461)	d	462)	a	463)	b	464)	c
465)	b	466)	c	467)	c	468)	a
469)	c	470)	b	471)	a	472)	d
473)	b	474)	a	475)	b	476)	b
477)	d	478)	a	479)	c	480)	b
481)	c	482)	b	483)	d	484)	a
485)	a	486)	c	487)	a	488)	c
489)	a	490)	d	491)	c	492)	a
493)	a	494)	a	495)	c	496)	c
497)	c	498)	a	499)	d	500)	c
501)	b	502)	a	503)	b	504)	d
505)	b	506)	d	507)	b	508)	c
509)	b	510)	b	511)	b	512)	c
513)	b	514)	a	515)	a	516)	c
517)	b	518)	c	519)	d	520)	a
521)	b	522)	b	523)	a	524)	c
525)	c	526)	c	527)	a	528)	d
529)	a	530)	b	531)	b	532)	c
533)	c	534)	b	535)	c	536)	a
537)	a	538)	c	539)	d	540)	d
541)	a	542)	d	543)	b	544)	d
545)	b	546)	d	547)	b	548)	c
549)	d	550)	c	551)	d	552)	a
553)	b	554)	d	555)	b	556)	d

557)	b	558)	b	559)	a	560)	b
561)	b	562)	a	563)	c	564)	d
565)	c	566)	b	567)	c	568)	c
569)	a	570)	b	571)	d	572)	a
573)	a	574)	c	575)	b	576)	b
577)	d	578)	d	579)	a	580)	c
581)	b	582)	c	583)	c	584)	a
585)	a	586)	d	587)	d	588)	b
589)	b	590)	a	591)	b	592)	c
593)	d	594)	c	595)	c	596)	b
597)	b	598)	a	599)	b	600)	d
601)	c	602)	c	603)	c	604)	b
605)	a	606)	a	607)	d	608)	c
609)	a	610)	b	611)	c	612)	c
613)	c	614)	c	615)	c	616)	d
617)	d	618)	a	619)	a	620)	b
621)	c	622)	b	623)	c	624)	b
625)	a	626)	c	627)	c	628)	a
629)	b	630)	a	631)	c	632)	c
633)	c	634)	a	635)	c	636)	a
637)	d	638)	b	639)	b	640)	c
641)	a	642)	a	643)	c	644)	a
645)	c	646)	d	647)	c	648)	c
649)	a						

5
LAWS OF MOTION
HINTS AND SOLUTIONS

1 **(d)**
Force applied against friction
$f_k = \mu_k R = \mu_k mg = \frac{1}{3} \times 60 \times 9.8$
$= 196$ N

2 **(c)**
Tension the string $= m(g + a) =$ Breaking force
$\Rightarrow 20(g + a) = 25 \times g \Rightarrow a = g/4$
$= 2.5\ m/s^2$

3 **(c)**
Force on the football,
$F = \frac{md\ v}{dt}$
$F = \frac{m(v_2 - v_1)}{dt} = \frac{0.5(10 - 0)}{1/50} = 250$ N

4 **(a)**
Mass of the person $M = 80$ kg
Mass of the parachute $m = 5$ kg
$\therefore$ Total mass of the system $= M + m =$ 85kg
Downward acceleration $a = 2.8$ ms^{-2}
Let upward force $= F$
Applying Newton's II law of motion to this system
$F = (m + M)g - (m + M)a$
or $F = (m + M)(g - a)$
$F = 85(9.8 - 2.8)$N
$= 85 \times 7$ N
$F = 595$ N

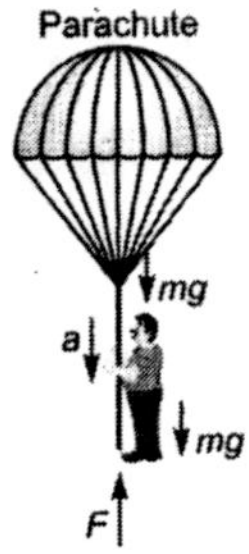

5 **(a)**
Figure shows free body diagram of the block

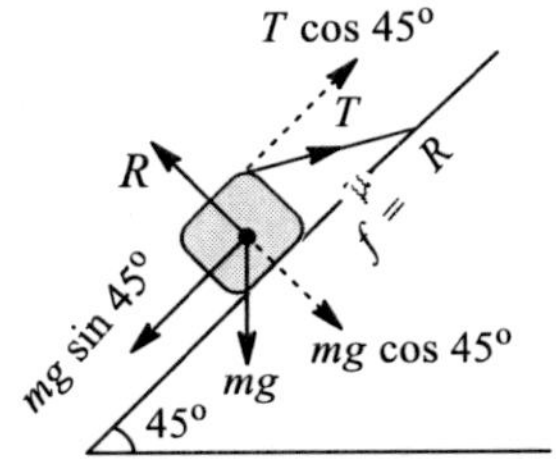

For equilibrium, along the place
$\mu R + T\cos 45° = mg\sin 45°$
$\mu R + \frac{T}{\sqrt{2}} = \frac{mg}{\sqrt{2}}$...(i)
For equilibrium, in direction perpendicular to inclined plane,
$R = T\sin 45° = mg\cos 45°$
$= \frac{T}{\sqrt{2}} + \frac{mg}{\sqrt{2}}$
Put in Eq. (i), $\frac{\mu}{\sqrt{2}}(T + mg) = -\frac{1}{\sqrt{2}}(mg - T)$
$\mu(50+15\times 10) = (15\times 10 - 50)$
$\mu = \frac{100}{200} = \frac{1}{2}$

6 **(c)**
Acceleration $a = \frac{\Delta v}{\Delta t} = \frac{v}{m} \cdot \frac{\Delta m}{\Delta t}$
$= \frac{50}{2} \times 0.1 = 2.5$ ms^{-2}

7 **(b)**
$T_C = 100$ g

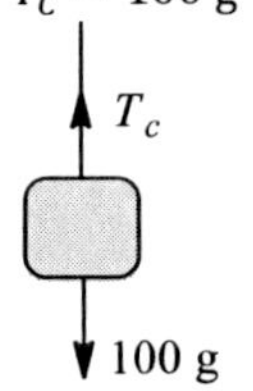

8 **(a)**
Since there in no resultant external force, linear momentum of the system remains constant

11 **(a)**
For body of mass 6 kg
$T = 6g = 6 \times 9.8 = 58.8$ N
For body of mass 4 kg
$T - T_1 = 4g = 4 \times 9.8 = 39.2$ N
$T_1 = T - 39.2$
$= 58.8 - 39.2 = 19.6$ N

12 **(c)**
Due to acceleration in forward direction, vessel is in an accelerated frame therefore a Pseudo force will be exerted in backward direction. Therefore water will be displaced in backward direction

13 **(d)**
$v = u - at \Rightarrow u - \mu gt = 0 \therefore \mu = \frac{u}{gt}$
$= \frac{6}{10 \times 10} = 0.06$

15 **(d)**

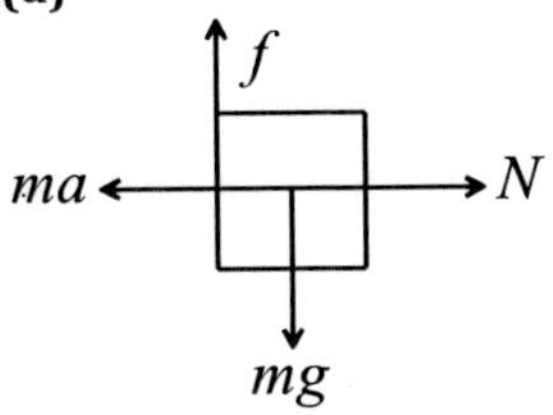

Here $f = mg$ and $N = m\alpha$ but $f \leq \mu N$
So $mg \leq \mu m\alpha \Rightarrow \alpha \geq \frac{g}{\mu}$

17 **(a)**
For equilibrium of street light,

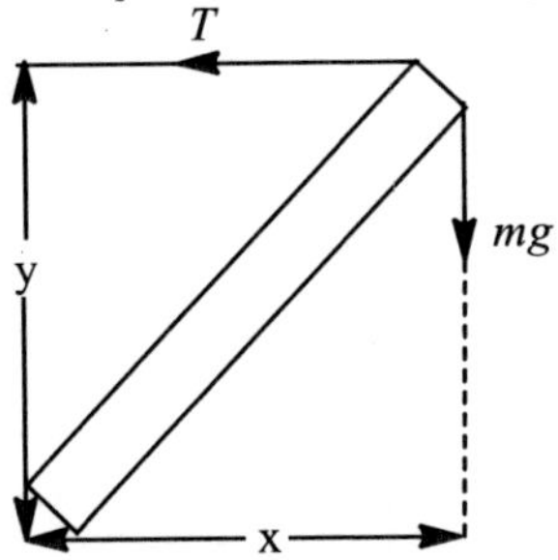

$mg \times x = T \times y$ or $T = \frac{mg\,x}{y}$
For $T$to be minimum$y$ should be maximum.
Hence, pattern A is more sturdy.

18 **(a)**
$F = \mu R = 0.3 \times 250 = 75\ N$

19 **(b)**
In one dimensional elastic collision when two particles of same mass colloid with each other their velocity are interchange *i.e.* change in momentum of particle
$|0 - mu| = mu$
But impulse = change in momentum
$\Rightarrow \frac{1}{2} \times F_0 \times T = mu$
$\therefore F_0 = \frac{2mu}{T}$

21 **(b)**
$T = \frac{2m_1m_2}{m_1 + m_2}g = \frac{2 \times 4 \times 2 \times 10}{4 + 2}$
$= \frac{160}{6} = 26.6 = 27\text{ N}$
Total downward thrust on the pulley
$= 2T = 2 \times 27 = 54\text{ N}$

22 **(c)**
Change of momentum $\mathbf{F}\,\Delta t = m\Delta\mathbf{v}$
$\Rightarrow \mathbf{F} = \frac{m\Delta\mathbf{v}}{t}$
By doing so time of change in momentum increases and impulsive force on knees decreases.

23 **(b)**
The acceleration of a rocket is given by
$a = \frac{v}{m}\left(\frac{\Delta m}{\Delta t}\right) - g$
$= \frac{400}{100}\left[\frac{5}{1}\right] - 10$
$= 20 - 10 = 10\text{ m/s}^2$

24 **(a)**
For limiting condition $\mu = \frac{m_B}{m_A + m_C} \Rightarrow 0.2 = \frac{5}{10 + m_C}$
$\Rightarrow 2 + 0.2m_C = 5 \Rightarrow m_C = 15\ kg$

25 **(a)**
Let m_1, m_2 be mass of bullet and gun and u_1, u_2 be their initial velocities and v_1 and $v_2$be their final velocities. From law of conservation of momentum.
$m_1u_1 + m_2u_2 = m_1v_1 + m_2v_2$
Given, $m_1 = 5\text{ g} = 5 \times 10^{-3}\text{ kg}, m_2 = 5\text{ kg}$
$u_1 = 0, u_2 = 0$ (at rest), $v_1 = 500\text{ ms}^{-1}$
$\therefore\ m_1 \times 0 + m_2 \times 0$
$= 5 \times 10^{-3} \times 500 + 5 \times v_2$
$\Rightarrow\ v_2 = -\frac{2.5}{5} = -0.5\text{ ms}^{-1}$

26 **(b)**
As the mass of 10 kg has acceleration 12 m/s^2 therefore it apply 120N force on mass 20kg in a backward direction
$\therefore$ Net forward force on 20 kg mass = $200 - 120 = 80N$
$\therefore$ Acceleration $= \frac{80}{20} = 4\ m/s^2$

27 **(b)**

Now, atB

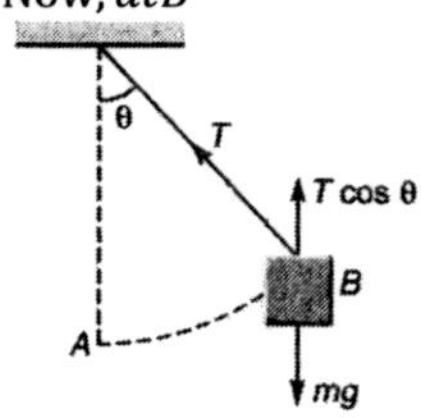

In equilibrium,

$T\cos\theta = mg$

$\Rightarrow \cos\theta = \dfrac{150\times 9.8}{2940}$

$\Rightarrow \cos\theta = 0.5 \Rightarrow \theta = 60°$

28 **(d)**

For upper half

$v^2 = u^2 + 2al/2 = 2(g\sin\theta)l/2$
$= gl\sin\theta$

For lower half

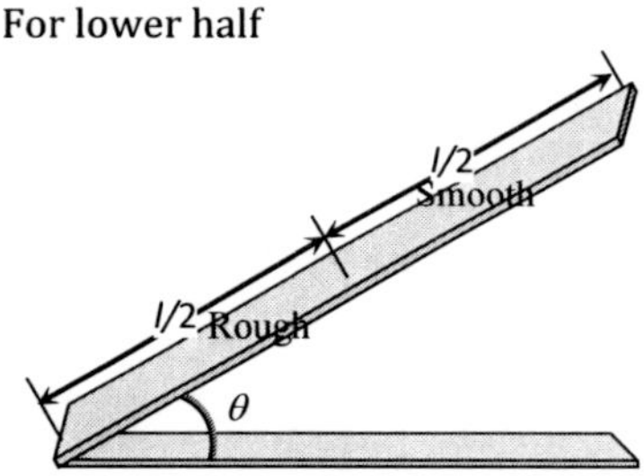

$\Rightarrow 0 = u^2 + 2g(\sin\theta - \mu\cos\theta)\dfrac{l}{2}$

$\Rightarrow -gl\sin\theta = gl(\sin\theta - \mu\cos\theta)$

$\Rightarrow \mu\cos\theta = 2\sin\theta \Rightarrow \mu = 2\tan\theta$

30 **(d)**

$\vec{p}(t) = A(\hat{\imath}\cos kt - \hat{\jmath}\sin kt)$

$\vec{F} = \dfrac{d}{dt}(\vec{p}(t)) = Ak(-\hat{\imath}\sin kt - \hat{\jmath}\cos kt)$

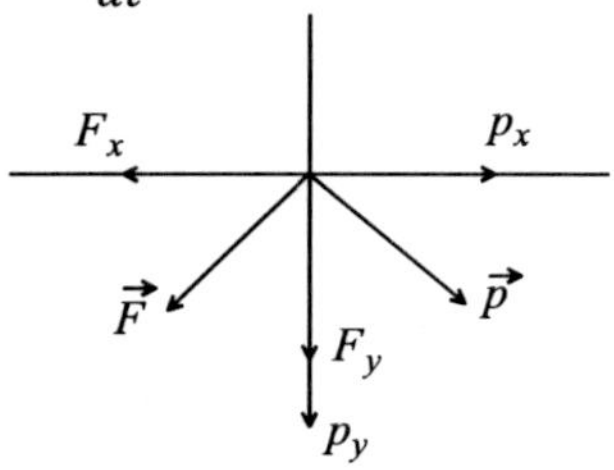

$\vec{F}.\vec{p} = A^2k(-\cos kt\sin kt$
$+ \sin kt\cos kt) = 0$

$\therefore$ The momentum and force are perpendicular to each other at 90°

31 **(a)**

When the lift moves upwards, the apparent weight,

$= m(g+a)$. Hence reading of spring balance increases

32 **(b)**

The acceleration of the centre of mass of the block,

$= \dfrac{g}{2}\sqrt{(\sin\angle ABC)^2 + (\sin\angle ABC)^2}$

$= \dfrac{g}{2}\sqrt{\sin^2 30° + \sin^2 60°}$

$= \dfrac{g}{2}\sqrt{(0.5)^2 + (0.866)^2} = \dfrac{g}{2}$

33 **(a)**

For angle of repose,

Friction = Component of weight along the plane

$= mg\sin\theta = 2\times 9.8\times\sin 45°$
$= 19.6\sin 45°$

34 **(a)**

Initially under equilibrium of mass m

$T = mg$

Now, the string is cut. Therefore, $T = mg$force is decreased on mass m upward and downwards on mass $2m$.

$\therefore\ a_m = \dfrac{mg}{m} = g(downwards)$

$anda_{2m} = \dfrac{mg}{2m} = \dfrac{g}{2}\ (upwards)$

35 **(d)**

Here friction force provides centripetal force so

$f = m\,\omega^2\,r$ but $f \le \mu mg$

So $m\omega^2 r \le \mu mg \Rightarrow r \le \frac{\mu g}{\omega^2}$

36 **(b)**

Horizontal velocity of apple will ren ain same but due to retardation of train, velocity of train and hence velocity of boy $w.r.t.$ ground decreases, so apple falls away from the hand of boy in the direction of motion of the train

37 **(a)**

Given, $m_1 = 1$kg, $m_2 = 2$ kg and g
$= 10$m/s^2

$a = \left(\dfrac{m_2 - m_1}{m_1 + m_2}\right)g$

$= \left(\dfrac{2-1}{1+2}\right)10 = \dfrac{10}{3}$

$S = \dfrac{1}{2}at^2$

$= \dfrac{1}{2}\times\dfrac{10}{3}\times 4 = \dfrac{20}{3}$

$m = \dfrac{2\times\frac{20}{3} - 1\times\frac{20}{3}}{3} = \dfrac{20}{9}$

39 **(b)**

If a large force F acts for a short time dt the impulse imparted I is

$I = F.dt, = \frac{dp}{dt}.dt$

$I = dp$ =change in momentum

40 **(b)**

Acceleration $(a) = g(\sin\theta - \mu\cos\theta)$ and $s = l$

$v = \sqrt{2as} = \sqrt{2gl(\sin\theta - \mu\cos\theta)}$

41 **(c)**

Various forces acting on the ball are as shown in figure. The three concurrent forces are in equilibrium. Using Lami's theorem

$\frac{T_1}{\sin 150°} = \frac{T_2}{\sin 120°} = \frac{10}{\sin 90°}$

$\frac{T_1}{\sin 30°} = \frac{T_2}{\sin 60°} = \frac{10}{1}$

$\therefore T_1 = 10\sin 30°$
$= 10 \times 0.5 = 5\text{N}$

$T_2 = 10\sin 60°$

and $= 10 \times \frac{\sqrt{3}}{2} = 5\sqrt{3}\text{N}$

42 **(c)**

As $\vec{v} = 5t\hat{i} + 2t\hat{j} \therefore \vec{a} = a_x\hat{i} + a_y\hat{j} = 5\hat{i} + 2\hat{j}$

$F = ma_x\hat{i} + m(g + a_y)\hat{j}$

$\therefore |\vec{F}| = m\sqrt{a_x^2 + (g + a_y)^2} = 26\ N$

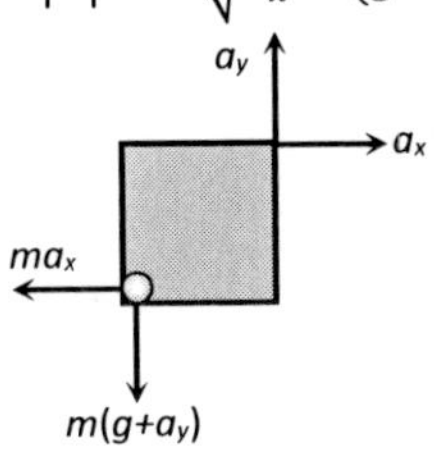

43 **(a)**

The system may be represented as follow

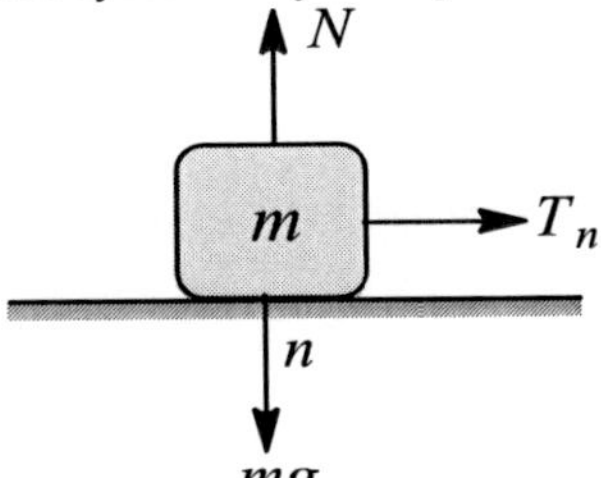

From the force diagram,

$Mg - T = Ma$ ...(i)

and $T = nma$ ...(ii)

From Eqs. (i) and (ii),

$a = \frac{Mg}{nm + M}$

The fore diagram of nth block is shown in figure

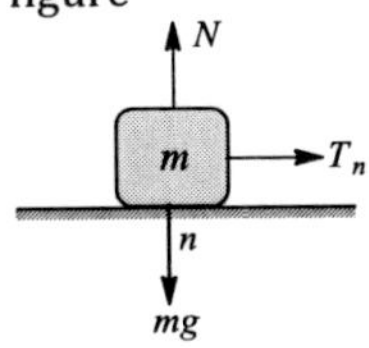

From the figure,

$T_n = ma = \frac{mMg}{nm + M}$

44 **(d)**

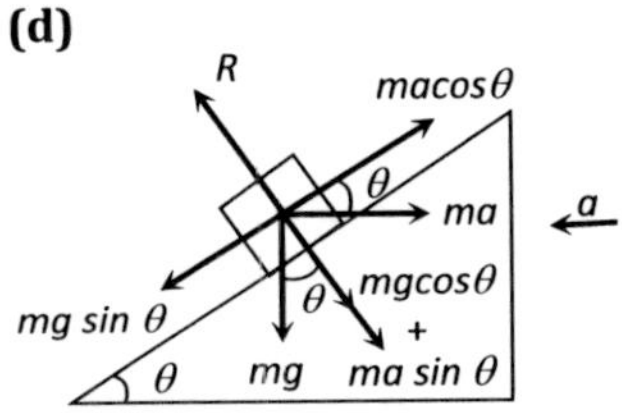

When the whole system is accelerated towards left then pseudo force (ma) works on a block towards right

For the condition of equilibrium

$mg\sin\theta = ma\cos\theta \Rightarrow a = \frac{g\sin\theta}{\cos\theta}$

$\therefore$ Force exerted by the wedge on the block

$R = mg\cos\theta + ma\sin\theta$

$R = mg\cos\theta + m\left(\frac{g\sin\theta}{\cos\theta}\right)\sin\theta$
$= \frac{mg(\cos^2\theta + \sin^2\theta)}{\cos\theta}$

$R = \frac{mg}{\cos\theta}$

45 **(b)**

Opposite force causes retardation

46 **(a)**

Slope of surface should change from one constant value (non zero) to another constant value (non zero) in terms of sign because force is constant piecewise

47 **(a)**

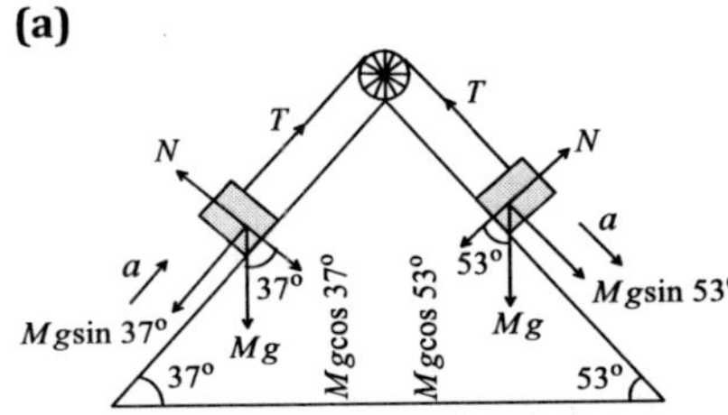

Let T be the tension in the string

Let a be acceleration of the system

The equations of motion are

$Ma = Mg \sin 53° - T$ ...(i)

And $Ma = T - Mg \sin 37°$ ...(ii)

Adding (i) and (ii, we get

$$a = \frac{Mg(\sin 53° - \sin 37°)}{2M}$$

$= g \cos 45° \sin 8°$

$\left[\because \sin A - \sin B = 2\cos\left(\frac{A+B}{2}\right)\sin\left(\frac{A-B}{2}\right)\right]$

$= 10 \times \frac{1}{\sqrt{2}} \times 0.139$

$= 0.98\, ms^{-2} \approx 1\, ms^{-2}$

48 **(c)**

$v = 3t^2 - 75; a = 6t$

$F = ma = 36t$. At $t = 4$ s, $F = 144$ N

49 **(c)**

Direction of second force should be at 180°

50 **(a)**

Here, mass of the body

$m = 0.05\, kg$

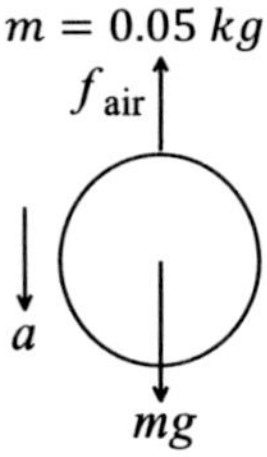

Acceleration $a = 9.5\, ms^{-2}, g = 9.8\, m/s^{-2}$

$\therefore mg - f_{air} = ma \Rightarrow f_{air} = m(g - a)$

$= 0.05(9.8 - 9.5) = 0.015\, N$

51 **(b)**

The accelerating force of the rocket

$= \text{upward thrust} = \frac{\Delta m}{\Delta t} \cdot v$

Given, $\frac{\Delta m}{\Delta t} = 50 \times 10^{-3} \text{kgs}^{-1}$,

$v = 400 \text{ ms}^{-1}$

So, accelerating force $= 50 \times 10^3 \times 400 = 20$ N

52 **(c)**

Sliding friction is greater than rolling friction

53 **(d)**

Given that, $u = 10 \text{ ms}^{-1}, \frac{dm}{dt} = 2 \text{ kgs}^{-1}$

Total mass of the truck, $M = (100 + 100)\text{kg} = 200$ kg

We know that,

$$F = \frac{u\, dm}{dt}$$

or $F = 10 \times 2 = 20$ N

or $Ma = 20$ N

or $200a = 20$ N

or $a = \frac{20}{200} ms^{-2} = \frac{1}{10} ms^{-2}$

now, we know that,

$a = \frac{v - u}{t}$ $(u = 10ms^{-2}, t = 50s)$

$\Rightarrow \frac{1}{10} = \frac{v - 10}{50} \Rightarrow v = 15 \text{ ms}^{-1}$

54 **(a)**

Form conservation of linear momentum

$m_1 v_1 = (m_1 + m_2)v$

or $1000 \times 50 = (1000 + 250)v$

$\Rightarrow v = 40 \text{ ms}^{-1}$

55 **(b)**

Let a be the acceleration of each block. Then,

$T_3 = (m_1 + m_2 + m_3)a$ .... (i)

and $T_2 = (m_1 + m_2)a$ .... (ii)

from Eq. (i) and (ii), we get

$T_2 = \left(\frac{m_1 + m_2}{m_1 + m_2 + m_3}\right) \times T_3$

$= \left(\frac{10 + 6}{10 + 6 + 4}\right) \times 40 = 32$ N

56 **(a)**

$\mu = \tan\theta\left(1 - \frac{1}{n^2}\right) = \tan\theta\left(1 - \frac{1}{2^2}\right)$

$= \frac{3}{4}\tan\theta$

57 **(b)**

Thrust force by rocket

$F_t = v_r\left(-\frac{dm}{dt}\right)$ (upwards)

Weight of the rocket

$w = mg$ (downwards)

Net force on the rocket

$F_{net} = f_t - w$

$\Rightarrow ma = v_r\left(\frac{-dm}{dt}\right) - mg$

$\Rightarrow ma = v_r\left(\frac{-dm}{dt}\right) = \frac{m(g + a)}{v_r}$

$\therefore$ Rate of the ejected per second

$= \frac{5000(10 + 20)}{800} = \frac{5000 \times 30}{800}$

$= 187.5 \text{ kgs}^{-1}$

58 **(d)**

Acceleration $= \frac{P}{M+m}$

Force on the block = Mass of block × acceleration

$= \frac{MP}{M + m}$

59 **(c)**
Maximum force by surface when friction works
$F = \sqrt{f^2 + R^2} = \sqrt{(\mu R)^2 + R^2}$
$= R\sqrt{\mu^2 + 1}$
Minimum force $= R$ where there is no friction
Hence ranging from R to $R\sqrt{\mu^2 + 1}$
We get, $Mg \le F \le Mg\sqrt{\mu^2 + 1}$

60 **(c)**
If 1 N and 2 N act in the same direction and 3 N acts in opposite direction, equilibrium is possible

61 **(a)**
$$v = \sqrt{\mu g r} \Rightarrow r = \frac{v^2}{\mu g} = \frac{100}{0.5 \times 10} = 20$$

62 **(d)**
Reading of spring balance $= T = \frac{2m_1 m_2 g}{m_1 + m_2} (N)$

64 **(c)**
Mass of rope, $m = 0.1\ kg, \theta = 10°$

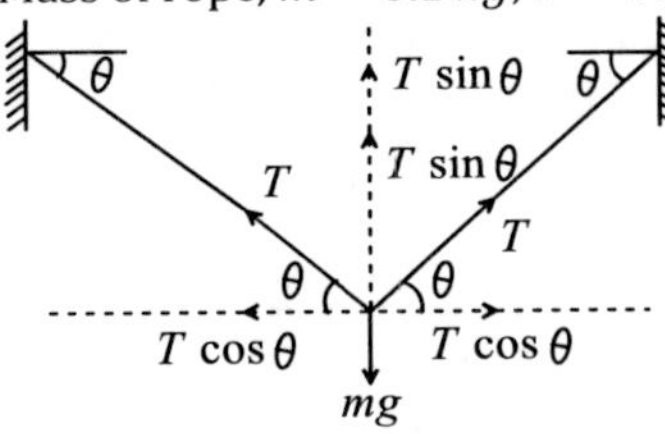

From figure, $2T \sin\theta = mg$
$$\Rightarrow T = \frac{mg}{2\sin\theta} = \frac{0.1 \times 9.8}{2 \sin 10°} = 2.82\ N$$

65 **(a)**
$$T = M \times a = M \times \left(\frac{F}{m + M}\right)$$

66 **(c)**
$F = \frac{dp}{dt}$, so the force is maximum when slope of graph is maximum

69 **(c)**
Mass of the packet $M = w/g$
For packet $F - w = Ma$
or $F - w = \left(\frac{w}{g}\right) \times 2g$
or $F = 3w$

70 **(b)**
$A + B = 18$ ...(i)
$12 = \sqrt{A^2 + B^2 + 2AB\cos\theta}$ ...(ii)
$\tan\alpha = \frac{B\sin\theta}{A + B\cos\theta} = \tan 90° \Rightarrow \cos\theta = -\frac{A}{B}$
...(iii)
By solving (i), (ii) and (iii), $A = 13\ N$ and $B = 5N$

71 **(a)**
$$F = \frac{W}{\mu} \therefore W = \mu F = 0.2 \times 10 = 2N$$

72 **(b)**
Since, force needed to overcome frictional force

73 **(a)**
Acceleration $= \frac{m_2}{m_1 + m_2} \times g = \frac{1}{2+1} \times 9.8 =$
$3.27\ m/s^2$
and $T = m_1 a = 2 \times 3.27 = 6.54\ N$

74 **(c)**
Acceleration of the car $= \frac{\text{Force on the car}}{\text{Mass of the car}}$
$$= \frac{mnv}{M} = \frac{0.01 \times 10 \times 500}{2000} \text{ms}^{-2}$$
$$= \frac{5}{200} \text{ms}^{-2}$$
$$= \frac{1}{40} \text{ms}^{-2}$$

75 **(d)**
Given that, $\frac{dm}{dt} = 0.1\ \text{kgs}^{-1}$;
mass of the rocket=100 kg
and $v = 1\ \text{kms}^{-1} = 1000\ \text{ms}^{-1}$
Thrust on the rocket, $F = v\frac{dm}{dt}$
$= 1000 \times 0.1$
Now, $F = Ma$
$$\therefore\ a = \frac{1000 \times 0.1}{100} = 1\ \text{ms}^{-2}$$

76 **(d)**
Using law of conservation of momentum, we get
$100 \times v = 0.25 \times 100 \Rightarrow v = 0.25\ m/s$

77 **(a)**
$$m = \frac{F}{a} = \frac{\sqrt{6^2 + 8^2 + 10^2}}{1} = \sqrt{200}$$
$$= 10\sqrt{2}\ kg$$

78 **(a)**
$F - F_1 = (dm)a$

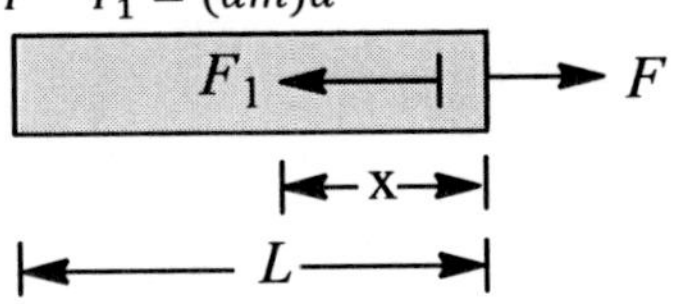

Acceleration of rope $a = \frac{F}{m}$
$$dm = \frac{m}{L}x$$
$$F - F_1 = \frac{mx}{L} \times \frac{F}{m}$$
$F - F_1 = \frac{Fx}{L}$ or $F_1 = \frac{F(L-x)}{L}$

79 **(c)**
When a person of mass m is placed on a weighing machine which is placed in a lift, then actual weight of the person is mg. This acts on a weighing machine which offers a reaction R given by the reading of weighing machine. This reaction exerted by the surface of contact on the person is the apparent weight of the person. When lift is accelerating upwards at the rate of a, then reaction

$R - mg = ma$

$\therefore \ R = m(g + a)$

Hence, apparent weight is greater than actual weight.

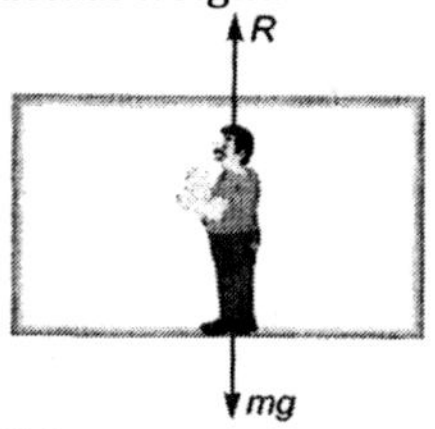

80 **(b)**
If v is common velocity of the block and movable wedge, then applying the principle o conservation of linear momentum we get,

$mu + 0 = (m + nm)v$

$v = \frac{mu}{m(1+n)} = \frac{u}{1+n}$

This infact, can be taken as velocity of centre of mass of the block and wedge

$ie, v_{CM} = v = \frac{u}{1+n}$

Applying the principle of conservation of energy

$\frac{1}{2}mu^2 = mgh + \frac{1}{2}m(1+n)\frac{u^2}{(1+n)^2}$

or $u^2 = 2gh + \frac{u^2}{1+n}$

$u = \sqrt{2gh\left(1 + \frac{1}{n}\right)}$

81 **(d)**
$F = mg \sin 30° = 50\ N = 5kg - wt$

82 **(b)**
$T_1 = 0.25 \times 100$ N

or $T_1 = 25$ N

83 **(c)**
Net force on the body = Applied force − Friction

$ma = F - \mu_k mg \Rightarrow \mu_k = \frac{F - ma}{mg}$

$= \frac{129.4 - 10 \times 10}{10 \times 9.8}$

$= 0.3$

84 **(c)**
$T = m(g + a) = 1000(9.8 + 1)$

$= 10800\ N$

85 **(d)**
In the frame of wedge, the force diagram of block is shown in figure. From free body diagram of wedge,

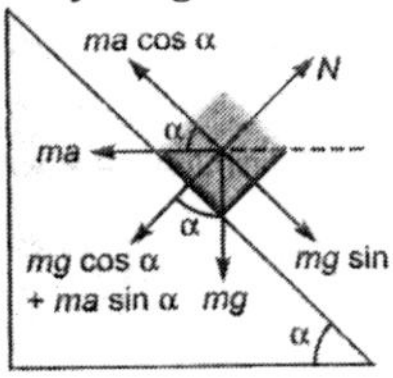

For block remain stationary,

$ma \cos \alpha = mg \sin \alpha$

$a = g \tan \alpha$

86 **(d)**
The coin will not slide if

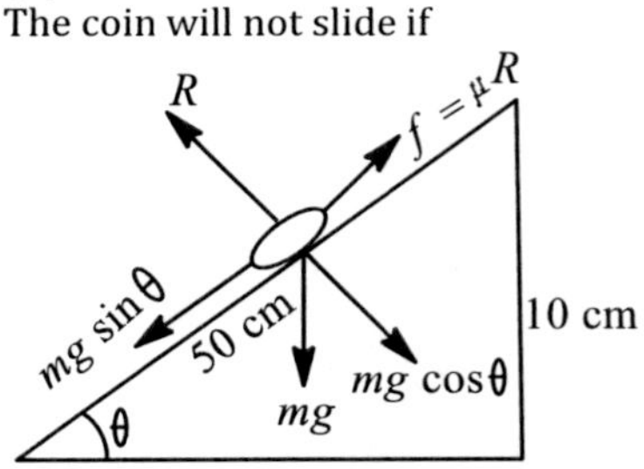

$mg \sin \theta = \mu R$

$ie, \ mg \sin \theta = \mu\, mg \cos \theta$ (as $R = mg \cos \theta$)

$ie, \tan \theta = \mu$

Here, $\tan \theta = \frac{10}{\sqrt{(50)^2 - (10)^2}} = 0.2$

$\therefore \ \mu = 0.2$

87 **(c)**
For given condition we can apply direct formula

$l_1 = \left(\frac{\mu}{\mu + 1}\right) l$

88 **(b)**
$F_k = \mu_k R = \mu_k mg \cos \theta$

$F_k = 1.7 \times 0.1 \times 10 \times \cos 30°$

$= 1.7 \times \frac{\sqrt{3}}{2} N$

89 **(c)**
Impulse $= \Delta P = m(V_f - V_i) = 0.4[1 - (-1)] = 0.8\ Ns$

90 **(a)**

Since acceleration of lift is zero

91 **(b)**

$F = \frac{m(v-u)}{t} = \frac{0.15[20-(-10)]}{0.1} = \frac{0.15\times 30}{0.1} =$ 45N

92 **(d)**

Force $F = \frac{dp}{dt}$

$= v\left[\frac{dM}{dt}\right]$

$= \alpha v^2$

$\therefore \quad a = \frac{F}{M} = \frac{\alpha v^2}{M}$

93 **(a)**

Here, thrust force is responsible to accelerate the rocket, so initial thrust of the blast.

$= ma$

$= 3.5 \times 10^4 \times 10$

$= 3.5 \times 10^5$ N

94 **(c)**

$5N$ force will not produce any tension in spring without support of other $5N$ force. So here the tension in the spring will be $5\,N$ only

95 **(c)**

$s = (v^2 - u^2)/2a$ is independent of mass

Hence both will stop at the same distance

96 **(d)**

$T = \frac{2m_1m_2}{m_1+m_2}g = \frac{2\times 10\times 6}{10+6}\times 9.8$

$= 73.5\ N$

97 **(d)**

Force acting on plate, $F = \frac{dp}{dt} = v\left(\frac{dm}{dt}\right)$

Mass of water reaching the plate per $sec = \frac{dm}{dt}$

$= Av\rho = A(v_1+v_2)\rho = \frac{V}{v_2}(v_1+v_2)\rho$

$(v = v_1 + v_2 =$ velocity of water coming out of jet $w.r.t.$ plate)

$\left[A = \text{Area of cross section of jet} = \frac{V}{v_2}\right]$

$\therefore F = \frac{dm}{dt}v = \frac{V}{v_2}(v_1+v_2)\rho \times (v_1+v_2)$

$= \rho\left[\frac{V}{v_2}\right](v_1+v_2)^2$

98 **(b)**

Tension between m_2 and m_3 is given by

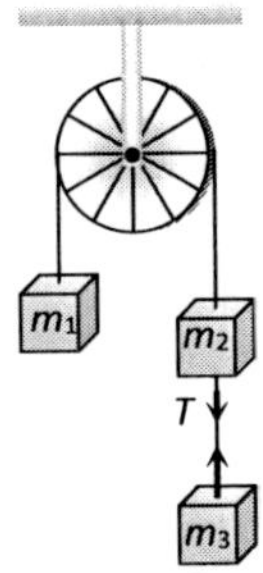

$T = \frac{2m_1m_3}{m_1+m_2+m_3}\times g$

$= \frac{2\times 2\times 2}{2+2+2}\times 9.8 = 13\ N$

99 **(d)**

As by an internal force momentum of the system can not be changed

100 **(a)**

Retardation in upward motion = $g(\sin\theta + \mu\cos\theta)$

$\therefore$ Force required just to move up $F_{up} = mg(\sin\theta + \mu\cos\theta)$

Similarly for down ward motion $a = g(\sin\theta - \mu\cos\theta)$

$\therefore$ Force required just to prevent the body sliding down

$F_{dn} = mg(\sin\theta - \mu\cos\theta)$

According to problem $F_{up} = 2F_{dn}$

$\Rightarrow mg(\sin\theta + \mu\cos\theta) = 2mg(\sin\theta - \mu\cos\theta)$

$\Rightarrow \sin\theta + \mu\cos\theta = 2\sin\theta - 2\mu\cos\theta$

$\Rightarrow 3\mu\cos\theta = \sin\theta \Rightarrow \tan\theta\ 3\mu$

$\Rightarrow \theta = \tan^{-1}(3\mu) = \tan^{-1}(3\times 0.25)$

$= \tan^{-1}(0.75) = 36.8°$

102 **(a)**

Angle of repose $\alpha = \tan^{-1}(\mu) = \tan^{-1}(0.8) = 38.6°$

Angle of inclined plane is given $\theta = 30°$. It means block is at rest therefore,

Static friction = component of weight in downward direction = $mg\sin\theta =$ $10\ N \quad \therefore m = \frac{10}{g\times \sin 30°} = 2\ kg$

103 **(d)**

$T = \frac{2\times m_Bm_C}{m_A+m_B+m_C}\times g = \frac{2\times 1\times 5}{3+1+5}\times g$

$= \frac{10}{9}g$

104 **(a)**

Free body diagram (FBD) of the block (shown by a dot) is shown in figure.

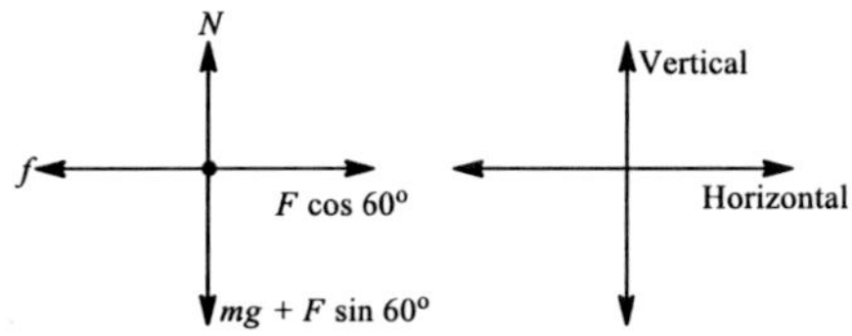

For vertical equilibrium of the block,

$N = mg + F\sin 60° = \sqrt{3}\,g + \sqrt{3}\frac{F}{2}$ ....(i)

For no motion, force of friction

$f \geq F\cos 60°$

or $°\mu N \geq F\cos 60°$

or $\frac{1}{2\sqrt{3}}\left(\sqrt{3}g + \frac{\sqrt{3}\,F}{2}\right) \geq \frac{F}{2}$

or $g \geq \frac{F}{2}$ or $F \leq 2\,g$ or 20 N

Therefore, maximum value of F is 20 N.

105 **(d)**

Effective value of acceleration due to gravity in the lift = g − a

Acceleration down the inclined plane

Using, $s = ut + \frac{1}{2}at^2$, we get

$l = \frac{1}{2}(g - a)\sin\theta t^2$, we get

$t = \sqrt{\frac{2l}{(g-a)\sin\theta}}$

106 **(b)**

Applying conservation of momentum,

$Mv = m \times 0 + (M - m)V$

or $V = \frac{M}{M-m}v$

107 **(b)**

Let a be the acceleration of each block and T_1 and T_2 be the tensions, in the two strings as shown in figure.

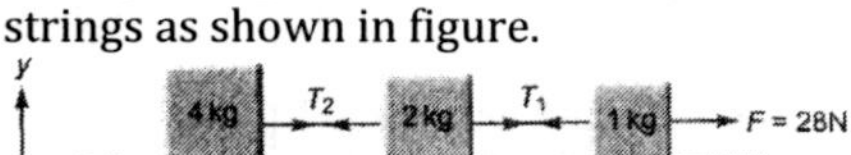

Using, $\sum F_x = ma_x$

or $28 = (4 + 2 + 1)a$

or $a = \frac{28}{7} = 4\ \text{ms}^{-2}$

108 **(d)**

$v = u - at \Rightarrow t = \frac{u}{a}$ [As $v = 0$]

$t = \frac{u \times m}{F} = \frac{30 \times 1000}{5000} = 6\ sec$

109 **(d)**

Velocity after 10 sec. $= \frac{1}{m}$

Area enclosed between $F - t$ graph and time axis

Area $= \frac{1}{2}\left[\frac{1}{2} \times 2 \times 10 + 2 \times 10 + \frac{1}{2} \times 30 \times 2 + \frac{1}{2} \times 4 \times 20\right]$

$= 50\ m/s$

110 **(d)**

In this case the internal force is applied on the system, so he will not succeed. According to Newton's law the state of a body can only be changed if some external force is applied on it.

111 **(a)**

For a smooth plane, $v = \sqrt{2g\sin\theta.s}$ and for a rough plane,

$\frac{v}{n} = \sqrt{2g(\sin\theta - \mu\cos\theta).s}$

$\therefore\ n = \sqrt{\frac{\sin\theta}{\sin\theta - \mu\cos\theta}}$ or $n^2 = \frac{\sin\theta}{\sin\theta - \mu\cos\theta}$

$\Rightarrow\ (n^2 - 1)\sin\theta = n^2\mu\cos\theta$

or $\mu = \left(\frac{n^2-1}{n^2}\right)\tan\theta = \tan\theta\left(1 - \frac{1}{n^2}\right)$

112 **(d)**

$T = mg = 50 \times 10^{-3} \times 10 = 0.5\ N$

113 **(d)**

$\mu_s = \frac{\text{Length of the chain hanging from the tal}}{\text{Length of the chain lying on the table}}$

$= \frac{l/3}{l - l/3} = \frac{l/3}{2l/3} = \frac{1}{2}$

114 **(c)**

As $2m > m$

The acceleration of the two block system will be as shown in figure.

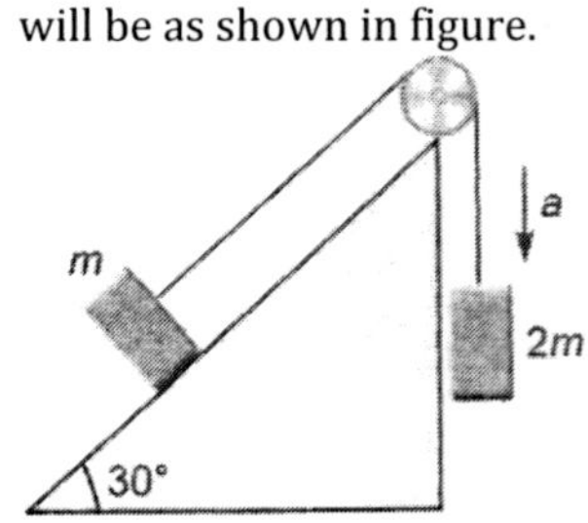

The equation for block of mass $2m$ is

$2mg - T = 2ma$ ...(i)

Similarly, for block of mass m

$T - mg\sin 30° = ma$...(ii)

From Eqs. (i) and (ii), we have

$2mg - \frac{mg}{2} = 3ma$

$\Rightarrow\ a = \frac{g}{2}$

115 **(d)**

Let, R be the normal contact force by wall on the block.

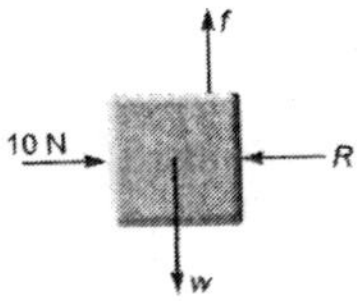

$R = 10$ N

$f_L = w$ and $f = \mu R$

$\therefore \ \mu R = w$

$\Rightarrow \quad w = 0.2 \times 10 = 2$ N

116 **(d)**

$$m_G v_G = m_B v_B \Rightarrow v_B = \frac{m_G v_G}{m_B} = \frac{1 \times 5}{10 \times 10^{-3}} = 500\ m/s$$

117 **(a)**

Let length of chain be l and mass m. Let a part x of chain can hang over one edge of table having coefficient of friction

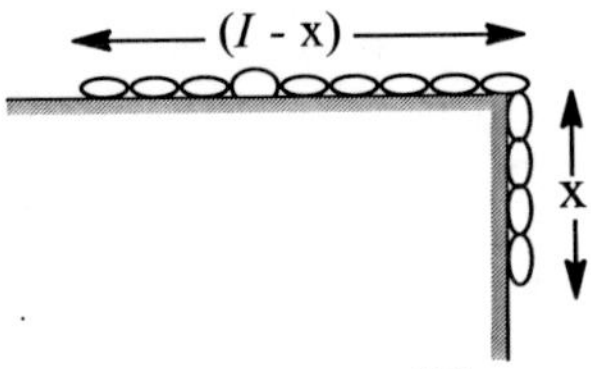

$\therefore$ Pulling force, $F = \frac{mx}{l}$g

and friction force, $f = \mu N = \mu \frac{m}{l}(l - x)$g

For equilibrium, $F = f$, hence

$$\frac{mx}{l}.\text{g} = \mu \frac{m}{l}(l - x)\text{g} = 0.25\frac{m}{l}(l - x)\text{g}$$

$\Rightarrow x = \frac{l}{5}$ or $\frac{x}{l} = \frac{1}{5} = 20\%$

118 **(b)**

$$v = \sqrt{\mu r g} = \sqrt{0.4 \times 30 \times 9.8} = 10.84\ m/s$$

119 **(c)**

Let smaller force be F_1. Resultant R of the forces is at 90° to AB,

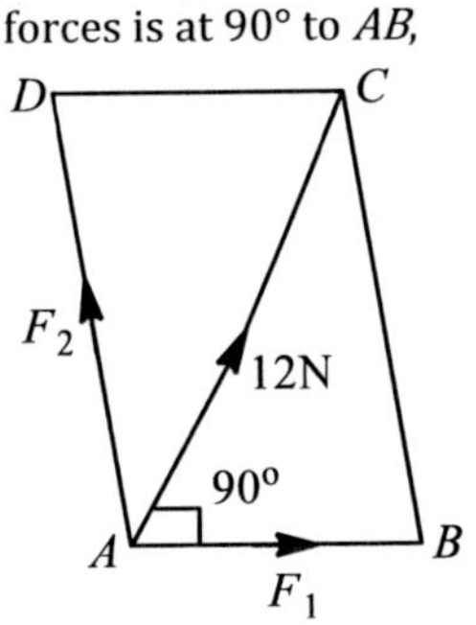

$\therefore R^2 + F_1^2 = F_2^2$ in ΔABC

or $(12)^2 = F_2^2 - F_1^2$ ...... (i)

or $144 = (F_2 - F_1)(F_2 + F_1)$

but $F_1 + F_2 = 18$ N (given) ....(ii)

$\therefore F_2 - F_1 = \frac{144}{18} = 8$ ..... (iii)

From Eqs. (ii) and (iii), $F_1 = 5$, $F_2 = 13$

Hence, forces are 5 N and 13 N.

120 **(d)**

Tension in the string, $T = \frac{m_1 m_2}{m_1 + m_2} g$

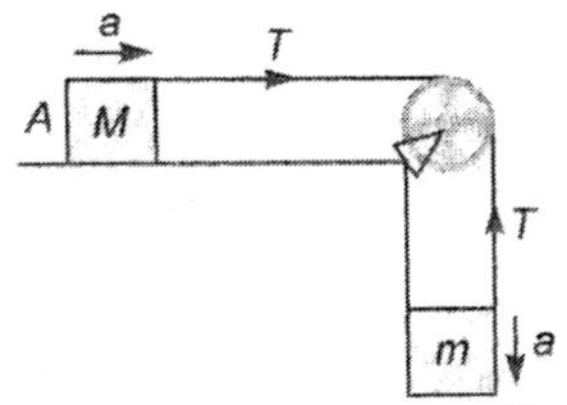

Here, tension $T = \frac{Mm}{M + m} g$

121 **(b)**

Acceleration produced in jet = $\frac{\text{Change in velocity}}{\text{Time}}$

$$a = \frac{(10^3 - 0)}{10} = 100\ m/s^2$$

$$\therefore \text{Mass} = \frac{\text{Force}}{\text{Acceleration}} = \frac{10^5}{10^2} = 10^3\ kg$$

122 **(a)**

Net force on the particle is zero so the $\vec{v}$ remains unchanged

123 **(c)**

According to principle of conservation of linear momentum $1000 \times 50 = 1250 \times v \Rightarrow v = 40\ km/hr$

124 **(a)**

In case of upward motion

$F = m(g + a)$

$= 60(9.8 + 4.9)$

$= 60(14.7)$

$= 882$ kg

125 **(d)**

Force $= m\left(\frac{dv}{dt}\right) = \frac{0.25 \times [(10) - (-10)]}{0.01} = 25 \times 20 = 500\ N$

126 **(d)**

Net force = Applied force − Friction force

$ma = 24 - \mu\, mg = 24 - 0.4 \times 5 \times 9.8$

$= 24 - 19.6$

$$\Rightarrow a = \frac{4.4}{5} = 0.88\ m/s^2$$

127 **(a)**

When friction absent

$a_1 = g \sin\theta$

$\therefore\ s_1 = \frac{1}{2}a_2 t_2^2$ ....(i)

When friction in present

$a_2 = g\sin\theta - \mu_k g\cos\theta$

$\therefore\ s_2 = \frac{1}{2}a_2 t_2^2$ ....(ii)

From Eqs. (i) and (ii), we have

$\frac{1}{2}a_1 t_1^2 = \frac{1}{2}a_2 t_2^2$

or $a_1 t_1^2 = a_2(nt_1)^2 \quad (\because\ t_2 = nt_2)$

or $a_1 = n^2 a_2$

or $\frac{a_2}{a_1} = \frac{g\sin\theta - \mu_k g\cos\theta}{g\sin\theta} = \frac{1}{n^2}$

or $\frac{g\sin 45° - \mu_k g\cos 45°}{g\sin 45°} = \frac{1}{n^2}$

or $1 - \mu_k = \frac{1}{n^2}$

or $\mu_k = 1 - \frac{1}{n^2}$

128 **(a)**

Reaction on m is mg. Maximum friction force by m on M is μmg. So, the force on M is μmg forward

Acceleration $= \frac{\mu mg}{M}$

129 **(d)**

When lift falls with acceleration (a) or rises with retardation $(-a)$, then a person apparently loses weight.

$R - mg = -ma$

$\Rightarrow\ R = m(g - a)$

In the given case scale reading changes from 60 kg to 50 kg for a while and then comes back to 60 kg mark. It happens while the lift in motion upwards suddenly stops.

130 **(a)**

To blocks of masses M and m are connected by a string passing over a pulley. Let T is the tension in the string and a is the downward acceleration. Applying II law of motion for block of mass M.

$T = ma$ ...(i)

For block of mass m

$ma = mg - T$ ...(ii)

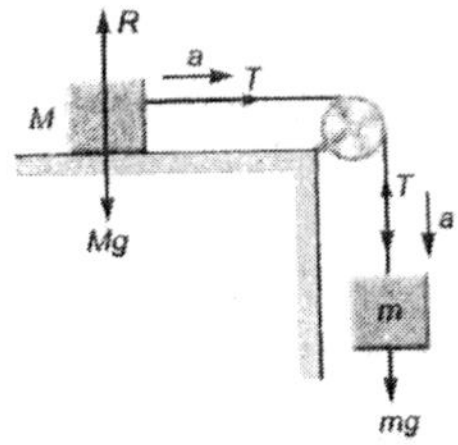

From Eqs. (i) and (ii), we get

$ma = mg - Ma$

or $a = \frac{mg}{m + M}$

131 **(b)**

$\uparrow 3mg$

$\boxed{A}\quad \boxed{B}$

$2mg \downarrow \quad \downarrow mg$

$a_A = g/2$

$a_B = g$

132 **(b)**

When false balance has equal arms then,

$W = \frac{X+Y}{2}$

133 **(b)**

$F = \frac{udm}{dt} = m(g + a)$

$\Rightarrow \frac{dm}{dt} = \frac{m(g+a)}{u} = \frac{5000 \times (10 + 20)}{800}$

$= 187.5\ kg/s$

134 **(b)**

Combined momentum $= 2p\hat{\mathbf{i}} + p\hat{\mathbf{j}}$

Magnitude of combined momentum

$= \sqrt{(2p)^2 + p^2} = \sqrt{5p^2} = \sqrt{5}p$

This must be equal to the momentum of the third part

135 **(d)**

From conservation of momentum

$MV + (4mv) = 0$

$\Rightarrow\ V = -\frac{4mv}{M}$

$= -\frac{4 \times 35 \times 10^{-3} \times 400}{20}$

$= -2.8\ \text{ms}^{-1}$

Force applied on the rifle,

$F = \frac{MV}{t} = -\frac{20 \times 2.8}{1} = -56\ \text{N}$

136 **(b)**

Since downward force along the inclined plane

$= mg\sin\theta = 5 \times 10 \times \sin 30° = 25N$

137 **(c)**

$v = 108 \times \frac{5}{18}\text{ms}^{-1} = 30\ \text{ms}^{-1}$

Momentum $= 120\times 30$ kg ms^{-1} (initial)
Final momentum $= 0$
Force $= \frac{\text{Change in momentum}}{\text{Time}}$
$= \frac{120\times30}{20} = 360$ N

140 **(d)**
If the applied force is less than limiting friction between block A and B , then whole system move with common acceleration

$i.e.\, a_A = a_B = \dfrac{F}{m_A + m_B}$

but the applied force increases with time, so when it becomes more than limiting friction between A and B, block B starts moving under the effect of net force $F - F_k$ Where F_k = kinetic friction between block A and B

$\therefore$ Acceleration of block B, $a_B = \frac{F-F_k}{m_B}$

As F is increasing with time so a_B will increase with time

Kinetic friction is the cause of motion of block A

$\therefore$ Acceleration of block A, $a_A = \frac{F_k}{m_A}$

It is clear that $a_B > a_A$. $i.e.$ graph (d) correctly represents the variation in acceleration with time for block A and B

141 **(a)**
$T\cos\theta = mg$
$T\sin\theta = ma$
$\tan\theta = \dfrac{a}{g}$

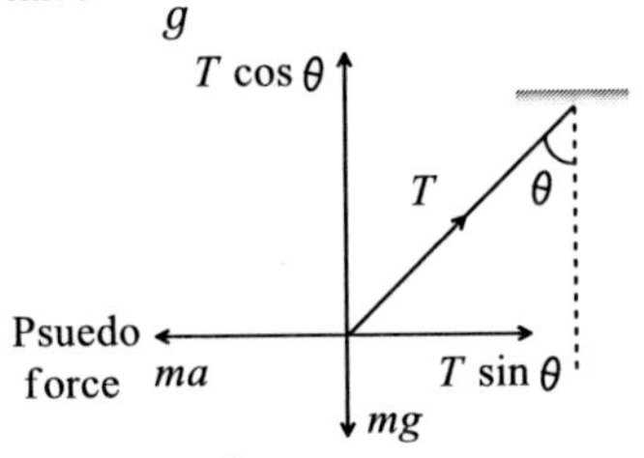

$\therefore \theta = \tan^{-1}(a/g)$

142 **(b)**
From Newton's second law

$F = n\cdot\left(\dfrac{\Delta p}{\Delta t}\right)$

Here, F = Force, n = number of bullets fired per second.

$\dfrac{\Delta p}{\Delta t}$ = rate of change of momentum of one bullet.

$\Rightarrow F = n\left[\dfrac{mv - 0}{\Delta t}\right]$

$\Rightarrow F = n\times\dfrac{mv}{\Delta t}$

Hence, $F = 144$ N, $m = 40$g
$= 40\times10^3$kg
and $v = 1200$ ms^{-1}, $\Delta t = 1$ s.

$\therefore 144 = n\times\dfrac{40\times10^{-3}\times1200}{1}$

$or n = \dfrac{144}{4\times12} \Rightarrow n = 3$

143 **(b)**
The maximum tension $= 30\times10\,N = 300\,N$
$T - mg = ma$
$300 - 10\times10 = 10a$
$\Rightarrow a = 20\,m/s^2$ [Maximum value]
When the mass is raised, $u = 0, a = 20\,m/s^2$
$S = 10m, t = ?$
$10 = \dfrac{1}{2}\times 20.t^2 \Rightarrow t = 1s$

144 **(c)**
Given $T_1 = (M_1 + M_2 + M_3)(g + a)$
$= (5+2+3)(9.8+2) = 118$ N
$T_2 = (M_2 + M_3)(g + a) = (2+3)(9.8+2) = 59$ N
$T_3 = M_3(g+a) = 3(9.8+2) = 35.5$ N

147 **(c)**
Man is sliding down the telegraphic pole with acceleration $g/4$. So,

$mg - F = \dfrac{mg}{4}$

$\Rightarrow F = mg - \dfrac{mg}{4}$

$\Rightarrow F = \dfrac{3mg}{4}$

$\Rightarrow F = \dfrac{3w}{4}$

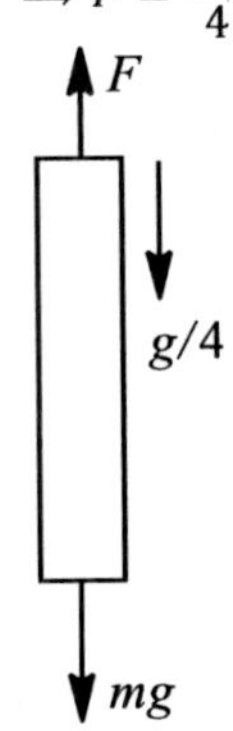

149 **(a)**
$R = mg + F_2\cos\theta, f = \mu R$
$f = \mu(mg + F_2\cos\theta)$

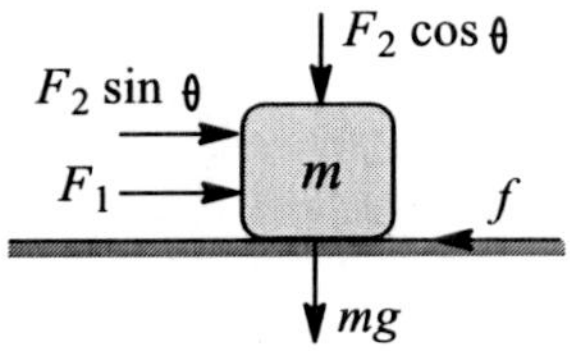

Also, $f = F_1 + F_2 \sin\theta$
Equating, $\mu(mg + F_2 \cos\theta)$
$= F_1 + F_2 \sin\theta$
or $\mu = \frac{F_1+F_2 \sin\theta}{mg+F_2 \cos\theta}$

150 **(b)**

$F = mg(\sin\theta + \mu\cos\theta)$
$= 10 \times 9.8(\sin 30° + 0.5 \cos 30°)$
$= 91.4\ N$

151 **(b)**

$$a = \frac{\text{Applied force} - \text{Kinetic friction}}{\text{mass}}$$
$$= \frac{100 - 0.5 \times 10 \times 10}{10} = 5\ m/s^2$$

152 **(b)**

Work done in max extension = stored $P.E.$

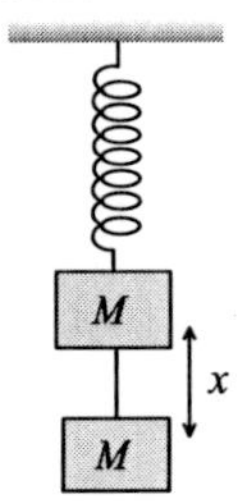

$$\Rightarrow Mg \times x = \frac{1}{2}kx^2$$
$$\Rightarrow x = \frac{2Mg}{k}$$

153 **(b)**

Applying law of conservation of liner momentum, *ie*,
$m_1u_1 + m_2u_2 = m_1v_1 + m_2v_2$
Here, $m_1 = 10\text{ g} = 10^{-2}\text{ kg}, m_2 = 5\text{ kg}$
$u_1 = 300\text{ ms}^{-1}, u_2 = 0$
$v_1 = 0, v_2 = ?$
$\therefore\ 10^{-2} \times 300 + 5 \times 0 = 10^{-2} \times 0 + 5v_2$
or $5v_2 = 3$
or $v_2 = \frac{3}{2}\text{ms}^{-1}$
$= 60\text{ cms}^{-1}$

154 **(d)**

The effective acceleration of ball observed by observer on earth = $(a - a_0)$
As $a_0 < a$, hence net acceleration is in downward direction

155 **(c)**

If T is tension in each part of the string holding mass $\sqrt{2}m$, then in equilibrium,
$T\cos\theta + T\cos\theta = \sqrt{2}\,mg$
$2T\cos\theta = \sqrt{2}\,mg$
But $T = mg$; $\therefore 2mg\cos\theta = \sqrt{2}mg$
$\cos\theta = \frac{1}{\sqrt{2}}$
$\theta = 45°$

156 **(c)**

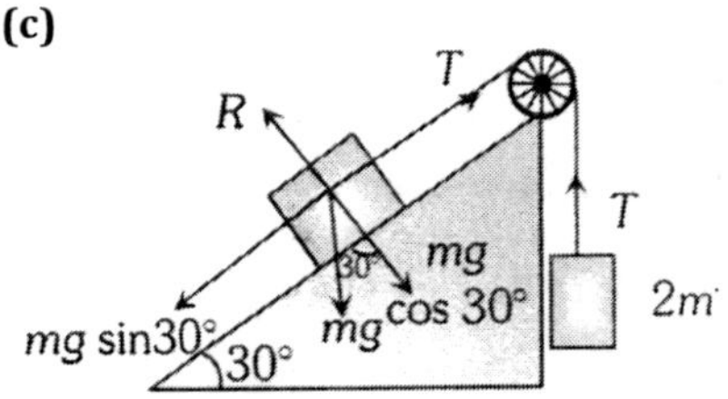

$2mg - T = 2ma$ ...(i)
$T - mg\sin 30° = ma$ ...(ii)
(i) + (ii) gives,
$2mg - \frac{mg}{2} = 3ma \Rightarrow a = \frac{g}{2}$

157 **(a)**

The net electromagnetic force = $\sqrt{N^2 + f^2}$
But $N = mg, f = \mu mg$
Force $= mg\sqrt{1 + \mu^2}$

158 **(a)**

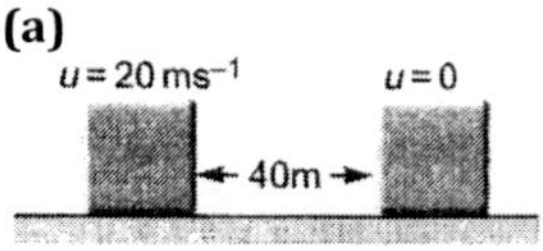

From equation of motion
$v^2 = u^2 - 2as$
Where v is final velocity, u the initial velocity, a acceleration and s the displacement.
Given, $v = 0, u = 20\text{ ms}^{-1}, s = 40\text{ m}$
$\therefore\ 0 = (20)^2 - 2 \times 40 \times a$
$\Rightarrow\ a = 5\text{ ms}^{-2}$
Kinetic (or dynamic) friction occurs when two objects are moving relative to each other and rub together. It is given by
$$\mu_x = \frac{a}{g} = \frac{5}{10} = 0.5$$

159 **(c)**

The trolley shall move backwards to conserve momentum. The backward momentum would be shared by both the trolley and man
Applying conservation of momentum

$60 \times 1(240+60)v$ or $60 = 300v$

or $v = \frac{60}{300} = \frac{1}{5}\text{ms}^{-1} = 0.2\ \text{ms}^{-1}$

Speed of man w.r.t. ground

$= (1-0.2)\text{ms}^{-1} = 0.8\ \text{ms}^{-1}$

Displacement of man $= 0.8 \times 4\text{m} = 3.2\text{m}$

161 **(a)**

There is no friction between the body B and surface of the table. If the body B is pulled with force F then

$F = (m_A + m_B)a$

Due to this force upper body A will feel the pseudo force in a backward direction

$f = m_A + a$

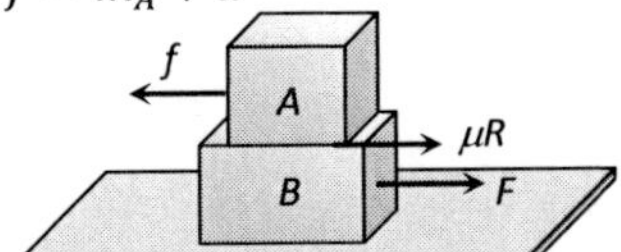

But due to friction $A$and B, body will not move. The body A will start moving when pseudo force is more than friction force

$i.e.$ for slipping, $m_A a = \mu\, m_A g$

$\therefore a = \mu g$

163 **(b)**

$P = f_{ms} = \mu_s mg$

When the body starts moving with acceleration a, then

$P - f_k = ma$

$\mu_s mg - \mu_k mg = ma$ or $a = (\mu_0 - \mu_k)g$

or $a = (0.5 - 0.4)10$

$= 0.1 \times 10\text{ms}^{-2} = 1\text{ms}^{-2}$

164 **(c)**

The friction force

$F = \mu R$

Here, R= Normal reaction

So, the coefficient of limiting friction

$\mu = \frac{F}{R}$

165 **(a)**

Block Bwill come to rest, it force applied to it will vanish due to frictional force acting between block $B$and surface ie, force applied = frictional force

ie, $\mu mg = ma$

or$\mu mg = m\left(\frac{v}{t}\right)$

or $t = \frac{v}{\mu g}$

166 **(c)**

Let P force is acting at an angle 30° with the horizontal

For the condition of motion $F = \mu R$

$P\cos 30° = \mu(mg - P\sin 30°)$

$\Rightarrow P\frac{\sqrt{3}}{2} = \frac{1}{\sqrt{3}}\left(100 - P\frac{1}{2}\right) \Rightarrow \frac{3P}{2}$

$= \left(100 - \frac{P}{2}\right)$

$\Rightarrow 2P = 100 \therefore P = 50\ N$

167 **(b)**

$a = \frac{m_2}{m_1 + m_2} \times g = \frac{5}{4+5} \times 9.8 = \frac{49}{9}$

$= 5.44\ m/s^2$

169 **(d)**

Pressure

$p = \frac{F}{A} = \frac{n\{mv - (mv)\}}{A} = \frac{2mnv}{A}$

$= \frac{2 \times 10^{-3} \times 10^4 \times 10^2}{10^{-4}} = 2 \times 10^7\text{Nm}^{-2}$

170 **(b)**

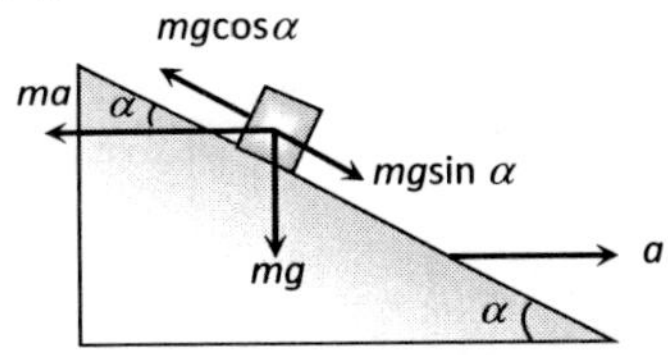

Let the mass of a block is m. It will remains stationary if forces acting on it are in equilibrium $i.e$, $ma\cos\alpha = mg\sin\alpha \Rightarrow a = g\tan\alpha$

Here ma = Pseudo force on block , mg = Weight

171 **(c)**

$F_l = \mu_s R = 0.4 \times mg = 0.4 \times 10 = 4N$ $i.e.$ minimum $4N$ force is required to start the motion of a body. But applied force is only $3N$. So the block will not move

172 **(a)**

For the limiting condition upward friction force between board and block will balance the weight of the block

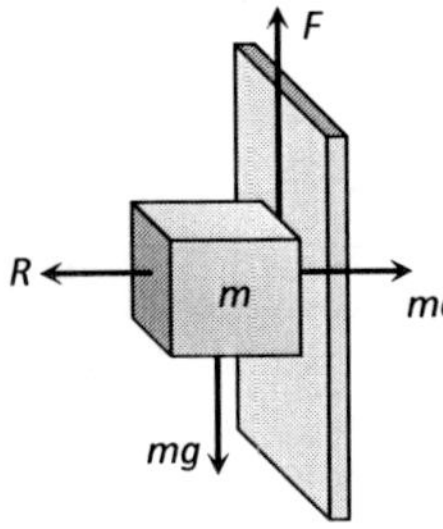

$i.e.F > mg$

$\Rightarrow \mu(R) > mg$

$\Rightarrow \mu(ma) > mg$

$\Rightarrow \mu > \frac{g}{a}$

173 **(b)**

As the spring balance are massless therefore both the scales read $M\ kg$ each

174 **(d)**

Impulse = Change in momentum

$F \times t = m(v - u)$

$F \times 0.4 = 80(5 - 0) \Rightarrow F = \frac{80\times5}{0.4} =$ 1000N

175 **(c)**

$T \sin 30° = 2$ kg-wt

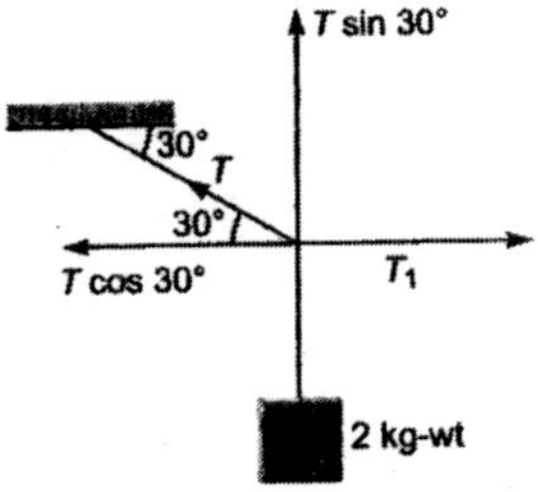

$\Rightarrow\ T = 4$ kg-wt

$T_1 = T \cos 30° = 4 \cos 30° = 2\sqrt{3}$

176 **(c)**

It works on the principle of conservation of momentum

177 **(d)**

$$\mu = \frac{F}{R} = \frac{F}{mg} = \frac{98}{100 \times 9.8} = \frac{1}{10} = 0.1$$

179 **(a)**

$$a = \left[\frac{m_1 - m_2}{m_1 + m_2}\right] g = \left[\frac{5 - 4.8}{5 + 4.8}\right] \times 9.8$$
$$= 0.2\ m/s^2$$

180 **(d)**

$20 - T = 6a$

$T - 10 = 4a$

$\therefore\ a = 1$ and $T = 14$ N

181 **(b)**

For body $A, T = M_1 a = 7a$

For body $B, M_2 g - T = 3a$

$3g - 7a = 3a$

$10a = 3g$

$$a = \frac{3g}{10} = \frac{3 \times 10}{10} = 3\ \text{ms}^{-2}$$

182 **(a)**

Tension, thrust, air resistance, weight are all common forces in mechanics where as impulse is not a force

Impulse = Force × Time duration

183 **(d)**

$T = 2\pi\sqrt{\frac{l}{g}}$. T will decrease, If g increases

It is possible when rocket moves up with uniform acceleration

184 **(b)**

Wheel of a car moving in backward direction so, friction force acts in forward direction

185 **(b)**

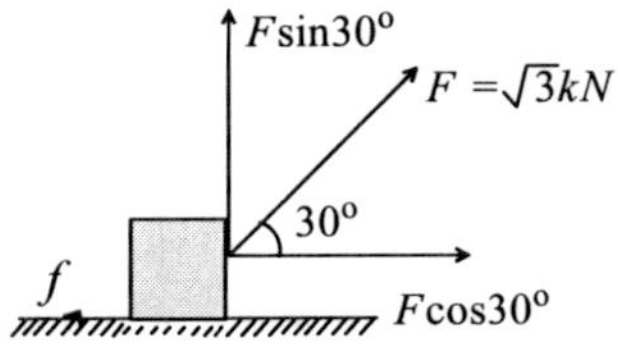

The component of applied force F in the direction of motion is $F \cos 30°$. The work done by applied force is

$$W = (F \cos 30°)S = \sqrt{3} \times 10^3 \times \frac{\sqrt{3}}{2} \times 10J$$
$$= 15 \times 10^3 J = 15kJ$$

186 **(d)**

Momentum acquired = Area of force – time graph

$$= \frac{1}{2} \times (2) \times (10) + 4 \times 10$$
$$= 50 \text{ N-S}$$

187 **(c)**

$$v = \sqrt{\mu g r} = \sqrt{0.8 \times 9.8 \times 15}$$
$$= 10.84\ m/s$$

188 **(b)**

The cloth can be pulled out without dislodging the dishes from the table due to law of inertia, which is Newton's first law. While, the statement II is true, but it is Newton's third law.

189 **(c)**

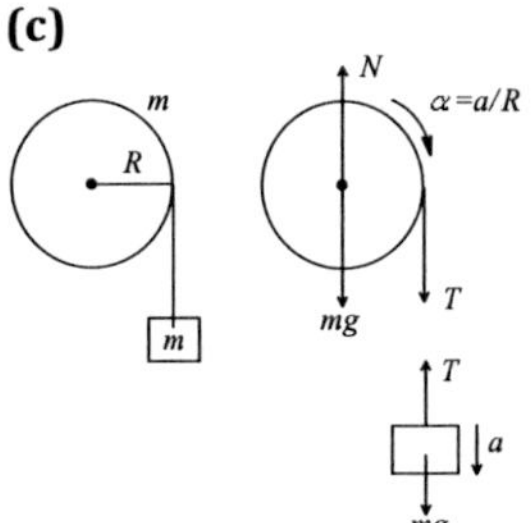

$mg - T = ma$

$$TR = \frac{mR^2\alpha}{2}$$

$$T = \frac{mR\alpha}{2} = \frac{ma}{2}$$

$$mg - \frac{ma}{2} = ma$$

$$\frac{3ma}{2} = mg$$

$$a = \frac{2g}{3}$$

190 **(c)**

Impulse = Force × Time = $50 \times 10^{-5} \times 3$ $= 1.5 \times 10^{-3} N\text{-}s$

191 **(a)**

$$F_{av} = \frac{\Delta p}{\Delta t} = \frac{mv - (-mv)}{\Delta t} = \frac{2mv}{\Delta t}$$
$$= \frac{2 \times 0.5 \times 2}{10^{-3}}$$
$$= 2000N$$

192 **(d)**

If a constant force **F** is applied on a body for a short interval of time Δt, then the impulse of this force is $F \times \Delta t$

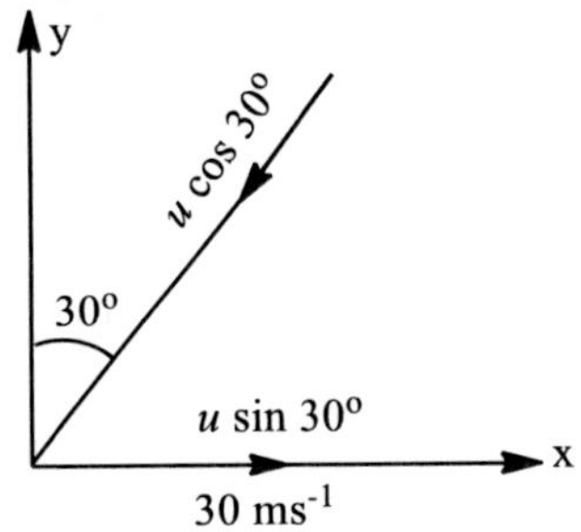

Since, impulse = change in momentum (Δp)

$\therefore\ F \times \Delta t = \Delta p$

$\Rightarrow\ F = \dfrac{\Delta p}{\Delta t}$

Change in x-direction

$$F = \frac{m[30 - (-15 \sin 30°)]}{0.01}$$
$$F = \frac{0.1 \times 37.5}{0.01} = 375 \text{ N}$$

193 **(c)**

Mass measured by physical balance remains unaffected due to variation in acceleration due to gravity

194 **(d)**

$t = \frac{v}{a} \Rightarrow t \propto \frac{1}{a}$ (v is same)

$$\Rightarrow \frac{t_1}{t_2} = \frac{a_2}{a_1} = \frac{m_1}{m_2} = \frac{3}{5}\left[\because a \propto \frac{1}{m}, F \text{ is same}\right]$$

195 **(d)**

Momentum acquired = Area of force - time graph

$= \frac{1}{2} \times (2) \times (10) + 4 \times 10 = 10 + 40 =$ $50\ N\text{-}s$

196 **(a)**

When the string C is stretched slowly, the tension in A is greater than that of C, because of the weight mg and the former reaches breaking point earlier

198 **(b)**

Refer to the free body diagram of painter

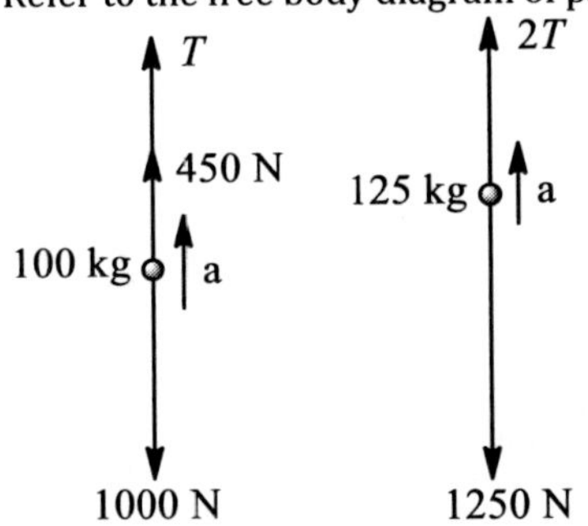

$T + 540 - 1000 = 100a$

or $\ T - 550 = 100a$ ...(i)

Refer to the free body diagram of the system

$2T - 1250 = 125a$ ....(ii)

From Eqs. (ii) and (i),

$2(550 + 100a) - 1250 = 125a$

or $\ 75a = 1250 - 1100 = 150$

or $\ a = \frac{150}{75} \text{ms}^{-2} = 2 \text{ ms}^{-2}$

199 **(b)**

From geometry, it is clear that $x = l \cos \theta$ and $y = l \sin \theta$

$\therefore\ v_x = \frac{dx}{dv} = -l \sin \theta \frac{d\theta}{dt}$ and $v_y = \frac{dy}{dt} =$ $l \cos \theta \frac{d\theta}{dt}$

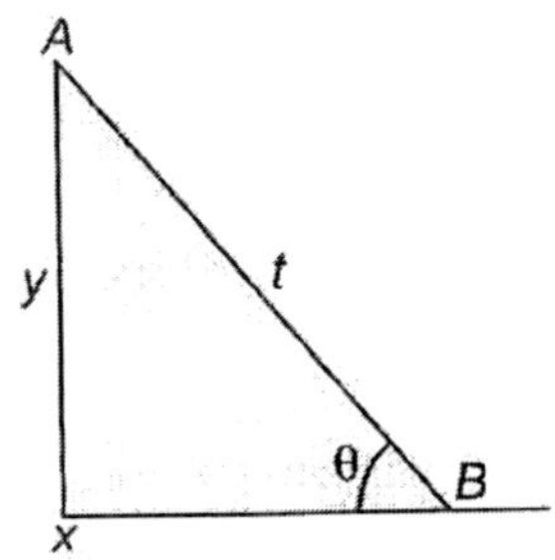

$\therefore\ \frac{v_y}{v_x} = -\cot \theta$ or $\ v_y = -v_x \cot \theta$

Since, $v_x = -3 \text{ ms}^{-1}$, hence

$$v_y = -(-3) \cot 60° = 3 \times \frac{1}{\sqrt{3}} = \sqrt{3} \text{ms}^{-1}$$

200 **(d)**

$m = 5kg, a = 5\ m/s^2$

$\because F \leq fr_{\max} \leq \mu N$

$F = ma = 5 \times 5 = 25N$

$\therefore fr = F \Rightarrow fr = 25N$

201 **(c)**

$mg \sin \theta - T = ma$

$T = ma$

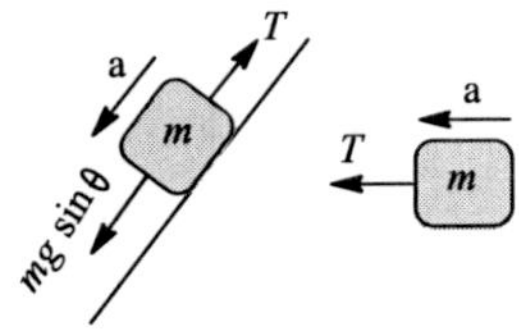

$\therefore\ mg\sin\theta - T = T$

or $2T = mg\sin\theta$ or $T = \frac{1}{2}mg\sin\theta$

202 **(d)**

The force is constant, *i.e.*, the acceleration is constant. The distance is decreasing with time. Therefore, it is retardation. 1st second $u.1 - (1/2)a.1 = 5m$ ...(i)

3rd second $u.3 - (1/2)a.9 - (u.2 - (1/2)a.4) = 3m$

$\therefore u.1 - (1/2)a.5 = 3m$ ...(ii)

(i) - (ii) gives $(1/2)a.4 = -2 \Rightarrow a = -1m/s^2$

$\therefore$ Constant force $= ma = 4 \times 1 = 4\ N$

[This is retarding force as a is negative]

203 **(d)**

$$P = \frac{F}{A} = \frac{n[mv - (-mv)]}{A} = \frac{2mnv}{A}$$

$$= \frac{2 \times 10^{-3} \times 10^4 \times 10^2}{10^{-4}} = 2 \times 10^7\ N/m^2$$

204 **(b)**

$0.05v = (0.450 + 0.05)V$

or $V = \frac{0.05}{0.50}v = \frac{v}{10}$

Using $v^2 - u^2 = 2as$, we get

$$0^2 - \left(\frac{v}{10}\right)^2 = -2 \times 10 \times 1.8$$

or $\left(\frac{v}{10}\right)^2 = 36$ or $v = 60\ \text{ms}^{-1}$

205 **(a)**

Acceleration

$$a = \frac{F}{m} = \frac{4}{20} = \frac{1}{5}\text{ms}^{-2}$$

Distance covered by body in 3rd second

$$= \frac{1}{2} \times \frac{1}{5} \times (2 \times 3 - 1) = \frac{5}{10} = \frac{1}{2}\text{m}$$

$$\therefore\ W = 4 \times \frac{1}{2} = 2\text{J}$$

206 **(b)**

From the graph, it is a straight line so, uniform motion. Because of impulse direction of velocity changes as can be seen from the slope of the graph.

Initial velocity $\mathbf{v}_1 = \frac{2}{2} = 1\ \text{ms}^{-1}$

Final velocity $\mathbf{v}_2 = \frac{-2}{2} = -1\ \text{ms}^{-1}$

$\mathbf{p}_i = m\mathbf{v}_1 = 0.4$ Ns

$\mathbf{p}_f = m\mathbf{v}_2 = -0.4$ Ns

$\mathbf{J} = \mathbf{p}_f - \mathbf{p}_i = -0.4 - 0.4$

$= -0.8\ Ns\ \ (\boldsymbol{J} = impulse)$

$\therefore\ |\mathbf{J}| = 0.8$ Ns

207 **(d)**

Let acceleration of system (rope + block) is a which acting along the direction of applied force. Then

= 1/3

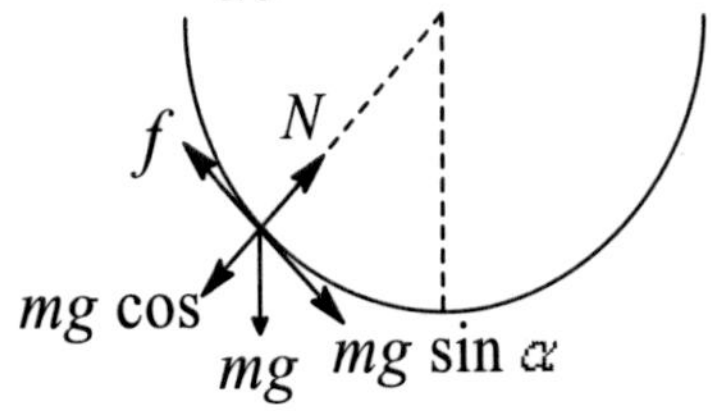

Where, T is the required parameter.

For block $T = Ma$

$$\Rightarrow\ T = \frac{MP}{M+m}$$

208 **(c)**

$$v = u + \frac{F}{m}t = 10 + \left(\frac{1000 - 500}{1000}\right) \times 10$$

$$= 15\ m/s$$

209 **(a)**

$$F = u\left(\frac{dm}{dt}\right)$$

$$= 20 \times \frac{50}{60}$$

$= 16.66$ N

210 **(b)**

$$\tan\alpha = \frac{2F\sin\theta}{F + 2F\cos\theta} = \infty\ (\text{as}\ \alpha = 90°)$$

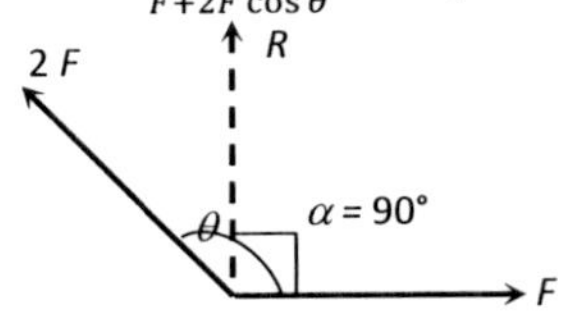

$\Rightarrow F + 2F\cos\theta = 0$

$$\Rightarrow \cos\theta = -\frac{1}{2}$$

$\theta = 120°$

211 **(d)**

$N\sin\theta = mg$

$N\cos\theta = ma$

$$\tan\theta = \frac{g}{a}$$

$$\cos\theta = \frac{a}{8} = \tan(90° - \theta) - \frac{dy}{dx} = 2k\ x$$

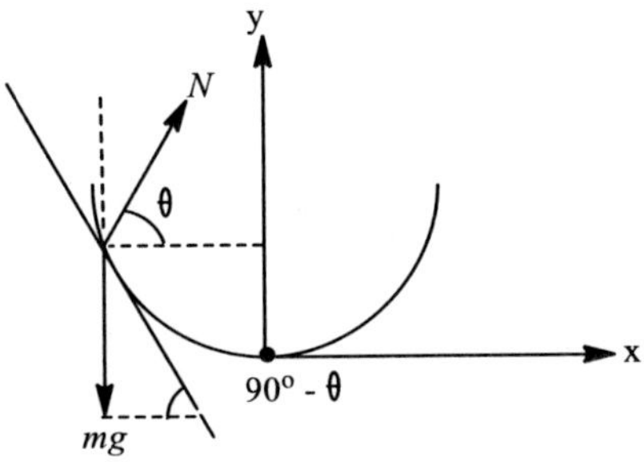

$\therefore x = \dfrac{a}{2k\, g}$

212 **(a)**

When the system accelerates upwards, the effective value of acceleration due to gravity is given by

$f' = g + a = g + g = 2g$

Now, $T = \frac{2(m_1)(m_2)}{m_1+m_2}(2g)$ or $T = \frac{4m_1m_2}{m_1+m_2}g$

213 **(d)**

Acceleration of the system

$$a = \frac{F}{\left(m + \frac{m}{3}\right)}$$

$\therefore$ Tension in the middle of the rope

$$T = \left(m + \frac{m}{6}\right)a$$

or $T = \dfrac{7m}{6} \cdot \dfrac{3F}{4m}$

$= \dfrac{7}{8}F$

214 **(c)**

Let the ball starts moving with velocity 'u' and it reaches upto maximum height $H_{\max}$, then

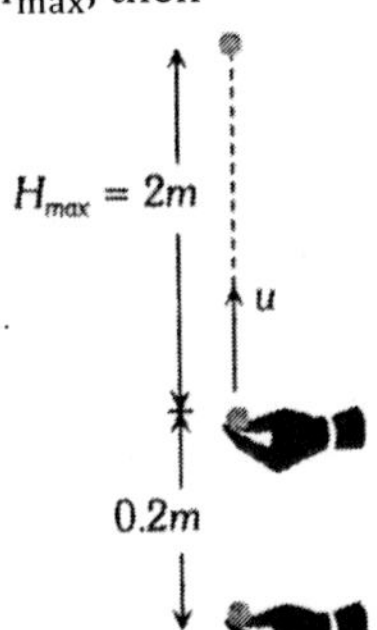

From $H_{\max} = \dfrac{u^2}{2g}$

$u = \sqrt{2g\,(H_{\max})}$

$= \sqrt{2 \times 10 \times 2} = 2\sqrt{10}\, m/s$

This velocity is supplied to the ball by the hand and initially the hand was at rest, it acquires this velocity in distance of $0.2\, meter$

$$\therefore a = \frac{u^2}{2s} = \frac{40}{2 \times 0.2} = 100\, m/s^2$$

So upward force on the ball $F = m(g + a)$

$= 0.2(10 + 100) = 0.2 \times 110 = 22N$

215 **(d)**

$$T_2 = (m_1 + m_2) \times \frac{T_3}{m_1 + m_2 + m_3}$$

$$= \frac{(10 + 6) \times 40}{20}$$

$$= 32\, N$$

216 **(d)**

As is clear from figure

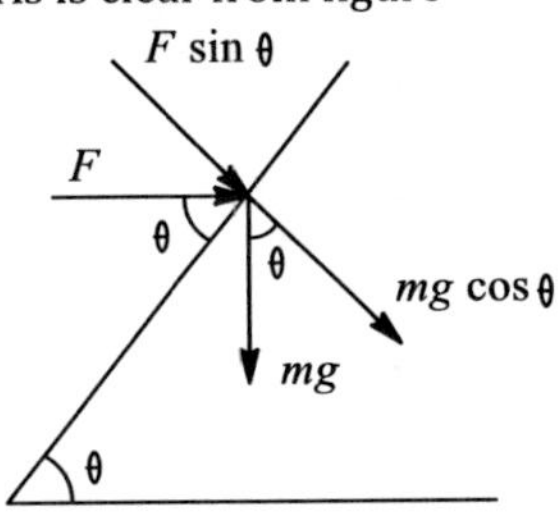

$R = mg \sin\theta + F \sin\theta$

$R = mg\cos\theta + F\sin\theta$

217 **(d)**

The situation is as shown in the figure

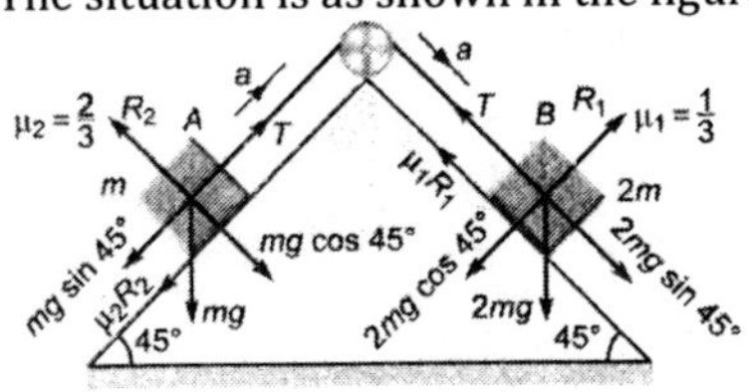

The equation of motion for body B.

$2mg \sin 45° - \mu_1 R_1 - T_1 - T_2 = 2ma$

$2mg \sin 45° - \dfrac{1}{3} 2mg \cos 45° - T = 2ma$

$\Rightarrow \quad 2mg \times \dfrac{1}{\sqrt{2}} - \dfrac{1}{3} 2mg \times \dfrac{1}{\sqrt{2}} - T$

$= 2ma \quad \ldots.(i)$

In the problem as $(m_B - m_A)g\cos\theta = (mg/\sqrt{2})$ is lesser than $(\mu_B m_B + \mu_A m_A)g\cos\theta = (4mg/3\sqrt{2})$ the masses will not move and hence.

Acceleration of B = acceleration of $A = 0$.

218 **(c)**

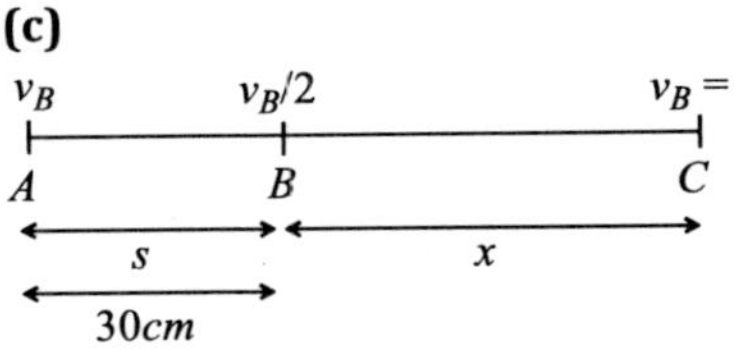

Let bullet is fired with velocity v_B at point A and its velocity becomes half when it travels a distance s and reaches at point B. When it reaches at point C, it comes to rest and travels a distance x

From A to B, using, $v^2 - u^2 = 2as$

$$\Rightarrow \left(\frac{v_B}{2}\right)^2 - v_B^2 = 2as \Rightarrow \frac{v_B^2}{4} - v_B^2 = 2as$$

$$\Rightarrow \frac{-3v_B^2}{4} = 2as \Rightarrow a = \frac{-3v_B^2}{8s}$$

$\therefore$ From B to C, using $v^2 - u^2 = 2as$

$$0^2 - \left(\frac{v_B}{2}\right)^2 = 2as \Rightarrow \frac{-v_B^2}{4} = 2\left(\frac{-3v_B^2}{8s}\right)x$$

$$\Rightarrow x = \frac{8s}{4 \times 6} = \frac{8 \times 30}{24} = 10\ cm$$

219 **(a)**

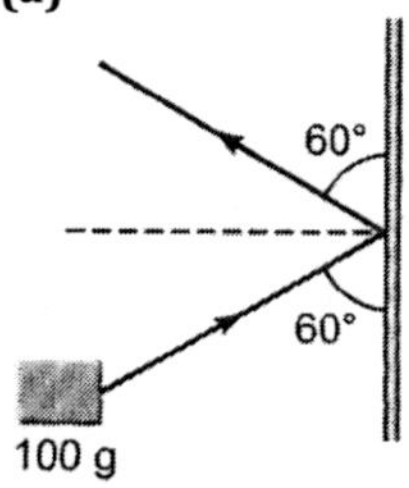

Change in the velocity $= v \sin \theta - (-v \sin \theta) = 2 \sin \theta$

Change in the momentum

$\Delta p = 2\ m\ v \sin \theta$

$\therefore$ Force applied $F = \frac{\Delta p}{\Delta t}$

$$= \frac{2 \times 100 \times 10^{-3} \times 5 \sin \theta\ 60°}{2 \times 10^{-3}}$$

$$= 100 \times 5 \times \frac{\sqrt{3}}{2}$$

$= 250\sqrt{3}$ N (To the right)

220 **(d)**

$$F = \frac{dp}{dt} = \frac{d}{dt}(a + bt^2) = 2bt \ \therefore F \propto t$$

221 **(d)**

$mg \sin \theta = ma$

$\therefore \ a = g \sin \theta$

Where a is long the inclined plane.

$\therefore$ Vertical component of acceleration is $g \sin^2\theta$.

$\therefore$ Relative vertical acceleration of A with respect to B is

$$g(\sin^2 60° - \sin^2 30°) = \frac{g}{2} = 4.9\ ms^{-2}$$

(in vertical direction)

222 **(d)**

Heavier gas will acquire largest momentum $i.e.$ Argon

223 **(c)**

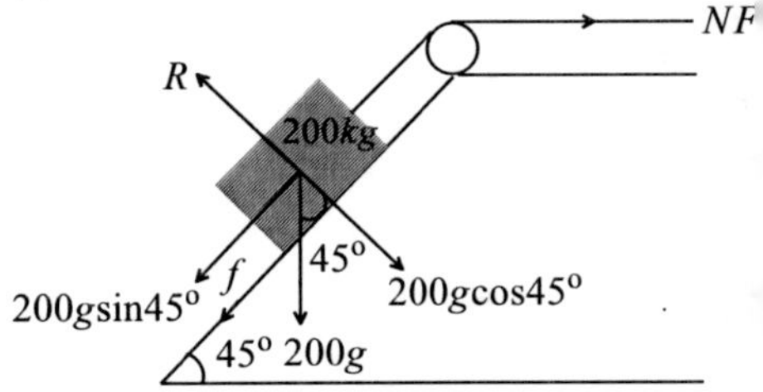

Here, mass of the block, $m = 200\ kg$

Coefficient of static friction, $\mu_s = 0.5 = \frac{1}{2}$

Angle of incline plane, $\theta = 45°$

Maximum force that each man can apply, $F = 500\ N$

Let N number of man are required for the block to just start moving up the plane

$NF = mg \sin \theta + f$

$= mg \sin \theta + \mu_s R$

$= mg \sin \theta + \mu_s\ mg \cos \theta$

$= mg\ [\sin \theta + \mu_s \cos \theta]$

$$= 200 \times 10 \left[\sin 45° + \frac{1}{2}\cos 45°\right]$$

$$= 200 \times 10 \left[\frac{1}{\sqrt{2}} + \frac{1}{2\sqrt{2}}\right] = \frac{200 \times 10 \times 3}{2\sqrt{2}}$$

$$\therefore N = \frac{200 \times 10 \times 3}{2\sqrt{2} \times 500} = 5$$

224 **(a)**

Let v be the velocity of the boat with respect to the water, then from conservation of linear momentum.

$(200 + 50)v + 50 \times 2$

$= 50 \times 0 + 200 \times 0$

$250v = -100$

$$v = -\frac{100}{250} = -\frac{2}{5}\ ms^{-1}$$

225 **(a)**

Upthrust on the body $= v\sigma g$. For freely falling body effective g becomes zero. So up thrust becomes zero

226 **(a)**

Applied force $= 2.5\ N$

Limiting friction $= \mu mg = 0.4 \times 2 \times 9.8 = 7.84\ N$

For the given condition applied force is very smaller than limiting friction

$\therefore$ Static friction on a body = Applied force $= 2.5N$

227

(b)
$$a=\left(\frac{m_1-m_2}{m_1+m_2}\right)g \Rightarrow \frac{g}{8}=\left(\frac{m_1-m_2}{m_1+m_2}\right)g$$
$$\Rightarrow \frac{m_1}{m_2}=\frac{9}{7}$$

228 **(c)**

Initially particle was at rest. By the application of force its momentum increases

Final momentum of the particle = Area of F-t graph

$\Rightarrow mu$ = Area of semi circle

$$mu=\frac{\pi r^2}{2}=\frac{\pi r_1 r_2}{2}=\frac{\pi(F_0)(T/2)}{2}\Rightarrow u$$
$$=\frac{\pi F_0 T}{4m}$$

229 **(d)**

Gravitational field is a conservative field. Therefore work done in moving a particle from A to B is independent of path chosen

230 **(a)**

During collision of ball with the wall horizontal momentum changes (vertical momentum remains constant)

$$\therefore F=\frac{\text{Change in horizontal momentum}}{\text{Time of contact}}$$

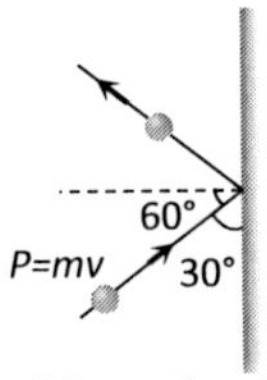

$$=\frac{2P\cos\theta}{0.1}=\frac{2mv\cos\theta}{0.1}$$
$$=\frac{2\times 0.1\times 10\times\cos 60^\circ}{0.1}=10\ N$$

231 **(b)**

From the formula

$F=ma$

$100=5a$

$\Longrightarrow\ a=20\ \text{cms}^{-2}$

from equation of motion

$v=u+at=0+20\times 10$

$v=200\ \text{cms}^{-1}$

232 **(a)**

$Mg'=Mg+M(4g)$ or $Mg'=5\ Mg$

233 **(c)**

$F\sin 30^\circ+N=Mg$

The block lifts when $N=0$

$\therefore\ F=\frac{10\times 10}{\frac{1}{2}}=200\text{N}$

234 **(d)**

u = velocity of bullet

$\frac{dm}{dt}$ = Mass of bullet fired per second by the gun

$\frac{dm}{dt}$ = Mass of one bullet $(m_B)\times$ Bullets fired per sec (N)

Maximum force that man can exert $F=u\left(\frac{dm}{dt}\right)$

$\therefore F=u\times m_B\times N$

$$\Rightarrow N=\frac{F}{m_B\times u}=\frac{144}{40\times 10^{-3}\times 1200}=3$$

235 **(c)**

Here, $m=20\ kg$, $u=20\ m/s, v=5m/s, s=100m$

Using $v^2=u^2+2as$

$$F=ma=m\times\left(\frac{v^2-u^2}{2s}\right)$$

Substituting the values, we get

$$F=20\times\left[\frac{(5)^2-(20)^2}{2\times 100}\right]=-37.5\ N$$

236 **(d)**

Equations of motion are

$m_1a=T-m_1g$

and $m_2a=m_2g-T$

$\Longrightarrow\ 8a=T-8g$ ....(i)

$12a=12g-T$ ....(ii)

Form Eqs. (i) and (ii), we get

$a=\frac{g}{5}=2\ \text{m/s}^2$

Substituting the value of a in Eq.(i)

We get $T=96$ N

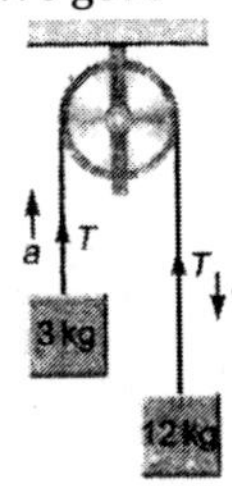

237 **(a)**

Work done = Force × displacement = $\mu\, mg\times(v\times t)$

$W=(0.2)\times 2\times 9.8\times 2\times 5\ joule$

Heat generated $Q=\frac{W}{J}=\frac{0.2\times 2\times 9.8\times 2\times 5}{4.2}=$ $9.33\ cal$

238 **(b)**

Using the relation

$$\frac{mv^2}{r}=\mu R, R=mg$$

$\frac{mv^2}{r} = \mu mg$ or $v^2 = \mu rg$

or $v^2 = 0.6 \times 150 \times 10$

or $v = 30 \text{ ms}^{-1}$

239 **(d)**

$T = mg = 5 \times 10^{-3} \times 10$

$= 0.5 \text{ N}$

240 **(a)**

$f = \mu mg = 0.8 \times 4 \times 10 = 32 \text{ N}$

Applied force $F < f$ therefore, answer will be (a)

241 **(d)**

According to Newton's second law of motion,

$F = \frac{\Delta p}{\Delta t}$

Also, $F = ma$

$\therefore \; ma = \frac{p_2 - p_1}{\Delta t}$

or $a = \frac{m_2 - (-mv_1)}{m\Delta t}$

or $a = \frac{v_2 + v_1}{\Delta t}$

$\therefore \; a = \frac{\sqrt{2 \times 10 \times 20} + \sqrt{2 \times 10 \times 5}}{0.02}$

or $a = \frac{20 + 10}{0.02} = 1500 \text{ ms}^{-2}$

242 **(d)**

$R = m(g + a) = m(g + g) = 2mg$

243 **(d)**

To find the acceleration of the block we will need the tension in the string that can be obtained by considering the pulley as the system.

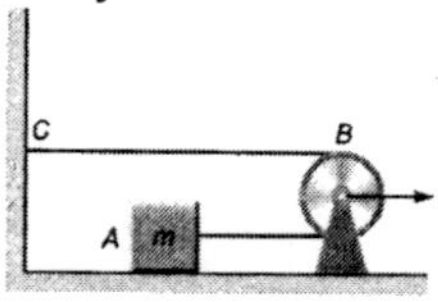

The force acting on the pulley are

(i) F towards right

(ii) T towards left by the portion BC of the string and

(iii) T towards left by the portion BA of the string.

Since there is no vertical motion.

Then vertical force if any, add to zero.

As the pulley is massless , the equation of motion is

$F - 2T = 0$ or $T = \frac{F}{2}$

The acceleration of the block is

$a = \frac{T}{m} = \frac{F}{2m}$

244 **(b)**

$F = 2T\cos\theta$

$\Rightarrow T = \frac{F}{2\cos\theta}$

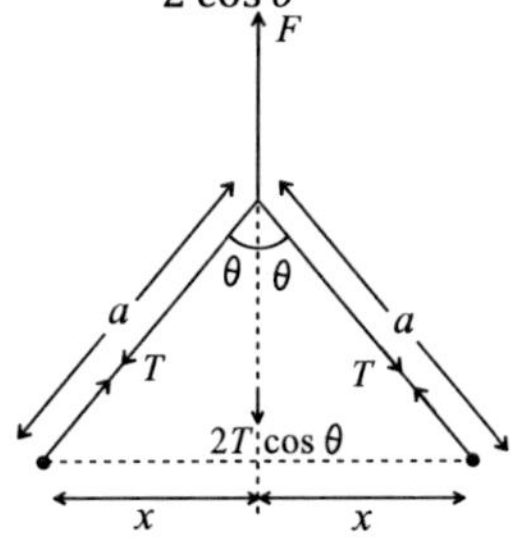

Magnitude of acceleration of the particle

$= \frac{T\sin\theta}{m}$

$= \frac{F\tan\theta}{2m} = \frac{F}{2m}\frac{x}{\sqrt{a^2 - x^2}}$

245 **(b)**

$8x = wy$

$\frac{x}{y} = \frac{w}{8}$ ...(i)

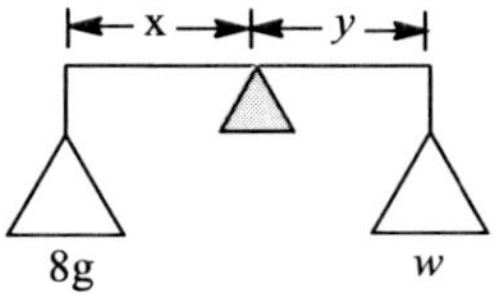

$wx = 18y$

$\frac{x}{y} = \frac{18}{w}$ ...(ii)

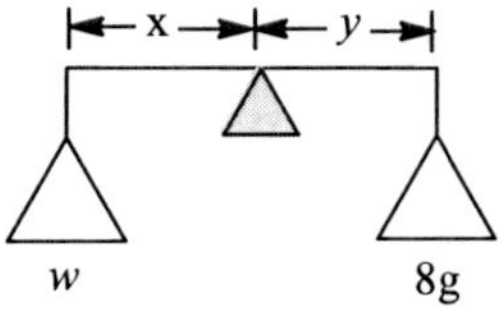

Dividing Eq. (i) by Eq. (ii)

$\frac{\frac{x}{y}}{\frac{x}{y}} = \frac{\frac{w}{8}}{\frac{18}{w}}$

$\Rightarrow w = \sqrt{18 \times 8} = 12\text{g}$

246 **(b)**

$f = \mu R = \mu mg | m$ is mass of the combination

$f = 0.5 \times 10 \times 10 \text{ N} = 50 \text{ N}$

So, a force of 10 N is unable to start the motion of the system. There is no relative motion between A and B

247 **(c)**

As external force is zero

$m_1v_1 + m_2v_2 = 0$

$v_2 = \frac{-m_1v_1}{m_2} = -\frac{60 \times 0.4}{30} = -0.8 \text{ ms}^{-1}$

It means that boy moves with speed $0.8\ ms^{-1}$ opposite to velocity of man.
So, relative velocity of man and boy is
$v_r = 0.8 + 0.4 = 1.2\ ms^{-1}$
Hence, separation $d = v_r t = 1.2 \times 5 =$ 6m

248 **(c)**

Force on the football $F = m\frac{dv}{dt}$

$$F = \frac{m(v_2 - v_1)}{dt} = \frac{0.5 \times (10 - 0)}{1/50}$$
$$= 250N$$

249 **(d)**

$T = 1 \times a$ or $1.5 = a$
$F = (1 + 22)\ kg \times 1.5\ ms^{-2} = 4.5\ N$

250 **(a)**

Let the mass of block be m

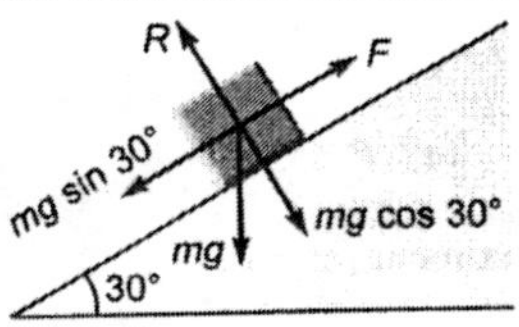

Friction force is rest position
$F = mg \sin 30°$
(This is static frictional force and may be less than the limiting frictional force)

$$10 = m \times 10 \times \frac{1}{2}$$
$$\therefore\ m = \frac{2 \times 10}{10} = 2\ kg$$

251 **(d)**

Applying law of conservation of momentum,

$\frac{m}{20} v = \left(m + \frac{m}{20}\right) V$ or $V = \frac{v}{20} \times \frac{20}{21} = \frac{v}{21}$

252 **(c)**

For given condition $s \propto \frac{1}{m^2}$

$$\therefore \frac{s_2}{s_1} = \left(\frac{m_1}{m_2}\right)^2 = \left(\frac{200}{300}\right)^2$$
$$\Rightarrow s_2 = s_1 \times \frac{4}{9} = 36 \times \frac{4}{9} = 16\ m$$

253 **(d)**

We know that,
$\mu = \tan\theta$
$\therefore = \tan 30° = \frac{1}{\sqrt{3}}$

254 **(a)**

As shown in figure, component of weight ($mg \sin\theta$) is always down the inclined plane, whether the cylinder is following up or it is rolling down. Therefore, for no slipping, sense of angular acceleration must be the same in both the cases.
Therefore, force of friction (f) acts up the inclined plane in both the cases

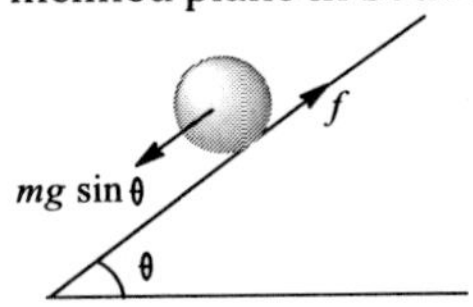

255 **(b)**

Change of momentum of one bullet is mv
Time for 1 bullet $= \frac{1}{n}$
$\therefore$ Force = time rate of change of momentum
$$= \frac{mv}{1/n} = mnv$$

256 **(d)**

Force $2mg$applied at the free end of the string acts on mass m. Therefore, its acceleration
$$a = \frac{\text{Force}}{\text{mass}}$$
$$= \frac{2mg}{m} = 2g$$

257 **(a)**

If μ is the mass/length, then
Weight of hanging length $= \mu l g$
Weight of chain on table $= \mu(L - l)g$
$R = \mu(L - l)g$
$f = \mu_s R = \mu_s \mu(L - l)g$
Equating, $\mu_s \mu(L - l)g = \mu l g$ or $\mu_s = \frac{l}{L-l}$

258 **(b)**

Let contact force = R

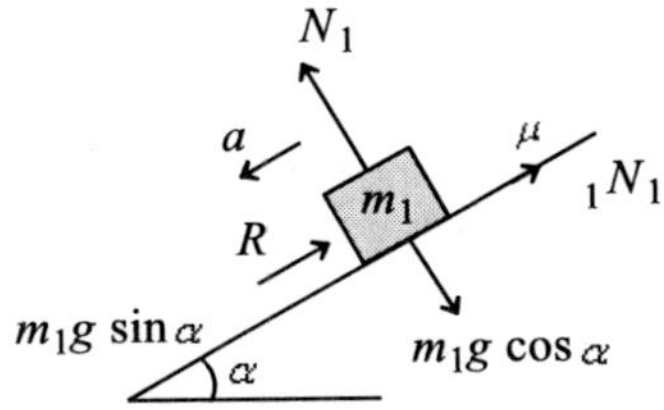

$$m_1 g \sin\alpha - R - \mu_1 N_1 = m_1 a \quad ...(i)$$
$$R = \frac{(\mu_2 - \mu_1) m_1 m_2 g \cos\alpha}{(m_1 + m_2)}$$
$$\Rightarrow R = \frac{0.1 \times 8 \times 9.8 \times \sqrt{\frac{3}{2}}}{6} \quad ...(ii)$$

On simplify equation (i) and (ii),
$a = 2.646\ m/s^2$

259 **(c)**

Acceleration of system, $a = \frac{F}{m + M}$

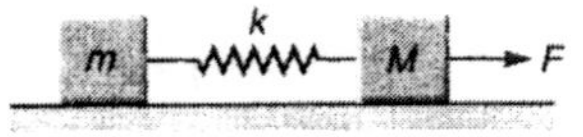

So, force acting on mass

$$m = ma = \frac{mF}{m+M}$$

260 **(d)**

Kinetic energy being a scalar quantity, hence measured from different inertial frame gives the same value, while the other three being vector quantities their values vary.

261 **(a)**

Weight of the disc will be balanced by the force applied by the bullet on the disc in vertically upward direction

$F = nmv = 40 \times 0.05 \times 6 = Mg$

$$\Rightarrow M = \frac{40 \times 0.05 \times 6}{10} = 1.2\ kg$$

262 **(c)**

Initially due to upward acceleration apparent weight of the body increases but then it decreases due to decrease in gravity

263 **(d)**

$$S = \frac{u^2}{2\mu g} = \frac{m^2u^2}{2\mu g m^2} = \frac{P^2}{2\mu m^2 g}$$

264 **(b)**

$$s = \frac{u^2}{2\mu g} = \frac{(20)^2}{2\times0.5\times10} = 40\ m,$$

$$72\ km/hr = 72 \times \frac{5}{18} = 20\ m/s$$

265 **(a)**

$\vec{v} = \vec{u} + \vec{a}t \therefore v$

$= \sqrt{u^2 + a^2t^2 + 2u\ at\cos\theta}$

v

$= \sqrt{200 + 100 + 2 \times 10\sqrt{2} \times 10 \times \cos 13\text{5}}$

$= 10\ m/s$

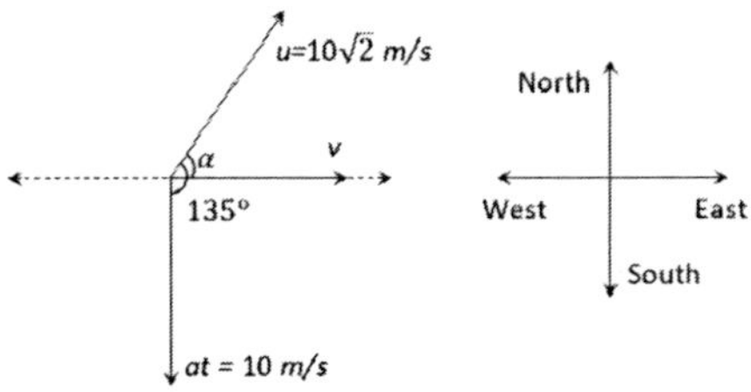

$$\tan\alpha = \frac{at\sin\theta}{u + at\cos\theta} \frac{10\sin 135°}{10\sqrt{2} + 10\cos 135°}$$

$= 1 \therefore \alpha = 45°$

i.e., resultant velocity is 10 m/s towards East

266 **(c)**

Refer to the free body diagram of block B

$5g - T = 5a$ or $T = 5g - 5a$

Refer to the free body diagram of block A

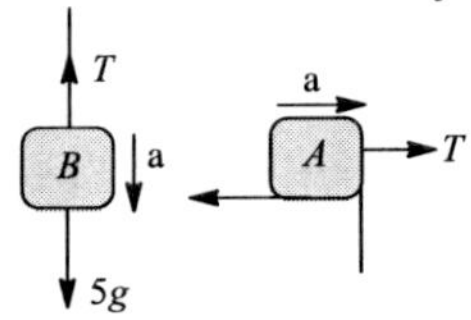

$T - f = 5a$

or $5g - 5a - 0.5 \times 5 \times g = 5a$

or $10a = 2.5g = 2.5 \times 9.8$

or $a = 0.25 \times 9.8\ ms^{-2} = 2.45\ ms^{-2}$

Again, $T = (5 \times 9.8 - 5 \times 2.45)$N

$= (49 - 12.5)$ N=36.75 N

267 **(a)**

Here, Mass of the box, $M = 40\ kg$

Acceleration of the truck, $a = 2\ ms^{-2}$

Distance of the box from the rear end ,

$d = 5m$

Coefficient of friction between the box and the surface below it, $\mu = 0.15$

The various forces acting on the block are as shown in the figure

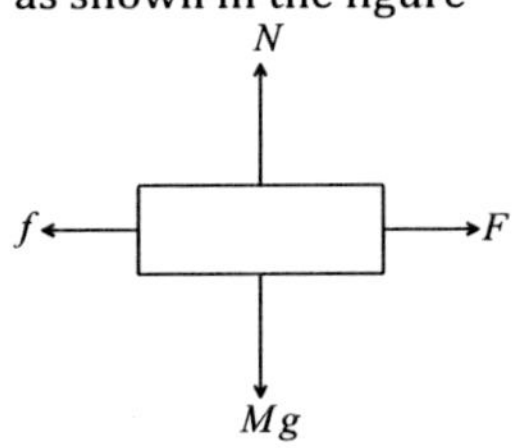

As the truck moves in forward direction with the acceleration $a = 2ms^{-2}$, the box experiences a force F in the backward direction and it is given by

$f = Ma = (40\ kg) \times (2\ ms^{-2}) = 80\ N$ in backward direction

Under the action of this force, the box will tend to move toward the rear end of the truck . As it does so, its motion will be opposed by the force of friction which acts in the forward direction and it is given by

$f = \mu N = \mu Mg = 0.15 \times 40 \times 10 = 60\ N$

The acceleration of the box relative to the truck toward the rear end is, $a_1 = \frac{F-f}{M} = \frac{80\ N - 60\ N}{40\ kg} = 0.5\ ms^{-1}$

Let t be the time taken for the box to fall off the truck

Using, $S = ut + \frac{1}{2}at^2$, we get , $d = 0 \times t + \frac{1}{2}a_1 t^2 [\because u = 0]$

$$5 = \frac{1}{2} \times 0.5 \times t^2, t = \sqrt{\frac{2 \times 5}{0.5}} = \sqrt{20}\ s$$

During this time, the truck covers a distance x

Using $S = ut + \frac{1}{2}at^2$

We get $x = 0 \times t + \frac{1}{2} \times 2 \times (\sqrt{20})^2 [\because u = 0]$

$x = 20\ m$

268 **(a)**

On moving upward,

weight, $w = m(g + a)$

$= 5(10 + 0.25)$

$= 5 \times 10.25 = 51.25$ N

269 **(b)**

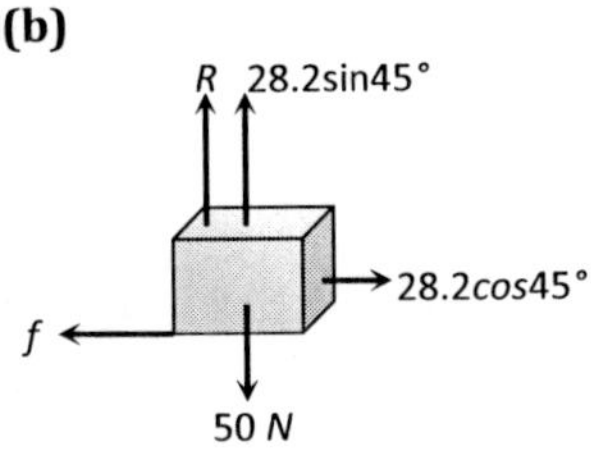

Frictional force $= f = 28.2 \cos 45° = 28.2 \times \frac{1}{\sqrt{2}} = 20N$

Normal reaction $R = 50 - 28.2 \sin 45° = 30\ N$

270 **(a)**

Equation for the given system

$10\ g - T = 10a$ ...(i)

$T - 6g = 6a$ ....(ii)

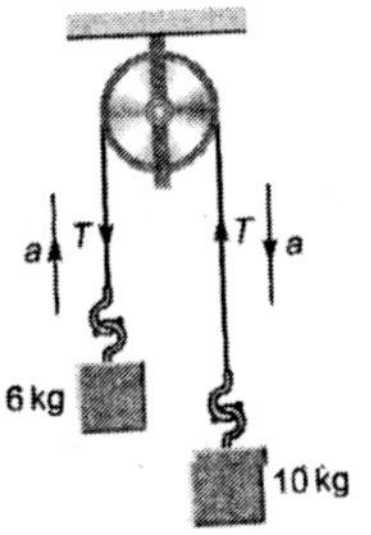

From Eqs. (i) and (ii)

$T = 75$ N

271 **(a)**

$W = \mu mgS = 0.2 \times 50 \times 9.8 \times 1 = 98\ J$

272 **(b)**

Rate of flow will be more when lift will move in upward direction with some acceleration because the net downward pull will be more and vice-versa

$F_{\text{upward}} = m(g + a)$ and $F_{\text{downward}} = m(g - a)$

273 **(b)**

An inertial frame is a non accelerated frame, so option (b) is correct

274 **(a)**

In stationary position,

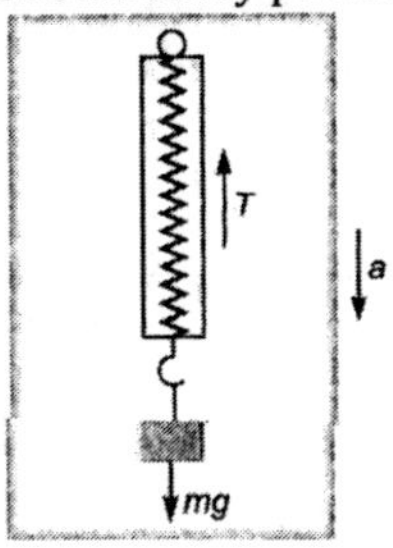

Spring balancing reading

$= mg = 49$

$m = \frac{49}{9.8} = 5$kg

When lift moves downward,

$mg - T = ma$

Reading of balance

$T = mg - ma$

$= 5(9.8 - 5) = 5 \times 4.8 = 24.0$ N

275 **(c)**

$$m\frac{dv}{dt} = F_0 e^{-bt} \Rightarrow \frac{dv}{dt} = \frac{F_0}{m}e^{-bt}$$

$$\Rightarrow \int_0^v dv = \frac{F_0}{m}\int_0^t e^{-bt}dt$$

$$\Rightarrow v = \frac{F_0}{m}\left[\frac{e^{-bt}}{-b}\right]_0^t$$

$$\Rightarrow v = \frac{F_0}{mb}(1 - e^{-bt})$$

276 **(d)**

By law of conservation of linear momentum.

$m_1\mathbf{v}_1 + m_2\mathbf{v}_2 + m_3\mathbf{v}_3 = 0$

Here $m_1 = m_2 = m_3 = 1$ kg,

$\mathbf{v}_1 = 3\,\hat{\mathbf{i}}, \mathbf{v}_2 = 4\,\hat{\mathbf{j}}$

$\therefore\ 3\,\hat{\mathbf{i}} + 4\,\hat{\mathbf{j}} + \mathbf{u}_3 = 0$

The average force acting on the third piece is

$$F = \frac{m\,\mathbf{v}_3}{t}$$

$$= \frac{1 \times -(3\,\hat{\mathbf{i}} + 4\,\hat{\mathbf{j}})}{10^4}\text{N}$$

$$= -(3\,\hat{\mathbf{i}} + 4\,\hat{\mathbf{j}}) \times 10^4\text{N}$$

277 **(c)**

From Newton's second law of motion,

Force = mass × acceleration

Acceleration along y-axis is

$$a_y = -\frac{5}{5}\text{ms}^{-2} = -1\text{ms}^{-2}$$

Form equation of motion

$$v_y = u_y + a_y t$$

Where v_y is initial velocity along y-axis, t the time and v_y the final velocity along y-axis.

Given, $v_y = 0, u_y = 40\text{ ms}^{-1}$,

$$a_y = -1\text{ ms}^{-1}$$

$\therefore\ 0 = 40 - 1 \times t$

$\Rightarrow t = 40\text{ s}$

278 **(b)**

Force,

$$F = \frac{m(u-v)}{t} = \frac{150 \times 10^{-3}(20 - 0}{0.1}$$

$= 30\text{ N}$

279 **(b)**

Due to Newton's third law

280 **(d)**

$$\tan\theta = \frac{5}{5} = 1$$

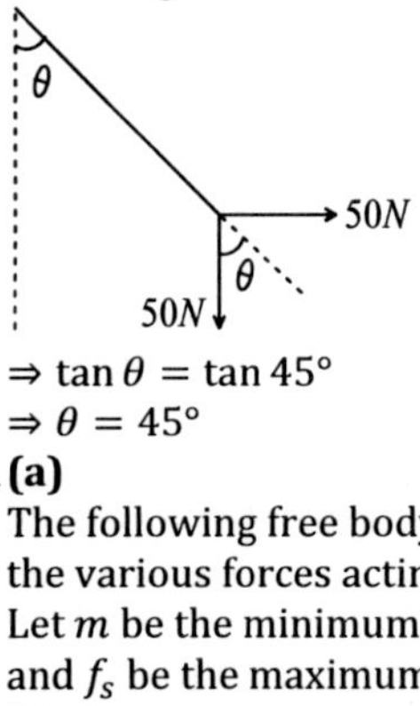

$\Rightarrow \tan\theta = \tan 45°$

$\Rightarrow \theta = 45°$

281 **(a)**

The following free body diagram shows the various forces acting on the system. Let m be the minimum mass of block C and f_s be the maximum value of static friction.

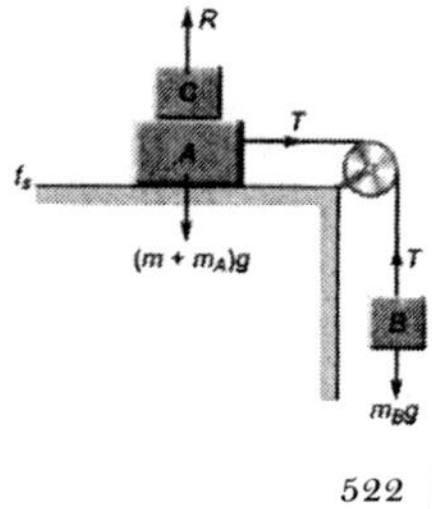

For block A

$R = (m + m_A)g, f_s = T$

$\therefore\ \mu(m + m_A)g = T$ ...(i)

For block B

$T = m_B g$ ...(ii)

From Eqs. (i) and (ii), we get

$$m = \frac{m_B - \mu\, m_A}{\mu}$$

$$m = \frac{10 - 0.4 \times 15}{0.4} = 10\text{ kg}$$

282 **(d)**

As the springs are light in weight, therefore, tension in both springs will be same. So, both springs will show same reading 8 kg.

283 **(b)**

Initially, the weight of load L is the force on the system of mass 8 kg

Acceleration $= \frac{2\times10}{8} = \frac{20}{8}$ units

Towards the end, force $= (2 + 1) \times 10\text{N}$ $= 30\text{N}$

So, acceleration now is $\frac{30}{8}$ units

284 **(d)**

Coefficent of friction $\mu = \frac{F}{R}$

$$= \frac{mg/3}{2mg/3} = \frac{1}{2}$$

285 **(c)**

If momentum remains constant, the force will be zero because

$$F = \frac{dp}{dt}$$

286 **(d)**

Momentum of the piece

$$p = \sqrt{p_x^2 + p_y^2}$$

$$= \sqrt{(16)^2 + (12)^2}$$

$= 20$

$$v = \frac{p}{m} = \frac{20}{0.5} = 40\text{ m/s}$$

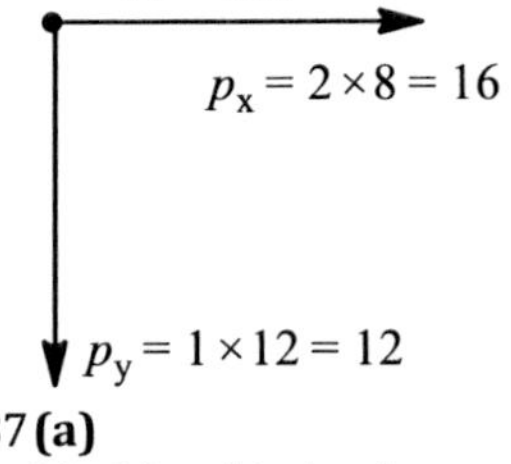

287 **(a)**

Limiting friction between block and slab

$= \mu_s m_A g$

$= 0.6 \times 10 \times 9.8 = 58.8\ N$

But applied force on block A is 100 N. So the block will slip over a slab
Now kinetic friction works between block and slab
$F_k = \mu_k m_A g = 0.4 \times 10 \times 9.8 = 39.2\ N$
This kinetic friction helps to move the slab
$\therefore$ Acceleration of slab $= \frac{39.2}{m_B} = \frac{39.2}{40}$
$= 0.98\ m/s^2$

288 **(a)**
From force diagram shown in figure,

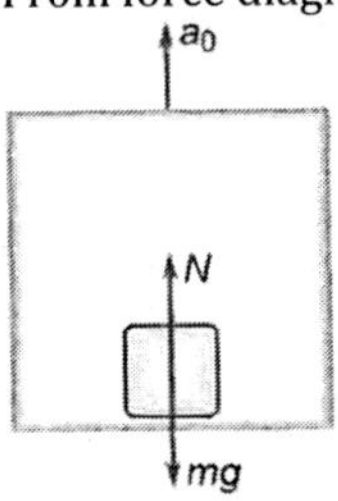

$N - mg = ma_0$
or $120 - mg = 2m$
$\therefore m = \dfrac{100}{10+2}$
$= 10$ kg

289 **(c)**
$T = \dfrac{F(L-x)}{L} = \dfrac{5(5-1)}{5} = 4N$

290 **(c)**
$F = v\frac{dm}{dt} = \frac{4\times5}{2.5}$ dyne $= 8$ dyne

291 **(d)**
Force on the car
$F = \mu R$
or $ma = \mu mg\ \ (\because R = mg)$
or $a = \mu g$
Now from 2nd equation of motion
$s = ut + \dfrac{1}{2}at^2$
or $s = 0 + \frac{1}{2}at^2\ \ (\because u = 0)$
or $t = \sqrt{\frac{2s}{\mu g}}$
$\therefore t = \sqrt{\dfrac{2s}{\mu g}}$
or $t \propto \frac{1}{\sqrt{\mu}}$

292 **(c)**
$T = \dfrac{T_0}{[1-(v^2/c^2)]^{1/2}}$
By substituting $T_0 = 1$ day and $T = 2$ days we get
$v = 2.6 \times 10^8\ ms^{-1}$

293 **(a)**
Change in momentum

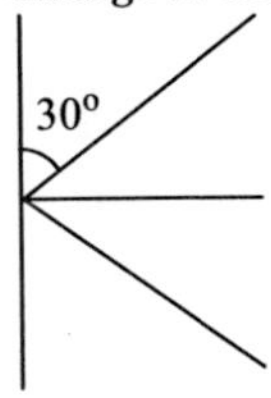

$\Delta p = 2\, mu \sin 30°$
$= 2 \times 0.1 \times 10 \times \dfrac{1}{2}$
$= 1$ kg-ms^{-1}
$F_{av} = \frac{\Delta p}{\Delta t} = \frac{1}{0.1} = 10$N

294 **(b)**
For a given mass $P \propto v$. If the momentum is constant then its velocity must be constant

295 **(c)**
Impulse, $I = F \times \Delta t = 50 \times 10^{-5} \times 3 = 1.5 \times 10^{-3}\ N - s$

296 **(d)**
Acceleration $= \frac{500}{25} = 20\ m/s^2$ in both the cases

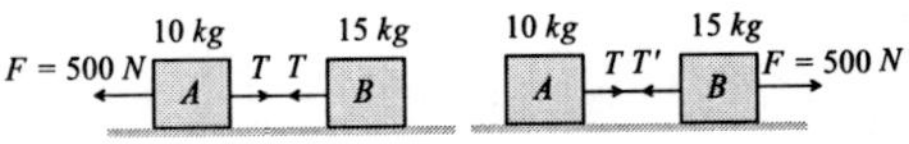

In fig 1, $T = 500 - 10 \times 20 = 300\ N$
In fig 2, $T'' = 500 - 15 \times 20 = 200\ N$

297 **(c)**
Component of weight mg in a direction perpendicular to the inclined plane is $mg \cos\theta$. According to Newton's third law of motion, the force exerted by the plane on the block is $mg \cos\theta$

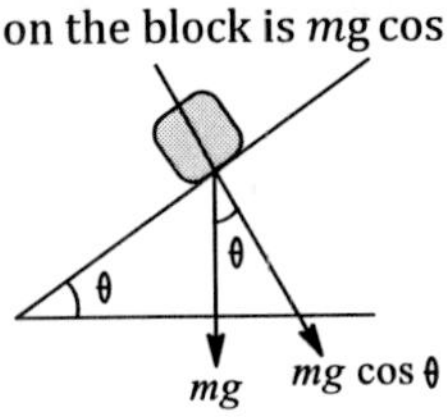

298 **(a)**
Since, $F = \dfrac{\Delta p}{\Delta t}$
or $\Delta p = F\,\Delta t$
We can say that momentum between 0 to 7 s is equal to the vector area enclosed by the force-time graph from 0 to 7 s. So,
Change in linear momentum
= vector area of triangle OAB + vector area of square BCDE+ vector area of

triangle EFG + vector area of square *GHIJ* + vector area of triangle *JKL*

$$= \left[\frac{1}{2} \times 1 \times (-1)\right] + [2 \times 2] + \left[\frac{1}{2} \times 2 \times (-2)\right] + [1 \times 1] + \left[\frac{1}{2} \times 1 \times (-1)\right]$$

$$= -\frac{1}{2} + 4 - 2 + 1 - \frac{1}{2} = 2 \text{ Ns}$$

299 **(b)**

$3mg = ma + ma$

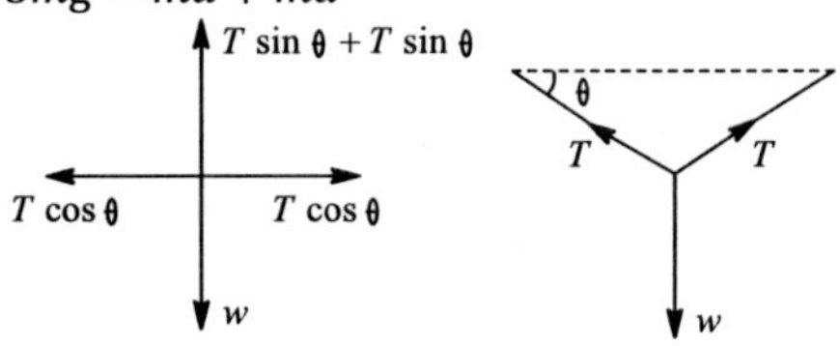

or $2mg = ma$ or $a = 2g$

300 **(a)**

Download acceleration

$a = g(\sin\theta - \mu\cos\theta)$

$$= g\left[\sin 30 - \frac{1}{\sqrt{3}}\cos 30\right]$$

$$= g\left[\frac{1}{2} - \frac{1}{\sqrt{3}} \times \frac{\sqrt{3}}{2}\right] = 0$$

301 **(a)**

$m_1 = 1000\ kg.\ v_1 = 50\ m/s, m_2 = 1250\ kg, v_2 = ?$

From conservation law of linear momentum

$m_1 v_1 = m_2 v_2$

$\Rightarrow 1000 \times 50 = 1250 \times v_2 \Rightarrow v_2 = 40\ m/s$

302 **(c)**

$500g - T = 500a$

$T - 100g\sin 30° - T' = 100a$

or $T - T' - 50g = 100a$

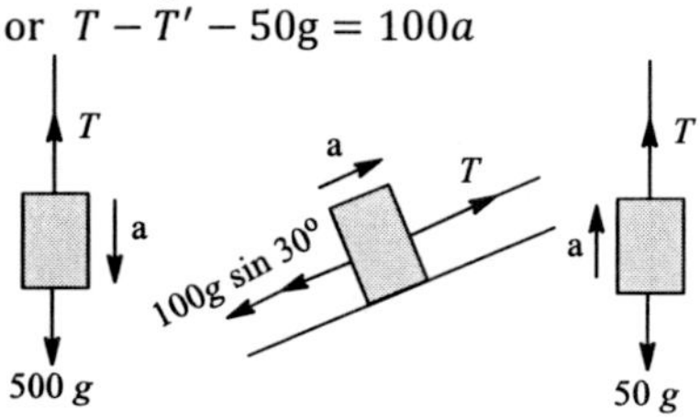

Again, $T' - 50g = 50a$

From Eqs. (ii) and (iii)

$T - 100g = 150a$

Adding Eqs. (i) and (iv), $400g = 650\ a$ or

$a = \frac{400g}{650} = \frac{8g}{13}$

This acceleration is downwards

303 **(b)**

Kinetic energy $= 10\ J$

$\Rightarrow \frac{1}{2}mv^2 = 10 \Rightarrow v^2 = 4$

From third equation of motion $v^2 = u^2 + 2as$

$4 = 0 + 2 \times a \times 2$

$\Rightarrow a = 1\ m/s^2$

$\because F_s = F - ma = 20 - 5 \times 1 = 15\ N$

304 **(c)**

Here, $\sin\theta = \frac{1}{l}$

Let required acceleration of inclined plane be a for the object to remain stationary relative to incline, we have

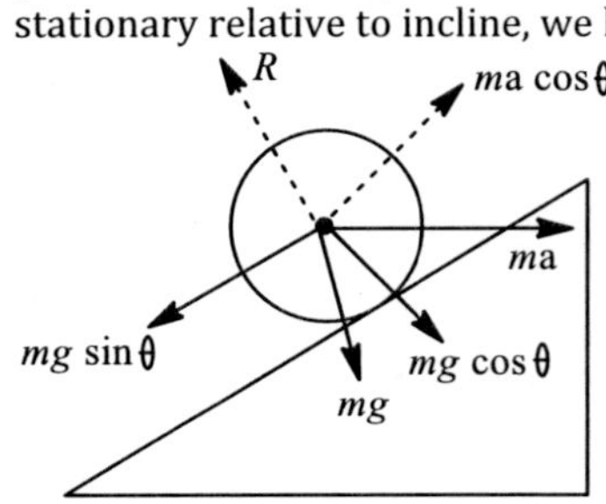

$ma\cos\theta = mg\sin\theta$

$$a = g\tan\theta = g\frac{1}{\sqrt{l^2 - 1}}$$

305 **(b)**

$l = \frac{1}{2}g\cos 60°\ t_1^2$ ....(i)

$l\cos\theta = \frac{1}{2}gt_2^2$ ...(ii)

$\frac{t_1^2}{t_2^2} = \frac{1}{\cos^2 60°}$

$= \frac{1}{4}$

$t_1 : t_2 = 2:1$

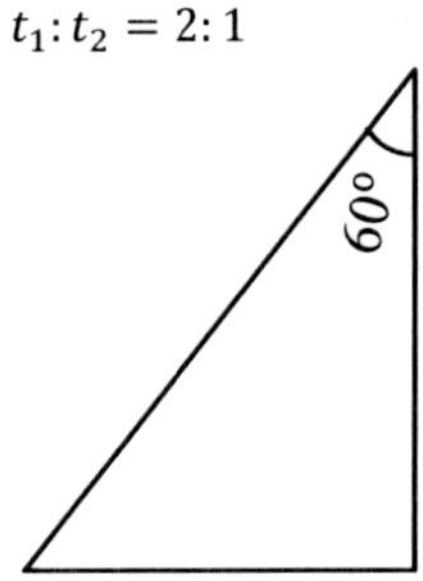

306 **(c)**

Normal reaction $R = mg - P\sin 30° = mg - \frac{P}{2}$

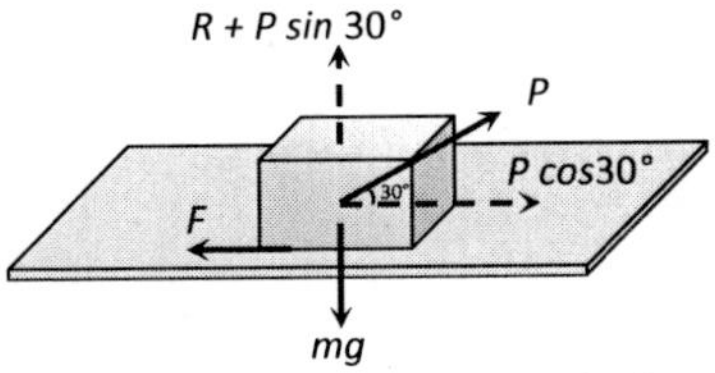

∴ Limiting friction between body and surface is given by, $F = \mu R = \mu\left(mg - \frac{P}{2}\right)$

307 **(a)**

As F_2 and F_3 are mutually perpendicular, their resultant $= \sqrt{F_2^2 + F_3^2}$

As particle is stationary under F_1, F_2, F_3, therefore, $=\sqrt{F_2^2 + F_3^2}$ must be equal and opposite to F_1

308 **(d)**

The situation is shown in figure. At initial time, the ball is at P, then under the action of a force (exerted by hand) from P to A and then from A to B, let acceleration of ball during PA is a ms^{-2}[assumed to be constant] in upward direction and velocity of ball at A is v ms^{-1}.

Then for PA, $v^2 = 0^2 + 2a \times 0.2$

For AB, $0 = v^2 - 2 \times g \times 2$

$\Rightarrow \quad v^2 = 2g \times 2$

From above equation,

$a = 10g = 100 \text{ ms}^{-2}$

Then for PA, FBD of ball is

$F - mg = ma$ [F is the force exerted by hand on ball]

$\Rightarrow \; F = m(g + a) = 0.2\,(11g)$

$= 22$ N

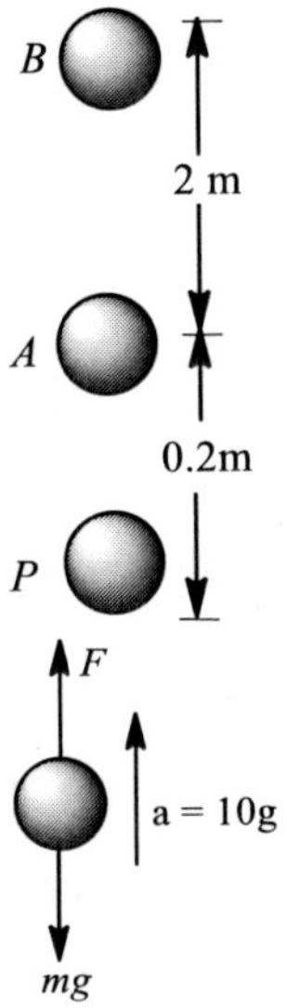

309 **(c)**

$$W = \mu\, mg \cos\theta\, S = 0.5 \times 1 \times 9.8 \times \frac{1}{2} \times 1$$
$$= 2.45\, J$$

310 **(c)**

When force F is applied on $2m$ from left, contact force,

$F_1 = \dfrac{m}{m + 2m} F = \dfrac{F}{3}$

When force F is applied on m from right, contact force

$F_2 = \dfrac{2m}{m + 2m} F = \dfrac{2F}{3}$

∴ $F_1 : F_2 = 1 : 2$

311 **(d)**

In terms of three significant figure

Momentum $P = mv = 3.513 \times 5.00 = 17.6$

312 **(d)**

Force equation for 'M'

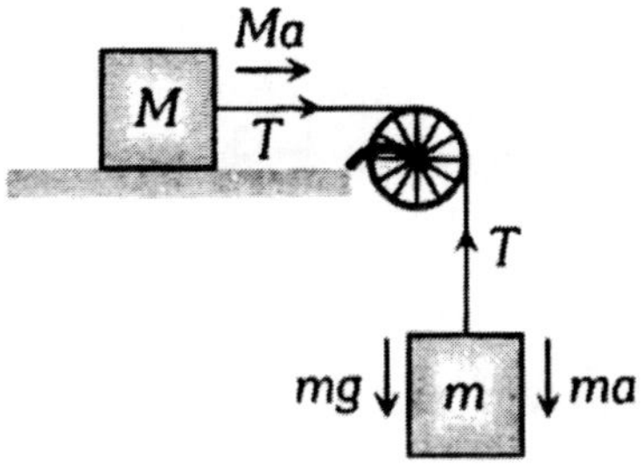

$Ma = T$...(i)

Force equation for, m

$mg - T = ma$ ...(ii)

On solving (i) and (ii)

$T = \left(\dfrac{Mm}{M + m}\right) g$

314 **(d)**

We know that $a = \mu g = 0.2 \times 9.8 = 1.96\, m/s^2$

315 **(b)**

Velocity $u = 72 \text{ kmh}^{-1} = 20 \text{ ms}^{-1}$

$a = \mu g = 0.5 \times 10 \text{ ms}^{-2}$

From $\quad v^2 = u^2 - 2as$

∴ $(0)^2 = (20)^2 - 2 \times 0.5 \times 10 \times s$

∴ $s = \dfrac{20 \times 20}{2 \times 0.5 \times 10}$ or $s = 40$m

316 **(d)**

$$\text{Force } \mathbf{F} = \frac{d\mathbf{p}}{dt} = -k\,A\sin(kt)\hat{\mathbf{i}} - k\,A\cos(kt)\hat{\mathbf{j}}$$

$\mathbf{p} = A\cos(kt)\hat{\mathbf{i}} - A\sin(kt)\hat{\mathbf{j}}$

Since, $\quad \mathbf{F}\cdot\mathbf{p} = 0$

∴ Angle between $\mathbf{F}$ and $\mathbf{p}$ should be 90°

317 **(c)**

Resistive force $= \dfrac{\text{initial kinetic energy}}{\text{stopping distance}}$

$= \dfrac{\frac{1}{2} \times 0.05 \times (80)^2}{0.40} = 400\text{N}$

318 **(c)**

Block A moves with velocity $0.15 ms^{-1}$ compresses the spring which pushes B towards right. A goes on compressing the spring till the velocity acquired by B becomes equal to the velocity of A. Let this velocity be v. This state occurs when the spring is in a state of maximum compression. Let x be the maximum compression in this stage According to the law of conservation of linear momentum, we get

$m_A u = (m_A + m_B)v$

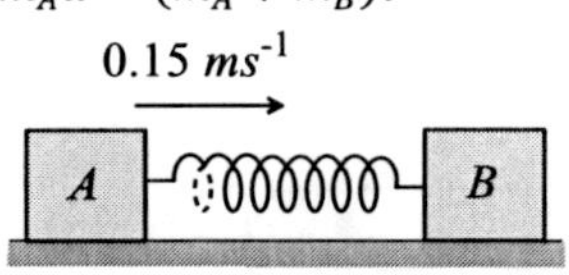

$\Rightarrow v = \dfrac{m_A u}{m_A + m_B} = \dfrac{2 \times 0.15}{2+3} = 0.06\, ms^{-1}$

According to the conservation of energy, we get

$\frac{1}{2} m_A u^2 = \frac{1}{2}(m_A + m_B)v^2 + \frac{1}{2}kx^2$

$\frac{1}{2} m_A u^2 - \frac{1}{2}(m_A + m_B)v^2 = \frac{1}{2}kx^2$

$\frac{1}{2} \times 2 \times (0.15)^2 - \frac{1}{2}(2+3)(0.06)^2 = \frac{1}{2}\, kx^2$

$0.0225 - 0.009 = \frac{1}{2}kx^2 \Rightarrow 0.0135 = \frac{1}{2}kx^2$

$\Rightarrow x = \sqrt{\dfrac{0.027}{k}} = \sqrt{\dfrac{0.027}{10.8}} = 0.05\, m$

319 **(a)**

Relative vertical acceleration of A with respect to B

$= g(\sin^2 60° - \sin^2 30°)$

$= 9.8\left(\frac{3}{4} - \frac{1}{4}\right) = 4.9\, m/s^2$

321 **(d)**

Net force in forward direction = Accelerating force + Friction

$= ma + \mu\, mg = m(a + \mu\, g)$

$= (1500 + 500)(1 + 0.2 \times 10)$

$= 2000 \times 3 = 6000\, N$

322 **(c)**

For $W, 2W, 3W$ apparent weight will be zero because the system is falling freely. So the distances of the weight from the rod will be same

323 **(d)**

At the highest point, velocity of shell before explosion

$= v \cos\theta$

$\therefore$ Momentum of shell before explosion

$= mv\cos\theta$

After explosion momentum

$= -\frac{m}{2} v\cos\theta + \frac{m}{2}V$

According to the law of conservation of momentum

$mv\cos\theta = -\dfrac{mv\cos\theta}{2} + \dfrac{m}{2}V$

$\therefore\ V = 3v\cos\theta$

326 **(d)**

Three guns are fired towards the centre of circle as shown in figure.

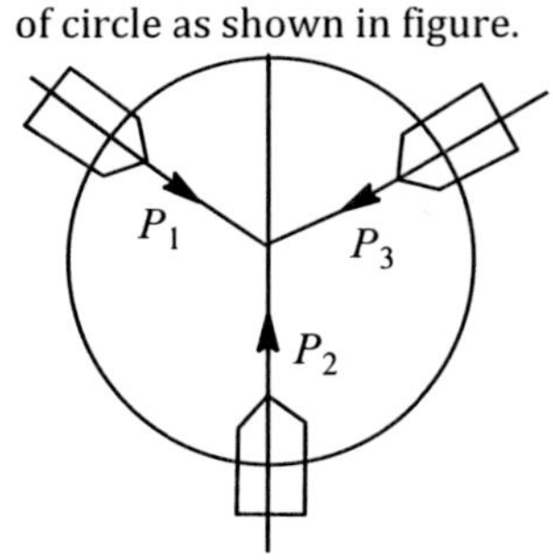

Since, total final momentum is zero, and no external force is acting on the system, so total initial momentum should be also zero.

So, $\mathbf{p}_1 + \mathbf{p}_2 + \mathbf{p}_3 = 0$

Three vectors, which are at an angle of 120° leads to zero resultant if and only if they have same magnitude.

So, $4.5v_1 = 2.5 \times 575 = 4.5v_2$

After solving, we will get v_1 and v_2 come out be 320 ms^{-1}.

327 **(d)**

Net download force = Weight – Friction

$\therefore ma = 25 \times 9.8 - 2 \Rightarrow a = \dfrac{25 \times 9.8 - 2}{25}$

$= 9.72\, m/s^2$

328 **(a)**

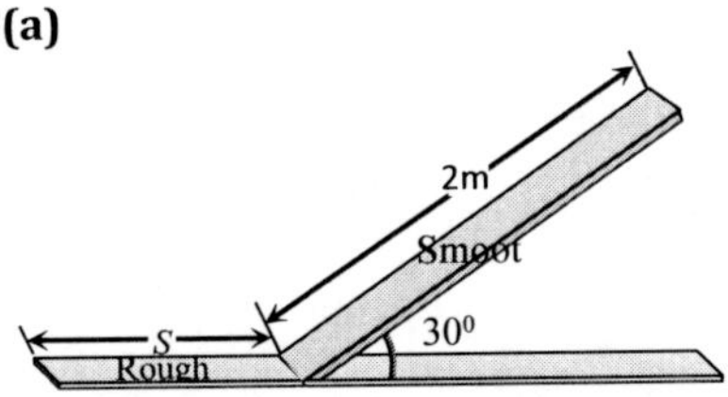

$v^2 = u^2 + 2as = 0 + 2 \times g \sin 30 \times 2$

$\Rightarrow v = \sqrt{20}$

Let it travel distance 'S' before coming to rest

$S = \frac{v^2}{2\mu g} = \frac{20}{2 \times 0.25 \times 10} = 4m$

329 **(d)**

Coefficient of friction $\mu = \tan\theta\left[1 - \frac{1}{2^2}\right]$

Here,, $\theta = 45°$ and $n = 2$

$\therefore \quad \mu = \tan 45°\left[1 - \frac{1}{2^2}\right]$

$= 1 - \frac{1}{4}$

$= \frac{3}{4} = 0.75$

331 **(d)**

Pseudo force on the block $= m \times 4$ N (backward)

Force of friction $= 0.4 \times m \times 10$ N (forward)

Equating, $m \times 4 = 0.4 \times m \times 10 = 4m$

Clearly the equation holds good for all values of m

332 **(c)**

We know that in the given condition $s \propto \frac{1}{m^2}$

$\therefore \frac{s_2}{s_1} = \left(\frac{m_1}{m_2}\right)^2 \Rightarrow s_2 = \left(\frac{m_1}{m_2}\right)^2 \times s_1$

333 **(b)**

Net force on mass m, $ma = F - T \;\therefore a = \frac{F-T}{m}$

334 **(c)**

Coefficient of friction = Tangent of angle of repose

$\therefore \mu = \tan\theta$

335 **(b)**

From $s = ut + \frac{1}{2}at^2 = 0 + \frac{1}{2}at^2$

$t = \sqrt{\frac{2s}{a}}$

For smooth plane, $a = g\sin\theta$

For rough plane, $a' = g \quad (\sin\theta - \mu\cos\theta)$

$\therefore \quad t' = \sqrt{\frac{2s}{g\,(\sin\theta - \mu\cos\theta)}} = nt$

$= n\sqrt{\frac{2s}{g\sin\theta}}$

$\therefore \; n^2 g(\sin\theta - \mu\cos\theta) = g\sin\theta$

When $\theta = 45°, \sin\theta = \cos\theta = 1/\sqrt{2}$

Solving we get $\mu = 1 - \frac{1}{n^2}$

336 **(a)**

The nail has fallen through a total vertical distance of $(h + x)$. Hence loss in tis potential energy $= mg(h + x)$. If average retarding force exerted by sand on nail is f, then work done $W = -Fx$

For equilibrium,

$mg(h + x) = Fx$ or $F = mg\frac{h+x}{x} = mg\left(\frac{h}{x} + 1\right)$

337 **(b)**

The pressure on the rear side would be more due to fictitious force (acting in the opposite direction of acceleration) on the rear face. Consequently the pressure in the front side would be lowered

338 **(a)**

$F = m\left(\frac{v-u}{t}\right) = \frac{5(65 - 15) \times 10^{-2}}{0.2}$

$= 12.5\ N$

339 **(c)**

Impulse = Change in momentum = $m(v_2 - v_1)$ ...(i)

Again impulse = Area between the graph and time axis

$= \frac{1}{2} \times 2 \times 4 + 2 \times 4 + \frac{1}{2}(4 + 2.5) \times 0.5 + 2 \times 2.5$

$= 4 + 8 + 1.625 + 5 = 18.625$ ...(ii)

From (i) and (ii), $m(v_2 - v_1) = 18.625$

$\Rightarrow v_2 = \frac{18.625}{m} + v_1 = \frac{18.625}{2} + 5$

$= 14.25\ m/s$

340 **(a)**

When

$P = mg\,(\sin\theta - \mu\cos\theta)$

$f = \mu mg\cos\theta$(upwards)

when $P = mg\sin\theta$

$f = 0$

and when $P = mg(\sin\theta + \mu\cos\theta)$

$f = \mu\, mg\cos\theta$ (downwards)

Hence, friction is first positive, then zero and then negative.

341 **(b)**

We know $s = \frac{u^2}{2\mu g}$ $\therefore$ $\mu = \frac{u^2}{2gs} = \frac{(6)^2}{2\times10\times9} =$ 0.2

342 **(d)**

Distance travelled by the lift

= Area under velocity time graph

$= \left(\frac{1}{2} \times 2 \times 3.6\right) + (8 \times 3.6)$
$\quad + \left(\frac{1}{2} \times 2 \times 3.6\right)$
$= 36\, m$

343 **(c)**

Force $= v\frac{dm}{dt}$

Force $= v\, M$ $\quad$ [$\therefore$ Force $= M\, v$ newton]

344 **(a)**

Work done against gravity $= mgh = 2 \times 10 \times 10 = 200\, J$

Work done against friction = (Total work done − work done against gravity) $= 300 - 200 = 100\, J$

345 **(a)**

The situation is shown in figure.

At an angle of 60°.

$T\cos\theta = mg$

$T = \frac{mg}{\cos\theta}$

$= \frac{10g}{\cos 60°}$

$= \frac{10}{1/2}$ kg-wt

20 kg-wt

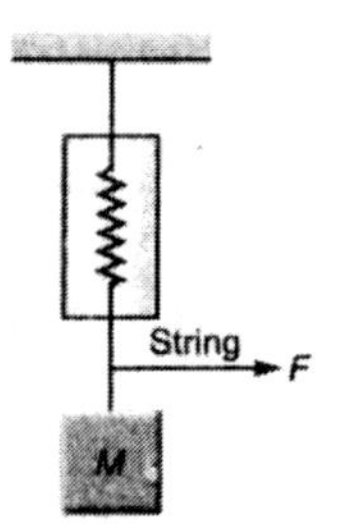

346 **(a)**

From acting on block are shown in adjoining figure

As the block does not move, hence

$F\cos 60° = f = \mu N = \mu(Mg + F\sin 60°)$

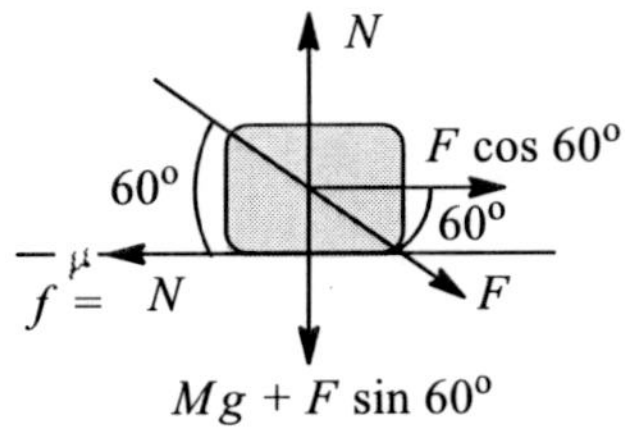

$\therefore F\frac{1}{2} = \frac{1}{2\sqrt{3}}\left(\sqrt{3}\times 10 + F\cdot\frac{\sqrt{3}}{2}\right)$

On simplification, we get $F = 20$ N

347 **(d)**

Initial velocity of steel ball $u = 14\ \text{ms}^{-1}$

Mass of steel ball $m = 0.5$ kg

Force acting on the steel ball from opposite direction.

$F = -1200$ N

Time duration for collision,

$t = 25\ \text{ms} = 25\times10^{-3}$ s

Let final velocity of ball$= v$

Applying Newton's II law of motion to steel ball *ie*,

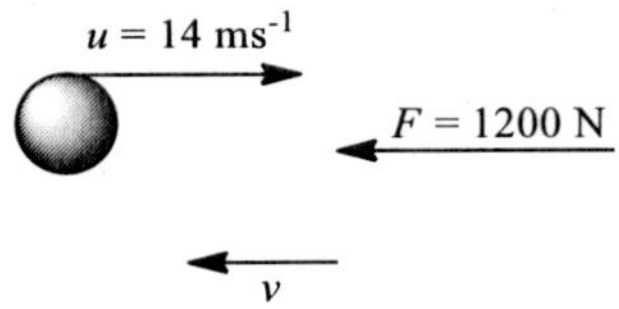

$F = \frac{p_2 - p_1}{t}$

$\Rightarrow\ -1200 = \frac{mv - mu}{t}$

$\Rightarrow\ \frac{-1200}{m} + u = v$

$\Rightarrow\ v = \frac{-1200\times25\times10^{-3}}{0.5} = -60\ \text{ms}^{-1}$

$v = -60 + 14\ \text{ms}^{-1}$

$= -46\ \text{ms}^{-1}$

[−ve sign shows that now ball will move in the direction of force]

348 **(d)**

Wall applies 2 forces of the block (i) normal reaction, $R = 12$ N, and (ii) frictional force, $f_2 = mg = 5$ N tangentially upward

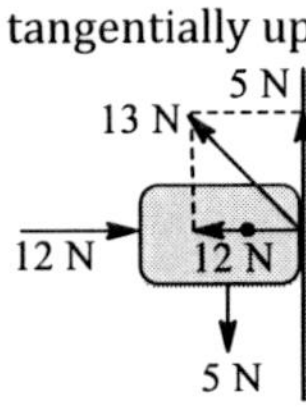

$\therefore$ Total force exerted by wall on block

$F = \sqrt{N^2 + f_s^2} = \sqrt{(12)^2 + (5)^2} = 13$N

349 **(d)**

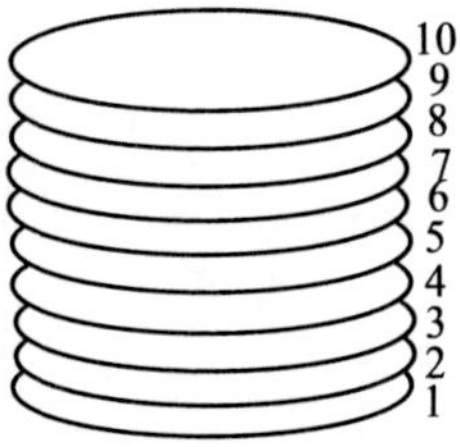

(a) Is correct 6th coin has four coins on its top which exert a force $4mg$ on it

(b) Is correct. 7th coin has three coins, placed over it. Thus 7th coin exerts a force $4mg$ on 6th coin (downwards)

(c) Is correct. As what is explained in (b), the reaction of 6th coin on the 7th coin is $4mg$ (upwards)

(d) Is wrong 10th coin, which is the topmost coin, experiences a reaction force of mg (upwards) from all the coins below it

350 **(d)**

Applying the law of conservation of linear momentum, we get

$$0.5 \times v = \sqrt{(2 \times 8)^2 + (1 \times 12)^2}$$
$$= \sqrt{256 + 144} = \sqrt{400}$$
$$0.5\, v = 20 \Rightarrow v = \frac{20}{0.5} = 40\ ms^{-1}$$

351 **(a)**

Newton's second and third laws lead to the conservation of linear momentum

352 **(b)**

Weight of chain on table $= mg - \frac{1}{4}\, mg = \frac{3}{4} mg$

For maximum possible friction,

$$\frac{3}{4} \mu mg = \frac{1}{4} mg$$
$$\therefore \mu = \frac{1}{3}$$

353 **(d)**

$$F_1 = mg(\sin\theta + \mu\cos\theta)$$
$$F_2 = mg(\sin\theta + \mu\cos\theta)$$
$$\frac{F_1}{F_2} = \frac{\sin\theta + \mu\cos\theta}{\sin\theta - \mu\cos\theta}$$
$$= \frac{\tan\theta + \mu}{\tan\theta - \mu}$$
$$= \frac{2\mu + \mu}{2\mu - \mu}$$
$$= 3$$

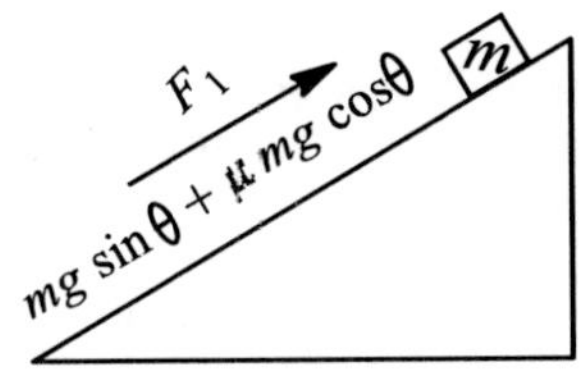

354 **(b)**

For the given condition, Static friction = Applied force = Weight of body = $2 \times 10 = 20\ N$

355 **(a)**

Making FBD of block with respect to disc

Let A be the acceleration of block with respect to disc

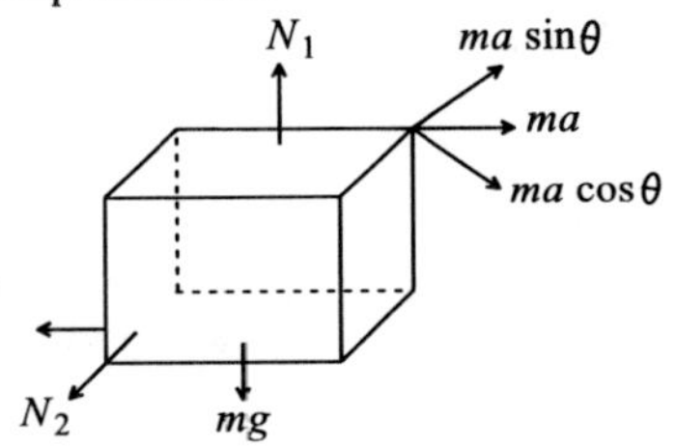

$$N_1 = mg$$
$$N_2 = m\, a \sin\theta$$
$$A = \frac{m\, a\cos\theta - \mu N_2 - \mu N_1}{m} = 10 m/s^2$$

356 **(a)**

Weight of the disc will be balanced by the force applied by the bullet on the disc in vertically upward direction

$$F = n\, m\, v$$
$$= 40 \times 0.05 \times 6 = Mg$$
$$M = \frac{40 \times 0.05 \times 6}{16}$$
$$= 1.2\ \text{kg}$$

357 **(b)**

Horizontal component of 10N is 10 cos 60° *ie* $10 \times \frac{1}{2}$ N or 5 N. It is balanced by 5 N force along x-axis

Y-component of 10 N force is 10 sin 60 N *ie*, $10 \times \frac{\sqrt{3}}{2}$N or $5\sqrt{3}$N

$$a_y = \frac{5\sqrt{3}\text{N}}{\sqrt{3}\ \text{kg}} = 5\ \text{ms}^{-2}$$

358 **(b)**

$$F = \sqrt{(F)^2 + (F)^2 + 2F.F\cos\theta} \Rightarrow \theta = 120°$$

359 **(b)**

$$a = \frac{m_2}{m_1 + m_2} g = \frac{3}{7+3} 10 = 3\ m/s^2$$

360 **(a)**

Let the tension in the wire be T. The equations of motion of the two locks are,
$T - 10 = 1\,a$
and $20 - T = 2\,a$
Eliminating a form these equations,
$T = \left(\frac{40}{3}\right)$ N
Stress, $T = \frac{\left(\frac{40}{3}\right)}{\pi\, r^2}$
If the minimum radius needed to avoid breakings is r.
$2 \times 10^9 = \frac{\left(\frac{40}{3}\right)}{\pi\, r^2}$
Solving this,
$r = 4.6 \times 10^{-5}$ m

361 **(a)**

Force equilibrium of system, $F_1 = \sqrt{F_2^2 + F_3^2}$ [As $\theta = 90°$]
In the absence of force F_1, Acceleration $= \frac{\text{Net Force}}{\text{Mass}}$
$= \frac{\sqrt{F_2^2 + F_3^2}}{m} = \frac{F_1}{m}$

362 **(c)**

Minimum force required to move a body up a rough inclined plane
$F_1 = mg(\sin\theta + \mu\cos\theta)$
Minimum force required to prevent the body from sliding down the rough inclined plane.
$F_2 = \mu\, mg\cos\theta$
According to question
$F_1 = 3\,F_2$
$\therefore\ mg(\sin\theta + \mu\cos\theta) = 3(\mu\, mg\cos\theta)$
$\sin\theta + \mu\cos\theta = 3\mu\cos\theta$
$\sin\theta = 2\mu\cos\theta$
$\tan\theta = 2\mu = 2 \times \frac{1}{2\sqrt{3}} = \frac{1}{\sqrt{3}} = \tan 30°$
$\theta = 30°$

363 **(b)**

Newton second law
$F = ma \Rightarrow 6 = (7+5)a\,;\ a = \frac{1}{2}\frac{m}{s^2};\ F' \rightarrow 5\,kg$
Now, $F' = 5 \times \frac{1}{2} = 2.5\ N$

364 **(d)**

Angular momentum is an axial vector, so its direction is along the axis, perpendicular to the plane of motion which is not changing because of change of speed. Therefore, the direction of angular momentum remains the same and its magnitude may vary

365 **(c)**

Impulse = Area between force and time graph and it is maximum for graph (III) and (IV)

366 **(c)**

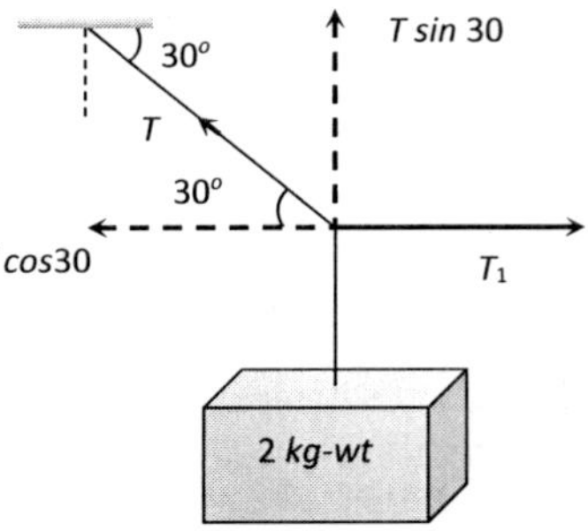

$T\sin 30 = 2\,kg\,wt$
$\Rightarrow T = 4\,kg\,wt$
$T_1 = T\cos 30°$
$= 4\cos 30°$
$= 2\sqrt{3}$

367 **(c)**

From geometry $l^2 = x^2 + y^2$ but y is constant, hence differentiating, we have,
$2l\frac{dl}{dt} = 2x\frac{dx}{dt}$
But $\frac{dl}{dt} = v$. Hence horizontal velocity of block,
$v_x = \frac{dx}{dt}$

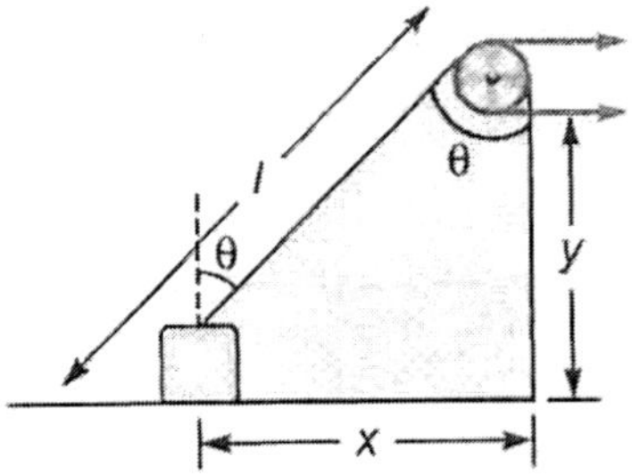

$\Rightarrow lv = x.v_x$ or $v_x = \frac{l.v}{x} = \frac{v}{\sin\theta}$

368 **(c)**

As the spring balances are massless therefore the reading of both balance should be equal

369 **(c)**

Gas will come out with sufficient speed in forward direction, so reaction of this forward force will change the reading of the spring balance

370 **(a)**

We can realise the situation as shown.

Maximum retardation

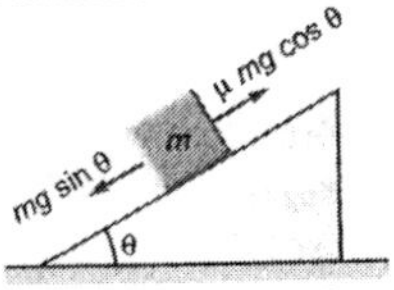

$$a = \frac{f_k - mg\sin\theta}{m}$$

$$= \frac{\mu_k\, mg\cos\theta - mg\sin\theta}{m}$$

$$= \frac{\frac{1}{\sqrt{3}} mg\cos 30° - mg\sin 30°}{m} \quad (\because\ \theta = 30°)$$

$$= \frac{\frac{1}{\sqrt{3}} g \times \frac{\sqrt{3}}{2} - g \times \frac{1}{2}}{1} = 0$$

Hence, under the effect of kinetic friction between block and inclined plane, acceleration of block is zero.

371 **(c)**

For accelerated upward motion

$R = m\,(g + a) = 80\,(10 + 5) = 1200\ N$

372 **(a)**

Coefficient of friction $\mu_s = \frac{F_1}{R} = \frac{75}{mg} = \frac{75}{20\times 9.8} = 0.38$

373 **(a)**

For equilibrium of the block

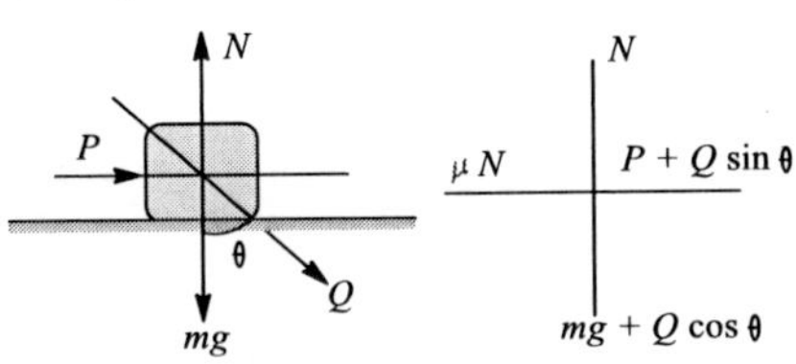

$N = mng + Q\cos\theta$

$Q\sin\theta + P = \mu N$

$Q\sin\theta + P = \mu(mg + Q\cos\theta)$

$$\mu = \left(\frac{Q\sin\theta + P}{mg + Q\cos\theta}\right)$$

374 **(d)**

Here : Mass of ship $m = 2\times 10^7\ kg$,

Force $F = 25\times 10^5 N$

Displacement $s = 25\ m$

According to the Newton's second law of motion

$F = ma$

$$\Rightarrow a = \frac{F}{m} = \frac{25\times 10^5}{2\times 10^7} = 12.5\times 10^{-2}\ m/s^2$$

The relation for final velocity is

$$v^2 = u^2 + 2as \Rightarrow v^2 = 0 + 2 \times (12.5\times 10^{-2}) \times 25$$

$\Rightarrow v = \sqrt{6.25} = 2.5\ m/s$

375 **(a)**

Resultant acceleration $(a)_R = \sqrt{2}a - a$

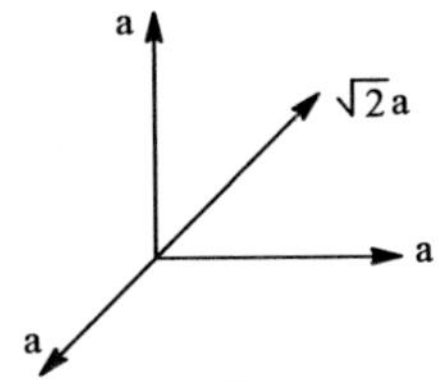

$(a)_R = (\sqrt{2} - 1)a$

376 **(b)**

$$F = \frac{W}{\mu} = \frac{1\times 9.8}{0.2} = 49\ N$$

377 **(a)**

$$F = u\left(\frac{dm}{dt}\right) = 20 \times \frac{50}{60} = 16.66\ N$$

378 **(a)**

On releasing, the motion of the system will be according to figure,

$m_1 g - T = m_1 a$ ....(i)

and $T - m_2 g = m_2 a$ ...(ii)

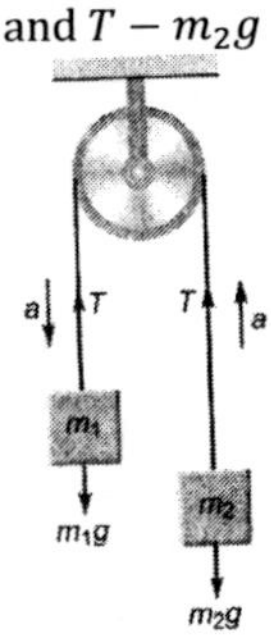

On solving:

$$a = \left(\frac{m_1 - m_2}{m_1 + m_2}\right) g \quad \text{.... (iii)}$$

Here, $m_1 = 5$ kgi, $m_2 = 4.8$ kg, $g = 9.8\ ms^{-2}$

$$\therefore\quad a = \left(\frac{5 - 4.8}{5 + 4.8}\right) \times 9.8$$

$$= \frac{0.2}{9.8} \times 9.8 = 0.2\ ms^{-2}$$

379 **(c)**

For minimum mass of m, mass M breaks off contact when elongation in spring is maximum

At the time of break off, block A is at lowest position and its speed is zero. At an instant t_1

$mg - kx = ma$

$v\dfrac{dv}{dx} = \dfrac{mg - kx}{m}$

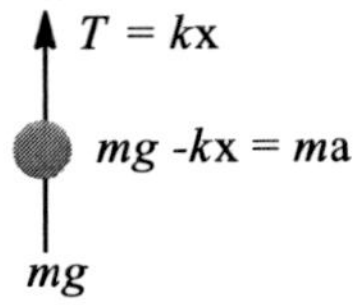

$$\int_0^0 vdv = \int_0^x \left(g - \frac{k}{m}x\right) dx$$

Where x_0 is maximum elongation is spring

$0 = gx_0 - \dfrac{kx_0^2}{2m}$

$x = \dfrac{2mg}{k}$

At the time of break off of block B

$Mg = kx_0$

$Mg = 2mg$

$m = \dfrac{M}{2}$

380 **(d)**

$T = mg + ma$

$2000g = 1000g + 1000a$

or $a = g$

Direction is upward

Now, $0^2 - 2.5^2 = -2 \times 10 \times s$

or $s = \frac{2.5\times2.5}{20}$

$= \dfrac{625}{100 \times 20} = \dfrac{25}{80}\text{m} = \dfrac{5}{16}\text{m}$

381 **(d)**

Because if the angle of inclination is equal to or more than angle of repose then box will automatically slide down the plane

382 **(d)**

Initial thrust $mg + ma = m(g + a)$

$= 10^5(10 + 5)\text{N}$

$= 1'5 \times 10^5\text{N} = 1.5 \times 10^6\text{N}$

383 **(d)**

In the given condition the required centripetal force is provided by frictional force between the road and tyre

$\dfrac{mv^2}{R} = \mu mg$

$\therefore v = \sqrt{\mu Rg}$

384 **(b)**

As weight $= 9.8\ N$ $\therefore$ Mass $= 1\ kg$

Acceleration $= \frac{\text{Force}}{\text{Mass}} = \frac{5}{1} = 5\ m/s^2$

385 **(b)**

Force exerted by the ball

$\Rightarrow F = m\left(\dfrac{dv}{dt}\right) = 0.15 \times \dfrac{20}{0.1} = 30\ N$

386 **(b)**

$2mnv = Mg$

$v = \dfrac{Mg}{2mn}$

or $= \frac{10\times980}{2\times5\times10}\text{cms}^{-1}$

$= \dfrac{9800}{100}\text{cms}^{-1}$

$= 98\ \text{cms}^{-1}$

387 **(b)**

Here, mass of bullet $m = 10g = \frac{10}{1000}\ kg$

Mass of ice, $M = 5kg$

According to the conservation of linear momentum, we get

$m \times 300 + M \times 0 = m \times 0 + mv$

$\Rightarrow \dfrac{10}{1000} \times 300 + M \times 0 = 5v$

$\therefore v = \dfrac{3}{5} = 0.6\ m/s = 60\ cm/s$

388 **(d)**

Particle will move with uniform velocity due to inertia

389 **(d)**

$\mu_s = \dfrac{m_B}{m_A} \Rightarrow 0.2 = \dfrac{m_B}{2} \Rightarrow m_B = 0.4\ kg$

390 **(b)**

As the amplitude is increased, the maximum acceleration of the platform (along with coin as long as they doesn't get separated) increases

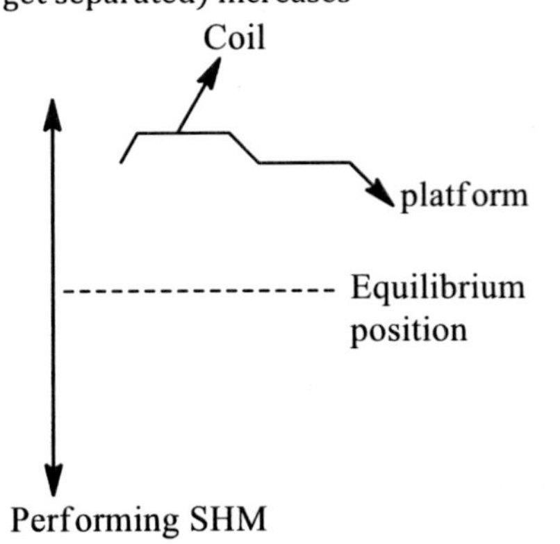

if we draw the FBD for coin at the extreme position as shown

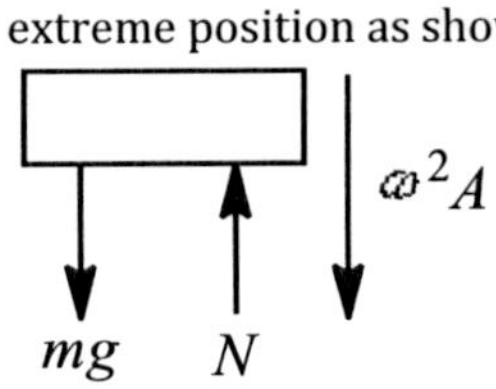

then from Newton's law, $mg - N = m\omega^2 A$

For loosing contact with the platform, $N = 0$

So, $A = \frac{g}{\omega^2}$

392 **(c)**

From the figure,

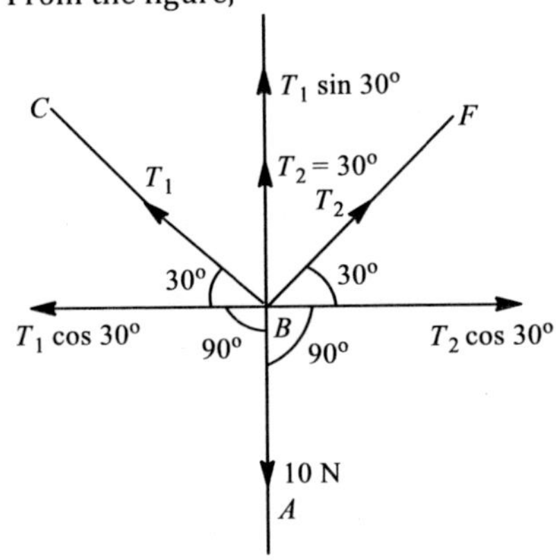

$T_1 \cos 30° = T_2 \cos 30°$

$\therefore\ T_1 = T_2 = T$(Let)

Again, $T_1 \sin 30° + T_2 \sin 30° = 10$

$2T \sin 30° = 10$

$2T\frac{1}{2} = 10 \quad \Rightarrow T = 10\text{ N}$

$\therefore$ Tension in section BC and BF are 10 N and 10 N.

393 **(c)**

$$l = l_0\sqrt{1 - \frac{v^2}{c^2}} = 1\sqrt{1 - \left(\frac{2.7 \times 10^8}{3 \times 10^8}\right)^2}$$

$$\Rightarrow l = 0.44\ m$$

394 **(a)**

By drawing the free body diagram of the block for critical condition

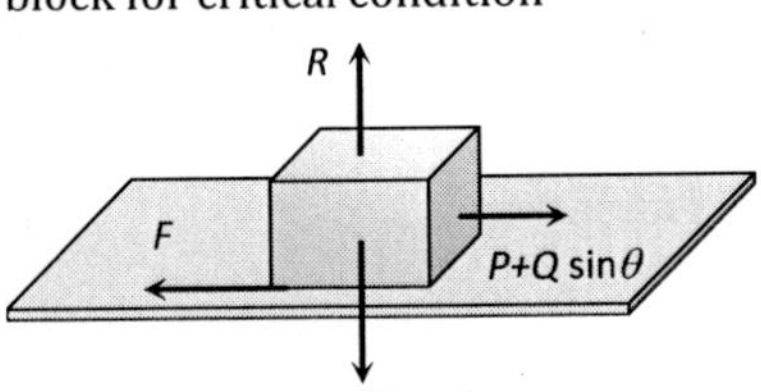

$F = \mu R \Rightarrow P + Q \sin\theta$
$= \mu\,(mg + Q\cos\theta)$

$\therefore \mu = \frac{P + Q\sin\theta}{mg + Q\cos\theta}$

395 **(a)**

$\vec{F} = \frac{\Delta\vec{p}}{\Delta t} \Rightarrow \Delta t = \frac{|\Delta\vec{p}|}{|\vec{F}|} = \frac{0.4}{2} = 0.2\ s$

396 **(c)**

$|\hat{\mathbf{j}}| = |\Delta\,\vec{\mathbf{p}}| = 0.2(20 + 15)\text{N s} = 7\text{N s}$

397 **(b)**

$F = \frac{d}{dt}(p)$

$F = \frac{d}{dt}(a + bt + ct^2)$ or $F = b + 2ct$

Clearly, the force is time-dependent

398 **(b)**

In this case, one 2 $kg\ \omega t$ on the left will act as the support for the spring balance. Hence its reading will be 2 kg

400 **(b)**

$a = \mu g = 5$

$v^2 = u^2 + 2as$

$0 = 2^2 + 2 \times (5)s$

$s = -\frac{2}{5}$ w.r.t. belt

Or distance $= 0.4\ m$

401 **(b)**

We know that $F = v\frac{dm}{dt}$

$\frac{dm}{dt}$ is increased by a factor of 4 and v is doubled

So, F is increased by a factor of 8

402 **(c)**

Distance travelled by the body in n^{th} second is given by

$S_n = u + \frac{a}{2}(2n - 1)$

$5 = u + \frac{a}{2}(2 \times 1 - 1)$

$5 = u + \frac{a}{2} \quad \ldots..(i)$

$2 = u + \frac{a}{2}(2 \times 3 - 1)$

$2 = u + \frac{5}{2}a \ldots.(ii)$

Solving Eqs. (i) and (ii), we get

$a = -\frac{6}{4}\text{ms}^{-2}$

ie, body is decelerating

Given, mass = 4kg

$\therefore F = m \times a = 4 \times \frac{6}{4} = 6\text{ N}$

403 **(d)**

Let x be distance from A to O, and L the total length of string. Then , ratio of moment is

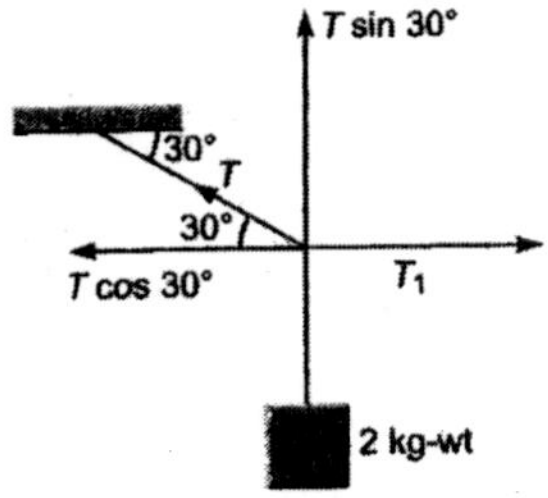

$$\frac{m_1}{m_2}=\frac{(L-x)}{L}=\frac{3-2}{3}=\frac{1}{3}$$

404 **(c)**

$$F=v\frac{dm}{dt}=10\times 5\text{ N}=50\text{ N}$$

405 **(c)**

Applying Lami's theorem

$$\frac{T_1}{\sin(90°+\beta)}=\frac{T_2}{\sin(90°+\alpha)}=\frac{mg}{\sin[180°-(\alpha+\beta)]}$$

or $\frac{T_1}{\cos\beta}=\frac{T_2}{\cos\alpha}=\frac{mg}{\sin(\alpha+\beta)}$

$\therefore\ T_1=\frac{mg\cos\beta}{\sin(\alpha+\beta)};T_2=\frac{mg\cos\alpha}{\sin(\alpha+\beta)}$

406 **(d)**

$$T=(M+m)(g+a)$$
$$=(940+60)(10+1)$$
$$=11000\ N$$

407 **(d)**

According to law of conservation of momentum the third piece has momentum

$=1\times-(3i+4\hat{j})kg\ ms^{-1}$

Impulse = Average force × time

$\Rightarrow$ Average force $=\frac{\text{Impulse}}{\text{time}}$

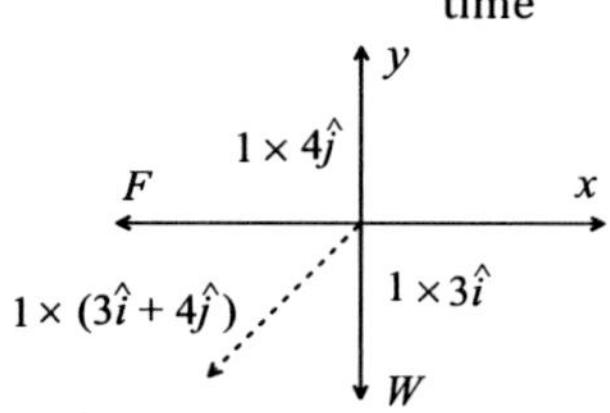

$$=\frac{\text{Change in momentum}}{\text{time}}$$
$$=\frac{-(3\hat{\imath}+4\hat{\jmath})kg\ ms^{-1}}{10^{-4}s}$$
$$=-(3\hat{\imath}+4\hat{\jmath})\times 10^4\ N$$

408 **(c)**

$$T'=(m_1+m_2)\times\frac{T}{m_1+m_2+m_3}$$

409 **(a)**

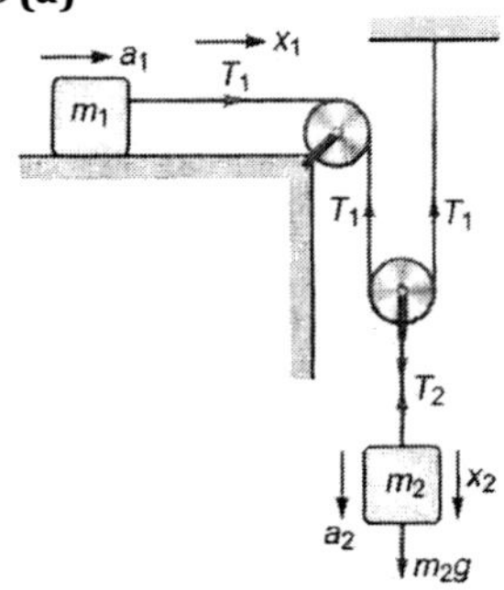

From force diagram,

$T_1=m_1a_1$ ...(i)

$T_2=2T_1$...(ii)

$m_2g-T_2=m_2a_1$

$m_2g-2T_2=m_2a_1$ ...(iii)

Total work done by tensions should be zero,

$\therefore T_1x_1-T_2x_2=0$

or $T_1x_1=T_2x_2$ or $T_1x_2=2T_1x_2$

or $x_1=2x_2$ or $\frac{d^2x_1}{dt^2}=\frac{2d^2x_2}{dt^2}$

$\therefore a_1=2a_2$...(iv)

After solving Eqs. (i), (iii) and (iv),

$$a_2=\frac{m_2g}{4m_1+m_2}$$

410 **(a)**

Minimum force required to move the block

$=\mu R=\mu mg=0.4\times 2\times 10=8\text{ N}$

Since the force applied is only 2.8 N, the block fails to move and static fraction = applied force = 2.8 N

411 **(b)**

For discussion of motion of the point mass m, we assume that observer is situated at the triangular block. The force diagram of point mass m is shown in figure

From force diagram,

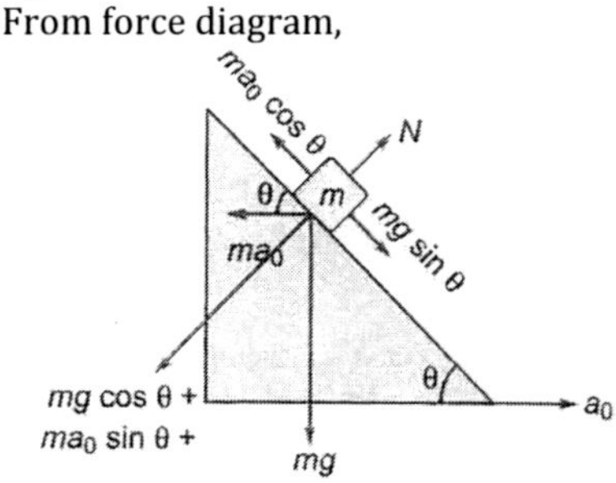

$mg\sin\theta-ma_0\cos\theta=ma$

$\therefore a=g\sin\theta-a_0\cos\theta$

412 **(a)**

For equilibrium, $mg.x = T \times y \Rightarrow T = \frac{mg.x}{y}$

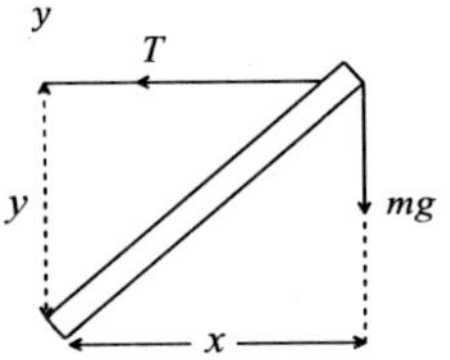

For T to be minimum y should be maximum

413 **(b)**

Kinetic energy required by body

= (Total work done on the body)

− (work against friction)

$= F \times S - \mu mgS = 25 \times 10 - 0.2 \times 5 \times 10 \times 10$

$= 250 - 100 = 150\, Joule$

414 **(d)**

Time taken by the bullet and ball to strike the ground is

$$t = \sqrt{\frac{2h}{g}} = \sqrt{\frac{2 \times 5}{10}} = 1 \text{ s}$$

Let v_1 and v_2 are the velocities of ball and bullet after collision. Then applying

$x = vt$

We have, $20 = v_1 \times 1$

or $v_1 = 20$ m/s

$100 = v_2 \times 1$ or $v_2 = 100$ m/s

Now, from conservation of linear momentum before and after collision we have,

$0.01v = (0.2 \times 20) + (0.01 \times 100)$

On solving, we get

$v = 500$ m/s

415 **(a)**

$S_{\text{Horizontal}} = ut = 1.5 \times 4 = 6m$

$S_{\text{Vertical}} = \frac{1}{2}at^2 = \frac{1}{2}\frac{F}{m}t^2 = \frac{1}{2} \times 1 \times 16 = 8\, m$

$S_{\text{Net}} = \sqrt{6^2 + 8^2} = 10\, m$

416 **(d)**

The fictitious force will act downwards. So the reading of spring balance will increase. In case of physical balance, the fictitious force will act on both the pans, so the equilibrium is not affected

417 **(a)**

$ma = mg \sin\theta - f$

or $f = mg\sin\theta - ma$

$= 8\left[10 \times \frac{1}{2} - 0.4\right]$ N $= 8\times 4.6$N $= 36.8$ N

418 **(b)**

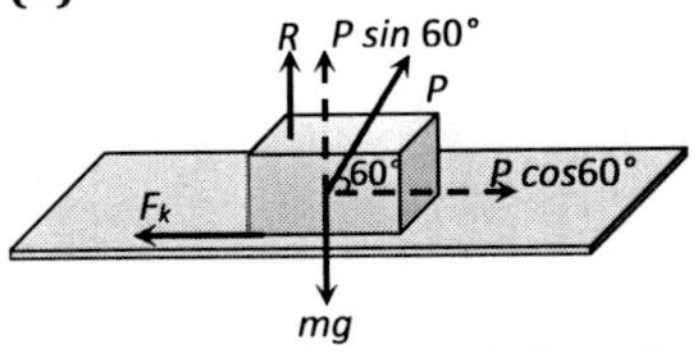

Let body is dragged with force P, making an angle 60° with the horizontal

F_k =Kinetic friction in the motion= $\mu_k R$

From the figure $F_k = p\cos 60°$ and $R = mg - P\sin 60°$

$\therefore P\cos 60° = \mu_k(mg - P\sin 60°)$

$$\Rightarrow \frac{P}{2} = 0.5\left(60 \times 10 - \frac{P\sqrt{3}}{2}\right)$$

$\Rightarrow P = 315.1N$

$\therefore K_k = P\cos 60° = \frac{315.1}{2}N$

Work done $= F_k \times s = \frac{315.1}{2} \times 2 =$ 315 $Joule$

419 **(b)**

For rotational equilibrium about point "P",

$$mg\sin\theta\left(\frac{b}{2}\right) = mg\cos\theta\left(\frac{a}{2}\right)$$

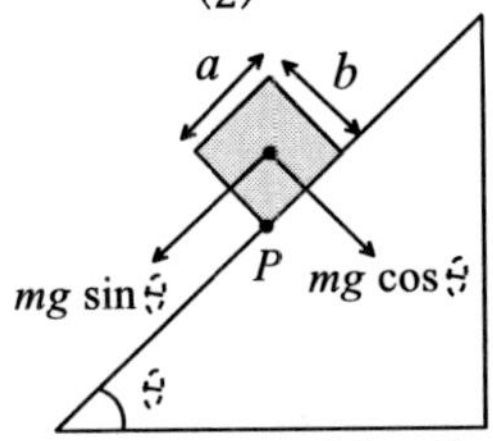

$\Rightarrow \tan\theta = \frac{a}{b} = \frac{10}{15} = \frac{2}{3}$

$\Rightarrow \theta = 33.69°$

i.e., toppling starts at $\theta = 33.69°$

and angle of repose $= \tan^{-1}(\mu) = \tan^{-1}(\sqrt{3}) = 60°$

It mean the block will remain at rest on the plane up to certain angle θ and then it will topple

420 **(d)**

Gravity is the force by which earth attracts all bodies towards itself. It is same for all bodies and does not depend upon the mass of body.

421 **(a)**

$\tan 30° = \frac{1}{\sqrt{3}} = \frac{x}{10}$

$x = \frac{10}{\sqrt{3}}$

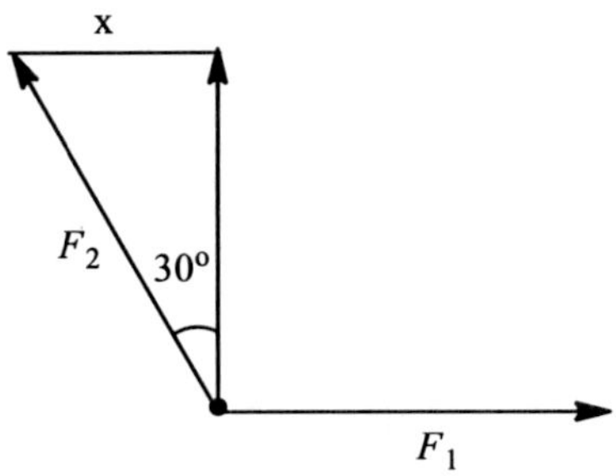

422 **(b)**

We can realize the situation as shown in figure. Let the direction of mass m makes angle θ with the z-axis. Resolve momentum $6.5m$ along x and y –axis and equate.

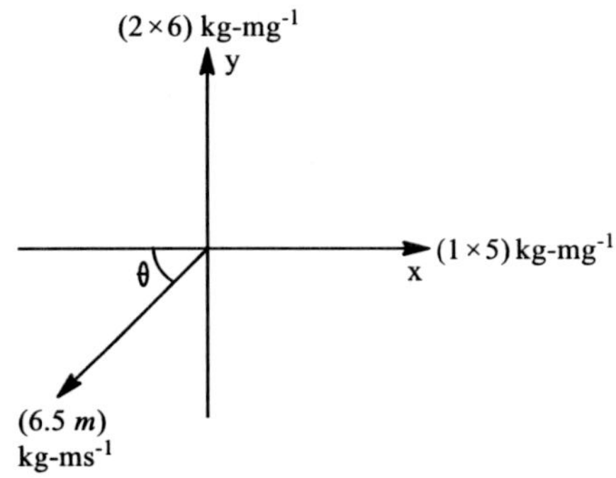

Therefore, $6.5m\cos\theta = 5\times 1$

and $6.5\,m\sin\theta = 6\times 2$

or $(6.5m)^2 = (5)^2 + (12)^2 = (13)^2$

or $m = \frac{13}{6.5} = 2$ kg

therefore, total mass of the shell

$= 1 + 2 + 2 = 5$ kg

423 **(c)**

$5g - T_2 = 5a$ ....(i)

$T_2 - T_1 - 3g = 3a$ ...(ii)

$T_1 - g = a$ ...(iii)

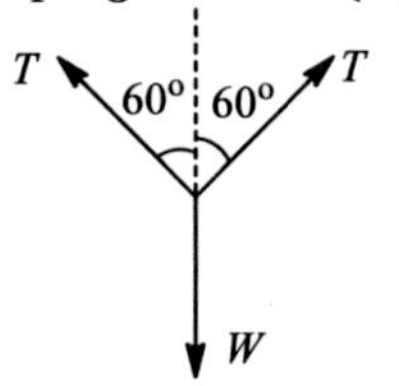

Adding Eqs (i) and (iii),

$-T_2 + T_1 + 4g = 6a$

Adding this to Eq. (ii), we get

$g = 9a$ or $a = \frac{g}{9}$

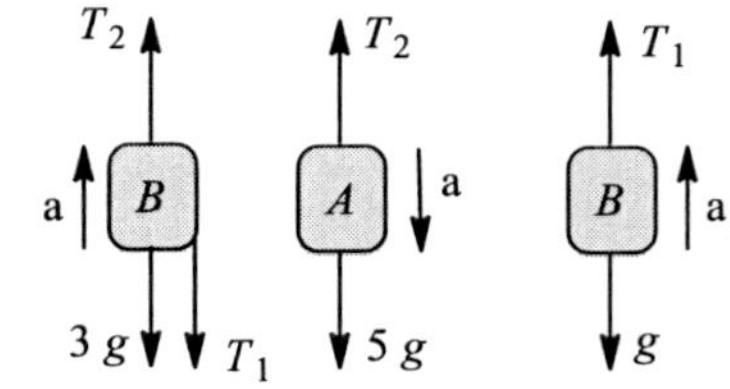

424 **(a)**

Acceleration of combined system,

$a = \frac{m_1 - m_2}{m_1 + m_2}.g$

$= \frac{3-2}{3+2}\times 9.8 = 1.96\text{ ms}^{-2}$

Vertically upward velocity of 2 kg mass at the time breaking of string,

$v = at = 5\times 1.96 = 9.8\text{ ms}^{-2}$

After breaking of string, mass m_2 moves under gravity and go further higher through a height h, where final velocity is zero. Hence

$(0)^2 - (9.8)^2 = 2\times(-9.8)\times h$ or $h =$ 4.9 m

425 **(c)**

At 11th second lift is moving upward with acceleration

$a = \frac{0 - 3.6}{2} = -1.8\,m/s^2$

Tension in rope, $T = m(g - a)$

$= 1500(9.8 - 1.8) = 12000\,N$

426 **(b)**

$R' = m(g + a) = m(g + g)$

$= 2mg$ N $= 2m$ kg-wt

$= 2$ kg-wt $[\because m = 1]$

427 **(c)**

Force applied by engine $= 6m$

When two cars are pulled,

$(m + m)a = 6m$

$2ma = 6m$

$a = 3\text{ ms}^{-2}$

428 **(a)**

Since, $P = (M + m)a$

Now as in free body diagram of block,

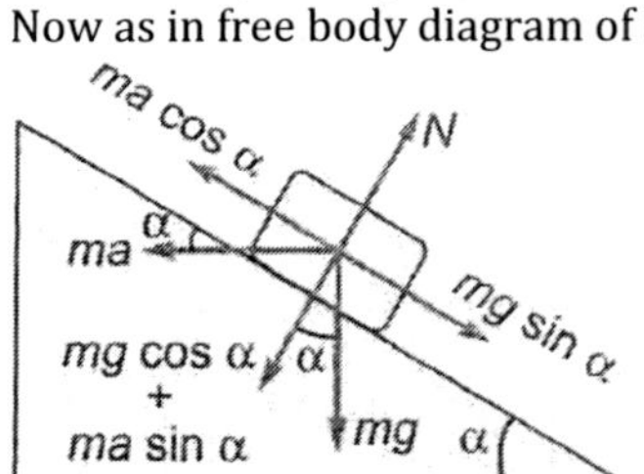

$ma\cos\alpha = mg\sin\alpha$

$\therefore\ a = g\frac{\sin\alpha}{\cos\alpha} = g\tan\alpha$ or $P = (M + m)g\tan\alpha$

429 **(c)**

Reading =Weight of cage + Reaction by bird

$= 20 + 0.5(10 + 2) = 26\ N$

430 **(b)**

$\frac{mg}{m(g-a)} = \frac{3}{2} \Rightarrow a = g/3$

431 **(a)**

$F_l = \mu mg = 0.6 \times 1 \times 9.8 = 5.88\ N$

Pseudo force on the block $= ma = 1 \times 5 = 5\ N$

Pseudo is less then limiting friction hence static force of friction $= 5\ N$

432 **(c)**

Momentum of one piece $= \frac{M}{4} \times 3$

Momentum of the other piece $= \frac{M}{4} \times 4$

$\therefore$ Resultant momentum $= \sqrt{\frac{9M^2}{16} + M^2} = \frac{5M}{4}$

The third piece should also have the same momentum

Let its velocity be v, then

$\frac{5M}{4} = \frac{M}{2} \times v \Rightarrow v = \frac{5}{2} = 2.5\ m/sec$

433 **(b)**

In the given system,

$a = \frac{m_1 - m_2}{m_1 + m_2} = \frac{g}{8}$

$\therefore \frac{m_1 - m_2}{m_1 + m_2} = \frac{1}{8}$

$8m_1 - 8m_2 = m_1 + m_2$

$7m_1 = 9m_2$

$\frac{m_1}{m_2} = \frac{9}{7}$

434 **(a)**

Force $= \frac{p}{t} = \frac{N\,m(v-u)}{t}$

$= m\,N\,v \quad (\text{as } u = 0, t = 1\text{ s})$

435 **(c)**

$a = \frac{m_2 - m_1}{m_1 + m_2} g = \frac{10-5}{10+5} g = \frac{g}{3}$

436 **(a)**

Friction is the retarding force for the block

$F = ma = \mu R = \mu mg$

Retardation $a = \mu g$

From first equation to motion $v = u - at$

$\Rightarrow 0 = V - \mu g \times t \Rightarrow t = \frac{V}{g\mu}$

437 **(a)**

$F = m\left(\frac{dv}{dt}\right) = \frac{100 \times 5}{0.1} = 5000\ N$

438 **(c)**

From force diagram shown in figure,

$T = mg$..(i)

and $2T\cos\theta = \sqrt{2}\,mg$ ... (ii)

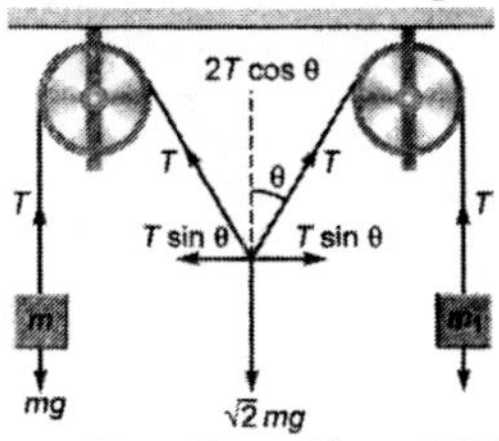

Combing Eqs. (i) and (ii), we have

$2mg\cos\theta = \sqrt{2}\,mg$

or $\cos\theta = \frac{1}{\sqrt{2}}$

or $\theta = 45°$

439 **(a)**

The bullets are initially at rest

Change of momentum per second $= mvN$

Where N is the number of bullets fired per second

440 **(b)**

$R + F\sin 60° = mg$ or $R = mg - \frac{\sqrt{3}F}{2}$

$F\cos 60° = f = \mu R$

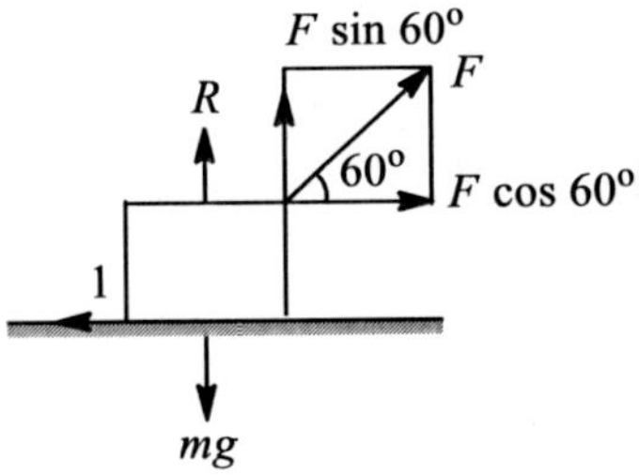

or $\frac{F}{2} = 0.5\left[1 \times 10 - \frac{\sqrt{3}F}{2}\right]$

or $F + \frac{\sqrt{3}F}{2} = 10$ or $F = \frac{20}{2+\sqrt{3}}$

or $F = \frac{20}{3.732}$ N= 5.36 N

441 **(a)**

$T_1 = m(g+a) = 1 \times \left(g + \frac{g}{2}\right) = \frac{3g}{2}$

$T_2 = m(g-a) = 1 \times \left(g - \frac{g}{2}\right) = \frac{g}{2} \therefore \frac{T_1}{T_2} = \frac{3}{1}$

442 **(c)**

$F = \frac{dp}{dt} = v\left(\frac{dM}{dt}\right) = \alpha v^2 \therefore a = \frac{F}{M} = \frac{\alpha v^2}{M}$

443 **(d)**

law of conservation of momentum gives

$m_1 v_1 = m_2 v_2$

$\Rightarrow \frac{m_1}{m_2} = \frac{v_2}{v_1}$

But, $m = \frac{4}{3}\pi r^3 \rho$

or $m \propto r^3$

$\therefore \frac{m_1}{m_2} = \frac{r_1^3}{r_2^3} = \frac{v_2}{v_1}$

$\Rightarrow \frac{r_1}{r_2} = \left(\frac{1}{2}\right)^{1/3}$

$\therefore r_1 : r_2 = 1 : 2^{1/3}$

444 **(d)**

For moving on circular path without slipping, centripetal force must equal frictional force

That is,

$\frac{mv^2}{r} = \mu m g$

$\Rightarrow mr\omega^2 = \mu m g [\because v = r\omega]$

$\Rightarrow r\omega^2 = \mu g$

$\therefore \omega = \sqrt{\frac{\mu g}{r}} = \sqrt{\frac{0.5 \times 9.8}{10}} = 0.7\ rad/s$

445 **(b)**

Here the tension in the cord is given by

$T = mg + ma$

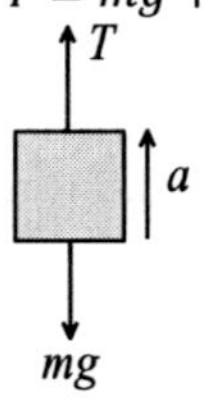

Here upwards acceleration = a,

Mass of sphere = m,

$T = 4mg$

$\Rightarrow 4mg = mg + ma$

$\Rightarrow 3mg = ma \Rightarrow a = 3g$

446 **(a)**

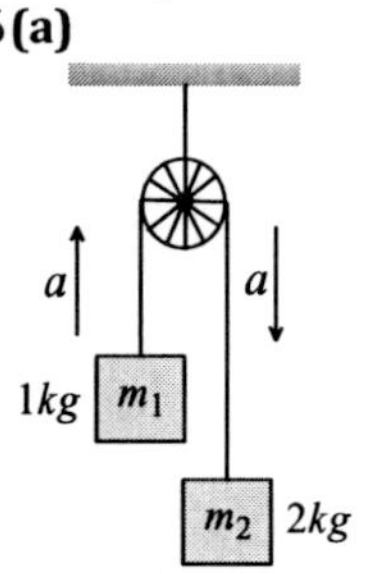

Here, $m_1 = 1kg, m_2 = 2kg$

The acceleration of the system is

$$a = \frac{(m_2 - m_1)g}{m_1 + m_2} = \frac{(2-1)g}{1+2} = \frac{g}{3} = \frac{10}{3}$$

Acceleration of the centre of mass is

$$a_{cm} = \frac{m_1 a_1 + m_2 a_2}{m_1 + m_2} = \frac{1(-a) + 2(a)}{1+2}$$

$$= \frac{1\left(\frac{-g}{3}\right) + 2\left(\frac{g}{3}\right)}{3}$$

$$= \frac{g}{9} = \frac{10}{9}$$

The distance travelled by the centre of mass in two seconds is

$$S = \frac{1}{2}a_{cm}t^2 = \frac{1}{2} \times \frac{10}{9} \times (2)^2 = \frac{20}{9}m$$

447 **(c)**

Given that,

$v_1 = 10\ ms^{-1};\ m_1 = 10\ kg$

$v_2 = 0;\ m_2 = 9\ kg$

$v_3 = v;\ m_3 = 1\ kg$

According to conservation of momentum

$m_1 v_1 = m_2 v_2 + m_3 v_3$

$10 \times 10 = 9 \times 0 + 1 \times v \Rightarrow v = 100\ ms^{-1}$

448 **(b)**

$T = m(g + a) \Rightarrow 28000 = 2000(10 + a) \Rightarrow a = 4\ m/s^2$, upwards

450 **(b)**

As masses are equal, the two pieces must move in opposite direction with the same velocity to conserve initial linear momentum, which is zero

451 **(d)**

Resultant force is zero, as three forces acting on the particle can be represented in magnitude and direction by three sides of a triangle in order. Hence, by Newton's 2nd law$\left(\mathbf{F} = m\frac{dv}{dt}\right)$, particle velocity ($\mathbf{v}$) will be same.

452 **(b)**

Tension is rope, $T <$ Breaking load, $\frac{2}{3}$mg

$\therefore m(g - a) < \frac{2}{3}mg$ or $a > \frac{g}{3}$

453 **(b)**

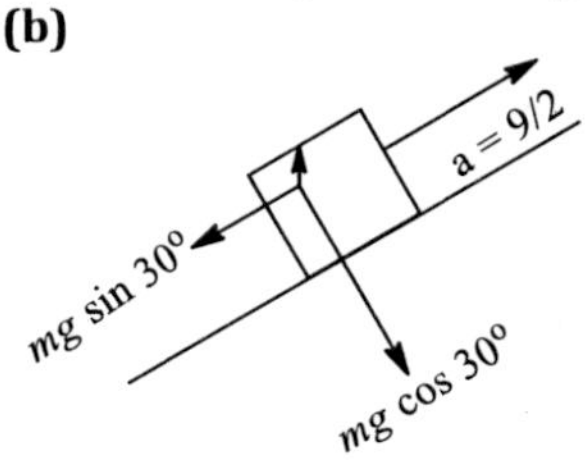

$T\sin\theta - mg\sin\theta = ma$

$T\sin\theta = mg\sin\theta + \frac{mg}{2}$ ...(i)

$T\cos\theta = mg\cos\theta$ ...(ii)
Dividing Eq. (i) by Eq. (ii), we get
$\tan\theta = \frac{2}{\sqrt{3}}$

454 **(a)**
The mass m is not moving with respect to the lift and also has no tendency to move. Hence, friction force acting on it is equal to zero

455 **(c)**
$v_1 = 10\,m/s$; $m_1 = 10\,kg$
$v_2 = 0$; $m_2 = 9\,kg$
$v_3 = v$; $m_3 = 1\,kg$
According to conservation of momentum
$m_1v_1 = m_2v_2 + m_3v_3$
$10 \times 10 = 9 \times 0 + 1 \times v; v = 100\,m/s$

456 **(c)**
$\frac{dM}{dt} = 0.1kg/s, v_{\text{gas}} = 50\,m/s,$
Mass of the rocket $= 2\,kg. Mv =$ constant
$-v\frac{dM}{dt} + M\frac{dv}{dt} = 0 \therefore \frac{dv}{dt} = \frac{1}{M}v\frac{dM}{dt}$
$\Rightarrow$ Acceleration $= \frac{1}{2} \times 50 \times 0.1 = 2.5\,m/s^2$

457 **(d)**
$F = mnv = 150 \times 10^{-3} \times 20 \times 800$
$= 2400\,N$

458 **(a)**

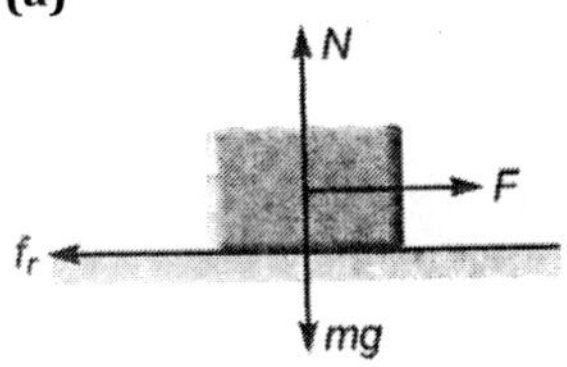

Given that, $\mu = 0.1$ and $g = 9.8\text{ ms}^{-2}$
So, $F = f_r = \mu N = \mu mg$
$= 0.1 \times 1 \times 9.8 = 0.98$ N

459 **(a)**

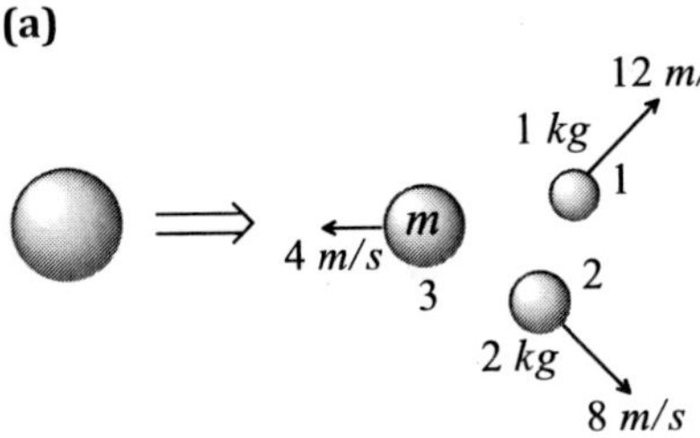

According to conservation of linear momentum
$p_3 = \sqrt{p_1^2 + p_2^2}$
$\Rightarrow m \times 4 = \sqrt{(1 \times 12)^2 + (2 \times 8)^2} = 20$
$\Rightarrow m = 5\,kg$

460 **(d)**
Condition of sliding is
$mg\sin\theta > \mu\, mg\cos\theta$
or $\tan\theta > \mu$
or $\tan\theta > \sqrt{3}$ ...(i)
condition of toppling is

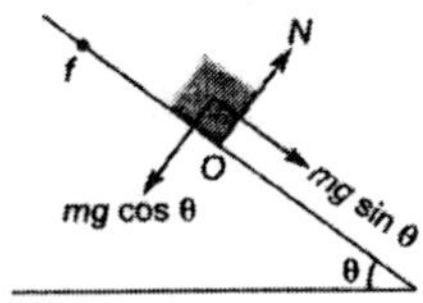

Torque of $mg\sin\theta$about $O >$ torque of $mg\cos\theta$about.
$\therefore \quad (mg\sin\theta)\left(\frac{15}{2}\right) > (mg\cos\theta)\left(\frac{10}{2}\right)$
or $\tan\theta > \frac{2}{3}$ .... (ii)
With increase in value of θ, condition of sliding is satisfied first.

461 **(d)**
The acceleration of the system is

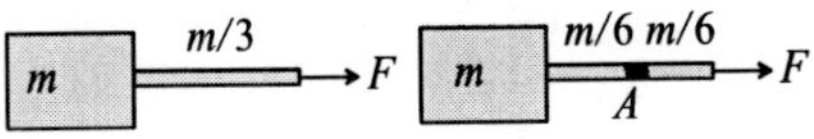

$a = \frac{F}{m+\frac{m}{3}} = \frac{3F}{4m}$ ...(i)
The tension in the middle of the rope ($i.e.$, at point A) is
$T = \left(m + \frac{m}{6}\right)a = \frac{7m}{6} \times \frac{3F}{4m} \Rightarrow T = \frac{7F}{8}$
[Using (i)]

462 **(a)**
The free body diagram showing the various forces acting on the pulley mass are as follows

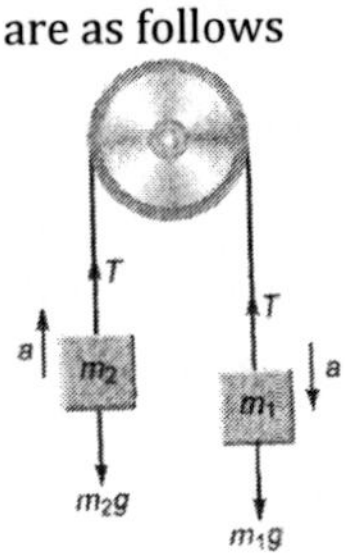

Equating the vertical forces, we have
$m_1g - T = m_1a$ ...(i)
$T - m_2g = m_2a$ ...(ii)
From Eqs. (i) and (ii), we get
$a = \frac{m_1g - m_2g}{m_1 + m_2}$ .... (iii)
The acceleration of centre of mass is
$a_{\text{CM}} = \frac{m_1a - m_2a}{m_1 + m_2}$

Putting the value of a from Eq. (iii), we get

$$a_{CM} = \frac{(m_1 - m_2)^2}{(m_1 + m_2)^2} g$$

463 **(b)**

$R^2 = (3P)^2 + (2P)^2 + 2 \times 3P \times 2P \times \cos\theta$ ...(i)

$(2R)^2 = (6P)^2 + (2P)^2 + 2 \times 6P \times 2P \times \cos\theta$ ...(ii)

By solving (i) and (ii), $\cos\theta = -1/2 \Rightarrow \theta = 120°$

464 **(c)**

$v_r \frac{dm}{dt} = m_0 g$ or $\frac{dm}{dt} = \frac{m_0 g}{v_r}$

or $\frac{dm}{dt} = \frac{5\times10^5}{10\times10^3}$ kg s^{-1} = 500 kg s^{-1}

466 **(c)**

Here applied horizontal force F acts as normal reaction. For holding the block Force of friction = Weight of block

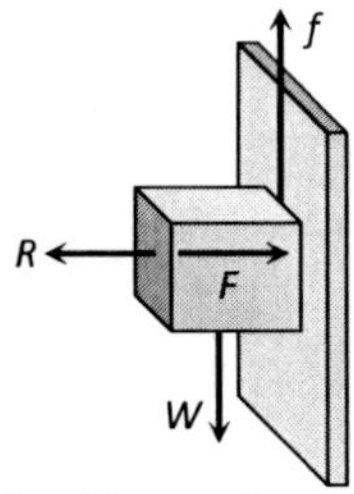

$f = W \Rightarrow \mu R = W$

$\Rightarrow \mu F = W$

$\Rightarrow F = \frac{W}{\mu}$

As $\mu < 1$ $\therefore F > W$

467 **(c)**

The limiting force of friction is

$f_s = 10\ N$

As $F = 8N < f_s$, therefore, block does not move. As static friction is a self adjusting force, therefore the frictional force on the block is 8 N

468 **(a)**

$F = f_r = \mu N = \mu mg = 0.1 \times 1 \times 9.8 = 0.98\ N$

(Assuming that the value of $\mu = 0.1$ is the coefficient of static friction)

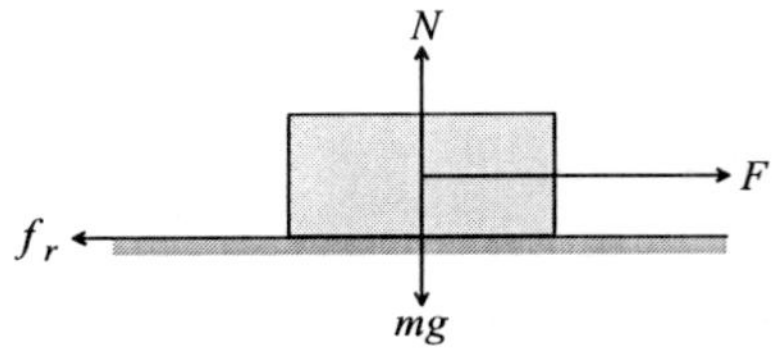

469 **(c)**

Total force required

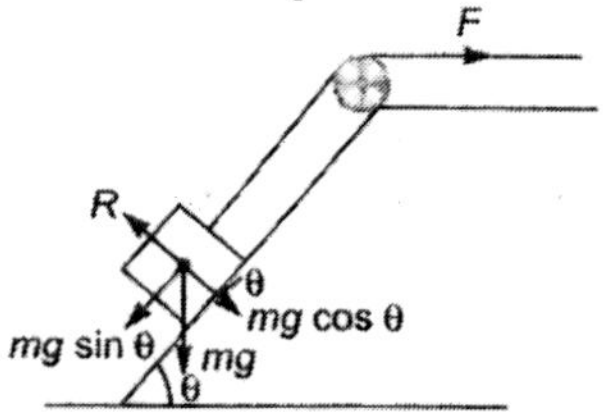

$F = mg\sin\theta + f$

$= mg\sin\theta + \mu_s R$

$= mg\sin\theta + \mu_s mg\cos\theta$

$= mg(\sin\theta + \mu_s\cos\theta)$

$= 200 \times 10(\sin 45° + 0.5\cos 45°)$

$F = \frac{200 \times 10 \times 3}{2\sqrt{2}} = 5$

The number of men required will be

$\frac{200 \times 10 \times 3}{500 \times 2\sqrt{2}} = 5$

470 **(b)**

If f_1 is the contact force in the first case, then

$f_1 = (2m)$ (acceleration)

$= 2m \times \frac{P}{5m} = \frac{2P}{5}$

If f_2 is the contact force in the second case, then

$f_2 = (3m)$ (acceleration)

$= 3m \times \frac{P}{5m} = \frac{3P}{5}$

Required ratio is $\frac{2P}{5} : \frac{3P}{5}$ or $2 : 3$

Note that the total acceleration of the system is the same as the acceleration of mass $2m$ or mass $3m$

471 **(a)**

Change in momentum = Impulse

$\Rightarrow \Delta p = F \times \Delta t \Rightarrow \frac{\Delta p}{F} = \frac{125}{250} = 0.5$ sec

472 **(d)**

$u = 250\ m/s, v = 0, s = 0.12\ metre$

$F = ma = m\left(\frac{u^2 - v^2}{2s}\right)$

$= \frac{20 \times 10^{-3} \times (250)^2}{2 \times 0.12}$

$\therefore F = 5.2 \times 10^3 N$

473 **(b)**

Velocity between $t = 0$, and $t = 2\ sec$

$\Rightarrow v_i = \frac{dx}{dt} = \frac{4}{2} = 2\ m/s$

Velocity at $t = 2$ sec, $v_f = 0$

Impulse = Change in momentum = $m(v_f - v_i)$

$= 0.1(0 - 2) = -0.2\ kg\ m\ sec^{-1}$

474 **(a)**

For solving the problem, we assume that observer is situated in the frame of pulley (non-inertial reference frame)

$m_1 g = w_1$

$m_2 g = w_2$

From force diagram,

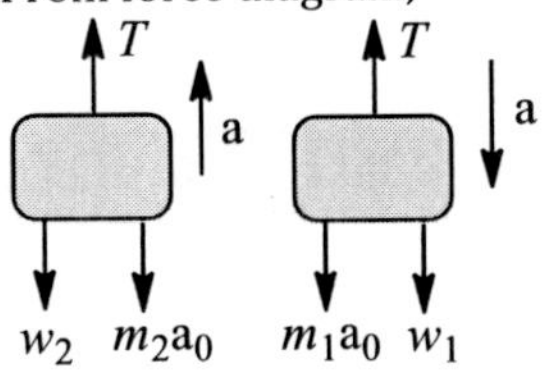

$T - m_2 a_0 - w_2 = m_2 a$

or $T - m_2 g - w_2 = m_2 a$ $(\because a_0 = g)$

or $T - 2w_2 = m_2 a$...(i)

From force diagram,

$m_1 a_0 + w_1 - T = m_1 a$

or $m_1 a_0 + w_1 - T = m_1 a$

or $2w_1 - T = m_1 a$ ...(ii) $(\because a_0 = g)$

From Eqs. (i) and (ii),

$T = \frac{4w_1 w_2}{w_1 + w_2}$

475 **(b)**

$F = m \times a$, If force is constant then $a \propto \frac{1}{m}$

So If mass is doubled then acceleration becomes half

476 **(b)**

As both the balls are of same size, force of buoyancy on each is same. Therefore, in equilibrium,

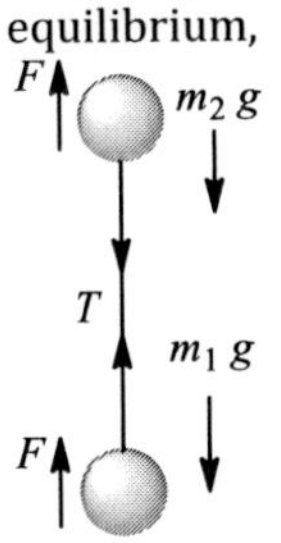

$F + F = m_1 g + m_2 g$ or $F = (m_1 + m_2)\frac{g}{2}$

Considering the equilibrium of lower ball,

$T + F = m_1 g$

$T = m_1 g - F$

$T = m_1 g - (m_1 + m_2)\frac{g}{2}$

$T = (m_1 - m_2)\frac{g}{2}$

477 **(d)**

Tangent of angle of repose = coefficient of static friction

$\mu_s = \tan\theta$

$\mu_s \approx \frac{10}{50}$

≈ 0.2

478 **(a)**

In figure O is the centre of the bowl of radius r. The insect will crawl (from B to A) till component of its weight (mg) along the bowl is balanced by the force of limiting friction (f)

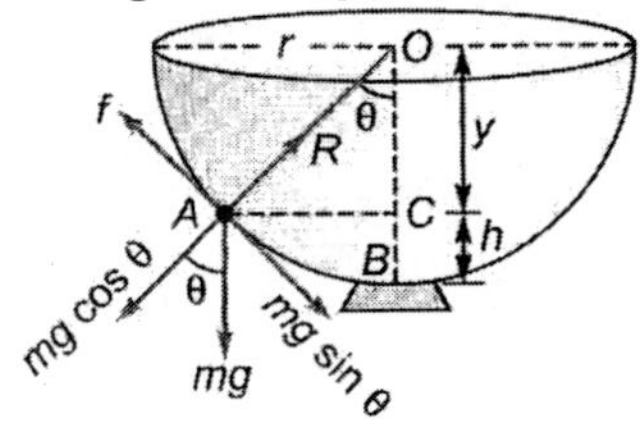

$ie, mg\sin\theta = f = \mu R = \mu\, mg\cos\theta$

or $\mu = \tan\theta = \frac{AC}{OC}$

or $= \frac{\sqrt{OA^2 - OC^2}}{OC} = \frac{\sqrt{r^2 - y^2}}{y}$

or $\mu^2 = \frac{r^2 - y^2}{y^2}$

$\mu^2 y^2 + y^2 = r^2$

$y = \frac{r}{\sqrt{\mu^2 + 1}}$

$h = BC = OB - OC = r - y$

$= r - \frac{r}{\sqrt{\mu^2 + 1}} = r\left[1 - \frac{1}{\sqrt{\mu^2 + 1}}\right]$

479 **(c)**

Here B is implying A but A is not implying B, as kinetic energy of system of particles is zero means speed of each and every particles is zero, which says the momentum of every particle is zero. But statement A means linear momentum of system of particle is zero, which may the true even if particles have equal and opposite momentums and hence, having non-zero KE.

480 **(b)**

When block R collides with block P, it transfers its momentum to block P, due to which it moves towards blocks Q. The spring connecting blocks P and Q gets compressed, which will push the block Q outwards. Due to outward motion of the block Q, the spring gets stretched, the block Q is pulled back. The spring gets compressed, it pushes the block P towards left and so on

482 **(b)**

The work done by the force is $F\cos 37°$, Where $F\cos 37° = f = \mu N$

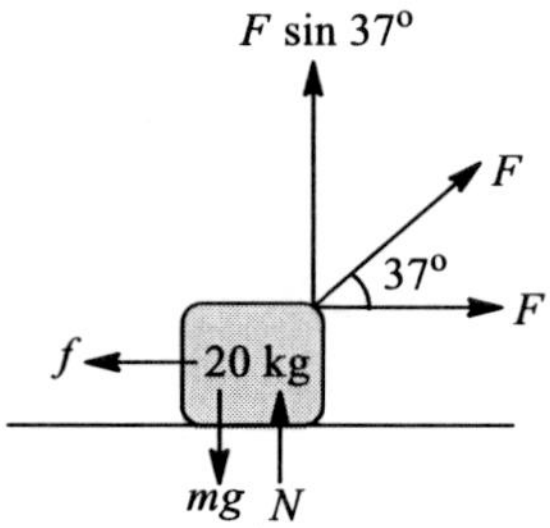

In this case, $N = mg - F\sin 37°$,

So that, $F = \frac{\mu mg}{(\cos 37° + \mu \sin 37°)}$

Here, $\mu = 0.40$ and $m = 20$ kg

$\therefore F = 75.4$ N

Hence, $W = (75.4\cos 37°)(8.0) = 482$ J

483 **(d)**

Work done by friction can be positive, negative and zero depending upon the situation

484 **(a)**

FBD of mass 2 kg FBD of mass $4kg$

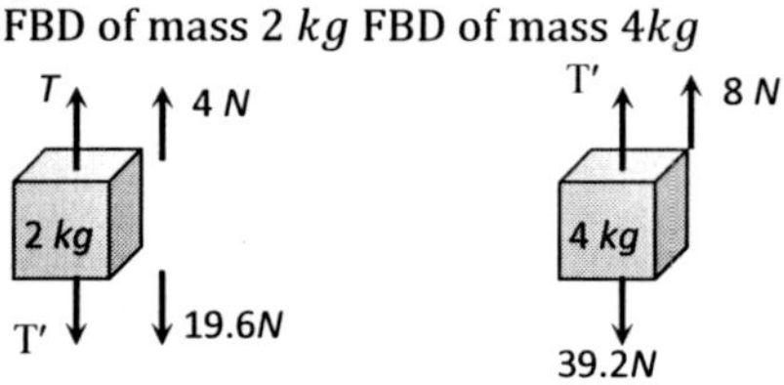

$TT' - 19.6 = 4$ ... (i)

$T' - 39.2 = 8$ ... (ii)

From (ii), $T' = 47.2\ N$

And substituting T' in (i), we get

$T = 4 + 19.6 + 47.2 \Rightarrow T = 70.8\ N$

485 **(a)**

$T = m_1 g\sin\theta - m_1 a$

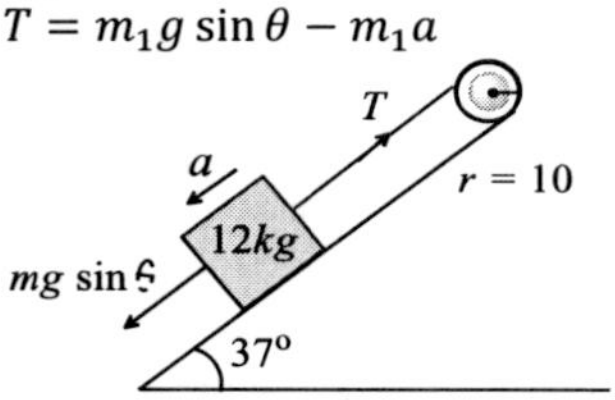

$T = 12 \times 10\sin 37 - 12 \times 2$

$T = 120 \times 0.6018 - 24$

$T = 72.21 - 24 = 48.21 = 48$

486 **(c)**

$2mg\cos\theta = \sqrt{2}mg$

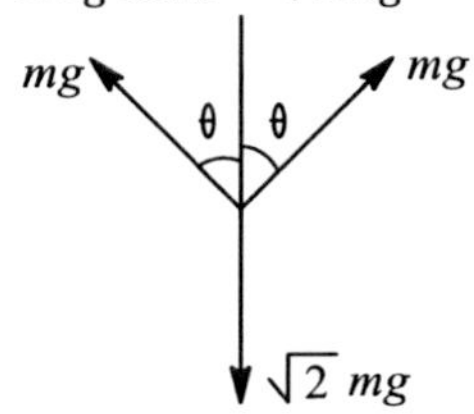

or $\cos\theta = \frac{1}{\sqrt{2}}$ or $\theta = 45°$

487 **(a)**

Relating force $F = ma = \mu R = \mu mg$

$\therefore a = \mu g$

Now from equation of motion $v^2 = u^2 - 2as$

$$\Rightarrow 0 = u^2 - 2as \Rightarrow s = \frac{u^2}{2a} = \frac{u^2}{2\mu g}$$

$$\therefore s = \frac{v_0^2}{2\mu g}$$

488 **(c)**

As the elevator is going down with decreasing speed, so acceleration is upward duration, Let it is a

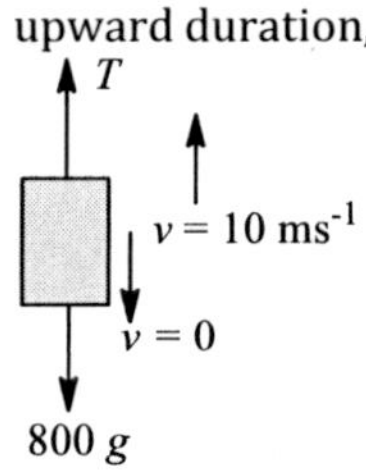

$T - 800g = 800a$,

$T = 800(g + a)$

From $v^2 = u^2 - 2as$,

$\therefore a = 2\text{ms}^{-2}$

$\therefore T = 800(10 + 2)$,

$\therefore T = 9600$ N

489 **(a)**

$T_2 - T_1 = m_2 a$

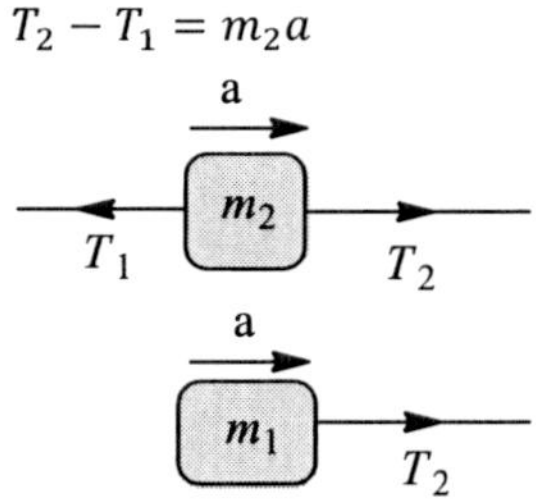

Dividing, $\frac{T_2 - T_1}{T_1} = \frac{m_2}{m_1}$

or $\frac{T_2}{T_1} = \frac{m_2}{m_1} + 1 = \frac{m_2 + m_1}{m_1}$

or $\frac{T_2}{T_1} = \frac{m_1}{m_1 + m_2}$

490 **(d)**

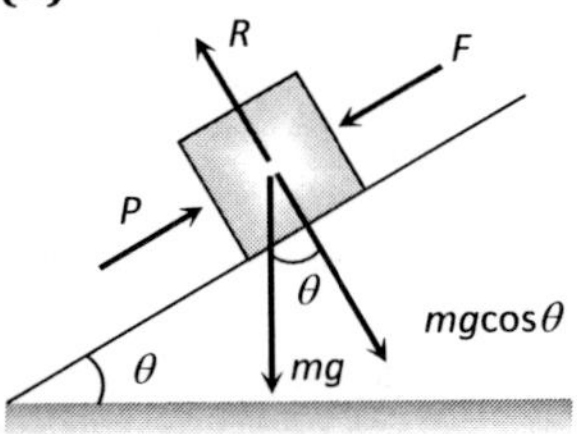

Net force along the plane
$= P - mg\sin\theta = 750 - 500 = 250\ N$
Limiting friction $= F_1 = \mu_s R =$
$\mu_s mg\cos\theta$
$= 0.4 \times 102 \times 9.8 \times \cos 30 = 346\ N$
As net external force is less than limiting friction therefore friction on the body will be 250 N

491 **(c)**
$f_{ms} = 0.4t \times 10\text{N} = 4\text{ N}$
The applied force is less than f_{ms}. So, the block would not move

492 **(a)**
$$T = \frac{2m_1m_2}{(m_1+m_2)}(g+a) = \frac{2m_1m_2(g+g)}{m_1+m_2}$$
$$\Rightarrow T = \frac{4m_1m_2}{m_1+m_2}g = \frac{4\omega_1\omega_2}{\omega_1+\omega_2}$$

493 **(a)**
Downward retardation means upward acceleration
$g' = g + a$
Now, $t = \sqrt{\frac{2L}{g'\sin\theta}}$ or $t = \sqrt{\frac{2L}{(g+a)\sin\theta}}$

494 **(a)**
The various forces acting on the block are as shown
As the truck moves in forward direction with acceleration 2 m/s^2, the box experiences a force F in backward direction,

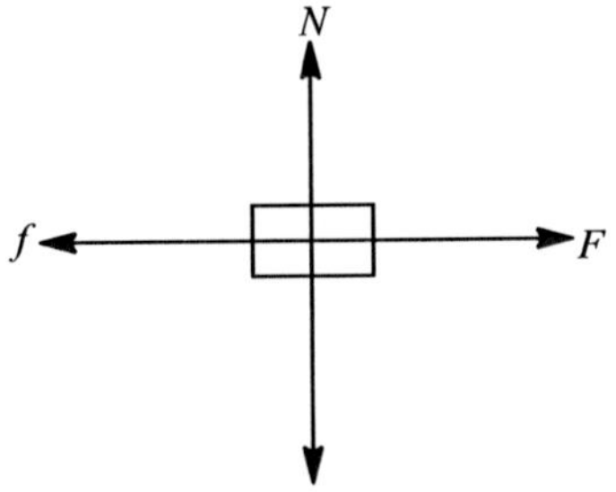

$F = ma = 40 \times 2 = 80$ N
in backward direction.

Its motion will be opposed by force of friction
$f = \mu N = \mu mg = 0.15 \times 40 \times 10 = 60\text{N}$
The acceleration of the box relative to the truck toward the rear end is
$$a = \frac{F-f}{m} = \frac{80-60}{40} = 0.5\ m/s^2$$
If t be the time taken by the box to fall off the truck
$$s = ut + \frac{1}{2}at^2$$
$$5 = 0 + \frac{1}{2} \times 0.5 \times t^2$$
$t = \sqrt{20}\ s$
During this time distance covered by truck
$$x = 0 \times t + \frac{1}{2} \times 2 \times (\sqrt{20})^2 = 20\text{ m}$$

495 **(c)**
The frictional force $F$is given by
$F = \mu_k mg$
Given, $\mu_k = 0.4, m = 2$ kg,
$g = 10\text{ ms}^{-2}$
$\therefore F = 0.4 \times 2 \times 10$
$F = 8\text{N}$

496 **(c)**
Here, , $m = 2$ kg, $\mu_s = 0.54, F = 2.8$ N, $g = 10\text{ ms}^{-2}$
Limiting force of friction,

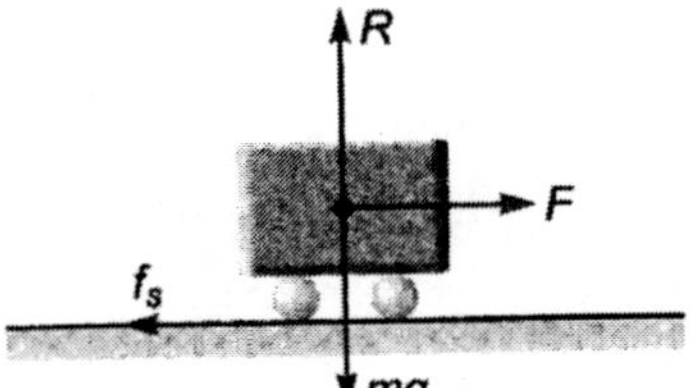

$f_s = \mu_s R = \mu_s mg = 0.54 \times 2 \times 10$
$= 10.8$ N
As $F < f_s$, therefore, the block dose not move. As static friction is itself an adjusting friction. Hence, frictional force between the block and the floor will be 2.8 N

497 **(c)**
Distance travelled in t^{th} second is,
$$s_t = u + at - \frac{1}{2}a$$
Given, $u = 0$
$$\therefore \frac{s_n}{s_n+1} = \frac{an - \frac{1}{2}a}{a(n+1) - \frac{1}{2}a} = \frac{2n-1}{2n+1x}$$

498 **(a)**

The coin falls behind him it means the velocity of train was increasing otherwise the coin fall directly into the hands of thrower

499 **(d)**

For block to continue motion on belt, acceleration $a = +\mu g = 0.2 \times 10 = 2\ \text{ms}^{-2}$

$\therefore$ Velocity of belt = Velocity of block after $4\ \text{s} = 2 \times 4$

$= 8\ \text{ms}^{-1}$

500 **(c)**

Acceleration $= \frac{(m_2 - m_1)}{(m_2 + m_1)} g$

$= \frac{4-3}{4+3} \times 9.8 = \frac{9.8}{7} = 1.4\ m/\text{sec}^2$

501 **(b)**

$v = u + at = 0 + \left(\frac{F}{m}\right) t = \left(\frac{100}{5}\right) \times 10$

$= 200\ cm\ /sec$

502 **(a)**

Force exerted by ball on wall

= rate of change in momentum of ball

$= \frac{mv - (-mv)}{t} = \frac{2mv}{t}$

503 **(b)**

From the relation, acceleration

$a = \frac{F}{m_1 + m_2 + m_3} \Rightarrow a = \frac{40}{10 + 6 + 4}$

$= 2\text{ms}^{-2}$

$\therefore 40 - T = 10 \times 2$

$T = 20\ \text{N}$

504 **(d)**

$s = \frac{u^2}{2\mu g} = \frac{(100)^2}{2 \times 0.5 \times 10} = 1000\ m$

505 **(b)**

When the lift is stationary $W = mg$

$\Rightarrow 49 = m \times 9.8 \Rightarrow m = 5\ kg$

When the lift is moving downward with an acceleration

$R = m(9.8 - a) = 5[9.8 - 5] = 24\ N$

506 **(d)**

Momentum acquired by the particle is numerically equal to area enclosed between the F-t curve and time axis. For the given diagram area in upper half is positive and in lower half is negative (and equal to upper half), so net area is zero. Hence the momentum acquired by the particle will be zero

507 **(b)**

Weight of body $= 2 \times 10 = 20\ \text{N}$

This force has the tendency to move the block, so friction force = 20 N.

508 **(c)**

Newton's first law of motion defines the inertia of body. It states that every body has a tendency to remain in its state (either rest or motion) due to its inertia

509 **(b)**

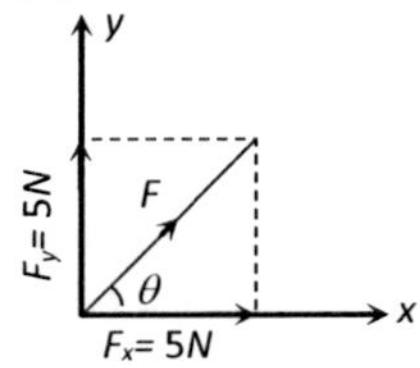

$|\vec{F}| = \sqrt{5^2 + 5^2} = 5\sqrt{2}\ N$

And $\tan\theta = \frac{5}{5} = 1$

$\Rightarrow \theta = \pi/4$

510 **(b)**

The various forces acting on the block are as shown

From Newton's law

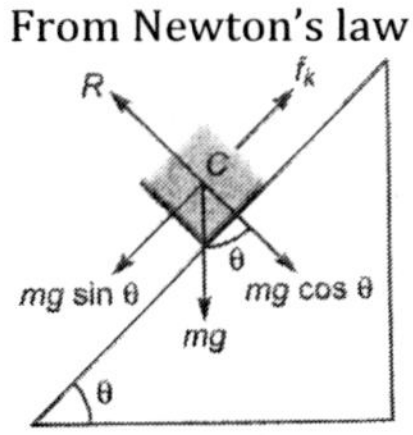

$mg \sin\theta - f = ma$ ....(i)

Where f is frictional force and a the acceleration downwards.

Since, there is no motion perpendicular to surface, we have

$R - mg\cos\theta = 0$

$\Rightarrow\ R = mg\cos\theta$ ....(ii)

Also, $f = \mu R = \mu\, mg \cos\theta$

Putting the value in Eq. (i) we get

$mg\sin\theta - \mu\, mg\cos\theta = ma$

$\Rightarrow\ a = g\sin\theta - \mu\, g\cos\theta$

Now, velocity at bottom

$v^2 = u^2 - 2as$

Since, $v = 0$

$\therefore u = \sqrt{2as}$

Given, $s = l$, $a = g \sin\theta - g\mu\cos\theta$

$\therefore u = \sqrt{2l(g\sin\theta - g\mu\cos\theta)}$

$u = \sqrt{2gl(\sin\theta - \mu\cos\theta)}$

511 **(b)**

$\vec{F} = m\vec{a}$

512 **(c)**

For the smooth portion BC,
$u = 0, s = l, a = g\sin\phi$
$u = ?$
From $v^2 - u^2 = 2as$
$v^2 - 0 = 2g\sin\phi \times l$
For the rough portion CO
$u = v = \sqrt{2g\sin\phi \cdot l}$
$v = 0, a = g(\sin\phi - \mu\cos\phi)$
$s = l$
From $v^2 - u^2 = 2as$
$0 - 2gl\sin\phi = 2g(\sin\phi - \mu\cos\phi)l$
$-\sin\phi = \sin\phi - \mu\cos\phi$
$\mu\cos\phi = 2\sin\phi$
$\mu = 2\tan\phi$

513 **(b)**

Mass of each bullet$(m) = 1\text{ g} = 0.001\text{ kg}$
Velocity of bullet $(v) = 10\text{ ms}^{-1}$
Applied force $(F) = 5$ g-wt.
$= \frac{5}{1000} \times 10\text{ N}$
$= 0.05\text{ N}$
Let n bullets are fired per second, then
Force= rate of change of linear momentum
ie, $F = n \times mv$
$\therefore$ Number of bullets fired per second
$n = \frac{F}{mv}$
$= \frac{0.05}{0.001 \times 10} = 5$

514 **(a)**

$F = u\left(\frac{dm}{dt}\right) \Rightarrow \frac{dm}{dt} = \frac{F}{u} = \frac{210}{300} = 0.7\ kg/s$

515 **(a)**

$F_l \propto R \therefore F_l \propto m$ *i.e.* limiting friction depends upon the mass of body. So,
$\Rightarrow (F_l)' = \frac{3}{2} \times F_l = \frac{3}{2} \times 19.6 = 29.4\ N$

516 **(c)**

By drawing the free body diagram of point B

Let the tension in the section BC and BF are T_1 and T_2 respectively

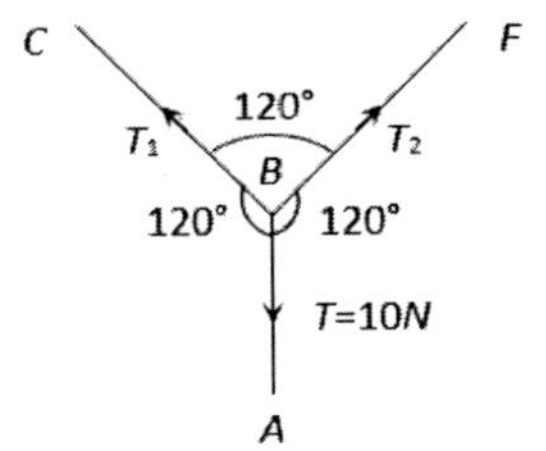

From Lami's theorem

$$\frac{T_1}{\sin 120°} = \frac{T_2}{\sin 120°} = \frac{T}{\sin 120°}$$

$\Rightarrow T = T_1 = T_2 = 10\ N$

517 **(b)**

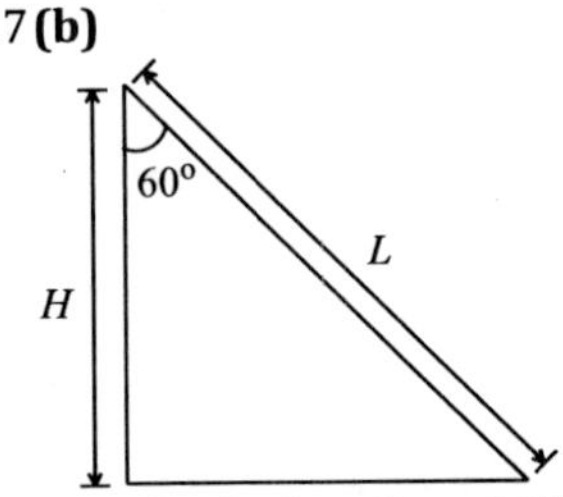

Let L be the length and H be height of the inclined plane respectively
Acceleration of the block slide down the smooth incline plane is
$a = g\cos 60°$
$\therefore L = \frac{1}{2}g\cos 60°\, t_1^2 \quad [\because u = 0]$ ...(i)
Acceleration of another block dropped vertically down from the same inclined plane is
$a = g$

$\therefore H = \frac{1}{2}at_2^2 = \frac{1}{2}gt_2^2 [\because u = 0]$

From figure,

$\cos 60° = \frac{H}{L} \Rightarrow H = L\cos 60°$

$\therefore L\cos 60° = \frac{1}{2}gt_2^2$ ...(ii)

Divide (i) by (ii), we get

$\frac{t_1^2 \cos 60°}{t_2^2} = \frac{1}{\cos 60°}$

$\frac{t_1^2}{t_2^2} = \frac{1}{\cos^2 60°} = \frac{4}{1} \Rightarrow \frac{t_1}{t_2} = \frac{2}{1}$

518 **(c)**

Acceleration of the system $= \frac{F}{M+m}$ and

Force on the block $m = Kx = ma = \frac{mF}{m+M}$

519 **(d)**

$f_{ms} = 0.6 \times 10 \times 9.8$ N

$= 58.8$ N

Since the applied force is greater than f_{ms} therefore the block will be in motion.

So, we should consider f_k

$f_k = 0.4 \times 10 \times 9.8$ N or $f_k = 4 \times 9.8$ N

This would cause acceleration of 40 kg block

Acceleration $= \frac{4 \times 9.8 \text{ N}}{40 \text{ kg}} = 0.98 \text{ ms}^{-2}$

520 **(a)**

The arrangement in shown in figure

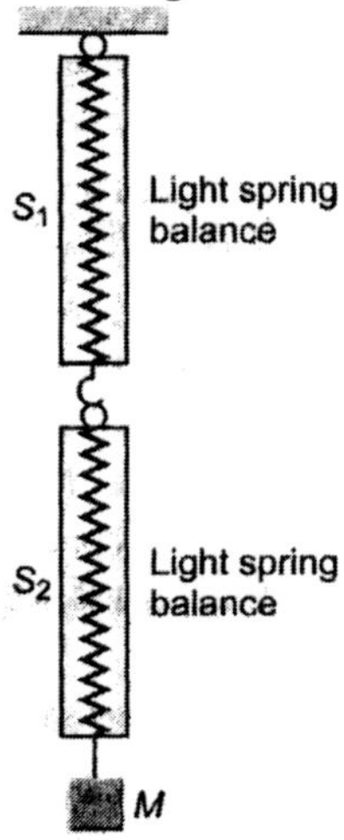

Now, draw the free body diagram of the spring balances and block.

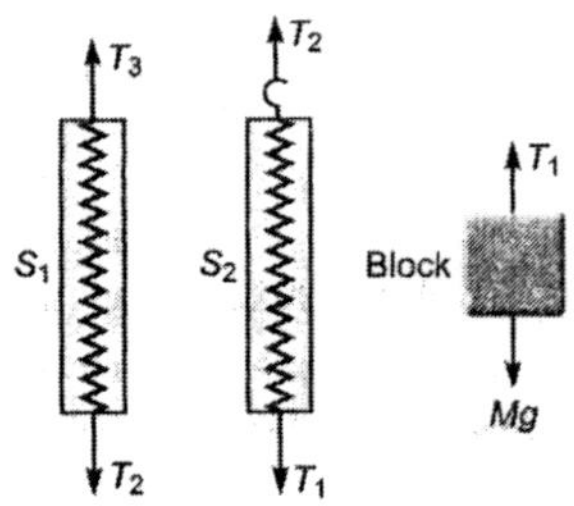

For equilibrium of block, $T_1 = Mg$

Where, T_1 = Reading of S_2

For equilibrium of $S_2, T_2 = T_1$

Where, T_2 = Reading of S_1

For equilibrium of $S_1, T_2 = T_3$

Hence, $T_1, = T_2 = Mg$

So, both scales read M kg.

522 **(b)**

$T = \frac{2 \times 3 \times 5}{3+5}$ kgf $= \frac{15}{4}$ kgf

Pressure on the pulley $= 2T = 2 \times \frac{15}{4}$ kgf

$= 7.5$ kgf

523 **(a)**

Acceleration of block in a stationary lift = $g\sin\theta$

If lift is descending with acc. then it will be $(g - a)\sin\theta$ but in the problem acceleration = $-a$ (retardation)

$\therefore$ Acceleration of block = $[g - (-a)]\sin\theta = (g + a)\sin\theta$

524 **(c)**

Apparent weight of ball

$w' = w - R$

$R = ma$ acts upward $= ma$

$w' = mg - ma = m(g - a)$

Hence, apparent acceleration in the lift is $g - a$. Now if the man is standing stationary on the ground, then the apparent acceleration of the falling ball is g.

525 **(c)**

Here, $\sin\theta = \frac{1}{l}$

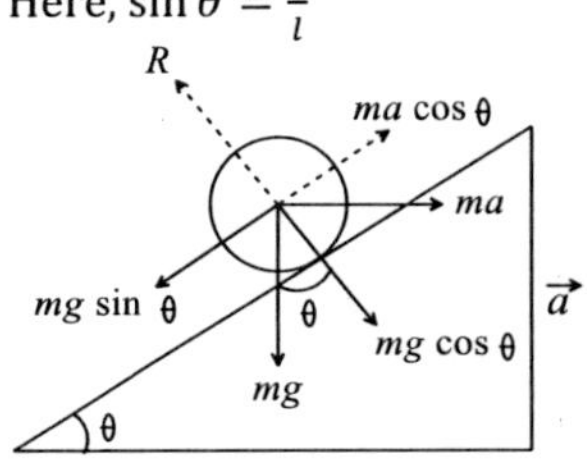

Let required acceleration of inclined plane to be a for the object to remain stationary relative to incline, we have

$ma\cos\theta = mg\sin\theta$

$a = g\tan\theta = g\frac{1}{\sqrt{l^2-1}}$

526 **(c)**

There is a static friction between tyres and road, so frictional force cause the retardation in velocity of a automobile.

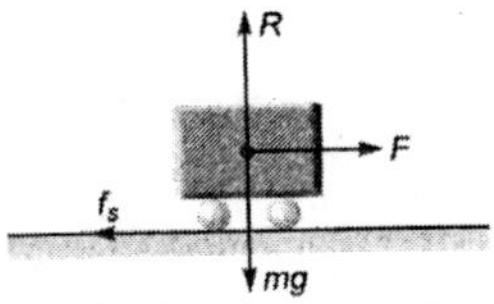

Free body diagram of automobile is shown. From Newton's third law

$F = f_s = \mu R = \mu mg$

Where m is the mass of automobile.

Also, $F = ma$

$ma = \mu mg$

$\Rightarrow a = \text{retardation} = \mu g = 0.5\ g$

Let automobile stops at a distance x, then from equation of motion

$v^2 = v^2 - 2\ ax$

Given,, $v = 0, u = 72\ \text{kmh}^{-1}$

$= 72 \times \frac{5}{18}\text{ms}^{-1}$

$= 20\ \text{ms}^{-1}$,

$g = 9.8\ \text{ms}^{-2}$

$\therefore\ 0^2 = (20)^2 - 2 \times 0.5 \times 9.8x$

$\Rightarrow\ x = \frac{20 \times 20}{2 \times 0.5 \times 9.8} = 40.8\ \text{m}$

527 **(a)**

For jumping the presses the spring platform, so the reading of spring balance increases first and finally it becomes zero

528 **(d)**

The linear momentum of the particle will be

$p = \sqrt{p_x^2 + p_y^2} = 2\sqrt{\cos^2 t + \sin^2 t} = 2$ unit

It the clear that p is constant hence, the angle between **F** and **p**is 90°.

529 **(a)**

Coefficient of friction is constant for two given surfaces in contact. It does not depend upon the weight or normal reaction

530 **(b)**

Surfaces always slide over each other

532 **(c)**

The value measured by O_1 in $N_1 = mg$ because acceleration of body with respect to O_1 is zero

The value measured by O_2 is

$N_2 - mg = ma_0$

$\therefore\ N_2 = m(g + a_0)$

So, $N_1 \neq N_2$

533 **(c)**

Impulse = Force × time $= m\ a\ t$

$= 0.15 \times 20 \times 0.1 = 0.3\ N - s$

534 **(b)**

$u = 100\ m/s, v = 0, s = 0.06\ m$

Retardation $= a = \frac{u^2}{2s} = \frac{(100)^2}{2 \times 0.06} = \frac{1 \times 10^6}{12}$

$\therefore$ Force $= ma = \frac{5 \times 10^{-3} \times 1 \times 10^6}{12} = \frac{5000}{12} =$

$417\ N$

535 **(c)**

Acceleration $a = \frac{1}{m}\left(\frac{-dm}{dt}\right)v_r = \frac{1}{1}\left(\frac{1}{60}\right) \times$

$2400 = 40\ ms^{-2}$

536 **(a)**

As is clear from figure

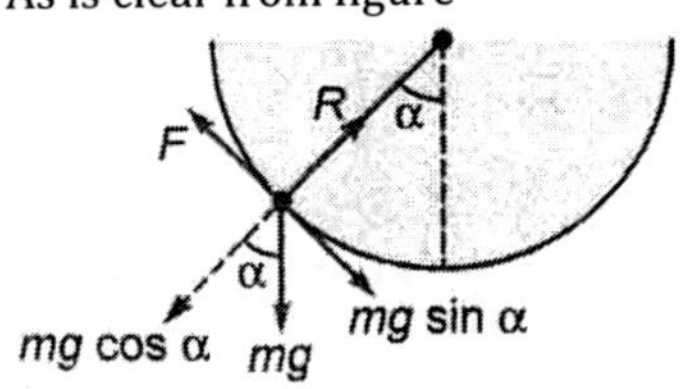

$F = mg\sin\alpha$

$R = mg\cos\alpha$

$\frac{F}{R} = \tan\alpha$

$ie\ \ \mu = \tan\alpha = \frac{1}{3}$

$\therefore\ \cot a = 3$

537 **(a)**

Net frictional force between block and surface is

$F = \mu R = 0.5 \times 10 \times 10 = 5$ N

Applied force is 10 N and it is less than 50 N.

$\therefore$ System is at rest and no friction between A and B.

538 **(c)**

Resultant force,

$F = \sqrt{F_1^2 + F_2^2 + 2F_1F_2\cos\theta}$

$= \sqrt{(3)^2 + (4)^2 + 2 \times 3 \times 4\cos 90°}$

$= 5$ N

539 **(d)**

$K = \frac{F}{x}$ and increment in length is proportional the original length $i.e. x \propto l \therefore K \propto \frac{1}{l}$

It means graph between K and l should be hyperbolic in nature

540 **(d)**

Net downward acceleration = $\frac{\text{Weight−Friction force}}{\text{Mass}}$

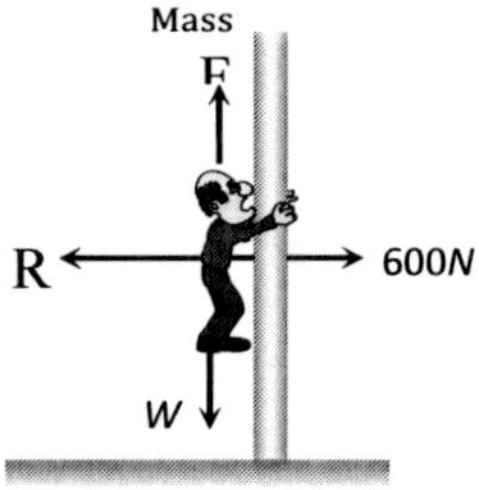

$= \frac{(mg - \mu R)}{m}$

$= \frac{60 \times 10 - 0.5 \times 600}{60}$

$= \frac{300}{60} = 5\, m/s^2$

542 **(d)**

Since action and reaction acts in opposite direction on same line, hence angle between them is 180°

543 **(b)**

Resolve momentum 6.5 $m$along x and y axes and equate

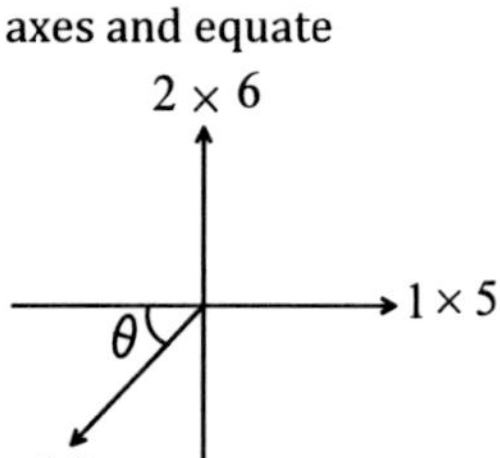

$\therefore 6.5\, m \cos\theta = 5 \times 1$

and $6.5\, m \sin\theta = 6 \times 2$

$\Rightarrow (6.5\, m)^2 = (5)^2 + (12)^2$

$\Rightarrow 6.5m = 13 \Rightarrow m = 2\, kg$

$\therefore$ Total mass = $1 + 2 + 2 = 5kg$

544 **(d)**

Common acceleration, $a = \frac{40}{10+6+4} \text{ms}^{-2} = 2\text{ms}^{-2}$

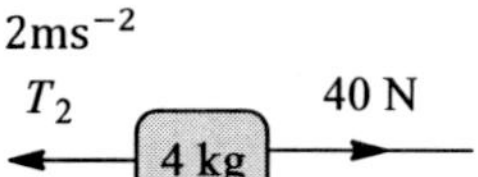

Now, $40 - T_2 = 4 \times 2$ or $T_2 = (40 - 8)$N= 32 N

545 **(b)**

Force on particle at 20 cm away $F = kx$

$F = 15 \times 0.2 = 3N$ [As $k = 15\, N/m$]

$\therefore$Acceleration = $\frac{\text{Force}}{\text{Mass}} = \frac{3}{0.3} = 10\, m/s^2$

546 **(d)**

Horizontal velocity of ball and person are same as both will cover equal horizontal distance in a given interval of time and after following the parabolic path the ball falls exactly in the hand which threw it up

547 **(b)**

Let the minimum mass of $B$is M_B.

Force applied by it $F = M_B g$

Friction force on block A

$f = \mu M_A g$

For motion to start

$M_B g = \mu M_A g$

$M_B = 0.22 \times 10$

$= 2.2$ kg

548 **(c)**

Sand is used to increase the friction

550 **(c)**

$\vec{F}\Delta t = m\Delta\vec{v} \Rightarrow F = \frac{m\Delta\vec{v}}{\Delta t}$

By doing so time of change in momentum increases and impulsive force on knees decreases

551 **(d)**

The various forces acting on the book are as shown in the figure

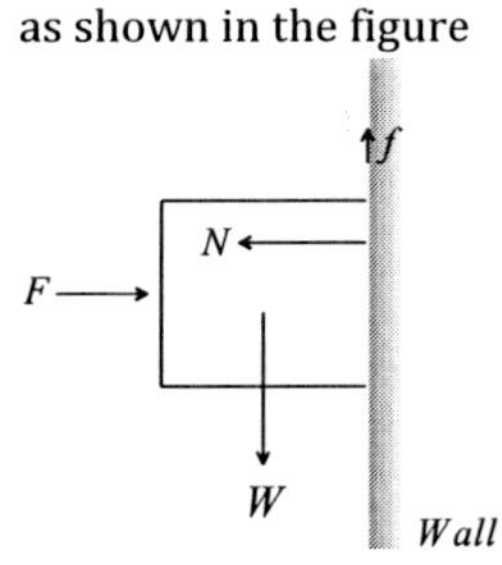

Where, f = Frictional force

W = Weight of a book

F = Applied force

N = Normal reaction

From figure,

The direction of force of friction between the wall and the book is upwards

552 **(a)**

The correct surface profile will be (a), because slope of surface should change from one constant value (non-zero) in terms of sign because force is constant picewise.

553 **(b)**

Spring balance reading in terms of kgf

$$= \frac{4m_1m_2}{m_1+m_2} = \frac{4\times5\times1}{6} = \frac{10}{3}$$

This is less than 6 kgf

554 **(d)**

From 0 to T area is +ive and from T to$2T$, areais − ive. So, net area is zero. Hence, there is no change in momentum.

555 **(b)**

$$F = u\left(\frac{dm}{dt}\right) = 400 \times 0.05 = 20\ N$$

556 **(d)**

Special theory of relativity is based on two postulates

(i) All laws of physics are the same in all inertial reference frames

(ii) The speed of light in vacuum has the same value in all inertial frames, regardless of the velocity of the observer or the velocity of source emitting the light

557 **(b)**

Acceleration of the body down the rough inclined plane = $g \sin\theta$

∴ Force applied on spring balance

$= mg\sin\theta = 5\times10\times\sin30°$

$= 5\times10\times\frac{1}{2} = 25$ N

558 **(b)**

Impulse is given by the product of force and time. Form Newton's second law

$$F = ma = m\frac{\Delta v}{\Delta t}$$

$\Rightarrow\ F\Delta t = m\Delta v$

= change in the momentum of the body.

559 **(a)**

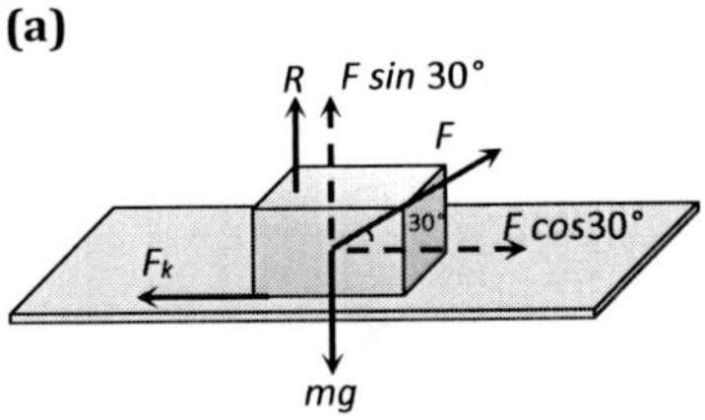

Kinetic friction $= \mu_k R = 0.2(mg - F\sin30°)$

$$= 0.2\left(5\times10 - 40\times\frac{1}{2}\right) = 0.2(50-20) = 6N$$

Acceleration of the block = $\frac{F\cos30° - \text{Kinetic friction}}{\text{mass}}$

$$= \frac{40\times\frac{\sqrt{3}}{2} - 6}{5} = 5.73\ m/s^2$$

560 **(b)**

When two blocks performs simple harmonic motion together then at the extreme position (at amplitude = A)

Restoring force $F = KA + 2ma$

$$\Rightarrow a = \frac{KA}{2m}$$

There will be no relative motion between P and Q if pseudo force on block P is less than or just equal to limiting friction between P and Q

$i.e.\ m = \left(\frac{KA}{2m}\right)$ = Limiting friction

∴ Maximum friction = $\frac{KA}{2}$

561 **(b)**

$$a_{cm} = \left(\frac{m_1-m_2}{m_1+m_2}\right)^2 g = \left(\frac{3m-m}{3m+m}\right)^2 g = \frac{g}{4}$$

562 **(a)**

If A is climbing with constant velocity, then

$T' = 5g + T$ and $T = 2g$

$T' = 5g + 2g = 7g = 7\times10N=70N$

Suppose A is climbing with acceleration a such that $T = 30$ N

That $T = 30$ N

$T - 2a = 2a$

$30 - 2\times10 = 2a$

or $a = 5\ \text{ms}^{-2}$

Again, $T' - T - 5g = 5a$ or $T' = T + 5g + 5a$

or $T' = (30+50+25)$ N= 105 N

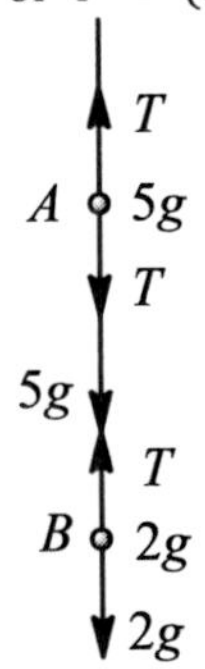

563 **(c)**

$u_y = 40\ m/s, F_y = -5\ N, m = 5\ kg$

So $a_y = \frac{F_y}{m} = -1\ m/s^2$ (As $v = u + at$)

∴ $v_y = 40 - 1\times t = 0 \Rightarrow t = 40\ sec$

564 **(d)**

The three forces acting on the mass at location A have been shown in figure. Since the mass is in equilibrium, therefore, the three forces acting on the mass must be represented by the three sides of a triangle taken in one order.

Hence

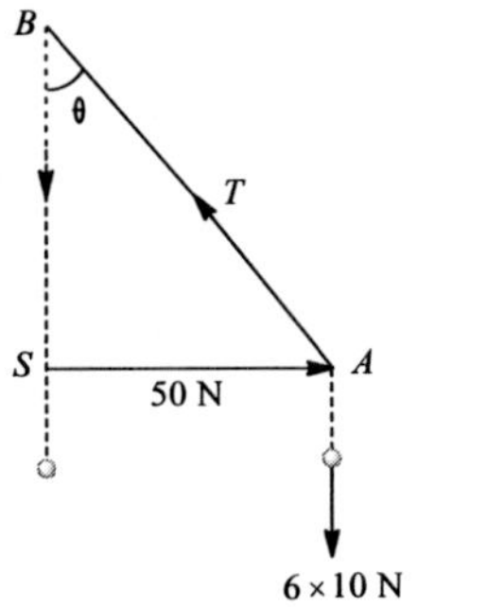

$\frac{50}{SA} = \frac{6\times10}{SB}$ or $\frac{SA}{SB} = \frac{50}{60} = \frac{5}{6}$

or $\tan\theta = \frac{SA}{SB} = \frac{5}{6} = 0.8333$

$= \tan 40°$

$\therefore \theta = 40°$

565 **(c)**

When the bird files, it pushes air down to balance its weight. So the weight of the bird and closed cage assembly remains unchanged

567 **(c)**

Effective upward force = 310 – mg

$= 310 - 24 \times 9.8 = 74.8$ N

Upward acceleration

$\therefore a = 74.8/24 = 3.12\ \text{ms}^{-2}$

As $s = ut + \frac{1}{2}at^2$

$4.6 = 0 + \frac{1}{2} \times 3.12 \times t^2$

or $t^2 = \frac{4.6}{1.56} = 2.95$

or $t = \sqrt{2.95} = 1.7$s

568 **(c)**

$F = 600 - 2 \times 10^5 t = 0 \Rightarrow t$
$= 3 \times 10^{-3}$sec

Impulse $I = \int_0^t F\,dt = \int_0^{3\times10^{-3}} (600 - 2 \times 10^5 t)dt$

$= [600t - 10^5 t^2]_0^{3\times10^{-3}} = 0.9\ N \times \sec$

569 **(a)**

$\mu = \tan\theta\left(1 - \frac{1}{n^2}\right) = 1 - \frac{1}{n^2}$ [As $\theta = 45°$]

570 **(b)**

Angular frequency of the system,

$\omega = \sqrt{\frac{k}{m+m}} = \sqrt{\frac{k}{2m}}$

Maximum acceleration of the system will be, $\omega^2 A$or $\frac{kA}{2m}$. This acceleration to the lower block is provided by friction.

Hence, $f_{\max} = ma_{\max}$

$= m\omega^2 A = m\left(\frac{k\,A}{2m}\right) = \frac{k\,A}{2}$

571 **(d)**

$\Delta p = 2mv = 2 \times 0.25 \times 10 = 5$ kg m/s

$F = \frac{\Delta p}{\Delta t} = \frac{5}{0.01} = 500$ N

572 **(a)**

Mass of the rocket $m = 1000$ kg

$\frac{\Delta m}{\Delta t} = 4\ \text{kgs}^{-1}$, $\quad v = 3000\ \text{ms}^{-1}$

Thrust on the rocket

$F = -v\frac{\Delta m}{\Delta t}$

$= -3000 \times 4$

$= -12000$ N

(Negative sign indicates that thrust applied in a direction opposite to the direction of escaping gas)

573 **(a)**

Mass of ball $m = 1.5$ kg

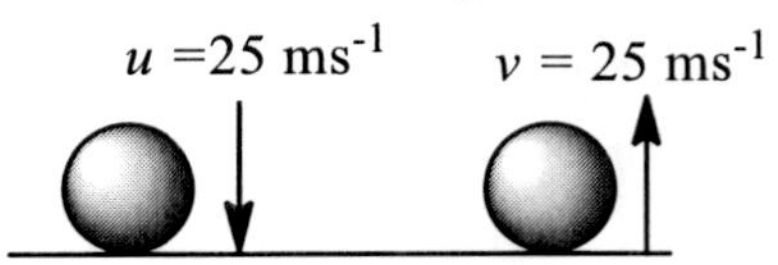

Speed of ball at the time of hitting,

$u = 25\ \text{ms}^{-1}$

Speed of ball while rebounding

$v = 15\ \text{ms}^{-1}$

Duration of contact with floor $t = 0.03$ s

Let force exerted by the ball on floor = F

Applying Newton's II law of motion

$F = \frac{\Delta p}{\Delta t}$

$F = \frac{mv - mu}{t} = \frac{1.5 \times 15 \times (-25)}{0.03}$

$= \frac{1.5(15 + 25)}{0.03} = 50 \times 40N$

$F = 2000$ N

574 **(c)**

$\mu = \tan\theta\left(1 - \frac{1}{n^2}\right)$

$\theta = 45°$ and $n = 2$ [Given]

$\therefore \mu = \tan 45°\left(1 - \frac{1}{2^2}\right) = 1 - \frac{1}{4} = \frac{3}{4}$
$= 0.75$

575 **(b)**

For A, $T = f = 2mg$

$2mg - mg = ma_1$

$\therefore\ a_1 = g$

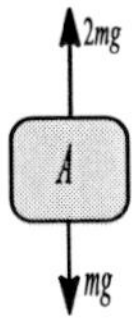

For B,
From force diagram shown in figure,

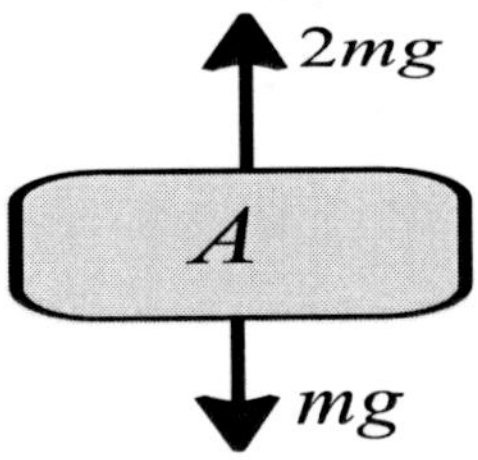

$2mg - mg = 3ma_2$
$a_2 = \frac{g}{3}$
For C,
$\therefore\ 2mg - mg = 2ma_3$
$\therefore\ a_3 = \frac{g}{2}$

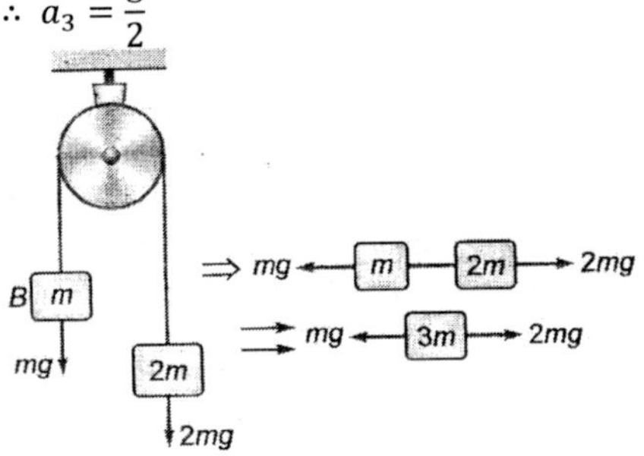

So, $a_1 > a_3 > a_2$

576 **(b)**
Force, $F = (M\text{ kg s}^{-1})(v\text{ m s}^{-1})$
$= Mv\text{ kg m s}^{-2} = Mv\text{ N}$

577 **(d)**
The stopping distance, $S \propto u^2 (\because v^2 = u^2 - 2as)$
$\Rightarrow \frac{S_2}{S_1} = \left(\frac{u_2}{u_1}\right)^2 = \left(\frac{120}{60}\right)^2 = 4$
$\Rightarrow S_2 = 4 \times S_1 = 4 \times 20 = 80\ m$

578 **(d)**
$R = m(g - a) = m(10 - 10) =$zero

579 **(a)**
Retardation of train $= \frac{36\text{ kmh}^{-1}}{5\ s}$
$= \frac{35 \times \frac{5}{18}\text{ms}^{-1}}{5s} = 2\text{ ms}^{-2}$
It acts in the backward direction
Fictitious force on suitcase $= 2m$ N,
Where m is the mass of suitcase
It acts in the forward direction
Due to this force, the suitcase has a tendency to slide forward. If suitcase is not to slide, then $2m =$ force f of friction
or $2m = \mu mg$ or $\mu = \frac{2}{g} = \frac{2}{9.8} = \frac{20}{98} = \frac{10}{49}$

580 **(c)**
$$v_G = \frac{m_B v_B}{m_G} = \frac{0.2 \times 5}{1} = 1\ m/s$$

581 **(b)**
Momentum (p) is defined as product of mass (m) and velocity (v)
$p = mv$
$p_1 = mv_1,\ \ p_2 = m(2v_1)$
$\therefore \frac{p_1}{p_2} = \frac{m\,v_1}{2m\,v_1} = \frac{1}{2}$
$\Rightarrow\ \ p_2 = 2\,p_1$
Which implies that momentum is doubled?

582 **(c)**
Let coefficient of friction isμ, and then retardation will be μg.
From equation of motion, $v = u + at$
$\Rightarrow\ \ 0 = 6 - \mu\,g \times 10$
$\Rightarrow\ \ \mu = \frac{6}{100} = 0.06$

583 **(c)**
As $m_1{:}m_2{:}m_3 = 1{:}1{:}3$
and momentum is conserved,
$\therefore\ \sqrt{p_1^2 + p_2^2 + p_3^2} = 3v_3$
$\sqrt{1 \times 39^2 + 1 \times 39^2} = 3v_3$
$39\sqrt{2} = 3v_3$
$v_3 = \frac{39\sqrt{2}}{3} = 13\sqrt{2}\text{ms}^{-1}$

584 **(a)**
For equilibrium of forces, the resultant of two (smaller) forces should be equal and opposite to third one

585 **(a)**
According to Newton's second law:
Force = rate of change of linear momentum

586 **(d)**
If rope of lift breaks suddenly, acceleration becomes equal to g so that tension, $T = m(g - g) = 0$

587 **(d)**
Weight of the body = 64 N
So mass of the body $m = 6.4\ kg, \mu_s = 0.6, \mu_k = 0.4$

Net acceleration = $\frac{\text{Applied force} - \text{Kinetic friction}}{\text{Mass of the body}}$

$= \frac{\mu_s mg - \mu_k mg}{m} = (\mu_s - \mu_k)g$

$= (0.6 - 0.4)g = 0.2g$

588 **(b)**

When the lift is moving upward with constant velocity then,

$R = mg \therefore F = \mu R = \mu mg$

589 **(b)**

Limiting friction $F_1 = \mu_s R = 0.5 \times (5) = 2.5\ N$

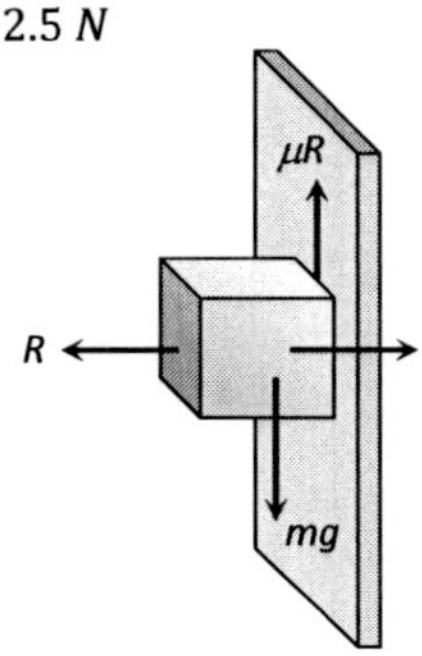

Since downward force is less than limiting friction therefore block is at rest so the static force of friction will work on it

F_s = downward force = Weight

$= 0.1 \times 9.8 = 0.98\ N$

590 **(a)**

On the system of particle if,

$\sum \mathbf{F}_{\text{ext}} = 0$

then $\mathbf{P}_{\text{system}}$ = constant

No other conclusions can be drawn.

591 **(b)**

$u = 2 m/s, v = 0, t = 10$ sec

$\therefore a = \frac{v-u}{t} = \frac{0-2}{10} = -\frac{2}{10} = -\frac{1}{5}$

$= -0.2\ m/s^2$

$\therefore$Friction force$= ma = 1 \times (-0.2) = -0.2\ N$

592 **(c)**

If man slides down with some acceleration, then its apparent weight decreases. For critical condition rope can bear only 2/3 of his weight. If a is the minimum acceleration, then tension in the rope= $m(g-a)$ breaking strength

$\Rightarrow m(g-a) = \frac{2}{3} mg$

$\Rightarrow a = g - \frac{2g}{3} = \frac{g}{3}$

594 **(c)**

$F_{\text{max}} = 5 + 10 = 15N$ and $F_{\text{min}} = 10 - 5 = 5\ N$

Range of resultant $5 \le F \le 15$

595 **(c)**

Here, $\mu = 0.8$. Let F be horizontal force that the body is applying on the pole. The various forces are acting on the boy as shown in the figure

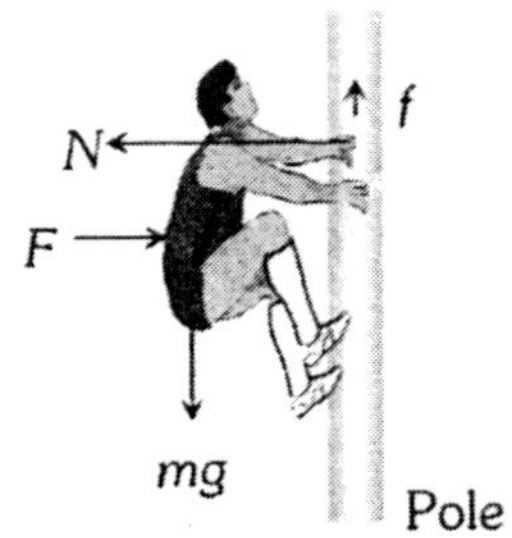

Frictional force,

$f = \mu N = mg$

$N = \frac{mg}{\mu} = \frac{40 \times 10}{0.8} = 5000N \Rightarrow F = N$

$= 500\ N$

598 **(a)**

$\Delta P = p_i - p_f = mv - (-mv) = 2\ mv$

599 **(b)**

Frictional force is independent of the area of surfaces in contact, as long as normal reaction between the two bodies in contact remains the same

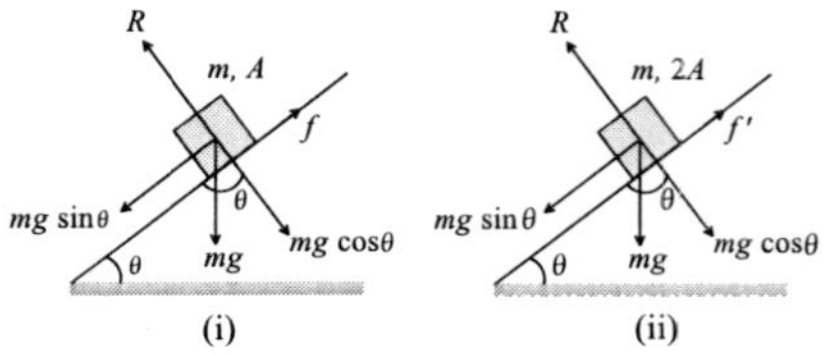

When an iron block of mass m with surface area A is placed on an inclined plane inclined at an angle θ, it will experience a frictional force f as shown in figure (i)

$f = \mu R = \mu mg \cos\theta$ ...(i)

When an iron block of mass m with surface area $2A$ is placed on an inclined plane inclined at an angle θ, it will experience a frictional force f'' as shown in figure (ii)

$f'' = \mu R = \mu mg \cos\theta$ ...(ii)

From (i) and (ii), we get $f'' = f$

600 **(d)**

Are under the $F-t$cure=change in momentum

or $\frac{1}{2}\times 2\times(10)+2\times 10+\frac{1}{2}(10+20)$
$\times 2+\frac{1}{2}\times 4\times 20$
$=m(v-u)$

or $10+20+30+40=2(v-0)$

or $100=2v$

or $v=50\text{ ms}^{-1}$

601 **(c)**

When lift is at rest, $T=2\pi\sqrt{l/g}$

If acceleration becomes $g/4$ then

$$T'=2\pi\sqrt{\frac{l}{g/4}}=2\pi\sqrt{\frac{4l}{g}}=2\times T$$

602 **(c)**

If monkey move downward with acceleration a then its apparent weight decreases. In that condition

Tension in string $=m(g-a)$

This should not be exceed over breaking strength of the rope *i.e.* $360\geq m(g-a)\Rightarrow 360\geq 60(10-a)\Rightarrow a\geq 4\,m/s^2$

603 **(c)**

The reading of balance A will decrease due to the upward thrust caused b buoyancy. The upthrust will be equal to the weight of water displaced. The net downward force due to mass immersed in water will add to effective weight of the system. So, the reading of balance B will increase

604 **(b)**

Resultant force $F_{\text{net}}=\sqrt{F_1^2+F_2^2+2F_1F_2\cos\theta}$

$=\sqrt{(10)^2+(10)^2+2\times 10\times 10\times\cos 60^\circ}$

$=\sqrt{100+100+100}=10\sqrt{3}$

Mass of the body =10 kg

$\therefore$ Acceleration $=\frac{\text{force}}{\text{mass}}$

$=\frac{10\sqrt{3}}{10}=\sqrt{3}\text{ms}^{-2}$

605 **(a)**

$v=\sqrt{\mu rg}=\sqrt{0.5\times 500\times 10}=50\,m/s$

606 **(a)**

Here, initial velocity of passenger train $u=v_1$; final velocity $v=v_2, a=-a$, distance $s=?$

As $v^2=u^2+2as$, so$v_2^2=v_1^2+2(-a)s$

or $s=(v_1^2-v_2^2)/2a$

607 **(d)**

Refer to the free-body diagrams

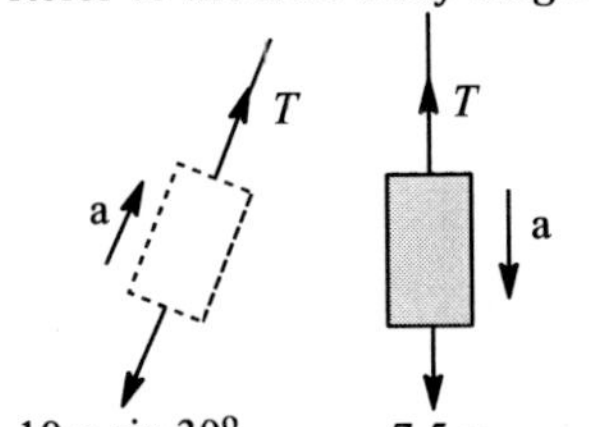

$T-10\text{g}\sin 30^\circ=10a$or $T-5\text{g}=10a$

Again , $7.5-T=7.5\alpha$

Adding, $2.5\text{g}=17.5\alpha$

or $\alpha=\frac{25\text{g}}{175}=\frac{\text{g}}{7}$

609 **(a)**

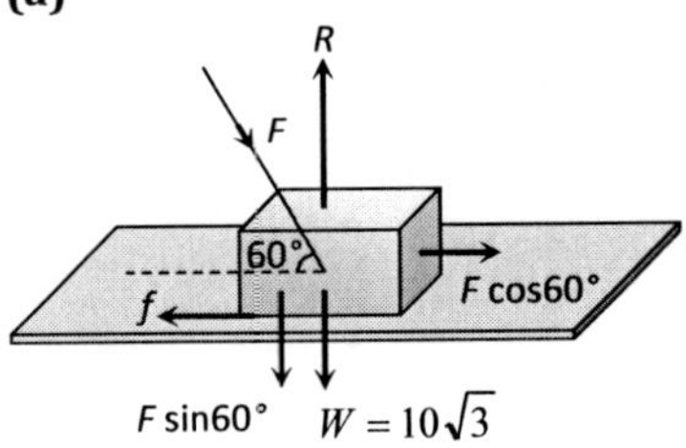

$f=\mu R\Rightarrow F\cos 60^\circ=\mu(W+F\sin 60^\circ)$

Substituting $\mu=\frac{1}{2\sqrt{3}}$ & $W=10\sqrt{3}$

We get $F=20\,N$

610 **(b)**

When a sudden jerk is given to C, an impulsive tension exceeding the breaking tension develops in C first, which breaks before this impulse can reach A as a wave through block

611 **(c)**

Force applied by engine= 6 m

When two cars are pulled,

$(m+m)a=6\,m$

or $2ma=6m$ or $a=3\text{ ms}^{-2}$

612 **(c)**

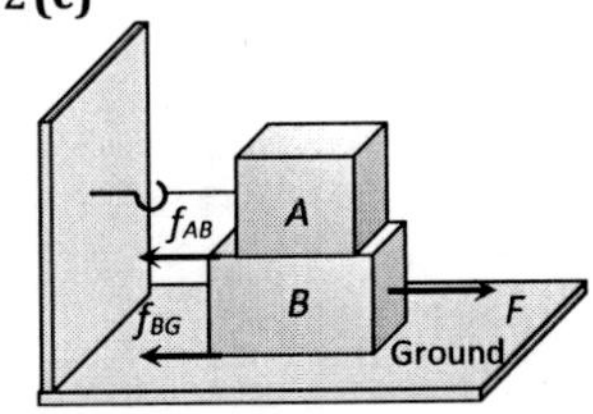

$F=f_{AB}+f_{BG}=\mu_{AB}m_a g+\mu_{BG}(m_A+m_B)g$

$=0.2\times 100\times 10+0.3(300)\times 10$

$=200+900=1100N$

613 **(c)**

Swimming is a result of pushing water in the opposite direction of the motion

614 **(c)**

Retardation $= g(\sin 60° + \mu \cos 60°)$

$= 10\left(\frac{\sqrt{3}}{2} + \mu\frac{1}{2}\right) = 5(\sqrt{3} + \mu)$

$v = u - at$

$0 = 20 - 5(\sqrt{3} + \mu) \times 2$

$\mu = 2 - 1.732 \approx 0.27$

615 **(c)**

$T_1 = \left(\frac{m_2 + m_3}{m_1 + m_2 + m_3}\right) g = \frac{3 + 5}{2 + 3 + 5} \times 10$
$= 8\ N$

616 **(d)**

Let T be the tension in the string. Since the system is in equilibrium, therefore from figure

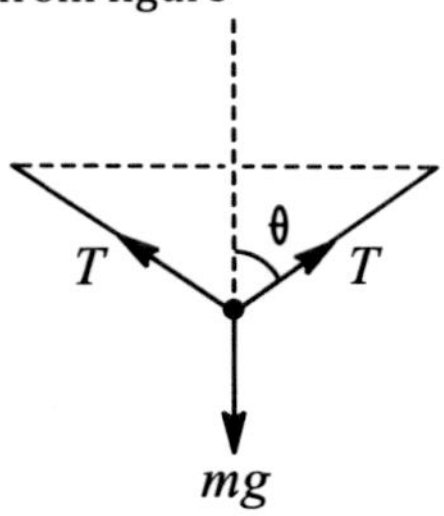

$2T \cos \theta = mg$

or $T = mg/2\ \cos\theta$

The string will be straight if $\theta = 90°$

$\therefore\ T = mg/2\ \cos 90° = mg/2(0) = \infty$

617 **(d)**

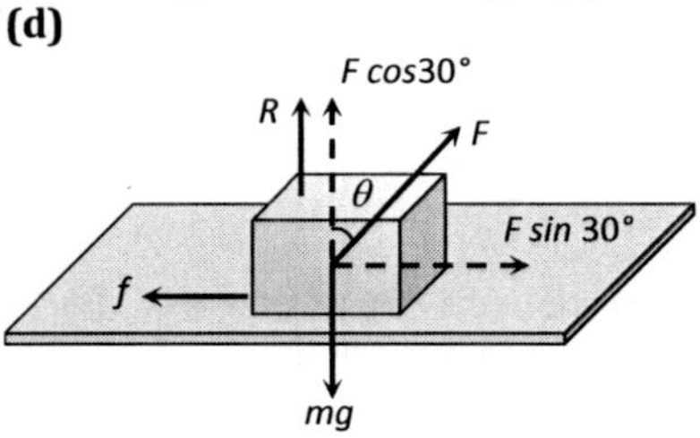

For limiting condition $f = \mu R$

$F \sin 30° = \mu(mg - F \cos\ 30°)$,

By solving, $F = 294.3\ N$

618 **(a)**

Given that $\vec{P} = P_x i + P_y \hat{j} = 2 \cos t\, \hat{\imath} + 2 \sin t\, \hat{j}$

$\therefore \vec{F} = \frac{d\vec{p}}{dt} = -2 \sin t\, \hat{\imath} + 2 \cos t\, \hat{j}$

Now, $\vec{F}.\vec{p} = 0 i.e.$ angle between $\vec{F}$ and $\vec{P}$ is 90°

619 **(a)**

$\mu = \tan 30° = \frac{1}{\sqrt{3}}$

620 **(b)**

$dp = F \times dt = 10 \times 10 = 100\ kg\ m/s$

621 **(c)**

Thrust $F = u\left(\frac{dm}{dt}\right) = 5 \times 10^4 \times 40 = 2 \times 10^6 N$

622 **(b)**

$T = 2\pi\sqrt{\frac{l}{g}}$ and $T' = 2\pi\sqrt{\frac{l}{4g/3}}$

[As $g' = g + a = g + \frac{g}{3} = \frac{4g}{3}$]

$\therefore T' = \frac{\sqrt{3}}{2} T$

623 **(c)**

Initial thrust must be

$m[g + a] = 3.5 \times 10^4(10 + 10)$
$= 7 \times 10^5\ N$

624 **(b)**

Friction between block A and block B and between block B and surface will oppose the P.

$\therefore P = F_{AB} + F_{BS}$

$\mu_{AB} m_a g + \mu_{BS}(m_A + m_B)g$

$= 0.25 \times 100 \times 10 + \frac{1}{2}(100 + 200) \times 10$
$= 1250$ N

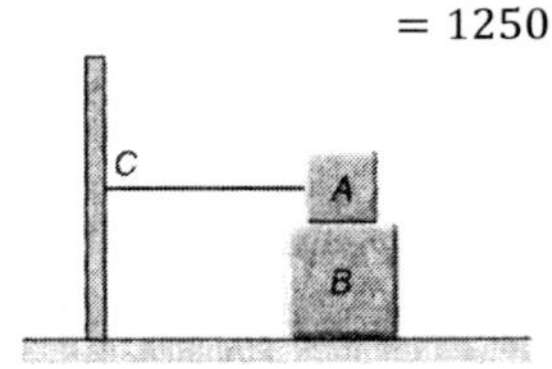

625 **(a)**

Velocity by which the ball hits the bat

$v_1 = \sqrt{2gh_1} = \sqrt{2 \times 10 \times 5}$ or $\vec{v_1} = +10 m/s = 10\ m/s$

Velocity of rebound

$v_2 = \sqrt{2gh_2} = \sqrt{2 \times 10 \times 20} = 20\ m/s$ or $\vec{v_2} = -20\ m/s$

$F = m\frac{dv}{dt} = \frac{m(\vec{v_1} - \vec{v_2})}{dt} = \frac{0.4(-10 - 20)}{dt}$
$= 100\ N$

By solving $dt = 0.12$ sec

628 **(a)**

Let R be the reaction of base of lift and g the acceleration due to gravity, acting downwards. Their resultant provides the net acceleration to the lift. Therefore,

$R - mg = ma$

$\Rightarrow\ R = m(g + a)$

Also, $R = mg'$

$\Rightarrow\ g' = g + a$

Therefore, net acceleration increases and hence, reading indicated by spring balance will increase.

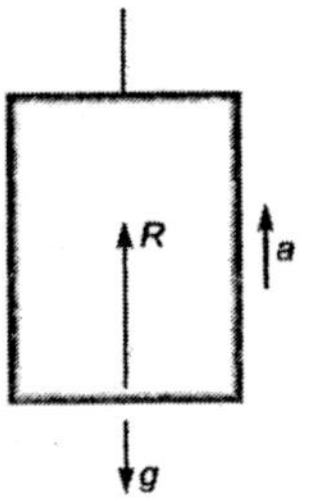

629 **(b)**

Density of cork $= d$, Density of water $= \rho$
Resultant upward force on cork $= V(\rho - d)g$

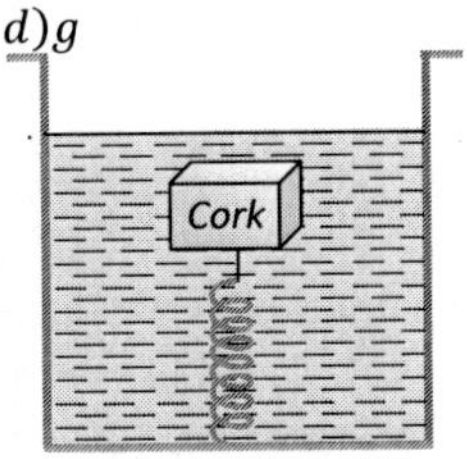

This causes elongation in the spring.
When the lift moves down with acceleration a, the resultant upward force on cork $= V(\rho - d)(g - a)$ which is less than the previous value. So the elongation decreases

630 **(a)**

Initial velocity of ball $= v$
When it strikes the wall normally and reflected back, then final velocity $= -v$
Change in velocity $= v - v = 2v$
Force exerted by the ball on the wall is given by Newton's second law, *ie*,
$F = ma$
$= \frac{m\,\Delta v}{\Delta t}$
$= \frac{m(2v)}{t} = \frac{2\,mv}{t}$

631 **(c)**

For upward acceleration apparent weight $= m(g + a)$
If lift suddenly stops during upward motion then apparent weight $= m(g - a)$ because instead of acceleration, we will consider retardation
In the problem it is given that scale reading initially was 60 kg and due to sudden jerk reading decreasing and finally comes back to the original mark $i.e.$, 60 kg
So, we can conclude that lift was moving upward with constant speed and suddenly stops

632 **(c)**

Resultant downward force along the incline
$= mg\,(\sin\theta - \mu\cos\theta)$
Normal reaction $= mg\cos\theta$
Given : $mg\cos\theta = 2\,mg(\sin\theta - \mu\cos\theta)$
By solving $\theta = 45°$

633 **(c)**

Two masses are moving with equal kinetic energy.
$\frac{1}{2}Mv_1^2 = \frac{1}{2}4Mv_2^2$
or $\frac{v_1}{v_2} = 2$
The ratio of linear momentum is
$\frac{p_1}{p_2} = \frac{Mv_1}{4Mv_2}$
or $\frac{p_1}{p_2} = \frac{1}{4}\left(\frac{v_1}{v_2}\right)$
or $\frac{p_1}{p_2} = \frac{2}{4} = \frac{1}{2}$
or $p_1 : p_2 = 1 : 2$

634 **(a)**

Acceleration, $a = \frac{M_1 - M_2}{M_1 + M_2}g$
$= \frac{M - \frac{M}{2}}{M + \frac{M}{2}}g = \frac{\frac{M}{2}}{\frac{3M}{2}}g = \frac{g}{3}$

637 **(d)**

$\frac{dm}{dt} = 0.1\,kg\,/\,\text{sec}$; Mass of the rocket $=$ 100 kg
$v = 1\,km/\text{sec} = 1000\,m/\text{sec}$
$F = \frac{d(mv)}{dt} = m\frac{dv}{dt} - v\frac{dm}{dt} = 0$ as the mass is decreasing
$100a - 1000 \times 0.1 = 0$
$a = +1\,m/s^2$

639 **(b)**

$\mu = \tan(\text{Angle of repose}) = \tan 60°$
$= 1.732$

640 **(c)**

Apparent weight of the man, $R = m(g + a)$
$= m(g + 4g) = 5mg$

641 **(a)**

Since the given system is in equilibrium therefore tension in the string is 5g
When we consider the combination of A and C , then

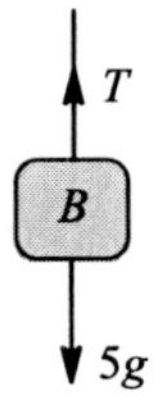

$T = \mu R$

or $T = 0.2(10 + \text{mass of C})g$

or $5g = 0.2(10 + \text{mass of C})g$

or mass of $C = 15$ kg

642 **(a)**

$a = g(\sin\theta - \mu\cos\theta)$

$= 9.8\,(\sin 45° - 0.5\cos 45°)$

$= \frac{4.9}{\sqrt{2}}\ m/sec^2$

643 **(c)**

Mass measured by physical balance remains unaffected due to variation in acceleration due to gravity

646 **(d)**

Rate of flow of water $\frac{V}{t} = \frac{10\ cm^3}{\sec} = 10 \times 10^{-6}\frac{m^3}{\sec}$

Density of water $\rho = \frac{10^3 kg}{m^3}$

Cross-sectional area of pipe $A = \pi(0.5 \times 10^{-3})^2$

Force $= m\frac{dv}{dt} = \frac{mv}{t} = \frac{V\rho v}{t} = \frac{\rho V}{t} \times \frac{V}{At} = \left(\frac{V}{t}\right)^2 \frac{\rho}{A}\left(\because v = \frac{V}{At}\right)$

By substituting the value in the above formula we get $F = 0.127\ N$

647 **(c)**

Given, $F = 600 - 2 \times 10^5\ t = 0$

$\Rightarrow\ t = 3 \times 10^{-3}$s

Impulese $I = \int_0^t F \cdot dt$

$= \int_0^{3\times10^{-3}} (600 - 2 \times 10^3\ t)dt$

$= [600\,t - 10^5 t^2]_0^{3\times10^{-3}} = 0.9$ Ns

648 **(c)**

Rate of change of momentum of the bullet in forward direction = Force required to hold the gun

$F = nmv = 4 \times 20 \times 10^{-3} \times 300 = 24\ N$

649 **(a)**

In Case of projectile motion at the highest point

$(v)_{\text{vertical}} = 0$ and $(v)_{\text{horizontal}} = v\cos\theta$

The initial linear momentum of the system will be $mv\cos\theta$. Now as force of blasting is internal and force of gravity is vertical

So linear momentum of the system along horizontal is conserved

$p_1 + p_2 = mv\cos\theta$

$m_1v_1 + m_2v_2 = mv\cos\theta$

But it is given that $m_1 = m_2 = \frac{m}{2}$ and as one part retraces its path,

$v_1 = -v\cos\theta$

$\therefore \frac{1}{2}m(-v\cos\theta) + \frac{1}{2}mv^2 = mv\cos\theta$

or $v_2 = 3v\cos\theta$

8
GRAVITATION

SINGLE CORRECT ANSWER TYPE

1. Radius of orbit of satellite of earth is R. Its kinetic energy is proportional to
   a) $\frac{1}{R}$ b) $\frac{1}{\sqrt{R}}$ c) R d) $\frac{1}{R^{3/2}}$
2. A satellite is to revolve round the earth in a circle of radius 8000 km. The speed at which this satellite be projected into an orbit, will be
   a) 3 km/s b) 16 km/s c) 7.15 km/s d) 8 km/s
3. Time speed of revolution of a nearest satellite around a planet of radius $R$is T. Period of revolution around another planet, whose radius is 3R but having same density is
   a) T b) $3T$ c) $9T$ d) $3\sqrt{3}T$
4. The acceleration due to gravity increase by 0.5 % when we go from the equator to the poles. What will be the time period of the pendulum at the equator which beats seconds at the poles?
   a) 1.950 s b) 1.995 s c) 2.050 s d) 2.005 s
5. Two identical thin rings each of radius R are coaxially placed at a distance R. If the rings have a uniform mass distribution and each has mass m_1 and m_2 respectively, then the work done in moving a mass m from centre of one ring to that of the other is

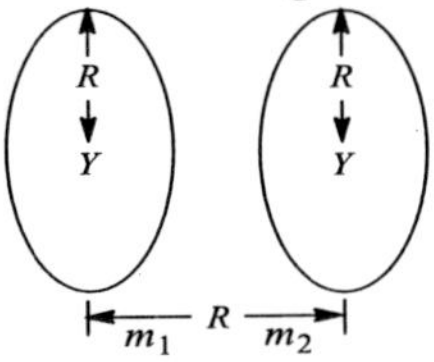

   a) $\frac{Gmm_1(\sqrt{2}+1)}{m_2R}$
   b) $\frac{Gm(m_1-m_2)(\sqrt{2}+1)}{\sqrt{2}R}$
   c) $\frac{Gm\sqrt{2}(m_1+m_2)}{R}$
   d) Zero
6. Kepler's second law regarding constancy of aerial velocity of a planet is consequence of the law of conservation of
   a) Energy b) Angular momentum c) Linear momentum d) None of these
7. The periodic time of a communication satellite is
   a) 6 *hours* b) 12 *hours* c) 18 *hours* d) 24 *hours*
8. The curves for potential energy (U) and kinetic energy (E_k) of a two particle system are shown in figure. At what points the system will be bound

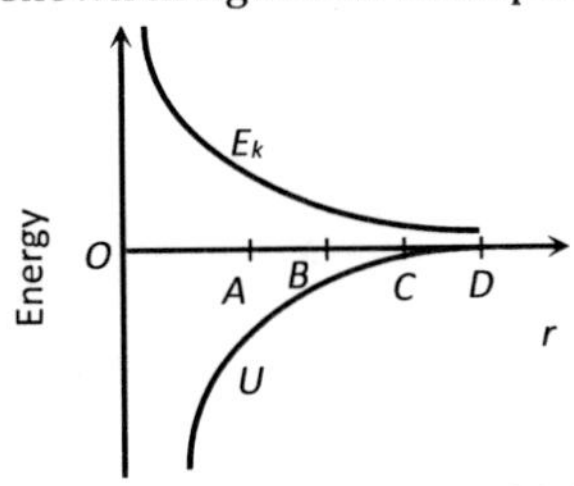

   a) Only at point D b) Only at point A c) At point D and A d) At points A, B and C

9. The acceleration of a body due to the attraction of the earth (radius R) at a distance $2R$ from the surface of the earth is (g = acceleration due to gravity at the surface of the earth)

a) $\frac{g}{9}$ b) $\frac{g}{3}$ c) $\frac{g}{4}$ d) g

10. Two equal mass m and m are hung from balance whose scale pans differ in vertical height by h. Calculate the error in weighing. If any, in terms of density of earth ρ.

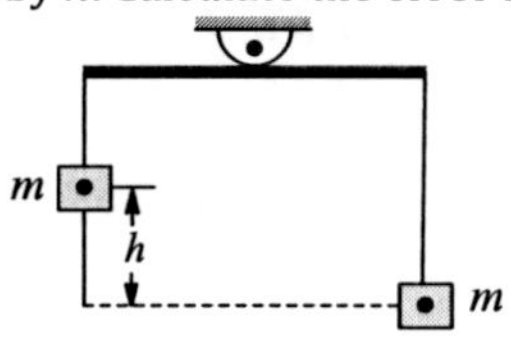

a) $\frac{2}{3}\pi\rho R^3 Gm$ b) $\frac{8}{3}\pi\rho Gmh$ c) $\frac{8}{3}\pi\rho R^3 Gm$ d) $\frac{4}{3}\pi\rho Gm^2 h$

11. Weight of $1kg$ becomes 1/6 on moon. If radius of moon is $1.768 \times 10^6\ m$, then the mass of moon will be

a) $1.99 \times 10^{30} kg$ b) $7.56 \times 10^{22} kg$ c) $5.98 \times 10^{24} kg$ d) $7.65 \times 10^{22} kg$

12. A satellite of mass m is placed at a distance r from the centre of earth (mass M). The mechanical energy of the satellite is

a) $-\frac{GMm}{r}$ b) $\frac{GMm}{r}$ c) $\frac{GMm}{2r}$ d) $-\frac{GMm}{2r}$

13. A satellite S is moving in an elliptical orbit around earth. The mass of the satellite is very small compared to the mass of the earth?

a) The acceleration of S is always directed towards the centre of the earth

b) The angular momentum of S about the centre of the earth changes in direction but its magnitude remains constant

c) The total mechanical energy of S varies periodically with time

d) The linear momentum of S remains constant in magnitude

14. The escape velocity of a body on the surface of the earth is $11.2\ km/s$. If the earth's mass increases to twice its present value and the radius of the earth becomes half, the escape velocity would become

a) $5.6\ km/s$ b) $11.2\ km/s$ (remain unchanged)

c) $22.4\ km/s$ d) $44.8\ km/s$

15. A geostationary satellite is orbiting the earth at a height of $5R$ above the surface of the earth, R being the radius of the earth. The time period of another satellite in hours at a height of $2R$ from the surface of the earth is

a) 5 b) 10 c) $6\sqrt{2}$ d) $\frac{6}{\sqrt{2}}$

16. The escape velocity of a sphere of mass m from earth having mass M and radius R is given by

a) $\sqrt{\frac{2GM}{R}}$ b) $2\sqrt{\frac{GM}{R}}$ c) $\sqrt{\frac{2GMm}{R}}$ d) $\sqrt{\frac{GM}{R}}$

17. Gravitational potential on the surface of earth is (M =mass of the earth, R = radius of earth)

a) $-GM/2R$ b) $-gR$ c) gR d) GM/R

18. If orbital velocity of planet is given by $v = G^a M^b R^c$, then

a) $a = 1/3, b = 1/3, c = -1/3$ b) $a = 1/2, b = 1/2, c = -1/2$

c) $a = 1/2, b = -1/2, c = 1/2$ d) $a = 1/2, b = -1/2, c = -1/2$

19. If both the masses and radius o the earth, each decreases by 50%, the acceleration due to gravity would

a) Remain same b) Decrease by 50% c) Decrease by 100% d) Increase by 100%

20. A spherical planet for out in space has a mass M_0 and diameter D_0. A particle of mass m falling freely new the surface of this planet will experience an acceleration due to gravity which is equal to
a) GM_0/D_0^2 b) $4mGM_0/D_0^2$ c) $4GM_0/D_0^2$ d) GmM_0/D_0^2
21. Kepler discovered
a) Laws of motion b) Laws of rotational motion
c) Laws of planetary motion d) Laws of curvilinear motion
22. If r represents the radius of the orbit of a satellite of mass m moving around a planet of mass M, the velocity of the satellite is given by
a) $v^2 = g\frac{M}{r}$ b) $v^2 = \frac{GMm}{r}$ c) $v = \frac{GM}{r}$ d) $v^2 = \frac{GM}{r}$
23. The ratio $\frac{g}{g_h}$, where g and g_h are the accelerations due to gravity at the surface of the earth and at a height h above the earth's surface respectively, is
a) $\left(1+\frac{h}{R}\right)^2$ b) $\left(1+\frac{R}{h}\right)^2$ c) $\left(\frac{R}{h}\right)^2$ d) $\left(\frac{h}{R}\right)^2$
24. To an astronaut in a spaceship, the sky appears
a) Black b) White c) Green d) Blue
25. In planetary motion the areal velocity of position vector of a planet depends on angular velocity (ω) and the distance of the planet from sun (r). If so the correct relation for areal velocity is
a) $\frac{dA}{dt} \propto \omega r$ b) $\frac{dA}{dt} \propto \omega^2 r$ c) $\frac{dA}{dt} \propto \omega r^2$ d) $\frac{dA}{dt} \propto \sqrt{\omega r}$
26. Earth binds the atmosphere because of
a) Gravity b) oxygen between earth and atmosphere
c) Both (a) and (b) d) None of the above
27. A point mass is placed inside a thin spherical shell of radius R and mass M at a distance $R/2$ from the centre of the shell. The gravitational force exerted by the shell on the point mass is
a) $\frac{GM}{2R^2}$ b) $-\frac{GM}{2R^2}$ c) Zero d) $\frac{GM}{4R^2}$
28. A man weighs 80 kg on earth surface. The height above ground where he will weigh 40kg, is (radius of earth is 6400 km)
a) 0.31 times r b) 0.41 times r c) 0.51 times r d) 0.61 times r
29. A particle falls towards earth from infinity. It's velocity on reaching the earth would be
a) Infinity b) $\sqrt{2gR}$ c) $2\sqrt{gR}$ d) Zero
30. What is the height the weight of body will be the same as at the same depth from the surface of the earth? Radius of earth is R
a) $\frac{R}{2}$ b) $\sqrt{5}R - R$ c) $\frac{\sqrt{5}R - R}{2}$ d) $\frac{\sqrt{3}R - R}{2}$
31. The time period of geostationary satellite at a height 36000 km is 24 h. A spy satellite orbits earth at a height 6400km. What will be the time period of sky satellite? (Radius of earth = 6400 km)
a) 5 h b) 4 h c) 3 h d) 12 h
32. The acceleration to gravity at a height 1/20th of the radius of the earth above the earth surface is 9ms^{-2}. Its value at a point at an equal distance below the surface of the earth in ms^{-2} is about below the surface of the earth in ms^{-2} is about
a) 8.5 b) 9.5 c) 9.8 d) 11.5
33. Two satellite A and B go round a planet orbits having radii $4R$ and R, respectivly. If the speed of satellite A is $3v$, then speed of satellite B is
a) $\frac{3v}{2}$ b) $\frac{4v}{2}$ c) $6v$ d) $12v$
34. An asteroid of mass m is approaching earth, initially at a distance of 10 R_e with speed v_i. It hits the earth with a speed v_f (R_e and M_e are radius and mass of earth), then

a) $v_f^2 = v_i^2 + \frac{2Gm}{M_e R}\left(1 - \frac{1}{10}\right)$ b) $v_f^2 = v_i^2 + \frac{2GM_e}{R_e}\left(1 + \frac{1}{10}\right)$

c) $v_f^2 = v_i^2 + \frac{2GM_e}{R_e}\left(1 - \frac{1}{10}\right)$ d) $v_f^2 = v_i^2 + \frac{2Gm}{R_e}\left(1 - \frac{1}{10}\right)$

35. The distance of a geo-stationary satellite from the centre the earth (Radius $R = 6400\ km$) is nearest to
a) 5 R b) 7 R c) 10 R d) 18 R
36. A satellite moves in elliptical orbit about a planet. The maximum and minimum velocities of satellites are $3 \times 10^4 \text{ms}^{-1}$ and $1 \times 10^3 \text{ms}^{-1}$ respectively. What is the minimum distance of satellite from planet, if maximum distance is 4×10^4km?
a) 4×10^3km b) 3×10^3km c) $4/3 \times 10^3$km d) 1×10^3km
37. Assuming earth to be a sphere of radius R, if $g_{30°}$ is value of acceleration due to gravity at latitude of 30° and g at the equator, the value of $g - g_{30°}$ is
a) $\frac{1}{4}\omega^2 R$ b) $\frac{3}{4}\omega^2 R$ c) $\omega^2 R$ d) $\frac{1}{2}\omega^2 R$
38. Two satellites of earth, S_1 and S_2, are moving in the same orbit. The mass of S_1 is four times the mass of S_2. Which one of the following statements is true?
a) The time period of S_1 is four times that of S_2
b) The potential energies of earth and satellite in the two cases are equal
c) S_1 and S_2 are moving with the same speed
d) The kinetic energies of the two satellites are equal
39. The ratio of acceleration due to gravity at a height h above the surface of the earth and at a depth h below the surface of the earth for $h <<$ radius of earth
a) Is constant
b) Increases linearly with h
c) Decreases linearly with h
d) Decreases parabolically with h
40. The weight of an object in the coal mine, sea level, at the top of the mountain are W_1, W_2 and W_3 respectively, then
a) $W_1 < W_2 > W_3$ b) $W_1 = W_2 = W_3$ c) $W_1 < W_2 < W_3$ d) $W_1 > W_2 > W_3$
41. As we go from the equator to the poles, the value of g
a) Remains the same b) Decreases
c) Increases d) Decreases upto a latitude of 45°
42. Spot the *wrong* statement:
The acceleration due to gravity $'g'$ decreases if
a) We go down from the surface of the earth towards its centre
b) We go up from the surface of the earth
c) We go from the equator towards the poles on the surface of the earth
d) The rotational velocity of the earth is increased
43. The earth revolves about the sun in an elliptical orbit with mean radius $9.3 \times 10^7 m$ in a period of 1 year. Assuming that there are no outside influences
a) The earth's kinetic energy remains constant b) The earth's angular momentum remains constant
c) The earth's potential energy remains constant d) All are correct
44. The escape velocity of a body from earth's surface is v_e. The escape velocity of the same body from a height equal to $7R$ from earth's surface will be
a) $\frac{v_e}{\sqrt{2}}$ b) $\frac{v_e}{2}$ c) $\frac{v_e}{2\sqrt{2}}$ d) $\frac{v_e}{4}$
45. The escape velocity from the earth is 11 kms^{-1}. The escape velocity from a planet having twice the radius and the same mean density as the earth would be
a) 5.5 kms^{-1} b) 11 kms^{-1} c) 15.5 kms^{-1} d) 22 kms^{-1}

46. If density of earth increased 4 times and its radius become half of what it is, our weight will

a) Be four times its present value
b) Be doubled
c) Remain same
d) Be halved

47. The time period of a geostationary satellite is

a) 12 hours
b) 24 hours
c) 6 hours
d) 48 hours

48. The weight of a body on surface of earth is 12.6 N. When it is raised to a height half the radius of earth its weight will be

a) 2.8 N
b) 5.6 N
c) 12.5 N
d) 25. 2N

49. Imagine a light plant revolving around a very massive star in circular orbit of radius r with a period of revolution T. If the gravitational force of attraction between the planet and the star is proportional to $r^{-5/2}$. Then the correct relation is

a) $T^2 \propto r^{5/2}$
b) $T^2 \propto r^{7/2}$
c) $T \propto r^{5/2}$
d) $T^2 \propto r^{7/2}$

50. If V, R and g denote respectively the escape velocity from the surface of the earth radius of the earth, and acceleration due to gravity, then the correct equation is

a) $V = \sqrt{gR}$
b) $V = \sqrt{\frac{4}{3}gR^3}$
c) $V = R\sqrt{g}$
d) $V = \sqrt{2gR}$

51. In an elliptical orbit under gravitational force, in general

a) Tangential velocity is constant
b) Angular velocity is constant
c) Radial velocity is constant
d) Areal velocity is constant

52. A thin uniform annular disc (see figure) of mass M has outer radius $4R$ and inner radius $3R$. The work required to take a unit mass from point P on its axis to infinity is

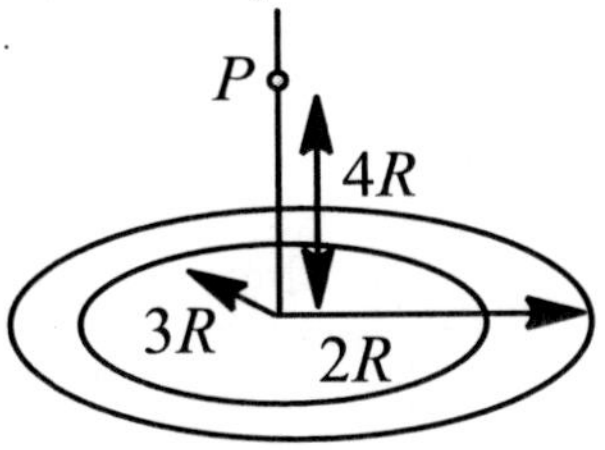

a) $\frac{2GM}{7R}(4\sqrt{2} - 5)$
b) $-\frac{2GM}{7R}(4\sqrt{2} - 5)$
c) $\frac{GM}{4R}$
d) $\frac{2GM}{5R}(\sqrt{2} - 1)$

53. If then radius of earth R, then the height h at which the value of g becomes one-fourth, will be

a) $\frac{R}{8}$
b) $\frac{3R}{8}$
c) $\frac{3R}{4}$
d) $\frac{R}{2}$

54. Two small and heavy spheres, each of mass M, are placed a distance r apart on a horizontal surface. The gravitational potential at the mid-point on the line joining the centre of the spheres is

a) Zero
b) $-\frac{GM}{r}$
c) $-\frac{2GM}{r}$
d) $-\frac{4GM}{r}$

55. If the mass of moon is $\frac{1}{90}$ of earth^' smass, its radius is $\frac{1}{3}$ of earth^' sradius and if g is acceleration due to gravity on earth, then the acceleration due to gravity on moon is..

a) $\frac{g}{3}$
b) $\frac{g}{90}$
c) $\frac{g}{10}$
d) $\frac{g}{9}$

56. Which of the following graphs represents the motion of a planet moving about the sun

a)
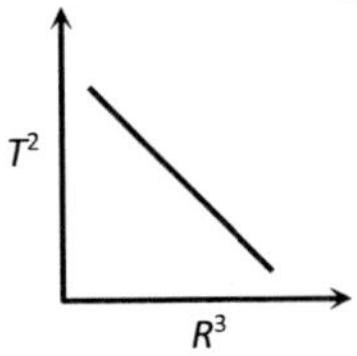

b)
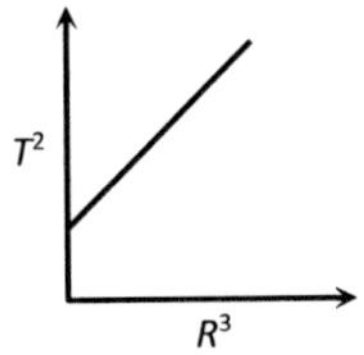

c)
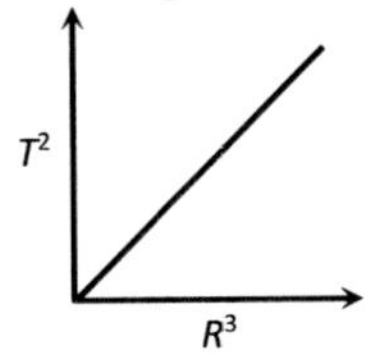

d)
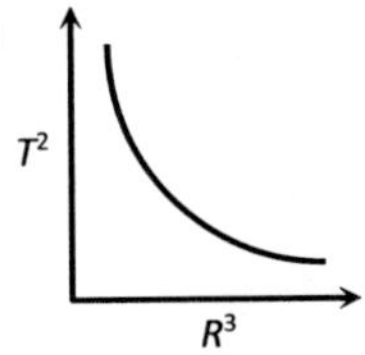

57. The escape velocity for a body projected vertically upwards from the surface of earth is 11 kms^{-1}. If the body is projected at an angle of 45° with the vertical, the escape velocity will be

a) $11\sqrt{2}kms^{-1}$ b) 22 kms^{-1} c) 11 kms^{-1} d) $11/\sqrt{2}ms^{-1}$

58. A rocket is launched with velocity 10 km/s. If radius of earth is R, then maximum height attained by it will be

a) $2R$ b) $3R$ c) $4R$ d) $5R$

59. The mass of the earth is 6.00×10^{22} kg. The constant of gravitation g = $6.67 \times 10^{-11} Nm^2kg^{-2}$. The potential energy of the system is -7.73×10^{28} J. The mean distance between earth and moon is

a) $3.80 \times 10^8 m$ b) $3.37 \times 10^6 m$ c) $7.60 \times 10^4 m$ d) $1.90 \times 10^2 m$

60. Gravitational mass is proportional to gravitational

a) Field b) Force c) Intensity d) All of these

61. The acceleration due to gravity is g at a point distant r from the centre of earth of radius R. If $r < R$, then

a) $g \propto r$ b) $g \propto r^2$ c) $g \propto r^{-1}$ d) $g \propto r^{-2}$

62. In the following four periods

(i) Time of revolution of a satellite just above the earth's surface (T_{st})

(ii) Period of oscillation of mass inside the tunnel bored along the diameter of the earth (T_{ma})

(iii) Period of simple pendulum having a length equal to the earth's radius in a uniform field of $9.8N/kg(T_{sp})$

(iv) Period of an infinite length simple pendulum in the earth's real gravitational filed (T_{is})

a) $T_{st} > T_{ma}$ b) $T_{ma} > T_{st}$

c) $T_{sp} > T_{is}$ d) $T_{st} = T_{ma} = T_{sp} = T_{is}$

63. The height from the earth surface at which the value of acceleration due to gravity reduces to $1/4^{th}$ of its value at earth's surface (assume earth to be sphere of radius 6400 km)

a) 6400 km b) 2649 km c) 2946 km d) 1600 km

64. A space ship moves from earth to moon and back. The greatest energy required for the space ship is to overcome the difficulty in

a) Entering the earth's gravitational field
b) Take off from earths field
c) Take off from lunar surface
d) Entering the moon's lunar surface

65. If the mass of earth is 80 times of that of a planet and diameter is double that of planet and $'g'$ on earth is $9.8m/s^2$, then the value of $'g'$ on that planet is

a) 4.9 m/s^2 b) 0.98 m/s^2 c) 0.49 m/s^2 d) 49 m/s^2

66. An iron ball and a wooden ball of the same radius are released from a height $'h'$ in vacuum. The time taken by both of them to reach the ground is

a) Unequal b) Exactly equal c) Roughly equal d) Zero

67. If gravitational force on a body of mass 1.5 kg at point is 45N, then the intensity of the gravitational field at that point is

a) 67.5 N kg^{-1} b) 45 N kg^{-1} c) 30 N kg^{-1} d) 15 N kg^{-1}

68. A man inside an artificial satellite feels weightlessness because the force of attraction due to earth is

a) Zero at that place
b) Is balanced by the force of attraction due to moon
c) Equal to the centripetal force
d) Non-effective due to particular design of the satellite

69. A geostationary satellite is orbiting the earth at a height of $6R$ above the surface of the earth; R being the radius of the earth. What will be the time period of another satellite at a height 2.5 R from the surface of the earth?
   a) $6\sqrt{2}$ h b) $6\sqrt{2.5}$ h c) $6\sqrt{3}$ h d) 12 h
70. If the radius of the earth shrinks by 1%, its mass remaining same, the acceleration due to gravity on the surface of earth will
   a) Decrease by 2% b) Decrease by 0.5% c) Increase by 2% d) Increase by 0.5%
71. If a body describes a circular motion under inverse square field, the time taken to complete one revolution T is related to the radius of the of the circular orbit is
   a) $T \propto r$ b) $T \propto r^2$ c) $T^2 \propto r^3$ d) $T \propto r^4$
72. Which of the following graphs between the square of the time period and cube of the distance of the planet from the sun is correct?
   a) 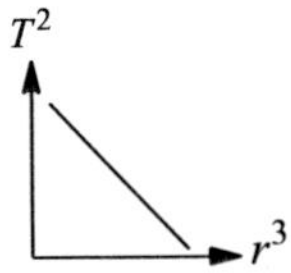

   b) 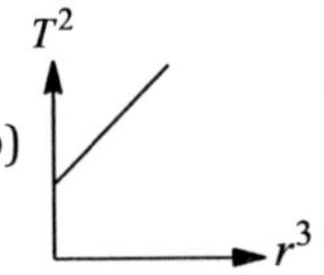

   c) 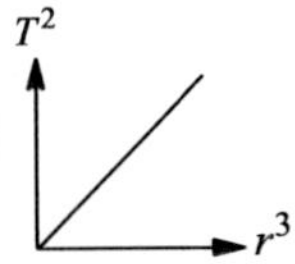

   d) 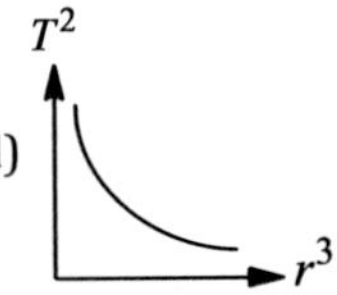

73. Two identical solid copper spheres of radius R are placed in contact with each other. The gravitational attraction between them is proportional to
   a) R^2 b) R^{-2} c) R^4 d) R^{-4}
74. If the density of a small planet is the same as that of earth, while the radius of the planet is 0.2 times that of the earth, the gravitational acceleration of the surface of that planet is
   a) 0.2 g b) 0.4 g c) 2 g d) 4 g
75. Acceleration due to gravity on moon is 1/6 of the acceleration due to gravity on earth. If the ratio of densities of earth (ρ_e) and moon (ρ_m) is $\left(\frac{\rho_e}{\rho_m}\right) = \frac{5}{3}$ then radius of moon (R_m) in terms of R_e will be
   a) $\frac{5}{18} R_e$ b) $\frac{1}{6} R_e$ c) $\frac{3}{18} R_e$ d) $\frac{1}{2\sqrt{3}} R_e$
76. Gravitational field is
   a) Conservative b) Non-conservative c) Electromagnetic d) Magnetic
77. The mass of a planet is six times that of the earth. The radius of the planet is twice that of the earth. If the escape velocity from the earth is v, then the escape velocity from the planet is
   a) $\sqrt{3}v$ b) $\sqrt{2}v$ c) v d) $\sqrt{5}v$
78. Acceleration due to gravity is maximum at (R is the radius of earth)
   a) A height $\frac{R}{2}$ from the earth's surface b) The centre of the earth
   c) The surface of the earth d) A depth $\frac{R}{2}$ from the earth's surface
79. An artificial satellite is revolving round the earth in a circular orbit. Its velocity is half the escape velocity. Its height from earth's surface is
   a) 6400 km b) 12800 km c) 3200 km d) 1600 km
80. A satellite is revolving round the earth in an orbit of radius r with time period T. If the satellite is revolving round the earth in an orbit of radius $r + \Delta r (\Delta r \ll r)$ with time period $T + \Delta T (\Delta T \ll T)$ then
   a) $\frac{\Delta T}{T} = \frac{3}{2}\frac{\Delta r}{r}$ b) $\frac{\Delta T}{T} = \frac{2}{3}\frac{\Delta r}{r}$ c) $\frac{\Delta T}{T} = \frac{\Delta r}{r}$ d) $\frac{\Delta T}{T} = -\frac{\Delta r}{r}$
81. Assume that the acceleration due to gravity on the surface of the moon is 0.2 times the acceleration due to gravity on the surface of the earth. If R_e is the maximum range of a projectile on the earth's surface, what is the maximum range on the surface of the moon for the same velocity of projection
   a) 0.2 R_e b) 2 R_e c) 0.5 R_e d) 5 R_e

82. The escape velocity of a body from the earth is v_e. If the radius of earth contracts to 1/4th of its value, keeping the mass of the earth constant, the escape velocity will be
a) Doubled b) Halved c) Tripled d) Unaltered

83. The ratio of the distances of two planets from the sun is 1.38. The ratio of their period of revolution around the sun is
a) 1.38 b) $1.38^{3/2}$ c) $1.38^{1/2}$ d) 1.38^3

84. A planet in a distant solar system is 10 times more massive than the earth and its radius is 10 times smaller. Given that the escape velocity from the earth is 11kms^{-1}, the escape velocity from the surface of the planet would be
a) 1.1 kms^{-1} b) 11 kms^{-1} c) 110 kms^{-1} d) 0.11 kms^{-1}

85. A satellite moves in a circle around the earth. The radius of this circle is equal to one-half of the radius of the moon's orbit. The satellite completes one revolution in
a) $\frac{1}{2}$ lunar month b) $\frac{2}{3}$ lunar month c) $2^{-3/2}$ lunar month d) $2^{3/2}$ lunar month

86. A body is released from a point distance r from the centre of earth. If R is the earth and $r > R$, then the velocity of the body at the time of striking the earth will be
a) $\sqrt{gR}$ b) $\sqrt{2gR}$ c) $\sqrt{\frac{2gR}{r-R}}$ d) $\sqrt{\frac{2gR(r-R)}{r}}$

87. Two equal masses m and m are hung from a balance whose scale pans differ in height by h. If ρis the mean density of earth, then the error in weighing is
a) Zero b) $4\pi G\rho mh/3$ c) $8\pi G\rho mh/3$ d) $2\pi G\rho mh/3$

88. A 20 cm long capillary tube is dipped in water. The water risesupto 8 cm. if entire arrangement is put in a freely falling elevator lengths of water column in the capillary tube will be
a) 4 cm b) 8 cm c) 10 cm d) 20 cm

89. A satellite revolves around the earth in an elliptical orbit. Its speed
a) Is the same at all points in the orbit
b) Is greatest when it is closest to the earth
c) Is greatest when it is farthest from the earth
d) Goes on increasing or decreasing continuously depending upon the mass of the satellite

90. A geostationary satellite
a) Revolves about the polar axis
b) Has a time period less than that of the near earth satellite
c) Moves faster than a near earth satellite
d) Is stationary in the space

91. A shell of mass M and radius R has a point mass m placed at a distance r from its centre.

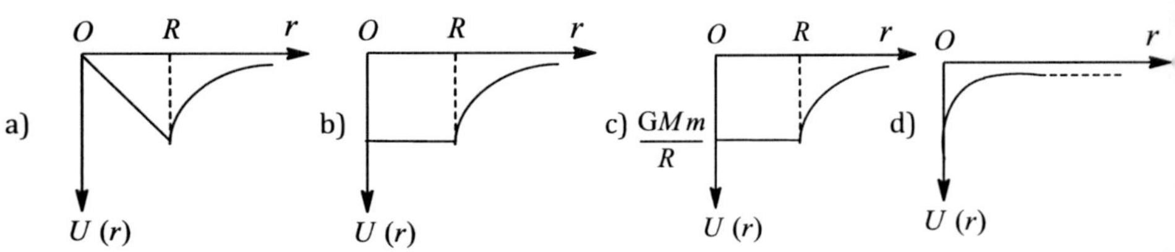

92. The angular velocity of the earth with which it has to rotate so that acceleration due to gravity on 60° latitude becomes zero is (Radius of earth = 6400 km. At the poles $g = 10\ ms^{-2}$)
a) $2.5 \times 10^{-3} rad/s$ b) $5.0 \times 10^{-1} rad/s$ c) $1 \times 10^{1} rad/s$ d) $7.8 \times 10^{-2} rad/s$

93. In a satellite, if the time of revolution is T, then KE is proportional to
a) $\frac{1}{T}$ b) $\frac{1}{T^2}$ c) $\frac{1}{T^3}$ d) $T^{-2/3}$

94. Three identical bodies of mass M are located at the vertices of an equilateral triangle of side L. They revolve under the effect of mutual gravitational force in a circular orbit,

circumscribing the triangle while preserving the equilateral triangle. Their orbital velocity is

a) $\sqrt{\frac{GM}{L}}$ b) $\sqrt{\frac{3GM}{2L}}$ c) $\sqrt{\frac{3GM}{L}}$ d) $\sqrt{\frac{2GM}{3L}}$

95. Imagine a new planet having the same density as that of earth but it is 3 times bigger than the earth in size. If the acceleration due to gravity on the surface of earth is g and that on the surface of the new planet is g', then

a) $g' = 2g$ b) $g' = 3g$ c) $g' = 4g$ d) $g' = 5g$

96. An artificial satellite of the earth moves at an altitude to $h = 670$ km along a circular orbit. The velocity of the satellite is

a) 7.5 kms^{-1} b) 8.5 kms^{-1} c) 11.2 kms^{-1} d) 4.5 kms^{-1}

97. Escape velocity of a body of 1 kg mass on a planet is 100 m/sec. Gravitational Potential energy of the body at the Planet is

a) $-5000\,J$ b) $-1000\,J$ c) $-2400\,J$ d) $5000\,J$

98. A particle of mass m is placed at the centre of a uniform spherical shell of mass 3 m and radius R. The gravitational potential on the surface of the shell is

a) $-\frac{Gm}{R}$ b) $-\frac{3Gm}{R}$ c) $-\frac{4Gm}{R}$ d) $-\frac{2Gm}{R}$

99. The work that must be done in lifting a body of weight P from the surface of the earth to a height h is

a) $\frac{PRh}{R-h}$ b) $\frac{R+h}{PRh}$ c) $\frac{PRh}{R+h}$ d) $\frac{R-h}{PRh}$

100. Two spherical planets A and B have same mass but densities in the ratio 8:1. For these planets, the ratio of acceleration due to gravity at the surface of A to its value at the surface of B is

a) 1 : 4 b) 1 : 2 c) 4 : 1 d) 8 : 1

101. A satellite which is geostationary in a particular orbit is taken to another orbit. Its distance from the centre of earth in new orbit is 2 times that of the earlier orbit. The time period in the second orbit is

a) 4.8 *hours* b) $48\sqrt{2}$ *hours* c) 24 *hours* d) $24\sqrt{2}$ *hours*

102. An object weighs 10N at the north-pole of the earth. In a geostationary satellite distance $7R$ from the centre of earth (of radius R) what will be its true weight?

a) 3 N b) 5 N c) 2 N d) 0.2 N

103. The value of g decreases inside the surface of earth because

a) A force of upward attraction is applied by the shell of earth above
b) The shell of earth above exerts no net force
c) The distance from the centre of the earth decreases
d) The density of the material at the centre of the earth is very small

104. What is the escape velocity for a body on the surface of a planet on which the acceleration due to gravity is $(3.1)^2 ms^{-2}$ and whose radius is 8100 km

a) 2790 $km.s^{-1}$ b) 27.9 $km.s^{-1}$ c) $\frac{27.9}{\sqrt{5}} km.s^{-1}$ d) $27.9\sqrt{5} km.s^{-1}$

105. If satellite is shifted towards the earth. Then time period of satellite will be

a) Increase b) Decrease c) Unchanged d) Nothing can be said

106. A spherically symmetric gravitational system of particles has a mass density

$$\rho = \begin{cases} \rho_0 \text{ for } r \le R \\ 0 \text{ for } r > R \end{cases}$$

where ρ_0 is a constant. A test mass can undergo circular motion under the influence of the gravitational field of particles. Its speed v as a function of distance $r(0 < r < \infty)$ from the centre of the system is represented by

a) 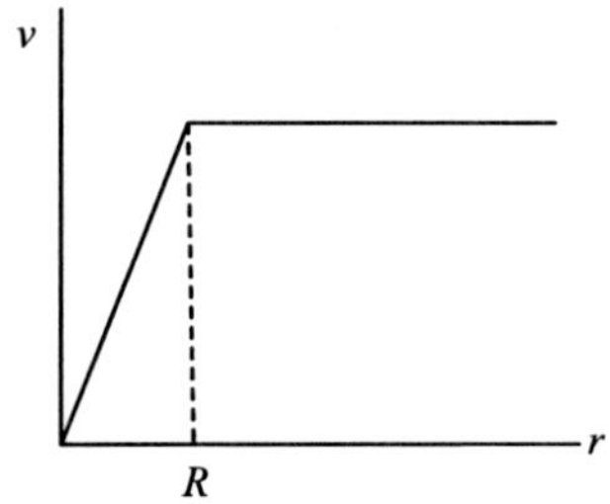

b) 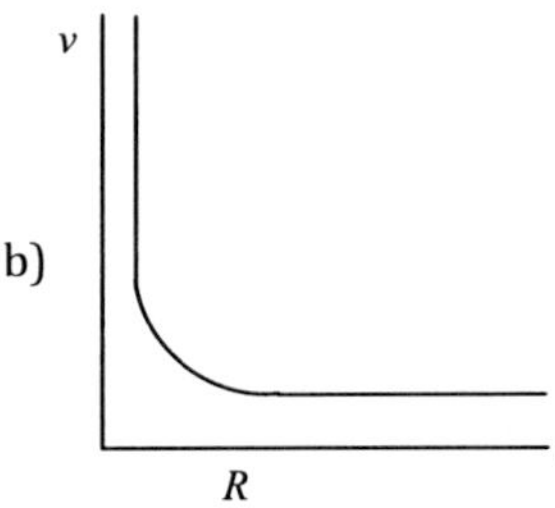

c) 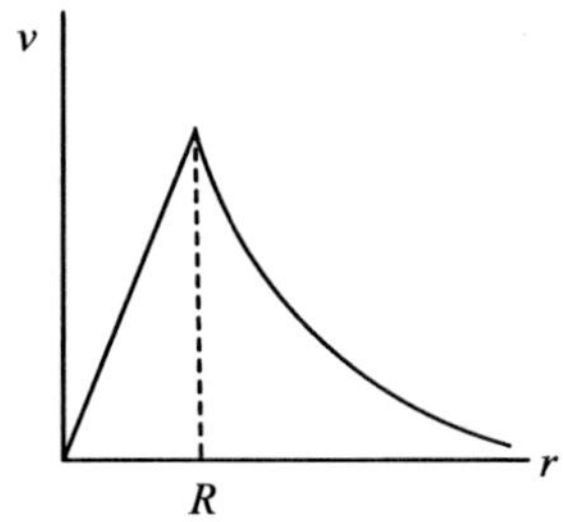

d) 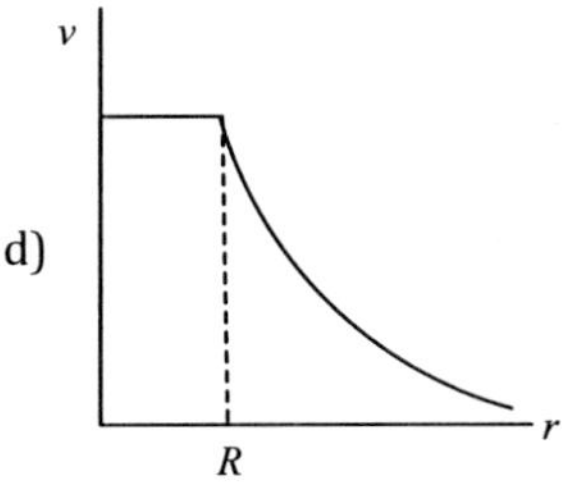

107. If the angular speed of the earth is doubled, the value of acceleration due to gravity (g) at the north pole
a) Doubles b) Becomes half c) Remains same d) Becomes zero

108. One can easily "weight the earth" by calculating the mass of earth using the formula (in usual notation)
a) $\frac{G}{g}R_E^2$ b) $\frac{g}{G}R_E^2$ c) $\frac{g}{G}R_E$ d) $\frac{G}{g}R_E^3$

109. Pick out the most correct statement with reference to earth satellites
a) Geostationary satellites are used for remote sensing
b) Polar satellites are used for telecommunications
c) INSAT group of satellites belong to geostationary satellites
d) Polar satellites are at a height of about 36,000 km

110. The radius of orbit of a planet is two times that of the earth. The time period of planet is
a) 4.2 *years* b) 2.8 *years* c) 5.6 *years* d) 8.4 *yeasrs*

111. The earth (mass $= 6 \times 10^{24}$kg) revolves around the sun with angular velocity 2×10^{-7} rads^{-1} in a circular orbit of radius 1.5×10^8km. The force exerted by the sun on the earth in newton is
a) Zero b) 18×10^{25} c) 27×10^{39} d) 36×10^{21}

112. If the Earth losses its gravity, then for a body
a) Weight becomes zero, but not the mass b) Mass becomes zero, but not the weight
c) Both mass and weight become zero d) Neither mass nor weight become zero

113. The mass of the moon is about 1.2% of the mass of the earth. Compared to the gravitational force the earth exerts on the moon, the gravitational force the moon exerts on earth
a) Is the same b) Is smaller c) Is greater d) Varies with its phase

114. Select the proper graph between the gravitational potential (v_g) due to hollow sphere and distance (r) from its centre

a) 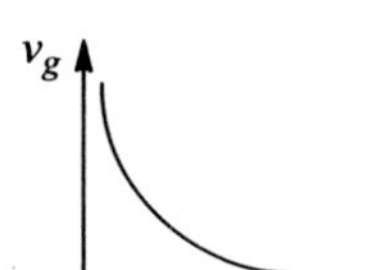

b) 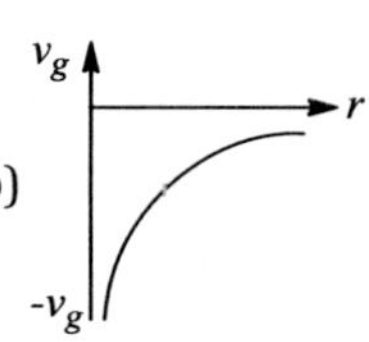

c) 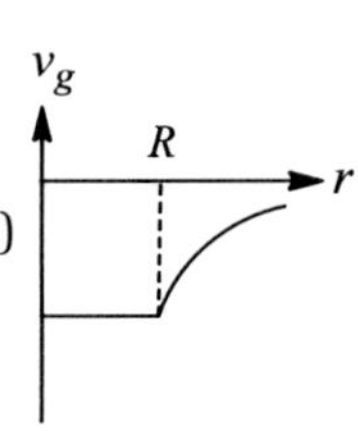

d) 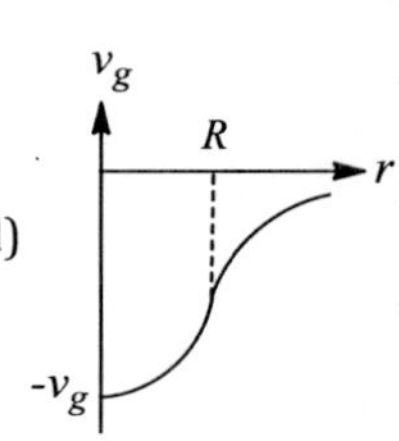

115. A person sitting on a chair in a satellite feels weightless because
a) The earth dose not attract the object in a satellite
b) The normal force by the chair on the person balances the earth's attraction
c) The normal force is zero
d) The person in satellite is not accelerated

116. A missile is launched with a velocity less than the escape velocity. The sum of its kinetic and potential energy is
a) Positive
b) Negative
c) Zero
d) May be positive or negative depending upon its initial velocity

117. The ratio of the radii of the planets P_1 and P_2 is a. The ratio of their acceleration due to gravity is b. The ratio of the escape velocities from them will be
a) ab
b) $\sqrt{ab}$
c) $\sqrt{a/b}$
d) $\sqrt{b/a}$

118. Weight of a body is maximum at
a) Moon
b) Poles of earth
c) Equator of earth
d) Centre of earth

119. Three weights w, $2w$ and $3w$ are connected to identical spring suspended from a rigid horizontal rod. The assembly of the rod and weights fall freely. The positions of the weight from the rod are such that
a) $3w$ will be farthest
b) w will be farthest
c) All will be at the same distance
d) $2w$ will be farthest

120. Potential energy of a satellite having mass $'m'$ and rotating at a height of $6.4 \times 10^6 m$ from the earth surface is
a) $-0.5\, mgR_e$
b) $-mgR_e$
c) $-2\, mgR_e$
d) $4\, mgR_e$

121. The kinetic energy needed to project a body of mass m from the earth surface (radius R) to infinity is
a) $mgR/2$
b) $2\, mgR$
c) mgR
d) $mgR/4$

122. The ratio of radii of earth to another planet is $\frac{2}{3}$ and the ratio of their mean densities is $\frac{4}{5}$. If an astronaut can jump to a maximum height of 1.5 m on the earth, with the same effort, the maximum height he can jump on the planet is
a) 1 m
b) 0.8 m
c) 0.5 m
d) 1.25 m

123. There are two planets. The ratio of radius of the two planets is K but ratio of acceleration due to gravity of both planets is g. What will be the ratio of their escape velocity
a) $(Kg)^{1/2}$
b) $(Kg)^{-1/2}$
c) $(Kg)^2$
d) $(Kg)^{-2}$

124. The binding energy of a satellite of mass m in a orbit of radius r is (R = radius of earth, g = acceleration due to gravity)
a) $\frac{mgR^2}{r}$
b) $\frac{mgR^2}{2r}$
c) $-\frac{mgR^2}{r}$
d) $-\frac{mgR^2}{2r}$

125. Two satellites of mass m and $9m$ are orbiting a planet in orbit of radius R. Their periods of revolution will be in the ratio of
a) 1:3
b) 1:1
c) 3:1
d) 9:1

126. At what depth below the surface of the earth, acceleration due to gravity g will be half its value 1600 km above the surface of the earth
a) $4.2 \times 10^6 m$
b) $3.19 \times 10^6 m$
c) $1.59 \times 10^6 m$
d) None of these

127. If ρ is the density of the planet, the time period of nearby satellite is given by
a) $\sqrt{\frac{4\pi}{3G\rho}}$
b) $\sqrt{\frac{4\pi}{G\rho}}$
c) $\sqrt{\frac{3\pi}{G\rho}}$
d) $\sqrt{\frac{\pi}{G\rho}}$

128. Two spheres of mass m and M are situated in air and the gravitational force between them is F. The space around the masses is now filled with a liquid of specific gravity 3. The gravitational force will now be
a) F
b) $\frac{F}{3}$
c) $\frac{F}{9}$
d) $3F$

129. For a body to escape from earth, angle at which it should be fired is?

a) 45° b) > 45° c) < 45° d) any angle

130. If $g \propto \frac{1}{R^3}$ (instead of $\frac{1}{R^2}$), then the relation between time period of a satellite near earth's surface and radius R will be

a) $T^2 \propto R^3$ b) $T \propto R^2$ c) $T^2 \propto R$ d) $T \propto T$

131. If the force inside the earth surface varies as x^n, where r is the distance of body from the centre of earth, then the value of n will be

a) -1 b) -2 c) 1 d) 2

132. The time period of a satellite of earth is 5h. If the separation between the earth and the satellite is increased to 4 times the previous value, the new time period will become

a) 10 h b) 18 h c) 40 h d) 20 h

133. The escape velocity from the earth is 11 kms^{-1}. The escape velocity from a planet having twice the radius and same mean density as that of earth is

a) 5.5 kms^{-1} b) 11 kms^{-1} c) 22 kms^{-1} d) None of these

134. If the value of g acceleration due to gravity at earth surface is 10 ms^{-2}. Its value in ms^{-2} at the centre of the earth, which is assumed to be a sphere of radius R metre and uniform mass density is

a) 5 b) $10/R$ c) $10/2R$ d) Zero

135. If the earth rotates faster than its present speed, the weight of an object will

a) Increase at the equator but remain unchanged at the poles
b) Decrease at the equator but remain unchanged at the poles
c) Remain unchanged at the equator but decrease at the poles
d) Remain unchanged at the equator but increase at the poles

136. The distance of neptune and saturn from sun are nearly 10^{13} and $10^{12} m$ respectively. Assuming that they move in circular orbits, their periodic times will be in the ratio

a) $\sqrt{10}$ b) 100 c) $10\sqrt{10}$ d) $1/\sqrt{10}$

137. At a distance 320 km above the surface of earth, the value of acceleration due to gravity will be lower than its value on the surface of the earth by nearly (radius of earth = 6400 km)

a) 2% b) 6% c) 10% d) 14%

138. If an object of mass m is taken from the surface of earth (radius R) to a height $2R$, then the work done is

a) $2mgR$ b) mgR c) $\frac{2}{3}mgR$ d) $\frac{3}{2}mgR$

139. A thief stole a box full of valuable articles of weight w and while carrying it on his head jumped down from a wall of height h from the ground. Before he reaches the ground, he experienced a load

a) Zero b) $w/2$ c) w d) $2\,w$

140. The gravitational potential energy of a body of mass 'm' at the earth's surface is $-mgR_e$. Its gravitational potential energy at a height R_e from the earth's surface will be (Here R_e is the radius of the earth)

a) $-2mgR_e$ b) $2mgR_e$ c) $\frac{1}{2}mgR_e$ d) $-\frac{1}{2}mgR_e$

141. An astronaut orbiting the earth in a circular orbit 120 km above the surface of earth, gently drops a spoon out of space-ship. The spoon will

a) Fall vertically down to the earth b) Move towards the moon
c) Will move along with space-ship d) Will move in an irregular way then fall down to earth

142. Two bodies of masses m_1 and m_2 are initially at rest at infinite distance apart. They are then allowed to move towards each other under mutual gravitational attraction. Their relative velocity of approach at a separation distance r between them is

a) $\left[2G\frac{(m_1 - m_2)}{r}\right]^{1/2}$ b) $\left[\frac{2G}{r}(m_1 + m_2)\right]^{1/2}$ c) $\left[\frac{r}{2G(m_1 m_2)}\right]^{1/2}$ d) $\left[\frac{2G}{r}m_1 m_2\right]^{1/2}$

143. In the solar system, which is conserved

a) Total Energy b) K.E. c) Angular Velocity d) Linear Momentum

144. The total energy of a circularly orbiting satellite is

a) Twice the kinetic energy of the satellite b) Half the kinetic energy of the satellite
c) Twice the potential energy of the satellite d) Half the potential energy of the satellite

145. The time period of a simple pendulum on a freely moving artificial satellite is

a) Zero b) 2 sec c) 3 sec d) Infinite

146. A satellite is revolving around the planet. The gravitational force between them varies with $R^{-5/2}$, where R is the radius of the satellite. The square of the time period T will be directly proportional to

a) R^3 b) $R^{7/2}$ c) $R^{3/2}$ d) $R^{5/7}$

147. A body is orbiting very close to the earth's surface with kinetic energy KE. The energy required to completely escape from it is

a) KE b) 2 KE c) $\frac{KE}{2}$ d) $\frac{3KE}{2}$

148. A man is standing on an international space station, which is orbiting earth at an altitude 520 km with a constant speed 7.6 km/s. If the man's weight is 50 kg, his acceleration is

a) $7.6km/s^2$ b) $7.6m/s^2$ c) $8.4m/s^2$ d) $10m/s^2$

149. Three or two planets. The ratio of radius of the two planets is K but ratio of acceleration due to gravity of both planets is g. What will be the ratio of their escape velocity?

a) $(Kg)^{1/2}$ b) $(Kg)^{-1/2}$ c) $(Kg)^2$ d) $(Kg)^{-2}$

150. A planet revolves around the sun in an elliptical orbit. The linear speed of the planet will be maximum at

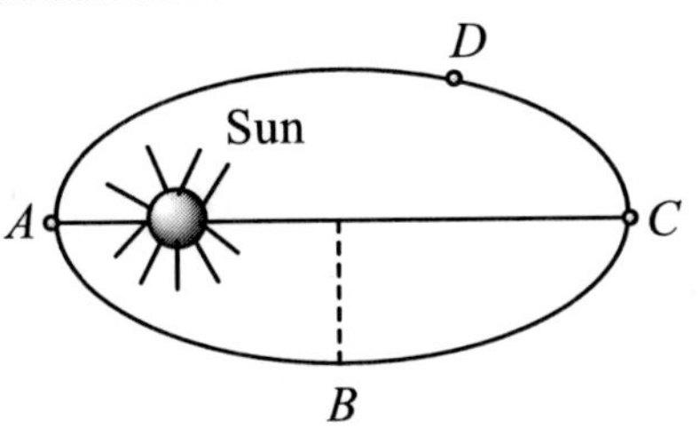

a) D b) B c) A d) C

151. If a man weighs 90 kg on the surface of earth, the height above the surface of the earth of radius R, where the weight is 30 kg is

a) 0.73 R b) $R/\sqrt{3}$ c) $R/3$ d) $\sqrt{3}R$

152. Correct form of gravitational law is

a) $F = -\frac{Gm_1m_2}{r^2}$ b) $\vec{F} = -\frac{Gm_1m_2}{r^2}$ c) $\vec{F} = -\frac{Gm_1m_2}{r^3}\hat{r}$ d) $\vec{F} = -\frac{Gm_1m_2\vec{r}}{r^3}$

153. At what height above the earth's surface does the force of gravity decrease by 10%? The radius of the earth is 6400 km?

a) 345.60 km b) 687.20 km c) 1031.8 km d) 12836.80 km

154. The satellite of mass m revolving in a circular orbit of radius r around the earth has kinetic energy E. Then its angular momentum will be

a) $\sqrt{\frac{E}{mr^2}}$ b) $\frac{E}{2mr^2}$ c) $\sqrt{2\,Emr^2}$ d) $\sqrt{2\,Emr}$

155. A satellite of mass m is circulation around the earth with constant angular velocity. If radius of the orbit is R_0 and mass of the earth M, the angular momentum about the centre of the earth is

a) $m\sqrt{GMR_0}$ b) $M\sqrt{GMR_0}$ c) $m\sqrt{\frac{GM}{R_0}}$ d) $M\sqrt{\frac{GM}{R_0}}$

156. Two satellite $A$and B, ratio of masses 3 : 1 are in circular orbits of radii r and $4r$. Then ratio of total mechanical energy of A to B is

a) 1 : 3 b) 3 : 1 c) 3 : 4 d) 12 : 1

157. The required kinetic energy of an object of mass m, so that it may escape, will be

a) $\frac{1}{4}mgR$ b) $\frac{1}{2}mgR$ c) mgR d) $2mgR$

158. The acceleration due to gravity on a planet is 1.96 ms^{-2}. If it is safe to jump from a height of 3 m on the earth, the corresponding height on the planet will be

a) 3 m b) 6 m c) 9 m d) 15 m

159. If g_e, g_h and g_d be the accelerations due to gravity at earth's surface, a height h and at depth d respectively. Then

a) $g_e > g_h > g_d$ b) $g_e > g_h < g_d$ c) $g_e < g_h < g_d$ d) $g_e < g_h > g_d$

160. The radius of a planet is 1/4 of earth's radius and its acceleration due to gravity is double that of earth's acceleration due to gravity. How many times will the escape velocity at the planet's surface be as compared to its value on earth's surface

a) $\frac{1}{\sqrt{2}}$ b) $\sqrt{2}$ c) $2\sqrt{2}$ d) 2

161. The mass of diameter of a planet are twice those of earth. The period of oscillation of pendulum on this planet will be (if it is a second's pendulum on earth)

a) $\frac{1}{\sqrt{2}}s$ b) $2\sqrt{2}\ s$ c) $2s$ d) $\frac{1}{2}s$

162. At what temperature, the hydrogen molecule will escape from earth's surface?

a) 10^1K b) 10^2 K c) 10^3 K d) 10^4 K

163. If the radius of a planet is R and its density is ρ, the escape velocity from its surface will be

a) $v_e \propto \rho R$ b) $v_e \propto \sqrt{\rho R}$ c) $v_e \propto \frac{\sqrt{\rho}}{R}$ d) $v_e \propto \frac{1}{\sqrt{\rho R}}$

164. A body is orbiting around earth at a mean radius which is two times as greater as parking orbit of a satellite, the period of body is

a) 4 days b) 16 days c) $2\sqrt{2}$ days d) 64 days

165. If M is the mass of the earth and R its radius, the ratio of the gravitational acceleration and the gravitational constant is

a) $\frac{R^2}{M}$ b) $\frac{M}{R^2}$ c) MR^2 d) $\frac{M}{R}$

166. What would be the velocity of earth due to rotation about its own axis so that the weight at equator become 3/5 of initial value. Radius of earth on equator is 6400 km

a) $7.4 \times 10^{-4} rad/sec$ b) $6.7 \times 10^{-4} rad/sec$ c) $7.8 \times 10^{-4} rad/sec$ d) $8.7 \times 10^{-4} rad/sec$

167. The rotation period of an earth satellite close to the surface of the earth is 83 min. the satellite in a orbit at a distance of three times earth radii from its surface will be

a) 83 min b) $83 \times \sqrt{8}$ min c) 664 min d) 249 min

168. The escape velocity on earth is

a) 1.12 kms^{-1} b) 11.2 ms^{-1} c) 11.2 kmh^{-1} d) 11.2 kms^{-1}

169. In a gravitational field, at a point where the gravitational potential is zero

a) The gravitational field is necessarily zero b) The gravitational field is not necessarily zero c) Nothing can be said definitely about the gravitational field d) None of these

170. Orbital velocity of earth's satellite near the surface is 7 km/s. When the radius of the orbit is 4 times than that of earth's radius, then orbital velocity in that orbit is

a) 3.5 km/s b) 7 km/s c) 72 km/s d) 14 km/s

171. The time period T of the moon of planet Mars ($mass\ M_m$) is related to its orbital radius R(G = Gravitational constant)as

a) $T^2 = \frac{4\pi^2 R^3}{GM_m}$ b) $T^2 = \frac{4\pi^2 GR^3}{M_m}$ c) $T^2 = \frac{2\pi R^3 G}{M_m}$ d) $T^2 = 4\pi M_m GR^3$

172. A body is projected vertically upwards from the surface of a planet of radius R with a velocity equal to half the escape velocity for that planet. The maximum height attained by the body is

a) $R/3$ b) $R/2$ c) $R/4$ d) $R/5$

173. If mass of a body is M on the earth surface, then the mass of the same body on the moon surface is

a) $M/6$ b) Zero c) M d) None of these

174. A synchronous satellite goes around the earth once in every 24 h. What is the radius of orbit of the synchronous satellite in terms of the earth's radius (Given mass of the earth, $m_e = 5.98 \times 10^{24} kg$. radius of earth, $r_e = 6.37 \times 10^6 m$, Universal constant of gravitation, $G = 6.67 \times 10^{-11} Nm^2/kg^2$)

a) $2.4r_e$ b) $3.6r_e$ c) $4.8r_e$ d) $6.6r_e$

175. Four particles each of mass M, are located at the vertices of a square with side L. The gravitational potential due to this at the centre of the square is

a) $-\sqrt{32}\frac{GM}{L}$ b) $-\sqrt{64}\frac{GM}{L^2}$ c) Zero d) $\sqrt{32}\frac{GM}{L}$

176. An astronaut on a strange planet finds that acceleration due to gravity is twice as that on the surface of earth. Which of the following could explain this

a) Both the mass and radius of the planet are half as that of earth
b) Radius of the planet is half as that of earth, but the mass is the same as that of earth
c) Both the mass and radius of the planet are twice as that of earth
d) Mass of the planet is half as that of earth, but radius is same as that of earth

177. If g is the acceleration due to gravity on earth's surface, the gain of the potential energy of an object of mass m raised from the surface of the earth to a height equal to the radius R of the earth is

a) $2mgR$ b) mgR c) $\frac{1}{2}mgR$ d) $\frac{1}{4}mgR$

178. An object weighs 72 N on earth. Its weight at a height of $R/2$ from earth is

a) 32 N b) 56 N c) 72 N d) Zero

179. A point mass m is placed inside a spherical shell of radius R and mass M. at a distance $R/2$ from the centre of the shell. The gravitational force exerted by the shell on the point mass is

a) $\frac{GMm}{R^2}$ b) $-\frac{GMm}{R^2}$ c) Zero d) $4\frac{GMm}{R^2}$

180. Two particles of equal mass m go around a circle of radius R under the action of their mutual gravitational attraction. The speed of each particle with respect to their center of mass is

a) $\sqrt{\frac{Gm}{R}}$ b) $\sqrt{\frac{Gm}{4R}}$ c) $\sqrt{\frac{Gm}{3R}}$ d) $\sqrt{\frac{Gm}{2R}}$

181. A straight rod of length L extends from $x = a$ to $x = L + a$. Find the gravitational force it, exerts on a point mass m at $x = 0$ if the linear density of rod $\mu = A + Bx^2$

a) $Gm\left[\frac{A}{a} + BL\right]$ b) $Gm\left[A\left(\frac{1}{a} - \frac{1}{a+L}\right) + BL\right]$

c) $Gm\left[BL + \frac{A}{a+L}\right]$ d) $Gm\left[BL - \frac{A}{a}\right]$

182. The escape velocity of an object from the earth depends upon the mass of the earth (M), its mean density, (ρ), its radius (R) and the gravitational constant (G). Thus the formula for escape velocity is

a) $v = R\sqrt{\frac{8\pi}{3}G\rho}$ b) $v = M\sqrt{\frac{8\pi}{3}GR}$ c) $v = \sqrt{2GMR}$ d) $v = \sqrt{\frac{2GM}{R^2}}$

183. Two spheres of radius r and $2r$ are touching each other. The fore of attraction between them is proportional to

a) r^6 b) r^4 c) r^2 d) r^{-2}

184. If the radius of the earth were to shrink by two percent, its mass remaining the same, the acceleration due to gravity on the earth's surface would

a) Decrease by 2% b) Increase by 2% c) Increase by 4% d) Decrease by 4%

185. A projectile is projected with velocity kv_e in vertically upward direction from the ground into the space. (v_e is escape velocity and $k < 1$). If resistance is considered to be negligible then the maximum height from the centre of earth to which it can go, will be : (R = radius of earth)

a) $\frac{R}{k^2+1}$ b) $\frac{R}{k^2-1}$ c) $\frac{R}{1-k^2}$ d) $\frac{R}{k+1}$

186. Mass of moon is 7.34×10^{22} kg. If the acceleration due to gravity on the moon is $1.4\ \text{ms}^{-2}$, the radius of the moon is($G = 6.667 \times 10^{11}\text{Nm}^2\text{kg}^{-2}$)

a) 0.56×10^4m b) 1.87×10^6m c) 1.92×10^6m d) 1.01×10^8m

187. Choose the correct statement from the following :
Weightlessness of an astronaut moving in a satellite is a situation of

a) Zero g b) No gravity c) Zero mass d) Free fall

188. The distance between centre of the earth and moon is 384000 km. If the mass of the earth is 6×10^{24} kg and $G = 6.66 \times 10^{-11} Nm^2/kg^2$. The speed of the moon is nearly

a) 1 km/sec b) 4 km/sec c) 8 km/sec d) 11.2 km/sec

189. If Gravitational constant is decreasing with time, what will remain unchanged in case of a satellite orbiting around earth

a) Time period b) Orbiting radius c) Tangential velocity d) Angular velocity

190. Two spherical bodies of mass M and $5M$ and radii R and $2R$ respectively are released in free space with initial separation between their centres equal to $12R$. If they attract each other due to gravitational force only, then the distance covered by the smaller body just before collision is

a) 2.5 R b) 4.5 R c) 7.5 R d) 1.5 R

191. A body weighs w newton at the surface of the earth. Its weight at a height equals to half the radius of the earth, will be

a) $\frac{w}{2}$ b) $\frac{2w}{3}$ c) $\frac{4w}{9}$ d) $\frac{8w}{27}$

192. Suppose the law of gravitational attraction suddenly changes and becomes an inverse cube law i.e. $F \propto 1/r^3$, but still remaining a central force. Then

a) Keplers law of areas still holds
b) Keplers law of period still holds
c) Keplers law of areas and period still hold
d) Neither the law of areas, nor the law of period still holds

193. A planet moves around the sun. At a given point P, it is closest from the sun at a distance d_1 and has a speed v_1. At another point Q, when it is farthest from the sun at a distance d_2, its speed will be

a) $\frac{d_1^2 v_1}{d_2^2}$ b) $\frac{d_2 v_1}{d_1}$ c) $\frac{d_1 v_1}{d_2}$ d) $\frac{d_2^2 v_1}{d_1^2}$

194. A body of mass m rises to a height $h = R/5$ from the surface of earth, where R is the radius of earth. If g is the acceleration due to gravity at the surface of earth, the increase in potential energy is

a) $(4/5)mgh$ b) $(5/6)mgh$ c) $(6/7)mgh$ d) mgh

195. Orbital velocity of an artificial does not depend upon

a) Mass of the earth b) Mass of the satellite

c) Radius of the earth
d) Acceleration due to gravity

196. A body has weight 90 kg on the earth's surface, the mass of the moon is 1/9 that of the earth's mass and its radius is 1/2 that of the earth's radius. On the moon the weight of the body is

a) 45 kg
b) 202.5 kg
c) 90 kg
d) 40 kg

197. A research satellite of mass 200 kg circles the earth in an orbit of average radius $3R/2$ where R is the radius of the earth. Assuming the gravitational pull on a mass of 1 kg on the earth's surface to be 10 N, the pull on the satellite will be

a) 880 N
b) 889 N
c) 890 N
d) 892 N

198. A pendulum clock is set to give correct time at the sea level. This clock is moved to hill station at an altitude of 2500 m above the sea level. In order to keep correct time of the hill station, the length of the pendulum

a) Has to be reduced
b) Has to be increased
c) Needs no adjustment
d) Needs no adjustment but its mass has to be increased

199. The mean radius of the earth is R, its angular speed on its own axis is ω and the acceleration due to gravity at earth's surface is g. The cube of the radius of the orbit of a geostationary satellite will be

a) R^2g/ω
b) $R^2\omega^2/g$
c) Rg/ω^2
d) R^2g/ω^2

200. The Earth is assumed to be a sphere of radius R. A platform is arranged at a height R from the surface of the Earth. The escape velocity of a body from this platform is fv, where v is its escape velocity from the surface of the Earth. The value of f is

a) $\frac{1}{3}$
b) $\frac{1}{2}$
c) $\sqrt{2}$
d) $\frac{1}{\sqrt{2}}$

201. A satellite is revolving around the earth with a kinetic energy E. The minimum addition of kinetic energy needed to make it escape from its orbit is

a) $2E$
b) $\sqrt{E}$
c) $E/2$
d) E

202. How high a man be able to jump on surface of a planet of radius 320 km, but having density same as that of the earth if he jumps 5 m on the surface of the earth (Radius of earth = 6400 km)

a) 60 m
b) 80 m
c) 100 m
d) 120 m

203. The change in the value of g at a height h above the surface of the earth is the same as at a depth d below the surface of earth. When both d and h are much smaller than the radius of earth, then which one of the following is correct?

a) $d = \frac{h}{2}$
b) $d = \frac{3h}{2}$
c) $d = 2h$
d) $d = h$

204. A particle is projected vertically upwards from the surface of earth (radius R_e) with a kinetic energy equal to half of the minimum value needed for it to escape. The height to which it rises above the surface of earth is

a) R_e
b) $2R_e$
c) $3R_e$
d) $4R_e$

205. The ratio of the radii of planets A and B is k_1 and ratio of acceleration due to gravity on them is k_2. The ratio of escape velocities from them will be

a) k_1k_2
b) $\sqrt{k_1k_2}$
c) $\sqrt{\frac{k_1}{k_2}}$
d) $\sqrt{\frac{k_2}{k_1}}$

206. $P$is point at a distance r from the centre of a solid sphere of radius r. The variation of gravitational potential at P (ie, V) and distance r from the centre of sphere is represented by the curve.

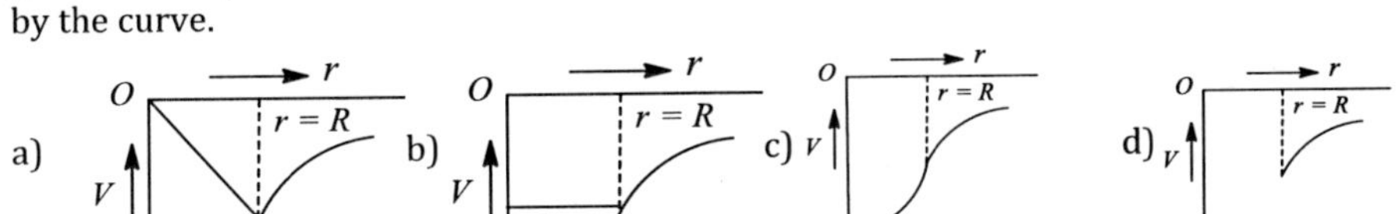

207. If the escape velocity of a planet is 3 times that of the earth and its radius is 4 times that of the earth, then the mass of the planet is (Mass of the earth $= 6 \times 10^{24} kg$)

a) $1.62 \times 10^{22} kg$ b) $0.72 \times 10^{22} kg$ c) $2.16 \times 10^{26} kg$ d) $1.22 \times 10^{22} kg$

208. Two identical satellite A and B are circulating round the earth at the height of R and $2R$ respectively. (where R is radius of the earth). The ratio of kinetic energy of A to that of B is

a) $\frac{1}{2}$ b) $\frac{2}{3}$ c) 2 d) $\frac{3}{2}$

209. What is the binding energy of earth-sun system neglecting the effect of other planets and satellites? (Mass of earth $M_e = 6 \times 10^{24}$kg, mass of the sun $M_x = 2 \times 10^{30}$kg^{-2})

a) 8.8×10^{10} J b) 8.8×10^{3} J c) 5.2×10^{33} J d) 2.6×10^{33} J

210. Which one of the following statements regarding artificial satellite of the earth is incorrect

a) The orbital velocity depends on the mass of the satellite

b) A minimum velocity of 8 km/sec is required by a satellite to orbit quite close to the earth

c) The period of revolution is large if the radius of its orbit is large

d) The height of a geostationary satellite is about 36000 km from earth

211. There are two bodies of masses 100,000 kg and 1000 kg separated by a distance of 1 m. At what distance (in metre) from the smaller body, the intensity of gravitational field will be zero?

a) 1/9 b) 1/10 c) 1/11 d) 10/11

212. The centripetal force acting on a satellite orbiting round the earth and the gravitational force of earth acting on the satellite both equal F. The net force on the satellite is

a) Zero b) F c) $F\sqrt{2}$ d) $2\ F$

213. The angular velocity of rotation of star (of mass M and radius R) at which the matter start to escape from its equator will be

a) $\sqrt{\frac{2GM^2}{R}}$ b) $\sqrt{\frac{2GM}{g}}$ c) $\sqrt{\frac{2GM}{R^3}}$ d) $\sqrt{\frac{2GR}{M}}$

214. Two bodies of masses m and $4m$ are placed at a distance r. The gravitational potential at a point on the line joining them where the gravitational field is zero is

a) $-\frac{4Gm}{r}$ b) $-\frac{6Gm}{r}$ c) $-\frac{9Gm}{r}$ d) zero

215. A body is acted upon by a force towards a point. The magnitude of the force is inversely proportional to the square of the distance. The path of body will be

a) Ellipse b) Hyperbola c) Circle d) Parabola

216. Three equal masses of $1kg$ each are placed at the vertices of an equilateral triangle PQR and a mass of $2kg$ is placed at the centroid O of the triangle which is at a distance of $\sqrt{2}\ m$ from each of the vertices of the triangle. The force, in newton, acting on the, mass of $2kg$ is

a) 2 b) $\sqrt{2}$ c) 1 d) Zero

217. The change in potential energy when a body of mass m is raised to a height nR from the centre of earth (R= radius of earth)

a) $mgR\frac{(n-1)}{n}$ b) $nmgR$ c) $mgR\frac{n^2}{n^2+1}$ d) $mgR\frac{n}{n+1}$

218. If three particles each of mass M are placed at the three corners of an equilateral triangle of side a, the forces exerted by this system on another particle of mass M placed (i) at the mid point of a side and (ii) at the centre of the triangle are respectively

a) 0, 0 b) $\frac{4GM^2}{3a^2}, 0$ c) $0, \frac{4GM^2}{3a^2}$ d) $\frac{3GM^2}{a^2}, \frac{GM^2}{a^2}$

219. How much energy will be necessary for making a body of 500 kg escape from the earth [$g = 9.8\ ms^{-2}$, radius of earth $= 6.4 \times 10^6$m]

a) About 9.8×10^6J b) About 6.4×10^8J c) About 3.1×10^{10}J d) About 27.4×10^{12}J

220. Two satellites S_1 and S_2 revolve around a planet in coplanar circular orbits in the same sense. Their periods of revolution are 1 h and 8 h respectively. The radius of orbit of S_1 is 10^4 km. When S_2 is closest to S_1, the speed of S_2 relative to S_1 is

a) $\pi \times 10^4$ kmh^{-1} b) $2\pi \times 10^4$ kmh^{-1} c) $3\pi \times 10^4$ kmh^{-1} d) $4\pi \times 10^4$ kmh^{-1}

221. The tidal waves in the sea are primarily due to

a) The gravitational effect of the moon on the earth
b) The gravitational effect of the sun on the earth
c) The gravitational effect of venus on the earth
d) The atmospheric effect of the earth itself

222. If mass of earth is M, radius is R and gravitational constant is G, then work done to take 1 kg mass from earth surface to infinity will be

a) $\sqrt{\frac{GM}{2R}}$ b) $\frac{GM}{R}$ c) $\sqrt{\frac{2GM}{R}}$ d) $\frac{GM}{2R}$

223. Which of the following graphs represents correctly the variation of the intensity of gravitational field (I) with the distance (r) from the centre of a spherical shell of mass M and radius a?

a) 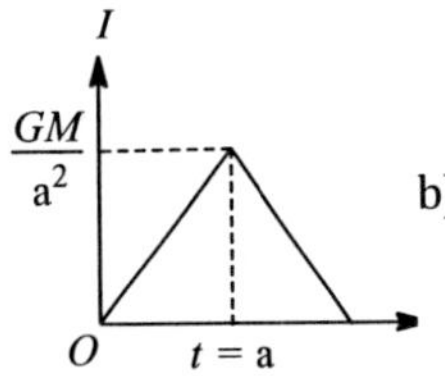

b) 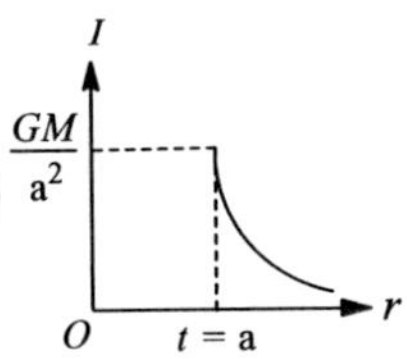

c) 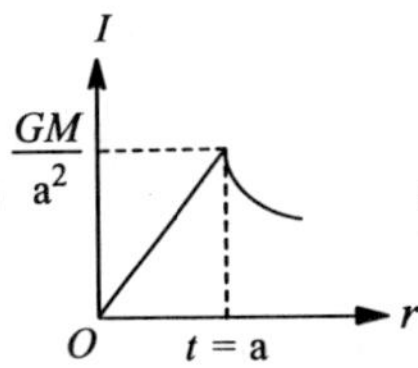

d) 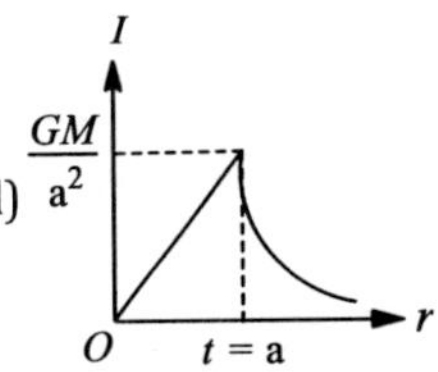

224. The orbital angular momentum of a satellite revolving at a distance r from the centre is L. If the distance is increased to $16r$, then the new angular momentum will be

a) $16\,L$ b) $64\,L$ c) $\frac{L}{4}$ d) $4\,L$

225. A body of mass m is moved to a height h equal to the radius of the earth. The increase in potential energy is

a) $2\,mgR$ b) mgR c) $mgR/2$ d) $mgR/4$

226. The ratio of acceleration due to gravity at a height $3R$ above earth's surface to the acceleration due to gravity on the surface of the earth is ($R=$ radius of earth)

a) $\frac{1}{9}$ b) $\frac{1}{4}$ c) $\frac{1}{16}$ d) $\frac{1}{3}$

227. The mean radius of the earth's orbit round the sun is 1.5×10^{11}. The mean radius of the orbit of mercury round the sun is $6 \times 10^{10} m$. The mercury will rotate around the sun in

a) A year b) Nearly 4 years c) Nearly $\frac{1}{4}$ year d) 2.5 years

228. The mass of the moon is 1/8 of the earth but the gravitational pull is 1/6 of the earth. It is due to the fact that

a) Moon is the satellite of the earth
b) The radius of the earth is 8.6 the moon
c) The radius of the earth is $\sqrt{8/6}$ of the moon
d) The radius of the moon is 6/8 of the earth

229. Where will it be profitable to purchase 1 kilogram sugar

a) At poles b) At equator c) At 45° latitude d) At 40° latitude

230. An artificial satellite is moving in a circular orbit around the earth with a speed equal to half the magnitude of escape velocity from the earth. The height of the satellite above the earth's surface will be

a) 6000 km b) 5800 km c) 7500 km d) 6400 km

231. Three particles each of mass m rotate in a circle of radius r with uniform angular speed ω under their mutual gravitational attraction. If at any instant the points are on the vertex of an equilateral of side L, then angular velocity ω is

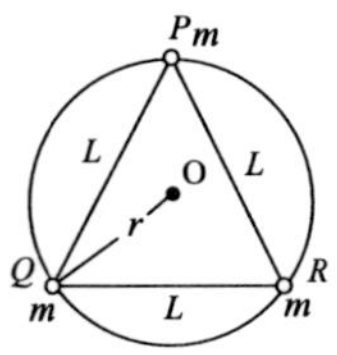

a) $\sqrt{\frac{2Gm}{L^3}}$ b) $\sqrt{\frac{3Gm}{L^3}}$ c) $\sqrt{\frac{5Gm}{L^3}}$ d) $\sqrt{\frac{Gm}{L^3}}$

232. A solid sphere of mass M and radius R has a spherical gravity of radius $\frac{R}{2}$ such that the centre of cavity is at distance $R/2$ from the centre of the sphere. A point mass m is placed inside the cavity at a distance $R/4$ from the centre of sphere. The gravitational pull between the sphere and the point mass m is

a) $\frac{11GMm}{R^2}$ b) $\frac{14GMm}{R^2}$ c) $\frac{GMm}{2R^2}$ d) $\frac{GMm}{R^2}$

233. The changes in potential energy when a body of mass m is raised to a height nR from earth's surface is (R = radius of the earth)

a) $mgR\frac{n}{(n-1)}$ b) mgR c) $mgR\frac{n}{(n+1)}$ d) $mgR\frac{n^2}{(n^2+1)}$

234. A man starts walking from a point on the surface of earth (assumed smooth) and reaches diagonally opposite point. What is the work done by him?

a) Zero b) Positive c) Negative d) Nothing can be said

235. The radius of the earth is R. The height of a point vertically above the earth's surface at which acceleration due to gravity becomes 1% of its value at the surface is

a) 8 R b) 9 R c) 10 R d) 20 R

236. The gravitational potential difference between the surface of a planet and a point 20 m above it is 14 $\mathrm{Jkg^{-1}}$. The work done in moving a 2.0 kg mass by 8.0 m on a slop of 60° from the horizontal is equal to

a) 7 J b) 9.6 J c) 16 J d) 32 J

237. The radius of the earth is about 6400 km and that of the mars is 3200 km. The mass of the earth is about 10 times the mass of the mars. An object weighs 200 N on the surface of earth, its weight on the surface of mars will be

a) 8 N b) 20 N c) 40 N d) 80 N

238. At what distance from the centre of the earth, the value of acceleration due to gravity g will be half that on the surface (R = radius of earth)

a) 2 R b) R c) 1.414 R d) 0.414 R

239. Which of the following astronomer first proposed that sun is static and earth rounds sun

a) Copernicus b) Kepler c) Galileo d) None

240. The gravitational field in a region is given by $\vec{\mathbf{I}} = (4\hat{\mathbf{i}} + \hat{\mathbf{j}})\mathrm{Nkg^{-1}}$. Work done by this field is zero when a particle is moved along the line

a) $x + y = 6$ b) $x + 4y = 6$ c) $y + 4x = 6$ d) $x - y = 6$

241. A sphere of mass M and radius R_2 has a concentric cavity of radius R_1 as shown in figure. The force F exerted by the sphere on a particle of mass m located at a distance r from the centre of sphere varies as ($0 \le r \le \infty$)

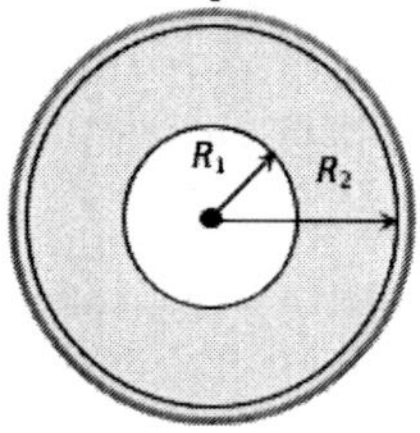

a) 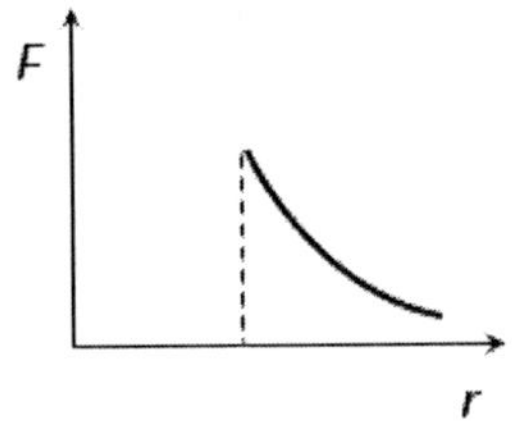

b) 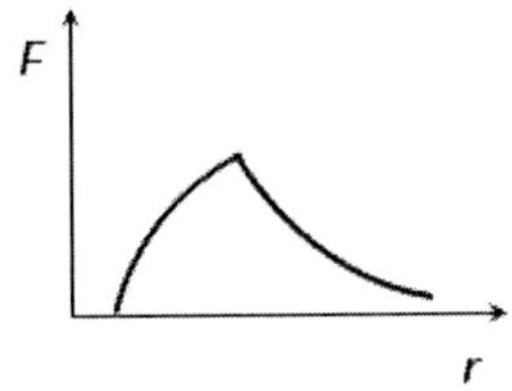

c) 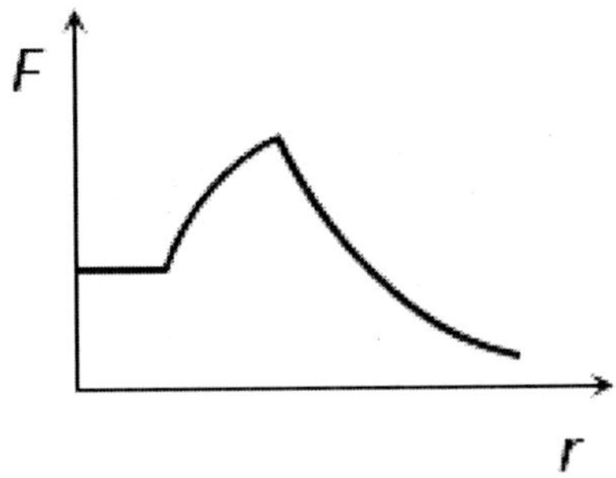

d) 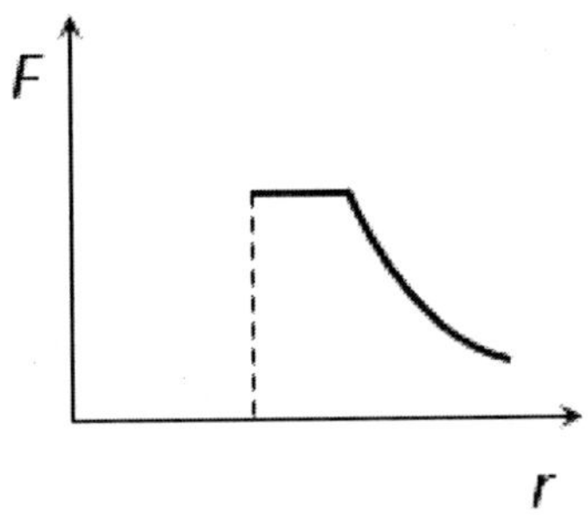

242. $R$is the radius of the earth and ω is its angular velocity and g_p is the value of g at the poles. The effective value of g at the latitude $\lambda = 60°$ will be equal to

a) $g_p - \frac{1}{4}R\omega^2$ b) $g_p - \frac{3}{4}R\omega^2$ c) $g_p - R\omega^2$ d) $g_p + \frac{1}{4}R\omega^2$

243. The escape velocity of a particle of mass m varias as

a) m^2 b) m c) m^0 d) m^{-1}

244. The value of g on the surface of earth is smallest at the equator because

a) The centripetal force is maximum at equator
b) The centripetal force is least at equator
c) The angular speed of earth is maximum at equator
d) The angular speed of earth is least at equator

245. The gravitational potential energy of a body of mass m at a distance r from the centre of the earth is U. What is the weight of the body at this distance?

a) U b) Ur c) $\frac{U}{r}$ d) $\frac{U}{2r}$

246. If g is the acceleration due to gravity at the earth's surface and r is the radius of the earth, the escape velocity for the body to escape out of earth's gravitational field is

a) gr b) $\sqrt{2\,gr}$ c) g/r d) r/g

247. At what height in km over the earth's pole the free fall acceleration decreases by one percent? (Assume the radius of the earth to be 6400 km)

a) 32 b) 64 c) 80 d) 1.253

248. According to Kepler's law T^2 is proportional to

a) R^3 b) R^2 c) R d) R^{-1}

249. The height at which the weight of a body becomes $1/16^{\text{th}}$, its weight on the surface of earth (radius R), is

a) $5R$ b) $15R$ c) $3R$ d) $4R$

250. Let g be the acceleration due to gravity at earth's surface and K be the rotational kinetic energy of the earth. Suppose the earth's radius decreases by 2% keeping all other quantities same, then

a) g decreases by 2% and K decreases by 4% b) g decreases by 4% and K increases by 2%
c) g increases by 4% and K increases by 4% d) g decreases by 4% and K increases by 4%

251. A geostationary satellite is revolving around the earth. To make it escape from gravitational field of earth, its velocity must be increased

a) 100% b) 41.4% c) 50% d) 59.6%

252. The correct graph representing the variation of total energy (E) kinetic energy (K) and potential energy (U) of a satellite with its distance from the centre of earth is

a) 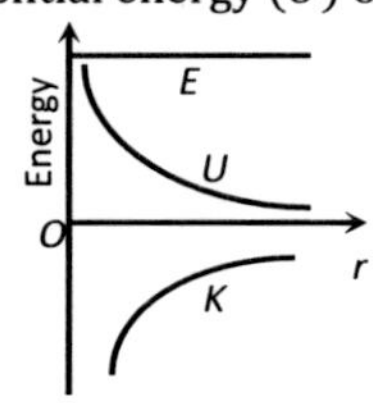

b) 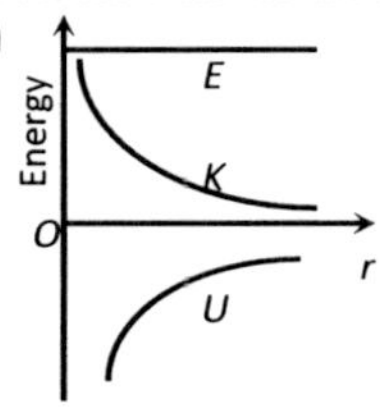

c) 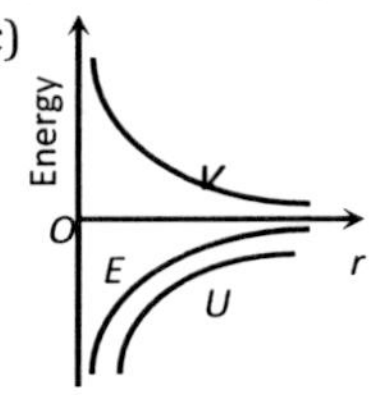

d) 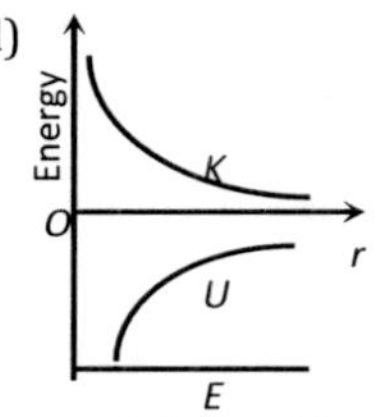

253. Two identical satellites are at R and $7R$ away from earth surface, the wrong statement is (R =Radius of earth)

a) Ratio of total energy will be y
b) Ratio of kinetic energies will be y
c) Ratio of potential energies will be y
d) Ratio of total energy will be y but ratio of potential and kinetic energy will be z

254. If the earth suddenly shrinks (without changing mass) to half of its present radius, the acceleration due to gravity will be

a) $g/2$ b) $4g$ c) $g/4$ d) $2g$

255. The gravitational force F_g between two objects does not depends on

a) Sum of the masses b) Product of the masses
c) Gravitational constant d) Distance between the masses

256. A point $P(R\sqrt{3},0,0)$ lies on the axis of a ring of mass M and radius R. The ring is located in $y-z$ plane with its centre at origin O. A small particle of mass m starts from P and reaches O under gravitational attraction only. Its speed at O will be

a) $\sqrt{\frac{GM}{R}}$ b) $\sqrt{\frac{Gm}{R}}$ c) $\sqrt{\frac{GM}{2R}}$ d) $\sqrt{\frac{Gm}{2R}}$

257. In the previous question, the angular speed of S_2 as actually observed by an astronaut is S_1

a) $\frac{\pi}{2}\,\text{radh}^{-1}$ b) $\pi\,\text{radh}^{-1}$ c) $\frac{2\pi}{3}\,\text{radh}^{-1}$ d) $\frac{\pi}{3}\,\text{radh}^{-1}$

258. The mass of the moon is 7.34×10^{22} kg and the radius is 1.74×10^{6}m. the value of gravitational field intensity will be

a) $1.45\ \text{Nkg}^{-1}$ b) $1.55\ \text{Nkg}^{-1}$ c) $1.7\ \text{Nkg}^{-1}$ d) $1.62\ \text{Nkg}^{-1}$

259. A body of mass $m\ kg$. starts falling from a point $2R$ above the Earth's surface. Its kinetic energy when it has fallen to a point $'R'$ above the Earth's surface [R-Radius of Earth, M-Mass of Earth, G-Gravitational Constant]

a) $\frac{1}{2}\frac{GMm}{R}$ b) $\frac{1}{6}\frac{GMm}{R}$ c) $\frac{2}{3}\frac{GMm}{R}$ d) $\frac{1}{3}\frac{GMm}{R}$

260. If acceleration due to gravity on the surface of a planet is two times that on surface of earth and its radius is double that of earth. Then escape velocity from the surface of that planet in comparison to earth will be

a) $2\ v_e$ b) $3\ v_e$ c) $4\ v_e$ d) None of these

261. Planetary system in the solar system describes

a) Conservation of energy b) Conservation of linear momentum
c) Conservation of angular momentum d) None of these

262. The ratio of the radius of the earth to that of the moon is 10. The ratio of acceleration due to gravity on the earth and on the moon is 6. The ratio of the escape velocity from the earth's surface to that from the moon is

a) 10 b) 6 c) Nearly 8 d) 1.66

263. The depth d at which the value of acceleration due to gravity becomes $1/n$ times the value of the surface, is [R = radius of the earth]

a) $\frac{R}{n}$ b) $R\left(\frac{n-1}{n}\right)$ c) $\frac{R}{n^2}$ d) $R\left(\frac{n}{n+1}\right)$

264. If v_e and v_o represent escape velocity and orbital velocity of a satellite corresponding to a circular orbit of radius R, then

a) $v_e = v_o$ b) $\sqrt{2}v_o = v_e$
c) $v_e = v_o/\sqrt{2}$ d) v_e and v_o are not related

265. If the distance between two masses is doubled, the gravitational attraction between them

a) Is doubled b) Becomes four times
c) Is reduced to half d) Is reduced to a quarter

266. If W_1, W_2 and W_3 represent the work done in moving a particle from A to B along three different paths 1,2 and 3 respectively (as shown) in a gravitational field of point mass m, then

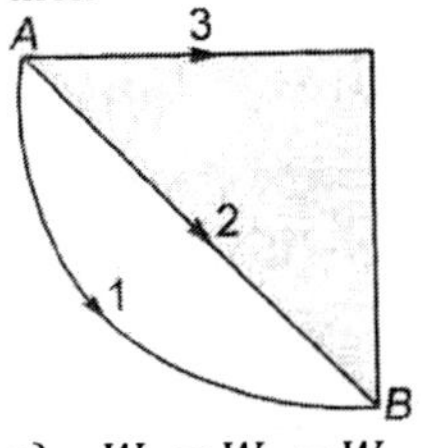

a) $W_1 = W_2 = W_3$ b) $W_1 > W_2 > W_3$ c) $W_1 > W_2 < W_3$ d) $W_1 < W_3 < W_2$

267. A satellite whose mass is M, is revolving in circular orbit of radius r around the earth. Time of revolution of satellite is

a) $T \propto \frac{r^5}{GM}$ b) $T \propto \sqrt{\frac{r^3}{GM}}$ c) $T \propto \sqrt{\frac{r}{GM^2/3}}$ d) $T \propto \sqrt{\frac{r^3}{GM^1/4}}$

268. Escape velocity on a planet is v_e. If radius of the planet remains same and mass becomes 4 times, the escape velocity becomes

a) $4v_e$ b) $2v_e$ c) v_e d) $\frac{1}{2}v_e$

269. If g is the acceleration due to gravity on the surface of the earth, the gain in potential energy of an object of mass m raised from the earth's surface to a height equal to the radius R of the earth is

a) $\frac{mgR}{4}$ b) $\frac{mgR}{2}$ c) mgR d) $2mgR$

270. A geostationary satellite is orbiting the earth at the height of 6 R above the surface of earth, R being radius of earth. The time period of another satellite at a height of 2.5 R from the surface of earth, is

a) 10 h
b) $\frac{6}{\sqrt{2}}$ h
c) 6 h
d) $6\sqrt{2}$ h

271. A satellite moves round the earth in a circular orbit of radius R making 1 rev/day. A second satellite moving in a circular orbit, moves round the earth ones in 8 days. The radius of the orbit of the second satellite is

a) 8 R
b) 4 R
c) 2 R
d) R

272. A satellite in launched in a circular orbit of radius R around the earth. A second satellite is launch in to an orbit of radius 1.01 R. The period of second satellite is longer than the first one (approximately) by

a) 1.5 %
b) 0.5%
c) 3%
d) 1%

273. Gravitational acceleration on the surface of a planet is $\frac{\sqrt{6}}{11}g$, where g is the gravitational acceleration on the surface of earth. The average mass density of the planet is $\frac{2}{3}$ times that of the earth. If the escape speed on the surface of the earth is taken on be 11 kms^{-1}, the escape speed on the surface of the planet in kms^{-1} will be

a) 5
b) 7
c) 3
d) 11

274. A solid sphere is of density ρ and radius R. The gravitational field at a distance r from the centre of the sphere, where $r < R$, is

a) $\frac{\rho\pi G R^3}{r}$
b) $\frac{4\pi G\rho r^2}{3}$
c) $\frac{4\pi G\rho R^3}{3r^2}$
d) $\frac{4\pi G\rho r}{3}$

275. A body revolved around the sun 27 times faster than the earth. What is the ratio of their radii

a) 1/3
b) 1/9
c) 1/27
d) 1/4

276. Satellite A and B are revolving around the orbit of earth. The mass of A is 10 times of mass of B. The ratio of time period $\left(\frac{T_A}{T_B}\right)$ is

a) 10
b) 1
c) $\frac{1}{5}$
d) $\frac{1}{10}$

277. A mass M is split into two parts m and $(M - m)$, which are then separated by a certain distance. The ratio m/M which maximizes the gravitational force between the parts is

a) 1 : 4
b) 1 : 2
c) 4 : 1
d) 2 : 1

278. The mass and radius of the sun are $1.99 \times 10^{30} kg$ and $R = 6.96 \times 10^8$m. The escape velocit of a rocket from the Sun is

a) 11.2 km/s
b) 2.38 km/s
c) 59/5 km/s
d) 618 km/s

279. The height at which the acceleration due to gravity becomes $\frac{g}{9}$ (where g = the acceleration due to gravity on the surface of the earth) in terms of R, the radius of the earth, is

a) $2R$
b) $\frac{R}{\sqrt{3}}$
c) $\frac{R}{2}$
d) $\sqrt{2}R$

280. The escape velocity for the earth is v_e. The escape velocity for a planet whose radius is four times and density is nine times that of the earth, is

a) $36v_e$
b) $12v_e$
c) $6v_e$
d) $20v_e$

281. The ratio of the radius of a planet 'A' to that of planet 'B' is 'r'. The ratio of acceleration due to gravity on the planets is 'x'. The ratio of the escape velocities from the two planets is

a) xr
b) $\sqrt{\frac{r}{x}}$
c) $\sqrt{rx}$
d) $\sqrt{\frac{x}{r}}$

282. A satellite of the earth is revolving in a circular orbit with a uniform speed v. If the gravitational force suddenly disappears, the satellite will

a) Continue to move with velocity v along the original orbit
b) Move with a velocity v, tangentially to the original orbit

c) Fall down with increasing velocity
d) Ultimately come to rest somewhere on the original orbit

283. If the earth stops rotating, the value of g at the equator

a) increases b) decreases c) no effect d) None of these

284. If a planet of given density were made larger (keeping its density unchanged) its force of attraction for an object on its surface would increase because of increased mass of the planet but would decrease because of larger separation between the centre of the planet and its surface. Which effect would dominate?

a) Increase in mass b) Increase in radius
c) Both affect the attraction equally d) None of the above

285. If different planets have the same density but different radii, then the acceleration due to gravity on the surface of the planet is related to the radius (R) of the planet as

a) $g \propto R^2$ b) $g \propto R$ c) $g \propto \frac{1}{R^2}$ d) $g \propto \frac{1}{R}$

286. A mass of 6×10^{24}kg is to be compressed in a sphere in such a way that the escape velocity from the sphere is 3×10^8m/s. What should be the radius of the sphere? ($G = 6.67 \times 10^{-11}$N– m^2/kg^2)

a) 9 km b) 9 m c) 9 mc d) 9 mm

287. The gravitational force between two point masses m_1 and m_2 at separation r is given by

$F = k\frac{m_1 m_2}{r^2}$

The constant k

a) Depends on system of units only b) Depends on medium between masses only
c) Depends on both (a) and (b) d) Is independent of both (a) and (b)

288. For a satellite moving in an orbit around the earth, the ratio of kinetic energy to potential energy is

a) 2 b) $\frac{1}{2}$ c) $\frac{1}{\sqrt{2}}$ d) $\sqrt{2}$

289. A satellite A of mass m is at a distance of r from the centre of the earth. Another satellite B of mass $2m$ is at a distance of $2r$ from the earth's centre. Their time periods are in the ratio of

a) 1 : 2 b) 1 : 16 c) 1 : 32 d) $1 : 2\sqrt{2}$

290. The largest and the shortest distance of the earth from the sun are r_1 and r_2, its distance from the sun when it is at the perpendicular to the major axis of the orbit drawn from the sun

a) $\frac{r_1 + r_2}{4}$ b) $\frac{r_1 r_2}{r_1 + r_2}$ c) $\frac{2r_1 r_2}{r_1 + r_2}$ d) $\frac{r_1 + r_2}{3}$

291. The total energy of satellite moving with an orbital velocity v around the earth is

a) $\frac{1}{2}mv^2$ b) $-\frac{1}{2}mv^2$ c) mv^2 d) $\frac{3}{2}mv^2$

292. The potential energy of gravitational interaction of a point mass m and a thin uniform rod of mass M and length l, if they are located along a straight line at distance$a$ from each other is

a) $U = \frac{GMm}{a}\log_e\left(\frac{a+l}{a}\right)$ b) $U = GMm\left(\frac{1}{a} - \frac{1}{a+l}\right)$
c) $U = -\frac{GMm}{l}\log_e\left(\frac{a+l}{a}\right)$ d) $U = -\frac{GMm}{a}$

293. Three identical bodies of mass M are located at the vertices of an equilateral triangle of side L. They revolve under the effect of mutual gravitational force in a circular orbit, circumscribing the triangle while preserving the equilateral triangle. Their orbital velocity is

a) $\sqrt{\frac{GM}{L}}$ b) $\sqrt{\frac{3GM}{2L}}$ c) $\sqrt{\frac{3GM}{L}}$ d) $\sqrt{\frac{2GM}{3L}}$

294. If the diameter of mars is 6760 km and mass one-tenth that of earth. The diameter of earth is 12742 km. If acceleration due to gravity on earth is 9.8 ms^{-2}, the acceleration due to gravity on mass is

a) 34.8 ms^{-2} b) 2.84 ms^{-2} c) 3.48 ms^{-2} d) 28.4 ms^{-2}

295. A clock S is based on oscillation of a spring and clock P is based on pendulum motion. Both clock run at the same rate on earth. On a planet having the same density as earth but twice the radius,

a) S will run faster than P
b) P will run faster than S
c) Both will run at the same rate as on the earth
d) Both will run at the same rate which will be different from that on the earth

296. The correct option is

a) The time taken in travelling DAB is less than that for BCD
b) The time taken in travelling DAB is greater than that for BCD
c) The time taken in travelling $CDAD$ is less than that for ABC
d) The time taken in travelling CDA is greater than that for ABC

297. In the above Question find apparent weight of the object?

a) 3 N b) Zero c) 2 N d) 0.2 N

298. The earth revolves round the sun in one year. If the distance between them becomes double, the new period of revolution will be

a) $1/2$ *year* b) $2\sqrt{2}$ *years* c) 4 *years* d) 8 *years*

299. At what height h above earth, the value of g becomes $g/2$? (R = radius of earth)

a) $3R$ b) $\sqrt{2}\,R$ c) $(\sqrt{2}-1)R$ d) $\frac{1}{\sqrt{2}}R$

300. When of the following graphs correctly represents the variation of g on earth?

a) 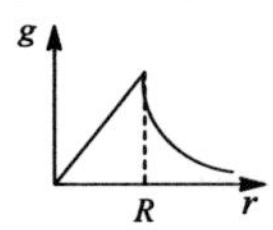

b) 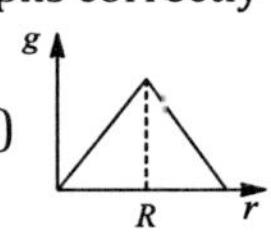

c) 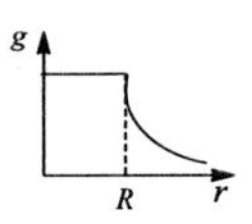

d) 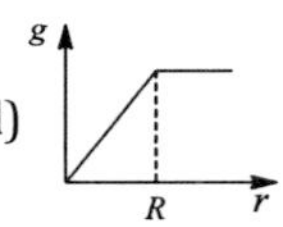

301. Rockets are launched in Eastward direction to take advantage of

a) The clear sky on Eastesn side b) The thinner atmosphere on this side
c) Earth's rotation d) Earth's tilt

302. If the diameter of mass is 6760 km and mass one-tenth that of earth. The diameter of earth is 12742 km. If acceleration due to gravity on earth is 9.8 ms^{-2}, the acceleration due to gravity on mass is

a) 34.8 ms^{-2} b) 2.48 ms^{-2} c) 3.48 ms^{-2} d) 28.4 ms^{-2}

303. An artificial satellite is moving in a circular orbit around the Earth. The height of the satellite above the surface of Earth is R. Suppose the satellite is stopped suddenly in its orbit and allowed to fall freely. On reaching Earth, its speed will be

a) $\sqrt{gR}$ b) $2\sqrt{gR}$ c) $3\sqrt{gR}$ d) $5\sqrt{gR}$

304. A planet moving along an elliptical orbit is closest to the sun at a distance r_1 and farthest away at a distance of r_2. If v_1 and v_2 are the linear velocities at these points respectively. Then the ratio $\frac{v_1}{v_2}$ is

a) $\frac{r_1}{r_2}$ b) $\left(\frac{r_1}{r_2}\right)^2$ c) $\frac{r_2}{r_1}$ d) $\left(\frac{r_2}{r_1}\right)^2$

305. A spherical hollow is made in a lead sphere of radius R such that its surface touches the outside surface of the lead sphere and passes through the centre. The mass of the lead sphere before hollowing was M. The force of attraction that this sphere would exert on a particle of mass m which lies at a distance $d(> R)$ from the centre of the lead sphere on the straight line joining the centres of the sphere and the hollow is

a) $\frac{GM\,m}{d^2}$ b) $\frac{GM\,m}{8d^2}$

c) $\frac{GM\,m}{d^2}\left[1+\frac{1}{8\left(1+\frac{R}{2d}\right)}\right]$ d) $\frac{GM\,m}{d^2}\left[1-\frac{1}{8\left(1-\frac{R}{2d}\right)^2}\right]$

306. A planet has twice the radius but the mean density is $\frac{1}{4}$ th as comparsed to earth. What is the ratio of escape velocity from earth to that from the planet?
a) 3:1 b) 1:2 c) 1:1 d) 2:1

307. A clock S is based on oscillation of a spring and a clock P is based on pendulum motion. Both clocks run at the same rate on earth. On a planet having the same density as earth but twice the radius
a) S will run faster than P
b) P will run faster than S
c) They will both run at the same rate as on the earth
d) None of these

308. Select the correct statement from the following
a) The orbital velocity of a satellite increases with the radius of the orbit
b) Escape velocity of a particle from the surface of the earth depends on the speed with which it is fired
c) The time period of satellite does not depend on the radius of the orbit
d) The orbital velocity is inversely proportional to the square root of the radius of the orbit

309. A body of mass 500 g is thrown upward with a velocity 20ms^{-1} and reaches back to the surface of a planet after 20 s. Then the weight of the body on that planet is
a) 2 N b) 4 N c) 5 N d) 1 N

310. A planet of mass m moves around the sun of mass M in an elliptical orbit. The maximum and minimum distance of the planet from the sun are $r^1$and r^2, respectively. The time period of the planet is proportional to
a) (r_1+r_2) b) $(r_1+r_2)^{1/2}$ c) $(r_1-r_2)^{3/2}$ d) $(r_1+r_2)^{3/2}$

311. Halley's comet has a period of 76, had distance of closest approach to the sun equal to 8.9×10^{10}m. the comet's farthest distance from the sun if the mass of sun is 2×10^{30} kg and $G=6.67\times 10^{11}$ in MKS units is
a) 2×10^{12}m b) 2.7×10^{13}m c) 5.3×10^{12}m d) 5.3×10^{13}m

312. If the earth were to spin faster, acceleration due to gravity at the poles
a) increase
b) decreases
c) remain the same
d) depends on how fast it spins

313. Which force in nature exits every where
a) Nuclear force
b) Electromagnetic force
c) Weak force
d) Gravitation

314. Kepler's second law states that the straight line joining the planet to the sun sweeps out equal times. This statement is equivalent to saying that
a) Total acceleration is zero
b) Tangential acceleration is zero
c) Longitudinal acceleration is zero
d) Radial acceleration is zero

315. Which of the following statement about the gravitational constant is true?
a) It is a force
b) It has no unit
c) It has same value in all system of units
d) It does not depend on the nature of the medium in which the bodies are kept

316. A satellite is orbiting around the earth with orbital radius R and time period T. The quantity which remain constant is
a) T/R b) T^2/R c) T^2/R^2 d) T^2/R^3

317. Assuming earth to be a sphere of a uniform density, what is the value of gravitational acceleration in a min 100 km below the earth' surface (Given $R=6400\ km$)
a) 9.66 m/s^2 b) 7.64 m/s^2 c) 5.06 m/s^2 d) 3.10 m/s^2

318. The atmosphere is held to the earth by

a) Winds　b) Gravity　c) Clouds　d) None of the above

319. The diagram showing the variation of gravitational potential of earth with distance from the centre of earth is

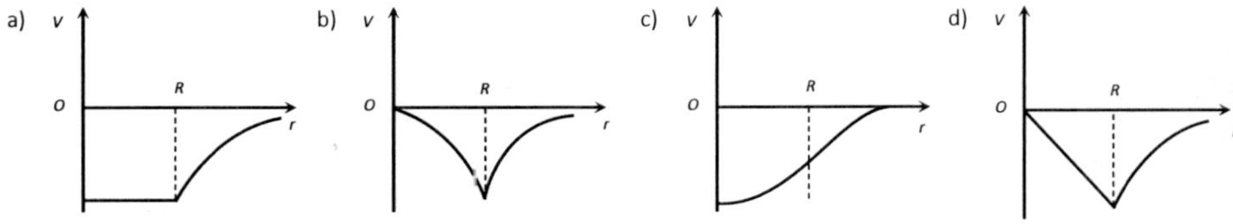

320. The escape velocity of an object on a planet whose g value is 9 times on earth and whose radius is 4 times that of earth in km/s is

a) 67.2　b) 33.6　c) 16.8　d) 25.2

321. Infinite number of masses, each 1 kg, are placed along the x-axis at $x = \pm 1\text{m}, \pm 2\text{m}, \pm 4\text{m}, \pm 8\text{m}, \pm 16\text{m}$ ..... The magnitude of the resultant gravitational potential in terms of gravitational constant G at the origin $(x = 0)$ is

a) $G/2$　b) G　c) $2G$　d) $4G$

322. There is a mine of depth about 2.0 km. In this mine the conditions as compared to those at the surface are

a) Lower air pressure, higher acceleration due to gravity

b) Higher air pressure, lower acceleration due to gravity

c) Higher air pressure, higher acceleration due to gravity

d) Lower air pressure, lower acceleration due to gravity

323. A satellite of mass m is orbiting close to the surface of the earth (Radius R = 6400 km) has a kinetic energy k. The corresponding kinetic energy of the satellite to escape from the earth's gravitational field is

a) K　b) $2K$　c) mgR　d) mK

324. What is the intensity of gravitational field at the centre of a spherical shell

a) Gm/r^2　b) g　c) Zero　d) None of these

325. Two particles each of mass m are moving around a circle of radius R due to their mutual gravitational force of attraction, velocity of each particle is

a) $v = \sqrt{\frac{Gm}{2R}}$　b) $v = \sqrt{\frac{Gm}{R}}$　c) $v = \sqrt{\frac{Gm}{4R}}$　d) None of these

326. Average density of the earth

a) does not depend on g　b) is a complex function of g

c) is directly proportional to g　d) is inversely proportional g

327. Venus looks brighter than other planets because

a) It is heavier than other planets　b) It has higher density than other planets

c) It is closer to the earth than other planets　d) It has no atmosphere

328. A spherical planet has a mass M_P and diameter D_P. A particle of mass m falling freely near the surface of this planet will experience an acceleration due to gravity, equal to

a) $4GM_P/D_P^2$　b) GM_Pm/D_P^2　c) GM_P/D_P^2　d) $4GM_Pm/D_P^2$

329. If the radius of earth decreases by 1% and its mass remains same, then the acceleration due to gravity

a) increases by 1%　b) decreases by 1%　c) increases by 2%　d) decreases by 2%

330. Suppose the gravitational force varies inversely as the nth power of distance. Then the time period of a planet in circular orbit of radius R around the sun will be proportional to

a) $R^{\left(\frac{n+1}{2}\right)}$　b) $R^{\left(\frac{n-1}{2}\right)}$　c) R^n　d) $R^{\left(\frac{n-2}{2}\right)}$

331. The additional kinetic energy to be provided to a satellite of mass m revolving around a planet of mass M, to transfer it from a circular orbit of radius R_1 to another of radius $R_2 (R_2 > R_1)$ is

a) $GmM\left(\frac{1}{R_1^2}-\frac{1}{R_2^2}\right)$ b) $GmM\left(\frac{1}{R_1}-\frac{1}{R_2}\right)$ c) $2GmM\left(\frac{1}{R_1}-\frac{1}{R_2}\right)$ d) $\frac{1}{2}GmM\left(\frac{1}{R_1}-\frac{1}{R_2}\right)$

332. If the radius of earth's orbit is made 1/4th, then duration of an year will become

a) 8 times b) 4 times c) 1/8 times d) 1/4 times

333. A particle of mass M is situated at the centre of a spherical shell of same mass and radius a. The magnitude of the gravitational potential at a point situated at $a/2$ distance from the centre, will be

a) $\frac{4GM}{a}$ b) $\frac{GM}{a}$ c) $\frac{2GM}{a}$ d) $\frac{3GM}{a}$

334. Read the following statements

S_1 : An object shall weigh more at pole than at equator when weighed by using a physical balance

S_2 : It shall weigh the same at pole and equator when weighed by using a physical balance

S_3 : It shall weigh the same at pole and equator when weighed by using a spring balance

S_4 : It shall weigh more at the pole than at equator when weighed using a spring balance

Which of the above statements is/are correct

a) S_1 and S_2 b) S_1 and S_4 c) S_2 and S_3 d) S_3 and S_4

335. The distance between the earth and the moon is 3.85×10^8m. At what distance from the earth's centre, the intensity of gravitational field will be zero? The masses of earth and moon are 5.98×10^{24} kg and 7.35×10^{22} kg respectively

a) 3.47×10^8m b) 0.39×10^8m c) 1.82×10^8m d) None of these

336. A satellite of mass m revolves around the earth of radius R at a height x from its surface. If g is the acceleration due to gravity on the surface of the earth, the orbital speed the satellite is

a) gx b) $\frac{gR}{R-x}$ c) $\frac{gR^2}{R+x}$ d) $\left(\frac{gR^2}{R+x}\right)^{1/2}$

337. The gravitational field due to a mass distribution is $1 = \frac{C}{x^2}$ in x direction. Hence C is constant. Taking the gravitational potential to be zero at infinity, potential at x is

a) $\frac{2C}{x}$ b) $\frac{C}{x}$ c) $\frac{2C}{x^2}$ d) $\frac{C}{2x^2}$

338. Reason of weightlessness in a satellite is

a) Zero gravity b) Centre of mass
c) Zero reaction force by satellite surface d) None

339. The relay satellite transmits the T.V. programme continuously from one part of the world to another because its

a) Period is greater than the period of rotation of the earth
b) Period is less than the period of rotation of the earth about its axis
c) Period has no relation with the period of the earth about its axis
d) Period is equal to the period of rotation of the earth about its axis

340. A comet of mass m moves in a highly elliptical orbit around the sun of mass M. The maximum and minimum distances of the comet from the centre of the sun are r_1 and r_2 respectively. The magnitude of angular momentum of the comet with respect to the centre of sun is

a) $\left[\frac{GMr_1}{(r_1+r_2)}\right]^{1/2}$ b) $\left[\frac{GMmr_1}{(r_1+r_2)}\right]^{1/2}$ c) $\left(\frac{2Gm^2r_1r_2}{r_1+r_2}\right)^{1/2}$ d) $\left(\frac{2GMm^2r_1r_2}{r_1+r_2}\right)^{1/2}$

341. Sun is about 330 times heavier and 100 times bigger in radius than earth. The ratio of mean density of the sun to that of earth is

a) 3.3×10^{-6} b) 3.3×10^{-4} c) 3.3×10^{-2} d) 1.3

342. Astronaut is in a stable orbit around the earth when he weighs a body of mass 5 kg. What is reading of spring balance?
a) Spring will not be extended
b) Spring will be extended according to Hook's law
c) Less than 5 kg-wt
d) More than 5 kg-wt

343. If the change in the value of $'g'$ at a height h above the surface of the earth is the same as at a depth x below it, then (both x and h being much smaller than the radius of the earth)
a) $x = h$ b) $x = 2h$ c) $x = \frac{h}{2}$ d) $x = h^2$

344. A satellite is launched into a circular orbit of radius R around the earth. A second satellite is launched into an orbit of radius $4R$. The ratio of their respective periods is
a) 4:1 b) 1:8 c) 8:1 d) 1:4

345. When earth moves around the sun, the quantity which remains constant is
a) Angular velocity b) Kinetic energy c) Potential energy d) Areal velocity

346. If distance between earth and sun become four times, then time period becomes
a) 4 times b) 8 times c) 1/4 times d) 1/8 times

347. The escape velocity for a body projected vertically upwards from the surface of the earth is 11.2 kms^{-1}. If the body is projected in a direction making an angle of 45° with the vertical, the escape velocity will be
a) 11.2 kms^{-1} b) $11.2 \times \sqrt{2}$ kms^{-1} c) 11.2×2 kms^{-1} d) $11.2/\sqrt{2}$ kms^{-1}

348. The period of revolution of planet A around the sun is 8 times that B. The distance of a from the sun is how many times greater than that of B from the sun?
a) 2 b) 3 c) 4 d) 5

349. What does not change in the field of central force
a) Potential energy b) Kinetic energy c) Linear momentum d) Angular momentum

350. If a planet of given density were made larger its force of attraction for an object on its surface would increase because of planet's greater mass but would decease because of the greater distance from the object to the centre of the planet. Which effect predominate?
a) Increases in mass b) Increase in radius
c) Both affect attraction equally d) None of the above

351. The potential energy of 4-particalse each of mass 1 kg placed at the four vertices of a square of side length 1 m is
a) +4.0 G b) −7.5 G c) −5.4 G d) +6.3 G

352. In the above problem, the ratio of the time duration of his jump on the moon to that of his jump on the earth is
a) 1 : 6 b) 6 : 1 c) $\sqrt{6} : 1$ d) $1 : \sqrt{6}$

353. A mass m is placed at a point B in the gravitational field of mass M. When the mass m is brought from B to near point A, its gravitational potential energy will
a) Remain unchanged b) Increase c) Decrease d) Become zero

354. An artificial of mass $'m'$ revolves around the earth near to its surface then its binding energy is [R_e, g are radius and acceleration due to gravity respectively of the earth]
a) $\frac{1}{2}mg\,R_e$ b) $-\frac{1}{2}mg\,R_e$ c) $mg\,R_e$ d) $-mg\,R_e$

355. Energy required to move a body of mass m from an orbit of radius 2 $R$to $3R$ is
a) $GMm/12R^2$ b) $GMm/3R^2$ c) $GMm/8R$ d) $GMm/6R$

356. The distance of a planet from the sun is 5 times, the distance between the earth and the sun. the time period of the planet is
a) $6^{3/2}T$ yr b) $5^{3/2}T$ yr c) $5^{3/1}T$ yr d) $5^{1/2}T$ yr

357. The orbital velocity of the planet will be maximum at

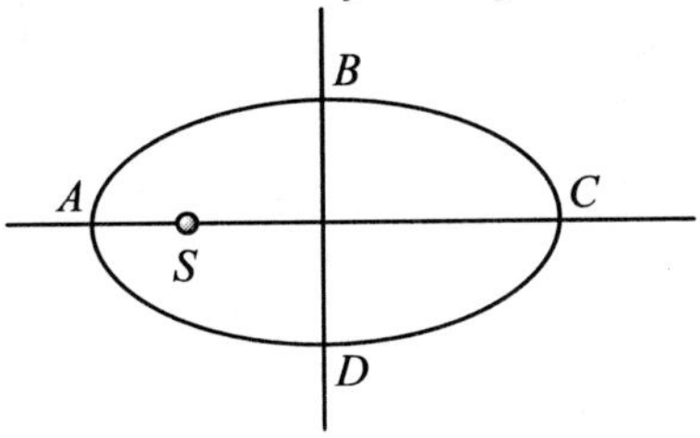

a) A b) B c) C d) D

358. Acceleration due to gravity g for a body of mass m on earth's surface is proportional to (Radius of earth = R, mass of earth = M)

a) M/R^2 b) m^0 c) mM d) $1/R^{3/2}$

359. The mass of the moon is $\frac{1}{81}$ of the earth but the gravitational pull is $\frac{1}{6}$ of the earth. It is due to the fact that

a) The radius of the moon is $\frac{81}{6}$ of the earth b) The radius of the earth is $\frac{9}{\sqrt{6}}$ of the moon

c) Moon is the satellite of the earth d) None of the above

360. Two planets have radii r_1 and r_2 and densities d_1 and d_2 respectively. Then the ratio of acceleration due to gravity on them will be

a) $r_1d_1 : r_2d_2$ b) $r_1d_2 : r_2d_1$ c) $r_1^2d_1 : r_2^2d_2$ d) $r_1 : r_2$

361. The escape velocity from earth is v_{es}. A body is projected with velocity $2v_{es}$ with what constant velocity will it move in the inter planetary space

a) v_{es} b) $3v_{es}$ c) $\sqrt{3}v_{es}$ d) $\sqrt{5}v_{es}$

362. The moon's radius is 1/4 that of the earth and its mass is 1/80 times that of the earth. If g represents the acceleration due to gravity on the surface of the earth, that on the surface of the moon is

a) $g/4$ b) $g/5$ c) $g/6$ d) $g/8$

363. When a satellite going round the earth in a circular orbit of radius r and speed v loses some of its energy, then r and v change as

a) r and v both will increase b) r and v both will decrease

c) r will decrease and v will increase d) r will decrease and v will decrease

364. A particle of mass 10 g is kept on the surface of a uniform sphere of mass 100 kg and radius 10 cm. Find the work to be done against the gravitational force between them, to take the particle far away from the sphere.
(You may take $G = 6.67 \times 10^{-11}\,\text{Nm}^2\text{kg}^2$)

a) 13.34×10^{-10}J b) 3.33×10^{-10}J c) 6.67×10^{-9}J d) 6.67×10^{-10}J

365. Two equal masses m and m are hung from a balance whose scale pan differs in vertical height by $h/2$. The error in weighing in terms of density of the earth ρ is

a) $\frac{1}{3}\pi G\rho mh$ b) $\pi G\rho mh$ c) $\frac{4}{3}\pi G\rho mh$ d) $\frac{8}{3}G\rho mh$

366. In a certain region of space, the gravitational field is given by $-k/r$, where r is the distance and k is a constant. If the gravitational potential at $r = r_0$ be V_0, then what is the expression for the gravitational potential V?

a) $k\log(r/r_0)$ b) $k\log(r_0/r)$ c) $V_0 + k\log(r/r_0)$ d) $V_0 + k\log(r_0/r)$

367. The density of earth in terms of acceleration due to gravity (g), radius of earth (R) and universal gravitational constant (G) is

a) $\frac{4\pi RG}{3g}$ b) $\frac{3\pi RG}{4g}$ c) $\frac{4g}{3\pi RG}$ d) $\frac{3g}{4\pi RG}$

368. The velocity with which is projectile must be fired so that it escapes earth's gravitation does not depend on

a) Mass of the earth b) Mass of the projectile

c) Radius of the projectile's orbit d) Gravitational constant

369. Two planets have the same average density but their radii are R_1 and R_2. If acceleration due to gravity on these planets be g_1 and g_2 respectively, then

a) $\frac{g_1}{g_2}=\frac{R_1}{R_2}$ b) $\frac{g_1}{g_2}=\frac{R_2}{R_1}$ c) $\frac{g_1}{g_2}=\frac{R_1^2}{R_2^2}$ d) $\frac{g_1}{g_2}=\frac{R_1^3}{R_2^3}$

370. A solid sphere of uniform density and radius r applies a gravitational force of attraction equal to F_1 on a particle placed at P, distance $2R$ from the centre O of the sphere. A spherical cavity of radius $R/2$ is now made in the sphere as shown in figure. The sphere with cavity now applied an gravitational force F_2 on same particle placed at P. The ratio F_2/F_1 will be

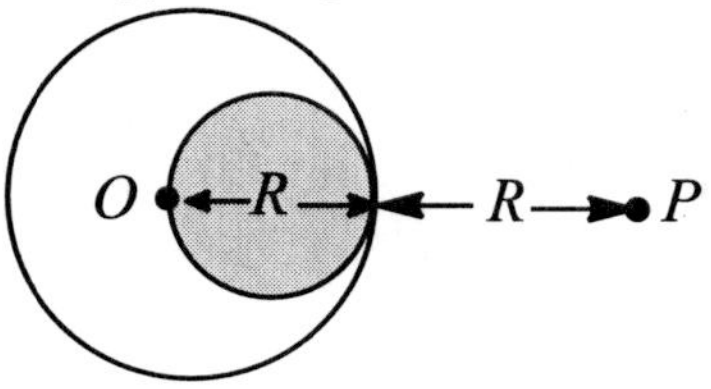

a) 1/2 b) 7/9 c) 3 d) 7

371. The height at which the acceleration due to gravity decreases by 36% of its value on the surface of the earth. (The radius of the earth is R)

a) $\frac{R}{6}$ b) $\frac{R}{4}$ c) $\frac{R}{2}$ d) $\frac{2}{3}R$

372. If a planet was suddenly stopped in its orbit, k suppose to be circular, find how much time will it take in falling onto the sun?

a) $\sqrt{2}/8$ times the period of the planet's revolution
b) $4\sqrt{2}$ times the period of the planet's revolution
c) $3\sqrt{2}$ times the period of the planet's revolution
d) 9 times the period of the planet's revolution

373. A body is projected upwards with a velocity of 4×11.2 kms^{-1} from the surface of earth. What will be the velocity of the body when it escapes from the gravitational pull of earth?

a) 11.2 kms^{-1} b) 2×11.2 kms^{-1} c) 3×11.2 kms^{-1} d) $\sqrt{15} \times 11.2$ kms^{-1}

374. A body falls freely under gravity. Its speed is v when it has lost an amount U of the gravitational energy. Then its mass is

a) $\frac{Ug}{v^2}$ b) $\frac{U^2}{g}$ c) $\frac{2U}{v^2}$ d) $2Ugv^2$

375. If the density of the earth is doubled keeping radius constant, find the new acceleration due to gravity? ($g = 9.8$ m/s^2)

a) 9.8 m/s^2 b) 19.6 m/s^2 c) 4.9 m/s^2 d) 39.2 m/s^2

376. Three particles each of mass m are kept at verities of an equilateral triangle of side L. The gravitational field at centre due to these particle is

a) Zero b) $\frac{3GM}{L^2}$ c) $\frac{9GM}{L^2}$ d) $\frac{12}{\sqrt{3}}\frac{GM}{L^2}$

377. The acceleration due to gravity on the planet A is 9 times the acceleration due to gravity on planet B. A man jumps to height of 2 m on the surface of A. What is the height of jump by the same person on the planet B?

a) 6 m b) $\frac{3}{2}$ m c) 2/9 m d) 18 m

378. A satellite is moving with a constant speed v in a circular orbit about the earth. An object of mass m is ejected from the satellite such that it just escapes from the gravitational pull of the earth. At the time of its ejection, the kinetic energy of the object is

a) $\frac{1}{2}mv^2$ b) mv^2 c) $\frac{3}{2}mv^2$ d) $2mv^2$

379. Two planets of radii in the ratio 2:3 are made from the material of density in the ratio 3:2. Then, the ratio of acceleration due to gravity $\frac{g_1}{g_2}$ at the surface of the two planets will be

a) 1 b) 2.25 c) 4/9 d) 0.12

380. A particle of mass m is located at a distance r from the centre of a shell of mass M and radius R. The force between the shell and mass is $F(r)$. The plot of $F(r)$ *versus* r is

a) 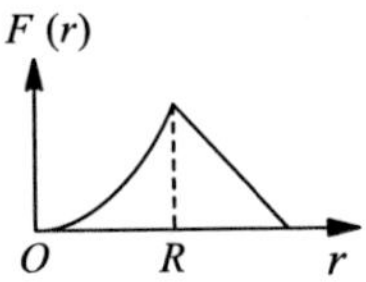
b) 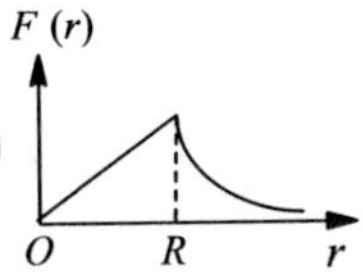
c) 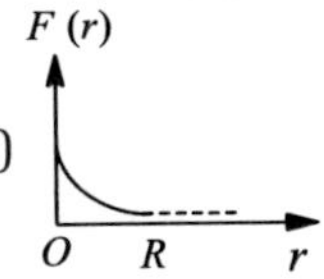
d) 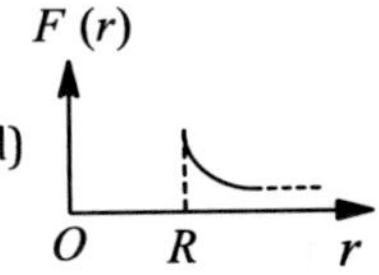

381. The condition for a uniform spherical mass m of radius r to be a black hole is [G = gravitational constant and g = acceleration due to gravity]

a) $(2Gm/r)^{1/2} \leq c$ b) $(2Gm/r)^{1/2} = c$ c) $(2Gm/r)^{1/2} \geq c$ d) $(gm/r)^{1/2} \geq c$

382. If mass of a satellite is doubled and time period remain constant the ratio of orbit in the two cases will be

a) 1 : 2 b) 1 : 1 c) 1 : 3 d) None of these

383. Periodic time of a satellite revolving above Earth's surface at a height equal to R, radius of Earth, is
[g is acceleration due to gravity at Earth's surface]

a) $2\pi\sqrt{\frac{2R}{g}}$ b) $4\sqrt{2\pi}\sqrt{\frac{R}{g}}$ c) $2\pi\sqrt{\frac{R}{g}}$ d) $8\pi\sqrt{\frac{R}{g}}$

384. An earth satellite of mass m revolves in a circular orbit of a height h from the surface of the earth. $R$is the radius of the earth and g is acceleration due to gravity at the surface of the earth. The velocity of the satellite in the orbit is given by

a) $\frac{gR^2}{R+h}$ b) gR c) $\frac{gR}{R+h}$ d) $\sqrt{\frac{gR^2}{R+h}}$

385. At what depth below the surface of the earth, the value of g is the same as that at a height of 5 km?

a) 1.25 km b) 2.5 km c) 5 km d) 10 km

386. Two bodies of masses 2kg and 8kg are separated by a distance of 9 m. the point where the resultant gravitational field intensity is zero is at a distance of

a) 4.5 m from each mass b) 6 m from 2 kg c) 6 m from 8 kg d) 2.5 m from 2 kg

387. A satellite is launched into a circular orbit of radius 'R' around earth while a second satellite is launched into an orbit of radius 1.02 R. The percentage difference in the time periods of the two satellites is

a) 0.7 b) 1.0 c) 1.5 d) 3

388. Which of the following statements is correct in respect of a geostationary satellite

a) It moves in a plane containing the Greenwich meridian
b) It moves in a plane perpendicular to the celestial equatorial plane
c) Its height above the earth's surface is about the same as the radius of the earth
d) Its height above the earth's surface is about six times the radius of the earth

389. If g is the acceleration due to gravity on the surface of earth, its value at a height equal to double the radius of earth is

a) g b) $\frac{g}{2}$ c) $\frac{g}{3}$ d) $\frac{g}{9}$

390. A spaceship is launched into a circular orbit close to earth's surface. The additional velocity that should be imparted to the spaceship in the orbit to overcome the gravitational pull is (Radius of earth = 6400 km and g = 9.8 ms^{-2})

a) 11.2 kms^{-1} b) 8 kms^{-1} c) 3.2 kms^{-1} d) 1.5 kms^{-1}

391. If r denotes the distance between the sun and the earth, then the angular momentum of the earth around the sun is proportional to

a) $r^{3/2}$ b) r c) $\sqrt{r}$ d) r^2

392. Escape velocity on the surface of earth is 11.2 km/s. Escape velocity from a planet whose mass is the same as that of earth and radius 1/4 that of earth is

a) 2.8 km/s b) 15.6 km/s c) 22.4 km/s d) 44.8 km/s

393. A satellite is orbiting around the earth. By what percentage should we increase its velocity, so as to enable it escape away from the earth?

a) 41.4% b) 50% c) 82.8% d) 100%

394. The escape velocity from the earth is 11.2 kms^{-1}. The escape velocity from a planet having twice the radius and the same mean density is (in kms^{-1})

a) 11.2 b) 5.6 c) 15 d) 22.4

395. In a certain region of space gravitational field is given by $I(Kr)$. Taking the reference point to be at $r = V_0$, find the potential.

a) $K \log \frac{r}{r_0} + V_0$ b) $K \log \frac{r_0}{r} + V_0$ c) $K \log \frac{r}{r_0} - V_0$ d) $\log \frac{r}{r_0} - V_0 r$

396. The escape velocity of projectile on the earth's surface is 11.2 kms^{-1}. A body is projected out with thrice this speed. The speed of the body for away from the earth will be

a) 22.4 kms^{-1} b) 31.7 kms^{-1} c) 33.6 kms^{-1} d) None of these

397. The change in the gravitational potential energy when a body mass m is raised to a height nR above the surface of the earth is (here R is the radius of the earth)

a) $\left(\frac{n}{n+1}\right) mgR$ b) $\left(\frac{n}{n-1}\right) mgR$ c) $nmgR$ d) $\frac{mgR}{n}$

398. Mass M is divided into two parts xM and $(1-x)M$. For a given separation, the value of x for which the gravitational attraction between the two pieces becomes maximum is

a) $\frac{1}{2}$ b) $\frac{3}{5}$ c) 1 d) 2

399. A satellite is moving around the earth with speed v in a circular orbit of radius r. If the orbit radius is decreased by 1%, its speed will

a) Increase by 1% b) Increase by 0.5% c) Decrease by 1% d) Decrease by 0.5%

400. If the height of a satellite from the earth is negligible in comparison to the radius of the earth R, the orbital velocity of the satellite is

a) gR b) $gR/2$ c) $\sqrt{g/R}$ d) $\sqrt{gR}$

401. The escape velocity for a body of mass 1 kg from the earth's surface is 11.2 kms^{-1}. The escape velocity for a body of mass 100 kg would be

a) $11.2 \times 10^2 kms^{-1}$ b) 112 kms^{-1} c) 11.2 kms^{-1} d) $11.2 \times 10^{-2} kms^{-1}$

402. Weight of a body of mass m decreases by 1% when it is raised to height h above the earth's surface. If the body is taken on a depth h in a mine, change in its weight is

a) 0.5% decrease b) 2% decrease c) 0.5% increase d) 1% increase

403. An earth satellite is moved from one stable circular orbit to farther stable circular orbit. Which one of the following quantities increase?

a) Linear orbit speed b) Gravitational force
c) Centripetal acceleration d) Gravitational potential energy

404. g_e and g_p denote the acceleration due to gravity on the surface of the earth and another planet whose mass and radius are twice to that of the earth, then

a) $g_p = \frac{g_e}{2}$ b) $g_p = g_e$ c) $g_p = 2g_e$ d) $g_p = \frac{g_e}{\sqrt{2}}$

405. Who among the following gave first the experimental value of G

a) Cavendish b) Copernicus c) Brook Teylor d) None of these

406. Assuming the earth to have a constant density, point out which of the following curves show the variation of acceleration due to gravity from the centre of earth to the points far away from the surface of earth

a) 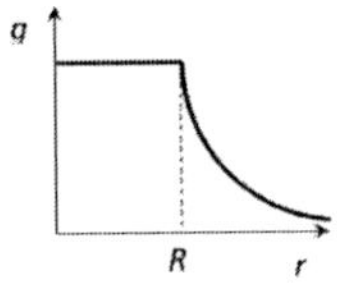

b) 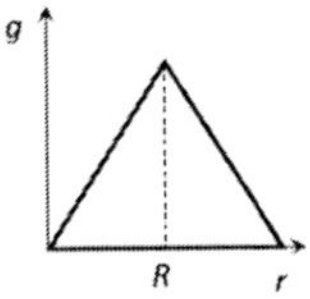

c) 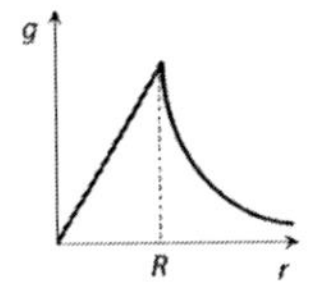

d) None of these

407. The masses of two planets are in the ratio 1:2. Their radii are in the ratio 1:2. The acceleration due to gravity on the planets are in the ratio

a) 1:2 b) 2:1 c) 3:5 d) 5:3

408. According to Kelper's law of planetary motion if T represent time period and r is orbital radius, then for two planets these are related as

a) $\left(\frac{T_1}{T_2}\right)^3 = \left(\frac{r_1}{r_2}\right)^3$ b) $\left(\frac{T_1}{T_2}\right)^{\frac{3}{2}} = \frac{r_1}{r_2}$ c) $\left(\frac{T_1}{T_2}\right)^4 = \left(\frac{r_1}{r_2}\right)^3$ d) $\left(\frac{T_1}{T_2}\right)^2 = \left(\frac{r_1}{r_2}\right)^3$

409. The maximum and minimum distances of a comet from the sun are $8 \times 10^{12} m$ and $1.6 \times 10^{12} m$. If its velocity when nearest to the sun is 60 m/s, what will be its velocity in m/s when it is farthest

a) 12 b) 60 c) 112 d) 6

410. A satellite is placed in a circular orbit around earth at such a height that it always remains stationary with respect to earth surface. In such case, its height from the earth surface is

a) 32000 km b) 36000 km c) 6400 km d) 4800 km

411. Out of the following, the only correct statement about satellites is

a) A satellite cannot move in a stable orbit in a plane passing through the earth's centre
b) Geostationary satellites are launched in the equatorial plane
c) We can use just one geostationary satellite for global communication around the globe
d) The speed of satellite increases with an increase in the radius of its orbit

412. Acceleration due to gravity is g on the surface of the earth. Then the value of the acceleration due to gravity at a height of 32 km above earth's surface is (Assume radius of earth to be 6400 km)

a) 0.99 g b) 0.8 g c) 1.01 g d) 0.9 g

413. In the above problem, if the shell is replaced by a sphere of same mass and radius then the graph of $F(r)$ *versus* r will be

a) 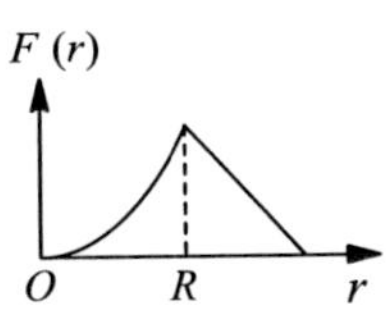

b) 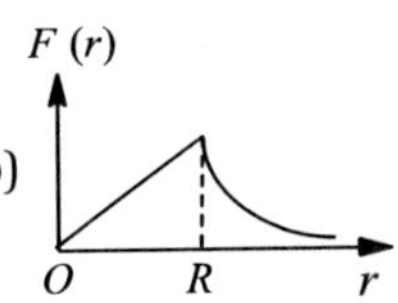

c) 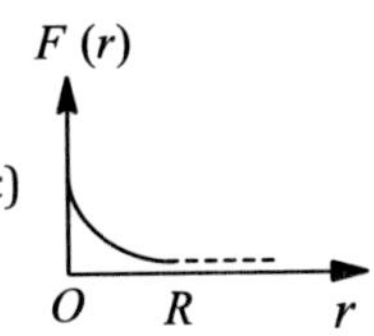

d) 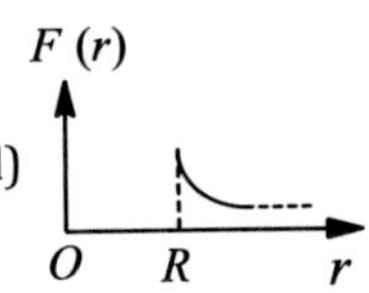

414. The orbital velocity of an artificial satellite in a circular orbit just above the earth's surface is V. For a satellite orbiting at an altitude of half of the earth's radius, the orbital velocity is

a) $\frac{3}{2}V$ b) $\sqrt{\frac{3}{2}}V$ c) $\sqrt{\frac{2}{3}}V$ d) $\frac{2}{3}V$

415. A body is projected with velocity of 2×11.2 km/s from the form the surface of earth. The velocity of the body when it escapes the gravitational pull of earth is

a) $\sqrt{3} \times 11.2$ km/s b) 11.2 km/s c) $\sqrt{2} \times 11.2$ km/s d) 0.5×11.2 km/s

416. The escape velocity of a body on the earth's surface is v_e. A body is thrown up with a speed $\sqrt{5}v_e$. Assuming that the sun and planets do not influence the motion of the body, velocity of the body at infinite distance is

a) Zero b) v_e c) $\sqrt{2}v_e$ d) $2v_e$

417. The period of a satellite in a circular orbit of radius R is T, the period of another satellite in a circular orbit of radius $4R$ is

a) $4T$ b) $T/4$ c) $8T$ d) $T/8$

418. Two point masses A and B having masses in the ratio 4:3 are separated by a distance of 1 m. When another point mass C of mass M is placed in between $A$and B, the force between A and C is $\frac{1}{3}$rd of the force between B and C. Then the distance of C from A is

a) $\frac{2}{3}$ m b) $\frac{1}{3}$ m c) $\frac{1}{4}$ m d) $\frac{2}{7}$ m

419. A rocket is sent vertically up with a velocity v less than the escape velocity from the earth. Taking M and r as the mass and radius of earth, the maximum height h attained by the rocket is given by the following expression

a) $v^2R^2/(2GR - Mv)$ b) $v^2R^2/(2GR + v^2R)$

c) $v^2R^2/(2GR - v^2R)$ d) $v^2R^2/(2GRv + RM)$

420. A particle of mass m is thrown upwards from the surface of the earth, with a velocity u. The mass and the radius of the earth are, respectively, M and R. G is gravitational constant and g is acceleration due to gravity on the surface of the earth. The minimum value of u so that the particle does not return back to earth is

a) $\sqrt{2gR^2}$ b) $\sqrt{\frac{2GM}{R^2}}$ c) $\sqrt{\frac{2GM}{R}}$ d) $\sqrt{\frac{2gM}{R^2}}$

421. At the surface of a certain planet, acceleration due to gravity is one-quarter of that on earth. If a brass ball is transported to this planet, then which one of the following statements is not correct

a) The mass of the brass ball on this planet is a quarter of its mass as measured on earth
b) The weight of the brass ball on this planet is a quarter of the weight as measured on earth
c) The brass ball has the same mass on the other planet as on earth
d) The brass ball has the same volume on the other planet as on earth

422. Distance between the centres of two stars is 10 a. The masses of these stars are M and 16 M and their radii a and $2a$ respectively. A body of mass m is fired straight from the surface of the larger star towards the smaller star. The minimum initial speed for the body to reach the surface of smaller star is

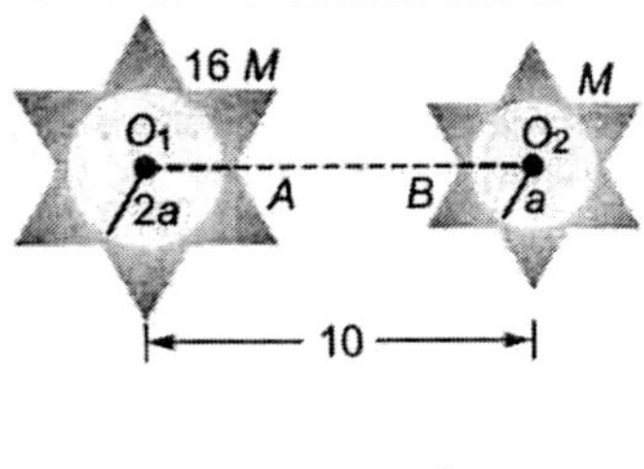

a) $\frac{2}{3}\sqrt{\frac{GM}{a}}$ b) $\frac{3}{2}\sqrt{\frac{5GM}{a}}$ c) $\frac{2}{3}\sqrt{\frac{5GM}{a}}$ d) $\frac{3}{2}\sqrt{\frac{GM}{a}}$

423. At a given place where, acceleration due to gravity is g ms^{-2}, a sphere of lead of density d kgm^{-2} is gently released in a column of liquid of density ρkgm^{-3}. If $d > \rho$, the sphere will
   a) Fall vertically with an acceleration of g ms^{-2}
   b) Fall vertically with no acceleration
   c) Fall vertically with an acceleration $g\left(\frac{d-\rho}{d}\right)$
   d) Fall vertically with an acceleration ρ/d

424. The orbital speed of Jupiter is
   a) Greater than the orbital speed of earth
   b) Less than the orbital speed of earth
   c) Equal to the orbital speed of earth
   d) Zero

425. A body is at rest on the surface of the earth. Which of the following statement is correct?
   a) No force is acting on the body
   b) Only weight of the body acts on it
   c) Net downward force is equal to the net upward force
   d) None of the above statement is correct

426. A body is taken to a height of nR from the surface of the earth. The ratio of the acceleration due to gravity on the surface to that at the altitude is
   a) $(n+1)^2$ b) $(n+1)^{-2}$ c) $(n+1)^{-1}$ d) $(n+1)$

427. If suppose moon is suddenly stopped and then released (given radius of moon is one-fourth the radius of earth)and the acceleration of moon with respect to earth is 0.0027 ms^{-2}), then the acceleration of the moon just before striking the earth's surface is (Take g = 10 ms^{-2})
   a) 0.0027 ms^{-2} b) 5.0 ms^{-2} c) 6.4 ms^{-2} d) 10 ms^{-2}

428. The effect of rotation of the earth on the value of acceleration due to gravity is
   a) g is maximum at the equator and maximum at the poles
   b) g is minimum at the equator and maximum at the poles
   c) g is maximum at the both poles
   d) g is minimum at the both poles

429. The period of a planet around sun is 27 times that of earth. The ratio of radius of planet's orbit to the radius of earth's orbit is
   a) 4 b) 9 c) 64 d) 27

430. The value of gon the earth's surface is 980 cms^{-2}. Its value at a height of 64 km from the earth's surface is
   a) 960.40 cms^{-2} b) 984.90 cms^{-2} c) 982.45 cms^{-2} d) 977.55 cms^{-2}

431. What should be the angular speed of earth in rad^{-1} so that a body 5kg weighs zero at the equator? (Take g = 10 ms^{-2} and radius of earth = 6400 km)
   a) 1/1600 b) 1/800 c) 1/400 d) 1/80

432. The acceleration due to gravity about the earth's surface would be half of its value on the surface of the earth at an altitude of ($R = 4000$ *mile*)
   a) 1200 *mile* b) 2000 *mile* c) 1600 *mile* d) 4000 *mile*

433. An artificial satellite moving in circle orbit around the earth has a total (kinetic + potential) energy E_0. Its potential energy and kinetic energy respectively are
   a) $2E_0$ and $-2E_0$ b) $-2E_0$ and $-3E_0$ c) $2E_0$ and $-E_0$ d) $-2E_0$ and $-E_0$

434. Two identical trains P and Q move with equal speeds on parallel tracks along the equator. P moves from east to west and Q from west to east
   a) Data is sufficient to arrive at a conclusion
   b) Both exert equal force on track
   c) Train Q exerts force on track
   d) Train P exerts greater force on track

435. A person will get more quantity of matter in kg-wt at
   a) Poles b) at latitude of 60° c) Equator d) Satellite

436. A satellite with kinetic energy E_k is revolving round the earth in a circular orbit. How much more kinetic energy should be given to it so that it may just escape into outer space

a) E_k b) $2\ E_k$ c) $\frac{1}{2}E_k$ d) $3\ E_k$

437. Force of gravity is least of

a) The equator b) The poles c) A point in between equator and any pole d) None of these

438. The speed of earth's rotation about its axis is ω. Its speed is increased to x times to make the effective acceleration due to gravity equal to zero at the equator, then x is around($g = 10 ms^{-2}, R = 6400$ km)

a) 1 b) 8.5 c) 17 d) 34

439. A body weighs 700 $g\ wt$ on the surface of the earth. How much will it weigh on the surface of a planet whose mass is $\frac{1}{7}$ and radius is half that of the earth

a) 200 $g\ wt$ b) 400 $g\ wt$ c) 50 $g\ wt$ d) 300 $g\ wt$

440. The escape velocity of a planet having mass 6 times and radius 2 times as that of earth is

a) $\sqrt{3}V_e$ b) $3\ V_e$ c) $\sqrt{2}V_e$ d) $2\ V_e$

441. The bodies situated on the surface of earth at its equator, becomes weightless, when the earth has KE about it axis

a) mgR b) $2\ mgR/5$ c) $MgR/5$ d) $5MgR/2$

442. The depth from the surface of the earth of radius R at which the acceleration due to gravity will be 75% of the value on the surface of the earth is

a) $R/4$ b) $R/2$ c) $3R/4$ d) $R/8$

443. Two balls, each of radius R, equal mass and density are placed in contact, then the force of gravitation between them is proportional to

a) $F \propto \frac{1}{R^2}$ b) $F \propto R$ c) $F \propto R^4$ d) $F \propto \frac{1}{R}$

444. If a planet consists of a satellite whose mass and radius were both half that of the earth, the acceleration due to gravity at its surface would be (g on earth $= 9.8\ m/sec^2$)

a) $4.9m/sec^2$ b) $8.9m/sec^2$ c) $19.6m/sec^2$ d) $29.4m/sec^2$

445. At what height from the ground will the value of $'g'$ be the same as that in 10 km deep mine below the surface of earth

a) 20 km b) 10 km c) 15 km d) 5 km

446. The mass of a spaceship is 1000kg. It is to be launched from the earth's surface out into free space. The value of $'g'$ and $'R'$ (radius of earth) are $10m/s^2$ and 6400km respectively

a) 6.4×10^{11} $Joules$ b) 6.4×10^8 $Joules$ c) 6.4×10^9 $Joules$ d) 6.4×10^{10} $Joules$

447. The acceleration due to gravity on a planet is same as that on earth and its radius is four times that of earth. What will be the value of escape velocity on that planet if it is v_e on earth

a) v_e b) $2v_e$ c) $4v_e$ d) $\frac{v_e}{2}$

448. The gravitational field due to a mass distribution is $I = k/x^3$ in the x-direction (k is a constant). Taking the gravitational potential to be zero at infinity, its value at a distance $x/\sqrt{2}$ is

a) k/x b) $k/2x$ c) k/x^2 d) $k/2x^2$

449. A uniform ring of mass M and radius r is placed directly above a uniform sphere of mass $8M$ and of same radius R. The centre of the ring is at a distance of $d = \sqrt{3}R$ from the centre of the sphere. The gravitational attraction beween the sphere and the ring is

a) $\frac{GM^2}{R^2}$ b) $\frac{3GM^2}{2R^2}$ c) $\frac{2GM^2}{\sqrt{2}R^2}$ d) $\frac{\sqrt{3}GM^2}{R^2}$

450. A small satellite is revolving near earth's surface. Its orbital velocity will be nearly

a) 8 km/sec b) 11.2 km/sec c) 4 km/sec d) 6 km/sec

451. A satellite in a circular orbit of radius R has a period of 4 h. Another satellite with orbital radius $3R$ around the same planet will have a period (in hour)

a) 16 b) 4 c) $4\sqrt{27}$ d) $4\sqrt{8}$

452. What will be the acceleration due to gravity at height h if $h >> R$. Where R is radius of earth and g is acceleration due to gravity on the surface of earth

a) $\frac{g}{\left(1+\frac{h}{R}\right)^2}$ b) $g\left(1-\frac{2h}{R}\right)$ c) $\frac{g}{\left(1-\frac{h}{R}\right)^2}$ d) $g\left(1-\frac{h}{R}\right)$

453. Imagine a light planet revolving around a very massive star in a circular orbit of radius r with a period of revolution T. If the gravitational force of attraction between the planet and the star is proportional to $R^{-3/2}$, then T^2 is proportional to

a) R^3 b) $R^{5/2}$ c) $R^{3/2}$ d) $R^{7/2}$

454. The period of moon's rotation around the earth is nearly 29 *days*. If moon's mass were 2 fold, its present value and all other things remained unchanged, the period of moon's rotation would be nearly

a) $29\sqrt{2}$ *days* b) $29\sqrt{2}$ *days* c) 29×2 *days* d) 29 *days*

455. Which is constant for a satellite in orbit

a) Velocity b) Angular momentum c) Potential energy d) Acceleration

456. Two bodies of masses 100 kg and 1000 kg are separated by distance of 1 m. What is the intensity of gravitational field at the mid point of the line joining them?

a) $6.6 \times 10^{-11} Nm^2kg^{-2}$ b) $2.4 \times 10^{-8} Mkg^{-1}$ c) $2.4 \times 10^{-7} Nkg^{-1}$ d) $2.4 \times 10^{-6} Nkg^{-1}$

457. What will be the effect on the weight of a body placed on the surface of earth, if earth suddenly starts rotating with half of its angular velocity of rotation?

a) No effect
b) Weight will increase
c) Weight will decrease
d) Weight will become zero

458. One goes from the centre of the earth to a distance two third the radius of the earth, where will the acceleration due to gravity be the greatest?

a) At the centre of the earth
b) At a height half the radius of the earth
c) At a height one-third the radius of the earth
d) At a height two-third the radius of the earth

459. The gravitational attraction between the two bodies increases when their masses are

a) Reduced and distance is reduced b) Increased and distance is reduced
c) Reduced and distance is increased d) Increased and distance is increased

460. A particle of mass m is placed inside a spherical shell, away from its centre. The mass of the shell is M

a) The particle will move towards the centre if $m < M$, and away from the centre if $m > M$
b) The particle will move towards the centre
c) The particle will oscillate about the centre of shell
d) The particle will remain stationary

461. Two planets of mean distance d_1 and d_2 from the sun and their frequencies are n_1 and n_2 respectively then

a) $n_1^2d_1^2 = n_2d_2^2$ b) $n_2^2d_2^3 = n_1^2d_1^3$ c) $n_1d_1^2 = n_2d_2^2$ d) $n_1^2d_1 = n_2^2d_2$

462. If satellite is revolving around a planet of mass M in an elliptical orbit of semi-major axis a, find the orbital speed of the satellite when it is at a distance r from the focus

a) $v^2 = GM\left[\frac{2}{r}-\frac{1}{a}\right]$ b) $v^2 = GM\left[\frac{2}{r^2}-\frac{1}{a}\right]$ c) $v^2 = GM\left[\frac{2}{r^2}-\frac{1}{a^2}\right]$ d) $v^2 = G\left[\frac{2}{r}-\frac{1}{a}\right]$

463. Two stars of mass m_1 and m_2 are parts of a binary system. The radii of their orbits are r_1 and r_2 respectively, measured from the C.M. of the system. The magnitude of gravitational force m_1 exerts on m_2 is

a) $\frac{m_1 m_2 G}{(r_1+r_2)^2}$ b) $\frac{m_1 G}{(r_1+r_2)^2}$ c) $\frac{m_2 G}{(r_1+r_2)^2}$ d) $\frac{(m_1+m_2)}{(r_1+r_2)^2}$

464. If the earth were to suddenly contract to $\frac{1}{n}$ th of its present radius without any change in its mass, the duration of the new day will be nearly

a) $\frac{24}{n}$ h b) $24n$ h c) $\frac{24}{n^2}$ h d) $24n^2$h

465. A man can jump to a height of 1.5 m on a planet A. What is the height he may be able to jump on another planet whose density and radius are, respectively, one-quarter and one-third that of planet A

a) 1.5 m b) 15 m c) 18 m d) 28 m

466. The value of escape velocity on a certain planet is 2 km s^{-1}. Then, the value of orbital speed for a satellite orbiting close to its surface is

a) 112 kms^{-1} b) 1 kms^{-1} c) $\sqrt{2}$ Kms^{-1} d) $2\sqrt{2}$ kms^{-1}

467. The maximum vertical distance through which a full dressed astronaut can jump on the earth is 0.5 m. Estimate the maximum vertical distance through which he can jump on the moon, which has a mean density 2/3rd that of earth and radius one quarter that of the eart

a) 1.5 m b) 3 m c) 6 m d) 7.5 m

468. Assuming that the earth is a sphere of radius R_E with uniform density, the distance from its centre at which the acceleration due to gravity is equal to $\frac{g}{3}$ (g is the acceleration due to gravity on the surface of earth) is

a) $\frac{R_E}{3}$ b) $\frac{2R_E}{3}$ c) $\frac{R_E}{2}$ d) $\frac{R_E}{4}$

469. Radius of earth is around 6000 km. The weight of body of height of 6000 km from earth surface becomes

a) Half b) One-fourth c) One third d) No change

470. The mass of the moon is 1/81 of earth's mass and its radius 1/4th that of the earth. If the escape velocity from the earth's surface is 11.2 kms^{-1}, its value for the moon will be

a) 0.15 kms^{-1} b) 5 kms^{-1} c) 2.5 kms^{-1} d) 0.5 kms^{-1}

471. Geostationary satellite

a) Falls with g towards the earth b) Has period of 24 hrs
c) Has equatorial orbit d) Above all correct

472. If a new planet is discovered rotating around the sum with the orbital radius double that of earth, then what will be its time period (in earth's days)?

a) 1032 b) 1023 c) 1024 d) 1043

473. The mass of a planet that has a moon whose time period and orbital radius are T and R respectively can be written as

a) $4\pi^2 R^3 G^{-1} T^{-2}$ b) $8\pi^2 R^3 G^{-1} T^{-2}$ c) $12\pi^2 R^3 G^{-1} T^{-2}$ d) $16\pi^2 R^3 G^{-1} T^{-2}$

474. Choose the correct statement from the following. The radius of the orbit of a geostationary satellite depends upon

a) Mass of the satellite, its time period and the gravitational constant
b) Mass of the satellite, mass of the earth and the gravitational constant
c) Mass of the earth, mass of the satellite, time period of the satellite and the gravitational constant
d) Mass of the earth, time period of the satellite and the gravitational constant

475. The diameters of two planets are in the ratio 4:1 and their mean densities in the ratio 1:2. The acceleration due to gravity on the planets will be in ratio

a) 1 : 2 b) 2 : 3 c) 2 : 1 d) 4 : 1

476. Given radius of Earth $'R'$ and length of a day $'T'$ the height of a geostationary satellite is [G-Gravitational Constant, M-Mass of Earth]

a) $\left(\frac{4\pi^2 GM}{T^2}\right)^{1/3}$ b) $\left(\frac{4\pi GM}{R^2}\right)^{1/3} - R$ c) $\left(\frac{GMT^2}{4\pi^2}\right)^{1/3} - R$ d) $\left(\frac{GMT^2}{4\pi^2}\right)^{1/3} + R$

477. According to Kepler, the period of revolution of a planet (T) and its mean distance from the sun (r) are related by the equation

a) $T^3r^3 = \text{constant}$ b) $T^2r^{-3} = \text{constant}$ c) $Tr^3 = \text{constant}$ d) $T^2r = \text{constant}$

478. Gas escapes from the surface of a planet because it acquires an escape velocity. The escape velocity will depend on which of the following factors :

I. Mass of the planet

II. Mass of the particle escaping

III. Temperature of the planet

IV. Radius of the planet

Select the correct answer from the codes given below :

a) I and II b) II and IV c) I and IV d) I, III and IV

479. The escape velocity of a projectile from the earth is approximately

a) 11.2 m/sec b) 112 km/sec c) 11.2 km/sec d) 11200 km/sec

480. Escape velocity on the earth

a) Is less than that on the moon b) Depends upon the mass of the body c) Depends upon the direction of projection d) Depends upon the height from which it is projected

481. The weight of an astronaut, in an artificial satellite revolving around the earth, is

a) Zero b) Equal to that on the earth c) More than that on the earth d) Less than that on the earth

482. The acceleration due to gravity becomes $\left(\frac{g}{2}\right)$ (g = acceleration due to gravity on the surface of the earth) at a height equal to

a) $4R$ b) $\frac{R}{4}$ c) $2R$ d) $\frac{R}{2}$

483. The orbit of geostationary satellite is circular, the time period of satellite depends on

(i) mass of the satellite

(ii) mass of the earth

(iii) radius of the orbit

(iv) height of the satellite from the surface of earth

Which of the following correct?

a) (i) only b) (i) and (ii) c) (i), (ii) and (iii) d) (ii), (iii) and (iv)

484. If suddenly the gravitational force of attraction between earth and a satellite revolving around it becomes zero, then the satellite will

a) Continue to move in its orbit with same velocity

b) Move tangentially to the original orbit with the same velocity

c) Become stationary in its orbit

d) Move towards the earth

485. The value of $'g'$ at a particular point is 9.8 m/s^2. Suppose the earth suddenly shrinks uniformly to half its present size without losing any mass. The value of $'g'$ at the same point (assuming that the distance of the point from the centre of earth does not shrink) will now be

a) 4.9 m/sec^2 b) 3.1 m/sec^2 c) 9.8 m/sec^2 d) 19.6 m/sec^2

486. A spring balance is graduated on sea level. If a body is weighed with this balance at consecutively increasing heights from earth's surface, the weight indicated by the balance

a) Will go on increasing continuously b) Will go on decreasing continuously c) Will remain same d) Will first increase and then decrease

487. The acceleration due to gravity near the surface of a planet of radius R and density d is proportional to

a) $\frac{d}{R^2}$ b) dR^2 c) dR d) $\frac{d}{R}$

488. 320 km above the surface of earth, the value of acceleration due to gravity is nearly 90% of its value on the surface of the earth. Its value will be 95% of the value on the earth's surface

a) Nearly 160 km below the earth's surface
b) Nearly 80 km below the earth's surface
c) Nearly 640 km below the earth's surface
d) Nearly 320 km below the earth's surface

489. In some region, the gravitational field is zero. The gravitational potential in this region

a) Must be variable b) Must be constant c) Cannot be zero d) Must be zero

490. A body of weight 500 N on the surface of the earth. How much would it weigh half-way below the surface of the earth?

a) 125 N b) 250 N c) 500 N d) 1000 N

491. LANDSAT series of satellites move in near polar orbits at an altitude of

a) 3600 km b) 3000 km c) 918 km d) 512 km

492. For the moon to cease to remain the earth's satellite, its orbital velocity has to increase by a factor of

a) 2 b) $\sqrt{2}$ c) $1/\sqrt{2}$ d) $\sqrt{3}$

493. The force of gravitation is

a) Repulsive b) Electrostatic c) Conservative d) Non-conservative

494. Two astronauts have deserted their space ships in a region of space far from the gravitational attraction of any other body. Each has a mass of 100 kg and they are 100 m apart. They are initially at rest relative to one another. How long will it be before the gravitational attraction brings them 1 cm closer together?

a) 2.52 days b) 1.41 days c) 0.70 days d) 0.41 days

495. Distance of geostationary satellite from the surface of earth $radius(R_e = 6400\ km)$ in terms of R_e is

a) 13.76 R_e b) 10.76 R_e c) 6.56 R_e d) 2.56 R_e

496. The gravitational force between a point like mass M and an infinitely long, thing rod of linear mass density perpendicular to distance L from M is

a) $\frac{MG\lambda}{L}$ b) $\frac{1}{2}\frac{MG\lambda}{L}$ c) $\frac{2MG\lambda}{L^2}$ d) Infinite

497. If the radius of the earth were to shrink by 1% its mass remaining same, the acceleration due to gravity on the earth's surface would

a) Decrease by 2% b) Remain unchanged c) Increase by 2% d) Become zero

498. For a body lying on the equator to appear weightless, what should be the angular speed of the earth?(Take g = $10 ms^{-2}$; radius of earth = 6400 km)

a) 0.125 $rads^{-1}$ b) 1.25 $rads^{-1}$ c) $1.25 \times 10^{-3} rads^{-1}$ d) $1.25 \times 10^{-2} rads^{-1}$

499. Two metallic spheres each of mass M are suspended by two strings each of length L. The distance between the upper ends of strings is L. The angle which the strings will make with the vertical due to mutual attraction of the spheres is

a) $\tan^{-1}\left[\frac{GM}{gL}\right]$ b) $\tan^{-1}\left[\frac{GM}{2gL}\right]$ c) $\tan^{-1}\left[\frac{GM}{gL^2}\right]$ d) $\tan^{-1}\left[\frac{2GM}{gL^2}\right]$

500. Hubble's law states that the velocity with which milky ways is moving away from the earth is proportional to

a) Square of the distance of the milky way from the earth
b) Distance of milky way from the earth
c) Mass of the milky way
d) Product of the mass of the milky way and its distance from the earth

501. If the moon is to escape from the gravitational field of the earth forever, it will require a velocity

a) $11.2 kms^{-1}$ b) Less than $11.2\ kms^{-1}$
c) Slightly more than $111.2\ kms^{-1}$ d) $22.4\ kms^{-1}$

502. The time period of an earth satellite in circular orbit is independent of

a) The mass of the satellite
b) Radius of its orbit
c) Both the mass and radius of the orbit
d) Neither the mass of the satellite nor the radius of its orbit

503. The orbital speed of an artificial satellite very close to the surface of the earth is V_o. Then the orbital speed of another artificial satellite at a height equal to three times the radius of the earth is

a) $1\ V_o$ b) $2\ V_o$ c) $0.5\ V_o$ d) $4\ V_o$

504. A particle is fired vertically upwards from the surface of earth and reaches a height 6400 km. The initial velocity of the particle is ($R = 6400$ km, $g = 10\text{ms}^{-2}$)

a) $11.2\ \text{ms}^{-1}$ b) $8\ \text{kms}^{-1}$ c) $3.2\ \text{kms}^{-1}$ d) None of these

505. The masses and radii of the earth and moon are M_1, R_1 and M_2, R_2 respectively. Their centres are distance d apart. The minimum velocity with which a particle of mass m shoule be projected from a point midway between their centres so that it escapes to infinity is

a) $2\sqrt{\frac{G}{d}(M_1+M_2)}$ b) $2\sqrt{\frac{2G}{d}(M_1+M_2)}$ c) $2\sqrt{\frac{Gm}{d}(M_1+M_2)}$ d) $2\sqrt{\frac{Gm(M_1+M_2)}{d(R_1+R_2)}}$

506. Where can a geostationary satellite be installed

a) Over any city on the equator
b) Over the north or south pole
c) At height R above earth
d) At the surface of earth

507. At some point the gravitational potential and also the gravitational field due to earth is zero. The speed is

a) On earth's surface
b) Below earth's surface
c) At a height R_e from earth's surface (R_e= radius of the earth)
d) At infinity

8
GRAVITATION
ANSWER KEY

1)	a	2)	c	3)	a	4)	d
5)	b	6)	b	7)	d	8)	d
9)	a	10)	b	11)	d	12)	d
13)	a	14)	c	15)	c	16)	a
17)	b	18)	b	19)	d	20)	c
21)	c	22)	d	23)	a	24)	a
25)	c	26)	a	27)	c	28)	b
29)	b	30)	c	31)	b	32)	b
33)	c	34)	c	35)	b	36)	c
37)	b	38)	c	39)	c	40)	a
41)	c	42)	c	43)	b	44)	c
45)	d	46)	b	47)	b	48)	b
49)	b	50)	d	51)	d	52)	a
53)	b	54)	d	55)	c	56)	c
57)	c	58)	c	59)	a	60)	d
61)	a	62)	c	63)	a	64)	b
65)	c	66)	b	67)	c	68)	c
69)	a	70)	c	71)	c	72)	c
73)	c	74)	a	75)	a	76)	a
77)	a	78)	c	79)	a	80)	a
81)	d	82)	a	83)	b	84)	c
85)	c	86)	d	87)	c	88)	d
89)	b	90)	a	91)	b	92)	a
93)	d	94)	a	95)	b	96)	a
97)	a	98)	c	99)	c	100)	c
101)	b	102)	d	103)	b	104)	b
105)	b	106)	c	107)	c	108)	b
109)	c	110)	b	111)	d	112)	a
113)	a	114)	c	115)	c	116)	b
117)	b	118)	b	119)	c	120)	a
121)	c	122)	c	123)	a	124)	b

125)	b	126)	a	127)	c	128)	a
129)	d	130)	b	131)	c	132)	c
133)	c	134)	d	135)	b	136)	c
137)	c	138)	c	139)	a	140)	d
141)	c	142)	b	143)	a	144)	d
145)	d	146)	b	147)	a	148)	c
149)	a	150)	c	151)	a	152)	d
153)	a	154)	c	155)	a	156)	d
157)	c	158)	d	159)	b	160)	a
161)	b	162)	d	163)	b	164)	c
165)	b	166)	c	167)	c	168)	d
169)	a	170)	a	171)	a	172)	a
173)	c	174)	d	175)	a	176)	a
177)	c	178)	a	179)	c	180)	b
181)	b	182)	a	183)	d	184)	c
185)	c	186)	b	187)	d	188)	a
189)	c	190)	c	191)	c	192)	d
193)	c	194)	b	195)	b	196)	d
197)	b	198)	a	199)	d	200)	d
201)	d	202)	c	203)	c	204)	a
205)	b	206)	c	207)	c	208)	d
209)	d	210)	a	211)	c	212)	b
213)	c	214)	c	215)	a	216)	d
217)	a	218)	b	219)	c	220)	a
221)	a	222)	b	223)	b	224)	d
225)	c	226)	c	227)	c	228)	c
229)	b	230)	d	231)	b	232)	b
233)	c	234)	a	235)	b	236)	b
237)	d	238)	c	239)	a	240)	c
241)	b	242)	a	243)	c	244)	c
245)	c	246)	b	247)	a	248)	a
249)	c	250)	c	251)	b	252)	c
253)	d	254)	b	255)	a	256)	a
257)	d	258)	d	259)	b	260)	a
261)	c	262)	c	263)	b	264)	b
265)	d	266)	a	267)	b	268)	b

269)	b	270)	d	271)	b	272)	a
273)	c	274)	d	275)	b	276)	b
277)	b	278)	d	279)	a	280)	b
281)	c	282)	b	283)	a	284)	b
285)	b	286)	d	287)	a	288)	b
289)	d	290)	c	291)	b	292)	c
293)	a	294)	c	295)	b	296)	b
297)	b	298)	b	299)	c	300)	a
301)	c	302)	c	303)	a	304)	c
305)	d	306)	c	307)	b	308)	d
309)	d	310)	d	311)	c	312)	c
313)	d	314)	b	315)	d	316)	d
317)	a	318)	b	319)	c	320)	a
321)	c	322)	b	323)	b	324)	c
325)	c	326)	b	327)	c	328)	a
329)	c	330)	a	331)	d	332)	c
333)	d	334)	d	335)	a	336)	d
337)	b	338)	c	339)	d	340)	d
341)	b	342)	a	343)	b	344)	b
345)	d	346)	b	347)	a	348)	b
349)	d	350)	a	351)	c	352)	b
353)	c	354)	a	355)	d	356)	b
357)	c	358)	a	359)	b	360)	a
361)	c	362)	b	363)	c	364)	d
365)	c	366)	c	367)	d	368)	b
369)	a	370)	b	371)	b	372)	a
373)	d	374)	c	375)	b	376)	a
377)	d	378)	b	379)	a	380)	d
381)	c	382)	b	383)	b	384)	d
385)	d	386)	c	387)	d	388)	d
389)	d	390)	c	391)	c	392)	c
393)	a	394)	d	395)	a	396)	b
397)	a	398)	a	399)	b	400)	d
401)	c	402)	a	403)	d	404)	a
405)	a	406)	c	407)	b	408)	d
409)	a	410)	b	411)	b	412)	a

413)	b	414)	c	415)	a	416)	d
417)	c	418)	a	419)	c	420)	c
421)	a	422)	b	423)	c	424)	b
425)	c	426)	a	427)	c	428)	a
429)	b	430)	a	431)	b	432)	c
433)	c	434)	d	435)	c	436)	a
437)	a	438)	c	439)	b	440)	a
441)	c	442)	a	443)	c	444)	c
445)	d	446)	d	447)	b	448)	c
449)	d	450)	a	451)	c	452)	a
453)	b	454)	d	455)	b	456)	c
457)	b	458)	d	459)	b	460)	d
461)	b	462)	a	463)	a	464)	c
465)	c	466)	c	467)	b	468)	a
469)	b	470)	c	471)	d	472)	a
473)	a	474)	d	475)	c	476)	c
477)	b	478)	c	479)	c	480)	d
481)	a	482)	b	483)	d	484)	c
485)	c	486)	b	487)	c	488)	d
489)	b	490)	b	491)	c	492)	b
493)	c	494)	b	495)	c	496)	c
497)	c	498)	c	499)	c	500)	b
501)	a	502)	a	503)	c	504)	b
505)	a	506)	a	507)	d		

8
GRAVITATION
HINTS AND SOLUTIONS

1 **(a)**

$$K.E. = \frac{GMm}{2R}$$

2 **(c)**

$$v_0 = \sqrt{\frac{GM}{r}} = \sqrt{\frac{gR^2}{r}} = \sqrt{\frac{10 \times (64 \times 10^5)^2}{8000 \times 10^3}}$$

$= 71.5 \times 10^2 m/s = 7.15\ km/s$

3 **(a)**

Time period of satellite which is very near to planet

$$T = 2\pi\sqrt{\frac{R^3}{GM}} = 2\pi\sqrt{\frac{R^3}{G\frac{4}{3}\pi R^3 \rho}} \therefore T \propto \sqrt{\frac{1}{\rho}}$$

*i.e.*time period of nearest satellite does not depends upon the radius of planet, it only depends upon the density of the planet.

In the problem, density is same so time period will be same

4 **(d)**

$$T = 2\pi\sqrt{\frac{l}{g}}$$

$$\frac{\Delta T}{T} = \frac{\Delta g}{2g}$$

or $\Delta T = -\frac{\Delta g}{2g} \times T = -\frac{1}{2} \times \left(\frac{-0.5}{100}\right) \times 2 =$ $+0.005$ s

$\therefore$ Time period at equator

$= 2 + 0.005 = 2.005$ s

5 **(b)**

V_A =(Potential at A due to A) + (Potential at A due to B)

$$\Rightarrow V_A = -\frac{Gm_1}{R} - \frac{Gm_2}{\sqrt{2}R}$$

Similarly,

V_B =(Potential at B due to A) + (Potential at B due to B)

$$\Rightarrow V_B = -\frac{Gm_2}{R} - \frac{Gm_1}{\sqrt{2}R}$$

Since, $W_{A\to B} = m(V_B - V_A) \Rightarrow W_{A\to B}$

$$= \frac{Gm(m_1 - m_2)(\sqrt{2} - 1)}{\sqrt{2}R}$$

6 **(b)**

$\frac{dA}{dt} = \frac{L}{2m} =$ constant

8 **(d)**

The system will be bound at points where total energy is negative.In the given curve at point A, B and C the P.E. is more than K.E.

9 **(a)**

$$g' = g\left(\frac{R}{R+h}\right)^2 = g\left(\frac{R}{R+2R}\right)^2 = \frac{g}{9}$$

10 **(b)**

As with height g varies as

$$g'' = \frac{g}{[1 + h/R]^2} = g\left[1 - \frac{2h}{R}\right]$$

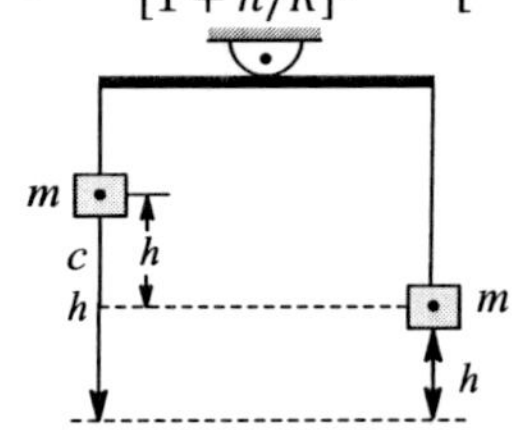

and in according with figure $h_1 > h_2$, so W_1 will be lesser than W_2 and

$$W_2 - W_1 = mg_2 - mg_1 = 2mg\left[\frac{h_1}{R} - \frac{h_2}{R}\right]$$

or $W_2 - W_1 = 2m\frac{GM}{R^2}\frac{h}{R}$

$$\left[\text{as} g = \frac{GM}{R^2} \text{and} (h_1 - h_2) = h\right]$$

or $W_2 - W_1 = \frac{2mhG}{R^3}\left(\frac{4}{3}\pi R^3 \rho\right)$

$$= \frac{8}{3}\pi\rho Gmh \left[as\ M = \frac{4}{3}\pi R^3 \rho\right]$$

11 **(d)**

$g_m = \frac{GM_m}{R_m^2}$ and $g_m = \frac{g_e}{6} = \frac{9.8}{6} m/s^2 =$ $1.63\ m/s^2$

Substituting $R_m = 1.768 \times 10^6 m$, $g_m =$ $1.63\ m/s^2$

and $G = 6.67 \times 10^{-11} N\text{-}m^2/kg^2$ We get $M_m = 7.65 \times 10^{22} kg$

13 **(a)**

When a satellite is moving in on elliptical orbit, it's angular momentum ($= \vec{\mathbf{r}} \times \vec{\mathbf{p}}$) about the centre of earth dos not change its direction. The linear momentum ($= m\vec{\mathbf{v}}$) does not remain constant as velocity of satellite is not constant. The total mechanical energy of S is constant at all locations.

The acceleration of S(=centripetal acceleration) is always directed towards the centre of earth

14 **(c)**

$$v_e = \sqrt{\frac{2GM}{R}} \therefore v_e \propto \sqrt{\frac{M}{R}}$$

If M becomes double and R becomes half then escape velocity becomes two times

15 **(c)**

$$\frac{T_1^2}{T_2^2} = \frac{R_1^3}{R_2^3} = \frac{(6R)^3}{(3R)^3} = 8$$

$$\frac{24 \times 24}{T_2^2} = 8$$

$$T_2^2 = \frac{24 \times 24}{8}$$

$$T_2^2 = 72$$

$$T_2^2 = 36 \times 2$$

$$T_2 = 6\sqrt{2}$$

16 **(a)**

Escape velocity does not depend on the mass of the projectiles

17 **(b)**

Gravitational potential at a point on the surface of earth

$$V = \frac{-GM}{R} = \frac{-gR^2}{R} = -gR$$

18 **(b)**

$$v = \sqrt{\frac{GM}{R}} = G^{1/2}M^{1/2}R^{-1/2}$$

19 **(d)**

Here, $g = GM/R$ and $g' = \frac{G(M/2)}{(R/2)^2} = \frac{2GM}{R^2} = 2g$

$\therefore$ % increase in $g = \left(\frac{g'-g}{g}\right) \times 100$

$$= \left(\frac{2g - g}{g}\right) \times 100 = 100\%$$

20 **(c)**

$$g = \frac{GM}{R^2} = \frac{GM_0}{(D_0/2)^2} = \frac{4GM_0}{D_0^2}$$

23 **(a)**

The value of acceleration due to gravity g at height h above the surface of earth is

$$g_h = \frac{g}{\left(1 + \frac{h}{R}\right)^2}$$

Where R is radius of earth.

$$\therefore \quad \frac{g}{g_h} = \left(1 + \frac{h}{R}\right)^2$$

25 **(c)**

$$\frac{dA}{dt} = \frac{L}{2m} \Rightarrow \frac{dA}{dt} \propto vr \propto \omega r^2$$

26 **(a)**

Earth is surrounded by an atmosphere of gases (air). The reason is that in earth's atmosphere the average thermal velocity of even the highest molecules at the maximum possible temperature is small compared to escape velocity which in turn depends upon gravity ($v_e = \sqrt{gR_e}$). Therefore, the molecules of gases cannot escape from the earth. Hence, an atmosphere exists around the earth.

27 **(c)**

Gravitational field inside hollow sphere will be zero

28 **(b)**

Weight of body at height above the earth's surface is

$$w' = \frac{w}{\left(1 + \frac{h}{r}\right)^2}$$

$$\Rightarrow \quad 40 = \frac{80}{\left(1 + \frac{h}{r}\right)^2}$$

$$\Rightarrow \quad h = 0.41r$$

29 **(b)**

This should be equal to escape velocity $i.e., \sqrt{2gR}$

30 **(c)**

$$\frac{gR^2}{(R + h)^2} = g\left(1 - \frac{h}{R}\right)$$

or $\left(1 - \frac{h}{R}\right)\left(1 + \frac{h^2}{R^2} + \frac{2h}{R}\right) = 1$

or $\frac{h^3}{R^3} + \frac{h^2}{R^2} - \frac{h}{R} = 0$

or $\frac{h}{R}\left(\frac{h^2}{R^2} + \frac{h}{R} - 1\right) = 0$

or $\frac{h}{R} = \frac{-1 \pm \sqrt{1+4}}{2} = \frac{\sqrt{5}-1}{2}$

or $h = \frac{\sqrt{5}R - R}{2}$

31 **(b)**

From Kepler's third law of planetary motion

$T^2 \propto R^3$

$$\therefore \quad \frac{T_2^2}{T_1^2} = \frac{R_2^3}{R_1^3}$$

or $\frac{T_2^2}{(24)^2} = \left(\frac{6400 + 6400}{36000 + 6400}\right)^3$

or $T_2^2 = (24)^2 \times \left(\frac{16}{53}\right)^3$

$\Rightarrow \quad T_2 = 4$ h

32 **(b)**

Given, $g_h = 9 = \frac{gR^2}{(R+R/20)^2} = \frac{20\times20}{21\times21}g$

or $g = \frac{9\times21\times21}{20\times20}$

Now, $g_d = g\left(1-\frac{d}{R}\right)$

$= \frac{9\times21\times21}{20\times20}\left[1-\frac{R/20}{R}\right] = 9.5\text{ms}^{-2}$

33 **(c)**

According to Kepler's third law, we have $T^2 \propto R^3$

Hence, $\frac{T_A^2}{T_B^2} = \left(\frac{4R}{R}\right)^3 = \frac{64}{1}$

or $\frac{T_A}{T_B} = \frac{8}{1}$

or $\frac{2\pi\omega_B}{2\pi\omega_A} = \frac{8}{1}$

or $\frac{v_B\times4R}{R\times v_A} = \frac{8}{1}$

or $\frac{v_B}{3v} = 2$

or $v_B = 6v$

34 **(c)**

Applying law of conservation of energy for asteroid at a distance 10 R_e and at earth's surface.

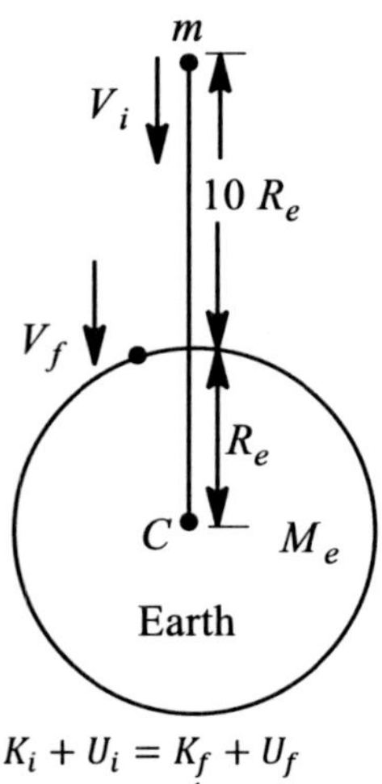

$K_i + U_i = K_f + U_f$ .... (i)

Now, $K_f = \frac{1}{2}mv_i^2$ and $U_i = -\frac{GM_em}{10R_e}$

$K_f = \frac{1}{2}mv_f^2$ and $U_f = -\frac{GM_em}{R_e}$

Substituting these values in Eq. (i), we get

$\frac{1}{2}mv_i^2 - \frac{GM_em}{10R_e} = \frac{1}{2}mv_f^2 - \frac{GM_em}{R_e}$

$\Rightarrow \frac{1}{2}mv_f^2 = \frac{1}{2}mv_f^2 + \frac{GM_em}{R_e} - \frac{GM_em}{10R_e}$

$\Rightarrow v_f^2 = v_i^2 + \frac{2GM_e}{R_e} - \frac{2GM_e}{10R_e}$

$\therefore v_f^2 = v_i^2 + \frac{GM_em}{R_e}\left(1-\frac{1}{10}\right)$

35 **(b)**

$6R$ from the surface of earth and $7R$ from the centre

36 **(c)**

If no external torque acts on a system, then angular momentum of the system does not change.

ie, If $\tau = 0$

$\Rightarrow \frac{dL}{dt} = 0$

$\therefore L = \text{constant}$

Hence, $mv_{\max}r_{\min} = mv_{\min}r_{\max}$

$\Rightarrow r_{\min} = \frac{v_{\min}\times r_{\max}}{v_{\max}}$

$= \frac{1\times10^3\times4\times10^4}{3\times10^4}$

$= \frac{4}{3}\times10^3\text{km}$

37 **(b)**

The value of acceleration due to gravity at latitude λ is given by

$g_\lambda = g - R\omega^2\cos^2\lambda$

$\therefore g - g_\lambda = R\omega^2\cos^2\lambda$

At λ = 30°,

$g - g_{30°} = R\omega^2\cos^2 30°$

$= R\omega^2\left(\frac{\sqrt{3}}{2}\right)^2$

$= \frac{3}{4}R\omega^2$

38 **(c)**

When two satellite of earth are moving in same orbit, then time period of both are equal. From Kepler's third law

$T^2 \propto r^3$

Time period is independent of mass, hence their time periods will be equal. The potential energy and kinetic energy are mass dependent, hence the PE and KE of satellites are not equal.

But, if they are orbiting in a same orbit, then they have equal orbital speed.

39 **(c)**

Acceleration due to gravity at height h,

$g_1 = g\left(1-\frac{2h}{R}\right)$

Acceleration due to gravity at depth h,

$g_1 = g\left(1-\frac{h}{R}\right)$

$$\therefore \frac{g_1}{g_2} = \frac{1 - 2h/R}{1 - h/R} = \left(1 - \frac{2h}{R}\right)\left(1 - \frac{h}{R}\right)^{-1}$$
$$= \left(1 - \frac{h}{R}\right)$$

$\therefore \frac{g_1}{g_2}$ decreases linearly with h

40 **(a)**
Because value of g decreases when we move either in coal mine or at the top of mountain

42 **(c)**
Value of g decreases when we go from poles to equator

43 **(b)**
Kinetic and potential energies varies with position of earth $w.r.t.$ sun. Angular momentum remains constant every where

44 **(c)**
$v_e \propto \frac{1}{\sqrt{r}}$ where r is a position of body from the surface

$$\frac{v_1}{v_2} = \sqrt{\frac{r_1}{r_2}} = \sqrt{\frac{R + 7R}{R}} \Rightarrow v_2 = \frac{v_1}{2\sqrt{2}}$$

45 **(d)**

Escape velocity $v_e = \sqrt{\frac{2GM}{R}}$

$$v_e = \sqrt{\frac{2G\,\frac{4}{3}\pi R^3 \times d}{R}}$$

$$\sqrt{2G\,\frac{4}{3}\pi R^3 \times d} = R\sqrt{\frac{8}{3}\pi G d}$$

where d = mean density of earth

$\because\ v_e \propto R\sqrt{d}$

$$\therefore\ \frac{v_e}{v_p} = \frac{R_e}{R_p}\sqrt{\frac{d_e}{d_p}}$$

$$= \frac{R_e}{2R_e}\sqrt{\frac{d_e}{d_e}}$$

$= v_p = 2v_e$

$= 2 \times 11 = 22\text{kms}^{-1}$

46 **(b)**
$g \propto \rho R$

48 **(b)**
Weight of body on the surface of earth
$mg = 12.6$ N
At height h, the value of g' is given by

$$g' = g\frac{R^2}{(R+h)^2}$$

Now, $\quad h = \frac{R}{2}$

$$\therefore \quad g' = g\left(\frac{R}{R + (R/2)}\right)^2 = g\frac{4}{9}$$

Weight at height $h = mg\frac{4}{9}$

$$= 12.6 \times \frac{4}{9} = 5.6 \text{ N}$$

49 **(b)**
Since, gravitation provides centripetal force

$\frac{mv^2}{r} = \frac{k}{r^{5/2}}$ $ie, v^2 = \frac{k}{mr^{3/2}}$

So that $T = \frac{2\pi r}{v} = \sqrt{\frac{mr^{3/2}}{k}}$ $ie, T^2 = \frac{4\pi^2 m}{k}r^{7/2}$

$\therefore T^2 \propto r^{7/2}$

51 **(d)**
From Kepler's second law of planetary motion, a line joining any planet to the sun sweeps out equal areas in equal times, that is, the aerial velocities of the planet remains constant dA= area of the curved triangle SAB

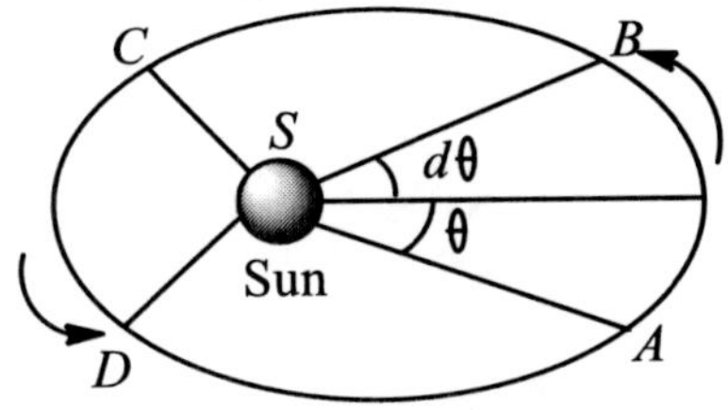

$$\approx \frac{1}{2}(AB \times SA)$$
$$\approx \frac{1}{2}(rd\theta \times r)$$
$$\approx \frac{1}{2}r^2 d\theta$$

Thus, the areal (instantaneous) velocity of the planet is

$$\frac{dA}{dt} = \frac{1}{2}r^2\frac{d\theta}{dt} = \frac{1}{2}r^2\omega = \text{constant}$$

whereω is angular speed of planet and r its radius.

52 **(a)**

$$W = \Delta U = U_f - U_i = U_\infty - U_P$$
$$= -U_P = -mV_P$$
$$= -V_P(\text{as } m = 1)$$

Potential at point P will be obtained by in integration as given below. Let dM be the mass of small rings as shown

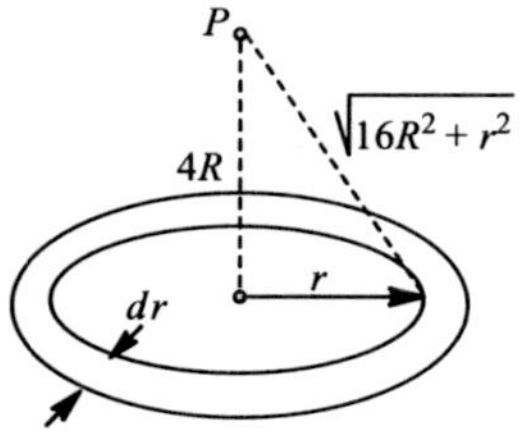

$$dM = \frac{M}{\pi(4R)^2 - \pi(3R)^2}(2\pi r)dr$$
$$= \frac{2Mr\,dr}{7R^2}$$
$$dV_P = -\frac{G.dM}{\sqrt{16R^2+r^2}}$$
$$= -\frac{2GM}{7R^2}\int_{3R}^{4R}\frac{r}{\sqrt{16R^2+r^2}}\cdot dr$$
$$= -\frac{2GM}{7R}(4\sqrt{2}-5)$$
$$\therefore \quad W = +\frac{2GM}{7R}(4\sqrt{2}-5)$$

53 **(b)**

The value of acceleration due to gravity at height h above the surface of the earth is given by

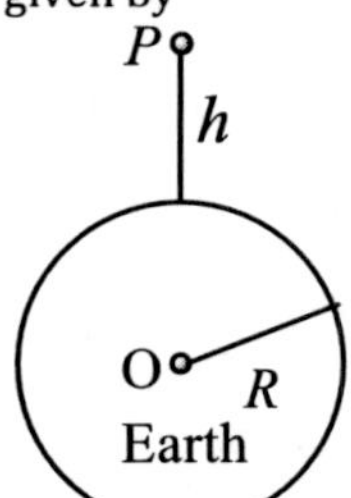

$$g' = \frac{g}{\left(1+\frac{h}{R}\right)^2}$$
$$g' = g\left(1+\frac{h}{R}\right)^{-2} = g\left(1-\frac{2h}{R}\right)$$
Given, $\quad g' = \frac{g}{4}$
$$\frac{g}{4} = g\left(1-\frac{2h}{R}\right)$$
$$\Rightarrow \quad \frac{1}{4} = 1-\frac{2h}{R}$$
$$\Rightarrow \quad \frac{2h}{R} = \frac{3}{4}$$
$$\Rightarrow \quad h = \frac{3R}{8}$$

54 **(d)**

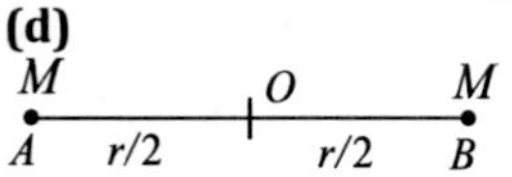

Gravitational potential of A at $O = -\frac{GM}{r/2} = -\frac{2GM}{r}$

For B, potential at $O = -\frac{GM}{r/2} = -\frac{2GM}{r}$

$\therefore$ Total potential $= -\frac{4GM}{r}$

55 **(c)**

Acceleration due to gravity on moon g_m
$$= \frac{G \times M/90}{(R/3)^2} = \frac{1}{10}g$$

56 **(c)**

Kepler's law $T^2 \propto R^3$

57 **(c)**

The escape velocity is independent of angle of projection, hence, it will remain same *ie.* 11 kms^{-1}.

58 **(c)**

If the body is projected with velocity $v(v < v_e)$ then height up to where it rises,
$$h = \frac{R}{\frac{v_e^2}{v^2}-1}$$
$$\Rightarrow h = \frac{R}{\left(\frac{11.2}{10}\right)^2 - 1} = 4R \text{ (approx.)}$$

59 **(a)**

$U = \frac{-GMm}{r}$ or $r = \frac{-GMm}{U}$

$$r = \frac{-6.67\times10^{-11}\times6\times10^{24}\times7.4\times10^{22}}{-7.79\times10^{38}}$$
$$= 3.8\times10^8\text{m}$$

61 **(a)**

Inside the earth $g' = \frac{4}{3}\pi\rho G r \;\therefore g' \propto r$

62 **(c)**

(i) $T_{st} = 2\pi\sqrt{\frac{(R+h)^3}{GM}} = 2\pi\sqrt{\frac{R}{g}}$[As $h << R$ and $GM = gR^2$]

(ii) $T_{ma} = 2\pi\sqrt{\frac{R}{g}}$

(iii) $T_{sp} = 2\pi\sqrt{\frac{1}{g\left(\frac{1}{l}+\frac{1}{R}\right)}} = 2\pi\sqrt{\frac{R}{2g}}$ [As $l = R$]

(iv) $T_{is} = 2\pi\sqrt{\frac{R}{g}}$[As $l = \infty$]

63 **(a)**

$$g' = \frac{g}{\left(1+\frac{h}{R}\right)^2}$$
$$\frac{g}{4} = \frac{g}{\left(1+\frac{h}{R}\right)^2}$$

$1+\frac{h}{R}=2 \Rightarrow \frac{h}{R}=1$
$\Rightarrow h=R$
$\therefore h=6400km$

64 **(b)**
When the spaceship is to take off, gravitational pull of earth requires more energy to be spent to overcome it

65 **(c)**

$$g_p=g_e\left(\frac{M_p}{M_e}\right)\left(\frac{R_e}{R_p}\right)^2=9.8\left(\frac{1}{80}\right)(2)^2$$
$=9.8/20=0.49m/s^2$

66 **(b)**
Time of decent $t=\sqrt{\frac{2h}{g}}$. In vacuum no other force works except gravity so time period will be exactly equal

67 **(c)**
$F=mI$
$\therefore I=\frac{F}{m}=\frac{45}{1.5}=30\text{N kg}^{-1}$

69 **(a)**
According to Kepler's law of periods
$T^2 \propto a^3[a=\text{semi-major axis}]$
Here, in case I a is $7R$ as satellite is $6R$ above the earth and for a geostationary satellite $T=24$ h
$\therefore \quad (24)^2 \propto (7R)^3 \quad ....(i)$
Similarly for case II
$T^2 \propto (3.5R)^3 \quad ....(ii)$
Dividing Eq. (i) by Eq. (ii), we get
$\frac{(24)^2}{T^2}=\frac{(7R)^3}{(3.5R)^3}$
$\Rightarrow \quad T^2=\frac{(24)^2}{8}$
or $\quad T=6\sqrt{2}\text{h}$

70 **(c)**
$g=\frac{GM}{r^2}$
$\therefore \log g=\log G+\log M-2\log r$
Differentiating both sides w.r.t. t
$\frac{1}{g}=\frac{dg}{dt}=0-2\times\frac{1}{r}\frac{dr}{dt}\left(\frac{dr}{dt}\times 100=-1\right)$
$\Rightarrow \frac{1}{g}\left(\frac{dg}{dt}\times 100\right)=-2\times\frac{1}{r}\left(\frac{dr}{dt}\times 100\right)$
$\Rightarrow \frac{dg}{dt}\times 100=-2\times(-1)=2$
$\therefore$ g increasing by 2%

72 **(c)**
According to kepler's third law $T^2 \propto r^3$;At $r=0, T=0$. It shows that the graph between T^2 and r^2 is a straight line passing through origin

73 **(c)**

$$F=\frac{G\times m\times m}{(2R)^2}=\frac{G\times\left(\frac{4}{3}\pi R^3\rho\right)^2}{4R^2}$$
$$=\frac{4}{3}\pi^2\rho^2R^4$$
$\therefore F\propto R^4$

74 **(a)**
Let R be the radius of earth and ρ its density, then since shape of earth is assumed spherical
we have

Mass of earth = volume × density

$$M=\frac{4}{3}\pi R^3\times\rho \quad ....(i)$$

The acceleration due to gravity which arises in the body due to gravitational force of attraction is given by

$$g=\frac{GM}{R^2} \quad ....(ii)$$

Putting the value of M from Eq.(i), we get

$$g=\frac{G\frac{4}{3}\pi R^3\rho}{R^2}=G\frac{4}{3}\pi R\rho \quad ....(iii)$$

Given, $\rho_p=\rho, R_p=0.2R_e$
$\therefore g_p=G\frac{4}{3}\pi R_p\rho_p=G\times\frac{4}{3}\pi\times 0.2R\rho$
$=0.2g$

75 **(a)**
$g=\frac{4}{3}\pi G\rho R \Rightarrow g\propto\rho R \Rightarrow \frac{g_e}{g_m}=\frac{\rho_e}{\rho_m}\times\frac{R_e}{R_m}$
$\Rightarrow \frac{6}{1}=\frac{5}{3}\times\frac{R_e}{R_m} \Rightarrow R_m=\frac{5}{18}R_e$

77 **(a)**

Escape velocity, $v_e=\sqrt{\frac{2GM_e}{R_e}}$

Given, $M_p=6M_e, R_p=2R_e$

$\therefore \quad v_p=\sqrt{\frac{2G\cdot 6M_e}{(2R_e)}}=\sqrt{3}\,v_e$

78 **(c)**

Acceleration due to gravity at a height h above the earth's surface is $g_h = \frac{g}{\left(1+\frac{h}{R}\right)^2}$

Where g is the acceleration due to gravity on the earth's surface

At $h = \frac{R}{2}, g_h = \frac{g}{\left(1+\frac{R}{2R}\right)^2} = \frac{4g}{9}$

At $h = R, g_h = \frac{g}{\left(1+\frac{R}{R}\right)^2} = \frac{g}{4}$

Acceleration due to gravity at a depth $d$below the earth's surface is $g_d = g\left(1 - \frac{d}{R}\right)$

At $d = \frac{R}{2}, g_d = g\left(1 - \frac{2}{2R}\right) = \frac{g}{2}$

At the centre of earth, $d = R$

$g_d = g\left(1 - \frac{R}{R}\right) = 0$

Thus, the acceleration due to gravity is maximum on the earth's surface

79 **(a)**

$$v = \sqrt{\frac{GM}{R+h}} = \frac{1}{2}\sqrt{\frac{2GM}{R}}$$

$\Rightarrow 4R = 2(R+h) \Rightarrow h = R = 6400\ km$

80 **(a)**

Since, $T^2 = kr^3$

$$\Rightarrow \frac{2\Delta T}{T} = \frac{3\Delta r}{r} \Rightarrow \frac{\Delta T}{T} = \frac{3}{2}\frac{\Delta r}{r}$$

81 **(d)**

Range of projectile $R = \frac{u^2 \sin 2\theta}{g}$

If u and θ are constant then $R \propto \frac{1}{g}$

$$\frac{R_m}{R_e} = \frac{g_e}{g_m} \Rightarrow \frac{R_m}{R_e} = \frac{1}{0.2} \Rightarrow R_m = \frac{R_e}{0.2} \Rightarrow R_m = 5R_e$$

82 **(a)**

Escape velocity $v_e = \sqrt{\frac{2GM}{R}}$

If $R' = \frac{R}{4}$

$$v'_e = 2\sqrt{\frac{2GM}{R}}$$

Since, G and M are constant hence, $v'_e = 2v_e$

84 **(c)**

Mass of planet, $M_p = 10M_e$, where M_e is mass of earth. Radius of planet,

$R_p = \frac{R_e}{10}$, where $R_e$is radius of earth.

Escape speed is given by,

$$v = \sqrt{\frac{2GM}{R}}$$

So, for planet $v_p = \sqrt{\frac{2\,G \times M_p}{R_p}}$

$$= \sqrt{\frac{100 \times 2\,GM_e}{R_e}}$$

$$= 10 \times v_e$$
$$= 10 \times 11\text{kms}^{-1}$$
$$= 110\text{kms}^{-1}$$

85 **(c)**

The period of revolution of a satellite at a height h from the surface of earth is given by

$$T = 2\pi\sqrt{\frac{(R_e+h)^3}{gR_e^2}}$$

Given, $T_m = 1$ lunar month,

$$\therefore \quad T_{\text{sat}} = 2\pi\sqrt{\frac{\left(R+\frac{h}{2}\right)^2}{gR^2}}$$

$$\Rightarrow \quad T_{\text{sat}} = \frac{1}{2^{3/2}}$$

$T_{\text{moon}} = 2^{-3/2}$ lunar month

86 **(d)**

Using law of conservation of energy

$$-\frac{GMm}{r} = \frac{1}{2}mv^2 - \frac{GMm}{R}$$

$$\frac{v^2}{2} = \frac{GM}{R} - \frac{GM}{r}$$

$$= GM\left(\frac{r-R}{rR}\right)$$

$$= gR\left(\frac{r-R}{r}\right)$$

$$v = \sqrt{2gR(r-R)/r}$$

87 **(c)**

Error in weighing

$= mg - mg' = mg - mg(1 - 2h/R)$

$= mg2h/R = \frac{m2hg}{R}$

$= \frac{m2h}{R} \times \frac{G\frac{4}{3}\pi R^2 \rho}{R^2} = \frac{8\pi G\rho mh}{3}$

88 **(d)**

Water fills the tube entirely in gravity less condition.

91 **(b)**

$$U_{(r)} = \begin{cases} -\frac{GMm}{r}, r \geq R \\ -\frac{GMm}{R}, r < R \end{cases}$$

92 **(a)**

$$g' = g - \omega^2 R\cos^2\lambda \Rightarrow 0$$
$$= g - \omega^2 R\cos^2 60°$$
$$0 = g - \frac{\omega^2 R}{4} \Rightarrow \omega = 2\sqrt{\frac{g}{R}} = \frac{1}{400}\frac{rad}{sec}$$
$$= 2.5 \times 10^{-3}\frac{rad}{sec}$$

93 **(d)**

Velocity of satellite $v = \sqrt{\frac{GM}{r}}$

$$\text{KE} \propto v^2 \propto \frac{1}{r}$$

and $T^2 \propto r^3$

$$\text{KE} \propto T^{-2/3}$$

94 **(a)**

$$\frac{GMM}{L^2} = \frac{MV^2}{L}$$
$$\Rightarrow V = \sqrt{\frac{GM}{L}}$$

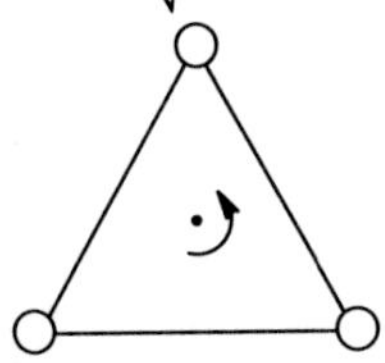

95 **(b)**

The acceleration due to gravity on the new planet can be using the relation

$$g = \frac{GM}{R^2} \quad \text{.... (i)}$$

but $M = \frac{4}{3}\pi R^3\rho$, ρ being density.

Thus, Eq. (i) becomes

$$\therefore \quad g = \frac{G \times \frac{4}{3}\pi R^3\rho}{R^2}$$
$$= G \times \frac{4}{3}\pi R\rho$$
$$\Rightarrow \quad g \propto R$$
$$\therefore \quad \frac{g'}{g} = \frac{R'}{R}$$
$$\Rightarrow \quad \frac{g'}{g'} = \frac{3R}{R} = 3$$
$$\Rightarrow \quad g' = 3g$$

96 **(a)**

For the satellite to move along closed orbit (a circle with a radius $R + h$) it should be acted upon by a force directed towards the centre. In this case, this is the force of earth's attraction. According to Newton's Second law

$$\frac{mv^2}{R+h} = \frac{GMm}{(R+h)^2}$$

At the earth's surface, $\frac{GMm}{R^2} = mg$

Therefore, $v = \sqrt{\frac{gR^2}{R+h}} = 7.5\ kms^{-1}$

97 **(a)**

$$v_e = \sqrt{\frac{2GM}{R}} = 100 \Rightarrow \frac{GM}{R} = 5000$$

Potential energy $U = -\frac{GMm}{R} = -5000\ J$

98 **(c)**

Gravitational potential on the surface of the shell is

$V =$ Gravitational potential due to particle (V_1)

+ Gravitational potential due to shell particle (V_2)

$$= -\frac{Gm}{R} + \left(-\frac{G3m}{R}\right) = -\frac{4Gm}{R}$$

99 **(c)**

Force on the body $= \frac{GMm}{x^2}$

To move it by a small distance dx,

Work done $= F\,dx = \frac{GMm}{x^2}dx$

Total work done $= GMm\int_R^{R+h}\frac{dx}{x^2} = \left[\frac{-GMm}{x}\right]_R^{R+h}$

$$= GMm\left[\frac{1}{R} - \frac{1}{R+h}\right]$$
$$= \left[\frac{(R+h)-R}{R(R+h)}\right] = \frac{GMmh}{R(R+h)}$$
$$\frac{GM}{R^3} \times \frac{mhR}{R+h} = \frac{gmhR}{R+h} = \frac{PRh}{R+h}$$

100 **(c)**

Mass of two planets is same, so

$$\frac{4}{3}\pi R_1^3\rho_1 = \frac{4}{3}\pi R_2^3\rho_2$$

or $\frac{R_1}{R_2} = \left(\frac{\rho_2}{\rho_1}\right)^{1/3} = \left(\frac{1}{8}\right)^{1/3} = \frac{1}{2}$

$$\frac{g_1}{g_2} = \frac{GM/R_1^2}{GM/R_2^2} = \left(\frac{R_2}{R_1}\right)^2 = (2)^2 = 4$$

101 **(b)**

$T \propto r^{3/2}$. If r becomes double then time period will become $(2)^{3/2}$ times

So new time period will be $24 \times 2\sqrt{2}\ hr\ i.e. T = 48\sqrt{2}$

102 **(d)**

The true weight of a body is given by mg and with height g decrease

So, $\frac{W_s}{W_E} = \frac{mg'}{mg} = \frac{1}{[1+(h/R)]^2}\left[\text{as} g' = \frac{g}{[1+(h/R)]^2}\right]$

But here, $h = 7R - R = 6R,\ ie, h/R = 6$

So, $W_S = \frac{W_E}{(1+6)^2} = \frac{10}{49} = 0.2\text{N}$

103 **(b)**

Consider that the earth is sphere of radius R and mass M. Then, value of acceleration due

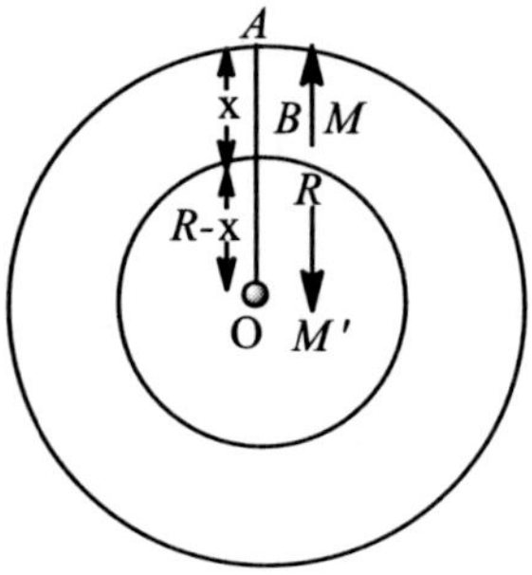

to gravity at the point A on the surface of earth is given by

$$g = \frac{GM}{R^2}$$

If ρ is density of the material of earth, then

$$M = \frac{4}{3}\pi R^3 \rho$$

$$\therefore \quad g = \frac{G \times \frac{4}{3}\pi R^3 \rho}{R^2}$$

or $\quad g = \frac{4}{3}\pi G R \rho$

Let g' be acceleration due to gravity at the point B at a depth x below the surface of earth. A body at point B will experience force only due to the portion of the earth of radius $OB(= R - x)$. The outer spherical shell, whose thickness is x, will not exert any force on body at point B.

104 **(b)**

$$v = \sqrt{\frac{2GM}{R}} = \sqrt{2gR}$$

$= \sqrt{2 \times (3.1)^2 \times 8100 \times (10)^3}$

$= 27.9\ km/sec^{-1}$

105 **(b)**

$T^2 \propto r^3$

106 **(c)**

For $r \leq R$:

$$\frac{mv^2}{r} = \frac{Gmm'}{r^2}$$

Here, $\quad m' = \left(\frac{4}{3}\pi r^3\right)\rho_0$

Substituting in Eq. (i) we get

$$v \propto r$$

ie, $v-r$ graph is a straight line passing through orgine.

For $r > R$:

$$\frac{mv^2}{r} = \frac{G\,m\left(\frac{4}{3}\pi R^3\right)\rho_0}{r^2}$$

or $\quad v \propto \frac{1}{\sqrt{r}}$

The corresponding $v-r$ graph will be as shown in option (c).

107 **(c)**

Acceleration due to gravity at poles is independent of the angular speed of earth

108 **(b)**

$mg = \frac{GM_E m}{R_E^2}$; where M_E and R_E is the mass and radius of the earth respectively. $M_E = \frac{g}{G}R_E^2$

110 **(b)**

$$T_2 = T_1\left(\frac{R_2}{R_1}\right)^{3/2} = 1 \times (2)^{3/2} = 2.8\ year$$

111 **(d)**

$F = mR\,\omega^2$

$= 6 \times 10^{24} \times (1.5 \times 10^{11})(2 \times 10^{-7})^2$

$= 36 \times 10^{21}\text{N}$

113 **(a)**

Force between earth and moon $F = \frac{Gm_m m_e}{r^2}$

This amount of force, both earth and moon will exert on each other *i.e.* they exert same force on each other

114 **(c)**

Gravitational potential at a pint outside the sphere $V_g = \frac{-GM}{r}$. But V_s is same at a point inside the hollow sphere as on the surface of sphere. Hence, graph (c) is correct.

115 **(c)**

A person feels weightlessness in satellite orbit because he is in free fall along with the satellite and experiences no force of support from the satellite. The perception of weight comes from the support force exerted on one by the floor, a chair etc. If that support is removed and one is in free fall, we feel no experience of weight.

116 **(b)**

If missile is launched with escape velocity, then it will escape from the gravitational field and at infinity its total energy becomes zero

But if the velocity of projection is less than escape velocity then sum of energies will be negative. This shows that attractive force is working on the missile

117 **(b)**

$$\frac{(v_e)p_1}{(v_e)p_2} = \frac{\sqrt{2g_1R_1}}{\sqrt{2g_1R_2}} = \sqrt{\frac{g_1}{g_2} \times \frac{R_1}{R_2}} = \sqrt{ab}$$

119 **(c)**

For $w2w$, $3w$ apparent weight will be zero because the system is falling freely. So the distances of the weights from the rod will be same.

120 **(a)**

$$\text{Potential energy} = \frac{-GMm}{r} = \frac{GMm}{R_e+h} = \frac{-GMm}{2R_e}$$

$$= -\frac{gR_e^2m}{2R_e} = -\frac{1}{2}mgR_e = -0.5mgR_e$$

121 **(c)**

$$\frac{1}{2}mv_e^2 = \frac{1}{2}m\,2gR = mgR$$

122 **(c)**

Given $\quad \frac{R_e}{R_p} = \frac{2}{3}$

$$\frac{d_e}{d_p} = \frac{4}{5}$$

As $\quad MG = gR_e^2$

and $\quad M = d_e \times \frac{4}{3}\pi R_e^3$

$$d_e \times \frac{4}{3}\pi R_e^3 \times G = g_eR_e^2$$

or $d_e \times \frac{4}{3}\pi R_e \times G = g_e \quad \ldots.(i)$

Similarly for planet

$d_p \times \frac{4}{3}\pi R_pG = g_p \quad \ldots.(ii)$

Dividing Eq. (i) by Eq. (ii), we get

$$\frac{g_e}{g_p} = \frac{R_e}{R_p} \times \frac{d_e}{d_p}$$

$$\frac{g_e}{g_p} = \frac{2}{3} \times \frac{4}{5} = \frac{8}{15} = 0.5$$

123 **(a)**

$$v = \sqrt{2gR} \therefore \frac{v_1}{v_2} = \sqrt{\frac{g_1}{g_2} \times \frac{R_1}{R_2}} = \sqrt{g \times K}$$

$$= (Kg)^{1/2}$$

124 **(b)**

The energy required to remove the satellite from its orbit around the earth to infinity is called binding energy of the satellite. It is equal to negative of total mechanical energy of satellite in its orbit.

Thus, binding energy $= -E = \frac{GMm}{2r}$

but, $\quad g = \frac{GM}{R^2}$

$\Rightarrow \quad GM = gR^2$

$\therefore \quad \text{BE} = \frac{gmR^2}{2r}$

125 **(b)**

Time period is independent of mass. Therefore their periods of revolution will be same.

126 **(a)**

Radius of earth $R = 6400\ km \therefore h = \frac{R}{4}$

Acceleration due to gravity at a height h

$$g_h = g\left(\frac{R}{R+h}\right)^2 = g\left(\frac{R}{R+\frac{R}{4}}\right)^2 = \frac{16}{25}g$$

At depth $'d'$ value of acceleration due to gravity

$g_d = \frac{1}{2}g_h$ (According to problem)

$$\Rightarrow g_d = \frac{1}{2}\left(\frac{16}{25}\right)g \Rightarrow g\left(1-\frac{d}{R}\right) = \frac{1}{2}\left(\frac{16}{25}\right)g$$

By solving we get $d = 4.3 \times 10^6 m$

127 **(c)**

Time period of nearby satellite

$$T = 2n\sqrt{\frac{r^3}{GM}}$$

$$= 2\pi\sqrt{\frac{R^3}{GM}}$$

$$= \frac{2\pi(R^3)^{1/2}}{\left[G\,\frac{4}{3}\pi R^3\rho\right]^{1/2}} = \sqrt{\frac{3\pi}{G\rho}}$$

128 **(a)**

Gravitational force dsesnot depend on the medium.

129 **(d)**

The body can be fired at any angle because the energy is sufficient to take the body out of the gravitational field of earth

130 **(b)**

Gravitational force provides the required centripetal force

$$m\omega^2R = \frac{GMm}{R^3} \Rightarrow \frac{4\pi^2}{T^2} = \frac{GM}{R^4} \Rightarrow T \propto R^2$$

131 **(c)**

Force acting on a body of mass M at a point at depth d. Inside the earth is

$F = mg' = mg\left(1 - \frac{d}{R}\right)$

$= \frac{mGM}{R^2}\left(\frac{R-d}{R}\right) = \frac{GM\,m}{R^3}r \quad (\because R - d = r)$

So, $F \propto r$; Given $F \propto r^n$

$n = 1$

132 **(c)**

According to Kepler's law

$T^2 \propto r^3$

or $\quad 5^2 \propto r^3 \quad$ ... (i)

and $\quad (T')^2 \propto (4r)^3 \quad$ ... (ii)

From Eqs.(i) and (ii), we have

$\frac{25}{(T')^2} = \frac{r^3}{64r^3}$

$$T = \sqrt{1600} = 40\text{h}$$

133 **(c)**

Escape velocity,

$v_{\text{ascape}} = \sqrt{\frac{2\,GM}{R}}$

$$= R\sqrt{\frac{8}{3}\pi G\rho}$$

$\therefore v_e \propto R$ if $\rho =$ constant.

Since the planet is having double radius in comparision to earth, therefore escape velocity becomes twice *ie*, 22 kms^{-1}.

134 **(d)**

Let the gravitational force on a body mass m at O due to moon of mass M and earth of mass $8/M$ be zero, where $EO = x$ and $MO = (r - x)$. Then,

$\frac{G81M \times m}{x^2} = \frac{GM\,m}{(r-x)^2}$

or $\frac{81}{x^2} = \frac{1}{(r-x)^2}$

or $\frac{9}{x} = \frac{1}{(r-x)}$

On solving; $x = 9r/10$

135 **(b)**

$g' = g - \omega^2 R\cos^2\lambda$

Rotation of the earth results in the decreased weight apparently. This decrease in weight is not felt at the poles as the angle of latitude is 90°

136 **(c)**

$\frac{T_1}{T_2} = \left(\frac{R_1}{R_2}\right)^{3/2} = \left(\frac{10^{13}}{10^{12}}\right)^{3/2} = (1000)^{1/2}$

$$= 10\sqrt{10}$$

137 **(c)**

$\frac{g'}{g} = 1 - \frac{2h}{R} = 1 - \frac{2 \times 320}{6400} = 1 - \frac{1}{10} = \frac{9}{10}$

$\therefore$ % decrease in g $= \left(\frac{g-g'}{g}\right) \times 100$

$= \frac{1}{100} \times 100 = 10\%$

138 **(c)**

Work done

$$W = \Delta U = \frac{mgh}{1 + \frac{-h}{R}}$$

Substituting $R = \frac{h}{L}$ we get

$$\Delta U = \frac{mg \times 2R}{1+2}$$

$$\Delta U = \frac{2mgR}{3}$$

139 **(a)**

When the thief with box on his head jumped down from a wall, he along with box is falling down with acceleration due to gravity, so the apparent weight of box becomes zero, (because, $R = mg - mg = 0$), so he experiences no load till he reaches the ground

140 **(d)**

$\Delta U = U_2 - U_1 = \frac{mgh}{1 + \frac{h}{R_e}} = \frac{mgR_e}{1 + \frac{R_e}{R_e}} = \frac{mgR_e}{2}$

$\Rightarrow U_2 - (-mgR_4) = \frac{mgR_e}{2} \Rightarrow U_2$

$$= -\frac{1}{2}mgR_e$$

141 **(c)**

The velocity of the spoon will be equal to the orbital velocity when dropped out of the space-ship

142 **(b)**

Let velocities of these masses at r distance from each other be v_1 and v_2 respectively

By conservation of momentum

$m_1v_1 - m_2v_2 = 0$

$\Rightarrow m_1v_1 = m_2v_2 \quad$ ...(i)

By conservation of energy

Change in $P.E. =$ change in $K.E.$

$\frac{Gm_1m_2}{r} = \frac{1}{2}m_1v_1^2 + \frac{1}{2}m_2v_2^2$

$\Rightarrow \frac{m_1^2v_1^2}{m_1} + \frac{m_2^2v_2^2}{m_2} = \frac{2Gm_1m_2}{r}$ ...(ii)

On solving equation (i) and (ii)

$v_1 = \sqrt{\frac{2Gm_2^2}{r(m_1+m_2)}}$ and $v_2 = \sqrt{\frac{2Gm_1^2}{r(m_1+m_2)}}$

$\therefore v_{\text{app}} = |v_1| + |v_2| = \sqrt{\frac{2G}{r}(m_1 + m_2)}$

144 **(d)**

Kinetic energy of the satellite is $K = \frac{GMm}{2r}$ ...(i)

Potential energy of the satellite is $U = -\frac{GMm}{r}$ ...(ii)

Total energy of the satellite is $E = -\frac{GMm}{2r}$ ...(iii)

Divide (iii) by (i), we get $\frac{E}{K} = -1$ or $E = -K$

Divide (iii) by (ii), we get $\frac{E}{U} = \frac{1}{2}$ or $E = \frac{U}{2}$

145 **(d)**

Time period of simple pendulum $T = 2\pi\sqrt{\frac{1}{g'}}$

In artificial satellite $g' = 0 \therefore T =$ infinite

146 **(b)**

Gravitational force provides the required centripetal force *ie*,

$$m\omega^2 R = \frac{GMm}{R^{\frac{5}{2}}}$$

$$\Rightarrow \quad \frac{m4\pi^2}{T^2} = \frac{GMm}{R^{\frac{7}{2}}}$$

$$\Rightarrow \quad T^2 \propto R^{7/2}$$

147 **(a)**

The energy given to the body so as to completely escape from its orbit is equal to its kinetic energy KE.

148 **(c)**

Acceleration due to gravity at an altitude h is

$g_h = \frac{gR_e^2}{(R_e+h)^2}$; where R_e is the radius of the earth

$$g_h = \frac{9.8m/s^2 \times (6400 \times 10^3 m)^2}{(6400 \times 10^3 m + 520 \times 10^3 m)^2} = 8.4m/s^2$$

149 **(a)**

Escape velocity $v = \sqrt{2gR}$

$$\therefore \quad \frac{v_1}{v_2} = \sqrt{\frac{g_1}{g_2} \times \frac{R_1}{R_2}} = \sqrt{g \times K} = (Kg)^{1/2}$$

150 **(c)**

From Kepler's second law of planetary motion, the linear speed of a planet is maximum, when its distance from the sun is least, *ie*, at point A.

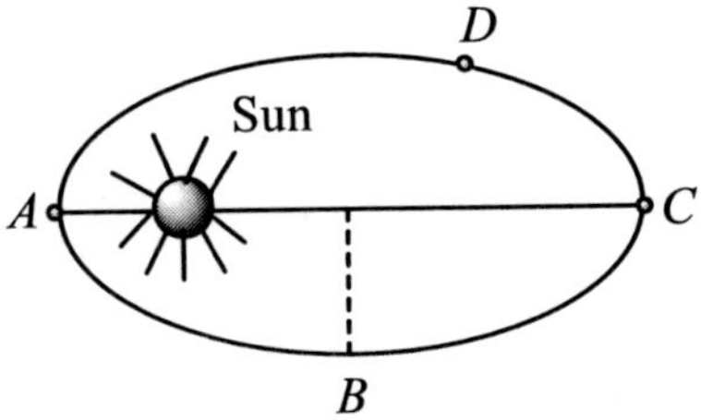

151 **(a)**

Given, $\frac{mg'}{mg} = \frac{30}{90}$ or $\frac{g'}{g} = \frac{1}{3}$

Now, $g' = g\frac{R^2}{(R+h)^2}$ or $\frac{g'}{g} = \frac{R^2}{(R+h)^2} = \frac{1}{3}$

or $\frac{R}{R+h} = \frac{1}{\sqrt{3}}$ or $(R+h) = \sqrt{3}R$

or $h = (\sqrt{3}-1)R = 0.73R$

153 **(a)**

$$g' = g - \frac{10g}{100} - \frac{90}{100}g$$

$$g' = g\frac{R^2}{(R+h)^2} \text{ or } \frac{9}{10} = \frac{R^2}{(R+h)^2}$$

or $\frac{3}{\sqrt{10}} = \frac{R}{R+h}$

or $h = (\sqrt{10}-3)R/3$

$$\frac{(\sqrt{10}-3) \times 6400}{3} = 345.60 \text{ km}$$

154 **(c)**

If m is the mass and v is the orbital velocity of the satellite, then kinetic energy.

$$E = \frac{1}{2}mv^2$$

or $\quad Em = \frac{1}{2}m^2v^2$

or $\quad m^2v^2 = 2Em$

or $\quad mv = \sqrt{2Em}$ ....(i)

If r is the radius of the orbit of the satellite, then its angular momentum

$$L = mvr$$

Using Eq. (i),

$$L = (\sqrt{2Em})r = \sqrt{2Emr^2}$$

155 **(a)**

Angular momentum = Mass × Orbital velocity × Radius

$$= m \times \left(\sqrt{\frac{GM}{R_0}}\right) \times R_0 = m\sqrt{GMR_0}$$

156 **(d)**

Total mechanical energy of satellite

$$E = \frac{-GMm}{2r} \Rightarrow \frac{E_A}{E_B} = \frac{m_A}{m_B} \times \frac{r_B}{r_A} \Rightarrow \frac{3}{1} \times \frac{4r}{r} = \frac{12}{1}$$

157 **(c)**

$$\text{kinetic energy} = \frac{1}{2}mv_e^2$$
$$= \frac{1}{2}m \times 2gR$$
$$= mgR$$

158 **(d)**

A person is safe, if his velocity while reaching the surface of moon from a height h' is equal to its velocity while falling from height h on earth. So

$\sqrt{2g'h'} = \sqrt{2gh}$

or $h' = gh/g' = 9.8 \times 3/1.96 = 15\text{m}$

159 **(b)**

When going above at a height h or at a depth d below earth's surface, in any case acceleration due to gravity decrease. Therefore,

$g_e > g_h$ and $g_e > g_d$

Moreover $g_h < g_d$, if $h = d$.

160 **(a)**

$$v = \sqrt{2gR} \Rightarrow \frac{v_p}{v_e} = \sqrt{\frac{g_p}{g_e} \times \frac{R_e}{R_p}} = \sqrt{2 \times \frac{1}{4}}$$
$$= \frac{1}{\sqrt{2}}$$

$$\therefore v_p = \frac{v_e}{\sqrt{2}}$$

161 **(b)**

Gravity, $g = \frac{GM}{R^2}$

$$\therefore \frac{g_{earth}}{g_{planet}} = \frac{M_e}{M_p} \times \frac{R_p^2}{R_e^2}$$

$$\Rightarrow \frac{g_e}{g_p} = \frac{2}{1}$$

Also, $T \propto \frac{1}{\sqrt{g}} \Rightarrow \frac{T_e}{T_p} = \sqrt{\frac{g_p}{g_e}}$

$$\Rightarrow \frac{2}{T_p} = \sqrt{\frac{1}{2}} \Rightarrow T_p = 2\sqrt{2}s$$

162 **(d)**

As we know gas molecules cannot escape from earth's atmosphere because their root mean square velocity is less than escape velocity at earth's surface. If we fill this requirement, then gas molecules can escape from earth's atmosphere.

ie, $v_{\text{rms}} = v_{\text{es}}$

or $\sqrt{\frac{3RT}{M}} = \sqrt{2gR_e}$

or $$T = \frac{2MgR_e}{3R} \quad \text{....(i)}$$

Given, $M = 2 \times 10^{-3}\text{kg}$,

$g = 9.8\ \text{ms}^{-2}$

$R_e = 6.4 \times 10^6\text{m}, R = 8.31\ \text{Jmol}^{-1}\text{–}\ \text{K}^{-1}$

Substituting in Eq. (i), we have

$$T = \frac{2 \times 2 \times 10^{-3} \times 9.8 \times 6.4 \times 10^6}{3 \times 8.31}$$
$$= 10^4\text{K}$$

163 **(b)**

$$v_e = R\sqrt{\frac{8}{3}G\pi\rho} \quad \therefore v_e \propto R\sqrt{\rho}$$

164 **(c)**

From Kepler 's third law of planetary motion,

$T^2 \propto a^3$

Given,

$T_1 = 1\text{day}$ (geostationary)

$a_1 = a, a_2 = 2a$

$$\therefore \frac{T_1^2}{T_2^2} = \frac{a_1^3}{a_2^3}$$

$$\Rightarrow T_2^2 = \frac{a_2^3}{a_1^3}T_1^2 = \frac{(2a)^3}{a^3} \times 1 = 8$$

$$\Rightarrow T_2 = 2\sqrt{2}\text{days}$$

165 **(b)**

Gravitational acceleration is given by

$$g = \frac{GM}{R^2}$$

where G = gravitational constant

$$\therefore \frac{g}{G} = \frac{M}{R^2}$$

166 **(c)**

Weight of the body at equator $= \frac{3}{5}$ of initial weight

$\therefore g' = \frac{3}{5}g$ (because mass remains constant)

$$g' = g - \omega^2 R\cos^2\lambda \Rightarrow \frac{3}{5}g$$
$$= g - \omega^2 R\cos^2(0°)$$

$$\Rightarrow \omega^2 = \frac{2g}{5R} \Rightarrow \omega = \sqrt{\frac{2g}{5R}}$$
$$= \sqrt{\frac{2 \times 10}{5 \times 6400 \times 10^3}}$$
$$= 7.8 \times 10^{-4}\frac{rad}{sec}$$

167 **(c)**

$$T_2 = T_1\left(\frac{r_2}{r_1}\right)^{2/3} = 83\left(\frac{R+3R}{R}\right)^{3/2}$$
$= 83 \times 8 = 664$ min

169 **(a)**

$$I = \frac{-dV}{dx}$$
If $V = 0$ then gravitational field is necessarily zero

170 **(a)**

$v \propto \frac{1}{\sqrt{r}}$. If orbital radius becomes 4 times then orbital velocity will becomes half, $i.e. \frac{7}{2} = 3.5\ km/s$

171 **(a)**

Time period, $T = \frac{2\pi R}{\sqrt{\frac{GM_m}{R}}} = \frac{2\pi R^{3/2}}{\sqrt{GM_m}}$

Where the symbols have their meaning as given in the question

Squaring both sides, we get
$$T^2 = \frac{4\pi^2 R^3}{GM_m}$$

172 **(a)**

If body is projected with velocity $v (v < v_e)$ then

Height up to which it will rise, $h = \frac{R}{\frac{v_e^2}{v^2}-1}$

$v = \frac{v_e}{2}$ (Given) $\therefore h = \frac{R}{\left(\frac{(v_e)}{v_e/2}\right)^2 - 1} = \frac{R}{4-1} = \frac{R}{3}$

173 **(c)**

Mass does not vary from place to place

174 **(d)**

Time period (T) of a synchronous satellite around the earth is given by
$$T^2 = \frac{4\pi^2 r^3}{Gm_e} \Rightarrow r = \left(\frac{T^2 Gm_e}{4\pi^2}\right)^{1/3}$$
Substituting the given values, we get
$$r = \left[\frac{(24 \times 60 \times 60)^2 + 6.67 \times 10^{-11} \times 5.98 \times 10^{24}}{4 \times \frac{22}{7} \times \frac{22}{7}}\right]^{1/3}$$
$r = 42.08 \times 10^6 m$
$$\therefore \frac{r}{r_e} = \frac{42.08 \times 10^6 m}{6.37 \times 10^6 m} = 6.6 \Rightarrow r = 6.6 r_e$$

175 **(a)**

Potential at the centre due to single mass $= \frac{-GM}{L/\sqrt{2}}$

Potential at the centre due to all four masses
$$= -4\frac{GM}{L/\sqrt{2}} = -4\sqrt{2}\frac{GM}{L}$$
$$= -\sqrt{32} \times \frac{GM}{L}$$

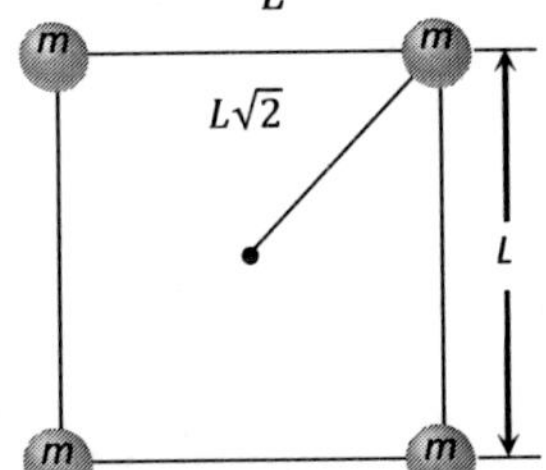

176 **(a)**

Acceleration due to gravity on the surface of the earth is
$$g_e = \frac{GM_e}{R_e^2}$$
Where M_e and R_e are the mass and the radius of the earth respectively

Acceleration due to gravity on the surface of the planetis $g_p = \frac{GM_p}{R_p^2}$

Where M_p and R_p be the mass and the radius of the planet respectively

If both mass and radius of the planet are half as that of the earth, then
$$g_p = \frac{G(M_e/2)}{(R_e/2)^2} = 2\frac{GM_e}{R_e^2} = 2g_e$$

177 **(c)**

The potential energy of an object at the surface of the earth
$$U_1 = -\frac{GMm}{R} \quad ....(i)$$
The potential energy of the subject at a height $h = R$ from the surface of the earth
$$U_2 = -\frac{GMm}{R+h} = -\frac{GMm}{R+R} \quad ...(ii)$$
Hence, the gain in potential energy of the object
$$\Delta U = U_2 - U_1$$
$$\Delta U = -\frac{GMm}{R+R} + \frac{GMm}{R}$$
$$\Delta U = -\frac{GMm}{2R} + \frac{GMm}{R}$$
$$\Delta U = \frac{1}{2}\frac{GMm}{R}$$
But we know that $GM = gR^2$

Hence, $\Delta U = \frac{1}{2}\frac{gR^2 m}{R}$

or $\Delta U = \frac{1}{2}mgR$

178 **(a)**

$$g' = g\left(\frac{R}{R+h}\right)^2 = g\left(\frac{R}{R+\frac{R}{2}}\right)^2 = \frac{4}{9}g$$

$$\therefore W' = \frac{4}{9}\times W = \frac{4}{9}\times 72 = 32N$$

179 **(c)**

Gravitational field due to a spherical shell

At a point inside the shell, *i.e.*, $r < R$

$E_{\text{inside}} = 0$

∴ The gravitational force acting on a point mass m at a distance $R/2$ is

$F = mE_{\text{inside}} = 0$

180 **(b)**

Gravitational force provides necessary centripetal force

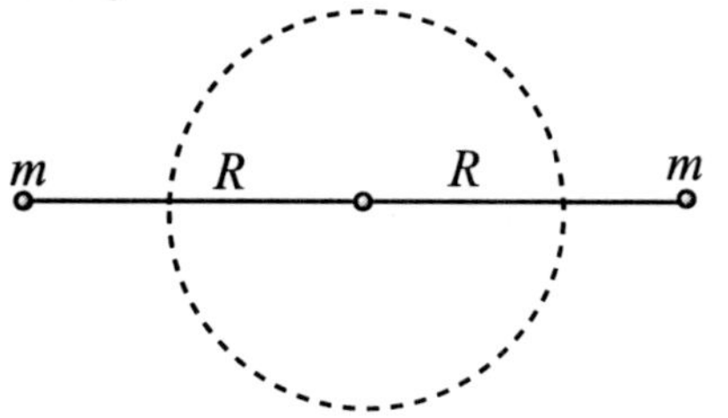

$$\frac{Gm^2}{(2R)^2} = \frac{mv^2}{R}$$

$$\Rightarrow \quad v = \sqrt{\frac{Gm}{4R}}$$

181 **(b)**

$$\therefore dF = \frac{Gm(\mu dx)}{x^2}$$

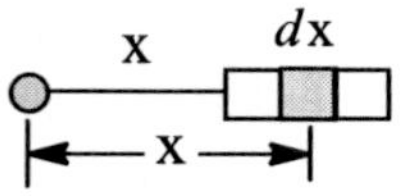

$$F = Gm\int_a^{a+L}(A+Bx)\frac{dx}{x^2}$$

$$F = Gm\left[A\left(\frac{1}{a}-\frac{1}{a+L}\right)+BL\right]$$

183 **(d)**

$$F = \frac{Gm_1m_2}{(r+2r)^2} = \frac{Gm_1m_2}{9r^2}, ie, F \propto r^{-2}$$

Note that $F \propto r^4$ by taking $m = \frac{4}{3}\pi r^4 \rho$ and then

$$F \propto \frac{r^3r^3}{r^2}, ie, F \propto r^4$$

is not correct because the gravitational law obeys inverse square law and is not related with densities

184 **(c)**

$g = \frac{GM}{R^2}$; If R decreases then g increases.

Taking logarithm of both the sides;

$\log g = \log G + \log M = -2\log R$

Differentiating it we get; $\frac{dg}{g} = 0 + 0 - \frac{2dR}{R}$

$$= -2\left(\frac{-2}{100}\right) = \frac{4}{100}$$

∴ % increase in $g = \frac{dg}{g}\times 100 = \frac{4}{100}\times 100 = 4\%$

185 **(c)**

Kinetic energy = Potential energy

$$\frac{1}{2}m(kv_e)^2 = \frac{mgh}{1+\frac{h}{R}} \Rightarrow \frac{1}{2}mk^2 2gR = \frac{mgh}{1+\frac{h}{R}}$$

$$\Rightarrow h = \frac{Rk^2}{1-k^2}$$

Height of Projectile from the earth's surface = h

Height from the centre $r = R + h = R + \frac{Rk^2}{1-k^2}$

By solving $r = \frac{R}{1-k^2}$

186 **(b)**

$$g = \frac{GM}{R^2} \text{ or } R = \sqrt{\frac{GM}{g}}$$

$$= \sqrt{6.67\times10^{-11}\times7.34\times10^{22}/1.4}$$
$$= 1.87\times10^6 \text{ m}$$

188 **(a)**

$$v = \sqrt{\frac{GM}{r}} = \sqrt{\frac{6.67\times10^{-11}\times6\times10^{24}}{384000\times10^3}}$$
$$= 1\ km/s$$

189 **(c)**

$T^2 = \frac{4\pi^2}{GM}r^3$. If G is variable then time period, angular velocity and orbital radius also changes accordingly

190 **(c)**

Let at O there will be a collision. If smaller sphere moves x distance to reach at O, then bigger sphere will move a distance of $(9R - x)$

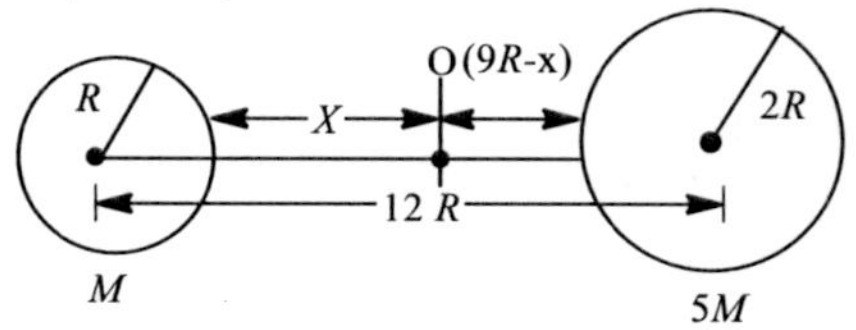

$$a_{\text{small}} = \frac{F}{M} = \frac{G\times 5M}{(12R-x)^2}$$

$$a_{big} = \frac{F}{5M} = \frac{GM}{(12R-x)^2}$$

$$x = \frac{1}{2}a_{small}t^2$$

$$= \frac{1}{2}\frac{G \times 5M}{(12R-x)^2} \quad \text{... (i)}$$

$$(9R - x) = \frac{1}{2}a_{big}t^2$$

$$= \frac{1}{2}\frac{GM}{(12R-x)^2}t^2 \quad \text{... (ii)}$$

Thus, dividing Eq. (i) by Eq. (ii), we get

$$\therefore \quad \frac{x}{9R-x} = 5$$

$$\Rightarrow \quad x = 45R - 5x$$

$$\Rightarrow \quad 6x = 45R$$

$$\Rightarrow \quad x = 7.5\,R$$

191 **(c)**

The value of acceleration due to gravity at a height h above the earth's surface is given by

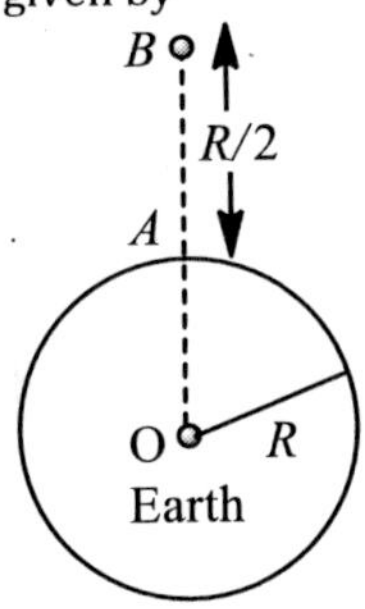

$$g' = \frac{g}{\left(1+\frac{h}{R}\right)^2}$$

where R is radius of earth.

When $\quad h = \frac{R}{2}$

$$g' = \frac{g}{\left(1+\frac{h}{R}\right)^2} = \frac{4g}{9}$$

Hence, weight $w' = mg' = \frac{4}{9}mg = \frac{4}{9}w.$

193 **(c)**

Angular momentum remains constant

$$mv_1d_1 = mv_2d_2 \Rightarrow v_2 = \frac{v_1d_1}{d_2}$$

194 **(b)**

Gravitational force on a body at a distance x from the centre of earth $F = \frac{GMm}{x^2}$

Work done,

$$W = \int_R^{R+h} F\,dx = \int_R^{R+h} \frac{GM\,m}{x^2}dx$$

$$= GMm\left[-\frac{1}{x}\right]_R^{R+h} = mgR^2\left[\frac{1}{R} - \frac{1}{R+h}\right]$$

This work done appears as increase in potential energy

$$\Delta E_p = mgR^2\left[\frac{1}{R} - \frac{1}{R+h}\right]$$

$$= mg(5h)^2\left[\frac{1}{5h} - \frac{1}{6h}\right] = \frac{5}{6}mgh$$

195 **(b)**

$$v = \sqrt{\frac{GM}{r}}$$

196 **(d)**

Acceleration due to gravity on earth is given by $g = \frac{GM}{R^2}$

$\left(\text{Here, } M_m = \frac{M_e}{9}, R_m = \frac{R_e}{2}\right)$

Hence, $\quad \frac{g_e}{g_m} = \frac{M_e}{M_m} \times \frac{R_m^2}{R_e^2}$

$$= \frac{9M_e}{M_e} \times \left(\frac{R_e}{2R_e}\right)^2$$

or $\frac{g_e}{g_m} = \frac{9}{4}$

So, $\frac{g_m}{g_e} = \frac{4}{9}$

$\because$ Weight of body on moon

$= $ weight of body on earth $\times g_m/g_e$

$$= 90 \times \frac{4}{9} = 90 \times \frac{4}{9} = 40\text{kg}$$

197 **(b)**

$$g' = g\left(\frac{R}{R+h}\right)^2 = g\left(\frac{R}{(3R/2)}\right)^2 = \frac{4}{9}g \quad [g = 10m/sec^2]$$

$$\therefore W' = \frac{4}{9} \times mg = \frac{4 \times 200 \times 10}{9} = 889\,N$$

198 **(a)**

$T = 2\pi\sqrt{\frac{l}{g}}$. At the hill g will decrease so to keep the time period same the length of pendulum has to be reduced

199 **(d)**

Orbital velocity $v_0 = \sqrt{\frac{GM}{r}} = \sqrt{\frac{gR^2}{r}}$ and

$v_0 = r\omega$

This gives $r^3 = \frac{R^2g}{\omega^2}$

200 **(d)**

Escape velocity of the body from the surface of earth is $v = \sqrt{2gR}$

Escape velocity of the body from the platform

Potential energy + Kinetic energy = 0

$$\Rightarrow -\frac{GMm}{2R}+\frac{1}{2}mv_p^2=0 \Rightarrow v_p=\sqrt{\frac{GM}{R^2}.R}$$

$$=\sqrt{gR}$$

$$=\frac{1}{\sqrt{2}}\sqrt{2gR}=\frac{1}{\sqrt{2}};\ \therefore f=\frac{1}{\sqrt{2}}$$

201 **(d)**

Kinetic energy of satellite in its orbit

$$E=\frac{1}{2}mv_o^2$$

or $$E=\frac{1}{2}m\left(\frac{GM}{r}\right)=\frac{GMm}{2r}$$

kinetic energy at escape velocity

$$E'=\frac{1}{2}mv_e^2$$

$$=\frac{1}{2}m\left(\frac{2GM}{r}\right)=\frac{GMm}{r}$$

$$=2E$$

Therefore, additional kinetic energy required

$$=2E-E=E$$

202 **(c)**

For earth, $g=\frac{GM}{R^2}=\frac{4}{3}\pi R\rho G$

For the planet, $g_1=\frac{GM_1}{R_1^2}=\frac{4}{3}\pi R_1\rho G$

$$\frac{g}{g_1}=\frac{R}{R_1}=\frac{6400}{320}=20$$

Let h and h_1 be the distance upto which the man can jump on surface of the earth and planet, then

$mgh=mg_1h_1$

$$\therefore\ h_1=\frac{g}{g_1}h=20\times5=100\text{ m}$$

203 **(c)**

Acceleration due to gravity at height h

$$g_h=g\left(1-\frac{2h}{R}\right) \quad \text{....(i)}$$

and depth d

$$g_d=g\left(1-\frac{d}{R}\right) \quad \text{....(ii)}$$

From Eq. (i) and (ii),

$$g\left(1-\frac{2h}{R}\right)=g\left(1-\frac{d}{g}\right)$$

$$\Rightarrow \quad 2h=d$$

204 **(a)**

$$(\text{KE})_{\text{escape}}=\frac{1}{2}m\left(\sqrt{\frac{2GM}{R_e}}\right)^2=\frac{GMm}{R_e}$$

$$(\text{KE})_{\substack{\text{body}\\\text{initially}}}=\frac{1}{2}\frac{GMm}{R_e}$$

By law of conservation of energy

$$\begin{pmatrix}\text{Total}\\\text{mechanical}\\\text{energy}\end{pmatrix}=\begin{pmatrix}\text{Total final}\\\text{mechanical}\\\text{energy}\end{pmatrix}$$

$$(\text{KE + PE})_{\text{surface}}=(\text{KE + PE})_{\text{at height }h}$$

$$\Rightarrow \quad \frac{1}{2}\frac{GMM}{R_e}-\frac{GMm}{R_e}=0-\frac{GMm}{R_e+h}$$

($\therefore$ velocity at maximum height is zero)

$\Rightarrow v=R_e$

205 **(b)**

$$v=\sqrt{2gR}\Rightarrow\frac{v_A}{v_B}=\sqrt{\frac{g_A}{g_B}\times\frac{R_A}{R_B}}=\sqrt{k_1\times k_2}$$

$$=\sqrt{k_1k_2}$$

206 **(c)**

$V_P=-\frac{GM}{2R^3}(3R^2-r^2)$ inside the sphere

and $V_P=-\frac{GM}{r}$

outside the sphere

207 **(c)**

Escape velocity of the planet is $v_p=\sqrt{\frac{2GM_P}{R_P}}$

Where M_p and R_p be the mass and radius of the planet respectively

Escape velocity of the earth is $v_e=\sqrt{\frac{2GM_e}{R_e}}$

Where M_e and R_e be the mass and radius of the earth respectively

According to given problem, $v_p=3v_e$ and $R_p=4R_e$

$$\therefore\ \sqrt{\frac{2GM_p}{4R_e}}=3\sqrt{\frac{2GM_e}{R_e}}\Rightarrow\frac{M_p}{4R_e}=\frac{9M_e}{R_e}$$

$$\Rightarrow M_p=36M_e=36\times6\times10^{24}kg$$

$$=216\times10^{24}kg=2.16\times10^{26}kg$$

208 **(d)**

$$\frac{K_A}{K_B}=\frac{r_B}{r_A}=\left(\frac{R+h_B}{R+h_A}\right)=\left(\frac{R+2R}{R+R}\right)=\frac{3}{2}$$

209 **(d)**

Binding energy of the system $=\frac{GM_eM_s}{2r}$

$$=\frac{6.6\times10^{-11}\times6\times10^{24}\times2\times10^{30}}{2\times1.5\times10^{11}}$$

$=2.6\times10^{33}$J

211 **(c)**

Let x be the distance of point from the smaller body where gravitational intensity is zero.

$$\therefore \frac{Gm_1}{(1-x)^2} = \frac{Gm_2}{x^2}$$

or $\frac{x}{1-x} = \sqrt{\frac{m_2}{m_1}} = \sqrt{\frac{1000}{100{,}000}} = \frac{1}{10}$

or $10x = -x$

or $x = (1/11)\text{m}$

212 **(b)**

Actually gravitational force provides the centripetal force

213 **(c)**

Escape velocity $v = \sqrt{\frac{2GM}{R}}$

If star rotates with angular velocity ω

Then $\omega = \frac{v}{R} = \frac{1}{R}\sqrt{\frac{2GM}{R}} = \sqrt{\frac{2GM}{R^3}}$

214 **(c)**

Let gravitation field is zero at P as shown in figure.

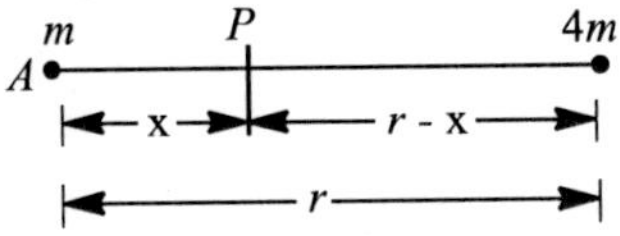

$$\therefore \quad \frac{Gm}{x^2} = \frac{G(4m)}{(r-x)^2}$$

$$\Rightarrow \quad 4x^2 = (r-x)^2$$

$$\Rightarrow \quad 2x = r - x$$

$$\Rightarrow \quad x = \frac{r}{3}$$

$$\therefore \quad V_p = \frac{Gm}{x} - \frac{G(4m)}{r-x}$$

$$= -\frac{3Gm}{r} - \frac{6Gm}{r} = -\frac{9Gm}{r}$$

215 **(a)**

When a body is acted on by the force towards a point and the magnitude of force is inversely proportional to the square of distance. It means it obeys inverse square law and represents ellipse, for example path of the planet around the sun and the force acts between sun and planet proportional to $\frac{1}{r^2}$

216 **(d)**

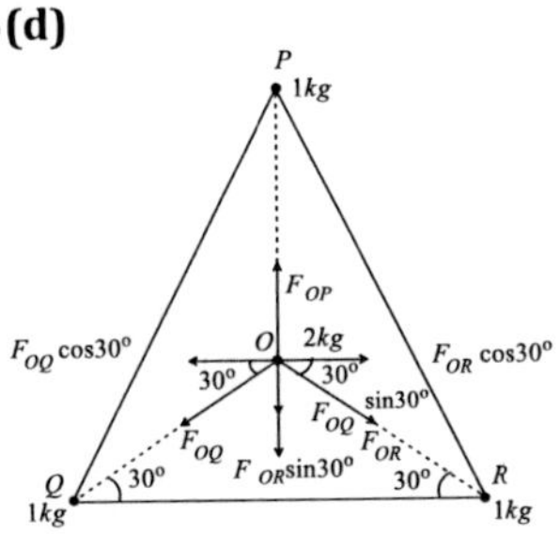

Here, $OP = OQ = OR = \sqrt{2}\ m$

The gravitational force on mass 2 kg at O due to mass

1 kg at P is $F_{OP} = \frac{G\times2\times1}{(\sqrt{2})^2} = G$ along OP

The gravitational force on mass 2 kg at O due to mass

1 kg at Q is $F_{OQ} = \frac{G\times2\times1}{(\sqrt{2})^2} = G$ along OQ

The gravitational force on mass 2 kg at O due to mass

1 kg at R is $F_{OR} = \frac{G\times2\times1}{(\sqrt{2})^2} = G$ along OR

Resolve forces F_{OQ} and F_{OR} into two rectangular components

$F_{OQ}\cos 30°$ and $F_{OR}\cos 30°$ are equal in magnitude of equal and opposite direction

$= F_{OP} - (F_{OQ}\sin 30° + F_{OR}\sin 30°)$

$= G - \left(G \times \frac{1}{2} + G \times \frac{1}{2}\right) = G - G = \text{Zero } N$

217 **(a)**

The change in potential energy in gravitational field is given by $\Delta E = GMm\left(\frac{1}{r_1} - \frac{1}{r_2}\right)$

In this problem; $r_1 = R$ and $r_2 = nR$

$$\Delta E = GMm\left(\frac{1}{R} - \frac{1}{nR}\right)$$

$$= \frac{GMm}{R}\left(\frac{n-1}{n}\right)$$

$$= mgR\left(\frac{n-1}{n}\right)\left(\because g = \frac{Gm}{R^2}\right)$$

218 **(b)**

(i)Gravitational force on the particle placed at mid point D of side BC of length a is

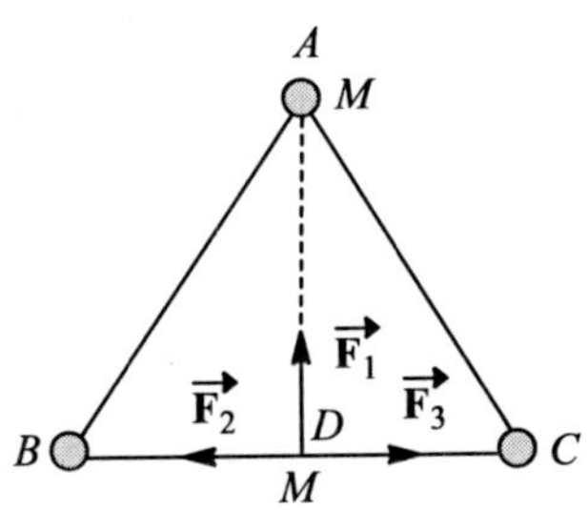

$$\vec{F} = \vec{F}_1 + \vec{F}_2 + \vec{F}_3$$

Here, $\vec{F}_2 = -\vec{F}_3$

$$\therefore \vec{F} = \vec{F}_1 + 0 = \vec{F}_1$$

or $F = F_1 = \frac{GMM}{[AD]^2} = \frac{GM^2}{(3a^2/4)} = \frac{4GM^2}{3a^2}$

(ii)gravitational force on the particle placed at the point O, *ie* the intersection of three medians is

$$\vec{\mathbf{F}} = \vec{\mathbf{F}}_1 + \vec{\mathbf{F}}_2 + \vec{\mathbf{F}}_3 = \vec{\mathbf{0}} \quad \text{or} \quad \mathrm{F} = 0$$

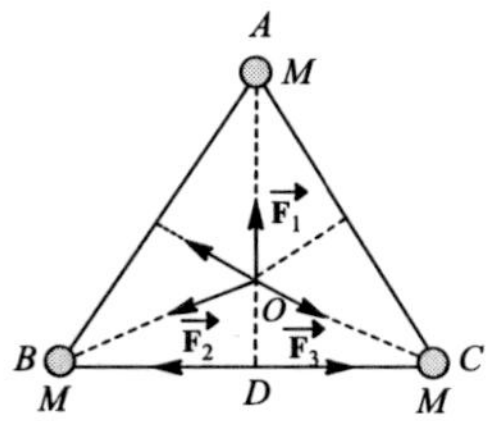

Since, the resultant of $\vec{\mathbf{F}}_2$ and $\vec{\mathbf{F}}_3$ is equal and opposite to $\vec{\mathbf{F}}_1$

219 **(c)**

Potential energy of a body at the surface of the earth

$PE = -\frac{GMm}{R} = -\frac{9R^2M}{R} = -mgR$

$= 500 \times 9.8 \times 6.4 \times 10^6$

$= -3.6 \times 10^{10}\mathrm{J}$

So, if we give this amount of energy in the form of kinetic energy then body escape from the earth

220 **(a)**

$\frac{mv^2}{R} = \frac{RMm}{R^2} \Rightarrow v^2 = \frac{GM}{R}$

$v = \frac{2\pi R}{T} \Rightarrow v^2 = \frac{4\pi^2R^2}{T^2} = \frac{GM}{R}$

$\therefore T^2 = \frac{4\pi^2R^3}{GM}$

If T_1 and T_2 are the time periods for satellite S_1 and S_2 respectively

$\left(\frac{T_1}{T_2}\right)^2 = \left(\frac{R_1}{R_2}\right)^3 \Rightarrow R_2 = \left(\frac{T_2}{T_1}\right)^{2/3} R_1$

$T_1 = 1\text{ h}, T_2 = 8\text{ h} = 10^4\text{km}$

$R_2 = \left(\frac{8}{1}\right)^{3/2} \times 10^4\text{km} = 4 \times 10^4 \text{ km}$

$v_1 = \frac{2\pi R_1}{T_1} = \frac{2\pi \times 10^4}{1} = 2\pi \times 10^4\text{kmh}^{-1}$

$v_2 = \frac{2\pi R_2}{T_2} = \frac{2\pi \times 4 \times 10^4}{8}$

$= \pi \times 10^4\text{kmh}^{-1}$

Relative velocity of S_2 with respect to S_1 is

$v = v_2 - v_1(\pi \times 10^4 - 2\pi \times 10^4)\text{ kmh}^{-1}$

$|v| = \pi \times 10^4\text{kmh}^{-1}$

222 **(b)**

Potential energy of the 1 kg mass which is placed at the earth surface $= -\frac{GM}{R}$

Its potential energy at infinite $= 0$

$\therefore$ Work done = change in potential energy $= \frac{GM}{R}$

223 **(b)**

Intensity of gravitational field at a point inside the spherical shell is zero and outside the shell is $I \propto 1/r^2$

224 **(d)**

$L = mvr = m\left(\sqrt{\frac{GM}{r}}\right)r = m\sqrt{GMr} \therefore L \propto \sqrt{r}$

225 **(c)**

Increase in potential energy,

$\Delta U = \frac{GMm}{(R+R)} - \left(-\frac{GMm}{R}\right)$

$= \frac{1}{2}\frac{GMm}{R} = \frac{1}{2}\left(\frac{GM}{R^2}\right)mR = \frac{1}{2}\text{mg}R$

226 **(c)**

The value of acceleration due to gravity changes with height (*ie*, altitude). If g' is the acceleration due to gravity at a point, at height h above the surface of earth, then

$g' = \frac{GM}{(R+h)^2}$

but, $g = \frac{\text{GM}}{R^2}$

$\therefore \frac{g'}{g} = \frac{GM}{(R+h)^2} \times \frac{R^2}{GM} = \frac{R^2}{(R+h)^2}$

Here, $\text{g}' = \frac{GM}{(R+h)^2} = \frac{GM}{(R+3R)^2}$

$= \frac{GM}{(4R)^2} = \frac{GM}{16R^2} = \frac{g_e}{16}$

227 **(c)**

$\frac{T_{\text{mercury}}}{T_{\text{earth}}} = \left(\frac{r_{\text{mercury}}}{r_{\text{earth}}}\right)^{3/2}$

$= \left(\frac{6 \times 10^{10}}{1.5 \times 10^{11}}\right)^{3/2} = \frac{1}{4}$

(approx.)

$\therefore T_{\text{mercury}} = \frac{1}{4}\text{year}$

228 **(c)**

$$\frac{g_m}{g_e} = \frac{G(M/8)}{GM/R_e^2} = \frac{R_e^2}{8R_m^2}; \;...\text{(i)}$$

Given, $\frac{mg_m}{mg_e} = \frac{1}{6}$

or $\frac{g_m}{g_e} = \frac{1}{6}$ ...(ii)

From Eqs. (i) and (ii); $\frac{R_e^2}{8R_m^2} = \frac{1}{6}$

or $R_e = \sqrt{8/6}R_m$

229 **(b)**

Weight is least at the equator

230 **(d)**

Given that, the orbital velocity of satellite $= \dfrac{\text{escape velocity}}{2}$

$$\Rightarrow \quad v_o = \frac{v_e}{2} \quad ....(i)$$

But we know that,

$$v_o = \sqrt{\frac{gR^2}{R+h}} \text{ and } v_e = \sqrt{2gR}$$

On putting these values in Eq. (i)

$$\sqrt{\frac{gR^2}{R+h}} = \frac{\sqrt{2gR}}{2}$$

On squaring both sides, we obtain

$$\frac{gR^2}{R+h} = \frac{2gR}{4}$$

or $2gR^2 = gR(R+h)$

or $2R = R + h$ or $R = h$

or $h = R = 6400$ km

231 **(b)**

$$a\frac{Gm^2}{L^2}\cos 30° = m\omega^2 r = \frac{m\omega^2 L}{\sqrt{3}} \quad \therefore r = \frac{L}{\sqrt{3}} \therefore \omega = \sqrt{\frac{3Gm}{L^3}}$$

232 **(b)**

To calculate the force of attraction on the point mass m we should calculate the force due to the solid sphere and subtract from this the force which the mass of the hollow sphere would have exerted on m *ie*,

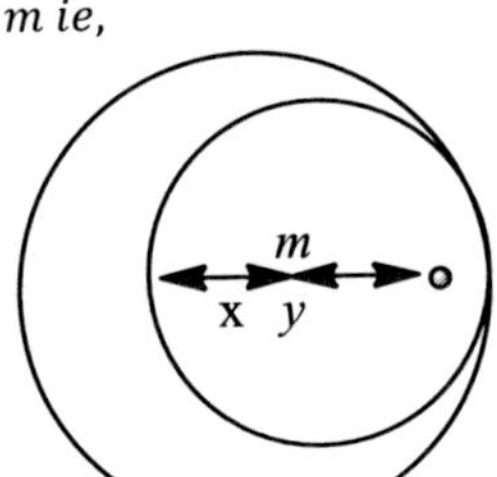

$$F = \frac{GmM}{x^2} - \frac{GmM'}{y^2}$$

$[x = R/4, x + y = R/2]$

$$M = \left(\frac{4}{3}\right)\pi R^3 \rho$$

and $M' = \frac{4}{3}\pi\left(\frac{R}{2}\right)^3 \rho = \frac{M}{8}$

$$F = \frac{GMm}{(R/4)^2} - \frac{Gm(M/8)}{(R/4)^2} = \frac{14GmM}{R^2}$$

233 **(c)**

Change in potential energy

$$\Delta U = U_2 - U_1$$

$$\therefore \quad \Delta U = -\frac{GMm}{(R+nR)} + \frac{GMm}{R}$$

or $\Delta U = -\frac{GMm}{R(1+n)} + \frac{GMm}{R}$

or $\Delta U = \frac{GMm}{R}\left[-\frac{1}{1+n} + 1\right]$

or $\Delta U = \frac{(R^2g)m}{R} \times \frac{n}{(1+n)}\left[\because g = \frac{GM}{R^2}\right]$

or $\Delta U = mgR\left(\frac{n}{n+1}\right)$

234 **(a)**

The earth behaves for all external points as if its mass M were concentrated at its centre. When man of mass m walks from a point on earth's surface and reaches diagonally opposite point, then gravitational potential energy given by

$$U = -\frac{GMm}{R}$$

Will remain same.

Hence, no work is done by the man against gravity.

235 **(b)**

Acceleration due to gravity at a height h from earth's surface

$$g' = \frac{GM}{(R+h)^2}$$

Since, $g' = \frac{g}{100}$

or $\frac{g}{100} = \frac{GM}{(R+h)^2}$

or $\frac{(R+h)^2}{100} = \frac{GM}{g}$

or $\frac{(R+h)^2}{100} = R^2 \left[\because g = \frac{GM}{R^2}\right]$

or $R + h = 10R$

$\Rightarrow \quad h = 9R$

236 **(b)**

Gravitational intensity,

$I=\frac{dV}{dx}=\frac{14}{20}=0.7\ \text{Nkg}^{-1}$

Acceleration due to gravity,

$g=I=0.7\ \text{Nkg}^{-1}$

Work done under this field in displacing a body on a slope of 60° through a distance s

$=m(g\sin 60°)s$

$=2\times(0.7\times\sqrt{3}/2)\times 8=9.6\ \text{J}$

237 **(d)**

Weight on mars $=mg'=\frac{mG(m/10)}{(R/2)^2}$

$=m\times\frac{4}{10}mg=\frac{4}{10}\times 200=80\ \text{N}$

238 **(c)**

$g'=g\left(\frac{R}{R+h}\right)^2\Rightarrow\frac{1}{\sqrt{2}}=\frac{R}{R+h}$

$\Rightarrow R+h=\sqrt{2}R\Rightarrow h=(\sqrt{2}-1)R$

$=0.414R$

Hence, distance from centre $=R+0.414\,R=1.414\,R$

240 **(c)**

Work done by the gravitational field is zero, when displacement is perpendicular to gravitational field. Here, gravitational field, $\vec{\mathbf{I}}=4\hat{\mathbf{i}}+\hat{\mathbf{j}}$. if θ_1 is the angle which makes with positive x-axis, then

$\tan\theta_1=\frac{1}{4}$ or $\theta_1=\tan^{-1}\left(\frac{1}{4}\right)=14°6'$

If θ_2 is the angle which the line $y+4x=6$ makes with positive x-axis, then $\theta_2=\tan^{-1}(-4)=75°56'$ so $\theta_1+\theta_2=90°$

ie, the line $y+4x=6$ is perpendicular to I

241 **(b)**

$F=0$ when $0\le r\le R_1$

Because intensity is zero inside the cavity

F increase when $R_1\le r\le R_2$

$F\propto\frac{1}{r^2}$ when $r>R_2$

242 **(a)**

$g=g_p-R\omega^2\cos^2\lambda$

$=g_p-\omega^2R\cos^2 60°$

$=g_p-\frac{1}{4}R\omega^2$

243 **(c)**

The escape velocity of a particle

$v_e=\sqrt{2gR}$

Hence, the escape velocity is independent of mass of the particle.

244 **(c)**

Due to rotation of earth the effective acceleration due to gravity $g'=g-R\omega^2\cos^2\lambda$.

For a given point on the surface of earth g decreases as ω increases. The angular speed of earth is maximum at equator hence, the value of g on the surface of the earth is smallest.

245 **(c)**

Gravitational potential energy, $U=\frac{GMm}{r}$

or $U=\frac{GMm}{r^2}\times r$

or $U=g\times mr$

or $U=(mg)r$

or $mg=\frac{U}{r}$

247 **(a)**

At a height h, (Taking $h<<R$) from the surface of earth

$g_h=g\left(1-\frac{2h}{R}\right)$ or $\frac{g_h}{g}=1-\frac{2h}{R}=\frac{90}{100}$

or $\frac{2h}{R}=1-\frac{99}{100}=\frac{1}{100}$

or $g=\frac{R}{100}=\frac{6400}{200}=32\ \text{km}$

249 **(c)**

$g'=\frac{g}{\left(1+\frac{h}{R}\right)^2}$

$\frac{g}{16}=\frac{g}{\left(1+\frac{h}{R}\right)^2}$

$\left(1+\frac{h}{R}\right)^2=16$

$1+\frac{h}{R}=4$

$\frac{h}{R}=3$

$h=3R$

250 **(c)**

$g=\frac{GM}{R^2}$ and $K=\frac{L^2}{2I}$

If mass of the earth and its angular momentum remains constant then $g\propto\frac{1}{R^2}$ and $K\propto\frac{1}{R^2}$

i.e., if radius of earth decreases by 2% then g and K both increases by 4%

251 **(b)**

The period of revolution of geostationary satellite is the same as that of the earth.

Orbital velocity $v_o=\sqrt{gR_e}$

Escape velocity $v_e=\sqrt{2gR_e}$

whereR_e is radius of earth.

$$\% \text{ increase} = \frac{v_e - v_o}{v_o} \times 100$$

% increase

$$= \frac{\sqrt{2gR_e} - \sqrt{gR_e}}{\sqrt{gR_e}} \times 100$$

$$= (\sqrt{2} - 1) \times 100$$

$$= (1.141 - 1) \times 100$$

$$= 41.4\%$$

252 **(c)**

$U = \frac{-GMm}{r}, K = \frac{GMm}{2r}$ and $E = \frac{-GMm}{2r}$

For a satellite U, K and E varies with r and also U and E remains negative whereas K remain always positive

253 **(d)**

Orbital radius of satellites $r_1 = R + R = 2R$

$r_2 = R + 7R = 8R$

$U_1 = -\frac{GMm}{r_1}$ and $U_2 = -\frac{GMr}{r_2}$

$K_1 = \frac{GMm}{2r_1}$ and $K_2 = \frac{GMm}{2r_2}$

$E_1 = \frac{GMm}{2r_1}$ and $E_2 = \frac{GMm}{2r_2}$

$$\therefore \frac{U_1}{U_2} = \frac{K_1}{K_2} = \frac{E_1}{E_2} = 4$$

254 **(b)**

$g = \frac{GM}{R^2}$. If radius shrinks to half of its present value then g will becomes four times

256 **(a)**

$$\begin{pmatrix}\text{Total}\\\text{mechanical}\\\text{energy}\end{pmatrix}_P = \begin{pmatrix}\text{Total final}\\\text{mechanical}\\\text{energy}\end{pmatrix}_O$$

$$\Rightarrow \frac{1}{2}m(0)^2 - \frac{GMM}{\sqrt{(\sqrt{3}R)^2 R^2}} = \frac{1}{2}mv^2 - \frac{GMm}{R}$$

$$\Rightarrow -\frac{GMm}{2R} - \frac{1}{2}mv^2 - \frac{GMm}{R}$$

$$\Rightarrow v = \sqrt{\frac{GM}{R}}$$

257 **(d)**

$$\omega = \frac{|v|}{R_2 - R_1} = \frac{\pi \times 10^4}{4 \times 10^4 - 1 \times 10^4} = \frac{\pi}{3} \text{ radh}^{-1}$$

258 **(d)**

Gravitational field intensity

$$I = \frac{GM}{R^2} = \frac{6.6 \times 10^{-11} \times 7.34 \times 10^{22}}{(1.74 \times 10^6)^2}$$

$= 1.62 \text{ Nkg}^{-1}$

259 **(b)**

Potential energy $U = \frac{-GMm}{r} = -\frac{GMm}{R+h}$

$U_{initial} = -\frac{GMm}{3R}$ and $U_{final} = -\frac{GMm}{2R}$

Loss in PE = gain in $KE = \frac{GMm}{2R} - \frac{GMm}{3R} = \frac{GMm}{6R}$

260 **(a)**

$v = \sqrt{2gR}$. If acceleration due to gravity and radius of the planet, both are double that of earth then escape velocity will be two times, *i.e.* $v_p = 2v_e$

262 **(c)**

$$\frac{v_e}{v_m} = \sqrt{\frac{g_e}{g_m}\frac{R_e}{R_m}} = \sqrt{6 \times 10} = \sqrt{60} = 8$$

(nearly)

263 **(b)**

$$g' = g\left(1 - \frac{d}{R}\right) \Rightarrow \frac{g}{n} = g\left(1 - \frac{d}{R}\right) \Rightarrow d = \left(\frac{n-1}{n}\right)R$$

264 **(b)**

$v_e = \sqrt{2gR}$ and $v_0 = \sqrt{gR}$ $\therefore$ $\sqrt{2}v_0 = v_e$

265 **(d)**

$F \propto \frac{1}{r^2}$. If r becomes double then F reduces to $\frac{F}{4}$

266 **(a)**

Since the gravitational field is conservative field, hence, the work done in taking a particle from one point to another in a gravitational field is path independent

268 **(b)**

$v_e = \sqrt{\frac{2GM}{R}} \Rightarrow v_e \propto \sqrt{M}$ if R = constant

If the mass of the planet becomes four times then escape velocity will become 2 times

269 **(b)**

$$\Delta U = \frac{mgh}{1 + \frac{h}{R}} = \frac{mgh}{1 + \frac{R}{R}} = \frac{mgR}{2}$$

270 **(d)**

By Kepler's law $T^2 \propto R^3$

Hence, $\left(\frac{T_2}{T_1}\right)^2 = \left(\frac{R_2}{R_1}\right)^3$

$= \left(\frac{2.5R + R}{6R + R}\right)^3$

$= \left(\frac{1}{2}\right)^3$

$T_2 = \frac{T_1}{(2)^{3/2}}$

For a geostationary satellite

$T_1 = 24\text{ h}$

So, $T_2 = \frac{24}{2\sqrt{2}} = 6\sqrt{2}\text{ h}$

271 **(b)**

From Kepler's third law of planetary motion

$T^2 \propto R^3$

Given, $T_1 = 1, T_2 = 8, R_1 = R$

$\therefore \quad \frac{T_1^2}{T_2^2} = \frac{R_1^3}{R_2^3}$

$R_2^3 = R_1^3 \frac{T_2^2}{T_1^2}$

$R_2^3 = R_1^3 \times (8)^2$

$R_2^3 = R^3 \times (2^3)^2$

$\Rightarrow \quad R_2 = R \times 4 = 4R$

272 **(a)**

From Kepler's law, $T^2 \propto R^3$

or $\left(\frac{T_2}{T_1}\right)^2 = \left(\frac{R_2}{R_1}\right)^3 = \left(\frac{1.01\,R}{R}\right)^3$

$= (1 + 0.01)^3$

or $\frac{T_2}{T_1} = (1 + 0.01)^{3/2}$

$= 1 + \left(\frac{3}{2} \times 0.01\right)$

or $\frac{T_2 - T_1}{T_1} = \frac{1.5}{100} = 1.5\%$

273 **(c)**

$$g = \frac{GM}{R^2} = \frac{G\left(\frac{4}{3}\pi R^3\right)\rho}{R^2}$$

or $g \propto \rho R$

or $R \propto \frac{g}{\rho}$

Now escape velocity, $v_e = \sqrt{2gR}$

or $v_e \propto \sqrt{gR}$

or $v_e \propto \sqrt{g \times \frac{g}{\rho}} \propto \sqrt{\frac{g^2}{\rho}}$

$\therefore (v_e)_{\text{panet}} = (11\text{ ms}^{-1})\sqrt{\frac{6}{121} \times \frac{3}{2}}$

$= 3\text{km s}^{-1}$

274 **(d)**

When $r < R$, Gravitational field intensity,

$I = \frac{GM}{R^3}r = \frac{Gr}{R^3}\left(\frac{4}{3}\pi R^3 \rho\right) = \frac{4\pi G\rho r}{3}$

275 **(b)**

$\omega_{\text{body}} = 27\omega_{\text{earth}}$

$T^2 \propto r^3 \Rightarrow \omega^2 \propto \frac{1}{r^3} \Rightarrow \omega \propto \frac{1}{r^{3/2}} \therefore r \propto \frac{1}{\omega^{2/3}}$

$\Rightarrow \frac{r_{\text{body}}}{r_{\text{earth}}} = \left(\frac{\omega_{\text{earth}}}{\omega_{\text{body}}}\right)^{2/3} = \left(\frac{1}{27}\right)^{2/3} = \frac{1}{9}$

276 **(b)**

Time period of satellite

$$T = 2\pi\sqrt{\frac{(R_e + h)^2}{gR_e^2}}$$

Where, R_e = Radius of earth,

h = Height from earth surface .

Time period not depend on mass. So, time period of both satellite will be equal.

277 **(b)**

$F = \frac{Gm(M-m)}{x^2}$; For maxima,

$\frac{dF}{dm} = \frac{G}{x^2}(M - 2m) = 0$

or $\frac{m}{M} = \frac{1}{2}$

278 **(d)**

$v_e = \sqrt{\frac{2GM}{R}}$

$= \sqrt{\frac{2 \times 6.67 \times 10^{-11} \times 1.99 \times 10^{30}}{6.96 \times 10^8}}$

$= 618\ km/sec$

279 **(a)**

$g' = \frac{GM}{(R+h)^2}$, acceleration due to gravity at heigth$h$

$\Rightarrow \frac{g}{9} = \frac{GM}{R^2} \cdot \frac{R^2}{(R + h)^2}$

$= g\left(\frac{R}{R + h}\right)^2$

$\Rightarrow \frac{1}{9} = \left(\frac{R}{R + h}\right)^2$

$\Rightarrow \frac{R}{R + h} = \frac{1}{3}$

$\Rightarrow \quad 3R = R + h$

$\Rightarrow \quad 2R = h$

280 **(b)**

$v \propto R\sqrt{\rho} \therefore \frac{v_p}{v_e} = \frac{R_p}{R_e} \times \sqrt{\frac{\rho_p}{\rho_e}} = 4 \times \sqrt{9}$
$= 12$

$\Rightarrow v_p = 12v_e$

281 **(c)**

$v_e = \sqrt{2gR} \Rightarrow \frac{v_A}{v_B} = \sqrt{\frac{g_A}{g_B} \times \frac{R_A}{R_B}} = \sqrt{x \times r}$

$\therefore \frac{v_A}{v_B} = \sqrt{rx}$

282 **(b)**

Due to inertia of direction

283 **(a)**

The earth possesses rotational motion about an axis through its poles. The value of acceleration due to gravity at a place (at given latitude) is affected due to its rotational motion. If earth ceases to rotate, the weight of body at equator will increase. However, there will be no effect on the weight at poles. The effect of rotation of the earth on acceleration due to gravity is to decrease its value. Therefore, if the earth stops rotating, the value of g will increase.

284 **(b)**

The relation between mass and density of earth is given form Newton's law of gravitation, according to which

$M_e = \frac{gR_e^2}{G}$

whereM_e is mass of earth, G the gravitational constant, R_e the radius of earth and g the acceleration due to gravity.

Also, mass = volume × density

$g = \frac{G \times \text{volume} \times \text{density}}{R^2}$

Assuming spherical shape of earth volume

$= \frac{4}{3}\pi R^3$

$g = G \times \frac{4}{3}\frac{\pi R^3}{R^2}\rho$

$\Rightarrow \quad g = G \cdot \frac{4}{3}\pi R\rho$

Hence, increases in radius would dominate.

285 **(b)**

$g = \frac{GM}{R^2} = \frac{G}{R^2}\frac{4}{3}\pi R^3\rho = \frac{4}{3}\pi G\rho R \ ie, g = R$

286 **(d)**

Let escape velocity bev_e, then kinetic energy is

$= \frac{1}{2}mv_e^2 \quad \text{.... (i)}$

and escape energy $= +\frac{GM_e m}{R_e} \quad \text{... (ii)}$

Equating Eqs. (i) and (ii), we get

$\frac{1}{2}m_e^2 = \frac{GM_e m}{R_e}$

$\Rightarrow \quad v_e = \sqrt{\frac{2GM_e}{R_e}}$

$\Rightarrow \quad R = \frac{2GM_e}{v_e^2}$

Given, $G = 6.67 \times 10^{-11}\text{N–m}^2/\text{kg}$,
$M_e = 6 \times 10^{24}\text{kg}, v_e = 3 \times 10^8\text{m/s}^2$

$R = \frac{2 \times 6.67 \times 10^{-11} \times 6 \times 10^{24}}{(3 \times 10^8)^2}$

$R = 8.89 \times 10^{-3}$

$R \approx 9 \times 10^{-3}\text{m} = 9\text{mm}$

287 **(a)**

k represents gravitational constant which depends only on the system of units

288 **(b)**

For a moving satellite kinetic energy = $\frac{GMm}{2r}$

Potential energy = $\frac{-GMm}{r} \Rightarrow \therefore$

$\frac{\text{Kinetic energy}}{\text{Potential energy}} = \frac{1}{2}$

289 **(d)**

Mass of the satellite does not affect the time period

$\frac{T_A}{T_B} = \left(\frac{r_1}{r_2}\right)^{3/2} = \left(\frac{r}{2r}\right)^{3/2} = \left(\frac{1}{8}\right)^{1/2} = \frac{1}{2\sqrt{2}}$

290 **(c)**

The earth moves around the sun is elliptical path, so by using the properties of ellipse

$r_1 = (1+e)a$ and $r_2 = (1-e)a$

$\Rightarrow a = \frac{r_1+r_2}{2}$ and $r_1r_2 = (1-e^2)a^2$

Where a = semi major axis

b = semi minor axis

e = eccentricity

Now required distance = semi latusrectum= $\frac{b^2}{a}$

$$= \frac{a^2(1-e^2)}{a} = \frac{(r_1 r_2)}{(r_1+r_2)/2} = \frac{2r_1 r_2}{r_1+r_2}$$

291 **(b)**

Let a satellite is revolving around earth with orbital velocity v. The gravitational potential energy of satellite is

$$U = -\frac{GM_e m}{R_e} \quad \text{... (i)}$$

The kinetic energy of satellite is

$$K = \frac{1}{2}\frac{GM_e m}{R_e} \quad \text{... (ii)}$$

∴ Total energy of satellite

$$E = U + K$$
$$= -\frac{GM_e m}{R_e} + \frac{1}{2}\frac{GM_e m}{R_e}$$
$$= -\frac{1}{2}\frac{GM_e m}{R_e} \quad \text{...(iii)}$$

But we know that necessary centripetal force to the satellite is provided by the gravitational force. *ie*,

$$\frac{mv^2}{R_e} = \frac{GM_e m}{R_e^2}$$

or $\quad mv^2 = \frac{GM_e m}{R_e} \quad$ ... (iv)

Hence, from Eqs. (iii) and (iv), we get

$$E = -\frac{1}{2}mv^2$$

292 **(c)**

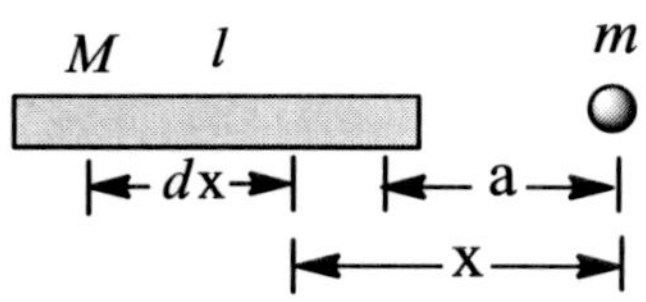

$$\Rightarrow dU = \frac{Gm\left(\frac{M}{l}dx\right)}{x}$$

$$\Rightarrow U\int dU = \frac{GmM}{l}\int_a^{a+l}\frac{dx}{x}$$

$$\Rightarrow U = -\frac{GmM}{l}\log_e\left(\frac{a+l}{a}\right)$$

293 **(a)**

Given, $F_1 = F_2 = F$ and $\theta = 60°$

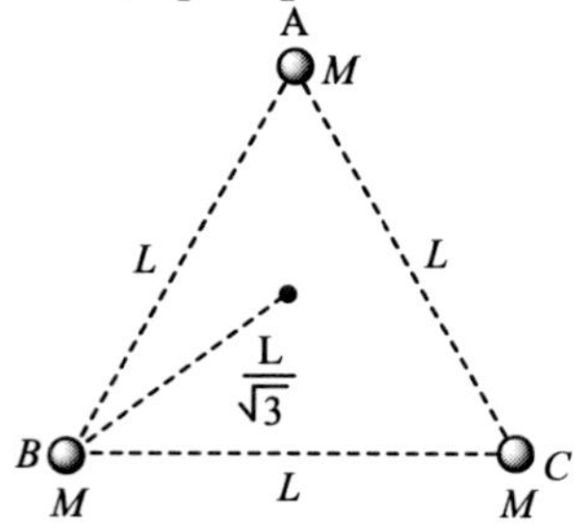

Resultant force $= \sqrt{3}F$

∴ Force on mass at A due to mass at B and

$$= \sqrt{3}\left(\frac{GM^2}{L^2}\right)$$

Centripetal force for circumscribing the triangle in a circular orbit is provided by mutual gravitational interaction.

ie, $\quad \frac{Mv^2}{(L/\sqrt{3})} = \sqrt{3}\left(\frac{GM^2}{L^2}\right)$

or $v = \sqrt{\frac{GM}{L}}$

294 **(c)**

$$g = \frac{GM}{R^2}$$

So, $\frac{g_M}{g_E} = \left(\frac{M_M}{M_E}\right) \times \left(\frac{R_E}{R_M}\right)^2 = \frac{1}{10} \times \left(\frac{12742}{6760}\right)^2$

$\therefore \frac{g_M}{g_E} = 0.35 \Rightarrow g_M = 9.8 \times 0.35$

$$= 3.48 \text{ ms}^{-2}$$

295 **(b)**

$$g = \frac{GM}{R^2} = \frac{G \times \frac{4}{3}\pi R^3 \rho}{R^2} = \frac{4}{3}\pi G\rho R \quad ie, g \propto R$$

For pendulum clock, g will increase on the planet, so time period will decrease. But for spring clock, it will not change. Hence, P will run faster than S

296 **(b)**

From Kepler's third law of planetary motion also known as law of periods

$T^2 = kr^3$

Where T is time period and r the mean distance from the sun. Hence, greater is the distance of planet from sun, greater is its period of revolution.

297 **(b)**

If g is the acceleration due to gravity of earth at the position of satellite, the apparent weight of a body in the satellite will be

$W_{app} = m(g' - a)$

But as satellite is a freely falling body, *ie*, $g' = a$

So, $W_{app} = 0$

298 **(b)**

$$\frac{T_2}{T_1} = \left(\frac{r_2}{r_1}\right)^{3/2} = (2)^{3/2} = 2\sqrt{2} \Rightarrow T_2$$
$$= 2\sqrt{2} \text{ } years$$

299 **(c)**

The value of acceleration due to gravity at height h (when h is not negligible as compared to R)

$$g' = g\frac{R^2}{(R+h)^2}$$

Here, $g' = \frac{g}{2}$

$$\therefore \quad \frac{g}{2} = g\frac{R^2}{(R+h)^2}$$

or $\frac{1}{2} = \frac{R^2}{(R+h)^2}$

or $\sqrt{\frac{1}{2}} = \frac{R}{R+h}$

or $R + h = \sqrt{2}\,R$

$\therefore \quad h = (\sqrt{2}-1)R$

300 **(a)**

Below the surface of the earth g $\propto r$ and above the surface of earth g $\propto 1/r^2$.

Therefore, the graph (a) is correct

301 **(c)**

Launching the rocket in the direction of earth's rotation allows it to exploit the earth's rotational velocity *ie*, launching it from West to East. (It gains speed from velocity addition with the earth's rotational velocity.)

302 **(c)**

Acceleration due to gravity, $g = \frac{GM}{R^2}$

$$\frac{g_M}{g_E} = \left(\frac{M_M}{M_E}\right) \times \left(\frac{R_E}{R_M}\right)^2$$

$$= \frac{1}{10} \times \left(\frac{12742}{6760}\right)^2$$

$$\frac{g_M}{g_E} = 0.35$$

$$g_M = 9.8 \times 0.35 = 3.48 \text{ ms}^{-2}$$

303 **(a)**

Using conservation of energy.

$$\frac{1}{2}mv^2 - \frac{GMm}{R} = 0 - \frac{GMm}{2R}$$

or $\frac{1}{2}mv^2 = \frac{GMm}{R} - \frac{GMm}{2R}$

or $v^2 = 2\frac{GM}{R}\left[1 - \frac{1}{2}\right]$

or $v^2 = \frac{GM}{R}$

But $gR^2 = GM$

$\therefore \quad v = \sqrt{\frac{gR^2}{R}}$

or $v = \sqrt{gR}$

304 **(c)**

$v_1r_1 = v_2r_2$[$\because$ angular momentum is constant]

305 **(d)**

Gravitational force between sphere of mass M and the particle of mass m at B is

$$F_1 = \frac{GM\,m}{d^2}$$

If M_1 is the mass of the removed part of sphere, then

$$M_1 = \frac{4}{3}\pi(R/2)^3\rho = \frac{1}{8}\left(\frac{4}{3}\pi R^2\rho\right) = \frac{M}{8}$$

Gravitational force between the removed part and the particle of mass m at B is

$$F_2 = \frac{GM_1m}{(d-R/2)^2} = \frac{G(M/8)m}{(d-R/2)^2}$$

$$= \frac{GM\;\;m}{8(d-R/2)^2}$$

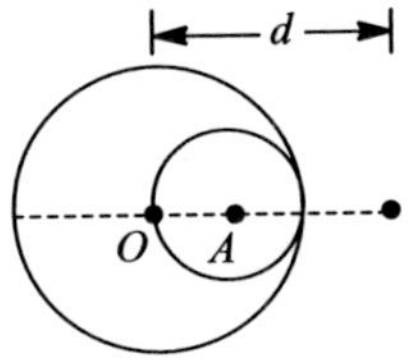

$\therefore$ Required force,

$$F = F_1 - F_2 = \frac{GM\,m}{d^2} - \frac{GMm}{8[d-(R/2)]^2}$$

$$= \frac{GM\,m}{d^2}\left[1 - \frac{1}{8\left(1-\frac{R}{2d}\right)^2}\right]$$

306 **(c)**

The maximum velocity with which a body must be projected in the atmosphere, so as to enable it to just overcome the gravitational pull, is known as escape velocity.

Escape velocity from earth's surface is

$$v_{es} = \sqrt{\frac{2GM_e}{R_e}}$$

$$= \sqrt{\frac{2G\cdot\frac{4}{3}\pi R_e^3 d_e}{R_e}}\left(\therefore M = \frac{4}{3}\pi R_e^3 d_e\right)$$

or $v_{es} \propto \sqrt{d_e} \times R_e$ .... (i)

similarly, for a planet

$v'_{es} \propto \sqrt{d_p} \times R_p$ ... (ii)

So, $\frac{v_{es}}{v'_{es}} = \left(\frac{d_e}{d_p}\right)^{1/2} \times \frac{R_e}{R_p}$

Given, $d_p = \frac{1}{4}d_e, R_p = 2R_e$

$$\frac{v_{es}}{v_{es}} = \left(\frac{\frac{d_e}{d_e}}{4}\right)^{\frac{1}{21}} \times \frac{R_e}{2R_e}$$

$$= (4)^{1/2} \times \frac{1}{2}$$

$$= 2 \times \frac{1}{2} = 1$$

So, $\frac{v_{es}}{v'_{es}} = 1:1$

307 **(b)**

$g = \frac{4}{3}\pi\rho GR$. If density is same then $g \propto R$

According to problem $R_p = 2R_e \therefore g_p = 2g_e$

For clock P (based on pendulum motion)

$T = 2\pi\sqrt{\frac{l}{g}}$

Time period decreases on planet so it will run faster because $g_p > g_e$

For clock S (based on oscillation of spring) $T = 2\pi\sqrt{\frac{m}{k}}$

So it does not change

308 **(d)**

$$v_0 = \sqrt{\frac{GM}{r}}$$

309 **(d)**

Here, $u = 20\ \text{ms}^{-1}, m = 500\text{g} = 0.5\text{kg}, t = 20\ \text{s}$

Using Newton's equation of motion

$$s = ut + \frac{1}{2}gt^2$$

$$0 = 20 \times 20 + \frac{1}{2}(-g)(20)^2$$

or $g = 2\ \text{ms}^{-2}$

$\therefore$ Weight of body on planet $= mg$

$$= 0.5 \times 2 = 1\ \text{N}$$

310 **(d)**

According to Kepler's third law, T^2 is proportional to cube of semi-major axis of the elliptical orbit.

Semi– major axis $= \frac{r_1 + r_2}{2}$

$\therefore$ $T^2 \propto \left[\frac{r_1 + r_2}{2}\right]^3$

or $T \propto (r_1 + r_2)^{3/2}$

311 **(c)**

It is self-evident that the orbit of the comet is elliptic with sun begin at one of the focus. Now, as for elliptic orbits, according to kepler's third law,

$$T^2 = \frac{4\pi^2 a^3}{GM} \Rightarrow a = \left(\frac{T^2 GM}{4\pi^2}\right)^{1/3}$$

$$a = \left[\frac{(76 \times 3.14 \times 10^7) \times 6.67 \times 10^{-11} \times 2 \times 10^{10}}{4\pi^2}\right]^{1/3}$$

But in case of ellipse,

$2a = r_{\min} + r_{\max}$

$\therefore r_{\max} = 2a - r_{\min}$

$= 2 \times 2.7 \times 10^{12} - 8.9 \times 10^{10}$

$\cong 5.3 \times 10^{12}\text{m}$

312 **(c)**

The variation of g with angular velocity (ω) is given by

$g' = g - R\omega^2$

If earth were to spin faster, that is angular velocity increases, then except at poles, the weight of bodies will decrease at all places.

314 **(b)**

Earth and all other planets move around the sun under the effect of gravitational force. This force always acts along the line joining the centre of the planet and the sun and is directed towards the sun. In other words, a planet moves around the sun under the effect of a purely radial force. Therefore, areal velocity of the planet must always remain constant.

$\therefore$ $\frac{\Delta \mathbf{A}}{\Delta t} = \frac{\mathbf{L}}{2m} =$ a constant vector

Therefore, Kepler's 2nd law is the consequence of the principle of conservation of angular momentum (L)

$\tau = 0$

Now, $\tau = I\alpha$

$\therefore$ $I\alpha = 0$ or $\alpha = 0$

or $\alpha_T = r\alpha = 0$

ie, tangential acceleration is zero.

316 **(d)**

From Kepler's third law of planetary motion,

$T^2 \propto R^3$

$\Rightarrow$ $\frac{T^2}{R^3} =$ constant

317 **(a)**

$$g' = g\left(1 - \frac{d}{R}\right) = 9.8\left(1 - \frac{100}{6400}\right) = 9.66 m/s^2$$

319 **(c)**

$$V_{in} = \frac{-GM}{2R}\left[3 - \left(\frac{r}{R}\right)^2\right], V_{\text{surface}} = \frac{-GM}{R}, V_{out} = \frac{-GM}{r}$$

320 **(a)**

$$\frac{v_p}{v_e} = \sqrt{\frac{g_p}{g_e} \times \frac{R_p}{R_e}} = \sqrt{9 \times 4} = 6 \therefore v_p = 6 \times v_e = 67.2 km/s$$

321 **(c)**

Gravitational potential

$$V = GM\left(\frac{1}{r_1} + \frac{1}{r_2} + \frac{1}{r_3} + \cdots\right)$$
$$= G \times 1\left(\frac{1}{1} + \frac{1}{2} + \frac{1}{4} + \frac{1}{8} + \frac{1}{16} + \cdots.\right)$$
$$= G\left(\frac{1}{1 - 1/2}\right)\left(\because \text{sum of GP} = \frac{a}{1-r}\right)$$
$$= 2G$$

322 **(b)**

Below the sea level the pressure is increasing with depth in mine due to presence of atmospheric air there. The acceleration due to gravity below the surface of the earth decreases with the distance from the surface of the earth as $g' = g\left(1 - \frac{d}{R}\right)$

323 **(b)**

$$K = \frac{GMm}{2R}$$

Escape velocity $V_e = \sqrt{\frac{2GM}{R}}$

Kinetic energy to escape $(K') = \frac{1}{2}m \times 2\frac{GM}{R}$

$K' = 2K$

325 **(c)**

Here the force of attraction between them provides the necessary centripetal force

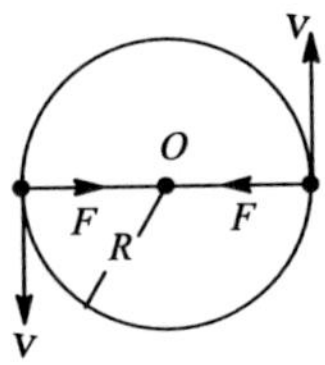

$$\therefore \frac{mv^2}{R} = \frac{Gm^2}{(4R)^2}$$

$$\therefore v = \sqrt{\frac{Gm}{4R}}$$

326 **(b)**

Acceleration due to gravity $g = \frac{GM}{R^2}$, $M = \left(\frac{4}{3}\pi R^3\right)\rho$

$$\therefore \quad g = \frac{4G}{3}\frac{\pi R^3}{R^2}\rho$$

$$\Rightarrow \quad g = \left(\frac{4G\pi R}{3}\right)\rho \quad (\rho = \text{average density})$$

$\Rightarrow g \propto \rho$ or $\rho \propto g$

328 **(a)**

Gravitational attraction force on particle B

$$F_g = \frac{GM_P m}{(D_P/2)^2}$$

Acceleration of particle due to gravity $a = \frac{F_g}{m} = \frac{4GM_P}{D_P^2}$

329 **(c)**

If M be mass of earth and R its radius, the acceleration due to gravity is given by

$$g = \frac{GM}{R^2} \quad \ldots.(i)$$

Where, G is gravitational constant.

Given, $R = 0.99R$

$$\therefore \quad g' = \frac{GM}{(0.99R)^2} \quad \ldots(ii)$$
$$= 1.02\left(\frac{GM}{R^2}\right)$$

From Eq. (i), we get

$g' = 1.02g$

Hence, acceleration due to gravity increases by

$g' - g = 1.02 - 1 = 0.02g$

Hence, percentage increases =2%.

330 **(a)**

The necessary centripetal force required for a planet to move round the sun = gravitational force exerted on it ie, $\frac{mv^2}{R} = \frac{GM_e m}{R^n}$

$$v = \left(\frac{GM_e}{R^{n-1}}\right)^{1/2}$$

Now, $T = \frac{2\pi R}{v} = 2\pi R \times \left(\frac{R^{n-1}}{GM_e}\right)^{1/2}$

$\Rightarrow$ $= 2\pi\left(\frac{R^2 \times R^{n-1}}{GM_e}\right)^{1/2}$

$$= 2\pi\left(\frac{R^{(n+1)/2}}{(GM_e)^{1/2}}\right)$$

$\Rightarrow$ $T \propto R^{(n+1)/2}$

331 **(d)**

$$-\frac{GMm}{2R_1} + KE = -\frac{GMm}{2R_2}$$

$$KE = \frac{GMm}{2}\left[\frac{1}{R_1} - \frac{1}{R_2}\right]$$

332 **(c)**

$$T_2 = T_1\left(\frac{r_2}{r_1}\right)^{3/2} = T_1\left(\frac{1}{4}\right)^{3/2} = \frac{1}{8} \text{ times}$$

333 **(d)**

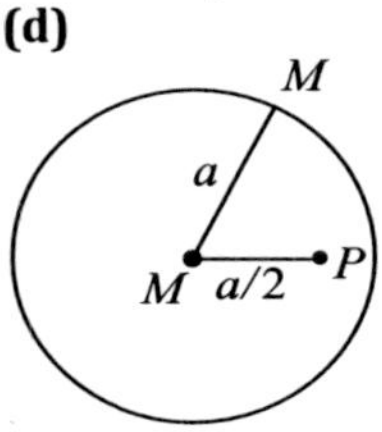

$$V_p = V_{\text{sphere}} + V_{\text{partical}}$$

$$= \frac{GM}{a} + \frac{GM}{a/2} = \frac{3GM}{a}$$

334 **(d)**

S_2 is correct because whatever be the g, the same force is acting on both the pans. Using a spring balance, the value of g is greater at the pole. Therefore mg at the pole is greater. S_4 is correct. S_2 and S_4 are correct

335 **(a)**

$$\frac{GM_e}{x^2} = \frac{GM_m}{(r-x)^2}$$

or $\frac{r-x}{x} = \sqrt{\frac{M_m}{M_e}} = \sqrt{\frac{7.35\times10^{22}}{5.98\times10^{24}}}$

or $r = 0.11x + x = 1.11x$

$x = r/1.11 = 3.85 \times 10^8/1.11$

$= 3.47 \times 10^8$ m

336 **(d)**

The gravitational force exerted on satellite at a height x is

$$F_G = \frac{GM_e m}{(R+x)^2}$$

where M_e = mass of earth.

Since, gravitational force provides the necessary centripetal force, so,

$$\frac{GM_e m}{(R+x)^2} = \frac{mv_o^2}{(R+x)}$$

where v_o is orbital speed of satellite.

$\Rightarrow$ $\frac{GM_e m}{(R+x)} = mv_o^2$

$\Rightarrow$ $\frac{gR^2 m}{(R+x)}$

$$= mv_o^2 \left(\because g = \frac{GM_e}{R^2}\right)$$

$$\Rightarrow v_o = \sqrt{\left[\frac{gR^2}{(R+x)}\right]} = \left[\frac{gR^2}{(R+x)}\right]^{1/2}$$

337 **(b)**

$$V = -\int_{\infty}^{x} I\,dx = -\int_{\infty}^{x} \frac{C}{x^2}\,dx = \frac{C}{x}$$

339 **(d)**

Telecommunication satellites are geostationary satellite

340 **(d)**

The gravitational force of sun on comet is radial, hence angular momentum is constant over the entire orbit. Using law of conservation of angular momentum, at locations A and B

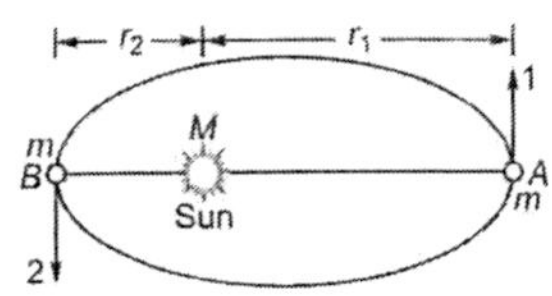

$L = mv_1r_1 = mv_2r_2$ or $v_2 = \frac{v_1r_1}{r_2}$ ...(i)

Using the principle of conservation of total energy at A and B

$$\frac{1}{2}mv_1^2 - \frac{GMm}{r_1} = \frac{1}{2}m\,v_2^2 - \frac{GMm}{r_2}$$

or $v_2^2 - v_1^2 = 2GM\left(\frac{1}{r_2} - \frac{1}{r_1}\right)$...(ii)

Putting the values from Eq. (i) in Eq. (ii) and solving, we get

$$v_1 = \left[\frac{2GMr_2}{r_1(r_1+r_2)}\right]^{1/2}$$

$$\therefore L = mv_1r_1 = m\left[\frac{2GMr_1r_2}{(r_1+r_2)}\right]^{1/2}$$

341 **(b)**

As mass, $M = \frac{4}{3}\pi R^2\rho$

or $\rho = \frac{3M}{4\pi R^3}$

$$\therefore \frac{\rho_s}{\rho_s} = \frac{M_s}{M_e} \times \frac{R_e^3}{R_s^3} = 330 \times \left(\frac{1}{100}\right)^3$$
$$= 3.3 \times 10^{-4}$$

342 **(a)**

As there is no gravity in space so spring will not be extened.

343 **(b)**

The value of g at the height h from the surface of earth

$$g' = g\left(1 - \frac{2h}{R}\right)$$

The value of g at depth x below the surface of earth

$$g' = g\left(1 - \frac{x}{R}\right)$$

These two are given equal, hence $\left(1 - \frac{2h}{R}\right) = \left(1 - \frac{x}{R}\right)$

On solving, we get $x = 2h$

344 **(b)**

We have

$T^2 \propto R^3$

$R_1 = r$

and $R_2 = 4r$

$$\therefore \quad \frac{T_1^2}{T_2^2} = \frac{(r)^3}{(4r)^3}$$

$$\text{or} \quad \frac{T_1}{T_2} = \frac{1}{8}$$

345 **(d)**

When earth moves round the sun then according to Kepler's second law, the radius vector drawn from the sun to earth, sweeps out equal areas in equal time, *ie*, its areal velocity (or the area swept out by it per unit time) is constant. While in such motion, angular velocity, kinetic energy and potential energy change.

346 **(b)**

According to Kelper's third law (law of periods),

we have $\quad T^2 \propto R^3$

where T is time taken by the planet to go once around the sun and R is semi-major axis (distance) of the elliptical orbit.

$$\therefore \quad T^3 = k\,R^3 \quad \text{...(i)}$$

Where k is constant of proportionality.

When R becomes 4 times let time period be T'.

$$\therefore \quad T'^2 = k(4R)^3 \quad \text{... (ii)}$$

$$\therefore \quad \frac{T^2}{T'^2} = \frac{1}{64}$$

$$\text{or} \quad \frac{T}{T'} = \frac{1}{8}$$

or $T' = 8T$

So, time period becomes 8 times of previous value.

347 **(a)**

Escape velocity of a body from the surface of earth is 11.2 kms^{-1} which is independent of the angle of project

348 **(b)**

As $T^2 \propto r^3$,

so, $\frac{T_A^2}{T_B^2} = \frac{r_A^3}{r_B^3}$

or $\frac{r_A}{r_B} = \left(\frac{T_A}{T_B}\right)^{2/3} = (8)^{2/3} = 4$

or $r_A = 4r_B$;

so $r_A - r_B = 4r_B - r_B = 3r_B$

349 **(d)**

For central force, torque is zero

$\because \tau = \frac{dL}{dt} = 0 \Rightarrow L = \text{constant}$

i.e. Angular momentum is constant

350 **(a)**

Let R be the original radius of a planet. Then attraction on a body of mass m placed on its surface will be

$$F = \frac{GM\,m}{R^2}$$

If size of the planet is made double *ie*, $R' = 2R$, then mass of the planet becomes

$$M' = \frac{4}{3}\pi(2R)^3\rho = 8 \times \frac{3}{4}\pi R^2\rho = 8M$$

New force $F' = -\frac{GM'm}{R'^2} = -\frac{G8M\times m}{(2R)^2} = 2F$

ie, force of attraction increases due to the increase in mass of the planet

351 **(c)**

Here, $\quad AB = BC = CD = DA = 1\text{m}$

$$BD = AC$$
$$= \sqrt{1^2 + 1^2}$$
$$= \sqrt{2}\text{m}$$

Total potential energy

$$U = \left[\frac{-G \times 1 \times 1}{AB}\right] + \left[\frac{-G \times 1 \times 1}{BC}\right] + \left[\frac{-G \times 1 \times 1}{CD}\right] + \left[\frac{-G \times 1 \times 1}{DA}\right] + \left[\frac{-G \times 1 \times 1}{BD}\right] + \left[\frac{-G \times 1 \times 1}{AC}\right]$$

$$= 4 \times \left[\frac{-G \times 1 \times 1}{1}\right] + 2\left[\frac{-G \times 1 \times 1}{\sqrt{2}}\right]$$
$$= -5.4G$$

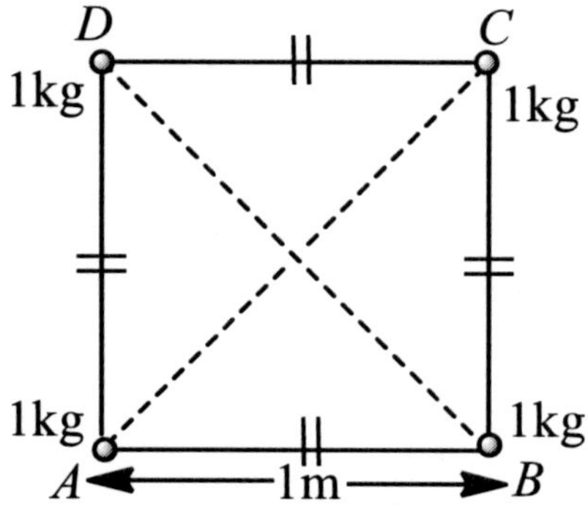

352 **(b)**

$\frac{g_e}{g_m} = \frac{R_e\rho_e}{R_m\rho_m} = \frac{2}{3} \times \frac{4}{1} = 6$ or $g_m = \frac{g_e}{6}$

For motion on earth, using the relation,

$s = ut + \frac{1}{2}at^2$

We have, $\frac{1}{2} = 0 + \frac{1}{2} \times 9.8r^2$ or $t = 1/\sqrt{9.8}s$

For motion on moon, $3 = 0 + \frac{1}{2}(9.8/6)t_1^2$

or $t_1 = 6\sqrt{9.8}s \therefore \frac{t_1}{t} = 6$ or $t_1 = 6t$

353 **(c)**

Gravitational potential energy of a body in the gravitational field, $E = \frac{-GM\,m}{r}$.

When r decreases negative value of E increase *ie*, E decreases

354 **(a)**

Binding energy $= |E|$

$= \frac{1}{2}\frac{GMm}{R_e} = \frac{1}{2}gmR_e$

355 **(d)**

Gravitational potential energy of body will be

$$E = \frac{GM_em}{r}$$

At $\quad r = 2R,$

$E_1 = -\frac{GM_em}{(2R)}$

At $\quad r = 3R$

$E_2 = -\frac{GM_em}{(3R)}$

Energy required to move a body of mass m from on orbit of radius $2R$ to $3R$ is

$$\Delta E = \frac{GM_em}{R}\left[\frac{1}{2} - \frac{1}{3}\right] = \frac{GM_em}{6R}$$

356 **(b)**

From Kepler's third law of planetary motion

$T^2 \propto R^3$

Given, $R_1 = R, \quad R_2 = 5R$

$\therefore \quad \frac{T_1^2}{T_2^2} = \frac{R^3}{(5R)^3}$

$\Rightarrow \quad \frac{T_1}{T_2} = \frac{1}{(5)^{3/2}}$

$T_2 = 5^{3/2}T_1$

$\therefore \quad T_2 = 5^{\frac{3}{2}}T \quad [\because T_1 = T]$

357 **(c)**

Let m be mass of planet and M that of sun, r the radius between the two. Let the planet be moving with velocity v_o, then Gravitational force = centripetal force

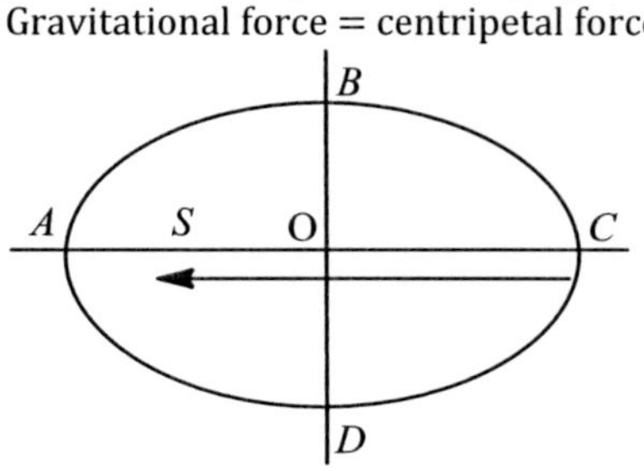

$\frac{GMm}{r^2} = \frac{mv_o^2}{r}$

$\Rightarrow \quad v_o = \sqrt{\frac{GM}{r}}$

$\Rightarrow \quad v_o \propto \frac{1}{\sqrt{r}}$

Hence, larger the distance, smaller the orbital velocity. At point C distance from sun is maximum, hence orbital velocity is lowest. At point A distance from sun is minimum, hence orbital velocity is maximum.

358 **(a)**

According of law of gravitation, the force of attraction acting on the body due to earth is given by

$$F = G\frac{Mm}{R^2} \quad \text{....(i)}$$

The acceleration due to gravity g in the body arises due to the force F from Newton's second law of motion, we have

$$F = mg \quad \text{...(ii)}$$

From Eqs. (i) and (ii), we get

$$mg = G\frac{Mm}{R^2}$$

$$\Rightarrow \quad g = \frac{GM}{R^2}$$

359 **(b)**

Gravitational pull depends upon the acceleration due to gravity on that planet

$M_m = \frac{1}{81}M_e, g_m = \frac{1}{6}g_e$

$$g = \frac{GM}{R^2} \Rightarrow \frac{R_e}{R_m} = \left(\frac{M_e}{M_m} \times \frac{g_m}{g_e}\right)^{1/2}$$

$$= \left(81 \times \frac{1}{6}\right)^{1/2}$$

$$\therefore R_e = \frac{9}{\sqrt{6}}R_m$$

360 **(a)**

The relation between density (d) and acceleration due to gravity (g) is

$$d = \frac{3g}{4\pi R_e G}$$

$$\therefore \quad \frac{d_1}{d_2} = \frac{g_1}{r_1} \times \frac{r_2}{g_2}$$

$$\Rightarrow \quad \frac{g_1}{g_2} = \frac{d_1 r_1}{d_2 r_2}$$

361 **(c)**

Velocity of body in inter planetary space

$v' = \sqrt{v^2 - v_{es}^2}$

Where v_{es}= escape velocity and

v = velocity of projection

$$\therefore v' = \sqrt{(2v_{es})^2 - v_{es}^2} = \sqrt{3v_{es}^2} \Rightarrow v'$$

$$= \sqrt{3}v_{es}$$

362 **(b)**

Using $g = \frac{GM}{R^2}$ we get $g_m = g/5$

363 **(c)**

$B.E. = -\frac{GMm}{r}$. If $B.E.$ decreases then r also decreases and v increases as $v \propto \frac{1}{\sqrt{r}}$

364 **(d)**

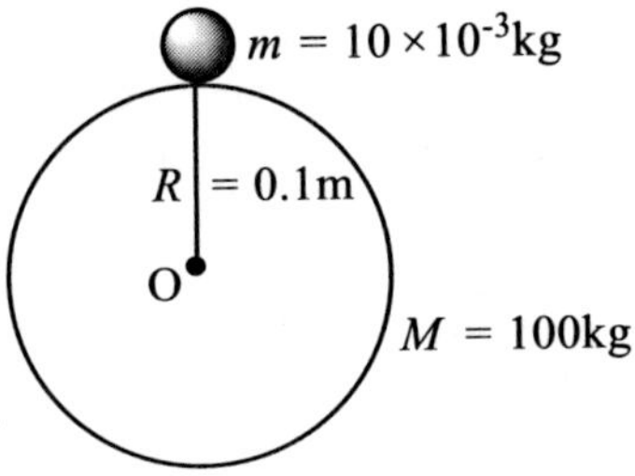

$$U_i = -\frac{GMm}{r}$$

$$U_i = \frac{6.67 \times 10^{-11} \times 100 \times 10^{-2}}{0.1}$$

$$U_i = -\frac{6.67 \times 10^{-11}}{0.1}$$

$$= -6.67 \times 10^{-10}\text{J}$$

We know

$$\therefore \quad W = \Delta U$$

$$= U_f - U_i \quad (\because U_f = 0)$$

$$\therefore \quad W = U_i = 6.67 \times 10^{-10}\text{J}$$

365 **(c)**

Error in weight = difference in weight at two different heights

$$= mg\left[1 - \frac{2h_1}{R}\right] - mg\left[1 - \frac{2h_2}{R}\right]$$

$$= \frac{2mg}{R}(h_2 - h_1) = \frac{2m}{R} \times \frac{GM}{R^2} \times \frac{h}{2}$$

[where, $h_2 - h_1 = h_2$]

$$= \frac{2m}{R^3} \times G \times \frac{4}{3}\pi R^2 \rho \times \frac{h}{2} = \frac{4}{3}\pi G m \rho h$$

366 **(c)**

Here, $I = \frac{dV}{dr} = -k/r$

or $dV = k\frac{dr}{r}$

Integrating it, we get

$$\int_{V_0}^{v} dV = \int_{r_0}^{r} k\frac{dr}{r}$$

or $V = V_0 + k \log r/r_0$

367 **(d)**

Acceleration due to gravity

$$g = \frac{GM}{R^2} = \frac{G}{R^2} \times \frac{4}{3}\pi R^3 \rho$$

$$\therefore \quad \rho = \frac{3g}{4\pi GR}$$

369 **(a)**

$g = \frac{4}{3}\pi\rho GR$. If ρ = constant then $\frac{g_1}{g_2} = \frac{R_1}{R_2}$

370 **(b)**

Gravitational force due to solid sphere,

$F_1 = \frac{GM\,m}{(2R)^2}$,

where$M$ and m are mass of the solid sphere and particle respectively and R is the radius of the sphere. The gravitational force on particle due to sphere with cavity = force due to solid sphere creating cavity, assumed to be present above at that position

$$ie, F_2 = \frac{GM\,m}{4R^2} - \frac{G(M/8)m}{(3R/2)^2} = \frac{7}{36}\frac{GM\,m}{R^2}$$

So, $\frac{F_2}{F_1} = \frac{7GM\,m}{36R^2} / \left(\frac{GM\,m}{4R^2}\right) = \frac{7}{9}$

371 **(b)**

The value of acceleration due to gravity at a height h reduces to

$= 100 - 36 = 64\% = \frac{64}{100}g$

$\therefore \frac{64}{100}g = \frac{gR^2}{(R+h)^2}$

or $\frac{8}{10} = \frac{R}{R+h}$ or $h = \frac{R}{4}$

372 **(a)**

If the mass of sun is M and radius of the planet's orbit is r,

then as $v_0 = \sqrt{GM/r}$

$T = \frac{2\pi r}{v_0} = 2\pi r\sqrt{\frac{r}{GM}}, ie, T^2 = \frac{4\pi^2 r^2}{GM}$ ...(i)

Now, if the planet (When stopped in the orbit) has velocity v when it is at a distance x from the sun, by conservation of mechanical energy,

$\frac{1}{2}mv^2 + \left(-\frac{GMm}{x}\right) = 0 - \frac{GMm}{r}$

or $\left(-\frac{dx}{dt}\right)^2 = \frac{2GM}{r}\left[\frac{r-x}{x}\right]$,

$ie, -\frac{dx}{dt} = \sqrt{\frac{2GM}{r}}\sqrt{\frac{(r-x)}{x}}$

or $\int_0^t dt = -\sqrt{\frac{r}{2GM}} \times \int_r^0 \left[\frac{x}{(r-x)}\right] dx$

Substituting $x\ r \sin^2\theta$ and solving the RHS,

$T = \sqrt{\frac{r}{2GM}} \times \left(\frac{\pi r}{2}\right)$

In the light of Eq. (i) reduces to

$t = \frac{1}{\sqrt{4}\sqrt{2}}T, ie, t = \left(\frac{\sqrt{2}}{8}\right)T$

373 **(d)**

The minimum velocity of projection to achieve escape velocity can be calculated as

$$\text{Intial KE} = \frac{1}{2}mv^2$$
$$= \frac{1}{2} \times m(4 \times 11.2)^2$$
$$= 16 \times \frac{1}{2}mv_e^2$$

As $\frac{1}{2}mv_e^2$ energy is used up in coming out from the gravitational pull of the earth, so final KE should be $15 \times \frac{1}{2}mv_e^2$

Hence, $\frac{1}{2}mv_e^2 = 15 \times \frac{1}{2}mv_e^2$

$\therefore$ $v'^2 = 15v_e^2$

or $v' = \sqrt{15}v_e$

$= \sqrt{15} \times 11.2\text{kms}^{-1}$

374 **(c)**

U = Loss in gravitational energy = gain in K.E.

So, $U = \frac{1}{2}mv^2 \Rightarrow m = \frac{2U}{v^2}$

375 **(b)**

Acceleration due to gravity $g = \frac{4}{3}\pi\rho GR$

or $g \propto \rho$

$\therefore \frac{g_1}{g_2} = \frac{\rho_1}{\rho_2}$

$\frac{g_1}{g_2} = \frac{\rho}{2\rho}[\because \rho_2 = 2\rho]$

$g_2 = g_1 \times 2 = 9.8 \times 2$

$g_2 = 19.6\ \text{m/s}^2$

376 **(a)**

The net force acting on a unit mass placed at O due to three equal masses M at verities A, B and C is the gravitational field intensity at point O. The gravitational force on the particle placed at the point of intersection of three medians.

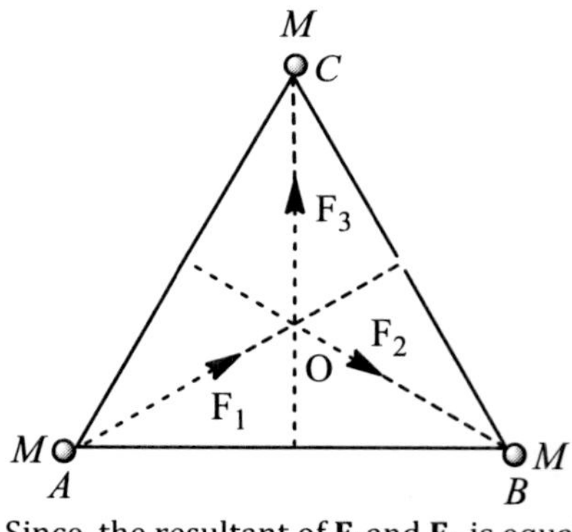

Since, the resultant of $\mathbf{F}_1$ and $\mathbf{F}_2$ is equal and opposite to $\mathbf{F}_3$.

377 **(d)**

It is given that, acceleration due to gravity on planet A is 9 times the acceleration due to gravity on planet B *ie*,

$g_A = 9g_B$ ... (i)

From third equation of motion

$v^2 = 2gh$

At planet A, $h_A = \frac{v^2}{2g_A}$ ... (ii)

At planet B, $h_B = \frac{v^2}{2g_B}$ ... (iii)

Dividing Eq. (ii) by Eq. (iii), we have

$\frac{h_A}{h_B} = \frac{g_B}{g_A}$

From Eq. (i), $g_A = 9g_B$

$\therefore \quad \frac{h_A}{h_B} = \frac{g_B}{9g_B} = \frac{1}{9}$

or $h_B = 9h_A = 9 \times 2 = 18$ m ($\therefore h_A = 2$m)

378 **(b)**

In circular orbit of a satellite of potential energy

$= -2 \times$ (kinetic energy)

$= -2 \times \frac{1}{2} m^v = -mv^2$

Just to escape from the gravitational pull, its total mechanical energy should be zero. Therefore , its kinetic energy should be $+mv^2$

379 **(a)**

The acceleration due to gravity (g) is given by

$$g = \frac{GM}{R^2}$$

where$M$ is mass, G the gravitational constant and R the radius.

Since, planets have a spherical shape

$$V = \frac{4}{3}\pi r^3$$

Also, mass (M) = volume(V) $\times$ density(ρ)

$$g = \frac{G\frac{4}{3}\pi R^3 \rho}{R^2}$$

$$\Rightarrow \quad g = \frac{4G\pi\rho R}{3}$$

Given, $R_1 : R_2 = 2 : 3$

$\rho_1 : \rho_2 = \frac{3}{2}$

$$\therefore \quad \frac{g_1}{g_2} = \frac{\rho_1 R_1}{\rho_2 R_2} = \frac{3}{2} \times \frac{2}{3} = 1$$

380 **(d)**

$$F = \left\{\frac{GMm}{r^2}\right\}, r \geq R$$

381 **(c)**

Escape velocity for that body $v_e = \sqrt{\frac{2Gm}{r}}$

v_e should be more than or equal to speed of light

$i.e. \sqrt{\frac{2Gm}{r}} \geq c$

382 **(b)**

Mass of satellite does not affect its orbital radius

383 **(b)**

$$T = 2\pi\sqrt{\frac{(R+h)^3}{gR^2}} = 2\pi\sqrt{\frac{(2R)^3}{gR^2}}$$

$$= 4\sqrt{2}\pi\sqrt{\frac{R}{g}}$$

384 **(d)**

$$v_0 = \sqrt{\frac{GM}{r}} = \sqrt{\frac{gR^2}{R+h}}$$

385 **(d)**

Acceleration due to gravity at depth d below the surface earth

$$g_d = g\left(1 - \frac{d}{R}\right)$$

Acceleration due to gravity at height h from the surface of the earth

$$g_h = g\left(g - \frac{2h}{R}\right)$$

Given $\quad g_h = g_d$

$\therefore \quad \frac{2h}{R} = \frac{d}{R}$

$d = 2h$

$d = 10$ km

386 **(c)**

If x is the distance of point on the line joining the two masses from mass $m_2{}'$ where gravitational field intensity is zero, then

$\frac{Gm}{(r-x)^2} = \frac{Gm_2}{x^2}$ or $\frac{2}{(9-x)^2} = \frac{8}{x^2}$

or $\frac{1}{9-x} = \frac{2}{x}$

On solving, $x = 6$m

387 **(d)**

% change in $T = \frac{3}{2}$(% change in R) $= \frac{3}{2} \times$ (2)% = 3%

389 **(d)**

Acceleration due to gravity at a height h form the surface of the earth

$$g' = g\frac{1}{\left(1 + \frac{h}{R}\right)^2}$$

Given, $\quad h = 2R$

$\therefore \quad g' = g\frac{1}{(1+2)^2}$

or $g' = \frac{g}{9}$

390 **(c)**

For orbiting the earth close to its surface

$$= \frac{mv^2}{R} = \frac{GMm}{R^2}, ie, v_0 = \sqrt{\frac{GM}{R}} = \sqrt{gR}$$

$\therefore v_0 = \sqrt{(9.8 \times 6.4 \times 10^6)} = 8 \text{ kms}^{-1}$

For escaping from close to the surface of earth,

$$\frac{GMm}{R} = \frac{1}{2}mv_t^2, v_e = \sqrt{\frac{2GM}{R}} = \sqrt{2gR}$$

$$v_e = \sqrt{2} \times v_0 = 1.41 \times 8\text{kms}^{-1} = 11.2 \text{ kms}^{-1}$$

$\therefore$ the additional velocity to be imparted to the orbiting satellite for escaping is $11.2 - 8 = 3.2 \text{ kms}^{-1}$

391 **(c)**

Angular momentum of the earth around the sun is

$L = M_E v_0 r$

$$\Rightarrow L = M_E\sqrt{\frac{GM_s}{r}}r \quad \left(\because v_0 = \sqrt{\frac{GM_s}{r}}\right)$$

$\Rightarrow L = [M_E^2 GM_s r]^{1/2}$

Where, M_E = Mass of the earth

M_s = Mass of the sun

r = Distance between the sun and the earth

$\therefore L \propto \sqrt{r}$

392 **(c)**

$v_e \propto \frac{1}{\sqrt{R}}$. If R becomes $\frac{1}{4}$ then v_e will be 2 times

393 **(a)**

The velocity with which satellite is orbiting around the earth is the orbital velocity (v_o) and that required to escape out of gravitational pull of earth is the escape velocity (v_e).

We know that

$v_e = \sqrt{2gR}$ and $v_o = \sqrt{gR}$

$\therefore$ Increase in velocity required

$$= \frac{v_e - v_o}{v_o} = \frac{\sqrt{2gR} - \sqrt{gR}}{\sqrt{gR}}$$

$$= \sqrt{2} - 1 = 0.414$$

Percent increase in velocity required $= 0.414 \times 100 = 41.4\%$

394 **(d)**

Escape velocity from the earth $(v_e) = 11.2 \text{ kms}^{-1}$

Let the mass, radius and density of earth be M, R and ρ respectively and for given planet mass, radius and density are M', R' and ρ', respectively.

$\therefore$ Escape velocity from the earth

$$v_e = \sqrt{\frac{2G \times \left(\frac{4}{3}\pi R^3 \rho\right)}{R}}$$

$$v_e = \sqrt{\frac{8G\pi R^2 \rho}{3}} \quad \text{.... (i)}$$

Similarly, escape velocity from the given planet

$$v'_e = \sqrt{\frac{8G\pi R'^2 \rho}{3}} \quad \text{.... (ii)}$$

Dividing Eq. (i) by Eq. (ii) we get

$$\frac{v_e}{v'_e} = \sqrt{\frac{8G\pi R^2\rho}{3}} \times \sqrt{\frac{3}{8G\pi R'^2 \rho}} = \sqrt{\frac{R^2}{R'^2}}$$

or $\frac{11.2}{v'_e} = \frac{R}{R'}$

or $\frac{11.2}{v'_e} = \frac{R}{2R}$

$\therefore \quad v'_e = 22.4 \text{ kms}^{-1}$

395 **(a)**

We know that intensity is negative gradient of potential,

$ie, m\, I = -(dV/dr)$ and as here $I = -(k/r)$, so

$$\frac{dV}{dr} = \frac{k}{r}, ie, \int_0^v dV = k\int_{r_0}^{r} \frac{dr}{r}$$

or $V - V_0 = k \log \frac{r}{r_0}$ so, $V = k \log \frac{r}{r_0} + V_0$

396 **(b)**

By the law of conservation of energy

$(U + K)_{\text{surface}} = (U + K)_\infty$

$$\Rightarrow -\frac{GMm}{R} + \frac{1}{2}m(3v_e)^2 = 0 + \frac{1}{2}mv^2$$

$$\Rightarrow -\frac{GM}{R} + \frac{9v_e^2}{2} = \frac{1}{2}v^2$$

Since, $v_e^2 = \frac{2GM}{R}$

$$\therefore -\frac{v_e^2}{2} + \frac{9v_e^2}{2} = \frac{1}{2}v^2 \Rightarrow v^2 = 8v_e^2$$

$$v = 2\sqrt{2}v_e = 2\sqrt{2} \times 11.2 = 31.7\text{kms}^{-1}$$

397 **(a)**

Gravitational potential energy of mass m at any point at a distance r from the centre of earth is

$$U = -\frac{GMm}{r}$$

At the surface of earth $r = R$

$$\therefore U_s = -\frac{GMm}{R} = -mgR \quad \left(\because g = \frac{GM}{R^2}\right)$$

At the height $h = nR$ from the surface of earth

$r = R + h = R + nR = R(1+n)$

$\therefore U_h = -\frac{GMm}{R(1+n)} = -\frac{mgR}{(1+n)}$

Change in gravitational potential energy is

$\Delta U = U_h - U_s = -\frac{mgR}{(1+n)} - (-mgR)$

$= -\frac{mgR}{1+n} + mgR = mgR\left(1 - \frac{1}{1+n}\right)$

$= mgR\left(\frac{n}{1+n}\right)$

398 **(a)**

$F \propto xm \times (1-x)m = xm^2(1-x)$

For maximum force $\frac{dF}{dx} = 0$

$\Rightarrow \frac{dF}{dx} = m^2 - 2xm^2 = 0$

$\Rightarrow x = 1/2$

399 **(b)**

$v \propto \frac{1}{\sqrt{r}}$

% increase in speed = 1/2 (% decrease in radius)

$= 1/2(1\%) = 0.5\%$

i.e. speed will increase by 0.5%

401 **(c)**

At a certain velocity of projection of the body will go out of the gravitational field of earth and never to return to earth. The initial velocity is called escape velocity

$v_e = \sqrt{2gR}$

Where g is acceleration due to gravity and R the radius. As is clear from above formula, that escape velocity dose not depends upon mass of body hence, it will be same for a body of 100kg as for 1kg body.

402 **(a)**

At an altitude h the acceleration due to gravity is

$g' = g\left(1 - \frac{2h}{R_e}\right)$

or $\quad mg' = mg\left(1 - \frac{2h}{R_e}\right)$

ie, $\quad w' = w\left(1 - \frac{2h}{R_e}\right)$

$\frac{99}{100}w = w\left(1 - \frac{2h}{R_e}\right)$

ie, $\quad h = 0.005R_e$

At point below the surface of earth at depth h. The weight of body given by

$w' = w\left(1 - \frac{2h}{R_e}\right)$

$\frac{w'}{w} = 0.995$

$\%\Delta w = \frac{(1-0.995)w}{w} \times 100$

$\%\Delta w = 0.5\%$(decreases)

403 **(d)**

Gravitational potential energy is given as

$U = -\frac{GMm}{r}$

$U_1 = -\frac{GMm}{r_1}, U_2 = -\frac{GMm}{r_2}$

As $r_2 > r_1$, hence,

$U_1 - U_2 = GMm\left[\frac{r_2 - r_1}{r_1 r_2}\right]$ is positie

ie, $\quad U_1 > U_2$

or $\quad U_2 < U_1$

*ie,*gravitational potential energy increases.

404 **(a)**

Acceleration due to gravity is given by

$g = \frac{GM}{R^2}$

where$G$ is gravitational constant.

For earth: $g_e = \frac{GM_e}{R_e^2}$

For planet: $g_p = \frac{GM_p}{R_p^2}$

Therefore, $\quad \frac{g_e}{g_p} = \frac{GM_e/R_e^2}{GM_p/R_p^2}$

or $\frac{g_e}{g_p} = \frac{M_e}{M_p} \times \frac{R_p^2}{R_e^2}$ ... (i)

Given, $M_p = 2M_e, \quad R_p = 2R_e$

Putting the values in the Eq. (i), we obtain

$\frac{g_e}{g_p} = \frac{M_e}{2M_e} \times \frac{(2R_e)^2}{R_e^2} = \frac{1}{2} \times \frac{4}{1} = 2$

$\therefore \quad g_p = \frac{g_e}{2}$

406 **(c)**

$g \propto r$ (if $r < R$) and $g \propto \frac{1}{r^2}$ (if $r > R$)

407 **(b)**

Acceleration due to gravity on earth is given by

$g = \frac{GM_e}{R_e^2}$

or $\quad g \propto \frac{M_e}{R_e^2}$

Hence, $\frac{g_{p_1}}{g_{p_2}} = \frac{M_{p_1}}{M_{p_2}} \times \left(\frac{R_{p_1}}{R_{p_2}}\right)^2$ ... (i)

Given, $\frac{M_{p_1}}{M_{p_2}} = \frac{1}{2}$ and $\frac{R_{p_1}}{R_{p_2}} = \frac{1}{2}$

Substituting the given value in Eq. (i), we get

$$\frac{g_{p_1}}{g_{p_2}} = \frac{1}{2} \times \left(\frac{2}{1}\right)^2 = \frac{2}{1}$$

$\therefore \quad g_{p_1} : g_{p_2} = 2 : 1$

408 **(d)**

According to Kepler's law $T^2 \propto r^3$

$$\Rightarrow \quad \left(\frac{T_1}{T_2}\right)^2 = \left(\frac{r_1}{r_2}\right)^3$$

409 **(a)**

By conservation of angular momentum

$mvr =$ constant

$v_{\min} \times r_{\max} = v_{\max} \times r_{\min}$

$$\therefore v_{\min} = \frac{60 \times 1.6 \times 10^{12}}{8 \times 10^{12}} = \frac{60}{5} = 12\ m/s$$

410 **(b)**

$$h = \left(\frac{T^2 R^2}{4\pi^2}\right)^{1/3} - R$$

$$= \left[\frac{(24 \times 60 \times 60)^2 \times (6.4 \times 10^6)^2 \times 9.8}{4 \times (22/7)^2}\right]^{1/3}$$

$- 6.4 \times 10^6$

$= 3.6 \times 10^7 \text{m} = 36000$ km

411 **(b)**

Orbital speed, $v_0 = \sqrt{\frac{GM}{r}}$; so speed of satellite decreases with the increase in the radius of its orbit. We need more than one satellite for global communication. For stable orbit it must pass through the centre of earth. So, only (b) is correct

412 **(a)**

Acceleration due to gravity at height h is

$$g' = \frac{g}{\left(1 + \frac{h}{R_e}\right)^2} = \frac{g}{\left(1 + \frac{32}{6400}\right)^2} = 0.99\ g$$

413 **(b)**

$F(r) = \begin{cases} \frac{GMm}{r^2} \\ \frac{4\pi G\rho rm}{3}, r < R \end{cases}$ (where ρ is density of sphere)

414 **(c)**

$$v = \sqrt{\frac{GM}{R}} = V,$$

$$v' = \sqrt{\frac{GM}{(R + R/2)}}$$

$$= \sqrt{\frac{2}{3}\frac{GM}{R}} = \sqrt{\frac{2}{3}}V$$

415 **(a)**

$$\text{KE} = \frac{1}{2}mv^2 - \frac{1}{2}m(11.2)^2$$

$$= \frac{1}{2}m(2 \times 11.2)^2 - \frac{1}{2}m(11.2)^2$$

$$\frac{1}{2}mv^2 = 3 \times \frac{1}{2}m \times (11.2)^2$$

$$v = \sqrt{3} \times 11.2$$

416 **(d)**

Since, velocity of projection (v)is greater than the escape velocity (v_e),therefore at infinite distance the body moves with a velocity

$$v' = \sqrt{v^2 - v_e^2}$$

$$\therefore \quad v' = \sqrt{(\sqrt{5}v_e)^2 - v_e^2} = 2v_e$$

417 **(c)**

$$\frac{T_1}{T_2} = \left(\frac{R_1}{R_2}\right)^{3/2} = \left(\frac{R}{4R}\right)^{3/2} \Rightarrow T_2 = 8T_1$$

418 **(a)**

Let a point mass C is placed at a distance of x m from the point mass A as shown in the figure

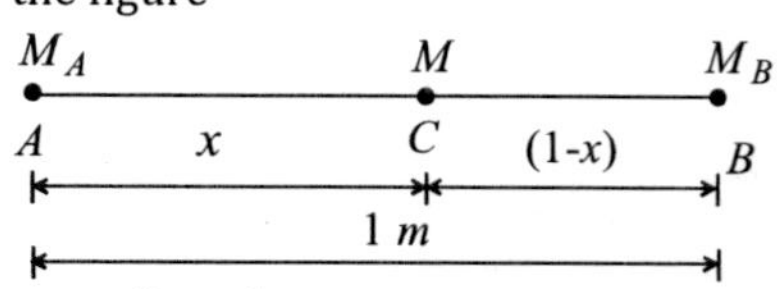

Here, $\frac{M_A}{M_B} = \frac{4}{3}$, Force between A and C is

$F_{AC} = \frac{GMM_A}{x^2}$ ...(i)

Force between B and C is

$F_{BC} = \frac{GMM_B}{(1-x)^2}$ ...(ii)

According to given problem $F_{AC} = \frac{1}{3}F_{BC}$

$\therefore \frac{GM_AM}{x^2} = \frac{1}{3}\left(\frac{GM_BM}{(1-x)^2}\right)$ [Using (i) and (ii)]

$$\frac{M_A}{x^2} = \frac{M_B}{3(1-x)^2} \text{ or } \frac{M_A}{M_B} = \frac{x^2}{3(1-x)^2}$$

$$\Rightarrow \frac{4}{3} = \frac{x^2}{3(1-x)^2} \text{ or } 4 = \frac{x^2}{(1-x)^2}$$

Or $2 = \frac{x}{1-x}$ or $2 - 2x = x$

$3x = 2$ or $x = \frac{2}{3}m$

419 **(c)**

If m is the mass of racket, M that of earth and R is the radius of earth, then gravitational potential energy of racket near the surface of earth

$$U_1 = \frac{GMm}{R}$$

Gravitational potential energy of racket at a height h from earth's surface

$$U_2 = -\frac{GMm}{(R+h)}$$

Increase in gravitational potential energy of racket

$$\Delta U = U_2 - U_1 = -\frac{GMm}{R+h} + \frac{GMm}{R}$$

or $$\Delta U = \frac{GMmh}{(R+h)R}$$

If v is the escape velocity of racket, then

$$\Delta U = \frac{1}{2}mv^2$$

$$\Rightarrow \quad \frac{1}{2}mv^2 = \frac{GMmh}{(R+h)R}$$

$$\Rightarrow \quad mv^2R^2 + mv^2Rh = 2GMm\,h$$

$$\Rightarrow \quad v^2R^2 = (2GM - v^2R)h$$

$$\therefore \quad h = \frac{v^2R^2}{2GM - v^2R}$$

420 **(c)**

$$GM = gR^2$$

$$u = \sqrt{2gR} = \sqrt{2\frac{GM}{R^2}R} = \sqrt{\frac{2GM}{R}}$$

421 **(a)**

Mass of the ball always remain constant. It does not depend upon the acceleration due to gravity

422 **(b)**

First we have to find a point where the resultant field due to both is zero. Let the point P be at a distance x from centre of bigger star.

$$\Rightarrow \frac{G(16M)}{x^2} = \frac{GM}{(10a-x)^2} \Rightarrow x = 8a(\text{from } O_1)$$

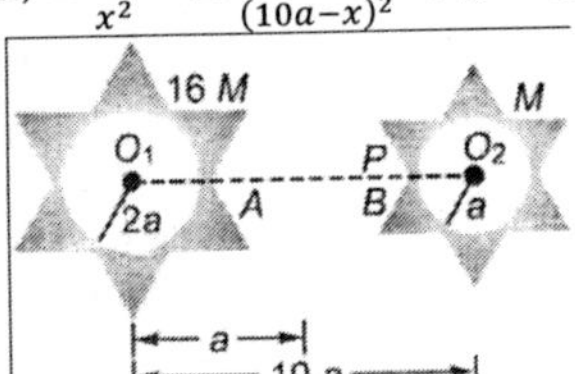

ie, once the body reaches P, the gravitational pull of attraction due to M takes the lead to make m move towards it automatically as the gravitational pull of attraction due to 16 Mvanishes *ie*, a minimum KE or velocity has to be imparted to m from surface of $16M$ such that it is just able to overcome the gravitational pull of $16M$. By law of conservation of energy.

(Total mechanical energy at A) = (Total mechanical energy at P)

$$\Rightarrow \frac{1}{2}mv_{min}^2 + \left[\frac{G(16M)m}{2a} - \frac{GMm}{8a}\right]$$

$$= 0 + \left[\frac{GMm}{2a} - \frac{G(16M)m}{8a}\right]$$

$$\Rightarrow \frac{1}{2}mv_{\min}^2 = \frac{GMm}{8a}(45) \Rightarrow v_{\min}$$

$$= \frac{3}{2}\sqrt{\frac{5GM}{a}}$$

423 **(c)**

When a sphere of mass m is released in a liquid, it falls vertically down with acceleration $= \frac{mg - F_B}{m}$

$$\frac{\frac{4}{3}\pi r^3 dg - \frac{4}{3}\pi r^3 \rho g}{\frac{4}{3}\pi r^2 d} = \frac{(d-\rho)g}{d}$$

424 **(b)**

Orbital radius of Jupiter > Orbital radius of Earth

$\frac{v_J}{v_e} = \frac{r_e}{r_J}$. As $r_J > r_e$ therefore $v_J < v_e$

426 **(a)**

Acceleration due to gravity at a height above the earth surface

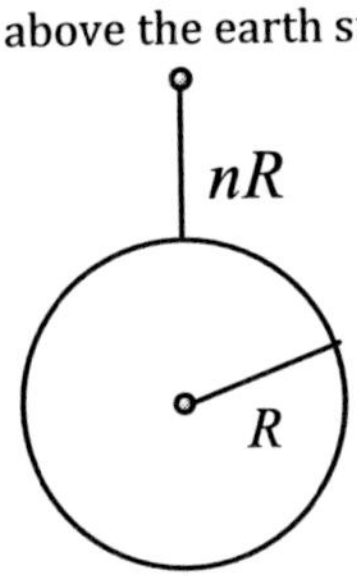

$$g' = g\left(\frac{R}{R+h}\right)^2$$

$$\frac{g}{g'} = \left(\frac{R+h}{R}\right)^2$$

$$\frac{g}{g'} = \left(\frac{R+nR}{R}\right)^2$$

$$\frac{g}{g'} = (1+n)^2$$

427 **(c)**

Just before striking, the distance between the centre of earth and moon is,

$$r = R_e + \frac{R_e}{4} = \frac{5R_e}{4}$$

So, acceleration of moon at this moment is

$$a = \frac{GM_e}{(5R_e/4)^2} = \frac{16}{25} \times 10 = 6.4 \text{ ms}^{-2}$$

428 **(a)**

As $g' = g - \omega^2 R \cos^2 \lambda$

The latitude at point on the surface of the earth is defined as the angle, which the line joining that point to the centre of earth makes with equatorial plane. It is denoted by λ. For the poles $\lambda = 90°$ and for equator $\lambda = 0°$.

(i) Substituting $\lambda = 90°$ in the above expression, we get

$g_{\text{pole}} = g - \omega^2 R\cos^2 90°$

$\therefore \quad g_{\text{pole}} = g$

ie,there is no effect of rotational motion of the earth on the value of g at the poles.

(ii) Substituting $\lambda = 0°$ in the above expression, we get

$g_{\text{equator}} = g - \omega^2 R\cos^2 0°$

$\therefore \quad g_{\text{equator}} = g - \omega^2 R$

ie, the effect of rotation of the earth on the value of g at the equator is maximum.

429 **(b)**

From kepler's third law of planetary motion:

$T^2 \propto R^3$

Given, $\quad T_p = 27T_{e'}$

$$\frac{T_e^2}{T_p^2} = \frac{R_e^3}{R_p^3}$$

$$\frac{T_e^2}{(27T_e)^2} = \frac{R_e^3}{R_p^3}$$

$$\frac{R_p}{R_e} = (27)^{1/2}$$

$$\frac{R_p}{R_e} = 3^2$$

$$R_p = 9R_e$$

430 **(a)**

$$g' = \frac{gR^2}{(R+h)^2}$$

$$= 980 \times \left(\frac{6400}{6400+64}\right)^2 = 960 \text{cms}^{-2}$$

431 **(b)**

Hence, $g' = g - R\omega^2 = 0$;

$\omega = \sqrt{g/R} = \sqrt{10/(6400 \times 10^3)} = 1/800$

432 **(c)**

$$\frac{g'}{g} = \left(\frac{R}{R+h}\right)^2 \Rightarrow \frac{1}{2} = \left(\frac{R}{R+h}\right)^2 \Rightarrow \frac{1}{2}$$

$$= \left(\frac{4000}{4000+h}\right)^2$$

By solving we get $h = 1656.85 \, mile \approx 1600 \, mile$

433 **(c)**

$\text{KE} = \frac{GMm}{2r} = -E_0$, and

$$\text{PE} = -\frac{GMm}{r} = 2E_0$$

$$\Rightarrow \text{TE} = \text{KE} + \text{PE} = -\frac{GMm}{2r} = E_0$$

434 **(d)**

Since, earth from west to east, so train Q has effectively more angular velocity in comparison to train P and hence, experiences a greater centrifugal force directed radially outwards. So, train Q will exert a lesser force on track Q in comparison to train P. Hence, P exerts greater force on track

435 **(c)**

The earth is not a solid sphere but is somewhat flattened at the poles and bulged at equator, its equatorial radius is 21 km larger than its polar radius, since,

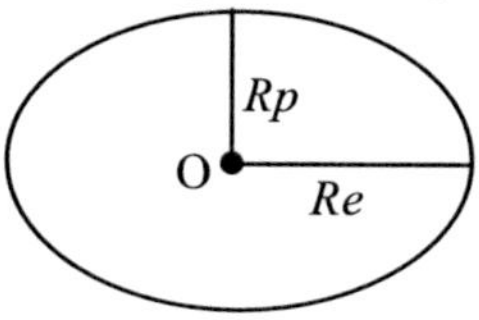

$$g = \frac{GM}{R^2}$$

Hence, value of g is least at equator and maximum at poles. Also, $W = mg$, therefore a person will get more quantity of matter in kg-wt at equator.

436 **(a)**

Binding energy = − kinetic energy

And if this amount of energy (E_K) given to satellite then it will escape into outer space

438 **(c)**

At equator, $g' = g - R\omega^2$. When angular velocity be

$\omega' (= x\omega)$, then, $0 = g - R\omega'^2$ or $\omega' = \sqrt{g/R} = x\omega$

or $x = (\sqrt{g/R})/\omega$

or $x = \frac{\sqrt{10/(6.4\times10^6)}}{2\pi} \times 24 \times 60 \times 60 = 17$

439 **(b)**

We know that $g = \frac{GM}{R^2}$

On the planet $g_p = \frac{GM/7}{R^2/4} = \frac{4}{7}g$

Hence weight on the planet $= 700 \times \frac{4}{7} =$ 400 $gm\ wt$

440 **(a)**

$$\frac{v_p}{v_e} = \sqrt{\frac{M_p}{M_e} \times \frac{R_e}{R_p}} = \sqrt{6 \times \frac{1}{2}} = \sqrt{3} \therefore v_p = \sqrt{3}v_e$$

441 **(c)**

When there is a weightlessness in the body at the equator, then $g' = r - R\omega^2 = 0$

or $\omega = \sqrt{g/R}$ and linear velocity $= \omega R = (\sqrt{g/R})R = \sqrt{gR}$

$\therefore$ KE of rotation of earth $= \frac{1}{2}I\omega^2$

$$= \frac{1}{2} \times \frac{2}{5}MR^2 \times \omega^2$$

$$= \frac{2}{5}M(\omega R)^2 = \frac{1}{5}MgR$$

442 **(a)**

At depth d from the surface of the earth.

$$g' = g\left(1 - \frac{d}{R}\right)$$

Given, $g' = \frac{75}{100}g = \frac{3}{4}g$

Then, $\frac{3g}{4} = g\left(1 - \frac{d}{R}\right)$

On solving, $d = R/4$

443 **(c)**

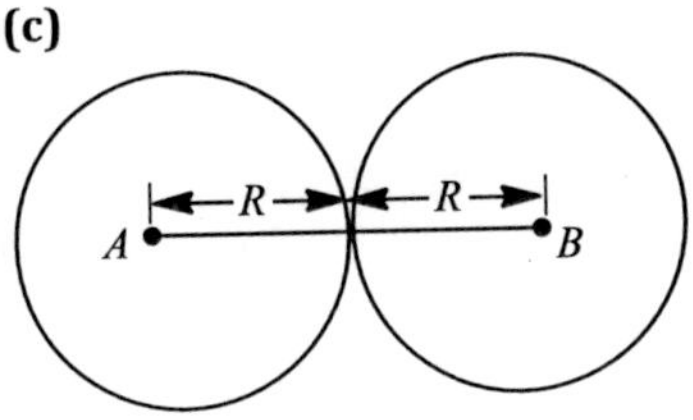

Let masses of two balls are $m_1 = m_2 = m$ (given) and the density be ρ.

Distance between their centres $= AB = 2R$

Thus, the magnitude of the gravitational force F that two balls separated by a distance $2R$ exert on each other is

$$F = G\frac{(m)(m)}{(2R)^2}$$

$$= G\frac{m^2}{4R^2} = G\frac{\left(\frac{4}{3}\pi R^3 \rho\right)^2}{4R^2}$$

$$\therefore \quad F \propto R^4$$

444 **(c)**

$$g = \frac{GM}{R^2} \therefore g \propto \frac{M}{R^2}$$

According to problem $M_p = \frac{M_e}{2}$ and $R_p = \frac{R_e}{2}$

$$\therefore \frac{g_p}{g_e} = \left(\frac{M_p}{M_e}\right)\left(\frac{R_e}{R_p}\right)^2 = \left(\frac{1}{2}\right) \times (2)^2 = 2$$

$$\Rightarrow g_p = 2g_e = 2 \times 9.8 = 19.6\ m/s^2$$

445 **(d)**

$$g'\left(1 - \frac{d}{R}\right) = g'\left(1 - \frac{2h}{R}\right)$$

d = depth of mine

h = height from surface

$$\therefore g'\left(1 - \frac{d}{R}\right) = g'\left(1 - \frac{2h}{R}\right)$$

$\Rightarrow d = 2h$

$\Rightarrow 10 = 2h$

$\Rightarrow h = 5km$

446 **(d)**

$$W = 0 - \left(-\frac{GMm}{R}\right) = \frac{GMm}{R}$$

$$= gR^2 \times \frac{m}{R} = mgR$$

$= 1000 \times 10 \times 6400 \times 10^3$

$= 64 \times 10^9 J$

$= 6.4 \times 10^{10} J$

447 **(b)**

$$v = \sqrt{2gR} \Rightarrow \frac{v_p}{v_e} = \sqrt{\frac{g_p}{g_e} \times \frac{R_p}{R_e}} = \sqrt{1 \times 4} = 2$$

$\therefore v_p = 2v_e$

448 **(c)**

$dV = -Edx$

or $V = -\int_{\infty}^{x/\sqrt{2}} Edx = -\int_{\infty}^{x/\sqrt{2}} kx^{-3}dx = k/x^2$

449 **(d)**

From the figure the gravitational intensity due to the ring at a distance $d = \sqrt{3}R$ on its axis is

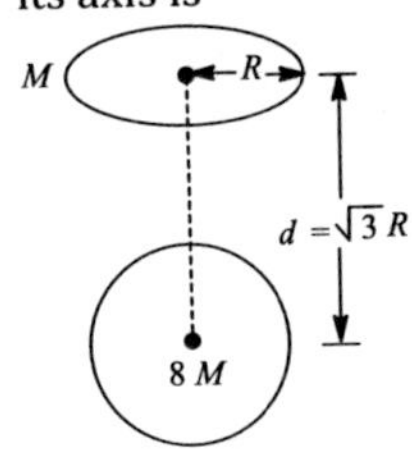

$$I=\frac{GM}{(d^2+R^2)^{3/2}}=\frac{GM\times\sqrt{3}R}{(3R^2+R^2)^{3/2}}$$
$$=\frac{\sqrt{3}GM}{8R^2}$$

Force on sphere = $(8M)I=(8M)\times\frac{\sqrt{3}GM}{8R^2}$

$$=\frac{\sqrt{3}GM^2}{R^2}$$

451 **(c)**

According to Kepler's third law

$T^2\propto R^3$

$$\Rightarrow\quad \frac{T_2}{T_1}=\left(\frac{R_2}{R_1}\right)^{3/2}$$
$$\therefore\quad \frac{T_2}{T_1}=\left(\frac{3R}{R}\right)^{3/2}$$
$$\Rightarrow\quad \frac{T_2}{T_1}=\sqrt{27}$$

$\therefore\ T_2=\sqrt{27}T_1=\sqrt{27}\times 4=4\sqrt{27}\text{h}$

452 **(a)**

$$g'=g\left(\frac{R}{R+h}\right)^2=\frac{g}{\left(1+\frac{h}{R}\right)^2}$$

453 **(b)**

Gravitational force $\left(=\frac{GM\,m}{R^{3/2}}\right)$ provides the necessary

centripetal force $(ie, m\,R\,\omega^2)$

So, $\frac{GM\,m}{R^{3/2}}=mR\omega^2=mR\left(\frac{2\pi}{T}\right)^2=\frac{4\pi^2 mR}{T^2}$

or $T^2=\frac{4\pi^2R^{5/4}}{GM}ie, T^2\propto R^{5/2}$

454 **(d)**

Time period does not depends upon the mass of satellite

455 **(b)**

Angular momentum is conserved in central field

456 **(c)**

Resultant gravitational intensity at a mid-point on the line joining the two bodies is

$$I=\frac{Gm_2}{(r/2)^2}-\frac{Gm_1}{(r/2)^2}=\frac{4G}{r^2}(m_2-m_1)$$
$$=\frac{4\times 6.6\times 10^{-11}}{1^2}(1000-100)$$
$$=2.4\times 10^{-7}\text{Mkg}^{-1}$$

457 **(b)**

The value of g at latitude $\lambda$is ;$g'=g-R\omega'^2\cos^2\lambda$. If earth stops rotating, $\omega=0$; $g'=g$. It means the weight of body will increase

458 **(d)**

The acceleration due to gravity at a depth d inside the earth is

$$g'=g\left(1-\frac{d}{R}\right)=g\left(\frac{R-d}{R}\right)=g\frac{r}{R}$$

where, $R-d=r$ = distance of a place from the centre of earth, therefore, $g'\propto r$

459 **(b)**

$F=Gm_1m_2/r^2$, thus on increasing masses and reducing distance r, force of gravitational attraction F will increase

460 **(d)**

The gravitational intensity at a point inside the spherical shell is zero

461 **(b)**

$\frac{T^2}{R^3}=\frac{T^2}{d^3}=\frac{1}{n^2d^3}$ = constant

$\therefore n_1^2d_1^3=n_2^2d_2^3$ [where n = frequency]

462 **(a)**

As in case of elliptic orbit of a satellite mechanical energy

$E=-(GMm/2a)$ remains constant, at any position of satellite in the orbit,

KE + PE $=-\frac{GMm}{2a}$ ...(i)

Now, if at position r, v is the orbital speed of satellite

KE $=\frac{1}{2}mv^2$ and PE $=-\frac{GMm}{r}$ ...(ii)

So, from Eqs. (i) and (ii), we have

$$\frac{1}{2}mv^2-\frac{GMm}{r}=-\frac{GMm}{2a}, ie, v^2=GM\left[\frac{2}{r}-\frac{1}{a}\right]$$

463 **(a)**

Both the stars with same angular velocity ω around the centre of mass (CM) in their respective orbits as shown in figure

The magnitude of gravitational force m_1 exerts on m_2 is $|F|=\frac{Gm_1m_2}{(r_1+r_2)^2}$

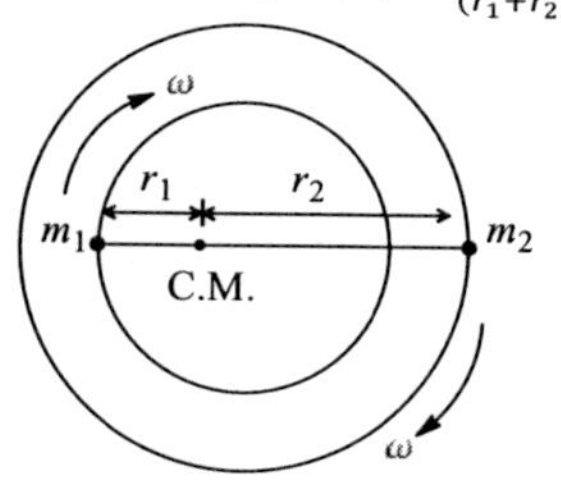

464 **(c)**

$I_1\omega_1=I_2\omega_2$

$$\frac{2}{5}MR^2\left(\frac{2\pi}{T_1}\right)=\frac{2}{5}M\cdot\frac{R^2}{n^2}\left(\frac{2\pi}{T_2}\right)$$
$$T_2=\frac{T_1}{n^2}=\frac{24}{n^2}$$

465 **(c)**

$H = \frac{u^2}{2g} \Rightarrow H \propto \frac{1}{g} \Rightarrow \frac{H_B}{H_A} = \frac{g_A}{g_B}$

Now $g_B = \frac{g_A}{12}$ as $g \propto \rho R$

$\therefore \frac{H_B}{H_A} = \frac{g_A}{g_B} = 12$

$\Rightarrow H_B = 12 \times H_A = 12 \times 1.5 = 18m$

466 **(c)**

The orbital velocity of satellite close to the earth is

$v_0 = \sqrt{gR_e}$ .... (i)

whereR_e is radius of the earth. The escape velocity for a body thrown from the earth's surface is

$v_e = \sqrt{gR_e}$ .... (ii)

Thus, $\frac{v_o}{v_e} = \frac{\sqrt{gR_e}}{\sqrt{2gR_e}} = \frac{1}{\sqrt{2}}$

$\therefore v_e = \sqrt{2}v_o$

or $v_o = \frac{v_e}{\sqrt{2}} = \frac{2}{\sqrt{2}} = \sqrt{2}\text{kms}^{-1}$

467 **(b)**

Here to point 7 of problem Solving skills

$[\frac{h_1}{h_2} = \frac{g_2}{g_1}$ or$h_2 = \frac{h_1 g_1}{g_2} = \frac{0.5 \times g}{g/6} = 3.0]$

Energy spent $= mgh_e = mg_m h_m$

or $h_m = g_e h_e / g_m$ ...(i)

$= \left(\frac{G\frac{4}{3}\pi R_e^3 \rho / R_e^2) h_e}{(G\frac{4}{3}\pi R_m^2 \rho_m / R_m^2)}\right)$

$= \left(\frac{G\frac{4}{3}\pi R_e^3 \rho / R_e^2) h_e}{(G\frac{4}{3}\pi R_m^2 \rho_m / R_m^2)}\right)$

$= \frac{R_e}{R_m} \times \frac{\rho_e}{\rho_m} \times h_e = \frac{3}{2} \times \frac{4}{1} \times 0.5 = 3\text{m}$

469 **(b)**

$g' = g\left(\frac{R}{R+h}\right)^2 \Rightarrow$ when $h = R$ then $g' = \frac{g}{4}$

So the weight of the body at this height will become one-fourth

470 **(c)**

$\frac{g_m}{g_e} = \frac{M_m}{M_e} \times \left(\frac{R_e}{R_m}\right) = \frac{1}{81} \times (4)^2 = \frac{16}{81}$

$g_m = \frac{16}{81} g_e$

$\therefore v_e = \sqrt{2g_e R_e}$

$= \sqrt{2 \times 9.8 \times (6400 \times 1000)}$

$\approx 11.2\ \text{kms}^{-1}$

$v_m = \sqrt{2g_m R_m} = \sqrt{2 \times \frac{16}{81} g_e \times \frac{1}{4} R_e}$

$= \frac{2}{9}\sqrt{2g_e R_e} = \frac{2}{9} \times 11 \approx 2.5\ \text{kms}^{-1}$

472 **(a)**

From Kepler's third law of planetary motion

$T^2 \propto R^3$

Given, $R_p = 2R_e$

$\therefore \frac{T_e^2}{T_p^2} = \frac{R_e^3}{R_p^3}$

$\Rightarrow \frac{T_e^2}{T_p^2} = \frac{R_e^3}{(2R_e)^3}$

$\Rightarrow \frac{T_e}{T_p} = \left(\frac{1}{2}\right)^{3/2}$

$\Rightarrow T_p = 2\sqrt{2}T_e$

Since, $T_e = 365$ days

$= 1$ year, we have

$T_p = 2\sqrt{2} \times 365$ days

$T_p = 1032.37$

$T_p = 1032$ days.

473 **(a)**

$m\omega^2 R = \frac{GMm}{R^2} \Rightarrow \left(\frac{2\pi}{T}\right)^2 R = \frac{GM}{R^2} \Rightarrow M$

$= \frac{4\pi^2 R^3}{GT^2}$

474 **(d)**

$T = 2\pi\sqrt{\frac{r^3}{GM}} \Rightarrow r^3 = \frac{GMT^2}{4\pi^2} \Rightarrow r$

$= \left[\frac{GMT^2}{4\pi^2}\right]^{1/3}$

475 **(c)**

$g = \frac{4}{3}G\pi R\rho \Rightarrow \frac{g_1}{g_2} = \frac{\rho_1 R_1}{\rho_2 R_2} = \frac{1}{2} \times \frac{4}{1} = \frac{2}{1}$

476 **(c)**

$T = 2\pi\sqrt{\frac{r^3}{GM}} \Rightarrow T^2 = \frac{4\pi^2}{GM}(R+h)^3$

$\Rightarrow R + h = \left[\frac{GMT^2}{4\pi^2}\right]^{1/3} \Rightarrow h$

$= \left[\frac{GMT^2}{4\pi^2}\right]^{\frac{1}{3}} - R$

477 **(b)**

$\frac{T^2}{r^3} =$ constant $\Rightarrow T^2 r^{-3} =$ constant

478 **(c)**

$v_e = \sqrt{\frac{2GM}{R}}$ $i.e.$ escape velocity depends upon the mass and radius of the planet

480 **(d)**

$$v_e = \sqrt{\frac{2GM}{(R+h)}}$$

482 **(b)**

The acceleration due to gravity

$$g = \frac{GM}{R^2}$$

At a height h above the earth's surface, the acceleration due to gravity is

$$g' = \frac{GM}{(R+h)^2}$$

$$\therefore \quad \frac{g}{g'} = \left(1+\frac{h}{R}\right)^{-2} = \left(1+\frac{h}{R}\right)^{2}$$

$$\frac{g'}{g} = \left(1+\frac{h}{R}\right)^{-2} = \left(1-\frac{2h}{R}\right)$$

but $\quad g' = \frac{g}{2}$ (given)

$$\therefore \quad \frac{g/2}{g} = 1 - \frac{2h}{R}$$

$$\frac{2h}{R} = \frac{1}{2}$$

$$h = \frac{R}{4}$$

483 **(d)**

Time period of satellite

$T \propto \frac{1}{M^{1/2}}$, where M is mass of earth.

$\propto (R+h)^{3/2}$ where R is radius of the orbit, h is the height of satellite from the earth's surface.

484 **(c)**

When gravitational force becomes zero, then centripetal force on satellite becomes zero and therefore, the satellite will become stationary in its orbit.

485 **(c)**

$g = \frac{GM}{r^2}$. Since M and r are constant, so

$g = 9.8 m/s^2$

486 **(b)**

Because value of g decreases with increasing height

487 **(c)**

$g = \frac{4}{3}\pi\rho GR \Rightarrow g \propto dR (\rho = d$ given in the problem)

488 **(d)**

At height h', $\frac{g'}{g} = 1 - \frac{2h}{R} = \frac{90}{100}$

or $\frac{2h}{R} = 1 - \frac{90}{100} = \frac{10}{100} = \frac{1}{10}$

or $R = 20h = 20 \times 320 = 6400\ km$

At dept d, $\frac{g'}{g} = 1 - \frac{d}{R} = \frac{95}{100}$

or $\frac{d}{R} = 1 - \frac{95}{100} = \frac{5}{100} = \frac{1}{20}$

or $d = \frac{R}{20} = \frac{6400}{20} = 320$ km

489 **(b)**

$I = \frac{-dV}{dr}$. If $I = 0$ then $V =$ constant

490 **(b)**

Weight on surface of earth, $mg = 500$N and weight below the surface of earth at

$$d = \frac{R}{2}$$

$$mg' = mg\left(1 - \frac{d}{R}\right)$$

$$= mg\left(1 - \frac{1}{2}\right)$$

$$= \frac{mg}{2} = 250\text{N}$$

491 **(c)**

Landsats 1 through 3 operated in a near polar orbit at an altitude of 920 km with an 18 day repeat coverage cycle. These satellites circled the earth every 103 min completing 14 orbits a day.

492 **(b)**

$v_e = \sqrt{2}v_o$, $i.e.$ if the orbital velocity of moon is increased by factor of $\sqrt{2}$ then it will escape out from the gravitational field of earth

494 **(b)**

Here, $m_1 = m_2 = 100$ kg ; $r = 100$m

Acceleration of first astronaut,

$$a_1 = \frac{Gm_1m_2}{r^2} \times \frac{1}{m_1} = \frac{Gm_1}{r^2}$$

Acceleration of second astronaut,

$$a_2 = \frac{Gm_1m_2}{r^2} \times \frac{1}{m_2} = \frac{Gm_2}{r^2}$$

Net acceleration of approach

$$a = a_1 + a_2 = \frac{Gm_2}{r^2} + \frac{Gm_1}{r^2} = \frac{2Gm_1}{r^2}$$

$$= \frac{2 \times (6.67 \times 10^{-11}) \times 100}{(100)^2}$$

$$= 2 \times 6.67 \times 10^{-13}\text{ms}^{-2}$$

As $\quad s = \frac{1}{2}at^2$

$\therefore t = \left(\frac{2s}{a}\right)^{1/2} = \left[\frac{2\times(1/100)}{2\times 6.67\times 10^{-13}}\right]^{1/2}$ second

On solving we get $t = 1.41$ days

496 **(c)**

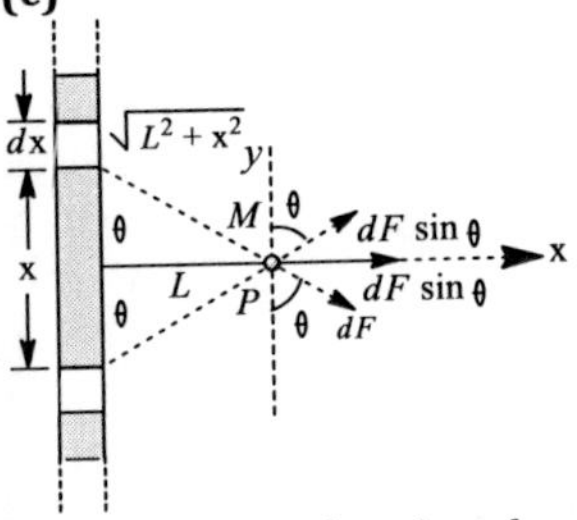

Let the mass M be placed symmetrically

$\Rightarrow F_{net}$

$$= \int_{-\infty}^{\infty} dF \sin\theta = \int_{-\infty}^{\infty} \frac{GM(\lambda dx)}{X^2 + L^2} \frac{L}{\sqrt{X^2 + L^2}}$$

$$\Rightarrow F_{net} = GM\lambda L \int_{-\infty}^{\infty} \frac{dx}{(X^2 + L^2)^{3/2}}$$

$$\Rightarrow F_{net} = \frac{GM\lambda L}{L^2}(2)$$

$$\Rightarrow F_{net} = \frac{2GM\lambda}{L^2}$$

497 **(c)**

$$g = \frac{GM}{R^2}; g' = \frac{GM}{R'^2} = \frac{GM(100)^2}{(99)^2 R^2}$$

% increase in $g = \frac{(g' - g) \times 100}{g}$

$$= \left(\frac{g'}{g} - 1\right) \times 100 = \left[\left(\frac{100}{99}\right)^2 - 1\right] \times 100$$

$$\left[\left(1 + \frac{1}{99}\right)^2 - 1\right] \times 100 \approx 2\%$$

498 **(c)**

At equator, $g' = g - R\omega^2 = 0$ or $\omega = \sqrt{g/R}$

or $\omega = \sqrt{10/(6.4 \times 10^6)} = 1.25 \times 10^{-3}$ rads^{-1}

499 **(c)**

The metallic spheres will be at positions as shown in the figure.

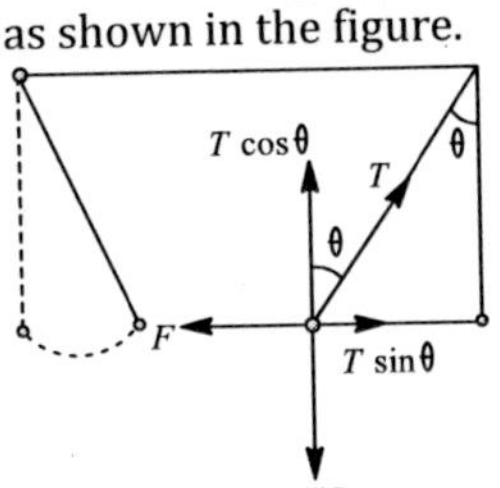

$$T \sin\theta = F = \frac{GM \times M}{L^2} = \frac{GM^2}{L^2}$$

$$T \cos\theta = Mg$$

$$\therefore \quad \tan\theta = \frac{GM}{gL^2}$$

or $\quad \theta = \tan^{-1}\left(\frac{GM}{gL^2}\right)$

501 **(a)**

The escape velocity at the surface of earth is 11.2 kms^{-1}

502 **(a)**

Time period of satellite

$$T = 2\pi \sqrt{\frac{(R + h)^3}{GM_e}}$$

where $R + h$ = orbital radius of satellite,

M_e = mass of earth.

Thus, time period does not depend on mass of satellite.

503 **(c)**

$v \propto \frac{1}{\sqrt{r}}$, If $r = R$ then $v = V_0$

If $r = R + h = R + 3R = 4R$ then $v = \frac{V_0}{2} = 0.5V_0$

504 **(b)**

If a body is projected from the surface of earth with a velocity v and reaches a height h, applying conservation of energy (relative to surface of earth)

$$\frac{1}{2}mv^2 = \frac{mgh}{[1 + (h/R)]}$$

$h = R = 6400$ km, $g = 10$ms^{-2}

So, $v^2 = gh$ *ie*, $v = \sqrt{10 \times 6400 \times 10^3} =$ 8 kms^{-1}

505 **(a)**

Gravitational potential at mid point

$$V = \frac{-GM_1}{d/2} + \frac{-GM_2}{d/2}$$

Now, $PE = m \times V = \frac{-2Gm}{d}(M_1 + M_2)$

(m = mass of particle)

So, for projecting particle from mid point to infinity

$KE = |PE|$

$$\Rightarrow \frac{1}{2}mv^2 = \frac{2\,Gm}{d}(M_1 + M_2) \Rightarrow v$$

$$= 2\sqrt{\frac{G(M_1 + M_2)}{d}}$$

507 **(d)**

$V = \frac{-GM}{r}$ and $I = \frac{GM}{r^2}$

$V = 0$ and $I = 0$ at $r = \infty$

NOTES